Progress in Mathematics
Volume 253

Algebraic Geometry and Number Theory

In Honor of Vladimir Drinfeld's 50th Birthday

Victor Ginzburg
Editor

Birkhäuser
Boston • Basel • Berlin

Victor Ginzburg
University of Chicago
Department of Mathematics
Chicago, IL 60637
U.S.A.
ginzburg@math.uchicago.edu

Mathematics Subject Classification (2000): 03C60, 11F67, 11M41, 11R42, 11S20, 11S80, 14C99, 14D20, 14G20, 14H70, 14N10, 14N30, 17B67, 20G42, 22E46 (primary); 05E15, 11F23, 11G45, 11G55, 11R39, 11R47, 11R58, 14F20, 14F30, 14H40, 14H42, 14K05, 14K30, 14N35, 22E67, 37K20, 53D17 (secondary)

Library of Congress Control Number: 2006931530

ISBN-10: 0-8176-4471-7 eISBN-10: 0-8176-4532-2
ISBN-13: 978-0-8176-4471-0 eISBN-13: 978-0-8176-4532-8

Printed on acid-free paper.

(JLS)

9 8 7 6 5 4 3 2 1

www.birkhauser.com

Vladimir Drinfeld

Preface

Vladimir Drinfeld's many profound contributions to mathematics reflect breadth and great originality. The ten research articles in this volume, covering a diversity of topics predominantly in algebra and number theory, reflect Drinfeld's vision in significant areas of mathematics, and are dedicated to him on the occasion of his 50th birthday.

The paper by Goncharov and Fock is devoted to the study of cluster varieties and their quantizations. This subject has its origins in the work of Fomin and Zelevinsky on cluster algebras and total positivity on the one hand, and, on the other hand, on various attempts to understand Kashiwara's theory of crystals and quantizations of moduli spaces of curves.

Starting with a split semisimple real Lie group G with trivial center, Goncharov and Fock define a family of varieties with additional structures called *cluster* $\mathcal{X}$-*varieties*. These varieties have a natural Poisson structure. The authors define a Poisson map from a cluster variety to the group G equipped with the standard Poisson–Lie structure as defined by V. Drinfeld. The map is birational and thus provides G with canonical rational coordinates. Further, Goncharov and Fock show how to construct complicated cluster $\mathcal{X}$-varieties from more elementary ones using an amalgamation procedure. This is used, in particular, to produce canonical (Darboux) coordinates for the Poisson structure on a Zariski open subset of the group G.

Some of the cluster varieties are very closely related to the double Bruhat cells studied by A. Berenshtein, S. Fomin, and A. Zelevinisky. On the other hand, the results of the paper play a key role in describing the cluster structure of the moduli spaces of local systems on surfaces, as studied by Goncharov and Fock in an earlier work.

The important role of Drinfeld's ideas—indeed, one of the central themes of his research—is evident in the paper by Frenkel and Gaitsgory, which is devoted to the (local) geometric Langlands correspondence from the point of view of D-modules and the representation theory of affine Kac–Moody algebras.

Let $\mathfrak{g}$ be a simple complex Lie algebra and G a connected algebraic group with Lie algebra $\mathfrak{g}$. The affine Kac–Moody algebra $\widehat{\mathfrak{g}}$ is the universal central extension of the formal loop agebra $\mathfrak{g}((t))$. Representations of $\widehat{\mathfrak{g}}$ have a parameter, an invariant bilinear form on $\mathfrak{g}$, which is called the *level*. Representations corresponding to the bilinear form that is equal to minus one-half of the Killing form are called representations of *critical level*. Such representations can be realized in spaces of global sections of twisted D-modules on the quotient of the loop group $G((t))$ by its "compact" subgroup K equal to $G[[t]]$, or to the Iwahori subgroup I.

This work by Frenkel and Gaitsgory is the first in a series of papers devoted to the study of the categories of representations of the affine Kac–Moody algebra $\widehat{\mathfrak{g}}$ of the critical level and D-modules on $G((t))/K$ from the point of view of a geometric version of the local Langlands correspondence. The local Langlands correspondence sets up a relation between two different types of data. Roughly speaking, the first data consist of the equivalence classes of homomorphisms from the Galois group of a local non-archimedean field $\mathbb{K}$ to $G(\mathbb{C})^\vee$, the Langlands dual group of G. The second data consist of the isomorphism classes of irreducible smooth representations, denoted by π, of the locally compact group $G(\mathbb{K})$.

A naive analogue of this correspondence in the geometric situation seeks to assign to a $G(\mathbb{C})^\vee$-local system on the formal punctured disc a representation of the formal loop group $G((t))$. However, the authors show that in contrast to the classical setting, this representation of $G((t))$ should be defined not on a vector space, but on a *category*, as explained in the paper.

In the contribution by Ihara, and in the closely related appendix by Tsfasman, the authors study the ζ-function $\zeta_{\mathbb{K}}(s)$ of a global field $\mathbb{K}$. Specifically, they are interested in the so-called Euler–Kronecker constant $\gamma_{\mathbb{K}}$, a real number attached to the power series expansion of the ζ-function at the point $s = 1$. In the special case of the field $\mathbb{K} = \mathbb{Q}$ of rational numbers, this constant reduces to the Euler constant

$$\gamma_{\mathbb{Q}} = \lim_{n\to\infty}\left(1 + \frac{1}{2} + \frac{1}{3} + \cdots + \frac{1}{n} - \log n\right).$$

The constant $\gamma_{\mathbb{K}}$ plays an important role in analytic number theory. On the other hand, for $\mathbb{K} = \mathbb{F}_q(X)$, the field of rational functions on a complete algebraic curve X over a finite field, the corresponding Euler–Kronecker constant is closely related to the number of $\mathbb{F}_q$-rational points of X.

Ihara addresses the question of how negative the constant $\gamma_{\mathbb{K}}$ may be, depending on the field $\mathbb{K}$. In the number field case, this happens when $\mathbb{K}$ has *many* primes with small norm. In the function field case, there are known towers of curves with many $\mathbb{F}_q$-rational points; the author studies the behavior of $\gamma_{\mathbb{K}}$ using the generalized Riemann hypothesis. In this way, he obtains very interesting explicit estimates of $\gamma_{\mathbb{K}}$. For instance, in the case $\mathbb{K} = \mathbb{F}_q(X)$ Ihara establishes an upper bound

$$\gamma_{\mathbb{K}} \leq 2\log((g-1)\log q) + \log q,$$

where g denotes the genus of the curve X. He also obtains similar estimates for the lower bound.

Hrushovski and Kazhdan in their paper lay the foundations of integration theory over, not necessarily locally compact, valued fields of residue characteristic zero. A valued field is a field $\mathbb{K}$ equipped with a "ring of integers" $\mathcal{O} \subset \mathbb{K}$, satisfying the property that $\mathbb{K} = \mathcal{O} \cup (\mathcal{O} \setminus \{0\})^{-1}$. In particular, the authors obtain new and base-field independent foundations for integration over local fields of large residue characteristic, extending results of Denef, Loeser, and Cluckers.

The work of Hrushovski and Kazhdan is on the border of logic and algebraic geometry. Their methods involve an analysis of definable sets. Specifically, they obtain a precise description of the Grothendieck semigroup of definable sets in terms of related groups over the residue field and value group. This yields new invariants of all definable bijections, as well as invariants of measure-preserving bijections. Their results are intended to be applied to the construction of Hecke algebras associated with reductive groups over a not necessarily locally compact valued field. In the case of a two-dimensional local field, the corresponding Hecke algebra is expected to be closely related to the double affine Hecke algebra introduced by Cherednik.

Kisin's paper is devoted to p-adic algebraic geometry and number theory. This subject is rapidly developing at this point in time. Following the ideas of Berger and Breuil, Kisin gives a new classification of crystalline representations. The objects involved may be viewed as local, characteristic 0 analogues of the "shtukas" introduced by Drinfeld. Kisin also gives a classification of p-divisible groups and finite flat group schemes, conjectured by Breuil. Furthermore, he shows that a crystalline representation with Hodge–Tate weights 0, 1 arises from a p-divisible group—a result conjectured by Fontaine.

Let k be a perfect field of characteristic $p > 0$, $W = W(k)$ its ring of Witt vectors, $\mathbb{K}_0 = W(k)[\frac{1}{p}]$, and $\mathbb{K} : \mathbb{K}_0$ a finite totally ramified extension. Breuil proposed a new classification of p-divisible groups and finite flat group schemes over the ring of integers $O_{\mathbb{K}}$ of $\mathbb{K}$. For p-divisible groups and $p > 2$, this classification was established in an earlier paper by Kisin, who also used a variant of Breuil's theory to describe flat deformation rings, and thereby establish a modularity lifting theorem for Barsotti–Tate Galois representations.

In the present paper, the author generalizes Breuil's theory to describe crystalline representations of higher weight or, equivalently, their associated weakly admissible modules.

Krichever's paper analyzes deep and important relations between the theory of integrable systems and the Riemann–Schottky problem. The Riemann–Schottky problem on the characterization of the Jacobians of curves among abelian varieties is more than 120 years old. Quite a few geometrical characterizations of the Jacobians have been found. None of them, however, provides an explicit system of equations for the image of the Jacobian locus in the projective space under the level-2 theta imbedding.

The link of this problem to integrable systems was first discovered in the 1980s. Specifically, T. Shiota established the first effective solution of the Riemann–Schottky problem, known as Novikov's conjecture. The conjecture says the following: *An indecomposable principally polarized abelian variety (X, θ) is the Jacobian of a curve of a genus g if and only if there exist g-dimensional vectors $U \neq 0$, V, W such that the function*

$$u(x, y, t) = -2\wp_x^2 \ln \theta(Ux + Vy + Wt + Z)$$

is a solution of the Kadomtsev–Petviashvilii (KP) equation

$$3u_{yy} = (4u_t + 6uu_x - u_{xxx})_x.$$

(Here $\theta(Z) = \theta(Z|B)$ is the Riemann theta function.)

In the present paper, Krichever proves that an indecomposable principally polarized abelian variety X is the Jacobian of a curve if and only if there exist vectors $U \neq 0$, V such that the roots $x_i(y)$ of the theta functional equation $\theta(Ux + Vy + Z) = 0$ satisfy the equations of motion of the *formal infinite-dimensional Calogero–Moser system*.

The main goal of Laumon's paper is to identify the fibers of the affine Springer resolution for the group GL_n with coverings of compactified Jacobians of projective singular curves. This work is part of a more general project of obtaining a geometric version of the "Fundamental Lemma" that appears in Langlands' works on automorphic forms.

Let F be a local non-archimedean field of equal characteristic, let $\mathcal{O}_F$ be its ring of integers, and let k be the residue field. Let E be a finite-dimensional F-vector space. The author considers the affine Grassmannian formed by $\mathcal{O}_F$-lattices M in E.

Given a regular semisimple and topologically nilpotent endomorphism γ of E, one defines the affine Springer fiber, X_γ, as the closed reduced subscheme of the affine Grassmannian formed by the γ-stable lattices $M \subset E$. Kazhdan and Lusztig have shown that X_γ is a scheme, locally of finite type over k. Moreover, this scheme comes equipped with a natural free action of an abelian algebraic group Λ_γ such that the quotient $Z_\gamma = X_\gamma/\Lambda_\gamma$ is a projective k-scheme.

In his paper, the author attaches to γ a projective algebraic curve C_γ over k with a single singular point such that the completed local ring at this point is isomorphic to $\mathcal{O}_F[\gamma] \subset F[\gamma] \subset \mathrm{Aut}_F(E)$. Furthermore, the author relates the varieties X_γ and Z_γ with the compactified Jacobian of the curve C_γ. This allows him to reprove some irreducibility results about compactified Jacobians due to Altman and Kleiman. In addition, the techniques developed in the paper provide an approach to an important "purity conjecture" concerning the cohomology of certain affine Springer fibers, due to Goresky, Kottwitz, and MacPherson.

The goal of the work of Manin presented in this volume is to study properties of the iterated integrals of modular forms in the upper half-plane. This setting generalizes simultaneously the theory of modular symbols and that of multiple zeta values. Multiple zeta values are the numbers given by the k-multiple Dirichlet series

$$\zeta(m_1, \ldots, m_k) = \sum_{0<n_1<\cdots<n_k} \frac{1}{n_1^{m_1} \cdots n_k^{m_k}} \tag{0.1}$$

or, equivalently, by the m-multiple iterated integrals $m = m_1 + \cdots + m_k$,

$$\zeta(m_1, \ldots, m_k) = \int_0^1 \frac{dz_1}{z_1} \int_0^{z_1} \frac{dz_2}{z_2} \int_0^{z_2} \cdots \int_0^{z_{m_k-1}} \frac{dz_{m_k}}{1 - z_{m_k}} \cdots . \tag{0.2}$$

Multiple zeta values are interesting because they and their generalizations appear in many different contexts involving mixed Tate motives, deformation quantization (Kontsevich), knot invariants, etc.

Multiple zeta values satisfy certain combinatorial relations, called double-shuffle relations. The relations in question can be succinctly written in terms of formal generating series for (regularized) iterated integrals (0.2). Such integrals appeared more than 15 years ago in the celebrated work by Drinfeld on what is nowadays known as *the Drinfeld associator*. However, the question about interdependence of (double-) shuffle and associator relations does not seem to be settled at the moment.

In the paper, the author defines 1-forms of modular and cusp modular type and studies iterated integrals and the total Mellin transform for families of such forms. The functional equation for the total Mellin transform is deduced. This result extends the classical functional equation for L series. The author also introduces an iterated modular symbol as a certain noncommutative 1-cohomology class of the relevant subgroup of the modular group. The paper establishes some analogues of the classical identity (0.1) = (0.2) but different from it in two essential respects. First, iterated integrals are only linear combinations of certain multiple Dirichlet series. Second, the identities obtained in the paper involve integrals which are *not* of the usual type,

$$\sum_{0<n_1<\cdots<n_k} \frac{a_{1,n_1}\cdots a_{n,n_k}}{n_1^{m_1}\cdots n_k^{m_k}};$$

in fact, their coefficients depend on pairwise differences $n_j - n_i$.

In the paper by Eskin and Okounkov, the authors prove that natural generating functions for enumeration of branched coverings of the pillowcase orbifold are level-2 quasimodular forms. This gives us a way to compute the volumes of the strata of the moduli space of quadratic differentials.

Consider a complex torus $T^2 = \mathbb{C}/L$, where $L \subset \mathbb{C}$ is a lattice. Its quotient

$$P = T^2/\pm 1$$

by the automorphism $z \mapsto -z$ is a sphere with four $(\mathbb{Z}/2)$-orbifold points, which is sometimes called the *pillowcase* orbifold. The map $T^2 \to P$ is essentially the Weierstraß $\wp$-function. The quadratic differential $(dz)^2$ on T^2 descends to a quadratic differential on P. Viewed as a quadratic differential on the Riemann sphere, $(dz)^2$ has simple poles at corner points.

Let μ be a partition and ν a partition of an even number into *odd* parts. The authors are interested in enumeration of degree $2d$ maps

$$\pi : C \to P \tag{0.3}$$

with the following ramification data. Viewed as a map to the sphere, π has profile $(\nu, 2^{d-|\nu|/2})$ over $0 \in P$ and profile (2^d) over the other three corners of P. Additionally, π has the profile $(\mu_i, 1^{2d-\mu_i})$ over some $\ell(\mu)$ given points of P and unramified elsewhere. Here $\ell(\mu)$ is the number of parts in μ.

The paper by Schechtman may be viewed as a continuation of the work by Gorbunov, Malikov, and Schechtman on the chiral de Rham complex. Specifically, the paper in the volume introduces a certain chiral analogue of the third Chern–Simons class of a vector bundle.

Let X be a smooth variety over a field k of characteristic zero, and write $\Omega^{\cdot}_X$ for the de Rham complex of X. Associated with any vector bundle E on X, one has the corresponding Chern classes

$$c_i^{DR}(E) \in H^{2i}(X, \Omega_X^{\cdot}), \quad i \geq 1,$$

respectively, the Chern–Simons classes

$$c_i^{CS}(E) \in H^i(X, \Omega_X^{[i,2i-1\rangle}),$$

where

$$\Omega_X^{[i,2i-1\rangle} = \sigma_{\geq i}\tau_{\leq 2i-1}\Omega_X[i] : \Omega_X^i \to \cdots \to \Omega_X^{2i-1,\mathrm{cl}},$$

and where $\Omega_X^{j,\mathrm{cl}} \subset \Omega_X^j$ stands for the subsheaf of closed forms. These classes are related via the canonical morphism

$$H^i(X, \Omega_X^{[i,2i-1\rangle}) \to H^{2i}(X, \Omega_X^{\cdot}),$$

which sends the class $c_i^{CS}(E)$ to $c_i^{DR}(E)$. One also defines the corresponding "Chern character" by setting $\mathrm{ch}_1 = c_1$, $\mathrm{ch}_2 = c_1^2 - \frac{c_2}{2}$, etc.

The present paper addresses the problem of giving explicit de Rham representatives for the classes $\mathrm{ch}_i^{CS}(T_X)$ for $i = 1, 2, 3$, where T_X denotes the tangent sheaf on X. Writing a de Rham representative for $\mathrm{ch}_1^{CS}(T_X)$ involves a choice of flat connection on $\det T_X$. Similarly, it is shown in the paper that the data required for writing de Rham representatives for the classes $\mathrm{ch}_i^{CS}(T_X)$, $i = 2, 3$ involve three maps

$$\gamma : \mathcal{O}_X \otimes_k T_X \to \Omega_X^1, \qquad \langle , \rangle : S^2 T_X \to \mathcal{O}_X, \quad \text{and} \quad c : \Lambda^2 T_X \to \Omega_X^1.$$

These maps must satisfy certain identities.

It turns out that exactly the same data of three maps γ, $\langle , \rangle$, and c appears in the theory of *vertex algebras*. Specifically, Gorbunov, Malikov, and Schechtman have given a mathematical definition of a special class of vertex algebras, called sheaves of chiral differential operators. The main result of the present paper by Schechtman says that the sheaves of chiral differential operators on a manifold X form a *gerbe* over the complex $\Omega_X^{[1,2\rangle}$, and the caracteristic class of that gerbe is equal to $\mathrm{ch}_2^{CS}(T_X)$. This provides "la raison d'être" for the appearance of the class $\mathrm{ch}_2^{CS}(T_X)$.

Victor Ginzburg
University of Chicago

June 2006

A glimpse into the life and work of V. Drinfeld

Volodya[1] Drinfeld was born in 1954 in Kharkov, Ukraine. He graduated from Moscow State University in 1974 at the age of 20, and defended his Ph.D. thesis in 1978. His vision of mathematics was, to a great extent, influenced by Yu. I. Manin, his advisor, and by the Algebraic Geometry Seminar ("Manin's Seminar") that functioned with regularity at Moscow State University for about two decades.

Because of his Jewish origin and the absence of a Moscow "propiska,"[2] the Soviet system made it extremely difficult for the talented Drinfeld, in spite of his obvious mathematical achievements, to get any reasonable job in mathematics in Moscow.

Therefore, after receiving his Ph.D., Drinfeld went to Ufa, a town in the Ural Mountains, where he taught mathematics at a local university. Later, he moved back to his native city, Kharkov, where he lived with his family until after the collapse of the Soviet Union.

It was in Kharkov where Drinfeld learned that he was to be awarded the Fields Medal, which he received at the Kyoto International Congress of Mathematicians (ICM) in 1990. In 1998, Drinfeld left Kharkov. Not long after migrating to the United States, he became a Distinguished Service Professor at the University of Chicago.

Almost immediately upon his arrival in Chicago, Drinfeld and A. Beilinson jointly organized the Geometric Langlands Seminar. Following, perhaps, in the tradition of the famous Gelfand Seminar in Moscow, the Geometric Langlands Seminar now runs regularly on Mondays from 4:30PM until both the speaker and the participants are completely exhausted.

In the course of his mathematical career, Drinfeld has worked on many different subjects, but his most fundamental contributions are in two fields. The first pertains to Drinfeld's fascination with quantum groups, which were discovered by the Leningrad school—L. D. Faddeev's students and collaborators. Drinfeld's outstanding contributions revitalized the entire subject. With his celebrated talk at the Berkeley ICM

[1] Volodya is the Russian diminutive for "Vladimir."

[2] People in the Soviet Union had their addresses written in their passports, the so-called "propiska." A person was not allowed to get a job in any town different from the one indicated in the propiska. Changing one's propiska was close to impossible.

in 1986, Drinfeld effectively "played a decisive role in the crystallization of this new domain." To be sure, we do not discount the articles of M. Jimbo and others who drew attention of this field to so many mathematicians today.

The second major contribution of Drinfeld is in the area known as the "Langlands program." Although the program itself was launched by Langlands in the late 1960s and early 1970s, it was Drinfeld who contributed crucial geometric insights. Drinfeld himself proved the Langlands conjecture in the special case of the group GL_2 over function fields; this, together with his achievements in quantum groups, earned him the Fields Medal. Drinfeld's ideas have been extended to the GL_n case by L. Lafforgue (2001), and a geometric refinement of this result was proved by D. Gaitsgory shortly afterwards. The general case of the Langlands conjecture still remains wide open.

Almost as important as Drinfeld's own works were the remarks and ideas that he generously shared, either during private discussions or in his letters to other mathematicians. For instance, in a (widely circulated) letter to V. Schechtman, Drinfeld outlined his vision of deformation theory, emphasizes the role of DG-algebras, Maurer–Cartan equations, and stacks. All of these have later found their place in M. Kontsevich's approach to deformation theory via A_∞-algebras.

As another example, one may cite a classic work of Deligne and Lusztig that to a large extent owes its existence to a remark made by Drinfeld to T. Springer in a private conversation. In that remark, Drinfeld sketched a geometric construction of representations of the groups $SL_2(\mathbb{F}_q)$ in terms of what is nowadays known as Deligne–Lusztig varieties.

In the same spirit, I remember how Volodya once asked me, while walking in the corridor of Moscow State University sometime around 1987, whether or not the convolution of two spherical perverse sheaves on the loop grassmannian was again a perverse sheaf. The following day, I told him that this was indeed true and could be deduced from Lusztig's results on Hecke algebras. In this way, thanks to his question, Drinfeld effectively created the theory of geometric Satake isomorphism.

I would like to finish with a couple of examples that show, I believe, that many of Drinfeld's insights are still awaiting "discovery." One such example is related to symplectic reflection algebras, a notion introduced by P. Etingof and myself in 2002. After having worked on the subject for several years, we discovered (in January 2005) that the definition of symplectic reflection algebras was essentially contained in two lines of Drinfeld's paper "Degenerate Affine Hecke Algebras and Yangians," written 15 years earlier! Although the paper itself is very well known, it seems nobody has read those two lines of Drinfeld's very densely written text carefully enough.

The second example is equally amazing. I was preparing for a course on representation theory, which I teach regularly in Chicago. Volodya mentioned to me that he had some old notes with exercises on representation theory, written for his students in Kharkov back in the 1980s. As usual, Volodya's notes were very systematic; they contained both the exercises and the solutions. Somewhere in the middle of the notes, I found a digression on "q-analogues" that contained computations equivalent, essentially, to the important geometric construction of the quantum group discovered by Beilinson, Lusztig, and MacPherson 10 years later!

We wish Volodya Drinfeld many more years of good health and inspirational mathematics which have contributed so much to so many of us from all over the mathematical world.

Victor Ginzburg
University of Chicago

June 2006

Contents

Algebraic Geometry and Number Theory

Pillowcases and quasimodular forms

Alex Eskin* and Andrei Okounkov†

* Department of Mathematics
University of Chicago
5734 South University Avenue
Chicago, IL 60637
USA
eskin@math.uchicago.edu
† Department of Mathematics
Princeton University
Fine Hall
Washington Road
Princeton, NJ 08544
USA
okounkov@math.princeton.edu

To Vladimir Drinfeld on his 50th birthday.

Summary. We prove that natural generating functions for enumeration of branched coverings of the pillowcase orbifold are level 2 quasimodular forms. This gives a way to compute the volumes of the strata of the moduli space of quadratic differentials.

Subject Classifications: Primary 14N10, 14N30. Secondary 11F23, 14N35.

1 Introduction

1.1 Pillowcase covers and quadratic differentials

Consider a complex torus $\mathbb{T}^2 = \mathbb{C}/L$, where $L \subset \mathbb{C}$ is a lattice. Its quotient

$$\mathfrak{P} = \mathbb{T}^2/\pm$$

by the automorphism $z \mapsto -z$ is a sphere with four $(\mathbb{Z}/2)$-orbifold points which is sometimes called the *pillowcase* orbifold. The map $\mathbb{T}^2 \to \mathfrak{P}$ is essentially the Weierstraß $\wp$-function. The quadratic differential $(dz)^2$ on $\mathbb{T}^2$ descends to a quadratic differential on $\mathfrak{P}$. Viewed as a quadratic differential on the Riemann sphere, $(dz)^2$ has simple poles at corner points.

Let μ be a partition and ν a partition of an even number into *odd* parts. We are interested in enumeration of degree $2d$ maps

$$\pi : \mathcal{C} \to \mathfrak{P} \tag{1}$$

with the following ramification data. Viewed as a map to the sphere, π has profile $(\nu, 2^{d-|\nu|/2})$ over $0 \in \mathfrak{P}$ and profile (2^d) over the other three corners of $\mathfrak{P}$. Additionally, π has profile $(\mu_i, 1^{2d-\mu_i})$ over some $\ell(\mu)$ given points of $\mathfrak{P}$ and is unramified elsewhere. Here $\ell(\mu)$ is the number of parts in μ. This ramification data determines the genus of $\mathcal{C}$ by

$$\chi(\mathcal{C}) = \ell(\mu) + \ell(\nu) - |\mu| - |\nu|/2.$$

In principle, one could allow more general ramifications over 0 and the nonorbifold points, but this more general problem is readily reduced to the one above.[1]

Pulling back $(dz)^2$ via π gives a quadratic differential on $\mathcal{C}$ with zeros of multiplicities $\{\nu_i - 2\}$ and $\{2\mu_i - 2\}$. The periods of this differential, by construction, lie in a translate of a certain lattice. The enumeration of covers π is thus related to lattice point enumeration in the natural strata of the *moduli space of quadratic differentials*. In particular, the $d \to \infty$ asymptotics gives the volumes of these strata. These volumes are of considerable interest in ergodic theory, in particular in connection with billiards in rational polygons; see [6, 18]. Their computation was the main motivation for the present work.

A different way to compute the volume of the principal stratum was found by M. Mirzakhani [19].

1.2 Generating functions

1.2.1

Two covers $\pi_i : \mathcal{C}_i \to \mathfrak{P}$, $i = 1, 2$, are identified if there is an isomorphism $f : \mathcal{C}_1 \to \mathcal{C}_2$ such that $\pi_1 = f \circ \pi_2$. In particular, associated to every cover π is a finite group $\mathrm{Aut}(\pi)$. This group is trivial for most connected covers; see, e.g., [7, Section 3.1]. We form the generating function

$$Z(\mu, \nu; q) = \sum_{\pi} \frac{q^{\deg \pi}}{|\mathrm{Aut}(\pi)|}, \tag{2}$$

where π ranges over all inequivalent covers (1) with ramification data μ and ν as above. Note that the degree of any such π is even.

In particular, for $\mu = \nu = \emptyset$ any connected cover has the form

$$\pi : \mathbb{T}^2 \xrightarrow{\pi'} \mathbb{T}^2 \to \mathbb{T}^2/\pm$$

with π' unramified. We have $|\mathrm{Aut}(\pi)| = 2|\mathrm{Aut}(\pi')|$ corresponding to the lift of $\pm$. From the enumeration of possible π' we obtain,

[1] From first principles, the count of the branched coverings does not change if one replaces two ramification conditions by the product of the corresponding conjugacy classes in the class algebra of the symmetric group. In this way, one can generate complicated ramifications from simpler ones.

$$Z(\emptyset, \emptyset; q) = \prod_n (1 - q^{2n})^{-1/2}.$$

By definition, we set

$$Z'(\mu, \nu; q) = \frac{Z(\mu, \nu; q)}{Z(\emptyset, \emptyset; q)}. \tag{3}$$

This enumerates covers without unramified connected components. By the usual inclusion-exclusion, one can extract from (3) a generating function for connected covers. This generating function for connected covers will be denoted by $Z^\circ(\mu, \nu; q)$.

1.2.2

Recall the classical level 1 Eisenstein series

$$E_{2k}(q) = \frac{\zeta(1-2k)}{2} + \sum_{n=1}^{\infty} \left(\sum_{d|n} d^{2k-1} \right) q^n, \quad k = 1, 2, \ldots.$$

The algebra they generate is called the algebra $\mathcal{QM}(\Gamma(1))$ of *quasimodular forms* for $\Gamma(1) = SL_2(\mathbb{Z})$; see [16] and also below in Section 3.3.7. It is known that E_2, E_4, and E_6 are free commutative generators of $\mathcal{QM}(\Gamma(1))$. The algebra $\mathcal{QM}(\Gamma(1))$ is naturally graded by weight, where wt $E_{2k} = 2k$. Clearly, for any integer N, $E_{2k}(q^N)$ is a quasimodular form of weight $2k$ for the group

$$\Gamma_0(N) = \left\{ \begin{pmatrix} a & b \\ c & d \end{pmatrix} \middle| \, c \equiv 0 \bmod N \right\} \subset SL_2(\mathbb{Z}).$$

The quasimodular forms that will appear in this paper will typically be inhomogeneous, so instead of weight grading we will only keep track of the corresponding filtration. We define the weight of a partition by

$$\text{wt}\, \mu = |\mu| + \ell(\mu).$$

The main result of this paper is the following.

Theorem 1. *The series $Z'(\mu, \nu; q)$ is a polynomial in $E_2(q^2)$, $E_2(q^4)$, and $E_4(q^4)$ of weight* wt $\mu + |\nu|/2$.

Several explicit examples of the forms $Z'(\mu, \nu; q)$ are given in the appendix.

1.2.3

Quasimodular forms occur in nature, for example, as coefficients of the expansion of the odd genus 1 theta-function

$$\vartheta(x) = (x^{1/2} - x^{-1/2}) \prod_{i=1}^{\infty} \frac{(1 - q^i x)(1 - q^i / x)}{(1 - q^i)^2}$$

at the origin $x = 1$. The techniques developed below give a certain formula for (3) in terms of derivatives of $\vartheta(x)$ at $x = \pm 1$, from which the quasimodularity follows.

1.2.4

The following discussion closely parallels the corresponding discussion for the case of holomorphic differentials in [7, Section 1.2].

Let $\mathsf{Q}(\mu, \nu)$ denote the moduli space of pairs (Σ, ϕ), where ϕ is a quadratic differential on a curve Σ with zeroes of multiplicities $\{\nu_i - 2, 2\mu_i - 2\}$. Note that we allow $\nu_i = 1$; hence our quadratic differentials can have simple poles. For $(\Sigma, \phi) \in \mathsf{Q}(\mu, \nu)$, let $\widetilde{\Sigma}$ denote the double cover of Σ on which the differential

$$\omega = \sqrt{\phi}$$

is well defined. The pair $(\widetilde{\Sigma}, \omega)$ belongs to the corresponding space of holomorphic differentials with zeroes of multiplicity

$$\{\nu_i - 1, \mu_i - 1, \mu_i - 1\}.$$

By construction, Σ is the quotient of $\widetilde{\Sigma}$ by an involution σ. Let P denote the set of zeroes of ω; it is clearly stable under σ. Then σ acts as an involution on the relative homology group $H_1(\widetilde{\Sigma}, P, \mathbb{Z})$. Let H^- denote the subspace of $H_1(\widetilde{\Sigma}, P, \mathbb{Z})$ on which σ acts as multiplication by -1. Choose a basis $\{\gamma_1, \ldots, \gamma_n\}$ for H^-, and consider the period map $\Phi : \mathsf{Q}(\mu, \nu) \to \mathbb{C}^n$ defined by

$$\Phi(\Sigma, \phi) = \left(\int_{\gamma_1} \omega, \ldots, \int_{\gamma_n} \omega \right).$$

It is known [18] that $\Phi(\Sigma, \phi)$ is a local coordinate system on $\mathsf{Q}(\mu, \nu)$ and, in particular, $n = \dim_{\mathbb{C}} H^- = \dim_{\mathbb{C}} \mathsf{Q}(\mu, \nu)$.

Pulling back the Lebesgue measure from $\mathbb{C}^n$ yields a well-defined measure on $\mathsf{Q}(\mu, \nu)$. However, this measure is infinite since ϕ can be multiplied by any complex number. Thus we define $\mathsf{Q}_1(\mu, \nu)$ to be the subset satisfying

$$\mathrm{Area}(\widetilde{\Sigma}) \equiv \frac{\sqrt{-1}}{2} \int_{\widetilde{\Sigma}} \omega \wedge \overline{\omega} = 2.$$

As in the case of holomorphic differentials, the area function is a quadratic form in the local coordinates on $\mathsf{Q}(\mu, \nu)$, and thus the image under Φ of $\mathsf{Q}_1(\mu, \nu)$ can be identified with an open subset of a hyperboloid in $\mathbb{C}^n$.

Now let $E \subset \mathsf{Q}_1(\mu, \nu)$ be a set lying in the domain of a coordinate chart, and let $C\Phi(E) \subset \mathbb{C}^n$ denote the cone over $\Phi(E)$ with vertex 0. Then we can define a measure ρ on $\mathsf{Q}_1(\mu, \nu)$ via

$$\rho(E) = \mathrm{vol}(C\Phi(E)),$$

where vol is the Lebesgue measure. The proof of [7, Proposition 1.6] shows the analogue

$$\rho(\mathsf{Q}_1(\mu, \nu)) = \lim_{D \to \infty} D^{-\dim_{\mathbb{C}} \mathcal{Q}(\mu,\nu)} \sum_{d=1}^{2D} \mathrm{Cov}_d^0(\mu, \nu),$$

where $\mathrm{Cov}_d^0(\mu, \nu)$ is the number of inequivalent degree d connected covers $C \to \mathfrak{P}$. Thus, the volume $\rho(\mathsf{Q}_1(\mu, \nu))$ can be read off from the $q \to 1$ asymptotics of the connected generating function $Z^\circ(\mu, \nu; q)$.

Note that the moduli spaces $\mathsf{Q}(\mu, \nu)$ may be disconnected. Ergodic theory applications require the knowledge of volumes of each connected component. Fortunately, connected components of $\mathsf{Q}(\mu, \nu)$ have been classified by E. Lanneau [17] and these spaces turn out to be connected except for hyperelliptic components (whose volume can be computed separately) and finitely many sporadic cases.

1.2.5

The modular transformation

$$q = e^{2\pi i \tau} \mapsto e^{-2\pi i/\tau}$$

relates $q = 0$ and $q = 1$ and thus gives an easy handle on the $q \to 1$ asymptotics of (3). This gives an asymptotic enumeration of pillowcase covers and hence computes the volume of the moduli spaces of quadratic differentials.

1.2.6

In spirit, Theorem 1 is parallel to the results of [1, 8, 13]; see also [2, 3, 5] for earlier results in the physics literature. The main novelty is the occurrence of quasimodular forms of higher level. One might speculate whether similar lattice point enumeration in the space of Nth order differentials leads to level N quasimodular forms. Those spaces, however, do not admit an $SL_2(\mathbb{R})$-action and a natural interpretation of their volumes is not known.

1.2.7

The following enumerative problem is naturally a building block of the enumerative problem that we consider. Consider branched covers of the sphere ramified over 3 points $0, 1, \infty$ with profile $(\nu, 2^{d-|\nu|/2})$, (2^d), and μ, respectively, where μ is an arbitrary partition of $2d$.

The preimage of the segment $[0, 1]$ on the sphere is a graph $\mathcal{G}$ on a Riemann surface (also known as a *ribbon graph*) with many 2-valent vertices (that can be ignored) and a few odd valent vertices (namely, with valencies ν_i). The complement of $\mathcal{G}$ is a union of $\ell(\mu)$ disks (known as *cells*) with perimeters $2\mu_i$ in the natural metric on $\mathcal{G}$. The asymptotic enumeration of such combinatorial objects is, almost by definition, given by integrals of ψ-classes against Kontsevitch's combinatorial cycles in $\overline{\mathcal{M}}_{g,\ell(\mu)}$; see [15]. There is a useful expression for these integrals in terms of Schur Q-functions obtained in [4, 11]. In fact, our original approach to the results presented in this paper was based on these ideas.

While the proof that we give here is more direct, it is still interesting to investigate the connection with combinatorial classes further, especially since a natural geometric interpretation of combinatorial classes is still missing. Perhaps the Gromov–Witten theory of the orbifold $\mathfrak{P}$ is the natural place to look for it. This will be further discussed in [22].

2 Character sums

2.1 Characters of near-involutions

2.1.1

There is a classical way to enumerate branched coverings in terms of irreducible characters, which is reviewed, for example, in [10] or in [21]. Specialized to our case, it gives

$$Z(\mu, \nu; q) = \sum_{\lambda} q^{|\lambda|/2} \left(\frac{\dim \lambda}{|\lambda|!}\right)^2 \mathbf{f}_{\nu,2,2,\ldots}(\lambda)\mathbf{f}_{2,2,\ldots}(\lambda)^3 \prod_i \mathbf{f}_{\mu_i}(\lambda) \tag{4}$$

where summation is over all partitions, $\dim \lambda$ is the dimension of the corresponding representation of the symmetric group, and $\mathbf{f}_\eta(\lambda)$ is the *central character* of an element with cycle type η in the representation λ. Recall that the sum of all permutations with cycle type η acts as a scalar operator in any representation λ and, by definition, this number is $\mathbf{f}_\eta(\lambda)$. In (4), as usual, we abbreviate $\mathbf{f}_{k,1,1,\ldots}$ to $\mathbf{f}_k$.

2.1.2

A lot is known about the characters of the symmetric group $S(2d)$ in the situation when the representation is arbitrary but the support of the permutation is bounded by some number independent of d. In particular, explicit formulas exist for the functions $\mathbf{f}_k$.

Understanding the function $\mathbf{f}_{\nu,2,2,\ldots}$ is the key to evaluation of (4). That is, we must study characters of permutations that are a product of a permutation with finite support and a fixed-point-free involution. We call such permutations *near-involutions*.

2.1.3

By a result of Kerov and Olshanski [14], the functions $\mathbf{f}_k$ belong to the algebra Λ^* generated by

$$\mathbf{p}_k(\lambda) = (1 - 2^{-k})\zeta(-k) + \sum_i \left[(\lambda_i - i + \frac{1}{2})^k - \left(-i + \frac{1}{2}\right)^k \right]; \tag{5}$$

moreover, $\mathbf{f}_k$ has weight $k+1$ in the weight filtration on Λ^* defined by setting

$$\operatorname{wt} \mathbf{p}_k = k + 1.$$

The functions $\mathbf{p}_k$ are central characters of certain distinguished elements in the group algebra of symmetric group known as *completed cycles*. See [21] for the discussion of the relation between $\mathbf{p}_k$ and $\mathbf{f}_k$ from the viewpoint of Gromov–Witten theory.

2.1.4

Our next goal is to generalize the results of [14] to characters of near-involutions. This will require enlarging the algebra of functions. In addition to the polynomials $\mathbf{p}_k$, we will need quasi-polynomial functions $\bar{\mathbf{p}}_k$ defined in (6) below.

It is convenient to work with the generating function

$$\mathbf{e}(\lambda, z) \stackrel{\text{def}}{=} \sum_i e^{z(\lambda_i - i + \frac{1}{2})} = \frac{1}{z} + \sum_k \mathbf{p}_k(\lambda) \frac{z^k}{k!}.$$

By definition, set

$$\begin{aligned} \bar{\mathbf{p}}_k(\lambda) &= i k! [z^k] \mathbf{e}(\lambda, z + \pi i) \qquad (6) \\ &= \sum_i \left[(-1)^{\lambda_i - i + 1} \left(\lambda_i - i + \frac{1}{2} \right)^k - (-1)^{-i+1} \left(-i + \frac{1}{2} \right)^k \right] + \text{const}, \end{aligned}$$

where the constant terms are determined by the expansion

$$\sum_k \frac{z^k}{k!} \bar{\mathbf{p}}_k(\emptyset) = \frac{1}{e^{z/2} + e^{-z/2}}.$$

Up to powers of 2, they are Euler numbers.

2.1.5

Define

$$\bar{\Lambda} = \mathbb{Q}[\mathbf{p}_k, \bar{\mathbf{p}}_k]_{k \geq 1}.$$

Setting

$$\text{wt}\, \bar{\mathbf{p}}_k = k$$

gives the algebra $\bar{\Lambda}$ the weight grading. Note that if f is homogeneous, then

$$f(\lambda') = (-1)^{\text{wt} f} f(\lambda), \qquad (7)$$

where λ' denotes the conjugate partition.

2.1.6

In the definition of $\bar{\Lambda}$, we excluded the function

$$\bar{\mathbf{p}}_0(\lambda) = \frac{1}{2} + \sum_i [(-1)^{\lambda_i - i + 1} - (-1)^{-i+1}],$$

which measures the difference between the number of even and odd numbers among $\{\lambda_i - i + 1\}$, also known as the *2-charge* of a partition λ.

Every partition λ uniquely defines two partitions α and β, known as its *2-quotients*, such that

$$\left\{\lambda_i - i + \frac{1}{2}\right\} = \left\{2\left(\alpha_i - i + \frac{1}{2}\right) + \bar{\mathbf{p}}_0(\lambda)\right\} \sqcup \left\{2\left(\beta_i - i + \frac{1}{2}\right) - \bar{\mathbf{p}}_0(\lambda)\right\}.$$

A partition λ will be called *balanced* if $\bar{\mathbf{p}}_0(\lambda) = \frac{1}{2}$.

Several constructions related to 2-quotients will play an important role in this paper. A modern review of these ideas can be found, for example, in [9]. In particular, it is known that the character $\chi^{\lambda}_{2,2,\dots}$ of a fixed-point free involution in the representation λ vanishes unless λ is balanced, in which case

$$\left|\chi^{\lambda}_{2,2,\dots}\right| = \binom{|\lambda|/2}{|\alpha|,\,|\beta|} \dim\alpha \dim\beta. \tag{8}$$

It follows that only balanced partitions contribute to the sum (4).

2.1.7

For a balanced partition λ, define

$$\mathbf{g}_\nu(\lambda) = \frac{\mathbf{f}_{(\nu,2,2,\dots)}(\lambda)}{\mathbf{f}_{(2,2,\dots)}(\lambda)}. \tag{9}$$

We will prove that this function lies in $\bar{\Lambda}$ in the following sense.

Theorem 2. *The ratio* (9) *is the restriction of a unique function* $\mathbf{g}_\nu \in \bar{\Lambda}$ *of weight* $|\nu|/2$ *to the set of balanced partitions.*

Several examples of the polynomials $\mathbf{g}_\nu$ can be found in the appendix.

2.1.8

In view of Theorem 2, it is natural to introduce the *pillowcase weight*

$$\mathsf{w}(\lambda) = \left(\frac{\dim\lambda}{|\lambda|!}\right)^2 \mathbf{f}_{2,2,\dots}(\lambda)^4.$$

Theorem 1 follows from (4), Theorem 2, and the following result.

Theorem 3. *For any* $F \in \bar{\Lambda}$*, the average*

$$\langle F\rangle_{\mathsf{w}} = \frac{1}{Z(\emptyset,\emptyset;q)} \sum_\lambda q^{|\lambda|}\mathsf{w}(\lambda)F(\lambda) \tag{10}$$

is a polynomial in $E_2(q^2)$*,* $E_2(q^4)$*, and* $E_4(q^4)$ *of weight* $\operatorname{wt} F$*.*

Note that if F is homogeneous of odd weight, then $\langle F\rangle_{\mathsf{w}} = 0$. This can be seen directly from (7). Also note that (10) will *not* in general be of pure weight even if F is a monomial in the generators $\mathbf{p}_k$ and $\bar{\mathbf{p}}_k$. This contrast with [1, 8] hints to the existence of a better set of generators of the algebra $\bar{\Lambda}$. Probably such generators are related to descendents of orbifold points in the Gromov–Witten theory of $\mathfrak{P}$.

2.1.9

It will be convenient to work with the following generating functions for the sums (10):

$$F(x_1, \ldots, x_n) = \left\langle \prod \mathbf{e}(\lambda, \ln x_i) \right\rangle_{\mathsf{W}} . \tag{11}$$

The function (11) will be called the *n-point function.*

2.2 Proof of Theorem 2

2.2.1

In the proof of theorems 2 and 3 it will be very convenient to use the fermionic Fock space formalism. This formalism is standard and [12, 20] can be recommended as a reference. A quick review of these techniques can be found, for example, in [21, Section 2]. We follow the notation of [21].

2.2.2

By definition, the space $\Lambda^{\frac{\infty}{2}}{}_0 V$ is spanned by the infinite wedge products

$$v_\lambda = \underline{\lambda_1 - \tfrac{1}{2}} \wedge \underline{\lambda_2 - \tfrac{3}{2}} \wedge \underline{\lambda_3 - \tfrac{5}{2}} \wedge \cdots , \tag{12}$$

where $\underline{k}$, $k \in \mathbb{Z} + \frac{1}{2}$, is a basis of the underlying space V and λ is a partition. The subscript 0 in $\Lambda^{\frac{\infty}{2}}{}_0 V$ refers to the charge zero condition: the ith factor in (12) is $\underline{-i + \frac{1}{2}}$ for all sufficiently large i.

There is a natural projective representation of the Lie algebra $\mathfrak{gl}(V)$ on $\Lambda^{\frac{\infty}{2}}{}_0 V$. For us, the following elements of $\mathfrak{gl}(V)$ will be especially important:

$$\mathcal{E}_k[f(x)]\underline{i} = f\left(i - \frac{k}{2}\right)\underline{i - k}, \tag{13}$$

where f is a function on the real line. To define the action of $\mathcal{E}_0[f(x)]$ on $\Lambda^{\frac{\infty}{2}}{}_0 V$ one needs to regularize the infinite sum $\sum_{i<0} f(\frac{1}{2} - i)$. This regularization is the source of the central extension in the $\mathfrak{gl}(V)$ action. When f is an exponential as in

$$\mathcal{E}_k(z) = \mathcal{E}_k[e^{zx}],$$

this infinite sum is a geometric series and thus has a natural regularization. By differentiation, this leads to the ζ-regularization for operators $\mathcal{E}_k[f]$ with a polynomial function f.

2.2.3

Other very useful operators are

$$\alpha_k = \mathcal{E}_k[1], \quad k \neq 0.$$

The operator H defined by

$$H v_\lambda = |\lambda| v_\lambda$$

is known as the energy operator. It differs only by a constant from the operator $\mathcal{E}_0[x]$. The operator H defines a natural grading on $\Lambda^{\frac{\infty}{2}}{}_0 V$ and $\mathfrak{gl}(V)$.

2.2.4

A function $F(\lambda)$ on partitions of n can be viewed as a vector

$$\sum_{|\lambda|=n} F(\lambda) v_\lambda \in \Lambda^{\frac{\infty}{2}}{}_0 V$$

of energy n. For example, the vectors

$$|\mu\rangle \stackrel{\text{def}}{=} \frac{1}{\mathfrak{z}(\mu)} \prod \alpha_{-\mu_i} v_\emptyset = \frac{1}{\mathfrak{z}(\mu)} \sum_\lambda \chi^\lambda_\mu v_\lambda, \tag{14}$$

where

$$\mathfrak{z}(\mu) = |\mathrm{Aut}\,\mu| \prod \mu_i,$$

correspond to irreducible characters normalized by the order of the centralizer.

2.2.5

The operator $\mathcal{E}_0(z)$ is the generating function

$$\mathcal{E}_0(z) = \mathcal{E}_0[e^{zx}] = \frac{1}{z} + \sum_k \frac{z^k}{k!} \mathcal{P}_k,$$

for the operators $\mathcal{P}_k$ acting by

$$\mathcal{P}_k v_\lambda = \mathbf{p}_k(\lambda) v_\lambda.$$

In parallel to (6), we define operators $\bar{\mathcal{P}}_k$ by

$$i\mathcal{E}_0(z + \pi i) = \sum_k \frac{z^k}{k!} \bar{\mathcal{P}}_k.$$

Translated into the operator language, the statement of Theorem 2 is the following: the orthogonal projection of $|\nu, 2^{d-|\nu|/2}\rangle$ onto the subspace spanned by the v_λ with λ balanced is a linear combination of vectors

$$\prod \mathcal{P}_{\mu_i} \prod \bar{\mathcal{P}}_{\bar{\mu}_i} |2^d\rangle \tag{15}$$

with

$$\mathrm{wt}\,\mu + |\bar{\mu}| \leq |\nu|/2$$

and coefficients independent of d.

2.2.6

Let us call the span of v_λ with λ balanced the balanced subspace of $\Lambda^{\frac{\infty}{2}}{}_0V$. A convenient orthogonal basis of it is provided by the vectors

$$|\rho;\bar{\rho}\rangle \overset{\text{def}}{=} \frac{1}{\mathfrak{z}(\rho)\mathfrak{z}(\bar{\rho})}\prod \alpha_{-\rho_i}\prod \bar{\alpha}_{-\bar{\rho}_i} v_{\emptyset}, \qquad \rho_i, \bar{\rho}_i \in 2\mathbb{Z}, \tag{16}$$

where the operators $\bar{\alpha}_k$ are defined by

$$\bar{\alpha}_k = i^{k+1}\mathcal{E}_k(\pi i) = \sum_n (-1)^{n+\frac{1}{2}} E_{n-k,n} + \frac{\delta_k}{2}, \tag{17}$$

the operators $E_{i,j}$ being the matrix units of $\mathfrak{gl}(V)$. From the commutation relations for the operators $\mathcal{E}_k(z)$, we compute

$$[\bar{\alpha}_k, \bar{\alpha}_m] = [(-1)^k - (-1)^m]\alpha_{k+m} + k(-1)^k\delta_{k+m}, \tag{18}$$

$$[\alpha_k, \bar{\alpha}_m] = [1-(-1)^k]\left(\bar{\alpha}_{k+m} + \frac{\delta_{k+m}}{2}\right). \tag{19}$$

In particular, when both k and m are even, all these operators commute apart from the central term in $[\bar{\alpha}_k, \bar{\alpha}_{-k}]$.

The adjoint of $\bar{\alpha}_k$ is

$$\bar{\alpha}_k^* = (-1)^k\bar{\alpha}_{-k},$$

which gives the inner products

$$\langle\rho;\bar{\rho}|\rho';\bar{\rho}'\rangle = \frac{\delta_{\rho,\rho'}\delta_{\bar{\rho},\bar{\rho}'}}{\mathfrak{z}(\rho)\mathfrak{z}(\bar{\rho})}, \tag{20}$$

provided all parts of all partitions in (20) are even. In particular, the vectors (16) are orthogonal. It is clear that they lie in the balanced subspace and their number equals the dimension of the space. Therefore, they form a basis.

2.2.7

The projection of $|\nu, 2^{d-|\nu|/2}\rangle$ onto the balanced subspace is given in term of inner products of the form

$$\langle\nu, 2^{d-|\nu|/2}|(\rho, 2^{d-|\rho|/2-|\bar{\rho}|/2}); \bar{\rho}\rangle$$

where all parts of ν are odd, all parts of ρ and $\bar{\rho}$ are even, and ρ has no parts equal to 2. From the commutation relations (18) and (19) we conclude that this inner product vanishes unless

$$\rho = \emptyset.$$

The nonvanishing inner products are

$$\langle \nu, 2^k | 2^k; \bar{\rho} \rangle = \frac{2^{\ell(\nu)-\ell(\bar{\rho})}}{2^k k! \mathfrak{z}(\nu) \mathfrak{z}(\bar{\rho})} \mathsf{c}(\nu, \bar{\rho}), \tag{21}$$

where the combinatorial coefficient $\mathsf{c}(\nu, \bar{\rho})$ equals the number of ways to represent the parts of $\bar{\rho}$ as sums of parts of ν. For example,

$$\mathsf{c}((3,1,1,1),(4,2)) = 3, \qquad \mathsf{c}((3,1,1,1),(6)) = 1.$$

2.2.8

The matrix elements

$$\left\langle 2^d \left| \prod \mathcal{P}_{\mu_i} \prod \bar{\mathcal{P}}_{\bar{\mu}_i} \right| (\rho, 2^{d-|\rho|/2-|\bar{\rho}|/2}); \bar{\rho} \right\rangle, \quad \rho_i \neq 2, \tag{22}$$

describe the decomposition of the vectors (15) in the basis (16). Since

$$\mathcal{P}_1 |2^d\rangle = \left(d - \frac{1}{24}\right) |2^d\rangle, \tag{23}$$

we can also assume that $\mu_i \neq 1$.

We claim that (22) vanishes unless

$$\operatorname{wt} \mu + |\bar{\mu}| \geq \operatorname{wt} \rho/2 + |\bar{\rho}|/2, \tag{24}$$

where $\rho/2$ is the partition with parts $\rho_i/2$ (recall that all parts of ρ are even).

2.2.9

The usual way to evaluate a matrix element like (22) is to use commutation relations to commute all lowering operators to the right until they reach the vacuum (which they annihilate) and, similarly, commute the raising operators to the left.

We will exploit the following property of the operators $\mathcal{P}_k$ and $\bar{\mathcal{P}}_k$: their commutator with enough operators of the form $\alpha_{-2\rho_i}$ and $\bar{\alpha}_{-2\bar{\rho}_i}$ vanishes. All such commutators have the form $\mathcal{E}_k[f]$ with $f(x) = (\pm 1)^x p(x)$, where $p(x)$ is a polynomial. Commutation with $\alpha_{-2\rho_i}$ takes a finite difference of $p(x)$; commutation with $\bar{\alpha}_{-2\bar{\rho}_i}$ additionally flips the sign of ± 1.

Since a $(k+1)$-fold finite difference of a degree k polynomial vanishes, the commutator of $\mathcal{P}_k$ with more than $k+1$ operators of the form $\alpha_{-2\rho_i}$ or $\bar{\alpha}_{-2\bar{\rho}_i}$ vanishes. In fact, a $(k+1)$-fold commutator may be nonvanishing only because of the central extension term. To pick up this central term, the total energy of all operators involved should be zero and the number of $\bar{\alpha}$s should be even. The same reasoning applies to $\bar{\mathcal{P}}_k$, but now the number of $\bar{\alpha}$s should be odd to produce a nontrivial $(k+1)$-fold commutator.

2.2.10

Now look at one of the raising operators involved in (22), say $\bar{\alpha}_{-\bar{\rho}_i}$. This operator commutes with α_2 and its adjoint annihilates the vacuum, so only the terms involving the commutator of $\bar{\alpha}_{-\bar{\rho}_i}$ with one of the $\mathcal{P}_{\mu_i}$ or $\bar{\mathcal{P}}_{\bar{\mu}_i}$ give a nonzero contribution to (22). The commutator $[\mathcal{P}_{\mu_i}, \bar{\alpha}_{-\bar{\rho}_i}]$ has energy $(-\rho_i)$ and so its adjoint again annihilates the vacuum. The same is true for the commutation with $\bar{\mathcal{P}}_{\bar{\mu}_i}$. To bring these commutators back to zero energy, one needs to commute it $\bar{\rho}_i/2$ times with α_2. Given the above bounds on how many commutators we can afford, this implies (24).

2.2.11

When the bound (24) is saturated, then a further condition

$$\ell(\rho) + \ell(\bar{\rho}) \geq \ell(\mu) + \ell(\bar{\mu})$$

is clearly necessary for nonvanishing of (22). The unique nonzero coefficient saturating both bounds corresponds to

$$\rho = 2\mu, \quad \bar{\rho} = 2\bar{\mu}.$$

Moreover, when divided by the norm squared of the vector $|(\rho, 2^{d-|\rho|/2-|\bar{\rho}|/2}); \bar{\rho}\rangle$, this coefficient is independent of d.

2.2.12

For general ρ and $\bar{\rho}$, the similarly normalized coefficient will be a polynomial in d of degree

$$\frac{1}{2}(\operatorname{wt}\mu + |\bar{\mu}| - \operatorname{wt}\rho/2 - |\bar{\rho}|/2) \tag{25}$$

because so many operators α_{-2} can commute with $\mathcal{P}_{\mu_i}$s or $\bar{\mathcal{P}}_{\bar{\mu}_i}$s instead of commuting directly with α_2s.

By induction on weight and length, we can express the basis vectors (16) in terms of (15) with $\mu_i \neq 1$ and coefficients being polynomial in d of degree at most minus the difference (25). By (23), to have d-dependent coefficients and $\mu_i \neq 1$ is the same as to allow $\mu_i = 1$ and make the coefficients independent of d. The bound of degree in d ensures that this transition preserves weight. This concludes the proof of Theorem 2.

3 Proof of Theorem 3

3.1 The pillowcase operator

3.1.1

Consider the operator

$$\mathfrak{W} = \exp\left(\sum_{n>0} \frac{\alpha_{-2n-1}}{2n+1}\right) \exp\left(-\sum_{n>0} \frac{\alpha_{2n+1}}{2n+1}\right). \tag{26}$$

Because this operator is normally ordered, its matrix elements $(\mathfrak{W}v, w)$ are well defined for any vectors v and w of finite energy. The relevance of this operator for our purposes lies in the following.

Theorem 4. *The diagonal matrix elements of $\mathfrak{W}$ are*

$$(\mathfrak{W}v_\lambda, v_\lambda) = \begin{cases} \mathsf{w}(\lambda), & \lambda \text{ is balanced,} \\ 0 & \text{otherwise.} \end{cases}$$

The proof of this theorem will occupy the rest of Section 3.1.

3.1.2

Let N be chosen so large that $\lambda_{2N+1} = 0$. Then because the operator (26) is a product of an upper unitriangular and lower unitriangular operator, the vectors $\underline{\lambda_i - i + \frac{1}{2}}$ with $i > 2N$ in

$$v_\lambda = \underline{\lambda_1 - \tfrac{1}{2}} \wedge \underline{\lambda_2 - \tfrac{3}{2}} \wedge \underline{\lambda_3 - \tfrac{5}{2}} \wedge \cdots$$

are inert bystanders for the evaluation of $(\mathfrak{W}v_\lambda, v_\lambda)$. The whole computation is therefore a computation of a matrix element of an operator in a finite exterior power of a finite dimensional vector space $V^{[N]}$ with basis

$$e_k = \underline{-2N + k + \tfrac{1}{2}}, \quad k = 0, \ldots, \lambda_1 + 2N - 1.$$

By definition, matrix elements of $\mathfrak{W}$ in exterior powers of $V^{[N]}$ are determinants of the matrix elements of $\mathfrak{W}$ acting on the space $V^{[N]}$ itself. The latter matrix elements are determined in the following.

Proposition 1. *We have*

$$\frac{(\mathfrak{W}e_k, e_l)}{\mathsf{b}(k)\mathsf{b}(l)} = \begin{cases} 1, & k \equiv l \equiv 0 \mod 2, \\ 0, & k \equiv l \equiv 1 \mod 2, \\ 2/(k-l), & \text{otherwise,} \end{cases} \tag{27}$$

where

$$\mathsf{b}(k) = \frac{k!}{2^k \lfloor k/2 \rfloor !^2}.$$

3.1.3

For the proof of Proposition 1, form the generating function

$$f(x, y) = \sum_{k,l} x^k y^l (\mathfrak{W} e_k, e_l).$$

From the equality

$$\exp\left(\sum_{n>0} \frac{x^{2n+1}}{2n+1}\right) = \sqrt{\frac{1+x}{1-x}}$$

and definitions, we compute

$$f(x, y) = \frac{1}{1-xy}\sqrt{\frac{1+x}{1-x}}\sqrt{\frac{1-y}{1+y}}.$$

The factorization

$$\left(x\frac{\partial}{\partial x} - y\frac{\partial}{\partial y}\right) f(x, y) = \frac{(x+y)(1+x)(1-y)}{(1-x^2)^{3/2}(1-y^2)^{3/2}}$$

by elementary binomial coefficient manipulations proves (27) for $k \neq l$. To compute the diagonal matrix elements observe that the above differential equation uniquely determines $f(x, y)$ from its values on the diagonal $x = y$. On the diagonal, the skew-symmetric terms in (27) cancel out and evaluation is immediate.

3.1.4

We now proceed to the computation of the matrix element $(\mathfrak{W} v_\lambda, v_\lambda)$. We have the following.

Proposition 2. *We have*

$$(\mathfrak{W} v_\lambda, v_\lambda) = \left(2^N \prod_{i=1}^{2N} \mathsf{b}(\lambda_i - i + 2N) \prod_{i<j\leq 2N} (\lambda_i - \lambda_j + j - i)^{(-1)^{\lambda_i - \lambda_j + j - i}}\right)^2,$$

provided λ *is balanced and* $(\mathfrak{W} v_\lambda, v_\lambda) = 0$ *otherwise.*

The proof of this proposition is the following. Observe that by Proposition 1 the matrix element $(\mathfrak{W} v_\lambda, v_\lambda)$ is a determinant of a $2N \times 2N$ block matrix in which the odd-odd block is identically zero, the even-even block is a rank 1 matrix with all elements equal to 1 and the off-diagonal blocks have the form $(\frac{2}{x_i - y_j})$, where $\{x_i\}$ and $\{y_i\}$ are the odd and even subsets of $\{\lambda_i - i + 2N\}$. Since the odd-odd block is identically zero, its size has to be $\leq N$ for the determinant to be nonvanishing. Similarly, if the size of the even-even block is larger than N, then the determinant is easily seen to vanish. It follows that both blocks have size N, which precisely means that the partition λ is balanced. It remains to use the Cauchy determinant

$$\det\left(\frac{1}{x_i+y_j}\right)=\frac{\prod_{i<j}(x_i-x_j)\prod_{i<j}(y_i-y_j)}{\prod(x_i+y_j)}$$

to finish the proof.

3.1.5

Note that decomposition of $\{\lambda_i - i + 2N\}$ into the even and odd subsets is the same as the 2-quotient construction from Section 2.1.6. Theorem 4 follows from formula (8) and the classical formula

$$\frac{\dim\lambda}{|\lambda|!}=\frac{\prod_{i<j\le N}(\lambda_i-\lambda_j+j-i)}{\prod(\lambda_i+N-i)!},$$

where N is any number such that $\lambda_{N+1}=0$.

3.1.6

It would be interesting to find an interpretation of the operator $\mathfrak{W}$ in conformal field theory. Note that

$$\exp\left(\sum_{n>0}\frac{z^{-2n-1}}{2n+1}\right)\exp\left(-\sum_{n>0}\frac{z^{2n+1}}{2n+1}\right)=\sqrt{\frac{1+z^{-1}}{1-z^{-1}}}\sqrt{\frac{1-z}{1+z}}$$

is the Wiener–Hopf factorization of the function taking the value $\mp i$ on the upper/lower half-plane.

3.2 Formula for the n-point function

3.2.1

Theorem 4 yields the following operator formula for the n-point function (11):

$$F(x_1,\ldots,x_n)=\frac{1}{Z(\emptyset,\emptyset;q)}\operatorname{tr} q^H\prod\mathcal{E}_0(\ln x_i)\mathfrak{W}, \tag{28}$$

where the trace is taken in the charge zero subspace of the infinite wedge and H is the energy operator

$$Hv_\lambda=|\lambda|v_\lambda.$$

We have the following expression for the operator $\mathcal{E}_0$ in terms of the fermionic currents:

$$\mathcal{E}_0(\ln x)=[y^0]\psi(xy)\psi^*(y),$$

where $[y^0]$ denotes the constant coefficient in the Laurent series expansion in the variable y. Therefore,

$$F(x_1, \ldots, x_n) = \frac{1}{Z(\emptyset, \emptyset; q)} \times [y_1^0 \cdots y_n^0] \operatorname{tr} q^H \psi(x_1 y_1)\psi^*(y_1) \cdots \psi(x_n y_n)\psi^*(y_n)\mathfrak{W}. \tag{29}$$

3.2.2

By the main result of [9], we have

$$\mathsf{w}(\lambda) \leq 1 \tag{30}$$

for any partition λ. In other words, all diagonal matrix elements of $\mathfrak{W}$ are bounded by 1. For the off-diagonal elements, we prove the following cruder bound.

Proposition 3. *Let* $M = \max\{|\lambda|, |\mu|\}$. *Then*

$$(\mathfrak{W} v_\lambda, v_\mu) \leq \exp\left(\frac{1}{2} \sum_{n=0}^{\lfloor \frac{M-1}{2} \rfloor} \frac{1}{2n+1}\right) \sim \text{const} \cdot M^{1/4}. \tag{31}$$

To see this note that

$$(\mathfrak{W} v_\lambda, v_\mu) = (\mathfrak{W}^{[M]} v_\lambda, v_\mu),$$

where $\mathfrak{W}^{[M]}$ is the truncated operator

$$\exp\left(\sum_{2n+1 \leq M} \frac{\alpha_{-2n-1}}{2n+1}\right) \exp\left(-\sum_{2n+1 \leq M} \frac{\alpha_{2n+1}}{2n+1}\right).$$

We claim that the operator $\mathfrak{W}^{[M]}$ is a multiple of a unitary operator. Indeed,

$$(\mathfrak{W}^{[M]*})^{-1} = \exp\left(-\sum_{n=0}^{\lfloor \frac{M-1}{2} \rfloor} \frac{1}{2n+1}\right) \mathfrak{W}^{[M]},$$

whence the result.

In fact, we will only use that (31) is bounded by a polynomial in the sizes of the partitions.

3.2.3

By normally ordering all fermionic operators in (29) and using the estimate (31) one sees that the trace converges if

$$|y_n/q| > |x_1 y_1| > |y_1| > \cdots > |x_n y_n| > |y_n| > 1. \tag{32}$$

3.2.4

The proof of the following identity is given in [12, Theorem 14.10]:

$$\psi(xy)\psi^*(y) = \frac{1}{x^{1/2}-x^{-1/2}} \times \exp\left(\sum_n \frac{(xy)^n - y^n}{n}\alpha_{-n}\right) \exp\left(\sum_n \frac{y^{-n}-(xy)^{-n}}{n}\alpha_n\right). \tag{33}$$

It allows to express the operator in (29) in terms of bosonic operators α_n.

With respect to the action of the operators α_n, the charge zero subspace of the infinite wedge space decomposes as the infinite tensor product

$$\Lambda^{\frac{\infty}{2}}_0 V \cong \bigotimes_{n=1}^{\infty} \bigoplus_{k=0}^{\infty} \alpha_{-n}^k v_\emptyset,$$

the distinguished vector in each factor being $v_\emptyset$. This gives a factorization of the trace in (29). The trace in each tensor factor is computed as follows:

$$\operatorname{tr} e^{A\alpha_{-n}} e^{B\alpha_n}\Big|_{\bigoplus_{k=0}^{\infty}\alpha_{-n}^k v_\emptyset} = \frac{1}{1-q^n}\exp\left(\frac{nABq^n}{1-q^n}\right).$$

For example, this shows that

$$\operatorname{tr} q^H \mathfrak{W} = (q^2)_\infty^{-1/2} = Z(\emptyset,\emptyset;q),$$

where

$$(a)_\infty = \prod_{n\ge 0}(1-aq^n),$$

and so the 0-point function is $F(\,) = 1$, as expected. For the n-point function this gives the following.

Theorem 5. *We have*

$$F(x_1,\dots,x_n) = \prod \frac{1}{\vartheta(x_i)} \times [y_1^0\cdots y_n^0] \prod_{i<j} \frac{\vartheta(y_i/y_j)\vartheta(x_iy_i/x_jy_j)}{\vartheta(x_iy_i/y_j)\vartheta(y_i/x_jy_j)} \prod_i \sqrt{\frac{\vartheta(-y_i)\vartheta(x_iy_i)}{\vartheta(y_i)\vartheta(-x_iy_i)}}, \tag{34}$$

where the series expansion is performed in the domain (32).

3.3 Quasimodular forms

3.3.1

In the computation of (34), we can assume that $1 < |x_i| \ll |q^{-1}|$ for all i and hence

$$|y_i| > |y_j| \prod |x_k|^{\pm 1} > |q y_i|, \quad i < j.$$

The series expansion in (34) can then be performed using the following elementary lemma.

Lemma 4. *We have*

$$\frac{1}{2\pi i} \oint_{|y|=c} \frac{dy}{y} \prod_{i=1}^{n} \frac{\vartheta(y/a_i)}{\vartheta(y/b_i)} = \left(1 - \prod \frac{a_i}{b_i}\right)^{-1} \sum_{i=1}^{n} \frac{\prod_j \vartheta(b_i/a_j)}{\prod_{j \neq i} \vartheta(b_i/b_j)}, \tag{35}$$

provided $c > |b_i| > |q|c$ *for* $i = 1, \ldots, n$.

This is obtained by computing the difference of $\oint_{|y|=c}$ and $\oint_{|y|=|q|c}$ as a sum of residues using

$$\vartheta'(1) = 1.$$

3.3.2

There are two obstacles to literally applying this lemma to the evaluation of (34). The first is the square roots in (34). However, we are ultimately interested in the expansion of (34) about $x_i = \pm 1$. The expansion of the integrand about $x_i = \pm 1$ contains no square roots, only the theta function and its derivatives. Formulas for integrating derivatives can be obtained from (35) by differentiating with respect to parameters.

3.3.3

The other issue is that at $x_i = 1$ the integrand is an elliptic function of the corresponding y_i, and so the left hand side of (35) gives infinity times zero. This can be circumvented, for example, by replacing each factor of x_i in the argument of each theta function an independent variable and specializing them all back to x_i only after integration. By l'Hôpital's rule, this will produce an additional differentiation any time we expand around $x_i = 1$ for some i.

3.3.4

In the end, we will get some rather complicated polynomial in theta functions and their derivatives evaluated at ± 1 divided by a power of $\vartheta(-1)$. This means that we will get a combination of Eisenstein series arising from

$$\ln \frac{z}{\vartheta(e^z)} = 2 \sum_{k \geq 1} \frac{z^{2k}}{(2k)!} E_{2k}(q), \tag{36}$$

and

$$\ln \frac{\vartheta(-e^z)}{\vartheta(-1)} = 2 \sum_{k \geq 1} \frac{z^{2k}}{(2k)!} [E_{2k}(q) - 2^{2k} E_{2k}(q^2)], \tag{37}$$

together with the product

$$\vartheta(-1) = 2i \left(\prod_n \frac{1+q^n}{1-q^n} \right)^2 = \frac{\eta(q^2)^2}{\eta(q)^4}. \tag{38}$$

Note that (38) has weight -1.

3.3.5

Without knowing the precise form of the answer, one can still make some qualitative observations about it.

Suppose we are interested in the coefficient of $z_1^{k_1} \cdots z_n^{k_n}$ in the expansion of

$$F(e^{z_1}, \dots, e^{z_r}, -e^{z_{r+1}}, \dots, -e^{z_n})$$

in powers of z_i. We claim that the weight of this coefficient is at most $\sum k_i + r$. Indeed, we from (36) and (37) we have

$$\operatorname{wt} \left(x \frac{d}{dx} \right)^k \vartheta(x) \Big|_{x=\pm 1} = k - 1.$$

This gives the following count for the weight:

$$n - n + \sum k_i + r,$$

where the first n is added because of the prefactor in (34), the second n is subtracted due to integration in y_i (which, by Lemma 4 changes the balance of θ-factors by 1), $\sum k_i$ is the number of times we need to differentiate the integrand, and, finally, r additional differentiations are needed for reasons explained in Section 3.3.3.

3.3.6

We further claim that (34) is, in fact, a polynomial in the coefficients of (36), (37), and

$$\frac{1}{\vartheta(-1)^2} = -\frac{1}{4} \frac{\eta(q)^8}{\eta(q^2)^4} = 2E_2(q) - 12E_2(q^2) + 16E_2(q^4). \tag{39}$$

First, observe only even powers of (38) appear in the answer. This is because the formula (34) has a balance of minus signs in the arguments of theta functions in the numerator and denominator. Every time we specialize y_i to one of the poles in (35), the balance of minus signs changes by an even number.

Inverse powers of (39) cannot appear in the answer because they grow exponentially as $q \to 1$ and there are no other exponentially large terms to cancel this growth out. The averages (10) may grow only polynomially as $q \to 1$ because of the bound (30).

3.3.7

Recall from [16] that a *quasimodular form* for a congruence subgroup $\Gamma \subset SL_2(\mathbb{Z})$ is, by definition, the holomorphic part of an almost holomorphic modular form for Γ. A function of $|q| < 1$ is called almost holomorphic if it is a polynomial in $(\ln|q|)^{-1}$ with coefficients in holomorphic functions of q. Quasimodular forms for Γ form a graded algebra denoted by $\mathcal{QM}(\Gamma)$. By a theorem of Kaneko and Zagier [13],

$$\mathcal{QM}(\Gamma) = \mathbb{Q}[E_2] \otimes \mathcal{M}(\Gamma).$$

In particular,

$$E_2(q), E_2(q^2), E_2(q^4) \in \mathcal{QM}(\Gamma_0(4)) \tag{40}$$

where

$$\Gamma_0(4) = \left\{ \begin{pmatrix} a & b \\ c & d \end{pmatrix} \middle| \, c \equiv 0 \bmod 4 \right\} \subset SL_2(\mathbb{Z}).$$

Hence all averages (10) lie in $\mathcal{QM}(\Gamma_0(4))$.

In fact, the series (40) generate the subalgebra $\mathcal{QM}_{2*}(\Gamma_0(4))$ of even weight quasimodular forms. This is because $\mathcal{M}_{2*}(\Gamma_0(4))$ is freely generated by two generators of weight two, for example, by $E_2^{\text{odd}}(q)$ and $E_2^{\text{odd}}(q^2)$, where

$$E_2^{\text{odd}}(q) = E_2(q) - 2E_2(q^2) = \frac{1}{24} + \sum_{n=1}^{\infty} \left(\sum_{d|n, d \text{ odd}} d \right) q^n.$$

3.3.8

Note that because $\mathsf{w}(\lambda) = 0$ for any partition λ of odd size, the series (10) is in fact a series in q^2. It follows that it is quasimodular with respect to a bigger group, namely

$$\begin{pmatrix} 1 & 0 \\ 0 & 2 \end{pmatrix} \Gamma_0(2) \begin{pmatrix} 1 & 0 \\ 0 & 2 \end{pmatrix}^{-1} \supset \Gamma_0(4).$$

In other words, (10) is, in fact, obtained by substituting $q \mapsto q^2$ into an element of $\mathcal{QM}(\Gamma_0(2))$. We have

$$\mathcal{M}(\Gamma_0(2)) = \mathbb{Q}[E_2^{\text{odd}}(q), E_4(q^2)]$$

and hence

$$\mathcal{QM}(\Gamma_0(2)) = \mathbb{Q}[E_2(q), E_2(q^2), E_4(q^2)].$$

This concludes the proof of Theorem 3.

Appendix A: Examples

In this appendix, we list some simple examples of the quasimodular forms $Z'(\mu, \nu; q)$ appearing in Theorem 1 and polynomials $\mathbf{g}_\nu$ from Theorem 2.

A.1 Quasimodular forms $Z'(\mu, \nu; q)$

$$Z'((1,1),(2)) = 20E_2(q^4)^2 - 20E_2(q^4)E_2(q^2) + 4E_2(q^2)^2 - \frac{5}{3}E_4(q^4).$$

$$\begin{aligned}Z'((3,1),(3)) = &-\frac{2112}{5}E_2(q^4)^3 + \frac{3888}{5}E_2(q^4)^2E_2(q^2)\\ &-\frac{2304}{5}E_2(q^4)E_2(q^2)^2 + \frac{384}{5}E_2(q^2)^3\\ &+48E_4(q^4)E_2(q^4) - 36E_4(q^4)E_2(q^2).\end{aligned}$$

$$\begin{aligned}Z'((3,3),(2)) = &\frac{1056}{5}E_2(q^4)^3 - \frac{1044}{5}E_2(q^4)^2E_2(q^2)\\ &+\frac{252}{5}E_2(q^4)E_2(q^2)^2 - \frac{12}{5}E_2(q^2)^3 - 24E_4(q^4)E_2(q^4)\\ &+3E_4(q^4)E_2(q^2) + \frac{15}{2}E_2(q^4)^2 - \frac{15}{2}E_2(q^4)E_2(q^2)\\ &+\frac{3}{2}E_2(q^2)^2 - \frac{5}{8}E_4(q^4).\end{aligned}$$

$$\begin{aligned}Z'((5,1),(2)) = &\frac{3520}{3}E_2(q^4)^3 - 1160E_2(q^4)^2E_2(q^2) + 280E_2(q^4)E_2(q^2)^2\\ &-\frac{40}{3}E_2(q^2)^3 - \frac{400}{3}E_4(q^4)E_2(q^4) + \frac{50}{3}E_4(q^4)E_2(q^2)\\ &+\frac{125}{3}E_2(q^4)^2 - \frac{125}{3}E_2(q^4)E_2(q^2) + \frac{25}{3}E_2(q^2)^2\\ &-\frac{125}{36}E_4(q^4).\end{aligned}$$

$$Z'((1,1,1,1),\emptyset) = \frac{1}{4}E_2(q^4) + \frac{1}{96}.$$

$$\begin{aligned}Z'((3,3,1,1),\emptyset) = &\frac{9}{256} - 12E_2(q^4)^2 + \frac{27}{2}E_2(q^4)E_2(q^2) - \frac{9}{4}E_2(q^2)^2\\ &+\frac{5}{4}E_4(q^4) + \frac{9}{16}E_2(q^4) + \frac{3}{8}E_2(q^2).\end{aligned}$$

$$\begin{aligned}Z'((5,1,1,1),\emptyset) = &\frac{125}{1152} - 10E_2(q^4)^2 + 15E_2(q^4)E_2(q^2) - \frac{5}{2}E_2(q^2)^2\\ &+\frac{55}{24}E_2(q^4) + \frac{5}{12}E_2(q^2).\end{aligned}$$

$$\begin{aligned}Z'((3,3,3,3),\emptyset) = &-\frac{24}{5}E_2(q^4)^3 - \frac{84}{5}E_2(q^4)^2E_2(q^2) + \frac{423}{20}E_2(q^4)E_2(q^2)^2\\ &-\frac{39}{10}E_2(q^2)^3 + E_4(q^4)E_2(q^4) + \frac{7}{4}E_4(q^4)E_2(q^2)\end{aligned}$$

$$
\begin{aligned}
&- \frac{33}{4}E_2(q^4)^2 + \frac{141}{16}E_2(q^4)E_2(q^2) - \frac{21}{32}E_2(q^2)^2 \\
&+ \frac{25}{32}E_4(q^4) + \frac{27}{256}E_2(q^4) + \frac{9}{32}E_2(q^2) + \frac{27}{2048}. \\
Z'((5,3,3,1),\emptyset) = {}& 132E_2(q^4)^3 - 708E_2(q^4)^2E_2(q^2) + 639E_2(q^4)E_2(q^2)^2 \\
&- 114E_2(q^2)^3 - 15E_4(q^4)E_2(q^4) + 55E_4(q^4)E_2(q^2) \\
&- 310E_2(q^4)^2 + \frac{1365}{4}E_2(q^4)E_2(q^2) - \frac{285}{8}E_2(q^2)^2 \\
&+ \frac{175}{6}E_4(q^4) + \frac{615}{64}E_2(q^4) + \frac{85}{8}E_2(q^2) + \frac{375}{512}.
\end{aligned}
$$

A.2 Polynomials $\mathbf{g}_\nu$

$$
\begin{aligned}
\mathbf{g}_{1,1} &= \frac{1}{2}\bar{\mathbf{p}}_1. \\
\mathbf{g}_{3,1} &= \frac{1}{6}\bar{\mathbf{p}}_1^2 + \frac{1}{6}\bar{\mathbf{p}}_2 - \frac{1}{2}\mathbf{p}_1. \\
\mathbf{g}_{3,3} &= -\frac{1}{54}\bar{\mathbf{p}}_1^3 + \frac{1}{18}\bar{\mathbf{p}}_1\bar{\mathbf{p}}_2 + \frac{1}{54}\bar{\mathbf{p}}_3 - \frac{1}{4}\mathbf{p}_2 + \frac{3}{16}\bar{\mathbf{p}}_1. \\
\mathbf{g}_{5,1} &= \frac{1}{30}\bar{\mathbf{p}}_1^3 + \frac{1}{10}\bar{\mathbf{p}}_1\bar{\mathbf{p}}_2 - \frac{1}{2}\bar{\mathbf{p}}_1\mathbf{p}_1 + \frac{1}{15}\bar{\mathbf{p}}_3 - \frac{1}{2}\mathbf{p}_2 + \frac{25}{24}\bar{\mathbf{p}}_1. \\
\mathbf{g}_{5,3} &= -\frac{1}{360}\bar{\mathbf{p}}_1^4 - \frac{1}{60}\bar{\mathbf{p}}_1^2\bar{\mathbf{p}}_2 - \frac{1}{12}\bar{\mathbf{p}}_1^2\mathbf{p}_1 + \frac{2}{45}\bar{\mathbf{p}}_3\bar{\mathbf{p}}_1 + \frac{25}{36}\bar{\mathbf{p}}_1^2 + \frac{1}{40}\bar{\mathbf{p}}_2^2 \\
&\quad - \frac{1}{12}\bar{\mathbf{p}}_2\mathbf{p}_1 + \frac{5}{8}\mathbf{p}_1^2 + \frac{1}{60}\bar{\mathbf{p}}_4 - \frac{1}{2}\mathbf{p}_3 + \frac{25}{36}\bar{\mathbf{p}}_2 - \frac{25}{12}\mathbf{p}_1. \\
\mathbf{g}_{1,1,1,1} &= -\frac{1}{24}\bar{\mathbf{p}}_1^2 + \frac{1}{12}\bar{\mathbf{p}}_2 + \frac{1}{96}. \\
\mathbf{g}_{3,1,1,1} &= \frac{1}{108}\bar{\mathbf{p}}_1^3 - \frac{1}{36}\bar{\mathbf{p}}_1\bar{\mathbf{p}}_2 - \frac{1}{4}\bar{\mathbf{p}}_1\mathbf{p}_1 + \frac{2}{27}\bar{\mathbf{p}}_3 + \frac{3}{8}\bar{\mathbf{p}}_1. \\
\mathbf{g}_{3,3,1,1} &= \frac{1}{216}\bar{\mathbf{p}}_1^4 - \frac{1}{12}\bar{\mathbf{p}}_1^2\mathbf{p}_1 + \frac{1}{108}\bar{\mathbf{p}}_3\bar{\mathbf{p}}_1 - \frac{1}{8}\mathbf{p}_2\bar{\mathbf{p}}_1 + \frac{9}{32}\bar{\mathbf{p}}_1^2 - \frac{1}{72}\bar{\mathbf{p}}_2^2 \\
&\quad - \frac{1}{12}\bar{\mathbf{p}}_2\mathbf{p}_1 + \frac{1}{8}\mathbf{p}_1^2 + \frac{1}{36}\bar{\mathbf{p}}_4 + \frac{9}{16}\bar{\mathbf{p}}_2 - \frac{3}{4}\mathbf{p}_1 + \frac{9}{256}. \\
\mathbf{g}_{3,3,3,1} &= \frac{1}{4860}\bar{\mathbf{p}}_1^5 + \frac{1}{486}\bar{\mathbf{p}}_1^3\bar{\mathbf{p}}_2 + \frac{1}{108}\bar{\mathbf{p}}_1^3\mathbf{p}_1 - \frac{5}{972}\bar{\mathbf{p}}_3\bar{\mathbf{p}}_1^2 - \frac{1}{24}\mathbf{p}_2\bar{\mathbf{p}}_1^2 - \frac{1}{96}\bar{\mathbf{p}}_1^3 \\
&\quad + \frac{1}{324}\bar{\mathbf{p}}_1\bar{\mathbf{p}}_2^2 - \frac{1}{36}\bar{\mathbf{p}}_1\bar{\mathbf{p}}_2\mathbf{p}_1 + \frac{1}{162}\bar{\mathbf{p}}_4\bar{\mathbf{p}}_1 - \frac{5}{972}\bar{\mathbf{p}}_3\bar{\mathbf{p}}_2 - \frac{1}{108}\bar{\mathbf{p}}_3\mathbf{p}_1 - \frac{1}{24}\mathbf{p}_2\bar{\mathbf{p}}_2 \\
&\quad + \frac{1}{8}\mathbf{p}_2\mathbf{p}_1 + \frac{31}{96}\bar{\mathbf{p}}_1\bar{\mathbf{p}}_2 - \frac{19}{32}\bar{\mathbf{p}}_1\mathbf{p}_1 + \frac{1}{4}\bar{\mathbf{p}}_3 - \mathbf{p}_2 + \frac{2}{405}\bar{\mathbf{p}}_5 + \frac{153}{128}\bar{\mathbf{p}}_1. \\
\mathbf{g}_{3,3,3,3} &= \frac{1}{29160}\bar{\mathbf{p}}_1^6 - \frac{1}{2916}\bar{\mathbf{p}}_3\bar{\mathbf{p}}_1^3 + \frac{1}{216}\mathbf{p}_2\bar{\mathbf{p}}_1^3 - \frac{1}{432}\bar{\mathbf{p}}_1^4 + \frac{1}{1944}\bar{\mathbf{p}}_1^2\bar{\mathbf{p}}_2^2 - \frac{1}{972}\bar{\mathbf{p}}_4\bar{\mathbf{p}}_1^2 \\
&\quad + \frac{1}{972}\bar{\mathbf{p}}_3\bar{\mathbf{p}}_1\bar{\mathbf{p}}_2 - \frac{1}{72}\mathbf{p}_2\bar{\mathbf{p}}_1\bar{\mathbf{p}}_2 - \frac{7}{288}\bar{\mathbf{p}}_1^2\bar{\mathbf{p}}_2 - \frac{1}{12}\bar{\mathbf{p}}_1^2\mathbf{p}_1 - \frac{1}{2916}\bar{\mathbf{p}}_2^3
\end{aligned}
$$

$$- \frac{1}{1944}\bar{\mathbf{p}}_3^2 - \frac{1}{216}\bar{\mathbf{p}}_3\mathbf{p}_2 + \frac{59}{864}\bar{\mathbf{p}}_3\bar{\mathbf{p}}_1 + \frac{1}{32}\mathbf{p}_2{}^2 - \frac{3}{64}\mathbf{p}_2\bar{\mathbf{p}}_1 + \frac{1}{1215}\bar{\mathbf{p}}_5\bar{\mathbf{p}}_1$$
$$+ \frac{231}{512}\bar{\mathbf{p}}_1^2 + \frac{1}{32}\bar{\mathbf{p}}_2^2 - \frac{1}{12}\bar{\mathbf{p}}_2\mathbf{p}_1 + \frac{3}{8}\mathbf{p}_1{}^2 + \frac{5}{144}\bar{\mathbf{p}}_4 - \frac{5}{12}\mathbf{p}_3 + \frac{1}{2916}\bar{\mathbf{p}}_6$$
$$+ \frac{129}{256}\bar{\mathbf{p}}_2 - \frac{9}{8}\mathbf{p}_1 + \frac{27}{2048}.$$

Acknowledgments. We thank H. Masur, M. Mirzakhani, N. Nekrasov, R. Pandharipande, and A. Zorich for discussions and the anonymous referee for helpful suggestions.

The authors were partially supported by the NSF and the Packard Foundation.

References

[1] S. Bloch and A. Okounkov, The character of the infinite wedge representation, *Adv. Math.*, **149**-1 (2000), 1–60.
[2] R. Dijkgraaf, Mirror symmetry and elliptic curves, in R. Dijkgraaf, C. Faber, and G. van der Geer, eds., *The Moduli Space of Curves*, Progress in Mathematics, Vol. 129, Birkhäuser Boston, Cambridge, MA, 1995, 149–163.
[3] R. Dijkgraaf, Chiral deformations of conformal field theories, *Nuclear Phys.* B, **493**-3 (1997), 588–612.
[4] Ph. Di Francesco, C. Itzykson, and J.-B. Zuber, Polynomial averages in the Kontsevich model, *Comm. Math. Phys.*, **151**-1 (1993), 193–219.
[5] M. Douglas, Conformal field theory techniques in large *N* Yang-Mills theory, in *Quantum Field Theory and String Theory* (*Cargèse*, 1993), NATO Advanced Science Institutes Series B: Physics, Vol. 328, Plenum, New York, 1995, 119–135.
[6] A. Eskin, H. Masur, and A. Zorich, Moduli spaces of abelian differentials: The principal boundary, counting problems, and the Siegel-Veech constants, *Publ. Math. Inst. Hautes Études Sci.*, **97** (2003), 61–179.
[7] A. Eskin and A. Okounkov, Asymptotics of numbers of branched coverings of a torus and volumes of moduli spaces of holomorphic differentials, *Invent. Math.*, **145**-1 (2001), 59–103.
[8] A. Eskin, A. Okounkov, and R. Pandharipande, The theta characteristic of a branched covering, math.AG/0312186, 2003.
[9] S. Fomin and N. Lulov, On the number of rim hook tableaux, *Zap. Nauchn. Sem. S.-Peterburg. Otdel. Mat. Inst. Steklov* (*POMI*), **223** (1995), 219–226 (in Russian); *J. Math. Sci.*, **87**-6 (1997), 4118–4123 (in English).
[10] G. Jones, Characters and surfaces: A survey, in *The Atlas of Finite Groups: Ten years On* (*Birmingham*, 1995), London Mathematical Society Lecture Note Series, Vol. 249, Cambridge University Press, Cambridge, UK, 1998, 90–118.
[11] T. Józefiak, Symmetric functions in the Kontsevich-Witten intersection theory of the moduli space of curves, *Lett. Math. Phys.*, **33**-4 (1995), 347–351.
[12] V. Kac, *Infinite Dimensional Lie Algebras*, Cambridge University Press, Cambridge, UK, 1990, 1995.
[13] M. Kaneko and D. Zagier, A generalized Jacobi theta function and quasimodular forms, in *The Moduli Space of Curves* (*Texel Island*, 1994), Progress in Mathematics, Vol. 129, Birkhäuser Boston, Cambridge, MA, 1995, 165–172.

[14] S. Kerov and G. Olshanski, Polynomial functions on the set of Young diagrams, *C. R. Acad. Sci. Paris Sér.* I *Math.*, **319**-2 (1994), 121–126.

[15] M. Kontsevich, Intersection theory on the moduli space of curves and the matrix Airy function, *Comm. Math. Phys.*, **147** (1992), 1–23.

[16] M. Kontsevich and A. Zorich, Connected components of the moduli spaces of abelian differentials with prescribed singularities, *Invent. Math.*, **113**-3 (2003), 631–678.

[17] E. Lanneau, *Classification des composantes connexes des strates de l'espace des modules des différentielles quadratiques*, Ph.D. thesis, Université Rennes I, Rennes, France, 2003.

[18] H. Masur and A. Zorich, Multiple saddle connections on flat surfaces and principal boundary of the moduli spaces of quadratic differentials, math.GT/0402197, 2004.

[19] M. Mirzakhani, http://math.harvard.edu/~mirzak/.

[20] T. Miwa, M. Jimbo, and E. Date, *Solitons: Differential Equations, Symmetries and Infinite-Dimensional Algebras*, Cambridge Tracts in Mathematics, Vol. 135. Cambridge University Press, Cambridge, UK, 2000.

[21] A. Okounkov and R. Pandharipande, Gromov-Witten theory, Hurwitz theory, and completed cycles, math.AG/0204305, 2002.

[22] A. Okounkov and R. Pandharipande, in preparation.

Cluster $\mathcal{X}$-varieties, amalgamation, and Poisson–Lie groups

V. V. Fock* and A. B. Goncharov†

* Institute for Theoretical and Experimental Physics
B. Cheremushkinskaya 25
117259 Moscow
Russia
fock@math.brown.edu
† Department of Mathematics
Brown University
Box 1917
Providence, RI 02906
USA
sasha@math.brown.edu

To Vladimir Drinfeld for his 50*th birthday.*

Subject Classifications: Primary 22E46, 20G42. Secondary 05E15, 53D17.

1 Introduction

1.1 Summary

In this paper, starting from a split semisimple real Lie group G with trivial center, we define a family of varieties with additional structures. We describe them as *cluster $\mathcal{X}$-varieties*, as defined in [FG2]. In particular they are Poisson varieties. We define canonical Poisson maps of these varieties to the group G equipped with the standard Poisson–Lie structure defined by V. Drinfeld in [D, D1]. One of them maps to the group birationally and thus provides G with canonical rational coordinates.

We introduce a simple but important operation of *amalgamation* of cluster varieties. Our varieties are obtained as amalgamations of certain *elementary cluster varieties* $\mathcal{X}_{\mathbf{J}(\alpha)}$, assigned to positive simple roots α of the root system of G. An elementary cluster variety $\mathcal{X}_{\mathbf{J}(\alpha)}$ is a split algebraic torus of dimension $r+1$, where r is the rank of G. Its cluster, and in particular Poisson structure, is described in a very simple way by the Cartan matrix for G. Since one of them is a Zariski open part of G, we can develop the Poisson–Lie group structure on G from scratch, without the

r-matrix formalism, getting as a benefit canonical (Darboux) coordinates for the Poisson structure on G. Some of our varieties are very closely related to the double Bruhat cells studied by A. Berenstein, S. Fomin, and A. Zelevinisky in [FZ, BFZ3, BZq].

Using quantization of cluster $\mathcal{X}$-varieties developed in [FG2, Section 4] we get as a byproduct a quantization (i.e., a noncommutative q-deformation) of our varieties. The quantum version of the operation of amalgamation generalizes the standard quantum group structure.

The results of this paper enter as a building block into a description of the cluster structure of the moduli spaces of local systems on surfaces studied in [FG1].

1.2 Description of the results

We start the paper with a brief recollection of the definition and properties of cluster $\mathcal{X}$-varieties. Let us briefly discuss some of their features, postponing the detailed discussion until Section 2.

1.2.1 Cluster $\mathcal{X}$-varieties

Cluster $\mathcal{X}$-varieties are determined by combinatorial data similarly (although differently in some details) to that used for the definition of cluster algebras in [FZI], that is, by a *cluster seed* $\mathbf{I}$, which is a quadruple (I, I_0, ε, d), where

(i) I is a finite set;
(ii) $I_0 \subset I$ is a subset;
(iii) ε is a matrix (ε_{ij}), where $i, j \in I$, such that $\varepsilon_{ij} \in \mathbb{Z}$ unless $i, j \in I_0$;
(iv) $d = \{d_i\}$, where $i \in I$, is a set of positive integers, such that the matrix $(\widehat{\varepsilon}_{ij}) = (\varepsilon_{ij} d_j)$ is skew-symmetric.

The elements of the set I are called *vertices*, the elements of I_0 are called *frozen vertices*.

Given a seed $\mathbf{I}$, every nonfrozen vertex $k \in I - I_0$ gives rise to a *mutation*, producing a new, mutated seed $\mu_k(\mathbf{I})$. Compositions of mutations are called *cluster transformations of seeds*.

Following [FG2, Section 2], we associate to a seed $\mathbf{I}$ a torus $\mathcal{X}_{\mathbf{I}} = (\mathbb{G}_m)^I$ with a Poisson structure given by

$$\{x_i, x_j\} = \widehat{\varepsilon}_{ij} x_i x_j$$

where $\{x_i | i \in I\}$ are the standard coordinates on the factors. We shall call it the *seed $\mathcal{X}$-torus*. Cluster transformations of seeds give rise to Poisson birational transformations between the seed tori, called cluster transformations. Gluing the seed $\mathcal{X}$-tori according to these birational transformations we get a scheme $\mathcal{X}_{|\mathbf{I}|}$ over $\mathbb{Z}$, called below a *cluster $\mathcal{X}$-variety*. (However, $\mathcal{X}_{|\mathbf{I}|}$ may not be a scheme of finite type, and thus $\mathcal{X}_{|\mathbf{I}|} \otimes \mathbb{Q}$ may not be an algebraic variety.)

In [FZI], the values ε_{ij} for $i \in I_0$ were not defined, so (ε_{ij}) was a rectangular matrix with integral entries. In our approach the frozen variables play an important role. The values ε_{ij}, when $i, j \in I_0$, are essential, and may not be integers. Let us elaborate on this point.

1.2.2 Amalgamation

We introduce operations of *amalgamation* and *defrosting* of seeds. The amalgamation of a collection of seeds $\mathbf{I}(s)$, parametrised by a set S, is a new seed $\mathbf{K} = (K, K_0, \varepsilon_{ij}, d)$. The set K is defined by gluing some of the frozen vertices of the sets $I(s)$. The frozen subset K_0 is obtained by gluing the frozen subsets $I_0(s)$. The rest of the data of $\mathbf{K}$ is also inherited from those of $\mathbf{I}(s)$. Defrosting simply shrinks the subset of the frozen vertices of $\mathbf{K}$, without changing the set K. One can defrost any subset of K such that $\varepsilon_{ij} \in \mathbb{Z}$ for any i, j from this subset. All seeds in our paper are obtained by amalgamation followed by defrosting of certain elementary seeds. All vertices of the elementary seeds are frozen.

In order to state our results we have to introduce some notation related to the group.

1.2.3 Notation

Let G be a semisimple adjoint Lie group of rank r. There is the following data associated to G: the set of positive simple roots Π, the Cartan matrix $C_{\alpha\beta} = 2\frac{(\alpha,\beta)}{(\alpha,\alpha)}$, where $\alpha, \beta \in \Pi$, and the multipliers $d^\alpha = (\alpha, \alpha)/2, \alpha \in \Pi$, such that the matrix $\widehat{C}_{\alpha\beta} = C_{\alpha\beta}d^\beta$ is symmetric. Let Π^- be the set of negative simple roots and let $\mathfrak{W}$ be the semigroup freely generated by Π and Π^-. Any element D of $\mathfrak{W}$ is thus a word $\mu_1 \cdots \mu_{l(D)}$ in the letters from the alphabet $\Pi \cup \Pi^-$, where $l(D)$ is its length. For $\alpha \in \Pi$ we shall denote by $\bar{\alpha}$ the opposite element from Π^-.

1.2.4 The braid and Hecke semigroups

Let $\mathfrak{B}$ be the quotient of the semigroup $\mathfrak{W}$ by

$$\alpha\bar{\beta} = \bar{\beta}\alpha, \tag{1}$$

$$\begin{array}{rcll} \alpha\beta\alpha = \beta\alpha\beta & \text{and} & \bar{\alpha}\bar{\beta}\bar{\alpha} = \bar{\beta}\bar{\alpha}\bar{\beta} & \text{if } C_{\alpha\beta} = C_{\beta\alpha} = -1, \\ \alpha\beta\alpha\beta = \beta\alpha\beta\alpha & \text{and} & \bar{\alpha}\bar{\beta}\bar{\alpha}\bar{\beta} = \bar{\beta}\bar{\alpha}\bar{\beta}\bar{\alpha} & \text{if } C_{\alpha\beta} = 2C_{\beta\alpha} = -2, \\ \alpha\beta\alpha\beta\alpha\beta = \beta\alpha\beta\alpha\beta\alpha & \text{and} & \bar{\alpha}\bar{\beta}\bar{\alpha}\bar{\beta}\bar{\alpha}\bar{\beta} = \bar{\beta}\bar{\alpha}\bar{\beta}\bar{\alpha}\bar{\beta}\bar{\alpha} & \text{if } C_{\alpha\beta} = 3C_{\beta\alpha} = -3. \end{array} \tag{2}$$

The semigroup $\mathfrak{B}$ is called the *braid semigroup*. We denote by $p : \mathfrak{W} \to \mathfrak{B}$ the canonical projection.

Another semigroup appropriate in our context is the further quotient of $\mathfrak{B}$ by the relations

$$\alpha^2 = \alpha; \quad \bar{\alpha}^2 = \bar{\alpha}, \tag{3}$$

denoted by $\mathfrak{H}$ and called the *Hecke semigroup*. It is isomorphic as a set to the square of the Weyl group of G. We call an element of $\mathfrak{W}$ or $\mathfrak{B}$ *reduced* if its length is minimal among the elements having the same image in $\mathfrak{H}$.

Now we are ready to discuss our main goals and results.

1.2.5 Cluster $\mathcal{X}$-varieties related to the braid semigroup and their properties

We will define a cluster $\mathcal{X}$-variety $\mathcal{X}_B$ associated to any element $B \in \mathfrak{B}$. For this purpose, given a $D \in \mathfrak{W}$, we will define a seed $\mathbf{J}(D) = (J(D), J_0(D), \varepsilon(D), d(D))$. We prove that the seeds corresponding to different elements of $p^{-1}(B)$ are related by cluster transformations. Moreover, we define the evaluation and multiplication maps

$$\mathrm{ev} : \mathcal{X}_B \to G, \qquad m : \mathcal{X}_{B_1} \times \mathcal{X}_{B_2} \to \mathcal{X}_{B_1 B_2}.$$

The correspondence $B \mapsto \mathcal{X}_B$ and the maps m and ev are to satisfy the following properties:

1. ev is a Poisson map.
2. m is an amalgamation followed by defrosting of cluster $\mathcal{X}$-varieties (and thus a Poisson map).
3. The multiplication maps are associative in the obvious sense.
4. Multiplication commutes with the evaluation, i.e., the following diagram is commutative:

$$\begin{array}{ccc} \mathcal{X}_{B_1} \times \mathcal{X}_{B_2} & \stackrel{(\mathrm{ev},\mathrm{ev})}{\longrightarrow} & G \times G, \\ \downarrow m & & \downarrow m, \\ \mathcal{X}_{B_1 B_2} & \stackrel{\mathrm{ev}}{\longrightarrow} & G. \end{array} \tag{4}$$

We would like to stress that the multiplication m is a projection with fibers of nonzero dimension.

1.2.6 Cluster $\mathcal{X}$-varieties related to the Hecke semigroup

Let $\pi : \mathfrak{B} \to \mathfrak{H}$ be the canonical projection of semigroups. Considered as a projection of sets it has a canonical splitting $s : \mathfrak{H} \to \mathfrak{B}$. Namely, for every $H \in \mathfrak{H}$ there is a unique reduced element $s(H)$ in $\pi^{-1}(H)$, the *reduced representative* of H in $\mathfrak{B}$. So given an element $H \in \mathfrak{H}$ there is a cluster variety $\mathcal{X}_{s(H)}$. Abusing notation, we will denote it by $\mathcal{X}_H$.

A rational map of cluster $\mathcal{X}$-varieties is a *cluster projection* if in a certain cluster coordinate system it is obtained by forgetting one or more cluster coordinates.

We show the following:

1. There is a canonical cluster projection $\pi : \mathcal{X}_B \to \mathcal{X}_{\pi(B)}$. By the very definition, it is an isomorphism if B is reduced.
2. There is a multiplication map $m_{\mathcal{H}} : \mathcal{X}_{H_1} \times \mathcal{X}_{H_2} \to \mathcal{X}_{H_1 H_2}$, defined as the composition

$$\mathcal{X}_{H_1} \times \mathcal{X}_{H_2} := \mathcal{X}_{s(H_1)} \times \mathcal{X}_{s(H_2)} \stackrel{m}{\longrightarrow} \mathcal{X}_{s(H_1)s(H_2)} \stackrel{\pi}{\longrightarrow} \mathcal{X}_{H_1 H_2}.$$

   So it is a composition of an amalgamation, defrosting, and cluster projection. It follows from (4) that the maps $m_{\mathcal{H}}$ and m are related by a commutative diagram

$$\begin{array}{ccc} \mathcal{X}_{B_1} \times \mathcal{X}_{B_2} & \stackrel{m}{\longrightarrow} & \mathcal{X}_{B_1 B_2}, \\ \downarrow \pi \times \pi & & \downarrow \pi, \\ \mathcal{X}_{H_1} \times \mathcal{X}_{H_2} & \stackrel{m_{\mathcal{H}}}{\longrightarrow} & \mathcal{X}_{H_1 H_2}, \end{array} \tag{5}$$

where the vertical maps are the canonical cluster projections.

3. The multiplication maps are associative in the obvious sense (see the remark below).
4. If $H \in \mathfrak{H}$, the map ev : $\mathcal{X}_H \hookrightarrow G$ is injective at the generic point. If H is the longest element of $\mathfrak{H}$, then the image of ev is Zariski dense in G. The map ev : $\mathcal{X}_B \to G$ is a composition

$$\mathcal{X}_B \stackrel{\pi}{\longrightarrow} \mathcal{X}_H \stackrel{\mathrm{ev}_{\mathcal{H}}}{\hookrightarrow} G, \quad H = \pi(B).$$

Remark 1. A part of the above data is axiomatized as follows. Given a semigroup $\mathfrak{S}$, we assign to every $s \in \mathfrak{S}$ an object $\mathcal{X}_s$ of a monoidal category $\mathcal{M}$ (e.g., the category of Poisson varieties with the product as monoidal structure), and for every pair $s, t \in \mathfrak{S}$ a canonical morphism $m_{s,t} : \mathcal{X}_s \times \mathcal{X}_t \longrightarrow \mathcal{X}_{st}$. They must satisfy an associativity constraint, i.e., for every $r, s, t \in \mathfrak{S}$, the following diagram is commutative:

$$\begin{array}{ccc} \mathcal{X}_r \times \mathcal{X}_s \times \mathcal{X}_t & \stackrel{\mathrm{Id} \times m_{s,t}}{\longrightarrow} & \mathcal{X}_r \times \mathcal{X}_{st}, \\ m_{r,s} \downarrow \times \mathrm{Id} & & \downarrow m_{r,st}, \\ \mathcal{X}_{rs} \times \mathcal{X}_t & \stackrel{m_{rs,t}}{\longrightarrow} & \mathcal{X}_{rst}. \end{array} \tag{6}$$

Remark 2. Recall Lusztig's coordinates on the group G [L1]. Let $E^{\alpha}(t)$ and $F^{\alpha}(t)$ be the two standard one-parameter subgroups corresponding to a simple root α. Denote by $X^{\alpha}(t)$ the element $E^{\alpha}(t)$ if $\alpha \in \Pi$ and $F^{\alpha}(t)$ if $\alpha \in \Pi^-$. A reduced decomposition of the longest element in $W \times W$ is encoded by a sequence $\alpha_1, \ldots, \alpha_{2m}$ of $2m$ elements of $\Pi \cup \Pi^-$, where $2m + r = \dim G$. There is a birational isomorphism

$$H \times \mathbb{G}_m^{2m} \longrightarrow G, \quad (H, t_1, \ldots, t_{2m}) \longmapsto H X^{\alpha_1}(t_1), \ldots, X^{\alpha_{2m}}(t_{2m}). \tag{7}$$

There are similar coordinates on all double Bruhat cells [FZ]. However, they are not cluster $\mathcal{X}$-coordinates. Our coordinates are related to them by monomial transformations.

1.2.7 Quantization

Cluster $\mathcal{X}$-varieties were quantized in [FG2, Section 4]. The operations of amalgamation, defrosting and cluster projection have straightforward generalizations to the quantum $\mathcal{X}$-varieties. Thus we immediately get q-deformations of the cluster $\mathcal{X}$-varieties considered above for the braid and Hecke semigroups.

We understood the category of quantum spaces as in loc. cit. So, in particular, a morphism of quantum cluster spaces $\mathcal{X}^q \to \mathcal{Y}^q$, by definition, is given by a compatible collection of morphisms of the corresponding quantum tori algebras going in the opposite direction, i.e., the $\mathcal{Y}$-algebras map to the corresponding $\mathcal{X}$-algebras.

It follows that the quantum spaces enjoy properties similar to the properties of their classical counterparts listed above:

1. There is a canonical projection $\pi : \mathcal{X}_B \to \mathcal{X}_{\pi(B)}$.
2. There are multiplication maps of quantum spaces for the braid and Hecke semigroups:

$$m^q : \mathcal{X}^q_{B_1} \times \mathcal{X}^q_{B_2} \to \mathcal{X}^q_{B_1 B_2}, \qquad m^q_{\mathcal{H}} : \mathcal{X}^q_{H_1} \times \mathcal{X}^q_{H_2} \to \mathcal{X}^q_{H_1 H_2}.$$

3. They are related by the q-version of the diagram (5), and satisfy the associativity constraints given by the q-versions of the diagram (6).

Remark. One can show that there is a quantum evaluation map to the quantum deformation [D1] of the algebra of regular functions of G. Unlike the other properties, this is not completely straightforward, and will be elaborated elsewhere.

1.2.8 Proofs

They are easy if the Dynkin diagram of G is simply-laced. The other cases are reduced to the rank two cases. The B_2 case can be done by unenlightening calculations. The G_2 case is considerably more difficult.

A more conceptual approach is provided by the operation of *cluster folding* [FG2], briefly reviewed in Section 3.6, which clarifies the picture in the B_2 case and seems to be indispensable in the G_2 case.

1.2.9 Cluster structures of moduli spaces of triples of flags of types A_3 and G_2

In the process of proof we work with cluster $\mathcal{X}$-varieties related to the moduli spaces $\mathrm{Conf}_3(\mathcal{B}_{B_2})$ and $\mathrm{Conf}_3(\mathcal{B}_{G_2})$ of configurations of triples of flags in the Lie groups of type B_2 and G_2, respectively. There is a canonical embedding $\mathrm{Conf}_3(\mathcal{B}_{B_2}) \hookrightarrow \mathrm{Conf}_3(\mathcal{B}_{A_3})$ provided by the folding of the latter. It allows us to reduce the study of the former to the study of the latter. In the two appendices we investigate the cluster structures of the moduli spaces $\mathrm{Conf}_3(\mathcal{B}_{A_3})$ and $\mathrm{Conf}_3(\mathcal{B}_{G_2})$ in detail. Here is what we learned.

Recall the moduli space $\mathcal{M}_{0,6}$ of configurations of 6 points on $\mathbb{P}^1$. It has an $\mathcal{X}$-cluster structure of finite type A_3. In Appendix B we construct an explicit birational isomorphism

$$\Phi : \mathrm{Conf}_3(\mathcal{B}_{A_3}) \stackrel{\sim}{\longrightarrow} \mathcal{M}_{0,6}$$

respecting the $\mathcal{X}$-cluster structures.

The investigation of the moduli space $\mathrm{Conf}_3(\mathcal{B}_{G_2})$ turned out to be a subject of independent interest, which reveals the following story, discussed in Appendix A.

We say that two seeds $\mathbf{I} = (I, I_0, \varepsilon, d)$ and $\mathbf{I}' = (I', I'_0, \varepsilon', d')$ are isomorphic if there is a set isomorphism $\varphi : I \to I'$, preserving the frozen vertices and the ε- and d-functions.

Definition 1.1. A cluster $\mathcal{X}$-variety is of ε-*finite* type if the set of isomorphism classes of its seeds is finite.

Any cluster $\mathcal{X}$-variety gives rise to an orbifold, called the *modular orbifold*; see [FG2, Section 2]. We recall its definition in Section **??**. Its dimension is the dimension of the cluster $\mathcal{X}$-variety minus one. In the ε-finite case the modular orbifold is glued from a finite number of simplices. It is noncompact, unless the cluster $\mathcal{X}$-variety is of finite cluster type. Here is the main result:

Theorem 1.2.

(a) *The cluster $\mathcal{X}$-variety corresponding to the moduli space* $\mathrm{Conf}_3(\mathcal{B}_{G_2})$ *is of ε-finite type. The number of isomorphism classes of its seeds is seven.*
(b) *The corresponding modular orbifold is a manifold. It is homeomorphic to $S^3 - L$, where L is a two-component link, and $\pi_1(S^3 - L)$ is isomorphic to the braid group of type G_2.*

It is well known that the complement to the discriminant variety of type G_2 in $\mathbb{C}^2$ is a $K(\pi, 1)$-space, where π is the braid group of type G_2. Its intersection with a sphere S^3 containing the origin has two connected components. We conjecture that it is isomorphic to $S^3 - L$.

The *mapping class group* of a cluster $\mathcal{X}$-variety was defined in loc. cit. It acts by automorphisms of the cluster $\mathcal{X}$-variety. It is always infinite if the cluster structure is of ε-finite, but not of finite type. Theorem 1.2 immediately implies the following.

Corollary 1.3. *The mapping class group of the cluster $\mathcal{X}$-variety corresponding to* $\mathrm{Conf}_3(\mathcal{B}_{G_2})$ *is an infinite quotient of the braid group of type G_2.*

Conjecturally it coincides with the braid group. This is the first example of an infinite mapping class group different from the mapping class groups of surfaces.

2 Cluster $\mathcal{X}$-varieties and amalgamation

In this section we recall some definitions from [FG2]. For the reader's convinience, we repeat, verbatim, the definition of the cluster seed from Section 1.2.1.

2.1 Basic definitions

A *cluster seed*, or just *seed*, **I** is a quadruple (I, I_0, ε, d), where

(i) I is a finite set;
(ii) $I_0 \subset I$ is a subset;
(iii) ε is a matrix (ε_{ij}), where $i, j \in I$, such that $\varepsilon_{ij} \in \mathbb{Z}$ unless $i, j \in I_0$;
(iv) $d = \{d_i\}$, where $i \in I$, is a set of positive integers, such that the matrix $(\widehat{\varepsilon}_{ij}) = (\varepsilon_{ij} d_j)$ is skew-symmetric.

The elements of the set I are called *vertices*, the elements of I_0 are called *frozen vertices*. The matrix ε is called a *cluster function*, the numbers $\{d_i\}$ are called *multipliers*, and the function d on I whose value at i is d_i is called a *multiplier function*. We omit $\{d_i\}$ if all of them are equal to one, and therefore the matrix ε is skew-symmetric, and we omit the set J_0 if it is empty.

The seed $\mathbf{I} = (I, I_0, \varepsilon, d)$ is called a *subseed* of the seed $\mathbf{I}' = (I', I_0', \varepsilon, d)$ if $I \subset I'$, $I_0 \subset I_0'$ and the functions ε and d are the restriction of ε' and d', respectively. In this case we denote $\mathbf{I}$ by $\mathbf{I}'|_I$.

Recall the multiplicative group scheme $\mathbb{G}_m$. It is defined as the spectrum of the ring $\mathbb{Z}[X, X^{-1}]$. The direct product of several copies of the multiplicative group is called a split algebraic torus, or simply a torus. Readers who are not used to the language of schemes may just fix, once and for all, a field K, and replace everywhere $\mathbb{G}_m$ by $K^\times$; indeed $\mathbb{G}_m(K) = K^\times$.

For a seed $\mathbf{I}$ we associate a torus $\mathcal{X}_\mathbf{I} = (\mathbb{G}_m)^I$ with a Poisson structure given by

$$\{x_i, x_j\} = \widehat{\varepsilon}_{ij} x_i x_j, \tag{8}$$

where $\{x_i | i \in I\}$ are the standard coordinates on the factors. We shall call it the *seed $\mathcal{X}$-torus*.

Let $\mathbf{I} = (I, I_0, \varepsilon, d)$ and $\mathbf{I}' = (I', I_0', \varepsilon', d')$ be two seeds, and $k \in I$. A *mutation in the vertex k* is an isomorphism $\mu_k : I \to I'$ satisfying the following conditions:

1. $\mu_k(I_0) = I_0'$,
2. $d'_{\mu_k(i)} = d_i$,
3. $$\varepsilon'_{\mu_k(i)\mu_k(j)} = \begin{cases} -\varepsilon_{ij} & \text{if } i = k \text{ or } j = k, \\ \varepsilon_{ij} + \varepsilon_{ik}\max(0, \varepsilon_{kj}) & \text{if } \varepsilon_{ik} \geq 0, \\ \varepsilon_{ij} + \varepsilon_{ik}\max(0, -\varepsilon_{kj}) & \text{if } \varepsilon_{ik} < 0. \end{cases}$$

A *symmetry* of a seed $\mathbf{I} = (I, I_0, \varepsilon, d)$ is an automorphism σ of the set I preserving the subset I_0, the matrix ε and the numbers d_i. In other words, it satisfies the following conditions:

1. $\sigma(I_0) = I_0$,
2. $d_{\sigma(i)} = d_i$,
3. $\varepsilon_{\sigma(i)\sigma(j)} = \varepsilon_{ij}$

Symmetries and mutations induce (rational) maps between the corresponding seed $\mathcal{X}$-tori, which are denoted by the same symbols μ_k and σ and given by the formulas

$$x_{\sigma(i)} = x_i$$

and

$$x_{\mu_k(i)} = \begin{cases} x_k^{-1} & \text{if } i = k, \\ x_i(1 + x_k)^{\varepsilon_{ik}} & \text{if } \varepsilon_{ik} \geq 0 \text{ and } i \neq k, \\ x_i(1 + (x_k)^{-1})^{\varepsilon_{ik}} & \text{if } \varepsilon_{ik} \leq 0 \text{ and } i \neq k. \end{cases}$$

A *cluster transformation* between two seeds (and between two seed $\mathcal{X}$-tori) is a composition of symmetries and mutations. If the source and the target of a cluster transformation coincide, we call this map a *cluster automorphism*. Two seeds are called *equivalent* if they are related by a cluster transformation. The equivalence class of a seed $\mathbf{I}$ is denoted by $|\mathbf{I}|$.

Thus we have defined two categories. The first one has cluster seeds as objects and cluster transformations as morphisms. The second one has seed $\mathcal{X}$-tori as objects and cluster transformations of them as morphisms. There is a canonical functor from the first to the second. The objects in these two categories are the same. However, there are more morphisms in the second category.

A *cluster $\mathcal{X}$-variety* is obtained by taking a union of all seed $\mathcal{X}$-tori related to a given seed $\mathbf{I}$ by cluster transformations, and gluing them together using the above birational isomorphisms. It is denoted by $\mathcal{X}_{|\mathbf{I}|}$. Observe that the cluster $\mathcal{X}$-varieties corresponding to equivalent seeds are isomorphic. Every particular seed $\mathcal{X}$-torus provide our cluster variety with a rational coordinate system. The corresponding rational functions are called *cluster coordinates*.

Since in what follows we shall extencively use compositions of mutations we would like to introduce a shorthand notation for them. Namely, we denote an expression $\mu_{\mu_i(j)}\mu_i$ by $\mu_j\mu_i$, $\mu_{\mu_{\mu_i(j)}\mu_i(k)}\mu_{\mu_i(j)}\mu_i$ by $\mu_k\mu_j\mu_i$, and so on. We will also say that two sequences of mutations are equivalent ($\stackrel{\sim}{=}$) if they coincide as maps between the $\mathcal{X}$-tori up to permutation of coordinates.

The cluster transformations have the following basic properties (see [FG2, Section 2]):

1. Every seed is related to other seeds by exactly $\sharp(I - I_0)$ mutations.
2. Cluster transformations form a groupoid. In particular the inverse of a mutation is a mutation: $\mu_k\mu_k = \text{id}$. Cluster automorphisms form a group called the *mapping class group*. The groups of cluster automorphisms of equivalent seeds are isomorphic.
3. Cluster transformations preserve the Poisson structure. In particular a cluster $\mathcal{X}$-manifold has a canonical Poisson structure and the automorphism group of this manifold acts on it by Poisson transformations.
4. Cluster transformations are given by rational functions with positive integral coefficients.
5. If $\varepsilon_{ij} = \varepsilon_{ji} = 0$, then $\mu_i\mu_j\mu_j\mu_i = \text{id}$.
6. If $\varepsilon_{ij} = -\varepsilon_{ji} = -1$, then $\mu_i\mu_j\mu_i\mu_j\mu_i \stackrel{\sim}{=} \text{id}$. (This is called the *pentagon relation*.)
7. If If $\varepsilon_{ij} = -2\varepsilon_{ji} = -2$, then $\mu_i\mu_j\mu_i\mu_j\mu_i\mu_j = \text{id}$.
8. If If $\varepsilon_{ij} = -3\varepsilon_{ji} = -3$, then $\mu_i\mu_j\mu_i\mu_j\mu_i\mu_j\mu_i = \text{id}$.

Conjecturally all relations between mutations are exhausted by properties 5–8.

2.2 Amalgamation

We start from the simplest example: the amalgamation of two seeds. Let $\mathbf{J} = (J, J_0, \varepsilon, d)$ and $\mathbf{I} = (I, I_0, \zeta, c)$ be two seeds and let L be a set embedded into both I_0 and J_0 in a such a way that for any $i, j \in L$, we have $c(i) = d(i)$. Then the amalgamation of $\mathbf{J}$ and $\mathbf{I}$ is a seed $\mathbf{K} = (K, K_0, \zeta, b)$, such that $K = I \cup_L J$, $K_0 = I_0 \cup_L J_0$ and

$$\zeta_{ij} = \begin{cases} 0 & \text{if } i \in I - L \text{ and } j \in J - L, \\ 0 & \text{if } i \in J - L \text{ and } j \in I - L, \\ \eta_{ij} & \text{if } i \in I - L \text{ or } j \in I - L, \\ \varepsilon_{ij} & \text{if } i \in J - L \text{ or } j \in J - L, \\ \eta_{ij} + \varepsilon_{ij} & \text{if } i, j \in L. \end{cases}$$

This operation induces a homomorphism $\mathcal{X}_{\mathbf{J}} \times \mathcal{X}_{\mathbf{I}} \to \mathcal{X}_{\mathbf{K}}$ between the corresponding seed $\mathcal{X}$-tori given by the rule

$$z_i = \begin{cases} x_i & \text{if } i \in I - L, \\ y_i & \text{if } i \in J - L, \\ x_i y_i & \text{if } i \in L. \end{cases} \tag{9}$$

It is easy to check that it respects the Poisson structure and commutes with cluster transformations, and thus is defined for the cluster $\mathcal{X}$-varieties, and not only for the seeds.

If there is a subset $L' \subset L$ such that $\varepsilon_{ij} + \eta_{ij} \in \mathbb{Z}$ when $i, j \in L$, then we can *defrost* the vertices of L', getting a new seed $(K, K_0 - L', \zeta, b)$. In this way we get a different cluster ensemble, since we can now mutate the elements of L' as well.

Now let us present the general definition. Let

$$\mathbf{I}(s) = (I(s), I_0(s), \varepsilon(s), d(s)), \quad s \in S$$

be a family of seeds parametrised by a set S. Let us glue the sets I^s in such a way that

(a) only frozen vertices can be glued;
(b) if $i \in I(s)$ and $j \in I(t)$ are glued, then $d(s)_i = d(t)_j$.

Let us denote by K the set obtained by gluing the sets $I(s)$.

Alternatively, the *gluing data* can be described by the following data:

(i) a set K;
(ii) a collection of injective maps $p_s : I(s) \hookrightarrow K$, $s \in S$, whose images cover K; two images may intersect only at the frozen elements;
(iii) the multiplier function on $\cup_{s \in S} I(s)$ descends to a function d on the set K.

In other words, there is a cover of the set K by the subsets $I(s)$, any two elements covering the same point are frozen, and the values of the multiplier functions at these elements coincide.

We identify $\varepsilon(s)$ with a function on the square of the image of the set $I(s)$, and denote by $\varepsilon(s)'$ its extension by zero to K^2. Then we set

$$\varepsilon := \sum \varepsilon(s)'. \tag{10}$$

There is a map $P : \cup_{s \in S} I(s) \to K$. We set

$$K_0 := P(\cup_{s \in S} I_0(s)).$$

There is a unique function c on K such that $p_s^* d = d(s)$ for any $s \in C$.

Definition 2.1. The seed $\mathbf{K} := (K, K_0, \varepsilon, d)$ is the amalgamation of the seeds $\mathbf{I}(s)$ with respect to the given gluing data (i)–(iii).

Lemma 2.2. *The amalgamation of seeds commutes with cluster transformations.*

Proof. Thanks to (ii), for any element $i \in I(s) - I_0(s)$ one has $|P^{-1}(p_s(i))| = 1$. Thus when we do a mutation in the direction $p_s(i)$, we can change the values of the cluster function only on the subset $p_s(I(s)^2)$. The lemma follows.

The amalgamation map of cluster $\mathcal{X}$-varieties

Let us consider the following map of $\mathcal{X}$-tori:

$$m : \prod_{s\in S} \mathcal{X}_{\mathbf{I(s)}} \to \mathcal{X}_{\mathbf{K}}, \qquad m^* x_i = \prod_{j\in P^{-1}(i)} x_j. \tag{11}$$

The following lemma is obvious.

Lemma 2.3. *The maps* (11) *commute with mutations, and thus give rise to a map of cluster $\mathcal{X}$-varieties, called the amalgamation map:*

$$m : \prod_{s\in S} \mathcal{X}_{|\mathbf{I}(s)|} \to \mathcal{X}_{|\mathbf{K}|}.$$

Defrosting

Let $L \subset K_0$. Assume that the function ε restricted to $L \times L - K_0 \times K_0$ takes values in $\mathbb{Z}$. Set $K_0' := K_0 - L$. Then there is a new seed $\mathbf{K}' := (K, K_0', \varepsilon, c)$. We say that *the seed $\mathbf{K}'$ is obtained from $\mathbf{K}$ by defrosting of L.* There is a canonical open embedding $\mathcal{X}_{|\mathbf{K}|} \hookrightarrow \mathcal{X}_{|\mathbf{K}'|}$.

Amalgamation followed by defrosting is the key operation which we use below. Abusing notation, sometimes one may refer to this operation simply as amalgamation. However, then the defrosted subset must be specified.

3 Cluster $\mathcal{X}$-varieties related to a group G

3.1 An example: Rational coordinates on PGL_2

Observe that one has

$$H(x) := \begin{pmatrix} x^{1/2} & 0 \\ 0 & x^{-1/2} \end{pmatrix} \overset{PGL(2,\mathbb{C})}{=} \begin{pmatrix} x & 0 \\ 0 & 1 \end{pmatrix}.$$

They are elements of a Cartan subgroup in $\mathrm{PGL}(2, \mathbb{C})$.

Consider the map $\mathrm{ev}_{\bar{\alpha}\alpha} : (\mathbb{C}^\times)^3 \to \mathrm{PGL}(2, \mathbb{C})$ given by

$$\begin{aligned}\mathrm{ev}_{\bar\alpha\alpha} : (x_0, x_1, x_2) &\mapsto H(x_0)\begin{pmatrix}1&0\\1&1\end{pmatrix}H(x_1)\begin{pmatrix}1&1\\0&1\end{pmatrix}H(x_2)\\ &= (x_0x_1x_2)^{-1/2}\begin{pmatrix}x_0x_2(1+x_1) & x_0\\ x_2 & 1\end{pmatrix}.\end{aligned}$$

Consider also another map $\mathrm{ev}_{\alpha\bar\alpha} : (\mathbb{C}^\times)^3 \to \mathrm{PGL}(2, \mathbb{C})$:

$$\begin{aligned}\mathrm{ev}_{\alpha\bar\alpha} : (y_0, y_1, y_2) &\mapsto H(y_0)\begin{pmatrix}1&1\\0&1\end{pmatrix}H(y_1)\begin{pmatrix}1&0\\1&1\end{pmatrix}H(y_2)\\ &= (y_0y_1y_2)^{-1/2}\begin{pmatrix}y_0y_1y_2 & y_0y_1\\ y_1y_2 & 1+y_1\end{pmatrix}.\end{aligned}$$

One can see that these maps enjoy the following properties:

1. $\mathrm{ev}_{\bar\alpha,\alpha}(x_0, x_1, x_2) = \mathrm{ev}_{\alpha,\bar\alpha}(x_0(1+x_1^{-1})^{-1}, x_1^{-1}, x_2(1+x_1^{-1})^{-1})$.
2. $\mathrm{ev}_{\alpha,\bar\alpha}(y_0, y_1, y_2) = \mathrm{ev}_{\bar\alpha,\alpha}(y_0(1+y_1), y_1^{-1}, y_2(1+y_1))$.
3. Both maps are open embeddings.
4. The standard Poisson–Lie bracket on the group $\mathrm{PGL}(2, \mathbb{C})$ reads as

$$\begin{aligned}&\{x_0, x_1\} = x_0x_1; \qquad &&\{x_2, x_1\} = x_2x_1; \qquad &&\{x_0, x_2\} = 0;\\ &\{y_0, y_1\} = -y_0y_1; \qquad &&\{y_2, y_1\} = -y_2y_1; \qquad &&\{y_0, y_1\} = 0.\end{aligned}$$

Therefore, $\mathrm{ev}_{\bar\alpha\alpha}$ and $\mathrm{ev}_{\alpha\bar\alpha}$ provide the group variety $\mathrm{PGL}(2, \mathbb{C})$ with two rational coordinate systems. The transition between these coordinates is given by a mutation and thus the union of the images of $\mathrm{ev}_{\bar\alpha,\alpha}$ and $\mathrm{ev}_{\alpha,\bar\alpha}$ is a cluster variety corresponding to two equivalent seeds $(\{0, 1, 2\}, \{0, 2\}, \varepsilon)$ and $(\{0, 1, 2\}, \{0, 2\}, \eta)$, where

$$\varepsilon = -\eta = \begin{pmatrix}0 & 1 & 0\\ -1 & 0 & 1\\ 0 & -1 & 0\end{pmatrix}$$

(columns correspond to the first index).

One can also consider the maps

$$\mathrm{ev}_\alpha, \mathrm{ev}_{\bar\alpha} : (\mathbb{C}^\times)^2 \to \mathrm{PGL}(2, \mathbb{C}) \quad \text{and} \quad \mathrm{ev}_\emptyset : \mathbb{C}^\times \to \mathrm{PGL}(2, \mathbb{C})$$

given by

$$\begin{aligned}\mathrm{ev}_\alpha : (z_0, z_1) &\mapsto H(z_0)\begin{pmatrix}1&1\\0&1\end{pmatrix}H(z_1) = (z_0z_1)^{-1/2}\begin{pmatrix}z_0z_1 & z_0\\ 0 & 1\end{pmatrix},\\ \mathrm{ev}_{\bar\alpha} : (w_0, w_1) &\mapsto H(w_0)\begin{pmatrix}1&0\\1&1\end{pmatrix}H(w_1) = (w_0w_1)^{-1/2}\begin{pmatrix}w_0w_1 & 0\\ w_1 & 1\end{pmatrix},\\ \mathrm{ev}_\emptyset : (t) &\mapsto t^{-1/2}\begin{pmatrix}t & 0\\ 0 & 1\end{pmatrix}.\end{aligned}$$

and satisfying the following properties:

6. The images of ev_α, $\mathrm{ev}_{\bar{\alpha}}$, $\mathrm{ev}_\emptyset$ and the union of the images of $\mathrm{ev}_{\bar{\alpha}\alpha}$ and $\mathrm{ev}_{\alpha\bar{\alpha}}$ are pairwise disjoint. The complement to their union in the whole group is of codimension two. It consists of the antidiagonal matrices.
7. The images of $\mathrm{ev}_\alpha, \mathrm{ev}_{\bar{\alpha}}, \mathrm{ev}_\emptyset$ are Poisson subvarieties with respect to the standard Drinfeld–Jimbo Poisson–Lie structure on $\mathrm{PGL}(2, \mathbb{C})$. The Poisson bracket is given by

$$\{z_0, z_1\} = z_0 z_1; \qquad \{w_0, w_1\} = w_0 w_1.$$

In the above constructions one can replace $\mathbb{C}^\times$ by $\mathbb{G}_m$, $\mathrm{PGL}(2, \mathbb{C})$ by the group scheme $\mathrm{PGL}(2)$, and upgrade all maps to the maps of the corresponding schemes.

Our aim now is a generalization of this picture in two directions. We consider split semisimple adjoint groups of higher ranks, and construct Poisson varieties which map to the group respecting the Poisson structure, but may not inject into the group.

Below we will give two alternative definitions of the seed $\mathbf{J}(D)$. The first one is computation-free: we define first the elementary seeds $\mathbf{J}(\alpha)$ corresponding to simple roots α, and then define $\mathbf{J}(D)$ as an amalgamated product of the elementary seeds $\mathbf{J}(\alpha_1), \ldots, \mathbf{J}(\alpha_n)$, where $D = \alpha_1 \ldots \alpha_n$, followed by defrosting of some of the frozen variables. It is presented in Section 2, and it is the definition which we use proving the main properties of our varieties in Appendix A. The second definition is given by defining directly all components of the seed; its most important part is an explicit formula for the function ε_{ij}. The second definition is given in Section 3.

3.2 The seed $\mathbf{J}(D)$

Let us assume first that $D = \alpha$ is a simple positive root. Then we set

$$J(\alpha) = J_0(\alpha) := (\Pi - \{\alpha\}) \cup \{\alpha'\} \cup \{\alpha''\},$$

where α' and α'' are certain new elements. There is a *decoration* map

$$\pi : J(\alpha) \longrightarrow \Pi,$$

which sends α' and α'' to α, and is the identity map on $\Pi - \alpha$.

The collection of multipliers $\{d^\alpha\}$ gives rise to a function $\mathcal{D}$ on the set Π: $\mathcal{D}(\alpha) := d^\alpha$. We define the multipliers for $\mathbf{J}(\alpha)$ as the function $\mathcal{D} \circ \pi$ on $J(\alpha)$.

Finally, the function $\varepsilon(\alpha)$ is defined as follows. Its entry $\varepsilon(\alpha)_{\beta\gamma}$ is zero unless one of the indices is decorated by α. Further,

$$\varepsilon(\alpha)_{\alpha'\beta} = \frac{C_{\alpha\beta}}{2}, \qquad \varepsilon(\alpha)_{\alpha''\beta} = -\frac{C_{\alpha\beta}}{2}, \qquad \varepsilon(\alpha)_{\alpha'\alpha''} = -1. \tag{12}$$

If $D = \bar{\alpha}$ is a negative simple root, we have a similar set

$$J(\bar{\alpha}) = J_0(\bar{\alpha}) := (\Pi^- - \{\bar{\alpha}\}) \cup \{\bar{\alpha}'\} \cup \{\bar{\alpha}''\},$$

a similar decoration $\pi : J(\bar{\alpha}) \longrightarrow \Pi$, and similar multipliers $\mathcal{D} \circ \pi$ on $J(\bar{\alpha})$. The cluster function is obtained by reversing the signs, using the obvious identification of $J(\bar{\alpha})$ and $J(\alpha)$: $\varepsilon(\bar{\alpha}) := -\varepsilon(\alpha)$:

$$\varepsilon(\bar{\alpha})_{\bar{\alpha}'\bar{\beta}} = -\frac{C_{\alpha\beta}}{2}, \qquad \varepsilon(\bar{\alpha})_{\bar{\alpha}''\bar{\beta}s} = \frac{C_{\alpha\beta}}{2}, \qquad \varepsilon(\bar{\alpha})_{\bar{\alpha}'\bar{\alpha}''} = 1.$$

The torus $\mathcal{X}_\alpha$ can also be defined as follows. Let U_α be the one-parameter unipotent subgroup corresponding to the root α. Then $\mathcal{X}_\alpha = H \times (U_\alpha - \{0\}) = HU_\alpha - H$.

The general case

Observe that the subset of elements decorated by a simple root has one element unless this root is α when there are two elements. There is a natural linear order on the subset of elements of $\mathbf{J}(\alpha)$ decorated by a given simple positive root γ: it is given by (α_-, α_+) in the only nontrivial case when $\alpha = \gamma$. So for a given simple positive root γ there are the *minimal* and the *maximal* elements decorated by γ.

Definition 3.1. Let $D = \alpha_1 \ldots \alpha_n \in \mathfrak{W}$. Then $J(D)$ is obtained by gluing the sets $J(\alpha_1), \ldots, J(\alpha_n)$ as follows. For every $\gamma \in \Pi$, and for every $i = 1, \ldots, n-1$, we glue the maximal γ-decorated element of $J(\alpha_i)$ and the minimal γ-decorated element of $J(\alpha_{i+1})$.

The seed $\widetilde{\mathbf{J}}(D)$ is the amalgamated product for this gluing data of the seeds $\mathbf{J}(\alpha_1), \ldots, \mathbf{J}(\alpha_n)$.

The seed $\widetilde{\mathbf{J}}(D)$ has frozen vertices only: $\widetilde{J}(D) = \widetilde{J}_0(D)$. To define the seed $\mathbf{J}(D)$ we will defrost some of them, making the set $J_0(D)$ smaller.

In Definition 3.1 we glue only the elements decorated by the same positive simple root. Thus the obtained set $J(D)$ has a natural decoration $\pi : J(D) \longrightarrow \Pi$, extending those of the subsets $J(\alpha_i)$. Moreover, for every $\gamma \in \Pi$, the subset of γ-decorated elements of $J(D)$ has a natural linear order, induced by the ones on $J(\alpha_i)$, and the linear order of the word D.

Definition 3.2. The subset $J_0(D)$ is the union, over $\gamma \in \Pi$, of the extremal (i.e., minimal and maximal) elements for the defined above linear order on the γ-decorated part of $J(D)$.

The seed $\mathbf{J}(D)$ is obtained from the seed $\widetilde{\mathbf{J}}(D)$ by reducing $\widetilde{J}_0(D)$ to the subset $J_0(D)$.

Observe that $\varepsilon_{\alpha\beta}$ is integral unless both α and β are in $J_0(D)$. Thus the integrality condition for $\varepsilon_{\alpha\beta}$ holds.

3.3 An alternative definition of the seed $\mathbf{J}(D)$

The sets $J_0(D)$ and $J(D)$

Given a positive simple root $\alpha \in \Pi$, denote by $n^\alpha(D)$ the number of occurrences of α and $\bar{\alpha}$ in the word D. We set

$$J^\alpha(D) := \{\tbinom{\alpha}{i} | \alpha \in \Pi, 0 \leq i \leq n^\alpha\}, \qquad J^\alpha_0(D) := \{\tbinom{\alpha}{0}\} \cup \{\tbinom{\alpha}{n_\alpha}\}.$$

Then $J(D)$ (respectively, $J_0(D)$) is the disjoint union of $J^\alpha(D)$ (respectively, $J_0^\alpha(D)$) for all $\alpha \in \Pi$. Observe that if a root α does not enter the word D then $J_0^\alpha(D) = J^\alpha(D)$ is a one-element set.

One can picture elements of the set $J(D)$ as associated to the intervals between walls made by $\alpha, \bar{\alpha}$ or the ends of the word D for some root α. If at least one wall is just the end D, the corresponding element of $J(D)$ belongs to $J_0(D)$. We shall denote these elements by braces connecting the walls with the name of the corresponding elements in the middle.

Example. Let $r = \operatorname{rk} G = 3$ and $\Pi = \{\alpha, \beta, \gamma\}$. Take $D = \alpha\bar{\beta}\bar{\alpha}\bar{\alpha}\beta$. Then

$$n^\alpha(D) = 3, \qquad n^\beta(D) = 2, \qquad n^\gamma(D) = 0,$$
$$J(D) = \{\tbinom{\alpha}{0}, \tbinom{\alpha}{1}, \tbinom{\alpha}{2}, \tbinom{\alpha}{3}, \tbinom{\beta}{0}, \tbinom{\beta}{1}, \tbinom{\beta}{2}, \tbinom{\gamma}{0}\},$$
$$J_0(D) = \{\tbinom{\alpha}{0}, \tbinom{\alpha}{3}, \tbinom{\beta}{0}, \tbinom{\beta}{2}, \tbinom{\gamma}{0}\}.$$

In brace notation, the set $J(D)$ can be shown as

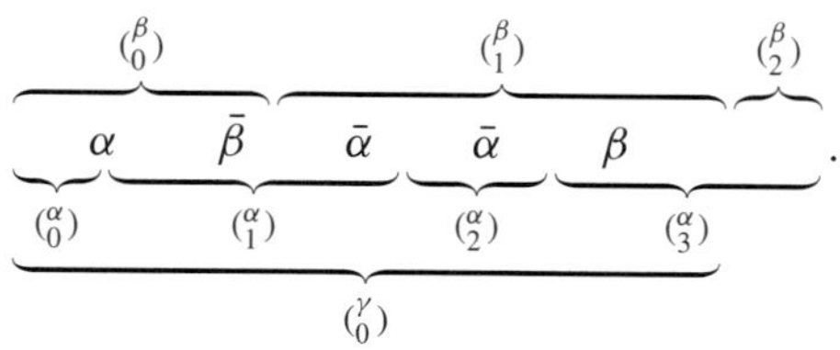

A description of ε and d

In order to give an explicit formula for the matrix $(\varepsilon_{\binom{\alpha}{i}\binom{\beta}{j}})$, we introduce more notation. Let $n^\alpha(k)$ be the number of letters α or $\bar{\alpha}$ among the first k letters of the word D. Let μ_k be the kth letter of D and let $\operatorname{sgn}(k)$ be $+1$ if $\mu_k \in \Pi$ and -1 otherwise. Let finally $|\mu_k| = \operatorname{sgn}(\mu_k)\mu_k \in \Pi$.

Definition 3.3. Let D be a word of $\mathfrak{W}$; then

- the multipliers are given by the rule $d_{\binom{\alpha}{i}}(D) = d^\alpha$;
- the integers $\widehat{\varepsilon}_{\binom{\alpha}{i}\binom{\beta}{j}}$ are defined by the formula

$$\sum_{\binom{\alpha}{i}\binom{\beta}{j}} \widehat{\varepsilon}_{\binom{\alpha}{i}\binom{\beta}{j}} \frac{\partial}{\partial \log x_i^\alpha} \wedge \frac{\partial}{\partial \log x_j^\beta}$$
$$= \frac{1}{2} \sum_{k=1}^{l(D)} \sum_\alpha \operatorname{sgn}(\mu_k) \widehat{C}_{\mu(k)\alpha} \frac{\partial}{\partial \log x_{n^\alpha(k)}^\alpha}$$
$$\wedge \left(\frac{\partial}{\partial \log x_{n^{|\mu_k|}(k)-1}^{|\mu_k|}} - \frac{\partial}{\partial \log x_{n^{|\mu_k|}(k)}^{|\mu_k|}} \right).$$

Remark. One can check that in the case when D is reduced, our function ε_{ij} is related to the cluster function b_{ij} defined in [BFZ3] for the corresponding double Bruhat cell as follows. Recall that b_{ij} is not defined if both i, j are frozen variables. Other than that, the values of b_{ij} turn out to be the same as for ε_{ij}.

It is easy to prove the following properties of the matrix ε:

1. $\varepsilon_{\binom{\alpha}{i}\binom{\beta}{j}}$ is integral unless both $\binom{\alpha}{i}$ and $\binom{\beta}{j}$ are in I_0.
2. For a given $\binom{\alpha}{i} \in I$ the number of $\binom{\beta}{j} \in I$ such that $\varepsilon_{\binom{\alpha}{i}\binom{\beta}{j}} \neq 0$ is no more than twice the number of $\beta \in \Pi$ such that $C_{\alpha\beta} \neq 0$. In particular, this number never exceeds 8.
3. The value of $\varepsilon_{\binom{\alpha}{i}\binom{\beta}{j}}$ is determined by the patterns of the walls in D corresponding to $\binom{\alpha}{i}$ and $\binom{\beta}{j}$. The list of all possibilities is too large to give explicitly, but we give just some of them—the patterns to the left and the corresponding values of $\varepsilon_{\binom{\alpha}{i}\binom{\beta}{j}}$ to the right (stars mean any roots or word ends compatible with the pattern):

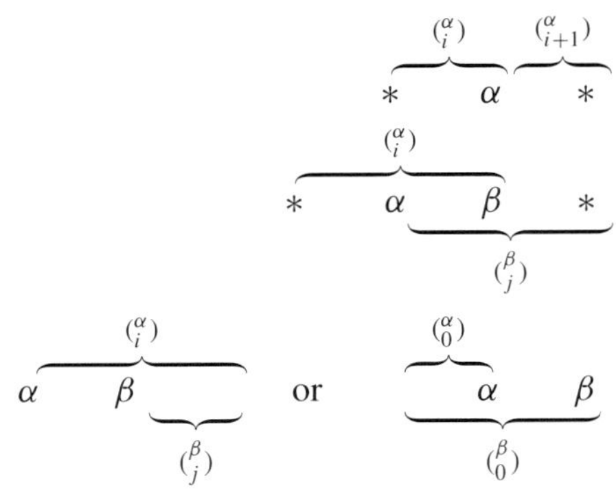

1

$C_{\alpha\beta}$

$C_{\alpha\beta}/2.$

4. If the word D consists of just one letter α, then $\varepsilon_{\binom{\alpha}{0}\binom{\beta}{0}} = C_{\alpha\beta}/2$, $\varepsilon_{\binom{\alpha}{1}\binom{\beta}{0}} = -C_{\alpha\beta}/2$. If the word D consists of just one letter $\bar{\alpha}$, then $\varepsilon_{\binom{\alpha}{0}\binom{\beta}{0}} = -C_{\alpha\beta}/2$, $\varepsilon_{\binom{\alpha}{1}\binom{\beta}{0}} = C_{\alpha\beta}/2$.

Example. For the word $D = \alpha\bar{\beta}\bar{\alpha}\bar{\alpha}\beta$ considered above, one can easily compute that all nonvanishing elements of ε are given by

$$
\begin{aligned}
\widehat{\varepsilon}_{\binom{\alpha}{0}\binom{\alpha}{1}} &= -\widehat{\varepsilon}_{\binom{\alpha}{1}\binom{\alpha}{2}} = -\widehat{\varepsilon}_{\binom{\alpha}{2}\binom{\alpha}{3}} = \widehat{C}_{\alpha\alpha}/2,\\
-\widehat{\varepsilon}_{\binom{\beta}{0}\binom{\beta}{1}} &= \widehat{\varepsilon}_{\binom{\beta}{1}\binom{\beta}{2}} = \widehat{C}_{\beta\beta}/2,\\
\widehat{\varepsilon}_{\binom{\alpha}{0}\binom{\beta}{0}} &= \widehat{\varepsilon}_{\binom{\alpha}{3}\binom{\beta}{2}} = \widehat{C}_{\alpha\beta}/2,\\
\widehat{\varepsilon}_{\binom{\alpha}{0}\binom{\gamma}{0}} &= -\widehat{\varepsilon}_{\binom{\alpha}{1}\binom{\gamma}{0}}/2 = \widehat{\varepsilon}_{\binom{\alpha}{3}\binom{\gamma}{0}} = \widehat{C}_{\alpha\gamma}/2,\\
-\widehat{\varepsilon}_{\binom{\beta}{0}\binom{\gamma}{0}} &= \widehat{\varepsilon}_{\binom{\beta}{1}\binom{\gamma}{0}}/2 = -\widehat{\varepsilon}_{\binom{\beta}{2}\binom{\gamma}{0}} = \widehat{C}_{\beta\gamma}/2.
\end{aligned}
$$

Proposition 3.4. *Definitions* 3.2 *and* 3.3 *are equivalent.*

Proof. Property 4 of the matrix ε tells us that the two definitions coincide for the elementary seeds $\mathbf{J}(\alpha)$. So it remains to check that the seed $\mathbf{J}(D)$ is the amalgamated product.

3.4 A map to the group

Recall the torus $\mathcal{X}_{\mathbf{J}(D)} = \mathbb{G}_m^{J(D)}$ and the natural coordinates $\{x_i^\alpha\}$ on it. Let us define the map $\mathrm{ev} : \mathcal{X}_{\mathbf{J}(D)} \to G$. In order to do this, we are going to construct a sequence of group elements, each of which is either a constant or depends on just one coordinate of $\mathcal{X}_{\mathbf{J}(D)}$. The product of the elements of the sequence will give the desired map.

Let $f_\alpha, h_\alpha, e_\alpha$ be Chevalley generators of the Lie algebra $\mathfrak{g}$ of G. They are defined up to an action of the Cartan subgroup H of G. Let $\{h^\alpha\}$ be another basis of the Cartan subalgebra defined by the property:

$$[h^\alpha, e_\beta] = \delta_\beta^\alpha e_j, \quad [h^\alpha, f_\beta] = -\delta_\beta^\alpha f_\beta.$$

This basis is related to the basis $\{h_\alpha\}$ via the Cartan matrix: $\sum_\beta C_{\alpha\beta} h^\beta = h_\alpha$.

Recall the lattice $X_*(H)$ of homomorphisms (cocharacters) $\mathbb{G}_m \to H$. The elements h_α and h^α give rise to cocharacters $H_\alpha, H^\alpha \in X_*(H)$, called the coroot and the coweight corresponding to the simple root α:

$$H_\alpha : \mathbb{G}_m \to H, \quad dH_\alpha(1) = h_\alpha, \qquad H^\alpha : \mathbb{G}_m \to H, \quad dH^\alpha(1) = h^\alpha.$$

One has $H^\alpha(x) = \exp(\log(x)h^\alpha)$. Let us introduce the group elements $\mathbf{E}^\alpha = \exp e_\alpha$, $\mathbf{F}^\alpha = \exp f_\alpha$.

Replace the letters in D by the group elements using the rule $\alpha \to \mathbf{E}^\alpha$, $\bar{\alpha} \to \mathbf{F}^\alpha$. For $\alpha \in \Pi$ and for any $\binom{\alpha}{i} \in J(D)$, insert $H^\alpha(x_i^\alpha)$ somewhere between the corresponding walls. The choice in placing every H is nonessential since it commutes with all Es and Fs unless they are marked by the same root.

In other words the sequence of group elements is defined by the following requirements:

- The sequence of $\mathbf{E}$s and $\mathbf{F}$s reproduce the sequence of letters in the word D.
- Any H depends on its own variable x_i^α.
- There is at least one $\mathbf{E}^\alpha$ or $\mathbf{F}^\alpha$ between any two H^αs.
- The number of H^αs is equal to the total number of $\mathbf{E}^\alpha$s and $\mathbf{F}^\alpha$s plus one.

Example. The word $\alpha\bar{\beta}\bar{\alpha}\bar{\alpha}\beta$ is mapped by ev to

$$H^\alpha(x_0^\alpha)H^\beta(x_0^\beta)\mathbf{E}^\alpha H^\alpha(x_1^\alpha)\mathbf{F}^\beta H^\beta(x_1^\beta)\mathbf{F}^\alpha H^\alpha(x_2^\alpha)\mathbf{F}^\alpha \mathbf{E}^\beta H^\alpha(x_3^\alpha)H^\beta(x_2^\beta)H^\gamma(x_0^\gamma).$$

3.5 The key properties of the spaces $\mathcal{X}_B$

We are going to show that the association of the seed $\mathcal{X}$-torus to the words is compatible with natural operations on the words.

Theorem 3.5. *Let A, B be arbitrary words and $\alpha, \beta \in \Pi$. Then there are the following rational maps commuting with the map* ev*:*

1. $\mathcal{X}_{\mathbf{J}(A\bar{\alpha}\beta B)} \to \mathcal{X}_{\mathbf{J}(A\beta\bar{\alpha}B)}$ *if* $\alpha \neq \beta$.
2. $\mathcal{X}_{\mathbf{J}(A\alpha\beta B)} \to \mathcal{X}_{\mathbf{J}(A\beta\alpha B)}$ *and* $\mathcal{X}_{\mathbf{J}(A\bar{\alpha}\bar{\beta}B)} \to \mathcal{X}_{\mathbf{J}(A\bar{\beta}\bar{\alpha}B)}$ *if* $C_{\alpha\beta} = 0$.
3. $\mathcal{X}_{\mathbf{J}(A\alpha\beta\alpha B)} \to \mathcal{X}_{\mathbf{J}(A\beta\alpha\beta B)}$ *and* $\mathcal{X}_{\mathbf{J}(A\bar{\alpha}\bar{\beta}\bar{\alpha}B)} \to \mathcal{X}_{\mathbf{J}(A\bar{\beta}\bar{\alpha}\bar{\beta}B)}$ *if* $C_{\alpha\beta} = -1$, $C_{\beta\alpha} = -1$.
4. $\mathcal{X}_{\mathbf{J}(A\alpha\beta\alpha\beta B)} \to \mathcal{X}_{\mathbf{J}(A\beta\alpha\beta\alpha B)}$ *and* $\mathcal{X}_{\mathbf{J}(A\bar{\alpha}\bar{\beta}\bar{\alpha}\bar{\beta}B)} \to \mathcal{X}_{\mathbf{J}(A\bar{\beta}\bar{\alpha}\bar{\beta}\bar{\alpha}B)}$ *if* $C_{\alpha\beta} = -2$, $C_{\beta\alpha} = -1$.
5. $\mathcal{X}_{\mathbf{J}(A\alpha\beta\alpha\beta\alpha\beta B)} \to \mathcal{X}_{\mathbf{J}(A\beta\alpha\beta\alpha\beta\alpha B)}$ *and* $\mathcal{X}_{\mathbf{J}(A\bar{\alpha}\bar{\beta}\bar{\alpha}\bar{\beta}\bar{\alpha}\bar{\beta}B)} \to \mathcal{X}_{\mathbf{J}(A\bar{\beta}\bar{\alpha}\bar{\beta}\bar{\alpha}\bar{\beta}\bar{\alpha}B)}$ *if* $C_{\alpha\beta} = -3$, $C_{\beta\alpha} = -1$.
6. $\mathcal{X}_{\mathbf{J}(A\alpha\alpha B)} \to \mathcal{X}_{\mathbf{J}(A\alpha B)}$ *and* $\mathcal{X}_{\mathbf{J}(A\bar{\alpha}\bar{\alpha}B)} \to \mathcal{X}_{\mathbf{J}(A\bar{\alpha}B)}$.
7. $\mathcal{X}_{\mathbf{J}(A)} \times \mathcal{X}_{\mathbf{J}(B)} \to \mathcal{X}_{\mathbf{J}(AB)}$.

Maps 1 *and* 2 *are isomorphisms. Maps* 3, 4, *and* 5 *are cluster transformations. (They are composions of* 1, 3, *and at least* 10 *mutations, respectively.) Map* 6 *is a composition of a cluster transformation and a projection along the coordinate axis. Map* 7 *is an amalgamated product. Map* ev *is a Poisson map. Maps* 1–7 *are also Poisson maps.*

Remark. It is not true that a mutation of a cluster seed $\mathbf{J}(D)$ is always a cluster seed corresponding to another word of the semigroup. For example, let $\Pi = \{\gamma, \Delta, \eta\}$ be the root system of type A_3 with $C_{\eta\gamma} = 0$. (It is convenient to use the capital letter Δ to distinguish the root which plays a special role below). Then $\mu_{\binom{\gamma}{1}}\mathbf{J}(\gamma\Delta\eta\gamma\Delta\gamma) = \mathbf{J}(\Delta\gamma\Delta\eta\Delta\gamma)$, but $\mu_{\binom{\Delta}{1}}\mathbf{J}(\gamma\Delta\eta\gamma\Delta\gamma)$ is a seed which does not correspond to any word.

Remark. There is a famous eight-term relation among the relations in the symmetric group. Namely, in the notations of the previous remark, we have

$$\begin{aligned}\gamma\Delta\eta\gamma\Delta\gamma &= \gamma\Delta\eta\gamma\Delta\gamma = \gamma\eta\Delta\eta\gamma\Delta \mathrel{\tilde{=}} \eta\gamma\Delta\gamma\eta\Delta = \eta\Delta\gamma\Delta\eta\Delta = \eta\Delta\gamma\eta\Delta\eta\\ &\mathrel{\tilde{=}} \eta\Delta\eta\gamma\Delta\eta = \Delta\eta\Delta\gamma\Delta\eta = \Delta\eta\gamma\Delta\gamma\eta \mathrel{\tilde{=}} \Delta\gamma\eta\Delta\eta\gamma\\ &= \Delta\gamma\Delta\eta\Delta\gamma = \gamma\Delta\gamma\eta\Delta\gamma = \gamma\Delta\eta\gamma\Delta\gamma.\end{aligned}$$

It is equivalent to the relation between mutations:

$$\mu_{\binom{\Delta}{1}}\mu_{\binom{\gamma}{2}}\mu_{\binom{\gamma}{1}}\mu_{\binom{\Delta}{1}}\mu_{\binom{\gamma}{2}}\mu_{\binom{\gamma}{1}}\mu_{\binom{\Delta}{1}}\mu_{\binom{\gamma}{2}} = \mathrm{id}\,.$$

It is an easy exercise to show that this relation is a corollary of properties 5 and 6 of mutations. Thus the eight-term relation can be reduced to pentagons in the cluster setting.

Proof. The proof of the last two statements follows immediately from the rest of the theorem. To prove the rest of the theorem, we define map 7 as the amalgamation map between the corresponding $\mathcal{X}$-varieties. Then it is sufficient to construct the other maps for the shortest word where the maps are defined, and then extend them using the multiplicativity property 7 to the general case. The claim that the evaluation map is Poisson will be proved in Section 3.8.

It is useful to recall an explicit description of the amalgamation map $\mathcal{X}_{\mathbf{J}(A)} \times \mathcal{X}_{\mathbf{J}(B)} \to \mathcal{X}_{\mathbf{J}(AB)}$. Let $\{x_i^\alpha\}$, $\{y_i^\alpha\}$ and $\{z_i^\alpha\}$ be the coordinates on $\mathcal{X}_{\mathbf{J}(A)}$, $\mathcal{X}_{\mathbf{J}(B)}$ and $\mathcal{X}_{\mathbf{J}(AB)}$, respectively. Then the map is given by the formula:

$$z_i^\alpha = \begin{cases} x_i^\alpha & \text{if } i < n^\alpha(A), \\ x_{n^\alpha(A)}^\alpha y_0^\alpha & \text{if } i = n^\alpha(A), \\ y_{i+n^\alpha(A)}^\alpha & \text{if } i > n^\alpha(A). \end{cases}$$

The crucial point is that, just by the construction, this map is compatible with the evaluation map ev to the group.

Maps 1–6 and their properties are deduced from Proposition 3.6 below. Some formulas of this proposition are equivalent to results available in the literature [L1, BZ, FZ], but are stated there in a different form. Our goal is to make apparent their cluster nature, i.e., to show that they transform as the $\mathcal{X}$-coordinates for cluster varieties. In the non-simply-laced cases these transformations are presented as compositions of several cluster transformations. The very existence of such presentations is a key new result.

Proposition 3.6. *There are the following identities between the generators* $\mathbf{E}^\alpha$, $H^\alpha(x)$, $\mathbf{E}^\alpha$:

$\underline{\alpha\alpha \to \alpha.}$ $\mathbf{E}^\alpha H^\alpha(x)\mathbf{E}^\alpha = H^\alpha(1+x)\mathbf{E}^\alpha H^\alpha(1+x^{-1})^{-1}$.

$\underline{\alpha\beta \to \beta\alpha.}$ *If* $C_{\alpha\beta} = 0$. *Then* $\mathbf{E}^\alpha\mathbf{E}^\beta = \mathbf{E}^\beta\mathbf{E}^\alpha$.

$\underline{\alpha\beta\alpha \to \beta\alpha\beta.}$ *If* $C_{\alpha\beta} = -1$, *then* $\mathbf{E}^\alpha H^\alpha(x)\mathbf{E}^\beta\mathbf{E}^\alpha = H^\alpha(1+x)H^\beta(1+x^{-1})^{-1}\mathbf{E}^\beta H^\beta(x)^{-1}\mathbf{E}^\alpha\mathbf{E}^\beta H^\alpha(1+x^{-1})^{-1}H^\beta(1+x)$.

$\underline{\alpha\beta\alpha\beta \to \beta\alpha\beta\alpha.}$ *If* $C_{\alpha\beta} = -2$, $C_{\beta\alpha} = -1$, *then* $\mathbf{E}^\alpha\mathbf{E}^\beta H^\alpha(x)H^\beta(y)\mathbf{E}^\alpha\mathbf{E}^\beta = H^\beta(a')H^\alpha(b')\mathbf{E}^\beta\mathbf{E}^\alpha H^\beta(y')H^\alpha(x')\mathbf{E}^\beta\mathbf{E}^\alpha H^\beta(q')H^\alpha(p')$, *where* a', b', x', y', p', q' *are rational functions of* x *and* y *given by* (17).

$\underline{\alpha\beta\alpha\beta\alpha\beta \to \beta\alpha\beta\alpha\beta\alpha.}$ *If* $C_{\alpha\beta} = -3$, $C_{\beta\alpha} = -1$, *then*

$$\begin{aligned}&\mathbf{E}^\alpha H^\alpha(x)\mathbf{E}^\beta H^\beta(y)\mathbf{E}^\alpha H^\alpha(z)\mathbf{E}^\beta H^\beta(w)\mathbf{E}^\alpha\mathbf{E}^\beta \\ &\quad = H^\beta(a')H^\alpha(b')\mathbf{E}^\beta H^\beta(y')\mathbf{E}^\alpha H^\alpha(x')\mathbf{E}^\beta H^\beta(w')\mathbf{E}^\alpha H^\alpha(z')\mathbf{E}^\beta H^\beta(q')\mathbf{E}^\alpha H^\alpha(p'),\end{aligned} \tag{13}$$

where a', b', x', y', z', w', p', q' *are rational functions of* x, y, z, w *given by* (24).

Further, one has

$\underline{\bar{\alpha}\alpha \to \alpha\bar{\alpha}.}$

$$\mathbf{F}^\alpha H^\alpha(x)\mathbf{E}^\alpha = \left(\prod_{\beta\neq\alpha} H^\beta(1+x)^{-C_{\alpha\beta}}\right) H^\alpha(1+x^{-1})^{-1}\mathbf{E}^\alpha H^\alpha(x^{-1})\mathbf{F}^\alpha H^\alpha(1+x^{-1})^{-1}.$$

$\underline{\bar\alpha\beta \to \beta\bar\alpha}$. $\mathbf{F}^\alpha\mathbf{E}^\beta = \mathbf{E}^\beta\mathbf{F}^\alpha$ *if* $\alpha \neq \beta$.

Applying the antiautomorphism of $\mathfrak{g}$ which acts as the identity on the Cartan subalgebra, and interchanges $\mathbf{F}^\alpha$ and $\mathbf{E}^\alpha$, we obtain similar formulas for $\bar\alpha\bar\alpha \to \bar\alpha$, $\bar\alpha\bar\beta \to \bar\beta\bar\alpha$, $\bar\alpha\bar\beta\bar\alpha \to \bar\beta\bar\alpha\bar\beta$ and $\bar\alpha\bar\beta\bar\alpha\bar\beta \to \bar\beta\bar\alpha\bar\beta\bar\alpha$. For example, if $C_{\alpha\beta} = C_{\beta\alpha} = -1$, then

$$\begin{aligned}&\mathbf{F}^\alpha H^\alpha(x)\mathbf{F}^\beta\mathbf{F}^\alpha\\ &\quad = H^\alpha(1+x^{-1})^{-1}H^\beta(1+x)\mathbf{F}^\beta H^\alpha(x)^{-1}\mathbf{F}^\alpha\mathbf{F}^\beta H^\alpha(1+x)H^\beta(1+x^{-1})^{-1}.\end{aligned}$$

Proposition 3.6 implies Theorem 3.5 minus part 5 and the "map ev is Poisson" part

Map 1 is a corollary of the obvious property $\underline{\bar\alpha\beta \to \beta\bar\alpha}$. Map 2 follows from $\underline{\alpha\beta \to \beta\alpha}$ and $\underline{\bar\alpha\bar\beta \to \bar\beta\bar\alpha}$. Map 6 follows from $\underline{\alpha\alpha \to \alpha}$ and $\underline{\bar\alpha\bar\alpha \to \bar\alpha}$. Each of these maps is obviously a composition of a mutation $\mu_{\binom{\alpha}{n^\alpha(A)+1}}$, or its $\bar\alpha$ version, and the projection along the corresponding coordinate. The maps 3 follow from $\underline{\alpha\beta\alpha \to \beta\alpha\beta}$ and $\underline{\bar\alpha\bar\beta\bar\alpha \to \bar\beta\bar\alpha\bar\beta}$ and are given by the mutations $\mu_{\binom{\alpha}{n^\alpha(A)+1}}$, or its $\bar\alpha$ version. Map 4 follows from $\underline{\alpha\beta\alpha\beta \to \beta\alpha\beta\alpha}$ and $\underline{\bar\alpha\bar\beta\bar\alpha\bar\beta \to \bar\beta\bar\alpha\bar\beta\bar\alpha}$ and can be easily shown to be given by composition of three mutations: $\mu_{\binom{\alpha}{n^\alpha(A)+1}}$, $\mu_{\binom{\beta}{n^\beta(A)+1}}$ and $\mu_{\binom{\alpha}{n^\alpha(A)+1}}$, or their bar counterparts. A conceptual proof explains this; see Section 3.7 below. However, it is not at all clear from the very complicated fromulas (24) why map 5 in Theorem 3.5 is a composition of mutations.

Proof of Proposition 3.6. This proof will occupy the end of this subsection and the next two subsections and will be combined with the proof of part 5 of Theorem 3.5, as well as a more conceptual proof of part 4.

Recall that there is a transposition antiautomorphism which interchanges $\mathbf{E}^\alpha$ and $\mathbf{F}^\alpha$ and does not change H^α. Thus the formulas $\underline{\bar\alpha\bar\alpha \to \bar\alpha}$, $\underline{\bar\alpha\bar\beta \to \bar\beta\bar\alpha}$, $\underline{\bar\alpha\bar\beta\bar\alpha \to \bar\beta\bar\alpha\bar\beta}$, and $\underline{\bar\alpha\bar\beta\bar\alpha\bar\beta \to \bar\alpha\bar\beta\bar\alpha\bar\beta}$ follow from the respective formulas for positive roots.

Let us introduce a more traditional generator $H_\alpha(x) = \exp(\log(x)h_\alpha)$.

- $\underline{\alpha\alpha \to \alpha}$. It is easy to show using computations with 2×2 matrices that

$$\mathbf{E}^\alpha H_\alpha(t)\mathbf{E}^\alpha = H_\alpha(1+t^2)^{1/2}\mathbf{E}^\alpha H_\alpha(1+t^{-2})^{-1/2}.$$

  Substituting $H_\alpha(t) = \prod_\beta H^\beta(t)^{C_{\alpha\beta}}$ and taking into account that $\mathbf{E}^\alpha$ and $\mathbf{H}^\beta$ commute when $\beta \neq \alpha$, $C_{\alpha\alpha} = 2$, and making the substitution $x = t^2$ one gets the identity $\underline{\alpha\alpha \to \alpha}$.
- $\underline{\bar\alpha\beta \to \beta\bar\alpha}$. The proof is similar to the previous one. It is based on the easily verifiable identity for 2×2 matrices

$$\mathbf{F}^\alpha H_\alpha(t)\mathbf{E}^\alpha = H_\alpha(1+t^{-2})^{-1/2}\mathbf{E}^\alpha H_\alpha(t)^{-1}\mathbf{F}^\alpha H_\alpha(1+t^{-2})^{-1/2}.$$

- $\alpha\beta\alpha \to \beta\alpha\beta$. This identity can be easily derived from the well-known identity, which is sufficient to check for SL_3:

$$e^{ae_\alpha}e^{be_\beta}e^{ce_\alpha} = e^{\frac{bc}{a+c}e_\beta}e^{(a+c)e_\alpha}e^{\frac{ab}{a+c}e_\beta}.$$

Taking into account that

$$H^\alpha(a)\mathbf{E}^\alpha H^\alpha(a)^{-1} = e^{ae_\alpha} \tag{14}$$

for any α and making the substitution $a/c \to x$, one obtains the desired identity.

3.6 Cluster folding

We start by recalling the notion of the *folding* of root systems. Let Π' and Π be two sets of simple roots corresponding to the root systems with the Cartan matrices C' and C, respectively. A surjective map $\pi : \Pi' \to \Pi$ is called folding if it satisfies the following properties:

1. $C_{\alpha'\beta'} = 0$ if $\pi(\alpha') = \pi(\beta')$, and $\alpha' \neq \beta'$.
2. $C_{\alpha,\beta} = \sum_{\alpha'\in\pi^{-1}(\alpha)} C_{\alpha'\beta'}$ if $\pi(\beta') = \beta$.

A folding induces an embedding (in the inverse direction) of the corresponding Lie algebras denoted by π^* and given by

$$\pi^*(h_\alpha) = \sum_{\alpha'\in\pi^{-1}(\alpha)} h'_{\alpha'}, \qquad \pi^*(e_\alpha) = \sum_{\alpha'\in\pi^{-1}(\alpha)} e'_{\alpha'}, \qquad \pi^*(e_{-\alpha}) = \sum_{\alpha'\in\pi^{-1}(\alpha)} e'_{-\alpha'},$$

where $\{h'_{\alpha'}, e'_{\alpha'}, e'_{-\alpha'}\}$ are the standard Chevalley generators of the Lie algebra $\mathfrak{g}'$ corresponding to the Cartan matrix C'.

A folding also induces maps between the corresponding Weyl groups, braid semigroups and braid groups, and Hecke semigroups, given by

$$\pi^*(\alpha) = \prod_{\alpha'\in\pi^{-1}(\alpha)} \alpha', \qquad \pi^*(\bar{\alpha}) = \prod_{\alpha'\in\pi^{-1}(\alpha)} \bar{\alpha}'.$$

(The order of the product does not matter since according to property 1 the factors commute.)

The main feature of the folding is that it gives embeddings of non-simply-laced Lie algebras and groups to the simply-laced ones. Namely, B_n is a folding of D_n, C_n is a folding of A_{2n-1}, F_4 is a folding of E_6 and G_2 is a folding of both B_3 and D_4. In these cases, the folding is provided by the action of a subgroup Γ of the automorphism group of the Dynkin diagram: one has $\Pi' := \Pi/\Gamma$, and the folding is the quotient map $\Pi \to \Pi/\Gamma$. On the level of Dynkin diagrams a folding corresponds precisely to the folding of the corresponding graph, thus explaining the origin of the name.

Following [FG2], we define a *folding of a cluster seed*.

Definition 3.7. A *folding* π of a cluster seed $\mathbf{J}' = (J', J'_0, \varepsilon', d')$ to a cluster seed $\mathbf{J} = (J, J_0, \varepsilon, d)$ is a surjective map $\pi : J' \to J$ satisfying the following conditions:

0. $\pi(J_0') = J_0', \pi(J' - J_0') = J - J_0$.
1. $\varepsilon'_{i'j'} = 0$ if $\pi(i') = \pi(j')$.
2. $\varepsilon_{ij} = \sum_{i' \in \pi^{-1} i} \varepsilon'_{i'j'}$, and all summands in this sum have the same signs or vanish.

A folding induces a map of the corresponding cluster tori $\pi^* : \mathcal{X}_{\mathbf{J}} \to \mathcal{X}_{\mathbf{J}'}$ by the formula $(\pi^*)^* x_{i'} = x_{\pi(i')}$. The main feature of this map is that it commutes with mutations in the following sense. To formulate it we need the following simple but basic fact.

Lemma 3.8. *If $\varepsilon_{kk'} = 0$, then mutations in the directions k and k' commute.*

Let $\mu_k : \mathbf{J} \to \mathbf{I}$ be a mutation, let $\pi : \mathbf{J}' \to \mathbf{J}$ be a folding, and let $\mathbf{I}' := (\prod_{k' \in \pi^{-1}(k)} \mu_{k'})\mathbf{J}'$. We define a map $\pi_k : \mathbf{I}' \to \mathbf{I}$ as the composition

$$\pi_k = \mu_k \pi \left(\prod_{k' \in \pi^{-1}(k)} \mu_{k'} \right)^{-1}.$$

The last factor in this formula is well defined since the mutations $\mu_{k'}$ commute thanks to Lemma 3.8 and the condition 1 of Definition 3.7. The map π_k is not always a folding (the condition 2 may not be satisfied), but if it is then, of course, $\mu_k \pi = \pi_k \prod_{k' \in \pi^{-1} k} \mu_{k'}$; furthermore, on the level of $\mathcal{X}$-tori we have $\pi^* \mu_k = \prod_{k' \in \pi^{-1} k} \mu_{k'} \pi^*$.

We would like to note that the folding map is not a Poisson map. However, it sends symplectic leaves to symplectic leaves, and, on being restricted to a symplectic leaf, multiplies the symplectic structure there by a constant.

The two foldings, of the Cartan matrices and of the cluster seeds, are closely related. Namely, let $\pi : \Pi' \to \Pi$ be a folding of the Cartan matrices. Denote by $\mathfrak{W}$ (respectively, $\mathfrak{W}'$) the free seimgroup generated by Π and $-\Pi$ (respectively, by Π' and $-\Pi'$). Let $D \in \mathfrak{W}$, and let $D' = \pi^*(D)$ be the image of D in the semigroup $\mathfrak{W}'$. The proof of the following proposition is rather straightforward and is thus left to the reader.

Proposition 3.9. *In the above notation, there is a natural map $\pi : \mathbf{J}(D') \to \mathbf{J}(D)$, which is a folding of cluster seeds. Moreover, the map of the corresponding seed $\mathcal{X}$-tori commutes with the evaluation map* ev *to the respective Lie groups.*

3.7 A proof of parts 4 and 5 of Theorem 3.5

Let us first prove the formula $\underline{\alpha\beta\alpha\beta \to \beta\alpha\beta\alpha}$. We need the folding map of the simple roots Π' of the Lie group A_3 to the simple roots Π of the Lie group B_2. Let $\Pi' = \{\gamma, \Delta, \eta\}$, $\Pi = \{\alpha, \beta\}$, $\pi(\gamma) = \pi(\eta) = \alpha$, $\pi(\Delta) = \beta$,

$$\begin{pmatrix} C'_{\gamma\gamma} & C'_{\gamma\Delta} & C'_{\gamma\eta} \\ C'_{\Delta\gamma} & C'_{\Delta\Delta} & C'_{\Delta\eta} \\ C'_{\eta\gamma} & C'_{\eta\Delta} & C'_{\eta\eta} \end{pmatrix} = \begin{pmatrix} 2 & -1 & 0 \\ -1 & 2 & -1 \\ 0 & -1 & 2 \end{pmatrix}, \qquad \begin{pmatrix} C_{\alpha\alpha} & C_{\alpha\beta} \\ C_{\beta\alpha} & C_{\beta\beta} \end{pmatrix} = \begin{pmatrix} 2 & -2 \\ -1 & 2 \end{pmatrix}.$$

There is the following sequence of relations in the braid group of type A_3:

$$\begin{aligned}\pi^*(\alpha\beta\alpha\beta) &= \gamma\eta\Delta\gamma\eta\Delta \,\tilde{=}\, \eta\gamma\Delta\gamma\eta\Delta = \eta\Delta\gamma\Delta\eta\Delta = \eta\Delta\gamma\eta\Delta\eta \,\tilde{=}\, \eta\Delta\eta\gamma\Delta\eta \\ &= \Delta\eta\Delta\gamma\Delta\eta = \Delta\eta\gamma\Delta\gamma\eta \,\tilde{=}\, \Delta\gamma\eta\Delta\gamma\eta = \pi^*(\beta\alpha\beta\alpha),\end{aligned} \tag{15}$$

which just shows that π^* is a semigroup homomorphism. Here $\tilde{=}$ stands for the elementary transformations provided by the relation $\gamma\eta = \eta\gamma$.

Hence we have

$$\begin{aligned}&\pi^*(\mathbf{E}^\alpha\mathbf{E}^\beta H^\alpha(x)H^\beta(y)\mathbf{E}^\alpha\mathbf{E}^\beta) \\ &\quad= \mathbf{E}^\gamma\mathbf{E}^\eta\mathbf{E}^\Delta H^\gamma(x)H^\eta(x)H^\Delta(y)\mathbf{E}^\gamma\mathbf{E}^\eta\mathbf{E}^\Delta \\ &\quad= \mathbf{E}^\eta\mathbf{E}^\gamma\mathbf{E}^\Delta H^\gamma(x)H^\eta(x)H^\Delta(y)\mathbf{E}^\gamma\mathbf{E}^\eta\mathbf{E}^\Delta \\ &\quad= H^\gamma(1+x)H^\Delta(1+x^{-1})^{-1}\mathbf{E}^\eta\mathbf{E}^\Delta H^\Delta(x^{-1})\mathbf{E}^\gamma\mathbf{E}^\Delta H^\Delta(z+xz) \\ &\qquad\cdot H^\eta(y)\mathbf{E}^\eta\mathbf{E}^\Delta H^\gamma(1+x^{-1})^{-1} \\ &\quad= \cdots = H^\eta(a')H^\gamma(a')H^\Delta(b')\mathbf{E}^\Delta\mathbf{E}^\gamma\mathbf{E}^\eta H^\gamma(x')H^\eta(x') \\ &\qquad\cdot H^\Delta(y')\mathbf{E}^\Delta\mathbf{E}^\gamma\mathbf{E}^\eta H^\eta(p')H^\gamma(p')H^\Delta(q') \\ &\quad= \pi^*(H^\beta(b')H^\alpha(a')\mathbf{E}^\beta\mathbf{E}^\alpha H^\beta(y')H^\alpha(x') \\ &\qquad\cdot\mathbf{E}^\beta\mathbf{E}^\alpha H^\beta(q')H^\alpha(p')),\end{aligned} \tag{16}$$

where the ellipsis $\cdots$ means repeated application of the formula $\underline{\alpha\beta\alpha \to \beta\alpha\beta}$ corresponding to the sequence of mutations $\mu_1^\gamma\mu_1^\eta\mu_1^\Delta\mu_1^\gamma$, and

$$\begin{aligned}a' &= \frac{1+x+2xy+xy^2}{1+x+xy}, & b' &= \frac{xy^2}{1+x+2xy+xy^2}, \\ p' &= 1+x+xy, & q' &= \frac{x(1+x+2xy+x^2y)}{(1+x+xy)^2}, \\ x' &= \frac{y}{1+x+2xy+xy^2}, & y' &= \frac{(1+x+xy)^2}{xy^2}.\end{aligned} \tag{17}$$

This proves the $\underline{\alpha\beta\alpha\beta \to \beta\alpha\beta\alpha}$ claim of Proposition 3.6. Further, from this we easily get, by adding pairs of elements of the Cartan group on both sides of (16), a birational transformation

$$\Psi_{B_2} : \mathbb{Q}(a,b,p,q,x,y) \longrightarrow \mathbb{Q}(a'',b'',p'',q'',x'',y''),$$

which reduces to (17) when $a=b=p=q=1$, and is determined by the formula

$$\begin{aligned}&\pi^*(H^\alpha(a)H^\beta(b)\mathbf{E}^\alpha\mathbf{E}^\beta H^\alpha(x)H^\beta(y)\mathbf{E}^\alpha\mathbf{E}^\beta H^\alpha(p)H^\beta(q)) \\ &\quad= \pi^*(H^\beta(b'')H^\alpha(a'')\mathbf{E}^\beta\mathbf{E}^\alpha H^\beta(y'')H^\alpha(x'')\mathbf{E}^\beta\mathbf{E}^\alpha H^\beta(q'')H^\alpha(p'')).\end{aligned} \tag{18}$$

To prove that map 4 from Theorem 3.5 is a cluster transformation of the original $\mathcal{X}$-torus, we need to show that there exists a sequence of mutations of the cluster seed $\mathbf{J}(\alpha\beta\alpha\beta)$ whose product is equal to the transformation Ψ_{B_2}.

Consider a cluster transformation $L_B : \mathcal{X}_{\alpha\beta\alpha\beta} \longrightarrow \mathcal{X}_{\beta\alpha\beta\alpha}$ given as a composition of three mutations:

$$L_B := \mu_{\binom{\alpha}{1}}\mu_{\binom{\beta}{1}}\mu_{\binom{\alpha}{1}}.$$

Let us show that the map Ψ_{B_2} is equal to the cluster transformation L_B.

The four nontrivial transformations in (15) give rise to a cluster transformation

$$L_A : \mathcal{X}_{\gamma\eta\Delta\gamma\eta\Delta} \longrightarrow \mathcal{X}_{\Delta\gamma\eta\Delta\gamma\eta}, \qquad L_A = \mu_{\binom{\gamma}{1}}\mu_{\binom{\eta}{1}}\mu_{\binom{\Delta}{1}}\mu_{\binom{\gamma}{1}}$$

given by composition of the corresponding sequence of four mutations. There is a diagram

$$\begin{array}{ccc} \mathcal{X}_{\alpha\beta\alpha\beta} & \hookrightarrow & \mathcal{X}_{\gamma\eta\Delta\gamma\eta\Delta}, \\ {\scriptstyle L_B}\downarrow & & \downarrow {\scriptstyle L_A}, \\ \mathcal{X}_{\beta\alpha\beta\alpha} & \hookrightarrow & \mathcal{X}_{\Delta\gamma\eta\Delta\gamma\eta}, \end{array} \tag{19}$$

where the horizontal arrows are the folding embeddings.

Lemma 3.10. *The diagram* (19) *is commutative.*

Proof. Consider the following cluster transformation, which is the image under the folding embedding of L_B:

$$\widehat{L}_B : \mathcal{X}_{\gamma\eta\Delta\gamma\eta\Delta} \longrightarrow \mathcal{X}_{\delta\gamma\eta\delta\gamma\eta}, \qquad \widehat{L}_B := \mu_{\binom{\eta}{1}}\mu_{\binom{\gamma}{1}}\mu_{\binom{\Delta}{1}}\mu_{\binom{\eta}{1}}\mu_{\binom{\gamma}{1}}. \tag{20}$$

It evidently makes the following diagram commutative:

$$\begin{array}{ccc} \mathcal{X}_{\alpha\beta\alpha\beta} & \hookrightarrow & \mathcal{X}_{\gamma\eta\Delta\gamma\eta\Delta}, \\ {\scriptstyle L_B}\downarrow & & \downarrow {\scriptstyle \widehat{L}_B}, \\ \mathcal{X}_{\beta\alpha\beta\alpha} & \hookrightarrow & \mathcal{X}_{\Delta\gamma\eta\Delta\gamma\eta}. \end{array} \tag{21}$$

It remains to show that the cluster transformations $\widehat{L}_B$ and L_A are equal. This can be done by computing the effect of the action of the latter sequence of mutations on the $\mathcal{X}$-torus, and checking that it coincides with the transformation Ψ_{B_2}. Another way is to use explicitly the pentagon relations. For the connoisseurs of the cluster varieties we give another proof just by drawning pictures. The seed $\mathbf{J}(\gamma\eta\Delta\gamma\eta\Delta)$ is of the finite type A_3; it has only a finite number of different seeds parametrized by triangulations of a hexagon. (See Appendix B, where we discuss this model and the isomorphism of the corresponding cluster variety with the configuration space of 6 points in $\mathbb{P}^1$ and of 3 flags in PGL_4.) In this framework mutations correspond to removing an edge of the triangulation and replacing it by another diagonal of the arising quadrilateral. Thus the first sequence of mutations corresponds to the sequence of triangualtions shown in Figure 1, while the second one is shown in Figure 2.

The lemma and hence part 4 of Theorem 3.5 are proved.

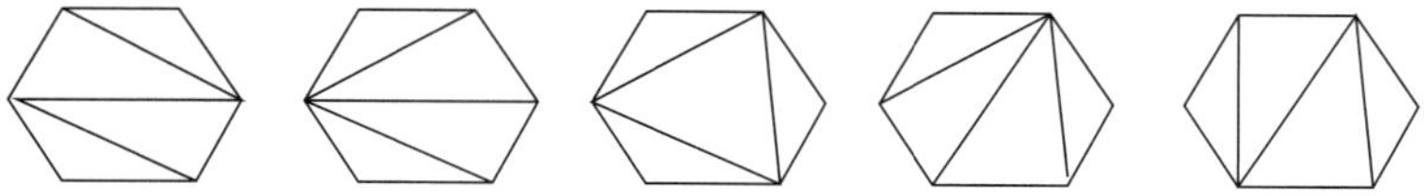

Fig. 1. The sequence of 4 mutations corresponding to L_A.

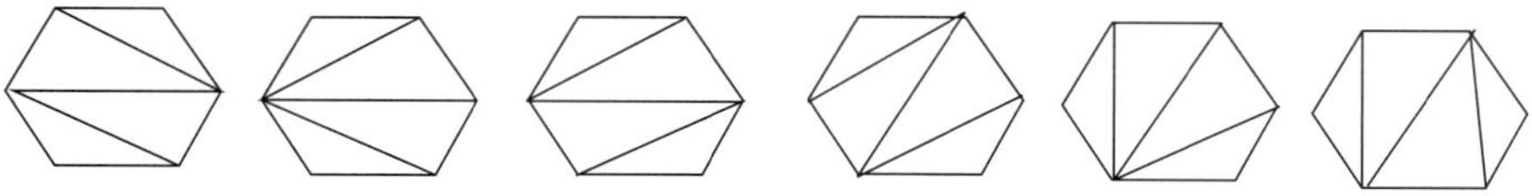

Fig. 2. The sequence of 5 mutations corresponding to $\widehat{L}_B$.

Now let us proceed to the proof of the formula $\alpha\beta\alpha\beta\alpha\beta \to \beta\alpha\beta\alpha\beta\alpha$. Consider the folding map π of the set of simple roots Π' of the Lie algebra D_4 to the set of simple roots Π of the Lie algebra G_2. Let $\Pi' = \{\gamma, \eta, \rho, \Delta\}$, $\Pi = \{\alpha, \beta\}$, $\pi(\gamma) = \pi(\eta) = \pi(\rho) = \alpha$, $\pi(\Delta) = \beta$.

$$\begin{pmatrix} C'_{\gamma\gamma} & C'_{\gamma\eta} & C'_{\gamma\rho} & C'_{\gamma\Delta} \\ C'_{\eta\gamma} & C'_{\eta\eta} & C'_{\eta\rho} & C'_{\eta\Delta} \\ C'_{\rho\gamma} & C'_{\rho\eta} & C'_{\rho\rho} & C'_{\rho\Delta} \\ C'_{\Delta\gamma} & C'_{\Delta\eta} & C'_{\Delta\rho} & C'_{\Delta\Delta} \end{pmatrix} = \begin{pmatrix} 2 & 0 & 0 & -1 \\ 0 & 2 & 0 & -1 \\ 0 & 0 & 2 & -1 \\ -1 & -1 & -1 & 2 \end{pmatrix}, \qquad \begin{pmatrix} C_{\alpha\alpha} & C_{\alpha\beta} \\ C_{\beta\alpha} & C_{\beta\beta} \end{pmatrix} = \begin{pmatrix} 2 & -3 \\ -1 & 2 \end{pmatrix}.$$

There is the following sequence of relations in the braid group of type D_4:

$$\begin{aligned}
&\pi^*(\alpha\beta\alpha\beta\alpha\beta) \\
&\quad = \gamma\eta\rho\Delta\gamma\eta\rho\Delta\gamma\eta\rho\Delta \stackrel{\sim}{=} \gamma\eta\rho\Delta\rho\eta\gamma\Delta\gamma\eta\rho\Delta = \gamma\eta\Delta\rho\Delta\eta\gamma\Delta\gamma\eta\rho\Delta \\
&\quad = \gamma\eta\Delta\rho\Delta\eta\Delta\gamma\Delta\eta\rho\Delta = \gamma\eta\Delta\rho\eta\Delta\eta\gamma\Delta\eta\rho\Delta \stackrel{\sim}{=} \gamma\eta\Delta\eta\rho\Delta\gamma\eta\Delta\eta\rho\Delta \\
&\quad = \gamma\Delta\eta\Delta\rho\Delta\gamma\eta\Delta\eta\rho\Delta = \gamma\Delta\eta\Delta\rho\Delta\gamma\Delta\eta\Delta\rho\Delta = \gamma\Delta\eta\Delta\rho\Delta\gamma\Delta\eta\rho\Delta\rho \\
&\quad = \gamma\Delta\eta\rho\Delta\rho\gamma\Delta\eta\rho\Delta\rho \stackrel{\sim}{=} \gamma\Delta\eta\rho\Delta\gamma\rho\Delta\rho\eta\Delta\rho = \gamma\Delta\eta\rho\Delta\gamma\Delta\rho\Delta\eta\Delta\rho \\
&\quad = \gamma\Delta\eta\rho\gamma\Delta\gamma\rho\Delta\eta\Delta\rho \stackrel{\sim}{=} \gamma\Delta\gamma\eta\rho\Delta\rho\gamma\Delta\eta\Delta\rho = \Delta\gamma\Delta\eta\rho\Delta\rho\gamma\Delta\eta\Delta\rho \\
&\quad = \Delta\gamma\Delta\eta\Delta\rho\Delta\gamma\Delta\eta\Delta\rho = \Delta\gamma\eta\Delta\eta\rho\Delta\gamma\Delta\eta\Delta\rho = \Delta\gamma\eta\Delta\eta\rho\Delta\gamma\eta\Delta\eta\rho \\
&\quad \stackrel{\sim}{=} \Delta\gamma\eta\Delta\rho\eta\Delta\eta\gamma\Delta\eta\rho = \Delta\gamma\eta\Delta\rho\Delta\eta\Delta\gamma\Delta\eta\rho = \Delta\gamma\eta\Delta\rho\Delta\eta\gamma\Delta\gamma\eta\rho \\
&\quad = \Delta\gamma\eta\rho\Delta\rho\eta\gamma\Delta\gamma\eta\rho \stackrel{\sim}{=} \Delta\gamma\eta\rho\Delta\gamma\eta\rho\Delta\gamma\eta\rho = \pi^*(\beta\alpha\beta\alpha\beta\alpha).
\end{aligned} \tag{22}$$

It shows that π^* is a homomorphism of semigroups. Here $\stackrel{\sim}{=}$ stands for the equalities which follow from the commutativity relations $\gamma\eta = \eta\gamma$, $\gamma\rho = \rho\gamma$ and $\eta\rho = \rho\eta$.

A computation similar to (16) is too long to write down here; it was done using a computer. The result is

$$\begin{aligned}
&\pi^*(\mathbf{E}^\alpha H^\alpha(x)\mathbf{E}^\beta H^\beta(y)\mathbf{E}^\alpha H^\alpha(z)\mathbf{E}^\beta H^\beta(w)\mathbf{E}^\alpha\mathbf{E}^\beta) \\
&\quad = \pi^*(H^\beta(a')H^\alpha(b')\mathbf{E}^\beta H^\beta(y')\mathbf{E}^\alpha H^\alpha(x')\mathbf{E}^\beta H^\beta(w')\mathbf{E}^\alpha H^\alpha(z') \\
&\qquad\quad \cdot \mathbf{E}^\beta H^\beta(q')\mathbf{E}^\alpha H^\alpha(p')).
\end{aligned} \tag{23}$$

where

$$
\begin{aligned}
a' &= \frac{xR_2}{R_3}, & b' &= R_3,\\
p' &= \frac{xyz^2w}{R_1}, & q' &= \frac{R_1^3}{R_4},\\
x' &= \frac{zR_1R_3}{R_4}, & y' &= \frac{yR_4}{R_2^3},\\
z' &= \frac{R_4}{xyz^2wR_2}, & w' &= \frac{wR_2^3}{R_3^3}
\end{aligned}
\tag{24}
$$

and

$$
\begin{aligned}
R_1 &= xyz^2w + 1 + x + yx + 2yxz + yxz^2,\\
R_2 &= y^2x^2z^3w + y^2x^2z^3 + 3y^2x^2z^2 + 3yx^2z + 2yxz + 3y^2x^2z + 1 + 2x + 2yx\\
&\quad + x^2 + 2yx^2 + y^2x^2,\\
R_3 &= 3x + 3x^2 + 3yx^2 + 3yx^2z + 1 + y^2x^3z^3w + y^2x^3z^3 + 3y^2x^3z^2 + 3yx^3z\\
&\quad + 3y^2x^3z + x^3 + 2yx^3 + y^2x^3,\\
R_4 &= 1 + 3x + 3y^2x^3z^4 + 12y^2x^3z + 6yx^3z + 18y^2x^3z^2 + 12y^2x^3z^3 + x^3\\
&\quad + 2y^2x^3z^3w + 3y^2x^2z^4w + 3y^2x^2z^3w + 3yx + 3y^2x^3z^4w + 6yxz\\
&\quad + 3x^2 + 6yx^2 + 12yx^2z + 3yxz^2 + 3y^2x^3 + 3yx^3 + 3yx^3z^2\\
&\quad + 2y^3x^3z^6w + 6y^3x^3z^5w + 6y^3x^3z^4w + y^3x^3 + 20y^3x^3z^3 + 6y^3x^3z^5\\
&\quad + 6y^3x^3z + 15y^3x^3z^4 + y^3x^3z^6 + 15y^3x^3z^2 + 2y^3x^3z^3w + y^3x^3z^6w^2\\
&\quad + 6yx^2z^2 + 12y^2x^2z + 18y^2x^2z^2 + 12y^2x^2z^3 + 3y^2x^2z^4 + 3y^2x^2.
\end{aligned}
$$

This proves the $\alpha\beta\alpha\beta\alpha\beta \to \beta\alpha\beta\alpha\beta\alpha$ claim of Proposition 3.6.

Just as in the B_2 case, to prove that map 5 from Theorem 3.5 is a cluster transformation, we need to show that there exists a cluster transformation of the seed $\mathbf{J}(\alpha\beta\alpha\beta\alpha\beta)$ which, being transformed by the folding map to the D_4 setup, equals the cluster transformation encoded in the sequence (22), and given explicitly as the left-hand side in the formula (25) below. To do this it is sufficient to show the following equality between two sequences of mutations:

$$
\begin{aligned}
&\mu_{\binom{\rho}{1}}\mu_{\binom{\rho}{2}}\mu_{\binom{\Delta}{1}}\mu_{\binom{\eta}{2}}\mu_{\binom{\eta}{1}}\mu_{\binom{\rho}{1}}\mu_{\binom{\gamma}{1}}\mu_{\binom{\gamma}{2}}\mu_{\binom{\rho}{2}}\mu_{\binom{\rho}{1}}\mu_{\binom{\Delta}{2}}\mu_{\binom{\eta}{2}}\mu_{\binom{\eta}{1}}\mu_{\binom{\Delta}{1}}\mu_{\binom{\gamma}{2}}\mu_{\binom{\rho}{1}}\\
&= \mu_{\binom{\Delta}{2}}\mu_{\binom{\rho}{1}}\mu_{\binom{\eta}{1}}\mu_{\binom{\gamma}{1}}\mu_{\binom{\Delta}{1}}\mu_{\binom{\Delta}{2}}\mu_{\binom{\rho}{2}}\mu_{\binom{\eta}{2}}\mu_{\binom{\gamma}{2}}\mu_{\binom{\Delta}{2}}\mu_{\binom{\rho}{1}}\mu_{\binom{\eta}{1}}\mu_{\binom{\gamma}{1}}\mu_{\binom{\rho}{2}}\mu_{\binom{\eta}{2}}\mu_{\binom{\gamma}{2}}\mu_{\binom{\Delta}{1}}\mu_{\binom{\Delta}{2}}.
\end{aligned}
\tag{25}
$$

Here the second composition is the image under the folding map of the sequence of mutations $\mu_{\binom{\beta}{2}}\mu_{\binom{\alpha}{1}}\mu_{\binom{\beta}{1}}\mu_{\binom{\beta}{2}}\mu_{\binom{\alpha}{2}}\mu_{\binom{\beta}{2}}\mu_{\binom{\alpha}{1}}\mu_{\binom{\alpha}{2}}\mu_{\binom{\beta}{1}}\mu_{\binom{\beta}{2}}$.

This was done by an explicit calculation of the action of the latter sequences of mutations on the $\mathcal{X}$-coordinates, performed by a computer, which showed that it coincides with the transformation given by (23)–(24). It would be interesting to find a proof which relates one of the sequences of mutations in the Lie group of type D_4 to the other by using the pentagon relations.

3.8 The evaluation map ev is a Poisson map

To prove this claim, it is sufficient to prove it in the simplest case.

Proposition 3.11. *Let α be a simple root. Then the evaluation map* ev : $\mathcal{X}_{\mathbf{J}(\alpha)} \hookrightarrow G$ *is a Poisson immersion. So its image is a Poisson subvariety of G, and the induced Poisson structure on $\mathcal{X}_{\mathbf{J}(\alpha)}$ coincides with the one* (8) *for the matrix ε given by* (12).

We deduce the general claim from the proposition by induction, using the following four facts: the multiplication map $G \times G \to G$ is a Poisson map for the standard Poisson structure on G, the evaluation map commutes with the multiplication, i.e., the diagram (4) is commutative, and the left vertical map in that diagram is a Poisson map, and a dominant map, i.e., its image is dense, and thus the induced map of functions is injective.

Proof of Proposition 3.11. The evaluation map is obviously an immersion in our case. Let us recall the standard Poisson structure on G. Let $R \subset \mathfrak{h}^*$ be the set of roots of the Lie algebra $\mathfrak{g}$ with Cartan subalgebra $\mathfrak{h}$, and let $R_+ \subset R$ be the subset of positive roots. The root decomposition of $\mathfrak{g}$ reads as $\mathfrak{g} = \oplus_{\beta \in R} \mathfrak{g}_\beta \oplus \mathfrak{h}$. Let $\{e_\alpha \in \mathfrak{g}_\alpha | \alpha \in R\}$ be a set of root vectors normalized so that $[e_\alpha, [e_{-\alpha}, e_\alpha]] = 2e_\alpha$. The vectors $e_{\pm\alpha}$, $\alpha \in R_+$, are defined by this condition uniquely up to rescaling $e_{\pm\alpha} \to \lambda_\alpha^{\pm 1} e_{\pm\alpha}$. Let $r \in \mathfrak{g} \otimes \mathfrak{g}$ be the standard r-matrix:

$$r = \sum_{\alpha \in R_+} d^\alpha e_\alpha \wedge e_{-\alpha} = \sum_{\alpha \in R_+} d^\alpha (e_\alpha \otimes e_{-\alpha} - e_{-\alpha} \otimes e_\alpha), \tag{26}$$

where as above $d^\alpha = \frac{(\alpha,\alpha)}{2}$. Observe that r does not depend on the choice of the vectors e_α satisfying the above normalization. The Poisson bracket on G is given by a bivector field $P = r^L - r^R$, where r^L (respectively, r^R) is the right-invariant (respectively, left-invariant) bivector field on G which equals r at the identity of G. If we identify the tangent space to G at a point $g \in G$ with $\mathfrak{g}$ using the right shift, the value $P(g)$ of the bivector field P at g is $P(g) = r - \mathrm{Ad}_g\, r$. We apply this formula in the special case when

$$g = \left(\prod_{\beta \in \Pi} H^\beta(x_0^\beta) \right) \mathbf{E}^\alpha H^\alpha(x_1^\alpha). \tag{27}$$

To make the computation we shall use the following formulas:

$$\mathrm{Ad}_{H^\alpha(x)}\, r = r, \qquad \mathrm{Ad}_{H^\alpha(x)}\, e_\alpha \wedge h_\alpha = x e_\alpha \wedge h_\alpha, \qquad \mathrm{Ad}_{\mathbf{E}^\alpha}\, r = r + d^\alpha e_\alpha \wedge h_\alpha. \tag{28}$$

The first two are obvious; for a proof of the third one see below. Using them, one easily derives

$$P(g) = r - \mathrm{Ad}_g\, r = d^\alpha x_0^\alpha h_\alpha \wedge e_\alpha. \tag{29}$$

So to find the Poisson bracket induced on $\mathcal{X}_\alpha$ we need to compute the right-invariant vector fields on $\mathcal{X}_\alpha$ corresponding to h_α and e_α. Obviously, h_α gives rise to a vector field $\sum_\beta C_{\alpha\beta}x_0^\beta \frac{\partial}{\partial x_0^\beta}$. The following computaiton, where g is from (27), shows that e_α gives rise to $\frac{\partial}{\partial x_0^\alpha} - \frac{x_1^\alpha}{x_0^\alpha}\frac{\partial}{\partial x_1^\alpha}$:

$$\begin{aligned}
e_\alpha g &= \frac{d}{dt}\exp te_\alpha g|_{t=0} = \frac{d}{dt}H^\alpha(t)\mathbf{E}^\alpha H^\alpha(t^{-1})g|_{t=0} \\
&= \frac{d}{dt}\left(\prod_{\beta\in\Pi-\{\alpha\}} H^\beta(x_0^\beta)\right) H^\alpha(t)\mathbf{E}^\alpha H^\alpha(t^{-1})H^\alpha(x_0^\alpha)\mathbf{E}^\alpha H^\alpha(x_1^\alpha)|_{t=0} \\
&= \left(\prod_{\beta\in\Pi-\{\alpha\}} H^\beta(x_0^\beta)\right)\frac{d}{dt}H^\alpha(x_0^\alpha+t)\mathbf{E}^\alpha H^\alpha(x_1^\alpha - tx_1^\alpha/x_0^\alpha)|_{t=0} \\
&= \left(\frac{\partial}{\partial x_0^\alpha} - \frac{x_1^\alpha}{x_0^\alpha}\frac{\partial}{\partial x_1^\alpha}\right)g.
\end{aligned}$$

Here we have used the formula $\underline{\alpha\alpha\to\alpha}$ from Proposition 3.6.

Substituting these expressions for the vector fields in (29), we get

$$\begin{aligned}
P &= -\sum_\beta d^\alpha C_{\alpha\beta}x_0^\beta x_0^\alpha\left(\frac{\partial}{\partial x_0^\alpha} - \frac{x_1^\alpha}{x_0^\alpha}\frac{\partial}{\partial x_1^\alpha}\right)\wedge\frac{\partial}{\partial x_0^\beta} \\
&= -\widehat{C}_{\alpha\beta}x_0^\alpha x_0^\beta\frac{\partial}{\partial x_0^\alpha}\wedge\frac{\partial}{\partial x_0^\beta} + \widehat{C}_{\alpha\beta}x_1^\alpha x_0^\beta\frac{\partial}{\partial x_1^\alpha}\wedge\frac{\partial}{\partial x_0^\beta},
\end{aligned} \tag{30}$$

which coincides with the expression given by (8) and (12).

Proof of the third formula in (28). Let $R(\alpha)\subset R$ be the root system for the Dynkin diagram obtained from the initial one by deleting the vertex corresponding to the simple positive root α. Let $R_+(\alpha)$ be the set of its positive roots. Let p be the projection of $\mathfrak{h}^*$ onto its quotient by the subspace spanned by α. Then $p(R)=R(\alpha)$. So we can rewrite the root decomposition as

$$\mathfrak{g} = \oplus_{\beta\in R(\alpha)}(\oplus_{\gamma\in p^{-1}(\beta)}\mathfrak{g}_\gamma)\oplus\mathfrak{h}(\alpha)\oplus i_\alpha(\mathfrak{sl}_2), \tag{31}$$

where $i_\alpha(\mathfrak{sl}_2)$ is the $\mathfrak{sl}(2)$-subalgebra spanned by $e_\alpha, e_{-\alpha}, h_\alpha=[e_\alpha,e_{-\alpha}]$, and $\mathfrak{h}(\alpha)\subset\mathfrak{h}$ is the kernel of α. Observe that the summands are $i_\alpha(\mathfrak{sl}_2)$-invariant.

Let us consider the quadratic Casimir:

$$t = \sum_{\beta\in R} d^\beta e_\beta\otimes e_{-\beta} + \frac{1}{2}\sum_{\beta\in\Pi} d^\beta h_\beta\otimes h^\beta \in \mathfrak{g}\otimes\mathfrak{g}.$$

It can be rewritten as

$$t = t_0 + \sum_{\beta\in R(\alpha)} t_\beta, \tag{32}$$

where

$$t_\beta = \sum_{\gamma\in p^{-1}(\beta)} d^\gamma e_\gamma \otimes e_{-\gamma}, \qquad t_0 = d^\alpha(e_\alpha\otimes e_{-\alpha} + e_{-\alpha}\otimes e_\alpha) + \frac{1}{2}\sum_{\beta\in\Pi} d^\beta h_\beta \otimes h^\beta.$$

Every term of the expression (32) is $i_\alpha(\mathfrak{sl}_2)$-invariant. The r-matrix (26) is decomposed in the same way:

$$r = r_0 + \sum_{\beta\in R_+(\alpha)} t_\beta - \sum_{\beta\in R_-(\alpha)} t_\beta,$$

where t_β are as above and $r_0 = d^\alpha e_\alpha \wedge e_{-\alpha}$. All the terms but the first one r_0 are $i_\alpha(\mathfrak{sl}_2)$-invariant and thus $\mathrm{Ad}_{\mathbf{E}^\alpha}(r) = r - r_0 + \mathrm{Ad}_{\mathbf{E}^\alpha}(r_0) = r - r_0 + \exp(\mathrm{ad}_{e_\alpha})(e_\alpha \wedge e_{-\alpha}) = r + d^\alpha e_\alpha \wedge h_\alpha$. We proved the third formula in (28), and hence Proposition 3.11. As was explained above, this implies that ev is a Poisson map in general.

Therefore, we have completed the proof of Theorem 3.5.

3.9 Duality conjectures and canonical bases

Let w_0 be the longest element of the Weyl group W_G of G. Let $\mathcal{X}_{G,w_0}$ be the corresponding cluster $\mathcal{X}$-variety. Recall that, given a seed $\mathbf{I}$, we defined in [FG2, Section 2] a positive space $\mathcal{A}_{|\mathbf{I}|}$ assigned to it. Let G^L be the Langlands dual group for G. Applying this construction to the seed corresponding to the element w_0 in G^L, we arrive at a positive space $\mathcal{A}_{G^L,w_0}$. Recall that for any semifield $\mathbb{F}$ and a positive space $\mathcal{X}$ there is a set $\mathcal{X}(\mathbb{F})$ of $\mathbb{F}$-points of $\mathcal{X}$ (loc. cit.). Recall the tropical semifield $\mathbb{Z}^t$: it is the set $\mathbb{Z}$ with the following semifield operations: the multiplication and division are given by the usual addition and subtraction in $\mathbb{Z}$, and the semifield addition is given by taking the maximum. Let $\mathcal{A}_{G^L,w_0}(\mathbb{Z}^t)$ be the set of $\mathbb{Z}^t$-points of the positive space $\mathcal{A}_{G^L,w_0}$.

Then, according to the duality conjecture from [FG2, Section 4], there should exist a basis in the algebra $\mathbb{Z}[\mathcal{X}_{G,w_0}]$ of regular functions on the variety $\mathcal{X}_{G,w_0}$, parametrised by the set $\mathcal{A}_{G^L,w_0}(\mathbb{Z}^t)$. Let us explain how it should be related to the (dual) canonical basis of Lusztig [L2].

The dual canonical basis is a basis of regular functions on the Borel subgroup B of G. Since $\mathcal{X}_{G,w_0}$ is birationally equivalent to B, the regular functions on the former are rational, but not necessarily regular, functions on B. Let us say that an element of our conjectural basis in $\mathbb{Z}[\mathcal{X}_{G,w_0}]$ is *regular* if it provides a regular function on B.

Conjecture 3.12. The regular elements of the conjectural basis in $\mathbb{Z}[\mathcal{X}_{G,w_0}]$ form a basis of the space of regular functions on B. Moreover, it coincides with Lusztig's dual canonical basis on B.

Let us try to determine when a rational function F on $\mathcal{X}_{G,w_0}$ is regular on B. Observe that $B = HU$, where U is the maximal unipotent in B. Clearly, F is regular on H. So it remains to determine when it is regular on U.

Pick a reduced decomposition of w_0. Let $(t_{i_1},\dots,t_{i_N})$, where $N = \dim U = l(w_0)$, be the corresponding Lusztig coordinates on B [L1], and $(x_{i_1},\dots,x_{i_{N+r}})$ the corresponding cluster $\mathcal{X}$-coordinates on $\mathcal{X}_{G,w_0}$. Choose coordinates $(h_1,\dots,h_r)$

on H. Evidently the t-coordinates are related to the (x, h)-coordinates by monomial transformations, and vice versa. In particular, any $F \in \mathbb{Z}[\mathcal{X}_{G,w_0}]$ is a Laurent polynomial in $(t_{i_1}, \ldots, t_{i_N})$.

Lemma 3.13. *An $F \in \mathbb{Z}[\mathcal{X}_{G,w_0}]$ is regular on B if and only if for any reduced decomposition of w_0 it is a polynomial in the corresponding coordinates $(t_{i_1}, \ldots, t_{i_N})$.*

Proof. The "only if" part is clear. Let us check the opposite. The subvarieties $\mathrm{ev}(\mathcal{X}_{G,w})$ in G, where $w \in W_G$, are of codimension $l(w_0) - l(w)$, and it follows from the Bruhat decomposition that the complement to their union is of codimension ≥ 2. Any irreducible component of divisors $\mathrm{ev}(\mathcal{X}_{G,w})$ is given by the equation $t_{i_k} = 0$ for a certain reduced decomposition of w_0 and certain k. The lemma follows.

3.10 Examples for PGL_m

Below we show how to visualize, in the case of PGL_m, the combinatorics of the cluster $\mathcal{X}$-coordinates by a wiring diagram. The wiring diagram language is well known [BFZ96]. Our goal is to show how it works for the $\mathcal{X}$-coordinates.

3.10.1

The $\mathcal{X}$-coordinates are assigned to the connected components of the complement to the wiring diagram, except the bottom and top components. The frozen $\mathcal{X}$-coordinates are assigned to the very left and right domains, i.e., to the domains which are not completely bounded by wires. The word itself is encoded by the wiring diagram as follows: we scan the diagram from the left to the right, and assign a generator for each vertex of the diagram: the generator σ_i is assigned to a vertex having $i - 1$ wires above it. See Figure 3, which illustrates the situation for the word $\sigma_3\sigma_1\sigma_2\sigma_1\sigma_3\sigma_2$ for PGL_4.

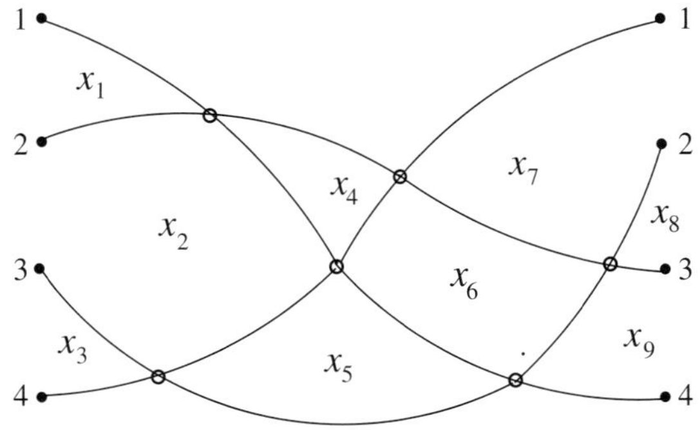

Fig. 3. The wiring diagram and $\mathcal{X}$-coordinates for the word $\sigma_3\sigma_1\sigma_2\sigma_1\sigma_3\sigma_2$.

The corresponding parametrization of the Borel subgroup of upper triangular 4×4 matrices is given by the product

$$H^3(x_3)H^1(x_1)\mathbf{E}^3\mathbf{E}^1H^2(x_2)H^1(x_4)H^3(x_5)\mathbf{E}^2\mathbf{E}^1H^2(x_6)\mathbf{E}^3H^1(x_7)\mathbf{E}^2H^2(x_8)H^3(x_9). \tag{33}$$

Here $\mathbf{E}^i$ is the elementary unipotent matrix corresponding to the ith simple positive root: it has 1s on the diagonal, and the only nonzero nondiagonal element is 1 at the entry $(i, i+1)$. Further, $H^j(t) = \text{diag}(\underbrace{t, \dots, t}_{j}, 1, \dots, 1)$ is the diagonal matrix corresponding to the jth simple coroot. The frozen variables are $x_1, x_2, x_3, x_7, x_8, x_9$. To record the expression (33), we scan the wiring diagram from the left to the right. The intersection points of wires provide the elementary matrices $\mathbf{E}^i$, while the domains contributes the Cartan elements $H^j(x)$. Observe that the order of factors in (33) is by no means uniquely determined: the Cartan elements commute, and some of them commute with some **E**s. The wiring diagram, considered modulo isotopy, encodes the element (33) in a more adequate way.

3.10.2

The Poisson structure tensor is encoded by the wiring diagram as follows. Take a (connected) domain of the wiring diagram corresponding to a nonfrozen coordinate x_0. It can be rather complicated, sharing boundary with many other domains; see Figure 4. However, it has two distinguished vertices, the very left and right ones, shown by circles. There are at most six outside domains sharing one of these two vertices. Let x_j be an $\mathcal{X}$-coordinate for the given wiring diagram. The Poisson bracket $\{x_0, x_j\} = \varepsilon_{0j}x_0x_j$ is nonzero if and only if x_j is assigned to one of those domains. One has $\varepsilon_{0j} = \pm 1$, and the sign is shown by arrows on the picture: $\varepsilon_{0j} = 1$ if and only if the arrow goes from x_0 to x_j. The Posson bracket between two frozen variables is obtained similarly, but the coefficient is divided by 2.

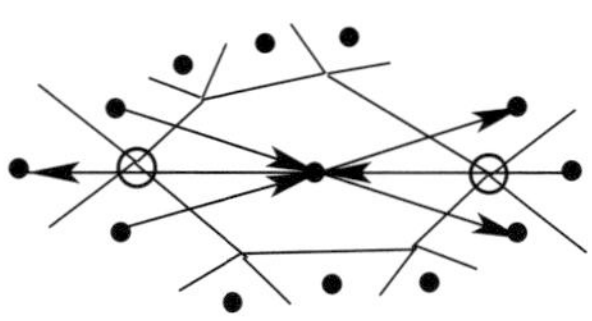

Fig. 4. Reading the Poisson tensor from a wiring diagram.

3.10.3

Yet another example, corresponding to a "standard" reduced decomposition of w_0 for PGL_m, is given on the left-hand side of Figure 5. The nonfrozen coordinates, shown by black points, give ries to coordinates on $H\backslash B/H$. Observe that there is a canonical birational isomorphism between $H\backslash B/H$ and the configuration space of triples of flags $\text{Conf}_3(\mathcal{B})$.

Using it, one can show that the nonfrozen coordinates in this case are identified with the canonical coordinates on the configuration space of triples of flags for PGL_m

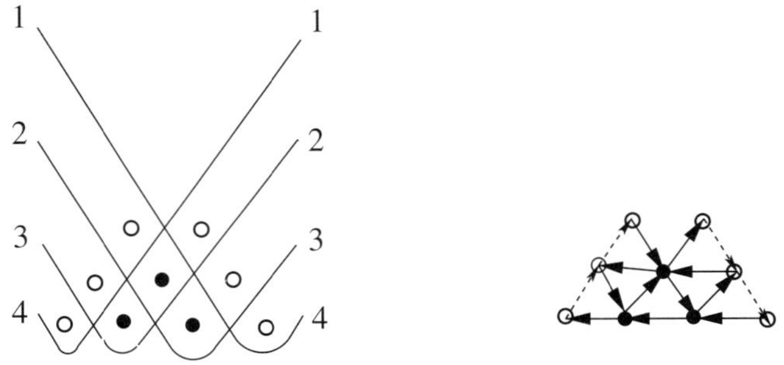

Fig. 5. The $\mathcal{X}$-coordinates for a standard reduced decomposition of w_0 for PGL_4.

introduced in [FG1, Section 9]. The right-hand side of Figure 5 shows the Poisson tensor. Its coefficients between the frozen coordinates are shown by dotted arrows.

4 Cluster $\mathcal{X}$-variety structure of partial flag varieties

Let P be a parabolic subgroup of G, and $P = M_P U_P$ its Levi decomposition. So G/P is a partial flag variety. Let w_0^G (respectively, w_0^M) be the longest element of the Weyl group of G (respectively, M_P). Write $w_0^G = w_0^M w_0^U$. Take a reduced decomposition $\widetilde{w}_0^U$ of w_0^U. It gives rise to a coordinate system on G/P as follows. Take the seed corresponding to $\widetilde{w}_0^U$ and the corresponding seed $\mathcal{X}$-torus $\mathcal{X}_{\widetilde{w}_0^U}$. The frozen part of the torus $\mathcal{X}_{\widetilde{w}_0^U}$ is a product $H_L \times H_R$ of two Cartan subgroups, called the left (H_L) and right (H_R) frozen Cartan subgroups. The canonical projection $\mathcal{X}_{\widetilde{w}_0^U} \to G/P$ provides a regular open embedding $H_L \backslash \mathcal{X}_{\widetilde{w}_0^U} \hookrightarrow U_P^{\mathrm{opp}} \hookrightarrow G/P$, where U_P^{opp} is the subgroup opposite to the unipotent radical U_P. It is the coordinate system on G/P corresponding to $\widetilde{w}_0^U$. It follows from Theorem 3.5 that the collection of coordinate systems on G/P corresponding to different reduced decompositions of w_0^U provides a set of cluster $\mathcal{X}$-coordinate systems.

Observe that P is a Poisson subgroup of G, so G/P has a natural Poisson structure. It follows from Theorem 3.5 that it coincides with the one given by the cluster $\mathcal{X}$-variety structure on G/P.

Example. For the Grassmannian $\mathrm{Gr}_k(n)$ of k-planes in an n-dimensional vector space the above construction provides a *canonical* coordinate system. Indeed, there is only one reduced decomposition of w_0^U for the Grassmannian. See Figure 6, where the case of $\mathrm{Gr}_3(6)$ is illustrated. The wiring diagram for the Grassmannian is on the right of the dotted vertical line. The frozen variables are shown by circles. The oriented graph providing the Poisson structure tensor is on the right. It is calculated using the recipe from Section 3.10. It coincides with the one defined in [GSV1].

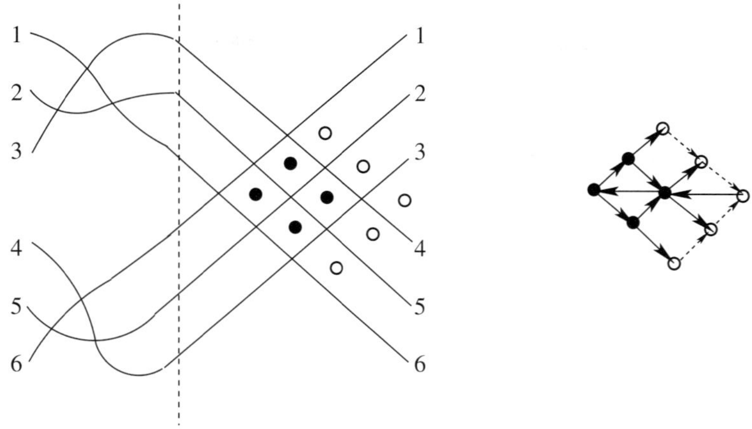

Fig. 6. The cluster $\mathcal{X}$-variety structure of $\mathrm{Gr}_3(6)$.

Appendix A: The braid group of type G_2 acts on triples of flags of type G_2

In this section we study the cluster $\mathcal{X}$-variety corresponding to the moduli space $\mathrm{Conf}_3(\mathcal{B}_{G_2})$ of configurations of triples of flags of type G_2. The combinatorial structure of a cluster $\mathcal{X}$-variety is reflected in the topology of the *modular orbifold*, defined in [FG2, Section 2]. Below we recall its definition. Then we determine the modular orbifold for the moduli space $\mathrm{Conf}_3(\mathcal{B}_{G_2})$, and compute its fundamental group, which turns out to be the braid group of type G_2.

A.1 The modular orbifold of a cluster $\mathcal{X}$-variety

It is constructed in three steps:

1. We assign to a seed $\mathbf{I} = (I, I_0, \varepsilon, d)$ a simplex $S_{\mathbf{I}}$ equipped with a bijection of the set of its codimension one faces with I. It induces a bijection between the set of its vertices and I: a vertex is labeled by the same element as the opposite codimension one face. Recall that an element $k \in I - I_0$ gives rise to a seed mutation $\mathbf{I} \to \mu_k(\mathbf{I})$. We glue the simplices $S_{\mathbf{I}}$ and $S_{\mu_k(\mathbf{I})}$ along their codimension one faces labeled by k, matching the vertices labeled by the same elements. We continue this process by making all possible mutations and gluing the corresponding simplices. In this way we get a simplicial complex $S_{|\mathbf{I}|}$.
2. We identify simplices corresponding to isomorphic seeds, getting a simplicial complex $\overline{S}_{|\mathbf{I}|}$.
3. We remove from $\overline{S}_{|\mathbf{I}|}$ certain faces of codimension ≥ 2 defined as follows. Recall that a seed $\mathbf{I}$ provides a torus $\mathcal{X}_{\mathbf{I}}$ with a coordinate system $\{x_i^{\mathbf{I}}\}$. Let $\varphi : \mathbf{I} \to \mathbf{I}'$ be a cluster transformation of seeds. We say that it is an $\mathcal{X}$-equivalence if the induced cluster transformation $\varphi_{\mathcal{X}} : \mathcal{X}_{\mathbf{I}} \to \mathcal{X}_{\mathbf{I}'}$ is an isomorphism respecting the coordinates: $\varphi_{\mathcal{X}}^* x_{\varphi(i)}^{\mathbf{I}'} = x_i^{\mathbf{I}}$. Let F be a face of $S_{|\mathbf{I}|}$. Consider the set of

$\mathcal{X}$-equivalence classes of seeds $\mathbf{I}'$ such that the simplex $S_{\mathbf{I}'}$ in $S_{|\mathbf{I}|}$ contains F. We say that F is of *infinite type* if this set is infinite. Removing from $\overline{S}_{|\mathbf{I}|}$ all faces of infinite type, we get the *modular orbifold* $M_{|\mathbf{I}|}$.

We proved (loc. cit.) that it is indeed an orbifold. Its dimension is the dimension of the cluster $\mathcal{X}$-variety minus one.

A.2 Main results

In the ε-finite case the modular orbifold is glued from a finite number of simplices. It is noncompact, unless the cluster $\mathcal{X}$-variety is of finite cluster type. In general it cannot be compactified by an orbifold, but sometimes it can. Here is the main result.

Theorem A.1.

(a) *The cluster $\mathcal{X}$-variety corresponding to the moduli space* $\mathrm{Conf}_3(\mathcal{B}_{G_2})$ *is of ε-finite type. The number of nonisomorphic seeds assigned to it is seven.*
(b) *The corresponding modular complex is homeomorphic to $S^3 - L$, where L is a link with two connected components, and $\pi_1(S^3 - L)$ is isomorphic to the braid group of type G_2.*

The *mapping class group* of a cluster $\mathcal{X}$-variety was defined in [FG2, Section 2]. It acts by automorphisms of the cluster $\mathcal{X}$-variety. It is always infinite if the cluster structure is of ε-finite, but not of finite, type. Theorem A.1 immediately implies the following.

Corollary A.2. *The mapping class group of this cluster $\mathcal{X}$-variety corresponding to* $\mathrm{Conf}_3(\mathcal{B}_{G_2})$ *is an infinite quotient of the braid group of type G_2.*

Remark. According to [FG2, Hypothesis 2.19], the modular complex of a cluster $\mathcal{X}$-variety is the classifying space (in general orbispace) for the corresponding mapping class group. This plus Theorem A.1 imply that the mapping class group should be isomorphic to the braid group of type G_2.

A.3 Proof of Theorem A.1

The cluster structure of the moduli space $\mathrm{Conf}_3(\mathcal{B}_{G_2})$ can be described by a seed with 4 vertices shown in Figure 7. The multipliers are equal to 3 for the top two vertices, and 1 for the bottom two.

We claim that mutating this seed we get exactly 7 different seeds shown in Figure 8, where we keep the same convention about the multipliers. To prove this claim we list below the 14 pairs of seeds from Figure 8 related by mutations.

Let us give an example explaining our notation. We denote by (a_k, b_k, c_k, d_k) the four vertices of seed number k in Figure 8 (k counts the seeds from the left to the right). The vertices are arranged as in Figure 7: (a_k, b_k) are the top two, and (c_k, d_k) the bottom vertices. The mutation λ_6 mutates the seed (a_2, b_2, c_2, d_2) at the vertex d_2, producing the seed (a_3, b_3, c_3, d_3). Forming the cluster/modular complex, we

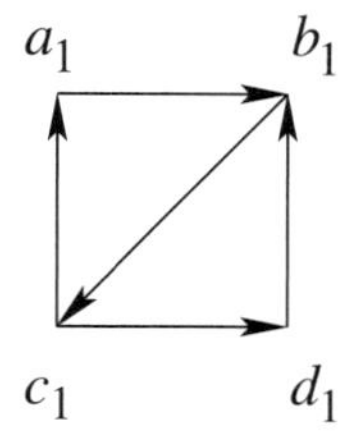

Fig. 7. The original seed.

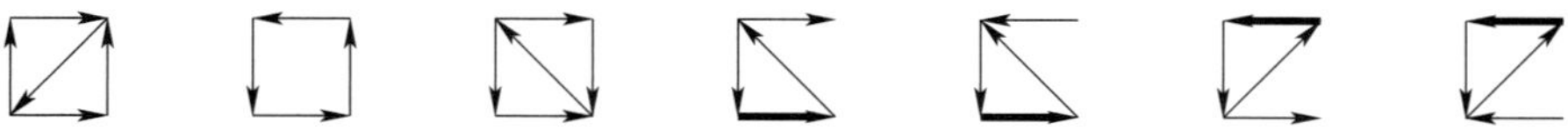

Fig. 8. The seven seeds.

glue the corresponding two tetrahedra so that the face (a_2, b_2, c_2) of the first is glued to the face (b_3, a_3, d_3) of the second, matching the ith vertices of these two triangles. We record this information as follows:

$$\lambda_6 : \begin{pmatrix} a_2 & b_2 & c_2 \\ b_3 & a_3 & d_3 \end{pmatrix}.$$

Below we list mutations $\lambda_1, \ldots, \lambda_{14}$, which together with their inverses give us all the mutations:

$$\lambda_1 : \begin{pmatrix} b_1 & c_1 & d_1 \\ b_2 & c_2 & d_2 \end{pmatrix}, \quad \lambda_2 : \begin{pmatrix} a_1 & c_1 & d_1 \\ b_4 & d_4 & c_4 \end{pmatrix}, \quad \lambda_3 : \begin{pmatrix} a_1 & b_1 & d_1 \\ a_7 & b_7 & d_7 \end{pmatrix}, \quad \lambda_4 : \begin{pmatrix} a_1 & b_1 & c_1 \\ b_2 & a_2 & d_2 \end{pmatrix},$$

$$\lambda_5 : \begin{pmatrix} a_2 & c_2 & d_2 \\ a_3 & c_3 & d_3 \end{pmatrix}, \quad \lambda_6 : \begin{pmatrix} a_2 & b_2 & c_2 \\ b_3 & a_3 & d_3 \end{pmatrix}, \quad \lambda_7 : \begin{pmatrix} b_3 & c_3 & d_3 \\ b_5 & d_5 & c_5 \end{pmatrix},$$

$$\lambda_8 : \begin{pmatrix} a_3 & b_3 & c_3 \\ a_6 & b_6 & d_6 \end{pmatrix}, \quad \lambda_9 : \begin{pmatrix} a_4 & c_4 & d_4 \\ a_5 & c_5 & d_5 \end{pmatrix}, \quad \lambda_{10} : \begin{pmatrix} a_4 & b_4 & d_4 \\ a_4 & b_4 & c_4 \end{pmatrix}, \quad \lambda_{11} : \begin{pmatrix} a_5 & b_5 & d_5 \\ a_5 & b_5 & c_5 \end{pmatrix},$$

$$\lambda_{12} : \begin{pmatrix} b_6 & c_6 & d_6 \\ a_6 & c_6 & d_6 \end{pmatrix}, \quad \lambda_{13} : \begin{pmatrix} a_6 & b_6 & c_6 \\ a_7 & b_7 & c_7 \end{pmatrix}, \quad \lambda_{14} : \begin{pmatrix} b_7 & c_7 & d_7 \\ a_7 & c_7 & d_7 \end{pmatrix}.$$

We present in Figure 9 the 1-skeleton of the simplicial complex dual to the modular complex. It has 7 vertices corresponding to the seven different seeds, and every vertex is connected with the four vertices related to it by mutations.

The gluing data for the edges

Here is how we glue the edges of seven tetrahedra forming the modular complex:

$$\begin{pmatrix} b_1 \\ d_1 \end{pmatrix} \xrightarrow{\lambda_1} \begin{pmatrix} b_2 \\ d_2 \end{pmatrix} \xrightarrow{\lambda_4^{-1}} \begin{pmatrix} a_1 \\ c_1 \end{pmatrix} \xrightarrow{\lambda_2} \begin{pmatrix} b_4 \\ d_4 \end{pmatrix} \xrightarrow{\lambda_{10}} \begin{pmatrix} b_4 \\ c_4 \end{pmatrix} \xrightarrow{\lambda_2^{-1}} \begin{pmatrix} a_1 \\ d_1 \end{pmatrix} \xrightarrow{\lambda_3} \begin{pmatrix} a_7 \\ d_7 \end{pmatrix}$$

$$\xrightarrow{\lambda_{14}^{-1}} \begin{pmatrix} b_7 \\ d_7 \end{pmatrix} \xrightarrow{\lambda_3^{-1}} \begin{pmatrix} b_1 \\ d_1 \end{pmatrix},$$

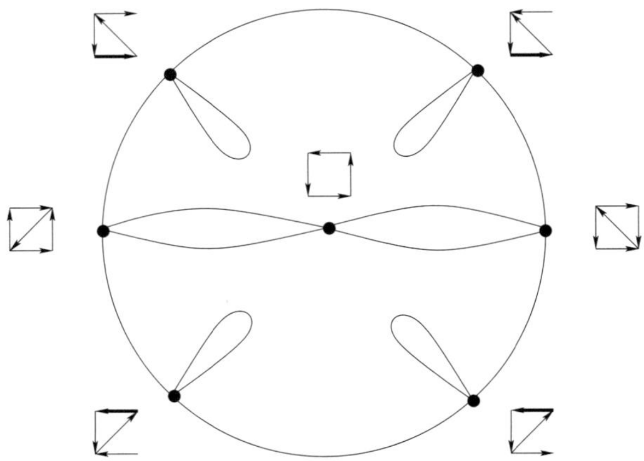

Fig. 9. The 1-skeleton of the dual to the modular complex.

$$\begin{pmatrix}a_2\\c_2\end{pmatrix} \xrightarrow{\lambda_5} \begin{pmatrix}a_3\\c_3\end{pmatrix} \xrightarrow{\lambda_8} \begin{pmatrix}a_6\\d_6\end{pmatrix} \xrightarrow{\lambda_{12}^{-1}} \begin{pmatrix}b_6\\d_6\end{pmatrix} \xrightarrow{\lambda_8^{-1}} \begin{pmatrix}b_3\\c_3\end{pmatrix} \xrightarrow{\lambda_7} \begin{pmatrix}b_5\\d_5\end{pmatrix} \xrightarrow{\lambda_{11}} \begin{pmatrix}b_5\\c_5\end{pmatrix}$$
$$\xrightarrow{\lambda_7^{-1}} \begin{pmatrix}b_3\\d_3\end{pmatrix} \xrightarrow{\lambda_6^{-1}} \begin{pmatrix}a_2\\c_2\end{pmatrix},$$
$$\begin{pmatrix}c_1\\d_1\end{pmatrix} \xrightarrow{\lambda_1} \begin{pmatrix}c_2\\d_2\end{pmatrix} \xrightarrow{\lambda_5} \begin{pmatrix}c_3\\d_3\end{pmatrix} \xrightarrow{\lambda_7} \begin{pmatrix}d_5\\c_5\end{pmatrix} \xrightarrow{\lambda_9^{-1}} \begin{pmatrix}d_4\\c_4\end{pmatrix} \xrightarrow{\lambda_2^{-1}} \begin{pmatrix}c_1\\d_1\end{pmatrix},$$
$$\begin{pmatrix}a_1\\b_1\end{pmatrix} \xrightarrow{\lambda_4} \begin{pmatrix}b_2\\a_2\end{pmatrix} \xrightarrow{\lambda_6} \begin{pmatrix}a_3\\b_3\end{pmatrix} \xrightarrow{\lambda_8} \begin{pmatrix}a_6\\b_6\end{pmatrix} \xrightarrow{\lambda_{13}} \begin{pmatrix}a_7\\b_7\end{pmatrix} \xrightarrow{\lambda_3^{-1}} \begin{pmatrix}a_1\\b_1\end{pmatrix},$$
$$\begin{pmatrix}b_1\\c_1\end{pmatrix} \xrightarrow{\lambda_1} \begin{pmatrix}b_2\\c_2\end{pmatrix} \xrightarrow{\lambda_6} \begin{pmatrix}a_3\\d_3\end{pmatrix} \xrightarrow{\lambda_5^{-1}} \begin{pmatrix}a_2\\d_2\end{pmatrix} \xrightarrow{\lambda_4^{-1}} \begin{pmatrix}b_1\\c_1\end{pmatrix},$$
$$\begin{pmatrix}a_4\\c_4\end{pmatrix} \xrightarrow{\lambda_9} \begin{pmatrix}a_5\\c_5\end{pmatrix} \xrightarrow{\lambda_{11}^{-1}} \begin{pmatrix}a_5\\d_5\end{pmatrix} \xrightarrow{\lambda_9^{-1}} \begin{pmatrix}a_4\\d_4\end{pmatrix} \xrightarrow{\lambda_{10}} \begin{pmatrix}a_4\\c_4\end{pmatrix},$$
$$\begin{pmatrix}a_6\\c_6\end{pmatrix} \xrightarrow{\lambda_{12}^{-1}} \begin{pmatrix}b_6\\c_6\end{pmatrix} \xrightarrow{\lambda_{13}} \begin{pmatrix}b_7\\c_7\end{pmatrix} \xrightarrow{\lambda_{14}} \begin{pmatrix}a_7\\c_7\end{pmatrix} \xrightarrow{\lambda_{13}^{-1}} \begin{pmatrix}a_6\\c_6\end{pmatrix},$$
$$\begin{pmatrix}c_6\\d_6\end{pmatrix} \xrightarrow{\lambda_{12}} \begin{pmatrix}c_6\\d_6\end{pmatrix}; \qquad \begin{pmatrix}c_7\\d_7\end{pmatrix} \xrightarrow{\lambda_{14}} \begin{pmatrix}c_7\\d_7\end{pmatrix};$$
$$\begin{pmatrix}a_4\\b_4\end{pmatrix} \xrightarrow{\lambda_{10}} \begin{pmatrix}a_4\\b_4\end{pmatrix}; \qquad \begin{pmatrix}a_5\\b_5\end{pmatrix} \xrightarrow{\lambda_{11}} \begin{pmatrix}a_5\\b_5\end{pmatrix}.$$

After the gluing, we get four vertices:

$$a_1 = a_2 = a_3 = a_6 = a_7 = b_1 = b_2 = b_3 = b_4 = b_5 = b_6 = b_7,$$
$$c_1 = c_2 = c_3 = c_4 = c_5 = d_1 = d_2 = d_3 = d_4 = d_5 = d_6 = d_7,$$
$$a_4 = a_5, \qquad c_6 = c_7.$$

Thus we have 7 tetrahedra, 14 faces, 11 triangles, and 4 vertices. So the Euler characteristic is 0.

Each of the edges λ_i, $i = 1, \dots, 14$, of the dual modular complex gives us a generator of the fundamental groupoid. Each of the 11 edges listed above of the original complex gives a relation in the fundamental groupoid.

To compute the fundamental group of the modular complex we use the following algorithm. Recall that a spanning tree of a graph is a maximal contractible subgraph of the graph. Clearly, a spanning tree contains all vertices of the graph. Let us shrink a spanning tree. Then every edge of the graph which does not belong to the spanning tree gives rise to a nontrivial loop on the quotient. These loops generate the fundamental group of the quotient based at the contracted spanning tree.

1. Choose a spanning tree of the dual to the modular complex.
2. Then the fundamental group has the following presentation:

Generators correspond to the edges of the dual modular complex which do not belong to the spanning tree. Relations correspond to the edges of the modular complex as follows: Take all triangles containing the given edge. A coorientation of this edge gives rise to a cyclic order of this set. Then the product of the corresponding generators, in an order compatible with the cyclic order, is a relation.

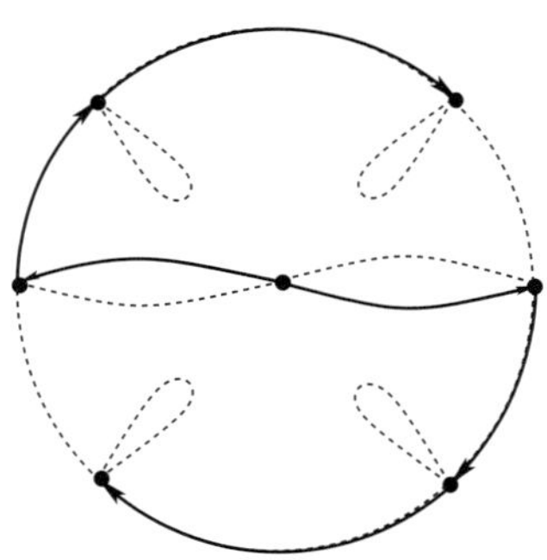

Fig. 10. A spanning tree.

Let us implement this algorithm. Choose a spanning tree shown by bold arcs in Figure 10. Then there are 8 generators, corresponding to mutations λ_3, λ_4, λ_5, λ_7, λ_{10}, λ_{11}, λ_{12}, λ_{14}. The relations can be read off from the gluing data of the edges:

$$\lambda_4^{-1}\lambda_{10}\lambda_3\lambda_{14}^{-1}\lambda_3^{-1} = 1, \qquad \lambda_5\lambda_{12}^{-1}\lambda_7\lambda_{11}\lambda_7^{-1} = 1, \qquad \lambda_5\lambda_7 = 1,$$
$$\lambda_4\lambda_3^{-1} = 1, \qquad \lambda_5^{-1}\lambda_4^{-1} = 1, \qquad \lambda_{11}^{-1}\lambda_{10} = 1, \qquad \lambda_{12}^{-1}\lambda_{14} = 1.$$

Let us simplify these equations. From the last five equations, we get

$$\rho = \lambda_3 = \lambda_4 = \lambda_5^{-1} = \lambda_7; \qquad a = \lambda_{10} = \lambda_{11}; \qquad x = \lambda_{12} = \lambda_{14}.$$

Substituting this into the first two equations we get an equivalent presentation of our group: the generators are a, x, β; they satisfy two relations:

$$\rho^{-1}\alpha\beta x^{-1}\rho^{-1} = 1, \qquad \rho^{-1}x^{-1}\rho a\rho^{-1} = 1. \tag{34}$$

Set $b = \rho x^{-1}\rho^{-1}$, i.e., $x^{-1} = \rho^{-1}b\rho$. Then the first equation in (34) is equivalent to $\beta = ab$. Thus the group is generated by a, b. The only relation comes from the second equation in (34). Therefore, substituting the above expressions for β and x^{-1}, we arrive at

$$(ab)^{-1}(ab)^{-1}b(ab)aba(ab)^{-1} = 1 \Leftrightarrow bababa = ababab, \tag{35}$$

which is the defining relation for the braid group of type G_2. The proof of Theorem A.1 is finished.

A.4 The action of the braid group of type G_2

This group is generated by two elements a, b subject to the single relation $ababab = bababa$. Let a be the composition of the three mutations at the vertices b_1, c_1, b_1, and b be the composition of the three mutations at the vertices c_1, b_1, c_1. Then one checks that they satisfy the above relation. So they are generators of the braid group of type G_2.

A.5 $S^3 - L$ and the discriminant variety for the Coxeter group of type G_2

Let W_{G_2} be the Coxeter group of type G_2. It is isomorphic to the dihedral group of order 12. It acts on the two-dimensional complex vector space V_2, the Cartan subalgebra of the complex Lie algebra of type G_2, equipped with a configuration of six one-dimensional subspaces, corresponding to the kernels of the roots. The group W_{G_2} acts freely on the complement V_2^{reg} to the union of these six lines. The quotient V_2^{reg}/W_{G_2} is known to be a $K(\pi, 1)$ space for $\pi = W_{G_2}$. The group $\mathbb{R}_+^*$ acts by dilatations on V_2^{reg}, commuting with the W_{G_2}-action. Hence $V_2^{\text{reg}}/R_+^* W_{G_2}$ is also a $K(\pi, 1)$ space for $\pi = W_{G_2}$.

Conjecture A.3. The space $V_2^{\text{reg}}/R_+^* W_{G_2}$ is homeomorphic to $S^3 - L$.

Here is evidence. The quotient of V_2 by the action of the group W_{G_2} is isomorphic to $\mathbb{C}^2$. The $\mathbb{R}_+^*$-action on V_2 descends to an $\mathbb{R}_+^*$-action on $V_2^{\text{reg}}/W_{G_2} = \mathbb{C}^2$ given by $t : (z_1, z_2) \longmapsto (t^2 z_1, t^6 z_2)$. Thus the quotient space $V_2^{\text{reg}} - \{0\}/W_{G_2}\mathbb{R}_+^*$ is a sphere S^3, given as the quotient of $\mathbb{C}^2 - \{0.0\}$ by the $(2, 6)$-weighted $\mathbb{R}_+^*$-action.

The intersection of each line with the unit sphere in V_2 is a circle. The action of the group W_{G_2} has two orbits on the set of the six lines, and hence on the set of six circles. So we get two circles in the quotient. They are the two connected components of the link L in S^3.

Appendix B: Cluster structure of the moduli of triples of flags in PGL$_4$

B.1 The cluster $\mathcal{X}$-structure for the moduli space $\mathcal{M}_{0,n+3}$

Recall that $\mathcal{M}_{0,n+3}$ is the moduli space of configurations of $n+3$ distinct points on $\mathbb{P}^1$. Following [FG1, Section 9], we provide it with a structure of cluster $\mathcal{X}$-variety as follows.

Let us consider an $(n+3)$-gon whose vertices are labeled by a configuration $(x_1, \ldots, x_{n+3})$ of points on $\mathbb{P}^1$, so that the cyclic structure on the points coincides with that of the vertices induced by the counterclockwise orientation of the $(n+3)$-gon. Then given a triangulation T of the $(n+3)$-gon, we define a rational coordinate system on $\mathcal{M}_{0,n+3}$ as follows. Recall the cross-ratio of four distinct points (x_1, x_2, x_3, x_4) on $\mathbb{P}^1$:

$$r^+(x_1, x_2, x_3, x_4) := \frac{(x_1 - x_2)(x_3 - x_4)}{(x_1 - x_4)(x_2 - x_3)}.$$

We assign to every (internal) edge of the triangulation T a rational function X_E^T on $\mathcal{M}_{0,n+3}$ as follows. We define a seed $\mathbf{I}^T = (I^T, I_0^T, \varepsilon_{ij}, d_i)$ as follows: I^T is the set of edges of T, I_0^T is empty, and $d_i = 1$.

Let (x_a, x_b, x_c, x_d) be the configuration of points assigned to the vertices of the 4-gon formed by the two triangles of the triangulation sharing the edge E, so that $E = x_b x_d$. Then

$$X_E^T(x_1, \ldots, x_{n+3}) := r^+(x_a, x_b, x_c, x_d).$$

Proposition B.1. *The above construction provides $\mathcal{M}_{0,n+3}$ with a structure of cluster $\mathcal{X}$-variety of finite type A_n.*

Proof. Take the snake triangulation shown in Figure 11. The corresponding ε_{ij}-function is the one assigned to the root system of type A_n. Let us change the triangulation by flipping a diagonal. Then it is easy to see that the resulting transformation of the coordinates is described by the cluster mutation corresponding to the flip [FG1, Section 9]. The proposition is proved.

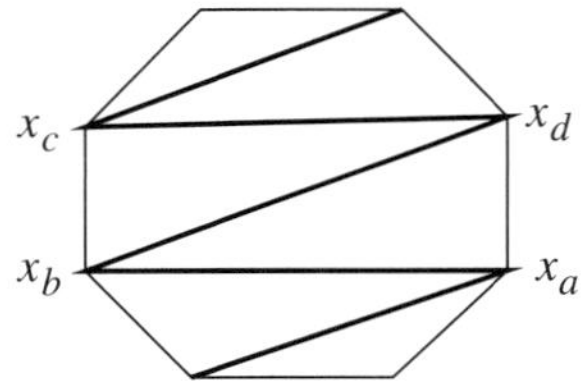

Fig. 11. A triangulation of the octagon provides a coordinate system on the moduli space $\mathcal{M}_{0,8}$.

B.2 The cluster $\mathcal{X}$-variety corresponding to the moduli space $\mathrm{Conf}_3(\mathcal{B}_{A_3})$ of triples of flags in PGL_4

Recall the rational coordinates on the moduli space $\mathrm{Conf}_3(\mathcal{B}_{A_3})$ defined in [FG1, Section 9.3]. A triple of vectors (a_1, a_2, a_3) in a four-dimensional vector space V_4 provides a flag $(A_0, A_1, A_2) := (a_1, a_1a_2, a_1a_2a_3)$ in $\mathbb{P}(V_4)$. Here A_i is the projectivization of the subspace spanned by $a_1, \dots, a_{i+1}$. Consider a triple of flags in $\mathbb{P}(V_4)$:

$$(A, B, C) = ((a_1, a_2, a_3), (b_1, b_2, b_3), (c_1, c_2, c_3)).$$

Let us choose a volume form $\omega \in \det V_4^*$. Then, for any four vectors (a, b, c, d), there is a determinant

$$\Delta(a, b, c, d) := \langle \omega, a \wedge b \wedge c \wedge d \rangle.$$

We define a rational function X_1 on $\mathrm{Conf}_3(\mathcal{B}_{A_3})$ as follows:

$$X_1(A, B, C) := -\frac{\Delta(a_1, a_2, a_3, b_1)\Delta(a_1, b_2, b_3, c_1)\Delta(a_1, c_1, c_2, a_2)}{\Delta(a_1, a_2, a_3, c_1)\Delta(a_1, b_1, b_2, a_2)\Delta(a_1, c_1, c_2, b_1)}$$

and the functions X_2 and X_3 are obtained by cyclic shifts:

$$X_2(A, B, C) := X_1(B, C, A), \qquad X_3(A, B, C) := X_1(C, A, B),$$

B.3 An isomorphism of cluster $\mathcal{X}$-varieties

Let us define a map

$$\Phi : \mathcal{M}_{0,6} \longrightarrow \mathrm{Conf}_3(\mathcal{B}_{A_3}).$$

Recall that a *normal curve* $N \subset \mathbb{P}^n$ is a curve of the minimal possible degree n which does not lie in a hyperplane. Any such curve is projectively equivalent to the image of the map $t \mapsto (1, t, t^2, \dots, t^n - 1)$. The group of projective transformations preserving a normal curve is isomorphic to PGL_2, that is, to the automorphism group of $\mathbb{P}^1$. For any $n+3$ generic points in $\mathbb{P}^n$ there exists a unique normal curve containing these points.

Let $(x_1, y_1, x_2, y_2, x_3, y_3)$ be a configuration of six distinct points on $\mathbb{P}^1$. We identify it with a configuration of points on a normal curve $N \subset \mathbb{P}^3$. Set

$$\begin{aligned}&\Phi(x_1, y_1, x_2, y_2, x_3, y_3)\\ &\quad = (X, Y, Z) := ((x_1, x_1y_1, y_3x_1y_1), (x_2, x_2y_2, y_1x_2y_2), (x_3, x_3y_3, y_2x_3y_3)).\end{aligned}$$

The inverse map Ψ is defined as follows. Let

$$(A, B, C) := (A_0, A_1, A_2), (B_0, B_1, B_2), (C_0, C_1, C_2) \tag{36}$$

be a triple of flags in $\mathbb{P}^3$. So A_0 is a point, A_1 is a line containing this point, and A_2 is a plane containing A_1. We assign to it the following collection of 6 points in $\mathbb{P}^3$:

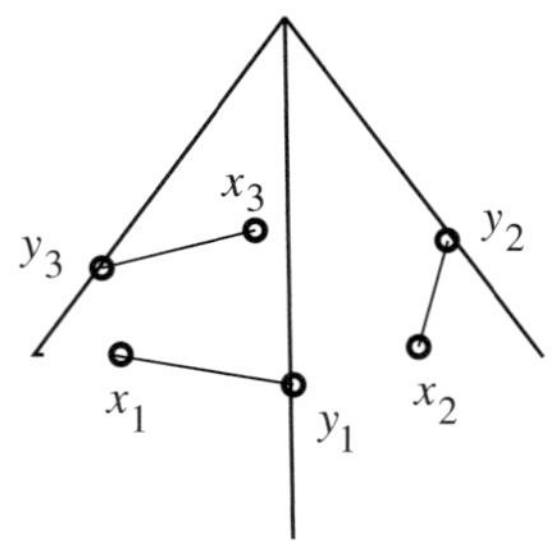

Fig. 12. A configuration of three flags in $\mathbb{P}^3$ from a configuration of six points on $\mathbb{P}^1$.

$$\Psi(A, B, C) := (A_0, A_1 \cap B_2, B_0, B_1 \cap C_2, C_0, C_1 \cap A_2).$$

Taking the unique normal curve passing through these points, we get a configuration of 6 points on $\mathbb{P}^1$. By the very definition, the compositions $\Psi \circ \Phi$ and $\Phi \circ \Psi$ are the identity maps.

Proposition B.2. *The map Φ provides an isomorphism of the cluster $\mathcal{X}$-varieties corresponding to the moduli spaces $\mathcal{M}_{0,6}$ and* $\mathrm{Conf}_3(\mathcal{B}_{A_3})$.

Proof. Observe that one has

$$\begin{aligned} &X_1\Phi(x_1, y_1, x_2, y_2, x_3, y_3) \\ &\quad = X_1(X, Y, Z) = -\frac{\Delta(x_1, y_1, y_3, x_2)\Delta(x_1, x_2, y_2, x_3)\Delta(x_1, x_3, y_3, y_1)}{\Delta(x_1, y_1, y_3, x_3)\Delta(x_1, x_2, y_2, y_1)\Delta(x_1, x_3, y_3, x_2)} \\ &\quad = \frac{\Delta(x_1, x_2, y_1, y_3)\Delta(x_1, x_2, x_3, y_2)}{\Delta(x_1, x_2, y_1, y_2)\Delta(x_1, x_2, y_3, x_3)} = r^+(y_1, y_3, x_3, y_2). \end{aligned}$$

Using the cyclic shifts, we get

$$X_2\Phi(x_1, y_1, x_2, y_2, x_3, y_3) = r^+(y_2, y_1, x_1, y_3),$$
$$X_2\Phi(x_1, y_1, x_2, y_2, x_3, y_3) = r^+(y_3, y_2, x_2, y_1).$$

These are the coordinates on $\mathcal{M}_{0,6}$ assigned to the triangulation of the hexagon shown in Figure 13. The proposition is proved.

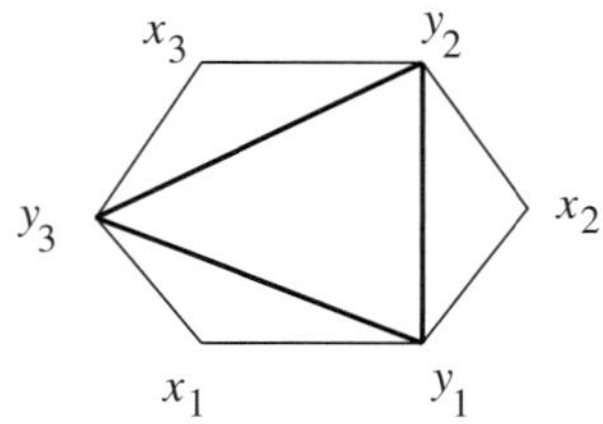

Fig. 13. A triangulation of the hexagon providing a coordinate system on $\mathcal{M}_{0,6}$.

Exercise. Using the above results, show that the cluster $\mathcal{X}$-variety corresponding to the moduli space $\mathrm{Conf}_3(\mathcal{B}_{B_2})$ of triples of flags in Sp_4 is of finite type B_2. *Hint*: Use the triangulations of the hexagon symmetric with respect to the center.

Acknowledgments. V.F. was supported by grants CRDF 2622 and 2660. He is grateful to the Mathematics Departments of the Université Lyon 1 and Brown University for their hospitality, and to Lacrimiora Iancu for a stimulating lecture.

A.G. was supported by NSF grant DMS-0400449. He is grateful to IHES for the hospitality.

References

[BFZ96] A. Berenstein, S. Fomin, and A. Zelevinsky, Parametrizations of canonical bases and totally positive matrices, *Adv. Math.*, **122**-1 (1996), 49–149.

[BFZ3] A. Berenstein, S. Fomin, and A. Zelevinsky, Cluster algebras III: Upper bounds and double Bruhat cells, *Duke Math. J.*, **126**-1 (2005), 1–52.

[BZ] A. Berenstein and A. Zelevinsky, Totally positivity in Schubert varieties, *Comm. Math. Helv.*, **72** (1997), 1–40.

[BZq] A. Berenstein and A. Zelevinsky, Quantum cluster algebras, math.QA/0404446, 2004; *Adv. Math.*, to appear.

[D] V. G. Drinfeld, Hamiltonian structures on Lie groups, Lie bialgebras and the geometric meaning of classical Yang-Baxter equations, *Dokl. Akad. Nauk SSSR*, **268**-2 (1983), 285–287.

[D1] V. G. Drinfeld, Quantum groups, in *Proceedings of the International Congress of Mathematicians*, Vols. 1–2, American Mathematical Society, Providence, RI, 1987, 798–820.

[FZ] S. Fomin and A. Zelevinsky, Double Bruhat cells and total positivity, *J. Amer. Math. Soc.*, **12**-2 (1999), 335–380.

[FZI] S. Fomin and A. Zelevinsky, Cluster algebras I: Foundations, *J. Amer. Math. Soc.*, **15**-2 (2002), 497–529.

[FG1] V. V. Fock and A. B. Goncharov, Moduli spaces of local systems and higher Teichmuller theory, math.AG/0311149, 2003.

[FG2] V. V. Fock and A. B. Goncharov, Cluster ensembles, quantization and the dilogarithm, math.AG/0311245, 2003.

[GSV1] M. Gekhtman, M. Shapiro, A. Vainshtein, Cluster algebras and Poisson geometry, *Moscow Math. J.*, **3** (2003), 899–934.

[L1] G. Lusztig, Total positivity in reductive groups, in *Lie Theory and Geometry*, Progress in Mathematics, Vol. 123, Birkhäuser Boston, Cambridge, MA, 1994, 531–568.

[L2] G. Lusztig, Canonical bases arising from quantized enveloping algebras, *J. Amer. Math. Soc.*, **3**-2 (1990), 447–498.

Local geometric Langlands correspondence and affine Kac–Moody algebras

Edward Frenkel* and Dennis Gaitsgory†

[1] Department of Mathematics
University of California at Berkeley
Berkeley, CA 94720
USA
frenkel@math.berkeley.edu

[2] Department of Mathematics
Harvard University
Cambridge, MA 02138
USA
gaitsgde@math.harvard.edu

Dedicated to Vladimir Drinfeld on his 50th birthday.

Subject Classifications: Primary 17B67, 14D20. Secondary 22E67, 11R39, 11G45.

0 Introduction

Let $\mathfrak{g}$ be a simple Lie algebra over $\mathbb{C}$ and G a connected algebraic group with Lie algebra $\mathfrak{g}$. The affine Kac–Moody algebra $\widehat{\mathfrak{g}}$ is the universal central extension of the formal loop agebra $\mathfrak{g}((t))$. Representations of $\widehat{\mathfrak{g}}$ have a parameter, an invariant bilinear form on $\mathfrak{g}$, which is called the level. Representations corresponding to the bilinear form which is equal to minus one half of the Killing form are called representations of *critical level*. Such representations can be realized in spaces of global sections of twisted D-modules on the quotient of the loop group $G((t))$ by its "open compact" subgroup K, such as $G[[t]]$ or the Iwahori subgroup I.

This is the first in a series of papers devoted to the study of the categories of representations of the affine Kac–Moody algebra $\widehat{\mathfrak{g}}$ of the critical level and D-modules on $G((t))/K$ from the point of view of a geometric version of the local Langlands correspondence. Let us explain what we mean by that.

0.1

First, we recall the classical setting of the local Langlands correspondence. Let $\hat{\mathcal{K}}$ be a non-archimedean local field such as $\mathbb{F}_q((t))$ and G a connected reductive algebraic group over $\hat{\mathcal{K}}$. The local Langlands correspondence sets up a relation between two different types of data. Roughly speaking, the first data are the equivalence classes of homomorphisms, denoted by σ, from the Galois group of $\hat{\mathcal{K}}$ (more precisely, its version, called the Weil–Deligne group) to $\check{G}$, the Langlands dual group of G. The second data are the isomorphism classes of irreducible smooth representations, denoted by π, of the locally compact group $G(\hat{\mathcal{K}})$ (we refer the reader to [Vog] for a precise formulation of this correspondence).

A naive analogue of this correspondence in the geometric situation is as follows. Since the geometric analogue of the Galois group is the fundamental group, the geometric analogue of a homomorphism from the Galois group of $\hat{\mathcal{K}}$ to $\check{G}$ is a $\check{G}$-local system on $\operatorname{Spec}\hat{\mathcal{K}}$. Now we wish to replace $\hat{\mathcal{K}} = \mathbb{F}_q((t))$ by $\mathbb{C}((t))$. Then $\operatorname{Spec}\mathbb{C}((t))$ is the formal punctured disc $\mathcal{D}^\times$. By a $\check{G}$-local system on $\mathcal{D}^\times$ we will always understand its de Rham version: a $\check{G}$-bundle on $\mathcal{D}^\times$ with a meromorphic connection that may have a pole of arbitrary order at the origin. By analogy with the classical local Langlands correspondence, we would like to attach to such a local system a representation of the formal loop group $G((t)) = G(\mathbb{C}((t)))$. However, we will argue in this paper that in contrast to the classical setting, this representation of $G((t))$ should be defined not on a vector space, but on a *category* (see Section 20.7 where the notion of a group acting on a category is spelled out).

Thus, to each $\check{G}$-local system σ we would like to attach an abelian category $\mathcal{C}_\sigma$ equipped with an action of the ind-group $G((t))$. This is what we will mean by a geometric local Langlands correspondence for the formal loop group $G((t))$. This correspondence may be viewed as a "categorification" of the classical local Langlands correspondence, in the sense that we expect the Grothendieck groups of the categories $\mathcal{C}_\sigma$ to "look like" irreducible smooth representations of $G(\hat{\mathcal{K}})$. At the moment we cannot characterize $\mathcal{C}_\sigma$ in local terms. Instead, we shall now explain how this local correspondence fits in with the pattern of the global geometric Langlands correspondence.

In the global geometric Langlands correspondence we start with a smooth projective connected curve X over $\mathbb{C}$ with distinct marked points $x_1, \dots, x_n$. Let σ^{glob} be a $\check{G}$-local system on $\overset{\circ}{X} = X \backslash \{x_1, \dots, x_n\}$, i.e., a $\check{G}$-bundle on $X \backslash \{x_1, \dots, x_n\}$ with a connection which may have poles of arbitrary order at the points $x_1, \dots, x_n$. Let $\operatorname{Bun}_G^{x_1,\dots,x_n}$ be the moduli stack classifying G-bundles on X with the full level structure at $x_1, \dots, x_n$ (i.e., trivializations on the formal discs $\mathcal{D}_{x_i}$ around x_i). Let $\mathfrak{D}(\operatorname{Bun}_G^{x_1,\dots,x_n})$-mod be the category of D-modules on $\operatorname{Bun}_G^{x_1,\dots,x_n}$. One defines, as in [BD1], the Hecke correspondence between $\operatorname{Bun}_G^{x_1,\dots,x_n}$ and $\overset{\circ}{X} \times \operatorname{Bun}_G^{x_1,\dots,x_n}$ and the notion of a Hecke "eigensheaf" on $\operatorname{Bun}_G^{x_1,\dots,x_n}$ with the "eigenvalue" σ^{glob}.

The *Hecke correspondence* is the following moduli space:

$$\mathcal{H} = \{(\mathcal{P}, \mathcal{P}', x, \phi) \mid \mathcal{P}, \mathcal{P}' \in \operatorname{Bun}_G^{x_1,\dots,x_n}, x \in \overset{\circ}{X}, \phi : \mathcal{P}|_{X \backslash x} \xrightarrow{\sim} \mathcal{P}'|_{X \backslash x}\}.$$

It is equipped with the projections

$$\begin{array}{ccc} & \mathcal{H} & \\ \overleftarrow{h} \swarrow & & \searrow \overrightarrow{h} \\ \mathrm{Bun}_G^{x_1,\dots,x_n} & & \overset{\circ}{X} \times \mathrm{Bun}_G^{x_1,\dots,x_n} \end{array}$$

where $\overleftarrow{h}(\mathcal{P}, \mathcal{P}', x, \phi) = \mathcal{P}$ and $\overrightarrow{h}(\mathcal{P}, \mathcal{P}', x, \phi) = (x, \mathcal{P}')$. The fiber of $\mathcal{H}$ over $(x, \mathcal{P}')$ is isomorphic to $\mathrm{Gr}_x^{\mathcal{P}'}$, the twist of the affine Grassmannian $\mathrm{Gr}_x = G(\hat{\mathcal{K}}_x)/G(\hat{\mathcal{O}}_x)$ by the $G(\hat{\mathcal{O}}_x)$–torsor of trivializations of $\mathcal{P}'|_{\mathcal{D}_x}$ (here we denote by $\hat{\mathcal{O}}_x$ and $\hat{\mathcal{K}}_x$ the completed local ring of X at x and its field of fractions, respectively). The stratification of Gr_x by $G(\hat{\mathcal{O}}_x)$–orbits induces a stratification of $\mathcal{H}$. The strata are parametrized by the set of isomorphism classes of irreducible representations of the Langlands dual group $\check{G}$. To each such isomorphism class V therefore corresponds an irreducible D-module on $\mathcal{H}$ supported on the closure of the orbit labeled by V. We denote it by $\mathcal{F}_V^{\mathrm{glob}}$.

One defines the Hecke functors H_V, $V \in \mathrm{Irr}(\mathcal{R}\mathrm{ep}(\check{G}))$ from the derived category of D-modules on $\mathrm{Bun}_G^{x_1,\dots,x_n}$ to the derived category of D-modules on $\overset{\circ}{X} \times \mathrm{Bun}_G^{x_1,\dots,x_n}$ by the formula

$$H_V(\mathcal{F}) = \overrightarrow{h}_!(\overleftarrow{h}^*(\mathcal{F}) \otimes \mathcal{F}_V^{\mathrm{glob}}).$$

A D-module on $\mathrm{Bun}_G^{x_1,\dots,x_n}$ is called a *Hecke eigensheaf with eigenvalue* σ^{glob} if we are given isomorphisms

$$H_V(\mathcal{F}) \simeq V_{\sigma^{\mathrm{glob}}} \boxtimes \mathcal{F} \tag{0.1}$$

of D-modules on $\overset{\circ}{X} \times \mathrm{Bun}_G^{x_1,\dots,x_n}$ which are compatible with the tensor product structure on the category of representations of $\check{G}$ (here $V_{\sigma^{\mathrm{glob}}}$ is the associated vector bundle with a connection on $\overset{\circ}{X}$ corresponding to σ^{glob} and V).

The aim of the global geometric Langlands correspondence is to describe the category $\mathfrak{D}(\mathrm{Bun}_G^{x_1,\dots,x_n})^{\mathrm{Hecke}}_{\sigma^{\mathrm{glob}}}$-mod of such eigensheaves.

For example, if there are no marked points, and so σ^{glob} is unramified everywhere, it is believed that this category is equivalent to the category of vector spaces, provided that σ^{glob} is sufficiently generic. In particular, in this case $\mathfrak{D}(\mathrm{Bun}_G)^{\mathrm{Hecke}}_{\sigma^{\mathrm{glob}}}$-mod should contain a unique, up to isomorphism, irreducible object, and all other objects should be direct sums of its copies. The irreducible Hecke eigensheaf may be viewed as a geometric analogue of an unramified automorphic function from the classical global Langlands correspondence. This Hecke eigensheaf has been constructed by A. Beilinson and V. Drinfeld in [BD1] in the case when σ^{glob} has an additional structure of an "oper."

In order to explain what we expect from the category $\mathfrak{D}(\mathrm{Bun}_G^{x_1,\dots,x_n})^{\mathrm{Hecke}}_{\sigma^{\mathrm{glob}}}$-mod when the set of marked points is nonempty, let us revisit the classical situation. Denote by $\mathbb{A}$ the ring of adeles of the field of rational functions on X. Let $\pi_{\sigma^{\mathrm{glob}}}$ be an irreducible automorphic representation of the adelic group $G(\mathbb{A})$ corresponding to σ^{glob} by the classical global Langlands correspondence. Denote by $(\pi_{\sigma^{\mathrm{glob}}})_{x_1,\dots,x_n}$ the subspace of $\pi_{\sigma^{\mathrm{glob}}}$ spanned by vectors unramified away from $x_1, \dots, x_n$. Then

$(\pi_{\sigma^{\text{glob}}})_{x_1,\dots,x_n}$ is a representation of the locally compact group $\prod_{i=1,\dots,n} G(\hat{\mathcal{K}}_{x_i})$ (here $\hat{\mathcal{K}}_{x_i}$ denotes the local field at x_i). A basic compatibility between the local and global classical Langlands correspondences is that this representation should be isomorphic to the tensor product of local factors

$$\bigotimes_{i=1,\dots,n} \pi_{\sigma_i},$$

where π_{σ_i} is the irreducible representation of $G(\hat{\mathcal{K}}_i)$, attached via the local Langlands correspondence to the restriction σ_i of σ^{glob} to the formal punctured disc around x_i.

In the geometric setting we view the category $\mathfrak{D}(\text{Bun}_G^{x_1,\dots,x_n})^{\text{Hecke}}_{\sigma^{\text{glob}}}$-mod as a "categorification" of the representation $(\pi_{\sigma^{\text{glob}}})_{x_1,\dots,x_n}$. Based on this, we expect that there should be a natural functor

$$\bigotimes_{i=1,\dots,n} \mathcal{C}_{\sigma_i} \to \mathfrak{D}(\text{Bun}_G^{x_1,\dots,x_n})^{\text{Hecke}}_{\sigma^{\text{glob}}}\text{-mod}, \tag{0.2}$$

relating the local and global categories. Moreover, we expect this functor to be an equivalence when σ^{glob} is sufficiently generic. This gives us a basic compatibility between the local and global geometric Langlands correspondences.

0.2

How can we construct the categories $\mathcal{C}_\sigma$ and the corresponding functors to the global categories? At the moment we see two ways to do that. In order to explain them, we first illustrate the main idea on a toy model.

Let G be a split reductive group over $\mathbb{Z}$, and B a Borel subgroup. A natural representation of the finite group $G(\mathbb{F}_q)$ is realized in the space of complex (or $\overline{\mathbb{Q}}_\ell$-) valued functions on the quotient $G(\mathbb{F}_q)/B(\mathbb{F}_q)$. We can ask what is the "correct" analogue of this representation when we replace the field $\mathbb{F}_q$ by the complex field and the group $G(\mathbb{F}_q)$ by $G(\mathbb{C})$. This may be viewed as a simplified version of our quest, since instead of considering $G(\mathbb{F}_q((t)))$ we now look at $G(\mathbb{F}_q)$.

The quotient $G(\mathbb{F}_q)/B(\mathbb{F}_q)$ is the set of $\mathbb{F}_q$-points of the algebraic variety G/B defined over $\mathbb{Z}$ called the flag variety of G. Let us recall the Grothendieck *faisceaux-fonctions* dictionary: if $\mathcal{F}$ is an ℓ-adic sheaf on an algebraic variety V over $\mathbb{F}_q$ and x is an $\mathbb{F}_q$-point of V, then we have the Frobenius conjugacy class Fr_x acting on the stalk $\mathcal{F}_x$ of $\mathcal{F}$ at x. Hence, we can define a $\overline{\mathbb{Q}}_\ell$-valued function $\mathtt{f}_q(\mathcal{F})$ on the set of $\mathbb{F}_q$-points of X, whose value at x is $\text{Tr}(\text{Fr}_x, \mathcal{F}_x)$. We also obtain in the same way a function on the set $V(\mathbb{F}_{q^n})$ of $\mathbb{F}_{q^n}$-points of V for $n > 1$. This passage from ℓ-adic sheaves to functions satisfies various natural properties. This construction identifies the Grothendieck group of the category of ℓ-adic sheaves on V with a subgroup of the direct product of the spaces of functions on $V(\mathbb{F}_{q^n})$, $n > 0$ (see [Lau]). Therefore, the category of ℓ-adic sheaves (or its derived category) may be viewed as a categorification of this space of functions.

This suggests that in order to pass from $\mathbb{F}_q$ to $\mathbb{C}$ we first need to replace the notion of a function on $(G/B)(\mathbb{F}_q)$ by the notion of an ℓ-adic sheaf on the variety $(G/B)_{\mathbb{F}_q} = G/B \underset{\mathbb{Z}}{\otimes} \mathbb{F}_q$.

Next, we replace the notion of an ℓ-adic sheaf on G/B, considered as an algebraic variety over $\mathbb{F}_q$, by a similar notion of a constructible sheaf on $(G/B)_{\mathbb{C}} = G/B \underset{\mathbb{Z}}{\otimes} \mathbb{C}$ which is an algebraic variety over $\mathbb{C}$. The group $G_{\mathbb{C}}$ naturally acts on $(G/B)_{\mathbb{C}}$ and hence on this category. We shall now apply two more metamorphoses to this category.

Recall that for a smooth complex algebraic variety V we have a Riemann–Hilbert correspondence which is an equivalence between the derived category of constructible sheaves on V and the derived category of D-modules on V that are holonomic and have regular singularities. Thus, over $\mathbb{C}$ we may pass from constructible sheaves to D-modules. Generalizing this, we consider the category of all D-modules on the flag variety $(G/B)_{\mathbb{C}}$. This category carries a natural action of $G_{\mathbb{C}}$.

Let us also recall that by taking global sections we obtain a functor from the category of D-modules on $(G/B)_{\mathbb{C}}$ to the category of $\mathfrak{g}$-modules. Moreover, A. Beilinson and J. Bernstein have proved [BB] that this functor is an equivalence between the category of D-modules on $(G/B)_{\mathbb{C}}$ and the category of $\mathfrak{g}$-modules on which the center of the universal enveloping algebra $U(\mathfrak{g})$ acts through the augmentation character. Observe that the latter category also carries a natural $G_{\mathbb{C}}$-action that comes from the adjoint action of $G_{\mathbb{C}}$ on $\mathfrak{g}$.

We arrive at the following conclusion: a meaningful geometric analogue of the notion of representation of $G(\mathbb{F}_q)$ is that of a *category* equipped with an action of $G_{\mathbb{C}}$. In particular, an analogue of the space of functions on $G(\mathbb{F}_q)/B(\mathbb{F}_q)$ is the category $\mathfrak{D}((G/B)_{\mathbb{C}})$-mod, which can be also realized as the category of $\mathfrak{g}$-modules with a fixed central character.

Our challenge is to find analogues of the above two categories in the case when the reductive group G is replaced by its loop group $G((t))$. The exact relation between them will be given by a loop group analogue of the Beilinson–Bernstein equivalence, and will be in itself of great interest to us.

As the previous discussion demonstrates, one possibility is to consider representations of the complex loop group $G((t))$ on various categories of D-modules on the ind-schemes $G((t))/K$, where K is an "open compact" subgroup of $G((t))$, such as $G[[t]]$ or the Iwahori subgroup I (the preimage of a Borel subgroup $B \subset G$ under the homomorphism $G[[t]] \to G$). The other possibility is to consider various categories of representations of the Lie algebra $\mathfrak{g}((t))$, or of its universal central extension $\widehat{\mathfrak{g}}$, because the group $G((t))$ still acts on $\widehat{\mathfrak{g}}$ via the adjoint action.

0.3

To explain the main idea of this paper, we consider an important example of a category of D-modules which may be viewed as a "categorification" of an irreducible *unramified* representation of the group $G(\hat{\mathcal{K}})$, where $\hat{\mathcal{K}} = \mathbb{F}_q((t))$. We recall that a representation π of $G(\hat{\mathcal{K}})$ is called unramified if it contains a nonzero vector v such that $G(\hat{\mathcal{O}})v = v$, where $\hat{\mathcal{O}} = \mathbb{F}_q[[t]]$. The spherical Hecke algebra $H(G(\hat{\mathcal{K}}), G(\hat{\mathcal{O}}))$

of bi-$G(\hat{\mathcal{O}})$-invariant compactly supported functions on $G(\hat{\mathcal{K}})$ acts on the subspace spanned by such vectors.

The Satake isomorphism identifies $H(G(\hat{\mathcal{K}}), G(\hat{\mathcal{O}}))$ with the representation ring $\mathrm{Rep}(\check{G})$ of finite-dimensional representations of the Langlands dual group $\check{G}$ [Lan]. This implies that equivalence classes of irreducible unramified representations of $G(\hat{\mathcal{K}})$ are parameterized by semisimple conjugacy classes in the Langlands dual group $\check{G}$. This is, in fact, a baby version of the local Langlands correspondence mentioned above, because a semisimple conjugacy class in $\check{G}$ may be viewed as an equivalence class of unramified homomorphisms from the Weil group $W_{\hat{\mathcal{K}}}$ to $\check{G}$ (i.e., one that factors through the homomorphism $W_{\hat{\mathcal{K}}} \to W_{\mathbb{F}_q} \simeq \mathbb{Z}$).

For a semisimple conjugacy class γ in $\check{G}$ denote by π_γ the corresponding irreducible unramified representation of $G(\hat{\mathcal{K}})$. It contains a unique, up to scalars, vector v_γ such that $G(\hat{\mathcal{O}})v_\gamma = v_\gamma$. It also satisfies the following property. For a finite-dimensional representation V of $\check{G}$ denote by F_V the element of $H(G(\hat{\mathcal{K}}), G(\hat{\mathcal{O}}))$ corresponding to $[V] \in \mathrm{Rep}(\check{G})$ under the Satake isomorphism. Then we have $F_V \cdot v_\gamma = \mathrm{Tr}(\gamma, V)v_\gamma$ (to simplify our notation, we omit a q-factor in this formula).

Now we embed π_γ into the space of locally constant functions on $G(\hat{\mathcal{K}})/G(\hat{\mathcal{O}})$, by using matrix coefficients, as follows:

$$u \in \pi_\gamma \mapsto f_u, \qquad f_u(g) = \langle u, gv_\gamma \rangle,$$

where $\langle , \rangle$ is an invariant bilinear form on π_γ. Clearly, the functions f_u are right $G(\hat{\mathcal{O}})$-invariant and satisfy the condition

$$f \star F_V = \mathrm{Tr}(\gamma, V) f, \tag{0.3}$$

where $\star$ denotes the convolution product. Let $C(G(\hat{\mathcal{K}})/G(\hat{\mathcal{O}}))_\gamma$ be the space of locally constant functions on $G(\hat{\mathcal{K}})/G(\hat{\mathcal{O}})$ satisfying (0.3). We have constructed an injective map $\pi_\gamma \to C(G(\hat{\mathcal{K}})/G(\hat{\mathcal{O}}))_\gamma$, and one can show that for generic γ it is an isomorphism.

Thus we obtain a realization of irreducible unramified representations of $G(\hat{\mathcal{K}})$ in functions on the quotient $G(\hat{\mathcal{K}})/G(\hat{\mathcal{O}})$. According to the discussion in the previous subsection, a natural complex geometric analogue of the space of functions on $G(\hat{\mathcal{K}})/G(\hat{\mathcal{O}})$ is the category of (right) D-modules on $G((t))/G[[t]]$. The latter has the structure of an ind-scheme over $\mathbb{C}$, which is called the affine Grassmannian and is denoted by Gr_G.

The classical Satake isomorphism has a categorical version due to Lusztig, Drinfeld, Ginzburg, and Mirković–Vilonen (see [MV]) which may be formulated as follows: the category of $G[[t]]$-equivariant D-modules on Gr_G, equipped with the convolution tensor product, is equivalent to the category $\mathcal{R}\mathrm{ep}(\check{G})$ of finite-dimensional representations of $\check{G}$ as a tensor category. For a representation V of $\check{G}$ let $\mathcal{F}_V$ be the corresponding D-module on Gr_G. A D-module $\mathcal{F}$ on Gr_G satisfies the geometric analogue of the property (0.3) if we are given isomorphisms

$$\alpha_V : \mathcal{F} \star \mathcal{F}_V \xrightarrow{\sim} \underline{V} \otimes \mathcal{F}, \qquad V \in \mathrm{Ob}\, \mathcal{R}\mathrm{ep}(\check{G}) \tag{0.4}$$

satisfying a natural compatibility with tensor products. In other words, observe that we now have two monoidal actions of the tensor category $\mathcal{R}ep(\check{G})$ on the category $\mathfrak{D}(\mathrm{Gr}_G)$-mod of right D-modules on Gr_G: one is given by tensoring D-modules with $\underline{V}$, the vector space underlying a representation V of $\check{G}$, and the other is given by convolution with the D-module $\mathcal{F}_V$. The collection of isomorphisms α_V in (0.4) should give us an isomorphism between these two actions applied to the object $\mathcal{F}$.

Let $\mathfrak{D}(\mathrm{Gr}_G)^{\mathrm{Hecke}}$-mod be the category whose objects are the data $(\mathcal{F}, \{\alpha_V\})$, where $\mathcal{F}$ is a D-module on Gr_G and $\{\alpha_V\}$ are the isomorphisms (0.4) satisfying the above compatibility. This category carries a natural action of the loop group $G((t))$ that is induced by the (left) action of $G((t))$ on the Grassmannian Gr_G. We believe that the category $\mathfrak{D}(\mathrm{Gr}_G)^{\mathrm{Hecke}}$-mod, together with this action of $G((t))$, is the "correct" geometric analogue of the unramified irreducible representations of $G(\mathbb{F}_q((t)))$ described above. Therefore, we propose

$$\mathcal{C}_{\sigma_0} \simeq \mathfrak{D}(\mathrm{Gr}_G)^{\mathrm{Hecke}}\text{-mod}, \tag{0.5}$$

where σ_0 is the trivial $\check{G}$-local system on $\mathcal{D}^\times$. This is our simplest example of the conjectural categories $\mathcal{C}_\sigma$, and indeed its Grothendieck group "looks like" an unramified irreducible representation of $G(\hat{\mathcal{K}})$.

0.4

Next, we attempt to describe the category $\mathcal{C}_{\sigma_0}$ in terms of representations of the affine Kac–Moody algebra $\widehat{\mathfrak{g}}$. Since the affine analogue of the Beilinson–Bernstein equivalence is a priori not known, the answer is not as obvious as in the finite-dimensional case. However, the clue is provided by the Beilinson–Drinfeld construction of the Hecke eigensheaves.

The point of departure is a theorem of [FF3] which states that the completed universal enveloping algebra of $\widehat{\mathfrak{g}}$ at the critical level has a large center. More precisely, according to [FF3], it is isomorphic to the algebra of functions of the affine ind-scheme $\mathrm{Op}_{\check{\mathfrak{g}}}(\mathcal{D}^\times)$ of $\check{\mathfrak{g}}$-*opers* over the formal punctured disc (where $\check{\mathfrak{g}}$ is the Langlands dual of the Lie algebra $\mathfrak{g}$). Thus each point $\chi \in \mathrm{Op}_{\check{\mathfrak{g}}}(\mathcal{D}^\times)$ defines a character of the center, and hence the category $\widehat{\mathfrak{g}}_{\mathrm{crit}}\text{-mod}_\chi$ of discrete $\widehat{\mathfrak{g}}$-modules of critical level on which the center acts according to the character χ.

We recall that a $\check{\mathfrak{g}}$-oper (on a curve or on a disc) is a $\check{G}$-local system plus some additional data. This notion was introduced in [DS, BD1] (see Section 1.1 for the definition). Thus we have a natural forgetful map $\mathrm{Op}_{\check{\mathfrak{g}}}(\mathcal{D}^\times) \to \mathrm{LocSys}_{\check{G}}(\mathcal{D}^\times)$, where $\mathrm{LocSys}_{\check{G}}(\mathcal{D}^\times)$ is the stack of $\check{G}$-local systems on $\mathcal{D}^\times$.[1] In this subsection we will restrict our attention to those opers which extend regularly to the formal disc $\mathcal{D}$; they correspond to points of a closed subscheme of regular opers $\mathrm{Op}^{\mathrm{reg}}_{\check{\mathfrak{g}}} \subset \mathrm{Op}_{\check{\mathfrak{g}}}(\mathcal{D}^\times)$. In particular, the local systems on $\mathcal{D}^\times$ defined by such opers are unramified, i.e., they extend to local systems on $\mathcal{D}$, which means that they are isomorphic to the trivial

[1] Note that it is not an algebraic stack, but in this paper we will work with its substacks which are algebraic.

local system (noncanonically, since the group $\check{G}$ acts by automorphisms of the trivial local system).

For a fixed point $x \in X$ Beilinson and Drinfeld construct a local-to-global functor $\widehat{\mathfrak{g}}_{\mathrm{crit}}$-mod $\to \mathfrak{D}(\mathrm{Bun}_G^x)$-mod as a Beilinson–Bernstein-type localization functor by realizing Bun_G^x as the quotient $G((t))/G(X-x)$.

Given a regular oper on the formal disc $\mathcal{D}$ around x, consider the restriction of this localization functor to the category $\widehat{\mathfrak{g}}_{\mathrm{crit}}\text{-mod}_\chi$. It was shown in [BD1] that the latter functor is nonzero if and only if χ extends to an oper on the global curve X, and in that case it gives rise to a functor

$$\widehat{\mathfrak{g}}_{\mathrm{crit}}\text{-mod}_\chi \to \mathfrak{D}(\mathrm{Bun}_G^x)^{\mathrm{Hecke}}_{\sigma^{\mathrm{glob}}}\text{-mod},$$

where σ^{glob} is the $\check{G}$-local system on X corresponding to the above oper.

This construction, combined with (0.2), suggests that for every regular oper χ on $\mathcal{D}$ we should have an equivalence of categories

$$\mathcal{C}_{\sigma_0} \simeq \widehat{\mathfrak{g}}_{\mathrm{crit}}\text{-mod}_\chi. \tag{0.6}$$

Thus we have two conjectural descriptions of the category $\mathcal{C}_{\sigma^0}$: one is given by (0.5), and the other by (0.6). Comparing the two, we obtain a conjectural analogue of the Beilinson–Bernstein equivalence for the affine Grassmannian:

$$\mathfrak{D}(\mathrm{Gr}_G)^{\mathrm{Hecke}}\text{-mod} \simeq \widehat{\mathfrak{g}}_{\mathrm{crit}}\text{-mod}_\chi \tag{0.7}$$

for any $\chi \in \mathrm{Op}^{\mathrm{reg}}_{\check{\mathfrak{g}}}$. In fact, as we shall see later, we should have an equivalence as in (0.7) for every trivialization of the local system on $\mathcal{D}$ corresponding to the oper χ. In particular, the group of automorphisms of such a local system, which is noncanonically isomorphic to $\check{G}$, should act on the category $\widehat{\mathfrak{g}}_{\mathrm{crit}}\text{-mod}_\chi$ by automorphisms. In a sense, it is this action that replaces the Satake parameters of irreducible unramified representations of $G(\hat{\mathcal{K}})$ in the geometric setting.

Let us note that the equivalence conjectured in (0.7) does not explicitly involve the Langlands correspondence. Thus our attempt to describe the simplest of the categories $\mathcal{C}_\sigma$ has already paid dividends: it has led us to a formulation of Beilinson–Bernstein-type equivalence for Gr_G.

It is instructive to compare it with the Beilinson–Bernstein equivalence for a finite-dimensional flag variety $(G/B)_{\mathbb{C}}$, which says that the category of D-modules on $(G/B)_{\mathbb{C}}$ is equivalent to the category of $\mathfrak{g}$-modules with a fixed central character. Naively, one might expect that the same pattern holds in the affine case as well, and the category $\mathfrak{D}(\mathrm{Gr}_G)$-mod is equivalent to the category of $\widehat{\mathfrak{g}}_{\mathrm{crit}}$-modules with a fixed central character. However, in contrast to the finite-dimensional case, the category $\mathfrak{D}(\mathrm{Gr}_G)$-mod carries an additional symmetry, namely, the monoidal action of the category $\mathcal{R}\mathrm{ep}(\check{G})$ (which can be traced back to the action of the spherical Hecke algebra in the classical setting). The existence of this symmetry means that, unlike the category $\widehat{\mathfrak{g}}_{\mathrm{crit}}\text{-mod}_\chi$, the category $\mathfrak{D}(\mathrm{Gr}_G)$-mod is a $\check{G}$-equivariant category (in other words, $\mathfrak{D}(\mathrm{Gr}_G)$-mod is a category over the stack $\mathrm{pt}/\check{G}$; see below). From the point of view of Langlands correspondence, this equivariant structure is related to the fact that $\check{G}$ is the group of automorphisms of the trivial local system $\check{G}$. In order to obtain a Beilinson–Bernstein-type equivalence, we need to deequivariantize this category and replace it by $\mathfrak{D}(\mathrm{Gr}_G)^{\mathrm{Hecke}}$-mod.

0.5

Our next goal is to try to understand in similar terms what the categories $\mathcal{C}_\sigma$ look like for a general local system σ. Unfortunately, unlike the unramified case, we will not be able to construct them directly as categories of D-modules on some homogeneous space of $G((t))$. The reason for this can be traced to the classical picture. If σ is ramified, then the corresponding irreducible representation π of the group $G(\mathbb{F}_q((t)))$ does not contain nonzero vectors invariant under $G(\mathbb{F}_q[[t]])$, but it contains vectors invariant under a smaller compact subgroup $K \subset G(\mathbb{F}_q[[t]])$. As in the ramified case, we can realize π, by taking matrix coefficients, in the space of functions on $G(\mathbb{F}_q((t)))/K$ with values in the space π^K of K-invariant vectors in π satisfying a certain Hecke property. However, unlike the case of unramified representations, π^K generically has dimension greater than one. When we pass to the geometric setting, we need, roughly speaking, to find a proper "categorification" not only for the space of functions on $G(\mathbb{F}_q((t)))/K$ (which is the category of D-modules on the corresponding ind-scheme, as explained above), but also for π^K and for the Hecke property. In the case when σ is tamely ramified, we can take as K the Iwahori subgroup I. Then the desired categorification of π^I and the Hecke property can be constructed following R. Bezrukavnikov's work [Bez], as we will see below. This will allow us to relate the conjectural category $\mathcal{C}_\sigma$ to the category of D-modules on $G((t))/I$. But we do not know how to do that for more general local systems.

Therefore, we try first to describe the categories $\mathcal{C}_\sigma$ in terms of the category of representations of $\widehat{\mathfrak{g}}$ at the critical level rather than categories of D-modules on homogeneous spaces of $G((t))$.

A hint is once again provided by the Beilinson–Drinfeld construction of Hecke eigensheaves from representations of $\widehat{\mathfrak{g}}$ at the critical level described above, because it may be applied in the ramified situation as well. Extending (0.6), we conjecture that for *any* oper χ on $\mathcal{D}^\times$ and the corresponding local system σ, we have an equivalence of categories

$$\mathcal{C}_\sigma \simeq \widehat{\mathfrak{g}}_{\mathrm{crit}}\text{-mod}_\chi \tag{0.8}$$

equipped with an action of $G((t))$. This statement implies, in particular, that the category $\widehat{\mathfrak{g}}_{\mathrm{crit}}$-mod$_\chi$ depends not on the oper χ, but only on the underlying local system! This is in itself a deep conjecture about representations of $\widehat{\mathfrak{g}}$ at the critical level.

At this point, in order to elaborate more on what this conjecture implies and to describe the results of this paper, we will need to discuss a more refined version of the local geometric Langlands correspondence indicated above. For that we have to use the notion of an abelian or a triangulated category over a stack. In the abelian case this is an elementary notion, introduced, e.g., in [Ga1]. It amounts to a sheaf (in the faithfully-flat topology) of abelian categories over a given stack $\mathcal{Y}$. When $\mathcal{Y}$

is an affine scheme $\operatorname{Spec}(A)$, this amounts to the notion of A-linear abelian category. In the triangulated case, some extra care is needed, and we refer the reader to [Ga2] for details. The only property of this notion needed for the discussion that follows is that whenever $\mathcal{C}$ is a category over $\mathcal{Y}$ and $\mathcal{Y}' \to \mathcal{Y}$ is a map of stacks, we can form the base-changed category $\mathcal{C}' = \mathcal{C} \underset{\mathcal{Y}}{\times} \mathcal{Y}'$; in particular, for a point $y \in \mathcal{Y}(\mathbb{C})$ we have the category-fiber $\mathcal{C}_y$.

A refined version of the local geometric Langlands correspondence should attach to any $\mathcal{Y}$-family of $\check{G}$-local systems $\widetilde{\sigma}$ on $\mathcal{D}^\times$ a category $\mathcal{C}_{\widetilde{\sigma}}$ over $\mathcal{Y}$, equipped with an action of $G((t))$, in a way compatible with the above base change property. Such an assignment may be viewed as a category $\mathcal{C}$ over the stack $\operatorname{LocSys}_{\check{G}}(\mathcal{D}^\times)$ equipped with a "fiberwise" action of $G((t))$. Then the categories $\mathcal{C}_\sigma$ discussed above may be obtained as the fibers of $\mathcal{C}$ at $\mathbb{C}$-points σ of $\operatorname{LocSys}_{\check{G}}(\mathcal{D}^\times)$.

We shall now present a refined version of (0.8). Namely, although at the moment we cannot construct $\mathcal{C}$, the following meta-conjecture will serve as our guiding principle:

$$\widehat{\mathfrak{g}}_{\mathrm{crit}}\text{-mod} \simeq \mathcal{C} \underset{\operatorname{LocSys}_{\check{G}}(\mathcal{D}^\times)}{\times} \operatorname{Op}_{\check{\mathfrak{g}}}(\mathcal{D}^\times). \tag{0.9}$$

We will not even try to make this meta-conjecture precise in this paper. Instead we will derive from it some more concrete conjectures, and the goal of this paper will be to give their precise formulation and provide evidence for their validity.

0.6

Let us first revisit the unramified case discussed above. Since the trivial local system σ_0 has $\check{G}$ as the group of its automorphisms, we have a natural map from the stack $\mathrm{pt}/\check{G}$ to $\operatorname{LocSys}_{\check{G}}(\mathcal{D}^\times)$. Let us denote by $\mathcal{C}_{\mathrm{reg}}$ the base change of (the still conjectural category) $\mathcal{C}$ under the above map. Then, by definition, we have an equivalence:

$$\mathcal{C}_{\sigma_0} \simeq \mathcal{C}_{\mathrm{reg}} \underset{\mathrm{pt}/\check{G}}{\times} \mathrm{pt}. \tag{0.10}$$

Now observe that the geometric Satake equivalence of Section 0.3 gives us an action of the tensor category $\mathcal{R}\mathrm{ep}(\check{G})$ on $\mathfrak{D}(\mathrm{Gr}_G)$-mod, $V, \mathcal{F} \mapsto \mathcal{F} \star \mathcal{F}_V$. This precisely amounts to saying that $\mathfrak{D}(\mathrm{Gr}_G)$-mod *is a category over the stack* $\mathrm{pt}/\check{G}$. Moreover, we then have the following base change equivalence:

$$\mathfrak{D}(\mathrm{Gr}_G)^{\mathrm{Hecke}}\text{-mod} \simeq \mathfrak{D}(\mathrm{Gr}_G)\text{-mod} \underset{\mathrm{pt}/\check{G}}{\times} \mathrm{pt}. \tag{0.11}$$

Combining (0.10) and (0.11), we arrive at the following generalization of (0.5):

$$\mathcal{C}_{\mathrm{reg}} \simeq \mathfrak{D}(\mathrm{Gr}_G)\text{-mod}. \tag{0.12}$$

Let us now combine this with (0.9). Let us denote by $\widehat{\mathfrak{g}}_{\mathrm{crit}}\text{-mod}_{\mathrm{reg}}$ the subcategory of $\widehat{\mathfrak{g}}$-modules at the critical level on which the center acts in such a way that their scheme-theoretic support in $\operatorname{Op}_{\check{\mathfrak{g}}}(\mathcal{D}^\times)$ belongs to $\operatorname{Op}_{\check{\mathfrak{g}}}^{\mathrm{reg}}$. By the definition of the map $\mathrm{Op}_{\check{\mathfrak{g}}}(\mathcal{D}^\times) \to \mathrm{LocSys}_{\check{G}}(\mathcal{D}^\times)$, its restriction to $\mathrm{Op}^{\mathrm{reg}}_{\check{\mathfrak{g}}}$ factors through a map $\mathrm{Op}^{\mathrm{reg}}_{\check{\mathfrak{g}}} \to \mathrm{pt}/\check{G}$, which assigns to an oper χ the $\check{G}$-torsor on $\mathrm{Op}^{\mathrm{reg}}_{\check{\mathfrak{g}}}$ obtained by taking the fiber of χ at the origin in $\mathcal{D}$.

Thus, combining (0.9) with the identification

$$\widehat{\mathfrak{g}}_{\mathrm{crit}}\text{-mod}_{\mathrm{reg}} \simeq \widehat{\mathfrak{g}}_{\mathrm{crit}}\text{-mod} \underset{\mathrm{Op}_{\check{\mathfrak{g}}}(\mathcal{D}^\times)}{\times} \mathrm{Op}^{\mathrm{reg}}_{\check{\mathfrak{g}}}, \tag{0.13}$$

we obtain the following statement:

$$\widehat{\mathfrak{g}}_{\mathrm{crit}}\text{-mod}_{\mathrm{reg}} \simeq \mathfrak{D}(\mathrm{Gr}_G)\text{-mod} \underset{\mathrm{pt}/\check{G}}{\times} \mathrm{Op}^{\mathrm{reg}}_{\check{\mathfrak{g}}}, \tag{0.14}$$

By making a further base change with respect to an embedding of the point-scheme into $\mathrm{Op}^{\mathrm{reg}}_{\check{\mathfrak{g}}}$ corresponding to some regular oper χ, we obtain (0.7). Thus (0.14) is a family version of (0.7).

Let us now comment on one more aspect of the conjectural equivalence proposed in (0.14). With any category $\mathcal{C}$ acted on by $G((t))$ and an "open compact" subgroup $K \subset G((t))$ we can associate the category $\mathcal{C}^K$ of K-equivariant objects. Applying this to $\widehat{\mathfrak{g}}_{\mathrm{crit}}\text{-mod}_{\mathrm{reg}}$ we obtain the category consisting of those representations which are K-integrable (i.e., those, for which the action of $\mathrm{Lie}\, K$ may be exponentiated to that of K). In the case of $\mathfrak{D}(\mathrm{Gr}_G)$-mod we obtain the category of K-equivariant D-modules in the usual sense.

Let us take $K = G[[t]]$, and compare the categories obtained from the two sides of (0.14):

$$\widehat{\mathfrak{g}}_{\mathrm{crit}}\text{-mod}_{\mathrm{reg}}^{G[[t]]} \simeq \mathfrak{D}(\mathrm{Gr}_G)^{G[[t]]}\text{-mod} \underset{\mathrm{pt}/\check{G}}{\times} \mathrm{Op}^{\mathrm{reg}}_{\check{\mathfrak{g}}}. \tag{0.15}$$

However, the Satake equivalence mentioned above says that $\mathfrak{D}(\mathrm{Gr}_G)^{G[[t]]} \simeq \mathcal{R}\mathrm{ep}(\check{G})$, implying that the RHS of (0.15) is equivalent to the category of quasi-coherent sheaves on $\mathrm{Op}^{\mathrm{reg}}_{\check{\mathfrak{g}}}$:

$$\widehat{\mathfrak{g}}_{\mathrm{crit}}\text{-mod}_{\mathrm{reg}}^{G[[t]]} \simeq \mathrm{QCoh}(\mathrm{Op}^{\mathrm{reg}}_{\check{\mathfrak{g}}}). \tag{0.16}$$

The latter equivalence is not conjectural, but has already been established in [FG, Theorem 6.3] (see also [BD1]).

Thus we obtain a description of the category of modules at the critical level with a specified integrability property and a condition on the central character as a category of quasi-coherent sheaves on a scheme related to the Langlands dual group. Such a description is a prototype for the main conjecture of this paper, described below.

0.7

The main goal of this paper is to develop a picture similar to the one presented above, for *tamely ramified* local systems σ on $\mathcal{D}^\times$, i.e., those with regular singularity at the origin and unipotent monodromy. The algebraic stack classifying such local systems is isomorphic to $\mathcal{N}_{\check{G}}/\check{G}$, where $\mathcal{N}_{\check{G}} \subset \check{\mathfrak{g}}$ is the nilpotent cone. Let $\mathcal{C}_{\mathrm{nilp}}$ be the corresponding hypothetically existing category over $\mathcal{N}_{\check{G}}/\check{G}$ equipped with a fiberwise action of $G((t))$.

We shall first formulate a conjectural analogue of theorem (0.16) in this setup. As we will see, one essential difference from the unramified case is the necessity to consider derived categories.

Denote by $I \subset G[[t]]$ the Iwahori subgroup; it is the preimage of a once and for all fixed Borel subgroup of G under the homomorphism $G[[t]] \to G$. We wish to give a description of the I-monodromic part $D^b(\mathcal{C}_{\mathrm{nilp}})^{I,m}$ of the derived category $D^b(\mathcal{C}_{\mathrm{nilp}})$ that is similar in spirit to the one obtained in the unramified case (the notion of Iwahori-monodromic derived category will be introduced in Section 5.3).

In Section 2.13 we will introduce a subscheme $\mathrm{Op}_{\check{\mathfrak{g}}}^{\mathrm{nilp}} \subset \mathrm{Op}_{\check{\mathfrak{g}}}(\mathcal{D}^\times)$ of *opers with nilpotent singularities*. Note that $\mathrm{Op}_{\check{\mathfrak{g}}}^{\mathrm{nilp}}$ contains as a closed subscheme the scheme $\mathrm{Op}_{\check{\mathfrak{g}}}^{\mathrm{reg}}$ of regular opers. Denote by $\widehat{\mathfrak{g}}_{\mathrm{crit}}\text{-mod}_{\mathrm{nilp}}$ the subcategory of $\widehat{\mathfrak{g}}_{\mathrm{crit}}\text{-mod}$ whose objects are the $\widehat{\mathfrak{g}}_{\mathrm{crit}}$-modules whose scheme-theoretic support in $\mathrm{Op}_{\check{\mathfrak{g}}}(\mathcal{D}^\times)$ is contained in $\mathrm{Op}_{\check{\mathfrak{g}}}^{\mathrm{nilp}}$.

Let $\widetilde{\mathcal{N}}_{\check{G}}$ be the Springer resolution of $\mathcal{N}_{\check{G}}$. We will show in Section 2.13 that the composition $\mathrm{Op}_{\check{\mathfrak{g}}}^{\mathrm{nilp}} \to \mathrm{Op}_{\check{\mathfrak{g}}}(\mathcal{D}^\times) \to \mathrm{LocSys}_{\check{G}}(\mathcal{D}^\times)$ factors as

$$\mathrm{Op}_{\check{\mathfrak{g}}}^{\mathrm{nilp}} \xrightarrow{\mathrm{Res}^{\mathrm{nilp}}} \widetilde{\mathcal{N}}_{\check{G}}/\check{G} \to \mathcal{N}_{\check{G}}/\check{G} \hookrightarrow \mathrm{LocSys}_{\check{G}}(\mathcal{D}^\times).$$

The first map, denoted by $\mathrm{Res}^{\mathrm{nilp}}$, is smooth.

Then our Main Conjecture 6.2 describes the (bounded) derived category of $\widehat{\mathfrak{g}}_{\mathrm{crit}}\text{-mod}_{\mathrm{nilp}}$ as follows:

$$D^b(\widehat{\mathfrak{g}}_{\mathrm{crit}}\text{-mod}_{\mathrm{nilp}})^{I,m} \simeq D^b(\mathrm{QCoh}(\widetilde{\check{\mathfrak{g}}}/\check{G} \underset{\check{\mathfrak{g}}/\check{G}}{\times} \mathrm{Op}_{\check{\mathfrak{g}}}^{\mathrm{nilp}})), \tag{0.17}$$

where $\widetilde{\check{\mathfrak{g}}} \to \check{\mathfrak{g}}$ is Grothendieck's alteration. This is an analogue for nilpotent opers of theorem (0.16).

As will be explained below, the scheme $\widetilde{\check{\mathfrak{g}}}/\check{G} \underset{\check{\mathfrak{g}}/\check{G}}{\times} \mathrm{Op}_{\check{\mathfrak{g}}}^{\mathrm{nilp}}$, appearing in the RHS of (0.17), has a natural interpretation as the moduli space of Miura opers with nilpotent singularities (see Section 3). The main motivation for the above conjecture came from the theory of Wakimoto modules introduced in [FF2, F]. Namely, to each Miura oper with nilpotent singularity one can attach a Wakimoto module which is an object of the category $\widehat{\mathfrak{g}}_{\mathrm{crit}}\text{-mod}_{\mathrm{nilp}}^{I,m}$. Our (0.17) extends this "pointwise" correspondence to an equivalence of categories.

0.8

Next, we would like to formulate conjectures concerning $\mathcal{C}_{\mathrm{nilp}}$ that are analogous to (0.12) and (0.14), and relate them to conjecture (0.17) above.

The main difficulty is that we do not have an explicit description of $\mathcal{C}_{\text{nilp}}$ in terms of D-modules as the one for $\mathcal{C}_{\text{reg}}$, given by (0.12). Instead, we will be able to describe a certain base change of $\mathcal{C}_{\text{nilp}}$, suggested by the work of S. Arkhipov and R. Bezrukavnikov [Bez, AB].

Let $\text{Fl}_G = G((t))/I$ be the affine flag variety and the affine Grassmannian corresponding to G and $\mathfrak{D}(\text{Fl}_G)$-mod the category of right D-modules on Fl_G. The group $G((t))$ naturally acts on $\mathfrak{D}(\text{Fl}_G)$-mod. According to [AB], the triangulated category $D^b(\mathfrak{D}(\text{Fl}_G)\text{-mod})$ is a category over the stack $\widetilde{\mathcal{N}}_{\check{G}}/\check{G}$.

We propose the following conjecture, describing the hypothetically existing category $\mathcal{C}_{\text{nilp}}$, which generalizes (0.12):

$$D^b(\mathcal{C}_{\text{nilp}}) \underset{\mathcal{N}_{\check{G}}/\check{G}}{\times} \widetilde{\mathcal{N}}_{\check{G}}/\check{G} \simeq D^b(\mathfrak{D}(\text{Fl}_G)\text{-mod}). \tag{0.18}$$

Combining (0.18) with our meta-conjecture (0.9), we arrive at the statement

$$D^b(\widehat{\mathfrak{g}}_{\text{crit}}\text{-mod}_{\text{nilp}}) \simeq D^b(\mathfrak{D}(\text{Fl}_G)\text{-mod}) \underset{\widetilde{\mathcal{N}}_{\check{G}}/\check{G}}{\times} \text{Op}_{\check{\mathfrak{g}}}^{\text{nilp}} \tag{0.19}$$

(the RHS of the above equivalence uses the formalism of triangulated categories over stacks from [Ga2]). Note that conjecture (0.19) is an analogue for opers with nilpotent singularities of conjecture (0.14) for regular opers.

We would now like to explain the relation of conjectures (0.18) and (0.19) to the description of $D^b(\widehat{\mathfrak{g}}_{\text{crit}}\text{-mod}_{\text{nilp}})$ via quasi-coherent sheaves, given by conjecture (0.17) once we pass to the I-monodromic category.

We propose the following description of the category $D(\mathcal{C}_{\text{nilp}})^{I,m}$:

$$D^b(\mathcal{C}_{\text{nilp}})^{I,m} \simeq D^b(\text{QCoh}(\widetilde{\check{\mathfrak{g}}} \underset{\check{\mathfrak{g}}}{\times} \mathcal{N}_{\check{G}}/\check{G})). \tag{0.20}$$

Let us note that conjecture (0.20) is compatible with (0.18). Namely, by combining the two we obtain the following:

$$D^b(\mathfrak{D}(\text{Fl}_G)\text{-mod})^{I,m} \simeq D^b(\text{QCoh}(\widetilde{\check{\mathfrak{g}}} \underset{\check{\mathfrak{g}}}{\times} \widetilde{\mathcal{N}}_{\check{G}}/\check{G})). \tag{0.21}$$

However, this last statement is, in fact, a theorem, which is one of the main results of Bezrukavnikov's work [Bez].

Finally, combining (0.21) and (0.19), we arrive at the statement of conjecture (0.17), providing another piece of motivation for it, in addition to the one via Wakimoto modules given above.

0.9

The principal objective of our project is to prove conjectures (0.14) and (0.17). In the present paper we review some background material necessary to introduce the objects we are studying and formulate the above conjectures precisely. We also prove two results concerning the category of representations of affine Kac–Moody algebras at the critical level which provide us with additional evidence for the validity of these conjectures.

Our first result is Main Theorem 6.9, and it deals with a special case of Main Conjecture 6.2. Namely, in Section 6.7 we will explain that if $\mathcal{C}$ is a category endowed with an action of $G((t))$, the corresponding category $\mathcal{C}^{I,m}$ of Iwahori-monodromic objects admits a Serre quotient, denoted ${}^f\mathcal{C}^{I,m}$, by the subcategory consisting of the so-called partially integrable objects. (Its classical analogue is as follows: given a representation π of a locally compact group $G(\hat{\mathcal{K}})$, we first take the subspace π^I is Iwahori-invariant vectors, and then inside π^I we take the subspace of vectors corresponding to the sign character of the Iwahori–Hecke algebra.)

Performing this procedure on the two sides of (0.19), we should arrive at an equivalence of the corresponding triangulated categories:

$$ {}^f D^b(\widehat{\mathfrak{g}}_{\text{crit}}\text{-mod}_{\text{nilp}})^{I,m} \simeq {}^f D^b(\mathfrak{D}(\text{Fl}_G)\text{-mod})^{I,m} \underset{\widetilde{\mathcal{N}}_{\check{G}}/\check{G}}{\times} \text{Op}_{\check{\mathfrak{g}}}^{\text{nilp}}. \tag{0.22} $$

However, using Bezrukavnikov's result (see Theorem 6.8), the RHS of the above expression can be rewritten as $D^b(\text{QCoh}(\text{Spec}(h_0) \times \text{Op}_{\check{\mathfrak{g}}}^{\text{nilp}}))$, where h_0 is a finite-dimensional commutative algebra isomorphic to $H(\check{G}/\check{B}, \mathbb{C})$. The resulting description of ${}^f D^b(\widehat{\mathfrak{g}}_{\text{crit}}\text{-mod}_{\text{nilp}})^{I,m}$ is our Main Theorem 6.9. In fact, we show that at the level of quotient categories by partially integrable objects, the equivalence holds not only at the level of triangulated categories, but also at the level of abelian ones:

$$ {}^f\widehat{\mathfrak{g}}_{\text{crit}}\text{-mod}_{\text{nilp}}^{I,m} \simeq \text{QCoh}(\text{Spec}(h_0) \times \text{Op}_{\check{\mathfrak{g}}}^{\text{nilp}}). $$

We note that while we use [Bez] for motivational purposes, the proofs presented in this paper are independent of the results of [Bez].

Our second main result is Theorem 8.17. We construct a natural functor from the RHS of (0.14) to the LHS and prove that it is fully faithful at the level of derived categories.

0.10

Let us now describe the structure of the paper. It is logically divided into five parts.

Part I is a review of results concerning opers and Miura opers.

In Part II we discuss various categories of representations of affine Kac–Moody algebras at the critical level. We give more precise formulations of the conjectural equivalences that we mentioned above and the interrelations between them. In particular, we prove one of our main results, Theorem 8.17.

In Part III we review Wakimoto modules. We present a definition of Wakimoto modules by means of a kind of semi-infinite induction functor. We also describe various important properties of these modules.

In Part IV we prove Main Theorem 6.9, which establishes a special case of our conjectural equivalence of categories (0.17).

Part V is an appendix, most of which is devoted to the formalism of group action on categories.

Finally, a couple of comments on notation.

We will write $X \underset{Z}{\times} Y$ for the fiber product of schemes X and Y equipped with morphisms to a scheme Z. To distinguish this notation from the notation for associated fiber bundles, we will write $\mathcal{Y} \overset{K}{\times} \mathcal{F}_K$ for the fiber bundle associated to a principal K-bundle $\mathcal{F}_K$ over some base, where K is an algebraic group and $\mathcal{Y}$ is a K-space. We also denote this associated bundle by $\mathcal{Y}_{\mathcal{F}_K}$.

If a group G acts on a variety X, we denote by X/G the stack-theoretic quotient. If X is affine, we denote by $X//G$ the GIT quotient, i.e., the spectrum of the algebra of invariant functions. We have a natural morphism $X/G \to X//G$.

Part I: Opers and Miura Opers

We this part we collect the definitions and results on opers and Miura opers. As a mathematical object, opers first appeared in [DS], and their connection to representations of affine Kac–Moody algebras at the critical level was discovered in [FF3].

In Section 1 we recall the definition of opers following [BD1] and the explicit description of the scheme classifying them as a certain affine space.

In Section 2 we study opers on the formal punctured disc with a prescribed form of singularity at the closed point. After reviewing some material from [BD1], we show that the subscheme of opers with regular singularities and a specific value of the residue can be interpreted as a scheme of opers with nilpotent singularities, which we denote by $\mathrm{Op}_{\mathfrak{g}}^{\mathrm{nilp}}$. We show that the scheme $\mathrm{Op}_{\mathfrak{g}}^{\mathrm{nilp}}$ admits a natural secondary residue map to the stack $\mathfrak{n}/B \simeq \widetilde{\mathcal{N}}/G$.

In Section 3 we study Miura opers. The notion of Miura oper was introduced in [F], following earlier work of Feigin and Frenkel. By definition, a Miura oper on a curve X is an oper plus a reduction of the underlying G-local system to a Borel subgroup B^- opposite to the oper Borel subgroup B. The functor $\mathrm{MOp}_{\mathfrak{g}}(X)$ of Miura opers admits a certain open subfunctor, denoted by $\mathrm{MOp}_{\mathfrak{g},\mathrm{gen}}(X)$, that corresponds to generic Miura opers. The (D-) scheme classifying the latter is affine over X, and as was shown in [F], it is isomorphic to the (D-) scheme of connections on some fixed H-bundle over X, where H is the Cartan quotient of B.[2] The new results in this section are Proposition 3.10 which describes the forgetful map from generic Miura opers to opers over the locus of opers with regular singularities and Theorem 3.16 which describes the behavior of Miura opers and generic Miura opers over $\mathrm{Op}_{\mathfrak{g}}^{\mathrm{nilp}}$.

In Section 4 we introduce the isomonodromy groupoid over the ind-scheme $\mathrm{Op}_{\mathfrak{g}}(\mathcal{D}^\times)$ and its various subschemes. We recall the definition of Poisson structure on the space $\mathrm{Op}_{\mathfrak{g}}(\mathcal{D}^\times)$ of opers on the formal punctured disc introduced in [DS].

[2] The corresponding space of H-connections on the formal punctured disc and its map to the space of opers were introduced in [DS] as the phase space of the generalized mKdV hierarchy and the Miura transformation from this space to the phase space of the generalized KdV hierarchy.

Following [BD1] and [CHA], we interpret this Poisson structure as a structure of Lie algebroid on the cotangent sheaf $\Omega^1(\mathrm{Op}_{\mathfrak{g}}(\mathcal{D}^\times))$ and following [DS] we show that it is isomorphic to the Lie algebroid of the isomonodromy groupoid on the space of opers. The new results in this section concern the behavior of this algebroid along the subscheme $\mathrm{Op}_{\mathfrak{g}}^{\mathrm{nilp}}$.

1 Opers

1.1 Definition of opers

Throughout Part I (except in Section 1.11), we will assume that G is a simple algebraic group of *adjoint type*. Let B be a Borel subgroup and $N = [B, B]$ its unipotent radical, with the corresponding Lie algebras $\mathfrak{n} \subset \mathfrak{b} \subset \mathfrak{g}$. There is an open B-orbit $\mathbf{O} \subset \mathfrak{g}/\mathfrak{b}$ consisting of vectors which are invariant with respect to the radical $N \subset B$, and such that all of their negative simple root components, with respect to the adjoint action of $H = B/N$, are nonzero. This orbit may also be described as the B-orbit of the sum of the projections of simple root generators $f_\imath$ of any nilpotent subalgebra $\mathfrak{n}^-$, which is in generic position with respect to $\mathfrak{b}$, onto $\mathfrak{g}/\mathfrak{b}$. The torus $H = B/N$ acts simply transitively on $\mathbf{O}$, so $\mathbf{O}$ is an H-torsor. Note in addition that $\mathbf{O}$ is invariant with respect to the action of $\mathbb{G}_m$ on $\mathfrak{g}$ by dilations.

Let X be a smooth curve, or the formal disc $\mathcal{D} = \mathrm{Spec}(\hat{\mathcal{O}})$, where $\hat{\mathcal{O}}$ is a one-dimensional complete local ring, or the formal punctured disc $\mathcal{D}^\times = \mathrm{Spec}(\hat{\mathcal{K}})$, where $\hat{\mathcal{K}}$ is the field of fractions of $\hat{\mathcal{O}}$. We will denote by ω_X the canonical line bundle on X; by a slight abuse of notation we will identify it with the corresponding $\mathbb{G}_m$-torsor on X.

Suppose we are given a principal G-bundle $\mathcal{F}_G$ on X, together with a connection ∇ (automatically flat) and a reduction $\mathcal{F}_B$ of $\mathcal{F}_G$ to the Borel subgroup B of G. Then we define the relative position of ∇ and $\mathcal{F}_B$ (i.e., the failure of ∇ to preserve $\mathcal{F}_B$) as follows. Locally, choose any connection ∇' on $\mathcal{F}$ preserving $\mathcal{F}_B$, and take the difference $\nabla - \nabla' \in \mathfrak{g}_{\mathcal{F}_G} \simeq \mathfrak{g}_{\mathcal{F}_B}$. It is clear that the projection of $\nabla - \nabla'$ to $(\mathfrak{g}/\mathfrak{b})_{\mathcal{F}_B} \otimes \omega_X$ is independent of ∇'; we will denote it by $\nabla/\mathcal{F}_B$. This $(\mathfrak{g}/\mathfrak{b})_{\mathcal{F}_B}$-valued one-form on X is by definition the relative position of ∇ and $\mathcal{F}_B$.

Following Beilinson and Drinfeld (see [BD1, Section 3.1] and [BD2]), one defines a $\mathfrak{g}$-*oper* on X to be a triple $(\mathcal{F}_G, \nabla, \mathcal{F}_B)$, where $\mathcal{F}_G$ is a principal G-bundle $\mathcal{F}_G$ on X, ∇ is a connection on $\mathcal{F}_G$, and $\mathcal{F}_B$ is a B-reduction of $\mathcal{F}_G$ such that the one-form $\nabla/\mathcal{F}_B$ takes values in

$$\mathbf{O}_{\mathcal{F}_B,\omega_X} := \mathbf{O} \overset{B\times\mathbb{G}_m}{\times} (\mathcal{F}_B \times \omega_X) \subset (\mathfrak{g}/\mathfrak{b})_{\mathcal{F}_B} \otimes \omega_X.$$

Consider the H-bundle $\omega_X^{\check{\rho}}$ on X, induced from the line bundle ω_X by means of the homomorphism $\check{\rho} : \mathbb{G}_m \to H$. (The latter is well defined, since G was assumed to be of adjoint type.)

Lemma 1.2. *For an oper* $(\mathcal{F}_G, \nabla, \mathcal{F}_B)$*, the induced* H*-bundle* $\mathcal{F}_H := N\backslash\mathcal{F}_B$ *is canonically isomorphic to* $\omega_X^{\check{\rho}}$.

Proof. We have to show that for every simple root $\alpha_\imath : B \to \mathbb{G}_m$, the line bundle $\mathbb{C}^{\alpha_\imath}_{\mathcal{F}_B}$ is canonically isomorphic to ω_X.

Decomposing $\nabla/\mathcal{F}_B$ with respect to negative simple roots, we obtain for every $\imath$ a nonvanishing section of the line bundle

$$\mathbb{C}^{-\alpha_\imath}_{\mathcal{F}_B} \otimes \omega_X.$$

This provides the required identification. □

Here is an equivalent way to think about opers. Let us choose a trivialization of the B-bundle $\mathcal{F}_B$, and let ∇^0 be the tautological connection on it. Then an oper is given by a connection ∇ of the form

$$\nabla = \nabla^0 + \sum_\imath \phi_\imath \cdot f_\imath + \mathbf{q}, \tag{1.1}$$

where each $\phi_\imath$ is a nowhere vanishing one-form on X, and $\mathbf{q}$ is a $\mathfrak{b}$-valued one-form. If we change the trivialization of $\mathcal{F}_B$ by $\mathbf{g} : X \to B$, the connection will get transformed by the corresponding gauge transformation:

$$\nabla \mapsto \mathrm{Ad}_{\mathbf{g}}(\nabla) := \nabla^0 + \mathrm{Ad}_{\mathbf{g}}\left(\sum_\imath \phi_\imath \cdot f_\imath + \mathbf{q}\right) - \mathbf{g}^{-1} \cdot d(\mathbf{g}). \tag{1.2}$$

The following will be established in the course of the proof of Proposition 1.6.

Lemma 1.3. *If* $\mathrm{Ad}_{\mathbf{g}}(\nabla) = \nabla$, *then* $\mathbf{g} = 1$.

In a similar way one defines the notion of an R-family of opers on X, where R is an arbitrary commutative $\mathbb{C}$-algebra. We shall denote this functor by $\mathrm{Op}_{\mathfrak{g}}(X)$. For $X = \mathcal{D}$ (respectively, $X = \mathcal{D}^\times$) some extra care is needed when one defines the notion of R-family of bundles. To simplify the notation we will choose a coordinate t on $\mathcal{D}$, thereby identifying $\hat{\mathcal{O}} \simeq \mathbb{C}[[t]]$ and $\hat{\mathcal{K}} \simeq \mathbb{C}((t))$. Although this choice of the coordinate trivializes ω_X by means of dt, we will keep track of the distinction between functions and forms by denoting the $\hat{\mathcal{O}}$-module $\omega_{\mathcal{D}}$ (respectively, the $\hat{\mathcal{K}}$-vector space $\omega_{\mathcal{D}^\times}$) by by $\mathbb{C}[[t]]dt$ (respectively, $\mathbb{C}((t))dt$).

By definition, an R-family of G-bundles on $X = \mathcal{D}$ is a G-bundle on $\mathrm{Spec}(R[[t]])$, or what is the same, a compatible family of G-bundles on $\mathrm{Spec}(R[t]/t^i)$; such a family is always locally trivial in the étale topology on $\mathrm{Spec}(R)$.

An R-family of G-bundles on $\mathcal{D}^\times$ is a G-bundle on $\mathrm{Spec}(R((t)))$, which *we require to be locally trivial in the étale topology in* $\mathrm{Spec}(R)$.

Connections on the trivial R-family of G-bundles on $\mathcal{D}$ and $\mathcal{D}^\times$ are expressions of the form $\nabla^0 + \phi$, where ϕ is an element of $\mathfrak{g} \otimes R[[t]]dt$ and $\mathfrak{g} \otimes R((t))dt$, respectively. Gauge transformations are elements of $G(R[[t]])$ and $G(R((t)))$, respectively, and they act on connections by the formula (1.2).

Thus $\mathrm{Op}_{\mathfrak{g}}(\mathcal{D})$ and $\mathrm{Op}_{\mathfrak{g}}(\mathcal{D}^\times)$ are well defined as functors on the category of $\mathbb{C}$-algebras. Following [BD1, Section 3.1.10], we will prove below that these functors are representable by a scheme and ind-scheme, respectively.

1.4 D-scheme picture

When X is a curve of finite type, a natural way to think of $\mathfrak{g}$-opers on X is in terms of D-schemes. (We refer the reader to [CHA, Section 2.3] for the general discussion of D-schemes, and to [CHA, Section 2.6.8] for the discussion of opers in this context.)

Namely, let us notice that the notion of R-family of $\mathfrak{g}$-opers on X makes sense when R is a $\mathfrak{D}_X$-algebra, i.e., a quasi-coherent sheaf of algebras over X, endowed with a connection.

Repeating the argument of Proposition 1.6 (see below), one obtains that the above functor on the category of $\mathfrak{D}_X$-algebras is representable; the corresponding affine $\mathfrak{D}_X$-scheme, denoted $\mathrm{Op}_{\mathfrak{g}}(X)^{\mathfrak{D}}$, is isomorphic to the $\mathfrak{D}_X$-scheme of jets into a finite-dimensional vector space.

By definition, for a $\mathbb{C}$-algebra R we have

$$\mathrm{Op}_{\mathfrak{g}}(X)(R) \simeq \mathrm{Hom}_{\mathfrak{D}_X}(\mathrm{Spec}(R \otimes \mathcal{O}_X), \mathrm{Op}_{\mathfrak{g}}(X)^{\mathfrak{D}}). \tag{1.3}$$

If $\mathcal{D}$ is the formal neighborhood of a point $x \in X$ with a local coordinate t, the functors $\mathrm{Op}_{\mathfrak{g}}(\mathcal{D})$ and $\mathrm{Op}_{\mathfrak{g}}(\mathcal{D}^\times)$ are reconstructed as

$$\begin{aligned} &R \mapsto \mathrm{Hom}_{\mathfrak{D}_X}\left(\mathrm{Spec}(R[[t]]), \mathrm{Op}_{\mathfrak{g}}(X)^{\mathfrak{D}}\right) \quad \text{and} \\ &R \mapsto \mathrm{Hom}_{\mathfrak{D}_X}\left(\mathrm{Spec}\left(R((t))\right), \mathrm{Op}_{\mathfrak{g}}(X)^{\mathfrak{D}}\right), \end{aligned} \tag{1.4}$$

respectively.

In addition, one also has an isomorphism between the scheme $\mathrm{Op}_{\mathfrak{g}}(\mathcal{D})$ and the fiber of $\mathrm{Op}_{\mathfrak{g}}(X)^{\mathfrak{D}}$, regarded as a mere scheme over X, at $x \in X$.

1.5 Explicit description and canonical representatives

To analyze opers on $\mathcal{D}^\times$ (respectively, $\mathcal{D}$) more explicitly we will continue to use an identification $\mathcal{O} \simeq \mathbb{C}[[t]]$, and we will think of opers as equivalence classes of connections of the form

$$\nabla = \nabla^0 + \sum_\imath \phi_\imath(t)dt \cdot f_\imath + \mathbf{q}(t)dt, \tag{1.5}$$

where now $\phi_\imath$ and $\mathbf{q}$ are elements of $R((t))$ and $\mathfrak{b} \otimes R((t))$ (respectively, $R[[t]]$ and $\mathfrak{b} \otimes R[[t]]$), such that each $\phi_\imath$ is invertible. Two such connections are equivalent, if they can be conjugated one into another by a gauge transformation by an element of $\mathrm{Hom}(\mathrm{Spec}(R((t))), B)$ (respectively, $\mathrm{Hom}(\mathrm{Spec}(R[[t]]), B)$).

Let us observe that since $H \simeq B/N$ acts simply-transitively on $\mathbf{O}$, any connection as above can be brought to the form in which all the functions $\phi_\imath(t)$ are equal to 1. Moreover, this can be done uniquely, up to a gauge transformation by means of $\mathrm{Hom}(\mathrm{Spec}(R((t)), N)$.

The operator $\mathrm{ad}\,\check{\rho}$ defines the principal grading on $\mathfrak{b}$, with respect to which we have a direct sum decomposition $\mathfrak{b} = \bigoplus\limits_{d \geq 0} \mathfrak{b}_d$. Set

$$p_{-1} = \sum_{\imath} f_\imath;$$

we shall call this element the negative principal nilpotent.

Let p_1 be the unique element of $\mathfrak{n}$ such that $\{p_{-1}, 2\check{\rho}, p_1\}$ is an $\mathfrak{sl}_2$-triple. Let $V_{\text{can}} = \underset{d>0}{\oplus} V_{\text{can},d}$ be the space of ad p_1-invariants in $\mathfrak{n}$. The operator ad p_{-1} acts from $\mathfrak{b}_{d+1}$ to $\mathfrak{b}_d$ injectively for all $d \geq 0$, and we have $\mathfrak{b}_d = [p_{-1}, \mathfrak{b}_{d+1}] \oplus V_{\text{can},d}$.

We will call the $\mathbb{G}_m$-action on V_{can}, resulting from the above grading, "principal." We will call the $\mathbb{G}_m$-action on V_{can}, obtained by multiplying the principal one by the standard character, "canonical." Recall that by a theorem of Kostant, the map

$$V_{\text{can}} \overset{\mathbf{c} \mapsto p_{-1}+\mathbf{c}}{\longrightarrow} \mathfrak{g} \to \mathfrak{g}/\!/G \simeq \mathfrak{h}/\!/W \tag{1.6}$$

is an isomorphism. This map is compatible with the canonical $\mathbb{G}_m$-action on V_{can} and the action on $\mathfrak{h}/\!/W$, induced by the standard $\mathbb{G}_m$-action on $\mathfrak{h}$.

Proposition 1.6 ([DS]). *The gauge action of* $\text{Hom}(\text{Spec}(R(\!(t)\!)), B)$ *on the set of connections of the form* (1.5) *is free. Each gauge equivalence class contains a unique representative of the form*

$$\nabla = \nabla^0 + p_{-1}dt + \mathbf{v}(t)dt, \quad \mathbf{v}(t) \in V_{\text{can}} \otimes R(\!(t)\!). \tag{1.7}$$

As we shall see, the same assertion with the same proof is valid if we replace $R(\!(t)\!)$ by $R[[t]]$. In what follows we will refer to (1.7) as the canonical representative of an oper.

Proof. We already know that we can bring a connection (1.5) to the form

$$\nabla^0 + p_{-1}dt + \mathbf{q}(t)dt,$$

uniquely up to an element in $\text{Hom}(\text{Spec}(R(\!(t)\!)), N)$. We need to show now that there exists a unique element $u(t) \in \mathfrak{n} \otimes R(\!(t)\!)$ such that

$$\text{Ad}_{\exp(u(t))}\left(\nabla^0 + p_{-1}dt + \mathbf{q}(t)dt\right) = \nabla^0 + p_{-1}dt + \mathbf{v}(t)dt,$$

$\mathbf{v}(t) \in V_{\text{can}} \otimes R(\!(t)\!)$.

Let us now decompose the unknown element $u(t)$ as $\sum\limits_d u_d(t)$, where $u_d(t) \in \mathfrak{n}_d \otimes R(\!(t)\!)$, and we claim that we can find the elements $u_d(t)$ by induction d. Indeed, let us assume that $\mathbf{q}_{d'} \in V_{\text{can},d'}$ for $d' < d$. Then $u_{d+1}(t)$ must satisfy

$$[u_{d+1}(t), p_{-1}] + \mathbf{q}_d(t) \in V_{\text{can},d},$$

and this indeed has a unique solution. □

Corollary 1.7. *The set of R-families of opers on* $\mathcal{D}$ *and* $\mathcal{D}^\times$ *is isomorphic to* $V_{\text{can}} \otimes R[[t]]$ *and* $V_{\text{can}} \otimes R(\!(t)\!)$, *respectively. In particular, the functor* $\text{Op}_{\mathfrak{g}}(\mathcal{D})$ *(respectively,* $\text{Op}_{\mathfrak{g}}(\mathcal{D}^\times)$*) is representable by the scheme (respectively, ind-scheme), isomorphic to* $V_{\text{can}}[[t]]$ *(respectively,* $V_{\text{can}}(\!(t)\!)$*).*

We should note, however, that the isomorphisms $\mathrm{Op}_{\mathfrak{g}}(\mathcal{D}) \simeq V_{\mathrm{can}}[[t]]$ and $\mathrm{Op}_{\mathfrak{g}}(\mathcal{D}^\times) \simeq V_{\mathrm{can}}((t))$ are not canonical, since they depend on the choice of the coordinate t on $\mathcal{D}$.

By the very definition, on the scheme

$$\mathrm{Op}_{\mathfrak{g}}(\mathcal{D})\widehat{\times}\mathcal{D} := \mathrm{Spec}(\mathrm{Fun}\,\mathrm{Op}_{\mathfrak{g}}(\mathcal{D}))[[t]]$$

there exists a universal G-bundle $\mathcal{F}_{G,\mathrm{Op}_{\mathfrak{g}}(\mathcal{D})}$ with a reduction to a B-bundle $\mathcal{F}_{B,\mathrm{Op}_{\mathfrak{g}}(\mathcal{D})}$ and a connection $\nabla_{\mathrm{Op}_{\mathfrak{g}}(\mathcal{D})}$ in the $\mathcal{D}$-direction such that the triple

$$(\mathcal{F}_{G,\mathrm{Op}_{\mathfrak{g}}(\mathcal{D})}, \nabla_{\mathrm{Op}_{\mathfrak{g}}(\mathcal{D})}, \mathcal{F}_{B,\mathrm{Op}_{\mathfrak{g}}(\mathcal{D})})$$

is an $\mathrm{Op}_{\mathfrak{g}}(\mathcal{D})$-family of $\mathfrak{g}$-opers on $\mathcal{D}$. By the above, when we identify $\hat{\mathcal{O}} \simeq \mathbb{C}[[t]]$, the G-bundle $\mathcal{F}_{B,\mathrm{Op}_{\mathfrak{g}}(\mathcal{D})}$, and hence $\mathcal{F}_{G,\mathrm{Op}_{\mathfrak{g}}(\mathcal{D})}$, becomes trivialized. But this trivialization depends on the choice of the coordinate.

In what follows we will denote by $\mathcal{P}_{G,\mathrm{Op}_{\mathfrak{g}}(\mathcal{D})}$ (respectively, $\mathcal{P}_{B,\mathrm{Op}_{\mathfrak{g}}(\mathcal{D})}$) the restriction of $\mathcal{F}_{G,\mathrm{Op}_{\mathfrak{g}}(\mathcal{D})}$ (respectively, $\mathcal{F}_{B,\mathrm{Op}_{\mathfrak{g}}(\mathcal{D})}$) to the subscheme $\mathrm{Op}_{\mathfrak{g}}(\mathcal{D}) \subset \mathrm{Op}_{\mathfrak{g}}(\mathcal{D})\widehat{\times}\mathcal{D}$, corresponding to the closed point of $\mathcal{D}$. Note that $\mathcal{P}_{G,\mathrm{Op}_{\mathfrak{g}}(\mathcal{D})}$ can also be defined as the torsor of horizontal, with respect to the connection along $\mathcal{D}$, sections of $\mathcal{F}_{G,\mathrm{Op}_{\mathfrak{g}}(\mathcal{D})}$.

1.8 Action of $\mathrm{Aut}(\mathcal{D})$

Let $\mathrm{Aut}(\mathcal{D})$ (respectively, $\mathrm{Aut}(\mathcal{D}^\times)$) be the group scheme (respectively, group ind-scheme) of automorphisms of $\mathcal{D}$ (respectively, $\mathcal{D}^\times$).[3] Since $\mathrm{Op}_{\mathfrak{g}}(\mathcal{D})$ (respectively, $\mathrm{Op}_{\mathfrak{g}}(\mathcal{D}^\times)$) is canonically attached to $\mathcal{D}$ (respectively, $\mathcal{D}^\times$), it carries an action of $\mathrm{Aut}(\mathcal{D})$ (respectively, $\mathrm{Aut}(\mathcal{D}^\times)$); see Section 19.2 for the definition of the latter notion.

By transport of structure, the action of $\mathrm{Aut}(\mathcal{D})$ on $\mathrm{Op}_{\mathfrak{g}}(\mathcal{D})\widehat{\times}\mathcal{D}$ lifts onto $\mathcal{F}_{G,\mathrm{Op}_{\mathfrak{g}}(\mathcal{D})}$ and $\mathcal{F}_{B,\mathrm{Op}_{\mathfrak{g}}(\mathcal{D})}$. The interpretation of $\mathcal{P}_{G,\mathrm{Op}_{\mathfrak{g}}(\mathcal{D})}$ as the space of horizontal sections of $\mathcal{F}_{G,\mathrm{Op}_{\mathfrak{g}}(\mathcal{D})}$ implies that the action of $\mathrm{Aut}(\mathcal{D})$ on $\mathrm{Op}_{\mathfrak{g}}(\mathcal{D})$ lifts also onto the G-torsor $\mathcal{P}_{G,\mathrm{Op}_{\mathfrak{g}}(\mathcal{D})}$.

To a choice of a coordinate t on $\mathcal{D}$ there corresponds a homomorphism $\mathbb{G}_m \to \mathrm{Aut}(\mathcal{D})$ that acts by the "loop rotation," i.e., $t \mapsto c \cdot t$. We shall now describe the resulting action of $\mathbb{G}_m$ on $\mathrm{Op}_{\mathfrak{g}}(\mathcal{D}^\times)$ in terms of the isomorphism of Corollary 1.7.

Lemma 1.9.

(1) *The trivialization $\mathcal{P}_{G,\mathrm{Op}_{\mathfrak{g}}(\mathcal{D})} \simeq G \times \mathrm{Op}_{\mathfrak{g}}(\mathcal{D})$ corresponding to the given choice of a coordinate is compatible with the $\mathbb{G}_m$-action, via the homomorphism $\check{\rho} : \mathbb{G}_m \to G$.*

(2) *The action of $c \in \mathbb{G}_m$ on*

$$\mathrm{Op}_{\mathfrak{g}}(\mathcal{D}^\times) \simeq V_{\mathrm{can}}((t)) \simeq \bigoplus_d V_{\mathrm{can},d}((t))$$

[3] Note that $\mathrm{Aut}(\mathcal{D})$ is not reduced; see [BD1, Section 2.6.5].

is given by

$$\mathbf{v}_d(t) \in V_{\mathrm{can},d}(\!(t)\!) \mapsto c^{d+1} \cdot \mathbf{v}_d(c \cdot t).$$

Proof. By definition, the action of $c \in \mathbb{G}_m$ on a connection in the form (1.7) transforms it to

$$\nabla^0 + p_{-1}d(c \cdot t) + \mathbf{v}(c \cdot t)d(c \cdot t) = \nabla^0 + c \cdot p_{-1}dt + c \cdot \mathbf{v}(c \cdot t)dt. \tag{1.8}$$

In order to bring it back to the form (1.7), we need to apply a gauge transformation by means of the constant H-valued function $\check{\rho}(c)$. This implies point (1) of the lemma.

This gauge transformation transforms (1.8) to

$$\nabla^0 + p_{-1}dt + c \cdot \mathrm{Ad}_{\check{\rho}(c)}(\mathbf{v}(c \cdot t))dt.$$

This implies point (2) of the lemma, since $\mathrm{Ad}_{\check{\rho}(c)}(\mathbf{v}_d(c \cdot t)) = c^d \cdot \mathbf{v}_d(c \cdot t)$. □

1.10 Quasi-classics: The Hitchin space

Recall that the Hitchin space $\mathrm{Hitch}_{\mathfrak{g}}(X)$ corresponding to the Lie algebra $\mathfrak{g}$ and a curve X is a functor on the category of algebras that attaches to R the set of sections of the pull-back to $\mathrm{Spec}(R) \times X$ of the fiber bundle

$$(\mathfrak{h}/\!/W) \overset{\mathbb{G}_m}{\times} \omega_X,$$

where $\mathfrak{h}/\!/W := \mathrm{Spec}\,(\mathrm{Sym}(\mathfrak{h}^*)^W) \simeq \mathrm{Spec}\,(\mathrm{Sym}(\mathfrak{g}^*)^G)$ is endowed with canonical action of $\mathbb{G}_m$.

When $X = \mathcal{D}$ or $X = \mathcal{D}^\times$, in the above definition we replace $\mathrm{Spec}(R) \times X$ by $\mathrm{Spec}(R[[t]])$ and $\mathrm{Spec}\,(R(\!(t)\!))$, respectively.

For $X = \mathcal{D}$ the Hitchin space is a scheme, isomorphic to $\bigoplus\limits_d V_{\mathrm{can},d} \otimes \omega_{\mathcal{D}}^{\otimes d+1}$. For $X = \mathcal{D}^\times$ this is an ind-scheme, isomorphic to $\bigoplus\limits_d V_{\mathrm{can},d} \otimes \omega_{\mathcal{D}^\times}^{\otimes d+1}$. In particular, $\mathrm{Hitch}_{\mathfrak{g}}(\mathcal{D})$ (respectively, $\mathrm{Hitch}_{\mathfrak{g}}(\mathcal{D}^\times)$) has a natural structure of group scheme (respectively, group ind-scheme).

According to [BD1, Section 2.4.1], the natural map

$$\mathrm{Spec}(\mathrm{Sym}(\check{\mathfrak{g}}(\!(t)\!)/\check{\mathfrak{g}}[[t]])^{\check{\mathfrak{g}}[[t]]}) \to \mathrm{Hitch}_{\mathfrak{g}}(\mathcal{D})$$

is an isomorphism, where $\check{\mathfrak{g}}$ is the Langlands dual Lie algebra. This implies that the maps

$$\begin{aligned}\mathrm{Fun}\left(\mathrm{Hitch}_{\mathfrak{g}}(\mathcal{D}^\times)\right) &\to \left(\varprojlim_k \mathrm{Sym}\left(\check{\mathfrak{g}}(\!(t)\!)/t^k \cdot \check{\mathfrak{g}}[[t]]\right)\right)^{\check{\mathfrak{g}}(\!(t)\!)} \\ &\to \varprojlim_k \mathrm{Sym}\left(\check{\mathfrak{g}}(\!(t)\!)/t^k \cdot \check{\mathfrak{g}}[[t]]\right)^{\check{\mathfrak{g}}[[t]]}\end{aligned}$$

are also isomorphisms.

By Proposition 1.6 the scheme $\mathrm{Op}_{\mathfrak{g}}(\mathcal{D})$ (respectively, $\mathrm{Op}_{\mathfrak{g}}(\mathcal{D}^\times)$) is noncanonically isomorphic to $\mathrm{Hitch}_{\mathfrak{g}}(\mathcal{D})$ (respectively, $\mathrm{Hitch}_{\mathfrak{g}}(\mathcal{D}^\times)$). However, one can deduce from the proof (see [BD1, Section 3.10.11]) that $\mathrm{Op}_{\mathfrak{g}}(\mathcal{D})$ (respectively, $\mathrm{Op}_{\mathfrak{g}}(\mathcal{D}^\times)$) is *canonically* a torsor over $\mathrm{Hitch}_{\mathfrak{g}}(\mathcal{D})$ (respectively, $\mathrm{Hitch}_{\mathfrak{g}}(\mathcal{D}^\times)$).

In particular, the algebra $\mathrm{Fun}(\mathrm{Op}_{\mathfrak{g}}(\mathcal{D}))$ acquires a filtration, whose associated graded is $\mathrm{Fun}(\mathrm{Hitch}_{\mathfrak{g}}(\mathcal{D}))$. This filtration can also be defined as follows; see [BD1, Section 3.11.14]:

We claim that there exists a flat $\mathbb{G}_m$-equivariant family of schemes over $\mathbb{A}^1 \simeq \mathrm{Spec}(\mathbb{C}[\hbar])$, whose fiber over $1 \in \mathbb{A}^1$ is $\mathrm{Op}_{\mathfrak{g}}(\mathcal{D})$, and whose fiber over $0 \in \mathbb{A}^1$ is $\mathrm{Hitch}_{\mathfrak{g}}(\mathcal{D})$.

Indeed, this family is obtained from $\mathrm{Op}_{\mathfrak{g}}(\mathcal{D})$ by replacing the word "connection" by "$\hbar$-connection." The identification at the special fiber results from Kostant's theorem that the adjoint action of B on the preimage of $\mathbf{O}$ in $\mathfrak{g}$ is free, and the quotient projects isomorphically onto $\mathfrak{h}/\!/W$.

1.11 The case of groups of nonadjoint type

In the rest of the paper we will have to consider the case when the group G is not necessarily of adjoint type. Let $Z(G)$ be the center of G.

The notion of R-family of G-opers in this case is formally the same as in the adjoint case, i.e., a triple $(\mathcal{F}_G, \nabla, \mathcal{F}_B)$, where $\mathcal{F}_G$ is an R-family of G-bundle on X, $\mathcal{F}_B$ is its reduction to B, and ∇ is a connection on $\mathcal{F}_G$ in the X-direction, which satisfies the same condition on $\nabla/\mathcal{F}_B$.

We will denote the functor of R-families of G-opers on X by $\mathrm{Op}_G(X)$. The difference from the adjoint case is that now $\mathrm{Op}_G(\mathcal{D})$ is not representable by a scheme, but rather by a Deligne–Mumford stack, which is noncanonically isomorphic to $\mathrm{Op}_{\mathfrak{g}}(\mathcal{D}) \times \mathrm{pt}/Z(G)$; see [BD1, Section 3.4].

The following statement, established in [BD1, Section 3.4], will suffice for our purposes.

Lemma 1.12. *Every choice of the square root* $\omega_X^{\frac{1}{2}}$ *of the canonical bundle gives a map of functors* $\mathrm{Op}_{\mathfrak{g}}(X) \to \mathrm{Op}_G(X)$.

In particular, the lemma implies that for every choice of a square root of $\omega_{\mathcal{D}}$, there exists a canonically defined family of G-opers over $\mathrm{Op}_{\mathfrak{g}}(\mathcal{D})$. A similar statement holds for $\mathcal{D}$ replaced by $\mathcal{D}^\times$.

Proof. One only has to show how to lift the $B/Z(G)$-bundle $\mathcal{F}_{B/Z(G)}$ to a B-bundle. This is equivalent to lifting the H-bundle $\mathcal{F}_{H/Z(G)}$ to an H-bundle $\mathcal{F}_H$.

We set $\mathcal{F}_H$ to be the bundle induced by means of the homomorphism $2\check{\rho} : \mathbb{G}_m \to H$ from the line bundle $\omega_X^{\frac{1}{2}}$. By Lemma 1.2, it satisfies our requirement. □

2 Opers with singularities

2.1

For X a curve of finite type over $\mathbb{C}$ we shall fix $x \in X$ to be any closed point. For $X = \mathcal{D}$, we let x to be the unique closed point of $\mathrm{Spec}(\mathbb{C}[[t]])$. We shall now define the notion of $\mathfrak{g}$-*oper on* X *with singularity of order* k *at* x.

By definition, this is a triple $(\mathcal{F}_G, \nabla, \mathcal{F}_B)$, where $(\mathcal{F}_G, \mathcal{F}_B)$ are as in the definition of opers, but the connection ∇ on $\mathcal{F}_G$ is required to have a pole of order k such that

$$(\nabla - \nabla') \bmod \mathfrak{b}_{\mathcal{F}_B} \otimes \omega_X(k \cdot x) \in \mathbf{O}_{\mathcal{F}_B, \omega_X(k \cdot x)} \subset (\mathfrak{g}/\mathfrak{b})_{\mathcal{F}_B} \otimes \omega_X(k \cdot x), \qquad (2.1)$$

for any regular connection ∇' on $\mathcal{F}_G$ that preserves $\mathcal{F}_B$.

Again, if we trivialize $\mathcal{F}_B$ and choose a coordinate t near x, the set of opers with singularity of order k at x identifies with the set of equivalence classes of connections of the form

$$\nabla^0 + t^{-k} \left(\sum_\imath \phi_\imath(t) dt \cdot f_\imath + \mathbf{q}(t) dt \right), \qquad (2.2)$$

where $\phi_\imath(t)$ are nowhere vanishing functions on X, and $\mathbf{q}(t)$ is a $\mathfrak{b}$-valued function. Two such connections are equivalent if they are conjugate by means of an element of $\mathrm{Hom}(X, B)$. Equivalently, opers with singularity of order k at x comprise the set of $\mathrm{Hom}(X, N)$-equivalence classes of connections of the form

$$\nabla^0 + t^{-k} (p_{-1} dt + \mathbf{q}(t) dt)$$

for $\mathbf{q}(t)$ as above.

As in Lemma 1.2 one has the following.

Lemma 2.2. *For $(\mathcal{F}_G, \nabla, \mathcal{F}_B)$-an oper on X with singularity of order k at x, the H-bundle $\mathcal{F}_H = N \backslash \mathcal{F}_B$ is canonically isomorphic to $(\omega_X(x))^{\check{\rho}}$.*

One defines the notion of R-family of opers on X with singularity of order k at x in a straightforward way. The corresponding functor on the category of R-algebras will be denoted $\mathrm{Op}_\mathfrak{g}^{\mathrm{ord}_k}(X)$. We will be mainly concerned with the case when $X = \mathcal{D}$; if no confusion is likely to occur, we will denote the corresponding functor simply by $\mathrm{Op}_\mathfrak{g}^{\mathrm{ord}_k}$. We have an evident morphism of functors $\mathrm{Op}_\mathfrak{g}^{\mathrm{ord}_k} \to \mathrm{Op}_\mathfrak{g}(\mathcal{D}^\times)$, and below we will see (see Corollary 2.7) that $\mathrm{Op}_\mathfrak{g}^{\mathrm{ord}_k}$ is a representable by a closed subscheme of $\mathrm{Op}_\mathfrak{g}(\mathcal{D}^\times)$. Note that for $k = 0$ we recover $\mathrm{Op}_\mathfrak{g}(\mathcal{D})$, and we will often use the notation $\mathrm{Op}_\mathfrak{g}^{\mathrm{reg}}$ for it.

As in the case of usual opers, there exists a naturally defined functor $\mathrm{Op}_\mathfrak{g}^{\mathrm{ord}_k}(X)^{\mathfrak{D}}$ on the category of $\mathfrak{D}_X$-algebras. The functors $\mathrm{Op}_\mathfrak{g}^{\mathrm{ord}_k}(X)$ and $\mathrm{Op}_\mathfrak{g}^{\mathrm{ord}_k}$ are reconstructed by the analogues of (1.3) and (1.4), respectively. Corollary 2.7 implies that $\mathrm{Op}_\mathfrak{g}^{\mathrm{ord}_k}(X)^{\mathfrak{D}}$ is representable by an affine $\mathfrak{D}_X$-scheme, which over the curve $(X - x)$ is isomorphic to $\mathrm{Op}_\mathfrak{g}(X)^{\mathfrak{D}}$.

2.3 Changing k

Proposition 2.4 ([BD1], 4.3). *For every k there is a natural morphism of functors $\mathrm{Op}_{\mathfrak{g}}^{\mathrm{ord}_k} \to \mathrm{Op}_{\mathfrak{g}}^{\mathrm{ord}_{k+1}}$. We have*

$$\mathrm{Op}_{\mathfrak{g}}(\mathcal{D}^{\times}) \simeq \varinjlim_{k} \mathrm{Op}_{\mathfrak{g}}^{\mathrm{ord}_k}.$$

Proof. Given a triple $(\mathcal{F}_G, \nabla, \mathcal{F}_B) \in \mathrm{Op}_{\mathfrak{g}}^{\mathrm{ord}_k}(R)$, we define the corresponding

$$(\mathcal{F}'_G, \nabla', \mathcal{F}'_B) \in \mathrm{Op}_{\mathfrak{g}}^{\mathrm{ord}_{k+1}}(R)$$

as follows:

Let us choose (locally) a trivialization of $\mathcal{F}_B$, and let us apply the gauge transformation by means of $t^{\check{\rho}} \in H((t))$, where t is any uniformizer on $\mathcal{D}$. We thus obtain a different extension of $\mathcal{F}_B$ from $\mathcal{D}^{\times}$ to $\mathcal{D}$, and let it be our $\mathcal{F}'_B$. It is clear that $\mathcal{F}'_B$ is independent of both the choice of the trivialization and the coordinate.

Let $\mathcal{F}'_G$ be the induced G-bundle, and ∇' the resulting meromorphic connection on it. By (1.2), ∇' has the form required by the (2.2).

Now let $(\mathcal{F}_G, \nabla, \mathcal{F}_B)$ be an R-point of $\mathrm{Op}_{\mathfrak{g}}(\mathcal{D}^{\times})$, represented as a gauge equivalence class of some connection written in the form (1.5), $\phi_\imath(t) \in (R((t)))^{\times}$.

Consider the R-point of $H((t))$ equal to $(\prod_\imath (\check{\omega}_\imath)(\phi_\imath))^{-1} \cdot t^{k\cdot\check{\rho}}$, where each $\check{\omega}_\imath$ is regarded as a homomorphism $\mathbb{G}_m \to H$. It is clear from (1.2) that for k large enough, the resulting connection will be of the form (2.2). □

2.5 Description in terms of canonical representatives

By repeating the proof of Proposition 1.6, we obtain the following.

Lemma 2.6. *For every R-point of $\mathrm{Op}_{\mathfrak{g}}^{\mathrm{ord}_k}$, the canonical form of its image in $\mathrm{Op}_{\mathfrak{g}}(\mathcal{D}^{\times})(R)$ is such that each homogeneous component $\mathbf{v}_d(t)$ has a pole in t of order $\leq k \cdot (d+1)$.*

Corollary 2.7. *The morphisms of functors $\mathrm{Op}_{\mathfrak{g}}^{\mathrm{ord}_k} \to \mathrm{Op}_{\mathfrak{g}}^{\mathrm{ord}_{k+1}}$ and $\mathrm{Op}_{\mathfrak{g}}^{\mathrm{ord}_k} \to \mathrm{Op}_{\mathfrak{g}}(\mathcal{D}^{\times})$ are closed embeddings. The latter identifies with the subscheme*

$$\bigoplus_d t^{-k\cdot(d+1)} \cdot V_{\mathrm{can},d}[[t]] \subset V_{\mathrm{can},d}((t)).$$

Proof. Evidently, given a point of $\mathrm{Op}_{\mathfrak{g}}(\mathcal{D}^{\times})$, written in the canonical form (1.7) such that $t^{k\cdot(d+1)} \cdot \mathbf{v}_d(t) \in V_{\mathrm{can},d}[[t]]$, by applying the gauge transformation by means of $t^{k\cdot\check{\rho}}$, we bring it to the form (2.2). Thus we obtain the maps

$$\mathrm{Op}_{\mathfrak{g}}^{\mathrm{ord}_k} \rightleftarrows \bigoplus_d t^{-k\cdot(d+1)} \cdot V_{\mathrm{can},d}[[t]].$$

Finally, by induction on d it is easy to see that if some $\mathbf{g} \in B((t))(R)$ conjugates an R-point of $\mathrm{Op}_{\mathfrak{g}}^{\mathrm{ord}_k}$ to another point of $\mathrm{Op}_{\mathfrak{g}}^{\mathrm{ord}_k}$, then $\mathbf{g} \in B[[t]](R)$. □

Recall the ind-scheme $\mathrm{Hitch}_{\mathfrak{g}}(\mathcal{D}^\times)$, and let us denote by $\mathrm{Hitch}_{\mathfrak{g}}^{\mathrm{ord}_k}$ its subscheme corresponding to sections of $(\mathfrak{h}/\!/W) \overset{\mathbb{G}_m}{\times} \omega_{\mathcal{D}}(k \cdot x)$. This is a scheme, canonically isomorphic to $\bigoplus_d V_{\mathrm{can},d} \otimes (\omega_{\mathcal{D}}(k \cdot x))^{\otimes d+1}$, which gives it a structure of group scheme. Evidently, $\mathrm{Hitch}_{\mathfrak{g}}(\mathcal{D}^\times)$ is isomorphic to $\varinjlim_k \mathrm{Hitch}_{\mathfrak{g}}^{\mathrm{ord}_k}$. We also have an isomorphism:

$$\mathrm{Fun}(\mathrm{Hitch}_{\mathfrak{g}}^{\mathrm{ord}_k}) \simeq \mathrm{Sym}\left(\check{\mathfrak{g}}((t))/t^k \cdot \check{\mathfrak{g}}[[t]]\right)^{\check{\mathfrak{g}}[[t]]}.$$

As in the case of $k = 0$, the scheme $\mathrm{Hitch}_{\mathfrak{g}}^{\mathrm{ord}_k}$ acts simply transitively on $\mathrm{Op}_{\mathfrak{g}}^{\mathrm{ord}_k}$. Moreover,

$$\mathrm{Op}_{\mathfrak{g}}^{\mathrm{ord}_k} \simeq \mathrm{Op}_{\mathfrak{g}}(\mathcal{D}) \overset{\mathrm{Hitch}_{\mathfrak{g}}(\mathcal{D})}{\times} \mathrm{Hitch}_{\mathfrak{g}}^{\mathrm{ord}_k}.$$

This defines a filtration on the algebra $\mathrm{Fun}(\mathrm{Op}_{\mathfrak{g}}^{\mathrm{ord}_k})$, whose associated graded is $\mathrm{Fun}(\mathrm{Hitch}_{\mathfrak{g}}^{\mathrm{ord}_k})$. This filtration can be alternatively described by the deformation procedure mentioned at the end of Section 1.10.

2.8 Opers with regular singularities

In the context of the previous subsection let us set $k = 1$, in which case we will replace the superscript ord_1 by RS, and call the resulting scheme $\mathrm{Op}_{\mathfrak{g}}^{\mathrm{RS}}$ "the scheme of opers with regular singularities." The terminology is partly justified by the following assertion, which will be proved in the next section.

Proposition 2.9. *If a $\mathbb{C}$-point $(\mathcal{F}_G, \nabla, \mathcal{F}_B)$ of $\mathrm{Op}_{\mathfrak{g}}(\mathcal{D}^\times)$ has regular singularities as a G-bundle with connection, then it belongs to $\mathrm{Op}_{\mathfrak{g}}^{\mathrm{RS}}$.*

We claim now that there exists a canonical map $\mathrm{Res}^{\mathrm{RS}} : \mathrm{Op}_{\mathfrak{g}}^{\mathrm{RS}} \to \mathfrak{h}/\!/W$; see [BD1, Section 3.8.11]:

Recall first that if $(\mathcal{F}_G, \nabla)$ is an R-family of G-bundles on X with a connection that has a pole of order 1 at x, its residue (or polar part) is well defined as a section of $\mathfrak{g}_{\mathcal{P}_G}$, where $\mathcal{P}_G$ is the restriction of $\mathcal{F}_G$ to $\mathrm{Spec}(R) \times x \subset \mathrm{Spec}(R) \times X$. In other words, we obtain an R-point of the stack $\mathfrak{g}/G$.

Given an R-point $(\mathcal{F}_G, \nabla, \mathcal{F}_B) \in \mathrm{Op}_{\mathfrak{g}}^{\mathrm{RS}}$, we compose the above map with $\mathfrak{g}/\mathrm{Ad}(G) \to \mathfrak{h}/\!/W$. The resulting map $\mathrm{Spec}(R) \to \mathfrak{h}/\!/W$ is the map $\mathrm{Res}^{\mathrm{RS}}$.

Explicitly, to a connection written as

$$\nabla^0 + t^{-1}\left(\sum_\imath \phi_\imath(t)dt \cdot f_\imath + \mathbf{q}(t)dt\right), \tag{2.3}$$

we attach the projection to $\mathfrak{h}/\!/W$ of the element $\sum_\imath \phi_\imath(0) \cdot f_\imath + \mathbf{q}(0)$.

Let ϖ denote the tautological projection $\mathfrak{h} \to \mathfrak{h}/\!/W$. For $\check{\lambda} \in \mathfrak{h}$ we will denote by $\mathrm{Op}_{\mathfrak{g}}^{\mathrm{RS},\varpi(\check{\lambda})}$ the preimage under $\mathrm{Res}^{\mathrm{RS}}$ of the point $\varpi(\check{\lambda}) \in \mathfrak{h}/\!/W$. From the proof of Proposition 2.4 for $k=0$ we obtain that the subscheme $\mathrm{Op}_{\mathfrak{g}}(\mathcal{D}) =: \mathrm{Op}_{\mathfrak{g}}^{\mathrm{reg}} \subset \mathrm{Op}_{\mathfrak{g}}^{\mathrm{RS}}$ is contained in $\mathrm{Op}_{\mathfrak{g}}^{\mathrm{RS},\varpi(-\check{\rho})}$.

Let us now describe the map $\mathrm{Res}^{\mathrm{RS}}$ in terms of the isomorphism of Corollary 2.7.

Lemma 2.10. *The composition*

$$\bigoplus_d t^{-d-1} \cdot V_{\mathrm{can},d}[[t]] \simeq \mathrm{Op}_{\mathfrak{g}}^{\mathrm{RS}} \overset{\mathrm{Res}^{\mathrm{RS}}}{\to} \mathfrak{h}/\!/W$$

equals the map

$$\bigoplus_d t^{-d-1} \cdot V_{\mathrm{can},d}[[t]] \to \bigoplus_d V_{\mathrm{can},d} \simeq V_{\mathrm{can}} \simeq \mathfrak{h}/\!/W,$$

where the last arrow is given by (1.6), *and the first arrow is defined as follows:*

- *For $d \neq 1$, this is the projection on the top polar part.*
- *For $d=1$, this is the projection on the top polar part, followed by the affine shift by $\frac{p_1}{4}$.*

Proof. By the proof of Proposition 1.6, we have to check that for any $\mathbf{v}' \in \bigoplus_{d\neq 1} V_{\mathrm{can},d}$ and $\mathbf{v}'' \in V_{\mathrm{can},1}$, the elements of $\mathfrak{g}$ given by $p_{-1}+\mathbf{v}'+\mathbf{v}''-\check{\rho}$ and $p_{-1}+\mathbf{v}'+\mathbf{v}''+\frac{p_1}{4}$ project to the same element of $\mathfrak{h}/\!/W$. However, this follows from the fact that $\exp(\frac{p_1}{2})$ conjugates one to the other. □

Corollary 2.11. *Under the isomorphism of Corollary* 2.7 *the subscheme* $\mathrm{Op}_{\mathfrak{g}}^{\varpi(-\check{\rho})} \subset \mathrm{Op}_{\mathfrak{g}}^{\mathrm{RS}}$ *identifies with*

$$\bigoplus_d t^{-d} \cdot V_{\mathrm{can},d}[[t]] \subset \bigoplus_d t^{-d-1} \cdot V_{\mathrm{can},d}[[t]].$$

Of course, as in the case of Corollary 2.7, the isomorphism

$$\mathrm{Op}_{\mathfrak{g}}^{\varpi(-\check{\rho})} \simeq \bigoplus_d t^{-d} \cdot V_{\mathrm{can},d}[[t]]$$

depends on the choice of the coordinate t. Canonically, $\mathrm{Op}_{\mathfrak{g}}^{\varpi(-\check{\rho})}$ can be described in terms of the Hitchin space as follows.

Let us denote $\mathrm{Hitch}_{\mathfrak{g}}^{\mathrm{ord}_1}$ by $\mathrm{Hitch}_{\mathfrak{g}}^{\mathrm{RS}}$, and let us note that we have a natural homomorphism $\mathrm{Hitch}_{\mathfrak{g}}^{\mathrm{RS}} \to V_{\mathrm{can}}$. Let $\mathrm{Hitch}_{\mathfrak{g}}^{\mathrm{nilp}} \subset \mathrm{Hitch}_{\mathfrak{g}}^{\mathrm{RS}}$ be the preimage of 0. The algebra of functions on $\mathrm{Hitch}_{\mathfrak{g}}^{\mathrm{nilp}}$ also admits the description (see [F, Lemma 9.4])

$$\mathrm{Fun}(\mathrm{Hitch}_{\mathfrak{g}}^{\mathrm{nilp}}) \simeq \mathrm{Sym}\left(\check{\mathfrak{g}}((t))/\mathrm{Lie}(\check{I})\right)^{\check{I}},$$

where $\check{I} \subset \check{G}[[t]]$ is the Iwahori subgroup.

We have

$$\mathrm{Op}_{\mathfrak{g}}^{\varpi(-\check{\rho})} \simeq \mathrm{Op}_{\mathfrak{g}}(\mathcal{D}) \underset{\mathrm{Hitch}_{\mathfrak{g}}(\mathcal{D})}{\times} \mathrm{Hitch}_{\mathfrak{g}}^{\mathrm{nilp}}.$$

Now consider the gradings on the algebras $\mathrm{Fun}(\mathrm{Op}_{\mathfrak{g}}^{\mathrm{ord}_k})$, $\mathrm{Fun}(\mathrm{Op}_{\mathfrak{g}}^{\mathrm{RS}})$, and $\mathrm{Fun}(\mathrm{Op}_{\mathfrak{g}}^{\varpi(-\check{\rho})})$, coming from $\mathbb{G}_m \to \mathrm{Aut}(\mathcal{D})$, corresponding to some choice of a coordinate t on $\mathcal{D}$. From Lemma 1.9, we obtain the following.

Lemma 2.12.

(1) *The algebra* $\mathrm{Fun}(\mathrm{Op}_{\mathfrak{g}}^{\mathrm{RS}})$ *is nonpositively graded. The subalgebra consisting of degree* 0 *elements is identified with* $\mathrm{Fun}(\mathfrak{h}/\!/W)$ *under the map* $\mathrm{Res}^{\mathrm{RS}}$.

(2) *For every* $k \geq 2$, *the ideal of* $\mathrm{Fun}(\mathrm{Op}_{\mathfrak{g}}^{\mathrm{ord}_k}) \to \mathrm{Fun}(\mathrm{Op}_{\mathfrak{g}}^{\mathrm{RS}})$ *is freely generated by finitely many elements each having a positive degree.*

(3) *The algebra* $\mathrm{Fun}(\mathrm{Op}_{\mathfrak{g}}^{\varpi(-\check{\rho})})$ *is freely generated by elements of strictly negative degrees.*

2.13 Opers with nilpotent singularities

Let X and x be as above. We define a $\mathfrak{g}$-oper on X with a nilpotent singularity at x to be a triple $(\mathcal{F}_G, \nabla, \mathcal{F}_B)$, where $(\mathcal{F}_G, \mathcal{F}_B)$ are as in the definition of opers, and the connection ∇ has a pole of order 1 at x such that for some (or any) regular connection ∇' that preserves B, we have the following:

(i) $(\nabla - \nabla')$, which a priori is an element of $\mathfrak{g}_{\mathcal{F}_G} \otimes \omega_X(x)$, is, in fact, contained in $\mathfrak{b}_{\mathcal{F}_B} \otimes \omega_X(x) + \mathfrak{g}_{\mathcal{F}_G} \otimes \omega_X \subset \mathfrak{g}_{\mathcal{F}_G} \otimes \omega_X(x)$. Once this condition is satisfied, we impose the following two:

(ii) $(\nabla - \nabla') \bmod \mathfrak{g}_{\mathcal{F}_G} \otimes \omega_X$, which is an element of $\mathfrak{b}_{\mathcal{P}_B} \simeq \mathfrak{b}_{\mathcal{F}_B} \otimes \omega_X(x)/\mathfrak{b}_{\mathcal{F}_B} \otimes \omega_X$, must be contained in $\mathfrak{n}_{\mathcal{P}_B} \subset \mathfrak{b}_{\mathcal{P}_B}$, where $\mathcal{P}_B$ is the fiber of $\mathcal{F}_B$ at x.

(iii) $(\nabla - \nabla') \bmod \mathfrak{b}_{\mathcal{F}_B} \otimes \omega_X(x)$, which is a section of $(\mathfrak{g}/\mathfrak{b})_{\mathcal{F}_B} \otimes \omega_X$, must be contained in $\mathbf{O}_{\mathcal{F}_B, \omega_X} \subset (\mathfrak{g}/\mathfrak{b})_{\mathcal{F}_B} \otimes \omega_X$.

In other words, we are looking at gauge equivalence classes with respect to $\mathrm{Hom}(X, B)$ of connections of the form

$$\nabla^0 + \sum_\imath \phi_\imath \cdot f_\imath + \mathbf{q}, \tag{2.4}$$

where $\phi_\imath$ are as in (1.1), and $\mathbf{q}$ is a $\mathfrak{b}$-valued one-form on X with a pole of order 1 at x, whose residue belongs to $\mathfrak{n}$.

This definition makes sense for R-families, so we obtain a functor on the category of $\mathbb{C}$-algebras, which we will denote by $\mathrm{Op}_{\mathfrak{g}}^{\mathrm{nilp}}(X)$. For $X = \mathcal{D}$ we will denote the corresponding functor simply by $\mathrm{Op}_{\mathfrak{g}}^{\mathrm{nilp}}$.

As in the previous cases, one can define the functor $\mathrm{Op}_{\mathfrak{g}}^{\mathrm{nilp}}(X)^{\mathfrak{D}}$ on the category of $\mathfrak{D}_X$-algebras. Once we prove its representability (see below), this functor will be related to $\mathrm{Op}_{\mathfrak{g}}^{\mathrm{nilp}}(X)$, $\mathrm{Op}_{\mathfrak{g}}^{\mathrm{nilp}}$ and $\mathrm{Op}_{\mathfrak{g}}(\mathcal{D}^\times)$ in the same way as in the case of $\mathrm{Op}_{\mathfrak{g}}^{\mathrm{RS}}(X)^{\mathfrak{D}}$.

2.14

We have an evident morphism of functors $\mathrm{Op}_{\mathfrak{g}}^{\mathrm{nilp}} \to \mathrm{Op}_{\mathfrak{g}}(\mathcal{D}^{\times})$.

Theorem 2.15. *The above map is a closed embedding of functors, and an isomorphism onto* $\mathrm{Op}_{\mathfrak{g}}^{\mathrm{RS},\varpi(-\check{\rho})}$.

Since the assertion is local, a similar statement holds for any pair (X, x). Before proving this theorem let us make the following observation, which implies in particular that the map in question is injective at the level of $\mathbb{C}$-points.

Let $(\mathcal{F}_G, \nabla, \mathcal{F}_B)$ be a $\mathbb{C}$-point of $\mathrm{Op}_{\mathfrak{g}}(\mathcal{D}^{\times})$, and let us first regard it as a G-local system on $\mathcal{D}^{\times}$. Recall that if a local system $(\mathcal{F}_G, \nabla)$ on $\mathcal{D}^{\times}$ admits an extension to a bundle on $\mathcal{D}$ with a meromorphic connection with a pole of order 1 *and nilpotent residue*, then such extension is unique; we will refer to it as Deligne's extension.[4]

Thus a necessary condition for $(\mathcal{F}_G, \nabla, \mathcal{F}_B)$ to come from $\mathrm{Op}_{\mathfrak{g}}^{\mathrm{nilp}}$ is that it admits such an extension. Since the flag variety G/B is compact, the B-bundle $\mathcal{F}_B$, which is a priori defined on $\mathcal{D}^{\times}$, admits a unique extension to $\mathcal{D}$, compatible with the above extension of $\mathcal{F}_G$.

Having fixed this extension, our point comes from $\mathrm{Op}_{\mathfrak{g}}^{\mathrm{nilp}}$ if and only if conditions (i) and (iii) from the definition of opers with nilpotent singularities hold (condition (ii) is automatic from (i) and the nilpotency assumption on the residue).

2.16 Proof of Theorem 2.15

To an oper with nilpotent singularities, written in the form

$$\nabla^0 + \sum_{\imath} \phi_\imath(t)dt \cdot f_\imath + \frac{\mathbf{q}(t)}{t}dt, \tag{2.5}$$

$\phi_\imath(t) \in (R[[t]])^{\times}$, $\mathbf{q}(t) \in \mathfrak{b} \otimes R[[t]]$ with $\mathbf{q}(0) \in \mathfrak{n} \otimes R$, we associate a point of $\mathrm{Op}_{\mathfrak{g}}^{\mathrm{RS},\varpi(-\check{\rho})}$ by applying the gauge transformation by means of $t^{\check{\rho}}$.

The gauge action of $B[[t]](R)$ on connections of the form (2.5) gets transformed into the gauge action on connections with regular singularities by means of $\mathrm{Ad}_{t^{\check{\rho}}}(B[[t]])(R)$, which is a subgroup of $B[[t]]$. This shows that the map $\mathrm{Op}_{\mathfrak{g}}^{\mathrm{nilp}} \to \mathrm{Op}_{\mathfrak{g}}(\mathcal{D}^{\times})$ factors through $\mathrm{Op}_{\mathfrak{g}}^{\mathrm{RS},\varpi(-\check{\rho})}$.

To prove the theorem we must show that any connection written as

$$\nabla = \nabla^0 + \frac{p_{-1} - \check{\rho} + \mathbf{q}(t)}{t}dt, \tag{2.6}$$

with $\mathbf{q}(t) \in \mathfrak{b}[[t]]$ such that the image of $p_{-1} - \check{\rho} + \mathbf{q}(0)$ in $\mathfrak{h}/\!/W$ equals $\varpi(-\check{\rho})$ can be conjugated by means of $N[[t]]$ into a connection of similar form such that $t^{-d} \cdot \mathbf{q}_d(t) \in \mathfrak{b}_d[[t]]$, uniquely up to $\mathrm{Ad}_{t^{\check{\rho}}}(N[[t]])$.

[4] Such an extension exists if and only if $(\mathcal{F}_G, \nabla)$ has regular singularities, and when regarded analytically, has unipotent monodromy.

Note first of all that, by applying a gauge transformation by means of a *constant* loop into N, we can assume that $\mathbf{q}(0) = 0$. By induction on d we will prove the following statement:

Every connection as in (2.6) *can be conjugated by means of* $N[[t]]$ *to one which satisfies*

$$t^{-d'} \cdot \mathbf{q}_{d'}(t) \in \mathfrak{b}_{d'}[[t]] \quad \textit{for } d' \leq d, \tag{2.7}$$

and $t^{-d} \cdot \mathbf{q}_{d''}(t) \in \mathfrak{b}_{d''}[[t]]$ *for* $d'' \geq d$.

By the above, the statement holds for $d = 1$. To perform the induction step we will again use a *descending* inductive argument. We assume that ∇ satisfies (2.7), and that for some $k \geq d + 1$,

$$t^{-d-1} \cdot \mathbf{q}_{k'}(t) \in \mathfrak{b}_{k'}[[t]] \quad \text{for } k' \text{ satisfying } k' > k$$

and

$$t^{-d} \cdot \mathbf{q}_{k''}(t) \in \mathfrak{b}_{k''}[[t]] \quad \text{for } d + 1 \leq k'' \leq k.$$

We will show how to modify ∇ so that it continues to satisfy (2.7), and, in addition,

$$t^{-d-1} \cdot \mathbf{q}_{k'}(t) \in \mathfrak{b}_{k'}[[t]] \quad \text{for } k' \text{ satisfying } k' \geq k$$

and

$$t^{-d} \cdot \mathbf{q}_{k''}(t) \in \mathfrak{b}_{k''}[[t]] \quad \text{for } d + 1 \leq k'' < k.$$

Namely, we will replace ∇ by

$$\nabla' := \mathrm{Ad}_{\exp(t^d \cdot u_k)}(\nabla) = \nabla^0 + \frac{p_{-1} - \check{\rho} + \mathbf{q}'(t)}{t} dt$$

for a certain element $u_k \in \mathfrak{b}_k$.

For any such u_k the conditions involving $\mathbf{q}'_{k'}(t)$ for k' with either $k' < k$ or $k' > k$ hold automatically. The condition on $\mathbf{q}'_k(t)$ reads as follows:

$$-d \cdot u_k + [u_k, -\check{\rho}] = -t^{-d} \mathbf{q}_k(t) \bmod t. \tag{2.8}$$

However, $[u_k, -\check{\rho}] = k \cdot u_k$, and since $k > d$ the above condition is indeed solvable uniquely.

This finishes the proof of the fact that any connection as in (2.6) can be conjugated by means of $N[[t]]$ to one satisfying $t^{-d} \cdot \mathbf{q}_d(t) \in \mathfrak{b}_d[[t]]$. The uniqueness of the solution of (2.8) implies that the conjugation is unique modulo $\mathrm{Ad}_{t^{\check{\rho}}}(N[[t]])$. Thus the proof of Theorem 2.15 is complete. Let us note that the same argument proves the following generalization.

Proposition 2.17. *Let $\check{\lambda}$ be an antidominant coweight. Then the data of an R-point of $\mathrm{Op}_{\mathfrak{g}}^{\mathrm{RS},\varpi(\lambda-\rho)}$ are equivalent to the data of $B[[t]]$-conjugacy class of connections of the form*

$$\nabla^0 + \sum_\imath \phi_\imath(t)dt \cdot f_\imath + \frac{\mathbf{q}(t)}{t}dt,$$

where $\phi_\imath(t)$ are as in (2.5), *and $\mathbf{q}(t) \in \mathfrak{b} \otimes R[[t]]$ is such that $\mathbf{q}(0) \bmod \mathfrak{n} = \check{\lambda}$.*

2.18 The secondary residue map

Note that by definition the scheme $\mathrm{Op}_{\mathfrak{g}}^{\mathrm{nilp}} \widehat{\times} \mathcal{D}$ carries a universal oper with nilpotent singularities. Let us denote by $\mathcal{P}_{G,\mathrm{Op}_{\mathfrak{g}}^{\mathrm{nilp}}}$ (respectively, $\mathcal{P}_{B,\mathrm{Op}_{\mathfrak{g}}^{\mathrm{nilp}}}$) the resulting G-bundle (respectively, B-bundle) on $\mathrm{Op}_{\mathfrak{g}}^{\mathrm{nilp}}$. In particular, we obtain a map $\mathrm{Op}_{\mathfrak{g}}^{\mathrm{nilp}} \to \mathrm{pt}/B$.

By taking the residue of the connection (see Section 2.8), we obtain a map from $\mathrm{Op}_{\mathfrak{g}}^{\mathrm{nilp}}$ to the stack $\mathfrak{n}/B$, where B acts on $\mathfrak{n}$ by means of the adjoint action; we will denote this map by $\mathrm{Res}^{\mathrm{nilp}}$.

Lemma 2.19.

(1) *The map $\mathrm{Res}^{\mathrm{nilp}}$ is smooth. Moreover, the B-scheme $\mathfrak{n} \underset{\mathfrak{n}/B}{\times} \mathrm{Op}_{\mathfrak{g}}^{\mathrm{nilp}}$ can be represented as a product of an infinite-dimensional affine space by a finite-dimensional variety with a free action of B.*

(2) *We have a natural identification:*

$$\mathrm{pt}/B \underset{\mathfrak{n}/B}{\times} \mathrm{Op}_{\mathfrak{g}}^{\mathrm{nilp}} \simeq \mathrm{pt}/B \underset{\mathrm{pt}/G}{\times} \mathrm{Op}_G^{\mathrm{reg}},$$

where $\mathrm{pt}/B \to \mathfrak{n}/B$ corresponds to $0 \in \mathfrak{n}$, and the map $\mathrm{Op}_{\mathfrak{g}}^{\mathrm{reg}} := \mathrm{Op}_{\mathfrak{g}}(\mathcal{D}) \to \mathrm{pt}/G$ is given by $\mathcal{P}_{G,\mathrm{Op}_{\mathfrak{g}}(\mathcal{D})}$.

Proof. The second point of the lemma results from the definitions. To prove the first point, note that $\mathfrak{n} \underset{\mathfrak{n}/B}{\times} \mathrm{Op}_{\mathfrak{g}}^{\mathrm{nilp}}$ identifies with the quotient of the space of connections of the form (2.5) by gauge transformations by means of $B(\mathbb{C}[[t]])$. As in the proof of Proposition 1.6, we obtain that any such connection can be uniquely, up to the action of $B(t\mathbb{C}[[t]])$, brought into the form

$$\nabla^0 + \left(\sum_\imath a_\imath \cdot f_\imath + \frac{\mathbf{q}'}{t} + \mathbf{q}'' + t \cdot \mathbf{v}(t)\right) dt,$$

where $0 \neq a_\imath \in \mathbb{C}$, $\mathbf{q}' \in \mathfrak{n}$, $\mathbf{q}'' \in \mathfrak{b}$ and $\mathbf{v}(t) \in V_{\mathrm{can}}[[t]]$. This scheme projects onto the variety of expressions of the form

$$\sum_\imath a_\imath \cdot f_\imath + \frac{\mathbf{q}'}{t} + \mathbf{q}'',$$

on which B acts freely. □

2.20 Opers with an integral residue

For completeness, we shall now give a description of the scheme $\mathrm{Op}_{\mathfrak{g}}^{\mathrm{RS},\varpi(-\check{\lambda}-\check{\rho})}$ when $\check{\lambda}$ is an integral coweight with $\check{\lambda}+\check{\rho}$ dominant, similar to the one given by Theorem 2.15 in the case when $\check{\lambda}=0$.

Let $\mathfrak{J}$ be the subset of the set $\mathfrak{I}$ of vertices of the Dynkin diagram, corresponding to those simple roots, for which $\langle \alpha_\jmath, \check{\lambda}\rangle = -1$. Let $\mathfrak{p}_{\mathfrak{J}} \subset \mathfrak{g}$ be the corresponding standard parabolic subalgebra, $\mathfrak{n}_{\mathfrak{J}} \subset \mathfrak{n}$ its unipotent radical, and $\mathfrak{m}_{\mathfrak{J}}$ the Levi factor.

We introduce the notion of oper with $\check{\lambda}$-nilpotent singularity to be a triple $(\mathcal{F}_G, \nabla, \mathcal{F}_B)$ as in the definition of nilpotent opers, where conditions (i)–(iii) are replaced by the following ones:

(i) $(\nabla-\nabla') \bmod \mathfrak{b}_{\mathcal{F}_B} \otimes \omega_X(x)$, which is a section of $(\mathfrak{g}/\mathfrak{b})_{\mathcal{F}_B} \otimes \omega_X(x)$, must be contained in $\mathbf{O}_{\mathcal{F}_B} \overset{\mathbb{G}_m}{\times} \omega_X^{\check{\rho}}(-\check{\lambda}\cdot x)$.
(ii) $\mathrm{Res}(\nabla) := (\nabla-\nabla') \bmod \mathfrak{g}_{\mathcal{F}_G} \otimes \omega_X$, which is a priori an element of $\mathfrak{g}_{\mathcal{P}_B}$, is contained in $(\mathfrak{p}_{\mathfrak{J}})_{\mathcal{P}_B}$.
(iii) The image of $\mathrm{Res}(\nabla)$ under $(\mathfrak{p}_{\mathfrak{J}})_{\mathcal{P}_B} \to (\mathfrak{m}_{\mathfrak{J}})_{\mathcal{P}_B}$ is nilpotent.

As in the case of $\check{\lambda}=0$, this definition makes sense for R-families, where R is a $\mathbb{C}$-algebra or a $\mathfrak{D}_X$-algebra. We will denote by $\mathrm{Op}_{\mathfrak{g}}^{\check{\lambda},\mathrm{nilp}}$ the resulting functor for $X=\mathcal{D}$. Explicitly, an R-point of $\mathrm{Op}_{\mathfrak{g}}^{\check{\lambda},\mathrm{nilp}}$ is a $B[[t]](R)$-equivalence class of connections of the form

$$\nabla^0 + \sum_\imath t^{\langle \alpha_\imath, \check{\lambda}\rangle} \cdot \phi_\imath(t) dt \cdot f_\imath + \frac{\mathbf{q}(t)}{t} dt, \tag{2.9}$$

where $\phi_\imath(t)$ and $\mathbf{q}(t)$ are as in (2.5), subject to the condition that the element

$$\sum_{\jmath \in \mathfrak{J}} \phi_\jmath(0) + \mathbf{q}(0) \bmod \mathfrak{n}_{\mathfrak{J}} \in \mathfrak{m}_{\mathfrak{J}}$$

be nilpotent.

As in the case of $\check{\lambda}=0$, there exists a natural map of functors $\mathrm{Op}_{\mathfrak{g}}^{\check{\lambda},\mathrm{nilp}} \to \mathrm{Op}_{\mathfrak{g}}(\mathcal{D}^\times)$.

Theorem 2.21. *The above map is an isomorphism onto the subscheme* $\mathrm{Op}_{\mathfrak{g}}^{\mathrm{RS},\varpi(-\check{\lambda}-\check{\rho})}$.

The proof of this theorem repeats that of Theorem 2.15, where instead of the principal grading on $\mathfrak{n}$ we use the one defined by the adjoint action of $\check{\lambda}+\check{\rho}$.

Consider the subvariety of $\mathfrak{p}_{\mathfrak{J}}$, denoted $\mathbf{O}_{\mathfrak{J}}$, consisting of elements of the form

$$\sum_{\jmath \in \mathfrak{J}} c_\jmath \cdot f_\jmath + \mathbf{q}, \quad c_\jmath \neq 0, \quad \mathbf{q} \in \mathfrak{b},$$

that are nilpotent. We have a natural action of B on $\mathbf{O}_{\mathfrak{J}}$.

As in the case of nilpotent opers, i.e, $\check{\lambda} = 0$, there exists a natural smooth map

$$\mathrm{Res}^{\check{\lambda},\mathrm{nilp}} : \mathrm{Op}_{\mathfrak{g}}^{\check{\lambda},\mathrm{nilp}} \to \mathbf{O}_{\mathfrak{Z}}/B,$$

obtained by taking the polar part of a connection as in (2.9).

Finally, let us consider the case when $\check{\lambda}$ itself is dominant. In this case $\mathfrak{Z} = \emptyset$, and $\mathbf{O}_{\mathfrak{Z}} = \mathfrak{n}$. Let us denote by $\mathrm{Op}_{\mathfrak{g}}^{\check{\lambda},\mathrm{reg}}$ the preimage of $\mathrm{pt}/B \subset \mathfrak{n}/B$ under the map $\mathrm{Res}^{\check{\lambda},\mathrm{nilp}}$, where $\mathrm{pt} \to \mathfrak{n}$ corresonds to the point 0.

The scheme $\mathrm{Op}_{\mathfrak{g}}^{\check{\lambda},\mathrm{reg}}$ is the scheme of $\check{\lambda}$-opers introduced earlier by Beilinson and Drinfeld. As in the case of $\mathrm{Op}_{\mathfrak{g}}^{\check{\lambda},\mathrm{nilp}}$, we have the notion of (an R-family of) regular $\check{\lambda}$-opers over any curve. By definition, this is a triple $(\mathcal{F}_G, \nabla, \mathcal{F}_B)$, where $\mathcal{F}_G$ and ∇ are a principal G-bundle and a connection on it, defined on the entire X, and $\mathcal{F}_B$ is a reduction of $\mathcal{F}_G$ to B, such that $\nabla/\mathcal{F}_B$, as a section of $(\mathfrak{g}/\mathfrak{b})_{\mathcal{F}_B} \otimes \omega_X$, belongs to $\mathbf{O}_{\mathcal{F}_B} \overset{\mathbb{G}_m}{\times} \omega_X^{\check{\rho}}(-\check{\lambda} \cdot x)$.

3 Miura opers

3.1

Ler R be a $\mathfrak{D}_X$-algebra. Let us fix once and for all another Borel subgroup B^- of G which is in generic relative position with B. The definition of Miura opers given below uses B^-. However, the resulting scheme of Miura opers is defined canonically and is independent of this choice.

Following [F, Section 10.3], one defines a *Miura oper* over R to be a quadruple

$$(\mathcal{F}_G, \nabla, \mathcal{F}_B, \mathcal{F}_{B^-}),$$

where

- $(\mathcal{F}_G, \nabla, \mathcal{F}_B)$ are as in the definition of opers, i.e., $(\mathcal{F}_G, \mathcal{F}_B)$ is a G-bundle on $\mathrm{Spec}(R)$ with a reduction to B, and ∇ is a connection on $\mathcal{F}_G$ along X such that $\nabla/\mathcal{F}_B \in \mathbf{O}_{\mathcal{F}_B,\omega_X}$;
- $\mathcal{F}_{B^-}$ is a reduction of $\mathcal{F}_G$ to the opposite Borel subgroup B^- which is *preserved by the connection* ∇.

We will denote the functor of Miura opers on the category of $\mathfrak{D}_X$-algebras by $\mathrm{MOp}_{\mathfrak{g}}(X)^{\mathfrak{D}}$, and the resulting functor on the category of $\mathbb{C}$-algebras by $\mathrm{MOp}_{\mathfrak{g}}(X)$, i.e.,

$$\mathrm{MOp}_{\mathfrak{g}}(X)(R) := \mathrm{MOp}_{\mathfrak{g}}(X)^{\mathfrak{D}}(R \otimes \mathcal{O}_X).$$

Lemma 3.2. *The functor* $\mathrm{MOp}_{\mathfrak{g}}(X)^{\mathfrak{D}}$ *is representable by a* $\mathfrak{D}_X$*-scheme.*

Proof. Since the functor $\mathrm{Op}_{\mathfrak{g}}(X)^{\mathfrak{D}}$ is known to be representable, it suffices to show that the morphism $\mathrm{MOp}_{\mathfrak{g}}(X)^{\mathfrak{D}} \to \mathrm{Op}_{\mathfrak{g}}(X)^{\mathfrak{D}}$ is representable as well.

Consider another functor on the category of $\mathfrak{D}_X$-algebras that associates to R the set of quadruples $(\mathcal{F}_G, \nabla, \mathcal{F}_B, \mathcal{F}_{B^-})$, *but without the condition that $\mathcal{F}_{B^-}$ be compatible with the connection*. The latter functor is clearly representable over $\mathrm{Op}_{\mathfrak{g}}(X)^{\mathfrak{D}}$, and it contains $\mathrm{MOp}_{\mathfrak{g}}(X)^{\mathfrak{D}}$ as a closed subfunctor. □

We will denote by $\mathrm{MOp}_{\mathfrak{g}}(\mathcal{D})$ or $\mathrm{MOp}_{\mathfrak{g}}^{\mathrm{reg}}$ the resulting scheme of Miura opers over $\mathcal{D}$. Note, however, that since the flag variety G/B^- is nonaffine, the $\mathfrak{D}_X$-scheme $\mathrm{MOp}_{\mathfrak{g}}(X)$ is nonaffine over X. Hence, the functor $\mathrm{MOp}_{\mathfrak{g}}(\mathcal{D}^{\times})$ on the category of $\mathbb{C}$-algebras is ill-behaved; in particular, it cannot be represented by an ind-scheme.

We define the $\mathfrak{D}_X$-schemes $\mathrm{MOp}_{\mathfrak{g}}^{\mathrm{ord}_k}(X)^{\mathfrak{D}}$ (respectively, $\mathrm{MOp}_{\mathfrak{g}}^{\mathrm{RS}}(X)^{\mathfrak{D}}$, $\mathrm{MOp}_{\mathfrak{g}}^{\mathrm{nilp}}(X)^{\mathfrak{D}}$) to classify quadruples $(\mathcal{F}_G, \nabla, \mathcal{F}_B, \mathcal{F}_{B^-})$, where the first three pieces of data are as in the definition of $\mathrm{Op}_{\mathfrak{g}}^{\mathrm{ord}_k}(X)^{\mathfrak{D}}$ (respectively, $\mathrm{Op}_{\mathfrak{g}}^{\mathrm{RS}}(X)^{\mathfrak{D}}$, $\mathrm{Op}_{\mathfrak{g}}^{\mathrm{nilp}}(X)^{\mathfrak{D}}$), and $\mathcal{F}_{\mathcal{B}^-}$ is a reduction of the G-bundle $\mathcal{F}_G$, which is defined on the entire X, to the subgroup $\mathcal{F}_{B^-}$, compatible with the connection ∇. The last condition means, in the case of $\mathrm{Op}_{\mathfrak{g}}^{\mathrm{ord}_k}(X)^{\mathfrak{D}}$, that the operator $\nabla_{t^k \partial_t}$ preserves $\mathcal{F}_{B^-}$. For the other $\mathfrak{D}_X$-schemes this condition is defined similarly.

Each of these $\mathfrak{D}_X$-schemes is isomorphic to $\mathrm{MOp}_{\mathfrak{g}}(X)^{\mathfrak{D}}$ over the curve $(X - x)$.

We will denote the corresponding schemes for $X = \mathcal{D}$ simply by $\mathrm{MOp}_{\mathfrak{g}}^{\mathrm{ord}_k}$, $\mathrm{MOp}_{\mathfrak{g}}^{\mathrm{RS}}$ and $\mathrm{MOp}_{\mathfrak{g}}^{\mathrm{nilp}}$, respectively.

Note that there are *no* natural maps from $\mathrm{MOp}_{\mathfrak{g}}^{\mathrm{ord}_k}$ to $\mathrm{MOp}_{\mathfrak{g}}^{\mathrm{ord}_{k+1}}$ or from $\mathrm{MOp}_{\mathfrak{g}}^{\mathrm{nilp}}$ to $\mathrm{MOp}_{\mathfrak{g}}^{\mathrm{RS}}$.

3.3 Generic Miura opers

Following [F, Section 10.3], we shall say that a Miura oper $(\mathcal{F}_G, \nabla, \mathcal{F}_B, \mathcal{F}_{B^-})$ is *generic* if the given reductions of $\mathcal{F}_G$ to B and B^- are in generic relative position. More precisely, observe that given a G-bundle on a scheme X with two reductions to B and B^-, we obtain a morphism $X \to B\backslash G/B^-$. The Miura oper is called generic if this morphism takes values in the open part $B \cdot B^-$ of $B\backslash G/B^-$.

Lemma 3.4. *Let $(\mathcal{F}_G, \nabla, \mathcal{F}_B)$ be a $\mathbb{C}$-valued oper on $\mathcal{D}^{\times}$, and let $\mathcal{F}_{B^-}$ be any horizontal reduction of $\mathcal{F}_G$ to B^-. Then it is in generic relative position with respect to $\mathcal{F}_B$.*

Proof. The following short argument is due to Drinfeld. The G-bundle $\mathcal{F}_G$ can be assumed to be trivial, and we can think of $\mathcal{F}_B$ and $\mathcal{F}_{B^-}$ as two families of Borel subalgebras

$$\mathfrak{b}_1^- \subset \mathfrak{g}((t)) \supset \mathfrak{b}_2.$$

The connection on $\mathcal{F}_G$ has the form $\nabla^0 + \mathbf{q}(t)$, where $\mathbf{q}(t) \in \mathfrak{b}_1^-$.

Let $\mathfrak{h}' \subset \mathfrak{g}((t))$ be any Cartan subalgebra contained in both $\mathfrak{b}_1^-$ and $\mathfrak{b}_2$. Let us decompose $\mathbf{q}(t)$ with respect to the characters of $\mathfrak{h}'$, acting on $\mathfrak{g}((t))$, i.e., with respect to the roots.

Then, on the one hand, each $\mathbf{q}(t)_\alpha$ belongs to $\mathfrak{b}_1^-$. That is, if $\mathbf{q}(t)_\alpha \neq 0$, then α is positive with respect to $\mathfrak{b}_1^-$.

On the other hand, if $\alpha_\imath$ be a simple root of $\mathfrak{h}'$ with respect to $\mathfrak{b}_2$, then by the oper condition, $\mathbf{q}(t)_{-\alpha_\imath} \neq 0$. Hence, every positive simple root with respect to $\mathfrak{b}_2$ is negative with respect to $\mathfrak{b}_1^-$. This implies that $\mathfrak{b}_1^- \cap \mathfrak{b}_2 = \mathfrak{h}'$, i.e., the two reductions are in generic relative position. □

Evidently, generic Miura opers form an open $\mathfrak{D}_X$-subscheme of $\mathrm{MOp}_{\mathfrak{g}}(X)^{\mathfrak{D}}$; we will denote it by $\mathrm{MOp}_{\mathfrak{g},\mathrm{gen}}(X)^{\mathfrak{D}}$.

Lemma 3.5. *The $\mathfrak{D}_X$-scheme $\mathrm{MOp}_{\mathfrak{g},\mathrm{gen}}(X)^{\mathfrak{D}}$ is affine over X.*

Proof. We know that the $\mathfrak{D}_X$-scheme $\mathrm{Op}_{\mathfrak{g}}(X)^{\mathfrak{D}}$ is affine. Hence, it is sufficient to show that $\mathrm{MOp}_{\mathfrak{g},\mathrm{gen}}(X)^{\mathfrak{D}}$ is affine over $\mathrm{Op}_{\mathfrak{g}}(X)$.

By definition, $\mathrm{MOp}_{\mathfrak{g},\mathrm{gen}}(X)^{\mathfrak{D}}$ is a closed subfunctor of the functor that associates to a $\mathfrak{D}_X$-algebra R the set of quadruples $(\mathcal{F}_G, \nabla, \mathcal{F}_B, \mathcal{F}_{B^-})$, where $(\mathcal{F}_G, \nabla, \mathcal{F}_B)$ are as above, and $\mathcal{F}_{B^-}$ is a reduction of $\mathcal{F}_G$ to B^-, which is in generic relative position with $\mathcal{F}_B$, and *not necessarily compatible with the connection.*

Since the big cell $B^- \cdot 1 \subset G/B$ is affine, the latter functor is evidently affine over $\mathrm{Op}_{\mathfrak{g}}(X)^{\mathfrak{D}}$, implying our assertion. □

We will denote by $\mathrm{MOp}^{\mathrm{ord}_k}_{\mathfrak{g},\mathrm{gen}}(X)^{\mathfrak{D}}$ (respectively, $\mathrm{MOp}^{\mathrm{RS}}_{\mathfrak{g},\mathrm{gen}}(X)^{\mathfrak{D}}, \mathrm{MOp}^{\mathrm{nilp}}_{\mathfrak{g},\mathrm{gen}}(X)^{\mathfrak{D}}$) the corresponding open $\mathfrak{D}_X$-subscheme of $\mathrm{MOp}^{\mathrm{ord}_k}_{\mathfrak{g}}(X)^{\mathfrak{D}}$ (respectively, $\mathrm{MOp}^{\mathrm{RS}}_{\mathfrak{g}}(X)^{\mathfrak{D}}$, $\mathrm{MOp}^{\mathrm{nilp}}_{\mathfrak{g}}(X)^{\mathfrak{D}}$). We will denote by

$$\mathrm{MOp}^{\mathrm{reg}}_{\mathfrak{g},\mathrm{gen}} := \mathrm{MOp}_{\mathfrak{g},\mathrm{gen}}(\mathcal{D}), \quad \mathrm{MOp}^{\mathrm{nilp}}_{\mathfrak{g},\mathrm{gen}}, \quad \mathrm{MOp}^{\mathrm{RS}}_{\mathfrak{g},\mathrm{gen}}, \quad \mathrm{MOp}^{\mathrm{ord}_k}_{\mathfrak{g},\mathrm{gen}}$$

the corresponding open subschemes of

$$\mathrm{MOp}^{\mathrm{reg}}_{\mathfrak{g}} := \mathrm{MOp}_{\mathfrak{g}}(\mathcal{D}), \quad \mathrm{MOp}^{\mathrm{nilp}}_{\mathfrak{g}}, \quad \mathrm{MOp}^{\mathrm{RS}}_{\mathfrak{g}}, \quad \mathrm{MOp}^{\mathrm{ord}_k}_{\mathfrak{g}},$$

respectively. By Lemma 3.5, it makes sense also to consider the ind-scheme $\mathrm{MOp}_{\mathfrak{g},\mathrm{gen}}(\mathcal{D}^\times)$.

3.6 Miura opers and H-connections

We will now establish a crucial result that connects generic Miura opers with another, very explicit, $\mathfrak{D}_X$-scheme.

Consider the H-bundle $\omega_X^{\check{\rho}}$, and let $\mathrm{Conn}_H(\omega_X^{\check{\rho}})^{\mathfrak{D}}$ be the $\mathfrak{D}_X$-scheme of connections on it, i.e., it associates to a $\mathfrak{D}_X$-algebra R the set of connections on the pull-back of $\omega_X^{\check{\rho}}$ to $\mathrm{Spec}(R)$ along X. This is a principal homogeneous space with respect to the group $\mathfrak{D}_X$-scheme that associates to R the set of $\mathfrak{h}$-valued sections of the pull-back of ω_X to $\mathrm{Spec}(R)$. In particular, $\mathrm{Conn}_H(\omega_X^{\check{\rho}})^{\mathfrak{D}}$ is affine over X.

We will denote the resulting functor on $\mathbb{C}$-algebras by $\mathrm{Conn}_H(\omega_X^{\check{\rho}})$. For $X = \mathcal{D}$ (respectively, $\mathcal{D}^\times$) this functor is evidently representable by a scheme (respectively, ind-scheme), which we will denote by $\mathrm{Conn}_H(\omega_{\mathcal{D}}^{\check{\rho}}) =: \mathrm{Conn}_H(\omega_{\mathcal{D}}^{\check{\rho}})^{\mathrm{reg}}$ (respectively, $\mathrm{Conn}_H(\omega_{\mathcal{D}^\times}^{\check{\rho}})$). This scheme (respectively, ind-scheme) is a principal homogeneous space with respect to $\mathfrak{h} \otimes \omega_{\mathcal{D}}$ (respectively, $\mathfrak{h} \otimes \omega_{\mathcal{D}^\times}$).

Note that we have a natural map of $\mathfrak{D}_X$-schemes

$$\mathrm{MOp}_{\mathfrak{g},\mathrm{gen}}(X)^{\mathfrak{D}} \to \mathrm{Conn}_H(\omega_X^{\check{\rho}})^{\mathfrak{D}}. \tag{3.1}$$

Indeed, given an R-point of $\mathrm{MOp}_{\mathfrak{g},\mathrm{gen}}(X)^{\mathfrak{D}}$, let $\mathcal{F}'_H$ be the H-bundle with connection, induced by means of $\mathcal{F}_{B^-}$. However, the assumption that the Miura oper is generic implies that $\mathcal{F}'_H \simeq \mathcal{F}'_B \cap \mathcal{F}_{B^-} \simeq \mathcal{F}_H$, where $\mathcal{F}_H$ is the H-bundle induced by $\mathcal{F}_B$. Now let us recall that by Lemma 1.2, we have a canonical isomorphism $\mathcal{F}_H \simeq \omega_X^{\check{\rho}}$.

Proposition 3.7 ([F, Proposition 10.4]). *The map* (3.1) *is an isomorphism.*

Proof. We construct the inverse map $\mathrm{Conn}_H(\omega_X^{\check{\rho}})^{\mathfrak{D}} \to \mathrm{MOp}_{\mathfrak{g},\mathrm{gen}}(X)^{\mathfrak{D}}$ as follows. Recall first that the data of a G-bundle with two reductions to B and B^- in generic position is equivalent to a data of an H-bundle. Thus, from $\mathcal{F}_H := \omega_X^{\check{\rho}}$ we obtain the data $(\mathcal{F}_G, \mathcal{F}_B, \mathcal{F}_{B^-})$ from the Miura oper quadruple.

A connection on (the pull-back of) $\omega_X^{\check{\rho}}$ (to some $\mathfrak{D}_X$-scheme) induces a connection, that we will call ∇_H, on $\mathcal{F}_G$, compatible with both reductions. We produce the desired connection ∇ on $\mathcal{F}_G$ by adding to ∇_H the $\mathfrak{n}^-_{\mathcal{F}_H} \otimes \omega_X$-valued 1-form equal to $\sum_\imath \phi_\imath$, where each $\phi_\imath$ is the tautological trivialization of $(\mathfrak{n}^-_{-\alpha_\imath})_{\mathcal{F}_H} \otimes \omega_X$.

The resulting connection preserves $\mathcal{F}_{B^-}$ and satisfies the oper condition with respect to $\mathcal{F}_B$. Hence $(\mathcal{F}_G, \nabla, \mathcal{F}_B, \mathcal{F}_{B^-})$ is a generic Miura oper. Clearly, the two maps

$$\mathrm{MOp}_{\mathfrak{g},\mathrm{gen}}(X)^{\mathfrak{D}} \rightleftarrows \mathrm{Conn}_H(\omega_X^{\check{\rho}})^{\mathfrak{D}}$$

are mutually inverse. □

This proposition immediately implies the isomorphisms of $\mathfrak{D}_X$-schemes

$$\mathrm{Conn}_H(\omega_{\mathcal{D}}^{\check{\rho}})^{\mathrm{reg}} \simeq \mathrm{MOp}_{\mathfrak{g},\mathrm{gen}}^{\mathrm{reg}} \quad \text{and} \quad \mathrm{Conn}_H(\omega_{\mathcal{D}^\times}^{\check{\rho}}) \to \mathrm{MOp}_{\mathfrak{g},\mathrm{gen}}(\mathcal{D}^\times).$$

Let us denote by $(\mathrm{Conn}_H(\omega_X^{\check{\rho}})^{\mathrm{ord}_k})^{\mathfrak{D}}$ (respectively, $(\mathrm{Conn}_H(\omega_X^{\check{\rho}})^{\mathrm{RS}})^{\mathfrak{D}}$) the $\mathfrak{D}_X$-scheme of meromorphic connections on $\omega_X^{\check{\rho}}$ over $\mathcal{D}$ with poles of order $\leq k$ (respectively, ≤ 1). Each of these $\mathfrak{D}_X$-schemes is isomorphic to $\mathrm{Conn}_H(\omega_X^{\check{\rho}})^{\mathfrak{D}}$ over $(X-x)$. We will denote by $\mathrm{Conn}_H(\omega_{\mathcal{D}}^{\check{\rho}})^{\mathrm{RS}}$ and $\mathrm{Conn}_H(\omega_{\mathcal{D}}^{\check{\rho}})^{\mathrm{ord}_k}$ the resulting schemes of connections on $\mathcal{D}$.

Using the fact that connections on $\omega_X^{\check{\rho}}$ with a pole of order k, $k \geq 1$, are in a canonical bijection with those on $\omega_X^{\check{\rho}}(\check{\lambda} \cdot x)$ for any coweight $\check{\lambda}$, from the above proposition we obtain also the isomorphisms

$$\left(\mathrm{Conn}_H(\omega_X^{\check{\rho}})^{\mathrm{ord}_k}\right)^{\mathfrak{D}} \simeq \mathrm{MOp}_{\mathfrak{g},\mathrm{gen}}^{\mathrm{ord}_k}(X)^{\mathfrak{D}}$$

and

$$\left(\mathrm{Conn}_H(\omega_X^{\check{\rho}})^{\mathrm{RS}}\right)^{\mathfrak{D}} \simeq \mathrm{MOp}_{\mathfrak{g},\mathrm{gen}}^{\mathrm{RS}}(X)^{\mathfrak{D}},$$

implying that

$$\mathrm{Conn}_H(\omega_{\mathcal{D}}^{\check{\rho}})^{\mathrm{ord}_k} \simeq \mathrm{MOp}_{\mathfrak{g},\mathrm{gen}}^{\mathrm{ord}_k} \quad \text{and} \quad \mathrm{Conn}_H(\omega_{\mathcal{D}}^{\check{\rho}})^{\mathrm{RS}} \simeq \mathrm{MOp}_{\mathfrak{g},\mathrm{gen}}^{\mathrm{RS}}. \tag{3.2}$$

We call the composed map of $\mathfrak{D}_X$-schemes

$$\mathrm{Conn}_H(\omega_X^{\check{\rho}})^{\mathfrak{D}} \to \mathrm{MOp}_{\mathfrak{g},\mathrm{gen}}(X)^{\mathfrak{D}} \to \mathrm{Op}_{\mathfrak{g}}(X)^{\mathfrak{D}} \tag{3.3}$$

the *Miura transformation* and denote it by MT. By a slight abuse of notation, we will denote by the same symbol MT the corresponding maps

$$\mathrm{Conn}_H(\omega_{\mathcal{D}}^{\check{\rho}})^{\mathrm{RS}} \to \mathrm{Op}_{\mathfrak{g}}^{\mathrm{RS}}, \qquad \mathrm{Conn}_H(\omega_{\mathcal{D}}^{\check{\rho}})^{\mathrm{ord}_k} \to \mathrm{Op}_{\mathfrak{g}}^{\mathrm{ord}_k},$$
$$\mathrm{Conn}_H(\omega_{\mathcal{D}^\times}^{\check{\rho}}) \to \mathrm{Op}_{\mathfrak{g}}(\mathcal{D}^\times).$$

3.8 An application: Proof of Proposition 2.9

Let $(\mathcal{F}_G, \nabla, \mathcal{F}_B)$ be an oper on $\mathcal{D}^\times$ such that the G-bundle with connection $(\mathcal{F}_G, \nabla)$ has regular singularities, i.e., $\mathcal{F}_G$ can be extended to a G-bundle $\mathcal{F}'_G$ on $\mathcal{D}$, so that ∇ has a pole of order ≤ 1.

Then it is known that $(\mathcal{F}_G, \nabla)$ admits at least one horizontal reduction to B^-; call it $\mathcal{F}_{B^-}$. By Lemma 3.4, the quadruple $(\mathcal{F}_G, \nabla, \mathcal{F}_B, \mathcal{F}_{B^-})$ is a generic Miura oper on $\mathcal{D}^\times$.

By the compactness of G/B^-, the above reduction extends uniquely to the entire $\mathcal{D}$. The connection on the resulting B^--bundle $\mathcal{F}'_{B^-}$ has a pole of order ≤ 1. Hence, the connection on the H-bundle $\mathcal{F}'_H$, induced from $\mathcal{F}'_{B^-}$, also has a pole of order ≤ 1.

Therefore, the above point $(\mathcal{F}_G, \nabla, \mathcal{F}_B, \mathcal{F}_{B^-}) \in \mathrm{MOp}_{\mathfrak{g},\mathrm{gen}}(\mathcal{D}^\times)$, viewed as a point of $\mathrm{Conn}_H(\omega_{\mathcal{D}^\times}^{\check{\rho}})$, belongs to $\mathrm{Conn}_H(\omega_{\mathcal{D}}^{\check{\rho}})^{\mathrm{RS}}$. Hence, the triple $(\mathcal{F}_G, \nabla, \mathcal{F}_B)$, being the image of the above point under the map MT, belongs to $\mathrm{Op}_{\mathfrak{g}}^{\mathrm{RS}}$.

3.9 Miura opers with regular singularities

Consider the map $\mathrm{Conn}_H(\omega_{\mathcal{D}}^{\check{\rho}})^{\mathrm{RS}} \to \mathfrak{h}$ that assigns to a connection with a pole of order 1 its residue; we will denote it by $\mathrm{Res}^{\mathfrak{h}}$. For $\check{\lambda} \in \mathfrak{h}$ we will denote by $\mathrm{Conn}_H(\omega_{\mathcal{D}}^{\check{\rho}})^{\mathrm{RS},\check{\lambda}}$ the preimage of $\check{\lambda}$ under $\mathrm{Res}^{\mathfrak{h}}$.

A coweight $\check{\lambda}$ such that $\langle \alpha_\imath, \check{\lambda}\rangle \notin \mathbb{Z}^{<0}$ (respectively, $\notin \mathbb{Z}^{>0}$) for $\alpha \in \Delta^+$ will be called *dominant* (respectively, *antidominant*).

Proposition 3.10.

(1) *We have a commutative diagram,*

$$\begin{array}{ccc} \mathrm{Conn}_H(\omega_{\mathcal{D}}^{\check{\rho}})^{\mathrm{RS}} & \xrightarrow{\mathrm{MT}} & \mathrm{Op}_{\mathfrak{g}}^{\mathrm{RS}} \\ {\scriptstyle \mathrm{Res}^{\mathfrak{h}}}\downarrow & & {\scriptstyle \mathrm{Res}^{\mathrm{RS}}}\downarrow \\ \mathfrak{h} & \longrightarrow & \mathfrak{h}/\!/W, \end{array}$$

where the bottom arrow is $\check{\lambda} \mapsto \varpi(\check{\lambda} - \check{\rho})$.

(2) *If* $\check{\lambda}$ *is dominant with respect to* B, *then the map* $\mathrm{MT} : \mathrm{Conn}_H(\omega_{\mathcal{D}}^{\check{\rho}})^{\mathrm{RS},\check{\lambda}+\check{\rho}} \to \mathrm{Op}_{\mathfrak{g}}^{\mathrm{RS},\varpi(\check{\lambda})}$ *is an isomorphism.*

The rest of this subsection is devoted to the proof of this proposition. Part (1) follows from the construction:

Given a generic Miura oper with regular singularities $(\mathcal{F}_G, \nabla, \mathcal{F}_B, \mathcal{F}_{B^-})$, the induced H-bundle $\mathcal{F}_H$ is $\omega_X^{\check{\rho}}(\check{\rho} \cdot x)$, by Lemma 2.6. The polar part of ∇ is a section $\mathbf{q} \in \mathfrak{b}^-_{\mathcal{P}_{B^-}}$. Let $\check{\lambda}$ denote the projection of $\mathbf{q}$ onto $\mathfrak{b}^-_{\mathcal{P}_{B^-}}/\mathfrak{n}^-_{\mathcal{P}_{B^-}} \simeq \mathfrak{h}$, which equals the polar part of the connection on $\mathcal{F}_H$.

Then $\mathrm{Res}^{\mathrm{RS}}(\mathcal{F}_G, \nabla, \mathcal{F}_B)$ equals the projection of $\mathbf{q}$ under $\mathfrak{g}/B^- \to \mathfrak{g}/G \to \mathfrak{h}/\!/W$, and hence it equals $\varpi(\check{\lambda})$. It remains to notice that the resulting connection on $\omega_X^{\check{\rho}}$ has the polar part equal to $\check{\lambda} + \check{\rho}$.

To prove part (2), we will use the following general assertion.

Lemma 3.11. *Let* $(\mathcal{F}_G, \nabla)$ *be an* R*-family of* G*-connections on* $\mathcal{D}$ *with a pole of order* 1, *and let* $\mathcal{P}_G$ *be the fiber of* $\mathcal{F}_G$ *at the closed point of the disc. Let* $\mathcal{P}_{B^-}$ *be a reduction of* $\mathcal{P}_G$ *to* G *with the property that the residue* $\mathbf{q}$ *of* ∇, *which is a priori an element of* $\mathfrak{g}_{\mathcal{P}_G}$, *belongs to* $\mathfrak{b}^-_{\mathcal{P}_{B^-}}$. *Assume that the projection of* $\mathbf{q}$ *to* $\mathfrak{b}^-_{\mathcal{F}_{B^-}}/\mathfrak{n}^-_{\mathcal{F}_{B^-}} \simeq \mathfrak{h}$ *is constant and antidominant with respect to* B^-.

Then there exists a unique B^-*-reduction* $\mathcal{F}_{B^-}$ *of* $\mathcal{F}_G$, *which is compatible with* ∇ *and whose fiber at* x *equals* $\mathcal{P}_{B^-}$.

Let us first show how this lemma implies the proposition. Consider the subvariety

$$(p_{-1} + \mathfrak{b})^{\check{\lambda}} \subset (p_{-1} + \mathfrak{b}) \subset \mathfrak{g},$$

consisting of elements whose image in $\mathfrak{h}/\!/W$ equals $\check{\lambda}$. This is the N-orbit of the element $p_{-1} + \check{\lambda}$. We claim that each point of this orbit is contained in a *unique* Borel subalgebra of $\mathfrak{g}$ that is in generic relative position with $\mathfrak{b}$.

More precisely, consider the *Grothendieck alteration* $\widetilde{\mathfrak{g}} \to \mathfrak{g}$ defined as the subvariety of $\mathfrak{g} \times G/B^-$ consisting of the pairs

$$\widetilde{\mathfrak{g}} = \{\mathbf{q} \in \mathfrak{g}, \mathfrak{b}'^- \in G/B^- | \mathbf{q} \in \mathfrak{b}'^-\}. \tag{3.4}$$

Let $\widetilde{(p_{-1} + \mathfrak{b})}^{\check{\lambda}}$ be the scheme-theoretic intersection of the preimages of $(p_{-1} + \mathfrak{b})^{\check{\lambda}} \subset \mathfrak{g}$ and the big cell $B \cdot 1 \subset G/B^-$ in $\widetilde{\mathfrak{g}}$.

Lemma 3.12. *The projection* $\widetilde{(p_{-1} + \mathfrak{b})}^{\check{\lambda}} \to (p_{-1} + \mathfrak{b})^{\check{\lambda}}$ *is an isomorphism.*

Proof. The inverse map $(p_{-1}+\mathfrak{b})^{\check{\lambda}} \to \widetilde{(p_{-1}+\mathfrak{b})}^{\check{\lambda}}$ is obtained by conjugating the element $\{p_{-1}+\check{\lambda}, \mathfrak{b}^-\}$ by means of N. □

Let us denote by $\mathcal{F}_{G,\mathrm{Op}_{\mathfrak{g}}^{\mathrm{RS},\check{\lambda}}}$ (respectively, $\mathcal{P}_{G,\mathrm{Op}_{\mathfrak{g}}^{\mathrm{RS},\check{\lambda}}}$) the universal G-bundle with connection on $\mathrm{Op}_{\mathfrak{g}}^{\mathrm{RS},\check{\lambda}} \widehat{\times} \mathcal{D}$ (respectively, its restriction to $\mathrm{Op}_{\mathfrak{g}}^{\mathrm{RS},\check{\lambda}} \times x \subset \mathrm{Op}_{\mathfrak{g}}^{\mathrm{RS},\check{\lambda}} \widehat{\times} \mathcal{D}$). Let $\mathcal{F}_{B,\mathrm{Op}_{\mathfrak{g}}^{\mathrm{RS},\check{\lambda}}}$ and $\mathcal{P}_{B,\mathrm{Op}_{\mathfrak{g}}^{\mathrm{RS},\check{\lambda}}}$ be their reductions to B given by the oper structure.

From the above lemma we obtain that the G-bundle $\mathcal{P}_{G,\mathrm{Op}_{\mathfrak{g}}^{\mathrm{RS},\check{\lambda}}}$ over $\mathrm{Op}_{\mathfrak{g}}^{\mathrm{RS},\check{\lambda}}$ admits a unique reduction to B^- such that the polar part of the connection belongs to $\mathfrak{b}^-_{\mathcal{F}_{B^-}}$ and its image in $\mathfrak{h}$ equals $\check{\lambda}$. Moreover, the resulting B^--bundle $\mathcal{P}_{B^-,\mathrm{Op}_{\mathfrak{g}}^{\mathrm{RS},\check{\lambda}}}$ is in generic relative position with $\mathcal{P}_{B,\mathrm{Op}_{\mathfrak{g}}^{\mathrm{RS},\check{\lambda}}}$.

Note that if $\check{\lambda}$ is dominant with respect to B, then it is antidominant with respect to B^-. Hence, by our assumption on $\check{\lambda}$ and Lemma 3.11, the G-bundle $\mathcal{F}_{G,\mathrm{Op}_{\mathfrak{g}}^{\mathrm{RS},\check{\lambda}}}$ on $\mathrm{Op}_{\mathfrak{g}}^{\mathrm{RS},\check{\lambda}} \widehat{\times} \mathcal{D}$ admits a unique horizontal reduction to B^-. This reduction is automatically in generic position with $\mathcal{F}_{B,\mathrm{Op}_{\mathfrak{g}}^{\mathrm{RS},\check{\lambda}}}$, because this is so over the closed point $x \in \mathcal{D}$. Thus we have constructed the inverse map

$$\mathrm{Op}_{\mathfrak{g}}^{\mathrm{RS},\varpi(\check{\lambda})} \to \mathrm{Conn}_H(\omega_{\mathcal{D}}^{\check{\rho}})^{\mathrm{RS},\check{\lambda}+\check{\rho}}.$$

This map is evidently a left inverse of the map MT. The uniqueness assertion of Lemma 3.11, combined with Lemma 3.12, implies that it is also a right inverse. This completes the proof of part (2) of Proposition 3.10.

Let us now prove Lemma 3.11.

Proof (Drinfeld). With no loss of generality, we can assume that our G-bundle $\mathcal{F}_G$ is trivial, and the connection has the form $\nabla = \nabla^0 + \frac{\mathbf{q}(t)}{t}$, where $\mathbf{q}(t) \in \mathfrak{g}[[t]]$ and $\mathbf{q}(0) \in \mathfrak{b}^-$. We must show that there exists an element $\mathbf{g} \in \ker(G[[t]] \to G)$, unique modulo B^-, such that

$$\mathrm{Ad}_{\mathbf{g}}\left(\nabla^0 + \frac{\mathbf{q}(t)}{t}\right) =: \nabla' = \nabla^0 + \frac{\mathbf{q}'(t)}{t}$$

is such that $\mathbf{q}'(t) \in \mathfrak{b}^-[[t]]$.

Assume by induction that $\mathbf{q}(t) \bmod t^k \in \mathfrak{b}^-[t]/t^k$. We must show that there exists an element $u \in \mathfrak{g}$, unique modulo $\mathfrak{b}^-$, such that

$$t \cdot \left(\mathrm{Ad}_{\exp(t^k \cdot u)}\left(\frac{\mathbf{q}(t)}{t}\right) - k \cdot t^{k-1} \cdot u\right) \bmod t^{k+1} \in \mathfrak{b}^-[t]/t^{k+1}.$$

This can be rewritten as

$$k \cdot u + [\mathbf{q}_0, u] = \mathbf{q}_k.$$

However, this equation is indeed solvable uniquely in $\mathfrak{g}/\mathfrak{b}^-$, since by assumption, negative integers are not among the eigenvalues of the adjoint action of $\mathbf{q}_0$ on $\mathfrak{g}/\mathfrak{b}^-$. □

We will now describe the behavior of the map MT restricted to $\mathrm{Conn}_H(\omega_{\mathcal{D}}^{\check{\rho}})^{\mathrm{RS},\check{\mu}}$, for $\check{\mu}$ antidominant and integral. This is the case which is in a sense opposite to the one considered in Proposition 3.10(2).

Proposition 3.13. *Let $\check{\lambda}$ be a dominant integral weight. Then the image of the map* $\mathrm{MT}|_{\mathrm{Conn}_H(\omega_{\mathcal{D}}^{\check{\rho}})^{\mathrm{RS},-\check{\lambda}}}$ *belongs to the closed subscheme* $\mathrm{Op}_{\mathfrak{g}}^{\check{\lambda},\mathrm{reg}} \subset \mathrm{Op}_{\mathfrak{g}}^{\check{\lambda},\mathrm{nilp}} \simeq \mathrm{Op}_{\mathfrak{g}}^{\mathrm{RS},\varpi(-\check{\lambda}-\check{\rho})}$. *Moreover, we have a Cartesian square:*

$$\begin{array}{ccc} \mathrm{Conn}_H(\omega_{\mathcal{D}}^{\check{\rho}})^{\mathrm{RS},-\check{\lambda}} & \longrightarrow & (B^- \overset{\circ}{\backslash} G)/B \\ {\scriptstyle \mathrm{MT}}\downarrow & & \downarrow \\ \mathrm{Op}_{\mathfrak{g}}^{\check{\lambda},\mathrm{reg}} & \xrightarrow{\mathrm{Res}^{\check{\lambda},\mathrm{nilp}}} & \mathrm{pt}/B, \end{array} \tag{3.5}$$

where $B^- \overset{\circ}{\backslash} G$ denotes the open B-orbit in the flag variety $B^-\backslash G$.

Proof. Choosing a coordinate on $\mathcal{D}$, and thus trivializing ω_X, a point of $\mathrm{Conn}_H(\omega_{\mathcal{D}}^{\check{\rho}})^{\mathrm{RS},-\check{\lambda}}$ can be thought of as a connection on the trivial bundle of the form

$$\nabla^0 + \frac{\mathbf{q}(t)}{t}dt$$

with $\mathbf{q}(t) \in \mathfrak{h}[[t]]$ and $\mathbf{q}(0) = -\check{\lambda}$. The oper, corresponding to the Miura transformation of the above connection, equals

$$\nabla^0 + p_{-1}dt + \frac{\mathbf{q}(t)}{t}dt.$$

Conjugating this connection by means of $t^{-\check{\lambda}}$ we obtain a connection of the form (2.9). Let us denote by $(\mathcal{F}_G, \nabla, \mathcal{F}_B)$ the resulting point of $\mathrm{Op}_{\mathfrak{g}}^{\check{\lambda},\mathrm{reg}}$.

Note that the horizontal generic reduction to B^- of $\mathcal{F}_G$, which was defined over $\mathcal{D}^\times$, extends to one over $\mathcal{D}$. Indeed, under the above trivialization of $\mathcal{F}_G$, the reduction to B corresponds to the subgroup B itself, and the reduction to B^- corresponds to B^-, which are manifestly in the generic position.

This defines the upper horizontal map in (3.5). To show that this diagram is indeed Cartesian, it suffices to show that given a (R-) point $(\mathcal{F}_G, \nabla, \mathcal{F}_B)$ of $\mathrm{Op}_{\mathfrak{g}}^{\check{\lambda},\mathrm{reg}}$, any reduction to B^- of the fiber $\mathcal{P}_G$ at x of $\mathcal{F}_G$, which is in the generic position with respect $\mathcal{P}_B$ (the latter being the fiber of $\mathcal{F}_B$ at x), comes from a unique reduction of $\mathcal{F}_G$ to B^-. However, this immediately follows from Lemma 3.11, since $\check{\lambda}$ was assumed dominant with respect to B and hence antidominant with respect to B^-. □

3.14 Miura opers with nilpotent singularities

Let us observe that we have four geometric objects that may be called "Miura opers with nilpotent singularities"

$$\mathrm{MOp}_{\mathfrak{g}}^{\mathrm{nilp}}, \qquad \mathrm{MOp}_{\mathfrak{g}}^{\mathrm{RS}} \underset{\mathrm{Op}_{\mathfrak{g}}^{\mathrm{RS}}}{\times} \mathrm{Op}_{\mathfrak{g}}^{\mathrm{nilp}}, \qquad \mathrm{MOp}_{\mathfrak{g},\mathrm{gen}}^{\mathrm{RS}} \underset{\mathrm{Op}_{\mathfrak{g}}^{\mathrm{RS}}}{\times} \mathrm{Op}_{\mathfrak{g}}^{\mathrm{nilp}},$$

and

$$\mathrm{MOp}_{\mathfrak{g},\mathrm{gen}}(\mathcal{D}^\times) \underset{\mathrm{Op}_{\mathfrak{g}}(\mathcal{D}^\times)}{\times} \mathrm{Op}_{\mathfrak{g}}^{\mathrm{nilp}}.$$

The first three of the above objects are schemes, and the fourth is an ind-scheme. In this section we will study the relationship between them.

First, we have the following.

Lemma 3.15. *The sets of $\mathbb{C}$-points of the four objects above are in a natural bijection.*

Proof. In all the four cases the set in question classifies the data of an oper with a nilpotent singularity on $\mathcal{D}$, and its horizontal reduction to B^- over $\mathcal{D}^\times$ (which is necessarily generic by Lemma 3.4). □

We will establish the following.

Theorem 3.16. *There exist natural maps*

$$\begin{array}{ccccc}
\mathrm{MOp}_{\mathfrak{g},\mathrm{gen}}^{\mathrm{RS}} \underset{\mathrm{Op}_{\mathfrak{g}}^{\mathrm{RS}}}{\times} \mathrm{Op}_{\mathfrak{g}}^{\mathrm{nilp}} & \xrightarrow{\mathbf{1}} & \mathrm{MOp}_{\mathfrak{g},\mathrm{gen}}(\mathcal{D}^\times) \underset{\mathrm{Op}_{\mathfrak{g}}(\mathcal{D}^\times)}{\times} \mathrm{Op}_{\mathfrak{g}}^{\mathrm{nilp}} & \xrightarrow{\mathbf{2}} & \mathrm{MOp}_{\mathfrak{g}}^{\mathrm{nilp}} \\
\Big\downarrow{\scriptstyle \mathbf{3}} & & & & \\
\mathrm{MOp}_{\mathfrak{g}}^{\mathrm{RS}} \underset{\mathrm{Op}_{\mathfrak{g}}^{\mathrm{RS}}}{\times} \mathrm{Op}_{\mathfrak{g}}^{\mathrm{nilp}}, & & & &
\end{array}$$

which commute with the projection to $\mathrm{Op}_{\mathfrak{g}}^{\mathrm{nilp}}$, and which on the level of $\mathbb{C}$-points induce the bijection of Lemma 3.15. *Moreover, the map* **1** *is a closed embedding, the map* **2** *is formally smooth, and the map* **3** *is an isomorphism.*

The rest of this section is devoted to the proof of this theorem. Note, however, that the existence of the map **1** and the fact that it is a closed embedding is immediate from the fact that $\mathrm{MOp}_{\mathfrak{g},\mathrm{gen}}^{\mathrm{RS}} \to \mathrm{MOp}_{\mathfrak{g},\mathrm{gen}}(\mathcal{D}^\times)$ is a closed embedding.

Also, the map **3** comes from the tautological map $\mathrm{MOp}_{\mathfrak{g},\mathrm{gen}}^{\mathrm{RS}} \to \mathrm{MOp}_{\mathfrak{g}}^{\mathrm{RS}}$. Since the latter is an open embedding, the map **3** is one too. Since it induces a bijection on the set of $\mathbb{C}$-points by Lemma 3.15, we obtain that it is an isomorphism.

To construct the map **2** appearing in Theorem 3.16, we need to describe the corresponding schemes more explicitly. First, by (3.2), we have an isomorphism:

$$\mathrm{MOp}_{\mathfrak{g},\mathrm{gen}}^{\mathrm{RS}} \underset{\mathrm{Op}_{\mathfrak{g}}^{\mathrm{RS}}}{\times} \mathrm{Op}_{\mathfrak{g}}^{\mathrm{nilp}} \simeq \mathrm{Conn}_H(\omega_{\mathcal{D}}^{\check{\rho}})^{\mathrm{RS}} \underset{\mathrm{Op}_{\mathfrak{g}}^{\mathrm{RS}}}{\times} \mathrm{Op}_{\mathfrak{g}}^{\mathrm{nilp}},$$

and the latter identifies, by Theorem 2.15, with

$$\mathrm{Conn}_H(\omega_{\mathcal{D}}^{\check{\rho}})^{\mathrm{RS}} \underset{\mathfrak{h}}{\times} (\mathfrak{h} \underset{\mathfrak{h}/\!/W}{\times} \mathrm{pt}),$$

where $\mathrm{pt} \to \mathfrak{h}/\!/W$ corresponds to the point $\varpi(-\check{\rho})$. Hence, by Proposition 3.10, we obtain an isomorphism

$$\mathrm{MOp}^{\mathrm{RS}}_{\mathfrak{g},\mathrm{gen}} \underset{\mathrm{Op}^{\mathrm{RS}}_{\mathfrak{g}}}{\times} \mathrm{Op}^{\mathrm{nilp}}_{\mathfrak{g}} \simeq \underset{w \in W}{\cup} \mathrm{Conn}_H(\omega^{\check{\rho}}_{\mathcal{D}})^{\mathrm{RS},\check{\rho}-w(\check{\rho})}.$$

Since the map **1** in Theorem 3.16 is a closed embedding and an isomorphism at the level of $\mathbb{C}$-points, the ind-scheme $\mathrm{MOp}_{\mathfrak{g},\mathrm{gen}}(\mathcal{D}^\times) \underset{\mathrm{Op}_{\mathfrak{g}}(\mathcal{D}^\times)}{\times} \mathrm{Op}^{\mathrm{nilp}}_{\mathfrak{g}}$ also splits into connected components, numbered by elements of W; we will denote by $\mathrm{MOp}_{\mathfrak{g},\mathrm{gen}}(\mathcal{D}^\times)^w$ the component corresponding to a given $w \in W$.

Now let $\widetilde{\mathfrak{g}}$ be the Grothendieck alteration of $\mathfrak{g}$ defined in (3.4). Let $\widetilde{\mathfrak{n}}$ be the scheme-theoretic preimage of $\mathfrak{n} \subset \mathfrak{g}$ under the forgetful map $\widetilde{\mathfrak{g}} \to \mathfrak{g}$; this is a scheme acted on by B. Note that $\widetilde{\mathfrak{n}}$ is connected and nonreduced.

By Lemma 3.11, we have the following.

Corollary 3.17. *There exists a canonical isomorphism*

$$\mathrm{MOp}^{\mathrm{nilp}}_{\mathfrak{g}} \simeq \mathrm{Op}^{\mathrm{nilp}}_{\mathfrak{g}} \underset{\mathfrak{n}/B}{\times} \widetilde{\mathfrak{n}}/B.$$

Now let $\widetilde{\mathfrak{n}}^w$ be the subvariety of $\widetilde{\mathfrak{n}}$, obtained by requiring that the pair $(\mathbf{q} \in \mathfrak{g}, \mathfrak{b}'^-) \in \widetilde{\mathfrak{g}}$ be such that the Borel subalgebra $\mathfrak{b}'^-$ is in position w with respect to $\mathfrak{b}$, i.e., the corresponding point of G/B^- belongs to the B-orbit $B \cdot w^{-1} \cdot B^-$. This is a reduced scheme isomorphic to the affine space of dimension $\dim(\mathfrak{n})$. Let us denote by $\widetilde{\mathfrak{n}}^{w,\mathrm{th}}$ the formal neighborhood of $\widetilde{\mathfrak{n}}^w$ in $\widetilde{\mathfrak{n}}$, regarded as an ind-scheme. Clearly, the action of B on $\widetilde{\mathfrak{n}}$ preserves each $\widetilde{\mathfrak{n}}^w$, and

$$\widetilde{\mathfrak{n}}(\mathbb{C}) \simeq \underset{w \in W}{\cup} \widetilde{\mathfrak{n}}^w(\mathbb{C}).$$

Let us denote by $\mathrm{MOp}^{\mathrm{nilp},w}_{\mathfrak{g}}$ the subscheme of $\mathrm{MOp}^{\mathrm{nilp}}_{\mathfrak{g}}$ equal to $\mathrm{Op}^{\mathrm{nilp}}_{\mathfrak{g}} \underset{\mathfrak{n}/B}{\times} \widetilde{\mathfrak{n}}^w/B$ in terms of the isomorphism of Corollary 3.17. Let us denote by $\mathrm{MOp}^{\mathrm{nilp},w,\mathrm{th}}_{\mathfrak{g}}$ the ind-scheme $\mathrm{Op}^{\mathrm{nilp}}_{\mathfrak{g}} \underset{\mathfrak{n}/B}{\times} \widetilde{\mathfrak{n}}^{w,\mathrm{th}}/B$.

Theorem 3.18. *For every $w \in W$ there exists an isomorphism*

$$\mathrm{MOp}_{\mathfrak{g},\mathrm{gen}}(\mathcal{D}^\times)^w \simeq \mathrm{MOp}^{\mathrm{nilp},w,\mathrm{th}}_{\mathfrak{g}},$$

compatible with the forgetful map to $\mathrm{Op}^{\mathrm{nilp}}_{\mathfrak{g}}$ *and the bijection of Lemma* 3.15.

Clearly, Theorem 3.18 implies the remaining assertions of Theorem 3.16. In addition, by passing to reduced schemes underlying the isomorphism of Theorem 3.18, and using Lemma 2.19(1), we obtain the following.

Corollary 3.19. *There exists a canonical isomorphism* $\mathrm{Conn}_H(\omega^{\check{\rho}}_{\mathcal{D}})^{\mathrm{RS},\check{\rho}-w(\check{\rho})} \simeq \mathrm{MOp}^{\mathrm{nilp},w}_{\mathfrak{g}}$.

3.20 Proof of Theorem 3.18

We begin by constructing the map

$$\mathrm{MOp}_{\mathfrak{g},\mathrm{gen}}(\mathcal{D}^\times)^w \to \mathrm{MOp}_{\mathfrak{g}}^{\mathrm{nilp}}. \tag{3.6}$$

Given an R-point $(\mathcal{F}_G, \nabla, \mathcal{F}_B, \mathcal{F}_{B^-})$ of $\mathrm{MOp}_{\mathfrak{g},\mathrm{gen}}(\mathcal{D}^\times)^w$, let $(\mathcal{F}'_G, \mathcal{F}'_B)$ be an extension of the pair $(\mathcal{F}_G, \mathcal{F}_B)$ onto $\mathcal{D}$ such that the resulting triple $(\mathcal{F}'_G, \nabla, \mathcal{F}'_B)$ is a point of $\mathrm{Op}_{\mathfrak{g}}^{\mathrm{nilp}}$. Such an extension exists, according to Theorem 2.15. We claim that the reduction to B^- of $\mathcal{F}_G$, given by $\mathcal{F}_{B^-}$, gives rise to a reduction of $\mathcal{F}'_G$ to B^-:

Let us think of a reduction to B^- in the Plücker picture (see [FGV]). Let V^λ be the irreducible representation representation of $\mathfrak{g}$ with highest weight λ. Then our point of $\mathrm{MOp}_{\mathfrak{g},\mathrm{gen}}(\mathcal{D}^\times)^w$ gives rise to a system of meromorphic maps

$$V^\lambda_{\mathcal{F}'_G} \to \omega_X^{\langle \lambda, \check{\rho}\rangle},$$

for dominant weights λ, compatible with the (meromorphic) connections on the two sides. Note that the connection on $\omega_{\mathcal{D}}^{\check{\rho}}$, corresponding to $(\mathcal{F}_G, \nabla, \mathcal{F}_B, \mathcal{F}_{B^-})$, restricted to the subscheme

$$\mathrm{Spec}(R) \underset{\mathrm{MOp}_{\mathfrak{g},\mathrm{gen}}(\mathcal{D}^\times)^w}{\times} \mathrm{Conn}_H(\omega_{\mathcal{D}}^{\check{\rho}})^{\mathrm{RS}, \check{\rho} - w(\check{\rho})}$$

has the property that its pole is of order 1 and the residue equals $\check{\rho} - w(\check{\rho})$. We apply the following.

Lemma 3.21. *Let $(\mathcal{F}_H, \nabla_H)$ be an R-family of H-bundles with meromorphic connections on $\mathcal{D}$. Assume that there exists a quotient $R \twoheadrightarrow R'$ by a nilpotent ideal such that the connection on the resulting R'-family has a pole of order* 1 *and a fixed residue integral $\check{\lambda} \in \mathfrak{h}$. Then there exists a unique modification $\mathcal{F}'_H$ of $\mathcal{F}_H$ at x such that the resulting connection on $\mathcal{F}'_H$ is regular.*

The lemma produces an R-family of H-bundles $\mathcal{F}'_H$ with a regular connection, and a horizontal system of a priori *meromorphic* maps

$$\mathfrak{s}^\lambda : V^\lambda_{\mathcal{F}'_G} \to \mathbb{C}^\lambda_{\mathcal{F}'_H},$$

satisfying the Plücker equations. We claim that each of these maps $\mathfrak{s}^\lambda$ is, in fact, regular and surjective. This is a particular case of the following lemma.

Lemma 3.22. *Let $\mathcal{V}$ and $\mathcal{L}$ be R-families of vector bundles and line bundles on $\mathcal{D}$, respectively, both equipped with connections such that on $\mathcal{V}$ it has a pole of order* 1 *and nilpotent residue, and on $\mathcal{L}$ the connection is regular. Let $\mathcal{V} \to \mathcal{L}$ be a nonzero meromorphic map, compatible with the connections. Then this map is regular and surjective.*

Thus we obtain a horizontal reduction $\mathcal{F}'_{B^-}$ of $\mathcal{F}'_G$ to B^-, and the desired map in (3.6).

Consider the restriction of the map (3.6) to

$$\mathrm{Conn}_H(\omega_{\mathcal{D}}^{\check{\rho}})^{\mathrm{RS},\check{\rho}-w(\check{\rho})} \subset \mathrm{MOp}_{\mathfrak{g},\mathrm{gen}}(\mathcal{D}^\times)^w.$$

Since the former scheme is reduced and irreducible, the image of this map is contained in $\mathrm{MOp}_{\mathfrak{g}}^{\mathrm{nilp},w'}$ for some $w' \in W$. This implies that the map (3.6) itself factors through $\mathrm{MOp}_{\mathfrak{g}}^{\mathrm{nilp},w',\mathrm{th}}$ for the same w'.

We have to show that $w' = w$ and that the resulting map is an isomorphism. We claim that for this purpose it is sufficient to construct a map in the opposite direction

$$\mathrm{MOp}_{\mathfrak{g}}^{\mathrm{nilp},w} \to \mathrm{Conn}_H(\omega_{\mathcal{D}}^{\check{\rho}})^{\mathrm{RS},\check{\rho}-w(\check{\rho})}, \tag{3.7}$$

compatible with the identification of Lemma 3.15. This follows from the next observation.

Lemma 3.23. *Let $(\mathcal{F}_G, \nabla, \mathcal{F}_B)$ be an R-point of $\mathrm{Op}_{\mathfrak{g}}^{\mathrm{nilp}}$, and let R' be a quotient of R by a nilpotent ideal. Let $(\mathcal{F}'_G, \nabla', \mathcal{F}'_B, \mathcal{F}'_{B^-})$ be a lift of the induced R'-family to a point of $\mathrm{MOp}_{\mathfrak{g},\mathrm{gen}}(\mathcal{D}^\times)$. Then the sets of extensions of this lift to R-points of $\mathrm{MOp}_{\mathfrak{g},\mathrm{gen}}(\mathcal{D}^\times)$ and $\mathrm{MOp}_{\mathfrak{g}}^{\mathrm{nilp}}$ are in bijection.*

The lemma follows from the fact a deformation over a nilpotent base of a generic Miura oper remains generic.

Given an R-point of $\mathrm{MOp}_{\mathfrak{g}}^{\mathrm{nilp},w}$ and a dominant weight λ, consider the diagram

$$\omega_X^{\langle\lambda,\check{\rho}\rangle} \overset{\mathfrak{s}'}{\to} V^\lambda_{\mathcal{F}_G} \overset{\mathfrak{s}}{\to} \mathcal{L}, \tag{3.8}$$

where the map $\mathfrak{s}'$ corresponds to the reduction of $\mathcal{F}_G$ to B, and $\mathcal{L}$ is *some* line bundle on $\mathrm{Spec}(R[[t]])$ with a regular connection $\nabla_{\mathcal{L}}$ in the t-direction, and the map $\mathfrak{s}$ is a surjective bundle map, compatible with connections, corresponding to the reduction of $\mathcal{F}_G$ to B^-. We will denote by $\nabla(\partial_t)$ (respectively, $\nabla_{\mathcal{L}}(\partial_t)$) the action of the vector field ∂_t on $\mathcal{D}$ on sections of $V^\lambda_{\mathcal{F}_G}$ (respectively, $\mathcal{L}$), given by the connection.

To construct the map as in (3.7), we have to show that the composition $\mathfrak{s}' \circ \mathfrak{s}$ has a zero of order $\langle\lambda, \check{\rho} - w(\check{\rho})\rangle$. This is equivalent to the following: let $\mathbf{v}$ be a nonvanishing section of $\omega_X^{\langle\lambda,\check{\rho}\rangle}$, thought of as a section of $V^\lambda_{\mathcal{F}_G}$ by means of $\mathfrak{s}'$. We need to show that the section $\nabla_{\mathcal{L}}(\partial_t)^{n'}(\mathfrak{s}(\mathbf{v}))$ of $\mathcal{L}$ is regular and nonvanishing for $n' = n := \langle\lambda, \check{\rho} - w(\check{\rho})\rangle$, and has a zero at x if $n' < n$. Since the map $\mathfrak{s}$ is compatible with connections, we have to calculate $\mathfrak{s}(\nabla(\partial_t)^n(\mathbf{v}))$.

Let $F^j(V^\lambda)$ be the increasing B-stable filtration on V^λ, defined by the condition that a vector $v \in V^\lambda$ of weight λ' belongs to $F^j(V^\lambda)$ if and only if $\langle\lambda-\lambda', \check{\rho}\rangle \le j$. Let $F^j(V^\lambda_{\mathcal{F}_B})$ be the corresponding induced filtration on the vector bundle $V^\lambda_{\mathcal{F}_B} \simeq V^\lambda_{\mathcal{F}_G}$. Each successive quotient $F^j(V^\lambda_{\mathcal{F}_B})/F^{j-1}(V^\lambda_{\mathcal{F}_B})$ is isomorphic to

$$\bigoplus_{\lambda',\langle\lambda-\lambda',\check{\rho}\rangle=j} F^j(V^\lambda)/F^{j-1}(V^\lambda) \otimes \omega_{\mathcal{D}}^{\langle\lambda,\check{\rho}\rangle}.$$

By the condition on ∇,

$$\nabla(\partial_t)(F^j(V^\lambda_{\mathcal{F}_B})) \subset F^{j-1}(V^\lambda_{\mathcal{F}_B})(x) + F^{j+1}(V^\lambda_{\mathcal{F}_B}), \tag{3.9}$$

and the induced map

$$\nabla(\partial_t) : F^j(V^\lambda_{\mathcal{F}_B})/F^{j-1}(V^\lambda_{\mathcal{F}_B}) \to F^{j+1}(V^\lambda_{\mathcal{F}_B})/F^j(V^\lambda_{\mathcal{F}_B}) \tag{3.10}$$

comes from the map $F^j(V^\lambda)/F^{j-1}(V^\lambda) \to F^{j+1}(V^\lambda)/F^j(V^\lambda)$, given by p_{-1}. (The latter makes sense, since the vector field ∂_t trivializes the line bundle $\omega_{\mathcal{D}}$.)

Let us denote by n'' the maximal integer such that the composition $\mathfrak{s} \circ \mathfrak{s}'$ vanishes to order n'' along $\operatorname{Spec}(R) \times x \subset \operatorname{Spec}(R[[t]])$. By induction on j, from (3.9), we obtain that the map

$$\mathfrak{s} : F^j(V^\lambda_{\mathcal{F}_B}) \to \mathcal{L} \to \mathcal{L}_x$$

vanishes to order $n'' - j$, where $\mathcal{L}_x$ is the restriction of $\mathcal{L}$ to $\operatorname{Spec}(R) \times x$.

Assume first that $n'' < n$. Then, by the maximality assumption on n'', the image of $\mathfrak{s}(\nabla(\partial_t)^{n''}(\mathbf{v}))$ in $\mathcal{L}_x$ is nonzero. However, this is impossible since the composition

$$F^j(V^\lambda) \hookrightarrow V^\lambda \to (V^\lambda)_{\mathfrak{b}'^-}$$

vanishes for any $\mathfrak{b}'^- \in G/B^-$ in relative position w with respect to B and $j < \langle \lambda - w(\lambda), \check{\rho} \rangle = n$.

Thus $\mathfrak{s}(\nabla(\partial_t)^n(\mathbf{v}))$ is regular, and it remains to show that its image in $\mathcal{L}_x$ is nowhere vanishing. However, this follows from (3.10), since for a highest weight vector $v \in V^\lambda$ and n and $\mathfrak{b}'^-$ as above, the image of $p_{-1}^n(v)$ in $(V^\lambda)_{\mathfrak{b}'^-}$ is nonzero.

4 Groupoids and Lie algebroids associated to opers

4.1 The isomonodromy groupoid

Let us recall that a *groupoid* over a scheme S is a scheme $\mathcal{G}$ equipped with morphisms $l : \mathcal{G} \to S$, $r : \mathcal{G} \to S$, $m : \mathcal{G} \underset{r,S,l}{\times} \mathcal{G} \to \mathcal{G}$, an involution $\gamma : \mathcal{G} \to \mathcal{G}$, and a morphism $u : S \to \mathcal{G}$ that satisfy the following conditions:

- associativity: $m \circ (m \times \mathrm{id}) = m \circ (\mathrm{id} \times m)$ as morphisms $\mathcal{G} \underset{r,S,l}{\times} \mathcal{G} \underset{r,S,l}{\times} \mathcal{G} \to \mathcal{G}$;
- unit: $r \circ u = l \circ u = \mathrm{id}_S$.
- inverse: $l \circ \gamma = r$, $r \circ \gamma = l$, $m \circ (\gamma \times \mathrm{id}_{\mathcal{G}}) = u \circ r$, $m \circ (\mathrm{id}_{\mathcal{G}} \times \gamma) = u \circ l$.

If $S_1 \subset S$ is a subscheme, we will denote by $\mathcal{G}|_{S_1}$ the restriction of $\mathcal{G}$ to S_1, i.e., the subscheme of $\mathcal{G}$ equal to $(l \times r)^{-1}(S_1 \times S_1)$. This is a groupoid over S_1.

The normal sheaf to S inside $\mathcal{G}$ acquires a structure of *Lie algebroid*; we will denote it by $\mathfrak{G}$, and by **anch** the anchor map $\mathfrak{G} \to T(S)$, where $T(S)$ is the tangent algebroid of S. (We refer to [Ma] for more details on groupoids and Lie algebroids).

The notion of groupoid generalizes in a straightforward way to the case when both S and $\mathcal{G}$ are ind-schemes. However, to speak about a Lie algebroid attached to a Lie groupoid, we will need to assume that $\mathcal{G}$ is formally smooth over S (with respect to either, or equivalently, both projections). In this case $\mathfrak{G}$ will be a Tate vector bundle over S; we refer the reader to Section 19.2 for details.

We define the *isomonodromy groupoid* $\mathsf{Isom}_{\mathrm{Op}_\mathfrak{g}}$ over the ind-scheme $\mathrm{Op}_\mathfrak{g}(\mathcal{D}^\times)$. Points of the ind-scheme $\mathsf{Isom}_{\mathrm{Op}_\mathfrak{g}}$ over an algebra R are triples (χ, χ', ϕ), where $\chi = (\mathcal{F}_G, \nabla, \mathcal{F}_B)$ and $\chi' = (\mathcal{F}'_G, \nabla', \mathcal{F}'_B)$ are both R-points of $\mathrm{Op}_\mathfrak{g}(\mathcal{D}^\times)$, and ϕ is an isomorphism of G-bundles with connections $(\mathcal{F}_G, \nabla) \simeq (\mathcal{F}'_G, \nabla')$.

Explicitly, if χ and χ' are connections ∇ and ∇', respectively, on the trivial bundle, both of the form

$$\nabla^0 + p_{-1}dt + \phi(t)dt, \quad \phi(t) \in \mathfrak{b} \otimes R((t)), \tag{4.1}$$

then a point of $\mathsf{Isom}_{\mathrm{Op}_\mathfrak{g}}(R)$ over (χ, χ) is an element $\mathbf{g} \in G(R((t)))$ such that $\mathrm{Ad}_\mathbf{g}(\nabla) = \nabla'$. Two triples $(\chi_1, \chi'_1, \mathbf{g}_1)$ and $(\chi_2, \chi'_2, \mathbf{g}_2)$ are equivalent if there exist elements $\mathbf{g}, \mathbf{g}' \in N(R((t)))$ such that $\nabla_1 = \mathrm{Ad}_\mathbf{g}(\nabla_2)$, $\nabla'_2 = \mathrm{Ad}_{\mathbf{g}'}(\nabla'_1)$ and $\mathbf{g}_2 = \mathbf{g}' \cdot \mathbf{g}_1 \cdot \mathbf{g}$.

The morphisms l and r send (χ, χ', ϕ) to χ and χ', respectively. The morphism m sends the pair $(\chi, \chi', \phi), (\chi', \chi'', \phi')$ to $(\chi, \chi'', \phi' \circ \phi)$, the morphisms γ sends (χ, χ', ϕ) to (χ', χ, ϕ^{-1}) and the morphism $u : \mathrm{Op}_\mathfrak{g}(\mathcal{D}^\times) \to \mathsf{Isom}_{\mathrm{Op}_\mathfrak{g}}$ sends χ to $(\chi, \chi, \mathrm{id})$.

We call $\mathsf{Isom}_{\mathrm{Op}_\mathfrak{g}}$ the isomonodromy groupoid for the following reason. In the analytic context two connections on the trivial bundle on a punctured disc are called isomonodromic if they have the same monodromy and the Stokes data (in case of irregular singularity). In the case of connections on the formal punctured disc the appropriate analogue of the notion of isomonodromy is the notion of gauge equivalence of connections.

Proposition 4.2. *The groupoid* $\mathsf{Isom}_{\mathrm{Op}_\mathfrak{g}}$ *is formally smooth over* $\mathrm{Op}_\mathfrak{g}(D^\times)$.

4.3 Description of tangent space and proof of Proposition 4.2

Let $R' \to R$ be a homomorphism of rings such that its kernel $\mathbf{I}$ satisfies $\mathbf{I}^2 = 0$. Let $\chi' = (\mathcal{F}'_G, \nabla', \mathcal{F}'_B)$ be an R'-point of $\mathrm{Op}_\mathfrak{g}(\mathcal{D}^\times)$, and let $\chi = (\mathcal{F}_G, \nabla, \mathcal{F}_B)$ be the corresponding R-point. Let $\mathbf{g}$ be an automorphism of $\mathcal{F}_G$ such that the quadruple $(\mathcal{F}_G, \nabla, \mathcal{F}_B, \mathbf{g})$ is an R-point of $\mathsf{Isom}_{\mathrm{Op}_\mathfrak{g}}$ over χ. We need to show that it can be lifted to an R'-point $(\mathcal{F}'_G, \nabla', \mathcal{F}'_B, \mathbf{g}')$ of $\mathsf{Isom}_{\mathrm{Op}_\mathfrak{g}}$.

Since the ind-scheme $G((t))$ is formally smooth, we can always find some automorphism $\mathbf{g}'_1$ of $\mathcal{F}'_G$, lifting $\mathbf{g}$. To show the existence of the required lift we must find an element $\mathbf{u} \in \mathfrak{g}_{\mathcal{F}_G} \underset{R((t))}{\otimes} \mathbf{I}((t))$ such that the point $\mathbf{g}' = \mathbf{g}'_1 \cdot (1 + \mathbf{u})$ satisfies

$$\mathrm{Ad}_{\mathbf{g}'}(\nabla') - \nabla' \in \mathfrak{b}_{\mathcal{F}'_B} \underset{R'((t))}{\otimes} R'((t))dt.$$

By assumption, $\mathbf{q} := \mathrm{Ad}_{\mathbf{g}'_1}(\nabla') - \nabla'$ belongs to the subspace

$$\mathfrak{b}_{\mathcal{F}'_B} \underset{R'((t))}{\otimes} \mathbf{I}((t))dt \simeq \mathfrak{b}_{\mathcal{F}_B} \underset{R((t))}{\otimes} \mathbf{I}((t))dt.$$

Therefore, the desired element $\mathbf{u}$ must satisfy

$$\nabla(\mathbf{u}) = \mathbf{q} \in (\mathfrak{g}/\mathfrak{b})_{\mathcal{F}_B} \underset{R((t))}{\otimes} \mathbf{I}((t))dt.$$

Hence, it is sufficient to show that the map

$$\mathfrak{g}_{\mathcal{F}_B} \underset{R((t))}{\otimes} \mathbf{I}((t)) \overset{\nabla}{\to} \mathfrak{g}_{\mathcal{F}_B} \underset{R((t))}{\otimes} \mathbf{I}((t))dt \to (\mathfrak{g}/\mathfrak{b})_{\mathcal{F}_B} \underset{R((t))}{\otimes} \mathbf{I}((t))dt$$

is surjective. But this follows from the oper condition on $\nabla/\mathcal{F}_B$.[5]

Thus Proposition 4.2 is proved. In particular, the Lie algebroid $\mathsf{isom}_{\mathrm{Op}_\mathfrak{g}}$, corresponding to the groupoid $\mathsf{Isom}_{\mathrm{Op}_\mathfrak{g}}$, is well defined. Let us write down an explicit expression for $\mathsf{isom}_{\mathrm{Op}_\mathfrak{g}}$ and for the anchor map.

Since the ind-scheme $\mathrm{Op}_\mathfrak{g}(\mathcal{D}^\times)$ is reasonable and formally smooth, its tangent $T(\mathrm{Op}_\mathfrak{g}(\mathcal{D}^\times))$ is a Tate vector bundle. For an R-point $(\mathcal{F}_G, \nabla, \mathcal{F}_B)$ of $\mathrm{Op}_\mathfrak{g}(\mathcal{D}^\times)$, we have

$$T(\mathrm{Op}_\mathfrak{g}(\mathcal{D}^\times))|_{\mathrm{Spec}(R)} \simeq \mathrm{coker}(\nabla) : \mathfrak{n}_{\mathcal{F}_B} \to \mathfrak{b}_{\mathcal{F}_B} \underset{\mathbb{C}((t))}{\otimes} \mathbb{C}((t))dt, \tag{4.2}$$

and

$$\mathsf{isom}_{\mathrm{Op}_\mathfrak{g}}|_{\mathrm{Spec}(R)} \simeq \ker(\nabla) : (\mathfrak{g}/\mathfrak{n})_{\mathcal{F}_G} \to (\mathfrak{g}/\mathfrak{b})_{\mathcal{F}_B} \underset{\mathbb{C}((t))}{\otimes} \mathbb{C}((t))dt. \tag{4.3}$$

The anchor map **anch** : $\mathsf{isom}_{\mathrm{Op}_\mathfrak{g}} \to T(\mathrm{Op}_\mathfrak{g}(\mathcal{D}^\times))$ acts as follows:

$$\mathbf{u} \in \mathfrak{g}_{\mathcal{F}_G} \mapsto \nabla(\mathbf{u}) \in \mathfrak{b}_{\mathcal{F}_B} \underset{\mathbb{C}((t))}{\otimes} \mathbb{C}((t))dt.$$

Consider the cotangent sheaf $\Omega^1(\mathrm{Op}_\mathfrak{g}(\mathcal{D}^\times))$; this is also a Tate vector bundle on $\mathrm{Op}_\mathfrak{g}(\mathcal{D}^\times)$. From (4.2), we obtain that once we identify $\mathfrak{g}$ with its dual by means of any invariant form $\kappa : \mathfrak{g} \otimes \mathfrak{g} \to \mathbb{C}$, we obtain an isomorphism

$$\Omega^1(\mathrm{Op}_\mathfrak{g}(\mathcal{D}^\times)) \simeq \mathsf{isom}_{\mathrm{Op}_\mathfrak{g}}. \tag{4.4}$$

As we shall see in the next subsection, a choice of κ defines a Poisson structure on $\mathrm{Op}_\mathfrak{g}(\mathcal{D}^\times)$, and in particular makes $\Omega^1(\mathrm{Op}_\mathfrak{g}(\mathcal{D}^\times))$ into a Lie algebroid. We will show that the above identification of bundles is compatible with the Lie algebroid structure.

Remark 4.4. In the analytic context this Poisson structure is used to define the KdV flow on $\mathrm{Op}_\mathfrak{g}(\mathcal{D}^\times)$ as the system of evolution equations corresponding to a certain Poisson-commuting system functions on the space of opers. The isomorphism with $\mathsf{isom}_{\mathrm{Op}_\mathfrak{g}}$ implies in particular that the KdV flows preserve gauge equivalence classes.

[5] The above description makes it explicit that both $\Omega^1(\mathrm{Op}_\mathfrak{g}(\mathcal{D}^\times))$ and the conormal $N^*_{\mathrm{Op}_\mathfrak{g}(\mathcal{D}^\times)/\mathsf{Isom}_{\mathrm{Op}_\mathfrak{g}}}$ are Tate vector bundles on $\mathrm{Op}_\mathfrak{g}(\mathcal{D}^\times)$, i.e., we do not have to use the general Theorem 19.4 to prove this fact.

4.5 The Drinfeld–Sokolov reduction and Poisson structure

Consider the space of all connections on the trivial G-bundle on $\mathcal{D}^\times$, i.e., the space $\mathrm{Conn}_G(\mathcal{D}^\times)$ of operators of the form

$$\nabla^0 + \phi(t), \phi(t) \in \mathfrak{g} \otimes \omega_{\mathcal{D}^\times}. \tag{4.5}$$

This is an ind-scheme, acted on by the group $G((t))$ by gauge transformations. We can consider the natural isomonodromy ind-groupoid over $\mathrm{Conn}_G(\mathcal{D}^\times)$:

$$\mathsf{Isom}_{\mathrm{Conn}_G(\mathcal{D}^\times)} := \{\mathbf{g}, \nabla, \nabla' | \, \mathrm{Ad}_{\mathfrak{g}}(\nabla) = \nabla'\}, \quad l(\mathbf{g}, \nabla, \nabla') = \nabla, r(\mathbf{g}, \nabla, \nabla') = \nabla'.$$

Since $\mathsf{Isom}_{\mathrm{Conn}_G(\mathcal{D}^\times)} \simeq \mathrm{Conn}_G(\mathcal{D}^\times) \times G((t))$, it is formally smooth over $\mathrm{Conn}_G(\mathcal{D}^\times)$.

Let us choose a symmetric invariant form $\kappa : \mathfrak{g} \otimes \mathfrak{g} \to \mathbb{C}$, and let $\widehat{\mathfrak{g}}_\kappa$ be the corresponding Kac–Moody extension of $\mathfrak{g}((t))$. Using the form κ, we can identify the space $\mathrm{Conn}_G(\mathcal{D}^\times)$ with hyperplane in $\widehat{\mathfrak{g}}_\kappa^*$ equal to the preimage of $1 \in \mathbb{C}$ under the natural map $\widehat{\mathfrak{g}}_\kappa \to \mathbb{C}$. It is well known that under this identification the coadjoint action of $G((t))$ on $\widehat{\mathfrak{g}}_\kappa^*$ corresponds to the gauge action of $G((t))$ on $\mathrm{Conn}_G(\mathcal{D}^\times)$.

The space $\widehat{\mathfrak{g}}_\kappa^*$ carries a canonical Poisson structure, which induces a Poisson structure also on $\mathrm{Conn}_G(\mathcal{D}^\times)$.

Lemma 4.6. *We have a canonical isomorphism of Lie algebroids*

$$\Omega^1(\mathrm{Conn}_G(\mathcal{D}^\times)) \simeq \mathsf{isom}_{\mathrm{Conn}_G(\mathcal{D}^\times)}, \tag{4.6}$$

where $\mathsf{isom}_{\mathrm{Conn}_G(\mathcal{D}^\times)}$ *is the Lie algebroid of* $\mathsf{Isom}_{\mathrm{Conn}_G(\mathcal{D}^\times)}$.

Proof. We claim that (global sections of) both the LHS and the RHS identify with

$$\mathfrak{g}((t)) \overset{!}{\otimes} \mathrm{Fun}(\mathrm{Conn}_G(\mathcal{D}^\times))$$

with the natural bracket (we refer to Section 19.1, where the notation $\overset{!}{\otimes}$ is introduced).

The assertion concerning $\mathsf{isom}_{\mathrm{Conn}_G(\mathcal{D}^\times)}$ follows from the fact that $\mathsf{Isom}_{\mathrm{Conn}_G(\mathcal{D}^\times)}$ is the product of $\mathrm{Conn}_G(\mathcal{D}^\times)$ and the group $G((t))$ acting on it, and $\mathfrak{g}((t))$ is the Lie algebra of $G((t))$.

The assertion concerning $\Omega^1(\mathrm{Conn}_G(\mathcal{D}^\times))$ follows from the identification of $\mathrm{Conn}_G(\mathcal{D}^\times)$ with a hyperplane in $\widehat{\mathfrak{g}}_\kappa^*$, and the description of the Poisson structure on the dual space to a Lie algebra. □

For any group ind-subscheme $K \subset G((t))$ such that $\widehat{\mathfrak{g}}_\kappa$ is split over $\mathsf{k} \subset \mathfrak{g}((t))$, the map $\widehat{\mathfrak{g}}_\kappa^* \to \mathsf{k}^*$ is a moment map for the action of K on $\widehat{\mathfrak{g}}_\kappa^*$, and, in particular, on $\mathrm{Conn}_G(\mathcal{D}^\times)$.

We take $K = N((t))$, and we obtain a moment map

$$\mu : \mathrm{Conn}_G(\mathcal{D}^\times) \to (\mathfrak{n}((t)))^* \simeq \mathfrak{g}/\mathfrak{b} \otimes \omega_{\mathcal{D}^\times},$$

where we identify $\mathfrak{n}^* \simeq \mathfrak{g}/\mathfrak{b}$ using κ.

We have an identification

$$\mathrm{Op}_{\mathfrak{g}}(\mathcal{D}^\times) \simeq \left(\mu^{-1}(p_{-1}dt)\right)/N(\hat{\mathcal{K}}), \tag{4.7}$$

where the action of $N(\hat{\mathcal{K}})$ on $\mu^{-1}(p_{-1}dt)$ is free. It is in this fashion that $\mathrm{Op}_{\mathfrak{g}}(\mathcal{D}^\times)$ was originally introduced in [DS] and this is why this Hamiltonian reduction is called the Drinfeld–Sokolov reduction.

Lemma 4.7. *There exists a canonical isomorphism of Lie algebroids over* $\mathrm{Op}_{\mathfrak{g}}(\mathcal{D}^\times)$

$$\Omega^1(\mathrm{Op}_{\mathfrak{g}}(\mathcal{D}^\times)) \simeq \mathsf{isom}_{\mathrm{Op}_{\mathfrak{g}}}.$$

Proof. Note that the action of $N((t))$ on $\mathrm{Conn}_G(\mathcal{D}^\times)$ lifts naturally to an action of the group $N((t)) \times N((t))$ on $\mathsf{Isom}_{\mathrm{Conn}_G(\mathcal{D}^\times)}$. We have a canonical identification of $\mathsf{Isom}_{\mathrm{Op}_{\mathfrak{g}}}$ with the two-sided quotient of $\mathsf{Isom}_{\mathrm{Conn}_G(\mathcal{D}^\times)}$:

$$\mathsf{Isom}_{\mathrm{Op}_{\mathfrak{g}}} \simeq \left(((\mu \times \mu) \circ (l \times r))^{-1}\left((p_{-1} \cdot dt) \times (p_{-1} \cdot dt)\right)\right)/N((t)) \times N((t)). \tag{4.8}$$

Hence, $\mathsf{isom}_{\mathrm{Op}_{\mathfrak{g}}}$ is obtained as a reduction with respect to $N((t))$ of the Lie algebroid $\mathsf{isom}_{\mathrm{Conn}_G(\mathcal{D}^\times)}$. By the definition of the Poisson structure on $\mathrm{Op}_{\mathfrak{g}}(\mathcal{D}^\times)$, the Lie algebroid $\Omega^1(\mathrm{Op}_{\mathfrak{g}}(\mathcal{D}^\times))$ is the reduction of the Lie algebroid $\Omega^1(\mathrm{Conn}_G(\mathcal{D}^\times))$ on $\mathrm{Conn}_G(\mathcal{D}^\times)$.

Hence, the assertion of the lemma follows from (4.6). □

4.8 The groupoid and Lie algebroid over regular opers

Let S be an ind-scheme with a Poisson structure and $S_1 \subset S$ be a reasonable subscheme, which is coisotropic, i.e., the ideal $\mathbf{I} = \ker(\mathrm{Fun}(S) \to \mathrm{Fun}(S_1))$ satisfies $[\mathbf{I}, \mathbf{I}] \subset \mathbf{I}$. We will assume that both S and S_1 are formally smooth; we will also assume that the normal bundle $N_{S_1/S}$ (which by our assumption is discrete) is locally projective.[6]

In this case the conormal $N^*_{S_1/S}$ acquires a structure of Lie algebroid, and the sheaf $\Omega^1(S_1)$ is a module over it. Moreover, we have the commutative diagram

$$\begin{array}{ccccccccc} 0 & \longrightarrow & N^*_{S_1/S} & \longrightarrow & \Omega^1(S)|_{S_1} & \longrightarrow & \Omega^1(S_1) & \longrightarrow & 0 \\ & & \big\downarrow & & \big\downarrow{\scriptstyle \mathbf{anch}} & & \big\downarrow & & \\ 0 & \longrightarrow & T(S_1) & \longrightarrow & T(S)|_{S_1} & \longrightarrow & N_{S_1/S} & \longrightarrow & 0 \end{array} \tag{4.9}$$

such that the right vertical arrow is a map of modules over $N^*_{S_1/S}$.

We claim the following.

Lemma 4.9. *The subscheme* $\mathrm{Op}^{\mathrm{reg}}_{\mathfrak{g}} \subset \mathrm{Op}_{\mathfrak{g}}(\mathcal{D}^\times)$ *is coisotropic.*

[6] We do not know whether this follows directly from the formal smoothness assumption.

Proof. Consider the subscheme $\mathrm{Conn}_G^{\mathrm{reg}}$ of $\mathrm{Conn}_G(\mathcal{D}^\times)$ obtained by imposing the condition that $\phi(t)$ belongs to $\mathfrak{g}\otimes\omega_{\hat{\mathcal{O}}}$. It is coisotropic, since the corresponding ideal in $\mathrm{Fun}(\mathrm{Conn}_G(\mathcal{D}^\times))$ is generated by $\mathfrak{g}\otimes\omega_{\hat{\mathcal{O}}}\subset\widehat{\mathfrak{g}}_\kappa$, which is a subalgebra.

By Section 1.5, the scheme $\mathrm{Op}_{\mathfrak{g}}^{\mathrm{reg}}$ can be realized as

$$\left(\mu^{-1}(p_{-1}\cdot dt)\cap\mathrm{Conn}_G^{\mathrm{reg}}\right)/N[[t]],$$

which implies the assertion of the lemma. □

Let $\mathsf{Isom}^{\mathrm{reg}}_{\mathrm{Op}_{\mathfrak{g}}}$ be the groupoid over the scheme $\mathrm{Op}_{\mathfrak{g}}^{\mathrm{reg}}=\mathrm{Op}_{\mathfrak{g}}(D)$ whose R-points are triples (χ,χ',ϕ), where $\chi=(\mathcal{F}_G,\nabla,\mathcal{F}_B)$ and $\chi=(\mathcal{F}_G,\nabla,\mathcal{F}_B)$ are R-points of $\mathrm{Op}_{\mathfrak{g}}^{\mathrm{reg}}$ and ϕ is an isomorphism of R-families of G-bundles on $\mathcal{D}$ with connections $(\mathcal{F}_G,\nabla)\to(\mathcal{F}'_G,\nabla')$.

Recall now the principal G-bundle $\mathcal{P}_{G,\mathrm{Op}_{\mathfrak{g}}^{\mathrm{reg}}}$ over $\mathrm{Op}_{\mathfrak{g}}^{\mathrm{reg}}$ obtained by restriction to $\mathrm{Op}_{\mathfrak{g}}^{\mathrm{reg}}\times x$ from the tautological G-bundle $\mathcal{F}_G$ on $\mathrm{Op}_{\mathfrak{g}}^{\mathrm{reg}}\hat{\times}\mathcal{D}$. This G-bundle defines a map $\mathrm{Op}_{\mathfrak{g}}^{\mathrm{reg}}\to\mathrm{pt}/G$.

Lemma 4.10.

(1) *The natural map* $\mathsf{Isom}^{\mathrm{reg}}_{\mathrm{Op}_{\mathfrak{g}}}\to\mathsf{Isom}_{\mathrm{Op}_{\mathfrak{g}}}|_{\mathrm{Op}_{\mathfrak{g}}^{\mathrm{reg}}}$ *is an isomorphism.*

(2) *The groupoid* $\mathsf{Isom}^{\mathrm{reg}}_{\mathrm{Op}_{\mathfrak{g}}}$ *is naturally isomorphic to*

$$\mathrm{Op}_{\mathfrak{g}}^{\mathrm{reg}}\underset{\mathrm{pt}/G}{\times}\mathrm{Op}_{\mathfrak{g}}^{\mathrm{reg}}.$$

Proof. The assertion of the lemma amounts to the following. Let $S=\mathrm{Spec}(R)$ be an affine scheme and let $(\mathcal{F}_G,\nabla)$, $(\mathcal{F}'_G,\nabla')$ be two G-bundles on $\mathrm{Spec}(R[[t]])$ with a regular connection along t. Let $\mathcal{P}_G,\mathcal{P}'_G$ be their restrictions to $\mathrm{Spec}(R)$, respectively. Then the set of connection-preserving isomorphisms $\mathcal{F}_G\to\mathcal{F}'_G$ maps isomorphically to both the set of connection-preserving isomorphisms $\mathcal{F}_G|_{\mathcal{D}^\times}\to\mathcal{F}'_G|_{\mathcal{D}^\times}$ and the set of isomorphisms $\mathcal{P}_G\to\mathcal{P}'_G$. □

Let $\mathsf{isom}^{\mathrm{reg}}_{\mathrm{Op}_{\mathfrak{g}}}$ be the Lie algebroid of $\mathsf{Isom}^{\mathrm{reg}}_{\mathrm{Op}_{\mathfrak{g}}}$. Lemma 4.10(2) implies that $\mathsf{isom}^{\mathrm{reg}}_{\mathrm{Op}_{\mathfrak{g}}}$ is identified with the Atiyah algebroid $\mathrm{At}(\mathcal{P}_{G,\mathrm{Op}^{\mathrm{reg}}})$ of infinitesimal symmetries of the G-bundle $\mathcal{P}_{G,\mathrm{Op}^{\mathrm{reg}}}$. Therefore, it fits in the exact sequence

$$0\to\mathfrak{g}_{\mathrm{Op}_G^{\mathrm{reg}}}\to\mathsf{isom}^{\mathrm{reg}}_{\mathrm{Op}_{\mathfrak{g}}}\to T(\mathrm{Op}_{\mathfrak{g}}^{\mathrm{reg}})\to 0,$$

where $\mathfrak{g}_{\mathrm{Op}_G^{\mathrm{reg}}}:=\mathfrak{g}_{\mathcal{P}_{\mathrm{Op}_G^{\mathrm{reg}}}}$. In what follows we will denote by $\mathfrak{b}_{\mathrm{Op}_G^{\mathrm{reg}}}$ (respectively, $\mathfrak{n}_{\mathrm{Op}_G^{\mathrm{reg}}}$) the subbundle of $\mathfrak{g}_{\mathrm{Op}_G^{\mathrm{reg}}}$, corresponding to the reduction $\mathcal{P}_{B,\mathrm{Op}_G^{\mathrm{reg}}}$ of $\mathcal{P}_{G,\mathrm{Op}_G^{\mathrm{reg}}}$ to B.

Note that by Lemmas 4.7 and 4.10(1) we have a natural map of algebroids

$$N^*_{\mathrm{Op}_{\mathfrak{g}}^{\mathrm{reg}}/\mathrm{Op}_{\mathfrak{g}}(\mathcal{D}^\times)}\to\mathsf{isom}^{\mathrm{reg}}_{\mathrm{Op}_{\mathfrak{g}}}. \tag{4.10}$$

Following [BD1, Section 3.7.16], we have Proposition 4.11.

Proposition 4.11. *The map of* (4.10) *is an isomorphism.*

Proof. The assertion of the proposition amounts to the fact that the map

$$\Omega^1(\mathrm{Op}_{\mathfrak{g}}^{\mathrm{reg}}) \to N_{\mathrm{Op}_{\mathfrak{g}}^{\mathrm{reg}}/\mathrm{Op}_{\mathfrak{g}}(\mathcal{D}^\times)} \tag{4.11}$$

from (4.9) is an injective bundle map.

Since the scheme $\mathrm{Op}_{\mathfrak{g}}^{\mathrm{reg}}$ is smooth, for an R-point $(\mathcal{F}_G, \nabla, \mathcal{F}_B)$ of $\mathrm{Op}_{\mathfrak{g}}^{\mathrm{reg}}$, the restrictions of $T(\mathrm{Op}_{\mathfrak{g}}^{\mathrm{reg}})$ and $\Omega^1(\mathrm{Op}_{\mathfrak{g}}^{\mathrm{reg}})$ to $\mathrm{Spec}(R)$ can be canonically identified with

$$\mathrm{coker}(\nabla) : \mathfrak{n}_{\mathcal{F}_B} \to \mathfrak{b}_{\mathcal{F}_B} \underset{\mathbb{C}[[t]]}{\otimes} \mathbb{C}[[t]]dt$$

and

$$\ker(\nabla) : (\mathfrak{g}/\mathfrak{n})_{\mathcal{F}_B} \underset{\mathbb{C}[[t]]}{\otimes} (\mathbb{C}((t))/\mathbb{C}[[t]]) \to (\mathfrak{g}/\mathfrak{b})_{\mathcal{F}_B} \underset{\mathbb{C}[[t]]}{\otimes} (\mathbb{C}((t))dt/\mathbb{C}[[t]]dt),$$

respectively, where we have used the identification $\mathfrak{g}^* \simeq \mathfrak{g}$ given by κ.

Hence, the restriction of $N_{\mathrm{Op}_{\mathfrak{g}}^{\mathrm{reg}}/\mathrm{Op}_{\mathfrak{g}}(\mathcal{D}^\times)}$ to $\mathrm{Spec}(R)$ can be identified with

$$\mathrm{coker}(\nabla) : \mathfrak{n}_{\mathcal{F}_B} \underset{\mathbb{C}[[t]]}{\otimes} (\mathbb{C}((t))/\mathbb{C}[[t]]) \to \mathfrak{b}_{\mathcal{F}_B} \underset{\mathbb{C}[[t]]}{\otimes} (\mathbb{C}((t))dt/\mathbb{C}[[t]]dt),$$

and the map of (4.11) is given by

$$\mathbf{u} \in \mathfrak{g}_{\mathcal{F}_B} \underset{\mathbb{C}[[t]]}{\otimes} (\mathbb{C}((t))/\mathbb{C}[[t]]) \mapsto \nabla(\mathbf{u}) \in \mathfrak{b}_{\mathcal{F}_B} \underset{\mathbb{C}[[t]]}{\otimes} (\mathbb{C}((t))dt/\mathbb{C}[[t]]dt).$$

The injectivity of the map in question is now evident from the oper condition on $\nabla/\mathcal{F}_B$. □

Corollary 4.12. *The kernel and the cokernel of the anchor map*

$$\mathbf{anch} : \Omega^1(\mathrm{Op}_{\mathfrak{g}}(\mathcal{D}^\times))|_{\mathrm{Op}_{\mathfrak{g}}^{\mathrm{reg}}} \to T(\mathrm{Op}_{\mathfrak{g}}(\mathcal{D}^\times))|_{\mathrm{Op}_{\mathfrak{g}}^{\mathrm{reg}}}$$

are both isomorphic to $\mathfrak{g}_{\mathrm{Op}_G^{\mathrm{reg}}}$ *as* $N^*_{\mathrm{Op}_{\mathfrak{g}}^{\mathrm{reg}}/\mathrm{Op}_{\mathfrak{g}}(\mathcal{D}^\times)}$*-modules.*

Proof. The isomorphism concerning the kernel follows by combining Proposition 4.11 and Lemma 4.10. The isomorphism concerning the cokernel follows from the first one by a general D-scheme argument; see [CHA, Section 2.5.22].

Let us, however, reprove both isomorphisms directly. We have

$$\ker(\mathbf{anch}|_{\mathrm{Op}_{\mathfrak{g}}^{\mathrm{reg}}}) \simeq \ker(\nabla) : \mathfrak{g}_{\mathcal{F}_G} \underset{\mathbb{C}[[t[[}{\otimes} \mathbb{C}((t)) \to \mathfrak{g}_{\mathcal{F}_G} \underset{\mathbb{C}[[t]]}{\otimes} \mathbb{C}((t))dt,$$

which is easily seen to be identified with $\mathfrak{g}_{\mathrm{Op}_G^{\mathrm{reg}}}$.

The assertion concerning $\mathrm{coker}(\mathbf{anch}|_{\mathrm{Op}_{\mathfrak{g}}^{\mathrm{reg}}})$ follows by Serre duality on $\mathcal{D}^\times$. Indeed, the dual of $\Omega^1(\mathrm{Op}_{\mathfrak{g}}(\mathcal{D}^\times))$ is canonically isomorphic to $T(\mathrm{Op}_{\mathfrak{g}}(\mathcal{D}^\times))$, and under this isomorphism, the dual of the map **anch** goes to itself. Hence,

$$(\mathrm{coker}(\mathbf{anch}))^* \simeq \ker(\mathbf{anch}) \simeq \mathfrak{g}_{\mathrm{Op}_G^{\mathrm{reg}}},$$

which we identify with $\mathfrak{g}^*_{\mathrm{Op}_G^{\mathrm{reg}}}$ using the form κ. □

To summarize, we obtain the following commutative diagram:

$$\begin{array}{ccccccccc}
 & & 0 & & 0 & & & & \\
 & & \downarrow & & \downarrow & & & & \\
0 & \longrightarrow & \mathfrak{g}_{\mathrm{Op}_{\mathfrak{g}}^{\mathrm{reg}}} & \xrightarrow{\mathrm{id}} & \mathfrak{g}_{\mathrm{Op}_{\mathfrak{g}}^{\mathrm{reg}}} & \longrightarrow & 0 & & \\
 & & \downarrow & & \downarrow & & \downarrow & & \\
0 & \longrightarrow & N^*_{\mathrm{Op}_{\mathfrak{g}}^{\mathrm{reg}}/\mathrm{Op}_{\mathfrak{g}}(\mathcal{D}^\times)} & \longrightarrow & \Omega^1(\mathrm{Op}_{\mathfrak{g}}(\mathcal{D}^\times)) & \longrightarrow & \Omega^1(\mathrm{Op}_{\mathfrak{g}}^{\mathrm{reg}}) & \longrightarrow & 0 \\
 & & \downarrow & & \downarrow & & \downarrow & & \\
0 & \longrightarrow & T(\mathrm{Op}_{\mathfrak{g}}^{\mathrm{reg}}) & \longrightarrow & T(\mathrm{Op}_{\mathfrak{g}}(\mathcal{D}^\times))|_{\mathrm{Op}_{\mathfrak{g}}^{\mathrm{reg}}} & \longrightarrow & N_{\mathrm{Op}_{\mathfrak{g}}^{\mathrm{reg}}/\mathrm{Op}_{\mathfrak{g}}(\mathcal{D}^\times)} & \longrightarrow & 0 \\
 & & \downarrow & & \downarrow & & \downarrow & & \\
 & & 0 & \longrightarrow & \mathfrak{g}_{\mathrm{Op}_{\mathfrak{g}}^{\mathrm{reg}}} & \xrightarrow{\mathrm{id}} & \mathfrak{g}_{\mathrm{Op}_{\mathfrak{g}}^{\mathrm{reg}}}. & & \\
 & & & & \downarrow & & \downarrow & & \\
 & & & & 0 & & 0 & &
\end{array}$$

We will conclude this subsection by the following remark. Let $(\mathcal{F}_G, \nabla, \mathcal{F}_B)$ be an R-point of $\mathrm{Op}_{\mathfrak{g}}^{\mathrm{reg}}$, and let $\mathbf{u}$ be an element of $\mathfrak{b}_{\mathcal{F}_B} \underset{\mathbb{C}[[t]]}{\otimes} \mathbb{C}((t))dt$, giving rise to a section of $T(\mathrm{Op}_{\mathfrak{g}}(\mathcal{D}^\times))|_{\mathrm{Spec}(R)}$ by (4.2).

From the proof of Corollary 4.12, we obtain the following.

Lemma 4.13. *The image of* $\mathbf{u}$ *in* $N_{\mathrm{Op}_{\mathfrak{g}}^{\mathrm{reg}}/\mathrm{Op}_{\mathfrak{g}}(\mathcal{D}^\times)}/\,\mathsf{isom}_{\mathrm{Op}_{\mathfrak{g}}}^{\mathrm{reg}} \simeq \mathfrak{g}_{\mathrm{Op}_{\mathfrak{g}}^{\mathrm{reg}}}$ *equals the image of* $\mathbf{u}$ *under the composition*

$$\mathfrak{b}_{\mathcal{F}_B} \underset{\mathbb{C}[[t]]}{\otimes} \mathbb{C}((t))dt \to \mathfrak{g}_{\mathcal{F}_G} \underset{\mathbb{C}[[t]]}{\otimes} \mathbb{C}((t))dt \to H^0_{DR}(\mathcal{D}^\times, \mathfrak{g}_{\mathcal{F}_G}) \simeq \mathfrak{g}_{\mathrm{Op}_{\mathfrak{g}}^{\mathrm{reg}}}.$$

4.14 The groupoid and algebroid on opers with nilpotent singularities

Now consider the subscheme $\mathrm{Op}_{\mathfrak{g}}^{\mathrm{nilp}} \subset \mathrm{Op}_{\mathfrak{g}}(\mathcal{D}^\times)$. As in Lemma 4.9, it is easy to see that $\mathrm{Op}_{\mathfrak{g}}^{\mathrm{nilp}}$ is coisotropic since

$$\mathrm{Op}_{\mathfrak{g}}^{\mathrm{nilp}} \simeq \left(\mu^{-1}(p_{-1}dt) \cap \mathrm{Conn}_G^{\mathrm{nilp}}\right)/N[[t]],$$

where $\mathrm{Conn}_G^{\mathrm{nilp}}$ is the subscheme of $\mathrm{Conn}_G(\mathcal{D}^\times)$, consisting of connections as in (4.5), for which $\phi(t) \in \mathfrak{g}[[t]] + \mathfrak{n} \otimes t^{-1}\mathbb{C}[[t]]$, and the latter is the orthogonal complement to the Iwahori subalgebra in $\widehat{\mathfrak{g}}_\kappa$.

Let us consider the groupoid

$$\mathsf{Isom}_{\mathrm{Op}_{\mathfrak{g}}}^{\mathrm{nilp}} := \mathrm{Op}_{\mathfrak{g}}^{\mathrm{nilp}} \underset{\mathfrak{n}/B}{\times} \mathrm{Op}_{\mathfrak{g}}^{\mathrm{nilp}}$$

over $\mathrm{Op}_{\mathfrak{g}}^{\mathrm{nilp}}$, and let $\mathsf{isom}_{\mathrm{Op}_{\mathfrak{g}}}^{\mathrm{nilp}}$ be the corresponding Lie algebroid.

Lemma 4.15. *There exists a natural closed embedding* $\mathsf{Isom}_{\mathrm{Op}_{\mathfrak{g}}}^{\mathrm{nilp}} \to \mathsf{Isom}_{\mathrm{Op}_{\mathfrak{g}}}|_{\mathrm{Op}_{\mathfrak{g}}^{\mathrm{nilp}}}$.

Proof. The lemma is proved in the following general framework. Let $(\mathcal{F}_G, \nabla)$ and $(\mathcal{F}'_G, \nabla')$ be two R-families of bundles with connections on $\mathcal{D}$ with poles of order 1 and nilpotent residues. Let $\mathcal{P}_G$ and $\mathcal{P}'_G$ be the resulting G-bundles on $\mathrm{Spec}(R)$, and $\mathrm{Res}(\nabla)$ (respectively, $\mathrm{Res}(\nabla')$) be the residue, which is an element in $\mathfrak{g}_{\mathcal{P}_G}$ (respectively, $\mathfrak{g}_{\mathcal{P}'_G}$).

Then there is a bijection between the set of connection-preserving isomorphisms $\mathcal{F}_G \to \mathcal{F}'_G$ of bundles on $\mathrm{Spec}(R[[t]])$ and isomorphisms $\mathcal{P}_G \to \mathcal{P}'_G$ which map $\mathrm{Res}(\nabla)$ to $\mathrm{Res}(\nabla')$. □

Note, however, that unlike the case of regular opers, the map of Lemma 4.15 is *not* an isomorphism. Indeed, the restriction of $\mathsf{Isom}_{\mathrm{Op}_{\mathfrak{g}}}^{\mathrm{nilp}}$ to $\mathrm{Op}_{\mathfrak{g}}^{\mathrm{reg}}$ is $\mathrm{Op}_{\mathfrak{g}}^{\mathrm{reg}} \underset{\mathrm{pt}/B}{\times} \mathrm{Op}_{\mathfrak{g}}^{\mathrm{reg}}$, which is strictly contained in $\mathrm{Op}_{\mathfrak{g}}^{\mathrm{reg}} \underset{\mathrm{pt}/G}{\times} \mathrm{Op}_{\mathfrak{g}}^{\mathrm{reg}} \simeq \mathsf{Isom}_{\mathrm{Op}_{\mathfrak{g}}}^{\mathrm{reg}}$.

We shall now establish the following.

Proposition 4.16. *The map of* (4.9) *induces an isomorphism of Lie algebroids*

$$N^*_{\mathrm{Op}_{\mathfrak{g}}^{\mathrm{nilp}}/\mathrm{Op}_{\mathfrak{g}}(\mathcal{D}^\times)} \simeq \mathsf{isom}_{\mathrm{Op}_{\mathfrak{g}}}^{\mathrm{nilp}}.$$

Proof. For an R-point $(\mathcal{F}_G, \nabla, \mathcal{F}_B)$ of $\mathrm{Op}_{\mathfrak{g}}^{\mathrm{nilp}}$ let us describe the restrictions of $N^*_{\mathrm{Op}_{\mathfrak{g}}^{\mathrm{nilp}}/\mathrm{Op}_{\mathfrak{g}}(\mathcal{D}^\times)}$ and $\mathsf{isom}_{\mathrm{Op}_{\mathfrak{g}}}^{\mathrm{nilp}}$ to $\mathrm{Spec}(R)$ as subspaces of the restriction of $\Omega^1(\mathrm{Op}_{\mathfrak{g}}(\mathcal{D}^\times)) \simeq \mathsf{isom}_{\mathrm{Op}_{\mathfrak{g}}}$. We have

$$T(\mathrm{Op}_{\mathfrak{g}}^{\mathrm{nilp}})|_{\mathrm{Spec}(R)} = \mathrm{coker}(\nabla) : \mathfrak{n}_{\mathcal{F}_B} \to$$
$$\to \left(\mathfrak{b}_{\mathcal{F}_B} \underset{\mathbb{C}[[t]]}{\otimes} \mathbb{C}[[t]]dt + \mathfrak{n}_{\mathcal{F}_B} \underset{\mathbb{C}[[t]]}{\otimes} t^{-1}\mathbb{C}[[t]]dt\right).$$

Hence, $N_{\mathrm{Op}_{\mathfrak{g}}^{\mathrm{nilp}}/\mathrm{Op}_{\mathfrak{g}}(\mathcal{D}^\times)}|_{\mathrm{Spec}(R)}$ is isomorphic to the cokernel of ∇:

$$\mathfrak{n}_{\mathcal{F}_B} \underset{\mathbb{C}[[t]]}{\otimes} (\mathbb{C}((t))dt/\mathbb{C}[[t]]dt) \to$$

$$\to \mathfrak{b}_{\mathcal{F}_B} \underset{\mathbb{C}[[t]]}{\otimes} \mathbb{C}((t))dt/(\mathfrak{b}_{\mathcal{F}_B} \underset{\mathbb{C}[[t]]}{\otimes} \mathbb{C}[[t]]dt + \mathfrak{n}_{\mathcal{F}_B} \underset{\mathbb{C}[[t]]}{\otimes} t^{-1}\mathbb{C}[[t]]dt).$$

Finally,

$$N^*_{\mathrm{Op}_\mathfrak{g}^{\mathrm{nilp}} / \mathrm{Op}_\mathfrak{g}(\mathcal{D}^\times)}|_{\mathrm{Spec}(R)} \simeq \ker(\nabla) : \big((\mathfrak{g}/\mathfrak{n})_{\mathcal{F}_B}(-x) + (\mathfrak{b}/\mathfrak{n})_{\mathcal{F}_B}\big) \to (\mathfrak{g}/\mathfrak{b})_{\mathcal{F}_B} \underset{\hat{\mathcal{O}}}{\otimes} \omega_{\mathcal{D}}.$$

So we can identify $N^*_{\mathrm{Op}_\mathfrak{g}^{\mathrm{nilp}} / \mathrm{Op}_\mathfrak{g}(\mathcal{D}^\times)}|_{\mathrm{Spec}(R)}$ as a subset of $\Omega^1(\mathrm{Op}_\mathfrak{g}(\mathcal{D}^\times))|_{\mathrm{Spec}(R)}$ with

$$\{\mathbf{u} \in \mathfrak{g}_{\mathcal{F}_G}(-x) + \mathfrak{b}_{\mathcal{F}_B} \subset \mathfrak{g}_{\mathcal{F}_G} | \nabla(\mathbf{u}) \in \mathfrak{b}_{\mathcal{F}_B}(x) \underset{\hat{\mathcal{O}}}{\otimes} \omega_{\mathcal{D}}\}/\{\mathbf{u} \in \mathfrak{n}_{\mathcal{F}_B}\}.$$

The latter is easily seen to be the image of $\mathsf{isom}^{\mathrm{nilp}}_{\mathrm{Op}_\mathfrak{g}}|_{\mathrm{Spec}(R)}$ inside $\mathsf{isom}_{\mathrm{Op}_\mathfrak{g}}|_{\mathrm{Spec}(R)}$. □

We shall now study the behavior of the restriction of $\mathsf{isom}^{\mathrm{nilp}}_{\mathrm{Op}_\mathfrak{g}}$ to the subscheme $\mathrm{Op}_\mathfrak{g}^{\mathrm{reg}} \subset \mathrm{Op}_\mathfrak{g}^{\mathrm{nilp}}$. The above proposition combined with Lemma 2.19(2) implies the following.

Corollary 4.17. *The Lie algebroid $N^*_{\mathrm{Op}_\mathfrak{g}^{\mathrm{nilp}} / \mathrm{Op}_\mathfrak{g}(\mathcal{D}^\times)}$ preserves the subscheme $\mathrm{Op}_\mathfrak{g}^{\mathrm{reg}}$. The restriction $N^*_{\mathrm{Op}_\mathfrak{g}^{\mathrm{nilp}} / \mathrm{Op}_\mathfrak{g}(\mathcal{D}^\times)}|_{\mathrm{Op}_\mathfrak{g}^{\mathrm{reg}}}$ identifies with the Atiyah algebroid $\mathrm{At}(\mathcal{P}_{B,\mathrm{Op}_\mathfrak{g}^{\mathrm{reg}}})$ of the B-bundle $\mathcal{P}_{B,\mathrm{Op}_\mathfrak{g}^{\mathrm{reg}}}$, and we have a commutative diagram*

$$\begin{array}{ccccccccc}
0 & \longrightarrow & \mathfrak{b}_{\mathrm{Op}_\mathfrak{g}^{\mathrm{reg}}} & \longrightarrow & N^*_{\mathrm{Op}_\mathfrak{g}^{\mathrm{nilp}} / \mathrm{Op}_\mathfrak{g}(\mathcal{D}^\times)}|_{\mathrm{Op}_\mathfrak{g}^{\mathrm{reg}}} & \longrightarrow & T(\mathrm{Op}_\mathfrak{g}^{\mathrm{nilp}})|_{\mathrm{Op}_\mathfrak{g}^{\mathrm{reg}}} & & \\
 & & \downarrow & & \downarrow & & \uparrow & & \\
0 & \longrightarrow & \mathfrak{g}_{\mathrm{Op}_\mathfrak{g}^{\mathrm{reg}}} & \longrightarrow & N^*_{\mathrm{Op}_\mathfrak{g}^{\mathrm{reg}} / \mathrm{Op}_\mathfrak{g}(\mathcal{D}^\times)} & \longrightarrow & T(\mathrm{Op}_\mathfrak{g}^{\mathrm{reg}}) & \longrightarrow & 0.
\end{array}$$

Corollary 4.18. *The composition*

$$N_{\mathrm{Op}_\mathfrak{g}^{\mathrm{reg}} / \mathrm{Op}_\mathfrak{g}^{\mathrm{nilp}}} \to N_{\mathrm{Op}_\mathfrak{g}^{\mathrm{reg}} / \mathrm{Op}_\mathfrak{g}(\mathcal{D}^\times)} \to N_{\mathrm{Op}_\mathfrak{g}^{\mathrm{reg}} / \mathrm{Op}_\mathfrak{g}(\mathcal{D}^\times)}/\,\Omega^1(\mathrm{Op}_\mathfrak{g}^{\mathrm{reg}}) \simeq \mathfrak{g}_{\mathrm{Op}_\mathfrak{g}^{\mathrm{reg}}}$$

is an injective bundle map, and its image coincides with $\mathfrak{n}_{\mathrm{Op}_\mathfrak{g}^{\mathrm{reg}}} \subset \mathfrak{g}_{\mathrm{Op}_\mathfrak{g}^{\mathrm{reg}}}$.

Proof. We claim that it is enough to show that the natural surjection $N_{\mathrm{Op}_\mathfrak{g}^{\mathrm{reg}} / \mathrm{Op}_\mathfrak{g}(\mathcal{D}^\times)} \to N_{\mathrm{Op}_\mathfrak{g}^{\mathrm{nilp}} / \mathrm{Op}_\mathfrak{g}(\mathcal{D}^\times)}|_{\mathrm{Op}_\mathfrak{g}^{\mathrm{reg}}}$ fits into a commutative diagram with exact rows

$$\begin{array}{ccccccccc}
 & & \Omega^1(\mathrm{Op}_\mathfrak{g}^{\mathrm{nilp}})|_{\mathrm{Op}_\mathfrak{g}^{\mathrm{reg}}} & \xrightarrow{\mathbf{anch}} & N_{\mathrm{Op}_\mathfrak{g}^{\mathrm{nilp}} / \mathrm{Op}_\mathfrak{g}(\mathcal{D}^\times)}|_{\mathrm{Op}_\mathfrak{g}^{\mathrm{reg}}} & \longrightarrow & (\mathfrak{g}/\mathfrak{n})_{\mathrm{Op}_\mathfrak{g}^{\mathrm{reg}}} & \longrightarrow & 0 \\
 & & \downarrow & & \uparrow & & \uparrow & & \\
0 & \longrightarrow & \Omega^1(\mathrm{Op}_\mathfrak{g}^{\mathrm{reg}}) & \longrightarrow & N_{\mathrm{Op}_\mathfrak{g}^{\mathrm{reg}} / \mathrm{Op}_\mathfrak{g}(\mathcal{D}^\times)} & \longrightarrow & \mathfrak{g}_{\mathrm{Op}_\mathfrak{g}^{\mathrm{reg}}} & \longrightarrow & 0.
\end{array} \tag{4.12}$$

Indeed, this would imply that the map $N_{\mathrm{Op}_{\mathfrak{g}}^{\mathrm{reg}}/\mathrm{Op}_{\mathfrak{g}}^{\mathrm{nilp}}} \to \mathfrak{g}_{\mathrm{Op}_{\mathfrak{g}}^{\mathrm{reg}}}$ appearing in the corollary is a surjective bundle map onto $\mathfrak{n}_{\mathrm{Op}_{\mathfrak{g}}^{\mathrm{reg}}}$; hence it must be an isomorphism because of the equality of the ranks.

By Serre duality, the existence of the diagram (4.12) is equivalent to the diagram appearing in the previous corollary. □

Let us now consider the sequence of embeddings of schemes

$$\mathrm{Op}_{\mathfrak{g}}^{\mathrm{reg}} \hookrightarrow \mathrm{Op}_{\mathfrak{g}}^{\mathrm{nilp}} \hookrightarrow \mathrm{Op}_{\mathfrak{g}}^{\mathrm{RS}}.$$

By Theorem 2.15, the normal bundle $N_{\mathrm{Op}_{\mathfrak{g}}^{\mathrm{nilp}}/\mathrm{Op}_{\mathfrak{g}}^{\mathrm{RS}}}$ is canonically trivialized and its fiber is isomorphic to the tangent space to $\mathfrak{h}//W$ at the point $-\check{\rho}$; this tangent space is in turn canonically isomorphic to $\mathfrak{h}$.

Lemma 4.19. *The composition*

$$\mathfrak{h} \to N_{\mathrm{Op}_{\mathfrak{g}}^{\mathrm{nilp}}/\mathrm{Op}_{\mathfrak{g}}^{\mathrm{RS}}}|_{\mathrm{Op}_{\mathfrak{g}}^{\mathrm{reg}}} \to N_{\mathrm{Op}_{\mathfrak{g}}^{\mathrm{nilp}}/\mathrm{Op}_{\mathfrak{g}}(\mathcal{D}^\times)}|_{\mathrm{Op}_{\mathfrak{g}}^{\mathrm{reg}}} \twoheadrightarrow (\mathfrak{g}/\mathfrak{n})_{\mathrm{Op}_{\mathfrak{g}}^{\mathrm{reg}}}$$

equals the canonical map

$$\mathfrak{h} \simeq (\mathfrak{b}/\mathfrak{n})_{\mathrm{Op}_{\mathfrak{g}}^{\mathrm{reg}}} \hookrightarrow (\mathfrak{g}/\mathfrak{n})_{\mathrm{Op}_{\mathfrak{g}}^{\mathrm{reg}}}.$$

Proof. Let $(\mathcal{F}_G, \nabla, \mathcal{F}_B)$ be an R-point of $\mathrm{Op}_{\mathfrak{g}}^{\mathrm{reg}}$, written in the form $\nabla^0 + p_{-1}dt + \phi(t)dt$, $\phi(t) \in \mathfrak{b} \otimes R[[t]]$. Then by Proposition 2.17 the map

$$\mathfrak{h} \to N_{\mathrm{Op}_{\mathfrak{g}}^{\mathrm{nilp}}/\mathrm{Op}_{\mathfrak{g}}^{\mathrm{RS}}}|_{\mathrm{Spec}(R)} \to N_{\mathrm{Op}_{\mathfrak{g}}^{\mathrm{nilp}}/\mathrm{Op}_{\mathfrak{g}}(\mathcal{D})}|_{\mathrm{Spec}(R)}$$

can be realized by

$$\check{\lambda} \mapsto \frac{\check{\lambda}}{t} \in \mathfrak{b} \otimes \mathbb{C}((t))dt \subset T(\mathrm{Op}_{\mathfrak{g}}(\mathcal{D}^\times))|_{\mathrm{Spec}(R)}.$$

To prove the lemma it would be enough to show that the image of $\frac{\check{\lambda}}{t}$ under

$$T(\mathrm{Op}_{\mathfrak{g}}(\mathcal{D}^\times))|_{\mathrm{Spec}(R)} \to N_{\mathrm{Op}_{\mathfrak{g}}^{\mathrm{reg}}/\mathrm{Op}_{\mathfrak{g}}(\mathcal{D}^\times)} \to \mathfrak{g}_{\mathrm{Op}_{\mathfrak{g}}^{\mathrm{reg}}}$$

equals λ. But this follows from Lemma 4.13. □

4.20 The case of opers with an integral residue

For completeness, let us describe the behaviour of the groupoid $\mathsf{Isom}_{\mathrm{Op}_{\mathfrak{g}}}$ and the algebroid $\mathsf{isom}_{\mathrm{Op}_{\mathfrak{g}}}$, when restricted to the subscheme

$$\mathrm{Op}_{\mathfrak{g}}^{\check{\lambda},\mathrm{nilp}} \simeq \mathrm{Op}_{\mathfrak{g}}^{\mathrm{RS},\varpi(-\check{\lambda}-\check{\rho})} \subset \mathrm{Op}_{\mathfrak{g}}(\mathcal{D}^\times)$$

when $\check{\lambda} + \check{\rho}$ dominant and integral.

Recall that to $\check{\lambda}$ as above there corresponds a subset $\mathcal{J}$ of vertices of the Dynkin diagram, and a map

$$\mathrm{Res}^{\check{\lambda},\mathrm{nilp}} : \mathrm{Op}_{\mathfrak{g}}^{\check{\lambda},\mathrm{nilp}} \to \mathbf{O}_{\mathcal{J}}/B.$$

Let us denote by $\mathsf{Isom}_{\mathrm{Op}_{\mathfrak{g}}}^{\check{\lambda},\mathrm{nilp}}$ the groupoid

$$\mathsf{Isom}_{\mathrm{Op}_{\mathfrak{g}}}^{\check{\lambda},\mathrm{nilp}} := \mathrm{Op}_{\mathfrak{g}}^{\check{\lambda},\mathrm{nilp}} \underset{\mathbf{O}_{\mathcal{J}}/B}{\times} \mathrm{Op}_{\mathfrak{g}}^{\check{\lambda},\mathrm{nilp}},$$

and let $\mathsf{isom}_{\mathrm{Op}_{\mathfrak{g}}}^{\check{\lambda},\mathrm{nilp}}$ be the corresponding algebroid on $\mathrm{Op}_{\mathfrak{g}}^{\check{\lambda},\mathrm{nilp}}$.

As in the case of $\check{\lambda}$ there exists a natural closed embedding

$$\mathsf{Isom}_{\mathrm{Op}_{\mathfrak{g}}}^{\check{\lambda},\mathrm{nilp}} \hookrightarrow \mathsf{Isom}_{\mathrm{Op}_{\mathfrak{g}}} |_{\mathrm{Op}_{\mathfrak{g}}^{\check{\lambda},\mathrm{nilp}}}.$$

Repeating the proofs in the $\check{\lambda} = 0$ case, we obtain the following.

Proposition 4.21. *The subscheme* $\mathrm{Op}_{\mathfrak{g}}^{\check{\lambda},\mathrm{nilp}} \subset \mathrm{Op}_{\mathfrak{g}}(\mathcal{D}^{\times})$ *is coisotropic. The map* (4.9) *induces an isomorphism*

$$N^*_{\mathrm{Op}_{\mathfrak{g}}^{\check{\lambda},\mathrm{nilp}} / \mathrm{Op}_{\mathfrak{g}}(\mathcal{D}^{\times})} \simeq \mathsf{isom}_{\mathrm{Op}_{\mathfrak{g}}}^{\check{\lambda},\mathrm{nilp}}.$$

Let us consider a particular example of $\check{\lambda} = -\check{\rho}$. In this case $\mathcal{J} = \mathcal{I}$, and

$$\mathbf{O}_{\mathcal{J}}/B \simeq \mathrm{pt},$$

Therefore, the map

$$\mathsf{isom}_{\mathrm{Op}_{\mathfrak{g}}}^{\check{\rho},\mathrm{nilp}} \to T(\mathrm{Op}_{\mathfrak{g}}^{\check{\rho},\mathrm{nilp}})$$

is surjective. Therefore by Proposition 4.21, the map

$$N^*_{\mathrm{Op}_{\mathfrak{g}}^{\check{\lambda},\mathrm{nilp}} / \mathrm{Op}_{\mathfrak{g}}(\mathcal{D}^{\times})} \to T(\mathrm{Op}_{\mathfrak{g}}^{\check{\rho},\mathrm{nilp}}),$$

given by the Poisson structure, is surjective as well. By Serre duality, the map

$$\Omega^1(\mathrm{Op}_{\mathfrak{g}}^{\check{\lambda},\mathrm{nilp}}) \to N_{\mathrm{Op}_{\mathfrak{g}}^{\check{\lambda},\mathrm{nilp}} / \mathrm{Op}_{\mathfrak{g}}(\mathcal{D}^{\times})}$$

is injective. This means that the map $\mathsf{isom}_{\mathrm{Op}_{\mathfrak{g}}}^{-\check{\rho},\mathrm{nilp}} \hookrightarrow \mathsf{isom}_{\mathrm{Op}_{\mathfrak{g}}} |_{\mathrm{Op}_{\mathfrak{g}}^{-\check{\rho},\mathrm{nilp}}}$ is an isomorphism. In fact, it is easy to see that the map $\mathsf{Isom}_{\mathrm{Op}_{\mathfrak{g}}}^{-\check{\rho},\mathrm{nilp}} \hookrightarrow \mathsf{Isom}_{\mathrm{Op}_{\mathfrak{g}}} |_{\mathrm{Op}_{\mathfrak{g}}^{-\check{\rho},\mathrm{nilp}}}$ is an isomorphism.

Finally, let us consider the case of $\check{\lambda}$ which is integral and dominant. We have the subscheme $\mathrm{Op}_{\mathfrak{g}}^{\check{\lambda},\mathrm{reg}} \subset \mathrm{Op}_{\mathfrak{g}}^{\check{\lambda},\mathrm{nilp}}$, and we claim that the behavior of the groupoid $\mathsf{Isom}_{\mathrm{Op}_{\mathfrak{g}}}$ and the algebroid $\mathsf{isom}_{\mathrm{Op}_{\mathfrak{g}}}$, restricted to it, are the same as in the $\check{\lambda} = 0$ case. In particular, the analogues of Corollaries 4.17 and 4.18 hold, when we replace $\mathrm{Op}_{\mathfrak{g}}^{\mathrm{nilp}}$ by $\mathrm{Op}_{\mathfrak{g}}^{\check{\lambda},\mathrm{nilp}}$.

4.22 Grading on the Lie algebroid

Recall the action of the group scheme $\mathrm{Aut}(\mathcal{D})$ on the scheme $\mathrm{Op}_{\mathfrak{g}}(\mathcal{D})$ and the ind-scheme $\mathrm{Op}_{\mathfrak{g}}(\mathcal{D}^\times)$. It is is easy to see that this action lifts to a map from $\mathrm{Aut}(\mathcal{D})$ to the groupoids $\mathsf{Isom}_{\mathrm{Op}_{\mathfrak{g}}}$ and $\mathsf{Isom}^{\mathrm{reg}}_{\mathrm{Op}_{\mathfrak{g}}}$, respectively. In particular, we obtain a map

$$\mathrm{Der}(\hat{\mathcal{O}}) \simeq \mathrm{Lie}(\mathrm{Aut}(\mathcal{D})) \to \mathsf{isom}^{\mathrm{reg}}_{\mathrm{Op}_{\mathfrak{g}}}.$$

We choose a coordinate on $\mathcal{D}$ and consider two distinguished elements $L_0 = t\partial_t$ and $L_{-1} = \partial_t$ in $\mathrm{Der}(\hat{\mathcal{O}})$. The action of L_0 integrates to an action of $\mathbb{G}_m$, thus defining a grading on $\mathsf{isom}^{\mathrm{reg}}_{\mathrm{Op}_{\mathfrak{g}}}$. Recall also that this choice of a coordinate trivializes the B-bundle $\mathcal{P}_{B,\mathrm{Op}^{\mathrm{reg}}_{\mathfrak{g}}}$ on $\mathrm{Op}^{\mathrm{reg}}_{\mathfrak{g}}$.

Proposition 4.23.

(1) *The image of L_{-1} under*

$$\mathsf{isom}^{\mathrm{reg}}_{\mathrm{Op}_{\mathfrak{g}}} \simeq \mathrm{At}(\mathcal{P}_{G,\mathrm{Op}^{\mathrm{reg}}_{\mathfrak{g}}}) \to \mathrm{At}(\mathcal{P}_{G,\mathrm{Op}^{\mathrm{reg}}_{\mathfrak{g}}})/\,\mathrm{At}(\mathcal{P}_{B,\mathrm{Op}^{\mathrm{reg}}_{\mathfrak{g}}}) \simeq (\mathfrak{g}/\mathfrak{b})_{\mathrm{Op}^{\mathrm{reg}}_{\mathfrak{g}}}$$

identifies, under the trivialization of $(\mathfrak{g}/\mathfrak{b})_{\mathrm{Op}^{\mathrm{reg}}_{\mathfrak{g}}}$ corresponding to the above choice of a coordinate, with the element $p_{-1} \in \mathfrak{g}/\mathfrak{b}$.

(2) *Under the above trivialization of $\mathcal{P}_B$, the subspace $\mathfrak{g} \subset \mathfrak{g}_{\mathrm{Op}^{\mathrm{reg}}_{\mathfrak{g}}}$ is L_0-stable, and grading arising on it equals the one induced by* $\mathrm{ad}_{\check{\rho}}$.

Proof. The proof is essentially borrowed from [BD1, Proposition 3.5.18].

The action of L_{-1} on $\mathrm{Op}^{\mathrm{reg}}_{\mathfrak{g}} \hat{\times} \mathcal{D}$ lifts to the triple $(\mathcal{F}_{G,\mathrm{Op}^{\mathrm{reg}}_{\mathfrak{g}}}, \nabla_{\mathrm{Op}^{\mathrm{reg}}_{\mathfrak{g}}}, \mathcal{F}_{B,\mathrm{Op}^{\mathrm{reg}}_{\mathfrak{g}}})$ by definition. The lift of L_{-1} to the G-bundle $\mathcal{P}_{G,\mathrm{Op}^{\mathrm{reg}}_{\mathfrak{g}}}$ is obtained via the identification of the latter with the space of horizontal (with respect to ∇) sections of $\mathcal{F}_{G,\mathrm{Op}^{\mathrm{reg}}_{\mathfrak{g}}}$. This lift does not preserve the reduction of $\mathcal{P}_{G,\mathrm{Op}^{\mathrm{reg}}_{\mathfrak{g}}}$ to B; the resulting element in $\mathrm{At}(\mathcal{P}_{G,\mathrm{Op}^{\mathrm{reg}}_{\mathfrak{g}}})/\,\mathrm{At}(\mathcal{P}_{B,\mathrm{Op}^{\mathrm{reg}}_{\mathfrak{g}}})$, which is the element appearing in point (1) of the proposition, equals, by definition, the value of

$$\langle \nabla_{\mathrm{Op}^{\mathrm{reg}}_{\mathfrak{g}}}/\mathcal{F}_{B,\mathrm{Op}^{\mathrm{reg}}_{\mathfrak{g}}}, \partial_t \rangle \in (\mathfrak{g}/\mathfrak{b})_{\mathcal{F}_{B,\mathrm{Op}^{\mathrm{reg}}_{\mathfrak{g}}}}$$

at $\mathrm{Op}^{\mathrm{reg}}_{\mathfrak{g}} \times x \subset \mathrm{Op}^{\mathrm{reg}}_{\mathfrak{g}} \hat{\times} \mathcal{D}$.

When the triple $(\mathcal{F}_{G,\mathrm{Op}^{\mathrm{reg}}_{\mathfrak{g}}}, \nabla_{\mathrm{Op}^{\mathrm{reg}}_{\mathfrak{g}}}, \mathcal{F}_{B,\mathrm{Op}^{\mathrm{reg}}_{\mathfrak{g}}})$ is written as a connection on the trivial B-bundle in the form $\nabla^0 + p_{-1}dt + \mathbf{q}(t)dt$, $\mathbf{q}(t) \in \mathfrak{b}[[t]]$, the above value equals p_{-1}, as required.

The second point of the proposition follows immediately from Lemma 1.9(1). □

Part II: Categories of Representations

This part of the paper is devoted to the discussion of various categories of representations of affine Kac–Moody algebras of critical level.

In Section 5 we recall the results of [FF3, F] about the structure of the center of the completed universal enveloping algebra of $\widehat{\mathfrak{g}}$ at the critical level. According to [FF3, F], the spectrum of the center is identified with the space $\mathrm{Op}_{\check{\mathfrak{g}}}(\mathcal{D}^\times)$ of $\check{\mathfrak{g}}$-opers over the formal punctured disc. This means that the category $\widehat{\mathfrak{g}}_{\mathrm{crit}}$-mod "fibers" over the affine ind-scheme $\mathrm{Op}_{\check{\mathfrak{g}}}(\mathcal{D}^\times)$. Next, we introduce the categories of representations that we study in this project, and in Section 6 we formulate our Main Conjecture 6.2 and Main Theorem 6.9.

In Section 7 we collect some results concerning the structure of the category $\widehat{\mathfrak{g}}_{\mathrm{crit}}$-mod over its center. In particular, we discuss the various incarnations of the renormalized universal enveloping algebra at the critical level. This renormalization is a phenomenon that has to do with the fact that we are dealing with a one-parameter family of associative algebras (the universal enveloping of the Kac–Moody Lie algebra, depending on the level), which at some special point (the critical level) acquires a large center.

In Section 8 we discuss the subcategory $\widehat{\mathfrak{g}}_{\mathrm{crit}}$-$\mathrm{mod}_{\mathrm{reg}}$ of representations at the critical level, whose support over $\mathrm{Op}_{\check{\mathfrak{g}}}(\mathcal{D}^\times)$ belongs to the subscheme of regular opers. We study its relation with the category of D-modules on the affine Grassmannian $\mathrm{Gr}_G \simeq G((t))/G[[t]]$, and this leads us to Main Conjecture 8.11. We prove Theorem 8.17 which states that a natural functor in one direction is fully faithful at the level of derived categories. The formalism of convolution action, developed in [BD1, Section 7], and reviewed in Part V below, allows us to reduce this assertion to a comparison of self-Exts of a certain basic object in both cases. On one side the required computation of Exts was performed in [ABG], and on the other side it follows from the recent paper [FT].

Section 9 plays an auxiliary role: we give a proof of one of the steps in the proof of Theorem 8.17 mentioned above, by analyzing how the algebra of $G[[t]]$-equivariant self-Exts of the vacuum module $\mathbb{V}_{\mathrm{crit}}$ interacts with the G-equivariant cohomology of the point.

5 Definition of categories

5.1

Let $\mathfrak{g}$ be a simple finite-dimensional Lie algebra. For an invariant inner product κ on $\mathfrak{g}$ (which is unique up to a scalar) define the central extension $\widehat{\mathfrak{g}}_\kappa$ of the formal loop algebra $\mathfrak{g} \otimes \mathbb{C}((t))$ which fits into the short exact sequence

$$0 \to \mathbb{C}\mathbf{1} \to \widehat{\mathfrak{g}}_\kappa \to \mathfrak{g} \otimes \mathbb{C}((t)) \to 0.$$

This sequence is split as a vector space, and the commutation relations read

$$[x \otimes f(t), y \otimes g(t)] = [x, y] \otimes f(t)g(t) + \kappa(x, y) \cdot \mathrm{Res}(gdf) \cdot \mathbf{1}, \tag{5.1}$$

and $\mathbf{1}$ is a central element. The Lie algebra $\widehat{\mathfrak{g}}_\kappa$ is the *affine Kac–Moody algebra* associated to κ. We will denote by $\widehat{\mathfrak{g}}_\kappa$-mod the category of *discrete* representations of $\widehat{\mathfrak{g}}_\kappa$ (i.e., such that any vector is annihilated by $\mathfrak{g} \otimes t^n\mathbb{C}[[t]]$ for sufficiently large n), on which $\mathbf{1}$ acts as the identity.

Let $U_\kappa(\widehat{\mathfrak{g}})$ be the quotient of the universal enveloping algebra $U(\widehat{\mathfrak{g}}_\kappa)$ of $\widehat{\mathfrak{g}}_\kappa$ by the ideal generated by $(\mathbf{1} - 1)$. Define its completion $\widetilde{U}_\kappa(\widehat{\mathfrak{g}})$ as follows:

$$\widetilde{U}_\kappa(\widehat{\mathfrak{g}}) = \underset{\longleftarrow}{\lim} U_\kappa(\widehat{\mathfrak{g}})/U_\kappa(\widehat{\mathfrak{g}}) \cdot (\mathfrak{g} \otimes t^n\mathbb{C}[[t]]).$$

It is clear that $\widetilde{U}_\kappa(\widehat{\mathfrak{g}})$ is a topological algebra, whose discrete continuous representations are the same as objects of $\widehat{\mathfrak{g}}_\kappa$-mod.

The following theorem, due to [FF3, F], describes the center $Z_\kappa(\widehat{\mathfrak{g}})$ of $\widetilde{U}_\kappa(\widehat{\mathfrak{g}})$.

Let κ_{crit} be the *critical* inner product on $\mathfrak{g}$ defined by the formula

$$\kappa_{\text{crit}}(x, y) = -\frac{1}{2}\operatorname{Tr}(\operatorname{ad}(x) \circ \operatorname{ad}(y)).$$

Denote by $\check{G}$ the group of adjoint type whose Lie algebra $\check{\mathfrak{g}}$ is Langlands dual to $\mathfrak{g}$ (i.e., the Cartan matrix of $\check{\mathfrak{g}}$ is the transpose of that of $\mathfrak{g}$).

Theorem 5.2.

(1) $Z_\kappa(\widehat{\mathfrak{g}}) = \mathbb{C}$ *if* $\kappa \neq \kappa_{\text{crit}}$.
(2) $Z_{\text{crit}}(\widehat{\mathfrak{g}})$ *is isomorphic to the algebra* $\operatorname{Fun}(\operatorname{Op}_{\check{G}}(\mathcal{D}^\times))$ *of functions on the space of* $\check{G}$*-opers on the punctured disc* $\mathcal{D}^\times$.

From now on we will denote $Z_{\text{crit}}(\widehat{\mathfrak{g}})$ simply by $\mathfrak{z}_{\mathfrak{g}}$.

5.3

Let I be the Iwahori subgroup of the group $G[[t]]$, i.e., the preimage of a fixed Borel subgroup $B \subset G$ under the evaluation homomorphism $G[[t]] \to G$. Let $I^0 \subset I$ be the pro-unipotent radical of I. Noncanonically we have a splitting $\operatorname{Lie}(I) = \operatorname{Lie}(I^0) \oplus \mathfrak{h}$, where $\mathfrak{h}$ is a Cartan subalgebra of $\mathfrak{g}$.

Recall that an object $\mathcal{M} \in \mathfrak{g}_\kappa$-mod is called I-integrable (respectively, I^0-integrable) if the action of $\operatorname{Lie}(I) \subset \widehat{\mathfrak{g}}_\kappa$ (respectively, $\operatorname{Lie}(I^0)$) on $\mathcal{M}$ integrates to an action of the pro-algebraic group I (respectively, I^0). In the case of I^0 this condition is equivalent to saying that $\operatorname{Lie}(I^0)$ acts locally nilpotently, and in the case of I that, in addition, $\mathfrak{h}$ acts semisimply with eigenvalues corresponding to integral weights. (The latter condition is easily seen to be independent of the choice of the splitting $\mathfrak{h} \to \operatorname{Lie}(I)$).

Following the conventions of Section 20.3, we will denote the corresponding subcategories of $\widehat{\mathfrak{g}}_{\text{crit}}$-mod by $\widehat{\mathfrak{g}}_{\text{crit}}\text{-mod}^I$ and $\widehat{\mathfrak{g}}_{\text{crit}}\text{-mod}^{I^0}$, respectively. We will denote by $D(\widehat{\mathfrak{g}}_\kappa\text{-mod})^I$ and $D(\widehat{\mathfrak{g}}_\kappa\text{-mod})^{I^0}$ the corresponding triangulated categories; see Section 20.8.

Recall also that an object $\mathcal{M} \in \mathfrak{g}_\kappa$-mod is called I-monodromic if it is Lie(I^0)-integrable and $\mathfrak{h}$ acts locally finitely with generalized eigenvalues corresponding to integral weights. It is evident that a module $\mathcal{M}$ is I-monodromic if and only if it has an increasing filtration with successive quotients being I-integrable. We will denote the subcategory of monodromic modules by $\mathfrak{g}_\kappa\text{-mod}^{I,m}$. We will denote by $D(\widehat{\mathfrak{g}}_\kappa\text{-mod})^{I,m}$ the full subcategory of $D(\widehat{\mathfrak{g}}_\kappa\text{-mod})$ consisting of complexes with I-monodromic cohomology.

Let us note that the above notions make sense more generally for an arbitrary category $\mathcal{C}$ endowed with a Harish-Chandra action of I (see Section 20.7). Namely, we have the full subcategories

$$\mathcal{C}^I \subset \mathcal{C}^{I^0} \subset \mathcal{C}$$

along with the equivariant categories $D(\mathcal{C})^I$, $D(\mathcal{C})^{I,m}$, $D(\mathcal{C})^{I_0}$. Since the group I^0 is pro-unipotent, the functor

$$D^+(\mathcal{C})^{I_0} \to D^+(\mathcal{C})$$

is fully faithful and its image consists of complexes whose cohomologies are I^0-equivariant. We also introduce the I-monodromic category $\mathcal{C}^{I,m}$ as the full subcategory of $\mathcal{C}$ consisting of objects that admit a filtration whose subquotients belong to $\mathcal{C}^I$; we let $D(\mathcal{C})^{I,m}$ be the full subcategory of $D(\mathcal{C})$ which consists of complexes whose cohomologies belong to $\mathcal{C}^{I,m}$.

From now on let us take $\kappa = \kappa_{\text{crit}}$. Recall the subscheme $\text{Op}_{\check{\mathfrak{g}}}^{\text{nilp}} \subset \text{Op}_{\check{\mathfrak{g}}}(\mathcal{D}^\times)$; see Section 2.13. Let $\widehat{\mathfrak{g}}_{\text{crit}}\text{-mod}_{\text{nilp}} \subset \mathfrak{g}_{\text{crit}}\text{-mod}$ be the subcategory consisting of modules on which the action of the center $\mathfrak{Z}_{\mathfrak{g}} \simeq \text{Fun}(\text{Op}_{\check{\mathfrak{g}}}(\mathcal{D}^\times))$ factors through the quotient $\mathfrak{Z}_{\mathfrak{g}}^{\text{nilp}} := \text{Fun}(\text{Op}_{\check{\mathfrak{g}}}^{\text{nilp}})$. This is a category endowed with an action of $G((t))$ and, in particular, of I.

Our main object of study is the category $\widehat{\mathfrak{g}}_{\text{crit}}\text{-mod}_{\text{nilp}}^{I,m}$, where we follow the above conventions regarding the notion of the I-monodromic subcategory. In other words,

$$\widehat{\mathfrak{g}}_{\text{crit}}\text{-mod}_{\text{nilp}}^{I,m} = \widehat{\mathfrak{g}}_{\text{crit}}\text{-mod}^{I,m} \cap \widehat{\mathfrak{g}}_{\text{crit}}\text{-mod}_{\text{nilp}}.$$

The following will be established in Section 7.19.

Lemma 5.4. *The inclusion functor*

$$\widehat{\mathfrak{g}}_{\text{crit}}\text{-mod}^{I,m} \cap \widehat{\mathfrak{g}}_{\text{crit}}\text{-mod}_{\text{nilp}} \to \widehat{\mathfrak{g}}_{\text{crit}}\text{-mod}^{I^0} \cap \widehat{\mathfrak{g}}_{\text{crit}}\text{-mod}_{\text{nilp}}$$

is, in fact, an equivalence.

(In other words, any module in $\widehat{\mathfrak{g}}_{\text{crit}}$-mod, which is I^0-integrable, and on which the center acts via $\mathfrak{Z}_{\mathfrak{g}}^{\text{nilp}}$, is automatically I-monodromic.)

By the above lemma, the inclusion

$$D^+(\widehat{\mathfrak{g}}_{\text{crit}}\text{-mod}_{\text{nilp}})^{I,m} \hookrightarrow D^+(\widehat{\mathfrak{g}}_{\text{crit}}\text{-mod}_{\text{nilp}})^{I^0}$$

is an equivalence, and both these categories identify with the full subcategory of $D^+(\widehat{\mathfrak{g}}_{\text{crit}}\text{-mod}_{\text{nilp}})$, consisting of complexes whose cohomologies belong to $\widehat{\mathfrak{g}}_{\text{crit}}\text{-mod}^{I,m}_{\text{nilp}}$.

The following assertion seems quite plausible, but we are unable to prove it at the moment.

Conjecture 5.5. *The natural functor* $D(\widehat{\mathfrak{g}}_{\text{crit}}\text{-mod}^{I,m}_{\text{nilp}}) \to D^b(\widehat{\mathfrak{g}}_{\text{crit}}\text{-mod}_{\text{nilp}})^{I^0}$ *is an equivalence.*

We will not need it in what follows.

6 The main conjecture

6.1

We shall now formulate our main conjecture. Recall the scheme $\text{MOp}^{\text{nilp}}_{\check{\mathfrak{g}}}$; see Section 3.14. Let $D^b(\text{QCoh}(\text{MOp}^{\text{nilp}}_{\check{\mathfrak{g}}}))$ be the bounded derived category of quasi-coherent sheaves on $\text{MOp}^{\text{nilp}}_{\check{\mathfrak{g}}}$.

Our main conjecture is as follows.

Main Conjecture 6.2. *We have an equivalence of triangulated categories*

$$D^b(\widehat{\mathfrak{g}}_{\text{crit}}\text{-mod}_{\text{nilp}})^{I^0} \simeq D^b\left(\text{QCoh}(\text{MOp}^{\text{nilp}}_{\check{\mathfrak{g}}})\right).$$

In what follows we will provide some motivation for this conjecture. We will denote a functor establishing the conjectural equivalence $D^b(\widehat{\mathfrak{g}}_{\text{crit}}\text{-mod}_{\text{nilp}})^{I^0} \to D^b(\text{QCoh}(\text{MOp}^{\text{nilp}}_{\check{\mathfrak{g}}}))$ by F.

Note that both categories $D^b(\text{QCoh}(\text{MOp}^{\text{nilp}}_{\check{\mathfrak{g}}}))$ and $D^b(\widehat{\mathfrak{g}}_{\text{crit}}\text{-mod}_{\text{nilp}})^{I^0}$ come equipped with natural t-structures. The functor F will not be exact, but we expect it to be of bounded cohomological amplitude, and hence to extend to an equivalence of the corresponding unbounded derived categories.

Recall the ind-scheme $\text{MOp}_{\check{\mathfrak{g}},\text{gen}}(D^\times)$ from Section 3.3. Following [FF2, F], to a quasi-coherent sheaf $\mathcal{R}$ on $\text{MOp}_{\check{\mathfrak{g}},\text{gen}}(D^\times) \simeq \text{Conn}_{\check{H}}(\omega^{\rho}_{\mathcal{D}^\times})$ one can attach a Wakimoto module $\mathbb{W}^{w_0}_{\text{crit}}(\mathcal{R}) \in \widehat{\mathfrak{g}}_{\text{crit}}\text{-mod}$ (see Section 13.1 for a review of this construction).

If $\mathcal{R}$ is supported on the closed subscheme $\text{MOp}_{\check{\mathfrak{g}},\text{gen}}(D^\times) \underset{\text{Op}_{\check{\mathfrak{g}}}(\mathcal{D}^\times)}{\times} \text{Op}^{\text{nilp}}_{\check{\mathfrak{g}}}$, then it turns out that $\mathbb{W}^{w_0}_{\text{crit}}(\mathcal{R})$ belongs to the subcategory $\widehat{\mathfrak{g}}_{\text{crit}}\text{-mod}^{I,m}_{\text{nilp}}$. The main compatibility property that we expect from the functor F is that $\mathsf{F}(\mathbb{W}^{w_0}_{\text{crit}}(\mathcal{R}))$ will be isomorphic to the direct image of $\mathcal{R}$ under the morphism

$$\text{MOp}_{\check{\mathfrak{g}},\text{gen}}(D^\times) \underset{\text{Op}_{\check{\mathfrak{g}}}(\mathcal{D}^\times)}{\times} \text{Op}^{\text{nilp}}_{\check{\mathfrak{g}}} \to \text{MOp}^{\text{nilp}}_{\check{\mathfrak{g}}} \tag{6.1}$$

of Theorem 3.16.

In view of this requirement, the functor F^{-1}, inverse to F, should be characterized by the property that it extends the Wakimoto module construction from quasi-coherent sheaves on $\mathrm{MOp}_{\check{\mathfrak{g}},\mathrm{gen}}(D^\times) \underset{\mathrm{Op}_{\check{\mathfrak{g}}}(\mathcal{D}^\times)}{\times} \mathrm{Op}_{\check{\mathfrak{g}}}^{\mathrm{nilp}}$ to those on $\mathrm{MOp}_{\check{\mathfrak{g}}}^{\mathrm{nilp}}$. This was, in fact, the main motivation for Main Conjecture 6.2.

6.3

In this subsection we would like to explain a point of view on Conjecture 6.2 as a localization-type statement for affine algebras at the critical level that connects D-modules on the affine flag variety to $\widehat{\mathfrak{g}}_{\mathrm{crit}}$-mod.

This material will not be used in what follows, and for that reason we shall allow ourselves to appeal to some results and constructions that are not available in the published literature. One set of such results is Bezrukavnikov's theory of perverse sheaves on the affine flag scheme (see [Bez]) and another the formalism of triangulated categories over stacks (to be developed in [Ga2]).

Let Fl_G be the affine flag scheme corresponding to G, i.e., $\mathrm{Fl}_G \simeq G((t))/I$. Let $\mathfrak{D}(\mathrm{Fl}_G)$-mod denote the category of right D-modules on Fl. Let $\mathfrak{D}(\mathrm{Fl}_G)\text{-mod}^I$, $\mathfrak{D}(\mathrm{Fl}_G)\text{-mod}^{I^0}$ and $\mathfrak{D}(\mathrm{Fl}_G)\text{-mod}^{I,m}$ be the subcategories of I-equivariant, I^0-equivariant, and I-monodromic D-modules, respectively. One easily shows that the inclusion functor

$$\mathfrak{D}(\mathrm{Fl}_G)\text{-mod}^{I,m} \to \mathfrak{D}(\mathrm{Fl}_G)\text{-mod}^{I^0}$$

is, in fact, an equivalence of categories.

Let $D(\mathfrak{D}(\mathrm{Fl}_G)\text{-mod})^I$ and $D(\mathfrak{D}(\mathrm{Fl}_G)\text{-mod})^{I^0}$ denote the corresponding triangulated categories.

Recall the Grothendieck alteration $\widetilde{\check{\mathfrak{g}}} \to \check{\mathfrak{g}}$ from Section 3.14. Let $\widetilde{\mathcal{N}}_{\check{G}}$ be the Springer resolution of the nilpotent cone $\mathcal{N}_{\check{G}} \subset \check{\mathfrak{g}}$. Let $\mathrm{St}_{\check{G}}$ be the "thickened" Steinberg variety

$$\mathrm{St}_{\check{G}} := \widetilde{\check{\mathfrak{g}}} \underset{\check{\mathfrak{g}}}{\times} \widetilde{\mathcal{N}}_{\check{G}}.$$

Note that the scheme $\widetilde{\check{\mathfrak{n}}} := \check{\mathfrak{n}} \underset{\check{\mathfrak{g}}}{\times} \widetilde{\check{\mathfrak{g}}}$ introduced in Section 3.14 equals the preimage of $\check{G}/\check{B}^- \times \{\check{\mathfrak{b}}\}$ under the natural map $\mathrm{St}_{\check{G}} \to \check{G}/\check{B}^- \times \check{G}/\check{B}^-$, and we have natural isomorphisms of stacks:

$$\widetilde{\mathcal{N}}_{\check{G}}/\check{G} \simeq \check{\mathfrak{n}}/\check{B} \quad \text{and} \quad \mathrm{St}_{\check{G}}/\check{G} \simeq \widetilde{\check{\mathfrak{n}}}/\check{B}.$$

The next lemma ensures that the definition of the scheme $\widetilde{\check{\mathfrak{n}}}$ (and, hence, of $\mathrm{St}_{\check{G}}$) is not too naive, i.e., that we do not neglect lower cohomology.

Lemma 6.4. *The derived tensor product*

$$\mathrm{Fun}(\widetilde{\check{\mathfrak{g}}}) \overset{L}{\underset{\mathrm{Fun}(\check{\mathfrak{g}})}{\otimes}} \mathrm{Fun}(\check{\mathfrak{n}}) \in \mathrm{QCoh}(\widetilde{\check{\mathfrak{g}}})$$

is concentrated in cohomological dimension 0.

Proof. Consider the vector space $\check{\mathfrak{g}}/\check{\mathfrak{n}}$. It is enough to show that the composed map

$$\widetilde{\check{\mathfrak{g}}} \to \check{\mathfrak{g}} \to \check{\mathfrak{g}}/\check{\mathfrak{n}}$$

is flat near $0 \in \check{\mathfrak{g}}/\check{\mathfrak{n}}$. Since the varieties we are dealing with are smooth, it is enough to check that the dimension of the fibers is constant. The latter is evident. □

According to [AB], there exists a natural tensor functor

$$D^b\left(\mathrm{Coh}(\widetilde{\mathcal{N}}_{\check{G}}/\check{G})\right) \to D^b(\mathfrak{D}(\mathrm{Fl}_G)\text{-mod})^I.$$

In particular, using the convolution action of the monoidal category $D^b(\mathfrak{D}(\mathrm{Fl}_G)\text{-mod})^I$ on the entire $D^b(\mathfrak{D}(\mathrm{Fl}_G)\text{-mod})$, we obtain a monoidal action of $D^b(\mathrm{Coh}(\widetilde{\mathcal{N}}_{\check{G}}/\check{G}))$ on $D^b(\mathfrak{D}(\mathrm{Fl}_G)\text{-mod})$. This construction can be upgraded to a structure on $D^b(\mathfrak{D}(\mathrm{Fl}_G)\text{-mod})$ of triangulated category over the stack $\widetilde{\mathcal{N}}_{\check{G}}/\check{G}$, see [Ga2]. In particular, it makes sense to consider the base-changed triangulated category

$$D^b\left(\mathfrak{D}(\mathrm{Fl}_G)\text{-mod}\right) \underset{\widetilde{\mathcal{N}}_{\check{G}}/\check{G}}{\times} \mathrm{Op}^{\mathrm{nilp}}_{\check{\mathfrak{g}}}, \tag{6.2}$$

where we are using the map $\mathrm{Res}^{\mathrm{nilp}} : \mathrm{Op}^{\mathrm{nilp}}_{\check{\mathfrak{g}}} \to \check{\mathfrak{n}}/\check{B} \simeq \widetilde{\mathcal{N}}_{\check{G}}/\check{G}$.

A far-reaching generalization of Conjecture 6.2 is the following statement in the spirit of the localization theorem of [BB].

Conjecture 6.5. *There is an equivalence of triangulated categories*

$$D^b\left(\mathfrak{D}(\mathrm{Fl}_G)\text{-mod}\right) \underset{\widetilde{\mathcal{N}}_{\check{G}}/\check{G}}{\times} \mathrm{Op}^{\mathrm{nilp}}_{\check{\mathfrak{g}}} \simeq D^b(\widehat{\mathfrak{g}}_{\mathrm{crit}}\text{-mod}_{\mathrm{nilp}}).$$

A version of this conjecture concerning $\widehat{\mathfrak{g}}_{\mathrm{crit}}\text{-mod}_{\mathrm{reg}}$, rather than $\widehat{\mathfrak{g}}_{\mathrm{crit}}\text{-mod}_{\mathrm{nilp}}$, can be made precise without the machinery of categories over stacks, and it will be discussed in Section 8.

Let us explain the connection between the above Conjectures 6.5 and 6.2. Namely, we claim that the latter is obtained from the former by passing to the corresponding I^0-equivariant categories on both sides. In order to explain this, we recall the main result of Bezrukavnikov's theory.

Theorem 6.6. *There is a natural equivalence*

$$D^b\left(\mathfrak{D}(\mathrm{Fl}_G)\text{-mod}\right)^{I^0} \simeq D^b\left(\mathrm{Coh}\left(\mathrm{St}_{\check{G}}/\check{G}\right)\right).$$

This theorem implies that the base-changed category

$$D^b\left(\mathfrak{D}(\mathrm{Fl}_G)\text{-mod}\right)^{I_0} \underset{\widetilde{\mathcal{N}}_{\check{G}}/\check{G}}{\times} \mathrm{Op}^{\mathrm{nilp}}_{\check{\mathfrak{g}}}$$

is equivalent to

$$D^b\left(\mathrm{QCoh}(\mathrm{St}_{\check{G}}/\check{G} \underset{\widetilde{\mathcal{N}}_{\check{G}}/\check{G}}{\times} \mathrm{Op}^{\mathrm{nilp}}_{\check{\mathfrak{g}}})\right),$$

which by Corollary 3.17 is the same as $D^b(\mathrm{QCoh}(\mathrm{MOp}^{\mathrm{nilp}}_{\check{\mathfrak{g}}}))$.

6.7

We shall now formulate one of the main results of this paper, which amounts to an equivalence as in Main Conjecture 6.2, but at the level of certain quotient categories. This result provides us with the main supporting evidence for the validity of Main Conjecture 6.2. Before stating the theorem, let us give some motivation along the lines of Theorem 6.6.

Let $\mathcal{F}$ be an I^0-integrable D-module on Fl_G. We will say that it is partially integrable if $\mathcal{F}$ admits a filtration $\mathcal{F} = \underset{k\geq 0}{\cup} \mathcal{F}_k$ such that each successive quotient $\mathcal{F}_k/\mathcal{F}_{k-1}$ is equivariant with respect to a parahoric subalgebra $\mathfrak{p}^\iota = \mathrm{Lie}(I) + \mathfrak{sl}_2^\iota$ for some vertex of the Dynkin graph $\iota \in \mathcal{I}$.

Similarly, we will call an object $\mathcal{M}$ of $\widehat{\mathfrak{g}}_{\mathrm{crit}}\text{-mod}^{I^0}$ partially integrable if there exists a filtration $\mathcal{M} = \underset{k\geq 0}{\cup} \mathcal{M}_k$ such that for each successive quotient $\mathcal{M}_k/\mathcal{M}_{k-1}$ there exists a parahoric subalgebra $\mathfrak{p}^\iota$ as above such that its action integrates to an action of the corresponding pro-algebraic group. More generally, the notion of partial integrability makes sense in any category equipped with a Harish-Chandra action of $G((t))$ (see Section 22, where the latter notion is introduced).

In both cases it is easy to see that partially integrable objects form a Serre subcategory. Let ${}^f\mathfrak{D}(\mathrm{Fl}_G)\text{-mod}^{I^0}$ (respectively, ${}^f\widehat{\mathfrak{g}}_{\mathrm{crit}}\text{-mod}_{\mathrm{nilp}}^{I,m}$) denote the quotient category of $\mathfrak{D}(\mathrm{Fl}_G)\text{-mod}^{I^0}$ (respectively, $\widehat{\mathfrak{g}}_{\mathrm{crit}}\text{-mod}_{\mathrm{nilp}}^{I,m}$) by the subcategory of partially integrable objects. We will denote by ${}^fD(\mathfrak{D}(\mathrm{Fl}_G)\text{-mod})^{I^0}$ (respectively, ${}^fD^b(\widehat{\mathfrak{g}}_{\mathrm{crit}}\text{-mod}_{\mathrm{nilp}})^{I^0}$) the triangulated quotient categories by the subcategories consisting of objects whose cohomologies are partially integrable.

Let us now recall the statement from [Bez] that describes the category ${}^fD^b(\mathfrak{D}(\mathrm{Fl}_G)\text{-mod})^{I^0}$ in terms of quasi-coherent sheaves.

Let h_0 denote the algebra of functions on the scheme $\varpi^{-1}(0)$, where ϖ is the natural projection $\mathfrak{h}^* \to \mathfrak{h}^*/\!/W$. This is a nilpotent algebra of length $|W|$.

Recall also that $\mathfrak{h}^* \simeq \check{\mathfrak{h}}$. We have a natural map

$$\mathrm{St}_{\check{G}} \simeq \widetilde{\check{\mathfrak{g}}} \underset{\check{\mathfrak{g}}}{\times} \widetilde{\mathcal{N}}_{\check{G}} \to \check{\mathfrak{h}} \underset{\check{\mathfrak{h}}/\!/W}{\times} \widetilde{\mathcal{N}}_{\check{G}} \simeq \mathrm{Spec}(h_0) \times \widetilde{\mathcal{N}}_{\check{G}}.$$

Theorem 6.8. *There is a canonical equivalence*

$${}^fD^b\left(\mathfrak{D}(\mathrm{Fl}_G)\text{-mod}\right)^{I^0} \simeq D^b\left(\mathrm{QCoh}\left(\mathrm{Spec}(h_0) \times \widetilde{\mathcal{N}}_{\check{G}}/\check{G}\right)\right),$$

so that under the equivalence of Theorem 6.6 *the functor*

$$D^b\left(\mathfrak{D}(\mathrm{Fl}_G)\text{-mod}\right)^{I^0} \to {}^fD^b\left(\mathfrak{D}(\mathrm{Fl}_G)\text{-mod}\right)^{I^0}$$

corresponds to the direct image under the projection $\mathrm{St}_{\check{G}}/\check{G} \to \mathrm{Spec}(h_0) \times \widetilde{\mathcal{N}}_{\check{G}}/\check{G}$.

Combining this with Conjecture 6.5, we arrive at the following statement, which is proved in Part IV of this paper and is one of our main results.

Main Theorem 6.9. *We have an equivalence:*

$$ {}^f\mathsf{F} : {}^fD^b(\widehat{\mathfrak{g}}_{\mathrm{crit}}\text{-mod}_{\mathrm{nilp}})^{I^0} \to D^b\left(\mathrm{QCoh}\left(\mathrm{Spec}(h_0) \times \mathrm{Op}_{\check{\mathfrak{g}}}^{\mathrm{nilp}}\right)\right). $$

Moreover, this functor is exact in the sense that it preserves the natural t-structures on both sides.

7 Generalities on $\widehat{\mathfrak{g}}_{\mathrm{crit}}$-modules

7.1

Recall that the ind-scheme $\mathrm{Op}_{\check{\mathfrak{g}}}(\mathcal{D}^\times)$ contains the following subschemes:

$$ \mathrm{Op}_{\check{\mathfrak{g}}}^{\mathrm{reg}} \subset \mathrm{Op}_{\check{\mathfrak{g}}}^{\mathrm{nilp}} \subset \mathrm{Op}_{\check{\mathfrak{g}}}^{\mathrm{RS}} = \mathrm{Op}_{\check{\mathfrak{g}}}^{\mathrm{ord}_1} \subset \mathrm{Op}_{\check{\mathfrak{g}}}^{\mathrm{ord}_k} \quad \text{for } k \geq 1. $$

Let us denote by

$$ \mathfrak{z}_{\mathfrak{g}}^{\mathrm{ord}_k} \twoheadrightarrow \mathfrak{z}_{\mathfrak{g}}^{\mathrm{RS}} \twoheadrightarrow \mathfrak{z}_{\mathfrak{g}}^{\mathrm{nilp}} \twoheadrightarrow \mathfrak{z}_{\mathfrak{g}}^{\mathrm{reg}}, $$

respectively, the corresponding quotients of $\mathfrak{z}_{\mathfrak{g}} \simeq \mathrm{Fun}(\mathrm{Op}_{\check{\mathfrak{g}}}(\mathcal{D}^\times))$.

Let us denote by $\mathfrak{z}_{\mathfrak{g}}$-mod the category of discrete $\mathfrak{z}_{\mathfrak{g}}$-modules. By definition, any object of this category is a union of subobjects, each of which is acted on by $\mathfrak{z}_{\mathfrak{g}}$ via the quotient $\mathfrak{z}_{\mathfrak{g}}^{\mathrm{ord}_k}$ for some k.

Let ι^{reg} (respectively, ι^{nilp}, ι^{RS}, ι^{ord_k}) denote the closed embedding of $\mathrm{Spec}(\mathfrak{z}_{\mathfrak{g}}^{\mathrm{reg}})$ (respectively, $\mathrm{Spec}(\mathfrak{z}_{\mathfrak{g}}^{\mathrm{nilp}})$, $\mathrm{Spec}(\mathfrak{z}_{\mathfrak{g}}^{\mathrm{RS}})$, $\mathrm{Spec}(\mathfrak{z}_{\mathfrak{g}}^{\mathrm{ord}_k})$) into the ind-scheme $\mathrm{Spec}(\mathfrak{z}_{\mathfrak{g}})$, and let $\iota_!^{\mathrm{reg}}$ (respectively, $\iota_!^{\mathrm{nilp}}$, $\iota_!^{\mathrm{RS}}$, $\iota_!^{\mathrm{ord}_k}$) denote the corresponding direct image functor on the category of modules.

It is easy to see that at the level of derived categories we have well-defined right adjoint functors from $D^+(\mathfrak{z}_{\mathfrak{g}}\text{-mod})$ to $D^+(\mathfrak{z}_{\mathfrak{g}}^{\mathrm{reg}}\text{-mod})$, $D^+(\mathfrak{z}_{\mathfrak{g}}^{\mathrm{nilp}}\text{-mod})$, $D^+(\mathfrak{z}_{\mathfrak{g}}^{\mathrm{RS}}\text{-mod})$ and $D^+(\mathfrak{z}_{\mathfrak{g}}^{\mathrm{ord}_k}\text{-mod})$, denoted $\iota^{\mathrm{reg}!}$, $\iota^{\mathrm{nilp}!}$, $\iota^{\mathrm{nilp}!}$ and $\iota^{\mathrm{ord}_k!}$, respectively.

7.2

Let $\widehat{\mathfrak{g}}_{\mathrm{crit}}\text{-mod}_{\mathrm{reg}}$ (respectively, $\widehat{\mathfrak{g}}_{\mathrm{crit}}\text{-mod}_{\mathrm{nilp}}$, $\widehat{\mathfrak{g}}_{\mathrm{crit}}\text{-mod}_{\mathrm{RS}}$, $\widehat{\mathfrak{g}}_{\mathrm{crit}}\text{-mod}_{\mathrm{ord}_k}$) denote the subcategory of $\widehat{\mathfrak{g}}_{\mathrm{crit}}$-mod whose objects are modules on which $\mathfrak{z}_{\mathfrak{g}}$ acts through the corresponding quotient.

The following basic result was established in [BD1, Theorem 3.7.9].

Theorem 7.3. *The induced module* $\mathrm{Ind}_{t^k\mathfrak{g}[[t]]\oplus\mathbb{C}\mathbf{1}}^{\widehat{\mathfrak{g}}_{\mathrm{crit}}}(\mathbb{C})$ *belongs to* $\widehat{\mathfrak{g}}_{\mathrm{crit}}\text{-mod}_{\mathrm{ord}_k}$.

Here and below, when considering the induced modules such as $\mathrm{Ind}_{t^k\mathfrak{g}[[t]]\oplus\mathbb{C}\mathbf{1}}^{\widehat{\mathfrak{g}}_{\mathrm{crit}}}(\mathbb{C})$, we will assume that $\mathbf{1}$ acts as the identity. We will also need the following.

Lemma 7.4. *The module* $\mathrm{Ind}_{t^k\mathfrak{g}[[t]]\oplus\mathbb{C}\mathbf{1}}^{\widehat{\mathfrak{g}}_{\mathrm{crit}}}(\mathbb{C})$ *is flat over* $\mathfrak{z}_{\mathfrak{g}}^{\mathrm{ord}_k}$.

Proof. By construction, the PBW filtration on $\mathfrak{Z}_\mathfrak{g}$ induces a filtration on $\mathfrak{Z}_\mathfrak{g}^{\mathrm{ord}_k}$ such that

$$\mathrm{gr}(\mathfrak{Z}_\mathfrak{g}^{\mathrm{ord}_k}) \simeq \left(\mathrm{Sym}\left(\mathfrak{g}((t))/t^k\mathfrak{g}[[t]]\right)\right)^{G[[t]]}.$$

This filtration is compatible with the natural filtration on $\mathrm{Ind}^{\widehat{\mathfrak{g}}_{\mathrm{crit}}}_{t^k\mathfrak{g}[[t]]\oplus\mathbb{C}\mathbf{1}}(\mathbb{C})$, and it suffices to check the flatness on the associated graded level.

This reduces the assertion to showing that the algebra $\mathrm{Sym}\,(\mathfrak{g}((t))/t^k\mathfrak{g}[[t]])$ is flat over $(\mathrm{Sym}\,(\mathfrak{g}((t))/t^k\mathfrak{g}[[t]]))^{G[[t]]}$. However, the multiplication by t^{-k} reduces us to the situation when $k = 0$, in which case the required assertion is proved in [EF]. □

Let us denote by $\iota_!^{\mathrm{reg}}$ (respectively, $\iota_!^{\mathrm{nilp}}$, $\iota_!^{\mathrm{RS}}$, $\iota_!^{\mathrm{ord}_k}$) the evident functor from $\widehat{\mathfrak{g}}_{\mathrm{crit}}\text{-mod}_{\mathrm{reg}}$ (respectively, $\widehat{\mathfrak{g}}_{\mathrm{crit}}\text{-mod}_{\mathrm{nilp}}$, $\widehat{\mathfrak{g}}_{\mathrm{crit}}\text{-mod}_{\mathrm{RS}}$, $\widehat{\mathfrak{g}}_{\mathrm{crit}}\text{-mod}_{\mathrm{ord}_k}$) to $\widehat{\mathfrak{g}}_{\mathrm{crit}}\text{-mod}$. It is easy to show that each of these functors admits an adjoint, denoted $\iota^{\mathrm{reg}!}$ (respectively, $\iota^{\mathrm{nilp}!}$, $\iota^{\mathrm{RS}!}$, $\iota^{\mathrm{ord}_k!}$), defined on $D^+(\widehat{\mathfrak{g}}_{\mathrm{crit}}\text{-mod})$.

From Lemmas 7.4 and 23.8, we obtain the following.

Lemma 7.5. *The functor* $\iota^{\mathrm{reg}!} : D^+(\widehat{\mathfrak{g}}_{\mathrm{crit}}\text{-mod}) \to D^+(\widehat{\mathfrak{g}}_{\mathrm{crit}}\text{-mod}_{\mathrm{reg}})$ *commutes in the natural sense with the forgetful functors* $D^+(\widehat{\mathfrak{g}}_{\mathrm{crit}}\text{-mod}) \to D^+(\mathfrak{Z}_\mathfrak{g}\text{-mod})$ *and* $D^+(\widehat{\mathfrak{g}}_{\mathrm{crit}}\text{-mod}_{\mathrm{reg}}) \to D^+(\mathfrak{Z}_\mathfrak{g}^{\mathrm{reg}}\text{-mod})$, *and similarly for the* nilp, RS *and* ord_k *versions.*

7.6

Now let K be a group subscheme of $G[[t]]$. Following our conventions, we will denote by $\widehat{\mathfrak{g}}_{\mathrm{crit}}\text{-mod}^K$ (respectively, $\widehat{\mathfrak{g}}_{\mathrm{crit}}\text{-mod}^K_{\mathrm{reg}}$, $\widehat{\mathfrak{g}}_{\mathrm{crit}}\text{-mod}^K_{\mathrm{nilp}}$, $\widehat{\mathfrak{g}}_{\mathrm{crit}}\text{-mod}^K_{\mathrm{RS}}$, $\widehat{\mathfrak{g}}_{\mathrm{crit}}\text{-mod}^K_{\mathrm{ord}_k}$) the corresponding abelian categories of K-equivariant objects; see Section 20.7. We will denote by $D(\widehat{\mathfrak{g}}_{\mathrm{crit}}\text{-mod})^K$ (respectively, $D(\widehat{\mathfrak{g}}_{\mathrm{crit}}\text{-mod}_{\mathrm{reg}})^K$, $D(\widehat{\mathfrak{g}}_{\mathrm{crit}}\text{-mod}_{\mathrm{nilp}})^K$, $D(\widehat{\mathfrak{g}}_{\mathrm{crit}}\text{-mod}_{\mathrm{RS}})^K$, $D(\widehat{\mathfrak{g}}_{\mathrm{crit}}\text{-mod}_{\mathrm{ord}_k})^K$) the corresponding triangulated categories.

The functors $\iota_!^{\mathrm{reg}}$ (respectively, $\iota_!^{\mathrm{nilp}}$, $\iota_!^{\mathrm{RS}}$, $\iota_!^{\mathrm{ord}_k}$) extend to the K-equivariant setting in a straightforward way. By Proposition 23.14, we have the following.

Lemma 7.7. *There exist functors* $\iota^{\mathrm{reg}!} : D(\widehat{\mathfrak{g}}_{\mathrm{crit}}\text{-mod})^K \to D(\widehat{\mathfrak{g}}_{\mathrm{crit}}\text{-mod}_{\mathrm{reg}})^K$ *(respectively,* $\iota^{\mathrm{nilp}!} : D^+(\widehat{\mathfrak{g}}_{\mathrm{crit}}\text{-mod})^K \to D^+(\widehat{\mathfrak{g}}_{\mathrm{crit}}\text{-mod}_{\mathrm{nilp}})^K$, $\iota^{\mathrm{RS}!} : D^+(\widehat{\mathfrak{g}}_{\mathrm{crit}}\text{-mod})^K \to D^+(\widehat{\mathfrak{g}}_{\mathrm{crit}}\text{-mod}_{\mathrm{RS}})^K$, $\iota^{\mathrm{ord}_k!} : D^+(\widehat{\mathfrak{g}}_{\mathrm{crit}}\text{-mod})^K \to D^+(\widehat{\mathfrak{g}}_{\mathrm{crit}}\text{-mod}_{\mathrm{ord}_k})^K$) *that are right adjoint to the functors* $\iota_!^{\mathrm{reg}}$ *(respectively,* $\iota_!^{\mathrm{nilp}}$, $\iota_!^{\mathrm{RS}}$, $\iota_!^{\mathrm{ord}_k}$*), and which commute with the forgetful functors to the corresponding derived categories* $D^+(\widehat{\mathfrak{g}}_{\mathrm{crit}}\text{-mod})$, $D^+(\widehat{\mathfrak{g}}_{\mathrm{crit}}\text{-mod}_{\mathrm{reg}})$, $D^+(\widehat{\mathfrak{g}}_{\mathrm{crit}}\text{-mod}_{\mathrm{nilp}})$, $D^+(\widehat{\mathfrak{g}}_{\mathrm{crit}}\text{-mod}_{\mathrm{RS}})$, *and* $D^+(\widehat{\mathfrak{g}}_{\mathrm{crit}}\text{-mod}_{\mathrm{ord}_k})$.

This lemma implies that if $\mathcal{M}_1, \mathcal{M}_2$ are two objects of, say $\widehat{\mathfrak{g}}_{\mathrm{crit}}\text{-mod}^K_{\mathrm{reg}}$, then there exists a spectral sequence, converging to $\mathrm{Ext}^\bullet_{D(\widehat{\mathfrak{g}}_{\mathrm{crit}}\text{-mod})^K}(\iota_!^{\mathrm{reg}}(\mathcal{M}_1), \iota_!^{\mathrm{reg}}(\mathcal{M}_2))$, and whose second term $E_2^{p,q}$ is given by

$$\mathrm{Ext}^p_{D(\widehat{\mathfrak{g}}_{\mathrm{crit}}\text{-mod}_{\mathrm{reg}})^K}(\mathcal{M}_1, \mathcal{M}_2) \underset{\mathfrak{Z}_\mathfrak{g}^{\mathrm{reg}}}{\otimes} \Lambda^q(N_{\mathfrak{Z}_\mathfrak{g}^{\mathrm{reg}}/\mathfrak{Z}_\mathfrak{g}}), \tag{7.1}$$

where $N_{\mathfrak{Z}_\mathfrak{g}^{\mathrm{reg}}/\mathfrak{Z}_\mathfrak{g}}$ denotes the normal bundle to $\mathrm{Spec}(\mathfrak{Z}_\mathfrak{g}^{\mathrm{reg}})$ inside $\mathrm{Spec}(\mathfrak{Z}_\mathfrak{g})$. The same spectral sequence exists when we replace the index reg by either of nilp, RS, or ord_k.

7.8

We shall now recall a construction related to that of the *renormalized* universal enveloping algebra at the critical level, following [BD1, Section 5.6].

The main ingredient is the action of the algebra $\mathfrak{Z}_{\mathfrak{g}}$ on $\widetilde{U}_{\text{crit}}(\widehat{\mathfrak{g}})$ by *outer* derivations. Let us recall the construction:

Let us pick a nonzero (symmetric, invariant) pairing $\kappa : \mathfrak{g} \otimes \mathfrak{g} \to \mathbb{C}$, and using it construct a one-parameter deformation of the critical pairing: $\kappa_\hbar = \kappa_{\text{crit}} + \hbar \cdot \kappa$. We obtain a one-parameter family of topological associative algebras $\widetilde{U}_\hbar(\widehat{\mathfrak{g}})$. For an element $a \in \mathfrak{Z}_{\mathfrak{g}}$, and its lift $a_\hbar \in \widetilde{U}_\hbar(\widehat{\mathfrak{g}})$ and $b \in \widetilde{U}_\hbar(\widehat{\mathfrak{g}})$, the element $[a_\hbar, b] \in \widetilde{U}_\hbar(\widehat{\mathfrak{g}})$ is 0 modulo $\hbar$.

Hence, the operation $b \mapsto \frac{[a_\hbar, b]}{\hbar} \bmod \hbar$ is a derivation of $\widetilde{U}_{\text{crit}}(\widehat{\mathfrak{g}})$. It does not depend on the choice of the lifting $a_\hbar$ up to inner derivations. This construction has the following properties.

Lemma 7.9.

(a) *The constructed map* $\mathfrak{Z}_{\mathfrak{g}} \to \text{Der}^{\text{out}}(\widetilde{U}_{\text{crit}}(\widehat{\mathfrak{g}}))$ *is a derivation, i.e., it extends to a (continuous) map of (topological)* $\mathfrak{Z}_{\mathfrak{g}}$*-modules* $\Omega^1(\mathfrak{Z}_{\mathfrak{g}}) \to \text{Der}^{\text{out}}(\widetilde{U}_{\text{crit}}(\widehat{\mathfrak{g}}))$.

(b) *Each of the above derivations preserves the subalgebra* $\mathfrak{Z}_{\mathfrak{g}} \subset \widetilde{U}_{\text{crit}}(\widehat{\mathfrak{g}})$*, i.e.,* $\mathfrak{Z}_{\mathfrak{g}}$ *is a topological Poisson algebra and* $\Omega^1(\mathfrak{Z}_{\mathfrak{g}})$ *is an algebroid over* $\text{Spec}(\mathfrak{Z}_{\mathfrak{g}})$.

The following result, which relates the Poisson algebra structure on $\mathfrak{Z}_{\mathfrak{g}}$ with Langlands duality, is crucial for this paper:

Recall from Section 4.1 that $\mathsf{Isom}_{\text{Op}_{\check{\mathfrak{g}}}}$ denotes the groupoid over the ind-scheme $\text{Op}_{\check{\mathfrak{g}}}(\mathcal{D}^\times)$, whose fiber over $\chi, \chi' \in \text{Op}_{\check{\mathfrak{g}}}(\mathcal{D}^\times)$ is the scheme of isomorphisms of $\check{G}$-local systems on $\mathcal{D}^\times$, corresponding to χ and χ', respectively, and $\mathsf{isom}_{\text{Op}_{\check{\mathfrak{g}}}}$ denotes its algebroid. One of the key properties of the isomorphism Theorem 5.2, proved in [FF3, F], is that it respects the Poisson structures. In other words, in terms of the corresponding Lie algebroids (see Section 4.1), we have the following.

Theorem 7.10. *Under the isomorphism* $\mathfrak{Z}_{\mathfrak{g}} \simeq \text{Fun}(\text{Op}_{\check{\mathfrak{g}}}(D^\times))$*, we have a canonical identification of Lie algebroids* $\Omega^1(\mathfrak{Z}_{\mathfrak{g}}) \simeq \mathsf{isom}_{\text{Op}_{\check{\mathfrak{g}}}}$.

Let us now derive some consequences from the construction described above. By Lemma 4.9 and its variant for the nilp, RS and ord_k cases, we obtain the following.

Corollary 7.11. *The ideal of each of the quotient algebras* $\mathfrak{Z}_{\mathfrak{g}}^{\text{reg}}$ *(respectively,* $\mathfrak{Z}_{\mathfrak{g}}^{\text{nilp}}$, $\mathfrak{Z}_{\mathfrak{g}}^{\text{RS}}$, $\mathfrak{Z}_{\mathfrak{g}}^{\text{ord}_k}$*) is stable under the Poisson bracket, i.e.,* $N^*_{\mathfrak{Z}_{\mathfrak{g}}^{\text{reg}}/\mathfrak{Z}_{\mathfrak{g}}}$ *(respectively,* $N^*_{\mathfrak{Z}_{\mathfrak{g}}^{\text{nilp}}/\mathfrak{Z}_{\mathfrak{g}}}$, $N^*_{\mathfrak{Z}_{\mathfrak{g}}^{\text{RS}}/\mathfrak{Z}_{\mathfrak{g}}}$, $N^*_{\mathfrak{Z}_{\mathfrak{g}}^{\text{ord}_k}/\mathfrak{Z}_{\mathfrak{g}}}$*) is an algebroid over the corresponding algebra.*

Let us observe that for any $\widehat{\mathfrak{g}}_{\text{crit}}$-module $\mathcal{M}$ we obtain a map

$$\Omega^1(\mathfrak{Z}_{\mathfrak{g}}) \to \text{Ext}^1_{\widehat{\mathfrak{g}}_{\text{crit}}\text{-mod}}(\mathcal{M}, \mathcal{M}). \tag{7.2}$$

This map is functorial in the sense that for a morphism of $\widehat{\mathfrak{g}}_{\text{crit}}$-modules $\mathcal{M} \to \mathcal{M}'$, the two compositions

$$\Omega^1(\mathfrak{Z}_{\mathfrak{g}}) \to \mathrm{Ext}^1_{\widehat{\mathfrak{g}}_{\mathrm{crit}}\text{-mod}}(\mathcal{M}, \mathcal{M}) \to \mathrm{Ext}^1_{\widehat{\mathfrak{g}}_{\mathrm{crit}}\text{-mod}}(\mathcal{M}, \mathcal{M}')$$

and

$$\Omega^1(\mathfrak{Z}_{\mathfrak{g}}) \to \mathrm{Ext}^1_{\widehat{\mathfrak{g}}_{\mathrm{crit}}\text{-mod}}(\mathcal{M}', \mathcal{M}') \to \mathrm{Ext}^1_{\widehat{\mathfrak{g}}_{\mathrm{crit}}\text{-mod}}(\mathcal{M}, \mathcal{M}')$$

coincide.

The next series of remarks is stated for the subscheme $\mathrm{Spec}(\mathfrak{Z}_{\mathfrak{g}}^{\mathrm{reg}}) \subset \mathrm{Spec}(\mathfrak{Z}_{\mathfrak{g}})$; however, they equally apply to the cases when reg is replaced by either of nilp, RS or ord_k.

Note that the Poisson structure, viewed as a map $\Omega^1(\mathfrak{Z}_{\mathfrak{g}}) \to T(\mathfrak{Z}_{\mathfrak{g}})$, gives rise to a commutative diagram:

$$\begin{array}{ccccccccc} 0 & \longrightarrow & N^*_{\mathfrak{Z}_{\mathfrak{g}}^{\mathrm{reg}}/\mathfrak{Z}_{\mathfrak{g}}} & \longrightarrow & \Omega^1(\mathfrak{Z}_{\mathfrak{g}})|_{\mathrm{Spec}(\mathfrak{Z}_{\mathfrak{g}}^{\mathrm{reg}})} & \longrightarrow & \Omega^1(\mathfrak{Z}_{\mathfrak{g}}^{\mathrm{reg}}) & \longrightarrow & 0 \\ & & \downarrow & & \downarrow & & \downarrow & & \\ 0 & \longrightarrow & T(\mathfrak{Z}_{\mathfrak{g}}^{\mathrm{reg}}) & \longrightarrow & T(\mathfrak{Z}_{\mathfrak{g}})|_{\mathrm{Spec}(\mathfrak{Z}_{\mathfrak{g}}^{\mathrm{reg}})} & \longrightarrow & N_{\mathfrak{Z}_{\mathfrak{g}}^{\mathrm{reg}}/\mathfrak{Z}_{\mathfrak{g}}} & \longrightarrow & 0. \end{array}$$

Let $\mathcal{M}$ and $\mathcal{M}'$ be two objects of $\widehat{\mathfrak{g}}_{\mathrm{crit}}\text{-mod}_{\mathrm{reg}}$. Note that we have an exact sequence

$$\begin{aligned} 0 \to \mathrm{Ext}^1_{\widehat{\mathfrak{g}}_{\mathrm{crit}}\text{-mod}_{\mathrm{reg}}}(\mathcal{M}, \mathcal{M}') \to \mathrm{Ext}^1_{\widehat{\mathfrak{g}}_{\mathrm{crit}}\text{-mod}}(\mathcal{M}, \mathcal{M}') \to \\ \to \mathrm{Hom}(\mathcal{M}, \mathcal{M}') \underset{\mathfrak{Z}_{\mathfrak{g}}^{\mathrm{reg}}}{\otimes} N_{\mathfrak{Z}_{\mathfrak{g}}^{\mathrm{reg}}/\mathfrak{Z}_{\mathfrak{g}}} \to \mathrm{Ext}^2_{\widehat{\mathfrak{g}}_{\mathrm{crit}}\text{-mod}_{\mathrm{reg}}}(\mathcal{M}, \mathcal{M}'). \end{aligned}$$

It is easy to see that the composed map

$$\mathrm{Hom}(\mathcal{M}, \mathcal{M}') \underset{\mathfrak{Z}_{\mathfrak{g}}}{\otimes} \Omega^1(\mathfrak{Z}_{\mathfrak{g}}) \to \mathrm{Ext}^1_{\widehat{\mathfrak{g}}_{\mathrm{crit}}\text{-mod}}(\mathcal{M}, \mathcal{M}') \to \mathrm{Hom}(\mathcal{M}, \mathcal{M}') \underset{\mathfrak{Z}_{\mathfrak{g}}^{\mathrm{reg}}}{\otimes} N_{\mathfrak{Z}_{\mathfrak{g}}^{\mathrm{reg}}/\mathfrak{Z}_{\mathfrak{g}}}$$

comes from the map $\Omega^1(\mathfrak{Z}_{\mathfrak{g}}) \to N_{\mathfrak{Z}_{\mathfrak{g}}^{\mathrm{reg}}/\mathfrak{Z}_{\mathfrak{g}}}$ from the above commutative diagram. Thus we obtain the following commutative diagram:

$$\begin{array}{ccc} \mathrm{Hom}(\mathcal{M}, \mathcal{M}') \underset{\mathfrak{Z}_{\mathfrak{g}}^{\mathrm{reg}}}{\otimes} N^*_{\mathfrak{Z}_{\mathfrak{g}}^{\mathrm{reg}}/\mathfrak{Z}_{\mathfrak{g}}} & \longrightarrow & \mathrm{Ext}^1_{\widehat{\mathfrak{g}}_{\mathrm{crit}}\text{-mod}_{\mathrm{reg}}}(\mathcal{M}, \mathcal{M}') \\ \downarrow & & \downarrow \\ \mathrm{Hom}(\mathcal{M}, \mathcal{M}') \underset{\mathfrak{Z}_{\mathfrak{g}}^{\mathrm{reg}}}{\otimes} \Omega^1(\mathfrak{Z}_{\mathfrak{g}})|_{\mathrm{Spec}(\mathfrak{Z}_{\mathfrak{g}}^{\mathrm{reg}})} & \longrightarrow & \mathrm{Ext}^1_{\widehat{\mathfrak{g}}_{\mathrm{crit}}\text{-mod}}(\mathcal{M}, \mathcal{M}') \\ \downarrow & & \downarrow \\ \mathrm{Hom}(\mathcal{M}, \mathcal{M}') \underset{\mathfrak{Z}_{\mathfrak{g}}^{\mathrm{reg}}}{\otimes} \Omega^1(\mathfrak{Z}_{\mathfrak{g}}^{\mathrm{reg}}) & \longrightarrow & \mathrm{Hom}(\mathcal{M}, \mathcal{M}') \underset{\mathfrak{Z}_{\mathfrak{g}}^{\mathrm{reg}}}{\otimes} N_{\mathfrak{Z}_{\mathfrak{g}}^{\mathrm{reg}}/\mathfrak{Z}_{\mathfrak{g}}} \end{array} \tag{7.3}$$

and a natural map

$$\mathrm{Hom}(\mathcal{M}, \mathcal{M}') \underset{\mathfrak{Z}_{\mathfrak{g}}^{\mathrm{reg}}}{\otimes} \left(N_{\mathfrak{Z}_{\mathfrak{g}}^{\mathrm{reg}}/\mathfrak{Z}_{\mathfrak{g}}} / \Omega^1(\mathfrak{Z}_{\mathfrak{g}}^{\mathrm{reg}})\right) \to \mathrm{Ext}^2_{\widehat{\mathfrak{g}}_{\mathrm{crit}}\text{-mod}_{\mathrm{reg}}}(\mathcal{M}, \mathcal{M}'). \tag{7.4}$$

Let us now consider once again the family $\widetilde{U}_\hbar(\widehat{\mathfrak{g}})$, and inside $\widetilde{U}_\hbar(\widehat{\mathfrak{g}}) \underset{\mathbb{C}[[\hbar]]}{\otimes} \mathbb{C}((\hbar))$ consider the $\mathbb{C}[[\hbar]]$-subalgebra generated by $\widetilde{U}_\hbar(\widehat{\mathfrak{g}})$ and elements of the form

$$\frac{a_\hbar}{\hbar} \quad \text{for } a_\hbar \bmod \hbar \in \ker(\mathfrak{z}_{\mathfrak{g}} \to \mathfrak{z}_{\mathfrak{g}}^{\mathrm{reg}}).$$

Taking this algebra modulo $\hbar$, we obtain an algebra, denoted $U^{\mathrm{ren,reg}}(\widehat{\mathfrak{g}}_{\mathrm{crit}})$, and called the renormalized enveloping algebra at the critical level. The algebra $U^{\mathrm{ren,reg}}(\widehat{\mathfrak{g}}_{\mathrm{crit}})$ has a natural filtration, with the 0th term isomorphic to $\widetilde{U}_{\mathrm{crit}}(\widehat{\mathfrak{g}}) \underset{\mathfrak{z}_{\mathfrak{g}}}{\otimes} \mathfrak{z}_{\mathfrak{g}}^{\mathrm{reg}}$, and the first associated graded quotient isomorphic to

$$\left(\widetilde{U}_{\mathrm{crit}}(\widehat{\mathfrak{g}}) \underset{\mathfrak{z}_{\mathfrak{g}}}{\otimes} \mathfrak{z}_{\mathfrak{g}}^{\mathrm{reg}}\right) \underset{\mathfrak{z}_{\mathfrak{g}}^{\mathrm{reg}}}{\widehat{\otimes}} N^*_{\mathfrak{z}_{\mathfrak{g}}^{\mathrm{reg}}/\mathfrak{z}_{\mathfrak{g}}}.$$

Let $U^{\mathrm{ren,reg}}(\widehat{\mathfrak{g}}_{\mathrm{crit}})$-mod denote the category of (discrete) $U^{\mathrm{ren,reg}}(\widehat{\mathfrak{g}}_{\mathrm{crit}})$-modules. We have a tautological homomorphism $\widetilde{U}_{\mathrm{crit}}(\widehat{\mathfrak{g}}) \to U^{\mathrm{ren,reg}}(\widehat{\mathfrak{g}}_{\mathrm{crit}})$, whose restriction to $\mathfrak{z}_{\mathfrak{g}}$ factors through $\mathfrak{z}_{\mathfrak{g}}^{\mathrm{reg}}$; thus we have a restriction functor $U^{\mathrm{ren,reg}}(\widehat{\mathfrak{g}}_{\mathrm{crit}})\text{-mod} \to \widehat{\mathfrak{g}}_{\mathrm{crit}}\text{-mod}_{\mathrm{reg}}$. In addition, the adjoint action of the algebra $\widetilde{U}_{\mathrm{crit}}(\widehat{\mathfrak{g}}) \underset{\mathfrak{z}_{\mathfrak{g}}}{\otimes} \mathfrak{z}_{\mathfrak{g}}^{\mathrm{reg}}$ on itself extends to an action of the first term of the above-mentioned filtration on $U^{\mathrm{ren,reg}}(\widehat{\mathfrak{g}}_{\mathrm{crit}})$.

Now let $\mathcal{M}_\hbar$ be an $\hbar$-family of modules over $\widehat{\mathfrak{g}}_\hbar$ such that the action of $\mathfrak{z}_{\mathfrak{g}}$ on $\mathcal{M} := \mathcal{M}_\hbar/\hbar \cdot \mathcal{M}_\hbar$ factors through $\mathfrak{z}_{\mathfrak{g}}^{\mathrm{reg}}$. Then $\mathcal{M}$ is naturally acted on by $U^{\mathrm{ren,reg}}(\widehat{\mathfrak{g}}_{\mathrm{crit}})$. This construction provides a supply of objects of $U^{\mathrm{ren,reg}}(\widehat{\mathfrak{g}}_{\mathrm{crit}})$-mod.

Lemma 7.12. *Let $\mathcal{M}, \mathcal{M}'$ be $U^{\mathrm{ren,reg}}(\widehat{\mathfrak{g}}_{\mathrm{crit}})$-modules. Then*

(a) *The map* $\mathrm{Hom}_{\widehat{\mathfrak{g}}_{\mathrm{crit}}\text{-mod}}(\mathcal{M}, \mathcal{M}') \underset{\mathfrak{z}_{\mathfrak{g}}^{\mathrm{reg}}}{\otimes} N^*_{\mathfrak{z}_{\mathfrak{g}}^{\mathrm{reg}}/\mathfrak{z}_{\mathfrak{g}}} \to \mathrm{Ext}^1_{\widehat{\mathfrak{g}}_{\mathrm{crit}}\text{-mod}_{\mathrm{reg}}}(\mathcal{M}, \mathcal{M}')$ *vanishes.*

(b) *We have a natural action of the algebroid* $N^*_{\mathfrak{z}_{\mathfrak{g}}^{\mathrm{reg}}/\mathfrak{z}_{\mathfrak{g}}}$ *on* $\mathrm{Ext}^\bullet_{\widehat{\mathfrak{g}}_{\mathrm{crit}}\text{-mod}_{\mathrm{reg}}}(\mathcal{M}, \mathcal{M}')$.

Finally, let us note that the category of $U^{\mathrm{ren,reg}}(\widehat{\mathfrak{g}}_{\mathrm{crit}})$-modules carries a Harish-Chandra action of $G((t))$. In particular, if K is a group subscheme of $G[[t]]$, we can introduce the categories $U^{\mathrm{ren,reg}}(\widehat{\mathfrak{g}}_{\mathrm{crit}})\text{-mod}^K$ and $D(U^{\mathrm{ren,reg}}(\widehat{\mathfrak{g}}_{\mathrm{crit}}))^K$. In addition, analogues of the diagrams appearing above remain valid for

$$\mathrm{Ext}^\bullet_{\widehat{\mathfrak{g}}_{\mathrm{crit}}\text{-mod}_{\mathrm{reg}}}(\mathcal{M}, \mathcal{M}') \quad \text{and} \quad \mathrm{Ext}^\bullet_{\widehat{\mathfrak{g}}_{\mathrm{crit}}\text{-mod}}(\mathcal{M}, \mathcal{M}')$$

replaced by

$$\mathrm{Ext}^\bullet_{D(\widehat{\mathfrak{g}}_{\mathrm{crit}}\text{-mod}_{\mathrm{reg}})^K}(\mathcal{M}, \mathcal{M}') \quad \text{and} \quad \mathrm{Ext}^\bullet_{D(\widehat{\mathfrak{g}}_{\mathrm{crit}}\text{-mod})^K}(\mathcal{M}, \mathcal{M}'),$$

respectively.

7.13

For the rest of this section we will be concerned with the category $\widehat{\mathfrak{g}}_{\rm crit}\text{-mod}_{\rm ord_1}$, denoted also by $\widehat{\mathfrak{g}}_{\rm crit}\text{-mod}_{\rm RS}$.

Now consider the functor $\mathfrak{g}\text{-mod} \to \widehat{\mathfrak{g}}_{\rm crit}\text{-mod}$ given by

$$M \mapsto \operatorname{Ind}^{\widehat{\mathfrak{g}}_{\rm crit}}_{\mathfrak{g}[[t]]\oplus\mathbb{C}\mathbf{1}}(M), \tag{7.5}$$

where $\mathfrak{g}[[t]]$ acts on M via the evaluation map $\mathfrak{g}[[t]] \to \mathfrak{g}$ and $\mathbf{1}$ acts as the identity.

By definition.

$$\operatorname{Ind}^{\widehat{\mathfrak{g}}_{\rm crit}}_{\mathfrak{g}[[t]]\oplus\mathbb{C}\mathbf{1}}(U(\mathfrak{g})) \simeq \operatorname{Ind}^{\widehat{\mathfrak{g}}_{\rm crit}}_{t\mathfrak{g}[[t]]\oplus\mathbb{C}\mathbf{1}}(\mathbb{C}),$$

and by Theorem 7.3, this module belongs to $\widehat{\mathfrak{g}}_{\rm crit}\text{-mod}_{\rm RS}$. This implies that the module $\operatorname{Ind}^{\widehat{\mathfrak{g}}_{\rm crit}}_{\mathfrak{g}[[t]]\oplus\mathbb{C}\mathbf{1}}(M) \in \widehat{\mathfrak{g}}_{\rm crit}\text{-mod}_{\rm RS}$ for any M.

In what follows we will need the following technical assertions, in which we use the notion of quasi-perfectness introduced in Section 19.20.

Proposition 7.14.

(1) *Representations of the form* $\operatorname{Ind}^{\widehat{\mathfrak{g}}_{\rm crit}}_{\mathfrak{g}[[t]]\oplus\mathbb{C}\mathbf{1}}(M)$ *for* $M \in \mathfrak{g}\text{-mod}$ *are quasi-perfect as objects of* $D(\widehat{\mathfrak{g}}_{\rm crit}\text{-mod})$.

(2) *Any object* $\mathcal{M} \in \widehat{\mathfrak{g}}_{\rm crit}\text{-mod}_{\rm RS}$, *which is quasi-perfect in* $\widehat{\mathfrak{g}}_{\rm crit}$-mod, *is also quasi-perfect in* $\widehat{\mathfrak{g}}_{\rm crit}\text{-mod}_{\rm RS}$. *The same is true when the RS condition is replaced by any of* ord_k, nilp, *or* reg.

Proof. Since the induction functor is exact, by adjunction,

$$\operatorname{Hom}_{D(\mathfrak{g}_{\rm crit}\text{-mod})}(\operatorname{Ind}^{\widehat{\mathfrak{g}}_{\rm crit}}_{\mathfrak{g}[[t]]\oplus\mathbb{C}\mathbf{1}}(M), \mathcal{M}^\bullet_1) \simeq \operatorname{Hom}_{D(\mathfrak{g}[[t]]\text{-mod})}(M, \mathcal{M}^\bullet_1).$$

When $\mathcal{M}^\bullet_1$ is bounded from below the latter is computed by the standard cohomological complex of $\mathfrak{g}[[t]]$ (see Section 19.17), which manifestly commutes with direct sums. This proves the first point of the proposition.

The second point follows from Proposition 23.12. □

7.15

Denote by $\mathbb{M}_\lambda$ (respectively, $\mathbb{M}^\vee_\lambda$, $\mathbb{L}_\lambda$) the $\widehat{\mathfrak{g}}_{\rm crit}$-module induced from the Verma module M_λ (respectively, the contragredient Verma module $M^\vee_\lambda$, the irreducible module L_λ) with highest weight λ over $\mathfrak{g}$:

$$\mathbb{M}_\lambda = \operatorname{Ind}^{\widehat{\mathfrak{g}}_{\rm crit}}_{\mathfrak{g}[[t]]\oplus\mathbb{C}\mathbf{1}}(M_\lambda), \quad \mathbb{M}^\vee_\lambda = \operatorname{Ind}^{\widehat{\mathfrak{g}}_{\rm crit}}_{\mathfrak{g}[[t]]\oplus\mathbb{C}\mathbf{1}}(M^\vee_\lambda), \quad \mathbb{L}_\lambda = \operatorname{Ind}^{\widehat{\mathfrak{g}}_{\rm crit}}_{\mathfrak{g}[[t]]\oplus\mathbb{C}\mathbf{1}}(L_\lambda).$$

Recall that we have the natural residue map $\mathrm{Res}^{\rm RS} : \mathrm{Op}^{\rm RS}_{\check{\mathfrak{g}}} \to \check{\mathfrak{h}}/\!/W \simeq \mathfrak{h}^*/\!/W$. At the level of algebras of functions we therefore have a map

$$\mathrm{Res}^{\rm RS\,*} : \mathrm{Sym}(\mathfrak{h})^W \to \mathfrak{z}^{\rm RS}_{\mathfrak{g}}. \tag{7.6}$$

Thus, for every $M \in \mathfrak{g}\text{-mod}$ we obtain two a priori different actions of $\mathrm{Sym}(\mathfrak{h})^W$ on $\operatorname{Ind}^{\widehat{\mathfrak{g}}_{\rm crit}}_{\mathfrak{g}[[t]]\oplus\mathbb{C}\mathbf{1}}(M)$:

One action corresponds to the map $\mathrm{Res}^{\mathrm{RS}\,*}$ and the action of $\mathfrak{z}^{\mathrm{RS}}_{\mathfrak{g}}$ on objects of $\widehat{\mathfrak{g}}_{\mathrm{crit}}\text{-mod}_{\mathrm{RS}}$. Another action comes from the Harish-Chandra isomorphism

$$\mathrm{Sym}(\mathfrak{h})^W \simeq Z(U(\mathfrak{g})), \tag{7.7}$$

(which we normalize so that the central character of M_λ equals $\varpi(\lambda+\rho)$), the action of $Z(U(\mathfrak{g}))$ by endomorphisms on M, and, hence, the induced action on $\mathrm{Ind}^{\widehat{\mathfrak{g}}_{\mathrm{crit}}}_{\mathfrak{g}[[t]]\oplus\mathbb{C}\mathbf{1}}(M)$, by functoriality.

Let τ be the involution of $Z(U(\mathfrak{g}))$, induced by the anti-involution $x \mapsto -x$ of $U(\mathfrak{g})$. Alternatively, τ can be thought of as induced by the outer involution of $\mathfrak{g}$ that acts on the weights as $\lambda \mapsto -w_0(\lambda)$.

Proposition 7.16. *The above two actions of* $\mathrm{Sym}(\mathfrak{h})^W$ *on* $\mathrm{Ind}^{\widehat{\mathfrak{g}}_{\mathrm{crit}}}_{\mathfrak{g}[[t]]\oplus\mathbb{C}\mathbf{1}}(M)$ *differ by* τ.

Proof. It is enough to consider the universal example of $M = U(\mathfrak{g})$. In the course of the proof of the proposition we will essentially reprove Theorem 7.3.

Consider the grading on $\widehat{\mathfrak{g}}_{\mathrm{crit}}$ induced by the $\mathbb{G}_m$-action on $\mathcal{D}$ by loop rotations. Then all our objects, such as $\mathfrak{z}_{\mathfrak{g}}$, $\mathfrak{z}^{\mathrm{RS}}_{\mathfrak{g}}$ and $\mathrm{Ind}^{\widehat{\mathfrak{g}}_{\mathrm{crit}}}_{t\mathfrak{g}[[t]]\oplus\mathbb{C}\mathbf{1}}(\mathbb{C})$, acquire a natural grading; the degree i subspace will be denoted by the subscript i, i.e., $(\cdot)_i$.

Consider the ideal $\mathfrak{z}_{\mathfrak{g}}\cdot(\mathfrak{z}_{\mathfrak{g}})_{>0}$ in $\mathfrak{z}_{\mathfrak{g}}$ generated by elements of positive degree. From Section 1.9, we know that the quotient $\mathfrak{z}_{\mathfrak{g}}/\mathfrak{z}_{\mathfrak{g}}\cdot(\mathfrak{z}_{\mathfrak{g}})_{>0}$ is precisely $\mathfrak{z}^{\mathrm{RS}}_{\mathfrak{g}}$. Since the grading on $\mathrm{Ind}^{\widehat{\mathfrak{g}}_{\mathrm{crit}}}_{t\mathfrak{g}[[t]]\oplus\mathbb{C}\mathbf{1}}(\mathbb{C})$ is nonpositive and the module is generated by the subspace of degree 0, the above ideal annihilates this module.

Now consider the subalgebra of degree 0 elements $(\mathfrak{z}_{\mathfrak{g}}/\mathfrak{z}_{\mathfrak{g}}\cdot(\mathfrak{z}_{\mathfrak{g}})_{>0})_0 \subset \mathfrak{z}^{\mathrm{RS}}_{\mathfrak{g}}$. According to Section 1.9, it is isomorphic to $\mathrm{Sym}(\mathfrak{h})^W$ and the resulting embedding

$$\mathrm{Sym}(\mathfrak{h})^W \to \mathfrak{z}^{\mathrm{RS}}_{\mathfrak{g}} \tag{7.8}$$

is the homomorphism $\mathrm{Res}^{\mathrm{RS}\,*}$.

The action of $(\mathfrak{z}_{\mathfrak{g}}/\mathfrak{z}_{\mathfrak{g}}\cdot(\mathfrak{z}_{\mathfrak{g}})_{>0})_0$ on $\mathrm{Ind}^{\widehat{\mathfrak{g}}_{\mathrm{crit}}}_{t\mathfrak{g}[[t]]\oplus\mathbb{C}\mathbf{1}}(\mathbb{C})$ preserves the subspace of degree 0 elements. However, the latter subspace is isomorphic to $U(\mathfrak{g})$. Therefore, $(\mathfrak{z}_{\mathfrak{g}}/\mathfrak{z}_{\mathfrak{g}}\cdot(\mathfrak{z}_{\mathfrak{g}})_{>0})_0$ acts on $U(\mathfrak{g})$ commuting with both the left and right module structure; hence it comes from a homomorphism $(\mathfrak{z}_{\mathfrak{g}}/\mathfrak{z}_{\mathfrak{g}}\cdot(\mathfrak{z}_{\mathfrak{g}})_{>0})_0 \to Z(U(\mathfrak{g}))$.

It remains to compare the resulting homomorphism

$$\mathrm{Sym}(\mathfrak{h})^W \to (\mathfrak{z}_{\mathfrak{g}}/\mathfrak{z}_{\mathfrak{g}}\cdot(\mathfrak{z}_{\mathfrak{g}})_{>0})_0 \to Z(U(\mathfrak{g}))$$

with the Harish-Chandra isomorphism. This has been proved in [F, Section 12.6]. Let us repeat the argument for completeness:

It is enough to show that for any weight $\lambda \in \mathfrak{h}^*$, the two characters, corresponding to $\mathrm{Sym}(\mathfrak{h})^W$ acting in the two ways on the module $\mathbb{M}^\vee_\lambda$, coincide.

Let $\mathbb{W}^{w_0}_{\mathrm{crit},\lambda}$ be the Wakimoto module corresponding to the weight λ, as in Section 11.5. By Lemma 13.2, the character of $\mathrm{Sym}(\mathfrak{h})^W$, acting on $\mathbb{W}^{w_0}_{\mathrm{crit},\lambda}$ via (7.8), is given by $\varpi(-\lambda-\rho)$.

By Section 11.5, we have have a nontrivial homomorphism $\mathbb{M}^\vee_\lambda \to \mathbb{W}^{w_0}_{\mathrm{crit},\lambda}$, and hence the center $\mathfrak{z}_{\mathfrak{g}}$ acts on both modules by the same character. □

Recall that for $\chi \in \mathfrak{h}^*//W \simeq \check{\mathfrak{h}}//W$ we have a subscheme $\mathrm{Op}_{\check{\mathfrak{g}}}^{\mathrm{RS},\chi} \subset \mathrm{Op}_{\check{\mathfrak{g}}}^{\mathrm{RS}}$; if $\mu \in \mathfrak{h}^*$ is integral and antidominant, then $\mathrm{Op}_{\check{\mathfrak{g}}}^{\mathrm{RS},\varpi(\mu)} \simeq \mathrm{Op}_{\check{\mathfrak{g}}}^{-\mu-\rho,\mathrm{nilp}}$; if, moreover, $\mu+\rho$ is antidominant, then the latter scheme contains the subscheme $\mathrm{Op}_{\check{\mathfrak{g}}}^{-\mu-\rho,\mathrm{reg}}$.

Let us denote by $\mathfrak{Z}_{\mathfrak{g}}^{\mathrm{RS},\chi}$, $\mathfrak{Z}_{\mathfrak{g}}^{-\mu-\rho,\mathrm{nilp}}$ and $\mathfrak{Z}_{\mathfrak{g}}^{-\mu-\rho,\mathrm{reg}}$, respectively, the corresponding quotients of $\mathfrak{Z}_{\mathfrak{g}}$. Let $\widehat{\mathfrak{g}}_{\mathrm{crit}}\text{-mod}_{\mathrm{RS},\chi}$, $\widehat{\mathfrak{g}}_{\mathrm{crit}}\text{-mod}_{-\mu-\rho,\mathrm{nilp}}$, $\widehat{\mathfrak{g}}_{\mathrm{crit}}\text{-mod}_{-\mu-\rho,\mathrm{reg}}$ be the corresponding subcategories of $\widehat{\mathfrak{g}}_{\mathrm{crit}}$-mod. The general results stated in this section, concerning the behavior of $\widehat{\mathfrak{g}}_{\mathrm{crit}}\text{-mod}_{\mathrm{reg}}$, $\widehat{\mathfrak{g}}_{\mathrm{crit}}\text{-mod}_{\mathrm{nilp}}$, $\widehat{\mathfrak{g}}_{\mathrm{crit}}\text{-mod}_{\mathrm{RS}}$ and $\widehat{\mathfrak{g}}_{\mathrm{crit}}\text{-mod}_{\mathrm{ord}_k}$, are equally applicable to $\widehat{\mathfrak{g}}_{\mathrm{crit}}\text{-mod}_{\mathrm{RS},\chi}$, $\widehat{\mathfrak{g}}_{\mathrm{crit}}\text{-mod}_{-\mu-\rho,\mathrm{nilp}}$ and $\widehat{\mathfrak{g}}_{\mathrm{crit}}\text{-mod}_{-\mu-\rho,\mathrm{reg}}$.

From Proposition 7.16, we obtain the following.

Corollary 7.17. *The modules $\mathbb{M}_\lambda$, $\mathbb{M}_\lambda^\vee$ and $\mathbb{L}_\lambda$ belong to $\widehat{\mathfrak{g}}_{\mathrm{crit}}\text{-mod}_{\mathrm{RS},\varpi(-\lambda-\rho)}$.*

For a dominant integral weight λ, let V^λ be the corresponding irreducible finite-dimensional $\mathfrak{g}$-module. Let $\mathbb{V}_{\mathrm{crit}}^\lambda$ denote the corresponding induced module at the critical level. In Section 13.7 we will also establish the following.

Proposition 7.18. *The module $\mathbb{V}_{\mathrm{crit}}^\lambda$ belongs to $\widehat{\mathfrak{g}}_{\mathrm{crit}}\text{-mod}_{\lambda,\mathrm{reg}}$.*

7.19

Recall now that the subscheme $\mathrm{Op}_{\check{\mathfrak{g}}}^{\mathrm{nilp}} \subset \mathrm{Op}_{\check{\mathfrak{g}}}^{\mathrm{RS}}$ is the preimage of $\varpi(-\rho) \in \mathfrak{h}^*//W$ under the map $\mathrm{res} : \mathrm{Op}_{\check{\mathfrak{g}}}^{\mathrm{RS}} \to \mathfrak{h}^*//W$.

In particular, if we denote by $\mathcal{O}_0$ the subcategory of the usual category $\mathcal{O}$ corresponding to $\mathfrak{g}$-modules with central character equal to $\varpi(-\rho)$, we obtain that the induction (7.5) defines a functor $\mathcal{O}_0 \to \widehat{\mathfrak{g}}_{\mathrm{crit}}\text{-mod}_{\mathrm{nilp}}^{I,m}$. In particular, the modules $\mathbb{M}_{w(\rho)-\rho}$, $\mathbb{M}_{w(\rho)-\rho}^\vee$ for $w \in W$ all belong to $\widehat{\mathfrak{g}}_{\mathrm{crit}}\text{-mod}_{\mathrm{nilp}}^{I,m}$.

In what follows we will consider sections of right D-modules on the affine flag variety Fl_G. Instead of ordinary right D-modules, we will consider the ones twisted by a line bundle, which is the tensor product of the critical line bundle on Gr_G and the $G((t))$-equivariant line bundle, corresponding to the weight 2ρ (this choice is such that the twisting induced on $G/B \subset \mathrm{Fl}_G$ corresponds to *left* D-modules on G/B.)

We will denote the resulting category by $\mathfrak{D}(\mathrm{Fl}_G)_{\mathrm{crit}}$-mod, and by a slight abuse of language we will continue to call its objects D-modules. Of course, as an abstract category $\mathfrak{D}(\mathrm{Fl}_G)_{\mathrm{crit}}$-mod is equivalent to $\mathfrak{D}(\mathrm{Fl}_G)$-mod, but the functor of global sections is different. We have

$$R\Gamma : D^+(\mathfrak{D}(\mathrm{Fl}_G)_{\mathrm{crit}}\text{-mod}) \to D^+(\widehat{\mathfrak{g}}_{\mathrm{crit}}\text{-mod}).$$

In particular, $\Gamma(\mathrm{Fl}_G, \delta_{1_{\mathrm{Fl}_G}}) \simeq \mathbb{M}_{-2\rho}$.

As usual, if K is a subgroup of $G[[t]]$, we will denote by $\mathfrak{D}(\mathrm{Fl}_G)_{\mathrm{crit}}\text{-mod}^K$ the abelian category of K-equivariant D-modules, and by $D(\mathfrak{D}(\mathrm{Fl}_G)_{\mathrm{crit}}\text{-mod}^K)^K$ the corresponding triangulated category.

For $\mathcal{F}^\bullet \in D^+(\mathfrak{D}(\mathrm{Fl}_G)_{\mathrm{crit}}\text{-mod})$, we have

$$\Gamma(\mathrm{Fl}_G, \mathcal{F}^\bullet) \simeq \mathcal{F}^\bullet \star \mathbb{M}_{-2\rho}.$$

Hence, we obtain the following.

Corollary 7.20. *The functor of global sections gives rise to a functor*

$$D^+ (\mathfrak{D}(\mathrm{Fl}_G)_{\mathrm{crit}}\text{-mod}) \to D^+(\widehat{\mathfrak{g}}_{\mathrm{crit}}\text{-mod}_{\mathrm{nilp}}).$$

7.21

Let us now prove Lemma 5.4.

Proof. Let $\mathcal{M}$ be an I^0-integrable module. Then it admits a filtration whose subquotients are quotients of modules of the form $\mathrm{Ind}_{\mathfrak{g}[[t]]\oplus\mathbb{C}\mathbf{1}}^{\widehat{\mathfrak{g}}_{\mathrm{crit}}}(M)$, where M is an N-integrable $\mathfrak{g}$-module,

If we impose the condition that $\mathcal{M} \in \widehat{\mathfrak{g}}_{\mathrm{crit}}\text{-mod}_{\mathrm{nilp}}$, then by Proposition 7.16, we can assume that the above M has central character $\varpi(-\rho)$. But, as is well known, this implies that $M \in \mathcal{O}_0$. □

In addition, we have the following result.

Lemma 7.22. *Any object* $\mathcal{M} \in \widehat{\mathfrak{g}}_{\mathrm{crit}}\text{-mod}_{\mathrm{nilp}}^{I,m}$ *admits a nonzero map* $\mathbb{L}_{w(\rho)-\rho} \to \mathcal{M}$.

Proof. By definition, any $\mathcal{M}$ contains a vector annihilated by $\mathrm{Lie}(I^0)$, and which is an eigenvector of $\mathfrak{h}$. Hence, we have a nontrivial map $\mathbb{M}_\lambda \to \mathcal{M}$. By Proposition 7.16, λ must be of the form $w(\rho) - \rho$ for some $w \in W$.

The Verma module $M_{w(\rho)-\rho}$ admits a filtration whose subquotients are the irreducibles $L_{w'(\rho)-\rho}$, $w' \geq w$. Since the induction functor is exact, $\mathbb{M}_{w(\rho)-\rho}$ admits a filtration with subquotients isomorphic to $\mathbb{M}_{w'(\rho)-\rho}$.

Let w' be the maximal element such that the corresponding term of the filtration on $\mathbb{M}_{w(\rho)-\rho}$ maps nontrivially to $\mathcal{M}$. This gives the desired map. □

8 The case of regular opers

8.1

Recall that the preimage of $\mathrm{pt}/\check{B} \hookrightarrow \check{\mathfrak{n}}/\check{B}$ under $\mathrm{Op}_{\check{\mathfrak{g}}}^{\mathrm{nilp}} \to \check{\mathfrak{n}}/\check{B}$ is the scheme $\mathrm{Op}_{\check{\mathfrak{g}}}^{\mathrm{reg}}$ of regular $\check{G}$-opers on the disc $\mathcal{D}$. From the point of view of representations, the algebra $\mathfrak{Z}_{\mathfrak{g}}^{\mathrm{reg}} \simeq \mathrm{Fun}(\mathrm{Op}_{\check{\mathfrak{g}}}^{\mathrm{reg}})$ is characterized as follows. Let

$$\mathbb{V}_{\mathrm{crit}} \simeq \mathrm{Ind}_{\mathfrak{g}[[t]]\oplus\mathbb{C}\mathbf{1}}^{\widehat{\mathfrak{g}}_{\mathrm{crit}}}(\mathbb{C})$$

be the vacuum Verma module of critical level. According to [FF3, F], the action of the center $\mathfrak{Z}_{\mathfrak{g}} \simeq \mathrm{Fun}(\mathrm{Op}_{\check{\mathfrak{g}}}(\mathcal{D}^\times))$ on $\mathbb{V}_{\mathrm{crit}}$ factors through its quotient $\mathrm{Fun}(\mathrm{Op}_{\check{\mathfrak{g}}}^{\mathrm{reg}})$. Moreover, the latter algebra is isomorphic to the algebra of endomorphisms of $\mathbb{V}_{\mathrm{crit}}$.

In this section we will be concerned with the category $\widehat{\mathfrak{g}}_{\text{crit}}\text{-mod}_{\text{reg}}$ and its derived category $D(\widehat{\mathfrak{g}}_{\text{crit}}\text{-mod}_{\text{reg}})$. We will see that there are many parallels between the categories $D(\widehat{\mathfrak{g}}_{\text{crit}}\text{-mod}_{\text{reg}})$ and $D(\widehat{\mathfrak{g}}_{\text{crit}}\text{-mod}_{\text{nilp}})$, but the structure of the former is considerably simpler.

Let $\widehat{\mathfrak{g}}_{\text{crit}}\text{-mod}_{\text{reg}}^{I,m}$ denote the full subcategory of $\widehat{\mathfrak{g}}_{\text{crit}}\text{-mod}_{\text{reg}}$ equal to the intersection $\widehat{\mathfrak{g}}_{\text{crit}}\text{-mod}_{\text{reg}} \cap \widehat{\mathfrak{g}}_{\text{crit}}\text{-mod}^{I,m}$; we let $D^b(\widehat{\mathfrak{g}}_{\text{crit}}\text{-mod}_{\text{reg}})^{I^0} := D(\widehat{\mathfrak{g}}_{\text{crit}}\text{-mod}_{\text{reg}})^{I_0}$ denote the corresponding full triangulated category.

In this section we will formulate a conjecture that describes these categories in terms of D-modules on the affine Grassmannian.

8.2

Before stating the conjecture we would like to motivate it by Bezrukavnikov's theory in the spirit of Section 6.3. In this subsection the discussion will be informal.

Let $\text{Gr}_G = G(\!(t)\!)/G[[t]]$ be the affine Grassmannian of the group G. We will consider right D-modules on Gr_G and denote this category by $\mathfrak{D}(\text{Gr}_G)\text{-mod}$. As before, we have the subcategories $\mathfrak{D}(\text{Gr}_G)\text{-mod}^I$, $\mathfrak{D}(\text{Gr}_G)\text{-mod}^{I_0} \simeq \mathfrak{D}(\text{Gr}_G)\text{-mod}^{I,m}$ and the corresponding triangulated categories

$$D(\mathfrak{D}(\text{Gr}_G)\text{-mod})^I,\ D(\mathfrak{D}(\text{Gr}_G)\text{-mod})^{I_0} \subset D(\mathfrak{D}(\text{Gr}_G)\text{-mod}).$$

Consider the two categories appearing in Conjecture 6.5, and let us apply a further base change with respect to the map $\text{Op}_{\check{\mathfrak{g}}}^{\text{reg}} \to \text{Op}_{\check{\mathfrak{g}}}^{\text{nilp}}$. We obtain an equivalence:

$$D^b(\mathfrak{D}(\text{Fl}_G)\text{-mod}) \underset{\widetilde{\mathcal{N}}_{\check{G}}/\check{G}}{\times} (\text{Op}_{\check{\mathfrak{g}}}^{\text{nilp}} \underset{\text{Op}_{\check{\mathfrak{g}}}^{\text{nilp}}}{\times} \text{Op}_{\check{\mathfrak{g}}}^{\text{reg}}) \simeq D^b(\widehat{\mathfrak{g}}_{\text{crit}}\text{-mod}_{\text{nilp}}) \underset{\text{Op}_{\check{\mathfrak{g}}}^{\text{nilp}}}{\times} \text{Op}_{\check{\mathfrak{g}}}^{\text{reg}}. \quad (8.1)$$

The right-hand side is by definition equivalent to $D^b(\widehat{\mathfrak{g}}_{\text{crit}}\text{-mod}_{\text{reg}})$. The left-hand side can be rewritten as

$$\left(D^b(\mathfrak{D}(\text{Fl}_G)\text{-mod}) \underset{\widetilde{\mathcal{N}}_{\check{G}}/\check{G}}{\times} \text{pt}/\check{B}\right) \underset{\text{pt}/\check{B}}{\times} \text{Op}_{\check{\mathfrak{g}}}^{\text{reg}}.$$

The theory of spherical sheaves on the affine Grassmannian implies that $D^b(\mathfrak{D}(\text{Gr}_G)\text{-mod})$ is naturally *a category over the stack* $\text{pt}/\check{G}$ in the sense explained in Section 0.3. It follows from Bezrukavnikov's theory [Bez] that the categories $D^b(\mathfrak{D}(\text{Gr}_G)\text{-mod})$ and $D^b(\mathfrak{D}(\text{Fl}_G)\text{-mod})$ are related as follows:

$$D^b(\mathfrak{D}(\text{Fl}_G)\text{-mod}) \underset{\widetilde{\mathcal{N}}_{\check{G}}/\check{G}}{\times} \text{pt}/\check{B} \simeq D^b(\mathfrak{D}(\text{Gr}_G)\text{-mod}) \underset{\text{pt}/\check{G}}{\times} \text{pt}/\check{B}.$$

Hence, from (8.1) we obtain the following conjecture:

$$D^b(\mathfrak{D}(\text{Gr}_G)\text{-mod}) \underset{\text{pt}/\check{G}}{\times} \text{Op}_{\check{\mathfrak{g}}}^{\text{reg}} \simeq D^b(\widehat{\mathfrak{g}}_{\text{crit}}\text{-mod}_{\text{reg}}).$$

Our Conjecture 8.11 below reformulates the last statement in terms that do not require the formalism of categories over a stack.

8.3

Recall from Section 4.8 the groupoid $\mathsf{Isom}^{\text{reg}}_{\text{Op}_{\check{\mathfrak{g}}}}$ on $\text{Op}^{\text{reg}}_{\check{\mathfrak{g}}}$ and the corresponding Lie algebroid $\mathsf{isom}^{\text{reg}}_{\text{Op}_{\check{\mathfrak{g}}}}$, which is the Atiyah algebroid of the principal $\check{G}$-bundle $\mathcal{P}_{\check{G},\text{Op}^{\text{reg}}_{\check{\mathfrak{g}}}}$. For $V \in \mathcal{R}\text{ep}(\check{G})$ we will denote by $V_{\text{Op}^{\text{reg}}_{\check{\mathfrak{g}}}}$ the corresponding (projective) module over $\text{Fun}(\text{Op}^{\text{reg}}_{\check{\mathfrak{g}}})$.

Using Theorem 5.2 we can transfer these objects to $\text{Spec}(\mathfrak{z}^{\text{reg}}_{\mathfrak{g}})$, and we will denote them by $\mathsf{Isom}_{\text{Op}^{\text{reg}}_{\check{\mathfrak{g}}}}$, $\mathsf{isom}_{\text{Op}^{\text{reg}}_{\check{\mathfrak{g}}}}$, $\mathcal{P}_{\check{G},\mathfrak{z}^{\text{reg}}_{\mathfrak{g}}}$ and $V_{\mathfrak{z}^{\text{reg}}_{\mathfrak{g}}}$, respectively. From Theorem 7.10 and Section 4.8, we obtain the following.

Corollary 8.4.

(a) *Under the isomorphism* $\mathfrak{z}^{\text{reg}}_{\mathfrak{g}} \simeq \text{Fun}(\text{Op}^{\text{reg}}_{\check{\mathfrak{g}}})$*, we have a canonical identification of Lie algebroids* $N^*_{\mathfrak{z}^{\text{reg}}_{\mathfrak{g}}/\mathfrak{z}_{\mathfrak{g}}} \simeq \mathsf{isom}^{\text{reg}}_{\text{Op}_{\mathfrak{g}}}$.

(b) *We have a commutative diagram*

$$\begin{array}{ccccccccc}
 & & 0 & & 0 & & & & \\
 & & \downarrow & & \downarrow & & & & \\
 & & \check{\mathfrak{g}}_{\mathfrak{z}^{\text{reg}}_{\mathfrak{g}}} & \xrightarrow{\text{id}} & \check{\mathfrak{g}}_{\mathfrak{z}^{\text{reg}}_{\mathfrak{g}}} & \longrightarrow & 0 & & \\
 & & \downarrow & & \downarrow & & \downarrow & & \\
0 & \longrightarrow & N^*_{\mathfrak{z}^{\text{reg}}_{\mathfrak{g}}/\mathfrak{z}_{\mathfrak{g}}} & \longrightarrow & \Omega^1(\mathfrak{z}_{\mathfrak{g}})|_{\text{Spec}(\mathfrak{z}^{\text{reg}}_{\mathfrak{g}})} & \longrightarrow & \Omega^1(\mathfrak{z}^{\text{reg}}_{\mathfrak{g}}) & \longrightarrow & 0 \\
 & & \downarrow & & \downarrow & & \downarrow & & \\
0 & \longrightarrow & T(\mathfrak{z}^{\text{reg}}_{\mathfrak{g}}) & \longrightarrow & T(\mathfrak{z}_{\mathfrak{g}})|_{\text{Spec}(\mathfrak{z}^{\text{reg}}_{\mathfrak{g}})} & \longrightarrow & N_{\mathfrak{z}^{\text{reg}}_{\mathfrak{g}}/\mathfrak{z}_{\mathfrak{g}}} & \longrightarrow & 0 \\
 & & \downarrow & & \downarrow & & \downarrow & & \\
 & & 0 & \longrightarrow & \check{\mathfrak{g}}_{\mathfrak{z}^{\text{reg}}_{\mathfrak{g}}} & \xrightarrow{\text{id}} & \check{\mathfrak{g}}_{\mathfrak{z}^{\text{reg}}_{\mathfrak{g}}} & & \\
 & & & & \downarrow & & \downarrow & & \\
 & & & & 0 & & 0 & &
\end{array}$$

8.5

In what follows will work not with usual right D-modules on Gr_G, but rather with the D-modules twisted by the critical line bundle, as in Section 7.19. We will denote the corresponding category by $\mathfrak{D}_{\text{crit}}(\text{Gr}_G)$-mod. We have the following result, established in [FG].

Theorem 8.6. *The functor of global sections* $\Gamma : \mathfrak{D}_{\text{crit}}(\text{Gr}_G)\text{-mod} \to \widehat{\mathfrak{g}}_{\text{crit}}\text{-mod}$ *is exact and faithful. Moreover, it factors canonically through a functor* $\Gamma^{\text{ren}} : \mathfrak{D}_{\text{crit}}(\text{Gr}_G)\text{-mod} \to U^{\text{ren,reg}}(\widehat{\mathfrak{g}}_{\text{crit}})\text{-mod}$*, and the latter functor is fully faithful.*

Consider the category $\mathrm{Sph}_G := \mathfrak{D}_{\mathrm{crit}}(\mathrm{Gr}_G)\text{-mod}^{G[[t]]}$. According to the results of Lusztig, Drinfeld, Ginzburg, and Mirković–Vilonen (see [MV]), this is a tensor category under the convolution product, which is equivalent to the category $\mathcal{R}\mathrm{ep}(\check{G})$ of representations of the algebraic group $\check{G}$. For $V \in \mathcal{R}\mathrm{ep}(\check{G})$ we will denote by $\mathcal{F}_V$ the corresponding (critically twisted) $G[[t]]$-equivariant D-module on Gr_G.

Let us recall the basic result of [BD1, Sections 5.5 and 5.6] that describes global sections of the (critically twisted) D-modules $\mathcal{F}_V$.

Theorem 8.7.

(a) *We have a canonical isomorphism of $\widehat{\mathfrak{g}}_{\mathrm{crit}}$-modules*

$$\Gamma(\mathrm{Gr}_G, \mathcal{F}_V) \simeq \mathbb{V}_{\mathrm{crit}} \underset{\mathfrak{z}^{\mathrm{reg}}_{\mathfrak{g}}}{\otimes} V_{\mathfrak{z}^{\mathrm{reg}}_{\mathfrak{g}}},$$

compatible with the tensor product of representations.

(b) *The isomorphisms of (a) and that of Corollary* 8.4 *are compatible in the sense that the $N^*_{\mathfrak{z}^{\mathrm{reg}}_{\mathfrak{g}}/\mathfrak{z}_{\mathfrak{g}}}$-action on* $\mathrm{Hom}(\mathbb{V}_{\mathrm{crit}}, \Gamma(\mathrm{Gr}_G, \mathcal{F}_V))$, *coming from Theorem* 8.6 *and Section* 7.8, *corresponds to the canonical $\mathsf{isom}^{\mathrm{reg}}_{\mathrm{Op}_{\check{\mathfrak{g}}}}$-action on $V_{\mathrm{Op}^{\mathrm{reg}}_{\check{\mathfrak{g}}}}$.*

We can take the convolution product of any D-module on Gr_G with a spherical one. A priori, this will be a complex of D-modules on Gr_G, but as in [Ga] one shows that this is a single D-module. (Alternatively, this follows from the lemma below, using Theorem 8.6). Thus we obtain an action of the tensor category $\mathcal{R}\mathrm{ep}(\check{G})$ on $\mathfrak{D}_{\mathrm{crit}}(\mathrm{Gr}_G)$-mod:

$$\mathcal{F}, V \mapsto \mathcal{F} \star \mathcal{F}_V.$$

Lemma 8.8. *For $\mathcal{F} \in \mathfrak{D}_{\mathrm{crit}}(\mathrm{Gr}_G)$-mod and $V \in \mathcal{R}\mathrm{ep}(\check{G})$ we have a canonical isomorphism:*

$$\Gamma(\mathrm{Gr}_G, \mathcal{F} \star \mathcal{F}_V) \simeq \Gamma(\mathrm{Gr}_G, \mathcal{F}) \underset{\mathfrak{z}^{\mathrm{reg}}_{\mathfrak{g}}}{\otimes} V_{\mathfrak{z}^{\mathrm{reg}}_{\mathfrak{g}}}.$$

Proof. Let us recall the formalism of the convolution action (see Section 22.6). We have the functors

$$D^b(\mathfrak{D}_{\mathrm{crit}}(\mathrm{Gr}_G)\text{-mod}) \times D^b(\mathfrak{D}_{\mathrm{crit}}(\mathrm{Gr}_G)\text{-mod})^{G[[t]]} \to D^b(\mathfrak{D}_{\mathrm{crit}}(\mathrm{Gr}_G)\text{-mod})$$

and

$$D^b(\mathfrak{D}_{\mathrm{crit}}(\mathrm{Gr}_G)\text{-mod}) \times D^b(\widehat{\mathfrak{g}}_{\mathrm{crit}}\text{-mod}_{\mathrm{reg}})^{G[[t]]} \to D^b(\widehat{\mathfrak{g}}_{\mathrm{crit}}\text{-mod}_{\mathrm{reg}}),$$

which are intertwined by the functor Γ. Note that $\Gamma(\mathrm{Gr}_G, \delta_{1_{\mathrm{Gr}_G}}) \simeq \mathbb{V}_{\mathrm{crit}}$, and $\Gamma(\mathrm{Gr}_G, \mathcal{F}) \simeq \mathcal{F} \star \mathbb{V}_{\mathrm{crit}}$.

Hence, we have

$$\begin{aligned}\Gamma(\mathrm{Gr}_G, \mathcal{F} \star \mathcal{F}_V) &\simeq (\mathcal{F} \star \mathcal{F}_V) \star \mathbb{V}_{\mathrm{crit}} \simeq \mathcal{F} \star (\mathcal{F}_V \star \mathbb{V}_{\mathrm{crit}}) \simeq \mathcal{F} \star \Gamma(\mathrm{Gr}_G, \mathcal{F}_V) \\ &\simeq \mathcal{F} \star (\mathbb{V}_{\mathrm{crit}} \underset{\mathfrak{z}^{\mathrm{reg}}_{\mathfrak{g}}}{\otimes} V_{\mathfrak{z}^{\mathrm{reg}}_{\mathfrak{g}}}) \simeq (\mathcal{F} \star \mathbb{V}_{\mathrm{crit}}) \underset{\mathfrak{z}^{\mathrm{reg}}_{\mathfrak{g}}}{\otimes} V_{\mathfrak{z}^{\mathrm{reg}}_{\mathfrak{g}}} \\ &\simeq \Gamma(\mathrm{Gr}_G, \mathcal{F}) \underset{\mathfrak{z}^{\mathrm{reg}}_{\mathfrak{g}}}{\otimes} V_{\mathfrak{z}^{\mathrm{reg}}_{\mathfrak{g}}},\end{aligned}$$

where the second-to-last isomorphism is given by Theorem 8.7. □

8.9

After these preparations we introduce the category $\mathfrak{D}(\mathrm{Gr}_G)^{\mathrm{Hecke}}_{\mathrm{crit}}$-mod which is conjecturally equivalent to $\widehat{\mathfrak{g}}_{\mathrm{crit}}\text{-mod}_{\mathrm{reg}}$.

Its objects are (critically twisted) D-modules $\mathcal{F}$ on Gr_G, endowed with an action of the algebra $\mathfrak{z}^{\mathrm{reg}}_{\mathfrak{g}}$ by endomorphisms, and a family of functorial isomorphisms

$$\alpha_V : \mathcal{F} \star \mathcal{F}_V \simeq V_{\mathfrak{z}^{\mathrm{reg}}_{\mathfrak{g}}} \underset{\mathfrak{z}^{\mathrm{reg}}_{\mathfrak{g}}}{\otimes} \mathcal{F}, \qquad V \in \mathcal{R}\mathrm{ep}(\check{G}),$$

compatible with tensor products of representations in the sense that for $U, V \in \mathcal{R}\mathrm{ep}(\check{G})$ the diagram

$$\begin{array}{ccc}
(\mathcal{F} \star \mathcal{F}_U) \star \mathcal{F}_V & \longrightarrow & \mathcal{F} \star (\mathcal{F}_U \star \mathcal{F}_V) \\
\downarrow & & \downarrow \\
(U_{\mathfrak{z}^{\mathrm{reg}}_{\mathfrak{g}}} \underset{\mathfrak{z}^{\mathrm{reg}}_{\mathfrak{g}}}{\otimes} \mathcal{F}) \star \mathcal{F}_V & & (U_{\mathfrak{z}^{\mathrm{reg}}_{\mathfrak{g}}} \underset{\mathfrak{z}^{\mathrm{reg}}_{\mathfrak{g}}}{\otimes} V_{\mathfrak{z}^{\mathrm{reg}}_{\mathfrak{g}}}) \underset{\mathrm{Fun}(\mathrm{Op}^{\mathrm{reg}}_{\check{\mathfrak{g}}})}{\otimes} \mathcal{F} \\
\downarrow & & \downarrow \\
U_{\mathfrak{z}^{\mathrm{reg}}_{\mathfrak{g}}} \underset{\mathfrak{z}^{\mathrm{reg}}_{\mathfrak{g}}}{\otimes} (\mathcal{F} \star \mathcal{F}_V) & \longrightarrow & U_{\mathfrak{z}^{\mathrm{reg}}_{\mathfrak{g}}} \underset{\mathfrak{z}^{\mathrm{reg}}_{\mathfrak{g}}}{\otimes} (V_{\mathfrak{z}^{\mathrm{reg}}_{\mathfrak{g}}} \underset{\mathfrak{z}^{\mathrm{reg}}_{\mathfrak{g}}}{\otimes} \mathcal{F})
\end{array}$$

is commutative, and that α_V, for V being the trivial representation, is the identity map.

In fact, one can show as in [AG2] that it is sufficient to give a family of *morphisms* $\{\alpha_V\}$ satisfying the above conditions; the fact that they are isomorphisms is then automatic. Morphisms in this category are maps of D-modules that commute with the action of $\mathfrak{z}^{\mathrm{reg}}_{\mathfrak{g}}$ and the data of α_V.

Note that the category $\mathfrak{D}(\mathrm{Gr}_G)^{\mathrm{Hecke}}_{\mathrm{crit}}$-mod is precisely the category

$$\mathfrak{D}(\mathrm{Gr}_G)\text{-mod} \underset{\mathrm{pt}/\check{G}}{\times} \mathrm{Op}^{\mathrm{reg}}_{\check{\mathfrak{g}}}$$

introduced above.

Consider the groupoid $\mathsf{Isom}_{\mathrm{Op}^{\mathrm{reg}}_{\check{\mathfrak{g}}}}$ and note that the algebra $\mathrm{Fun}(\mathsf{Isom}_{\mathrm{Op}^{\mathrm{reg}}_{\check{\mathfrak{g}}}})$ is isomorphic to

$$\bigoplus_{V \in \mathrm{Irr}(\mathcal{R}\mathrm{ep}(\check{G}))} V_{\mathfrak{z}^{\mathrm{reg}}_{\mathfrak{g}}} \underset{\mathbb{C}}{\otimes} V^*_{\mathfrak{z}^{\mathrm{reg}}_{\mathfrak{g}}},$$

and the unit section corresponds to the map

$$V_{\mathfrak{z}^{\mathrm{reg}}_{\mathfrak{g}}} \underset{\mathbb{C}}{\otimes} V^*_{\mathfrak{z}^{\mathrm{reg}}_{\mathfrak{g}}} \to V_{\mathfrak{z}^{\mathrm{reg}}_{\mathfrak{g}}} \underset{\mathrm{Fun}(\mathfrak{z}^{\mathrm{reg}}_{\mathfrak{g}})}{\otimes} V^*_{\mathfrak{z}^{\mathrm{reg}}_{\mathfrak{g}}} \to \mathrm{Fun}(\mathfrak{z}^{\mathrm{reg}}_{\mathfrak{g}}).$$

Let us consider the space of global sections of an object $\mathcal{F} \in \mathfrak{D}(\mathrm{Gr}_G)^{\mathrm{Hecke}}_{\mathrm{crit}}$-mod. From Lemma 8.8 we obtain the following.

Lemma 8.10. *For an object* $\mathcal{F}$ *of* $\mathfrak{D}(\mathrm{Gr}_G)^{\mathrm{Hecke}}_{\mathrm{crit}}$-mod*, the action of* $\mathfrak{Z}^{\mathrm{reg}}_{\mathfrak{g}}$ *on* $\Gamma(\mathrm{Gr}_G, \mathcal{F})$ *by* $\widehat{\mathfrak{g}}_{\mathrm{crit}}$*-endomorphisms canonically extends to an action of* $\mathrm{Fun}(\mathsf{Isom}_{\mathrm{Op}^{\mathrm{reg}}_{\check{\mathfrak{g}}}})$.

Consider the functor $\Gamma^{\mathrm{Hecke}} : \mathfrak{D}(\mathrm{Gr}_G)^{\mathrm{Hecke}}_{\mathrm{crit}}\text{-mod} \to \widehat{\mathfrak{g}}_{\mathrm{crit}}\text{-mod}_{\mathrm{reg}}$ given by

$$\mathcal{F} \mapsto \Gamma(\mathrm{Gr}, \mathcal{F}) \underset{\mathrm{Fun}(\mathsf{Isom}_{\mathrm{Op}^{\mathrm{reg}}_{\check{\mathfrak{g}}}})}{\otimes} \mathfrak{Z}^{\mathrm{reg}}_{\mathfrak{g}},$$

where $\mathfrak{Z}^{\mathrm{reg}}_{\mathfrak{g}}$ is considered as a $\mathrm{Fun}(\mathsf{Isom}_{\mathrm{Op}^{\mathrm{reg}}_{\check{\mathfrak{g}}}})$-algebra via the unit section.

We propose the following.

Main Conjecture 8.11. *The above functor* Γ^{Hecke} *is exact and defines an equivalence of categories* $\mathfrak{D}(\mathrm{Gr}_G)^{\mathrm{Hecke}}_{\mathrm{crit}}\text{-mod} \to \widehat{\mathfrak{g}}_{\mathrm{crit}}\text{-mod}_{\mathrm{reg}}$.

Note that by definition, the category $\mathfrak{D}(\mathrm{Gr}_G)^{\mathrm{Hecke}}_{\mathrm{crit}}$-mod carries a Harish-Chandra action of $G(\!(t)\!)$ at the critical level. By construction, the functor Γ^{Hecke} preserves this structure. In particular, we can consider the subcategories of I^0-equivariant objects on both sides. As a consequence we obtain another conjecture.

Main Conjecture 8.12. *The category* $\widehat{\mathfrak{g}}_{\mathrm{crit}}\text{-mod}^{I,m}_{\mathrm{reg}}$ *is equivalent to*

$$\mathfrak{D}(\mathrm{Gr}_G)^{\mathrm{Hecke}}_{\mathrm{crit}}\text{-mod}^{I_0}.$$

8.13

We now present another way of formulating Main Conjecture 8.11. Recall from [Ga1] that if $\mathcal{Y}$ is an affine variety, $\mathcal{C}$ is a $\mathrm{Fun}(\mathcal{Y})$-linear abelian category and $\mathcal{G}_{\mathcal{Y}}$ is an affine groupoid over $\mathcal{Y}$, it then makes sense to speak about a lift of the $\mathcal{G}_{\mathcal{Y}}$-action on $\mathcal{Y}$ to $\mathcal{C}$.

We take $\mathcal{Y} = \mathrm{Spec}(\mathfrak{Z}^{\mathrm{reg}}_{\mathfrak{g}})$, $\mathcal{G}_{\mathcal{Y}} = \mathsf{Isom}_{\mathrm{Op}^{\mathrm{reg}}_{\check{\mathfrak{g}}}}$ and $\mathcal{C} = \widehat{\mathfrak{g}}_{\mathrm{crit}}\text{-mod}_{\mathrm{reg}}$. One can show that Main Conjecture 8.11 is equivalent to the following one.

Conjecture 8.14. *The action of* $\mathsf{Isom}^{\mathrm{reg}}_{\mathrm{Op}^{\mathrm{reg}}_{\check{\mathfrak{g}}}}$ *on* $\mathrm{Spec}(\mathfrak{Z}^{\mathrm{reg}}_{\mathfrak{g}})$ *lifts to an action on* $\widehat{\mathfrak{g}}_{\mathrm{crit}}\text{-mod}_{\mathrm{reg}}$ *in such a way that*

(1) *this structure commutes in the natural sense with the Harish-Chandra action of* $G(\!(t)\!)$ *on* $\widehat{\mathfrak{g}}_{\mathrm{crit}}\text{-mod}_{\mathrm{reg}}$;
(2) *the functor* Γ *establishes an equivalence between the category* $\mathfrak{D}(\mathrm{Gr}_G)_{\mathrm{crit}}$-mod *and the category of* $\mathsf{Isom}_{\mathrm{Op}^{\mathrm{reg}}_{\check{\mathfrak{g}}}}$*-equivariant objects in* $\widehat{\mathfrak{g}}_{\mathrm{crit}}\text{-mod}_{\mathrm{reg}}$.

Remark 8.15. If we had an action of $\mathsf{Isom}_{\mathrm{Op}^{\mathrm{reg}}_{\check{\mathfrak{g}}}}$ on $\widehat{\mathfrak{g}}_{\mathrm{crit}}\text{-mod}_{\mathrm{reg}}$, as conjectured above, then at the infinitesimal level we would have functorial maps

$$\mathsf{isom}^{\mathrm{reg}}_{\mathrm{Op}^{\mathrm{reg}}_{\check{\mathfrak{g}}}} \to \mathrm{Ext}^1_{\widehat{\mathfrak{g}}_{\mathrm{crit}}\text{-mod}_{\mathrm{reg}}}(\mathcal{M}, \mathcal{M}),$$

for any $\mathcal{M} \in \widehat{\mathfrak{g}}_{\mathrm{crit}}\text{-mod}_{\mathrm{reg}}$. However, the latter maps are known to exist, as follows from (7.3) in Section 7.8.

8.16

Although we are unable to prove Main Conjecture 8.11 at the moment, we will establish one result in its direction, which we will use later on.

Let us denote by $\mathrm{L}\,\Gamma^{\mathrm{Hecke}} : D^-(\mathfrak{D}(\mathrm{Gr}_G)^{\mathrm{Hecke}}_{\mathrm{crit}}\text{-mod}) \to D^-(\widehat{\mathfrak{g}}_{\mathrm{crit}}\text{-mod}_{\mathrm{reg}})$ the functor given by

$$\mathcal{F} \mapsto \Gamma(\mathrm{Gr}_G, \mathcal{F}) \underset{\mathrm{Fun}(\mathsf{Isom}_{\mathrm{Op}^{\mathrm{reg}}_{\check{\mathfrak{g}}}})}{\overset{L}{\otimes}} \mathfrak{Z}^{\mathrm{reg}}_{\mathfrak{g}},$$

where $\overset{L}{\otimes}$ is defined using a left resolution of $\mathfrak{Z}^{\mathrm{reg}}_{\mathfrak{g}}$ by projective $\mathrm{Fun}(\mathsf{Isom}_{\mathrm{Op}^{\mathrm{reg}}_{\check{\mathfrak{g}}}})$-modules.

One easily shows (and we will see this in the course of the proof of the next theorem) that $\mathrm{L}\,\Gamma^{\mathrm{Hecke}}$ is, in fact, the left derived functor of Γ^{Hecke}.

Theorem 8.17. *The functor* $\mathrm{L}\,\Gamma^{\mathrm{Hecke}}$, *restricted to* $D^b(\mathfrak{D}(\mathrm{Gr}_G)^{\mathrm{Hecke}}_{\mathrm{crit}}\text{-mod})$, *is fully faithful.*

Before giving the proof, we need some preparations.

8.18

Let us observe that the obvious forgetful functor

$$\mathfrak{D}(\mathrm{Gr}_G)^{\mathrm{Hecke}}_{\mathrm{crit}}\text{-mod} \to \mathfrak{D}(\mathrm{Gr}_G)_{\mathrm{crit}}\text{-mod}$$

admits a left adjoint, which we will denote by $\mathrm{Ind}^{\mathrm{Hecke}}$. Indeed, it is given by

$$\mathcal{F} \mapsto \bigoplus_{V \in \mathrm{Irr}(\mathcal{R}\mathrm{ep}(\check{G}))} (\mathcal{F} \star \mathcal{F}_{V^*}) \underset{\mathbb{C}}{\otimes} V_{\mathfrak{Z}^{\mathrm{reg}}_{\mathfrak{g}}}.$$

Evidently, we have the following.

Lemma 8.19.

$$\mathrm{L}\,\Gamma^{\mathrm{Hecke}}(\mathrm{Gr}_G, \mathrm{Ind}^{\mathrm{Hecke}}(\mathcal{F})) \simeq \Gamma^{\mathrm{Hecke}}(\mathrm{Gr}_G, \mathrm{Ind}^{\mathrm{Hecke}}(\mathcal{F})) \simeq \Gamma(\mathrm{Gr}_G, \mathcal{F}).$$

Therefore, Theorem 8.17 implies the following.

Theorem 8.20. *For*

$$\mathcal{F}^\bullet_1, \mathcal{F}^\bullet_2, \in D^b(\mathfrak{D}(\mathrm{Gr}_G)_{\mathrm{crit}}) \quad \textit{and} \quad \mathcal{M}^\bullet_i = \Gamma(\mathrm{Gr}_G, \mathcal{F}^\bullet_i) \in D^b(\widehat{\mathfrak{g}}_{\mathrm{crit}}\text{-mod}_{\mathrm{reg}}),$$

the map, given by the functor $\mathrm{L}\,\Gamma^{\mathrm{Hecke}}$,

$$R\,\mathrm{Hom}_{D(\mathfrak{D}(\mathrm{Gr}_G)_{\mathrm{crit}}\text{-mod})}\left(\mathcal{F}^\bullet_1, \bigoplus_{V \in \mathrm{Irr}(\mathcal{R}\mathrm{ep}(\check{G}))} (\mathcal{F}^\bullet_2 \star \mathcal{F}_{V^*}) \underset{\mathbb{C}}{\otimes} V_{\mathfrak{Z}^{\mathrm{reg}}_{\mathfrak{g}}}\right)$$
$$\to \mathrm{RHom}_{D(\widehat{\mathfrak{g}}_{\mathrm{crit}}\text{-mod}_{\mathrm{reg}})}(\mathcal{M}^\bullet_1, \mathcal{M}^\bullet_2)$$

is an isomorphism.

From this theorem we obtain that all

$$\mathrm{R}^i\,\mathrm{Hom}_{D(\widehat{\mathfrak{g}}_{\mathrm{crit}}\text{-mod}_{\mathrm{reg}})}(\Gamma(\mathrm{Gr}_G,\mathcal{F}_1^\bullet),\Gamma(\mathrm{Gr}_G,\mathcal{F}_2^\bullet)),$$

viewed as quasi-coherent sheaves on $\mathrm{Spec}(\mathfrak{z}_{\mathfrak{g}}^{\mathrm{reg}})$, are equivariant with respect to the groupoid $\mathsf{Isom}_{\mathrm{Op}^{\mathrm{reg}}_{\check{\mathfrak{g}}}}$. We claim that we know a priori that the above R^i Hom is acted on by the algebroid $\mathsf{isom}_{\mathrm{Op}^{\mathrm{reg}}_{\check{\mathfrak{g}}}} \simeq N^*_{\mathfrak{z}_{\mathfrak{g}}^{\mathrm{reg}}/\mathfrak{z}_{\mathfrak{g}}}$, and the map in Theorem 8.20 is compatible with the action of $N^*_{\mathfrak{z}_{\mathfrak{g}}^{\mathrm{reg}}/\mathfrak{z}_{\mathfrak{g}}}$. This follows from Lemma 7.12 and Theorem 8.7.

8.21 Proof of Theorem 8.17

It is clear that any object of $\mathfrak{D}(\mathrm{Gr}_G)^{\mathrm{Hecke}}_{\mathrm{crit}}$-mod admits a surjection from an object of the form $\mathrm{Ind}^{\mathrm{Hecke}}(\mathcal{F})$ for some $\mathcal{F}\in\mathfrak{D}(\mathrm{Gr}_G)_{\mathrm{crit}}$-mod. Therefore, any bounded from above complex in $\mathfrak{D}(\mathrm{Gr}_G)^{\mathrm{Hecke}}_{\mathrm{crit}}$-mod admits a left resolution by a complex consisting of objects of this form. Hence, it is sufficient to show that for $\mathcal{F}_1\in\mathfrak{D}(\mathrm{Gr}_G)_{\mathrm{crit}}$-mod and $\mathcal{F}_2^\bullet\in D^+(\mathfrak{D}(\mathrm{Gr}_G)^{\mathrm{Hecke}}_{\mathrm{crit}}\text{-mod})$ the map

$$\mathrm{RHom}_{D\left(\mathfrak{D}(\mathrm{Gr}_G)^{\mathrm{Hecke}}_{\mathrm{crit}}\text{-mod}\right)}\left(\mathrm{Ind}^{\mathrm{Hecke}}(\mathcal{F}_1),\mathcal{F}_2^\bullet\right) \tag{8.2}$$

$$\to \mathrm{RHom}_{D(\widehat{\mathfrak{g}}_{\mathrm{crit}}\text{-mod}_{\mathrm{reg}})}\left(\Gamma(\mathrm{Gr}_G,\mathcal{F}_1),\mathrm{L}\,\Gamma^{\mathrm{Hecke}}(\mathrm{Gr}_G,\mathcal{F}_2^\bullet)\right). \tag{8.3}$$

is an isomorphism. Note that by adjunction the LHS of the above formula is isomorphic to $\mathrm{RHom}_{D(\mathfrak{D}(\mathrm{Gr}_G)_{\mathrm{crit}}\text{-mod})}(\mathcal{F}_1,\mathcal{F}_2^\bullet)$, where we regard $\mathcal{F}_2^\bullet$ just as an object of $D^+(\mathfrak{D}(\mathrm{Gr}_G)_{\mathrm{crit}}\text{-mod})$.

Without loss of generality we can assume that $\mathcal{F}_1$ is finitely generated, and is equivariant with respect to some congruence subgroup $K\subset G[[t]]$. By Section 20.10, we can replace $\mathcal{F}_2^\bullet$ by $\mathrm{Av}_K(\mathcal{F}_2^\bullet)$, i.e., without a loss of generality, we can assume that $\mathcal{F}_2^\bullet$ is also K-equivariant.

We will use the Harish-Chandra action of $G((t))$ on $\mathfrak{D}(\mathrm{Gr}_G)^{\mathrm{Hecke}}_{\mathrm{crit}}$-mod and $\widehat{\mathfrak{g}}_{\mathrm{crit}}$-mod$_{\mathrm{reg}}$. Namely, we will interpret $\mathcal{F}_1$ as $\mathcal{F}_1\star\delta_{1_{\mathrm{Gr}_G}}\in\mathfrak{D}(\mathrm{Gr}_G)_{\mathrm{crit}}\text{-mod}^K$, and hence

$$\mathrm{Ind}^{\mathrm{Hecke}}(\mathcal{F}_1)\simeq\mathcal{F}_1\star(\mathrm{Ind}^{\mathrm{Hecke}}(\delta_{1_{\mathrm{Gr}_G}}))\in\mathfrak{D}(\mathrm{Gr}_G)^{\mathrm{Hecke}}_{\mathrm{crit}}\text{-mod}^K.$$

Similarly, $\Gamma(\mathrm{Gr}_G,\mathcal{F}_1)\simeq\mathcal{F}_1\star\mathbb{V}_{\mathrm{crit}}$.

Let $\widetilde{\mathcal{F}_1}$ be the dual D-module in $\mathfrak{D}(G((t))/K)_{\mathrm{crit}}\text{-mod}^{G[[t]]}$; see Section 22.22. Set

$$\mathcal{F}^\bullet := \widetilde{\mathcal{F}_1}\star\mathcal{F}_2\in D^+\left(\mathfrak{D}(\mathrm{Gr}_G)^{\mathrm{Hecke}}_{\mathrm{crit}}\text{-mod}\right)^{G[[t]]}.$$

By Section 22.22, we have

$$\mathrm{RHom}_{D(\mathfrak{D}(\mathrm{Gr}_G)_{\mathrm{crit}}\text{-mod})}(\mathcal{F}_1,\mathcal{F}_2^\bullet)\simeq\mathrm{RHom}_{D(\mathfrak{D}(\mathrm{Gr}_G)_{\mathrm{crit}}\text{-mod})^{G[[t]]}}\left(\delta_{1_{\mathrm{Gr}_G}},\mathcal{F}^\bullet\right)$$

and $\mathrm{RHom}_{D(\widehat{\mathfrak{g}}_{\mathrm{crit}}\text{-mod}_{\mathrm{reg}})}(\Gamma(\mathrm{Gr}_G,\mathcal{F}_1),\mathrm{L}\,\Gamma^{\mathrm{Hecke}}(\mathrm{Gr}_G,\mathcal{F}_2^\bullet))$ is isomorphic to

$$\mathrm{RHom}_{D(\widehat{\mathfrak{g}}_{\mathrm{crit}}\text{-mod}_{\mathrm{reg}})^{G[[t]]}}\left(\mathbb{V}_{\mathrm{crit}}, \mathrm{L}\,\Gamma^{\mathrm{Hecke}}(\mathrm{Gr}_G, \mathcal{F}^\bullet)\right).$$

Evidently, we can assume $\mathcal{F}^\bullet$ is an object, denoted $\mathcal{F}$, of the abelian category $\mathfrak{D}(\mathrm{Gr}_G^{\mathrm{Hecke}})_{\mathrm{crit}}\text{-mod}^{G[[t]]}$.

Since the category $\mathfrak{D}(\mathrm{Gr}_G)_{\mathrm{crit}}\text{-mod}^{G[[t]]}$ is equivalent to $\mathcal{R}\mathrm{ep}(\check{G})$, we obtain that the category $\mathfrak{D}(\mathrm{Gr}_G)^{\mathrm{Hecke}}_{\mathrm{crit}}\text{-mod}^{G[[t]]}$ is equivalent to the category of $\mathfrak{Z}_{\mathfrak{g}}^{\mathrm{reg}}$-modules, with the functor being given by

$$\mathcal{L} \mapsto \mathrm{Ind}^{\mathrm{Hecke}}(\delta_{1_{\mathrm{Gr}_G}}) \otimes \mathcal{L}.$$

Therefore, the D-module $\mathcal{F}$ above has such a form for some $\mathfrak{Z}_{\mathfrak{g}}^{\mathrm{reg}}$-module $\mathcal{L}$.

By a reg- and $G[[t]]$-equivariant version of Proposition 7.14, we can assume that $\mathcal{L}$ is finitely presented. Since $\mathfrak{Z}_{\mathfrak{g}}^{\mathrm{reg}}$ is a polynomial algebra, every finitely presented module admits a finite resolution by projective ones. This reduces us to the case when $\mathcal{L} = \mathfrak{Z}_{\mathfrak{g}}^{\mathrm{reg}}$. Thus we obtain that it is enough to show the following:

(*) *The map*

$$\mathrm{Ext}^\bullet_{D(\mathfrak{D}(\mathrm{Gr}_G)_{\mathrm{crit}}\text{-mod})^{G[[t]]}}(\delta_{1_{\mathrm{Gr}_G}}, \mathrm{Ind}^{\mathrm{Hecke}}(\delta_{1_{\mathrm{Gr}_G}})) \\ \to \mathrm{Ext}^\bullet_{D(\widehat{\mathfrak{g}}_{\mathrm{crit}}\text{-mod}_{\mathrm{reg}})^{G[[t]]}}(\mathbb{V}_{\mathrm{crit}}, \mathbb{V}_{\mathrm{crit}})$$

is an isomorphism.

To establish (*) we proceed as follows. It is known from [ABG, Theorem 7.6.1] that

$$\mathrm{Ext}^\bullet_{D(\mathfrak{D}(\mathrm{Gr})_{\mathrm{crit}}\text{-mod})^{G[[t]]}}\left(\delta_{1_{\mathrm{Gr}_G}}, \bigoplus_{V\in \mathrm{Irr}(\mathcal{R}\mathrm{ep}(\check{G}))} \mathcal{F}_V \underset{\mathbb{C}}{\otimes} V^*\right) \simeq \mathrm{Sym}^\bullet(\check{\mathfrak{g}}),$$

viewed as a graded algebra with an action of $\check{G}$, where the generators $\check{\mathfrak{g}} \subset \mathrm{Sym}^\bullet(\check{\mathfrak{g}})$ have degree 2.

Hence, the left-hand side in (*) is isomorphic to the graded algebra over $\mathfrak{Z}_{\mathfrak{g}}^{\mathrm{reg}}$ obtained from the $\check{G}$-torsor $\mathcal{P}_{\check{G},\mathfrak{Z}_{\mathfrak{g}}^{\mathrm{reg}}}$ and the $\check{G}$-algebra $\mathrm{Sym}^\bullet(\check{\mathfrak{g}})$, i.e.,

$$\mathcal{P}_{\check{G},\mathfrak{Z}_{\mathfrak{g}}^{\mathrm{reg}}} \overset{\check{G}}{\times} \mathrm{Sym}^\bullet(\check{\mathfrak{g}}) \simeq \mathrm{Sym}^\bullet_{\mathfrak{Z}_{\mathfrak{g}}^{\mathrm{reg}}}(\check{\mathfrak{g}}_{\mathfrak{Z}_{\mathfrak{g}}^{\mathrm{reg}}}). \tag{8.4}$$

Now we claim that the right-hand side in (*) is also isomorphic to the algebra appearing in (8.4).

Theorem 8.22. *There exists a canonical isomorphism of algebras*

$$\mathrm{Ext}^\bullet_{D(\widehat{\mathfrak{g}}_{\mathrm{crit}}\text{-mod}_{\mathrm{reg}})^{G[[t]]}}(\mathbb{V}_{\mathrm{crit}}, \mathbb{V}_{\mathrm{crit}}) \simeq \mathrm{Sym}^\bullet_{\mathfrak{Z}_{\mathfrak{g}}^{\mathrm{reg}}}(\check{\mathfrak{g}}_{\mathfrak{Z}_{\mathfrak{g}}^{\mathrm{reg}}}),$$

compatible with the action of $N^*_{\mathfrak{Z}_{\mathfrak{g}}^{\mathrm{reg}}/\mathfrak{Z}_{\mathfrak{g}}}$*, where the generators* $\check{\mathfrak{g}} \subset \mathrm{Sym}^\bullet(\check{\mathfrak{g}})$ *have degree* 2.

8.23 Proof of Theorem 8.22

From Corollary 8.4 and (7.4) we obtain a map

$$\check{\mathfrak{g}}_{\mathfrak{z}_{\mathfrak{g}}^{\mathrm{reg}}} \to \mathrm{Ext}^2_{D(\widehat{\mathfrak{g}}_{\mathrm{crit}}\text{-mod}_{\mathrm{reg}})^{G[[t]]}}(\mathbb{V}_{\mathrm{crit}}, \mathbb{V}_{\mathrm{crit}}), \tag{8.5}$$

compatible with the action of $N^*_{\mathfrak{z}_{\mathfrak{g}}^{\mathrm{reg}}/\mathfrak{z}_{\mathfrak{g}}}$. We are going to show that this map induces the isomorphism stated in the theorem. We will do it by analyzing the spectral sequence of Section 7.6.

Since the $\widehat{\mathfrak{g}}_{\mathrm{crit}}$-action on $\mathbb{V}_{\mathrm{crit}}$ can be canonically extended to an action of $U^{\mathrm{ren,reg}}(\widehat{\mathfrak{g}}_{\mathrm{crit}})$, from Lemma 7.12 and (7.3), we obtain a map

$$\Omega^1(\mathfrak{z}_{\mathfrak{g}}^{\mathrm{reg}}) \to \mathrm{Ext}^1_{D(\widehat{\mathfrak{g}}_{\mathrm{crit}}\text{-mod})^{G[[t]]}}(\mathbb{V}_{\mathrm{crit}}, \mathbb{V}_{\mathrm{crit}}).$$

We will use the following result of [FT].

Theorem 8.24. *The cup-product induces an isomorphism of algebras*

$$\Omega^\bullet(\mathfrak{z}_{\mathfrak{g}}^{\mathrm{reg}}) \to \mathrm{Ext}^\bullet_{D(\widehat{\mathfrak{g}}_{\mathrm{crit}}\text{-mod})^{G[[t]]}}(\mathbb{V}_{\mathrm{crit}}, \mathbb{V}_{\mathrm{crit}}).$$

Recall that ι^{reg} denotes the embedding $\mathrm{Spec}(\mathfrak{z}_{\mathfrak{g}}^{\mathrm{reg}}) \hookrightarrow \mathrm{Spec}(\mathfrak{z}_{\mathfrak{g}})$. Consider the object $\iota^{\mathrm{reg}!}(\mathbb{V}_{\mathrm{crit}}) \in D(\widehat{\mathfrak{g}}_{\mathrm{crit}}\text{-mod}_{\mathrm{reg}})^{G[[t]]}$; see Section 7.6. By loc. cit., the jth cohomology of this complex is isomorphic to $\mathbb{V}_{\mathrm{crit}} \underset{\mathfrak{z}_{\mathfrak{g}}^{\mathrm{reg}}}{\otimes} \Lambda^j(N_{\mathfrak{z}_{\mathfrak{g}}^{\mathrm{reg}}/\mathfrak{z}_{\mathfrak{g}}})$.

Consider the cohomological truncation of $\iota^!(\mathbb{V}_{\mathrm{crit}})$, leaving the segment in the cohomological degrees j and $j+1$. It gives rise to a map in the derived category

$$\phi_j : \mathbb{V}_{\mathrm{crit}} \underset{\mathfrak{z}_{\mathfrak{g}}^{\mathrm{reg}}}{\otimes} \Lambda^{j+1}(N_{\mathfrak{z}_{\mathfrak{g}}^{\mathrm{reg}}/\mathfrak{z}_{\mathfrak{g}}}) \to \mathbb{V}_{\mathrm{crit}} \underset{\mathfrak{z}_{\mathfrak{g}}^{\mathrm{reg}}}{\otimes} \Lambda^j(N_{\mathfrak{z}_{\mathfrak{g}}^{\mathrm{reg}}/\mathfrak{z}_{\mathfrak{g}}})[2]. \tag{8.6}$$

Lemma 8.25. *The map ϕ_j equals the composition*

$$\begin{aligned}\mathbb{V}_{\mathrm{crit}} \underset{\mathfrak{z}_{\mathfrak{g}}^{\mathrm{reg}}}{\otimes} \Lambda^{j+1}(N_{\mathfrak{z}_{\mathfrak{g}}^{\mathrm{reg}}/\mathfrak{z}_{\mathfrak{g}}}) &\to \mathbb{V}_{\mathrm{crit}} \underset{\mathfrak{z}_{\mathfrak{g}}^{\mathrm{reg}}}{\otimes} N_{\mathfrak{z}_{\mathfrak{g}}^{\mathrm{reg}}/\mathfrak{z}_{\mathfrak{g}}} \underset{\mathfrak{z}_{\mathfrak{g}}^{\mathrm{reg}}}{\otimes} \Lambda^j(N_{\mathfrak{z}_{\mathfrak{g}}^{\mathrm{reg}}/\mathfrak{z}_{\mathfrak{g}}}) \\ &\overset{\phi_1\otimes\mathrm{id}}{\to} \mathbb{V}_{\mathrm{crit}} \underset{\mathfrak{z}_{\mathfrak{g}}^{\mathrm{reg}}}{\otimes} \Lambda^j(N_{\mathfrak{z}_{\mathfrak{g}}^{\mathrm{reg}}/\mathfrak{z}_{\mathfrak{g}}})[2].\end{aligned}$$

By Section 7.6, we obtain a spectral sequence, converging to

$$\mathrm{Ext}^\bullet_{D(\widehat{\mathfrak{g}}_{\mathrm{crit}}\text{-mod})^{G[[t]]}}(\mathbb{V}_{\mathrm{crit}}, \mathbb{V}_{\mathrm{crit}}),$$

whose second term is given by

$$E_2^{i,j} = \mathrm{Ext}^i_{D(\widehat{\mathfrak{g}}_{\mathrm{crit}}\text{-mod}_{\mathrm{reg}})^{G[[t]]}}(\mathbb{V}_{\mathrm{crit}}, \mathbb{V}_{\mathrm{crit}}) \underset{\mathfrak{z}_{\mathfrak{g}}^{\mathrm{reg}}}{\otimes} \Lambda^j(N_{\mathfrak{z}_{\mathfrak{g}}^{\mathrm{reg}}/\mathfrak{z}_{\mathfrak{g}}}).$$

Note also that by Lemma 8.25, the differential in the above spectral sequence, which maps $E_2^{i-2,j+1} \to E_2^{i,j}$, can be expressed through the case when $j=0$ as

$$E_2^{i-2,j+1} \simeq E_2^{i-2,0} \underset{\mathfrak{z}_\mathfrak{g}^{\mathrm{reg}}}{\otimes} \Lambda^{j+1}(N_{\mathfrak{z}_\mathfrak{g}^{\mathrm{reg}}/\mathfrak{z}_\mathfrak{g}}) \to E_2^{i-2,0} \underset{\mathfrak{z}_\mathfrak{g}^{\mathrm{reg}}}{\otimes} N_{\mathfrak{z}_\mathfrak{g}^{\mathrm{reg}}/\mathfrak{z}_\mathfrak{g}} \underset{\mathfrak{z}_\mathfrak{g}^{\mathrm{reg}}}{\otimes} \Lambda^{j}(N_{\mathfrak{z}_\mathfrak{g}^{\mathrm{reg}}/\mathfrak{z}_\mathfrak{g}})$$
$$\simeq E_2^{i-2,1} \underset{\mathfrak{z}_\mathfrak{g}^{\mathrm{reg}}}{\otimes} \Lambda^{j}(N_{\mathfrak{z}_\mathfrak{g}^{\mathrm{reg}}/\mathfrak{z}_\mathfrak{g}}) \to E_2^{i,0} \underset{\mathfrak{z}_\mathfrak{g}^{\mathrm{reg}}}{\otimes} \Lambda^{j}(N_{\mathfrak{z}_\mathfrak{g}^{\mathrm{reg}}/\mathfrak{z}_\mathfrak{g}}) \simeq E_2^{i,j}.$$

Let us observe that the canonical map

$$\mathrm{Ext}^j_{D(\widehat{\mathfrak{g}}_{\mathrm{crit}}\text{-mod})^{G[[t]]}}(\mathbb{V}_{\mathrm{crit}}, \mathbb{V}_{\mathrm{crit}}) \to E_2^{0,j} \tag{8.7}$$

identifies by construction with the map $\Omega^j(\mathfrak{z}_\mathfrak{g}^{\mathrm{reg}}) \to \Lambda^j(N_{\mathfrak{z}_\mathfrak{g}^{\mathrm{reg}}/\mathfrak{z}_\mathfrak{g}})$ coming from (7.3); in particular, it is injective.

We will prove by induction on $i = 1, 2, \ldots$ the following statements:

(i) $E_2^{2i-1,0} = 0$,

(ii) $E_2^{2i,0} \simeq \mathrm{Sym}^i(\check{\mathfrak{g}}_{\mathfrak{z}_\mathfrak{g}^{\mathrm{reg}}})$ such that the differential $E_2^{2i-2,1} \to E_2^{2i,0}$ is identified with the map $\mathrm{Sym}^{i-1}(\check{\mathfrak{g}}_{\mathfrak{z}_\mathfrak{g}^{\mathrm{reg}}}) \underset{\mathfrak{z}_\mathfrak{g}^{\mathrm{reg}}}{\otimes} N_{\mathfrak{z}_\mathfrak{g}^{\mathrm{reg}}/\mathfrak{z}_\mathfrak{g}} \to \mathrm{Sym}^i(\check{\mathfrak{g}}_{\mathfrak{z}_\mathfrak{g}^{\mathrm{reg}}})$.

Note that item (i) above implies that $E_2^{2i-1,j} = 0$ for any j and that item (ii) implies that $E_2^{2i,j} \simeq \mathrm{Sym}^i(\check{\mathfrak{g}}_{\mathfrak{z}_\mathfrak{g}^{\mathrm{reg}}}) \underset{\mathfrak{z}_\mathfrak{g}^{\mathrm{reg}}}{\otimes} \Lambda^j(N_{\mathfrak{z}_\mathfrak{g}^{\mathrm{reg}}/\mathfrak{z}_\mathfrak{g}})$ such that the differential is identified with the Koszul differential

$$\mathrm{Sym}^{i-1}(\check{\mathfrak{g}}_{\mathfrak{z}_\mathfrak{g}^{\mathrm{reg}}}) \underset{\mathfrak{z}_\mathfrak{g}^{\mathrm{reg}}}{\otimes} \Lambda^j(N_{\mathfrak{z}_\mathfrak{g}^{\mathrm{reg}}/\mathfrak{z}_\mathfrak{g}}) \to \mathrm{Sym}^{i}(\check{\mathfrak{g}}_{\mathfrak{z}_\mathfrak{g}^{\mathrm{reg}}}) \underset{\mathfrak{z}_\mathfrak{g}^{\mathrm{reg}}}{\otimes} \Lambda^{j-1}(N_{\mathfrak{z}_\mathfrak{g}^{\mathrm{reg}}/\mathfrak{z}_\mathfrak{g}}).$$

Consider first the base of the induction, i.e., the case $i = 1$. In this case we know a priori that $E_2^{1,0} = 0$. We obtain that $\Omega^1(\mathfrak{z}_\mathfrak{g}^{\mathrm{reg}})$ maps isomorphically onto the kernel of the map $N_{\mathfrak{z}_\mathfrak{g}^{\mathrm{reg}}/\mathfrak{z}_\mathfrak{g}} \simeq E_2^{0,1} \to E_2^{2,0}$. In particular, the map of (8.5) $\check{\mathfrak{g}}_{\mathfrak{z}_\mathfrak{g}^{\mathrm{reg}}} \hookrightarrow E_2^{2,0}$ is injective. We claim that the latter map is surjective as well. Indeed, if it were not, the map in (8.7) would not be injective for $j = 2$.

Hence, the differential $E_2^{0,j} \to E_2^{2,j-1}$ does identify with the corresponding term of the Koszul differential. In particular, $\mathrm{Ext}^i_{D(\widehat{\mathfrak{g}}_{\mathrm{crit}}\text{-mod})^{G[[t]]}}(\mathbb{V}_{\mathrm{crit}}, \mathbb{V}_{\mathrm{crit}})$ maps isomorphically to $E_3^{0,j} \simeq \ker(E_2^{0,i} \to E_2^{2,i-1})$. This implies, in particular, that all the higher differentials $E_k^{0,j} \to E_k^{k,j-k-1}$ for $k \geq 3$ vanish.

Let us now perform the induction step. Observe that by the induction hypothesis, all the terms of the spectral sequence $E_k^{i',j}$ for $0 < i' \leq 2i - 2$ with $k \geq 3$ vanish. Therefore, the term $E_2^{2i+1,0}$ injects into

$$\ker\left(\mathrm{Ext}^{2i+1}_{D(\widehat{\mathfrak{g}}_{\mathrm{crit}}\text{-mod})^{G[[t]]}}(\mathbb{V}_{\mathrm{crit}}, \mathbb{V}_{\mathrm{crit}}) \to E_2^{0,2i+1}\right),$$

and as the latter map is injective, we obtain that $E_2^{2i+1,0} = 0$.

By a similar argument we obtain that $E_3^{2i,1} = E_3^{2i+2,0} = 0$. Hence, $E_2^{2i+2,0}$ identifies with

$$\operatorname{coker}\left(\operatorname{Sym}^{i-1}(\check{\mathfrak{g}}_{\mathfrak{z}_{\mathfrak{g}}^{\mathrm{reg}}}) \underset{\mathfrak{z}_{\mathfrak{g}}^{\mathrm{reg}}}{\otimes} \Lambda^2(N_{\mathfrak{z}_{\mathfrak{g}}^{\mathrm{reg}}/\mathfrak{z}_{\mathfrak{g}}}) \to \operatorname{Sym}^{i}(\check{\mathfrak{g}}_{\mathfrak{z}_{\mathfrak{g}}^{\mathrm{reg}}}) \underset{\mathfrak{z}_{\mathfrak{g}}^{\mathrm{reg}}}{\otimes} N_{\mathfrak{z}_{\mathfrak{g}}^{\mathrm{reg}}/\mathfrak{z}_{\mathfrak{g}}}\right)$$
$$\simeq \operatorname{Sym}^{i+1}(\check{\mathfrak{g}}_{\mathfrak{z}_{\mathfrak{g}}^{\mathrm{reg}}}).$$

To finish the proof of theorem it remains to remark that, by construction, the cup-product map

$$\operatorname{Ext}^2_{D(\widehat{\mathfrak{g}}_{\mathrm{crit}}\text{-mod}_{\mathrm{reg}})^{G[[t]]}}(\mathbb{V}_{\mathrm{crit}}, \mathbb{V}_{\mathrm{crit}}) \otimes \operatorname{Ext}^{2i}_{D(\widehat{\mathfrak{g}}_{\mathrm{crit}}\text{-mod}_{\mathrm{reg}})^{G[[t]]}}(\mathbb{V}_{\mathrm{crit}}, \mathbb{V}_{\mathrm{crit}})$$
$$\to \operatorname{Ext}^{2i+2}_{D(\widehat{\mathfrak{g}}_{\mathrm{crit}}\text{-mod}_{\mathrm{reg}})^{G[[t]]}}(\mathbb{V}_{\mathrm{crit}}, \mathbb{V}_{\mathrm{crit}})$$

is identified with the multiplication map $\check{\mathfrak{g}}_{\mathfrak{z}_{\mathfrak{g}}^{\mathrm{reg}}} \otimes \operatorname{Sym}^{i}(\check{\mathfrak{g}}_{\mathfrak{z}_{\mathfrak{g}}^{\mathrm{reg}}}) \to \operatorname{Sym}^{i+1}(\check{\mathfrak{g}}_{\mathfrak{z}_{\mathfrak{g}}^{\mathrm{reg}}})$.

8.26

Thus the two graded algebras appearing in (*) are abstractly isomorphic to one another. It remains to see that the existing map indeed induces an isomorphism. Since both algebras are freely generated by their degree 2 part, it is sufficient to show that the map

$$\check{\mathfrak{g}}_{\mathfrak{z}_{\mathfrak{g}}^{\mathrm{reg}}} \simeq \operatorname{Ext}^2_{D(\mathfrak{D}(\mathrm{Gr}_G)_{\mathrm{crit}}\text{-mod})^{G[[t]]}}\left(\delta_{1_{\mathrm{Gr}_G}}, \underset{V\in\operatorname{Irr}(\mathcal{R}ep(\check{G}))}{\oplus} \mathcal{F}_V \underset{\mathbb{C}}{\otimes} V^*_{\mathfrak{z}_{\mathfrak{g}}^{\mathrm{reg}}}\right) \tag{8.8}$$
$$\to \operatorname{Ext}^2_{D(\widehat{\mathfrak{g}}_{\mathrm{crit}}\text{-mod}_{\mathrm{reg}})^{G[[t]]}}(\mathbb{V}_{\mathrm{crit}}, \mathbb{V}_{\mathrm{crit}}) \simeq \check{\mathfrak{g}}_{\mathfrak{z}_{\mathfrak{g}}^{\mathrm{reg}}} \tag{8.9}$$

is an isomorphism. Since the map of Theorem 8.20 is compatible with the action of the algebroid $N^*_{\mathfrak{z}_{\mathfrak{g}}^{\mathrm{reg}}/\mathfrak{z}_{\mathfrak{g}}}$, and since $\check{\mathfrak{g}}_{\mathfrak{z}_{\mathfrak{g}}^{\mathrm{reg}}}$ is irreducible as a $N^*_{\mathfrak{z}_{\mathfrak{g}}^{\mathrm{reg}}/\mathfrak{z}_{\mathfrak{g}}}$-module, if the map in (8.8) were not an isomorphism, it would be zero. We claim that this leads to a contradiction:

Consider the canonical maps of $H^\bullet(\mathrm{pt}/G) \simeq H^\bullet_{G[[t]]}(\mathrm{pt})$ to both the LHS and RHS of (*). Note that we have a canonical identification

$$H^\bullet(\mathrm{pt}/G) \simeq \operatorname{Sym}^\bullet(\mathfrak{h}^*)^W \simeq \operatorname{Sym}^\bullet(\check{\mathfrak{h}})^W \simeq \operatorname{Sym}^\bullet(\check{\mathfrak{g}})^{\check{G}}.$$

By the construction of the isomorphism in [ABG, Theorem 7.6.1],

$$H^\bullet(\mathrm{pt}/G) \to \operatorname{Ext}^\bullet_{D(\mathfrak{D}(\mathrm{Gr})_{\mathrm{crit}}\text{-mod})^{G[[t]]}}\left(\delta_{1_{\mathrm{Gr}}}, \underset{V\in\operatorname{Irr}(\mathcal{R}ep(\check{G}))}{\oplus} \mathcal{F}_V \otimes V^*\right)$$

corresponds to the canonical embedding $\operatorname{Sym}^\bullet(\check{\mathfrak{g}})^{\check{G}} \to \operatorname{Sym}^\bullet(\check{\mathfrak{g}})$. Therefore, if the map of (*) was 0 on the generators, it would also annihilate the augmentation ideal in $H^\bullet(\mathrm{pt}/G)$. However, we have the following assertion.

Theorem 8.27. *The map*

$$H^\bullet(\mathrm{pt}/G) \simeq H^\bullet_{G[[t]]}(\mathrm{pt}) \to \operatorname{Ext}^\bullet_{D(\widehat{\mathfrak{g}}_{\mathrm{crit}}\text{-mod}_{\mathrm{reg}})^{G[[t]]}}(\mathbb{V}_{\mathrm{crit}}, \mathbb{V}_{\mathrm{crit}})$$

corresponds under the isomorphism of Theorem 8.22 *to the map*

$$H^\bullet(\mathrm{pt}/G) \simeq \mathrm{Sym}^\bullet(\check{\mathfrak{g}})^{\check{G}} \overset{\tau}{\simeq} \mathrm{Sym}^\bullet(\check{\mathfrak{g}})^{\check{G}} \to \mathcal{P}_{\check{G},\mathfrak{z}_{\mathfrak{g}}^{\mathrm{reg}}} \overset{\check{G}}{\times} \mathrm{Sym}^\bullet(\check{\mathfrak{g}}) \simeq \mathrm{Sym}^\bullet(\check{\mathfrak{g}}_{\mathfrak{z}_{\mathfrak{g}}^{\mathrm{reg}}}),$$

where τ *is as in Section* 7.15.

The proof of this theorem will be given in the next section.

9 A manipulation with equivariant cohomology: Proof of Theorem 8.27

9.1

We will consider the algebra of self-Exts of $\mathbb{V}_{\mathrm{crit}}$ in a category bigger than $\widehat{\mathfrak{g}}_{\mathrm{crit}}\text{-mod}_{\mathrm{reg}}$, namely, in the category $\widehat{\mathfrak{g}}_{\mathrm{crit}}\text{-mod}_{\mathrm{nilp}}$.

Let $\mathcal{P}_{\check{B},\mathrm{Op}_{\check{\mathfrak{g}}}^{\mathrm{nilp}}}$ be the canonical $\check{B}$-torsor on the scheme $\mathrm{Op}_{\check{\mathfrak{g}}}^{\mathrm{nilp}}$, and let $\mathcal{P}_{\check{B},\mathrm{Op}_{\check{\mathfrak{g}}}^{\mathrm{reg}}}$ be its restriction to $\mathrm{Op}_{\check{\mathfrak{g}}}^{\mathrm{reg}}$. We will denote by $\mathcal{P}_{\check{B},\mathfrak{z}_{\mathfrak{g}}^{\mathrm{nilp}}}$ and $\mathcal{P}_{\check{B},\mathfrak{z}_{\mathfrak{g}}^{\mathrm{reg}}}$ the corresponding $\check{B}$-torsors on $\mathrm{Spec}(\mathfrak{z}_{\mathfrak{g}}^{\mathrm{nilp}})$ and $\mathrm{Spec}(\mathfrak{z}_{\mathfrak{g}}^{\mathrm{reg}})$, respectively. If V is a representation of $\check{B}$ (in practice we will take $V = \check{\mathfrak{b}}, \check{\mathfrak{n}}, \check{\mathfrak{g}}/\check{\mathfrak{n}}$, etc.), we will denote by $V_{\mathfrak{z}_{\mathfrak{g}}^{\mathrm{nilp}}}$, $V_{\mathfrak{z}_{\mathfrak{g}}^{\mathrm{reg}}}$ the corresponding modules over $\mathfrak{z}_{\mathfrak{g}}^{\mathrm{nilp}}$ and $\mathfrak{z}_{\mathfrak{g}}^{\mathrm{reg}}$, respectively.

Recall that by Corollary 4.18, the image of the normal $N_{\mathrm{Op}_{\check{\mathfrak{g}}}^{\mathrm{reg}}/\mathrm{Op}_{\check{\mathfrak{g}}}^{\mathrm{nilp}}}$ in the quotient

$$N_{\mathfrak{z}_{\mathfrak{g}}^{\mathrm{reg}}/\mathfrak{z}_{\mathfrak{g}}}/\Omega^1(\mathfrak{z}_{\mathfrak{g}}^{\mathrm{reg}})$$

identifies with $\check{\mathfrak{n}}_{\mathfrak{z}_{\mathfrak{g}}^{\mathrm{reg}}} \subset \check{\mathfrak{g}}_{\mathfrak{z}_{\mathfrak{g}}^{\mathrm{reg}}}$. From the proof of Theorem 8.22 we obtain the following statement.

Lemma 9.2. *The natural map*

$$\mathrm{Ext}^\bullet_{D(\widehat{\mathfrak{g}}_{\mathrm{crit}}\text{-mod}_{\mathrm{reg}})^{G[[t]]}}(\mathbb{V}_{\mathrm{crit}}, \mathbb{V}_{\mathrm{crit}}) \to \mathrm{Ext}^\bullet_{D(\widehat{\mathfrak{g}}_{\mathrm{crit}}\text{-mod}_{\mathrm{nilp}})^{G[[t]]}}(\mathbb{V}_{\mathrm{crit}}, \mathbb{V}_{\mathrm{crit}})$$

induces an isomorphism

$$\mathrm{Sym}^\bullet((\check{\mathfrak{g}}/\check{\mathfrak{n}})_{\mathfrak{z}_{\mathfrak{g}}^{\mathrm{reg}}}) \simeq \mathrm{Ext}^\bullet_{D(\widehat{\mathfrak{g}}_{\mathrm{crit}}\text{-mod}_{\mathrm{nilp}})^{G[[t]]}}(\mathbb{V}_{\mathrm{crit}}, \mathbb{V}_{\mathrm{crit}}).$$

By the equivariance of the map in (*) with respect to the algebroid $N^*_{\mathfrak{z}_{\mathfrak{g}}^{\mathrm{reg}}/\mathfrak{z}_{\mathfrak{g}}}$, the image of $H^\bullet(\mathrm{pt}/G)$ in $\mathrm{Sym}^\bullet_{\mathfrak{z}_{\mathfrak{g}}^{\mathrm{reg}}}(\check{\mathfrak{g}}_{\mathfrak{z}_{\mathfrak{g}}^{\mathrm{reg}}})$ is a priori contained in the subalgebra $\mathrm{Sym}^\bullet(\check{\mathfrak{g}})^{\check{G}}$. Hence, it is sufficient to show that the composition

$$\mathrm{Sym}^\bullet(\check{\mathfrak{h}})^W \simeq H^\bullet(\mathrm{pt}/G) \to \mathrm{Ext}^\bullet_{D(\widehat{\mathfrak{g}}_{\mathrm{crit}}\text{-mod}_{\mathrm{nilp}})^{G[[t]]}}(\mathbb{V}_{\mathrm{crit}}, \mathbb{V}_{\mathrm{crit}}) \simeq \mathrm{Sym}^\bullet((\check{\mathfrak{g}}/\check{\mathfrak{n}})_{\mathfrak{z}_{\mathfrak{g}}^{\mathrm{reg}}})$$

equals the natural map

$$\mathrm{Sym}^\bullet(\check{\mathfrak{h}})^W \overset{\tau}{\simeq} \mathrm{Sym}^\bullet(\check{\mathfrak{h}})^W \to \mathrm{Sym}^\bullet(\check{\mathfrak{h}}) \to \mathrm{Sym}^\bullet((\check{\mathfrak{b}}/\check{\mathfrak{n}})_{\mathfrak{z}_{\mathfrak{g}}^{\mathrm{reg}}}) \hookrightarrow \mathrm{Sym}^\bullet((\check{\mathfrak{g}}/\check{\mathfrak{n}})_{\mathfrak{z}_{\mathfrak{g}}^{\mathrm{reg}}}).$$

9.3

Now consider the module $\mathbb{M}_0 \in \widehat{\mathfrak{g}}_{\mathrm{crit}}\text{-mod}^I_{\mathrm{nilp}}$. Since $\mathrm{Av}_{G[[t]]/I}(\mathbb{M}_0) \simeq \mathbb{V}_{\mathrm{crit}}$, by Section 22.21, we obtain an isomorphism

$$\mathrm{RHom}_{D(\widehat{\mathfrak{g}}_{\mathrm{crit}}\text{-mod}_{\mathrm{nilp}})^{G[[t]]}}(\mathbb{V}_{\mathrm{crit}}, \mathbb{V}_{\mathrm{crit}}) \simeq \mathrm{RHom}_{D(\widehat{\mathfrak{g}}_{\mathrm{crit}}\text{-mod}_{\mathrm{nilp}})^I}(\mathbb{M}_0, \mathbb{V}_{\mathrm{crit}}).$$

It is easy to see that the composition

$$\begin{aligned} H^\bullet(\mathrm{pt}\,/\,G[[t]]) &\to \mathrm{Ext}^\bullet_{D(\widehat{\mathfrak{g}}_{\mathrm{crit}}\text{-mod}_{\mathrm{nilp}})^{G[[t]]}}(\mathbb{V}_{\mathrm{crit}}, \mathbb{V}_{\mathrm{crit}}) \\ &\to \mathrm{Ext}^\bullet_{D(\widehat{\mathfrak{g}}_{\mathrm{crit}}\text{-mod}_{\mathrm{nilp}})^I}(\mathbb{M}_0, \mathbb{V}_{\mathrm{crit}}) \end{aligned}$$

equals the map

$$\begin{aligned} H^\bullet(\mathrm{pt}\,/\,G[[t]]) \to H^\bullet(\mathrm{pt}\,/\,I) &\to \mathrm{Ext}^\bullet_{D(\widehat{\mathfrak{g}}_{\mathrm{crit}}\text{-mod}_{\mathrm{nilp}})^I}(\mathbb{M}_0, \mathbb{M}_0) \\ &\to \mathrm{Ext}^\bullet_{D(\widehat{\mathfrak{g}}_{\mathrm{crit}}\text{-mod}_{\mathrm{nilp}})^I}(\mathbb{M}_0, \mathbb{V}_{\mathrm{crit}}). \end{aligned}$$

By Corollary 13.9, the module $\mathbb{M}_0$ is flat over $\mathfrak{z}^{\mathrm{nilp}}_{\mathfrak{g}}$. Hence, by Lemma 23.3

$$\mathrm{RHom}^\bullet_{D(\widehat{\mathfrak{g}}_{\mathrm{crit}}\text{-mod}_{\mathrm{nilp}})^I}(\mathbb{M}_0, \mathcal{M}) \simeq \mathrm{RHom}^\bullet_{D(\widehat{\mathfrak{g}}_{\mathrm{crit}}\text{-mod}_{\mathrm{reg}})^I}(\mathbb{M}_{0,\mathrm{reg}}, \mathcal{M})$$

for any $\mathcal{M} \in \widehat{\mathfrak{g}}_{\mathrm{crit}}\text{-mod}^I_{\mathrm{reg}}$, where $\mathbb{M}_{0,\mathrm{reg}} := \mathbb{M}_0 \underset{\mathfrak{z}^{\mathrm{nilp}}_{\mathfrak{g}}}{\otimes} \mathfrak{z}^{\mathrm{reg}}_{\mathfrak{g}}$.

Moreover, the map

$$H^\bullet(\mathrm{pt}\,/\,I) \to \mathrm{Ext}^\bullet_{D(\widehat{\mathfrak{g}}_{\mathrm{crit}}\text{-mod}_{\mathrm{nilp}})^I}(\mathbb{M}_0, \mathbb{M}_0) \to \mathrm{Ext}^\bullet_{D(\widehat{\mathfrak{g}}_{\mathrm{crit}}\text{-mod}_{\mathrm{nilp}})^I}(\mathbb{M}_0, \mathbb{V}_{\mathrm{crit}})$$

that appears above equals the map

$$\begin{aligned} H^\bullet(\mathrm{pt}\,/\,I) &\to \mathrm{Ext}^\bullet_{D(\widehat{\mathfrak{g}}_{\mathrm{crit}}\text{-mod}_{\mathrm{reg}})^I}(\mathbb{M}_{0,\mathrm{reg}}, \mathbb{M}_{0,\mathrm{reg}}) \to \\ &\to \mathrm{Ext}^\bullet_{D(\widehat{\mathfrak{g}}_{\mathrm{crit}}\text{-mod}_{\mathrm{reg}})^I}(\mathbb{M}_{0,\mathrm{reg}}, \mathbb{V}_{\mathrm{crit}}) \simeq \mathrm{Ext}^\bullet_{D(\widehat{\mathfrak{g}}_{\mathrm{crit}}\text{-mod}_{\mathrm{nilp}})^I}(\mathbb{M}_0, \mathbb{V}_{\mathrm{crit}}). \end{aligned}$$

Thus we obtain a commutative diagram

$$\begin{array}{ccc} H^\bullet(\mathrm{pt}\,/\,G[[t]]) & \longrightarrow & H^\bullet(\mathrm{pt}\,/\,I) \\ \downarrow & & \downarrow \\ \mathrm{Sym}^\bullet((\check{\mathfrak{g}}/\check{\mathfrak{n}})_{\mathfrak{z}^{\mathrm{reg}}_{\mathfrak{g}}}) & \xrightarrow{\sim} & \mathrm{Ext}^\bullet_{D(\widehat{\mathfrak{g}}_{\mathrm{crit}}\text{-mod}_{\mathrm{reg}})^I}(\mathbb{M}_{0,\mathrm{reg}}, \mathbb{V}_{\mathrm{crit}}), \end{array}$$

and it is easy to see that the resulting map

$$\mathrm{Sym}^\bullet(\check{\mathfrak{h}}) \simeq H^\bullet(\mathrm{pt}\,/\,I) \to \mathrm{Sym}^\bullet((\check{\mathfrak{g}}/\check{\mathfrak{n}})_{\mathfrak{z}^{\mathrm{reg}}_{\mathfrak{g}}})$$

is a homomorphism of algebras.

Therefore, it suffices to show that the map

$$\check{\mathfrak{h}} \simeq H^2(\mathrm{pt}\,/I) \to \mathrm{Ext}^2_{D(\widehat{\mathfrak{g}}_{\mathrm{crit}}\text{-mod}_{\mathrm{reg}})^I}(\mathbb{M}_{0,\mathrm{reg}}, \mathbb{M}_{0,\mathrm{reg}}) \to \\ \to \mathrm{Ext}^2_{D(\widehat{\mathfrak{g}}_{\mathrm{crit}}\text{-mod}_{\mathrm{reg}})^I}(\mathbb{M}_{0,\mathrm{reg}}, \mathbb{V}_{\mathrm{crit}}) \simeq (\check{\mathfrak{g}}/\check{\mathfrak{n}})_{\mathfrak{z}^{\mathrm{reg}}_{\mathfrak{g}}}$$

equals the negative of the tautological map $\check{\mathfrak{h}} \to (\check{\mathfrak{g}}/\check{\mathfrak{n}})_{\mathfrak{z}^{\mathrm{reg}}_{\mathfrak{g}}}$.

Let $\mathcal{M}$ be an arbitrary I-equivariant object of $\widehat{\mathfrak{g}}_{\mathrm{crit}}\text{-mod}_{\mathrm{reg}}$. One easily establishes the following compatibility of spectral sequences.

Lemma 9.4. *The composition*

$$\mathrm{Ext}^1_{\widehat{\mathfrak{g}}_{\mathrm{crit}}\text{-mod}}(\mathcal{M}, \mathcal{M}) \to \mathrm{Hom}(\mathcal{M}, \mathcal{M}) \underset{\mathfrak{z}^{\mathrm{reg}}_{\mathfrak{g}}}{\otimes} N_{\mathfrak{z}^{\mathrm{reg}}_{\mathfrak{g}}/\mathfrak{z}_{\mathfrak{g}}} \to \mathrm{Ext}^2_{D(\widehat{\mathfrak{g}}_{\mathrm{crit}}\text{-mod}_{\mathrm{reg}})^I}(\mathcal{M}, \mathcal{M})$$

equals the composition

$$\mathrm{Ext}^1_{\widehat{\mathfrak{g}}_{\mathrm{crit}}\text{-mod}}(\mathcal{M}, \mathcal{M}) \to \mathrm{Hom}(\mathcal{M}, \mathcal{M}) \otimes H^2(\mathrm{pt}\,/I) \to \mathrm{Ext}^2_{D(\widehat{\mathfrak{g}}_{\mathrm{crit}}\text{-mod}_{\mathrm{reg}})^I}(\mathcal{M}, \mathcal{M}).$$

Therefore, to complete the proof of Theorem 8.27, it is sufficient to construct a map $\mathfrak{h}^* \to \mathrm{Ext}^1_{\widehat{\mathfrak{g}}_{\mathrm{crit}}\text{-mod}}(\mathbb{M}_{0,\mathrm{reg}}, \mathbb{M}_{0,\mathrm{reg}})$ such that the composition

$$\mathfrak{h}^* \to \mathrm{Ext}^1_{\widehat{\mathfrak{g}}_{\mathrm{crit}}\text{-mod}}(\mathbb{M}_{0,\mathrm{reg}}, \mathbb{M}_{0,\mathrm{reg}}) \to \mathrm{Hom}(\mathbb{M}_{0,\mathrm{reg}}, \mathbb{M}_{0,\mathrm{reg}}) \otimes H^2(\mathrm{pt}\,/I)$$

comes from the natural isomorphism $\mathfrak{h}^* \to H^2(\mathrm{pt}\,/I)$, and the composition

$$\check{\mathfrak{h}} \simeq \mathfrak{h}^* \to \mathrm{Ext}^1_{\widehat{\mathfrak{g}}_{\mathrm{crit}}\text{-mod}}(\mathbb{M}_{0,\mathrm{reg}}, \mathbb{M}_{0,\mathrm{reg}}) \to \mathrm{Hom}(\mathbb{M}_{0,\mathrm{reg}}, \mathbb{M}_{0,\mathrm{reg}}) \underset{\mathfrak{z}^{\mathrm{reg}}_{\mathfrak{g}}}{\otimes} N_{\mathfrak{z}^{\mathrm{reg}}_{\mathfrak{g}}/\mathfrak{z}_{\mathfrak{g}}} \\ \to \mathrm{Hom}(\mathbb{M}_{0,\mathrm{reg}}, \mathbb{M}_{0,\mathrm{reg}}) \underset{\mathfrak{z}^{\mathrm{reg}}_{\mathfrak{g}}}{\otimes} (\check{\mathfrak{g}}/\check{\mathfrak{n}})_{\mathfrak{z}^{\mathrm{reg}}_{\mathfrak{g}}}$$

equals the negative of the embedding $\check{\mathfrak{h}} \to (\check{\mathfrak{b}}/\check{\mathfrak{n}})_{\mathfrak{z}^{\mathrm{reg}}_{\mathfrak{g}}}$.

9.5

The required map $\mathfrak{h}^* \to \mathrm{Ext}^1_{\widehat{\mathfrak{g}}_{\mathrm{crit}}\text{-mod}}(\mathbb{M}_{0,\mathrm{reg}}, \mathbb{M}_{0,\mathrm{reg}})$ is constructed as follows. By deforming the highest weight, we obtain the "universal" Verma module $U(\mathfrak{g}) \underset{U(\mathfrak{n})}{\otimes} \mathbb{C} =: M_{\mathrm{univ}}$, and the corresponding induced module $\mathbb{M}_{\mathrm{univ}}$ over $\widehat{\mathfrak{g}}_{\mathrm{crit}}$. In particular, we have a map $\mathfrak{h}^* \to \mathrm{Ext}^1(\mathbb{M}_0, \mathbb{M}_0)$.

Clearly, the composition $\mathfrak{h}^* \to \mathrm{Ext}^1(\mathbb{M}_0, \mathbb{M}_0) \to \mathrm{Ext}^1(\mathbb{M}_0, \mathbb{M}_{0,\mathrm{reg}})$ factors canonically through $\mathrm{Ext}^1_{\widehat{\mathfrak{g}}_{\mathrm{crit}}\text{-mod}}(\mathbb{M}_{0,\mathrm{reg}}, \mathbb{M}_{0,\mathrm{reg}})$.

The fact that the composition

$$\mathfrak{h}^* \to \mathrm{Ext}^1_{\widehat{\mathfrak{g}}_{\mathrm{crit}}\text{-mod}}(\mathbb{M}_{0,\mathrm{reg}}, \mathbb{M}_{0,\mathrm{reg}}) \to \mathrm{Hom}(\mathbb{M}_{0,\mathrm{reg}}, \mathbb{M}_{0,\mathrm{reg}}) \otimes H^2(\mathrm{pt}\,/I)$$

comes from $\mathfrak{h}^* \to H^2(\mathrm{pt}\,/I)$ follows from the corresponding property of the composition $\mathfrak{h}^* \to \mathrm{Ext}^1_{\mathfrak{g}\text{-mod}}(M_0, M_0) \to \mathrm{Hom}(M_0, M_0) \otimes H^2(\mathrm{pt}\,/I)$.

Consider the composition $\check{\mathfrak{h}} \simeq \mathfrak{h}^* \to \mathrm{Hom}(\mathbb{M}_{0,\mathrm{reg}}, \mathbb{M}_{0,\mathrm{reg}}) \underset{\mathfrak{z}_{\mathfrak{g}}^{\mathrm{reg}}}{\otimes} (\check{\mathfrak{g}}/\check{\mathfrak{n}})_{\mathfrak{z}_{\mathfrak{g}}^{\mathrm{reg}}}$. This map is equivariant with respect to the group $\mathrm{Aut}(\mathcal{D})$. In particular, if we choose a coordinate on $\mathcal{D}$, the above map has degree 0 with respect to the action of $\mathbb{G}_m$ by loop rotations. Since $\check{\mathfrak{h}}$ equals the degree 0 subspace of $\mathrm{Hom}(\mathbb{M}_{0,\mathrm{reg}}, \mathbb{M}_{0,\mathrm{reg}}) \underset{\mathfrak{z}_{\mathfrak{g}}^{\mathrm{reg}}}{\otimes} (\check{\mathfrak{g}}/\check{\mathfrak{n}})_{\mathfrak{z}_{\mathfrak{g}}^{\mathrm{reg}}}$ (see Section 4.22), we obtain that the map in question factors through *some* map $\check{\mathfrak{h}} \to \check{\mathfrak{h}}$.

To prove that the latter map is, in fact, the negative of the identity, we proceed as follows. By Section 2.13, we have an identification $\mathfrak{h}^* \otimes \mathfrak{z}_{\mathfrak{g}}^{\mathrm{nilp}} \simeq N_{\mathfrak{z}_{\mathfrak{g}}^{\mathrm{nilp}}/\mathfrak{z}_{\mathfrak{g}}^{\mathrm{RS}}}$. Moreover, by Lemma 4.19, the composition

$$\mathfrak{h}^* \otimes \mathfrak{z}_{\mathfrak{g}}^{\mathrm{reg}} \simeq N_{\mathfrak{z}_{\mathfrak{g}}^{\mathrm{nilp}}/\mathfrak{z}_{\mathfrak{g}}^{\mathrm{RS}}}|_{\mathrm{Spec}(\mathfrak{z}_{\mathfrak{g}}^{\mathrm{reg}})} \to N_{\mathfrak{z}_{\mathfrak{g}}^{\mathrm{nilp}}/\mathfrak{z}_{\mathfrak{g}}}|_{\mathrm{Spec}(\mathfrak{z}_{\mathfrak{g}}^{\mathrm{reg}})} \to (\check{\mathfrak{g}}/\check{\mathfrak{n}})_{\mathfrak{z}_{\mathfrak{g}}^{\mathrm{reg}}},$$

maps identically onto $\mathfrak{h}^* \otimes \mathfrak{z}_{\mathfrak{g}}^{\mathrm{reg}} \subset (\check{\mathfrak{g}}/\check{\mathfrak{n}})_{\mathfrak{z}_{\mathfrak{g}}^{\mathrm{reg}}}$.

Now, our assertion follows from the fact that the map

$$\mathfrak{h}^* \otimes \mathfrak{z}_{\mathfrak{g}}^{\mathrm{nilp}} \to \mathrm{Ext}^1(\mathbb{M}_0, \mathbb{M}_0) \to \mathrm{Hom}(\mathbb{M}_0, \mathbb{M}_0) \underset{\mathfrak{z}_{\mathfrak{g}}^{\mathrm{nilp}}}{\otimes} N_{\mathfrak{z}_{\mathfrak{g}}^{\mathrm{nilp}}/\mathfrak{z}_{\mathfrak{g}}}$$

equals the negative of

$$\begin{aligned}\mathfrak{h}^* \otimes \mathfrak{z}_{\mathfrak{g}}^{\mathrm{nilp}} \simeq N_{\mathfrak{z}_{\mathfrak{g}}^{\mathrm{nilp}}/\mathfrak{z}_{\mathfrak{g}}^{\mathrm{RS}}} &\overset{1\otimes \mathrm{id}}{\longrightarrow} \mathrm{Hom}(\mathbb{M}_0, \mathbb{M}_0) \underset{\mathfrak{z}_{\mathfrak{g}}^{\mathrm{nilp}}}{\otimes} N_{\mathfrak{z}_{\mathfrak{g}}^{\mathrm{nilp}}/\mathfrak{z}_{\mathfrak{g}}^{\mathrm{RS}}} \\ &\to \mathrm{Hom}(\mathbb{M}_0, \mathbb{M}_0) \underset{\mathfrak{z}_{\mathfrak{g}}^{\mathrm{nilp}}}{\otimes} N_{\mathfrak{z}_{\mathfrak{g}}^{\mathrm{nilp}}/\mathfrak{z}_{\mathfrak{g}}}\end{aligned}$$

by Proposition 7.16.

Part III: Wakimoto Modules

In this part we review the Wakimoto modules which were introduced for an arbitrary affine Kac–Moody algebra $\widehat{\mathfrak{g}}$ in [FF1, FF2, F] following the work of Wakimoto [W] in the case of $\widehat{\mathfrak{sl}}_2$. On the intuitive level, Wakimoto modules are sections of certain D-modules on the Iwahori orbits on the semi-infinite flag manifold $G((t))/B((t))$. The construction of [FF1, FF2, F] may be phrased in terms of a kind of semi-infinite induction functor, as we explain below. This approach to the Wakimoto modules is similar to the one discussed in [Ar, Vor, GMS]. It uses the formalism of chiral algebras, and in particular, the chiral algebra of differential operators on the group G. It also uses the language of semi-infinite cohomology, which was introduced by Feigin [Fe] and, in the setting of chiral algebras, by Beilinson and Drinfeld [CHA].

Let $\overset{\circ}{G}$ be the big cell $B \cdot w_0 \cdot B \subset G$, and for an arbitrary level κ we consider the chiral algebra $\mathfrak{D}^{\mathrm{ch}}(\overset{\circ}{G})_\kappa$ of chiral differential operators on it. In Section 10 we define the chiral algebra $\mathfrak{D}^{ch}(G/\overset{\circ}{N})_\kappa$ as a BRST reduction of $\mathfrak{D}^{ch}(\overset{\circ}{G})_\kappa$ with respect to $\mathfrak{n}((t))$. This chiral algebra can be thought of as governing D-modules on the big cell in $G((t))/N((t))$; we show that the natural homomorphism to it from the chiral algebra, corresponding to the Kac–Moody Lie algebra $\widehat{\mathfrak{g}}_\kappa$, coincides with the free field realization homomorphism of Feigin and Frenkel.

By construction, any chiral module over $\mathfrak{D}^{ch}(G/\overset{\circ}{N})_\kappa$ is a bimodule over $\widehat{\mathfrak{g}}_\kappa$ and the Heisenberg algebra $\widehat{\mathfrak{h}}_{-\kappa+\kappa_{\mathrm{crit}}}$. In Section 11, for any such module we define the induction functor from the category $\widehat{\mathfrak{h}}_{\kappa-\kappa_{\mathrm{crit}}}$-mod to $\widehat{\mathfrak{g}}_\kappa$-mod. The resulting $\widehat{\mathfrak{g}}_\kappa$-modules are by definition the Wakimoto modules. Thus Wakimoto modules can be viewed as induced from $\widehat{\mathfrak{h}}_{\kappa-\kappa_{\mathrm{crit}}}$ to $\widehat{\mathfrak{g}}_\kappa$ using certain bimodules.

In Section 12 we study cohomological properties of Wakimoto modules and, in particular, their behavior with respect to the convolution functors. The crucial result that we need below is Proposition 12.12, which states that Wakimoto modules are essentially invariant under convolution with "lattice" elements in the Iwahori–Hecke algebra.

In Section 13 we specialize to the case $\kappa = \kappa_{\mathrm{crit}}$. The crucial result here, due to [F], is that certain Wakimoto modules are isomorphic to Verma modules over $\widehat{\mathfrak{g}}_{\mathrm{crit}}$. This fact will allow us to obtain information about the structure of Verma modules that will be used in the subsequent sections.

10 Free field realization

In what follows we will use the language of chiral algebras on a curve X, developed in [CHA]. We will fix a point $x \in X$ and identify $\mathfrak{D}_X$-modules supported at this point with underlying vector spaces. We will identify the formal disc $\mathcal{D}$ with the formal neighborhood of x in X.

10.1

Let L be a Lie-* algebra, which we assume to be projective and finitely generated as a $\mathfrak{D}_X$-module. Recall that there exists a canonical Tate central extension of L, which is a Lie-* algebra $\widehat{L}^{\mathrm{Tate}}$

$$0 \to \omega_X \to \widehat{L}^{\mathrm{Tate}} \to L \to 0$$

(see [CHA, Section 2.7]). The key property of $\widehat{L}^{\mathrm{Tate}}$ is that if $\mathcal{M}$ is a chiral module over $\widehat{L}^{-\mathrm{Tate}}$ (here "$-$" signifies the Baer negative central extension), then we have a well-defined complex of $\mathfrak{D}_X$-modules, denoted $\mathfrak{C}^{\frac{\infty}{2}}(L, \mathcal{M})$, which we will refer to as the semi-infinite complex of $\mathcal{M}$ with respect to L. We will denote by $H^{\frac{\infty}{2}}(L, \mathcal{M})$ (respectively, $H^{\frac{\infty}{2}+i}(L, \mathcal{M})$) the 0th (respectively, ith) cohomology of this complex.

If $\mathcal{M}$ is supported at the point $x \in X$, by definition $\mathfrak{C}^{\frac{\infty}{2}}(L, \mathcal{M})$ is given by the semi-infinite complex of the Tate Lie algebra $H^0_{DR}(\mathcal{D}^\times, L)$ with respect to the lattice $H^0_{DR}(\mathcal{D}, L) \subset H^0_{DR}(\mathcal{D}^\times, L)$.

If $\mathcal{A}$ is a chiral algebra with a homomorphism $\widehat{L}^{-\mathrm{Tate}} \to \mathcal{A}$, then $\mathfrak{C}^{\frac{\infty}{2}}(L, \mathcal{A})$ has a natural structure of a DG chiral algebra.

Now let L' and L'' be two central extensions of L by ω_X, whose Baer sum is identified with $\widehat{L}^{-\text{Tate}}$, and let $\mathcal{M}$ and $\mathcal{M}'$ be L'- and L''-modules, respectively. Then $\mathcal{M} \otimes \mathcal{M}'$ is a module over $\widehat{L}^{-\text{Tate}}$, and in this case we will use the notation

$$\mathcal{M} \overset{\frac{\infty}{2}}{\underset{L}{\otimes}} \mathcal{M}' \quad \text{or} \quad \mathcal{M} \overset{\frac{\infty}{2}}{\underset{H^0_{DR}(\mathcal{D}^\times, L), H^0_{DR}(\mathcal{D}, L)}{\otimes}} \mathcal{M}'$$

instead of $\mathfrak{C}^{\frac{\infty}{2}}(L, \mathcal{M} \otimes \mathcal{M}')$. If the latter is acyclic away from cohomological degree 0 we will denote by the same symbol the corresponding 0th cohomology.

Finally, let $\mathfrak{h}$ be a finite-dimensional subspace in $H^0_{DR}(X, L)$. In this case, $\mathfrak{C}^{\frac{\infty}{2}}(L, \mathcal{M})$ admits a subcomplex $\mathfrak{C}^{\frac{\infty}{2}}(L; \mathfrak{h}, \mathcal{M})$ of relative cochains. We will sometimes also use the notation $\mathfrak{C}^{\frac{\infty}{2}}(L; \mathfrak{h}, \cdot)$ and $\cdot \overset{\frac{\infty}{2}}{\underset{L;\mathfrak{h}}{\otimes}} \cdot$.

10.2

Let $L_{\mathfrak{g}}$, $L_{\mathfrak{b}}$ and $L_{\mathfrak{n}}$ be the Lie-* algebras corresponding to the Lie algebras $\mathfrak{g}$, $\mathfrak{b}$ and $\mathfrak{n}$, respectively. For a level κ, we will denote by $L_{\mathfrak{g},\kappa}$ the corresponding Kac–Moody extension of $L_{\mathfrak{g}}$ by ω_X, and by $L_{\mathfrak{b},\kappa}$ the induced central extension of $L_{\mathfrak{b}}$. Let $\widehat{L}^{\text{Tate}}_{\mathfrak{b}}$ be the Tate extension of $L_{\mathfrak{b}}$, and let $\widehat{L}'_{\mathfrak{b},\kappa}$ be the Baer sum of $\widehat{L}^{\text{Tate}}_{\mathfrak{b}}$ and $L_{\mathfrak{b},\kappa'}$, where $\kappa' = -\kappa - 2\kappa_{\text{crit}}$; let $\widehat{L}_{\mathfrak{b},\kappa}$ be the Baer negative of $\widehat{L}'_{\mathfrak{b},\kappa}$.

Since $\kappa'|_{\mathfrak{n}} = 0$, the extension induced by $L_{\mathfrak{b},\kappa'}$ on $L_{\mathfrak{n}}$ is canonically trivialized. The extension induced by $\widehat{L}^{\text{Tate}}_{\mathfrak{b}}$ is also canonically trivialized, since $\mathfrak{n}$ is nilpotent. Hence, $\widehat{L}_{\mathfrak{b},\kappa}$ comes from a well-defined central extension $\widehat{L}_{\mathfrak{h},\kappa}$ of the commutative Lie-* algebra $L_{\mathfrak{h}}$. We will denote by $\widehat{L}'_{\mathfrak{h},\kappa}$ the Baer negative of $\widehat{L}_{\mathfrak{h},\kappa}$.

Note that when κ is integral, the above central extensions of Lie algebras $H^0_{DR}(\mathcal{D}^\times, ?)$ all come from the corresponding central extensions of loop groups.

We will denote by $\mathfrak{H}_\kappa$ (respectively, $\mathfrak{H}'_\kappa$) the reduced universal enveloping chiral algebra of $\widehat{L}_{\mathfrak{h},\kappa}$ (respectively, $\widehat{L}'_{\mathfrak{h},\kappa}$). We will denote by $\mathcal{A}_{\mathfrak{g},\kappa}$ the reduced universal enveloping chiral algebra of $L_{\mathfrak{g},\kappa}$.

Let $\mathcal{M}$ be a chiral $L_{\mathfrak{b},\kappa'}$-module. Since the Tate extension of $L_{\mathfrak{b}}$, induced by the adjoint action equals the extension induced by the adjoint action on $L_{\mathfrak{n}}$, the complex $\mathfrak{C}^{\frac{\infty}{2}}(L_{\mathfrak{n}}, \mathcal{M})$ carries a chiral action of $\widehat{L}'_{\mathfrak{b},\kappa}$. The resulting action of $L_{\mathfrak{n}} \subset \widehat{L}'_{\mathfrak{b},\kappa}$ on the individual semi-infinite cohomologies $H^{\frac{\infty}{2}+i}(L_{\mathfrak{n}}, \mathcal{M})$ is trivial. Hence, we obtain that each $H^{\frac{\infty}{2}+i}(L_{\mathfrak{n}}, \mathcal{M})$ is a chiral $\mathfrak{H}'_\kappa$-module. If $\mathcal{R}$ is an $\mathfrak{H}_\kappa$-module, regarded as a $\widehat{L}_{\mathfrak{b},\kappa}$-module, $\mathfrak{C}^{\frac{\infty}{2}}(L_{\mathfrak{b}}, \mathcal{M} \otimes \mathcal{R})$ makes sense. If we suppose, moreover, that $\mathfrak{C}^{\frac{\infty}{2}}(L_{\mathfrak{n}}, \mathcal{M})$ is acyclic away from degree 0, then

$$\mathfrak{C}^{\frac{\infty}{2}}(L_{\mathfrak{b}}, \mathcal{M} \otimes \mathcal{R}) \simeq H^{\frac{\infty}{2}}(L_{\mathfrak{n}}, \mathcal{M}) \overset{\frac{\infty}{2}}{\underset{L_{\mathfrak{h}}}{\otimes}} \mathcal{R}.$$

10.3

Recall now that for any level κ we can introduce the chiral algebra of differential operators (CADO) $\mathfrak{D}^{\text{ch}}(G)_\kappa$, which admits two mutually commuting homomorphisms

$$\mathfrak{l}_{\mathfrak{g}} : \mathcal{A}_{\mathfrak{g},\kappa} \to \mathfrak{D}^{\mathrm{ch}}(G)_\kappa \leftarrow \mathcal{A}_{\mathfrak{g},\kappa'} : \mathfrak{r}_{\mathfrak{g}}.$$

Let $\overset{\circ}{G}$ denote the open Bruhat cell $B \cdot w_0 \cdot B \subset G$, where w_0 is the longest element of the Weyl group. We will denote by $\overset{\circ}{G/N}$, $\overset{\circ}{G/B}$ the corresponding open subsets in G/N and G/B, respectively.

Let $\mathfrak{D}^{\mathrm{ch}}(\overset{\circ}{G})_\kappa$ be the induced CADO on $\overset{\circ}{G}$. Consider the chiral DG algebra $\mathfrak{C}^{\frac{\infty}{2}}(L_{\mathfrak{n}}, \mathfrak{D}^{\mathrm{ch}}(\overset{\circ}{G})_\kappa)$, where we take $L_{\mathfrak{n}}$ mapping to $\mathfrak{D}^{\mathrm{ch}}(\overset{\circ}{G})_\kappa$ via

$$\mathfrak{D}^{\mathrm{ch}}(\overset{\circ}{G})_\kappa \leftarrow \mathfrak{D}^{\mathrm{ch}}(G)_\kappa \overset{\mathfrak{r}_{\mathfrak{g}}}{\longleftarrow} \mathcal{A}_{\mathfrak{g},\kappa'} \leftarrow L_{\mathfrak{g},\kappa'} \leftarrow L_{\mathfrak{n}}.$$

Since $\overset{\circ}{G} \to \overset{\circ}{G/N}$ is a principal N-bundle, from [CHA, Section 2.8.16] we obtain the following.

Lemma 10.4. *The complex $\mathfrak{C}^{\frac{\infty}{2}}(L_{\mathfrak{n}}, \mathfrak{D}^{\mathrm{ch}}(\overset{\circ}{G})_\kappa)$ is acyclic away from degree zero, and the resulting chiral algebra is a CADO on $\overset{\circ}{G/N}$.*

Let us denote $H^{\frac{\infty}{2}}(L_{\mathfrak{n}}, \mathfrak{D}^{\mathrm{ch}}(\overset{\circ}{G})_\kappa)$ by $\mathfrak{D}^{\mathrm{ch}}(\overset{\circ}{G/N})_\kappa$. By construction, we have a homomorphism of chiral algebras

$$\mathfrak{D}^{\mathrm{ch}}(\overset{\circ}{G/N})_\kappa \leftarrow \mathfrak{H}'_\kappa,$$

which we will denote by $\mathfrak{r}_{\mathfrak{h}}$. We define the chiral algebra $\mathfrak{D}^{\mathrm{ch}}(\overset{\circ}{G/B})_\kappa$ as the Lie-* centralizer of $\mathfrak{H}'_\kappa$ in $\mathfrak{D}^{\mathrm{ch}}(\overset{\circ}{G/N})_\kappa$. The map $\mathfrak{l}_{\mathfrak{g}} : \mathcal{A}_{\mathfrak{g},\kappa} \to \mathfrak{D}^{\mathrm{ch}}(G)_\kappa$ induces a homomorphism

$$\mathfrak{l}_{\mathfrak{g}} : \mathcal{A}_{\mathfrak{g},\kappa} \to \mathfrak{D}^{\mathrm{ch}}(\overset{\circ}{G/B})_\kappa. \tag{10.1}$$

Again, by construction, we have a canonical map

$$\mathfrak{D}^{\mathrm{ch}}(\overset{\circ}{G/B})_\kappa \to \mathfrak{D}^{\mathrm{ch}}(\overset{\circ}{G/N})_\kappa \overset{\frac{\infty}{2}}{\underset{L_{\mathfrak{h}};\mathfrak{h}}{\otimes}} \mathfrak{H}_\kappa. \tag{10.2}$$

Lemma 10.5. *The map in* (10.2) *is an isomorphism.*

The proof will become clear from the discussion in the next section.

10.6

Note that $\mathfrak{D}^{\mathrm{ch}}(\overset{\circ}{G/B})_\kappa$ is not a CADO on $\overset{\circ}{G/B}$. We will now give a more explicit, even if less canonical, description of the chiral algebras $\mathfrak{D}^{\mathrm{ch}}(\overset{\circ}{G/N})_\kappa$, $\mathfrak{D}^{\mathrm{ch}}(\overset{\circ}{G/B})_\kappa$ and the free field realization homomorphism.

Let us choose a representative of w_0 in W and identify the variety $\overset{\circ}{G} \simeq N \cdot w_0 \cdot B$ with the product $N \times B$, endowed with the action on N on the left and of B on the right. Then $\mathfrak{D}^{\mathrm{ch}}(\overset{\circ}{G})_\kappa$ becomes a CADO on this group, isomorphic to $\mathfrak{D}^{\mathrm{ch}}(N) \otimes \mathfrak{D}^{\mathrm{ch}}(B)_{\kappa'}$. We will denote the existing maps

$$\mathcal{A}_{\mathfrak{n}} \to \mathfrak{D}^{\mathrm{ch}}(N) \subset \mathfrak{D}^{\mathrm{ch}}(\overset{\circ}{G})_\kappa \quad \text{and} \quad \mathcal{A}_{\mathfrak{b},\kappa'} \to \mathfrak{D}^{\mathrm{ch}}(B)_{\kappa'} \subset \mathfrak{D}^{\mathrm{ch}}(\overset{\circ}{G})_\kappa$$

by $\mathfrak{l}_{\mathfrak{n}}$ and $\mathfrak{r}_{\mathfrak{b}}$, respectively, and the "new" maps, as in [AG1],

$$\mathcal{A}_{\mathfrak{n}} \to \mathfrak{D}^{\mathrm{ch}}(N) \subset \mathfrak{D}^{\mathrm{ch}}(\overset{\circ}{G})_\kappa \quad \text{and} \quad \widehat{\mathcal{A}}_{\mathfrak{b},\kappa} \to \mathfrak{D}^{\mathrm{ch}}(B)_{\kappa'} \subset \mathfrak{D}^{\mathrm{ch}}(\overset{\circ}{G})_\kappa$$

by $\mathfrak{r}_{\mathfrak{n}}$, $\mathfrak{l}_{\mathfrak{b}}$, respectively, where $\widehat{\mathcal{A}}_{\mathfrak{b},\kappa}$ is the reduced chiral universal envelope of $\widehat{L}_{\mathfrak{b},\kappa}$.

Then

$$\mathfrak{D}^{\mathrm{ch}}(\overset{\circ}{G}/N)_\kappa \simeq \mathfrak{D}^{\mathrm{ch}}(N) \otimes \widehat{\mathfrak{D}}^{\mathrm{ch}}(H)_\kappa, \tag{10.3}$$

where $\widehat{\mathfrak{D}}^{\mathrm{ch}}(H)_\kappa$ is a CADO on H with the maps

$$\mathfrak{H}_\kappa \xrightarrow{\mathfrak{l}_{\mathfrak{h}}} \widehat{\mathfrak{D}}^{\mathrm{ch}}(H)_\kappa \xleftarrow{\mathfrak{r}_{\mathfrak{h}}} \mathfrak{H}'_\kappa.$$

As usual, the centralizer of $\mathfrak{H}'_\kappa$ in $\widehat{\mathfrak{D}}^{\mathrm{ch}}(H)_\kappa$ is $\mathfrak{H}_\kappa$, and we obtain that

$$\mathfrak{D}^{\mathrm{ch}}(\overset{\circ}{G}/B)_\kappa \simeq \mathfrak{D}^{\mathrm{ch}}(N) \otimes \mathfrak{H}_\kappa.$$

The above isomorphism makes the assertion of Lemma 10.5 manifest: indeed, it follows from the fact that $\widehat{\mathfrak{D}}^{\mathrm{ch}}(H)_\kappa \overset{\frac{\infty}{2}}{\underset{L_{\mathfrak{h}};\mathfrak{h}}{\otimes}} \mathfrak{H}_\kappa \simeq \mathfrak{H}_\kappa$; see Section 22.8.

Homomorphism (10.1) therefore gives rise to a homomorphism from the affine Kac–Moody algebra to the tensor product of the chiral algebras $\mathfrak{D}^{\mathrm{ch}}(N)$ and $\mathfrak{H}_\kappa$:

$$\mathcal{A}_{\mathfrak{g},\kappa} \to \mathfrak{D}^{\mathrm{ch}}(N) \otimes \mathfrak{H}_\kappa. \tag{10.4}$$

This is the *free field realization* homomorphism of [FF2, F].

The CADO $\mathfrak{D}^{\mathrm{ch}}(N)$ may be identified with what physicists call the free field $\beta\gamma$ system, and $\mathfrak{H}_\kappa$ is a twisted form of a Heisenberg algebra, which is also related to a free bosonic system. This is why the homomorphism (10.4) is referred to as the free field realization.

10.7

Let us now explain in what sense the homomorphism

$$\mathcal{A}_{\mathfrak{g},\kappa} \to \mathfrak{D}^{\mathrm{ch}}(\overset{\circ}{G}/B)_\kappa \simeq \mathfrak{D}^{\mathrm{ch}}(N) \otimes \mathfrak{H}_\kappa \tag{10.5}$$

above is an affine analogue (i.e., chiralization) of a well-known phenomenon for finite-dimensional Lie algebras. We will appeal to notations introduced in [AG1].

Consider the variety $\overset{\circ}{G} \simeq N \times B$ with an action of the Lie algebra $\mathfrak{g}$ on the left. This action defines a map $\mathfrak{g} \to T(\overset{\circ}{G}/N)$, whose image consists of vector fields, that are invariant with respect to the action of $H \simeq B/N$ on the right. Since the Lie algebra of such vector fields is isomorphic to $T(N) \oplus (\mathrm{Fun}(N) \otimes \mathfrak{h})$, we obtain a map

$$\mathfrak{g} \to T(N) \oplus (\mathrm{Fun}(N) \otimes \mathfrak{h})\,. \tag{10.6}$$

The restriction of this map to $\mathfrak{n} \subset \mathfrak{g}$ is the homomorphism $\mathfrak{l}_n \to T(N)$. The restriction to $\mathfrak{h} \subset \mathfrak{g}$ is the sum of two maps: one is $\mathfrak{h} \to T(N)$, corresponding to the natural adjoint of H on N, and the other is the identity map $\mathfrak{h} \to \mathfrak{h} \subset \mathrm{Fun}(N) \otimes \mathfrak{h}$, twisted by w_0.

The map of (10.6) can be chiralized in a straightforward way, and we obtain a map of Lie-* algebras

$$L_{\mathfrak{g}} \to \Theta(N) \oplus \left(\mathrm{Fun}(\mathrm{Jets}(N)) \otimes L_{\mathfrak{h}}\right), \tag{10.7}$$

where for an affine scheme Y, we denote by $\mathrm{Jets}(Y)$ the $\mathfrak{D}_X$-scheme of jets into Y, and $\Theta(Y)$ denotes the tangent algebroid on this $\mathfrak{D}_X$-scheme. By construction, we have the following.

Lemma 10.8. *The image of $L_{\mathfrak{g},\kappa} \subset \mathcal{A}_{\mathfrak{g},\kappa}$ under* (10.5) *belongs to*

$$\mathfrak{D}^{\mathrm{ch}}(N)^{\leq 1} \oplus \left(\mathrm{Fun}(\mathrm{Jets}(N)) \otimes (\mathfrak{H}_\kappa)^{\leq 1}\right),$$

where $(\cdot)^{\leq i}$ denotes the PBW filtration. The composition

$$\begin{aligned} L_{\mathfrak{g},\kappa} &\to \mathfrak{D}^{\mathrm{ch}}(N)^{\leq 1} \oplus \left(\mathrm{Fun}(\mathrm{Jets}(N)) \otimes (\mathfrak{H}_\kappa)^{\leq 1}\right) \\ &\to \left(\mathfrak{D}^{\mathrm{ch}}(N)^{\leq 1}/\mathfrak{D}^{\mathrm{ch}}(N)^{\leq 0}\right) \oplus \left(\mathrm{Fun}(\mathrm{Jets}(N)) \otimes \left((\mathfrak{H}_\kappa)^{\leq 1}/(\mathfrak{H}_\kappa)^{\leq 0}\right)\right) \\ &\simeq \Theta(N) \oplus \left(\mathrm{Fun}(\mathrm{Jets}(N)) \otimes L_{\mathfrak{h}}\right) \end{aligned}$$

factors through $L_{\mathfrak{g}}$ and equals the map of (10.7).

10.9

For the remainder of this section we will specialize to the case when $\kappa = \kappa_{\mathrm{crit}}$. The following basic fact is established in [CHA, Section 2.8.17].

Proposition 10.10. *The Lie-* algebra $\widehat{L}'_{\mathfrak{h},\kappa}$ is commutative if and only if $\kappa = \kappa_{\mathrm{crit}}$. In this case there is a canonical isomorphism*

$$\mathrm{Spec}(\mathfrak{H}_{\mathrm{crit}}) \simeq \mathrm{Conn}_{\check{H}}(\omega_X^{\rho})^{\mathfrak{D}},$$

respecting the torsor structure on both sides with respect to the D-scheme of $\mathfrak{h}^*$-values 1-forms on X.

Since $\mathfrak{H}'_{\text{crit}}$ is commutative, it is contained as a chiral subalgebra in $\mathfrak{D}^{\text{ch}}(\overset{\circ}{G/B})_{\text{crit}}$; moreover, from (10.2) we infer

$$\mathfrak{H}'_{\text{crit}} \simeq \mathfrak{z}(\mathfrak{D}^{\text{ch}}(\overset{\circ}{G/B})_{\text{crit}}). \tag{10.8}$$

Since $\mathfrak{H}_{\text{crit}}$ is commutative as well, from the isomorphism (10.2) we obtain that there exists a homomorphism (which is easily seen to be an isomorphism) from $\mathfrak{H}_{\text{crit}}$ to $\mathfrak{z}(\mathfrak{D}^{\text{ch}}(\overset{\circ}{G/B})_{\text{crit}})$.

Lemma 10.11. *The resulting homomorphism $\mathfrak{H}_{\text{crit}} \to \mathfrak{H}'_{\text{crit}}$ comes from the sign-inversion isomorphism $L_{\mathfrak{h},\text{crit}} \to L'_{\mathfrak{h},\text{crit}}$ of commutative Lie-* algebras.*

10.12

We will now study the homomorphism

$$\mathfrak{l}_{\mathfrak{g}} : \mathcal{A}_{\mathfrak{g},\text{crit}} \to \mathfrak{D}^{\text{ch}}(\overset{\circ}{G/B})_{\text{crit}}. \tag{10.9}$$

Proposition 10.13. *The centralizer of $\mathcal{A}_{\mathfrak{g},\text{crit}}$ in $\mathfrak{D}^{\text{ch}}(\overset{\circ}{G/B})_{\text{crit}}$ equals $\mathfrak{H}'_{\text{crit}}$.*

Proof. Since $\mathfrak{H}'_{\text{crit}}$ is the center of the chiral algebra $\mathfrak{D}^{\text{ch}}(\overset{\circ}{G/B})_{\text{crit}}$, the fact that it centralizes the image of $\mathcal{A}_{\mathfrak{g},\text{crit}}$ is evident.

To prove the inclusion in the opposite direction, we will establish a stronger fact. Namely, that the centralizer in $\mathfrak{D}^{\text{ch}}(\overset{\circ}{G/B})_{\text{crit}}$ of the image of $L_{\mathfrak{n}} + \mathfrak{h}$ is already contained in $\mathfrak{H}'_{\text{crit}}$.

Using the description of $\mathfrak{D}^{\text{ch}}(\overset{\circ}{G/B})_{\text{crit}}$ given in Section 10.6, we obtain that the centralizer of $\mathfrak{l}_{\mathfrak{n}}(\mathcal{A}_{\mathfrak{n}})$ in it equals $\mathfrak{r}_{\mathfrak{n}}(\mathcal{A}_{\mathfrak{n}}) \otimes \mathfrak{H}'_{\text{crit}}$, in the notation of loc. cit.

Now consider the action of $\mathfrak{h} \in \Gamma(X, \mathcal{A}_{\mathfrak{g},\text{crit}})$ on

$$\mathfrak{D}^{\text{ch}}(\overset{\circ}{G/N})_{\text{crit}} \simeq \mathfrak{D}^{\text{ch}}(N) \otimes \widehat{\mathfrak{D}}^{\text{ch}}(H)_{\text{crit}}.$$

By Section 10.7, this action decomposes as a tensor product of the natural adjoint action on $\mathfrak{D}^{\text{ch}}(N)$, and the action on $\widehat{\mathfrak{D}}^{\text{ch}}(H)_{\kappa}$ given by $\mathfrak{l}_{\mathfrak{h}}$, twisted by w_0. Since $\widehat{\mathfrak{D}}^{\text{ch}}(H)_{\kappa}$ is commutative, the resulting action of $\mathfrak{h}$ on $\mathfrak{D}^{\text{ch}}(\overset{\circ}{G/B})_{\text{crit}} \simeq \mathfrak{D}^{\text{ch}}(N) \otimes \mathfrak{H}'_{\text{crit}}$ is the adjoint action along the first factor.

This implies our assertion since $(\mathcal{A}_{\mathfrak{n}})^{\mathfrak{h}} \simeq \mathbb{C}$, as $\mathfrak{h}$ acts on $\mathfrak{n}$, and hence on $\mathcal{A}_{\mathfrak{n}}$, by characters, which belong to the positive span of Δ^+. □

10.14

Now consider the composition

$$\mathfrak{z}_{\mathfrak{g}} = \mathfrak{z}(\mathcal{A}_{\mathfrak{g},\text{crit}}) \to \mathfrak{D}^{\text{ch}}(\overset{\circ}{G/B})_{\text{crit}}.$$

From Proposition 10.13 we immediately obtain the following result.

Corollary 10.15. *The image of* $\mathfrak{z}_{\mathfrak{g}}$ *is contained in* $\mathfrak{z}(\mathfrak{D}^{\mathrm{ch}}(\overset{\circ}{G/B})_{\mathrm{crit}}) = \mathfrak{H}'_{\mathrm{crit}}$.

Thus we obtain a homomorphism of commutative chiral algebras

$$\mathfrak{z}_{\mathfrak{g}} \to \mathfrak{H}'_{\mathrm{crit}} \simeq \mathfrak{H}_{\mathrm{crit}}. \tag{10.10}$$

Let us now recall that ultimate form of the isomorphism statement of [FF3, F] (see [F, Theorem 11.3]).

Theorem 10.16. *There exists a canonical isomorphism of commutative chiral algebras* $\mathfrak{z}_{\mathfrak{g}} \simeq \mathrm{Fun}(\mathrm{Op}_{\check{\mathfrak{g}}}(X))$ *such that the diagram*

$$\begin{array}{ccc} \mathfrak{z}_{\mathfrak{g}} & \xrightarrow{\sim} & \mathrm{Fun}(\mathrm{Op}_{\check{\mathfrak{g}}}(X)^{\mathfrak{D}}) \\ \downarrow & & \downarrow{\scriptstyle \mathrm{MT}^*} \\ \mathfrak{H}_{\mathrm{crit}} & \xrightarrow{\sim} & \mathrm{Fun}(\mathrm{Conn}_{\check{H}}(\omega_X^{\rho})^{\mathfrak{D}}) \end{array}$$

is commutative, where the left vertical arrow is the map of (10.10)*, the right vertical arrow is the Miura transformation of* (3.3)*, and the bottom horizontal arrow is the isomorphism of Proposition* 10.10*, composed with the automorphism, induced by the automorphism* $\tau := \check{\lambda} \mapsto -w_0(\check{\lambda})$ *of* $\check{H}$*.*

Remark 10.17. The isomorphism between $\mathfrak{z}_{\mathfrak{g}}$ and $\mathrm{Fun}(\mathrm{Op}_{\check{\mathfrak{g}}}(X)^{\mathfrak{D}})$ in the above diagram differs from the isomorphism in the corresponding diagram of Theorem 11.3 of [F] by the automorphism of $\mathfrak{z}_{\mathfrak{g}}$ induced by the automorphism τ of the Dynkin diagram of the Lie algebra $\mathfrak{g}$. This automorphism takes the vertex i of the diagram to $\bar{i}$, where $\alpha_{\bar{i}} = -w_0(\alpha_i)$.

10.18

To conclude this section let us return to the setup of Section 10.7. Consider the map $\mathfrak{g} \to T(N)$, obtained by composing the map of (10.6) with the projection on the $T(N)$-factor.

It is well known that lifts of this map to a homomorphism of Lie algebras $\mathfrak{g} \to \mathfrak{D}(N)^{\leq 1}$, which on $\mathfrak{n} \subset \mathfrak{g}$ induce the map $\mathfrak{l}_{\mathfrak{n}} : \mathfrak{n} \to T(N)$, are classified by characters of $\mathfrak{h}$ (and correspond to G-equivariant twistings on G/B). We would like to establish an affine analogue of this statement.

The analogue of characters of $\mathfrak{h}$ will be played by the set of chiral algebra homomorphisms $\psi : \mathfrak{H}_{\mathrm{crit}} \to \mathcal{O}_X$. For any such ψ, the composition

$$\phi : L_{\mathfrak{g},\mathrm{crit}} \to \mathfrak{D}^{\mathrm{ch}}(N) \otimes \mathfrak{H}_{\mathrm{crit}} \to \mathfrak{D}^{\mathrm{ch}}(N)$$

is a Lie-* algebra homomorphism, satisfying the following:

- The image of ϕ belongs to $\mathfrak{D}^{\mathrm{ch}}(N)^{\leq 1}$,
- The composition $L_{\mathfrak{g},\mathrm{crit}} \to \mathfrak{D}^{\mathrm{ch}}(N) \to \Theta(N)$ equals the composition of the map (10.7), followed by the projection on the $\Theta(N)$-factor,
- The restriction of ϕ to $L_{\mathfrak{n}}$ equals $\mathfrak{l}_{\mathfrak{n}}$.

Proposition 10.19. *Let $L'_{\mathfrak{g}}$ be a central extension of $L_{\mathfrak{g}}$ by means of ω_X, split over $L_{\mathfrak{n}}$, and let $\phi : L'_{\mathfrak{g}} \to \mathfrak{D}^{ch}(N)$ be a homomorphism of Lie-* algebras, satisfying the three properties above. Then $L'_{\mathfrak{g}} \simeq L_{\mathfrak{g},\mathrm{crit}}$ and ϕ is obtained from some $\psi : \mathfrak{H}_{\mathrm{crit}} \to \mathcal{O}_X$ in the manner described above.*

Proof. First, since $L'_{\mathfrak{g}}$ splits over $L_{\mathfrak{n}}$, we obtain that as a $\mathfrak{D}_X$-module $L'_{\mathfrak{g}} \simeq L_{\mathfrak{g}} \oplus \omega_X$. Let us show that the bracket on $L'_{\mathfrak{g}}$ corresponds to the critical pairing. For this, it is sufficient to calculate the bracket on $L_{\mathfrak{h}} \subset L_{\mathfrak{g}}$. However, since $L_{\mathfrak{h}}$ is commutative, the latter bracket is independent of the choice of a pair $(L'_{\mathfrak{g}}, \phi)$. Hence, we may choose the pair $L'_{\mathfrak{g}} = L_{\mathfrak{g},\mathrm{crit}}$ and a homomorphism corresponding to some homomorphism $\psi : \mathfrak{H}_{\mathrm{crit}} \to \mathcal{O}_X$. In the latter case, our assertion is clear.

Consider the set of all homomorphisms of chiral algebras $\psi : \mathfrak{H}_{\mathrm{crit}} \to \mathcal{O}_X$. By definition, this is a torsor over $\Gamma(X, \omega_X \otimes \mathfrak{h}^*)$. Now consider the space of homomorphisms $\phi_{\mathfrak{b}} : L_{\mathfrak{b},\mathrm{crit}} \to \mathfrak{D}^{ch}(N)^{\leq 1}$, satisfying the same three conditions as ϕ. This set is also a torsor over $\Gamma(X, \omega_X \otimes \mathfrak{h}^*)$. Moreover, it is easy to see that the map $\psi \mapsto \phi \mapsto \phi|_{L_{\mathfrak{b}}} =: \phi_{\mathfrak{b}}$ is a map of torsors.

Hence, for any ϕ as in the proposition, there exists a ψ, such that the two homomorphism $L_{\mathfrak{g},\mathrm{crit}} \to \mathfrak{D}^{ch}(N)^{\leq 1}$ coincide, when restricted to $L_{\mathfrak{b},\mathrm{crit}}$. We claim that in this case the two homomorphisms in question coincide on the entire of $L_{\mathfrak{g},\mathrm{crit}}$.

Indeed, let ϕ_1 and ϕ_2 be two such homomorphisms. Then $\phi_1 - \phi_2$ is a map $L_{\mathfrak{g}}/L_{\mathfrak{b}} \to \mathrm{Fun}(\mathrm{Jets}(N)) \otimes \omega_X$. Let f be a section of $L_{\mathfrak{n}^-}$, and let e be a section of $L_{\mathfrak{n}}$ such that $[\mathsf{e}, \mathsf{f}] \in L_{\mathfrak{h}}$. We obtain that $[\phi_1(\mathsf{f}) - \phi_2(\mathsf{f}), \phi_{\mathfrak{b}}(\mathsf{e})] = 0$. Hence, the image of $\phi_1 - \phi_2$ consists of $L_{\mathfrak{n}}$-invariant sections of $\mathrm{Fun}(\mathrm{Jets}(N)) \otimes \omega_X$, and the latter subspace is ω_X.

Again, for f above, let h be a section of $L_{\mathfrak{h}}$ such that $[\mathsf{f}, \mathsf{h}] = c \cdot \mathsf{f}$, where c is a nonzero scalar. We obtain $[\phi_1(\mathsf{f}) - \phi_2(\mathsf{f}), \phi_{\mathfrak{b}}(\mathsf{h})] = c \cdot (\phi_1(\mathsf{f}) - \phi_2(\mathsf{f}))$. However, by the above, $\phi_1(\mathsf{f}) - \phi_2(\mathsf{f})$ is central. Hence, $c \cdot (\phi_1(\mathsf{f}) - \phi_2(\mathsf{f})) = 0$, implying our assertion. □

11 Construction of Wakimoto modules

11.1

Homomorphism (10.1) allows us to produce representations of $\mathcal{A}_{\mathfrak{g},\kappa}$, i.e., $\widehat{\mathfrak{g}}_\kappa$-modules, by restricting modules of $\mathfrak{D}^{ch}(\overset{\circ}{G/B})_\kappa$. This should be regarded as a chiral analogue of the construction of $\mathfrak{g}$-modules by taking sections of twisted D-modules on the big Schubert cell $\overset{\circ}{G/B}$.

In the applications, modules over $\mathfrak{D}^{ch}(\overset{\circ}{G/B})_\kappa$ that we will consider are obtained using (10.2), from pairs of modules: $\mathcal{M} \in \mathfrak{D}^{ch}(\overset{\circ}{G/N})_\kappa$-mod, and $\mathcal{R} \in \mathfrak{H}_\kappa$-mod by taking $\mathcal{M} \overset{\frac{\infty}{2}}{\underset{L_{\mathfrak{h}}}{\otimes}} \mathcal{R}$. Let us describe the examples of $\mathfrak{D}^{ch}(\overset{\circ}{G/B})_\kappa$-modules that we will consider.

11.2

First, note that if $\mathfrak{D}^{ch}(Y)$ is a CADO on (the scheme of jets corresponding to) a smooth affine X-scheme Y, any left D-module on the scheme $Y[[t]]$ gives rise to a chiral module over $\mathfrak{D}^{ch}(Y)$.

Indeed, if $\mathcal{F}$ is such a D-module, it (or, rather, the space of its global sections) is naturally a chiral module over Fun(Jets(Y)) and a Lie-* module over Θ_Y. In this case we can induce it and obtain a chiral module over $\mathfrak{D}^{ch}(Y)$.

If $Y' \subset Y$ is a smooth locally closed subvariety, let us denote by $\mathrm{Dist}_Y(Y')$ the left D-module of distributions on Y' (i.e., the $*$-extension of the D-module $\mathrm{Fun}(Y')$), and let $\mathrm{Dist}_{Y[[t]]}(\mathrm{ev}^{-1}(Y'))$ denote the corresponding left D-module on $Y[[t]]$, i.e.,

$$\mathrm{Dist}_{Y[[t]]}(\mathrm{ev}^{-1}(Y')) \simeq \mathrm{ev}^*(\mathrm{Dist}_Y(Y')).$$

Finally, let $\mathrm{Dist}^{ch}_Y(\mathrm{ev}^{-1}(Y'))$ denote the resulting $\mathfrak{D}^{ch}(Y)$-module.

Let us take $Y = \overset{\circ}{G}$ and for each element $w \in W$ consider

$$Y' = \mathrm{Ad}_{w_0 w^{-1}}(N) \cdot w_0 \cdot N \subset \overset{\circ}{G}.$$

For example, if $w = w_0$ we get the D-module of functions on $N \cdot w_0 \cdot N$, and if $w = 1$ we get the δ-function at $w_0 \cdot N$.

Note that $\mathrm{ev}^{-1}(N) = I^0$. Therefore, we obtain a left D-module

$$\mathrm{Dist}_{G[[t]]}(\mathrm{Ad}_{w_0 w^{-1}}(I^0) \cdot w_0 \cdot I^0)$$

on $G[[t]]$ and

$$\mathrm{Dist}^{ch}_G(\mathrm{Ad}_{w_0 w^{-1}}(I^0) \cdot w_0 \cdot I^0)_\kappa \in \mathfrak{D}^{ch}(\overset{\circ}{G})_\kappa\text{-mod}.$$

Consider the chiral $\mathfrak{D}^{ch}(\overset{\circ}{G/N})_\kappa$-module

$$\mathrm{Dist}^{ch}_{\overset{\circ}{G/N}}(\mathrm{ev}^{-1}(\mathrm{Ad}_{w_0 w^{-1}}(N) \cdot w_0))_\kappa := H^{\frac{\infty}{2}}(L_{\mathfrak{n}}, \mathrm{Dist}^{ch}_G(\mathrm{Ad}_{w_0 w^{-1}}(I^0) \cdot w_0 \cdot I^0)_\kappa). \tag{11.1}$$

In other words, $\mathrm{Dist}^{ch}_{\overset{\circ}{G/N}}(\mathrm{ev}^{-1}(\mathrm{Ad}_{w_0 w^{-1}}(N) \cdot w_0))_\kappa$ is obtained by the above construction for $Y = \overset{\circ}{G/N}$ and $Y' = \mathrm{Ad}_{w_0 w^{-1}}(N) \cdot w_0 \subset \overset{\circ}{G/N}$. From Section 10.6 we obtain that $\mathrm{Dist}^{ch}_{\overset{\circ}{G/N}}(\mathrm{ev}^{-1}(\mathrm{Ad}_{w_0 w^{-1}}(N) \cdot w_0))_\kappa$ is indeed acyclic away from degree 0.

Moreover, as a module over $H^0_{DR}(\mathcal{D}^\times, L_{\mathfrak{n}} \oplus \widehat{L}'_{\mathfrak{h},\kappa})$, it is isomorphic to

$$\mathrm{Dist}^{ch}_N(\mathrm{Ad}_{w_0 w^{-1}}(N) \cap N) \otimes \mathrm{Ind}^{H^0_{DR}(\mathcal{D}^\times, \widehat{L}'_{\mathfrak{h},\kappa})}_{t\mathfrak{h}[[t]] \oplus \mathbb{C}} (\mathrm{Fun}\,(H(t\mathbb{C}[[t]]))).$$

In particular, as an $H^0_{DR}(\mathcal{D}^\times, \widehat{L}'_{\mathfrak{h},\kappa})$-module, it is $H(t\mathbb{C}[[t]])$-integrable, and injective as an $H(t\mathbb{C}[[t]])$-representation. Furthermore, it is free over over $\mathfrak{h}[t^{-1}]$ for any choice of a splitting $\mathfrak{h}[t^{-1}] \to H^0_{DR}(\mathcal{D}^\times, \widehat{L}'_{\mathfrak{h},\kappa})$.

11.3

Now, for $w \in W$ and an $\mathfrak{H}_\kappa$-module $\mathcal{R}$ we define the (complex of) $\mathfrak{D}^{ch}(G\overset{\circ}{/}B)_\kappa$-modules

$$'\mathbb{W}^w_\kappa(\mathcal{R}) := \mathrm{Dist}^{ch}_{G\overset{\circ}{/}N}(\mathrm{ev}^{-1}(\mathrm{Ad}_{w_0 w^{-1}}(N)\cdot w_0))_\kappa \underset{\mathfrak{h}((t)),t\mathfrak{h}[[t]]}{\overset{\frac{\infty}{2}}{\otimes}} \mathcal{R}. \tag{11.2}$$

Note that by the $H(t\mathbb{C}[[t]])$-integrability of $\mathrm{Dist}^{ch}_{G\overset{\circ}{/}N}(\mathrm{ev}^{-1}(\mathrm{Ad}_{w_0 w^{-1}}(N)\cdot w_0))_\kappa$, we have

$$'\mathbb{W}^w_\kappa(\mathcal{R}) \simeq {'\mathbb{W}^w_\kappa}(\mathrm{Av}_{H(t\mathbb{C}[[t]])}(\mathcal{R})), \tag{11.3}$$

where $\mathrm{Av}_{H(t\mathbb{C}[[t]])}$ denotes the averaging functor with respect to $H(t\mathbb{C}[[t]])$; see Section 20.10. Therefore, with no restriction of generality, we can (and will) assume that $\mathcal{R}$ is $H(t\mathbb{C}[[t]])$-integrable. Under this assumption, as a $\mathfrak{n}((t))$-module

$$'\mathbb{W}^w_\kappa(\mathcal{R}) \simeq \mathrm{Dist}^{ch}_N(\mathrm{ev}^{-1}(\mathrm{Ad}_{w_0 w^{-1}}(N)\cap N)) \otimes \mathcal{R}. \tag{11.4}$$

In particular, it is acyclic away from degree 0.

We restrict $'\mathbb{W}^w_\kappa(\mathcal{R})$ to $\mathcal{A}_{\mathfrak{g},\kappa}$ and obtain an object of $\widehat{\mathfrak{g}}_\kappa$-mod, denoted by the same symbol. Note that when defining $'\mathbb{W}^w_\kappa(\mathcal{R})$, we can avoid mentioning the chiral algebra $\mathfrak{D}^{ch}(G\overset{\circ}{/}B)_\kappa$. Namely,

$$'\mathbb{W}^w_\kappa(\mathcal{R}) \simeq \left(\mathrm{Dist}^{ch}_G(\mathrm{Ad}_{w_0 w^{-1}}(I^0)\cdot w_0 \cdot I^0)_\kappa\right) \underset{\mathfrak{b}((t)),\mathfrak{n}[[t]]+t\mathfrak{h}[[t]]}{\overset{\frac{\infty}{2}}{\otimes}} \mathcal{R}.$$

The $\mathfrak{D}^{ch}(G)_\kappa$-module $\mathrm{Dist}^{ch}_G(\mathrm{Ad}_{w_0 w^{-1}}(I^0)\cdot w_0 \cdot I^0)_\kappa$ is by construction equivariant with respect to the group $\mathrm{Ad}_{w_0 w^{-1}}(I^0)$, when we think of the action on $G((t))$ on itself by left multiplication. Let $\mathrm{Dist}^{ch}_G(I^0 \cdot w \cdot I^0)_\kappa$ be the chiral $\mathcal{A}_{\mathfrak{g},\kappa}$-module, obtained from the module $\mathrm{Dist}^{ch}_G(\mathrm{Ad}_{w_0 w^{-1}}(I^0)\cdot w_0 \cdot I^0)_\kappa$ by applying the left shift by $w \cdot w_0$.

Set

$$\mathbb{W}^w_\kappa(\mathcal{R}) := \left(\mathrm{Dist}^{ch}_G(I^0 \cdot w \cdot I^0)_\kappa\right) \underset{\mathfrak{b}((t)),\mathfrak{n}[[t]]+t\mathfrak{h}[[t]]}{\overset{\frac{\infty}{2}}{\otimes}} \mathcal{R}. \tag{11.5}$$

This is what we will call the Wakimoto module of type w corresponding to the $\mathfrak{H}_\kappa$-module $\mathcal{R}$.

Tautologically, as a $\widehat{\mathfrak{g}}_\kappa$-module, $\mathbb{W}^w_\kappa(\mathcal{R})$ is obtained from $'\mathbb{W}^w_\kappa(\mathcal{R})$ by the automorphism Ad_{ww_0} of $\mathfrak{g}$, and it is I^0-equivariant. Note, however, that $\mathbb{W}^w_\kappa(\mathcal{R})$ does not come by restriction from a $\mathfrak{D}^{ch}(G\overset{\circ}{/}B)_\kappa$-module, unless $w = w_0$.

We have a description of $\mathbb{W}^w_\kappa(\mathcal{R})$ similar to (11.4), but with respect to the subalgebra $\mathfrak{n}^{ww_0}((t))$, where we set $\mathfrak{n}^w := \mathrm{Ad}_w(\mathfrak{n})$, $N^w = \mathrm{Ad}_w(N)$. Namely,

$$\mathbb{W}^w_\kappa(\mathcal{R}) \simeq \mathrm{Dist}^{ch}_{N^{ww_0}}\left(\mathrm{ev}^{-1}(N^{ww_0}\cap N)\right) \otimes \mathcal{R}. \tag{11.6}$$

11.4

Assume now that the $\mathfrak{H}_\kappa$-module $\mathcal{R}$ is $H[[t]]$-integrable. (Having already the assumption that it is $H(t\mathbb{C}[[t]])$-integrable, this amounts to requiring that $\mathfrak{h}$ acts semisimply with eigenvalues corresponding to integral weights.) We claim that in this case the module $\mathbb{W}^w_\kappa(\mathcal{R})$ is I-integrable.

Indeed, let us instead of $\mathrm{Dist}^{\mathrm{ch}}_G(I^0 \cdot w \cdot I^0)_\kappa$ take the chiral $\mathfrak{D}^{\mathrm{ch}}(G)_\kappa$-module

$$\mathrm{Dist}^{\mathrm{ch}}_G(I^0 \cdot w \cdot I)_\kappa.$$

As an object of $\mathfrak{D}^{\mathrm{ch}}(G)_\kappa$-mod, it is clearly I-integrable with respect to both left and right action of $G((t))$ on itself.

Consider $H^{\frac{\infty}{2}}(L_\mathfrak{n}, \mathrm{Dist}^{\mathrm{ch}}(I^0 \cdot w \cdot I)_\kappa)$. This is an $\mathfrak{H}'_\kappa$-module, which is $H[[t]]$-integrable and injective as an $H[[t]]$-module.

One easily checks that $\mathbb{W}^w_\kappa(\mathcal{R})$ is isomorphic to

$$H^{\frac{\infty}{2}}\left(L_\mathfrak{n}, \mathrm{Dist}^{\mathrm{ch}}(I^0 \cdot w \cdot I)_\kappa\right) \overset{\frac{\infty}{2}}{\underset{L_\mathfrak{h};\mathfrak{h}}{\otimes}} \mathcal{R},$$

which is manifestly I-integrable.

11.5

For a weight $\lambda \in \mathfrak{h}^*$ consider the 1-dimensional Lie-* module over $\widehat{L}_{\mathfrak{h},\kappa}$ corresponding to the character λ. Let us denote by π_λ the induced chiral module over $\mathfrak{H}_\kappa$.

For future use we introduce the notation

$$\mathbb{W}^w_{\kappa,\lambda} := \mathbb{W}^w_\kappa(\pi_{w^{-1}(\lambda+\rho)+\rho}). \tag{11.7}$$

Observe that the definition of $\mathbb{W}^w_{\kappa,\lambda}$ can be rewritten as

$$\left(H^{\frac{\infty}{2}}\left(\mathfrak{n}((t)), \mathfrak{n}[[t]], \mathrm{Dist}^{\mathrm{ch}}_G(I^0 \cdot w \cdot I^0)_\kappa\right)^{t\mathfrak{h}[[t]]}\right) \underset{\mathfrak{h}}{\otimes} \mathbb{C}^{w^{-1}(\lambda+\rho)+\rho}.$$

Let M^w_λ be the $\mathfrak{g}$-module equal to $\mathrm{Dist}_G(N \cdot w \cdot N) \underset{\mathfrak{b},\mathfrak{n}}{\otimes} \mathbb{C}^{w^{-1}(\lambda+\rho)+\rho}$. Note that when $w = 1$, M^w_λ is the Verma module M_λ, and when $w = w_0$, M^w_λ is the dual Verma $M^\vee_\lambda$. In general, M^w_λ always has highest weight λ, and it is characterized by the property that it is free with respect to the Lie subalgebra $\mathfrak{n}^{ww_0} \cap \mathfrak{n}^-$ and cofree with respect to $\mathfrak{n}^{ww_0} \cap \mathfrak{n}$.

Set $\mathbb{M}^w_{\kappa,\lambda} := \mathrm{Ind}^{\widehat{\mathfrak{g}}_\kappa}_{\mathfrak{g}[[t]]\oplus\mathbb{C}\mathbf{1}}(M^w_\lambda)$ be the induced $\widehat{\mathfrak{g}}_\kappa$-module. We claim that we always have a map

$$\mathbb{M}^w_{\kappa,\lambda} \to \mathbb{W}^w_{\kappa,\lambda}. \tag{11.8}$$

This amounts to constructing a map of $\mathfrak{g}[[t]]$-modules $M^w_\lambda \to \mathbb{W}^w_{\kappa,\lambda}$. We have

$$M^w_\lambda \hookrightarrow \mathrm{Dist}_{G[[t]]}(I^0 \cdot w \cdot I^0) \underset{\mathfrak{b}[[t]],\mathfrak{n}[[t]]+t\mathfrak{h}[[t]]}{\otimes} \mathbb{C}^{w^{-1}(\lambda+\rho)+\rho} \hookrightarrow$$

$$\left(\mathrm{Dist}^{\mathrm{ch}}_G(I^0 \cdot w \cdot I^0)_\kappa\right)^{\mathfrak{n}[[t]]+t\mathfrak{h}[[t]]} \underset{\mathfrak{h}}{\otimes} \mathbb{C}^{w^{-1}(\lambda+\rho)+\rho},$$

which maps to the required semi-infinite cohomology.

12 Convolution action on Wakimoto modules

12.1

In this section we will apply the formalism of convolution functors

$$\star : \mathfrak{D}(G/K)_\kappa\text{-mod}^{K'} \times D(\widehat{\mathfrak{g}}_\kappa\text{-mod})^K \to D(\widehat{\mathfrak{g}}_\kappa\text{-mod})^{K'},$$

where K, K' are subgroups of $G[[t]]$ to derive some additional properties of Wakimoto modules.

The subgroups that we will use will be either I^0 or $G^{(1)}$, the first congruence subgroup in $G[[t]]$, and if κ is integral, also I. When confusion is likely to occur, we will use the notation $\cdot \underset{K}{\star} \cdot$ to emphasize which equivariant derived category we are working in; see Section 22.6. We will identify D-modules on G (respectively, G/N, G/B) with the corresponding κ-twisted D-modules on $G((t))/G^{(1)}$ (respectively, $\widetilde{\mathrm{Fl}}_G = G((t))/I^0$, $\mathrm{Fl}_G = G((t))/I$).

Another two pieces of notation that we will need are as follows. If $\mathbf{g}$ is a point of $G((t))$, and $\mathcal{F}$ an object of an arbitrary category with a Harish-Chandra action of $G((t))$ at level κ, we will denote by $\delta_{\mathbf{g}} \star \mathcal{F}$ the twist of $\mathcal{F}$ by $\mathbf{g}$. If $\mathcal{F}$ is equivariant with respect to a congruence subgroup $K \subset G[[t]]$, then

$$\delta_{\mathbf{g}} \star \mathcal{F} \simeq \delta_{\mathbf{g}_{G((t))/K}} \underset{K}{\star} \mathcal{F},$$

where $\delta_{\mathbf{g}_{G((t))/K}}$ is the unique κ-twisted D-module on $G((t))/K$, whose !-fiber at the point $\mathbf{g}_{G((t))/K} \in G((t))/K$ is $\mathbb{C}$.

Let $\mathbf{U}$ be a pro-unipotent subgroup such that $\kappa|_{\mathrm{Lie}(\mathbf{U})}$ is trivial. Then for $\mathcal{F}$ as above, $\underline{\mathbb{C}}_{\mathbf{U}} \star \mathcal{F}$ will denote the same thing as $\mathrm{Av}_{\mathbf{U}}(\mathcal{F})$. In other words, if $\mathcal{F}$ is equivariant with respect to some unipotent $K \subset G[[t]]$ containing a congruence subgroup, and $\mathbf{U}' = \mathbf{U} \cap K$, then

$$\underline{\mathbb{C}}_{\mathbf{U}} \star \mathcal{F} \simeq \mathrm{Dist}_{G((t))/K}(\mathbf{U}/\mathbf{U}')_\kappa \underset{K}{\star} \mathcal{F} \otimes \det\left(\mathrm{Lie}(\mathbf{U})/\mathrm{Lie}(\mathbf{U}')[1]\right)^{\otimes -1},$$

where $\underline{\mathbb{C}}_{\mathbf{U}/\mathbf{U}'}$ denotes the cohomologically shifted D-module on $\mathbf{U}/\mathbf{U}'$, corresponding via Riemann–Hilbert to the *constant sheaf* on $\mathbf{U}/\mathbf{U}'$, and $\mathrm{Dist}_{G((t))/K}(\mathbf{U}/\mathbf{U}')_\kappa$ is the unique κ-twisted D-module on $G((t))/K$, supported on $\mathbf{U}/\mathbf{U}' \subset G((t))/K$, and whose !-restriction to this subscheme is $\mathrm{Fun}(\mathbf{U}/\mathbf{U}')$; see Section 21.6.

We will use the following observation.

Lemma 12.2. *Suppose that* $\mathbf{U}$ *contains two subgroups* $\mathbf{U}_1$ *and* $\mathbf{U}_2$ *such that the multiplication map defines an isomorphism* $\mathbf{U}_1 \times \mathbf{U}_2 \to \mathbf{U}$. *Then*

$$\underline{\mathbb{C}}_{\mathbf{U}} \star \mathcal{F} \simeq \underline{\mathbb{C}}_{\mathbf{U}_1} \star (\underline{\mathbb{C}}_{\mathbf{U}_2} \star \mathcal{F}).$$

For $\widetilde{w} \in W_{\mathrm{aff}}$ we will denote by $\widetilde{j}_{\kappa,\widetilde{w}}$ the unique κ-twisted I^0-equivariant D-module on $\widetilde{\mathrm{Fl}}_G$, supported on $I^0 \cdot \widetilde{w} \subset \widetilde{\mathrm{Fl}}_G$, whose !-restriction to this subscheme is isomorphic to $\mathrm{Fun}(I^0 \cdot \widetilde{w})$, as an I^0-equivariant quasi-coherent sheaf. Of course, the isomorphism class of this D-module depends on the choice of a representative of $\widetilde{w}$ in $G((t))$.

Since $\widetilde{j}_{\kappa,\widetilde{w}} \simeq \underline{\mathbb{C}}_{I^0} \star \delta_{\widetilde{w}_{G((t))/I^0}} \otimes \det\left(\mathrm{Lie}(I^0)/\,\mathrm{Lie}(I^0) \cap \mathrm{Ad}_{\widetilde{w}}(\mathrm{Lie}(I^0))[1]\right)$, from Lemma 12.2 we obtain the following.

Lemma 12.3. *For $\widetilde{w} \in W_{\mathrm{aff}}$ assume that I^0 can be written as a product of subgroups $\mathbf{U}_1 \cdot \mathbf{U}_2$ such that $\mathrm{Ad}_{\widetilde{w}^{-1}}(\mathbf{U}_2) \subset I^0$. Then for an I^0-equivariant object $\mathcal{F}$ of a category with a Harish-Chandra action of $G((t))$, we have a canonical isomorphism*

$$\widetilde{j}_{\kappa,\widetilde{w}} \underset{I^0}{\star} \mathcal{F} \simeq \mathrm{Av}_{\mathbf{U}_1}(\delta_{\widetilde{w}} \star \mathcal{F}) \otimes \det\left(\mathrm{Lie}(\mathbf{U}_1)/\,\mathrm{Lie}(\mathbf{U}_1) \cap \mathrm{Ad}_{\widetilde{w}}(\mathrm{Lie}(\mathbf{U}_1))[1]\right).$$

Suppose that κ is integral, i.e., comes from a group ind-scheme extension $\widehat{G((t))}$ of $G((t))$ split over $G[[t]]$; let us denote by $\mathcal{P}^\kappa$ the resulting line bundle on $\mathrm{Gr}_G = G((t))/G[[t]]$. In this case we will denote by $j_{\widetilde{w},*}$ (respectively, $j_{\widetilde{w},!}$) the I-equivariant κ-twisted D-modules on Fl_G given by the $*$-extension (respectively, !-extension) of the twisted right D-module on $I \cdot \widetilde{w} \subset \mathrm{Fl}_G$, corresponding to the restriction of the line bundle $\widehat{G((t))}/I \to G((t))/I$ to this subscheme. If κ is not integral the above I-equivariant D-modules still make sense for $w \in W$.

If $l(\widetilde{w}_1) + l(\widetilde{w}_2) = l(\widetilde{w}_1 \cdot \widetilde{w}_2)$, then

$$j_{\widetilde{w}_1,*} \underset{I}{\star} j_{\widetilde{w}_2,*} \simeq j_{\widetilde{w}_1\cdot\widetilde{w}_2,*} \quad \text{and} \quad j_{\widetilde{w}_1,!} \underset{I}{\star} j_{\widetilde{w}_2,!} \simeq j_{\widetilde{w}_1\cdot\widetilde{w}_2,!}.$$

Since the functor $j_{\widetilde{w},*} \underset{I}{\star} \cdot$ is right exact, the above isomorphism implies that the functor $j_{\widetilde{w},!} \underset{I}{\star} \cdot$, being its quasi-inverse, is left exact.

Let us observe that the definition of $j_{\widetilde{w},*}$ (respectively, $j_{\widetilde{w},!}$) is evidently independent of the choice of representatives $\widetilde{w}$ in $G((t))$. The direct image of $\widetilde{j}_{\kappa,\widetilde{w}}$ under $\widetilde{\mathrm{Fl}}_G \to \mathrm{Fl}_G$ is isomorphic to $j_{\widetilde{w},*} \otimes (\mathfrak{l}_\kappa^{\widetilde{w}})^{\otimes -1}$, where $\mathfrak{l}_\kappa^{\widetilde{w}}$ is the line defined as

$$\mathfrak{l}_\kappa^{\widetilde{w}} := \Gamma\left(I \cdot \widetilde{w},\, \Omega^{\mathrm{top}}(I^0 \cdot \widetilde{w}) \otimes \mathcal{P}^\kappa|_{I^0\cdot\widetilde{w}}\right)^{I_0}. \tag{12.1}$$

12.4

Let us first observe that for $w \in W$

$$\mathrm{Dist}_G^{\mathrm{ch}}(I^0 \cdot w \cdot I^0)_\kappa \simeq \widetilde{j}_{\kappa,w} \underset{I^0}{\star} \mathrm{Dist}_G^{\mathrm{ch}}(I^0)_\kappa.$$

Hence, we obtain the following.

Lemma 12.5. $\mathbb{W}_\kappa^w(\mathcal{R}) \simeq \widetilde{j}_{\kappa,w} \underset{I^0}{\star} \mathbb{W}_\kappa^1(\mathcal{R})$.

If $\mathcal{R}$ is integrable with respect to $H[[t]]$, the above lemma implies that

$$\mathbb{W}^w_\kappa(\mathcal{R}) \simeq j_{w,*} \star \mathbb{W}^1(\mathcal{R}), \tag{12.2}$$

which, in turn, implies that

$$j_{w,!} \star \mathbb{W}^{w^{-1}}(\mathcal{R}) \simeq \mathbb{W}^1(\mathcal{R}), \tag{12.3}$$

and if $l(w_1 \cdot w_2) = l(w_1) + l(w_2)$, then

$$j_{w_1,*} \star \mathbb{W}^{w_2}(\mathcal{R}) \simeq \mathbb{W}^{w_1 \cdot w_2}(\mathcal{R}). \tag{12.4}$$

12.6

For $w \in W$ recall that $\mathfrak{n}^w$ (respectively, $\mathfrak{b}^w$) denotes the subalgebra $\mathrm{Ad}_w(\mathfrak{n}) \subset \mathfrak{g}$ (respectively, $\mathrm{Ad}_w(\mathfrak{b}) \subset \mathfrak{g}$). Note that the Cartan quotient of $\mathfrak{b}^w$ is still canonically identified with $\mathfrak{h}$. For $w = w_0$ we will sometimes also write $\mathfrak{n}^-$, $\mathfrak{b}^-$.

Proposition 12.7. *For any chiral $\mathfrak{H}_\kappa$-module $\mathcal{R}$,*

$$\mathbb{W}^w_\kappa(\mathcal{R}) \simeq (\mathrm{Dist}^{\mathrm{ch}}_G(I^0)_\kappa) \overset{\frac{\infty}{2}}{\underset{\mathfrak{b}^w((t)),t\mathfrak{b}^w[[t]]+\mathfrak{n}\cap\mathfrak{n}^w}{\otimes}} \mathcal{R}.$$

Proof. It is enough to show that

$$H^{\frac{\infty}{2}}(\mathfrak{n}((t)), \mathfrak{n}[[t]], \mathrm{Dist}^{\mathrm{ch}}_G(I^0 \cdot w \cdot I^0)_\kappa) \simeq H^{\frac{\infty}{2}}(\mathfrak{n}^w((t)), t\mathfrak{n}^w[[t]] + \mathfrak{n}^w \cap \mathfrak{n}, \mathrm{Dist}^{\mathrm{ch}}_G(I^0)_\kappa),$$

in a way compatible with the $\mathfrak{H}_\kappa$-actions.

Again, we have $\mathrm{Dist}^{\mathrm{ch}}_G(I^0 \cdot w \cdot I^0)_\kappa \simeq \mathrm{Dist}^{\mathrm{ch}}_G(I^0)_\kappa \underset{I^0}{\star} \widetilde{j}_{\kappa,w}$, where we are using the action of $G((t))$ on itself by right translations. We have

$$I^0 = (I^0 \cap B^-[[t]]) \cdot (I^0 \cap N[[t]]).$$

By Lemma 12.3, we obtain that

$$\mathrm{Dist}^{\mathrm{ch}}_G(I^0 \cdot w \cdot I^0)_\kappa \simeq \mathrm{Dist}^{\mathrm{ch}}_G(I^0)_\kappa \star \delta_w \star \underline{\mathbb{C}}_{N[[t]]} \otimes \det(\mathfrak{n}/\mathfrak{n}^{w^{-1}} \cap \mathfrak{n}[1]).$$

Hence, by Section 22.15

$$\begin{aligned} &H^{\frac{\infty}{2}}(\mathfrak{n}((t)), \mathfrak{n}[[t]], \mathrm{Dist}^{\mathrm{ch}}_G(I^0 \cdot w \cdot I^0)_\kappa) \\ &\quad \simeq H^{\frac{\infty}{2}}(\mathfrak{n}((t)), \mathfrak{n}[[t]], \mathrm{Dist}^{\mathrm{ch}}_G(I^0)_\kappa \star \delta_w) \otimes \det(\mathfrak{n}/\mathfrak{n}^{w^{-1}} \cap \mathfrak{n}[1]), \end{aligned}$$

which, in turn, is isomorphic to

$$H^{\frac{\infty}{2}}(\mathfrak{n}^w((t)), \mathfrak{n}^w[[t]], \mathrm{Dist}^{\mathrm{ch}}_G(I^0)_\kappa) \otimes \det(\mathfrak{n}^w/\mathfrak{n}^w \cap \mathfrak{n}[1]).$$

The determinant line exactly accounts for the change of the lattice $\mathfrak{n}^w[[t]] \mapsto t\mathfrak{n}^w[[t]] + \mathfrak{n} \cap \mathfrak{n}^w$. □

As a corollary, we obtain the following characterization of the Wakimoto modules $\mathbb{W}^w_\kappa(\mathcal{R})$.

Corollary 12.8. *For an I^0-integrable $\widehat{\mathfrak{g}}_{\kappa'}$-module $\mathcal{M}$ and a chiral $\mathfrak{H}_\kappa$-module $\mathcal{R}$, we have a quasi-isomorphism*

$$\mathcal{M} \underset{\mathfrak{g}((t)),\mathrm{Lie}(I^0)}{\overset{\frac{\infty}{2}}{\otimes}} \mathbb{W}^w_\kappa(\mathcal{R}) \simeq \mathfrak{C}^{\frac{\infty}{2}}(\mathfrak{b}^w((t)), t\mathfrak{b}^w[[t]] + \mathfrak{n}\cap\mathfrak{n}^w, \mathcal{M}\otimes\mathcal{R}).$$

Proof. In view of Proposition 12.7, it suffices to show that for any $\mathcal{M}$ as in the proposition,

$$\mathcal{M} \underset{\mathfrak{g}((t)),\mathrm{Lie}(I^0)}{\overset{\frac{\infty}{2}}{\otimes}} \left(\mathrm{Dist}^{\mathrm{ch}}_G(I^0)_\kappa\right) \simeq \mathcal{M}$$

as $\widehat{\mathfrak{g}}_\kappa$-modules. But this follows from Section 22.8. □

12.9

We will now show that Wakimoto modules of type $w \cdot w_0$ are well behaved with respect to the functor of semi-infinite cohomology of the algebra $\mathfrak{n}^w((t))$. This is, in fact, a fundamental property of Wakimoto modules which was found in [FF2].

Namely, let $\mathcal{L}$ be a module over $\mathfrak{n}^w((t))$, on which the subalgebra

$$t\mathfrak{n}^w[[t]] + \mathfrak{n}\cap\mathfrak{n}^w$$

acts locally nilpotently. Let $\mathcal{R}$ be an $\mathfrak{H}_\kappa$-module, on which $t\mathfrak{h}[[t]]$ acts locally nilpotently. (By (11.3) the latter is not really restrictive.)

Proposition 12.10. *Under the above circumstances,*

$$\mathcal{L} \underset{\mathfrak{n}^w((t)),t\mathfrak{n}^w[[t]]+\mathfrak{n}\cap\mathfrak{n}^w}{\overset{\frac{\infty}{2}}{\otimes}} \mathbb{W}^{ww_0}_\kappa(\mathcal{R})$$

is canonically isomorphic to $\mathcal{L}\otimes\mathcal{R}$.

Proof. By (11.6), it suffices to show that

$$\mathcal{L} \underset{\mathfrak{n}^w((t)),t\mathfrak{n}^w[[t]]+\mathfrak{n}\cap\mathfrak{n}^w}{\overset{\frac{\infty}{2}}{\otimes}} \mathrm{Dist}^{\mathrm{ch}}_{N^w}\left(\mathrm{ev}^{-1}(N^w\cap N)\right) \simeq \mathcal{L}.$$

However, this readily follows from Corollary 22.14(2). □

12.11

For an integral coweight $\check{\lambda}$ let us consider the corresponding point $t^{\check{\lambda}} \in G((t))$. We will also think of $\check{\lambda}$ as an element of W_{aff} corresponding to this orbit. Note that if λ is dominant, the orbit of $I\cdot t^{\check{\lambda}} \subset \mathrm{Fl}_G$ has the property that under the projection $\mathrm{Fl}_G \to \mathrm{Gr}_G$ it maps one-to-one.

We have already established the transformation property of Wakimoto modules with respect to convolution with $\widetilde{j}_{\kappa,w}$ for $w \in W$, see Lemma 12.5. Now we would like to study their behavior with respect to convolution with $\widetilde{j}_{\kappa,\check{\lambda}}$.

Note that we have a natural adjoint action of $H((t))$ on $H^0_{DR}(\mathcal{D}^\times, \widehat{L}_{\mathfrak{h},\kappa})$, and similarly for the Baer negative extension. Thus we obtain that $H((t))$ acts on the categories $\mathfrak{H}_\kappa$-mod and $\mathfrak{H}'_\kappa$-mod. For $t^{\check{\lambda}} \in H((t))$ we will denote the corresponding functor by $\mathcal{R} \mapsto t^{\check{\lambda}} \star \mathcal{R}$.

The following property of Wakimoto modules will play a crucial role.

Proposition 12.12. *For a dominant λ we have*

$$\widetilde{j}_{\kappa,\check{\lambda}} \underset{I^0}{\star} \mathbb{W}^{w_0}_\kappa(\mathcal{R}) \simeq \mathbb{W}^{w_0}_\kappa(t^{w_0(\check{\lambda})} \star \mathcal{R}).$$

12.13 Proof of Proposition 12.12

Consider the subscheme $I^0 \cdot t^{\check{\lambda}} \cdot I^0 \subset G((t))$. Clearly, there exists a unique irreducible object of $\mathfrak{D}^{\mathrm{ch}}(G)_\kappa\text{-mod}^{I^0,I^0}$, supported on this subset. Let us denote it by $\mathrm{Dist}^{\mathrm{ch}}_G(I^0 t^{\check{\lambda}} I^0)_\kappa$. In particular, for $\check{\lambda} = 0$ we recover $\mathrm{Dist}^{\mathrm{ch}}_G(I^0)_\kappa$.

We have

$$\widetilde{j}_{\kappa,\check{\lambda}} \underset{I^0}{\star} \mathrm{Dist}^{\mathrm{ch}}_G(I^0)_\kappa \simeq \mathrm{Dist}^{\mathrm{ch}}_G(I^0 t^{\check{\lambda}} I^0)_\kappa \simeq \mathrm{Dist}^{\mathrm{ch}}_G(I^0)_\kappa \underset{I^0}{\star} \widetilde{j}_{\kappa',\check{\lambda}}.$$

Therefore, by Proposition 12.7, we have to show that

$$\left(\mathrm{Dist}^{\mathrm{ch}}_G(I^0)_\kappa \underset{I^0}{\star} \widetilde{j}_{\kappa',\check{\lambda}}\right) \overset{\frac{\infty}{2}}{\underset{\mathfrak{b}^-((t)),t\mathfrak{b}^-[[t]]}{\otimes}} \mathcal{R} \simeq \left(\mathrm{Dist}^{\mathrm{ch}}_G(I^0)_\kappa\right) \overset{\frac{\infty}{2}}{\underset{\mathfrak{b}^-((t)),t\mathfrak{b}^-[[t]]}{\otimes}} (t^{w_0(\check{\lambda})} \star \mathcal{R}).$$

Let us write $I^0_+ = I^0 \cap B[[t]]$ and $I^0_- = I^0 \cap N^-[[t]]$, and recall that the product map defines an isomorphism $I^0 = I^0_+ \cdot I^0_-$. Note also that $\mathrm{Ad}_{t^{\check{\lambda}}}(I^0_+) \subset I^0$.

Hence,

$$\begin{aligned}\mathrm{Dist}^{\mathrm{ch}}_G(I^0)_\kappa \underset{I^0}{\star} \widetilde{j}_{\kappa',\check{\lambda}} \simeq{}& \mathrm{Dist}^{\mathrm{ch}}_G(I^0)_\kappa \star \delta_{t^{\check{\lambda}}} \star \underline{\mathbb{C}}_{I^0_-} \\ &\otimes \det\left(t\mathfrak{n}^-[[t]]/t\mathfrak{n}^-[[t]] \cap \mathrm{Ad}_{t^{-\check{\lambda}}}(t\mathfrak{n}^-[[t]])[1]\right).\end{aligned}$$

Therefore, by Lemma 12.3 we obtain that

$$\begin{aligned}\left(\mathrm{Dist}^{\mathrm{ch}}_G(I^0)_\kappa \star \widetilde{j}_{\kappa',\lambda}\right) \overset{\frac{\infty}{2}}{\underset{\mathfrak{b}^-((t)),t\mathfrak{b}^-[[t]]}{\otimes}} \mathcal{R} \simeq{}& H^{\frac{\infty}{2}}\left(\mathfrak{n}^-((t)), t\mathfrak{n}^-[[t]], \left(\mathrm{Dist}^{\mathrm{ch}}_G(I^0)_\kappa \star \delta_{t^{\check{\lambda}}}\right)\right) \\ &\overset{\frac{\infty}{2}}{\underset{\mathfrak{h}((t)),t\mathfrak{h}[[t]]}{\otimes}} \mathcal{R} \otimes \det\left(t\mathfrak{n}^-[[t]]/\mathrm{Ad}_{t^{-\check{\lambda}}}(t\mathfrak{n}^-[[t]])[1]\right).\end{aligned}$$

For an $L'_{\mathfrak{b},\kappa}$-module $\mathcal{M}$ we have

$$H^{\frac{\infty}{2}}(\mathfrak{n}^-((t)), t\mathfrak{n}^-[[t]], t^{\check{\lambda}} \star \mathcal{M}) \simeq t^{\check{\lambda}} \star H^{\frac{\infty}{2}}(\mathfrak{n}^-((t)), \mathrm{Ad}_{t^{\check{\lambda}}}(t\mathfrak{n}^-[[t]]), \mathcal{M}),$$

as $\mathfrak{H}'_\kappa$-modules. Hence, the expression above can be rewritten as

$$\left(H^{\frac{\infty}{2}}\left(\mathfrak{n}^-((t)), t\mathfrak{n}^-[[t]], \mathrm{Dist}^{\mathrm{ch}}_G(I^0)_\kappa\right) \star t^{\check{\lambda}}\right) \overset{\frac{\infty}{2}}{\underset{\mathfrak{h}((t)), t\mathfrak{h}[[t]]}{\otimes}} \mathcal{R},$$

where we have absorbed the determinant line into changing the lattice

$$\mathrm{Ad}_{t^{-\check{\lambda}}}(t\mathfrak{n}^-[[t]]) \mapsto t\mathfrak{n}^-[[t]].$$

The latter can, in turn, be rewritten as

$$H^{\frac{\infty}{2}}\left(\mathfrak{n}^-((t)), t\mathfrak{n}^-[[t]], \mathrm{Dist}^{\mathrm{ch}}_G(I^0)_\kappa\right) \overset{\frac{\infty}{2}}{\underset{\mathfrak{h}((t)), t\mathfrak{h}[[t]]}{\otimes}} (t^{w_0(\check{\lambda})} \star \mathcal{R}),$$

which is what we had to show.[7]

13 Wakimoto modules at the critical level

13.1

In this section we will consider in more detail Wakimoto modules at the critical level. By Proposition 10.10 and using the isomorphism $\mathfrak{H}'_{\mathrm{crit}} \simeq \mathfrak{H}_{\mathrm{crit}}$, if $\mathcal{R}$ is a quasi-coherent sheaf on $\mathrm{Conn}_{\check{H}}(\omega^\rho_{\mathcal{D}^\times})$, we can define Wakimoto modules $\mathbb{W}^w_{\mathrm{crit}}(\mathcal{R})$ for $w \in W$.

Note that for any $\mathcal{R}$, the Wakimoto module $\mathbb{W}^w_{\mathrm{crit}}(\mathcal{R})$ carries an action of $\mathfrak{H}_{\mathrm{crit}}$ by transport of structure, and the isomorphism

$$\mathbb{W}^w_{\mathrm{crit}}(\mathcal{R}) \simeq \mathrm{Dist}^{\mathrm{ch}}_{N^{ww_0}}(\mathrm{ev}^{-1}(N^{ww_0} \cap N)) \otimes \mathcal{R} \tag{13.1}$$

of (11.6) is compatible with the $\mathfrak{H}_{\mathrm{crit}}$-actions.

In particular, from Theorem 10.16 combined with Proposition 3.10, we obtain the following result. Recall that a weight λ is called antidominant, if $\langle \lambda, \check{\alpha}\rangle \notin \mathbb{Z}^{>0}$ for any $\alpha \in \Delta^+$, or equivalently, if the intersection of the two sets $\{\lambda - \mathrm{Span}_{\mathbb{Z}^+}(\Delta^+)\}$ and $\{w(\lambda), w \in W\}$ consists only of the element α.

Corollary 13.2. *The action of the center* $\mathfrak{Z}_{\mathfrak{g}} \simeq \mathrm{Fun}(\mathrm{Op}_{\check{\mathfrak{g}}}(\mathcal{D}^\times))$ *on* $\mathbb{W}^w_{\mathrm{crit},\lambda}$ *factors through* $\mathfrak{Z}^{\mathrm{RS},\varpi(-\lambda-\rho)}_{\mathfrak{g}} \simeq \mathrm{Fun}(\mathrm{Op}^{\mathrm{RS},\varpi(-\lambda-\rho)}_{\check{\mathfrak{g}}})$. *Moreover, if* $w^{-1}(\lambda+\rho)$ *is dominant, then* $\mathbb{W}^w_{\mathrm{crit},\lambda}$ *is flat over* $\mathrm{Fun}(\mathrm{Op}^{\mathrm{RS},\varpi(-\lambda-\rho)}_{\check{\mathfrak{g}}})$.

Another useful observation is the following.

Proposition 13.3. *Let* $\mathcal{R}_1$ *and* $\mathcal{R}_2$ *be two* $\mathfrak{H}_{\mathrm{crit}}$*-modules, on which* $\mathfrak{h} \subset \Gamma(X, \mathfrak{H}_{\mathrm{crit}})$ *acts by the same scalar. Then for any* $w \in W$ *the map*

$$\mathrm{Hom}_{\mathfrak{H}_{\mathrm{crit}}}(\mathcal{R}_1, \mathcal{R}_2) \to \mathrm{Hom}_{\hat{\mathfrak{g}}_{\mathrm{crit}}}(\mathbb{W}^w_{\mathrm{crit}}(\mathcal{R}_1), \mathbb{W}^w_{\mathrm{crit}}(\mathcal{R}_2))$$

is an isomorphism.

---

[7] The replacement of $\check{\lambda}$ by $w_0(\check{\lambda})$ comes from the fact that the identifications $B/N \simeq H \simeq B^-/N^-$ differ by the automorphism w_0 of H.

Proof. Since $\mathcal{R}$ is a subspace of $\mathbb{W}^w_{\mathrm{crit}}(\mathcal{R})$, the fact that the map in question is injective is evident. Let us prove the surjectivity.

It will be more convenient to work with $'\mathbb{W}^w_{\mathrm{crit}}(\mathcal{R}_i)$ instead of $\mathbb{W}^w_{\mathrm{crit}}(\mathcal{R}_i)$, $i = 1, 2$. As in (13.1), we have an identification

$$'\mathbb{W}^w_{\mathrm{crit}}(\mathcal{R}_i) \simeq \mathrm{Dist}^{\mathrm{ch}}_N \left(\mathrm{ev}^{-1}(N \cap N^{w_0 w^{-1}})\right) \otimes \mathcal{R}_i,$$

respecting the actions of $\mathfrak{n}((t))$ and $\mathfrak{H}_{\mathrm{crit}}$.

Let us first analyze the space of endomorphisms $\mathrm{Dist}^{\mathrm{ch}}_N (\mathrm{ev}^{-1}(N \cap N^{w_0 w^{-1}}))$ as a $\mathfrak{n}((t))$-module. We obtain, as in Section 22.14, that the map $\mathfrak{r}_{\mathfrak{n}} : L_{\mathfrak{n}} \to \mathfrak{D}^{\mathrm{ch}}(N)$ has the property that the image of $U(\mathfrak{n}((t)))$ is dense in $\mathrm{End}_{\mathfrak{n}((t))} (\mathrm{Dist}^{\mathrm{ch}}_N (\mathrm{ev}^{-1}(N \cap N^{w_0 w^{-1}})))$.

By the assumption on the $\mathfrak{h}$-action, and arguing as in the proof of Proposition 10.13, we obtain that any map of vector spaces $'\mathbb{W}^w_{\mathrm{crit}}(\mathcal{R}_1) \to {}'\mathbb{W}^w_{\mathrm{crit}}(\mathcal{R}_2)$ compatible with the action of $\mathfrak{n}((t))$ and $\mathfrak{h} \subset \Gamma(X, \mathcal{A}_{\mathfrak{g},\mathrm{crit}})$ has the form

$$\mathrm{Id}_{\mathrm{Dist}^{\mathrm{ch}}_N \left(\mathrm{ev}^{-1}(N \cap N^{w_0 w^{-1}})\right)} \otimes \varphi,$$

where φ is some map $\mathcal{R}_1 \to \mathcal{R}_2$ as vector spaces. To prove that φ is a map of $\mathfrak{H}_{\mathrm{crit}}$-modules, we argue as follows:

Recall that for a $\widehat{\mathfrak{g}}_{\mathrm{crit}}$-module $\mathcal{M}$, the semi-infinite cohomology

$$H^{\frac{\infty}{2}}(\mathfrak{n}((t)), t\mathfrak{n}[[t]] + \mathfrak{n} \cap \mathfrak{n}^{w_0 w^{-1}}, \mathcal{M})$$

is naturally an $\mathfrak{H}'_{\mathrm{crit}}$-module. We will regard it as an $\mathfrak{H}_{\mathrm{crit}}$-module via the isomorphism $\mathfrak{H}_{\mathrm{crit}} \simeq \mathfrak{H}'_{\mathrm{crit}}$. Recall the isomorphism

$$\mathcal{R} \simeq H^{\frac{\infty}{2}}(\mathfrak{n}((t)), t\mathfrak{n}[[t]] + \mathfrak{n} \cap \mathfrak{n}^{w_0 w^{-1}}, {}'\mathbb{W}^w_{\mathrm{crit}}(\mathcal{R})) \tag{13.2}$$

given by Proposition 12.10. From Lemma 10.11 we obtain the following.

Lemma 13.4. *The isomorphism* (13.2) *respects the* $\mathfrak{H}_{\mathrm{crit}}$*-module structures.*

From the construction of the isomorphism of Proposition 12.10 it is easy to see that any map $'\mathbb{W}^w_{\mathrm{crit}}(\mathcal{R}_1) \to {}'\mathbb{W}^w_{\mathrm{crit}}(\mathcal{R}_2)$ of the form $\mathrm{Id} \otimes \phi$ induces on the left-hand side of (13.2) the endomorphism equal to φ. Hence, the above lemma implies that ϕ respects the $\mathfrak{H}_{\mathrm{crit}}$-actions. □

13.5

We will now recall a crucial result of [F] (see [F, Proposition 6.3 and Remark 6.4]) that establishes isomorphisms between Wakimoto modules and Verma modules. Note that in [F] the module $\mathbb{W}^{w_0}_{\mathrm{crit},\lambda}$ is denoted by W_{λ,κ_c}.

Proposition 13.6. *Let* λ *be such that* $\lambda + \rho$ *is antidominant. Then* $\mathbb{W}^{w_0}_{\mathrm{crit},\lambda} \simeq \mathbb{M}_\lambda$.

Proof. First, we claim that when λ is antidominant, the $\mathfrak{g}$-module $M_\lambda^{w_0}$ (see Section 11.5) is, in fact, isomorphic to the Verma module M_λ.

Indeed, let us first note that $M_\lambda^{w_0}$ has a vector of highest weight λ, i.e., there is a morphism $M_\lambda \to M_\lambda^{w_0}$. Now, it is well known that the antidominance condition on $\lambda+\rho$ implies that M_λ is irreducible, hence the above map is injective. The assertion now follows from the fact that the two modules have the same formal character. This is a prototype of the argument proving the proposition.

By Section 11.5 we have a map in one direction

$$\mathbb{M}_\lambda \to \mathbb{W}^{w_0}_{\mathrm{crit},\lambda} \tag{13.3}$$

and we claim that it is an isomorphism. We will regard both sides of (13.3) as modules over the Kac–Moody algebra $\mathbb{C}\cdot t\partial_t \ltimes \widehat{\mathfrak{g}}_{\mathrm{crit}}$, where we normalize the action of $t\partial_t$ so that it annihilates the generating vector in $\mathbb{M}_\lambda$, and it acts on $\mathbb{W}^{w_0}_{\mathrm{crit},\lambda}$ by loop rotation.

The map (13.3) clearly respects this action. Moreover, both sides have well-defined formal characters with respect to the extended Cartan subalgebra $\mathbb{C}\cdot t\partial_t \oplus \mathfrak{h} \oplus \mathbb{C}\mathbf{1}$, and a computation shows that these characters are equal. Therefore, the map (13.3) is surjective if and only if it is injective.

Suppose that the kernel of the map in question is nonzero. Let $v \in \mathbb{M}_\lambda$ be a vector of highest weight with respect to $\mathbb{C}\cdot t\partial_t \oplus \mathfrak{h} \oplus \mathbb{C}\mathbf{1}$; let us denote this weight by $\widehat{\mu}$.

Then the quotient $\mathbb{W}^{w_0}_{\mathrm{crit},\lambda}/\operatorname{Im}\mathbb{M}_\lambda$ also contains a vector, call it v', of weight $\widehat{\mu}$. Moreover, by assumption, v' projects nontrivially to the space of coinvariants $(\mathbb{W}^{w_0}_{\mathrm{crit},\lambda})_{\mathfrak{n}^-[t^{-1}]\oplus t^{-1}\mathfrak{b}[t^{-1}]}$.

However, from (13.1), it follows that the projection

$$\operatorname{Fun}(N[[t]]) \to \left(\mathbb{W}^{w_0}_{\mathrm{crit},\lambda}\right)_{t^{-1}\mathfrak{n}[t^{-1}]\oplus t^{-1}\mathfrak{h}[t^{-1}]}$$

is an isomorphism. Therefore, $\widehat{\mu}$ must be of the form

$$\widehat{\mu} := (-n, \lambda-\beta, -\check{h}), \quad n \in \mathbb{Z}^{\geq 0}, \quad \beta \in \operatorname{Span}^+(\Delta^+). \tag{13.4}$$

We will now use the Kac–Kazhdan theorem [KK] that describes the possible highest weights of submodules of a Verma module. This theorem says that there must exist a sequence of weights

$$(0, \lambda, -\check{h}) = \widehat{\mu_1}, \widehat{\mu_2}, \ldots, \widehat{\mu}_{n-1}, \widehat{\mu}_n = \widehat{\mu}$$

and a sequence of positive affine roots $\alpha_{\mathrm{aff},k}$ such that

$$\widehat{\mu}_{k+1} = \widehat{\mu}_k - b_k \cdot \alpha_{\mathrm{aff},k} \tag{13.5}$$

with $b_k \in \mathbb{Z}^{>0}$ and such that

$$b_k \cdot (\alpha_{\mathrm{aff},k}, \alpha_{\mathrm{aff},k}) = 2 \cdot (\alpha_{\mathrm{aff},k}, \widehat{\mu}_k + \rho_{\mathrm{aff}}),$$

where $(\cdot,\cdot)$ is the invariant inner product on the Kac–Moody algebra.

Let us write $\widehat{\mu}_k = (n_k, \mu_k, -\check{h})$ and

$$\begin{cases} \alpha_{\text{aff},k} = (m_k, \epsilon_k \cdot \alpha_k, 0), \ m_k \geq 0, \ \alpha_k \in \Delta^+, \ \epsilon_k = \pm 1 & \text{if } \alpha_{\text{aff},k} \text{ is real,} \\ \alpha_{\text{aff},k} = (m_k, 0, 0) & \text{if } \alpha_{\text{aff},k} \text{ is imaginary.} \end{cases}$$

In the latter case we obtain $\mu_{k+1} = \mu_k$. In the former case we have

$$b_k = \langle \widehat{\mu}_k + \rho_{\text{aff}}, \check{\alpha}_{\text{aff},k} \rangle,$$

and since $\rho_{\text{aff}} = (0, \rho, \check{h})$ we obtain that $b_k = \epsilon_k \cdot \langle \mu_k + \rho, \alpha_k \rangle$, implying that

$$(\mu_{k+1} + \rho) = s_{\alpha_k}(\mu_k + \rho),$$

regardless of the sign of ϵ_k.

In particular, we obtain that $(\lambda + \rho) - \beta$ belongs to the W-orbit of $\lambda + \rho$, but this contradicts the antidominance of $\lambda + \rho$. □

13.7

We will use the above proposition to derive information about the structure of other Wakimoto and Verma modules.

Corollary 13.8. *For λ such that $\lambda + \rho$ is antidominant and $w \in W$ we have an isomorphism*

$$\mathbb{W}^{ww_0}_{\text{crit}, w(\lambda+\rho)-\rho} \simeq \mathbb{M}_{w(\lambda+\rho)-\rho}.$$

Proof. Let us assume that λ is integral. In this case all $\mathbb{W}^w_{\text{crit},\lambda}$ and $\mathbb{M}_{w(\lambda+\rho)-\rho}$ are I-integrable, and we can use the convolution action of D-modules on $G/B \subset \text{Fl}_G$ to pass from one another.

(If λ is not integral, the proof is essentially the same, where instead of B-equivariant D-modules on G/B we will use λ-twisted D-modules and replace the B-equivariant category by a λ-twisted version.)

It is known that for λ antidominant, $j_{w,!} \underset{B}{\star} M_\lambda = M_{w(\lambda+\rho)-\rho}$. Hence, $j_{w,!} \underset{I}{\star} \mathbb{M}_\lambda = \mathbb{M}_{w(\lambda+\rho)-\rho}$. This implies the corollary in view of (12.3). □

Since for every weight λ' there exists an element of the Weyl group such that $\lambda' = w(\lambda + \rho) - \rho$ with λ antidominant, every Verma module $\mathbb{M}_{\lambda'}$ is isomorphic to an appropriate Wakimoto module. By combining this with Propositions 3.10 and 13.3 we obtain the following statement.

Corollary 13.9. *The module $\mathbb{M}_\lambda$ is flat over $\mathfrak{z}_{\mathfrak{g}}^{\text{RS},\varpi(-\lambda-\rho)}$. The map*

$$\mathfrak{z}_{\mathfrak{g}}^{\text{RS},\varpi(-\lambda-\rho)} \to \text{End}_{\widehat{\mathfrak{g}}_{\text{crit}}}(\mathbb{M}_\lambda)$$

is an isomorphism.

Let us give an additional proof of the second assertion.

Proof. As above, we can reduce the statement to the case when $\lambda + \rho$ is itself antidominant, and $\mathbb{M}_\lambda \simeq \mathbb{W}^{w_0}_{\mathrm{crit},\lambda}$.

In the latter case, we have to show that the embedding of $\pi_{w_0(\lambda)}$ into the subspace of vectors of weight λ in $(\mathbb{W}^{w_0}_{\mathrm{crit},\lambda})^{\mathrm{Lie}(I^0)}$ is an isomorphism.

Consider the bigger subspace $(\mathbb{W}^{w_0}_{\mathrm{crit},\lambda})^{\mathfrak{n}[[t]]}$. As in the proof of Proposition 10.13, this subspace is isomorphic to $\mathrm{Ind}^{\mathfrak{n}((t))}_{\mathfrak{n}[[t]]}(\mathbb{C}) \otimes \pi_{w_0(\lambda)}$, which implies that the vectors of weight λ belong to $1 \otimes \pi_{w_0(\lambda)}$. □

Next, we shall prove Proposition 7.18.

Proof. Consider the Wakimoto module $\mathbb{W}^{w_0}_{\mathrm{crit},\lambda}$, where λ is dominant. We claim that $\mathfrak{Z}_{\mathfrak{g}}$ acts on it via $\mathfrak{Z}^{\lambda,\mathrm{reg}}_{\mathfrak{g}}$. This follows by combining Proposition 3.13 with Theorem 10.16 and the fact that the isomorphism of Proposition 10.10 sends the chiral $\mathfrak{H}_{\mathrm{crit}}$-module π_μ to Fun $(\mathrm{Conn}_{\check{H}}(\omega^\rho_X)^{\mathrm{RS},\mu})$.

Composing the map (11.8) with the natural embedding $\mathbb{V}^\lambda_{\mathrm{crit}} \to \mathbb{M}^\vee_{\mathrm{crit},\lambda}$, we obtain a map $\mathbb{V}^\lambda \to \mathbb{W}^{w_0}_{\mathrm{crit},\lambda}$, which can be shown to be injective.[8] We obtain that the ideal in the center that annihilates $\mathbb{W}^{w_0}_{\mathrm{crit},\lambda}$, annihilates $\mathbb{V}^\lambda$ as well, which is what we had to show. □

13.10

Finally, let us derive a corollary of Proposition 12.12 at the critical level. In this case the adjoint action of $H((t))$ on $H^0_{DR}(\mathcal{D}^\times, \widehat{L}_{\mathfrak{h},\mathrm{crit}})$ is trivial, and hence we obtain the following.

Corollary 13.11. *For a dominant coweight $\check{\lambda}$ and an $\mathfrak{H}_{\mathrm{crit}}$-module $\mathcal{R}$, we have an isomorphism*

$$\widetilde{j}_{\check{\lambda}} \underset{I^0}{\star} \mathbb{W}^{w_0}_{\mathrm{crit}}(\mathcal{R}) \simeq \mathbb{W}^{w_0}_{\mathrm{crit}}(\mathcal{R}).$$

Suppose now that $\mathcal{R}$ in $H[[t]]$-integrable. In this case, we obtain that

$$\widetilde{j}_{\check{\lambda}} \underset{I^0}{\star} \mathbb{W}^{w_0}_{\mathrm{crit}}(\mathcal{R}) \simeq j_{\check{\lambda},*} \underset{I}{\star} \mathbb{W}^{w_0}_{\mathrm{crit}}(\mathcal{R}) \simeq \mathbb{W}^{w_0}_{\mathrm{crit}}(\mathcal{R}),$$

where both the LHS and RHS are canonically defined, i.e., are independent of the choice of a representative $t^{\check{\lambda}} \in G((t))$. However, the isomorphism between them, given by Corollary 13.11, does depend on this choice. In the remainder of this chapter, we will need a more precise version of the above result.

Corollary 13.12. *We have an isomorphism*

$$j_{\check{\lambda},*} \underset{I}{\star} \mathbb{M}_{-2\rho} \simeq \mathbb{M}_{-2\rho} \otimes \omega_x^{\langle -\rho,\check{\lambda}\rangle},$$

where ω_x is the fiber of ω_X at $x \in X$, compatible with the natural actions of Lie(Aut($\mathcal{D}$)) *on both sides.*

[8] We will supply a proof in the next paper in the series.

Note that $\langle -\rho, \check{\lambda}\rangle$, appearing in the corollary, may be a half-integer. In the above formula the expression $\omega_x^{\langle -\rho,\check{\lambda}\rangle}$ involves a choice of a square root of ω_X, as does the construction of the critical line bundle on $\mathrm{Gr}_{G_{ad}}$. However, the character of $\mathrm{Lie}(\mathrm{Aut}(\mathcal{D}))$ on $\omega_x^{\langle -\rho,\check{\lambda}\rangle}$ is, of course, independent of this choice.

Remark 13.13. By considering the action of the renormalized universal enveloping algebra as in Section 7.8, one shows that, more generally, there is an isomorphism

$$j_{\check{\lambda},*} \underset{I}{\star} \mathbb{W}^{w_0}_{\mathrm{crit},\mu} \simeq \mathbb{W}^{w_0}_{\mathrm{crit},\mu} \otimes \omega_x^{\langle \mu+\rho,\check{\lambda}\rangle},$$

compatible with the $\mathrm{Lie}(\mathrm{Aut}(\mathcal{D}))$-actions.

Proof. The existence of an isomorphism stated in the corollary follows by combining Proposition 13.6 and Corollary 13.11. By Proposition 13.3, we obtain that there exists a line, acted on by (a double cover of) $\mathrm{Aut}(\mathcal{D})$, and a canonical isomorphism

$$j_{\check{\lambda},*} \underset{I}{\star} \mathbb{M}_{-2\rho} \simeq \mathbb{M}_{-2\rho} \otimes \mathfrak{l},$$

compatible with the $\mathrm{Lie}(\mathrm{Aut}(\mathcal{D}))$-actions.

We have to show that the character of $\mathrm{Lie}(\mathrm{Aut}(\mathcal{D}))$, corresponding to $\mathfrak{l}$, equals that of $\omega_x^{\langle -\rho,\check{\lambda}\rangle}$. Let $t\partial_t \in \mathrm{Lie}(\mathrm{Aut}(\mathcal{D}))$ be the Euler vector field, corresponding to the coordinate t on $\mathcal{D}$. It suffices to show that the highest weight of $j_{\check{\lambda},*} \underset{I}{\star} \mathbb{M}_{-2\rho}$ with respect to $\mathbb{C}\cdot t\partial_t \oplus \mathfrak{h}$ equals $(-\langle \rho, \check{\lambda}\rangle, -2\rho)$.

The module in question identifies with $\Gamma(\mathrm{Fl}_G, j_{\check{\lambda},*})$. The highest weight line in $\Gamma(\mathrm{Fl}_G, j_{\check{\lambda},*})$ consists of I^0-invariant sections of this D-module, that are supported on the I-orbit of $t^{\check{\lambda}}$. Now the fact that $t\partial_t$ acts on this line by the character equal to $-\langle \rho, \check{\lambda}\rangle$ is a straightforward calculation, as in [BD1, Section 9.1]. □

Part IV: Proof of Main Theorem 6.9

The goal of this part is to prove Main Theorem 6.9:

There is an equivalence of categories

$${}^f\mathsf{F} : {}^f\widehat{\mathfrak{g}}_{\mathrm{crit}}\text{-mod}^{I,m}_{\mathrm{nilp}} \simeq \mathrm{QCoh}(\mathrm{Spec}(h_0) \times \mathrm{Op}^{\mathrm{nilp}}_{\check{\mathfrak{g}}}).$$

In Section 14 we introduce the module $\mathbf{\Pi}$ as induced from the big projective module Π over the finite-dimensional Lie algebra $\mathfrak{g}$. We first review the properties of Π, and the corresponding properties of $\mathbf{\Pi}$, related to the notion of partial integrability. The functor ${}^f\mathsf{F}$ is then defined as

$${}^f\mathsf{F}(M) = \mathrm{Hom}(\mathbf{\Pi}, M),$$

and we state Main Theorem 14.15 which asserts that this functor is exact.

As we shall see later (see Section 16), both Main Theorems 14.15 and 6.9 follow once we can compute $\mathrm{RHom}_{D^b(\widehat{\mathfrak{g}}_{\mathrm{crit}}\text{-mod}_{\mathrm{nilp}})^{I^0}}(\mathbf{\Pi}, \mathbb{M}_{w_0})$, where $\mathbb{M}_{w_0}$ is the corresponding Verma module over $\widehat{\mathfrak{g}}_{\mathrm{crit}}$. Also in Section 16 (see Proposition 16.2) we show that it is sufficient to compute $\mathrm{RHom}_{D^b(\widehat{\mathfrak{g}}_{\mathrm{crit}}\text{-mod}_{\mathrm{reg}})^{I^0}}(\mathbf{\Pi}_{\mathrm{reg}}, \mathbb{M}_{w_0,\mathrm{reg}})$, where $\mathbf{\Pi}_{\mathrm{reg}}$ and $\mathbb{M}_{w_0,\mathrm{reg}}$ are the restrictions of the corresponding modules to the subscheme $\mathrm{Op}_{\check{\mathfrak{g}}}^{\mathrm{reg}} \subset \mathrm{Op}_{\check{\mathfrak{g}}}^{\mathrm{nilp}}$.

The computation of $\mathrm{RHom}_{D^b(\widehat{\mathfrak{g}}_{\mathrm{crit}}\text{-mod}_{\mathrm{reg}})^{I^0}}(\mathbf{\Pi}_{\mathrm{reg}}, \mathbb{M}_{w_0,\mathrm{reg}})$ is carried out in Section 15. We reduce it to a calculation involving D-modules on the affine Grassmannian once we can identify $\mathbf{\Pi}_{\mathrm{reg}}$ as sections of some specific critically twisted D-module on Gr. The latter identification is given by Theorem 15.6. This theorem is proved in Section 17 by a rather explicit argument.

Having proved Main Theorem 6.9, we compare in Section 18 the functor $\mathrm{Hom}_{\widehat{\mathfrak{g}}_{\mathrm{crit}}\text{-mod}_{\mathrm{nilp}}^{I,m}}(\mathbf{\Pi}, \cdot)$ with the one given by semi-infinite cohomology with respect to the Lie algebra $\mathfrak{n}^-((t))$ against a nondegenerate character. We show that the two functors are isomorphic. We also express the semi-infinite cohomology of $\mathfrak{n}((t))$ with coefficients in a $\widehat{\mathfrak{g}}_{\mathrm{crit}}$-module of the form $\Gamma(\mathrm{Gr}_G, \mathcal{F})$, where $\mathcal{F}$ is a critically twisted D-module on Gr_G, in terms of the de Rham cohomologies of the restrictions of $\mathcal{F}$ to $N((t))$-orbits in Gr_G.

14 The module $\mathbf{\Pi}$

14.1

Recall from Section 7.19 that $\mathcal{O}_0$ denotes the subcategory of the category $\mathcal{O}$ of $\mathfrak{g}$-modules whose objects are modules with central character $\varpi(\rho)$. According to [BB], the functor of global sections induces an equivalence between the category of N-equivariant (or, equivalently, B-monodromic) left D-modules on G/B and $\mathcal{O}_0$.

To simplify our notation slightly, we will use the notation M_w instead of $M_{w(\rho)-\rho}$ and $M_w^\vee$ instead of $M_{w(\rho)-\rho}^\vee$. We will denote by L_w the irreducible quotient of M_w. By $\mathbb{M}_w$, $\mathbb{M}_w^\vee$ and $\mathbb{L}_w$ we will denote the corresponding induced representations of $\widehat{\mathfrak{g}}$ at the critical level.

By definition, objects of $\mathcal{O}_0$ are N-integrable, and the condition on the central character implies that they are, in fact, B-monodromic. Hence, every object $M \in \mathcal{O}_0$ carries an action of the commutative algebra $\mathfrak{h}$. This is the obstruction to being B-equivariant. (The notions of B-integrability (equivalently, B-equivariance), N-integrability and B-monodromicity are defined as their I- and I^0-counterparts and make sense in any category $\mathcal{C}$ with a Harish-Chandra action of G.)

Lemma 14.2. *For every $M \in \mathcal{O}_0$, the action of* $\mathrm{Sym}(\mathfrak{h})$ *on M factors through* $\mathrm{Sym}(\mathfrak{h}) \to h_0$.

The lemma follows, e.g., from the localization theorem of [BB]. Thus we obtain that the algebra h_0 maps to the center of $\mathcal{O}_0$. In fact, it follows from [Be] that h_0 is isomorphic to the center of $\mathcal{O}_0$.

As in Section 6.7, we will call an object $M \in \mathcal{O}_0$ partially integrable if it admits a filtration such that each successive quotient is integrable with respect to a parabolic subalgebra $\mathfrak{b} + \mathfrak{sl}_2^\iota$ for some $\iota \in \mathcal{I}$. This notion makes sense in an arbitrary category with a Harish-Chandra action of G.

We will denote by ${}^f\mathcal{O}_0$ the quotient abelian category of $\mathcal{O}_0$ by the subcategory of partially integrable objects. We will denote by $M \mapsto {}^f M$ the projection functor $\mathcal{O}_0 \to {}^f\mathcal{O}_0$.

Let Π be a "longest" indecomposable projective in $\mathcal{O}_0$. By definition,

$$\mathrm{Hom}(\Pi, L_{w_0}) = \mathbb{C}, \quad \mathrm{Hom}(\Pi, L_w) = 0 \quad \text{if } w \neq w_0.$$

Moreover, Π is known to be isomorphic (noncanonically) to its contragredient dual. We have the following result (see [BG]).

Lemma 14.3.

(1) *The map* $h_0 \to \mathrm{End}(\Pi)$ *is an isomorphism.*
(2) *The functor* $M \mapsto \mathrm{Hom}(\Pi, M)$ *induces an equivalence* ${}^f\mathcal{O}_0 \to h_0$-mod.

By construction, the image ${}^f\Pi$ of Π in ${}^f\mathcal{O}_0$ identifies with the free h_0-module with one generator. The maps $M_{w_0} \to L_{w_0} \to M_{w_0}^\vee$ induce isomorphisms ${}^f M_{w_0} \to {}^f L_{w_0} \to {}^f M_{w_0}^\vee$, and all identify with the trivial h_0-module $\mathbb{C}$.

We will now recall the construction of Π as

$$\Pi = \Gamma(G/B, \Xi), \tag{14.1}$$

where Ξ is a certain left D-module on G/B.

14.4

To describe Ξ we need to introduce some notation, which will also be used in what follows. Let $\psi : N^- \to \mathbb{G}_a$ be a nondegenerate character. By a slight abuse of notation, we will denote also by ψ its differential: $\mathfrak{n}^- \to \mathbb{C}$.

Let $\mathbf{e}^\psi$ denote the pull-back of the "e^x" D-module from $\mathbb{G}_a$ to N^-. This is a "character sheaf" in the sense of Section 20.20.

If N^- acts (in the Harish-Chandra sense) on a category $\mathcal{C}$, we will denote by $\mathcal{C}^{N^-,\psi}$ the corresponding (N^-, ψ)-equivariant category (see Section 20.20), and by $D(\mathcal{C})^{N^-,\psi}$ the corresponding triangulated category. Since N^- is unipotent, the natural forgetful functor $D(\mathcal{C})^{N^-,\psi} \to D(\mathcal{C})$ is fully faithful; see Section 20.20.

Following Section 20.20, we will denote by $\mathrm{Av}_{N^-,\psi}$ the functor

$$\mathcal{M} \mapsto \mathbf{e}^\psi \star \mathcal{M} \otimes \det(\mathfrak{n}^-[1])^{-1} : D(\mathcal{C}) \to D(\mathcal{C})^{N^-,\psi}.$$

This functor is the right adjoint and a left quasi-inverse to $D(\mathcal{C})^{N^-,\psi} \to D(\mathcal{C})$.

Lemma 14.5. *Suppose that $\mathcal{C}$ is endowed with a Harish-Chandra action of G, and let $\mathcal{M} \in \mathcal{C}^{B,m}$ be partially integrable; then* $\mathrm{Av}_{N^-,\psi}(\mathcal{M}) = 0$.

Proof. We can assume that $\mathcal{M}$ is an object of $\mathcal{C}$ integrable with respect to a parabolic subgroup P^ι for some $\iota \in \mathcal{I}$. Then the convolution $\mathbf{e}^\psi \star \mathcal{F}$ factors through the direct image of $\mathbf{e}^\psi$ under $N^- \hookrightarrow G \twoheadrightarrow G/P^\iota$, and the latter is clearly 0. □

For example, if N^- acts on a scheme Y, in this way we obtain the category of (N^-, ψ)-equivariant D-modules on Y. In other words, its objects are D-modules $\mathcal{F}$ on Y, together with an isomorphism

$$\mathrm{act}^*(\mathcal{F}) \simeq \mathbf{e}^\psi \boxtimes \mathcal{F} \in \mathfrak{D}(N^- \times Y)\text{-mod},$$

compatible with the restriction to the unit section and associative in the natural sense. One can show that in this case the functor

$$D(\mathfrak{D}(Y)\text{-mod}^{N^-,\psi}) \to D(\mathfrak{D}(Y)\text{-mod})^{N^-,\psi}$$

is an equivalence.

If we restrict ourselves to holonomic D-modules, or, rather, if we take the corresponding triangulated category (which, by definition, is the full subcategory of $D(\mathfrak{D}(Y)\text{-mod})$ consisting of complexes with holonomic cohomologies), then in addition to the functor $\mathcal{F} \mapsto \mathbf{e}^\psi \star \mathcal{F}$ we also have a functor

$$\mathcal{F} \mapsto \mathbf{e}^\psi \overset{!}{\star} \mathcal{M} : D(\mathfrak{D}(Y)_{\mathrm{hol}}\text{-mod}) \to D(\mathfrak{D}(Y)_{\mathrm{hol}}\text{-mod})^{N^-,\psi},$$

corresponding to taking the direct image with compact supports. This functor, tensored with $\det(\mathfrak{n}^-[1])$, is the left adjoint and a left quasi-inverse to the tautological functor $D(\mathfrak{D}(Y)_{\mathrm{hol}}\text{-mod})^{N^-,-\psi} \to D(\mathfrak{D}(Y)_{\mathrm{hol}}\text{-mod})$.

Proposition 14.6. *Suppose that Y is acted on by G. Then for $\mathcal{F} \in D(\mathfrak{D}(Y)_{\mathrm{hol}}\text{-mod})^{B,m}$ the canonical arrow: $\mathbf{e}^\psi \overset{!}{\star} \mathcal{F} \to \mathbf{e}^\psi \star \mathcal{F}$ is an isomorphism. In particular, the functor*

$$D(\mathfrak{D}(Y)_{\mathrm{hol}}\text{-mod})^{B,m} \to D(\mathfrak{D}(Y)_{\mathrm{hol}}\text{-mod}) \overset{\mathbf{e}^\psi \star \cdot}{\longrightarrow} D(\mathfrak{D}(Y)_{\mathrm{hol}}\text{-mod})^{N^-,-\psi}$$

is exact.

Proof. It is enough to analyze the functor $\mathcal{F} \mapsto \mathbf{e}^\psi \star \mathcal{F}$ on the subcategory $\mathfrak{D}(Y)_{\mathrm{hol}}\text{-mod}^B$.

The basic observation is that the D-module

$$\mathrm{Dist}_{G/B}(N^-, \psi) := \mathbf{e}^\psi \star \delta_{1_{G/B}} \in \mathfrak{D}(G/B)\text{-mod}^{B^-,\psi},$$

which is by definition the $*$-extension of $\mathbf{e}^\psi$ under $N^- \cdot 1_{G/B} \hookrightarrow G/B$, is clean. This means that the $*$-extension coincides with the !-extension, or, what is the same, that the arrow $\mathbf{e}^\psi \overset{!}{\star} \delta_{1_{G/B}} \to \mathbf{e}^\psi \star \delta_{1_{G/B}}$ is an isomorphism. (One easily shows this by observing that for any $\mathbf{g} \in G/B \setminus N^-$, the restriction of ψ to its stabilizer in N^- is nontrivial.) In particular, $\mathrm{Dist}_{G/B}(N^-, \psi)$ is the Verdier dual of $\mathrm{Dist}_{G/B}(N^-, -\psi)$.

Note that for $\mathcal{F} \in \mathfrak{D}(Y)_{\mathrm{hol}}\text{-mod}^B$,

$$\mathbf{e}^\psi \star \mathcal{F} \simeq \mathrm{Dist}_{G/B}(N^-, \psi) \underset{B}{\star} \mathcal{F}, \tag{14.2}$$

and similarly for $\mathbf{e}^\psi \overset{!}{\star} \mathcal{F}$. This establishes the assertion of the proposition. □

14.7

After these preliminaries, we are ready to introduce Ξ:

$$\Xi := \underline{\mathbb{C}}_N \overset{!}{\star} \mathrm{Dist}_{G/B}(N^-, \psi) \otimes \det(\mathfrak{n}[1]) \in D(\mathfrak{D}(G/B)\text{-mod})^N,$$

where

$$\mathcal{F} \mapsto \underline{\mathbb{C}}_N \overset{!}{\star} \mathcal{F} \otimes \det(\mathfrak{n}[1])^{\otimes 2}$$

is the functor $D(\mathfrak{D}(G/B)_{\mathrm{hol}}\text{-mod}) \to D(\mathfrak{D}(G/B)_{\mathrm{hol}}\text{-mod})^N$, left adjoint to the tautological functor $D(\mathfrak{D}(G/B)_{\mathrm{hol}}\text{-mod})^N \to D(\mathfrak{D}(G/B)_{\mathrm{hol}}\text{-mod})$. Explicitly, $\underline{\mathbb{C}}_N \overset{!}{\star} \cdot \otimes \det(\mathfrak{n}[1])$ is given by convolution with compact supports with the constant D-module on N.

Proposition 14.8.

(1) *The complex* Ξ *is concentrated in cohomological degree* 0.
(2) Ξ *is projective as an object of* $\mathfrak{D}(G/B)\text{-mod}^N$.
(3) Ξ *is noncanonically Verdier self-dual, i.e.,*

$$\Xi \simeq \underline{\mathbb{C}}_N \star \mathrm{Dist}_{G/B}(N^-, -\psi) \otimes \det(\mathfrak{n}[1])^{-1}.$$

(4) Ξ *is canonically independent of the choice of* ψ.

Proof. Consider the functor $\mathcal{F} \mapsto \mathrm{RHom}(\Xi, \mathcal{F})$ on the category $D(\mathfrak{D}(G/B)\text{-mod})^N$. We have

$$\mathrm{RHom}_{D(\mathfrak{D}(G/B)\text{-mod})^N}(\Xi, \mathcal{F}) \simeq \mathrm{RHom}_{D(\mathfrak{D}(G/B)\text{-mod})}(\mathbf{e}^{\psi} \overset{!}{\star} \delta_{1_{G/B}}, \mathcal{F}) \otimes \det(\mathfrak{n}[1]),$$

which, in turn, is isomorphic to

$$\mathrm{RHom}_{D(\mathfrak{D}(G/B)\text{-mod})}(\delta_{1_{G/B}}, \mathbf{e}^{\psi'} \star \mathcal{F}) \otimes \det(\mathfrak{n}[1]),$$

where $\psi' = -\psi$.

By Lemma 14.6, $\mathbf{e}^{\psi'} \star \mathcal{F}$ is concentrated in cohomological degree 0. Moreover, it is lisse near $1_{G/B}$. Hence, the above RHom is concentrated in cohomological degree 0.

Now, we will use the fact that $D(\mathfrak{D}(G/B)\text{-mod})^N$ is equivalent to the derived category of the abelian category $\mathfrak{D}(G/B)$-mod. Then the above property of RHom implies simultaneously assertions (1) and (2) of the proposition.

The above expression for $\mathrm{RHom}(\Xi, \mathcal{F})$ also implies that it is 0 if $\mathcal{F}$ is partially integrable, and $\mathrm{RHom}(\Xi, \delta_{1_{G/B}})$ is one-dimensional. This implies that Ξ corresponds to a projective cover of $\delta_{1_{G/B}} \in \mathfrak{D}(G/B)\text{-mod}^N$, i.e., $\Gamma(G/B, \Xi) \simeq \Pi$.

Since it is known that contravariant duality on $\mathcal{O}_0$ goes over to Verdier duality on $\mathfrak{D}(G/B)\text{-mod}^N$, assertion (3) of the proposition holds.

The fact that Ξ is *noncanonically* independent of the choice of ψ also follows, since we have shown that (14.1) is valid for any choice of ψ. To establish that it is canonically independent, we argue as follows:

Let ψ' be another nondegenerate character of N^-. Then there exists an element $\mathbf{h} \in H$, which, under the adjoint action of H on N^-, transforms ψ to ψ'.

Since Ξ is B-monodromic, we have a canonical isomorphism of D-modules $\mathbf{h}^*(\Xi) \simeq \Xi$. However, from the construction of Ξ, we have $\mathbf{h}^*(\Xi) \simeq \Xi'$, where the latter is the D-module constructed starting from ψ'. □

For any category $\mathcal{C}$ with a Harish-Chandra action of G, we can consider the functor

$$\mathcal{F} \mapsto \Xi \underset{B}{\star} \mathcal{F} : D(\mathcal{C})^B \to D(\mathcal{C})^N.$$

Proposition 14.9.

(1) *The above functor is exact, and it annihilates an object $\mathcal{F} \in \mathcal{C}$ if and only if $\mathcal{F}$ is partially integrable.*
(2) *For $\mathcal{M}_1, \mathcal{M}_2 \in D(\mathcal{C})^B$ we have a noncanonical but functorial isomorphism*

$$\mathrm{RHom}_{D(\mathcal{C})}(\Xi \star \mathcal{M}_1, \mathcal{M}_2) \simeq \mathrm{RHom}_{D(\mathcal{C})}(\mathcal{M}_1, \Xi \star \mathcal{M}_2).$$

Proof. Using Proposition 14.8(3), we can rewrite the functor in question as

$$\mathcal{F} \mapsto \underline{\mathbb{C}}_N \star \mathbf{e}^{\psi} \star \mathcal{F} \otimes \det(\mathfrak{n}[1])^{-1}.$$

Hence, the fact that it annihilates partially integrable objects follows from Lemma 14.5.

Recall that the object $\Pi \in \mathcal{O}_0$ is tilting, i.e., it admits two filtrations: one, whose successive quotients are isomorphic to Verma modules, and another, whose successive quotients are dual Verma modules. Hence, Ξ also admits such filtrations, with subquotients being $j_{w,!}$ and $j_{w,*}$, respectively. It is clear that convolution with the latter is right exact. The convolution with $j_{w,!}$, being a quasi-inverse of the convolution with $j_{w^{-1},*}$, is therefore left exact. This proves the exactness assertion of the proposition.

Finally, let us show that if $\mathcal{F}$ is not partially integrable, then $\Xi \star \mathcal{F} \neq 0$. Let ${}^f\mathcal{C}$ be the quotient category of $\mathcal{C}$ by the Serre subcategory of partially integrable objects. Let ${}^f\mathcal{F}$ be the image of $\mathcal{F}$ in ${}^f\mathcal{C}$.

We claim that the image of $\Xi \star \mathcal{F}$ in ${}^f\mathcal{C}$ is endowed with an increasing filtration of length $|W|$, whose subquotients are all isomorphic to ${}^f\mathcal{F}$. This follows from the existence of the filtration on Ξ by $j_{w,!}$: Indeed, the cokernel of the map $\delta_{1_{G/B}} \to j_{w,!}$ is partially integrable, hence ${}^f\mathcal{F} \to {}^f(j_{w,!} \underset{B}{\star} \mathcal{F})$ is an isomorphism.

Now let us prove assertion (2) of the proposition. By Section 22.22,

$$\mathrm{RHom}_{D(\mathcal{C})}(\Xi \underset{B}{\star} \mathcal{M}_1, \mathcal{M}_2) \simeq \mathrm{RHom}_{D(\mathcal{C})^B}(\mathcal{M}_1, \widetilde{\Xi} \star \mathcal{M}_2),$$

where $\widetilde{\Xi}$ is the corresponding dual D-module on $B\backslash G$. Since $\mathcal{M}_2$ was assumed B-equivariant,

$$\widetilde{\Xi} \star \mathcal{M}_2 \simeq (\widetilde{\Xi} \star \underline{\mathbb{C}}_B) \underset{B}{\star} \mathcal{M}_2 \otimes \det(\mathfrak{b}[1]).$$

Similarly, by the B-equivariance of $\mathcal{M}_1$,

$$\mathrm{RHom}_{D(\mathcal{C})}(\mathcal{M}_1, \Xi \star \mathcal{M}_2) \simeq \mathrm{RHom}_{D(\mathcal{C})^B}(\mathcal{M}_1, (\underline{\mathbb{C}}_B \star \Xi) \star \mathcal{M}_2).$$

Hence, it remains to see that

$$\underline{\mathbb{C}}_B \star \Xi \simeq \widetilde{\Xi} \star \underline{\mathbb{C}}_B \otimes \det(\mathfrak{b}[1]) \in D(\mathfrak{D}(G/B)\text{-mod})^B \simeq D(\mathfrak{D}(G)\text{-mod})^{B\times B}.$$

The left-hand side is isomorphic to $\mathrm{Av}_{B\times B}(\mathrm{Dist}_G(N^-, \psi))$ using Proposition 14.8(3), and using Proposition 14.8(4) the right-hand side is isomorphic to the same thing. □

14.10

Let us return to representations of affine algebras at the critical level. We define the module $\mathbf{\Pi} \in \widehat{\mathfrak{g}}_{\mathrm{crit}}$-mod as

$$\mathbf{\Pi} = \mathrm{Ind}^{\widehat{\mathfrak{g}}_{\mathrm{crit}}}_{\mathfrak{g}[[t]]\oplus\mathbb{C}\mathbf{1}}(\Pi). \tag{14.3}$$

By Section 7.19, $\mathbf{\Pi}$ belongs to $\widehat{\mathfrak{g}}_{\mathrm{crit}}\text{-mod}^{I,m}_{\mathrm{nilp}}$.

From the tilting property of Π, we obtain that $\mathbf{\Pi}$ admits two filtrations: one whose subquotients are modules of the form $\mathbb{M}_w$, and another, whose subquotients are of the form $\mathbb{M}^\vee_w$. Together with Corollary 13.9 this implies the following.

Corollary 14.11. *The module $\mathbf{\Pi}$ is flat over $\mathfrak{Z}^{\mathrm{nilp}}_{\mathfrak{g}}$.*

Using our conventions concerning twisted D-modules on Fl_G, we can rewrite the definition of $\mathbf{\Pi}$ as

$$\Gamma(\mathrm{Fl}_G, \Xi),$$

where we think of Ξ as living on Fl_G via $G/B \hookrightarrow \mathrm{Fl}_G$.

Note that for $\mathcal{M}^\bullet \in D(\widehat{\mathfrak{g}}_{\mathrm{crit}}\text{-mod})^I$, the convolution $\Xi \underset{I}{\star} \mathcal{M}^\bullet$ is tautologically the same as $\Xi \underset{B}{\star} \mathcal{M}^\bullet$, when we think of $\mathcal{M}^\bullet$ is a $\mathfrak{g}$-module via $\mathfrak{g} \hookrightarrow \widehat{\mathfrak{g}}_{\mathrm{crit}}$.

Proposition 14.12. *If an object $\mathcal{M}$ of $\widehat{\mathfrak{g}}_{\mathrm{crit}}\text{-mod}^{I,m}_{\mathrm{nilp}}$ is partially integrable, then*

$$\mathrm{RHom}_{D^b(\widehat{\mathfrak{g}}_{\mathrm{crit}}\text{-mod}_{\mathrm{nilp}})^{I^0}}(\mathbf{\Pi}, \mathcal{M}) = 0.$$

Proof. By Proposition 7.14, we can assume that $\mathcal{M}$ is I-integrable. In this case the assertion follows readily from Proposition 14.9. □

Obviously, the induction functor $\mathcal{O}_0 \to \widehat{\mathfrak{g}}_{\mathrm{crit}}\text{-mod}^{I,m}_{\mathrm{nilp}}$ descends to a well-defined functor ${}^f\mathcal{O}_0 \to {}^f\widehat{\mathfrak{g}}_{\mathrm{crit}}\text{-mod}^{I,m}_{\mathrm{nilp}}$. Let

$$\mathcal{M} \mapsto {}^f\mathcal{M}$$

denote the projection functor $\widehat{\mathfrak{g}}_{\mathrm{crit}}\text{-mod}^{I,m}_{\mathrm{nilp}} \to {}^f\widehat{\mathfrak{g}}_{\mathrm{crit}}\text{-mod}^{I,m}_{\mathrm{nilp}}$. In particular, we obtain the modules ${}^f\mathbb{M}_w$ and ${}^f\mathbf{\Pi}$ in ${}^f\widehat{\mathfrak{g}}_{\mathrm{crit}}\text{-mod}^{I,m}_{\mathrm{nilp}}$.

From Proposition 14.12 we obtain the following.

Corollary 14.13. *The map*

$$\mathrm{RHom}_{D^b(\widehat{\mathfrak{g}}_{\mathrm{crit}}\text{-mod}_{\mathrm{nilp}})^{I^0}}(\mathbf{\Pi}, \mathcal{M}) \to \mathrm{RHom}_{{}^fD^b(\widehat{\mathfrak{g}}_{\mathrm{crit}}\text{-mod}_{\mathrm{nilp}})^{I^0}}({}^f\mathbf{\Pi}, {}^f\mathcal{M})$$

is an isomorphism.

Since we have a surjection $\mathbf{\Pi} \to \mathbb{M}_{w_0}$, we also obtain the following.

Corollary 14.14. *If an object* $\mathcal{M}$ *of* $\widehat{\mathfrak{g}}_{\mathrm{crit}}\text{-mod}_{\mathrm{nilp}}^{I,m}$ *is partially integrable, then*

$$\mathrm{Hom}(\mathbb{M}_{w_0}, \mathcal{M}) = 0.$$

The main theorem in Part IV, from which we will derive Main Theorem 6.9, is the following.

Main Theorem 14.15. *For any object* $\mathcal{M}$ *of* $\widehat{\mathfrak{g}}_{\mathrm{crit}}\text{-mod}_{\mathrm{nilp}}^{I,m}$, *we have*

$$\mathrm{R}^i\,\mathrm{Hom}_{D^b(\widehat{\mathfrak{g}}_{\mathrm{crit}}\text{-mod}_{\mathrm{nilp}})^{I^0}}(\mathbf{\Pi}, \mathcal{M}) = 0 \quad \textit{for } i > 0.$$

15 The module $\mathbf{\Pi}_{\mathrm{reg}}$ via the affine Grassmannian

We proceed with the proof of Main Theorem 14.15.

15.1

Consider the quotient $\mathfrak{Z}_{\mathfrak{g}}^{\mathrm{reg}}$ of $\mathfrak{Z}_{\mathfrak{g}}^{\mathrm{nilp}}$. We will denote by $\mathbf{\Pi}_{\mathrm{reg}}$ and $\mathbb{M}_{w_0,\mathrm{reg}}$ the modules $\mathbf{\Pi} \underset{\mathfrak{Z}_{\mathfrak{g}}^{\mathrm{nilp}}}{\otimes} \mathfrak{Z}_{\mathfrak{g}}^{\mathrm{reg}}$ and $\mathbb{M} \underset{\mathfrak{Z}_{\mathfrak{g}}^{\mathrm{nilp}}}{\otimes} \mathfrak{Z}_{\mathfrak{g}}^{\mathrm{reg}}$, respectively. The goal of this section is to express these $\widehat{\mathfrak{g}}_{\mathrm{crit}}$-modules as sections of critically twisted D-modules on the affine Grassmannian.

Consider the element of the extended affine Weyl group equal to $w_0 \cdot \check{\rho} = -\check{\rho} \cdot w_0$. Let $j_{w_0\cdot\check{\rho},*}$ and $j_{w_0\cdot\check{\rho},!}$ denote the corresponding critically twisted D-modules on Fl_G.

Note that $w_0 \cdot \check{\rho}$ is minimal in its coset in $W\backslash W_{\mathrm{aff}}/W$, in particular, the orbit $I \cdot (w_0 \cdot \check{\rho}) \subset \mathrm{Fl}_G$ projects one-to-one under $\mathrm{Fl}_G \to \mathrm{Gr}_G$. Hence, $j_{w_0\cdot\check{\rho},!} \underset{I}{\star} \delta_{1_{\mathrm{Gr}_G}}$ is the D-module on Gr_G obtained as the extension by 0 from the Iwahori orbit of the element $t^{-\check{\rho}} \in \mathrm{Gr}_G$. Let us denote by $\mathrm{IC}_{w_0\cdot\check{\rho},\mathrm{Gr}_G}$ the intersection cohomology D-module corresponding to the above I-orbit.

We have the maps

$$j_{w_0\cdot\check{\rho},!} \underset{I}{\star} \delta_{1_{\mathrm{Gr}_G}} \twoheadrightarrow \mathrm{IC}_{w_0\cdot\check{\rho},\mathrm{Gr}_G} \hookrightarrow j_{w_0\cdot\check{\rho},*} \underset{I}{\star} \delta_{1_{\mathrm{Gr}_G}}, \tag{15.1}$$

such that the kernel of the first map and cokernel of the second map are supported on the closed subset $\overline{\mathrm{Gr}_G}^{\check{\rho}} - \mathrm{Gr}_G^{\check{\rho}} \subset \mathrm{Gr}_G$.

Proposition 15.2. *The maps*

$$\Xi \underset{I}{\star} j_{w_0\cdot\check{\rho},!} \underset{I}{\star} \delta_{1_{\mathrm{Gr}_G}} \to \Xi \underset{I}{\star} \mathrm{IC}_{w_0\cdot\check{\rho},\mathrm{Gr}_G} \to \Xi \underset{I}{\star} j_{w_0\cdot\check{\rho},*} \underset{I}{\star} \delta_{1_{\mathrm{Gr}_G}}$$

are isomorphisms.

The proposition follows from Proposition 14.9, using the following lemma.

Lemma 15.3. *Any I-monodromic D-module on Gr_G supported on $\overline{\mathrm{Gr}_G}^{\check{\rho}} - \mathrm{Gr}_G^{\check{\rho}}$ is partially integrable.*

Proof. Recall that the $G[[t]]$–orbits on Gr_G are labeled by the set of dominant coweights of G; for a coweight $\check{\lambda}$ we will denote by

$$\mathrm{Gr}_G^{\check{\lambda}} \overset{\mathrm{emb}_{\check{\lambda}}}{\hookrightarrow} \mathrm{Gr}_G$$

the embedding of the corresponding orbit. The quotient $G^{(1)}\backslash \mathrm{Gr}_G^{\check{\lambda}}$ is a G-homogeneous space, isomorphic to a partial flag variety. We identify it with G/P by requiring that the point $w_0 \cdot t^{\check{\lambda}} \in G((t))$ project to $1_{G/P} \subset G/P$; we have $P = B$ if and only if $\check{\lambda}$ is regular.

Note that the $G[[t]]$-orbits appearing in $\overline{\mathrm{Gr}_G}^{\check{\rho}} - \mathrm{Gr}_G^{\check{\rho}}$ all correspond to irregular $\check{\lambda}$. Therefore, it is enough to show that an irreducible I-equivariant D-module on $\mathrm{Gr}_G^{\check{\lambda}}$ with irregular $\check{\lambda}$ is partially integrable.

Any such D-module arises as a pull-back from an irreducible B-equivariant D-module on G/P for some parabolic P, strictly larger than B. By definition, irreducible B-equivariant D-modules on G/P are IC-sheaves of closures of B-orbits on G/P. So, it is enough to show that any such closure is stable under SL_2^{ι} for some $\iota \in \mathcal{J}$. But this is nearly evident:

The orbit of $1_{G/P}$ is clearly P-stable. Any other orbit corresponds to some element $w \in W$ of length more than 1. Hence, there exists a simple reflection s_ι such that $s_\iota \cdot w < w$. Then the orbit corresponding to $s_\iota \cdot w$ is contained in the closure of the one corresponding to w, and their union is SL_2^{ι} stable. □

Let us denote by $\pi_{\check{\lambda}}$ the map from $\mathrm{Gr}_G^{\check{\lambda}}$ to the corresponding partial flag variety G/P. The following lemma follows directly from definitions.

Lemma 15.4. *Let $\widetilde{w}$ be an element of W_{aff} which is minimal in its double coset $W\backslash W_{\mathrm{aff}}/W$, and $\check{\lambda}$ the corresponding dominant coweight. Assume that $\check{\lambda}$ is regular, and let $\mathcal{F}$ be a D-module on G/B. We have*

$$\mathcal{F} \star j_{\widetilde{w},!} \star \delta_{1_{\mathrm{Gr}_G}} \simeq (\mathrm{emb}_{\check{\lambda}})_! \circ \pi_{\check{\lambda}}^*(\mathcal{F}) \quad \textit{and} \quad \mathcal{F} \star j_{\widetilde{w},*} \star \delta_{1_{\mathrm{Gr}_G}} \simeq (\mathrm{emb}_{\check{\lambda}})_* \circ \pi_{\check{\lambda}}^*(\mathcal{F}).$$

Therefore, we can rewrite

$$\Xi \underset{I}{\star} j_{w_0\cdot\check{\rho},!} \underset{I}{\star} \delta_{1_{\mathrm{Gr}_G}} \simeq (\mathrm{emb}_{\check{\rho}})_! \circ \pi_{\check{\rho}}^*(\Xi)$$

and

$$\Xi \underset{I}{\star} j_{w_0\cdot\check{\rho},*} \underset{I}{\star} \delta_{1_{\mathrm{Gr}_G}} \simeq (\mathrm{emb}_{\check{\rho}})_* \circ \pi^*_{\check{\rho}}(\Xi).$$

Hence, the assertion of Proposition 15.2 can be reformulated as cleanness of the perverse sheaf $\pi^*_{\check{\rho}}(\Xi)$ on $\mathrm{Gr}^{\check{\rho}}_G$, i.e., that the map

$$(\mathrm{emb}_{\check{\rho}})_! \circ \pi^*_{\check{\rho}}(\Xi) \to (\mathrm{emb}_{\check{\rho}})_* \circ \pi^*_{\check{\rho}}(\Xi)$$

is an isomorphism.

15.5

Set

$$\mathbf{L}_{w_0} := \mathrm{IC}_{w_0\cdot\check{\rho},\mathrm{Gr}_G} \underset{G}{\star} \mathbb{V}_{\mathrm{crit}} = \Gamma(\mathrm{Gr}_G, \mathrm{IC}_{w_0\cdot\check{\rho},\mathrm{Gr}_G}).$$

A key result, from which we will derive the main theorem, is the following.

Theorem 15.6. *There exists a canonically defined map* $\mathbb{M}_{w_0,\mathrm{reg}} \otimes \omega_x^{\langle\rho,\check{\rho}\rangle} \to \mathbf{L}_{w_0}$ *such that*

(a) *The above map is surjective and its kernel is partially integrable.*
(b) *The induced map*

$$\begin{aligned}\mathbf{\Pi}_{\mathrm{reg}} \otimes \omega_x^{\langle\rho,\check{\rho}\rangle} &\simeq \Xi \underset{I}{\star} \mathbb{M}_{w_0,\mathrm{reg}} \otimes \omega_x^{\langle\rho,\check{\rho}\rangle} \\ &\to \Xi \underset{I}{\star} \mathrm{IC}_{w_0\cdot\check{\rho},\mathrm{Gr}_G} \underset{G}{\star} \mathbb{V}_{\mathrm{crit}} \simeq \Gamma(\mathrm{Gr}_G, \Xi \underset{I}{\star} \mathrm{IC}_{w_0\cdot\check{\rho},\mathrm{Gr}_G})\end{aligned}$$

is an isomorphism.

This theorem will be proved in Section 17. Let us now state a corollary of Theorem 15.6 that will be used in the proof of Main Theorem 14.15.

Corollary 15.7. *For any* $i > 0$, $\mathrm{R}^i\,\mathrm{Hom}_{D^b(\widehat{\mathfrak{g}}_{\mathrm{crit}}\text{-mod}_{\mathrm{reg}})^{I^0}}(\mathbf{\Pi}_{\mathrm{reg}}, \mathbb{M}_{w_0,\mathrm{reg}}) = 0$, *and the natural map* $\mathfrak{Z}^{\mathrm{reg}}_{\mathfrak{g}} \to \mathrm{Hom}(\mathbf{\Pi}_{\mathrm{reg}}, \mathbb{M}_{w_0,\mathrm{reg}})$ *is an isomorphism.*

Let us prove this corollary. By Proposition 14.12, it is sufficient to compute

$$\mathrm{RHom}_{D^b(\widehat{\mathfrak{g}}_{\mathrm{crit}}\text{-mod}_{\mathrm{nilp}})^{I^0}}\left(\Gamma(\mathrm{Gr}_G, \Xi \underset{I}{\star} \mathrm{IC}_{w_0\cdot\check{\rho},\mathrm{Gr}_G}), \Gamma(\mathrm{Gr}_G, \mathrm{IC}_{w_0\cdot\check{\rho},\mathrm{Gr}_G})\right).$$

By Theorem 8.20, the latter RHom is isomorphic to

$$\mathrm{RHom}_{D(\mathfrak{D}(\mathrm{Gr}_G)_{\mathrm{crit}}\text{-mod})}\left(\Xi \underset{I}{\star} \mathrm{IC}_{w_0\cdot\check{\rho},\mathrm{Gr}_G}, \bigoplus_{V\in\mathrm{Irr}(\mathcal{R}\mathrm{ep}(\check{G}))} \mathrm{IC}_{w_0\cdot\check{\rho},\mathrm{Gr}_G} \star \mathcal{F}_{V^*} \underset{\mathbb{C}}{\otimes} V_{\mathfrak{Z}^{\mathrm{reg}}_{\mathfrak{g}}}\right).$$

Let $I^{-,0}$ be the subgroup of $G[[t]]$ equal to the preimage of $N^- \subset G$ under the evaluation map. By composing with $\psi : N^- \to \mathbb{G}_a$, we obtain a character on $I^{-,0}$, denoted in the same way, and we can consider the category $\mathfrak{D}(\mathrm{Gr}_G)_{\mathrm{crit}}\text{-mod}^{I^{-,0},\psi}$ of $(I^{-,0},\psi)$-equivariant D-modules, and the corresponding triangulated category.

As in Section 14.4, the forgetful functor

$$D(\mathfrak{D}(\mathrm{Gr}_G)_{\mathrm{crit}}\text{-mod}^{I^{-,0},\psi}) \hookrightarrow D(\mathfrak{D}(\mathrm{Gr}_G)_{\mathrm{crit}}\text{-mod})$$

admits a right adjoint, which we will denote by $\mathrm{Av}_{I^{-,0},\psi}$, given by convolution with the corresponding D-module on $I^{-,0}$.

From Proposition 14.6 we obtain that the composition

$$D(\mathfrak{D}(\mathrm{Gr}_G)_{\mathrm{crit}}\text{-mod})^{I,m} \to D(\mathfrak{D}(\mathrm{Gr}_G)_{\mathrm{crit}}\text{-mod}) \to D(\mathfrak{D}(\mathrm{Gr}_G)_{\mathrm{crit}}\text{-mod})^{I^{-,0},\psi},$$

where the last arrow is the functor $\mathcal{F} \mapsto \mathrm{Av}_{I^{-,0},\psi}(\mathcal{F}) \otimes \det(\mathfrak{n}^-[1])^{-1}$, is exact, and essentially commutes with the Verdier duality on the holonomic subcategory.

By the construction of Ξ, we have that, for $\mathcal{F}_1 \in \mathfrak{D}(\mathrm{Gr}_G)_{\mathrm{crit}}\text{-mod}^I$ and $\mathcal{F}_2 \in \mathfrak{D}(\mathrm{Gr}_G)_{\mathrm{crit}}\text{-mod}^{I,m}$,

$$\begin{aligned} &\mathrm{RHom}_{D(\mathfrak{D}(\mathrm{Gr}_G)_{\mathrm{crit}}\text{-mod})}(\Xi \underset{I}{\star} \mathcal{F}_1, \mathcal{F}_2) \\ &\quad \simeq \mathrm{RHom}_{D(\mathfrak{D}(\mathrm{Gr}_G)_{\mathrm{crit}}\text{-mod})^{I^{-,0},\psi}} \left(\mathrm{Av}_{I^{-,0},\psi}(\mathcal{F}_1), \mathrm{Av}_{I^{-,0},\psi}(\mathcal{F}_2)\right). \end{aligned}$$

Using the exactness property of $\mathrm{Av}_{I^{-,0},\psi}$ mentioned above, Corollary 15.7 follows from the next general result.

Theorem 15.8. *For any two $\mathcal{F}'_1, \mathcal{F}'_2 \in \mathfrak{D}(\mathrm{Gr}_G)_{\mathrm{crit}}\text{-mod}^{I^{-,0},\psi}$ and $i > 0$,*

$$\mathrm{R}^i\,\mathrm{Hom}_{D(\mathfrak{D}(\mathrm{Gr}_G)_{\mathrm{crit}}\text{-mod})}(\mathcal{F}'_1, \mathcal{F}'_2) = 0.$$

The functor $\mathfrak{D}(\mathrm{Gr}_G)_{\mathrm{crit}}\text{-mod}^{G[[t]]} \to \mathfrak{D}(\mathrm{Gr}_G)_{\mathrm{crit}}\text{-mod}^{I^{-,0},\psi}$, given by

$$\mathcal{F} \mapsto \mathrm{Av}_{I^{-,0},\psi}(\mathrm{IC}_{w_0\cdot\check{\rho}} \star \mathcal{F}),$$

is an equivalence of abelian categories.

The proof of Theorem 15.8 is a word-for-word repetition of the proof of the main theorem of [FGV], using the fact that the combinatorics of I^0 (respectively, $I^{-,0}$) orbits on Gr_G is the same as that of $N((t))$ (respectively, $N^-((t))$) orbits. The main point is that any irreducible object of $\mathfrak{D}(\mathrm{Gr}_G)_{\mathrm{crit}}\text{-mod}^{I^{-,0},\psi}$ is a clean extension from a character sheaf on an orbit.

16 Proofs of the main theorems

In this section we will prove Main Theorem 14.15 and derive from it Main Theorem 6.9, assuming Theorem 15.6 (which is proved in the next section).

16.1

In Corollary 15.7 we computed the extensions between $\mathbf{\Pi}_{\mathrm{reg}}$ and $\mathbb{M}_{w_0,\mathrm{reg}}$ in the category $D^b(\widehat{\mathfrak{g}}_{\mathrm{crit}}\text{-mod}_{\mathrm{reg}})^{I^0}$. Now we use this result to compute the extensions between $\mathbf{\Pi}$ and $\mathbb{M}_{w_0}$ in the category $D^b(\widehat{\mathfrak{g}}_{\mathrm{crit}}\text{-mod}_{\mathrm{nilp}})^{I^0}$.

Proposition 16.2. *The morphism* $\mathfrak{Z}_{\mathfrak{g}}^{\mathrm{nilp}} \to \mathrm{Hom}_{D^b(\widehat{\mathfrak{g}}_{\mathrm{crit}}\text{-mod}_{\mathrm{nilp}})^{I^0}}(\mathbf{\Pi}, \mathbb{M}_{w_0})$ *is an isomorphism and* $R^i\,\mathrm{Hom}_{D^b(\widehat{\mathfrak{g}}_{\mathrm{crit}}\text{-mod}_{\mathrm{nilp}})^{I^0}}(\mathbf{\Pi}, \mathbb{M}_{w_0}) = 0$ *for* $i > 0$.

Proof. Let us note that for any two objects $\mathcal{M}_1^\bullet, \mathcal{M}_2^\bullet \in D^b(\widehat{\mathfrak{g}}_{\mathrm{crit}}\text{-mod}_{\mathrm{nilp}})$ the complex $\mathrm{RHom}_{D(\widehat{\mathfrak{g}}_{\mathrm{crit}}\text{-mod}_{\mathrm{nilp}})}(\mathcal{M}_1^\bullet, \mathcal{M}_2^\bullet)$ is naturally an object of $D^+(\mathfrak{Z}_{\mathfrak{g}}^{\mathrm{nilp}}\text{-mod})$. Recall also that $\mathfrak{Z}_{\mathfrak{g}}^{\mathrm{reg}}$, as a module over $\mathfrak{Z}_{\mathfrak{g}}^{\mathrm{nilp}}$, admits a finite resolution by finitely generated projective modules. Therefore, the functor

$$\mathcal{M}^\bullet \mapsto \mathcal{M}^\bullet \overset{L}{\underset{\mathfrak{Z}_{\mathfrak{g}}^{\mathrm{nilp}}}{\otimes}} \mathfrak{Z}_{\mathfrak{g}}^{\mathrm{reg}}$$

is well defined as a functor $D^b(\widehat{\mathfrak{g}}_{\mathrm{crit}}\text{-mod}_{\mathrm{nilp}}) \to D^b(\widehat{\mathfrak{g}}_{\mathrm{crit}}\text{-mod}_{\mathrm{nilp}})$.

Almost by definition we obtain the following result.

Lemma 16.3.

$$\left(\mathrm{RHom}_{D(\widehat{\mathfrak{g}}_{\mathrm{crit}}\text{-mod}_{\mathrm{nilp}})}(\mathcal{M}_1^\bullet, \mathcal{M}_2^\bullet)\right) \overset{L}{\underset{\mathfrak{Z}_{\mathfrak{g}}^{\mathrm{nilp}}}{\otimes}} \mathfrak{Z}_{\mathfrak{g}}^{\mathrm{reg}}$$

is isomorphic to

$$\mathrm{RHom}_{D(\widehat{\mathfrak{g}}_{\mathrm{crit}}\text{-mod}_{\mathrm{nilp}})}\left(\mathcal{M}_1^\bullet, (\mathcal{M}_2^\bullet \overset{L}{\underset{\mathfrak{Z}_{\mathfrak{g}}^{\mathrm{nilp}}}{\otimes}} \mathfrak{Z}_{\mathfrak{g}}^{\mathrm{reg}})\right).$$

Since $\mathbf{\Pi}$ is flat over $\mathfrak{Z}_{\mathfrak{g}}^{\mathrm{nilp}}$, by Lemma 23.3, we obtain that for any $\mathcal{M} \in \widehat{\mathfrak{g}}_{\mathrm{crit}}\text{-mod}_{\mathrm{reg}}$, we have the following.

Lemma 16.4.

$$\mathrm{RHom}_{D(\widehat{\mathfrak{g}}_{\mathrm{crit}}\text{-mod}_{\mathrm{nilp}})}(\mathbf{\Pi}, \mathcal{M}) \simeq \mathrm{RHom}_{D(\widehat{\mathfrak{g}}_{\mathrm{crit}}\text{-mod}_{\mathrm{reg}})}(\mathbf{\Pi}_{\mathrm{reg}}, \mathcal{M}).$$

By combining this with Corollary 15.7, we obtain that the natural map

$$\mathfrak{Z}_{\mathfrak{g}}^{\mathrm{reg}} \to \left(\mathrm{RHom}_{D(\widehat{\mathfrak{g}}_{\mathrm{crit}}\text{-mod}_{\mathrm{nilp}})}(\mathbf{\Pi}, \mathbb{M}_{w_0})\right) \overset{L}{\underset{\mathfrak{Z}_{\mathfrak{g}}^{\mathrm{nilp}}}{\otimes}} \mathfrak{Z}_{\mathfrak{g}}^{\mathrm{reg}}$$

is a quasi-isomorphism. We will now derive the assertion of Proposition 16.2 by a Nakayama lemma–type argument.

Consider the $\mathbb{G}_m$-action on $\widehat{\mathfrak{g}}_{\mathrm{crit}}$ coming from $\mathbb{G}_m \hookrightarrow \mathrm{Aut}(\mathcal{D})$. We obtain that $\mathbb{G}_m$ acts weakly on the categories $\widehat{\mathfrak{g}}_{\mathrm{crit}}\text{-mod}_{\mathrm{nilp}}$ and $\widehat{\mathfrak{g}}_{\mathrm{crit}}\text{-mod}$. Since the objects $\mathbf{\Pi}$ and $\mathbb{M}_{w_0}$ are $\mathbb{G}_m$-equivariant, the Ext groups

$$\mathrm{Ext}^i_{\widehat{\mathfrak{g}}_{\mathrm{crit}}\text{-mod}}(\mathbf{\Pi}, \mathbb{M}_{w_0}) \quad \text{and} \quad \mathrm{Ext}^i_{\widehat{\mathfrak{g}}_{\mathrm{crit}}\text{-mod}_{\mathrm{nilp}}}(\mathbf{\Pi}, \mathbb{M}_{w_0})$$

acquire an action of $\mathbb{G}_m$ by Lemma 20.6.

We claim that the grading arising on $\mathrm{Ext}^i_{\widehat{\mathfrak{g}}_{\mathrm{crit}}\text{-mod}_{\mathrm{nilp}}}(\mathbf{\Pi}, \mathbb{M}_{w_0})$ is bounded from above. First, let us note that the grading on $\mathrm{Ext}^i_{\widehat{\mathfrak{g}}_{\mathrm{crit}}\text{-mod}}(\mathbf{\Pi}, \mathbb{M}_{w_0})$ is nonpositive. This is evident since $\mathrm{Ext}^i_{\widehat{\mathfrak{g}}_{\mathrm{crit}}\text{-mod}}(\mathbf{\Pi}, \mathbb{M}_{w_0})$ are computed by the standard complex $\mathfrak{C}^\bullet(\mathfrak{g}[[t]], \mathrm{Hom}_{\mathbb{C}}(\Pi, \mathbb{M}_{w_0}))$, whose terms are nonpositively graded. Note also that the algebra $\mathfrak{z}^{\mathrm{nilp}}_{\mathfrak{g}}$ is nonpositively graded, and the grading on $N_{\mathfrak{z}^{\mathrm{nilp}}_{\mathfrak{g}}/\mathfrak{z}_{\mathfrak{g}}}$ is such that only finitely many free generators have positive degrees. Now, the spectral sequence of Section 7.6 implies by induction on i that the grading on $\mathrm{Ext}^i_{\widehat{\mathfrak{g}}_{\mathrm{crit}}\text{-mod}_{\mathrm{nilp}}}(\mathbf{\Pi}, \mathbb{M}_{w_0})$ is bounded from above.

Since the algebra $\mathrm{Fun}(\mathrm{Op}^{\mathrm{nilp}}_{\check{\mathfrak{g}}})$ is itself nonpositively graded, we deduce that $\mathrm{RHom}_{D(\widehat{\mathfrak{g}}_{\mathrm{crit}}\text{-mod}_{\mathrm{nilp}})}(\mathbf{\Pi}, \mathbb{M}_{w_0})$ can be represented by a complex of graded modules such that the grading on each term is bounded from above, and which lives in nonnegative cohomological degrees. Recall again that the ideal of $\mathrm{Fun}(\mathrm{Op}^{\mathrm{reg}}_{\check{\mathfrak{g}}})$ in $\mathrm{Fun}(\mathrm{Op}^{\mathrm{nilp}}_{\check{\mathfrak{g}}})$ is generated by a regular sequence of homogeneous negatively graded elements. The proof is concluded by the following observation.

Lemma 16.5. *Let*

$$Q^\bullet := Q^0 \to Q^1 \to \cdots \to Q^n \to \cdots$$

be a complex of graded modules over a graded algebra $A = \mathbb{C}[x_1, \ldots, x_n]$, *where* $\deg(x_i) < 0$ *such that the grading on each* Q^i *is bounded from above. Assume that* $Q^\bullet \overset{L}{\underset{A}{\otimes}} \mathbb{C}$ *is acyclic away from cohomological degree* 0. *Then* $Q^\bullet$ *is itself acyclic away from cohomological degree* 0. □

Corollary 16.6. *For any* $\mathfrak{z}^{\mathrm{nilp}}_{\mathfrak{g}}$*-module* $\mathcal{L}$

$$\mathrm{R}^i\,\mathrm{Hom}_{D^b(\widehat{\mathfrak{g}}_{\mathrm{crit}}\text{-mod}_{\mathrm{nilp}})^{I^0}}(\mathbf{\Pi}, \mathbb{M}_{w_0} \underset{\mathfrak{z}^{\mathrm{nilp}}_{\mathfrak{g}}}{\otimes} \mathcal{L}) = 0$$

for $i > 0$ *and is isomorphic to* $\mathcal{L}$ *for* $i = 0$.

Proof. Since any module $\mathcal{L}$ is a direct limit of finitely presented ones, by Proposition 7.14, we may assume that $\mathcal{L}$ is finitely presented. Since $\mathfrak{z}^{\mathrm{nilp}}_{\mathfrak{g}}$ is isomorphic to a polynomial algebra, any finitely presented module admits a finite resolution by projective ones:

$$\mathcal{L}_n \to \cdots \to \mathcal{L}_1 \to \mathcal{L}_0 \to \mathcal{L}.$$

Since $\mathbb{M}_{w_0}$ is flat over $\mathfrak{z}^{\mathrm{nilp}}_{\mathfrak{g}}$ (cf, Corollary 13.9), we obtain a resolution

$$\mathbb{M}_{w_0} \underset{\mathfrak{Z}_{\mathfrak{g}}^{\mathrm{nilp}}}{\otimes} \mathcal{L}_n \to \cdots \to \mathbb{M}_{w_0} \underset{\mathfrak{Z}_{\mathfrak{g}}^{\mathrm{nilp}}}{\otimes} \mathcal{L}_1 \to \mathbb{M}_{w_0} \underset{\mathfrak{Z}_{\mathfrak{g}}^{\mathrm{nilp}}}{\otimes} \mathcal{L}_0 \to \mathbb{M}_{w_0} \underset{\mathfrak{Z}_{\mathfrak{g}}^{\mathrm{nilp}}}{\otimes} \mathcal{L}.$$

Hence, we obtain a spectral sequence, converging to

$$\mathrm{R}^i \operatorname{Hom}_{D(\widehat{\mathfrak{g}}_{\mathrm{crit}}\text{-mod})^I}(\mathbf{\Pi}, \mathbb{M}_{w_0} \underset{\mathfrak{Z}_{\mathfrak{g}}^{\mathrm{nilp}}}{\otimes} \mathcal{L}),$$

whose first term $E_1^{i,j}$ is given by

$$\mathrm{R}^i \operatorname{Hom}_{D(\widehat{\mathfrak{g}}_{\mathrm{crit}}\text{-mod})^I}(\mathbf{\Pi}, \mathbb{M}_{w_0} \underset{\mathfrak{Z}_{\mathfrak{g}}^{\mathrm{nilp}}}{\otimes} \mathcal{L}_{-j}).$$

Since $\mathcal{L}_\bullet$ are projective, by Proposition 16.2, we obtain that $E_1^{i,j} = 0$ unless $i = 0$, and in the latter case, it is isomorphic to $\mathcal{L}_{-j}$, implying the assertion of the corollary. □

Corollary 16.7. *For any object L of $\widehat{\mathfrak{g}}_{\mathrm{crit}}\text{-mod}_{\mathrm{nilp}}^{I,m}$ and the $\mathfrak{Z}_{\mathfrak{g}}^{\mathrm{nilp}}$-module*

$$\mathcal{L} := \operatorname{Hom}(\mathbb{M}_{w_0}, L),$$

the kernel of the natural map

$$\mathbb{M}_{w_0} \underset{\mathfrak{Z}_{\mathfrak{g}}^{\mathrm{nilp}}}{\otimes} \mathcal{L} \to L$$

is partially integrable.

Proof. Let $\mathcal{M}$ be the kernel of $\mathbb{M}_{w_0} \underset{\mathfrak{Z}_{\mathfrak{g}}^{\mathrm{nilp}}}{\otimes} \mathcal{L} \to L$, and suppose that it is not partially integrable. Let $\mathcal{M}' \subset \mathcal{M}$ be the maximal partially integrable submodule. Consider the short exact sequence

$$0 \to \mathcal{M}' \to \mathcal{M} \to \mathcal{M}'' \to 0.$$

By Lemma 7.22, we have a nonzero map $\mathbb{L}_w \to \mathcal{M}''$ for some $w \in W$. We claim that w necessarily equals w_0.

Indeed, all modules $\mathbb{L}_w$ with $w \neq w_0$ are partially integrable, and we would obtain that the preimage in $\mathcal{M}$ of $\operatorname{Im}(\mathbb{L}_w)$ is again integrable, and is strictly bigger than $\mathcal{M}$.

Hence, we have a map $\mathbb{M}_{w_0} \to \mathcal{M}''$, and by composing, we obtain a map $\mathbf{\Pi} \to \mathcal{M}''$. Now, by Proposition 14.12, this maps lifts to a map $\mathbf{\Pi} \to \mathcal{M}$, i.e., $\operatorname{Hom}(\mathbf{\Pi}, \mathcal{M}) \neq 0$.

Now consider the exact sequence

$$0 \to \operatorname{Hom}(\mathbf{\Pi}, \mathcal{M}) \to \operatorname{Hom}(\mathbf{\Pi}, \mathbb{M}_{w_0} \underset{\mathfrak{Z}_{\mathfrak{g}}^{\mathrm{nilp}}}{\otimes} \mathcal{L}) \to \operatorname{Hom}(\mathbf{\Pi}, L).$$

By Proposition 16.6, the middle term is isomorphic to $\mathcal{L}$, and it maps injectively to $\mathrm{Hom}(\mathbf{\Pi}, L)$, since

$$\mathcal{L} \simeq \mathrm{Hom}(\mathbb{M}_{w_0}, L) \hookrightarrow \mathrm{Hom}(\mathbf{\Pi}, L),$$

which is a contradiction. □

Now we are able to prove Main Theorem 14.15.

Proof. Let $\mathcal{M}$ be an object of $\widehat{\mathfrak{g}}_{\mathrm{crit}}\text{-mod}^{I,m}_{\mathrm{nilp}}$. It admits a filtration $0 = \mathcal{M}_0 \subset \mathcal{M}_1 \subset \mathcal{M}_2 \cdots$ whose subquotients $\mathcal{M}_j/\mathcal{M}_{j-1}$ have the property that each is a quotient of the module $\mathbb{L}_w$ for some $w \in W$. By Proposition 7.14, to prove that $\mathrm{Ext}^i_{\widehat{\mathfrak{g}}_{\mathrm{crit}}\text{-mod}_{\mathrm{nilp}}}(\mathbf{\Pi}, \mathcal{M}) = 0$ for $i = 0$, by devissage, we can assume that $\mathcal{M}$ itself is a quotient of some $\mathbb{L}_w$.

If $w \neq w_0$, then $\mathcal{M}$ is partially integrable and the vanishing of Exts follows from Proposition 14.12. Hence, we can assume that $\mathcal{M}$ is a quotient of $\mathbb{L}_{w_0} = \mathbb{M}_{w_0}$. In this case, the assertion of the theorem follows from Corollary 16.6 combined with Corollary 16.7. □

16.8 Proof of Main Theorem 6.9

Now we derive Main Theorem 6.9 from Main Theorem 14.15. We define the functor $\widehat{\mathfrak{g}}_{\mathrm{crit}}\text{-mod} \to \mathfrak{z}^{\mathrm{nilp}}_{\mathfrak{g}} \otimes h_0\text{-mod}$ by

$$\mathcal{M} \mapsto \mathrm{Hom}(\mathbf{\Pi}, \mathcal{M}).$$

Composing with the forgetful functor $\widehat{\mathfrak{g}}_{\mathrm{crit}}\text{-mod}^{I,m}_{\mathrm{nilp}} \to \widehat{\mathfrak{g}}_{\mathrm{crit}}\text{-mod}$, we obtain a functor

$$\widehat{\mathfrak{g}}_{\mathrm{crit}}\text{-mod}^{I,m}_{\mathrm{nilp}} \to \mathfrak{z}^{\mathrm{nilp}}_{\mathfrak{g}} \otimes h_0\text{-mod}.$$

By Main Theorem 14.15, the latter functor is exact, and by Proposition 14.12 it factors through ${}^f\widehat{\mathfrak{g}}_{\mathrm{crit}}\text{-mod}^{I,m}_{\mathrm{nilp}}$. This defines the desired functor

$${}^f\mathsf{F} : {}^fD^b(\widehat{\mathfrak{g}}_{\mathrm{crit}}\text{-mod}_{\mathrm{nilp}})^{I^0} \to D^b(\mathfrak{z}^{\mathrm{nilp}}_{\mathfrak{g}} \otimes h_0\text{-mod}).$$

We define a functor ${}^f\mathsf{G} : \mathfrak{z}^{\mathrm{nilp}}_{\mathfrak{g}} \otimes h_0\text{-mod} \to {}^f\widehat{\mathfrak{g}}_{\mathrm{crit}}\text{-mod}^{I,m}_{\mathrm{nilp}}$ by

$$\mathcal{L} \mapsto {}^f\mathbf{\Pi} \underset{\mathfrak{z}^{\mathrm{nilp}}_{\mathfrak{g}} \otimes h_0}{\otimes} \mathcal{L}.$$

From Lemma 14.3 and Corollary 13.9 it follows that this functor is exact. We will denote by the same character the resulting functor

$$D^b(\mathfrak{z}^{\mathrm{nilp}}_{\mathfrak{g}} \otimes h_0) \to {}^fD^b(\widehat{\mathfrak{g}}_{\mathrm{crit}}\text{-mod}_{\mathrm{nilp}})^{I^0}.$$

For $\mathcal{L}^\bullet \in D^-(\mathfrak{z}^{\mathrm{nilp}}_{\mathfrak{g}} \otimes h_0\text{-mod})$ and $\mathcal{M}^\bullet \in {}^fD^b(\widehat{\mathfrak{g}}_{\mathrm{crit}}\text{-mod}_{\mathrm{nilp}})^{I^0}$, by Corollary 14.13 we have a natural isomorphism

$$\mathrm{Hom}_{D^b(\mathfrak{z}_{\mathfrak{g}}^{\mathrm{nilp}}\otimes h_0\text{-mod})}(\mathcal{L}, {}^f\mathsf{F}(\mathcal{M}^\bullet)) \simeq \mathrm{Hom}_{{}^fD^b(\widehat{\mathfrak{g}}_{\mathrm{crit}}\text{-mod}_{\mathrm{nilp}})^{I^0}}({}^f\mathsf{G}(\mathcal{L}), \mathcal{M}^\bullet).$$

Hence, G and F are mutually adjoint. Let us show that they are, in fact, mutually quasi-inverse.

Let us first show that the adjunction morphism $\mathrm{Id} \to {}^f\mathsf{F} \circ {}^f\mathsf{G}$ is an isomorphism. By exactness, it suffices to show that for a $\mathfrak{z}_{\mathfrak{g}}^{\mathrm{nilp}} \otimes h_0$-module $\mathcal{L}$, on which the action of h_0 is trivial, the map

$$\mathcal{L} \mapsto \mathrm{Hom}\left({}^f\mathbf{\Pi}, {}^f\mathbf{\Pi} \underset{\mathfrak{z}_{\mathfrak{g}}^{\mathrm{nilp}}\otimes h_0}{\otimes} \mathcal{L}\right) \tag{16.1}$$

is an isomorphism.

We have $\Pi \underset{h_0}{\otimes} \mathbb{C} \simeq M_1^\vee$, and hence $\mathbf{\Pi} \underset{h_0}{\otimes} \mathbb{C} \simeq \mathbb{M}_1^\vee$. Since the kernel of $\mathbb{M}_1^\vee \to \mathbb{M}_{w_0}$ is partially integrable, we obtain that

$${}^f\mathbf{\Pi} \underset{\mathfrak{z}_{\mathfrak{g}}^{\mathrm{nilp}}\otimes h_0}{\otimes} \mathcal{L} \simeq {}^f\mathbb{M}_{w_0} \underset{\mathfrak{z}_{\mathfrak{g}}^{\mathrm{nilp}}}{\otimes} \mathcal{L},$$

and the assertion follows from Corollary 16.6.

To show that the adjunction ${}^f\mathsf{G} \circ {}^f\mathsf{F} \to \mathrm{Id}$ is an isomorphism, by exactness, it is again sufficient to evaluate it on a single module $\mathcal{M}$. Since the functor ${}^f\mathsf{F}$ is faithful, it is enough to show that

$${}^f\mathsf{F} \circ {}^f\mathsf{G} \circ {}^f\mathsf{F}(\mathcal{M}) \to {}^f\mathsf{F}(\mathcal{M})$$

is an isomorphism. But we already know that ${}^f\mathsf{F}(\mathcal{M}) \to {}^f\mathsf{F} \circ {}^f\mathsf{G} \circ {}^f\mathsf{F}(\mathcal{M})$ is an isomorphism, and our assertion follows.

This completes the proof of Main Theorem 6.9 modulo Theorem 15.6. □

17 Proof of Theorem 15.6

17.1

Let us first construct the map

$$\mathbb{M}_{w_0,\mathrm{reg}} \otimes \omega_x^{\langle\rho,\check{\rho}\rangle} \to \Gamma(\mathrm{Gr}_G, \mathrm{IC}_{w_0\cdot\check{\rho},\mathrm{Gr}_G}), \tag{17.1}$$

whose existence is stated in Theorem 15.6.

Consider the $\widehat{\mathfrak{g}}_{\mathrm{crit}}$-module $\Gamma(\mathrm{Gr}_G, j_{w_0\cdot\check{\rho},*} \underset{I}{\star} \delta_{1,\mathrm{Gr}_G})$; it is equivariant with respect to the action of $\mathbb{G}_m$ acting by loop rotations. This module contains a unique line, corresponding to those sections of the twisted D-module $j_{w_0\cdot\check{\rho},*} \underset{I}{\star} \delta_{1,\mathrm{Gr}_G}$, which are scheme-theoretically supported on the closure of the I-orbit of the element $t^{-\check{\rho}} \in \mathrm{Gr}_G$, and which are I^0-invariant.

This line has weight $-2\check{\rho}$ with respect to $\mathfrak{h}$, and has the highest degree with respect to the $\mathbb{G}_m$-action. Moreover, a straightforward calculation (see [BD1, Section 9.1.13]) shows that this line can be canonically identified with $\omega_x^{\langle\rho,\check{\rho}\rangle}$. This defines a map

$$\mathbb{M}_{w_0,\mathrm{reg}} \otimes \omega_x^{\langle\rho,\check{\rho}\rangle} \to \Gamma(\mathrm{Gr}_G, j_{w_0\cdot\check{\rho},*} \underset{I}{\star} \delta_{1,\mathrm{Gr}_G}).$$

We claim that the above map factors through

$$\Gamma(\mathrm{Gr}_G, \mathrm{IC}_{w_0\cdot\check{\rho},\mathrm{Gr}_G}) \subset \Gamma(\mathrm{Gr}_G, j_{w_0\cdot\check{\rho},*} \underset{I}{\star} \delta_{1,\mathrm{Gr}_G}).$$

Indeed, by Lemma 15.3, the quotient module is partially integrable, and from Corollary 14.14 we obtain that it cannot be the target of a nonzero map from $\mathbb{M}_{w_0,\mathrm{reg}}$.

Proposition 17.2. *The map* $\mathbb{M}_{w_0,\mathrm{reg}} \otimes \omega_x^{\langle\rho,\check{\rho}\rangle} \to \Gamma(\mathrm{Gr}_G, \mathrm{IC}_{w_0\cdot\check{\rho},\mathrm{Gr}_G})$ *constructed above is surjective.*

The proof will be given at the end of this section. We will now proceed with the proof of Theorem 15.6.

17.3

We shall now construct a map

$$\Gamma(\mathrm{Gr}_G, \mathrm{IC}_{w_0\cdot\check{\rho},\mathrm{Gr}_G}) \to \mathbb{M}_{1,\mathrm{reg}} \otimes \omega_x^{\langle\rho,\check{\rho}\rangle}. \tag{17.2}$$

First, by Section 20.10, for any I-equivariant $\widehat{\mathfrak{g}}_{\mathrm{crit}}$-module $\mathcal{M}$,

$$\mathrm{Hom}(\mathbb{V}_{\mathrm{crit}}, \mathcal{M}) \simeq R^0\,\mathrm{Hom}_{\widehat{\mathfrak{g}}\text{-mod}^{G[[t]]}}(\mathbb{V}_{\mathrm{crit}}, \mathrm{Av}_{G[[t]]/I}(\mathcal{M})).$$

Applying this to $\mathcal{M} = \mathbb{M}_{w_0,\mathrm{reg}}$, we calculate

$$\mathrm{Av}_{G[[t]]/I}(\mathbb{M}_{w_0,\mathrm{reg}}) \simeq \mathrm{Av}_{G[[t]]/I}(\mathbb{M}_{w_0}) \overset{L}{\underset{\mathfrak{z}_{\mathfrak{g}}^{\mathrm{nilp}}}{\otimes}} \mathfrak{z}_{\mathfrak{g}}^{\mathrm{reg}} \simeq \mathbb{V}_{\mathrm{crit}}[-\dim(G/B)] \overset{L}{\underset{\mathfrak{z}_{\mathfrak{g}}^{\mathrm{nilp}}}{\otimes}} \mathfrak{z}_{\mathfrak{g}}^{\mathrm{reg}}. \tag{17.3}$$

Hence, the 0th cohomology of $\mathrm{Av}_{G[[t]]/I}(\mathbb{M}_{w_0,\mathrm{reg}})$ is isomorphic to

$$\mathrm{Tor}^{\mathfrak{z}_{\mathfrak{g}}^{\mathrm{nilp}}}_{\dim(G/B)}(\mathbb{V}_{\mathrm{crit}}, \mathfrak{z}_{\mathfrak{g}}^{\mathrm{reg}}) \simeq \mathbb{V}_{\mathrm{crit}} \otimes \mathrm{Tor}^{\mathfrak{z}_{\mathfrak{g}}^{\mathrm{nilp}}}_{\dim(G/B)}(\mathfrak{z}_{\mathfrak{g}}^{\mathrm{reg}}, \mathfrak{z}_{\mathfrak{g}}^{\mathrm{reg}}).$$

However, from Corollary 4.18 and Proposition 4.23(2), it follows that

$$\mathrm{Tor}^{\mathfrak{z}_{\mathfrak{g}}^{\mathrm{nilp}}}_{\dim(G/B)}(\mathfrak{z}_{\mathfrak{g}}^{\mathrm{reg}}, \mathfrak{z}_{\mathfrak{g}}^{\mathrm{reg}}) \simeq \Lambda^{\dim(G/B)}(N^*_{\mathfrak{z}_{\mathfrak{g}}^{\mathrm{reg}}/\mathfrak{z}_{\mathfrak{g}}^{\mathrm{nilp}}}) \simeq \mathfrak{z}_{\mathfrak{g}}^{\mathrm{reg}} \otimes \omega_x^{-\langle 2\rho,\check{\rho}\rangle}.$$

Hence, the above 0th cohomology is isomorphic to $\mathbb{V}_{\mathrm{crit}} \otimes \omega_x^{\langle 2\rho,\check{\rho}\rangle}$, and we obtain a map

$$\mathbb{V}_{\mathrm{crit}} \to \mathbb{M}_{w_0,\mathrm{reg}} \otimes \omega_x^{-\langle 2\rho,\check{\rho}\rangle}.$$

By applying the convolution $j_{w_0\cdot\check{\rho},*} \underset{I}{\star} \cdot$ to both sides we obtain a map

$$\Gamma(\mathrm{Gr}_G, j_{w_0\cdot\check{\rho},*} \underset{I}{\star} \delta_{1,\mathrm{Gr}_G}) \to j_{w_0\cdot\check{\rho},*} \underset{I}{\star} \mathbb{M}_{w_0,\mathrm{reg}} \otimes \omega_x^{\langle 2\rho,\check{\rho}\rangle}. \tag{17.4}$$

However, by (13.12),

$$\begin{aligned} j_{w_0\cdot\check{\rho},*} \underset{I}{\star} \mathbb{M}_{w_0,\mathrm{reg}} &\simeq j_{w_0,!} \underset{I}{\star} j_{w_0,*} \underset{I}{\star} j_{w_0\cdot\check{\rho},*} \underset{I}{\star} \mathbb{M}_{w_0,\mathrm{reg}} \simeq j_{w_0,!} \underset{I}{\star} j_{\check{\rho},*} \underset{I}{\star} \mathbb{M}_{w_0,\mathrm{reg}} \\ &\simeq j_{w_0,!} \underset{I}{\star} \mathbb{M}_{w_0,\mathrm{reg}} \otimes \omega_x^{-\langle\rho,\check{\rho}\rangle} \simeq \mathbb{M}_{1,\mathrm{reg}} \otimes \omega_x^{-\langle\rho,\check{\rho}\rangle}, \end{aligned}$$

and by composing with the embedding

$$\Gamma(\mathrm{Gr}_G, \mathrm{IC}_{w_0\cdot\check{\rho},\mathrm{Gr}_G}) \hookrightarrow \Gamma(\mathrm{Gr}_G, j_{w_0\cdot\check{\rho},*} \underset{I}{\star} \delta_{1,\mathrm{Gr}_G}),$$

we obtain the map of (17.2). By construction, this map respects the $\mathbb{G}_m$-action.

17.4

Now consider the composition

$$\mathbb{M}_{w_0,\mathrm{reg}} \otimes \omega_x^{\langle\rho,\check{\rho}\rangle} \to \Gamma(\mathrm{Gr}_G, \mathrm{IC}_{w_0\cdot\check{\rho},\mathrm{Gr}_G}) \to \mathbb{M}_{1,\mathrm{reg}} \otimes \omega_x^{\langle\rho,\check{\rho}\rangle}. \tag{17.5}$$

Lemma 17.5. *The resulting map $\mathbb{M}_{w_0,\mathrm{reg}} \to \mathbb{M}_{1,\mathrm{reg}}$ is a nonzero multiple of the canonical map, coming from the embedding $M_{w_0} \to M_1$.*

Proof. First, the map in question is nonzero by Proposition 17.2. Secondly, our map $\mathbb{M}_{w_0,\mathrm{reg}} \to \mathbb{M}_{1,\mathrm{reg}}$ respects the $\mathbb{G}_m$-action by loop rotations. Since $\mathbb{M}_{w_0,\mathrm{reg}}$ is generated by a vector of degree 0, and the subspace in $\mathbb{M}_{1,\mathrm{reg}}$ consisting of elements of degree 0 is isomorphic to the Verma module M_0, any map $\mathbb{M}_{w_0,\mathrm{reg}} \to \mathbb{M}_{1,\mathrm{reg}}$, compatible with the grading, is a scalar multiple of the canonical map. □

17.6

Let us now derive Theorem 15.6 from Lemma 17.5.

Let us apply the convolution $\Xi \underset{I}{\star} \cdot$ to the three terms appearing in (17.5). We obtain the maps

$$\mathbf{\Pi} \otimes \omega_x^{\langle\rho,\check{\rho}\rangle} \simeq \Xi \underset{I}{\star} \mathbb{M}_{w_0,\mathrm{reg}} \otimes \omega_x^{\langle\rho,\check{\rho}\rangle} \to \Gamma(\mathrm{Gr}_G, \Xi \underset{I}{\star} \mathrm{IC}_{w_0\cdot\check{\rho},\mathrm{Gr}_G}) \to \Xi \underset{I}{\star} \mathbb{M}_{1,\mathrm{reg}} \otimes \omega_x^{\langle\rho,\check{\rho}\rangle}. \tag{17.6}$$

However, the canonical map $M_{w_0} \to M_0$ has the property that its cokernel is partially integrable. Hence, the cone of the resulting map $\mathbb{M}_{w_0,\mathrm{reg}} \to \mathbb{M}_{1,\mathrm{reg}}$ is also partially integrable.

Hence, by Theorem 17.5 and Proposition 14.9, the composed map in (17.6) is an isomorphism. In particular, we obtain that $\mathbf{\Pi}$ is a direct summand of $\Gamma(\mathrm{Gr}_G, \Xi \underset{I}{\star} \mathrm{IC}_{w_0\cdot\check{\rho},\mathrm{Gr}_G})$.

Lemma 17.7. *The map*

$$\mathfrak{z}_{\mathfrak{g}}^{\mathrm{reg}} \otimes h_0 \to \mathrm{End}(\Gamma(\mathrm{Gr}_G, \Xi \underset{I}{\star} \mathrm{IC}_{w_0\cdot\check{\rho},\mathrm{Gr}_G}))$$

is an isomorphism.

Proof. By Theorem 8.20 the assertion of the lemma is equivalent to the fact that $h_0 \simeq \mathrm{End}(\Xi \underset{I}{\star} \mathrm{IC}_{w_0\cdot\check{\rho},\mathrm{Gr}_G})$, and $\mathrm{Hom}(\Xi \underset{I}{\star} \mathrm{IC}_{w_0\cdot\check{\rho},\mathrm{Gr}_G}, \Xi \underset{I}{\star} \mathrm{IC}_{w_0\cdot\check{\rho},\mathrm{Gr}_G} \underset{G[[t]]}{\star} \mathcal{F}_{V^{\check{\lambda}}}) = 0$ for $\check{\lambda} \neq 0$.

The former isomorphism follows from the fact that $h_0 \simeq \mathrm{End}(\Xi)$, combined with Proposition 15.2 and the fact that the projection $\mathrm{Gr}_G^{\check{\rho}} \to G/B$ is smooth with connected fibers.

To prove the vanishing for $\check{\lambda} \neq 0$, it is enough to show that

$$\mathrm{Hom}(\Xi \underset{I}{\star} \mathrm{IC}_{w_0\cdot\check{\rho},\mathrm{Gr}_G}, \mathrm{IC}_{w_0\cdot\check{\rho},\mathrm{Gr}_G} \underset{G[[t]]}{\star} \mathcal{F}_{V^{\check{\lambda}}}) = 0,$$

because modulo partially integrable objects, Ξ appearing on the right-hand side is an extension of several copies of $\delta_{1_{G/B}}$.

As in the proof of Corollary 15.7, the latter Hom is isomorphic to

$$R^0 \mathrm{Hom}_{D(\mathfrak{D}(\mathrm{Gr}_G)_{\mathrm{crit}}\text{-mod})^{I^{-,0},\psi}} \left(\mathrm{Av}_{I^{-,0},\psi}(\delta_{1_{\mathrm{Gr}_G}}), \mathrm{Av}_{I^{-,0},\psi}(\delta_{1_{\mathrm{Gr}_G}}) \underset{G[[t]]}{\star} \mathcal{F}_{V^{\check{\lambda}}} \right),$$

and the latter vanishes, according to Theorem 15.8. □

Thus we obtain that the ring $\mathrm{End}(\Gamma(\mathrm{Gr}_G, \Xi \underset{I}{\star} \mathrm{IC}_{w_0\cdot\check{\rho},\mathrm{Gr}_G}))$ has no idempotents. In particular, the map $\mathbf{\Pi} \otimes \omega_x^{\langle \rho,\check{\rho}\rangle} \to \Gamma(\mathrm{Gr}_G, \Xi \underset{I}{\star} \mathrm{IC}_{w_0\cdot\check{\rho},\mathrm{Gr}_G})$ is an isomorphism, establishing point (b) of Theorem 15.6.

Proposition 17.2 states that the map (17.1) is surjective. Thus it remains to show that the kernel of the map (17.1) is partially integrable. But this follows from point (b) and Proposition 14.9. Therefore we obtain point (a) of Theorem 15.6. This completes the proof of Theorem 15.6 modulo Proposition 17.2, which is proved in the next section. □

17.8 Proof of Proposition 17.2

The crucial fact used in the proof of this proposition is that the module $\mathbf{L}_{w_0}$ carries an action of the renormalized algebra $U^{\mathrm{ren,reg}}(\widehat{\mathfrak{g}}_{\mathrm{crit}})$; see Section 7.8. Moreover, as an object of the category $U^{\mathrm{ren,reg}}(\widehat{\mathfrak{g}}_{\mathrm{crit}})$-mod, the module $\mathbf{L}_{w_0}$ is irreducible, because the D-module $\mathrm{IC}_{w_0\cdot\check{\rho},\mathrm{Gr}_G}$ is irreducible, and the global sections functor $\mathfrak{D}(\mathrm{Gr}_G)\text{-mod} \to U^{\mathrm{ren,reg}}(\widehat{\mathfrak{g}}_{\mathrm{crit}})$-mod is fully faithful, according to [FG].

Recall that the algebra $U^{\mathrm{ren,reg}}(\widehat{\mathfrak{g}}_{\mathrm{crit}})$-mod is naturally filtered,

$$U^{\mathrm{ren,reg}}(\widehat{\mathfrak{g}}_{\mathrm{crit}})\text{-mod} = \bigcup_i (U^{\mathrm{ren,reg}}(\widehat{\mathfrak{g}}_{\mathrm{crit}})\text{-mod})^i,$$

so that

$$(U^{\mathrm{ren,reg}}(\widehat{\mathfrak{g}}_{\mathrm{crit}})\text{-mod})^0 \simeq \widetilde{U}_{\mathrm{crit}}(\widehat{\mathfrak{g}}) \underset{\mathfrak{z}_{\mathfrak{g}}}{\otimes} \mathfrak{z}_{\mathfrak{g}}^{\mathrm{reg}},$$

and $(U^{\mathrm{ren,reg}}(\widehat{\mathfrak{g}}_{\mathrm{crit}})\text{-mod})^1/(U^{\mathrm{ren,reg}}(\widehat{\mathfrak{g}}_{\mathrm{crit}})\text{-mod})^0$ is a free $(U^{\mathrm{ren,reg}}(\widehat{\mathfrak{g}}_{\mathrm{crit}})\text{-mod})_0$-module, generated by the algebroid $N^*_{\mathfrak{z}_{\mathfrak{g}}^{\mathrm{reg}}/\mathfrak{z}_{\mathfrak{g}}}$.

Let us denote by $(\mathbf{L}_{w_0})^0 \subset \mathbf{L}_{w_0}$ the image of the map $\mathbb{M}_{w_0,\mathrm{reg}} \otimes \omega_x^{\langle\rho,\check{\rho}\rangle} \to \mathbf{L}_{w_0}$, and we define the submodule $(\mathbf{L}_{w_0})^i$ inductively as the image of $(\mathbf{L}_{w_0})^{i-1}$ under the action of $(U^{\mathrm{ren,reg}}(\widehat{\mathfrak{g}}_{\mathrm{crit}})\text{-mod})^1$. In particular, we have surjective maps

$$N^*_{\mathfrak{z}_{\mathfrak{g}}^{\mathrm{reg}}/\mathfrak{z}_{\mathfrak{g}}} \otimes (\mathbf{L}_{w_0})^i/(\mathbf{L}_{w_0})^{i-1} \to (\mathbf{L}_{w_0})^{i+1}/(\mathbf{L}_{w_0})^i,$$

and, hence, also surjective maps

$$\left(N^*_{\mathfrak{z}_{\mathfrak{g}}^{\mathrm{reg}}/\mathfrak{z}_{\mathfrak{g}}}\right)^{\otimes i} \otimes \left(\mathbb{M}_{w_0,\mathrm{reg}} \otimes \omega_x^{\langle\rho,\check{\rho}\rangle}\right) \twoheadrightarrow (\mathbf{L}_{w_0})^i/(\mathbf{L}_{w_0})^{i-1},$$

and $\underset{i}{\cup}(\mathbf{L}_{w_0})^i = \mathbf{L}_{w_0}$. Our task is to show that $(\mathbf{L}_{w_0})^0 = (\mathbf{L}_{w_0})^1$, i.e., that $(\mathbf{L}_{w_0})^0$ is stable under the action of $(U^{\mathrm{ren,reg}}(\widehat{\mathfrak{g}}_{\mathrm{crit}})\text{-mod})^1$.

Lemma 17.9. *The module $\mathbf{L}_{w_0}$ has no partially integrable subquotients.*

Proof. First, let us show first that $\mathbf{L}_{w_0}$ has no partially integrable quotient modules. Suppose that $\mathcal{M}$ is such a quotient module. Let i be the minimal integer such that the projection $(\mathbf{L}_{w_0})^i \to \mathcal{M}$ is nonzero; by definition this projection factors through $(\mathbf{L}_{w_0})^i/(\mathbf{L}_{w_0})^{i-1}$. Hence, some element of $(N^*_{\mathfrak{z}_{\mathfrak{g}}^{\mathrm{reg}}/\mathfrak{z}_{\mathfrak{g}}})^{\otimes i}$ gives rise to a nontrivial map $\mathbb{M}_{w_0,\mathrm{reg}} \to \mathcal{M}$. But this is a contradiction, since $\mathbb{M}_{w_0}$, and hence $\mathbb{M}_{w_0,\mathrm{reg}}$, cannot map to any partially integrable module.

Now consider $\mathbf{L}_{w_0}$ as a graded module, i.e., as a module over $\mathbb{C} \cdot t\partial_t \ltimes \widehat{\mathfrak{g}}_{\mathrm{crit}}$. It is easy to see that a graded module admits no partially integrable subquotients as a $\widehat{\mathfrak{g}}_{\mathrm{crit}}$-module if and only if it has the same property with respect to $\mathbb{C} \cdot t\partial_t \ltimes \widehat{\mathfrak{g}}_{\mathrm{crit}}$.

As was remarked earlier, the commutative Lie algebra $\mathbb{C} \cdot t\partial_t \oplus \mathfrak{h}$ has finite-dimensional eigenspaces on the module $\Gamma(\mathrm{Gr}_G, j_{w_0\cdot\check{\rho},*} \underset{I}{\star} \delta_{1_{\mathrm{Gr}_G}})$; hence the same will be true for $\mathbf{L}_{w_0}$.

Consider the maximal $\mathbb{C} \cdot t\partial_t \ltimes \widehat{\mathfrak{g}}_{\mathrm{crit}}$-stable submodule of $\mathbf{L}_{w_0}$ that does not contain the highest weight line, and take the quotient. Since this quotient is generated by a vector of weight -2ρ with respect to $\mathfrak{h}$, it is not partially integrable.

Let $\mathcal{M}''$ be the maximal $\mathbb{C} \cdot t\partial_t \ltimes \widehat{\mathfrak{g}}_{\mathrm{crit}}$-stable quotient of $\mathbf{L}_{w_0}$ which admits no partially integrable subquotients. It is well defined due to the above finite-dimensionality property. It is nonzero, because we have just exhibited one such quotient.

Let $\mathcal{M}' := \ker(\mathbf{L}_{w_0} \to \mathcal{M}'')$, and assume that $\mathcal{M}' \neq 0$. As above, some section of $N^*_{\mathfrak{z}_{\mathfrak{g}}^{\mathrm{reg}}/\mathfrak{z}_{\mathfrak{g}}}$ induces a nonzero map of $\widehat{\mathfrak{g}}_{\mathrm{crit}}$-modules $\mathcal{M}' \to \mathcal{M}''$. Therefore, $\mathcal{M}'$ also admits a quotient which has no partially integrable subquotients. This contradicts the definition of $\mathcal{M}''$. Hence, $\mathbf{L}_{w_0}$ has no partially integrable subquotients. □

Let us continue viewing $\mathbf{L}_{w_0}$ as a graded module. For an integer n we will denote by $(\cdot)_n$ the subspace of elements of degree n. By Section 17.1, $(\mathbf{L}_{w_0})_{n+\langle\rho,\check{\rho}\rangle} = 0$ if $n > 0$, and $(\mathbf{L}_{w_0})_{\langle\rho,\check{\rho}\rangle}$ identifies with the Verma module M_{w_0}; in particular, it is contained in $(\mathbf{L}_{w_0})^0$.

Lemma 17.10. *The subspace* $(\mathbf{L}_{w_0})_{\langle\rho,\check{\rho}\rangle-1}$ *is also contained in* $(\mathbf{L}_{w_0})^0$.

Proof. Let $\mathsf{V} \subset \Gamma(\mathrm{Gr}_G, j_{w_0\cdot\check{\rho},*} \underset{I}{\star} \delta_{1_{\mathrm{Gr}_G}})$ be the subspace of sections scheme-theoretically supported on the I-orbit $I \cdot t^{-\check{\rho}} \subset \mathrm{Gr}_G$.

Let $\mathrm{Lie}(I^0)^- \subset \mathfrak{g}((t))$ be the subalgebra, opposite to $\mathrm{Lie}(I^0)$, i.e., the one spanned by $t^{-1}\mathfrak{g}([t^{-1}])$ and $\mathfrak{n}^{-1} \subset \mathfrak{g}$. The module $\Gamma(\mathrm{Gr}_G, j_{w_0\cdot\check{\rho},*} \underset{I}{\star} \delta_{1_{\mathrm{Gr}_G}})$ is generated from V by means of $\mathrm{Lie}(I^0)^- \subset \mathfrak{g}((t))$.

Hence, the subspace $\Gamma(\mathrm{Gr}_G, j_{w_0\cdot\check{\rho},*} \underset{I}{\star} \delta_{1_{\mathrm{Gr}_G}})_{\langle\rho,\check{\rho}\rangle-1}$ is the direct sum of $(\mathfrak{g}\otimes t^{-1}) \cdot U(\mathfrak{n}^-) \cdot \mathsf{V}_{\langle\rho,\check{\rho}\rangle}$ and $U(\mathfrak{n}^-) \cdot \mathsf{V}_{\langle\rho,\check{\rho}\rangle-1}$. Note that $\mathsf{V}_{\langle\rho,\check{\rho}\rangle}$ is the highest weight line in $\mathbf{L}_{w_0}$. Hence,

$$(\mathfrak{g}\otimes t^{-1}) \cdot U(\mathfrak{n}^-) \cdot \mathsf{V}_{\langle\rho,\check{\rho}\rangle} \subset (\mathbf{L}_{w_0})^0.$$

Therefore, it remains to show that $\mathsf{V}_{\langle\rho,\check{\rho}\rangle-1} \cap \mathbf{L}_{w_0}$ is contained in $(\mathbf{L}_{w_0})^0$. Suppose not, and consider the image of $\mathsf{V}_{\langle\rho,\check{\rho}\rangle-1} \cap \mathbf{L}_{w_0}$ in $(\mathbf{L}_{w_0})^1/(\mathbf{L}_{w_0})^0$. This is a subspace annihilated by $\mathfrak{g}(t\mathbb{C}[[t]])$, and stable under the $\mathfrak{b}$-action. Take some highest weight vector. It gives rise to a map $\mathbb{M}_w \to (\mathbf{L}_{w_0})^1/(\mathbf{L}_{w_0})^0$ for some element $w \in W$; moreover $w = w_0$ if and only if the above highest weight is -2ρ.

However, the algebra of functions on $I \cdot t^{-\check{\rho}}$ is generated by elements whose weights with respect to $\mathfrak{h}$ are in $\mathrm{Span}^+(\alpha_i) - 0$. Therefore, the above highest weight is different from -2ρ. Thus we obtain a nonzero map $\mathbb{M}_w \to (\mathbf{L}_{w_0})^1/(\mathbf{L}_{w_0})^0$ for $w \neq w_0$, where $\mathbb{M}_w$ is endowed with a $\mathbb{G}_m$-action such that its generating vector has degree $\langle\rho,\check{\rho}\rangle - 1$. But this leads to a contradiction.

By Lemma 17.9, the image of $\mathbb{M}_w$ in $(\mathbf{L}_{w_0})^1/(\mathbf{L}_{w_0})^0$ equals the image of the submodule $\mathbb{M}_{w_0} \subset \mathbb{M}_w$, as the quotient is partially integrable. Hence, $\mathbb{M}_w$ admits a quotient, which is simultaneously a quotient module of $\mathbb{M}_{w_0}$. However, this is impossible, since we are working with the *Kac–Moody* algebra $\mathbb{C}\cdot t\partial_t \ltimes \widehat{\mathfrak{g}}_{\mathrm{crit}}$, and it is known that for Kac–Moody algebras, Verma modules have simple and mutually nonisomorphic cosocles. □

Now we are ready to finish the proof of Proposition 17.2. Consider the nilp-version of the renormalized universal enveloping algebra at the critical level, $U^{\mathrm{ren,nilp}}(\widehat{\mathfrak{g}}_{\mathrm{crit}})$; see Section 7.8. We have a natural homomorphism $U^{\mathrm{ren,nilp}}(\widehat{\mathfrak{g}}_{\mathrm{crit}}) \to U^{\mathrm{ren,reg}}(\widehat{\mathfrak{g}}_{\mathrm{crit}})$.

Consider the $\hbar$-family of $\widehat{\mathfrak{g}}_\hbar$-modules equal to $\mathbb{M}_{-2\rho+\kappa_\hbar(\check{\rho},\cdot)}$. Its specialization at $\hbar = 0$ is the module $\mathbb{M}_{w_0}$; and hence it acquires a $U^{\mathrm{ren,nilp}}(\widehat{\mathfrak{g}}_{\mathrm{crit}})$-action.

Lemma 17.11. *The map* $\mathbb{M}_{w_0} \otimes \omega_x^{\langle\rho,\check{\rho}\rangle} \to \mathbf{L}_{w_0}$ *is compatible with the* $U^{\mathrm{ren,nilp}}(\widehat{\mathfrak{g}}_{\mathrm{crit}})$-*actions.*

Proof. This follows from the fact that the map $\mathbb{M}_{w_0} \otimes \omega_x^{\langle\rho,\check{\rho}\rangle} \to \Gamma(\mathrm{Gr}_G, j_{w_0\cdot\check{\rho},*} \underset{I}{\star} \delta_{1_{\mathrm{Gr}_G}})$, constructed in Section 17.1, deforms away from the critical level. □

By Theorem 7.10 and Corollary 4.18, we have a short exact sequence

$$0 \to N^*_{\mathfrak{z}^{\mathrm{nilp}}_{\mathfrak{g}}/\mathfrak{z}_{\mathfrak{g}}}|_{\mathrm{Spec}(\mathfrak{z}^{\mathrm{reg}}_{\mathfrak{g}})} \to N^*_{\mathfrak{z}^{\mathrm{reg}}_{\mathfrak{g}}/\mathfrak{z}_{\mathfrak{g}}} \to (\check{\mathfrak{g}}/\check{\mathfrak{b}})_{\mathfrak{z}^{\mathrm{reg}}_{\mathfrak{g}}} \to 0.$$

Let $L_{-1} = \partial_t$ be the renormalized Sugawara operator, which we view as an element of $(U^{\mathrm{ren,reg}}(\widehat{\mathfrak{g}}_{\mathrm{crit}}))^1$. By Proposition 4.23, the image of L_{-1} in $N^*_{\mathfrak{z}^{\mathrm{reg}}_{\mathfrak{g}}/\mathfrak{z}_{\mathfrak{g}}} \twoheadrightarrow (\check{\mathfrak{g}}/\check{\mathfrak{b}})_{\mathfrak{z}^{\mathrm{reg}}_{\mathfrak{g}}}$ is a principal nilpotent element. Hence, $N^*_{\mathfrak{z}^{\mathrm{nilp}}_{\mathfrak{g}}/\mathfrak{z}_{\mathfrak{g}}}|_{\mathrm{Spec}(\mathfrak{z}^{\mathrm{reg}}_{\mathfrak{g}})}$ and L_{-1} generate $N^*_{\mathfrak{z}^{\mathrm{reg}}_{\mathfrak{g}}/\mathfrak{z}_{\mathfrak{g}}}$ as an algebroid. This, in turn, implies that L_{-1} and $(U^{\mathrm{ren,nilp}}(\widehat{\mathfrak{g}}_{\mathrm{crit}}))^1 \underset{\mathfrak{z}^{\mathrm{nilp}}_{\mathfrak{g}}}{\otimes} \mathfrak{z}^{\mathrm{reg}}_{\mathfrak{g}}$ generate $(U^{\mathrm{ren,reg}}(\widehat{\mathfrak{g}}_{\mathrm{crit}}))^1$ as an algebroid over $(U^{\mathrm{ren,reg}}(\widehat{\mathfrak{g}}_{\mathrm{crit}}))^0$.

Thus, to prove Proposition 17.2, it remains to check that L_{-1} preserves $(\mathbf{L}_{w_0})^0$. Since L_{-1} normalizes $(U^{\mathrm{ren,reg}}(\widehat{\mathfrak{g}}_{\mathrm{crit}}))^0$, and since $(\mathbf{L}_{w_0})^0$ is generated over $(U^{\mathrm{ren,reg}}(\widehat{\mathfrak{g}}_{\mathrm{crit}}))^0$ by its highest weight line, it suffices to show that L_{-1} maps this highest weight line to $(\mathbf{L}_{w_0})^0$.

However, the image of the highest weight line under L_{-1} has degree $\langle \rho, \check{\rho} \rangle - 1$, and our assertion follows from Lemma 17.10. This completes the proof of Proposition 17.2. □

17.12

We conclude this section by the following observation.

Proposition 17.13. *For every $\chi \in \mathrm{Spec}(\mathfrak{z}^{\mathrm{reg}}_{\mathfrak{g}})$, the module $\mathbf{L}_{w_0} \underset{\mathfrak{z}^{\mathrm{reg}}_{\mathfrak{g}}}{\otimes} \mathbb{C}_\chi$ is irreducible.*

Proof. Let us observe that, on the one hand, Corollary 16.7 implies that the module $\mathbb{M}_{w_0} \underset{\mathfrak{z}^{\mathrm{reg}}_{\mathfrak{g}}}{\otimes} \mathbb{C}_\chi$ has a unique irreducible quotient, denoted $\mathbf{L}_{w_0,\chi}$, such that the kernel of the projection

$$\mathbb{M}_{w_0} \underset{\mathfrak{z}^{\mathrm{reg}}_{\mathfrak{g}}}{\otimes} \mathbb{C}_\chi \to \mathbf{L}_{w_0,\chi}$$

is partially integrable.

On the other hand, by Theorem 15.6, the above projection factors through

$$\mathbb{M}_{w_0} \underset{\mathfrak{z}^{\mathrm{reg}}_{\mathfrak{g}}}{\otimes} \mathbb{C}_\chi \twoheadrightarrow \mathbf{L}_{w_0} \underset{\mathfrak{z}^{\mathrm{reg}}_{\mathfrak{g}}}{\otimes} \mathbb{C}_\chi \to \mathbf{L}_{w_0,\chi}.$$

Thus we obtain a surjective map $\mathbf{L}_{w_0} \underset{\mathfrak{z}^{\mathrm{reg}}_{\mathfrak{g}}}{\otimes} \mathbb{C}_\chi \to \mathbf{L}_{w_0,\chi}$, whose kernel is partially integrable. However, by Lemma 17.9, we conclude that this map must be an isomorphism. □

18 Comparison with semi-infinite cohomology

18.1

Consider the group ind-scheme $N^-((t))$, and let Ψ_0 denote a nondegenerate character $N^-((t)) \to \mathbb{G}_a$ of conductor 0. This means that the restriction of Ψ_0 to $N^-[[t]]$ is trivial, and its restriction to $\mathrm{Ad}_{t^{\check{\alpha}_\iota}}(N^-[[t]]) \subset N^-((t))$ is nontrivial for each $\iota \in \mathcal{I}$. Note that to specify Ψ_0 one needs to make a choice: e.g., of a nonvanishing 1-form on $\mathcal{D}$, in addition to a choice of $\psi : N^- \to \mathbb{G}_a$.

For a coweight $\check{\lambda}$, let $\Psi_{\check{\lambda}}$ denote the character obtained as a composition

$$N^-((t)) \xrightarrow{\mathrm{Ad}(t^{\check{\lambda}})} N^-((t)) \xrightarrow{\Psi_0} \mathbb{G}_a.$$

Note that for $\check{\lambda} = -\check{\rho}$, this character is canonical, modulo a choice of ψ (the latter we will consider fixed).

We will identify N^- and N by means of conjugation by a chosen lift of the element $w_0 \in W$; and denote by the same symbol $\Psi_{\check{\lambda}}$ the corresponding character on $N((t))$. We will also use the same notation for the corresponding characters on the Lie algebras.

In this section we will study the semi-infinite cohomology of $\mathfrak{n}^-((t))$ twisted by the characters $\Psi_{\check{\lambda}}$ with coefficients in $\widehat{\mathfrak{g}}_{\mathrm{crit}}$-modules. The complex computing semi-infinite cohomology was introduced by Feigin [Fe]; the construction is recalled in Section 19.17. We denote it by

$$\mathcal{M} \mapsto \mathfrak{C}^{\frac{\infty}{2}}(\mathfrak{n}^-((t)), ?, \mathcal{M} \otimes \Psi_{\check{\lambda}}),$$

where ? stands for a choice of a lattice in $\mathfrak{n}^-((t))$. Its cohomology will be denoted by

$$H^{\frac{\infty}{2}+\bullet}(\mathfrak{n}^-((t)), \mathfrak{n}^-[[t]], \mathcal{M} \otimes \Psi_{\check{\lambda}}).$$

Proposition 18.2. *For* $\mathcal{M} \in D(\widehat{\mathfrak{g}}_{\mathrm{crit}}\text{-mod})^{I^0}$,

$$\mathfrak{C}^{\frac{\infty}{2}}(\mathfrak{n}((t)), \mathfrak{n}[[t]], \mathcal{M} \otimes \Psi_0) \simeq \mathfrak{C}^{\frac{\infty}{2}}\left(\mathfrak{n}^-((t)), t\mathfrak{n}^-[[t]], (\widetilde{j}_{w_0\cdot\check{\rho},*} \underset{I^0}{\star} \mathcal{M}) \otimes \Psi_{-\check{\rho}}\right).$$

We do not give the proof, since it essentially repeats the proof of Proposition 12.12. (In particular, the assertion is valid at any level κ.)

Another important observation (also valid at any level) is the following.

Lemma 18.3. *If* $\mathcal{M} \in \widehat{\mathfrak{g}}_{\mathrm{crit}}\text{-mod}^{I^0}$ *is partially integrable, then*

$$H^{\frac{\infty}{2}+\bullet}(\mathfrak{n}^-((t)), \mathfrak{n}^-[[t]], \mathcal{M} \otimes \Psi_{-\check{\rho}}) = 0.$$

Proof. We can assume that $\mathcal{M}$ is integrable with respect to $\mathfrak{sl}_2^\iota$ for some $\iota \in \mathcal{I}$. Let $f_\iota \in \mathfrak{n}^- \subset \mathfrak{n}^-((t))$ be the corresponding Chevalley generator. With no loss of generality, we can assume that $\Psi_{-\check{\rho}}(f_\iota) = 1$.

Consider the complex $\mathfrak{C}^{\frac{\infty}{2}}(\mathfrak{n}^-((t)), \mathfrak{n}^-[[t]], \mathcal{M} \otimes \Psi_{-\check{\rho}})$, and recall (see Section 19.17) that we have an action of $\mathfrak{n}^-((t))[1]$ on it by "annihilation operators," $x \mapsto i(x)$, and the action of $\mathfrak{n}^-((t))$ by Lie derivatives $x \mapsto \mathrm{Lie}_x$ such that

$$[d, i(x)] = \mathrm{Lie}_x + \mathrm{Id} \cdot \Psi_{-\check{\rho}}(x).$$

Hence, $i(f_\iota)$ defines a homotopy between the identity map on

$$\mathfrak{C}^{\frac{\infty}{2}}(\mathfrak{n}^-((t)), \mathfrak{n}^-[[t]], \mathcal{M} \otimes \Psi_{-\check{\rho}})$$

and the map given by Lie_{f_ι}. However, by assumption, the latter acts locally nilpotently, implying the assertion of the lemma. □

18.4

In the rest of this section we will collect several additional facts concerning the semi-infinite cohomology functor $H^{\frac{\infty}{2}+i}(\mathfrak{n}^-((t)), t\mathfrak{n}^-[[t]], ? \otimes \Psi_{-\check{\rho}})$. By Lemma 18.3, this functor, when restricted to $D^b(\widehat{\mathfrak{g}}_{\mathrm{crit}}\text{-mod}_{\mathrm{nilp}})^{I^0}$, factors through ${}^f D^b(\widehat{\mathfrak{g}}_{\mathrm{crit}}\text{-mod}_{\mathrm{nilp}})^{I^0}$.

Theorem 18.5. *The two functors* ${}^f D^b(\widehat{\mathfrak{g}}_{\mathrm{crit}}\text{-mod}_{\mathrm{nilp}})^{I^0} \to D(\mathrm{Vect})$

$$\mathcal{M}^\bullet \mapsto H^{\frac{\infty}{2}+\bullet}(\mathfrak{n}^-((t)), t\mathfrak{n}^-[[t]], \mathcal{M}^\bullet \otimes \Psi_{-\check{\rho}}) \quad \textit{and} \quad \mathcal{M}^\bullet \mapsto \mathrm{Hom}(\mathbf{\Pi}, \mathcal{M}^\bullet)$$

are isomorphic. In particular, for $0 \neq \mathcal{M} \in {}^f\widehat{\mathfrak{g}}_{\mathrm{crit}}\text{-mod}_{\mathrm{nilp}}^{I,m}$, *we have*

$$H^{\frac{\infty}{2}+i}(\mathfrak{n}^-((t)), t\mathfrak{n}^-[[t]], \mathcal{M} \otimes \Psi_{-\check{\rho}}) = 0 \quad \textit{for } i \neq 0$$

and

$$H^{\frac{\infty}{2}}(\mathfrak{n}^-((t)), t\mathfrak{n}^-[[t]], \mathcal{M} \otimes \Psi_{-\check{\rho}}) \neq 0.$$

The proof of the theorem is based on the following observation.

Lemma 18.6. *For any* $\mathfrak{Z}_{\mathfrak{g}}^{\mathrm{nilp}}$*-module* $\mathcal{L}$ *and* $w \in W$, *we have*

$$H^{\frac{\infty}{2}+i}\left(\mathfrak{n}^-((t)), t\mathfrak{n}^-[[t]], (\mathbb{M}_w \underset{\mathfrak{Z}_{\mathfrak{g}}^{\mathrm{nilp}}}{\otimes} \mathcal{L}) \otimes \Psi_{-\check{\rho}}\right) \simeq \begin{cases} \mathcal{L}, & i = 0, \\ 0, & i \neq 0. \end{cases}$$

Proof. Since the quotients $\mathbb{M}_1/\mathbb{M}_w$ are all partially integrable, we can assume that the element $w \in W$, appearing in the lemma, equals 1. In the latter case, the assertion follows from Proposition 12.10 and Corollary 13.8. □

Let us now prove Theorem 18.5.

Proof. In view of Main Theorem 6.9, to prove the theorem we have to establish an isomorphism

$$H^{\frac{\infty}{2}}(\mathfrak{n}^-((t)), t\mathfrak{n}^-[[t]], \mathbf{\Pi} \otimes \Psi_{-\check{\rho}}) \simeq \mathfrak{Z}_{\mathfrak{g}}^{\mathrm{nilp}} \otimes h_0.$$

Consider the filtration on $\mathbf{\Pi}$, induced by the tilting filtration on Π with quotients M_w. By Lemma 18.6, we obtain that $H^{\frac{\infty}{2}+i}(\mathfrak{n}^-((t)), t\mathfrak{n}^-[[t]], \mathbf{\Pi} \otimes \Psi_{-\check{\rho}}) = 0$ for $i \neq 0$, and that $H^{\frac{\infty}{2}}(\mathfrak{n}^-((t)), t\mathfrak{n}^-[[t]], \mathbf{\Pi} \otimes \Psi_{-\check{\rho}})$ has a filtration, with subquotients isomorphic to $\mathfrak{Z}_{\mathfrak{g}}^{\mathrm{nilp}}$.

Hence, it remains to show that $H^{\frac{\infty}{2}}(\mathfrak{n}^-((t)), t\mathfrak{n}^-[[t]], \mathbf{\Pi} \otimes \Psi_{-\check{\rho}})$ is flat as an h_0-module, where the action of h_0 is induced from the identification $h_0 \simeq \mathrm{End}(\Pi)$.

It suffices to check that

$$\left(H^{\frac{\infty}{2}}(\mathfrak{n}^-((t)), t\mathfrak{n}^-[[t]], \mathbf{\Pi} \otimes \Psi_{-\check{\rho}})\right) \overset{L}{\underset{h_0}{\otimes}} \mathbb{C} \simeq \mathfrak{Z}_{\mathfrak{g}}^{\mathrm{nilp}}.$$

By Lemma 14.3, ${}^f\Pi \overset{L}{\underset{h_0}{\otimes}} \mathbb{C} \simeq {}^f M_{w_0}$. Hence,

$$\left(H^{\frac{\infty}{2}}(\mathfrak{n}^-((t)), t\mathfrak{n}^-[[t]], \mathbf{\Pi} \otimes \Psi_{-\check{\rho}})\right) \overset{L}{\underset{h_0}{\otimes}} \mathbb{C} \simeq H^{\frac{\infty}{2}}(\mathfrak{n}^-((t)), t\mathfrak{n}^-[[t]], \mathbb{M}_{w_0} \otimes \Psi_{-\check{\rho}}),$$

and the assertion follows from Lemma 18.6. □

Corollary 18.7. *For any object $\mathcal{M}$ of $\widehat{\mathfrak{g}}_{\mathrm{crit}}$-$\mathrm{mod}_{\mathrm{nilp}}^{I,m}$ and a dominant coweight $\check{\lambda}$*

$$H^{\frac{\infty}{2}+i}(\mathfrak{n}((t)), \mathfrak{n}[[t]], \mathcal{M} \otimes \Psi_{\check{\lambda}}) = 0 \quad \text{for } i > 0.$$

Proof. We have

$$H^{\frac{\infty}{2}+i}(\mathfrak{n}((t)), \mathfrak{n}[[t]], \mathcal{M} \otimes \Psi_{\check{\lambda}}) \simeq H^{\frac{\infty}{2}}(\mathfrak{n}^-((t)), t\mathfrak{n}^-[[t]], \widetilde{j}_{\check{\lambda},*} \underset{I^0}{\star} \widetilde{j}_{w_0\cdot\check{\rho},*} \underset{I^0}{\star} \mathcal{M} \otimes \Psi_{-\check{\rho}}),$$

as in Proposition 18.2.

Now the assertion of the corollary follows from the fact that the functor $\mathcal{M} \mapsto \widetilde{j}_{\check{\lambda},*} \underset{I^0}{\star} \widetilde{j}_{w_0\cdot\check{\rho},*} \underset{I^0}{\star} \mathcal{M}$ is right exact. □

As another application, we give an alternative proof of the following result of [FB] (see Theorem 15.1.9).

Theorem 18.8. *The natural map $\mathfrak{Z}_{\mathfrak{g}}^{\mathrm{reg}} \to H^{\frac{\infty}{2}}(\mathfrak{n}((t)), \mathfrak{n}[[t]], \mathbb{V}_{\mathrm{crit}} \otimes \Psi_0)$ is an isomorphism, and all other cohomologies $H^{\frac{\infty}{2}+i}(\mathfrak{n}((t)), \mathfrak{n}[[t]], \mathbb{V}_{\mathrm{crit}} \otimes \Psi_0)$, $i \neq 0$, vanish.*

Proof. Consider the map

$$\mathbb{M}_{w_0,\mathrm{reg}} \otimes \omega_x^{\langle\rho,\check{\rho}\rangle} \to j_{w_0\cdot\check{\rho},*} \underset{I}{\star} \mathbb{V}_{\mathrm{crit}}$$

of Section 17.1. Its kernel and cokernel are partially integrable; hence it induces isomorphisms

$$H^{\frac{\infty}{2}+i}(\mathfrak{n}^-((t)), t\mathfrak{n}^-[[t]], \mathbb{M}_{w_0,\mathrm{reg}} \otimes \Psi_{-\check{\rho}})$$
$$\to H^{\frac{\infty}{2}+i}(\mathfrak{n}^-((t)), t\mathfrak{n}^-[[t]], j_{w_0\cdot\check{\rho},*} \underset{I}{\star} \mathbb{V}_{\mathrm{crit}} \otimes \Psi_{-\check{\rho}}).$$

By Lemma 18.6,

$$H^{\frac{\infty}{2}+i}(\mathfrak{n}^-((t)), t\mathfrak{n}^-[[t]], \mathbb{M}_{w_0,\mathrm{reg}} \otimes \Psi_{-\check{\rho}}) \simeq \begin{cases} \mathfrak{Z}_{\mathfrak{g}}^{\mathrm{reg}}, & i = 0, \\ 0, & i \neq 0, \end{cases}$$

we obtain that

$$H^{\frac{\infty}{2}+i}(\mathfrak{n}^-((t)), t\mathfrak{n}^-[[t]], j_{w_0\cdot\check{\rho},*} \underset{I}{\star} \mathbb{V}_{\mathrm{crit}} \otimes \Psi_{-\check{\rho}}) \simeq \begin{cases} \mathfrak{Z}_{\mathfrak{g}}^{\mathrm{reg}}, & i = 0, \\ 0, & i \neq 0. \end{cases}$$

Applying Proposition 18.2 for $\mathcal{M} = \mathbb{V}_{\mathrm{crit}}$, we obtain that

$$H^{\frac{\infty}{2}+i}(\mathfrak{n}((t)), \mathfrak{n}[[t]], \mathbb{V}_{\mathrm{crit}} \otimes \Psi_0) = 0 \quad \text{for } i \neq 0$$

and

$$H^{\frac{\infty}{2}}(\mathfrak{n}((t)), \mathfrak{n}[[t]], \mathbb{V}_{\mathrm{crit}} \otimes \Psi_0) \simeq \mathfrak{Z}_{\mathfrak{g}}^{\mathrm{reg}}.$$

Moreover, by unraveling the isomorphism of Proposition 18.2, we obtain that the above isomorphism coincides with the one appearing in the statement of the theorem. □

18.9

Let $\mathcal{F}$ be a critically twisted D-module on Gr_G. In this subsection we will express the semi-infinite cohomology

$$H^{\frac{\infty}{2}+\bullet}(\mathfrak{n}((t)), \mathfrak{n}[[t]], \Gamma(\mathrm{Gr}_G, \mathcal{F}) \otimes \Psi_0) \tag{18.1}$$

in terms of the de Rham cohomologies of $\mathcal{F}$ along the $N((t))$-orbits in Gr_G.

For a coweight $\check{\lambda}$, consider the $N((t))$-orbit of the point $t^{\check{\lambda}}$ on Gr_G; by pulling back $\mathcal{F}$, by Section 21.6, we obtain a D-module on $N((t))$. We will denote it by $\mathcal{F}|_{N((t))\cdot t^{\check{\lambda}}}$. If Ψ_0 is a nondegenerate character of conductor 0, we will denote by $H^\bullet(N((t)), \mathcal{F}|_{N((t))\cdot t^{\check{\lambda}}} \otimes \Psi_0)$ the resulting de Rham cohomology. Note that this cohomology vanishes automatically unless $\check{\lambda}$ is dominant, since otherwise Ψ_0 would be nontrivial on the stabilizer of $t^{\check{\lambda}} \in \mathrm{Gr}_G$.

By decomposing $\mathcal{F}$ in the derived category with respect to the stratification of Gr_G by $N((t)) \cdot t^{\check{\lambda}}$, using Section 22.15, we obtain that, as an object of the derived category of $\mathfrak{Z}_{\mathfrak{g}}^{\mathrm{reg}}$-modules, $\mathfrak{C}^{\frac{\infty}{2}+\bullet}(\mathfrak{n}((t)), \mathfrak{n}[[t]], \Gamma(\mathrm{Gr}_G, \mathcal{F}) \otimes \Psi_0)$ is a successive extension of complexes

$$H^{\frac{\infty}{2}+\bullet}\left(\mathfrak{n}((t)), \mathfrak{n}[[t]], \Gamma(\mathrm{Gr}_G, \delta_{t^{\check{\lambda}}}) \otimes \Psi_0\right) \otimes H^\bullet(N((t)), \mathcal{F}|_{N((t))\cdot t^{\check{\lambda}}} \otimes \Psi_0). \quad (18.2)$$

Note also that $\Gamma(\mathrm{Gr}_G, \delta_{t^{\check{\lambda}}})$ is isomorphic to the vacuum module, twisted by $t^{\check{\lambda}} \in T((t))$. Hence,

$$\begin{aligned} &H^{\frac{\infty}{2}+\bullet}\left(\mathfrak{n}((t)), \mathfrak{n}[[t]], \Gamma(\mathrm{Gr}_G, \delta_{t^{\check{\lambda}}}) \otimes \Psi_0\right) \\ &\simeq H^{\frac{\infty}{2}+\bullet}\left(\mathfrak{n}((t)), \mathrm{Ad}_{t^{\check{\lambda}}}(\mathfrak{n}[[t]]), \mathbb{V}_{\mathrm{crit}} \otimes \Psi_{\check{\lambda}}\right). \end{aligned} \quad (18.3)$$

We will prove the following.

Theorem 18.10.

(1) *For* $\mathcal{F} \in \mathfrak{D}(\mathrm{Gr}_G)_{\mathrm{crit}}$-mod, *there is a canonical direct sum decomposition*

$$\begin{aligned} &H^{\frac{\infty}{2}+\bullet}(\mathfrak{n}((t)), \mathfrak{n}[[t]], \Gamma(\mathrm{Gr}_G, \mathcal{F}) \otimes \Psi_0) \\ &\simeq \bigoplus_{\check{\lambda}} H^{\frac{\infty}{2}+\bullet}\left(\mathfrak{n}((t)), \mathfrak{n}[[t]], \Gamma(\mathrm{Gr}_G, \delta_{t^{\check{\lambda}}}) \otimes \Psi_0\right) \\ &\quad \otimes H^\bullet(N((t)), \mathcal{F}|_{N((t))\cdot t^{\check{\lambda}}} \otimes \Psi_0). \end{aligned}$$

(2) *The cohomology* $H^{\frac{\infty}{2}+i}(\mathfrak{n}((t)), \mathfrak{n}[[t]], \mathbb{V}_{\mathrm{crit}} \otimes \Psi_{\check{\lambda}})$ *vanishes unless* $\check{\lambda}$ *is dominant and* $i = 0$, *and in the latter case, it is canonically isomorphic to* $V^{\check{\lambda}}_{\mathfrak{z}^{\mathrm{reg}}_{\mathfrak{g}}}$.

The rest of this section is devoted to the proof of this theorem. Let us first prove point (2). The fact that the semi-infinite cohomology in question vanishes unless $\check{\lambda}$ is dominant follows by the same argument as in Lemma 18.3. Therefore, let us assume that $\check{\lambda}$ is dominant and consider the D-module $\mathcal{F}_{V^{\check{\lambda}}}$; see Section 8.5.

By the geometric Casselman–Shalika formula (see [FGV]),

$$H^\bullet(N((t)), \mathcal{F}_{V^{\check{\lambda}}}|_{N((t))\cdot t^{\check{\mu}}} \otimes \Psi_0) = 0$$

unless $\mu = \lambda$. Therefore, all terms with $\mu \neq \lambda$ in the spectral sequence (18.2) vanish. We obtain, therefore,

$$H^{\frac{\infty}{2}+\bullet}\left(\mathfrak{n}((t)), \mathfrak{n}[[t]], \Gamma(\mathrm{Gr}_G, \mathcal{F}_{V^{\check{\lambda}}}) \otimes \Psi_0\right) \simeq H^{\frac{\infty}{2}+\bullet}\left(\mathfrak{n}((t)), \mathfrak{n}[[t]], \mathbb{V}_{\mathrm{crit}} \otimes \Psi_{\check{\lambda}}\right).$$

But by Theorem 8.7, $\Gamma(\mathrm{Gr}_G, \mathcal{F}_{V^{\check{\lambda}}}) \simeq \mathbb{V}_{\mathrm{crit}} \underset{\mathfrak{z}^{\mathrm{reg}}_{\mathfrak{g}}}{\otimes} V^{\check{\lambda}}_{\mathfrak{z}^{\mathrm{reg}}_{\mathfrak{g}}}$. By combining this with Theorem 18.8, we obtain

$$\begin{aligned} &H^{\frac{\infty}{2}}\left(\mathfrak{n}((t)), \mathfrak{n}[[t]], \mathbb{V}_{\mathrm{crit}} \otimes \Psi_{\check{\lambda}}\right) \simeq V^{\check{\lambda}}_{\mathfrak{z}^{\mathrm{reg}}_{\mathfrak{g}}}, \\ &H^{\frac{\infty}{2}+i}\left(\mathfrak{n}((t)), \mathfrak{n}[[t]], \mathbb{V}_{\mathrm{crit}} \otimes \Psi_{\check{\lambda}}\right) = 0, \quad i \neq 0, \end{aligned} \quad (18.4)$$

as required.

To prove point (1), we need some preparations.

Proposition 18.11. *Suppose that $\mathcal{M}$ is an object of $\widehat{\mathfrak{g}}_{\text{crit}}\text{-mod}_{\text{reg}}$ that comes by restriction from a $U^{\text{ren,reg}}(\widehat{\mathfrak{g}}_{\text{crit}})$-module. Then all $H^{\frac{\infty}{2}+i}(\mathfrak{n}((t)), \mathfrak{n}[[t]], \mathcal{M} \otimes \Psi_0)$ are naturally modules over the algebroid $N^*_{\mathfrak{Z}^{\text{reg}}_{\mathfrak{g}}/\mathfrak{Z}_{\mathfrak{g}}}$.*

Proof. We will assume that Ψ_0 comes from a character of the Lie-* algebra $L_{\mathfrak{n}}$. In this case, the BRST complex $\mathfrak{C}^{\frac{\infty}{2}}(L_{\mathfrak{n}}, \mathcal{A}_{\mathfrak{g},\text{crit}} \otimes \Psi_0)$ is itself a DG-chiral algebra.

Let $\mathcal{A}_{\mathfrak{g},\hbar}$ be a 1-st order deformation of $\mathcal{A}_{\mathfrak{g},\text{crit}}$ away from the critical level; i.e., $\mathcal{A}_{\mathfrak{g},\hbar}$ is flat over $\mathbb{C}[\hbar]/\hbar^2$, and $\mathcal{A}_{\mathfrak{g},\hbar}/\hbar \simeq \mathcal{A}_{\mathfrak{g},\text{crit}}$.

Let us consider the DG-chiral algebra $\mathfrak{C}^{\frac{\infty}{2}}(L_{\mathfrak{n}}, \mathcal{A}_{\mathfrak{g},\hbar} \otimes \Psi_0)$. From Theorem 18.8, it follows that it is acyclic off cohomological degree 0; in particular, its 0th cohomology is $\mathbb{C}[\hbar]/\hbar^2$-flat.

This implies that any section $a \in \mathfrak{z}_{\mathfrak{g}}$, which we think of as a 0-cocycle in $\mathfrak{C}^{\frac{\infty}{2}}(L_{\mathfrak{n}}, \mathcal{A}_{\mathfrak{g},\text{crit}} \otimes \Psi_0)$, can be lifted to a 0-cocycle $a_\hbar \in \mathfrak{C}^{\frac{\infty}{2}}(L_{\mathfrak{n}}, \mathcal{A}_{\mathfrak{g},\hbar} \otimes \Psi_0)$.

We will think of $\frac{a_\hbar}{\hbar}$ an element of the Lie-* algebra $\mathcal{A}^\sharp_{\mathfrak{g}} \otimes \text{Cliff}(L_{\mathfrak{n}})$, where $\mathcal{A}^\sharp_{\mathfrak{g}}$ is as in [FG], and $\text{Cliff}(L_{\mathfrak{n}})$ is the Clifford chiral algebra, used in the definition of the BRST complex.

By the construction of $\mathcal{A}^\sharp_{\mathfrak{g}}$, for $\mathcal{M}$ satisfying the properties of the proposition, we have an action of $\mathcal{A}^\sharp_{\mathfrak{g}}$ on $\mathcal{M}$, and hence, an action of the Lie-* algebra $\mathcal{A}^\sharp_{\mathfrak{g}} \otimes \text{Cliff}(L_{\mathfrak{n}})$ on the complex $\mathfrak{C}^{\frac{\infty}{2}}(L_{\mathfrak{n}}, \mathcal{M} \otimes \Psi_0)$. By taking the Lie-* bracket with the above element $\frac{a_\hbar}{\hbar}$ we obtain an endomorphism of $\mathfrak{C}^{\frac{\infty}{2}}(L_{\mathfrak{n}}, \mathcal{M} \otimes \Psi_0)$, which commutes with the differential.

It is easy to see that for a different choice of $a_\hbar$ the corresponding endomorphisms of $\mathfrak{C}^{\frac{\infty}{2}}(L_{\mathfrak{n}}, \mathcal{M} \otimes \Psi_0)$ will differ by a coboundary. Thus we obtain a Lie-* action of $\mathfrak{z}_{\mathfrak{g}}$ on each $H^{\frac{\infty}{2}+i}(L_{\mathfrak{n}}, \mathcal{M} \otimes \Psi_0)$. One easily checks that this action satisfies the Leibniz rule with respect to the $\mathfrak{z}_{\mathfrak{g}}$-module structure on $H^{\frac{\infty}{2}+i}(L_{\mathfrak{n}}, \mathcal{M} \otimes \Psi_0)$, and hence extends to an action of the Lie-* algebroid $\Omega^1(\mathfrak{z}_{\mathfrak{g}})$. The latter is the same as an action of the $\mathfrak{Z}^{\text{reg}}_{\mathfrak{g}}$-algebroid $N^*_{\mathfrak{Z}^{\text{reg}}_{\mathfrak{g}}/\mathfrak{Z}_{\mathfrak{g}}}$. □

We are now ready to finish the proof of Theorem 18.10. By Section 7.8 and Proposition 18.11, the terms of the spectral sequence (18.2) are acted on by the algebroid $N^*_{\mathfrak{Z}^{\text{reg}}_{\mathfrak{g}}/\mathfrak{Z}_{\mathfrak{g}}}$.

It is easy to see that the $N^*_{\mathfrak{Z}^{\text{reg}}_{\mathfrak{g}}/\mathfrak{Z}_{\mathfrak{g}}}$-action on $H^{\frac{\infty}{2}}(\mathfrak{n}((t)), \mathfrak{n}[[t]], \mathbb{V}_{\text{crit}} \otimes \Psi_0)$ identifies via Theorem 18.8 with the canonical $N^*_{\mathfrak{Z}^{\text{reg}}_{\mathfrak{g}}/\mathfrak{Z}_{\mathfrak{g}}}$-action on $\mathfrak{Z}^{\text{reg}}_{\mathfrak{g}}$. Moreover, from Theorem 8.7(a) we obtain that the isomorphisms of (18.4) are compatible with the $N^*_{\mathfrak{Z}^{\text{reg}}_{\mathfrak{g}}/\mathfrak{Z}_{\mathfrak{g}}}$-action.

This implies the canonical splitting of the spectral sequence. Indeed, from Theorem 8.7(b) it is easy to derive that there are no nontrivial Hom's and Ext^1's between different $V^{\check{\lambda}}_{\mathfrak{Z}^{\text{reg}}_{\mathfrak{g}}}$, regarded as $N^*_{\mathfrak{Z}^{\text{reg}}_{\mathfrak{g}}/\mathfrak{Z}_{\mathfrak{g}}}$-modules.

Part V: Appendix

This part, which may be viewed as a user guide to [BD1, Section 7], reviews some technical material that we need in the main body of this paper.

In Section 19 we review some background material: the three monoidal structures on the category of topological vector spaces, the notion of a family of objects of an abelian category over a scheme or an ind-scheme, and the formalism of DG-categories.

In Section 20 we introduce the notion of action of a group scheme on an abelian category. In fact, there are two such notions that correspond to weak and strong actions, respectively. A typical example of a weak action is when a group H acts on a scheme S, and we obtain an action of H on the category QCoh_H of quasi-coherent sheaves. A typical example of a strong (equivalently, infinitesimally trivial or Harish-Chandra-type) action is when in the above situation we consider the action of H on the category $\mathfrak{D}(S)$-mod of D-modules on S. We also discuss various notions related to equivariant objects and the corresponding derived categories.

In Section 21 we make a digression and discuss the notion of D-module over a group ind-scheme. The approach taken here is different, but equivalent, to the one developed in [AG1] via chiral algebras.

In Section 22 we generalize the discussion of Section 20 to the case of group ind-schemes. The goal of this section is to show that if $\mathcal{C}$ is a category that carries a Harish-Chandra action of some group ind-scheme G, then at the level of derived categories we have an action of the monoidal category of D-modules over G on $\mathcal{C}$. This formalism was developed in [BD1, Section 7], and in this section we essentially repeat it.

Finally, Section 23 serves a purely auxiliary role: we prove some technical assertions concerning the behavior of an abelian category over its center provided that a certain flatness assumption is satisfied.

19 Miscellanea

Unless specified otherwise, the notation in this part will be independent of that of Parts I–IV. We will work over the ground field $\mathbb{C}$, and all additive categories will be assumed $\mathbb{C}$-linear. Unless specified otherwise, by tensor product, we will mean tensor product over $\mathbb{C}$.

If $\mathcal{C}$ is a category, and X_i is a directed system of objects in it, then following the notation of SGA 4(I) notation, we write "$\varinjlim$" X_i for the resulting object in $\mathrm{Ind}(\mathcal{C})$, thought of as a contravariant functor on $\mathcal{C}$. In contrast, $\varinjlim X_i$ will denote the object of $\mathcal{C}$ representing the functor $\mathrm{Hom}(\varinjlim X_i, ?)$ on $\mathcal{C}$, provided that it exists.

19.1 Topological vector spaces and algebras

In this subsection we will briefly review the material of [CHA1]. By a topological vector space we will mean a vector space over $\mathbb{C}$ equipped with a linear topology, assumed complete and separated. We will denote this category by $\mathcal{T}\mathrm{op}$; it is closed under projective and inductive limits (note that the projective limits commute with the forgetful functor to vector spaces, and inductive limits do not). Every such topological vector space $\mathbf{V}$ can be represented as $\lim\limits_{\longleftarrow} V^i$, where V^i are usual (i.e., discrete) vector spaces and the transition maps $V^j \to V^i$ are surjective.

For a topological vector space $\mathbf{V}$ represented as a projective limit as above, its dual $\mathbf{V}^*$ is by definition the object of $\mathcal{T}op$ equal to

$$\lim_{\longrightarrow}(V_i)^*,$$

where each $(V_i)^*$ is a dual of the corresponding discrete vector space V_i, endowed with the natural (pro-finite-dimensional) topology. It is easy to see that $\mathbf{V}^*$ is well defined, i.e., independent of the presentation of $\mathbf{V}$ as a projective limit.

A topological vector space $\mathbf{V}$ is said to be of Tate type if it can be written in the form $\mathbf{V}_1 \oplus \mathbf{V}_2$, where $\mathbf{V}_1$ is discrete and $\mathbf{V}_2$ is pro-finite dimensional. In this case $\mathbf{V}^*$ is also of Tate type, and the natural map $(\mathbf{V}^*)^* \to \mathbf{V}$ is an isomorphism.

Following [CHA1], we endow the category $\mathcal{T}op$ with three different monoidal structures:

$$\mathbf{V}_1, \mathbf{V}_2 \mapsto \mathbf{V}_1 \overset{*}{\otimes} \mathbf{V}_2, \mathbf{V}_1 \overset{\rightarrow}{\otimes} \mathbf{V}_2 \quad \text{and} \quad \mathbf{V}_1 \overset{!}{\otimes} \mathbf{V}_2.$$

They are constructed as follows. Let us write $\mathbf{V}_1 = \lim\limits_{\longleftarrow} V_1^i$, $\mathbf{V}_2 = \lim\limits_{\longleftarrow} V_2^j$. Then

$$\mathbf{V}_1 \overset{!}{\otimes} \mathbf{V}_2 = \lim_{\substack{\longleftarrow \\ i,j}} V_1^i \otimes V_2^j.$$

It is easy to see that this monoidal structure is, in fact, a tensor one.

To define $\mathbf{V}_1 \overset{\rightarrow}{\otimes} \mathbf{V}_2$, we proceed in two steps. If $\mathbf{V}_2 = V$ is discrete and equal to $\bigcup\limits_k V_k$, where V_k are finite dimensional, we set

$$\mathbf{V}_1 \overset{\rightarrow}{\otimes} V = \lim_{\substack{\longrightarrow \\ k}} \mathbf{V}_1 \otimes V_k,$$

where the inductive limit is taken in $\mathcal{T}op$. For an arbitrary $\mathbf{V}_2$ written as $\mathbf{V}_2 = \lim\limits_{\longleftarrow} V_2^j$, we set

$$\mathbf{V}_1 \overset{\rightarrow}{\otimes} \mathbf{V}_2 = \lim_{\substack{\longleftarrow \\ j}} (\mathbf{V}_1 \overset{\rightarrow}{\otimes} V_2^j).$$

Finally, $\mathbf{V}_1 \overset{*}{\otimes} \mathbf{V}_2$ is characterized by the property that $\mathrm{Hom}(\mathbf{V}_1 \overset{*}{\otimes} \mathbf{V}_2, V)$, where V is discrete, is the set of bilinear continuous maps $\mathbf{V}_1 \times \mathbf{V}_2 \to W$. This monoidal structure is also tensor in a natural way.

We have natural maps

$$\mathbf{V}_1 \overset{*}{\otimes} \mathbf{V}_2 \to \mathbf{V}_1 \overset{\rightarrow}{\otimes} \mathbf{V}_2 \to \mathbf{V}_1 \overset{!}{\otimes} \mathbf{V}_2,$$

where the first arrow is an isomorphism if $\mathbf{V}_2$ is discrete and the second one is an isomorphism if $\mathbf{V}_1$ is discrete.

Note also that for three objects $\mathbf{V}_1, \mathbf{V}_2, \mathbf{V}_3 \in \mathcal{T}op$ we have natural maps

$$(\mathbf{V}_1 \overset{!}{\otimes} \mathbf{V}_2) \overset{\rightarrow}{\otimes} \mathbf{V}_3 \to \mathbf{V}_1 \overset{!}{\otimes} (\mathbf{V}_2 \overset{\rightarrow}{\otimes} \mathbf{V}_3) \quad \text{and} \quad \mathbf{V}_1 \overset{\rightarrow}{\otimes} (\mathbf{V}_2 \overset{!}{\otimes} \mathbf{V}_3) \to (\mathbf{V}_1 \overset{\rightarrow}{\otimes} \mathbf{V}_2) \overset{!}{\otimes} \mathbf{V}_3$$

and hence the map

$$(\mathbf{V}_1 \overset{!}{\otimes} \mathbf{V}_2) \overset{*}{\otimes} \mathbf{V}_3 \to \mathbf{V}_1 \overset{!}{\otimes} (\mathbf{V}_2 \overset{*}{\otimes} \mathbf{V}_3). \tag{19.1}$$

By an action of a topological vector space $\mathbf{V}$ from a discrete vector space W_1 to a discrete vector space W_2 we will mean a map

$$\mathbf{V} \overset{\rightarrow}{\otimes} W_1 \simeq \mathbf{V} \overset{*}{\otimes} W_1 \to W_2.$$

The latter amounts to a compatible system of maps $V' \otimes W_1' \to W_2$, defined for every finite-dimensional subspace $W_1' \subset W_1$ for some sufficiently large discrete quotient V' of $\mathbf{V}$.

By definition, a topological associative algebra is an object $\mathbf{A} \in \mathcal{T}op$ endowed with an associative algebra structure with respect to the $\overset{\rightarrow}{\otimes}$ product. By construction, any such $\mathbf{A}$ can be represented as $\varprojlim_{\mathbf{I}} \mathbf{A}/\mathbf{I}$, where $\mathbf{I} \subset \mathbf{A}$ are open left ideals. A discrete module over a topological associative algebra $\mathbf{A}$ is a vector space V endowed with an associative action map $\mathbf{A} \overset{\rightarrow}{\otimes} V \to V$; we shall denote the category of discrete $\mathbf{A}$-modules by $\mathbf{A}$-mod.

A topological associative algebra is called commutative if the operation $\mathbf{A} \overset{\rightarrow}{\otimes} \mathbf{A} \to \mathbf{A}$ factors through $\mathbf{A} \overset{!}{\otimes} \mathbf{A} \to \mathbf{A}$ and the latter map is commutative (in the sense of the commutativity constraint for the $\overset{!}{\otimes}$ product). In this case $\mathbf{A}$ can be represented as $\varprojlim_{i} A_i$, where A_i are discrete commutative quotients of $\mathbf{A}$.

For a commutative associative topological algebra, by a topological $\mathbf{A}$-module we shall mean a topological vector space $\mathbf{V}$, endowed with an associative map $\mathbf{A} \overset{!}{\otimes} \mathbf{V} \to \mathbf{V}$ such that $\mathbf{V}$ is separated and complete in the topology defined by open $\mathbf{A}$-submodules. Any such $\mathbf{V}$ can be represented as $\varprojlim V^i$, with V_i being discrete $\mathbf{A}$-modules, on each of which $\mathbf{A}$ acts through a discrete quotient. If $f : \mathbf{A} \to \mathbf{B}$ is a homomorphism, we define $f^*(\mathbf{V})$ as $\varprojlim \mathbf{B} \underset{\mathbf{A}}{\otimes} V^i$.

Note that if we regard $\mathbf{A}$ as an associative topological algebra, a discrete $\mathbf{A}$-module is a topological $\mathbf{A}$-module in the above sense if and only if $\mathbf{A}$ acts on it through some discrete quotient.

A topological Lie algebra $\mathfrak{g}$ is a Lie algebra in the sense of the $\overset{*}{\otimes}$ structure. A discrete module over such $\mathfrak{g}$ is a vector space V endowed with a map $\mathfrak{g} \overset{*}{\otimes} V \to V$, which is compatible with the bracket on $\mathfrak{g}$ in a natural way.

Let $\mathbf{A}$ be a commutative associative topological algebra. A Lie algebroid over $\mathbf{A}$ is a topological Lie algebra $\mathfrak{g}$ endowed with a topological $\mathbf{A}$-module structure $\mathbf{A} \overset{!}{\otimes} \mathfrak{g} \to \mathfrak{g}$ and a Lie algebra action map $\mathfrak{g} \overset{*}{\otimes} \mathbf{A} \to \mathbf{A}$, which satisfy the usual compatibility conditions via (19.1).

19.2

Here we shall recall some notions related to infinite-dimensional vector bundles and ind-schemes, borrowed from [BD1] and [Dr2].

By an ind-scheme we will understand an ind-object in the category of schemes, which can be represented as $\mathcal{Y} := \text{``}\underset{\underset{i \in I}{\longrightarrow}}{\lim}\text{''}\mathcal{Y}_i$, where the transition maps $f_{i,j} : \mathcal{Y}_i \to \mathcal{Y}_j$ are closed embeddings. We will always assume that the indexing set I is countable.

A closed subscheme Z of $\mathcal{Y}$ is called reasonable if for every i, the ideal of the subscheme $Z \cap \mathcal{Y}_i$ of $\mathcal{Y}_i$ is locally finitely generated. The ind-scheme $\mathcal{Y}$ is called reasonable if it can be represented as an inductive limit of its reasonable subschemes (or, in other words, one can choose a presentation such that the ideal of $\mathcal{Y}_i$ in $\mathcal{Y}_j$ is locally finitely generated).

We shall say that $\mathcal{Y}$ is ind-affine if all the schemes $\mathcal{Y}_i$ are affine. In this case, if we denote by A_i the algebra of functions of $\mathcal{Y}_i$, we will write $\mathcal{Y} = \text{Spec}(\mathbf{A})$, where $\mathbf{A} = \underset{\longleftarrow}{\lim}\, A_i$ and $\mathbf{A} = \mathcal{O}_{\mathcal{Y}}$.

Assume that $\mathcal{Y} = G$ is ind-affine and is endowed with a structure of group ind-scheme. This amounts to a coassociative counital map $\mathcal{O}_G \to \mathcal{O}_G \overset{!}{\otimes} \mathcal{O}_G$. By definition, an action of G on a topological vector space $\mathbf{V}$ is a map

$$\mathbf{V} \to \mathcal{O}_G \overset{!}{\otimes} \mathbf{V},$$

such that the two morphisms

$$\mathbf{V} \rightrightarrows \mathcal{O}_G \overset{!}{\otimes} \mathcal{O}_G \overset{!}{\otimes} \mathbf{V}$$

coincide.

If $\mathbf{V}$ is an associative or Lie topological algebra, we define in an evident way what it means for an action to be compatible with the operation of product on $\mathbf{V}$.

Assume now that G is a group scheme $H = \text{Spec}(\mathcal{O}_H)$.

Lemma 19.3. *Every* $\mathbf{V}$*, acted on by* H*, can be written as* $\underset{\longleftarrow}{\lim}\, V_i$*, where* $V_i \in \mathcal{R}ep(H)$ *are quotients of* $\mathbf{V}$.

Proof. Let V be some discrete quotient of $\mathbf{V}$. We must show that we can find an H-stable quotient V' such that $\mathbf{V} \twoheadrightarrow V' \twoheadrightarrow V$. Consider the map

$$\mathbf{V} \to \mathcal{O}_H \overset{!}{\otimes} \mathbf{V} \to \mathcal{O}_H \otimes V,$$

Let $\mathbf{V}'$ be the kernel of this map; this is an open subspace in $\mathbf{V}$. The associativity of the action implies that $\mathbf{V}'$ is H-stable. Hence, $\mathbf{V}/\mathbf{V}'$ satisfies our requirements. □

Let $\mathcal{Y}$ be an ind-scheme. A topological $*$-sheaf on $\mathcal{Y}$ is a rule that assigns to a commutative algebra R and an R-point y of $\mathcal{Y}$ a topological R-module $\mathcal{F}_y$, and for a morphism of algebras $f : R \to R'$ an isomorphism $\mathcal{F}_y \simeq f^*(\mathcal{F}_{y'})$, where y' is the induced R'-point of $\mathcal{Y}$, compatible with two-fold compositions. Morphisms between topological $*$-sheaves are defined in an evident manner and we will denote the resulting category by $\mathrm{QCoh}_{\mathcal{Y}}^{\mathrm{top},*}$. The cotangent sheaf $\Omega^1(\mathcal{Y})$ is an example of an object of $\mathrm{QCoh}_{\mathcal{Y}}^{\mathrm{top},*}$.

We let $\mathrm{Tate}_{\mathcal{Y}}$ denote the full subcategory of $\mathrm{QCoh}_{\mathcal{Y}}^{\mathrm{top},*}$ formed by Tate vector bundles (i.e., those, for which each $\mathcal{F}_y$ is an R-module of Tate type); see [Dr2, Section 6.3.2]. The following basic result was established in [Dr2, Theorem 6.2].

Theorem 19.4. *Let $\mathcal{Y}_1 \to \mathcal{Y}_2$ be a formally smooth morphism between ind-schemes with $\mathcal{Y}_1$ being reasonable. Then the topological $*$-sheaf of relative differentials $\Omega^1(\mathcal{Y}_1/\mathcal{Y}_2)$ is a Tate vector bundle on $\mathcal{Y}_1$.*

Assume now that $\mathcal{Y}$ is affine and isomorphic to $\mathrm{Spec}(\mathbf{A})$ for a commutative associative topological algebra $\mathbf{A}$. In this case, the category $\mathrm{QCoh}_{\mathcal{Y}}^{\mathrm{top},*}$ is tautologically equivalent to that of topological $\mathbf{A}$-modules. We have the notion of Lie algebroid over $\mathcal{Y}$ (which is the same as a topological Lie algebroid over $\mathbf{A}$).

Now let $\mathcal{G}$ be an ind-groupoid over an ind-affine ind-scheme $\mathcal{Y}$, such that both (or, equivalently, one of the) projections $l, r : \mathcal{G} \rightrightarrows \mathcal{Y}$ is formally smooth. Then by the above theorem, the normal to $\mathcal{Y}$ in $\mathcal{G}$, denoted $N_{\mathcal{Y}/\mathcal{G}}$, which is by definition the dual of the restriction to $\mathcal{Y}$ of $\Omega^1(\mathcal{G}/\mathcal{Y})$ with respect to either of the projections, is a Tate vector bundle. The standard construction endows it with a structure of Lie algebroid.

19.5 A class of categories

Let $\mathcal{C}$ be an abelian category, and let $\mathrm{Ind}(\mathcal{C})$ denote its ind-completion. We will assume that $\mathcal{C}$ is closed under inductive limits, i.e., that the tautological embedding $\mathcal{C} \to \mathrm{Ind}(\mathcal{C})$ admits a right adjoint $\mathrm{limInd} : \mathrm{Ind}(\mathcal{C}) \to \mathcal{C}$, and that the latter functor is exact. In particular, it makes sense to tensor objects of $\mathcal{C}$ by vector spaces.

We shall say that an object $X \in \mathcal{C}$ is finitely generated (or compact) if the functor $\mathrm{Hom}(X, \cdot) : \mathcal{C} \to \mathrm{Vect}$ commutes with direct sums. Let us denote by $\mathcal{C}^c$ the full subcategory of $\mathcal{C}$ formed by compact objects. We will assume that $\mathcal{C}^c$ is equivalent to a small category (i.e., that isomorphism classes of compact objects in $\mathcal{C}$ form a set).

We shall say that $\mathcal{C}$ satisfies (*) if every object of $\mathcal{C}$ is isomorphic to the inductive limit of its compact subobjects.

Lemma 19.6. *Assume that $\mathcal{C}$ satisfies (*), and let G be a left exact contravariant functor $\mathcal{C}^c \to \mathrm{Vect}$. The following conditions are equivalent:*

(1) *G is representable by $X \in \mathcal{C}$.*
(2) *For an inductive system $\{X_i\} \in \mathcal{C}^c$, whenever $X := \varinjlim X_i$ belongs to $\mathcal{C}^c$, the natural map*

$$\mathsf{G}(X) \to \varprojlim \mathsf{G}(X_i)$$

is an isomorphism.

(3) G *extends to a functor* $\mathcal{C} \to \text{Vect}$ *such that for any inductive system* $\{X_i\} \in \mathcal{C}$, *the map*

$$\mathsf{G}\left(\varinjlim X_i\right) \to \varprojlim \mathsf{G}(X_i)$$

is an isomorphism.

We shall say that $\mathcal{C}$ satisfies (**) if there exists an exact and faithful covariant functor $\mathsf{F} : \mathcal{C} \to \text{Vect}$, which commutes with inductive limits.

The following is standard.

Lemma 19.7. *Assume that* $\mathcal{C}$ *satisfies* (*) *and* (**). *Then we have the following:*

(1) F *is representable by some* "$\varprojlim_i$"$X_i \in \text{Pro}(\mathcal{C}^c)$.

(2) *Assume that* F *has the following additional property. Whenever a system of maps* $\alpha_k : X \to Y_k$ *is such that for any nonzero subobject* $X' \subset X$ *not all maps* $\alpha_k|_{X'}$ *are zero, then the map*

$$\mathsf{F}(X) \to \prod_k \mathsf{F}(Y_k)$$

is injective.

Then the projective system $\{X_i\}$ *as above can be chosen so that all the transition maps* $X_{i'} \to X_i$ *are surjective.*

(3) *Under the assumption of (2), the functor* F *gives rise an to an equivalence* $\mathcal{C} \to$ $\mathbf{A}$-mod, *where* $\mathbf{A}$ *is the topological associative algebra* "$\varprojlim_i$"$\mathsf{F}(X_i) \simeq \text{End}(\mathsf{F})$.

19.8

If A is an associative algebra, we will denote by A-mod$\otimes\mathcal{C}$ the category whose objects are objects of $\mathcal{C}$, endowed with an action of A by endomorphisms, and morphisms being $\mathcal{C}$-morphisms, compatible with A-actions. This is evidently an abelian category.

If M is a left A-module and $X \in \mathcal{C}$, we produce an example of an object of A-mod $\otimes\, \mathcal{C}$ by taking $M \otimes X$.

Let M be a right A-module. We have a naturally defined right exact functor

$$A\text{-mod} \otimes \mathcal{C} \to \mathcal{C} : X \mapsto M \underset{A}{\otimes} X.$$

Lemma 19.9. *If* M *is a flat* (*respectively, faithfully-flat* A*-algebra*), *then the above functor is exact* (*respectively, exact and faithful*).

For the proof see [Ga1, Lemma 4 and Proposition 5].[9]

We will say that $X \in A$-mod $\otimes\, \mathcal{C}$ is A-flat if the functor $M \mapsto M \underset{A}{\otimes} X$: A^{op}-mod $\to \mathcal{C}$ is exact. The functor of tensor product can be derived in either (or both) arguments and we obtain a functor

[9] Whereas the first of the assertions of the lemma is obvious from Lazard's lemma, the second is less so, and it was pointed out to us by Drinfeld.

$$D^-(A^{\mathrm{op}}\text{-mod}) \times D^-(A\text{-mod} \otimes \mathcal{C}) \to D^-(\mathcal{C}).$$

If M is a left A-module and $X \in A\text{-mod} \otimes \mathcal{C}$, we define a contravariant functor on $\mathcal{C}$ by

$$Y \mapsto \mathrm{Hom}_{\mathcal{C}\otimes A\text{-mod}}(Y \otimes M, X).$$

This functor is representable by an object that we will denote by $\mathrm{Hom}_A(M, X)$. If M is finitely presented as an A-module, the functor $X \mapsto \mathrm{Hom}_A(M, X)$ commutes with inductive limits.

Let $\phi : A \to B$ be a homomorphism of algebras. We have a natural forgetful functor $\phi_* : B\text{-mod} \otimes \mathcal{C} \to A\text{-mod} \otimes \mathcal{C}$, and its left adjoint ϕ^*, given by tensor product with B, viewed as a right A-module. The right adjoint to ϕ_*, denoted $\phi^!$, is given by $X \mapsto \mathrm{Hom}_A(B, X)$.

19.10 Objects parameterized by a scheme

Assume now that A is commutative and set $S = \mathrm{Spec}(A)$. In this case we will use the notation $\mathrm{QCoh}_S \otimes \mathcal{C}$ for $A\text{-mod} \otimes \mathcal{C}$. We will think of objects of $\mathrm{QCoh}_S \otimes \mathcal{C}$ as families of objects of $\mathcal{C}$ over S.

For a morphism of affine schemes $f : S_1 \to S_2$ we have the direct and inverse image functors $f_*, f^* : \mathrm{QCoh}_{S_1} \otimes \mathcal{C} \rightleftarrows \mathrm{QCoh}_{S_2} \otimes \mathcal{C}$, with f^* being exact (respectively, exact and faithful) if f is, by Lemma 19.9.

The usual descent argument shows the following.

Lemma 19.11. *Let $S' \to S$ be a faithfully flat map. Then the category $\mathrm{QCoh}_S \otimes \mathcal{C}$ is equivalent to the category of descent data on $\mathrm{QCoh}_{S'} \otimes \mathcal{C}$ with respect to $S' \underset{S}{\times} S' \rightrightarrows S'$.*

This allows us to define the category $\mathrm{QCoh}_S \otimes \mathcal{C}$ for any separated scheme S. Namely, let S' be an affine scheme covering S. We introduce $\mathrm{QCoh}_S \otimes \mathcal{C}$ as the category of descent data on $\mathrm{QCoh}_{S'} \otimes \mathcal{C}$ with respect to $S' \underset{S}{\times} S' \rightrightarrows S'$. Lemma 19.11 above ensures that $\mathrm{QCoh}_S \otimes \mathcal{C}$ is well defined, i.e., is independent of the choice of S' up to a unique equivalence. (In fact, the same definition extends more generally to stacks algebraic in the faithfully-flat topology, for which the diagonal map is affine.) For a morphism of schemes $f : S_1 \to S_2$ we have the evidently defined direct and inverse image functors. If f is a closed embedding and the ideal of S_1 in S_2 is locally finitely generated, then we also have the functor $f^! : \mathrm{QCoh}_{S_2} \otimes \mathcal{C} \to \mathrm{QCoh}_{S_1} \otimes \mathcal{C}$, right adjoint to f_*.

If S_1 is a closed subscheme of S_2, we say that an object $X \in \mathrm{QCoh}_{S_2} \otimes \mathcal{C}$ is set-theoretically supported on S_1, if X can be represented as an inductive limit of its subobjects, each of which is the direct image of an object in some $\mathrm{QCoh}_{S_1'} \otimes \mathcal{C}$, where S_1' is a nilpotent thickening of S_1 inside S_2.

Suppose now that S is of finite type over $\mathbb{C}$. We will denote by $\mathfrak{D}(S)$-mod the category of right D-modules on S. We define the category $\mathfrak{D}(S)\text{-mod} \otimes \mathcal{C}$ as follows:

First, we assume that S is affine and smooth. Then $\mathfrak{D}(S)\text{-mod} \otimes \mathcal{C}$ is by definition the category $\Gamma(S, \mathfrak{D}(S))^{\mathrm{op}}\text{-mod} \otimes \mathcal{C}$.

If $S_1 \to S_2$ is a closed embedding of affine smooth schemes, we have an analogue of Kashiwara's theorem, saying that $\mathfrak{D}(S_1)\text{-mod} \otimes \mathcal{C}$ is equivalent to the subcategory of $\mathfrak{D}(S_2)\text{-mod} \otimes \mathcal{C}$, consisting of objects set-theoretically supported on S_1, when considered as objects of $\text{QCoh}_{S_2} \otimes \mathcal{C}$.

This allows us to define $\mathfrak{D}(S)\text{-mod} \otimes \mathcal{C}$ for any affine scheme of finite type, by embedding it into a smooth scheme. Finally, for an arbitrary S, we define $\mathfrak{D}(S)\text{-mod} \otimes \mathcal{C}$ using a cover by affine schemes, as above.

19.12

In this subsection we will assume that $\mathcal{C}$ satisfies (*). Let $\mathbf{V}$ be a topological vector space, and $X, Y \in \mathcal{C}$. An action $\mathbf{V} \times X \to Y$ is a map

$$\mathbf{V} \otimes X \to Y,$$

satisfying the following continuity condition: For every compact subobject $X' \subset X$, the induced map $\mathbf{V} \otimes X' \to Y$ factors through $V \otimes X' \to Y$, where V is a discrete quotient of $\mathbf{V}$.

If $X' \to X$ (respectively, $Y \to Y'$, $\mathbf{V}' \to \mathbf{V}$) is a map, and we have an action $\mathbf{V} \times X \to Y$, we produce an action $\mathbf{V} \times X' \to Y$ (respectively, $\mathbf{V} \times X \to Y'$, $\mathbf{V}' \times X \to Y$).

Note that if $\mathbf{V}$ is pro-finite dimensional, with the dual $\mathbf{V}^* \in \text{Vect}$, an action $\mathbf{V} \times X \to Y$ is the same as a map $X \to \mathbf{V}^* \otimes Y$.

Lemma-Construction 19.13. *Let $\mathbf{V}_2 \times X \to Y$ and $\mathbf{V}_1 \times Y \to Z$ be actions. Then we have an action*

$$(\mathbf{V}_1 \overset{\to}{\otimes} \mathbf{V}_2) \times X \to Z.$$

Proof. The construction immediately reduces to the case when $\mathbf{V}_2 = V_2$ is discrete, X is compact, and we have an action map $V_2 \otimes X \to Y$.

Then for every finite-dimensional subspace $V_2^k \subset V_2$ we can find a compact subobject $Y^k \subset Y$, such that $V_2^k \otimes X \to Y$ maps to Y^k and the action $\mathbf{V}_1 \otimes Y^k \to Z$ factors through a discrete quotient V_1^k of $\mathbf{V}_1$. Then

$$\bigcup_k \ker(\mathbf{V}_1 \to \mathbf{V}_1^k) \otimes V_2^k \subset \mathbf{V}_1 \overset{\to}{\otimes} V_2$$

is an open neighborhood of 0, and we have an action map

$$\left(\mathbf{V}_1 \overset{\to}{\otimes} V_2 / \bigcup_k \ker(\mathbf{V}_1 \to \mathbf{V}_1^k) \otimes V_2^k\right) \otimes X \simeq \varinjlim (V_1^k \otimes V_2^k) \otimes X \to Z. \qquad \square$$

We shall say that $\mathbf{V}$ acts on X if we are given a map $\mathbf{V} \times X \to X$. Objects of $\mathcal{C}$, acted on by $\mathbf{V}$ naturally form a category, which is abelian.

Let $\mathbf{A}$ be an associative topological algebra. We shall say that an object $X \in \mathcal{C}$ is acted on by $\mathbf{A}$ if we are given an action map $\mathbf{A} \times X \to X$ such that the two resulting action maps $(\mathbf{A} \overset{\to}{\otimes} \mathbf{A}) \times X \rightrightarrows X$ coincide. Objects of $\mathcal{C}$ acted on by $\mathbf{A}$ form a category, denoted $\mathbf{A}\text{-mod} \otimes \mathcal{C}$.

19.14 Objects of a category parameterized by an ind-scheme

In this subsection we retain the assumption that $\mathcal{C}$ satisfies (*). Let $\mathcal{Y}$ be an ind-scheme, $\mathcal{Y} = \bigcup_i \mathcal{Y}_i$. We introduce the category $\mathrm{QCoh}^*_{\mathcal{Y}} \otimes \mathcal{C}$ to have as objects collections $\{X_i \in \mathrm{QCoh}^*_{\mathcal{Y}_i} \otimes \mathcal{C}\}$ together with a compatible system of isomorphisms $f^*_{i,j}(X_j) \simeq X_i$, where $f_{i,j}$ is the map $\mathcal{Y}_i \to \mathcal{Y}_j$. Morphisms in the category are evident.

It is easy to see that this category is independent of the presentation of $\mathcal{Y}$ as an inductive limit. However, $\mathrm{QCoh}^*_{\mathcal{Y}} \otimes \mathcal{C}$ is, in general, not abelian.

Given an object of $\mathbf{X} \in \mathrm{QCoh}^*_{\mathcal{Y}} \otimes \mathcal{C}$ and a scheme S mapping to $\mathcal{Y}$, we have a well-defined object $\mathbf{X}|_S \in \mathrm{QCoh}_S \otimes \mathcal{C}$.

Assume now that $\mathcal{Y}$ is strict and reasonable. That is, the system $\mathcal{Y}_i$ can be chosen so that the maps $f_{i,j}$ are closed embeddings, and the ideal of $\mathcal{Y}_i$ inside $\mathcal{Y}_j$ is locally finitely generated.

We introduce the category $\mathrm{QCoh}^!_{\mathcal{Y}} \otimes \mathcal{C}$ as follows. Its objects are collections $\mathbf{X} := \{X_i \in \mathrm{QCoh}^*_{\mathcal{Y}_i} \otimes \mathcal{C}\}$ together with a compatible system of isomorphisms $X_i \simeq f^!_{i,j}(X_j)$. The morphisms in this category are evident.

Lemma 19.15. $\mathrm{QCoh}^!_{\mathcal{Y}} \otimes \mathcal{C}$ *is an abelian category.*

Proof. If $\alpha = \{\alpha_i : X_i \to X'_i\}$ is a morphism in $\mathrm{QCoh}^!_{\mathcal{Y}} \otimes \mathcal{C}$, its kernel is given by the system $\{\ker(\alpha_i)\}$. It is easy to see that the cokernel and image of this morphism are given by the systems that assign to each i,

$$\varinjlim_{j \geq i} f^!_{i,j}\left(\mathrm{coker}(\alpha_j)\right), \qquad \varinjlim_{j \geq i} f^!_{i,j}\left(\mathrm{Im}(\alpha_j)\right),$$

respectively. The fact that the axioms of an abelian category are satisfied is shown in the same way as in the case of $\mathcal{C} = \mathrm{Vect}$. □

Now let $\mathbf{A}$ be a commutative topological algebra. Then $\mathbf{A}$ can be represented as $\varprojlim A_i$, where the A_i are discrete commutative algebras. Assume, moreover, that we can find such a presentation that the ideal of A_i in each A_j, $j \geq i$ is finitely generated. Then $\mathcal{Y} := \varinjlim \mathrm{Spec}(A_i)$ is reasonable.

Lemma 19.16. *Under the above circumstances, the categories* $\mathrm{QCoh}^!_{\mathcal{Y}} \otimes \mathcal{C}$ *and* $\mathbf{A}$-mod $\otimes\, \mathcal{C}$ *are equivalent.*

Proof. The functor $\mathrm{QCoh}^!_{\mathcal{Y}} \otimes \mathcal{C} \to \mathbf{A}\text{-mod} \otimes \mathcal{C}$ is evident. Its right adjoint is defined as follows: given an object $X \in \mathbf{A}\text{-mod} \otimes \mathcal{C}$, represented as $\bigcup_i X_i$ with $X_i \in A_i\text{-mod} \otimes \mathcal{C}$, we define an object $\{X'_i\}$ in $\mathrm{QCoh}^!_{\mathcal{Y}} \otimes \mathcal{C}$ by setting

$$X'_i = \varinjlim_{j \geq i} f^!_{i,j}(X_j).$$

The fact that the adjunction morphisms are isomorphisms is shown as in the case $\mathcal{C} = \mathrm{Vect}$. □

Now let $\mathbf{X} = \{X_i\}$ be an object of $\mathrm{QCoh}^*_{\mathcal{Y}} \otimes \mathcal{C}$ such that each X_i is $\mathcal{Y}_i$-flat. Let M be an object in $\mathrm{QCoh}^!_{\mathcal{Y}} \otimes \mathbf{A}$-mod, where $\mathbf{A}$ is some topological algebra. We then have a well-defined tensor product

$$\mathbf{X} \underset{\mathcal{O}_{\mathcal{Y}}}{\otimes} M \in \mathrm{QCoh}^!_{\mathcal{Y}} \otimes (\mathbf{A}\text{-mod} \otimes \mathcal{C}).$$

The corresponding system assigns to every $\mathcal{Y}_i$ the object

$$\mathbf{X}|_{\mathcal{Y}_i} \underset{\mathcal{O}_{\mathcal{Y}_i}}{\otimes} M_i \in \mathrm{QCoh}_{\mathcal{Y}_i} \otimes (\mathbf{A}\text{-mod} \otimes \mathcal{C}).$$

Finally, let $\mathcal{Y}$ be a strict ind-scheme of ind-finite type. Proceeding as above, one defines the category $\mathfrak{D}(\mathcal{Y})^!\text{-mod}\otimes\mathcal{C}$ as the category of systems $\{X_i\} \in \mathfrak{D}(\mathcal{Y}_i)^!\text{-mod}\otimes\mathcal{C}$ with isomorphisms $X_i \simeq f^!_{i,j}(X_j)$.

If $\mathcal{Y}$ is formally smooth, we can also introduce the DG-category of $\Omega^\bullet_{\mathcal{Y}}$-modules with coefficients in $\mathcal{C}$, and we will have an equivalence between the corresponding derived category of $\Omega^\bullet_{\mathcal{Y}}\text{-mod} \otimes \mathcal{C}$ and the derived category of $\mathfrak{D}(\mathcal{Y})^!\text{-mod} \otimes \mathcal{C}$.

19.17 BRST complex

If $\mathfrak{g}$ is a topological Lie algebra, an action of $\mathfrak{g}$ on $X \in \mathcal{C}$ is an action map $\mathfrak{g} \times X \to X$ such that the difference of the two iterations

$$(\mathfrak{g} \overset{*}{\otimes} \mathfrak{g}) \times X \to (\mathfrak{g} \overset{\to}{\otimes} \mathfrak{g}) \times X \to X$$

equals the action induced by the Lie bracket $\mathfrak{g} \overset{*}{\otimes} \mathfrak{g} \to \mathfrak{g}$.

Assume now that $\mathfrak{g} \simeq \mathsf{k}$ is pro-finite dimensional. Then its action on X is the same as a coaction of the Lie coalgebra $\mathsf{k}^* \in \mathrm{Vect}$ on X, i.e., a map $a : X \to \mathsf{k}^* \otimes X$, satisfying the suitable axioms. In this case we can form a complex of objects of $\mathcal{C}$, called the standard complex, $\mathfrak{C}(\mathsf{k}, X)$:

As a graded object of $\mathcal{C}$, it is isomorphic to $\mathfrak{C}(\mathsf{k}, \mathcal{M}) := X \otimes \Lambda^\bullet(\mathsf{k}^*)$. Let us denote by i (respectively, i^*) the action of $\mathsf{k}[1]$ (respectively, $\mathsf{k}^*[-1]$) on $\mathfrak{C}(\mathsf{k}, X)$ by the "annihilation" (respectively, "creation" operators), and by Lie the diagonal action of k. Then the differential d on $\mathfrak{C}(\mathsf{k}, X)$ is uniquely characterized by the property that $[d, i] = \mathrm{Lie}$. We automatically obtain that

- $d^2 = 0$,
- The map $i^* : \Lambda^\bullet(\mathsf{k}^*) \otimes \mathfrak{C}(\mathsf{k}, X) \to \mathfrak{C}(\mathsf{k}, X)$ is a map of complexes, where $\Lambda^\bullet(\mathsf{k}^*)$ is endowed with a differential coming from the Lie cobracket.

If $X^\bullet$ is a complex of objects of $\mathcal{C}$, acted on by k, we will denote by $\mathfrak{C}(\mathsf{k}, X^\bullet)$ the complex associated to the corresponding bicomplex. It is clear that if $X^\bullet$ is bounded from below and acyclic, then $\mathfrak{C}(\mathsf{k}, X^\bullet)$ is acyclic as well. However, this would not be true if we dropped the boundedness from below assumption.

The above setup can be generalized as follows. Now let $\mathfrak{g}$ be a topological Lie algebra, which is of Tate type as a topological vector space. Let $\mathsf{Cliff}(\mathfrak{g}, \mathfrak{g}^*)$ be the (topological) Clifford algebra, constructed on $\mathfrak{g}$ and $\mathfrak{g}^*$; it is naturally graded, where the "creation" operators (i.e., elements of $\mathfrak{g}^*$) have degree 1, and the annihilation operators (i.e., elements of $\mathfrak{g}$) have degree -1. Let $\mathsf{Spin}(\mathfrak{g})$ be some fixed irreducible representation of $\mathsf{Cliff}(\mathfrak{g})$, equipped with a grading. (Of course, up to a grading shift and a noncanonical isomorphism, $\mathsf{Spin}(\mathfrak{g})$ is unique.)

Recall that the canonical (i.e., Tate) central extension $\mathfrak{g}_{\mathrm{can}}$ of $\mathfrak{g}$ is characterized by the property that the adjoint action of $\mathfrak{g}$ on $\mathsf{Cliff}(\mathfrak{g}, \mathfrak{g}^*)$ is inner via a homomorphism $\mathfrak{g}_{\mathrm{can}} \to \mathsf{Cliff}(\mathfrak{g}, \mathfrak{g}^*)$. We will denote by $\mathfrak{g}_{-\mathrm{can}}$ the Baer negative central extension.

Let $\mathcal{C}$ be as above, and let $X \in \mathcal{C}$ be acted on by $\mathfrak{g}_{-\mathrm{can}}$. Consider the graded object of $\mathcal{C}$ given by

$$\mathfrak{C}^{\frac{\infty}{2}}(\mathfrak{g}, X) := X \otimes \mathsf{Spin}(\mathfrak{g}).$$

As in the case in which $\mathcal{C} = \mathrm{Vect}$, one shows that $\mathfrak{C}^{\frac{\infty}{2}}(\mathfrak{g}, X)$ acquires a canonical differential d, characterized by the property that $[d, i] = \mathrm{Lie}$, where i denotes the action of $\mathfrak{g}$ on $\mathfrak{C}^{\frac{\infty}{2}}(\mathfrak{g}, X)$ via $\mathsf{Spin}(\mathfrak{g})$ by creation operators, and Lie is the diagonal action of $\mathfrak{g}$ on $X \otimes \mathsf{Spin}(\mathfrak{g})$. We have

- $d^2 = 0$
- The action i^* of $\mathfrak{g}^*$ is compatible with the differential $\mathfrak{g}^* \to \mathfrak{g}^* \overset{!}{\otimes} \mathfrak{g}^*$ given by the bracket.

If $X^\bullet$ is a complex of objects of $\mathcal{C}$, acted on by $\mathfrak{g}_{-\mathrm{can}}$, we will denote by $\mathfrak{C}^{\frac{\infty}{2}}(\mathfrak{g}, X^\bullet)$ the complex associated to the corresponding bicomplex.

Lemma 19.18. *Assume that $X^\bullet$ is bounded from below and is acyclic. Then $\mathfrak{C}^{\frac{\infty}{2}}(\mathfrak{g}, X^\bullet)$ is also acyclic.*

Proof. Let us choose a lattice $\mathsf{k} \subset \mathfrak{g}$; we can then realize $\mathsf{Spin}(\mathfrak{g})$ as $\mathsf{Spin}(\mathfrak{g}, \mathsf{k})$–the module generated by an element, annihilated by both $\mathsf{k} \subset \mathfrak{g} \subset \Lambda^\bullet(\mathfrak{g})$ and $(\mathfrak{g}/\mathsf{k})^* \subset \mathfrak{g}^* \subset \Lambda^\bullet(\mathfrak{g}^*)$.

In this case the complex $\mathfrak{C}^{\frac{\infty}{2}}(\mathfrak{g}, X^\bullet)$ acquires a canonical increasing filtration, indexed by the natural numbers, so that

$$\mathrm{gr}^i\left(\mathfrak{C}^{\frac{\infty}{2}}(\mathfrak{g}, X^\bullet)\right) \simeq \mathfrak{C}(\mathsf{k}, X^\bullet \otimes \Lambda^i(\mathfrak{g}/\mathsf{k}))[i].$$

This readily implies the assertion of the lemma. □

In what follows we will need to consider the following situation. Let $X^\bullet$ be a complex of objects of $\mathcal{C}$, endowed with two actions of $\mathfrak{g}_{-\mathrm{can}}$, denoted a and a', respectively. Then $X^\bullet \otimes \mathsf{Spin}(\mathfrak{g})$ acquires two differentials, d and d'.

Assume that there exists a self-anticommuting action

$$i_h : \mathfrak{g}[1] \times X^\bullet \to X^\bullet,$$

such that $a'(x) - a(x) = [d_X, i_h(x)]$, $[a'(x), i_h(y)] = i_h([x, y])$, $[a(x), i_h(y)] = 0$, where d_X is the differential on $X^\bullet$.

Lemma 19.19. *Under the above circumstances, there exists a graded automorphism of the complex $X^\bullet \otimes \mathsf{Spin}(\mathfrak{g})$ that intertwines d and d'.*

Proof. Let $\Lambda^i(\mathfrak{g})$ and $\Lambda^i(\mathfrak{g}^*)$ denote the !-completed exterior powers of $\mathfrak{g}$ and $\mathfrak{g}^*$, respectively.

For a natural number i consider the canonical element $\mathrm{id}_i \in \Lambda^i(\mathfrak{g}) \overset{!}{\otimes} \Lambda^i(\mathfrak{g}^*)$. We define the operator

$$T : X^\bullet \otimes \mathsf{Spin}(\mathfrak{g}) \to X^\bullet \otimes \mathsf{Spin}(\mathfrak{g})$$

by $\sum\limits_{i\in\mathbb{N}} (i_h \otimes i^*)(\mathrm{id}_i)$, where i^* and i_h denote the extension of the actions of $\mathfrak{g}^*[-1]$ and $\mathfrak{g}[1]$, respectively, to the exterior powers.

Clearly, T is a grading-preserving isomorphism, and

$$T \circ i \circ T^{-1} = i + i_h.$$

One easily shows that $d' = T^{-1} \circ d \circ T$. □

19.20 DG categories

We will adopt the conventions regarding DG categories from [Dr1]. Let $\mathbf{C}$ be a $\mathbb{C}$-linear DG category, which admits arbitrary direct sums.

For $X^\bullet, Y^\bullet \in \mathbf{C}$ we will denote by $\mathcal{H}om_{\mathbf{C}}(X^\bullet, Y^\bullet)$ the corresponding complex, and by $\mathrm{Hom}_{\mathbf{C}}(X^\bullet, Y^\bullet)$ its 0th cohomology. By definition, the homotopy category $Ho(\mathbf{C})$ has the same objects as $\mathbf{C}$, with the Hom space being $\mathrm{Hom}_{\mathbf{C}}(X^\bullet, Y^\bullet)$

We will assume that $\mathbf{C}$ is strongly pretriangulated, i.e., that it admits cones. In this case $Ho(\mathbf{C})$ is triangulated.

We will assume that $\mathbf{C}$ is equipped with a cohomological functor H to an abelian category $\mathcal{C}'$. We will denote by $D(\mathbf{C})$ the corresponding localized triangulated category, and we will assume that H defines a t-structure on $D(\mathbf{C})$. We will denote by $\mathrm{RHom}_{D(\mathbf{C})}(\cdot,\cdot)$ the resulting functor $D(\mathbf{C})^{\mathrm{op}} \times D(\mathbf{C}) \to D(\mathrm{Vect})$.

We will denote by $D^b(\mathbf{C})$ (respectively, $D^+(\mathbf{C})$, $D^-(\mathbf{C})$,) the subcategory consisting of objects $X^\bullet$ such that $\mathsf{H}(X^\bullet[i]) = 0$ for i away from a bounded interval (respectively, $i \ll 0$, $i \gg 0$.)

In what follows we will also use the following notion: We shall say that an object $X^\bullet \in D(\mathbf{C})$ is quasi-perfect if it belongs to $D^-(\mathbf{C})$, and the functor $Y \mapsto \mathrm{Hom}_{D(\mathbf{C})}(X^\bullet, Y[i])$ commutes with direct sums in the core of $\mathbf{C}$ (i.e., those objects $Y \in \mathbf{C}$ for which $\mathsf{H}(Y[j]) = 0$ for $j \neq 0$).

Lemma 19.21. *Let $X^\bullet \in D(\mathbf{C})$ be quasi-perfect and $Y^\bullet \in D^+(\mathbf{C})$. Let $\mathcal{K}^\bullet$ be a bounded from below complex of vector spaces. Then* $\mathrm{RHom}_{D(\mathbf{C})}(X^\bullet, Y^\bullet \otimes \mathcal{K}^\bullet)$ *is quasi-isomorphic to* $\mathrm{RHom}_{D(\mathbf{C})}(X^\bullet, Y^\bullet) \otimes \mathcal{K}^\bullet$ *in* $D(\mathrm{Vect})$.

The most typical example of this situation is, of course, when $\mathbf{C} = \mathbf{C}(\mathcal{C})$ is the category of complexes of objects of an abelian category $\mathcal{C}$, and H comes from an exact functor $\mathcal{C} \to \mathcal{C}'$. If $\mathcal{C}' \simeq \mathcal{C}$, then $D(\mathbf{C})$ will be denoted $D(\mathcal{C})$; this is the usual derived category of $\mathcal{C}$.

An example of a quasi-perfect object of $D(\mathcal{C})$ is provided by a bounded from above complex consisting of projective finitely generated objects of $\mathcal{C}$.

Let $\mathbf{C}_1, \mathbf{C}_2$ be two DG categories as above, and let $\mathsf{G} : \mathbf{C}_1 \to \mathbf{C}_2$ be a DG functor. We shall say that G is exact if it sends acyclic objects (in the sense of H_1) to acyclic ones (in the sense of H_2).

The following (evident) assertion will be used repeatedly.

Lemma 19.22. *Let $\mathsf{G} : \mathbf{C}_1 \to \mathbf{C}_2$ and $\mathsf{G}' : \mathbf{C}_2 \to \mathbf{C}_1$ be mutually adjoint exact functors. Then the induced functors $\mathsf{G}, \mathsf{G}' : D(\mathbf{C}_1) \rightleftarrows D(\mathbf{C}_2)$ are also mutually adjoint.*

Proof. Let G be the left adjoint of G'. Note first of all that the functors induced by G and G' between the homotopy categories $Ho(\mathbf{C}_1)$ and $Ho(\mathbf{C}_2)$ are evidently mutually adjoint.

Then for $X^\bullet \in D(\mathbf{C}_1)$, $Y^\bullet \in D(\mathbf{C}_2)$,

$$\mathrm{Hom}_{D(\mathbf{C}_1)}(X^\bullet, \mathsf{G}'(Y^\bullet)) = \varinjlim_{X'^\bullet \to X^\bullet} \mathrm{Hom}_{Ho(\mathbf{C}_1)}(X'^\bullet, \mathsf{G}'(Y^\bullet)) \tag{19.2}$$

and

$$\mathrm{Hom}_{D(\mathbf{C}_2)}(\mathsf{G}(X^\bullet), Y^\bullet) = \varinjlim_{Y^\bullet \to Y'^\bullet} \mathrm{Hom}_{Ho(\mathbf{C}_2)}(\mathsf{G}(X^\bullet), Y'^\bullet), \tag{19.3}$$

where in both cases the inductive limits are taken over quasi-isomorphisms, i.e., morphisms in the homotopy category that become isomorphisms in the quotient triangulated category.

By adjunction, we rewrite the expression in (19.3) as

$$\varinjlim_{Y^\bullet \to Y'^\bullet} \mathrm{Hom}_{Ho(\mathbf{C}_2)}(X^\bullet, \mathsf{G}'(Y'^\bullet)),$$

and we map it to (19.2) as follows. For a quasi-isomorphism $Y^\bullet \to Y'^\bullet$ the map $\mathsf{G}(Y^\bullet) \to \mathsf{G}(Y'^\bullet)$ is a quasi-isomorphism as well, and given a map $X^\bullet \to \mathsf{G}'(Y'^\bullet)$, we can find a quasi-isomorphism $X'^\bullet \to X^\bullet$, so that the diagram

$$\begin{array}{ccc} \mathsf{G}'(Y^\bullet) & \longrightarrow & \mathsf{G}'(Y'^\bullet) \\ \uparrow & & \uparrow \\ X'^\bullet & \longrightarrow & X^\bullet \end{array}$$

commutes in $Ho(\mathbf{C}_1)$. The above map $X' \to \mathsf{G}'(Y^\bullet)$ defines an element in (19.2).

One constructs the map from (19.2) to (19.3) in a similar way, and it is straightforward to check that the two are mutually inverse. □

20 Action of a group on a category

20.1 Weak action

Let $\mathcal{C}$ be an abelian category as in Section 19.5, and let H be an affine group scheme. We will say that H acts weakly on $\mathcal{C}$ if we are given a functor

$$\text{act}^* : \mathcal{C} \to \text{QCoh}_H \otimes \mathcal{C},$$

and two functorial isomorphisms related to it. The first isomorphism is between the identity functor on $\mathcal{C}$ and the composition $\mathcal{C} \overset{\text{act}^*}{\to} \text{QCoh}_H \otimes \mathcal{C} \to \mathcal{C}$, where the second arrow corresponds to the restriction to $1 \in H$.

To formulate the second isomorphism, note that from the existing data we obtain a natural functor $\text{act}^*_S : \text{QCoh}_S \otimes \mathcal{C} \to \text{QCoh}_{S\times H} \otimes \mathcal{C}$ for any affine scheme S.

The second isomorphism is between the two functors $\mathcal{C} \to \text{QCoh}_{H\times H} \otimes \mathcal{C}$ that correspond to the two paths of the diagram

$$\begin{array}{ccc} \mathcal{C} & \xrightarrow{\text{act}^*} & \text{QCoh}_H \otimes \mathcal{C} \\ {\scriptstyle \text{act}^*}\downarrow & & \downarrow{\scriptstyle \text{act}^*_H} \\ \text{QCoh}_H \otimes \mathcal{C} & \xrightarrow{\text{mult}^*} & \text{QCoh}^*_{H\times H} \otimes \mathcal{C}, \end{array} \tag{20.1}$$

where mult denotes the multiplication map $H \times H \to H$.

We assume that the above two isomorphisms of functors satisfy the usual compatibility conditions. We will refer to these isomorphisms as the unit and associativity constraint of the action, respectively.

Lemma 20.2. *The functor* act* *is exact and faithful. For* $X \in \mathcal{C}$, *the* $\mathcal{O}_H$*-family* $\text{act}^*(X)$ *is flat.*

Proof. First, the faithfulness of act* is clear, since the fiber at $1 \in H$ provides a left quasi-inverse $\text{QCoh}_H \otimes \mathcal{C} \to \mathcal{C}$.

Let S be a scheme equipped with a map $\phi : S \to H$. Note that we have a self-functor $\text{act}^*_\phi : \text{QCoh}_S \otimes \mathcal{C} \to \text{QCoh}_S \otimes \mathcal{C}$ given by

$$\text{QCoh}_S \otimes \mathcal{C} \overset{\text{act}^*_S}{\to} \text{QCoh}_S \otimes \text{QCoh}_H \otimes \mathcal{C} \overset{(\text{id}_S \times \phi)^*}{\to} \text{QCoh}_S \otimes \mathcal{C}.$$

This is an equivalence of categories and its quasi-inverse is given by $\text{act}^*_{\phi^{-1}}$, where $\phi^{-1} : S \to G$ is obtained from ϕ by applying the inversion on H. Note that act^*_ϕ is $\mathcal{O}_S$-linear.

Let us take $S = H$ and ϕ to be inversion map. Then the composition $\text{act}^*_\phi \circ \text{act}^* : \mathcal{C} \to \text{QCoh}_H \times \mathcal{C}$ is isomorphic to the functor $X \mapsto \mathcal{O}_H \otimes X$, which is evidently exact. Hence, act* is exact as well.

Similarly, to show that $\text{act}^*(X)$ is $\mathcal{O}_H$-flat, it suffices to establish the corresponding fact for $\text{act}^*_\phi \circ \text{act}^*(X)$, which is again evident. □

Here are some typical examples of weak actions:

(1) Let H act on an ind-scheme $\mathcal{Y}$. Then the category $\text{QCoh}^!_{\mathcal{Y}}$ carries a weak H-action.
(2) Let H act on a topological associative algebra $\mathbf{A}$ (see Section 19.2). Then the category $\mathbf{A}$-mod of discrete $\mathbf{A}$-modules carries a weak H-action.

20.3 Weakly equivariant objects

Let us denote by p^* the tautological functor

$$\mathcal{C} \to \mathrm{QCoh}_H \otimes \mathcal{C} : X \mapsto \mathcal{O}_H \otimes X,$$

where $\mathcal{O}_H$ is the algebra of functions on H.

We will say that an object $X \in \mathcal{C}$ is weakly H-equivariant if we are given an isomorphism

$$\mathrm{act}^*(X) \simeq p^*(X), \tag{20.2}$$

which is compatible with the associativity constraint of the H-action on $\mathcal{C}$.

Clearly, weakly H-equivariant objects of $\mathcal{C}$ form an abelian category, which we will denote by $\mathcal{C}^{w,H}$. For example, let us take $\mathcal{C}$ to be Vect—the category of vector spaces with the obvious, i.e., trivial, H-action. Then $\mathcal{C}^{w,H}$ is the category of H-modules, denoted H-mod, or $\mathcal{R}\mathrm{ep}(H)$.

Let X be an object of $\mathcal{C}^{w,H}$, and $V \in \mathcal{R}\mathrm{ep}(H)$. We define a new object $V * X \in \mathcal{C}^{w,H}$ to be $V \otimes X$ as an object of $\mathcal{C}$, but where the isomorphism $\mathrm{act}^*(V \otimes X) \to p^*(V \otimes X)$ is multiplied by the coaction map $V \to \mathcal{O}_H \otimes V$.

We have an obvious forgetful functor $\mathcal{C}^{w,H} \to \mathcal{C}$, and it admits a right adjoint, denoted Av_H^w, given by $X \mapsto p_*(\mathrm{act}^*(X))$. For $X \in \mathcal{C}^{w,H}$,

$$\mathrm{Av}_H^w(X) \simeq \mathcal{O}_H * X.$$

For two objects X_1, X_2 of $\mathcal{C}^{w,H}$ we define a contravariant functor $\underline{\mathrm{Hom}}_{\mathcal{C}}(X_1, X_2)$ of $\mathcal{R}\mathrm{ep}(H)$ by

$$\mathrm{Hom}_{\mathcal{R}\mathrm{ep}(H)}(V, \underline{\mathrm{Hom}}_{\mathcal{C}}(X_1, X_2)) = \mathrm{Hom}_{\mathcal{C}^{w,H}}(V * X_1, X_2).$$

It is easy to see that this functor is representable.

Lemma 20.4. *Let X_1 be finitely generated as an object of $\mathcal{C}$. Then the forgetful functor* $\mathcal{R}\mathrm{ep}(H) \to \mathrm{Vect}$ *maps* $\underline{\mathrm{Hom}}_{\mathcal{C}}(X_1, X_2)$ *to* $\mathrm{Hom}_{\mathcal{C}}(X_1, X_2)$.

Proof. We have the map

$$\begin{aligned}\mathrm{Hom}_{\mathcal{C}}(X_1, X_2) &\simeq \mathrm{Hom}_{\mathcal{C}^{w,H}}(X_1, \mathrm{Av}_H^w(X_2)) \\ &\to \mathrm{Hom}_{\mathcal{C}}(X_1, \mathrm{Av}_H^w(X_2)) \simeq \mathrm{Hom}_{\mathcal{C}}(X_1, \mathcal{O}_H \otimes X_2),\end{aligned}$$

and the latter identifies with $\mathcal{O}_H \otimes \mathrm{Hom}_{\mathcal{C}}(X_1, X_2)$, by the assumption on X_1.

This endows $\mathrm{Hom}_{\mathcal{C}}(X_1, X_2)$ with a structure of H-module. It is easy to see that it satisfies the required adjunction property. □

Note that since

$$\mathrm{Hom}_{\mathcal{C}}(X_1, X_2) \simeq \mathrm{Hom}_{\mathcal{C}^{w,H}}(X_1, \mathcal{O}_H * X_2)$$

and

$$\mathrm{Hom}_{\mathcal{C}^{w,H}}(X_1, X_2) \simeq \mathrm{Hom}_H(\mathbb{C}, \underline{\mathrm{Hom}}_{\mathcal{C}}(X_1, X_2)),$$

we obtain that X_1 is finitely generated as an object of $\mathcal{C}$ if and only if it is so in $\mathcal{C}^{w,H}$.

20.5

Let $\mathbf{C}(\mathcal{C}^{w,H})$ denote the DG category of complexes of objects of $\mathcal{C}^{w,H}$, and let $D(\mathcal{C}^{w,H})$ be the corresponding derived category. By Lemma 19.22, the forgetful functor $D(\mathcal{C}^{w,H}) \to D(\mathcal{C})$ admits a right adjoint given by $X^\bullet \mapsto \mathrm{act}^*(X^\bullet)$.

For $X_1^\bullet, X_2^\bullet \in D(\mathcal{C}^{w,H})$, we define a contravariant cohomological functor $\underline{\mathrm{RHom}}_{D(\mathcal{C})}(X_1^\bullet, X_2^\bullet)$ on $D(\mathcal{R}\mathrm{ep}(H))$ by

$$V^\bullet \mapsto \mathrm{Hom}_{D(\mathcal{C}^{w,H})}(V^\bullet * X_1^\bullet, X_2^\bullet).$$

It is easy to see that this functor is representable.

Lemma 20.6. *Assume that $X_1^\bullet$ is quasi-perfect and $X_2^\bullet \in D^+(\mathcal{C}^{w,H})$. Then the image of $\underline{\mathrm{RHom}}_{D(\mathcal{C})}(X_1^\bullet, X_2^\bullet)$ under the forgetful functor $D(\mathcal{R}\mathrm{ep}(H)) \to D(\mathrm{Vect})$ is quasi-isomorphic to* $\mathrm{RHom}_{D(\mathcal{C})}(X_1^\bullet, X_2^\bullet)$.

The proof repeats that of Lemma 20.4.

20.7 Infinitesimally trivial actions

Let $H^{(i)}$ be the ith infinitesimal neighborhood of 1 in H, so that $H^{(0)} = 1$ and $H^{(1)} = \mathrm{Spec}(\mathbb{C} \oplus \epsilon \cdot \mathfrak{h}^*)$, where $\epsilon^2 = 0$. Note that if H weakly acts on $\mathcal{C}$, the restriction to $H^{(1)}$ yields for every object $X \in \mathcal{C}$ a short exact sequence in $\mathcal{C}$.

$$0 \to \mathfrak{h}^* \otimes X \to X^{(1)} \to X \to 0,$$

where $X^{(1)} := \mathrm{act}^*(X)|_{H^{(1)}}$.

We will say that the action of H on $\mathcal{C}$ is infinitesimally trivial, or of Harish-Chandra type if we are given a functorial isomorphism

$$\mathrm{act}^*(X)|_{H^{(1)}} \simeq p^*(X)|_{H^{(1)}}, \tag{20.3}$$

such that two compatibility condition (see below) are satisfied.

The first condition is that the isomorphism (20.3) respects the identification of the restrictions of both sides to $1 \in H$ with X. (In view of this condition, the data of (20.3) amounts to a functorial splitting $X \to X^{(1)}$.)

To formulate the second condition, consider the map of schemes

$$(h, h_1) \overset{\mathrm{Ad}}{\mapsto} \mathrm{Ad}_h(h_1) : H \times H^{(1)} \to H^{(1)}.$$

From (20.1) and (20.3) we obtain two a priori different identifications

$$\mathrm{Ad}^*(X^{(1)}) \rightrightarrows \mathcal{O}_{H \times H^{(1)}} \otimes X \in \mathrm{QCoh}_{H \times H^{(1)}} \otimes \mathcal{C}.$$

Our condition is that these two identifications coincide.

Here are some typical examples of Harish-Chandra actions:

(1) Let $\mathcal{Y}$ be an ind-scheme of ind-finite type acted on by H. Then the category $\mathfrak{D}(\mathcal{Y})$-mod carries an H-action of Harish-Chandra type.
(2) Let $\mathbf{A}$ be a topological associative algebra, acted on by H, and assume that the derived action of $\mathfrak{h}$ on $\mathbf{A}$ is inner, i.e., comes from a continuous map $\mathfrak{h} \to \mathbf{A}$. Then the action of H on $\mathbf{A}$-mod is of Harish-Chandra type.

Now let X be an object of $\mathcal{C}^{w,H}$. Note that in this case we have two identifications between $\mathrm{act}^*(X)|_{H^{(1)}}$ and $p^*(X)|_{H^{(1)}}$. Their difference is a map

$$a^\sharp : X \mapsto \mathfrak{h}^* \otimes X,$$

compatible with the cobracket on $\mathfrak{h}^*$, i.e., an action of $\mathfrak{h}$ on X; see Section 19.1. We will call this map "the obstruction to strong equivariance."

We will say that an object $X \in \mathcal{C}^{w,H}$ is strongly H-equivariant (or simply H-equivariant) if the map $a^\sharp$ is zero. Strongly equivariant objects form a full subcategory in $\mathcal{C}^{w,H}$, which we will denote by $\mathcal{C}^H$.

Let us consider the example where $\mathcal{C} = \mathfrak{D}(\mathcal{Y})$-mod, where X is an ind-scheme of ind-finite type acted on by H. Then $\mathfrak{D}(\mathcal{Y})\text{-mod}^{w,H}$ is the usual category of weakly H-equivariant D-modules, and $\mathfrak{D}(\mathcal{Y})\text{-mod}^H$ is the category of strongly H-equivariant $\mathfrak{D}$-modules.

More generally, if $\mathcal{C} = \mathbf{A}$-mod, where $\mathbf{A}$ is a topological associative algebra, acted on by H, then $\mathbf{A}\text{-mod}^{w,H}$ consists of $\mathbf{A}$-modules endowed with an algebraic action of H, compatible with the action of H on $\mathbf{A}$. If the action of H on $\mathbf{A}$ is of Harish-Chandra type, and $X \in A\text{-mod}^{w,H}$, the map $a^\sharp : X \to \mathfrak{h}^* \otimes X$ corresponds to the difference of the two actions of $\mathfrak{h}$ on X.

20.8

Let $\mathbf{C}(\mathcal{C}^H)$ denote the DG category of complexes of objects of $\mathcal{C}^H$, and $D(\mathcal{C}^H)$ the corresponding derived category. We have a natural functor $D(\mathcal{C}^H) \to D(\mathcal{C})$, but in general it does not behave well. Following Beilinson, we will now introduce the "correct" triangulated category, along with its DG model, that corresponds to strongly H-equivariant objects of $\mathcal{C}$.

Let $\mathbf{C}(\mathcal{C})^H$ be the category whose objects are complexes $X^\bullet$ of objects of $\mathcal{C}^{w,H}$, endowed with a map of complexes

$$i^\sharp : X^\bullet \to \mathfrak{h}^*[-1] * X^\bullet,$$

such that the following conditions are satisfied:

- $i^\sharp$ is a map in $\mathcal{C}^{w,H}$.
- The iteration of $i^\sharp$, viewed as a map $X^\bullet \to \Lambda^2(\mathfrak{h}^*)[-2] * X^\bullet$, vanishes.
- The map $[d, i^\sharp] : X^\bullet \to \mathfrak{h}^* * X^\bullet$ equals the map $a^\sharp$.

For two objects $X_1^\bullet$ and $X_2^\bullet$ of $\mathbf{C}(\mathcal{C})^H$ we define $\mathcal{H}om^k_{\mathbf{C}(\mathcal{C})^H}(X_1^\bullet, X_2^\bullet)$ to be the subcomplex of $\mathcal{H}om^k_{\mathbf{C}(\mathcal{C}^{w,H})}(X_1^\bullet, X_2^\bullet)$ consisting of graded maps $X_1^\bullet \to X_2^\bullet[k]$ that preserve the data of $i^\sharp$. This defines on $\mathbf{C}(\mathcal{C})^H$ a structure of DG-category.

Note that the usual cohomology functor defines a cohomological functor $\mathbf{C}(\mathcal{C})^H \to \mathcal{C}^H$. We will denote by $D(\mathcal{C})^H$ the resulting localized triangulated category, which henceforth we will refer to as the "H-equivariant derived category of $\mathcal{C}$."

By construction, the truncation functors $\tau^{<0}$, $\tau^{>0}$ are well defined at the level of $\mathbf{C}(\mathcal{C})^H$. So objects of the subcategory $D^b(\mathcal{C})^H$ (respectively, $D^+(\mathcal{C})^H$, $D^-(\mathcal{C})^H$) can be realized by complexes in $\mathbf{C}(\mathcal{C})^H$ that are concentrated in finitely many cohomological degrees (respectively, cohomological degrees $\gg -\infty$, $\ll \infty$).

20.9 Examples

First, take $\mathcal{C}$ to be Vect, in which case $\mathcal{C}^{w,H}$ identifies with the category $\mathcal{R}\text{ep}(H)$, and $\mathcal{C}^H$ is the same as $\mathcal{R}\text{ep}(H/H^0)$, where $H^0 \subset H$ is the connected component of the identity of H.

We will denote the resulting DG category by $\mathbf{C}(\text{pt}/H)$ and the triangulated category by $D(\text{pt}/H)$. Note that $\mathbf{C}(\text{pt}/H)$ is the standard, i.e., Cartan, DG-model for the H-equivariant derived category of the point-scheme.

Consider the de Rham complex on H, denoted $\mathbf{DR}_H$. The multiplication on H endows $\mathbf{DR}_H$ with a structure of a DG coalgebra. The category $\mathbf{C}(\text{pt}/H)$ is tautologically the same as the category of DG comodules over $\mathbf{DR}_H$. In particular, $\mathbf{DR}_H$ itself is naturally an object of $\mathbf{C}(\text{pt}/(H \times H))$.

More generally, let $\mathcal{C}$ be $\mathfrak{D}(\mathcal{Y})$-mod for $\mathcal{Y}$ as above. In this case $\mathbf{C}(\mathfrak{D}(\mathcal{Y})\text{-mod})^H$ is the DG-model for the H-equivariant derived category on $\mathcal{Y}$ studied in [BD1], where it is shown that the corresponding equivariant derived category is equivalent to the category of [BL].

20.10 Averaging

Note that for any $\mathcal{C}$ with an infinitesimally trivial action of H we have a natural tensor product functor

$$V^\bullet, X^\bullet \mapsto V^\bullet * X^\bullet : \mathbf{C}(\text{pt}/H) \times \mathbf{C}(\mathcal{C})^H \to \mathbf{C}(\mathcal{C})^H,$$

which extends to a functor $D(\text{pt}/H) \times D(\mathcal{C})^H \to D(\mathcal{C})^H$.

We have a tautological forgetful functor $\mathbf{C}(\mathcal{C})^H \to \mathbf{C}(\mathcal{C}^{w,H})$, and we claim that it admits a natural right adjoint, described as follows.

We will regard $X^\bullet \in \mathbf{C}(\mathcal{C}^{w,H})$ as a complex of objects of $\mathcal{C}$, acted on by $\mathfrak{h}$ via $a^\sharp$, and we can form the standard complex

$$\mathfrak{C}(\mathfrak{h}, X^\bullet) := \Lambda^\bullet(\mathfrak{h}^*) * X^\bullet$$

(see Section 19.17). It is naturally an object of $\mathbf{C}(\mathcal{C}^{w,H})$. The action of the annihilation operators defines on $\mathfrak{C}(\mathfrak{h}, X^\bullet)$ the structure of an object in $\mathbf{C}(\mathcal{C})^H$.

The resulting functor $\mathbf{C}(\mathcal{C}^{w,H}) \to \mathbf{C}(\mathcal{C})^H$ is exact when restricted to $\mathbf{C}^+(\mathcal{C}^{w,H})$, and the corresponding functor $D^+(\mathcal{C})^H \to D^+(\mathcal{C}^{w,H})$ is the right adjoint to the tautological forgetful functor, by Lemma 19.22.

We will denote the composed functor

$$\mathbf{C}(\mathcal{C}) \xrightarrow{\mathrm{Av}_H^w} \mathbf{C}(\mathcal{C}^{w,H}) \to \mathbf{C}(\mathcal{C})^H$$

(and the corresponding functor $D^+(\mathcal{C}) \to D^+(\mathcal{C})^H$) by Av_H. This functor is the right adjoint to the forgetful functor $\mathbf{C}(\mathcal{C})^H \to \mathbf{C}(\mathcal{C})$ (respectively, $D(\mathcal{C})^H \to D(\mathcal{C})$).

Let us consider two examples:

(1) For $\mathcal{C} = \mathrm{Vect}$, we have $\mathrm{Av}_H(\mathbb{C}) \simeq \mathbf{DR}_H \in \mathbf{C}(\mathrm{pt}/H)$.

(2) Let $\mathcal{C} = \mathfrak{D}(\mathcal{Y})$-mod, where $\mathcal{Y}$ is an ind-scheme of ind-finite type, acted on by H. The resulting functor at the level of derived categories $D^+(\mathfrak{D}(\mathcal{Y})\text{-mod}) \to D^+(\mathfrak{D}(\mathcal{Y})\text{-mod})^H$ is the corresponding $*$-averaging functor:

$$\mathcal{F} \mapsto p_* \circ \mathrm{act}^!(\mathcal{F}),$$

where p and act are the two maps $H \times \mathcal{Y} \to \mathcal{Y}$.

20.11 The unipotent case

Assume now that H is connected. We claim that in this case $\mathcal{C}^H$ is a full subcategory of $\mathcal{C}$. Indeed, for a $\mathcal{C}$-morphism $\phi : X_1 \to X_2$ between objects of $\mathcal{C}^H$, in the diagram

$$\begin{array}{ccc} \mathrm{act}^*(X_1) & \xrightarrow{\mathrm{act}^*(\phi)} & \mathrm{act}^*(X_2) \\ \sim\uparrow & & \sim\uparrow \\ \mathcal{O}_H \otimes X_1 & \longrightarrow & \mathcal{O}_H \otimes X_2 \end{array}$$

the bottom arrow is necessarily of the form $\mathrm{id} \otimes \phi'$, since its derivative along H is 0, as follows from the condition that $a^\sharp|_{X_1} = a^\sharp|_{X_2} = 0$. Then the unit constraint forces $\phi' = \phi$.

For H connected let us denote by $D(\mathcal{C})_{\mathcal{C}^H}$ the full subcategory of $D(\mathcal{C})$ consisting of objects whose cohomologies belong to $\mathcal{C}^H$.

Proposition 20.12. *Suppose that the group scheme H is pro-unipotent. Then the functor $D^+(\mathcal{C})^H \to D^+(\mathcal{C})$ is fully faithful, and it induces an equivalence $D^+(\mathcal{C})^H \simeq D^+(\mathcal{C})_{\mathcal{C}^H}$.*

Proof. Since $\mathrm{Av}_H : D^+(\mathcal{C}) \to D^+(\mathcal{C})^H$ is the right adjoint to the functor in question, to prove fully faithfulness it suffices to show that the adjunction map gives rise to an isomorphism between the composition

$$D^+(\mathcal{C})^H \to D^+(\mathcal{C}) \overset{\mathrm{Av}_H}{\to} D^+(\mathcal{C})^H$$

and the identity functor.

For $X^\bullet \in \mathbf{C}^+(\mathcal{C})^H$, the object $\mathrm{Av}_H(X^\bullet)$ is isomorphic to the tensor product of complexes

$$\mathbf{DR}_H * X^\bullet,$$

and the adjunction map in question corresponds to the natural map $\mathbb{C} \to \mathbf{DR}_H$. The latter is a quasi-isomorphism since H was assumed pro-unipotent.

It remains to show that $D^+(\mathcal{C})^H$ maps essentially surjectively onto $D^+(\mathcal{C})_{\mathcal{C}^H}$. For that it is sufficient to show that for $X^\bullet \in D^+(\mathcal{C})_{\mathcal{C}^H}$, the second adjunction map $\mathrm{Av}_H(X^\bullet) \to X^\bullet$ is a quasi-isomorphism.

By devissage, we can assume that $X^\bullet$ is concentrated in one cohomological dimension. However, such an object is quasi-isomorphic (up to a shift) to an object from $\mathcal{C}^H$, which makes the assertion manifest. □

20.13 Equivariant cohomology

For $X_1^\bullet, X_2^\bullet \in \mathbf{C}(\mathcal{C})^H$ we define a contravariant functor

$$\underline{\mathcal{H}om}_{\mathcal{C}}(X_1^\bullet, X_2^\bullet) : \mathbf{C}(\mathrm{pt}/H) \to \mathbf{C}(\mathrm{Vect})$$

by

$$V^\bullet \mapsto \mathcal{H}om_{\mathbf{C}(\mathcal{C})^H}(V^\bullet * X_1^\bullet, X_2^\bullet).$$

This functor is easily seen to be representable. When $X_1^\bullet$ is bounded from above and consists of objects that are finitely generated, the forgetful functor $\mathbf{C}(\mathcal{C})^H \to \mathbf{C}(\mathrm{Vect})$ maps $\underline{\mathcal{H}om}_{\mathcal{C}}(X_1^\bullet, X_2^\bullet)$ to $\mathcal{H}om_{\mathbf{C}(\mathcal{C})}(X_1^\bullet, X_2^\bullet)$.

Similarly, for $X_1^\bullet \in, X_2^\bullet \in D(\mathcal{C})^H$ the cohomological functor

$$V^\bullet \mapsto \mathrm{RHom}_{D(\mathcal{C})^H}(V^\bullet * X_1^\bullet, X_2^\bullet)$$

is representable by some $\underline{\mathrm{RHom}}_{D(\mathcal{C})}(X_1^\bullet, X_2^\bullet) \in D^+(\mathcal{C})^H$. We have the following assertion, whose proof repeats that of Lemma 20.4.

Lemma 20.14. *If $X_1^\bullet$ is quasi-perfect as an object of $D(\mathcal{C})$ and $X_2^\bullet$ is bounded from below, then the forgetful functor $D^+(\mathrm{pt}/H) \to D^+(\mathrm{Vect})$ maps $\underline{\mathrm{RHom}}_{D(\mathcal{C})}(X_1^\bullet, X_2^\bullet)$ to $\mathrm{RHom}_{D(\mathcal{C})}(X_1^\bullet, X_2^\bullet)$.*

The last lemma gives rise to the Leray spectral sequence that expresses Exts in the H-equivariant derived category as equivariant cohomology with coefficients in usual Exts.

We will now recall an explicit way of computing Exts in the category $D(\mathrm{pt}/H)$, in a slightly more general framework. For what follows we will make the following additional assumption on H (satisfied in the examples of interest):

We will assume that the group scheme H is such that its unipotent radical H_u is of finite codimension in H. We will fix a splitting $H/H_u =: H_{\mathrm{red}} \to H$.

Let $\mathcal{C}$ be an abelian category with the trivial action of H. We will denote the resulting equivariant DG category by $\mathbf{C}(\mathrm{pt}/H \otimes \mathcal{C})$. It consists of complexes of objects of $\mathcal{C}$, endowed with an algebraic $\mathcal{O}_H$-action, and an action of $\mathfrak{h}[1]$, satisfying the usual axioms.

Consider the functor $X^\bullet \mapsto \underline{\mathcal{H}\mathrm{om}}_{\mathbf{C}(\mathrm{pt}/H)}(\mathbb{C}, X^\bullet) : \mathbf{C}(\mathrm{pt}/H \otimes \mathcal{C}) \to \mathbf{C}(\mathcal{C})$, given by

$$X^\bullet \mapsto (X^\bullet)^{H,\mathfrak{h}[1]}.$$

Consider the corresponding derived functor

$$\underline{\mathcal{RH}\mathrm{om}}_{D(\mathrm{pt}/H)}(\mathbb{C}, ?) : D(\mathrm{pt}/H \otimes \mathcal{C}) \to D(\mathcal{C}).$$

Let us show how to compute it explicitly.

Let $BH^\bullet$ (respectively, $EH^\bullet$) be the standard simplicial model for the classifying space of H (respectively, the principal H-bundle over it). Let us denote by $\mathbf{DR}_{EH^\bullet}$ the de Rham complex of $EH^\bullet$. The action of H on $EH^\bullet$ makes $\mathbf{DR}_{EH^\bullet}$ a comodule over $\mathbf{DR}_H$, i.e., an object of $\mathbf{C}(\mathrm{pt}/H)$. Since $EH^\bullet$ is contractible, $\mathbf{DR}_{EH^\bullet}$ is quasi-isomorphic to $\mathbb{C}$.

Lemma 20.15. *For $X^\bullet \in \mathbf{C}(\mathrm{pt}/H \otimes \mathcal{C})$, there is a natural quasi-isomorphism*

$$\underline{\mathcal{H}\mathrm{om}}_{\mathbf{C}(\mathrm{pt}/H)}(\mathbb{C}, \mathbf{DR}_{EH^\bullet} * X^\bullet) \simeq \underline{\mathcal{RH}\mathrm{om}}_{D(\mathrm{pt}/H)}(\mathbb{C}, X^\bullet).$$

Proof. We only have to check that whenever $X^\bullet \in \mathbf{C}(\mathrm{pt}/H \otimes \mathcal{C})$ is acyclic, then

$$(\mathbf{DR}_{EH^\bullet} * X^\bullet)^{H,\mathfrak{h}[1]}$$

is acyclic as well.

Note that the rows of the corresponding bicomplex are isomorphic to

$$(\mathbf{DR}_{H^n} * X^\bullet)^{H,\mathfrak{h}[1]} \simeq \mathbf{DR}_{H^{n-1}} \otimes X^\bullet.$$

In particular, they are acyclic if $X^\bullet$ is. In other words, we have to show that the corresponding spectral sequence is convergent.

Consider the maps $\mathbf{DR}_{H^n} \to \mathbf{DR}_{H^n_{\mathrm{red}}}$, corresponding to the splitting $H_{\mathrm{red}} \to H$. They induce a quasi-isomorphism

$$(\mathbf{DR}_{EH^\bullet} * X^\bullet)^{H,\mathfrak{h}[1]} \to (\mathbf{DR}_{EH^\bullet_{\mathrm{red}}} * X^\bullet)^{H_{\mathrm{red}},\mathfrak{h}_{\mathrm{red}}[1]}.$$

This reduces us to the case when H is finite-dimensional, for which the convergence of the spectral sequence is evident. □

As a corollary, we obtain that the functor $\underline{\mathcal{RH}\mathrm{om}}_{D(\mathrm{pt}/H)}(\mathbb{C}, ?)$ commutes with direct sums. We will sometimes denote the functor $\underline{\mathcal{RH}\mathrm{om}}_{D(\mathrm{pt}/H)}(\mathbb{C}, ?)$ by $H^\bullet_{\mathbf{DR}}(\mathrm{pt}/H, ?)$.

20.16 Harish-Chandra modules

Let $\mathfrak{g}$ be a Tate Lie algebra, acted on by H by endomorphisms, and equipped with a homomorphism $\mathfrak{h} \to \mathfrak{g}$, so that $(\mathfrak{g}, H)$ is a Harish-Chandra pair. Then the category $\mathfrak{g}$-mod is a category with an infinitesimally trivial action of H.

The abelian category $\mathfrak{g}\text{-mod}^H$ is the same as $(\mathfrak{g}, H)$-mod, i.e., the category of Harish-Chandra modules. For $M^\bullet \in \mathbf{C}(\mathfrak{g}\text{-mod})^H$ we will denote by $x \mapsto a(x)$ the action of $\mathfrak{g}$ on $M^\bullet$ and for $x \in \mathfrak{h}$, by $a^\flat(x)$ the action obtained by deriving the algebraic H-action on $M^\bullet$. (Then, of course, $a^\sharp(x) = a^\flat(x) - a(x) = [d, i^\sharp(x)]$).

Let $D(\mathfrak{g}\text{-mod})^H$ be the corresponding derived category, and $D((\mathfrak{g}, H)\text{-mod})$ be the naive derived category of the abelian category $(\mathfrak{g}, H)$-mod.

Proposition 20.17. *Assume that $\mathfrak{g}$ is finite-dimensional. Then the evident functor $D(\mathfrak{g}, H)\text{-mod} \to D(\mathfrak{g}\text{-mod})^H$ is an equivalence.*

Proof. We will construct a functor $\Phi : \mathbf{C}(\mathfrak{g}\text{-mod})^H \to \mathbf{C}((\mathfrak{g}, H)\text{-mod})$ that would be the quasi-inverse of the tautological embedding at the level of derived categories.

For $M^\bullet \in \mathbf{C}(\mathfrak{g}\text{-mod})$ consider the tensor product

$$U(\mathfrak{g}) \otimes \Lambda^\bullet(\mathfrak{g}) \otimes M^\bullet \tag{20.4}$$

with the standard differential, where $U(\mathfrak{g})$ is the universal enveloping algebra.

Assume now that $M^\bullet$ is, in fact, an object of $\mathbf{C}(\mathfrak{g}\text{-mod})^H$. Consider an action i^o of $\mathfrak{h}[1]$ on (20.4), given by by $i^o(x) \cdot (u \otimes \omega \otimes m) = u \otimes \omega \wedge x \otimes m + u \otimes \omega \otimes i^\sharp(x) \cdot m$. Consider also an $\mathfrak{h}$-action Lie^o, given by

$$\text{Lie}^o_x \cdot (u \otimes \omega \otimes m) = -u \cdot x \otimes \omega \otimes m + u \otimes \text{ad}_x(\omega) \otimes m + u \otimes \omega \otimes a^\flat(x)(m).$$

We have the usual relation $[d, i^o(x)] = \text{Lie}^o_x$, and set

$$\Phi(M^\bullet) := (U(\mathfrak{g}) \otimes \Lambda^\bullet(\mathfrak{g}) \otimes M^\bullet)_{\mathfrak{h}, \mathfrak{h}[1]} \simeq U(\mathfrak{g}) \underset{U(\mathfrak{h})}{\otimes} (\Lambda^\bullet(\mathfrak{g}) \underset{\Lambda^\bullet(\mathfrak{h})}{\otimes} M^\bullet).$$

This is a complex of $\mathfrak{g}$-modules via the $\mathfrak{g}$-action by the left multiplication on $U(\mathfrak{g})$, Moreover, we claim that the action of $\mathfrak{h} \subset \mathfrak{g}$ on $\Phi(M^\bullet)$ integrates to an H-action. This follows from the fact that the $a^\flat$-action of $\mathfrak{h}$ on $M^\bullet$ is integrable, and that the adjoint of $\mathfrak{h}$ on $\mathfrak{g}$ is integrable. Therefore, $\Phi(M^\bullet)$ is an object of $\mathbf{C}((\mathfrak{g}, H)\text{-mod})$.

It is easy to see that $\Phi : \mathbf{C}(\mathfrak{g}\text{-mod})^H \to \mathbf{C}((\mathfrak{g}, H)\text{-mod})$ is exact, and hence it gives rise to a functor at the level of derived categories.

Note that for any $M^\bullet \in \mathbf{C}(\mathfrak{g}\text{-mod})^H$ we have the natural maps

$$M^\bullet \leftarrow U(\mathfrak{g}) \otimes \Lambda^\bullet(\mathfrak{g}) \otimes M^\bullet \to \Phi(M^\bullet),$$

both being quasi-isomorphisms. This implies the statement of the proposition. □

20.18 Relative BRST complex

Assume now that H is such that the adjoint action of $\mathfrak{h}$ on $\mathsf{Cliff}(\mathfrak{g}, \mathfrak{g}^*)$ can be lifted to an algebraic action of H on $\mathsf{Spin}(\mathfrak{g})$. In particular, the canonical extension $\mathfrak{g}_{-\text{can}}$ splits over $\mathfrak{h}$, and the category $\mathfrak{g}_{-\text{can}}$-mod also acquires an infinitesimally trivial H-action.

For an object $M^\bullet \in \mathbf{C}(\mathfrak{g}_{-\text{can}}\text{-mod})^H$, consider the complex $\mathfrak{C}^{\frac{\infty}{2}}(\mathfrak{g}, M^\bullet)$, associated with the corresponding bicomplex. We claim that it is naturally an object of $\mathbf{C}(\text{pt}\,/H)$:

As a complex of vector spaces, it carries the diagonal action of the group scheme H (we will denote the action of its Lie algebra by $\mathrm{Lie}^\flat$) and an action, denoted $i^\flat$, of $\Lambda^\bullet(\mathfrak{h})$ defined as $i|_{\mathfrak{h}}+i^\sharp$. Let us show how to compute $H^\bullet_{DR}(\mathrm{pt}/H, \mathfrak{C}^{\frac{\infty}{2}}(\mathfrak{g}, M^\bullet)) \in D(\mathrm{Vect})$ (see Section 20.13).

For $M^\bullet$ as above, let us denote by $\mathfrak{C}^{\frac{\infty}{2}}(\mathfrak{g}; H_{\mathrm{red}}, M^\bullet)$ (respectively, $\mathfrak{C}^{\frac{\infty}{2}}(\mathfrak{g}; H, M^\bullet)$ the subcomplex of $\mathfrak{C}^{\frac{\infty}{2}}(\mathfrak{g}, M^\bullet)$, equal to $(\mathfrak{C}^{\frac{\infty}{2}}(\mathfrak{g}, M^\bullet))^{H_{\mathrm{red}}, \mathfrak{h}_{\mathrm{red}}[1]}$ (respectively, $(\mathfrak{C}^{\frac{\infty}{2}}(\mathfrak{g}, M^\bullet))^{H, \mathfrak{h}[1]}$).

Lemma 20.19.

(1) *The complex* $H^\bullet_{DR}(\mathrm{pt}/H, \mathfrak{C}^{\frac{\infty}{2}}(\mathfrak{g}, M^\bullet))$ *is quasi-isomorphic to* $\mathfrak{C}^{\frac{\infty}{2}}(\mathfrak{g}; H_{\mathrm{red}}, M^\bullet)$.
(2) *If each* M^i *as above is injective as an* H_u*-module, then the embedding*

$$\mathfrak{C}^{\frac{\infty}{2}}(\mathfrak{g}; H, M^\bullet) \hookrightarrow \mathfrak{C}^{\frac{\infty}{2}}(\mathfrak{g}; H_{\mathrm{red}}, M^\bullet)$$

is a quasi-isomorphism.

Proof. First, by Section 20.13, we can assume that $M^\bullet$ is bounded from below. Secondly, arguing as in Proposition 20.12, we can replace the original complex $M^\bullet$ by one which consists of modules that are injective over H_u (and hence over H).

Hence, it is sufficient to check that in this case

$$H^\bullet_{DR}\left(\mathrm{pt}/H, \mathfrak{C}^{\frac{\infty}{2}}(\mathfrak{g}, M^\bullet)\right) \leftarrow \mathfrak{C}^{\frac{\infty}{2}}(\mathfrak{g}; H, M^\bullet) \rightarrow \mathfrak{C}^{\frac{\infty}{2}}(\mathfrak{g}; H_{\mathrm{red}}, M^\bullet)$$

are quasi-isomorphisms.

Consider $\mathfrak{C}^{\frac{\infty}{2}}(\mathfrak{g}, M^\bullet)$ as a module over the Clifford algebra $\mathsf{Cliff}(\mathfrak{h})$, where the annihilation operators act by i^*, and the creation operators act by means of $i^\flat$. We obtain that

$$\mathfrak{C}^{\frac{\infty}{2}}(\mathfrak{g}, M^\bullet) \simeq \mathfrak{C}(\mathfrak{h}, M_1^\bullet),$$

for some complex $M_1^\bullet$ of H-modules, which, moreover, consist of injective objects.

Thus we have reduced the original problem to the case when $\mathfrak{g} = \mathfrak{h}$. In this case, by Lemmas 19.19 and 20.14,

$$H^\bullet_{DR}\left(\mathrm{pt}/H, \mathfrak{C}(\mathfrak{h}, M^\bullet)\right) \simeq \mathrm{RHom}_{D(H\text{-mod})}(\mathbb{C}, M^\bullet),$$

which is quasi-isomorphic to

$$\mathfrak{C}(\mathfrak{h}; H, M^\bullet) \simeq M^\bullet,$$

if $M^\bullet$ consists of injective H-modules.

Moreover, by the Hochshild–Serre spectral sequence,

$$\mathrm{RHom}_{D(H\text{-mod})}(\mathbb{C}, M^\bullet) \simeq \left(\mathrm{RHom}_{D(H_u\text{-mod})}(\mathbb{C}, M^\bullet)\right)^{H_{\mathrm{red}}} \simeq \mathfrak{C}(\mathfrak{h}; H_{\mathrm{red}}, M^\bullet). \quad \square$$

20.20 Variant: Equivariance against a character

Now let ψ be a homomorphism $H \to \mathbb{G}_a$; we will denote by the same character the resulting character on $\mathfrak{h}$. For a category $\mathcal{C}$ as above, we introduce the category $\mathcal{C}^{H,\psi}$ to be the full subcategory of $\mathcal{C}^{w,H}$ consisting of objects for which the map $a^\flat$ is given by the character ψ.

Let us consider the example when $\mathcal{C} = \mathfrak{D}(\mathcal{Y})$-mod. Let $\mathbf{e}^\psi$ be the pull-back of the Artin–Schreier D-module on $\mathbb{G}_a$ under ψ. Its fiber at $1 \in H$ is trivialized and it is a character sheaf in the sense that we have a canonical isomorphism $\text{mult}^*(\mathbf{e}^\psi) \simeq \mathbf{e}^\psi \boxtimes \mathbf{e}^\psi$, which is associative in the natural sense.

The category $\mathfrak{D}(\mathcal{Y})\text{-mod}^{H,\psi}$ consists of D-modules $\mathcal{F}$ on $\mathcal{Y}$, endowed with an isomorphism $\text{act}^*(\mathcal{F}) \simeq \mathbf{e}^\psi \boxtimes \mathcal{F} \in \mathfrak{D}(H \times \mathcal{Y})$-mod, satisfying the associativity and unit conditions.

Returning to the general situation, we introduce the category $\mathbf{C}(\mathcal{C})^{H,\psi}$ in the same way as $\mathbf{C}(\mathcal{C})^H$ with the only difference that we require that $[d, i^\sharp] = a^\sharp + \psi$. This is a DG-category with a cohomological functor to $\mathcal{C}^{H,\psi}$. We will denote by $D(\mathcal{C})^{H,\psi}$ the resulting triangulated category.

Much of the discussion about $\mathbf{C}(\mathcal{C})^H$ carries over to this situation. For example, we have the averaging functor $\text{Av}_{H,\psi} : \mathbf{C}(\mathcal{C}) \to \mathbf{C}(\mathcal{C})^{H,\psi}$, right adjoint to the forgetful functor. It is constructed as the composition of Av_H^w and the functor

$$X^\bullet \mapsto \mathfrak{C}(\mathfrak{h}, X^\bullet \otimes \mathbb{C}^\psi) : \mathbf{C}(\mathcal{C}^{w,H}) \to \mathbf{C}(\mathcal{C})^{H,\psi},$$

where $\mathbb{C}^\psi$ is the 1-dimensional representation of $\mathfrak{h}$ corresponding to the character ψ.

When H is pro-unipotent, one shows in the same way as above that the functor $D^+(\mathcal{C})^{H,\psi} \to D^+(\mathcal{C})$ is an equivalence onto the full subcategory consisting of objects whose cohomologies belong to $\mathcal{C}^{H,\psi}$.

21 *D*-modules on group ind-schemes

21.1

Let G be an affine reasonable group ind-scheme, as in [BD1]. In particular, its Lie algebra $\mathfrak{g}$ is a Tate vector space. We will denote by $\mathcal{O}_G$ the topological commutative algebra of functions on G.

The multiplication on G defines a map $\Delta_G : \mathcal{O}_G \to \mathcal{O}_G \overset{!}{\otimes} \mathcal{O}_G$. We will denote by Lie_l and Lie_r the two maps

$$\mathfrak{g} \overset{*}{\otimes} \mathcal{O}_G \to \mathcal{O}_G,$$

corresponding to the action of G on itself by left (respectively, right) translations.

In addition, we have the maps

$$\Delta_{\mathfrak{g}} : \mathfrak{g} \to \mathcal{O}_G \overset{!}{\otimes} \mathfrak{g}, \Delta_{\mathfrak{g}^*} : \mathfrak{g}^* \to \mathcal{O}_G \overset{!}{\otimes} \mathfrak{g}^*$$

that correspond to the adjoint and coadjoint actions of G on $\mathfrak{g}$ and $\mathfrak{g}^*$, respectively.

Let us denote by $T(G)$ (respectively, $T^*(G)$) the topological $\mathcal{O}_G$-module of vector fields (respectively, 1-forms) on G. It identifies in two ways with $\mathcal{O}_G \overset{!}{\otimes} \mathfrak{g}$ (respectively, $\mathcal{O}_G \overset{!}{\otimes} \mathfrak{g}^*$), corresponding to the realization of $\mathfrak{g}$ (respectively, $\mathfrak{g}^*$) as right or left invariant vector fields (respectively, 1-forms). Note that $T(G)$ is a topological Lie algebra and $T^*(G)$ is a module over it.

21.2

Following [AG1], we introduce the category of D-modules on G, denoted $\mathfrak{D}(G)$-mod, as follows:

Its objects are (discrete) vector spaces $\mathcal{M}$, endowed with an action

$$\mathcal{O} \overset{\rightarrow}{\otimes} \mathcal{M} \simeq \mathcal{O} \overset{*}{\otimes} \mathcal{M} \overset{m}{\to} \mathcal{M}$$

and a Lie algebra action

$$a_l : \mathfrak{g} \overset{\rightarrow}{\otimes} \mathcal{M} \simeq \mathfrak{g} \overset{*}{\otimes} \mathcal{M} \to \mathcal{M},$$

such that the two pieces of data are compatible in the sense of the action of $\mathfrak{g}$ on $\mathcal{O}_G$ by *left* translations in the following sense:

We need that the difference of the two arrows:

$$\mathfrak{g} \overset{*}{\otimes} \mathcal{O}_G \overset{*}{\otimes} \mathcal{M} \overset{\mathrm{id}_G \otimes m}{\longrightarrow} \mathfrak{g} \overset{*}{\otimes} \mathcal{M} \overset{a_l}{\to} \mathcal{M}$$

and

$$\mathfrak{g} \overset{*}{\otimes} \mathcal{O}_G \overset{*}{\otimes} \mathcal{M} \simeq \mathcal{O}_G \overset{*}{\otimes} \mathfrak{g} \overset{*}{\otimes} \mathcal{M} \overset{\mathrm{id}_{\mathcal{O}_G} \otimes a_l}{\longrightarrow} \mathcal{O}_G \overset{*}{\otimes} \mathcal{M} \overset{m}{\to} \mathcal{M}$$

equals

$$\mathfrak{g} \overset{*}{\otimes} \mathcal{O}_G \overset{*}{\otimes} \mathcal{M} \overset{\mathrm{Lie}_l}{\longrightarrow} \mathcal{O}_G \overset{*}{\otimes} \mathcal{M} \overset{m}{\to} \mathcal{M}.$$

Morphisms in $\mathfrak{D}(G)$-mod are maps of vector spaces $\mathcal{M}_1 \to \mathcal{M}_2$ that commute with the actions of $\mathfrak{g}$ and $\mathcal{O}_G$.

21.3 Action of the Tate canonical extension

Following Beilinson, we will show now that if $\mathcal{M}$ is an object of $\mathfrak{D}(G)$-mod, then the underlying vector space carries a canonical action of $\mathfrak{g}_{-\mathrm{can}}$, denoted a_r, which commutes with the original action of $\mathfrak{g}$, and which is compatible with the action of $\mathcal{O}_G$ via the action of $\mathfrak{g}$ on $\mathcal{O}_G$ by *right* translations.

Set $\mathcal{M}^{DR} = \mathcal{M} \otimes \mathsf{Spin}(\mathfrak{g})$. Let us denote by i_r and i_r^* the actions on it of $\Lambda^\bullet(\mathfrak{g})$ and $\Lambda^\bullet(\mathfrak{g}^*)$, both of which are subalgebras of $\mathsf{Cliff}(\mathfrak{g}, \mathfrak{g}^*)$.

From the definition of $\mathfrak{D}(G)$-mod it follows that i_r and i_r^* on $\mathcal{M}$ extend to actions of the odd topological vector spaces $T(G)$ and $T^*(G)$, identified with $\mathcal{O}_G \overset{!}{\otimes} \mathfrak{g}$ and

$\mathcal{O}_G \overset{!}{\otimes} \mathfrak{g}^*$ using left-invariant vector fields and forms, respectively. We will denote the resulting actions simply by i and i^*.

Using the map

$$\mathfrak{g} \xrightarrow{-\Delta_{\mathfrak{g}}} \mathcal{O}_G \overset{!}{\otimes} \mathfrak{g} \xrightarrow{\gamma \otimes \mathrm{id}_{\mathfrak{g}}} \mathcal{O}_G \overset{!}{\otimes} \mathfrak{g}, \tag{21.1}$$

(here γ is the inversion on G), we obtain a new action i_l of $\Lambda^\bullet(\mathfrak{g})$ on $\mathcal{M}^{DR}$. Similarly, we have a new action i_l^* of $\Lambda^\bullet(\mathfrak{g}^*)$ on $\mathcal{M}^{DR}$. Altogether, we obtain a new action of the Clifford algebra $\mathsf{Cliff}(\mathfrak{g}, \mathfrak{g}^*)$ on $\mathcal{M}^{DR}$.

We will denote by the symbol Lie_l the action of $\mathfrak{g}$ on $\mathcal{M}^{DR}$ coming from the action a_l of $\mathfrak{g}$ on $\mathcal{M}$. We claim that this action extends to an action of the Lie algebra $T(G)$, denoted simply by Lie.

First we define an action of the noncompleted tensor product $\mathcal{O}_G \otimes \mathfrak{g}$ on $\mathcal{M}^{DR}$. Namely, for $x \in \mathfrak{g}$, $f \in \mathcal{O}_G$ and $v \in \mathcal{M}$, we set

$$(f \otimes x) \cdot v = f \cdot \mathrm{Lie}_l(x) \cdot v + i_l^*(df) \cdot i_l(x) \cdot v. \tag{21.2}$$

Note that

$$f \cdot \mathrm{Lie}_l(x) \cdot v + i_l^*(df) \cdot i_l(x) \cdot v = \mathrm{Lie}_l(x) \cdot f \cdot v + i_l(x) \cdot i_l^*(df) \cdot v.$$

This property implies that the above action of $\mathcal{O}_G \otimes \mathfrak{g}$ extends to the action of $\mathcal{O}_G \overset{!}{\otimes} \mathfrak{g} \simeq T(G)$. Indeed, when x is contained in a deep enough neighborhood of zero, then both $\mathrm{Lie}_{l(x)}$ and $i_l(x)$ annihilate any given $v \in \mathcal{M}$. Similarly, if f is contained in a deep neighborhood of zero, then v is annihilated by both f and $i_l^*(df)$.

One readily checks that the above action is compatible with the Lie algebra structure on $T(G)$. In particular, using the map $-\Delta_{\mathfrak{g}} : \mathfrak{g} \to \mathcal{O}_G \overset{!}{\otimes} \mathfrak{g}$, i.e., the embedding of $\mathfrak{g}$ into $T(G)$ as left-invariant vector fields, we obtain a new action of the Lie algebra $\mathfrak{g}$ on $\mathcal{M}$. We will denote this action by Lie_r.

We have

- $[\mathrm{Lie}_r(x), i_l(y)] = 0$, $[\mathrm{Lie}_r(x), i_l^*(y^*)] = 0$ for $x, y \in \mathfrak{g}$, $y^* \in \mathfrak{g}^*$.
- $[\mathrm{Lie}_r(x), i_r(y)] = i_r([x, y])$, $[\mathrm{Lie}_r(x), i_r^*(y^*)] = i_r^*(\mathrm{ad}_x(y^*))$,
- $[\mathrm{Lie}_r(x), f] = \mathrm{Lie}_{r(x)}(f)$ for $f \in \mathcal{O}_G$.
- $[\mathrm{Lie}_r(x), a_l(y)] = 0$.

Finally, we are ready to define the action a_r of $\mathfrak{g}_{-\mathrm{can}}$ on $\mathcal{M}^{DR}$. Namely, a_r is the difference of Lie_r and the canonical $\mathfrak{g}_{\mathrm{can}}$-action on $\mathsf{Spin}(\mathfrak{g})$.

It is easy to see that a_r is indeed an action. Moreover,

- $[a_r(x'), f] = \mathrm{Lie}_{r(x)}(f)$, for $x' \in \mathfrak{g}_{-\mathrm{can}}$ and its image $x \in \mathfrak{g}$,
- $[a_r(x'), a_l(y)] = 0$,
- $[a_r(x'), i_r(y)] = 0$, $[a_r(x'), i_r(y^*)] = 0$.

The last property implies that the a_r-action of $\mathfrak{g}_{-\mathrm{can}}$ on $\mathcal{M}^{DR}$ preserves the subspace $\mathcal{M}$; i.e., we obtain an action of $\mathfrak{g}_{-\mathrm{can}}$ on $\mathcal{M}$ that satisfies the desired commutation properties.

When we view $\mathcal{M} \in \mathfrak{D}(G)$-mod as a $\mathfrak{g}_{-\text{can}}$-module via a_r, we obtain that $\mathcal{M}^{DR}$ is identified with $\mathfrak{C}^{\frac{\infty}{2}}(\mathfrak{g}, \mathcal{M})$, where $i = i_r$, $i^* = i_r^*$, Lie $=$ Lie$_r$. In particular, $\mathcal{M}^{DR}$ acquires a natural differential d.

From the above commutation properties, it follows that this differential satisfies

- $[d, i(\xi)] = \text{Lie}(\xi)$ for $\xi \in T(G)$,
- $[d, f] = i^*(df)$ for $f \in \mathcal{O}_G$.

Of course, $\mathcal{M}^{DR}$ depends on the choice of the Clifford module $\mathsf{Spin}(\mathfrak{g})$.

21.4

Note that the above construction can be inverted: we can introduce the category $\mathfrak{D}(G)$-mod to consist of $(\mathcal{O}_G, \mathfrak{g}_{-\text{can}})$-modules, where the two actions are compatible in the sense of the $\mathfrak{g}_{-\text{can}}$-action on $\mathcal{O}_G$ via right translations. In this case, the vector space underlying a representation automatically acquires an action of $\mathfrak{g}$, which commutes with the $\mathfrak{g}_{-\text{can}}$-action and is compatible with the action of $\mathcal{O}_G$ via left translations.

Let us also note that in the definition of $\mathfrak{D}(G)$-mod we could interchange the roles of left and right:

Let us call the category introduced above $\mathfrak{D}(G)_l$-mod, and let us define the category $\mathfrak{D}(G)_r$-mod to consist of $(\mathcal{O}_G, \mathfrak{g})$-modules, where the two actions are compatible via the action of $\mathfrak{g}$ on $\mathcal{O}_G$ by right translations. We claim that the categories $\mathfrak{D}(G)_l$-mod and $\mathfrak{D}(G)_r$-mod are equivalent, but this equivalence does not respect the forgetful functor to vector spaces.

This equivalence is defined as follows. For $\mathcal{M}_l \in \mathfrak{D}(G)_l$-mod, the actions i_l, i_l^* define a new action of $\mathsf{Cliff}(\mathfrak{g}, \mathfrak{g}^*)$ on $\mathcal{M}^{DR}$. We define an object of $\mathcal{M}_r \in \mathfrak{D}(G)_r$-mod by $\text{Hom}_{\mathsf{Cliff}(\mathfrak{g},\mathfrak{g}^*)}(\mathsf{Spin}(\mathfrak{g}), \mathcal{M}^{DR})$ with respect to this new action.

Explicitly, this can be reformulated as follows. Let G_{can} be the canonical (i.e., Tate) central extension of G. It can be viewed as a line bundle $\mathcal{P}_{\text{can}}$ over G, whose fiber at a given point $\mathbf{g} \in G$ is the relative determinant line $\det(\mathfrak{g}, \text{Ad}_{\mathbf{g}}(\mathfrak{g}))$. The action of $\mathfrak{g}$ on G be left (respectively, right) translations extends to an action of $\mathfrak{g}_{\text{can}}$ (respectively, $\mathfrak{g}_{-\text{can}}$) on $\mathcal{P}_{\text{can}}$.

Then

$$\mathcal{M}_r \simeq \mathcal{M}_l \underset{\mathcal{O}_G}{\otimes} \mathcal{P}_{\text{can}}^{-1},$$

as $\mathcal{O}_G$-modules, respecting both the a_l and a_r actions.

In what follows, unless stated otherwise, we will think of $\mathfrak{D}(G)$-mod in the $\mathfrak{D}(G)_l$-mod realization.

21.5

Let H be a group scheme, mapping to G. We claim that the category $\mathfrak{D}(G)$-mod carries a natural infinitesimally trivial action of H, corresponding to the action of H on G by left translations. (As we shall see later, this is a part of a more general structure, the latter being an action of the group ind-scheme $G \times G$ on $\mathfrak{D}(G)$-mod).

For $\mathcal{M} \in \mathfrak{D}(G)$-mod we set $\mathrm{act}_l^*(\mathcal{M})$ to be isomorphic to $\mathcal{O}_H \otimes \mathcal{M}$ as an $\mathcal{O}_H$-module. The action of $\mathcal{O}_G$ is given via

$$\mathcal{O}_G \xrightarrow{\Delta_G} \mathcal{O}_G \overset{!}{\otimes} \mathcal{O}_G \xrightarrow{\gamma \otimes \mathrm{id}} \mathcal{O}_H \overset{!}{\otimes} \mathcal{O}_G.$$

The action a_l of $\mathfrak{g}$ is given via the map

$$\mathfrak{g} \xrightarrow{\Delta_\mathfrak{g}} \mathcal{O}_G \overset{!}{\otimes} \mathfrak{g} \xrightarrow{\gamma \otimes \mathrm{id}} \mathcal{O}_G \overset{!}{\otimes} \mathfrak{g} \to \mathcal{O}_H \overset{!}{\otimes} \mathfrak{g},$$

where $\mathcal{O}_H \overset{!}{\otimes} \mathfrak{g}$ acts on $\mathcal{O}_H \otimes \mathcal{M}$ by $\mathrm{id} \otimes m$.

To construct the isomorphism $\mathrm{act}^*(\mathcal{M})|_{H^{(1)}} \simeq p^*(\mathcal{M})|_{H^{(1)}}$ we identify both sides with $\mathcal{M} \oplus \epsilon \cdot \mathfrak{h}^* \otimes \mathcal{M}$ as vector spaces, and the required isomorphism is given by the action of $\mathfrak{h}$ on $\mathcal{M}$, obtained by restriction from a_l.

Note that by construction, the action of $\mathfrak{g}_{-\mathrm{can}}$ on $\mathrm{act}_l^*(\mathcal{M}) \simeq \mathcal{O}_H \otimes \mathcal{M}$ is via its action on the second multiple.

Now let H' be another group scheme mapping to G, and let us assume that there exists a splitting $\mathfrak{h}' \to \mathfrak{g}_{\mathrm{can}}$. In this case, we claim that there exists another infinitesimally trivial action of H' on $\mathfrak{D}(G)$-mod, corresponding to the action of H on G by right translations:

For $\mathcal{M} \in \mathfrak{D}(G)$-mod, we define $\mathrm{act}_r^*(\mathcal{M})$ to be isomorphic to $\mathcal{M} \otimes \mathcal{O}_{H'}$ as an $\mathcal{O}_{H'}$-module and as a $\mathfrak{g}$-module. The action of $\mathcal{O}_G$ is given by the comultiplication map $\mathcal{O}_G \to \mathcal{O}_G \overset{!}{\otimes} \mathcal{O}_{H'}$. It is easy to see that the commutation relation is satisfied. The associativity and unit constraint are evident.

To construct the isomorphism $\mathrm{act}_r^*(\mathcal{M})|_{H'^{(1)}} \simeq p^*(\mathcal{M})|_{H'^{(1)}}$, note that both sides are identified with $\mathcal{M} \oplus \epsilon \cdot \mathfrak{h}'^* \otimes \mathcal{M}$ as $\mathfrak{g}$-modules. The required isomorphism is given by the action of $\mathfrak{h}'$ on $\mathcal{M}$, obtained by restriction from a_r. Again, it is easy to see that the axioms of Harish-Chandra action hold.

Let us note that the action a_r of $\mathfrak{g}_{-\mathrm{can}}$ on $\mathrm{act}_r^*(\mathcal{M}) \simeq \mathcal{M} \otimes \mathcal{O}_{H'}$ is given via the map

$$\mathfrak{g}_{-\mathrm{can}} \xrightarrow{\Delta_\mathfrak{g}} \mathfrak{g}_{-\mathrm{can}} \overset{!}{\otimes} \mathcal{O}_G \to \mathfrak{g}_{-\mathrm{can}} \overset{!}{\otimes} \mathcal{O}_{H'}.$$

Let us denote by $\mathfrak{D}(G)\text{-mod}^{l(H)}$ (respectively, $\mathfrak{D}(G)\text{-mod}^{r(H')}$) the corresponding categories of strongly equivariant objects of $\mathfrak{D}(G)$-mod. Moreover, it is easy to see that the actions of H and H' commute in the natural sense, i.e., we have an action of $H \times H'$ on $\mathfrak{D}(G)$-mod. We will denote the resulting category by $\mathfrak{D}(G)\text{-mod}^{l(H),r(H')}$.

21.6

Now let $K \subset G$ be a group subscheme such that the quotient G/K exists as a strict ind-scheme of ind-finite type (in this case it is formally smooth). We will call such a K "open compact."

We will choose a particular model for the module $\mathsf{Spin}(\mathfrak{g})$, denoted $\mathsf{Spin}(\mathfrak{g}, \mathsf{k})$ by letting it be generated by a vector $\mathbf{1} \in \mathsf{Spin}(\mathfrak{g})$, annihilated by $\mathsf{k} \oplus (\mathfrak{g}/\mathsf{k})^* \subset$

$\mathfrak{g} \oplus \mathfrak{g}^* \subset \mathsf{Cliff}(\mathfrak{g}, \mathfrak{g}^*)$. This $\mathsf{Spin}(\mathfrak{g}, \mathsf{k})$ carries a natural action of K, which gives rise to a splitting of G_{can} over K.

By the assumption on G/K, it makes sense to consider right D-modules on it; we will denote this category by $\mathfrak{D}(G/K)$-mod.

Proposition 21.7. *We have a canonical equivalence* $\mathfrak{D}(G)\text{-mod}^{r(K)} \simeq \mathfrak{D}(G/K)\text{-mod}$.

Proof. Let π denote the projection $G \to G/K$. For $\mathcal{F} \in \mathfrak{D}(G/K)$-mod, consider the $\mathcal{O}_G$-module $\mathcal{M} := \Gamma(G, \pi^*(\mathcal{F}))$.

For $x \in \mathfrak{g}$, the (negative of the) corresponding vector field acting on $\mathcal{F}$ gives rise to a map $a_l(x) : \mathcal{M} \to \mathcal{M}$, as a vector space, and these data satisfy the conditions for $\mathcal{M}$ to be a $\mathfrak{D}(G)$-module.

We claim that the action of the Lie algebra $\mathsf{k} \subset \mathfrak{g}_{-\text{can}}$ on $\mathcal{M}$, given by a_r, coincides with the natural action of k on $\pi^*(\mathcal{F})$ obtained by deriving the group action. This would imply that $\mathcal{M}$ is naturally an object of $\mathfrak{D}(G)\text{-mod}^{r(K)}$.

To prove the assertion, we can assume that $\mathcal{F}$ is an extension of a D-module on an affine ind-subscheme of G/K. Then it is sufficient to check that the subspace $\Gamma(G/K, \mathcal{F}) \subset \mathcal{M} \subset \mathcal{M}^{DR}$ is annihilated by the operators $\text{Lie}_{r(x)}$ for $x \in \mathsf{k}$. But this is straightforward from the construction.

Vice versa, let $\mathcal{M}$ be an object of $\mathfrak{D}(G)\text{-mod}^{r(K)}$, which we identify with the corresponding quasi-coherent sheaf on G. Consider the complex of sheaves $\pi_*(\mathcal{M}^{DR})$ on G/K; it carries an action of the operators $i_r(x)$, $\text{Lie}_{r(x)}$, $x \in \mathfrak{g}$. We set $\mathcal{F}^{DR}$ to be the subcomplex of $\pi_*(\mathcal{M}^{DR})$ annihilated by the above operators for $x \in \mathsf{k}$.

Set $\mathcal{F}$ to be the degree 0 part of $\mathcal{F}^{DR}$; it is easy to see that $\mathcal{F} \simeq (\pi_*(\mathcal{M}))^K$. The degree -1 part of $\mathcal{F}^{DR}$ is identified with $\mathcal{F} \underset{\mathcal{O}(G/K)}{\otimes} T(G/K)$, and the differential

$$d : (\mathcal{F}^{DR})^{-1} \to (\mathcal{F}^{DR})^0$$

defines on $\mathcal{F}$ a structure of a right $\mathfrak{D}$-module. Moreover, the entire complex $\mathcal{F}^{DR}$ is identified with the de Rham complex of $\mathcal{F}$. □

Let $\delta_{K,G}$ be the object of $\mathfrak{D}(G)\text{-mod}^{r(K)}$ corresponding to the delta-function $\delta_{1,G/K}$ under the equivalence of categories of Proposition 21.7. It can be constructed as $\text{Ind}_{\mathsf{k}}^{\mathfrak{g}}(\mathcal{O}_K)$ as a module over $\mathfrak{g}$ and $\mathcal{O}_G$. As a module over $\mathfrak{g}_{-\text{can}}$ it is also isomorphic to $\text{Ind}_{\mathsf{k}}^{\mathfrak{g}_{-\text{can}}}(\mathcal{O}_K)$.

More generally, let $\mathcal{L}$ be an object of $\text{QCoh}^!_{G/K}$. Let $\text{Ind}_{\mathcal{O}_{G/K}}^{\mathfrak{D}_{G/K}}(\mathcal{L})$ be the induced D-module. The corresponding object of $\mathfrak{D}(G)\text{-mod}^{r(K)}$, i.e., $\Gamma(G, \pi^*(\text{Ind}_{\mathcal{O}_{G/K}}^{\mathfrak{D}_{G/K}}(\mathcal{L})))$ can be described as follows:

Consider the $\mathcal{O}_G$-module $\Gamma(G, \pi^*(\mathcal{L}))$; it is acted on naturally by K. Consider the $\mathfrak{g}_{-\text{can}}$-module $\text{Ind}_{\mathsf{k}}^{\mathfrak{g}_{-\text{can}}}(\Gamma(G, \pi^*(\mathcal{L})))$. It is naturally acted on by $\mathcal{O}_G$, so that the actions of $\mathfrak{g}_{-\text{can}}$ and $\mathcal{O}_G$ satisfy the commutation relation with respect to the right action of G on itself. Hence, $\text{Ind}_{\mathsf{k}}^{\mathfrak{g}_{-\text{can}}}(\Gamma(G, \pi^*(\mathcal{L})))$ is an object of $\mathfrak{D}(G)\text{-mod}^{r(K)}$ and we have a natural isomorphism:

$$\Gamma\left(G, \pi^*\left(\text{Ind}_{\mathcal{O}_{G/K}}^{\mathfrak{D}_{G/K}}(\mathcal{L})\right)\right) \simeq \text{Ind}_{\mathsf{k}}^{\mathfrak{g}_{-\text{can}}}\left(\Gamma(G, \pi^*(\mathcal{L}))\right). \tag{21.3}$$

21.8 The bi-equivariant situation

Now let K_1, K_2 be two "open compact" subgroups of G. Note that we have a natural equivalence of categories

$$\mathfrak{D}(G/K_1)\text{-mod}^{K_2} \to \mathfrak{D}(G/K_2)\text{-mod}^{K_1} : \mathcal{F} \mapsto \mathcal{F}^{\text{op}}, \tag{21.4}$$

defined as follows.

Assume, without loss of generality, that $\mathcal{F}$ is supported on a closed K_2-invariant subscheme $\mathcal{Y} \in G/K_1$, and let $\mathcal{Y}^{\text{op}}$ be the corresponding K_1-invariant subscheme in $K_2\backslash G$. We can find a normal "open compact" subgroup $K_2' \subset K_2$ such that if we denote by $\mathcal{Y}'^{\text{op}}$ the preimage of $\mathcal{Y}^{\text{op}}$ in $K_2'\backslash G$, the projection

$$\pi_1 : \mathcal{Y}'^{\text{op}} \to \mathcal{Y}$$

is well defined (and makes $\mathcal{Y}'^{\text{op}}$ a torsor with respect to the corresponding smooth group scheme over $\mathcal{Y}$).

Consider $\pi_1^!(\mathcal{F})$. This is a D-module on $\mathcal{Y}'^{\text{op}}$, equivariant with respect to the action of the group $(K_2/K_2') \times K_1$ on this scheme. Hence, it gives rise to a K_1-equivariant D-module on $K_2\backslash G$. To obtain $\mathcal{F}^{\text{op}}$ we apply the involution $\mathbf{g} \mapsto \mathbf{g}^{-1} : K_2\backslash G \to G/K_2$.

Let us now describe what this equivalence looks like in terms of the equivalences

$$\mathfrak{D}(G/K_1)\text{-mod}^{K_2} \simeq \mathfrak{D}(G)\text{-mod}^{l(K_2),r(K_1)}$$

and

$$\mathfrak{D}(G/K_2)\text{-mod}^{K_1} \simeq \mathfrak{D}(G)\text{-mod}^{l(K_1),r(K_2)}.$$

First, the inversion on G defines an equivalence

$$\mathfrak{D}(G)_l\text{-mod}^{l(K_2),r(K_1)} \simeq \mathfrak{D}(G)_r\text{-mod}^{l(K_2),r(K_1)},$$

and the sought-for equivalence is obtained from the one above via

$$\mathfrak{D}(G)_r\text{-mod}^{l(K_2),r(K_1)} \simeq \mathfrak{D}(G)_l\text{-mod}^{l(K_2),r(K_1)}.$$

(The determinant line that played a role in the $\mathfrak{D}(G)_r$-mod $\simeq \mathfrak{D}(G)_l$-mod equivalence appears also in (21.4), when we descend *right* D-modules from $\mathcal{Y}'^{\text{op}}$ to $\mathcal{Y}^{\text{op}}$.)

21.9

Now consider the DG-category $\mathbf{C}(\mathfrak{D}(G)\text{-mod})^{r(K)}$; we claim that the construction in Proposition 21.7 generalizes to a functor $\mathbf{C}(\mathfrak{D}(G)\text{-mod})^{r(K)} \to \mathbf{C}(\mathfrak{D}(G/K)\text{-mod})$.

Indeed, for $\mathcal{M}^\bullet \in \mathbf{C}(\mathfrak{D}(G)\text{-mod})^{r(K)}$ let us denote by $(\mathcal{M}^\bullet)^{DR}$ the total complex of the corresponding bicomplex. We have several actions of $\Lambda^\bullet(\mathsf{k})$ on it; let $i_r^\flat$ be the sum of the one given by restricting the i_r-action of $\mathfrak{g}$ and $i^\sharp$. In addition, $(\mathcal{M}^\bullet)^{DR}$

carries a natural action of K. These two structures combine to that of an object of $\mathbf{C}(\mathrm{pt}/K)$.

Consider again the complex of sheaves $\pi_*((\mathcal{M}^\bullet)^{DR})$ on G/K. It carries an action of $\Lambda^\bullet(\mathsf{k})$ coming from $i_r^\flat$ and an action of the group scheme K. Define

$$(\mathcal{F}^\bullet)^{DR} := \mathcal{H}om_{\mathbf{C}(\mathrm{pt}/K)}\left(\mathbb{C}, \pi_*((\mathcal{M}^\bullet)^{DR})\right).$$

This is an $\Omega^\bullet_{G/K}$-module on in the terminology of [BD1].

Finally, we consider the functor $\mathbf{C}(\mathfrak{D}(G)\text{-mod})^{r(K)} \to \mathbf{C}(\mathfrak{D}(G/K)\text{-mod})$ given by

$$\mathcal{M}^\bullet \mapsto \mathrm{Ind}_{\mathcal{O}_{G/K}}^{\mathfrak{D}(G/K)}\left((\mathcal{F}^\bullet)^{DR}\right) \in \mathbf{C}(\mathfrak{D}(G/K)),$$

where $\mathrm{Ind}_{\mathcal{O}_{G/K}}^{\mathfrak{D}(G/K)}$ is the induction functor from $\Omega^\bullet_{G/K}$-modules to D-modules on G/K; see [BD1, Section 7.11.12].

Lemma 21.10. *The resulting functor*

$$\mathcal{M}^\bullet \mapsto \mathrm{Ind}_{\mathcal{O}_{G/K}}^{\mathfrak{D}(G/K)}\left((\mathcal{F}^\bullet)^{DR}\right) : \mathbf{C}(\mathfrak{D}(G)\text{-mod})^{r(K)} \to \mathbf{C}(\mathfrak{D}(G/K)\text{-mod})$$

is exact.

Proof. This follows from the fact that the functor

$$\mathcal{M}^\bullet \mapsto (\mathcal{F}^\bullet)^{DR},$$

viewed as a functor from $\mathbf{C}(\mathfrak{D}(G)\text{-mod})^{r(K)}$ to the DG category of (non-quasi-coherent) sheaves on G/K is exact. □

Thus we obtain a well-defined functor $D(\mathfrak{D}(G)\text{-mod})^{r(K)} \to D(\mathfrak{D}(G/K)\text{-mod})$.

Proposition 21.11. *The above functor* $D(\mathfrak{D}(G)\text{-mod})^{r(K)} \to D(\mathfrak{D}(G/K)\text{-mod})$ *is an equivalence. Its quasi-inverse is given by*

$$D(\mathfrak{D}(G/K)\text{-mod}) \to D(\mathfrak{D}(G)\text{-mod}^{r(K)}) \to D(\mathfrak{D}(G)\text{-mod})^{r(K)}.$$

As a corollary, we obtain that in this case the evident functor $D(\mathfrak{D}(G)\text{-mod}^{r(K)}) \to D(\mathfrak{D}(G/K)\text{-mod})$ is an equivalence.

Proof. The functor

$$\mathcal{F}^\bullet \mapsto \Gamma(G, \pi^*(\mathcal{F}^\bullet)) =: \mathcal{M}^\bullet \mapsto \mathrm{Ind}_{\mathcal{O}_{G/K}}^{\mathfrak{D}(G/K)}\left(\mathcal{H}om_{\mathbf{C}(\mathrm{pt}/K)}\left(\mathbb{C}, \pi_*((\mathcal{M}^\bullet)^{DR})\right)\right)$$

is isomorphic to the composition

$$\mathbf{C}(\mathfrak{D}(G/K)\text{-mod}) \xrightarrow{DR} \Omega^\bullet_{G/K}\text{-mod} \xrightarrow{\mathrm{Ind}_{\mathcal{O}_{G/K}}^{\mathfrak{D}(G/K)}} \mathbf{C}(\mathfrak{D}(G/K)\text{-mod}),$$

and hence, on the derived level, it induces a functor isomorphic to the identity.

Vice versa, for $\mathcal{M}^\bullet \in \mathbf{C}(\mathfrak{D}(G)\text{-mod}^{r(K)})$ we have a natural map

$$\pi^*\left(\operatorname{Ind}_{\mathcal{O}_{G/K}}^{\mathfrak{D}(G/K)}\left(\mathcal{H}om_{\mathbf{C}(\mathrm{pt}/K)}\left(\mathbb{C}, \pi_*((\mathcal{M}^\bullet)^{DR})\right)\right)\right) \to \mathcal{M}^\bullet, \tag{21.5}$$

and we claim that it is a quasi-isomorphism. This follows from the fact that as a complex of vector spaces, the LHS of (21.5) is naturally filtered, and the associated graded is isomorphic to

$$\left(\mathrm{Sym}(\mathfrak{g}/\mathfrak{k}) \otimes \Lambda^\bullet(\mathfrak{g}/\mathfrak{k})\right) \otimes \mathcal{M}^\bullet,$$

where the first multiple has the Koszul differential. □

We shall now establish the following.

Proposition 21.12. *For $\mathcal{M}^\bullet \in D^+(\mathfrak{D}(G)\text{-mod})^K$ and $\mathcal{F}^\bullet \in D^+(\mathfrak{D}(G/K)\text{-mod})$, corresponding to each other under the equivalence of Proposition* 21.11, *we have a canonical quasi-isomorphism*

$$H^\bullet_{DR}(G/K, \mathcal{F}^\bullet) \simeq \mathfrak{C}^{\frac{\infty}{2}}(\mathfrak{g}; K_{\mathrm{red}}, \mathcal{M}^\bullet).$$

Note that by Lemma 20.19(1),

$$\mathfrak{C}^{\frac{\infty}{2}}(\mathfrak{g}; K_{\mathrm{red}}, \mathcal{M}^\bullet) \simeq H^\bullet_{DR}\left(\mathrm{pt}/K, (\mathcal{M}^\bullet)^{DR}\right).$$

Proof. We can assume that the complex $\mathcal{M}^\bullet$ is such that each $\mathcal{M}^i$, as a K-equivariant $\mathcal{O}_G$-module, is of the form $\pi^*(\mathcal{L}^i)$, where $\mathcal{L}^i$ is a quasi-coherent sheaf on G/K, which is the direct image from an affine subscheme. Such an $\mathcal{L}$ is obviously loose in the sense of [BD1], i.e., it has the property that the higher cohomologies $H^i(G/K, \mathcal{L} \otimes \mathcal{L}^1)$ vanish for any quasi-coherent sheaf $\mathcal{L}^1$ on G/K.

Hence, the de Rham cohomology of $\mathcal{F}^\bullet$ can be computed as $\Gamma(G/K, (\mathcal{F}^\bullet)^{DR})$. Note that the latter complex can be identified by definition with

$$\mathfrak{C}^{\frac{\infty}{2}}(\mathfrak{g}; K, \mathcal{M}^\bullet).$$

Hence, the assertion of the proposition follows from Lemma 20.19(2). □

21.13 Variant: Central extensions and twisting

Now let $\mathfrak{g}'$ be a central extension of $\mathfrak{g}$ by means of $\mathbb{C}$. We will denote by $\mathfrak{g}'_{-\mathrm{can}}$ the Baer sum of $\mathfrak{g}_{-\mathrm{can}}$ and the Baer negative of $\mathfrak{g}'$.

We introduce the category $\mathfrak{D}(G)'$-mod to consist of (discrete) vector spaces $\mathcal{M}$, endowed with an action

$$\mathcal{O}_G \overset{*}{\otimes} \mathcal{M} \overset{m}{\to} \mathcal{M}$$

as before, and a Lie algebra action

$$a_l : \mathfrak{g}' \overset{*}{\otimes} \mathcal{M} \to \mathcal{M},$$

(such that, of course, $1 \in \mathbb{C} \subset \mathfrak{g}'$ acts as an identity), and such that the two pieces of data are compatible in same way as in the definition of $\mathfrak{D}(G)$-mod.

We claim that in this case, the vector space underlying an object $\mathcal{M} \in \mathfrak{D}(G)'$-mod carries a canonically defined action, denoted a_r, of $\mathfrak{g}'_{-\text{can}}$, which commutes with a_l, and which satisfies $[a_r(x'), f] = \text{Lie}_{r(x)}(f)$ for $x' \in \mathfrak{g}'_{-\text{can}}$ and $f \in \mathcal{O}_G$.

We construct a_r by the same method as in the case of $\mathfrak{g}' = \mathfrak{g}$. Namely, we tensor $\mathcal{M}$ by $\mathsf{Spin}(\mathfrak{g})$, and show that it carries an action of $T(G)' := \mathcal{O}_G \overset{!}{\otimes} \mathfrak{g}'$, from which we produce the desired a_r.

Note, however, that in this case $\mathcal{M} \otimes \mathsf{Spin}(\mathfrak{g})$ does not carry any differential.

Let us now assume that $\mathfrak{g}'$ is a scalar multiple of an extension induced by some central extension of G by means of $\mathbb{G}_m$. Let K be an "open compact" subgroup of K, and assume that G' splits over K. We can then consider the category $\mathfrak{D}(G/K)'$-mod of twisted D-modules on G/K.

In this case we also have a well-defined category $\mathfrak{D}(G)'\text{-mod}^K$ (along with its DG and triangulated versions $\mathbf{C}(\mathfrak{D}(G)'\text{-mod})^K$ and $D(\mathfrak{D}(G)'\text{-mod})^K$). Propositions 21.7 and 21.11 generalize to the twisted context in a straightforward way.

21.14 *D*-modules with coefficients in a category

Let $\mathcal{C}$ be an abelian category, satisfying assumption (*) of Section 19.5. Then it makes sense to consider the category $\mathfrak{D}(G)\text{-mod} \otimes \mathcal{C}$, and all the results of the present section carry over to this context.

In particular, for an "open compact" subgroup $K \subset G$ we can consider the category $\mathfrak{D}(G/K)\text{-mod} \otimes \mathcal{C}$ (see Section 19.14), and we have the analogues of Propositions 21.7 and 21.11.

22 Convolution

22.1 Action of group ind-schemes on categories

We will now generalize the contents of Sections 20.1 and 20.7 to the context of group ind-schemes. Let G be an affine reasonable group ind-scheme as above. Let $\mathcal{C}$ be a category satisfying assumption (*) of Section 19.5.

A weak action of G on a $\mathcal{C}$ is the data of a functor

$$\text{act}^* : \mathcal{C} \to \text{QCoh}^*_G \otimes \mathcal{C},$$

and two functorial isomorphisms as in Section 20.1.

For $X \in \mathcal{C}$ and a scheme S mapping to G we obtain a functor

$$X \mapsto \text{act}^*(X)|_S : \mathcal{C} \to \text{QCoh}_S \otimes \mathcal{C}.$$

The following assertion is proved as Lemma 20.2.

Lemma 22.2. *For any $S \to G$, the functor $X \mapsto \mathrm{act}^*(X)|_S$ is exact, and its image consists of $\mathcal{O}_S$-flat objects.*

Let $G^{(1)}$ be the first infinitesimal neighborhood of $1 \in G$. This is a formal scheme equal to $\mathrm{Spf}(\mathbb{C} \oplus \epsilon \cdot \mathfrak{g}^*)$. If G acts on $\mathcal{C}$ and $X \in \mathcal{C}$, we obtain an object

$$X^{(1)} := \mathrm{act}(X)|_{G^{(1)}} \in \mathrm{QCoh}^*_{G^{(1)}} \otimes \mathcal{C}.$$

We say that the action of G on $\mathcal{C}$ is of Harish-Chandra type if we are given a functorial identification between $X^{(1)}$ and $p^*(X)|_{G^{(1)}}$, satisfying the same compatibility conditions as in Section 20.7.

Now let $\mathfrak{g}'$ be a central extension of $\mathfrak{g}$ by means of $\mathbb{C}$. Let $G'^{(1)}$ be the formal scheme $\mathrm{Spf}(\mathbb{C} \oplus \epsilon \cdot \mathfrak{g}'^*)$. It projects onto $G^{(1)}$ and contains $\mathrm{Spec}(\mathbb{C} \oplus \epsilon \cdot \mathbb{C})$ as a closed subscheme.

We say that a G action on $\mathcal{C}$ is of twisted Harish-Chandra type relative to $\mathfrak{g}'$ if for every $X \in \mathcal{C}$ we have a functorial isomorphism between $\mathrm{act}^*(X)|_{G'^{(1)}}$ and $p^*(X)|_{G'^{(1)}}$ such that the induced map

$$\mathrm{act}^*(X)|_{G'^{(1)}}|_{\mathrm{Spec}(\mathbb{C}\oplus\epsilon\mathbb{C})} \simeq X \oplus \epsilon \cdot X \simeq p^*(X)|_{G'^{(1)}}|_{\mathrm{Spec}(\mathbb{C}\oplus\epsilon\mathbb{C})},$$

is the automorphism

$$\mathrm{id}_X \oplus \epsilon \cdot \mathrm{id}_X : X \oplus \epsilon \cdot X \to X \oplus \epsilon \cdot X,$$

and which satisfies the second compatibility as in the nontwisted case.

22.3 Example: $\mathfrak{g}$-modules

Let $\mathbf{A}$ be an associative topological algebra with an action of G (see Section 19.2). Then the category $\mathbf{A}$-mod carries a weak G-action.

If, in addition, we have a continuous map $\mathfrak{g}' \to \mathbf{A}$ that sends $1 \in \mathbb{C} \subset \mathfrak{g}'$ to the identity in $\mathbf{A}$ such that the commutator map $\mathfrak{g} \overset{*}{\otimes} \mathbf{A} \to \mathbf{A}$ is the dual of the map $\mathbf{A} \to \mathfrak{g}^* \overset{!}{\otimes} \mathbf{A}$, obtained by deriving the G-action, then the above action of G on $\mathbf{A}$-mod is of twisted Harish-Chandra type relative to $\mathfrak{g}'$.

We will consider some particular cases of this situation. The most basic example is $\mathcal{C} = \mathfrak{g}'$-mod:

Let M be a $\mathfrak{g}'$-module. We will denote by a the action map $\mathfrak{g}' \overset{*}{\otimes} M \to M$ and by $a^* : M \to \mathfrak{g}'^* \overset{!}{\otimes} M$ its dual. For $S \to G$, we set $\mathrm{act}^*_S(M)$ to be isomorphic to $\mathcal{O}_S \otimes M$ as an $\mathcal{O}_S$-module. The $\mathfrak{g}'$-action on it is given via the map

$$\mathfrak{g}' \overset{\Delta_{\mathfrak{g}}}{\to} \mathcal{O}_G \overset{!}{\otimes} \mathfrak{g}' \to \mathcal{O}_S \overset{!}{\otimes} \mathfrak{g}'.$$

and the action of the latter on $\mathcal{O}_S \otimes M$ by means of $\mathrm{id} \otimes m$.

The restriction of $\mathrm{act}^*(M)$ to $G'^{(1)}$ identifies as a $(\mathbb{C} \oplus \epsilon \cdot \mathfrak{g}'^*)$-module with the free module

$$M \oplus \epsilon \cdot \mathfrak{g}'^* \overset{!}{\otimes} M.$$

In terms of this identification, the $\mathfrak{g}'$-action is given by

$$x \otimes (v_1 + \epsilon \cdot v_2) \mapsto a(x \otimes v_1) + \epsilon \cdot \big((a \otimes \mathrm{id}_{\mathfrak{g}'^*})(\mathrm{ad}^*(x) \otimes v_1) + a(x \otimes v_2)\big),$$

where ad^* is the map $\mathfrak{g}' \to \mathfrak{g}' \overset{!}{\otimes} \mathfrak{g}'^*$, adjoint to the bracket.

We construct an isomorphism between $M^{(1)}$ and

$$p^*(M)|_{G'^{(1)}} \simeq M \oplus \epsilon \cdot \mathfrak{g}'^* \overset{!}{\otimes} M$$

as $\mathfrak{g}'$-modules using the map

$$v_1 + \epsilon \cdot v_2 \mapsto v_1 + \epsilon \cdot (a^*(v_1) + v_2).$$

The category $\mathfrak{g}'$-mod is universal in the following sense. Let $\mathcal{C}$ be an abelian category as above, endowed an action of G and a functor $\mathsf{F} : \mathcal{C} \to \mathrm{Vect}$, respecting the action in the natural sense.

Assume that the G action on $\mathcal{C}$ is of Harish-Chandra type relative to $\mathfrak{g}'$. Then the functor F naturally lifts to a functor $\mathcal{C} \to \mathfrak{g}'$-mod.

22.4 Example: *D*-modules on *G*

Now consider the category $\mathfrak{D}(G)'$-mod. We claim that it carries an action of G of twisted Harish-Chandra type relative to $\mathfrak{g}'$, corresponding to the action of G on itself by left translations.

Let $\mathcal{M}$ be an object of $\mathfrak{D}(G)'$-mod, and S a scheme mapping to G. We define $\mathrm{act}^*(\mathcal{M})|_S$ to be isomorphic to $\mathcal{O}_S \otimes \mathcal{M}$ as an $\mathcal{O}_S$-module. The action of $\mathcal{O}_G$ is given via the comultiplication map $\mathcal{O}_G \overset{\Delta_G}{\to} \mathcal{O}_G \overset{!}{\otimes} \mathcal{O}_G \to \mathcal{O}_S \overset{!}{\otimes} \mathcal{O}_G$. The action of a_l of $\mathfrak{g}'$ is also given via the map $\mathfrak{g}' \overset{\Delta_{\mathfrak{g}}}{\to} \mathcal{O}_G \overset{!}{\otimes} \mathfrak{g} \to \mathcal{O}_S \overset{!}{\otimes} \mathfrak{g}'$.

Note that the action of $\mathfrak{g}'_{-\mathrm{can}}$ on $\mathrm{act}^*(\mathcal{M})|_S \simeq \mathcal{O}_S \otimes \mathcal{M}$ is via the a_r-action on the second multiple.

The infinitesimal trivialization of this action is defined in the same way as for $\mathfrak{g}'$-mod via the map $a_l^* : \mathcal{M} \to \mathfrak{g}'^* \overset{!}{\otimes} \mathcal{M}$.

We will now define another action of G on $\mathfrak{D}(G)'$-mod, corresponding to the action of G on itself by right translations. It will be of twisted Harish-Chandra type relative to $\mathfrak{g}'_{-\mathrm{can}}$:

For $\mathcal{M}$ and S as above, we let $\mathrm{act}^*(\mathcal{M})|_S$ again be isomorphic to $\mathcal{M} \otimes \mathcal{O}_S$ as a $\mathcal{O}_S$-module, and the $\mathcal{O}_G$-action be given via the comultiplication map $\mathcal{O}_G \to \mathcal{O}_G \overset{!}{\otimes} \mathcal{O}_S$. The a_l-action of $\mathfrak{g}'$ is $a_l \otimes \mathrm{id}_{\mathcal{O}_S}$. The resulting a_r-action of $\mathfrak{g}'_{-\mathrm{can}}$ is then given by the map $\mathfrak{g}' \to \mathcal{O}_G \overset{!}{\otimes} \mathfrak{g}' \to \mathcal{O}_S \overset{!}{\otimes} \mathfrak{g}'$.

The infinitesimal trivialization of the right action is defined in the same way as for the category $\mathfrak{g}'_{-\mathrm{can}}$-mod using the map $a_r^* : \mathcal{M} \to {\mathfrak{g}'_{-\mathrm{can}}}^* \overset{!}{\otimes} \mathcal{M}$.

It is easy to see that the two actions of G on $\mathfrak{D}(G)'$-mod commute in the natural sense. Thus we obtain an action of $G \times G$ on $\mathfrak{D}(G)'$-mod, which is of twisted Harish-Chandra type relative to $\mathfrak{g} \oplus \mathfrak{g}'_{-\text{can}}$.

22.5 The twisted product

Let $\mathcal{C}$ be a category equipped with an action of G of twisted Harish-Chandra type with respect to $\mathfrak{g}'$. Let X be an object of $\mathcal{C}$ and $\mathcal{M} \in \mathfrak{D}(G)'$-mod. We will define an object $\mathcal{M} \widetilde{\boxtimes} X \in \mathfrak{D}(G)\text{-mod} \otimes \mathcal{C}$:

As an object of $\text{QCoh}^!_G \otimes \mathcal{C}$, it is isomorphic to

$$\mathcal{M} \underset{\mathcal{O}_G}{\otimes} \text{act}^*(X)$$

(see Section 19.14). The action of $\mathfrak{g}'$ on it is defined as follows.

Consider the ind-subscheme $G^{(1)} \times G \subset G \times G$, and let p_2 denote its projection on the second multiple. Let $\mathsf{k} \subset \mathfrak{g}$ be a lattice and $\text{Spec}(\mathbb{C} \oplus \epsilon \cdot \mathsf{k}^*)$ the corresponding subscheme of $G^{(1)}$.

We have to construct an isomorphism

$$\text{mult}^* \left(\mathcal{M} \underset{\mathcal{O}_G}{\otimes} \text{act}^*(X) \right) |_{\text{Spec}(\mathbb{C} \oplus \epsilon \cdot \mathsf{k}^*) \times G} \simeq p_2^* \left(\mathcal{M} \underset{\mathcal{O}_G}{\otimes} \text{act}^*(X) \right) |_{\text{Spec}(\mathbb{C} \oplus \epsilon \cdot \mathsf{k}^*) \times G} \tag{22.1}$$

of objects of $\text{QCoh}^!_{\text{Spec}(\mathbb{C} \oplus \epsilon \cdot \mathsf{k}^*) \times G} \otimes \mathcal{C}$, compatible with the identification

$$\text{mult}^* \left(\mathcal{M} \underset{\mathcal{O}_G}{\otimes} \text{act}^*(X) \right) |_{1 \times G} \simeq \mathcal{M} \underset{\mathcal{O}_G}{\otimes} \text{act}^*(X) \simeq p_2^* \left(\mathcal{M} \underset{\mathcal{O}_G}{\otimes} \text{act}^*(X) \right) |_{1 \times G}.$$

Let k' be the preimage of k in $\mathfrak{g}'$, and let $\text{Spec}(\mathbb{C} \oplus \epsilon \cdot \mathsf{k}'^*)$ be the preimage of $\text{Spec}(\mathbb{C} \oplus \epsilon \cdot \mathsf{k}^*)$ in $G'^{(1)}$. We have an isomorphism

$$\text{mult}^*(\mathcal{M})|_{\text{Spec}(\mathbb{C} \oplus \epsilon \cdot \mathsf{k}'^*) \times G} \simeq p_2^*(\mathcal{M})|_{\text{Spec}(\mathbb{C} \oplus \epsilon \cdot \mathsf{k}'^*) \times G}$$

in $\text{QCoh}^!_{\text{Spec}(\mathbb{C} \oplus \epsilon \cdot \mathsf{k}'^*) \times G} \otimes \mathcal{C}$, given by the a_l-action of $\mathfrak{g}'$ on $\mathcal{M}$. We also have an isomorphism

$$\begin{aligned} &\text{mult}^*(\text{act}^*(X))|_{\text{Spec}(\mathbb{C} \oplus \epsilon \cdot \mathsf{k}'^*) \times G} \\ &\quad \simeq \text{act}^*(\text{act}^*(X))|_{\text{Spec}(\mathbb{C} \oplus \epsilon \cdot \mathsf{k}'^*) \times G} \simeq p_2^*(\text{act}^*(X))|_{\text{Spec}(\mathbb{C} \oplus \epsilon \cdot \mathsf{k}'^*) \times G} \end{aligned}$$

in $\text{QCoh}^*_{\text{Spec}(\mathbb{C} \oplus \epsilon \cdot \mathsf{k}'^*) \times G} \otimes \mathcal{C}$ where the first arrow is the associativity constraint for the action, and second one is the infinitesimal trivialization.

Combining the two we obtain an isomorphism

$$\text{mult}^* \left(\mathcal{M} \underset{\mathcal{O}_G}{\otimes} \text{act}^*(X) \right) |_{\text{Spec}(\mathbb{C} \oplus \epsilon \cdot \mathsf{k}'^*) \times G} \simeq p_2^* \left(\mathcal{M} \underset{\mathcal{O}_G}{\otimes} \text{act}^*(X) \right) |_{\text{Spec}(\mathbb{C} \oplus \epsilon \cdot \mathsf{k}'^*) \times G}$$

in $\text{QCoh}^!_{\text{Spec}(\mathbb{C} \oplus \epsilon \cdot \mathsf{k}'^*) \times G} \otimes \mathcal{C}$, but it is easy to see that the two central extensions cancel out, and we obtain an isomorphism as in (22.1).

By construction, this system of isomorphisms is compatible for different choices of k. Thus we obtain an action of $\mathfrak{g}$, as a Tate vector space, on $\mathcal{M} \underset{\mathcal{O}_G}{\otimes} \text{act}^*(X)$, satisfying the desired commutation relation with $\mathcal{O}_G$. Moreover, from the axioms it follows that this action of $\mathfrak{g}$ is compatible with the Lie algebra structure. We will denote this action by $\widetilde{a}_l$.

Thus $\mathcal{M} \widetilde{\boxtimes} X$ is an object of $\mathfrak{D}(G)$-mod $\otimes\, \mathcal{C}$; in particular, it carries an action of $\mathfrak{g}'_{-\text{can}}$, denoted $\widetilde{a}_r$. Let us describe this action explicitly:

Let k be a lattice in $\mathfrak{g}$ as above, and let $\mathsf{k}_{-\text{can}}$ denote its preimage in $\mathfrak{g}_{-\text{can}}$. We have to construct

$$\text{mult}^* \left(\mathcal{M} \underset{\mathcal{O}_G}{\otimes} \text{act}^*(X)\right)|_{G\times \text{Spec}(\mathbb{C}\oplus\epsilon\cdot \mathsf{k}^*_{-\text{can}})} \simeq p_1^* \left(\mathcal{M} \underset{\mathcal{O}_G}{\otimes} \text{act}^*(X)\right)|_{G\times \text{Spec}(\mathbb{C}\oplus\epsilon\cdot \mathsf{k}^*_{-\text{can}})}$$

in $\text{QCoh}^!_{G\times \text{Spec}(\mathbb{C}\oplus\epsilon\cdot \mathsf{k}^*_{-\text{can}})} \otimes \mathcal{C}$. It is constructed as in the previous case, using the a_r-action of $\mathfrak{g}'_{-\text{can}}$ on $\mathcal{M}$.

Finally, let us note that we can consider an object of $\mathcal{C}$ given by

$$(\mathcal{M} \widetilde{\boxtimes} X)^{DR} \simeq \mathfrak{C}^{\frac{\infty}{2}}(\mathfrak{g}, \mathcal{M} \widetilde{\boxtimes} X)$$

that carries a canonical differential. We will denote by $\widetilde{\text{Lie}}_l$, $\widetilde{\text{Lie}}_r$, $\widetilde{i}_l$, $\widetilde{i}^*_l$, $\widetilde{i}_r$, $\widetilde{i}^*_r$ the corresponding structures on it.

22.6 Definition of convolution

Now let $K \subset G$ be an "open compact" group subscheme over which $\mathfrak{g}'$ (and hence also $\mathfrak{g}'_{-\text{can}}$) is split. Let X be an object of $\mathcal{C}^{w,K}$ (respectively, $\mathcal{C}^K$) and $\mathcal{M}$ be an object of $\mathfrak{D}(G)'$-mod$^{w,r(K)}$ (respectively, $\mathfrak{D}(G)'$-mod$^{r(K)}$). We claim that in this case $\mathcal{M} \widetilde{\boxtimes} X$ is naturally an object of $\mathfrak{D}(G)$-mod$^{w,r(K)} \otimes \mathcal{C}$ (respectively, $\mathfrak{D}(G)$-mod$^{r(K)} \otimes \mathcal{C}$). This follows from the description of the action $\widetilde{a}_r$ given above.

More generally, if $X^\bullet \in \mathbf{C}(\mathcal{C})^K$ and $\mathcal{M}^\bullet \in \mathbf{C}(\mathfrak{D}(G)'\text{-mod})^{r(K)}$, then the complex $\mathcal{M}^\bullet \widetilde{\boxtimes} X^\bullet$ is naturally an object of $\mathbf{C}(\mathfrak{D}(G)\text{-mod} \otimes \mathcal{C})^{r(K)}$. We will denote by $(\mathcal{M}^\bullet \widetilde{\boxtimes} X^\bullet)_{G/K}$ the resulting object of $\mathbf{C}(\mathfrak{D}(G/K)\text{-mod} \otimes \mathcal{C})$.

We define a functor

$$\mathbf{C}(\mathfrak{D}(G)'\text{-mod})^{r(K)} \times \mathbf{C}(\mathcal{C})^K \to \mathbf{C}(\mathcal{C})$$

by

$$\mathcal{M}^\bullet, X^\bullet \mapsto \mathfrak{C}^{\frac{\infty}{2}}(\mathfrak{g}; K_{\text{red}}, \mathcal{M}^\bullet \widetilde{\boxtimes} X^\bullet). \tag{22.2}$$

This functor is exact when restricted to $\mathbf{C}^+(\mathfrak{D}(G)'\text{-mod})^{r(K)} \times \mathbf{C}^+(\mathcal{C})^K$, and hence we obtain a functor $D^+(\mathfrak{D}(G)'\text{-mod})^{r(K)} \times D^+(\mathcal{C})^K \to D(\mathcal{C})$, denoted

$$\mathcal{M}^\bullet, X^\bullet \mapsto \mathcal{M}^\bullet \underset{K}{\star} X^\bullet.$$

By Lemma 20.19 and Proposition 21.12,

$$\mathcal{M}^\bullet \underset{K}{\star} X^\bullet \simeq H^\bullet_{DR}(G/K, (\mathcal{M}^\bullet \widetilde{\boxtimes} X^\bullet)_{G/K}). \tag{22.3}$$

Using the equivalence $D(\mathfrak{D}(G)'\text{-mod})^{r(K)} \simeq D(\mathfrak{D}(G/K)'\text{-mod})$ we obtain also a functor $D^+(\mathfrak{D}(G/K)'\text{-mod}) \times D^+(\mathcal{C})^K \to D(\mathcal{C})$, denoted

$$\mathcal{F}^\bullet, X^\bullet \mapsto \mathcal{F}^\bullet \underset{K}{\star} X^\bullet.$$

Let $H \subset G$ be another group subscheme, not necessarily "open compact," and consider the category $\mathbf{C}(\mathfrak{D}(G)'\text{-mod})^{l(H),r(K)}$. The above convolution functor is easily seen to give rise to an exact functor

$$\mathbf{C}^+(\mathfrak{D}(G)'\text{-mod})^{l(H),r(K)} \times \mathbf{C}^+(\mathcal{C})^K \to \mathbf{C}(\mathcal{C})^H,$$

and the corresponding functor

$$\mathbf{C}^+(\mathfrak{D}(G/K)'\text{-mod})^H \times \mathbf{C}^+(\mathcal{C})^K \to \mathbf{C}(\mathcal{C})^H.$$

Note however, that if we start with an object $\mathcal{F}^\bullet \in \mathbf{C}^+(\mathfrak{D}(G/K)'\text{-mod})^H$ that comes from an object in the naive subcategory $\mathbf{C}^+(\mathfrak{D}(G/K)'\text{-mod}^H)$, the convolution $\mathcal{F}^\bullet \star X^\bullet$ is defined only as an object of $\mathbf{C}(\mathcal{C})^H$ (and not of $\mathbf{C}(\mathcal{C}^H)$). This is one of the reasons why one should work with $\mathbf{C}(\mathcal{C})^H$, rather than with $\mathbf{C}(\mathcal{C}^H)$.

Let us denote by $\mathbf{C}^{bd}(\mathfrak{D}(G/K)\text{-mod})$ the subcategory of $\mathbf{C}(\mathfrak{D}(G/K)\text{-mod})$ that consists of bounded from below complexes supported on a finite-dimensional closed subscheme of G/K. Let $D^{bd}(\mathfrak{D}(G/K)\text{-mod})$ be the corresponding full subcategory of $D(\mathfrak{D}(G/K)\text{-mod})$.

Let $\mathbf{C}^{bd}(\mathfrak{D}(G)\text{-mod})^{r(K)}$ be the subcategory of $\mathbf{C}(\mathfrak{D}(G)\text{-mod})^{r(K)}$ consisting of bounded from below complexes, supported set-theoretically on a preimage of a finite-dimensional closed subscheme of G/K; let $D^{bd}(\mathfrak{D}(G)\text{-mod})^{r(K)}$ be the corresponding full subcategory of $D(\mathfrak{D}(G)\text{-mod})^{r(K)}$.

One easily shows that under the equivalence

$$D(\mathfrak{D}(G/K)\text{-mod}) \simeq D(\mathfrak{D}(G)\text{-mod})^{r(K)},$$

the subcategories $D^{bd}(\mathfrak{D}(G/K)\text{-mod})$ and $D^{bd}(\mathfrak{D}(G)\text{-mod})^{r(K)}$ correspond to one another.

Lemma 22.7. *For $\mathcal{M}^\bullet \in D^{bd}(\mathfrak{D}(G)\text{-mod})^{r(K)}$ and $X^\bullet \in D^+(\mathcal{C})^K$ (respectively, $X^\bullet \in D^b(\mathcal{C})^K$), the convolution $\mathcal{M}^\bullet \underset{K}{\star} X^\bullet$ belongs to $D^+(\mathcal{C})$ (respectively, $D^b(\mathcal{C})$).*

Proof. Under the assumptions of the lemma the $\mathcal{C}$-valued complex of D-modules $(\mathcal{M}^\bullet \widetilde{\boxtimes} X^\bullet)_{G/K}$ is quasi-isomorphic to one bounded from below (respectively, bounded) and supported on a finite-dimensional closed subscheme of G/K. Hence, its de Rham cohomology is bounded from below (respectively, bounded). □

22.8 Examples

Let us consider the basic example, when $\mathcal{M}$ is the $\mathfrak{D}(G)'$-module

$$\delta'_{K,G} \simeq \mathrm{Ind}^{\mathfrak{g}'}_{\mathsf{k}\oplus\mathbb{C}}(\mathcal{O}_K).$$

Note that $\delta'_{K,G} \in \mathbf{C}(\mathfrak{D}(G)'\text{-mod})^{l(K),r(K)}$.

Proposition 22.9. *For $X^\bullet \in \mathbf{C}(\mathcal{C})^K$, we have a canonical quasi-isomorphism in $\mathbf{C}(\mathcal{C})^K$:*

$$\delta'_{K,G} \underset{K}{\star} X^\bullet \simeq X^\bullet.$$

Proof. Note that we can regard $\mathcal{C}$ as a category, acted on by K (rather than G). In particular, it makes sense to consider $\delta_{K,K} \widetilde{\boxtimes} X^\bullet \in \mathbf{C}(\mathfrak{D}(K)\text{-mod} \otimes \mathcal{C})^{r(K)\times K}$.

Let us regard $\delta'_{K,G} \widetilde{\boxtimes} X^\bullet$ as an object of the categories $\mathbf{C}(\mathfrak{g}_{-\mathrm{can}}\text{-mod} \otimes \mathcal{C})^{K\times K}$ and $\mathbf{C}(\mathsf{k}\text{-mod} \otimes \mathcal{C})^{K\times K}$. We have a natural map $\delta_{K,K} \widetilde{\boxtimes} X^\bullet \to \delta'_{K,G} \widetilde{\boxtimes} X^\bullet$ in the latter category, and since as a $\mathfrak{g}'_{-\mathrm{can}}$-module $\delta'_{K,G} \simeq \mathrm{Ind}^{\mathfrak{g}'_{-\mathrm{can}}}_{\mathsf{k}\oplus\mathbb{C}}(\mathcal{O}_K)$, the latter map induces an isomorphism

$$\mathrm{Ind}^{\mathfrak{g}'_{-\mathrm{can}}}_{\mathsf{k}\oplus\mathbb{C}}\left(\delta_{K,K} \widetilde{\boxtimes} X^\bullet\right) \to \delta'_{K,G} \widetilde{\boxtimes} X^\bullet \in \mathbf{C}(\mathfrak{g}_{-\mathrm{can}}\text{-mod} \otimes \mathcal{C})^{K\times K}.$$

Hence, as objects of $\mathbf{C}(\mathcal{C})^K$,

$$\delta'_{K,G} \underset{K}{\star} X^\bullet := \mathfrak{C}^{\frac{\infty}{2}}(\mathfrak{g}; K_{\mathrm{red}}, \delta'_{K,G} \widetilde{\boxtimes} X^\bullet) \overset{\text{quasi-isom}}{\simeq} \mathfrak{C}(\mathsf{k}; K_{\mathrm{red}}, \delta_{K,K} \widetilde{\boxtimes} X^\bullet).$$

This reduces the assertion of the proposition to the case when $G = K$. Note that we have a natural map

$$X^\bullet \to \mathfrak{C}(\mathsf{k}; K_{\mathrm{red}}, \delta_{K,K} \widetilde{\boxtimes} X^\bullet),$$

and it is easily seen to be a quasi-isomorphism, since as objects of $\mathbf{C}(\mathcal{C})$,

$$\mathfrak{C}(\mathsf{k}; K_{\mathrm{red}}, \delta_{K,K} \widetilde{\boxtimes} X^\bullet) \simeq \mathrm{Av}_{K_u}(X^\bullet).$$ □

More generally, let $K' \subset K$ be a group subscheme, and let $\delta'_{K/K',G/K'}$ be the twisted D-module on G/K' equal to the direct image of $\mathcal{O}_{K/K'}$ under $K/K' \to G/K'$. Arguing as above, we obtain the following.

Lemma 22.10. *For $X^\bullet \in \mathbf{C}(\mathcal{C})^{K'}$,*

$$\delta'_{K/K',G/K'} \underset{K'}{\star} X^\bullet \simeq \mathrm{Av}_K(X^\bullet) \in D(\mathcal{C})^K.$$

Now let $\mathbf{g}$ be a point of G. For an object $X \in \mathcal{C}$ we will denote by $X^{\mathbf{g}}$ (or $\delta_{\mathbf{g},G} \star \mathcal{M}$) the twist of X by means of $\mathbf{g}$, i.e., the restriction of $\mathrm{act}^*(X)$ to $\mathbf{g}$.

Applying this to $\mathcal{F} \in \mathfrak{D}(G/K)'$-mod, we obtain a $\mathbf{g}$-translate of $\mathcal{F}$ with respect to the action of G on G/K. In particular, $(\delta_{1,G/K})^{\mathbf{g}} \simeq (\delta_{\mathbf{g},G/K})$. The following results from the definitions.

Lemma 22.11. *For* $\mathcal{F} \in \mathbf{C}(\mathfrak{D}(G/K)'\text{-mod})$, $X^\bullet \in \mathbf{C}(\mathcal{C})^K$,

$$\left(\mathcal{F}^\bullet \underset{K}{\star} X^\bullet\right)^{\mathbf{g}} \simeq (\mathcal{F}^\bullet)^{\mathbf{g}} \underset{K}{\star} X^\bullet.$$

In particular, for $X^\bullet$ as above,

$$(X^\bullet)^{\mathbf{g}} \simeq \delta_{\mathbf{g},G/K} \underset{K}{\star} X^\bullet.$$

Let $G_1 \subset G$ be a group subindscheme, let $K_1 = K \cap G_1$, let $\mathcal{F}_1^\bullet$ be an object of $\mathbf{C}(\mathfrak{D}(G_1/K_1)'\text{-mod}$, and let $\mathcal{F}^\bullet \in \mathbf{C}(\mathfrak{D}(G/K)'\text{-mod})$ be its direct image under $G_1/K_1 \to G/K$.

The action of G on $\mathcal{C}$ induces an action of G_1; hence, for $X^\bullet \in \mathbf{C}(\mathcal{C})^K$ it makes sense to consider the object $\mathcal{F}_1^\bullet \underset{K_1}{\star} X^\bullet \in \mathbf{C}(\mathcal{C})$.

Lemma 22.12. *For* $\mathcal{F}_1^\bullet \in \mathbf{C}(\mathfrak{D}(G_1/K_1)'\text{-mod}$ *and* $X^\bullet \in \mathbf{C}(\mathcal{C})^K$*, the objects* $\mathcal{F}_1^\bullet \underset{K_1}{\star} X^\bullet$ *and* $\mathcal{F}^\bullet \underset{K}{\star} X^\bullet$ *in* $\mathbf{C}(\mathcal{C})$ *are canonically quasi-isomorphic.*

Proof. Let $\mathcal{M}^\bullet$ (respectively, $\mathcal{M}_1^\bullet$) be the object of $\mathbf{C}(\mathfrak{D}(G)'\text{-mod})^K$ (respectively, $\mathbf{C}(\mathfrak{D}(G_1)'\text{-mod})^{K_1}$) corresponding to $\mathcal{F}^\bullet$ (respectively, $\mathcal{F}_1^\bullet$) under the equivalence of Proposition 21.7.

Let $(\mathcal{M}^\bullet \widetilde{\boxtimes} X^\bullet)_{G/K}$ (respectively, $(\mathcal{M}_1^\bullet \widetilde{\boxtimes} X^\bullet)_{G_1/K_1}$) be the corresponding objects of the categories $\mathbf{C}(\mathfrak{D}(G/K)'\text{-mod} \otimes \mathcal{C})$ and $\mathbf{C}(\mathfrak{D}(G_1/K_1)'\text{-mod} \otimes \mathcal{C})$, respectively.

The assertion now follows from the fact that $\mathbf{C}(\mathfrak{D}(G/K)'\text{-mod} \otimes \mathcal{C})$ is the direct image $\mathbf{C}(\mathfrak{D}(G_1/K_1)'\text{-mod} \otimes \mathcal{C})$ under $G_1/K_1 \hookrightarrow G/K$. □

Finally, let us consider the example in which $\mathcal{C} = \mathfrak{D}(\mathcal{Y})'$, where $\mathcal{Y}$ is a strict ind-scheme, acted on by G, and $\mathfrak{D}(\mathcal{Y})'$ is the category of twisted D-modules on $\mathcal{Y}$, compatible with a twisting on G.

Recall that in this case we have a functor

$$D^{bd}(\mathfrak{D}(G/K)') \times D^b(\mathfrak{D}(\mathcal{Y})'\text{-mod})^K \to D^b(\mathfrak{D}(\mathcal{Y})'\text{-mod}) \tag{22.4}$$

defined as follows:

Consider the ind-scheme $G \underset{K}{\times} \mathcal{Y}$, which maps to $\mathcal{Y}$ via the action map of G on $\mathcal{Y}$; this ind-scheme is equipped with a twisting, which is pulled back from the one on $\mathcal{Y}'$ using the above map.

For $\mathcal{F}_1 \in D^{bd}(\mathfrak{D}(G/K)')$, $\mathcal{F}_2 \in D^b(\mathfrak{D}(\mathcal{Y})'\text{-mod})^K$ one can form their twisted external product

$$\mathcal{F}_1 \widetilde{\boxtimes} \mathcal{F}_2 \in D^b(\mathfrak{D}(G \underset{K}{\times} \mathcal{Y})').$$

Then $\mathcal{F}_1 \underset{K}{\star} \mathcal{F}_2$ is the direct image of $\mathcal{F}_1 \widetilde{\boxtimes} \mathcal{F}_2$ under the above map $G \underset{K}{\times} \mathcal{Y} \to \mathcal{Y}$.

It follows immediately from the definitions, that the functor (22.4) is canonically isomorphic to the one given by (22.2).

22.13 Convolution action on Harish-Chandra modules

We shall now study a particular case of the above situation, when $\mathcal{C} = \mathfrak{g}'$-mod. First, for $\mathcal{M} \in \mathfrak{D}(G)'$-mod and $N \in \mathfrak{g}'$-mod let us describe the object $\mathcal{M}\,\widetilde{\boxtimes}\,N \in \mathfrak{D}(G)\text{-mod} \otimes \mathfrak{g}'$-mod more explicitly.

As a vector space $\mathcal{M}\,\widetilde{\boxtimes}\,N$ is isomorphic to $\mathcal{M} \otimes N$. We will denote by m the action of $\mathcal{O}_G$ on $\mathcal{M}$, and by a_l, a_r the actions of $\mathfrak{g}'$, $\mathfrak{g}'_{-\mathrm{can}}$ on it. We will denote by a the action of $\mathfrak{g}'$ on N.

Let $\widetilde{m}, \widetilde{a}_l, \widetilde{a}_r$ and $\widetilde{a}$ be the actions of $\mathcal{O}_G$, $\mathfrak{g}$, $\mathfrak{g}_{-\mathrm{can}}$ and $\mathfrak{g}'$, respectively, on $\mathcal{M} \otimes N$ defining on $\mathcal{M}\,\widetilde{\boxtimes}\,N$ a structure of object of $\mathfrak{D}(G)\text{-mod} \otimes \mathfrak{g}'$-mod. We have

- $\widetilde{m} = m \otimes \mathrm{id}$,
- $\widetilde{a} = (m \otimes a) \circ (\gamma \otimes \mathrm{id}_{\mathfrak{g}}) \circ \Delta_{\mathfrak{g}'}$,
- $\widetilde{a}_l = (a_l \otimes \mathrm{id}) - (m \otimes a) \circ (\gamma \otimes \mathrm{id}_{\mathfrak{g}}) \circ \Delta_{\mathfrak{g}'}$,
- $\widetilde{a}_r = a_r \otimes \mathrm{id} + \mathrm{id} \otimes a$.

More generally, if $\mathcal{M}^\bullet$ is an object of $\mathbf{C}(\mathfrak{D}(G)'\text{-mod})^{r(K)}$ and $N^\bullet$ is an object of $\mathbf{C}(\mathfrak{g}'\text{-mod})^K$, the twisted product $\mathcal{M}\,\widetilde{\boxtimes}\,N$ is naturally an object of $\mathbf{C}(\mathfrak{D}(G)\text{-mod})^{r(K)} \otimes \mathbf{C}(\mathfrak{g}'\text{-mod})$, where the algebraic action of K on $\mathcal{M}^\bullet \otimes N^\bullet$ is the diagonal one, and so is the action of $\mathsf{k}[1]$.

In this case the convolution $\mathcal{M}^\bullet \underset{K}{\star} N^\bullet$ is computed by means of the complex

$$\mathfrak{C}^{\frac{\infty}{2}}\left(\mathfrak{g}; K_{\mathrm{red}}, \mathcal{M}^\bullet \otimes N^\bullet\right),$$

with respect to the diagonal action of $\mathfrak{g}_{-\mathrm{can}}$. The $\mathfrak{g}'$-module structure on this complex is given by $\widetilde{a}$.

Note, however, that the above complex carries a different action of $\mathfrak{g}'$, namely one given by a_l. We will denote this other functor $\mathbf{C}(\mathfrak{D}(G)'\text{-mod})^{r(K)} \times \mathbf{C}(\mathfrak{g}'\text{-mod})^K \to \mathbf{C}(\mathfrak{g}'\text{-mod})$ by

$$\mathcal{M}^\bullet, N^\bullet \mapsto \mathcal{M}^\bullet \overset{\natural}{\underset{K}{\star}} N^\bullet.$$

Note that if $\mathcal{M} \in \mathbf{C}(\mathfrak{D}(G)'\text{-mod})^{H,r(K)}$ for some group scheme H, then $\mathcal{M}^\bullet \overset{\natural}{\underset{K}{\star}} N^\bullet$ is naturally an object of $\mathbf{C}(\mathfrak{g}'\text{-mod})^H$.

The two actions of $\mathfrak{g}'$ on $\mathfrak{C}^{\frac{\infty}{2}}(\mathfrak{g}; K_{\mathrm{red}}, \mathcal{M}^\bullet \otimes N^\bullet)$ are related by the formula

$$a_l - \widetilde{a} = \widetilde{a}_l = [d, \widetilde{i}_l],$$

where $\widetilde{i}_l$ is the action of the annihilation operators on

$$\mathfrak{C}^{\frac{\infty}{2}}\left(\mathfrak{g}; K_{\mathrm{red}}, \mathcal{M}^\bullet \otimes N^\bullet\right) \subset \mathfrak{C}^{\frac{\infty}{2}}\left(\mathfrak{g}, \mathcal{M}^\bullet \otimes N^\bullet\right) \simeq (\mathcal{M}\,\widetilde{\boxtimes}\,N)^{DR}.$$

Therefore, the cohomologies of $\mathcal{M}^\bullet \underset{K}{\star} N^\bullet$ and $\mathcal{M}^\bullet \overset{\natural}{\underset{K}{\star}} N^\bullet$ are isomorphic as $\mathfrak{g}'$-modules.

Corollary 22.14.

(1) *For* $N \in (\mathfrak{g}', K)$-mod, *the complex*

$$\mathfrak{C}^{\frac{\infty}{2}}\left(\mathfrak{g}; K_{\mathrm{red}}, \delta'_{K,G} \otimes N\right)$$

is acyclic away from degree 0.

(2) *When regarded as a* $\mathfrak{g}'$*-module via the* a_l*-action on* $\delta'_{K,G}$*, the above* 0*th cohomology is isomorphic to* N.

(3) *The image of* $U(\mathfrak{g}')$ *in* $\mathrm{End}_{\mathfrak{g}'_{-\mathrm{can}}}(\delta'_{K,G})$ *is dense.*

Proof. The first two points follows from Proposition 22.9 and the above comparison of $\mathcal{M} \overset{\natural}{\underset{K}{\star}} N^\bullet$ and $\mathcal{M}^\bullet \underset{K}{\star} N^\bullet$.

Let $U(\mathfrak{g}', \mathsf{k})$ be the topological algebra of endomorphisms of the forgetful functor

$$(\mathfrak{g}', K)\text{-mod} \to \mathrm{Vect}\,.$$

Evidently, the image of $U(\mathfrak{g}')$ in $U(\mathfrak{g}', \mathsf{k})$ is dense. We claim now that $U(\mathfrak{g}', \mathsf{k})$ is isomorphic to $\mathrm{End}_{\mathfrak{g}'_{-\mathrm{can}}}(\delta'_{K,G})$.

The map in one direction, i.e., $U(\mathfrak{g}', \mathsf{k}) \to \mathrm{End}_{\mathfrak{g}'_{-\mathrm{can}}}(\delta'_{K,G})$, is evident: given an element in $U(\mathfrak{g}', \mathsf{k})$, we obtain a *functorial* endomorphism of every vector space underlying an object of $\mathfrak{g}'$-mod; in particular $\delta'_{K,G}$. This endomorphism commutes with $\mathfrak{g}'$-mod-endomorphisms of $\delta'_{K,G}$, in particular, with the action of $\mathfrak{g}'_{-\mathrm{can}}$.

To construct the map in the opposite direction, note that an endomorphism of $\delta'_{K,G}$ as a $\mathfrak{g}'_{-\mathrm{can}}$-module defines an endomorphism of the functor

$$N \mapsto h^0\left(\mathfrak{C}^{\frac{\infty}{2}}\left(\mathfrak{g}; K_{\mathrm{red}}, \delta'_{K,G} \otimes N\right)\right) : \mathfrak{g}'\text{-mod} \to \mathrm{Vect},$$

and the latter is isomorphic to the forgetful functor. □

We will now study the behavior of Lie algebra cohomology under convolution. We shall first consider a technically simpler case, when we will consider D-modules on a group scheme H, mapping to G, such that $\mathfrak{g}'$ splits over $\mathfrak{h}$. Let $K'_H, K''_H \subset H$ be group subschemes of finite codimension.

Proposition 22.15. *For* $N^\bullet \in D^+(\mathfrak{g}\text{-mod})^{K'_H}$ *and* $\mathcal{M}^\bullet \in D^+(\mathfrak{D}(H))^{l(K'_H), r(K''_H)}$*, the complex* $\mathfrak{C}(\mathfrak{h}; K'_{H red}, \mathcal{M}^\bullet \underset{K'_H}{\star} N^\bullet)$ *is quasi-isomorphic to*

$$H^\bullet_{DR}(H'_K \backslash H, \mathcal{F}^\bullet) \otimes \mathfrak{C}(\mathfrak{h}; H''_{K\,\mathrm{red}}, \mathcal{M}^\bullet),$$

where $\mathcal{F}^\bullet$ *is the object of* $D^+(\mathfrak{D}(H'_K \backslash H))$*, corresponding to* $\mathcal{M}^\bullet$.

Proof. By Lemma 19.19,

$$\mathfrak{C}(\mathfrak{h}; K'_{H\,\mathrm{red}}, \mathcal{M}^\bullet \underset{K''_H}{\star} N^\bullet) \simeq \mathfrak{C}(\mathfrak{h}; K'_{H\,\mathrm{red}}, \mathcal{M}^\bullet \overset{\natural}{\underset{K''_H}{\star}} N^\bullet).$$

The latter, by definition, can be rewritten as

$$\mathfrak{C}(\mathfrak{h} \oplus \mathfrak{h}; K'_{H\,\mathrm{red}} \times K''_{H\,\mathrm{red}}, \mathcal{M}^\bullet \otimes N^\bullet),$$

where the action of the first copy of $\mathfrak{h}$ is via a_l on $\mathcal{M}$, and the action of the second copy is diagonal with respect to a_r and a. Hence, the above expression can be rewritten as

$$\mathfrak{C}(\mathfrak{h}; K''_{H\,\mathrm{red}}, \mathfrak{C}(\mathfrak{h}; K'_{H\,\mathrm{red}}, \mathcal{M}^\bullet) \otimes N^\bullet),$$

where the $\mathfrak{h}$-action on $\mathfrak{C}(\mathfrak{h}; K'_{H\,\mathrm{red}}, \mathcal{M}^\bullet) \otimes N^\bullet$ is the diagonal one with respect to the a_r-action on $\mathcal{M}^\bullet$ and the existing action on $N^\bullet$.

Applying again Lemma 19.19, we can replace the a_r-action on $\mathcal{M}^\bullet$ by the trivial one. Hence,

$$\mathfrak{C}(\mathfrak{h}; K''_{H\,\mathrm{red}}, \mathfrak{C}(\mathfrak{h}; K'_{H\,\mathrm{red}}, \mathcal{M}^\bullet) \otimes N^\bullet) \simeq \mathfrak{C}(\mathfrak{h}; K'_{H\,\mathrm{red}}, \mathcal{M}^\bullet) \otimes \mathfrak{C}(\mathfrak{h}; K''_{H\,\mathrm{red}}, N^\bullet),$$

which is what we had to show. □

We will now generalize the above proposition to the case of semi-infinite cohomology with respect to $\mathfrak{g}$.

Let $N_1^\bullet$ and $N_2^\bullet$ be objects of $D^+(\mathfrak{g}'_{-\mathrm{can}}\text{-mod})^{K'}$ and $D^+(\mathfrak{g}'\text{-mod})^{K''}$, respectively, for some "open compact" $K, K'' \subset G$. Let $\mathcal{M}^\bullet$ be an object of $D^+(\mathfrak{D}(G)'\text{-mod})^{l(K'),r(K'')}$, supported over a closed pro-finite-dimensional subscheme of G. In this case the convolution $\mathcal{M}^\bullet \underset{K''}{\star} N_2^\bullet$ makes sense as an object of $D^+(\mathfrak{g}'\text{-mod})^{K'}$. Similarly, we can consider the convolution "on the right"

$$N_1^\bullet \underset{K'}{\star} \mathcal{M}^\bullet \in D^+(\mathfrak{g}'_{-\mathrm{can}}\text{-mod})^{K'}.$$

Proposition 22.16. *Under the above circumstances,*

$$\mathfrak{C}^{\frac{\infty}{2}}\left(\mathfrak{g}; K'_{\mathrm{red}}, N_1^\bullet \otimes \left(\mathcal{M}^\bullet \underset{K''}{\star} N_2^\bullet\right)\right)$$

and

$$\mathfrak{C}^{\frac{\infty}{2}}\left(\mathfrak{g}; K''_{\mathrm{red}}, \left(N_1^\bullet \underset{K'}{\star} \mathcal{M}^\bullet\right) \otimes N_2^\bullet\right)$$

are quasi-isomorphic.

Proof. By symmetry, it would be sufficient to show that there exists a quasi-isomorphism between

$$\mathfrak{C}^{\frac{\infty}{2}}\left(\mathfrak{g}; K'_{\mathrm{red}}, N_1^\bullet \otimes \left(\mathcal{M}^\bullet \underset{K''}{\star} N_2^\bullet\right)\right) \tag{22.5}$$

and

$$\mathfrak{C}^{\frac{\infty}{2}}\left(\mathfrak{g} \oplus \mathfrak{g}; K'_{\mathrm{red}} \times K''_{\mathrm{red}}, N_1^\bullet \otimes \mathcal{M}^\bullet \otimes N_2^\bullet\right), \tag{22.6}$$

where the first copy of $\mathfrak{g}_{-\mathrm{can}}$ acts diagonally on $N_1^\bullet \otimes \mathcal{M}^\bullet$ (via the existing $\mathfrak{g}'_{-\mathrm{can}}$ action on $N_1^\bullet$ and the a_l-action on $\mathcal{M}^\bullet$) and the second copy acts diagonally on $\mathcal{M}^\bullet \otimes N_2^\bullet$ (via the a_r-action on $\mathcal{M}^\bullet$ and the existing $\mathfrak{g}'$-action on $N_2^\bullet$).

By Lemma 19.19, in (22.6) we can replace the action of the first copy of $\mathfrak{g}_{-\mathrm{can}}$, by one where the $\mathfrak{g}'$-action on $\mathcal{M}^\bullet \otimes N_2^\bullet$ is given by $\widetilde{a}$. The resulting expression will be equal to the one in (22.5) modulo the following complication:

To define (22.5) one has to replace $\mathfrak{C}^{\frac{\infty}{2}}(\mathfrak{g}; K''_{\mathrm{red}}, \mathcal{M}^\bullet \otimes N_2^\bullet)$ by a quasi-isomorphic complex, which is bounded from below. We have to show that taking $\mathfrak{C}^{\frac{\infty}{2}}(\mathfrak{g}; K'_{\mathrm{red}}, N_1^\bullet \otimes ?)$ survives this quasi-isomorphism.

Let us first consider a particular case in which $\mathcal{M}^\bullet$ is induced from an $\mathcal{O}_G$-module, i.e., has the form

$$\mathrm{Ind}_{\mathsf{k}''}^{\mathfrak{g}'_{-\mathrm{can}}}(\mathcal{L}^\bullet) \tag{22.7}$$

for some complex $\mathcal{L}^\bullet$ of K'-equivariant $\mathcal{O}_G$-modules. In this case we have a quasi-isomorphism

$$\mathfrak{C}(\mathsf{k}''; K''_{\mathrm{red}}, \mathcal{L}^\bullet \otimes N_2^\bullet) \to \mathfrak{C}^{\frac{\infty}{2}}(\mathfrak{g}; K''_{\mathrm{red}}, \mathcal{M}^\bullet \otimes N_2^\bullet)$$

of complexes of $\mathfrak{g}'$-modules. Moreover, the PBW filtration defines a filtration on the RHS, of which $\mathfrak{C}(\mathsf{k}''; K''_{\mathrm{red}}, \mathcal{L}^\bullet \otimes N_2^\bullet)$ is the first term, by $\mathfrak{g}'$-stable subcomplexes, all quasi-isomorphic to one another.

Since the functor $\mathfrak{C}^{\frac{\infty}{2}}(\mathfrak{g}; K'_{\mathrm{red}}, N_1^\bullet \otimes ?)$ commutes with direct limits, the required assertion about quasi-isomorphism holds.

The case of a general $\mathcal{M}^\bullet$ follows from the one considered above, since the assumption on $\mathcal{M}^\bullet$ implies that it can be represented by a complex associated with a bicomplex with finitely many rows, each of the form (22.7). □

22.17 Convolution action on $\mathfrak{D}(G)$-modules

Let us now consider the case when $\mathcal{C} = \mathfrak{D}(G)'$ with the action of G by left translations.

Given two objects $\mathcal{M}_1, \mathcal{M}_2 \in \mathfrak{D}(G)'$-mod let us first describe how $\mathcal{M}_1 \widetilde{\boxtimes} \mathcal{M}_2$ looks like as an object of $\mathfrak{D}(G)$-mod $\otimes$ $\mathfrak{D}(G)'$-mod.

By construction as a vector space $\mathcal{M}_1 \widetilde{\boxtimes} \mathcal{M}_2 \simeq \mathcal{M}_1 \otimes \mathcal{M}_2$. We will denote by a_l^1, a_l^2 (respectively, a_r^1, a_r^2, m^1, m^2) the actions of $\mathfrak{g}'$ (respectively, $\mathfrak{g}'_{-\mathrm{can}}, \mathcal{O}_G$) on $\mathcal{F}_1$ and $\mathcal{F}_2$, respectively. We will denote by $\widetilde{a}_l^1, \widetilde{a}_l^2, \widetilde{a}_r^1, \widetilde{a}_r^2, \widetilde{m}^1, \widetilde{m}^2$ the actions of $\mathfrak{g}, \mathfrak{g}', \mathfrak{g}_{-\mathrm{can}}, \mathfrak{g}'_{-\mathrm{can}}$, respectively on $\mathcal{F}_1 \widetilde{\boxtimes} \mathcal{F}_2$, corresponding to the $\mathfrak{D}(G)$-mod $\otimes$ $\mathfrak{D}(G)'$-mod-structure.

The action of $\mathcal{O}_G$, corresponding to the $\mathfrak{D}(G)$-mod-structure on $\mathcal{M}_1 \widetilde{\boxtimes} \mathcal{M}_2$ is via the first multiple in $\mathcal{M}_1 \otimes \mathcal{M}_2$, which we will denote by m. The action of $\mathcal{O}_G$, corresponding to the $\mathfrak{D}(G)'$-mod-structure, is via the comultiplication map $\Delta_G : \mathcal{O}_G \to \mathcal{O}_G \overset{!}{\otimes} \mathcal{O}_G$.

These actions are described as follows:

- $\widetilde{m}^1 = m^1 \otimes \mathrm{id}$,
- $\widetilde{m}^2 = (m^1 \otimes m^2) \circ (\gamma \otimes \mathrm{id}) \circ \Delta_G$,
- $\widetilde{a}_l^1 = a_l^1 \otimes \mathrm{id} - (m \otimes a_l^2) \circ (\gamma \otimes \mathrm{id}) \circ \Delta_{\mathfrak{g}'}$,

- $\widetilde{a}_l^2 = (m \otimes a_l^2) \circ (\gamma \otimes \mathrm{id}) \circ \Delta_{\mathfrak{g}'}$,
- $\widetilde{a}_r^1 = a_r^1 \otimes \mathrm{id} + \mathrm{id} \otimes a_l^2$,
- $\widetilde{a}_r^2 = \mathrm{id} \otimes a_r^2$.

If $\mathcal{M}_1^\bullet \in \mathbf{C}^+(\mathfrak{D}(G)')\text{-mod}^{r(K)}$ and $\mathcal{M}_2^\bullet \in \mathbf{C}^+(\mathfrak{D}(G)')\text{-mod}^{l(K)}$, the convolution $\mathcal{M}_1^\bullet \underset{K}{\star} \mathcal{M}_2^\bullet$ is computed by means of

$$\mathfrak{C}^{\frac{\infty}{2}}(\mathfrak{g}; K_{\mathrm{red}}, \mathcal{M}_1^\bullet \otimes \mathcal{M}_2^\bullet),$$

with respect to the diagonal (i.e., $\widetilde{a}_r^1 = a_r^1 + a_l^2$) action of $\mathfrak{g}_{-\mathrm{can}}$, and the actions of $\mathcal{O}_G$, $\mathfrak{g}'$, and $\mathfrak{g}'_{-\mathrm{can}}$, specified above.

Note, however, that the above complex carries a different $\mathfrak{D}(G)'$-module structure. Namely, the action of $\mathcal{O}_G$ is $(m_1 \otimes m_2) \circ (\gamma \otimes \mathrm{id}) \circ \Delta_G$ as before, and the action of $\mathfrak{g}'$ is a_l^1. In this case the action of $\mathfrak{g}'_{-\mathrm{can}}$ equals $(a_r^1 \otimes m) \circ \Delta_{\mathfrak{g}}$.

We will denote this new functor

$$\mathbf{C}^+(\mathfrak{D}(G)')\text{-mod}^{r(K)} \times \mathbf{C}^+(\mathfrak{D}(G)')\text{-mod}^{l(K)} \to \mathbf{C}(\mathfrak{D}(G)'\text{-mod}$$

by

$$\mathcal{M}_1^\bullet, \mathcal{M}_2^\bullet \mapsto \mathcal{M}_1^\bullet \overset{\natural}{\underset{K}{\star}} \mathcal{M}_2^\bullet.$$

Lemma 22.18. *For $\mathcal{M}_1^\bullet \in D^{bd}(\mathfrak{D}(G)'\text{-mod})^{r(K)}$, $\mathcal{M}_2^\bullet \in D^{bd}(\mathfrak{D}(G)'\text{-mod})^{l(K)}$, the objects*

$$\mathcal{M}_1^\bullet \underset{K}{\star} \mathcal{M}_2^\bullet, \mathcal{M}_1^\bullet \overset{\natural}{\underset{K}{\star}} \mathcal{M}_2^\bullet \in D^b(\mathfrak{D}(G)'\text{-mod})$$

are isomorphic.

Proof. From the assumption it follows that there exist "open compact" groups K', K'' such that $\mathcal{M}_1^\bullet \in D(\mathfrak{D}(G)'\text{-mod})^{l(K'),r(K)}$ and $\mathcal{M}_2^\bullet \in D(\mathfrak{D}(G)'\text{-mod})^{l(K),r(K'')}$.

As we saw above, the convolution $\mathcal{M}_1^\bullet \underset{K}{\star} \mathcal{M}_2^\bullet$ can be interpreted as an action of $\mathcal{F}_1^\bullet \in D^{bd}(\mathfrak{D}(G/K)'\text{-mod})^{K'}$, corresponding to $\mathcal{M}_1^\bullet$, on $\mathcal{F}_2^\bullet \in D^b(\mathfrak{D}(G/K'')'\text{-mod})^K$, corresponding to $\mathcal{M}_2^\bullet$. The result is an object in $D^b(\mathfrak{D}(G/K'')\text{-mod})^{K'}$.

However, this convolution can be rewritten also as an action of

$${}'\mathcal{F}_2^\bullet \in D^{bd}(\mathfrak{D}(K\backslash G)'\text{-mod})^{K''}$$

on

$${}'\mathcal{F}_1^\bullet \in D^b(\mathfrak{D}(K'\backslash G)'\text{-mod})^K,$$

with the result being in

$$D^b(\mathfrak{D}(K'\backslash G)'\text{-mod})^{K''} \simeq D^b(\mathfrak{D}(G/K'')\text{-mod})^{K'}.$$

The latter convolution is manifestly the same as

$$\mathcal{M}_1^\bullet \overset{\natural}{\underset{K}{\star}} \mathcal{M}_2^\bullet \in D^b(\mathfrak{D}(G)'\text{-mod})^{l(K'),r(K'')}. \qquad \square$$

22.19 Associativity of convolution

Now let $\mathcal{M}_1^\bullet$ be an object of

$$\mathbf{C}^{bd}(\mathfrak{D}(G)'\text{-mod})^{r(K)}, \qquad \mathcal{M}_2^\bullet \in \mathbf{C}^{bd}(\mathfrak{D}(G)'\text{-mod})^{l(K),r(K')},$$

and

$$X^\bullet \in \mathbf{C}^+(\mathcal{C})^{K'}$$

for a category $\mathcal{C}$ as above.

Proposition 22.20. *Under the above circumstances, there exists a canonical isomorphism in $D^+(\mathcal{C})$*

$$\left(\mathcal{M}_1^\bullet \underset{K}{\star} \mathcal{M}_2^\bullet\right) \underset{K'}{\star} X^\bullet \simeq \mathcal{M}_1^\bullet \underset{K}{\star} \left(\mathcal{M}_2^\bullet \underset{K'}{\star} X^\bullet\right),$$

compatible with three-fold convolutions.

The rest of this subsection is devoted to the proof of this proposition.

Consider the bigraded object of $\mathcal{C}$ given by

$$\left(\mathcal{M}_1^\bullet \boxtimes \mathcal{M}_2^\bullet\right) \underset{\mathcal{O}_{G\times G}}{\otimes} \text{mult}^*\left(\text{act}^*(X^\bullet)\right). \tag{22.8}$$

It carries two actions of the Lie algebra $\mathfrak{g}_{-\text{can}} \oplus \mathfrak{g}_{-\text{can}}$, corresponding to the two isomorphisms

$$\mathcal{M}_1^\bullet \widetilde{\boxtimes} \left(\mathcal{M}_2^\bullet \widetilde{\boxtimes} X^\bullet\right) \simeq \left(\mathcal{M}_1^\bullet \boxtimes \mathcal{M}_2^\bullet\right) \underset{\mathcal{O}_{G\times G}}{\otimes} \text{mult}^*\left(\text{act}^*(X^\bullet)\right) \simeq \left(\mathcal{M}_1^\bullet \widetilde{\boxtimes} \mathcal{M}_2^\bullet\right) \widetilde{\boxtimes} X^\bullet.$$

The action of the second copy of $\mathfrak{g}_{-\text{can}}$ is the same in the two cases. The difference of the actions of the first copy of $\mathfrak{g}_{-\text{can}}$ is given by the $\mathfrak{g}$-action, coming from its $\widetilde{a}_l$-action on $\mathcal{M}_2^\bullet \widetilde{\boxtimes} X^\bullet$.

Hence, by Lemma 19.19, the two complexes

$$\mathfrak{C}^{\frac{\infty}{2}}\left(\mathfrak{g} \oplus \mathfrak{g}; K \times K', \mathcal{M}_1^\bullet \widetilde{\boxtimes} \left(\mathcal{M}_2^\bullet \widetilde{\boxtimes} X^\bullet\right)\right)$$

and

$$\mathfrak{C}^{\frac{\infty}{2}}\left(\mathfrak{g} \oplus \mathfrak{g}; K \times K', \left(\mathcal{M}_1^\bullet \widetilde{\boxtimes} \mathcal{M}_2^\bullet\right) \widetilde{\boxtimes} X^\bullet\right)$$

are isomorphic.

As in the proof of Proposition 22.16, we have to show that the above complexes are isomorphic in the derived category to $\mathcal{M}_1^\bullet \underset{K}{\star} (\mathcal{M}_2^\bullet \underset{K'}{\star} X^\bullet)$ and $(\mathcal{M}_1^\bullet \underset{K}{\star} \mathcal{M}_2^\bullet) \underset{K'}{\star} X^\bullet$, respectively. This is done as in the proof of Proposition 22.16 by replacing $\mathcal{M}_1^\bullet$ and $\mathcal{M}_2^\bullet$ by appropriately chosen complexes, for which the above semi-infinite complexes can be represented as direct limits of quasi-isomorphic complexes, bounded from below.

22.21 An adjunction in the proper case

Now let $K_1, K_2 \in G$ be two "open compact" subgroups of G, and assume that G/K_1 is ind-proper. Let $\mathcal{F}$ be a finitely generated object of $\mathfrak{D}(G/K_1)'\text{-mod}^{K_2}$. As in Section 21.8, we have a well-defined object $\mathcal{F}^{\text{op}}$ in $\mathfrak{D}(G/K_2)''\text{-mod}^{K_1}$, where the superscript $''$ indicates the twisting opposite to $'$. Then the Verdier dual $\mathbb{D}(\mathcal{F}^{\text{op}})$ is an object of $\mathfrak{D}(G/K_2)'\text{-mod}^{K_1}$.

Proposition 22.22. *The functor*

$$D(\mathcal{C})^{K_1} \to D(\mathcal{C})^{K_2} : X_1^\bullet \mapsto \mathcal{F} \underset{K_1}{\star} X_1^\bullet$$

is left adjoint to the functor

$$D(\mathcal{C})^{K_2} \to D(\mathcal{C})^{K_1} : X_2 \mapsto \mathbb{D}(\mathcal{F}^{\text{op}}) \underset{K_2}{\star} X_2.$$

Proof. We need to construct the adjunction maps

$$\mathcal{F} \underset{K_1}{\star} (\mathbb{D}(\mathcal{F}^{\text{op}}) \underset{K_2}{\star} X_2^\bullet) \to X_2^\bullet \quad \text{and} \quad X_1^\bullet \to \mathbb{D}(\mathcal{F}^{\text{op}}) \underset{K_2}{\star} (\mathcal{F} \underset{K_1}{\star} X_1^\bullet),$$

such that the identities concerning the two compositions hold.

In view of Proposition 22.20, it would suffice to construct the maps

$$\mathcal{F} \underset{K_1}{\star} \mathbb{D}(\mathcal{F}^{\text{op}}) \to \delta_{1,G/K_2} \in D(\mathfrak{D}(G/K_2)\text{-mod})^{K_2}$$

and

$$\delta_{1,G/K_1} \to \mathbb{D}(\mathcal{F}^{\text{op}}) \underset{K_2}{\star} \mathcal{F} \in D(\mathfrak{D}(G/K_1)\text{-mod})^{K_1},$$

such that the corresponding identities hold.

By the definition of convolution, constructing these maps is equivalent to constructing morphisms

$$H^\bullet(G/K_1, \Delta^*_{G/K_1}(\mathcal{F} \boxtimes \mathbb{D}(\mathcal{F})) \to \mathbb{C} \in D(\text{pt}/K_2) \tag{22.9}$$

and

$$\mathbb{C} \to H^\bullet(G/K_2, \Delta^!_{G/K_2}(\mathcal{F}^{\text{op}} \boxtimes \mathbb{D}(\mathcal{F}^{\text{op}})) \in D(\text{pt}/K_1), \tag{22.10}$$

respectively, where $\Delta_{G/K}$ denotes the diagonal morphism $G/K \to G/K \times G/K$. (Note that in each of the cases, the pull-back of the corresponding twisted D-module on the product under the diagonal map is a nontwisted right D-module.)

The morphism in (22.10) follows from Verdier duality, and likewise for (22.9), using the fact that

$$H^\bullet(G/K_1, \cdot) \simeq H_c^\bullet(G/K_1, \cdot).$$

The fact that the identities concerning the compositions of adjunction maps hold is an easy verification. □

23 Categories over topological commutative algebras

23.1 The notion of a category flat over an algebra

Let $\mathcal{C}$ be an abelian category as in Section 19.5, satisfying assumption (**), and let Z be a commutative algebra, mapping to the center of $\mathcal{C}$. An example of this situation is when $\mathbf{A}$ is a topological algebra, Z is a (discrete) commutative algebra mapping to the center of $\mathbf{A}$ and $\mathcal{C} = \mathbf{A}$-mod. Then the functor F factors naturally through a functor $\mathsf{F}_Z : \mathcal{C} \to Z$-mod.

Note that we have a naturally defined functor $Z\text{-mod} \times \mathcal{C} \to \mathcal{C}$ given by

$$M, X \mapsto M \underset{Z}{\otimes} X.$$

This functor is right exact in both arguments. We have

$$\mathsf{F}_Z(M \underset{Z}{\otimes} X) \simeq M \underset{Z}{\otimes} \mathsf{F}_Z(X).$$

This shows, in particular, that if M is Z-flat, then the above functor of tensor product is exact in Z. We will denote by

$$M^\bullet, X^\bullet \mapsto M^\bullet \overset{L}{\underset{Z}{\otimes}} X^\bullet : D^-(Z\text{-mod}) \times D^-(\mathcal{C}) \to D^-(\mathcal{C})$$

the corresponding derived functor. We have

$$\mathsf{F}_Z(M^\bullet \overset{L}{\underset{Z}{\otimes}} X^\bullet) \simeq M^\bullet \overset{L}{\underset{Z}{\otimes}} \mathsf{F}_Z(X^\bullet).$$

It is easy to see that for a fixed $X^\bullet \in \mathbf{C}^-(\mathcal{C})$, the derived functor of

$$M^\bullet \mapsto M^\bullet \underset{Z}{\otimes} X^\bullet : \mathbf{C}^-(Z\text{-mod}) \to \mathbf{C}^-(\mathcal{C})$$

is isomorphic to $M^\bullet \overset{L}{\underset{Z}{\otimes}} X^\bullet$. However, this is not, in general, true for the functor

$$X^\bullet \mapsto M^\bullet \underset{Z}{\otimes} X^\bullet : \mathbf{C}^-(\mathcal{C}) \to \mathbf{C}^-(\mathcal{C})$$

for a fixed $M^\bullet$.

We shall say that an object $X \in \mathcal{C}$ is flat over Z if the functor

$$M \mapsto M \underset{Z}{\otimes} X : Z\text{-mod} \to \mathcal{C}$$

is exact. This is equivalent to $\mathsf{F}_Z(X)$ being flat as a Z-module.

We shall say that $\mathcal{C}$ is flat over Z if every object of X admits a surjection $X' \to X$ for X being flat over Z.

Consider the example of $\mathcal{C} = \mathbf{A}$-mod. Suppose there exists a family of open left ideals $\mathbf{I} \subset \mathbf{A}$ such that $\mathbf{A} \simeq \varprojlim \mathbf{A}/\mathbf{I}$, such that each $\mathbf{A}/\mathbf{I}$ is flat as a Z-module. Then $\mathcal{C}$ is flat over Z.

Lemma 23.2. *Let $\mathcal{C}$ be flat over Z, then for a fixed $M^\bullet \in \mathbf{C}^-(Z\text{-mod})$ the left derived functor of*

$$X^\bullet \mapsto M^\bullet \underset{Z}{\otimes} X^\bullet : \mathbf{C}^-(\mathcal{C}) \to \mathbf{C}^-(\mathcal{C})$$

is isomorphic to $M^\bullet \overset{L}{\underset{Z}{\otimes}} X^\bullet$.

Proof. By assumption, every object in $\mathbf{C}^-(\mathcal{C})$ admits a quasi-isomorphism from one consisting of objects that are Z-flat. Hence, it suffices to show that if $X^\bullet \in \mathbf{C}^-(\mathcal{C})$ consists of Z-flat objects, and $M^\bullet \in \mathbf{C}^-(Z\text{-mod})$ is acyclic, then $M^\bullet \underset{Z}{\otimes} X^\bullet$ is acyclic as well. However, this is evident from the definitions. □

If $\phi : Z \to Z'$ is a homomorphism, we will denote by $\mathcal{C}_{Z'}$ the base-changed category, i.e., one whose objects are $X \in \mathcal{C}$, endowed with an action of Z', such that the two actions of Z on X coincide. Morphisms in this category are $\mathcal{C}$-morphisms that commute with the action of Z'.

By construction, Z' maps to the center of $\mathcal{C}_{Z'}$. The composed functor $\mathcal{C}_{Z'} \to \mathcal{C} \overset{\mathsf{F}_Z}{\to} Z\text{-mod}$ factors naturally through Z'-mod.

The forgetful functor $\mathcal{C}_{Z'} \to \mathcal{C}$ admits a left adjoint ϕ^* given by $X \mapsto Z' \underset{Z}{\otimes} X$. Note that this functor sends Z-flat objects in $\mathcal{C}$ to Z'-flat objects in $\mathcal{C}_{Z'}$. In particular, if $\mathcal{C}$ is flat over Z, then so is $\mathcal{C}_{Z'}$ over Z'.

As in Lemmas 23.2 and 19.22, we obtain the following.

Lemma 23.3. *Assume that $\mathcal{C}$ is Z-flat. Then the right derived functor of ϕ^**

$$L\phi^* : D^-(\mathcal{C}) \to D^-(\mathcal{C}_{Z'})$$

is well defined and is the left adjoint to the forgetful functor $D(\mathcal{C}_{Z'}) \to D(\mathcal{C})$. Moreover,

$$\mathsf{F}_{Z'} \circ L\phi^*(X^\bullet) \simeq \mathsf{F}(X^\bullet) \overset{L}{\underset{Z}{\otimes}} Z'.$$

In particular, we obtain that if $\mathcal{C}$ is flat over Z and $X \in \mathcal{C}$ is Z-flat, then for $Y^\bullet \in \mathbf{C}(\mathcal{C}_{Z'})$,

$$\mathrm{RHom}_{D(\mathcal{C}_{Z'})}(Z' \underset{Z}{\otimes} X, Y^\bullet) \simeq \mathrm{RHom}_{D(\mathcal{C})}(X, Y^\bullet).$$

Now let N be a Z-module. For $Y \in \mathcal{C}$ we define the object $\underline{\mathrm{Hom}}_Z(N, Y)$ by

$$\mathrm{Hom}_{\mathcal{C}}(X, \underline{\mathrm{Hom}}_Z(N, Y)) := \mathrm{Hom}_{\mathcal{C}}(N \underset{Z}{\otimes} X, Y).$$

If N is finitely presented, we have

$$\mathsf{F}_Z(\underline{\mathrm{Hom}}_Z(N, Y)) \simeq \mathrm{Hom}_Z(N, \mathsf{F}_Z(Y)).$$

For Z' as above, which is finitely presented as a Z-module, we define the functor $\phi^! : \mathcal{C} \to \mathcal{C}_{Z'}$ to be the right adjoint of the forgetful functor $\mathcal{C}_{Z'} \to \mathcal{C}$. It is given by $X \mapsto \underline{\mathrm{Hom}}_Z(Z', X)$. By definition, it maps injective objects in $\mathcal{C}$ to injectives in $\mathcal{C}_{Z'}$.

We will denote by $R\phi^! : D^+(\mathcal{C}) \to D^+(\mathcal{C}_{Z'})$ the corresponding right derived functor. It is easily seen to be the right adjoint of the forgetful functor $\mathbf{C}(\mathcal{C}_{Z'}) \to \mathbf{C}(\mathcal{C})$.

Proposition 23.4. *Assume that $\mathcal{C}$ is flat over Z, and that Z' is perfect as an object of* $\mathbf{C}(Z\text{-mod})$. *Then*

$$R\phi^! \circ \mathsf{F}_Z \simeq \mathsf{F}_{Z'} \circ R\phi^! : D^+(\mathcal{C}) \to D^+(Z'\text{-mod}).$$

Proof. To prove the proposition it suffices to check that if $Y^\bullet \in \mathbf{C}^+(\mathcal{C})$ is a complex consisting of injective objects of $\mathcal{C}$, and $M^\bullet \in \mathbf{C}^b(Z\text{-mod})$ is a complex of finitely presented modules, quasi-isomorphic to a perfect one, then $\mathsf{F}_Z(\underline{\mathrm{Hom}}(M^\bullet, Y^\bullet))$ is quasi-isomorphic to $\mathrm{RHom}_{D(Z\text{-mod})}(M^\bullet, \mathsf{F}_Z(Y^\bullet))$.

If $M^\bullet$ is a bounded complex consisting of finitely generated projective modules, then the assertion is evident. Hence, it remains to show that if $M^\bullet$ is an acyclic complex of finitely presented Z-modules, and $Y^\bullet$ is as above, then $\underline{\mathrm{Hom}}(M^\bullet, Y^\bullet)$ is acyclic. By assumption on $\mathcal{C}$, it would suffice to check that for $X \in \mathcal{C}$ which is Z-flat,

$$\mathrm{Hom}_{\mathcal{C}}(X, \underline{\mathrm{Hom}}(M^\bullet, Y^\bullet)) \simeq \mathrm{Hom}_{\mathcal{C}}(M^\bullet \underset{Z}{\otimes} X, Y^\bullet)$$

is acyclic. By the flatness assumption on Y, the complex $M^\bullet \underset{Z}{\otimes} X$ is also acyclic, and hence our assertion follows from the injectivity assumption on $Y^\bullet$. □

Corollary 23.5. *If, under the assumptions of the proposition, $X' \in D(\mathcal{C}'_Z)$ is quasi-perfect, then it is quasi-perfect also as an object of $D(\mathcal{C})$.*

Proof. This follows from the fact that the functor

$$R\phi^! : D(Z\text{-mod}) \to D(Z'\text{-mod})$$

commutes with direct sums, and hence so does the functor

$$R\phi^! : D(\mathcal{C}) \to D(\mathcal{C}_{Z'}).$$

□

23.6 A generalization

Let $\mathcal{C}$ be as in the previous subsection, and assume in addition that it satisfies assumption (*) of Section 19.5. Let $\mathbf{Z}$ be a topological commutative algebra, which acts functorially on every object of $\mathcal{C}$. In this case we will say that $\mathbf{Z}$ maps to the center $\mathcal{C}$. The functor F naturally factors through a functor $\mathsf{F}_{\mathbf{Z}} : \mathcal{C} \to \mathbf{Z}\text{-mod}$.

For every discrete quotient Z of $\mathbf{Z}$, let $\mathcal{C}_Z$ be the subcategory of $\mathcal{C}$ consisting of objects on which $\mathbf{Z}$ acts via Z. If $\mathbf{Z} \twoheadrightarrow Z \twoheadrightarrow Z'$, then $\mathcal{C}_{Z'}$ is obtained from $\mathcal{C}_Z$ by the procedure described in the previous subsection.

We shall say that $\mathcal{C}$ is flat over $\mathbf{Z}$ if each $\mathcal{C}_Z$ as above is flat over Z. Equivalently, we can require that this happens for a cofinal family of discrete quotients Z of $\mathbf{Z}$. Henceforth, we will assume that $\mathcal{C}$ is flat over $\mathbf{Z}$.

In what follows we will make the following additional assumption on $\mathbf{Z}$, namely, that we can present $\mathbf{Z}$ as $\lim\limits_{\overleftarrow{i}} Z_i$, such that for $j \geq i$ the ideal of $\phi_{j,i} : Z_j \to Z_i$ is perfect as an object of $D(Z_j\text{-mod})$.

Recall that a discrete quotient Z of $\mathbf{Z}$ reasonable if for some (equivalently, any) index i such that $\mathbf{Z} \to Z$ factors through Z_i, the algebra Z is finitely presented as a Z_i-module. We shall call Z admissible if the finite-presentation condition is replaced by the perfectness one.

Let us call an object $M \in \mathbf{Z}\text{-mod}$ finitely presented if M belongs to some Z-mod and is finitely presented as an object of this category, if Z is reasonable. By the assumption on $\mathbf{Z}$, this condition does not depend on a particular choice if Z.

For a finitely presented $M \in \mathbf{Z}\text{-mod}$ and $X \in \mathcal{C}$ we define $\underline{\mathrm{Hom}}_{\mathbf{Z}}(M, X) \in \mathcal{C}$ as

$$\lim_{\overrightarrow{X_i}} \underline{\mathrm{Hom}}_{Z_i}(M, X_i),$$

where X_i runs over the set of subobjects of X that belong to $\mathcal{C}_{Z_i}$ for some discrete reasonable quotient Z_i of $\mathbf{Z}$. We have

$$\mathsf{F}_{\mathbf{Z}}(\underline{\mathrm{Hom}}_{\mathbf{Z}}(M, X)) \simeq \mathrm{Hom}_{\mathbf{Z}}(M, \mathsf{F}_{\mathbf{Z}}(X)).$$

Consider $M = Z$ for some reasonable quotient $\phi : \mathbf{Z} \to Z$. Then $X \mapsto \underline{\mathrm{Hom}}_{\mathbf{Z}}(Z, X)$ defines a functor $\mathcal{C} \to \mathcal{C}_Z$, which we will denote by $\phi^!$.

Lemma 23.7. *The functor $\phi^!$ is the right adjoint to the forgetful functor $\mathcal{C}_Z \to \mathcal{C}$.*

Proof. By assumption (*), it suffices to check that for every finitely generated object Y of $\mathcal{C}_Z$,

$$\mathrm{Hom}_{\mathcal{C}_Z}(Y, \underline{\mathrm{Hom}}_{\mathbf{Z}}(Z, X)) \simeq \mathrm{Hom}_{\mathcal{C}}(Y, X).$$

By the finite generation assumption, we reduce the assertion to the case when $X \in \mathcal{C}_{Z_i}$ for some $\mathbf{Z} \twoheadrightarrow Z_i \twoheadrightarrow Z$, considered in the previous subsection. □

Clearly, the functor $\phi^!$ maps injective objects in $\mathcal{C}$ to injectives in $\mathcal{C}_Z$. Let $R\phi^!$ denote the right derived functor of $\phi^!$. By the above, it is the right adjoint to the forgetful functor $D(\mathcal{C}_Z) \to D(\mathcal{C})$.

Proposition 23.8. *Assume that Z is admissible. Then we have an isomorphism of functors:*

$$\mathsf{F}_Z \circ R\phi^! \simeq R\phi^! \circ \mathsf{F}_{\mathbf{Z}} : D^+(\mathcal{C}) \to D^+(Z\text{-mod}).$$

Proof. As in the proof of Proposition 23.4, it suffices to show that if $X^\bullet \in \mathbf{C}^+(\mathcal{C})$ is a complex consisting of injective objects of $\mathcal{C}$, and $M^\bullet$ is a perfect object of $D(Z\text{-mod})$, then $\mathcal{H}om_{\mathbf{C}(\mathbf{Z}\text{-mod})}(M^\bullet, \mathsf{F}_{\mathbf{Z}}(X^\bullet))$ computes $\mathrm{RHom}_{D(\mathbf{Z}\text{-mod})}(M^\bullet, \mathsf{F}_{\mathbf{Z}}(X^\bullet))$.

By devissage, we can assume that $X^\bullet$ consists of a single injective object $X \in \mathcal{C}$. For every Z_i such that $\mathbf{Z} \overset{\phi_i}{\twoheadrightarrow} Z_i \twoheadrightarrow Z$, note that $\phi_i^!(X)$ is an injective object of $\mathcal{C}_{Z_i}$, and $X \simeq \lim\limits_{\overrightarrow{i}} X_i$.

Using Proposition 23.4, the assertion of the present proposition follows from the next lemma.

Lemma 23.9. *For $N^\bullet \in \mathbf{C}^+(\mathbf{Z}\text{-mod})$ and $N_i^\bullet := \phi_i^!(N^\bullet)$, the map*

$$\varinjlim_i \operatorname{Hom}_{D(Z_i\text{-mod})}(M^\bullet, N_i^\bullet) \to \operatorname{Hom}_{D(\mathbf{Z}\text{-mod})}(M^\bullet, N^\bullet)$$

is a quasi-isomorphism, provided that $M^\bullet \in D(Z\text{-mod})$ is perfect. □

Proof of the lemma. The proof follows from the next observation:

Let $P^\bullet \to M^\bullet$ be a quasi-isomorphism, where $P^\bullet \in \mathbf{C}^-(\mathbf{Z}\text{-mod})$. Then we can find a quasi-isomorphism $Q^\bullet \to P^\bullet$ such that $Q^\bullet \in \mathbf{C}^-(\mathbf{Z}\text{-mod})$ and for any integer i, the module Q^i is supported on some discrete quotient of $\mathbf{Z}$. □

Corollary 23.10. *The functor $R\phi^! : D^+(\mathcal{C}) \to D^+(\mathcal{C}_Z)$ commutes with uniformly bounded from below direct sums.*

Proof. This follows from the corresponding fact for the functor $R\phi^! : D^+(\mathbf{Z}\text{-mod}) \to D^+(Z\text{-mod})$. □

Proposition 23.11. *Let $X_1^\bullet$ be a quasi-perfect object of $\mathbf{C}(\mathcal{C}_Z)$ and $X_2^\bullet$ be an object of $\mathbf{C}^+(\mathcal{C}_Z)$ for some discrete quotient Z. Then*

$$\operatorname{Hom}_{D(\mathcal{C})}(X_1^\bullet, X_2^\bullet) \simeq \varinjlim_{Z_i} \operatorname{Hom}_{D(\mathcal{C}_{Z_i})}(X_1^\bullet, X_2^\bullet),$$

where the direct limit is taken over the indices i such that $\mathbf{Z} \to Z$ factors through Z_i.

Proof. We can find a system of quasi-isomorphisms $X_2^\bullet \to Y_i^\bullet$, where each $Y_i^\bullet \in \mathbf{C}(\mathcal{C}_{Z_i})$ consists of injective objects of $\mathcal{C}_{Z_i}$, and such that these complexes form a direct system with respect to the index i, and such that all $Y_i^\bullet$ are uniformly bounded from below.

By Proposition 23.8 and Corollary 23.10, $R\phi^!(X_2^\bullet)$ is given by the complex

$$\varinjlim_i \phi_i^!(Y_i^\bullet).$$

Then, by the quasi-perfectness assumption,

$$\operatorname{Hom}_{D(\mathcal{C})}(X_1^\bullet, X_2^\bullet) \simeq \operatorname{Hom}_{D(\mathcal{C}_Z)}(X_1^\bullet, R\phi^!(X_2^\bullet)) \simeq \varinjlim_i \operatorname{Hom}_{D(\mathcal{C}_Z)}(X_1^\bullet, \phi_i^!(Y_i^\bullet)).$$

By Proposition 23.4, the latter is isomorphic to

$$\varinjlim_i \operatorname{Hom}_{D(\mathcal{C}_Z)}(X_1^\bullet, R\phi_i^!(Y_i^\bullet)) \simeq \varinjlim_i \operatorname{Hom}_{D(\mathcal{C}_{Z_i})}(X_1^\bullet, Y_i^\bullet),$$

which is what we had to show. □

Finally, we will prove the following assertion.

Proposition 23.12. *Let $X^\bullet$ be an object of $\mathbf{C}^-(\mathcal{C}_Z)$, where Z is an admissible quotient of $\mathbf{Z}$. Then $X^\bullet$ is quasi-perfect as an object of $D(\mathcal{C}_Z)$ if and only if it is quasi-perfect as an object of $D(\mathcal{C})$.*

Proof. Since the functor $R\phi^! : D^+(\mathcal{C}) \to D^+(\mathcal{C}_Z)$ commutes with direct sums, the implication "quasi-perfectness in $D(\mathcal{C}_Z)$" $\to$ "quasi-perfectness in $D(\mathcal{C})$" is clear.

To prove the implication in the opposite direction, we proceed by induction. We assume that the functor

$$Y \mapsto \mathrm{Hom}_{D(\mathcal{C}_Z)}(X^\bullet, Y[i']) : \mathcal{C}_Z \to \mathrm{Vect}$$

commutes with direct sums for $i' < i$. This assumption is satisfied for some i, since $X^\bullet$ is bounded from above.

Let us show that in this case the functor $Y \mapsto \mathrm{Hom}_{D(\mathcal{C}_Z)}(X^\bullet, Y[i])$ also commutes with direct sums. For $\bigoplus_\alpha Y_\alpha \in \mathcal{C}_Z$ consider the exact triangle in $D^+(\mathcal{C}_Z)$:

$$\bigoplus_\alpha Y_\alpha \to R\phi^!(\bigoplus_\alpha Y_\alpha) \to \tau^{>0}\left(R\phi^!(\bigoplus_\alpha Y_\alpha)\right),$$

where τ is the cohomological truncation.

Consider the corresponding commutative diagram:

$$\begin{array}{ccc}
\mathrm{Hom}_{D(\mathcal{C})}(X^\bullet, \bigoplus_\alpha Y_\alpha[i-1]) & \longleftarrow & \bigoplus_\alpha \mathrm{Hom}_{D(\mathcal{C})}(X^\bullet, Y_\alpha[i-1]) \\
\downarrow & & \downarrow \\
\mathrm{Hom}_{D(\mathcal{C}_Z)}\left(X^\bullet, \tau^{>0}\left(R\phi^!(\bigoplus_\alpha Y_\alpha[i-1])\right)\right) & \longleftarrow & \bigoplus_\alpha \mathrm{Hom}_{D(\mathcal{C}_Z)}\left(X^\bullet, \tau^{>0}\left(R\phi^!(Y_\alpha[i-1])\right)\right) \\
\downarrow & & \downarrow \\
\mathrm{Hom}_{D(\mathcal{C}_Z)}(X^\bullet, \bigoplus_\alpha Y_\alpha[i]) & \longleftarrow & \bigoplus_\alpha \mathrm{Hom}_{D(\mathcal{C}_Z)}(X^\bullet, Y_\alpha[i]) \\
\downarrow & & \downarrow \\
\mathrm{Hom}_{D(\mathcal{C})}(X^\bullet, \bigoplus_\alpha Y_\alpha[i]) & \longleftarrow & \bigoplus_\alpha \mathrm{Hom}_{D(\mathcal{C})}(X^\bullet, Y_\alpha[i]) \\
\downarrow & & \downarrow \\
\mathrm{Hom}_{D(\mathcal{C}_Z)}\left(X^\bullet, \tau^{>0}\left(R\phi^!(\bigoplus_\alpha Y_\alpha[i])\right)\right) & \longleftarrow & \bigoplus_\alpha \mathrm{Hom}_{D(\mathcal{C}_Z)}\left(X^\bullet, \tau^{>0}\left(R\phi^!(Y_\alpha[i])\right)\right).
\end{array}$$

The horizontal arrows in rows 1 and 4 are isomorphisms since $X^\bullet$ is quasi-perfect in $D(\mathcal{C})$. The arrows in rows 2 and 5 are isomorphisms by the induction hypothesis. Hence the map in row 3 is an isomorphism, which is what we had to show. □

23.13 The equivariant situation

Assume now that the category $\mathcal{C}$ as in the previous subsection is equipped with an infinitesimally trivial action of a group scheme H. Assume that this action commutes with that of $\mathbf{Z}$. The latter means that for every $X \in \mathcal{C}$, the $\mathbf{Z}$-action on $\text{act}^*(X)$ by transport of structure coincides with the action obtained by regarding it merely as an object of $\mathcal{C}$. Then for every discrete quotient Z of $\mathbf{Z}$, the category $\mathcal{C}_Z$ carries an infinitesimally trivial action of H.

We have a functor $\phi^! : \mathbf{C}^+(\mathcal{C})^H \to \mathbf{C}^+(\mathcal{C}_Z)^H$; let $R\phi^! : D^+(\mathcal{C})^H \to D^+(\mathcal{C}_Z)^H$ be its right derived functor. (Below we will show that it is well defined.) We are going to prove the following.

Proposition 23.14. *$R\phi^! : D^+(\mathcal{C})^H \to D^+(\mathcal{C}_Z)^H$ is the right adjoint to the forgetful functor $D(\mathcal{C}_Z)^H \to D(\mathcal{C})^H$. Moreover, the diagram of functors*

$$\begin{array}{ccc} D^+(\mathcal{C})^H & \xrightarrow{R\phi^!} & D^+(\mathcal{C}_Z)^H \\ {\scriptstyle \mathsf{F}_{\mathbf{Z}}}\downarrow & & {\scriptstyle \mathsf{F}_Z}\downarrow \\ D^+(\mathbf{Z}\text{-mod}) & \xrightarrow{R\phi^!} & D^+(Z\text{-mod}) \end{array}$$

is commutative.

Proof. For any quasi-isomorphism $X^\bullet \to X_1^\bullet$ in $\mathbf{C}^+(\mathcal{C})^H$ we can find a quasi-isomorphism from $X_1^\bullet$ to a complex, associated with a bicomplex $X_2^{\bullet,\bullet}$, whose rows are uniformly bounded from below and have the form $\text{Av}_H(Y^\bullet)$, where $Y^\bullet \in \mathbf{C}^+(\mathcal{C})$ consists of injective objects.

By Proposition 23.8 and Corollary 23.10, if we assign to $X^\bullet$ the complex in $\mathbf{C}(\mathcal{C}_Z)^H$ associated with the bicomplex $\phi^!(X_2^{\bullet,\bullet})$, this is the desired right derived functor of $\phi^!$. It is clear from the construction that the diagram of functors

$$\begin{array}{ccc} D^+(\mathcal{C})^H & \xrightarrow{R\phi^!} & D^+(\mathcal{C}_Z)^H \\ \downarrow & & \downarrow \\ D^+(\mathcal{C}) & \xrightarrow{R\phi^!} & D^+(\mathcal{C}_Z), \end{array}$$

where the vertical arrows are the forgetful functors, is commutative.

Hence, it remains to show that $R\phi^!$ satisfies the desired adjointness property. By devissage, we are reduced to showing that for $Y^\bullet$ as above and $Y_1^\bullet \in \mathbf{C}(\mathcal{C}_Z)^H$,

$$\text{Hom}_{D(\mathcal{C})^H}(Y_1^\bullet, \text{Av}_H(Y^\bullet)) \simeq \text{Hom}_{D(\mathcal{C}_Z)^H}\left(Y_1^\bullet, \phi^!(\text{Av}_H(Y^\bullet))\right).$$

However, the LHS is isomorphic to $\text{Hom}_{D(\mathcal{C})}(Y_1^\bullet, Y^\bullet)$, and the RHS is isomorphic to

$$\text{Hom}_{D(\mathcal{C}_Z)^H}\left(Y_1^\bullet, \text{Av}_H(\phi^!(Y^\bullet))\right) \simeq \text{Hom}_{D(\mathcal{C}_Z)}(Y_1^\bullet, \phi^!(Y^\bullet)),$$

and, as we have seen above, $\phi^!(Y^\bullet) \to R\phi^!(Y^\bullet)$ is an isomorphism in $D^+(\mathcal{C}_Z)$. □

Acknowledgments. We thank A. Beilinson, R. Bezrukavnikov, and B. Feigin for useful discussions.

Both authors were supported by DARPA grant HR0011-04-1-0031, and E.F. was also supported by NSF grant DMS-0303529.

References

[Ar] S. Arkhipov, A new construction of the semi-infinite BGG resolution, q-alg/9605043, 1996.

[AB] A. Arkhipov and R. Bezrukavnikov, Perverse sheaves on affine flags and Langlands dual group, math.RT/0201073, 2002.

[ABG] S. Arkhipov, R. Bezrukavnikov, V. Ginzburg, Quantum groups, the loop Grassmannian, and the Springer resolution, *J. Amer. Math. Soc.*, **17** (2004), 595–678.

[AG1] S. Arkhipov and D. Gaitsgory, Differential operators on the loop group via chiral algebras, *Internat. Math. Res. Notices*, **2002**-4 (2002), 165–210.

[AG2] S. Arkhipov and D. Gaitsgory, Another realization of the category of modules over the small quantum group, *Adv. Math.*, **173** (2003) 114–143.

[BB] A. Beilinson and J. Bernstein, Localisation de g-modules, *C. R. Acad. Sci. Paris Sér.* I *Math.*, **292** (1981), 15–18.

[CHA] A. Beilinson and V. Drinfeld, *Chiral Algebras*, American Mathematical Society Colloquium Publications, Vol. 51, American Mathematical Society, Providence, RI, 2004.

[CHA1] A. Beilinson, Tensor products of topological vector spaces, addendum to [CHA].

[BD1] A. Beilinson and V. Drinfeld, *Quantization of Hitchin's Integrable System and Hecke Eigensheaves*, available online from http://www.math.uchicago.edu/~arinkin/langlands/.

[BD2] A. Beilinson and V. Drinfeld, Opers, math.AG/0501398, 2005.

[BG] A. Beilinson and V. Ginzburg, Wall-crossing functors and D-modules, *Representation Theory*, **3** (1999), 1–31.

[Be] J. Bernstein, Trace in categories, in *Operator Algebras, Unitary Representations, Enveloping Algebras, and Invariant Theory* (*Paris*, 1989), Progress in Mathematics, Vol. 92, Birkhäuser Boston, Cambridge, MA, 1990, 417–423.

[BL] J. Bernstein and V. Luntz, *Equivariant Sheaves and Functors*, Lecture Notes in Mathematics, Vol. 1578, Springer-Verlag, Berlin, 1994.

[Bez] R. Bezrukavnikov, unpublished manuscript.

[Dr1] V. Drinfeld, DG quotients of DG categories, *J. Algebra*, **272** (2004), 643–691.

[Dr2] V. Drinfeld, Infinite-dimensional vector bundles in algebraic geometry, in P. Etinghof, V. Retakh, and I. Singer, eds., *The Unity of Mathematics*, Birkhäuser Boston, Cambridge, MA, 2005, 263–304.

[DS] V. Drinfeld and V. Sokolov, Lie algebras and KdV type equations, *J. Soviet Math.*, **30** (1985), 1975–2036.

[EF] D. Eisenbud and E. Frenkel, Appendix to M. Mustaţă, "Jet schemes of locally complete intersection canonical singularities," *Invent. Math.*, **145**-3 (2001), 397–424.

[Fe] B. Feigin, The semi-infinite cohomology of Kac–Moody and Virasoro Lie algebras, *Russian Math. Survey*, **39**-2 (1984), 155–156.

[FF1] B. Feigin and E. Frenkel, A family of representations of affine Lie algebras, *Russian Math. Survey*, **43**-5 (1988), 221–222.

[FF2] B. Feigin and E. Frenkel, Affine Kac-Moody algebras and semi-infinite flag manifolds, *Comm. Math. Phys.*, **128** (1990), 161–189.

[FF3] B. Feigin and E. Frenkel, Affine Kac-Moody algebras at the critical level and Gelfand-Dikii algebras, in A. Tsuchiya, T. Eguchi, and M. Jimbo, eds., *Infinite Analysis*, Advanced Series in Mathematical Physics, Vol. 16, World Scientific, Singapore, 1992, 197–215.

[F] E. Frenkel, Wakimoto modules, opers and the center at the critical level, *Adv. Math.*, **195** (2005), 297–404.

[FB] E. Frenkel and D. Ben-Zvi, *Vertex Algebras and Algebraic Curves*, 2nd ed., Mathematical Surveys and Monographs, Vol. 88, American Mathematical Society, Providence, RI, 2004.

[FG] E. Frenkel and D. Gaitsgory, D-modules on the affine Grassmannian and representations of affine Kac-Moody algebras, *Duke Math. J.*, **125** (2004), 279–327.

[FGV] E. Frenkel, D. Gaitsgory, and K. Vilonen, Whittaker patterns in the geometry of moduli spaces of bundles on curves, *Ann. Math.*, **153** (2001), 699–748.

[FT] E. Frenkel and K. Teleman, Self-extensions of Verma modules and differential forms on opers, *Compositio Math.*, **142** (2006), 477–500.

[Ga] D. Gaitsgory, Construction of central elements in the affine Hecke algebra via nearby cycles, *Invent. Math.*, **144** (2001), 253–280.

[Ga1] D. Gaitsgory, The notion of category over an algebraic stack, math.AG/0507192, 2005.

[Ga2] D. Gaitsgory, Notes on Bezrukavnikov's theory, in preparation.

[GMS] V. Gorbounov, F. Malikov, and V. Schechtman, On chiral differential operators over homogeneous spaces, *Internat. J. Math. Math. Sci.*, **26** (2001), 83–106.

[KK] V. Kac and D. Kazhdan, Structure of representations with highest weight of infinite-dimensional Lie algebras, *Adv. Math.*, **34** (1979), 97–108.

[Ko] B. Kostant, Lie group representations on polynomial rings, *Amer. J. Math.*, **85** (1963), 327–402.

[Lan] R. P. Langlands, Problems in the theory of automorphic forms, in *Lectures in Modern Analysis and Applications* III, Lecture Notes in Mathematics, Vol. 170, Springer-Verlag, Berlin, New York, Heidelberg, 1970, 18–61.

[Lau] G. Laumon, Transformation de Fourier, constantes d'équations fonctionelles et conjecture de Weil, *Publ. I.H.E.S.*, **65** (1987), 131–210.

[Ma] K. Mackenzie, *Lie Groupoids and Lie Algebroids in Differential Geometry*, London Mathematical Society Lecture Note Series, Vol. 124, Cambridge University Press, Cambridge, UK, 1987.

[MV] I. Mirković and K. Vilonen, Geometric Langlands duality and representations of algebraic groups over commutative rings, math.RT/0401222, 2004.

[Vog] D. Vogan, Local Langlands correspondence, in *Representation Theory of Groups and Algebras*, Contemporary Mathematics, Vol. 145, American Mathematical Society, Providence, RI, 1993, 305–379.

[Vor] A. Voronov, Semi-infinite induction and Wakimoto modules, *Amer. J. Math.*, **121** (1999), 1079–1094.

[W] M. Wakimoto, Fock representations of affine Lie algebra $A_1^{(1)}$, *Comm. Math. Phys.*, **104** (1986), 605–609.

Integration in valued fields

Ehud Hrushovski* and David Kazhdan*

Institute of Mathematics
Hebrew University of Jerusalem
Givat Ram
Jerusalem, 91904
Israel
ehud@math.huji.ac.il, kazhdan@math.huji.ac.il

Summary. We develop a theory of integration over valued fields of residue characteristic zero. In particular, we obtain new and base-field independent foundations for integration over local fields of large residue characteristic, extending results of Denef, Loeser, and Cluckers. The method depends on an analysis of definable sets up to definable bijections. We obtain a precise description of the Grothendieck semigroup of such sets in terms of related groups over the residue field and value group. This yields new invariants of all definable bijections, as well as invariants of measure-preserving bijections.

Subject Classifications: Primary 03C60, 14C99, 11S80.

1 Introduction

Since Weil's *Foundations*, algebraic varieties have been understood independently of a particular base field; thus an algebraic group G exists prior to the abstract or topological groups of points $G(F)$, taken over various fields F. For Hecke algebras, or other geometric objects whose definition requires integration, no comparable viewpoint exists. One uses the topology and measure theory of each local field separately; since a field F has measure zero from the point of view of any nontrivial finite extension, at the foundational level there is no direct connection between the objects obtained over different fields. The main thrust of this paper is the development of a theory of integration over valued fields, which is geometric in the sense of Weil. At present the theory covers local fields of residue characteristic zero or, in applications, large positive residue characteristic.

Our approach to integration continues a line traced by Kontsevich, Denef–Loeser, and Loeser–Cluckers (cf. [7]). In integration over non-archimedean local fields there are two sources for the numerical values. The first is counting points of varieties over the residue field. Kontsevich explained that these numerical values can be replaced, with a gain of geometric information, by the isomorphism classes of the varieties themselves up to appropriate transformations, or more precisely by their classes in a certain Grothendieck ring. This makes it possible to understand geometrically the changes in integrals upon unramified base change. In this aspect our approach is very similar. The main difference is a slight generalization of the notion of variety over the residue field, which allows us to avoid what amounted to a choice of uniformizer in the previous theory.

The second source of numerical values is the piecewise linear geometry of the value group. We geometrize this ingredient, too, obtaining a theory of integration taking values in an entirely geometric ring, a tensor product of a Grothendieck ring of generalized varieties over the residue field, and a Grothendieck ring of piecewise linear varieties over the value group.

Viewed in this way, the integral is an invariant of measure-preserving definable bijections. We actually find all such invariants. In addition, we consider and determine all possible invariants of definable bijections; we obtain in particular two Euler characteristics on definable sets, with values in the Grothendieck group of generalized varieties over the residue field.

At the level of foundations, until an additive character is introduced, we are able to work with Grothendieck semigroups rather than with classes in Grothendieck groups.

1.1 The logical setting

Let L be a valued field, with valuation ring $\mathcal{O}_L$. $\mathcal{M}$ denotes the maximal ideal. We let $\mathrm{VF}^n(L) = L^n$. The notation VF^n is analogous to the symbol $\mathbb{A}^n$ of algebraic geometry, denoting affine n-space. Let $\mathrm{RV}^m(L) = L^*/(1+\mathcal{M})$, $\Gamma(L) = L^*/\mathcal{O}_L^*$, $\mathbf{k}(L) = \mathcal{O}_L/\mathcal{M}_L$. Let $\mathrm{rv} : \mathrm{VF} \to \mathrm{RV}$ and $\mathrm{val} : \mathrm{VF} \to \Gamma$ be the natural maps. The natural map $\mathrm{RV} \to \Gamma$ is denoted $\mathrm{val}_{\mathrm{rv}}$. The exact sequence

$$0 \to \mathbf{k}^* \to \mathrm{RV} \to \Gamma \to 0$$

shows that RV is, at first approximation, just a way to wrap together the residue field and value group.

We consider expressions of the form $h(x) = 0$ and $\mathrm{val}\, f(x) \geq \mathrm{val}\, g(x)$ where $f, g, h \in L[X]$, $X = (X_1, \ldots, X_n)$. A *semialgebraic formula* is a finite Boolean combination of such basic expressions. A semialgebraic formula ϕ clearly defines a subset $D(L)$ of $\mathrm{VF}^n(L)$. Moreover, if $f, g, h \in L_0[X]$, we obtain a functor $L \mapsto D(L)$ from valued field extensions of L_0 to sets. We will later describe more general definable sets; but for the time being take a *definable subset of* VF^n to be a functor $D = D_\phi$ of this form.

An intrinsic description of definable subsets of RV^m is given in Section 2.1. In particular, definable subsets of $(\mathbf{k}^*)^m$ coincide with constructible sets in the usual Zariski sense; while modulo $(\mathbf{k}^*)^m$, a definable set is a piecewise linear subset of Γ^m. The structure of arbitrary definable subsets of RV^m is analyzed in Section 3.3.

The advantages of this approach are identical to the benefits in algebraic geometry of working with arbitrary algebraically closed fields, over arbitrary base fields. One can use Galois theory to describe rational points over subfields. Since function fields are treated on the same footing, one has a mechanism to inductively reduce higher-dimensional geometry to questions in dimension one, and often, in fact, to dimension zero. (As in algebraic geometry, statements about fields, applied to generic points, can imply birational statements about varieties.)

1.2 Model theory

Since topological tools are no longer available, it is necessary to define notions such as dimension in a different way. The basic framework comes from [15]; we recall and develop it further in Sections 2 and 4. It is in many respects analogous to the o-minimal framework of [37], that has become well accepted in real algebraic geometry.

In addition, whereas in geometry all varieties are made as it were of the same material, here a number of rather different types of objects coexist, and the interaction between them must be clarified. In particular, the residue field and the value group are orthogonal in a sense that will be defined below; definable subsets of one can never be isomorphic to subsets of the other, unless both are finite. This orthogonality has an effect on definable subsets of VF^n in general; for example, closed disks behave very differently from open ones. Here we follow and further develop [16].

Note that the set of rational points of closed and open disks over discrete valuation rings, for instance, cannot be distinguished; as in rigid geometry, the geometric setting is required to make sense of the notions. Nevertheless, they have immediate consequences for local fields. As an example, we define the notion of a definable distribution; this is defined as a function on the space of polydisks with certain properties. Making use of model-theoretic properties of the space of polydisks, we show that any definable distribution agrees outside a proper subvariety with one obtained by integrating a function. This is valid over any valued field of sufficiently large residue characteristic. In particular, for large p, the p-adic Fourier transform of a rational polynomial is a locally constant function away from an exceptional subvariety, in the usual sense (Corollary 11.10). The analogue for $\mathbb{R}$ and $\mathbb{C}$ was proved by Bernstein using D-modules. For an individual $\mathbb{Q}_p$, the same result can be shown using Denef integration and a similar analysis of definable sets over $\mathbb{Q}_p$. These results were obtained independently by Cluckers and Loeser; cf. [8].

1.3 More general definable sets

Throughout the chapter, we discuss not semialgebraic sets, but definable subsets of a theory with the requisite geometric properties (called V-minimality). This includes also the rigid analytic structures of [23]. The adjective "geometrically" can be take to mean here "in the sense of the V-minimal theory."

While we work geometrically throughout the paper, the isomorphisms we obtain are canonical and so specialize to rational points over substructures. Thus a posteriori our results apply to definable sets over any Hensel field of large residue characteristic. See Section 12.

For model theorists, this systematic use of algebraically closed valued fields to apply to other Hensel fields is only beginning to be familar. As an illustration, see Proposition 12.9, where it is shown that after a little analysis of definable sets over algebraically closed valued fields, quantifier elimination for Henselian fields of residue characteristic zero becomes a consequence of Robinson's earlier quantifier elimination in the algebraically closed case.

A third kind of generalization is an a posteriori expansion of the language in the RV sort. Such an expansion involves loss of information in the integration theory, but is sometimes useful. For instance, one may want to use the Denef–Pas language, splitting the exact sequence into a product of residue field and value group. Another example occurs in Theorem 12.5, where it is explained, given a valued field whose residue field is also a valued field, what happens when one integrates twice. To discuss this, the residue field is expanded so as to itself become a valued field.

1.4 Generalized algebraic varieties

We now describe the basic ingredients in more detail. Let L_0 be a valued field with residue field $\mathbf{k}_0$ and value group A. For each point $\gamma \in \mathbb{Q} \otimes A$, we have one-dimensional $\mathbf{k}$-vector space

$$V_\gamma = \{0\} \cup \frac{\{x \in K : \mathrm{val}(x) = \gamma\}}{1 + \mathcal{M}}.$$

As discussed above, V_γ should be viewed as a functor $L \mapsto V_\gamma(L)$ on valued field extensions L of L_0, giving a vector space over the residue field functor. If $\gamma - \gamma' \in A$, then V_γ, V'_γ are definably isomorphic, so one essentially has V_γ for $\gamma \in (\mathbb{Q} \otimes A)/A$.

Fix $\bar{\gamma} = (\gamma_1, \ldots, \gamma_n)$, and $V_i = V_{\gamma_i}$, $V_{\bar{\gamma}} = \Pi_i V_{\gamma_i}$. A *$\bar{\gamma}$-polynomial* is a polynomial $H(X) = \sum a_\nu X^\nu$ with $\mathrm{val}_p(a_\nu) + \sum_i \nu(i)\gamma_i = 0$ for each nonzero term $a_\nu X^\nu$. The coefficients a_ν are described in Section 5.5; for the purposes of the introduction, and of Theorem 1.3 below, it suffices to think of integer coefficients. Such a polynomial clearly defines a function $H : V_{\bar{\gamma}} \to \mathbf{k}$. In particular, one has the set of zeroes $Z(H)$. The *generalized residue structure* RES_{L_0} is the residue field, together with the collection of one-dimensional vector spaces $V_\gamma (\gamma \in \mathbb{Q} \otimes A)$ over it, and the functions $H : V_{\bar{\gamma}} \to \mathbf{k}$ associated to each $\bar{\gamma}$-polynomial.

The intersection W of finitely many zero sets $Z(H)$ is called a *generalized algebraic variety over the residue field*. Given a valued field extensions L of L_0, we have the set of points $W(L) \subseteq V_{\bar{\gamma}}(L)$. When L is a local field, $W(L)$ is finite.

We will systematically use the Grothendieck group of generalized varieties over the residue field, rather than the usual Grothendieck group of varieties. They are fundamentally of a similar nature: base change to an algebraically closed value field makes them isomorphic. But the generalized residue field makes it possible to see canonically objects that are only visible after base change in the usual approach. One application is Theorem 1.3 below.

$K_+ \mathrm{RES}_{L_0}[n]$ denotes the Grothendieck group of generalized varieties of dimension $\leq n$; in the paper we will omit L_0 from the notation.

1.5 Rational polyhedra over ordered Abelian groups

Let A be an ordered Abelian group. A *rational polyhedron Δ over A* is given by an expression

$$\Delta = \{x : Mx \geq b\}$$

with $x = (x_1, \ldots, x_n)$, M a $k \times n$ matrix with rational coefficients, and $b \in A^k$. We view this as a functor $B \mapsto \Delta(B)$ on ordered Abelian group extensions B of A. This functor is already determined by its value at $B = \mathbb{Q} \otimes A$. In particular, when $A \leq \mathbb{Q}$, Δ is an ordinary rational polyhedron.

$K_+ \Gamma_A[n]$ is the semigroup generated by such polyhedra, up to piecewise $\mathrm{GL}_n(\mathbb{Z})$-transformations and A-translations; see Section 9. When A is fixed it is omitted from the notation.

In our applications, A will be the value group of a valued field L_0. If B is the value group of a valued field extension L, write $\Delta(L)$ for $\Delta(B)$.

1.6 The Grothendieck semiring of definable sets

Fix a base field L_0. The word "definable" will mean $\mathbf{T}_{L_0}$-definable, with $\mathbf{T}$ a fixed V-minimal theory. To have an example in mind one can read "semialgebraic over L_0" in place of "definable."

Let VF$[n]$ be the category of definable subsets X of n-dimensional algebraic varieties over L_0; a morphism $X \to X'$ is a definable bijection $X \to X'$ (see Definition 3.65 for equivalent definitions). K_+ VF$[n]$ denotes the Grothendieck semigroup, i.e., the set of isomorphism classes of VF$[n]$ with the disjoint sum operation. $[X]$ denotes the class of X in the Grothendieck semigroup.

We explain how an isomorphism class of VF$[n]$ is determined precisely by isomorphism classes of generalized algebraic varieties and rational polyhedra, whose dimensions add up to n.

If $X \subseteq \mathrm{RES}^m$ and $f : X \to \mathrm{RES}^n$ is a finite-to-one map, let

$$\mathbb{L}(X, f) = \mathrm{VF}^n \times_{\mathrm{rv}, f} X = \{(v_1, \ldots, v_n, x) : v_i \in \mathrm{VF}, x \in X, \mathrm{rv}(v_i) = f_i(x)\}.$$

The VF$[n]$-isomorphism class $[\mathbb{L}(X, f)]$ does not depend on f, and is also denoted $[\mathbb{L}X]$.

When S is a smooth scheme over $\mathcal{O}$, X a definable subset of $S(\mathbf{k})$, $\pi : S(\mathcal{O}) \to S(\mathbf{k})$ the natural reduction map, we have $[\mathbb{L}X] = [\pi^{-1}X]$.

We let RES$[n]$ be the category of pairs (X, f) as above; a morphism $(X, f) \to (X', f')$ is just a definable bijection $X \to X'$. Let K_+ RES$[*]$ be the direct sum of the Grothendieck semigroups K_+ RES$[n]$.

On the other hand, we have already defined $K_+ \Gamma[n]$. Let $K_+ \Gamma[*]$ be the direct sum of the $K_+ \Gamma[n]$. An element of $K_+ \Gamma[n]$ is represented by a definable $X \subseteq \Gamma[n]$. Let $\mathbb{L}X = \mathrm{val}^{-1}(X)$, $\mathbb{L}[X] = [\mathbb{L}X]$.

It is shown in Proposition 10.2 that the Grothendieck semiring of RV is the tensor product $K_+ \mathrm{RES}[*] \otimes K_+ \Gamma[*]$ over the semiring $K_+\Gamma^{\mathrm{fin}}$ of classes of finite subsets of Γ; see Section 9.

Note that $\mathbb{L}([1]_1) = \mathbb{L}([1]_0) + \mathbb{L}([(0, \infty)]_1)$, where $[1]_1 \in K_+ \mathrm{RES}[1]$, $[1]_0 \in K_+ \mathrm{RES}[0]$ are the classes of the singleton set 1, and $[(0, \infty)]_1$ is the class in $K_+ \Gamma[1]$ of the semi-infinite segment $(0, \infty)$. Indeed, $\mathbb{L}([1]_1)$ is the unit open ball around 1, $\mathbb{L}([1]_0)$ is the point $\{1\}$, while $\mathbb{L}([(0, \infty)]_1)$ is the unit open ball around 0, isomorphic by a shift to the unit open ball around 1. This is the one relation that cannot be understood in terms of the Grothendieck semiring of RV; it will be seen to correspond to the analytic summation of geoemtric series in the Denef theory. Let $\mathrm{I_{sp}}$ be the congruence on the ring $K_+ \mathrm{RES}[*] \otimes K_+ \Gamma[*]$ generated by $[1]_1 \sim [1]_0 + [(0, \infty)]_1$.

The following theorem summarizes the relation between definable sets in VF and in RV; it follows from Theorem 8.4 together with Proposition 10.2 in the text.

Theorem 1.1. $\mathbb{L}$ *induces a surjective homomorphism of filtered semirings*

$$K_+ \mathrm{RES}[*] \otimes K_+ \Gamma[*] \to K_+(\mathrm{VF}).$$

The kernel is precisely the congruence $\mathrm{I_{sp}}$.

The inverse isomorphism $K_+(\mathrm{VF}) \to K_+ \mathrm{RES}[*] \otimes K_+ \Gamma[*]/\mathrm{I_{sp}}$ can be viewed as a kind of Euler characteristic, respecting products and disjoint sums, and can be functorial in various other ways.

The values of this Euler characteristic are themselves geometric objects, both on the algebraic-geometry side (RES) and the combinatorial-analytic side (Γ). This is valuable for some purposes; in particular, it becomes clear that the isomorphism is compatible with taking rational points over Henselian subfields (cf. Proposition 12.6).

For other applications, however, it would be useful to obtain more manageable numerical invariants; for this purpose one needs to analyze the structure of $K_+ \Gamma[*]$. We do not fully do this here, but using a number of homomorphisms on $K_+ \Gamma[*]$, we obtain a number of invariants. In particular, using the $\mathbb{Z}$-valued Euler characteristics on $K\,\Gamma[*]$ (cf. Section 9 and [26, 20]), we obtain two homomorphisms on $K_+ \mathrm{VF}[n]$ essentially to $K\,\mathrm{RES}[n]$. The reason there are two rather than one has to do with Poincaré duality; see Theorem 10.5.

For instance, when F is a field of characteristic 0, we obtain an invariant of rigid analytic varieties over $F((t))$, with values in the Grothendieck ring $K(\mathrm{Var}_F)$ of algebraic varieties over F; and another in $K(\mathrm{Var}_F)[[\mathbb{A}_1]^{-1}]$ (Proposition 10.8). It is instructive to compare this with the invariant of [25], with values in $K(\mathrm{RES}[n])/[G_m]$.[1] Since any two closed balls are isomorphic, via additive translation and multiplicative contractions, all closed balls must have the same invariant. Working with a discrete value group tends to force $[G_m] = 0$, since it appears that a closed ball B_0 of valuation radius 0 equals G_m times a closed ball B_1 of valuation radius 1. Since our technology is based on divisible value groups, the "equation" $[B_0] = [B_1][G_m]$ is replaced for us by $[B_0] = [B_0^o][G_m]$, where B_0^o is the open ball of valuation radius 0. Though B_1 and B_0^0 have the same $F((t))$-rational points, they are geometrically distinct (cf. Lemma 3.46) and so no collapse takes place. See also Sections 12.6 and 12.6 for two previously known cases.

By such Euler characteristic methods we can prove a statement purely concerning algebraic varieties, partially answering a question of Gromov and Kontsevich [13, p. 121]. In particular, two elliptic curves with isomorphic complements in projective

[1] The setting is somewhat different: Loeser–Sebag can handle positive characteristic, too, but assume smoothness.

space were previously known to be isogenous, by zeta function methods; we show that they are isomorphic. This also follows from [22]; the method there requires strong forms of resolution of singularities. See Theorem 13.1.

1.7 Integration of forms up to absolute value

Over local fields, data for integration consists of a triple (X, V, ω), with X a definable subset of a smooth variety V and ω a volume form on V. We are interested in an integral of the form $\int_X |\omega|$, so that multiplication of ω by a function with norm 1 does not count as a change, nor does removing a subvariety of V of smaller dimension. Using an equivalent description of VF$[n]$, where the objects come with a distinguished finite-to-one map into affine space, we can represent an integrand as a pair (X, ω) with $X \in \mathrm{Ob}\,\mathrm{VF}[n]$ and ω a function from X into Γ. Isomorphisms are essential bijections, preserving the form up to a function of norm 1. See Definition 8.10 for a precise definition of this category, the category $\mu_\Gamma\mathrm{VF}[n]$.

Integration is intended to be an invariant of isomorphisms in this category. Thus we can find the integral if we determine all invariants. We do this in complete analogy with Theorem 1.1.

For $n \geq 0$ let $\Gamma[n]$ be the category whose objects are finite unions of rational polyhedra over the group A of definable points of Γ. A morphism $f : X \to Y$ of $\Gamma[n]$ is a bijection such that for some partition $X = \cup_{i=1}^k X_i$ into rational polyhedra, $f|X_i$ is given by an element of $\mathrm{GL}_n(\mathbb{Z}) \ltimes A^n$. Let $\mu\Gamma[n]$ be the category of pairs (X, ω), with X an object of $\Gamma[n]$, and $\omega : X \to \Gamma$ a piecewise affine map. A morphism $f : (X, \omega) \to (X', \omega')$ is a morphism $f : X \to X'$ of $\Gamma[n]$ such that $\sum_{i=1}^l x_i + \omega(x) = \sum_{i=1}^l x'_i + \omega'(x')$ whenever $(x'_1, \dots, x'_n) = f(x_1, \dots, x_n)$. Given $(X, \omega) \in \mathrm{Ob}\,\mu\Gamma[n]$, define $\mathbb{L}X$ as above, and adjoint the pullback of ω to obtain an object of $\mu_\Gamma\mathrm{VF}[n]$. This gives a homomorphism $K_+\,\mu\Gamma[n] \to K_+\,\mu_\Gamma\mathrm{VF}[n]$.

Theorem 1.2. *$\mathbb{L}$ induces a surjective homomorphism of filtered semirings*

$$K_+\,\mathrm{RES}[*] \otimes_{\mathbb{N}} K_+\,\mu\Gamma[*] \to K_+(\mu_\Gamma\mathrm{VF})[*].$$

The kernel is generated by the relations $p \otimes 1 = 1 \otimes [(\mathrm{val}_{\mathrm{rv}}(p), \infty)]$ *and* $1 \otimes a = {\mathrm{val}_{\mathrm{rv}}}^{-1}(a) \otimes 1$.

In the statement of the theorem, p ranges over definable points of RES (actually one value suffices), and a ranges over definable points of Γ.

This can also be written as

$$K_+\,\mathrm{RES}[*] \otimes_{K_+(\mu\Gamma^{\mathrm{fin}})} K_+\,\mu\Gamma[*]/\mathrm{I}_{\mathrm{sp}}^{\mu} \simeq K_+(\mu_\Gamma\mathrm{VF})[*],$$

where $K_+(\mu\Gamma^{\mathrm{fin}})$ is the subsemiringof subsets of $\mu\Gamma$ with finite support, and $\mathrm{I}_{\mathrm{sp}}^{\mu}$ is a semiring congruence defined similarly to I_{sp}. The base of the tensor leads to the identification of a point of Γ with with a coset of $\mathbf{k}^*$ in RES, while $\mathrm{I}_{\mathrm{sp}}^{\mu}$ identifies a point of RES with an infinite interval of Γ. The inverse isomorphism can be viewed as an integral.

We introduce neither additive nor multiplicative inverses in $K_+ \mathrm{RES}[*]$ formally, so that the target of integration is completely geometric.

We proceed to give an application of the first part of the theorem (the surjectivity) in terms of ordinary p-adic integration.

1.8 Integrals over local fields: Uniformity over ramified extensions

Let L be a local field, finite extension of $\mathbb{Q}_p$ or $\mathbb{F}_p((t))$. We normalize the Haar measure μ in such a way that the maximal ideal has measure 1, the norm by $|a| = \mu\{x : |x| < |a|\}$. Let RES_L be the generalized residue field, and Γ_L be the value group. We assume $\mathbb{Q}_p$ or $\mathbb{F}_p((t))$ has value group $\mathbb{Z}$, and identify Γ_L with a subgroup of $\mathbb{Q}$.

Given $c = (c_1, \dots, c_k) \in L^k$ and $s = (s_1, \dots, s_k) \in \mathbb{R}^k$ with $s_i \geq 1$, let $|c|^s = \Pi_{i=1}^k |c_i|^{s_i}$.

Let λ be a multiplicative character $\mathbb{R}^n \to \mathbb{R}^*$. Define

$$\mathrm{ev}_\lambda(\Delta(B)) = \sum_{b \in \Delta(B)} \lambda(b),$$

provided this sum is absolutely convergent. Given linear functions $h_0, \dots, h_k$ on $\mathbb{R}^n$ and $s_1, \dots, s_k \in \mathbb{R}$, let $\mathrm{ev}_{h,s,Q} = \mathrm{ev}_\lambda$, where $\lambda(x) = Q^{h_0(x)+\sum s_i h_i(x)}$.

Theorem 1.3. *Fix $n, d, k \in \mathbb{N}$. Let p be a large prime compared to n, d, k, and let $f \in \mathbb{Q}_p[X_1, \dots, X_n]^k$ have degrees $\leq d$. Then there exist finitely many generalized varieties X_i over* $\mathrm{RES}(\mathbb{Q}_p)$, *rational polyhedra Δ_i, $\gamma(i) \in \mathbb{Q}^{\geq 0}$, $n_i \in \mathbb{N}$, and linear functions $h_0^i, \dots, h_k^i$ with rational coefficients, such that for any finite extension L of $\mathbb{Q}_p$ with residue field $GF(q)$ and* $\mathrm{val}(L^*) = (1/r)\mathbb{Z}$, $\mathrm{val}(p) = 1$, *and any $s \in \mathbb{R}^k_{\geq 1}$,*

$$\int_{\mathcal{O}_L^n} |f|^s = \sum_i q^{r\gamma(i)} (q-1)^{n_i} |X_i(L)| \,\mathrm{ev}_{h^i,s,q^r}(\Delta_i(L)).$$

Note the following:

(1) $\Delta_i(L)$ depends only on the ramification degree r of L over $\mathbb{Q}_p$.
(2) The formula is a sum of nonnegative terms.
(3) $\mathrm{ev}_{h,s,q^r}(\Delta_i((1/r)\mathbb{Z}))$ can be written in closed form as a rational function of q^{rs}. This follows from Denef, who shows it for more general sets Δ_i definable in Pressburger arithmetic; such analytic summation is an essential component of his integration theory. Since it plays no role in our approach we leave the statement in geometric form.
(4) The generalized varieties X_i and polyhedra Δ_i are simple functions of the coefficients f. Here we wish to emphasize not this, but the uniformity of the expression over ramified extensions of $\mathbb{Q}_p$.

The proof follows Proposition 10.10. (It uses only the easy surjectivity in this proposition and Proposition 4.5.)

1.9 Bounded and unbounded sets

The isomorphism of semirings of Theorem 1.2 obviously induces an isomorphism of rings. However, introducing additive inverses loses information on the Γ side; the class of the interval $[0, 1)$ becomes 0, since $[0, \infty)$ and $[1, \infty)$ are isomorphic. The classical remedy is to cut down to bounded sets before groupifying. This presents no difficulty, since the isomorphism respects boundedness.

In higher-dimensional local fields, stronger notions of boundedness may be useful, such as those introduced by Fesenko. Since these questions are not entangled with the theory of integration, and can be handled a posteriori, we will deal with them in a future work.

Here we mention only that even if one insists on integrating all definable integrands, with no boundedness condition, into a ring, some but not all information is lost. This is due to the existence of Euler characteristics on Γ, and thus again to the fact that we work geometrically, with divisible groups, even if the base field has a discrete group. We will see (Lemma 9.12) that $K_+(\mu\Gamma[n])$ can be identified with the group of definable functions $\Gamma \to K_+(\Gamma[n])$. Applying an appropriate Euler characteristic reduces to the group of piecewise constant functions on Γ into $\mathbb{Z}$. Recombining with RES we obtain a consistent definition of an integral on unbounded integrands, compatible with measure-preserving maps, sums, and products, with values in $K(\mathrm{RES})[A]/[\mathbb{A}_1]_1\, K(\mathrm{RES})[A]$, where A is the group of definable points of Γ, and $[\mathbb{A}_1]_1$ is the class of the affine line. See Theorem 10.11.

1.10 Finer volumes

We also consider a finer category of definable sets with RV-volume forms. This means that a volume form ω is identified with $g\omega$ only when $g - 1 \in \mathcal{M}$; $\mathrm{val}(g) = 0$ does not suffice. We obtain an integral whose values themselves are definable sets with volume forms; in particular, including algebraic varieties with volume forms over the residue field.

Theorem 1.4. $\mathbb{L}$ *induces a surjective homomorphism of graded semirings*

$$K_+\, \mu\mathrm{RV}[*] \to K_+(\mu\mathrm{VF})[*].$$

The kernel is precisely the congruence $\mathrm{I}_{\mathrm{sp}}^{\mu}$.

μRV is the category of definable subsets of $\mu\mathrm{RV}^*$ enriched with volume forms; see Definition 8.13. Again, an isomorphism is induced in the opposite direction, that can be viewed as a motivic integral

$$\int : K_+(\mu\mathrm{VF})[*] \to K_+\, \mu\mathrm{RV}[*]/\mathrm{I}_{\mathrm{sp}}^{\mu}.$$

This allows an iteration of the integration theory, either with an integral of the same nature if the residue field is a valued field, or with a different kind of integral if, for instance, the residue field is $\mathbb{R}$.

1.11 Hopes

We mention three. Until now, a deep obstacle existed to extending Denef's theory to positive characteristic; namely, the theory was based on quantifier elimination for Hensel fields of residue characteristic 0, or for finitely ramified extensions of $\mathbb{Q}_p$, and it is known that no similar quantifier elimination is possible for $\mathbb{F}_p((t))$, if any is. On the other hand, Robinson's quantifier elimination is perfectly valid in positive characteristic. This raises hopes of progress in this direction, although other obstacles remain.

It is natural to think that the theory can be applied to higher-dimensional local fields; we will consider this in a future work.

Another important target is asymptotic integration over $\mathbb{R}$. Nonstandard extensions of $\mathbb{R}$ admit natural valued field structures. This is the basis of Robinson's nonstandard analysis. These valued fields have divisible value groups, and so previous theories of definable integration do not apply. The theory of this paper applies, however, and we expect that it will yield connections between p-adic integration and asymptotics of real integrals.

1.12 Organization of the paper

After recalling some basic model theory in Section 2, we proceed in Section 3 to V-minimal theories.

In Section 4 we show that any definable subset of VF^n admits a constructible bijection with some $\mathbb{L}(X, f)$. In fact, only a very limited class of bijections is needed; a typical one has the form $(x_1, x_2) \mapsto (x_1, x_2 + f(x_1, x_2))$, so it is clearly measure preserving. The proof is simple and brief, and uses only a little of the preceding material. We note here that for many applications this statement is already sufficient; in particular, it suffices to give the surjectivity in Theorems 1.1 and 1.2, and hence the application Theorem 1.3.

In Section 5 we return to the geometry of V-minimal structures, developing a theory of differentiation. We show the compatibility between differentiation in RV and in VF. This is needed for Theorem 1.4. Differentiation in VF involves much finer scales than in RV; in effect RV can only see distances measured by valuation 0, while the derivative in VF involves distances of arbitrarily large valuation. The proof uses a continuity argument with respect to dependence on scales. It fails in positive characteristic, in its present form.

Section 6 is devoted to showing that $\mathbb{L}$ yields a well-defined map $K_+(\mathrm{RV}) \to K_+(\mathrm{VF})$; in other words, not only objects, but also isomorphisms can be lifted.

Sections 7 and 8 investigate the kernel of $\mathbb{L}$ in Theorem 1.1. This is the most technical part of the paper, and we have not been able to give a proof as functorial as we would have liked. See Question 7.9.

In Section 9 we study the piecewise linear Grothendieck group; see the introduction to this section.

Section 10 decomposes the Grothendieck group of RV into the components RES and Γ, used througout this introduction.

Section 11 introduces an additive character, and hence the Fourier transform. The isomorphism of volumes given by Theorem 1.4 suffices for this extension; it is not necessary to redo the theory from scratch, but merely to follow through the functoriality.

Section 12 contains the extension to definable sets over Hensel fields mentioned above, and Section 13 gives the application to the Grothendieck group of varieties.

2 First-order theories

The bulk of this paper uses no deep results from logic beyond Robinson's quantifier elimination for the theory of algebraically closed valued fields [33]. However, it is imbued with a model-theoretic viewpoint. We will not explain the most basic notions of logic: language, theory, model. Let us just mention that a language consists of basic relations and function symbols, and formulas are built out of these, using symbols for Boolean operations and quantifiers (cf., e.g., [11] or [19], or the first section of [9]); but we attempt in this section to bridge the gap between these and the model-theoretic language used in the paper.

A language L consists of a family of "sorts" S_i, a collection of variables ranging over each sort, a set of relation symbols R_j, each intended to denote a subset of a finite product of sorts, and a set of function symbols F_k intended to denote functions from a given finite product of sorts to a given sort. From these, and the logical symbols &, $\neg$, $\forall$, $\exists$ one forms *formulas*. A sentence is a formula with no free variables (cf. [11]). A theory T is a set of sentences of L. A theory is called *complete* if for every sentence ϕ of L, either ϕ or its negation $\neg\phi$ is in T.

A *universe* M for the language L consists, by definition, of a set $S(M)$ for each sort S of L. An L-structure consists of such a universe, together with an interpretation of each relation and a function symbol of L. One can define the truth value of a sentence in a structure M; more generally, if $\phi(x_1, \dots, x_n)$ is a formula, with x_i a variable of sort S_i, then one defines the interpretation $\phi(M)$ of ϕ in M, as the set of all $d \in S_1(M) \times \cdots \times S_n(M)$ of which ϕ is true. If every sentence in T is true in M, one says that M is a model of T ($M \models T$). The fundamental theorem here is a consequence of Gödel's completeness theorem called the *compactness theorem*: a theory T has a model if every finite subset of T has a model.

The language L_{rings} of rings, for example, has one sort, three function symbols $+, \cdots, -$, two constants 0, 1; any ring is an L_{rings}-structure; one can obviously write down a theory T_{fields} in this language whose models are precisely the fields.

2.1 Basic examples of theories

We will work with a number of theories associated with valued fields:

(1) ACF, the theory of algebraically closed fields. The language is the language of rings $\{+, \cdot, -, 0, 1\}$, mentioned earlier. The theory states that the model is a field, and for each n, that every monic polynomial of degree n has a root. For instance, for $n = 2$,

$$(\forall u_1)(\forall u_0)(\exists x)(x^2 + u_1 x + u_0 = 0).$$

In addition, ACF(0) includes the sentence $1 + 1 \neq 0, 1 + 1 + 1 \neq 0, \ldots$. This theory is complete (Tarski–Chevalley). It will arise as the theory of the residue field of our valued fields.

(2) Divisible ordered Abelian groups (DOAG). The language consists of a single sort, a binary relation symbol $<$, a binary function symbol $+$, a unary function symbol $-$, and a constant symbol 0. The theory states that a model is an ordered Abelian group. In addition, there are axioms asserting divisibility by n for each n, for instance, $(\forall x)(\exists y)(y + y = x)$.

This is the theory of the value group of a model of ACVF.

(3) The RV sort (extension of (2) by (1)). The language has one official sort, denoted RV, and includes Abelian group operations $\cdot$, $/$ on RV, a unary predicate $\mathbf{k}^*$ for a subgroup, and an operation $+ : \mathbf{k}^2 \to \mathbf{k}$, where $\mathbf{k}$ is $\mathbf{k}^*$ augmented by a constant 0. Finally, there is a partial ordering; the theory states that $\mathbf{k}^*$ is the equivalence class of 1; that $\leq$ is a total ordering on $\mathbf{k}^*$-cosets, making $\mathrm{RV}/\mathbf{k}^* =: \Gamma$ a divisible ordered Abelian group, and that $(\mathbf{k}, +, \cdot)$ is an algebraically closed field. (We thus have an exact sequence $0 \to \mathbf{k}^* \to \mathrm{RV} \to \Gamma \to 0$, but we treat Γ as an imaginary sort.) This theory TRV is complete, too.

We will sometimes view RV as an autonomous structure but it will arise from an algebraically closed valued field, as in (5) below.

(4) Let $M \models \mathrm{TRV}$, and let A be a subgroup of $\Gamma(M)$. Within TRV_A we see an interpretation of ACF, namely, the algebraically closed field $\mathbf{k}$. In addition, for each $a \in A$, we have a one-dimensional $\mathbf{k}$-space, the fiber of RV lying over Γ augmented by 0. Collectively, the field $\mathbf{k}$ with this collection of vector spaces will be denoted RES.

(5) ACVF, the theory of algebraically closed valued fields. According to Robinson, the completions, denoted ACVF(q, p), are obtained by specifying the characteristic q and residue characteristic p. We will be concerned with ACVF(0, 0) in this paper. However, since any sentence of ACVF(0, 0) lies in ACVF$(0, p)$ for almost all primes p, the results will a posteriori apply also to valued fields of characteristic zero and large residue characteristic.

We will take ACVF(0, 0) to have two sorts, VF and $\mathrm{RV} = \mathrm{VF}^*/(1 + \mathcal{M})$. The language includes the language of rings (1) on the VF sort, the language (3) on the RV sort, and a function symbol rv for a function $\mathrm{VF}^* \to \mathrm{RV}$. Denote $\mathrm{rv}^{-1}(\mathrm{RV}^{\geq 0}) = \mathcal{O}$, $\mathrm{rv}^{-1}(0) = \mathcal{M}$.

The theory states that VF is a valued field, with valuation ring $\mathcal{O}$ and maximal ideal $\mathcal{M}$ such that $\mathrm{rv} : \mathrm{VF}^* \to \mathrm{RV}$ is a surjective group homomorphism, and the restriction to $\mathcal{O}$ (augmented by $0 \mapsto 0$) is a surjective ring homomorphism.

The structure that ACVF_A induces on Γ is of a uniquely divisible Abelian group, with constants for the elements of $\Gamma(A)$. Thus every definable subset of Γ is a finite union of points and open intervals (possibly infinite).

(6) Rigid analytic expansions (Lipshitz). The theory $\mathrm{ACVF}^{\mathrm{R}}$ of algebraically closed valued fields expanded by a family R of analytic functions. See [23] and [24].

Our theory of definable sets will be carried out axiomatically, and are thus also valid for these rigid analytic expansions.

A *definable set* D is not really a set, but a functor from the category of models of T to the category of sets of the form $M \mapsto \phi(M)$, where ϕ is a formula of L. Model theorists do not really distinguish between the definable set D and the formula ϕ defining it; we will usually refer to definable sets rather than to formulas. If $R \subseteq D \times D'$ and for any model $M \models T$, $R(M)$ is the graph of a function $D(M) \to D(M')$, we say R is a *definable function of* T. Similarly, we say D is *finite* if $D(M)$ is finite for any $M \models T$, etc. It follows from the compactness theorem that if D is finite, then for some integer m we have $|D(M)| \leq m$ for any $M \models T$. We sometimes write S^* to denote S^n for some unspecified n.

By a *map* between L-structures A, B we mean a family $f = (f_S)$ indexed by the sorts of L, with $f_S : S(A) \to S(B)$; one extends f to products of sorts by setting $f((x_1, \ldots, x_n)) = (f(x_1), \ldots, f(x_n))$. f is *an embedding of structures* if $f^{-1}R(B) = R(A)$ for any atomic formula R of L. Taking R to be the equality relation, this includes, in particular, the statement that each f_S is injective.

On occasion we will use ∞-definable sets. An ∞-definable set is a functor of the form $M \mapsto \cap\mathcal{D}$, where $\mathcal{D}$ is a given collection of definable sets. In a complete theory a definable set is determined by the value it has at a single model; this is, of course, false for ∞-definable sets.

We write $a \in D$ to mean $a \in D(M)$ for some $M \models T$. It is customary, since Shelah, to choose a single universal domain $\mathbb{U}$ embedding all "small" models, and let $a \in D$ mean $a \in D(\mathbb{U})$; we will not require this interpretation, but the reader is welcome to take it.

We will sometimes consider *imaginary sorts*. If D is a definable set, and E a definable equivalence relation on D, then D/E may be considered to be an *imaginary sort*; as a definable set it is just the functor $M \mapsto D(M)/E(M)$. A definable subset of a product $\Pi_{i=1}^{n} D_i/E_i$ of imaginary sorts (and ordinary sorts) is taken to be a subset whose preimage in $\Pi_{i=1}^{n} D_i$ is definable; the notion of a definable function is thus also defined. In this way, the imaginary sorts can be treated on the same footing as the others. The set of all elements of all imaginary sorts of a structure M is denoted M^{eq}. It is easy to construct a theory T^{eq} in a language L^{eq} whose category of models is (essentially) $\{M^{\mathrm{eq}} : M \models T\}$. See [35] and [31, Section 16d].

Given a definable set $D \subseteq S \times X$, where S, X are definable sets, and given $s \in S$, let $D(s) = \{x \in X : (s, x) \in D\}$. Thus D is viewed as a *family* of definable subsets of X, namely, $\{D(s) : s \in S\}$. If $s \neq s'$ implies $D(s) \neq D(s')$, we say that the parameters are *canonical*, or that s is a *code* for $D(s)$. In particular, if E is a definable equivalence relation, the imaginary elements a/E can be considered as codes for the classes of E.

T is said to *eliminate imaginaries* if every imaginary sort admits a definable injection into a product of some of the sorts of L. For instance, the theory of algebraically closed fields eliminates imaginaries. See [32] for an excellent exposition of these issues. We note that T admits elimination of imaginaries iff for any family $D \subseteq S \times X$

there exists a family $D' \subseteq S' \times X$ such that for any $t \in S$ there exists a *unique* $t' \in S'$ with $D(t) = D'(t')$.

(Recall that $t \in S$ means $t \in S(M)$ for some $M \models T$. The uniqueness of t' implies in this case that one can choose $t' \in S'(M)$, too.) In this case, we also say that t' is called a *canonical parameter* or *code* for $D(t)$.

Example 2.1. Let b be a nondegenerate closed ball in a model the theory ACVF of algebraically closed valued fields. Then $b = \{x : \mathrm{val}(x - c) \geq \mathrm{val}(c - c')\}$ for some elements $c \neq c'$ of the field. b is coded by $\bar{b} = (c, c')/E$, where $(c, c')E(d, d')$ iff $\mathrm{val}(c - c') = \mathrm{val}(d - d') \leq \mathrm{val}(c - d)$. However, we often fail to distinguish notationally between b and $\bar{b}$, and, in particular, we write $A(b) = A(\bar{b})$.

The only imaginary sorts that will really be essential for us are the sorts $\mathfrak{B}$ of closed and open balls. The closed balls around 0 can be identified with their radius, hence the valuation group $\Gamma(M) = \mathrm{VF}^*(M)/\mathcal{O}^*(M)$ of a valued field M is embedded as part of $\mathfrak{B}$.

Notation. Let $\mathfrak{B} = \mathfrak{B}^o \cup \mathfrak{B}^{\mathrm{cl}}$, the sorts of open and closed subballs of VF. Let $\Gamma^+ = \{\gamma \in \Gamma : \gamma \geq 0\}$.

$$\mathfrak{B}^{\mathrm{cl}} = \dot{\bigcup_{\gamma \in \Gamma}} \mathfrak{B}^{\mathrm{cl}}_{\gamma}, \qquad \mathfrak{B}^{\mathrm{cl}}_{\gamma} = \mathrm{VF}/\gamma\mathcal{O},$$

$$\mathfrak{B}^{o} = \dot{\bigcup_{\gamma \in \Gamma}} \mathfrak{B}^{o}_{\gamma}, \qquad \mathfrak{B}^{o}_{\gamma} = \mathrm{VF}/\gamma\mathcal{M}.$$

Here $\gamma\mathcal{M} = \{x \in \mathrm{VF} : \mathrm{val}(x) > \gamma\}$, $\gamma\mathcal{O} = \{x \in \mathrm{RES} : \mathrm{val}(x) \geq \gamma\}$. The elements of $\mathfrak{B}^{\mathrm{cl}}_{\gamma}$, $f\, B^o_{\gamma}$ will be referred to as *closed and open balls of valuative radius* γ; though this valuative definition of radius means that bigger balls have smaller radius. The word "distance" will be used similarly.

By a *thin annulus* we will mean a closed ball of valuative radius γ, with an open ball of valuative radius γ removed.

Fix a model M of T. A *substructure* A of M (written $A \leq M$) consists of a subset A_S of $S(M)$, for each sort S of L, closed under all definable functions of T. For example, the substructures of models of T_{fields} are the integral domains.

In general, the *definable closure* of a set $A_0 \subset M$ is the smallest substructure containing A_0; it is denoted $\mathrm{dcl}(A_0)$ or $\langle A_0 \rangle$. An element of $\langle A_0 \rangle$ can be written as $g(a_1, \ldots, a_n)$ with $a_i \in A_0$ and g a definable function; i.e., it is an element satisfying a formula $\phi(x, a_1, \ldots, a_n)$ of L_{A_0} in one variable that has exactly one solution in M. If A is a substructure, $\mathrm{dcl}(A \cup \{c\})$ is also denoted $A(c)$. These notions apply equally when A, c contain elements of the imaginary sorts. If B is contained in sorts $S_1, \ldots, S_n$, then $\mathrm{dcl}(B)$ is said to be an $S_1, \ldots, S_n$-generated substructure. In the special case of valued fields, where one of the sorts VF is the "main" valued field sort, a VF-generated structure will be said to be field-generated, or sometimes just "a field."

For any definable set D, we let $D(A)$ be the set of points of $D(M)$ with coordinates in A. If $S = D/E$ is an imaginary sort, $S(A)$ is the set of $a \in S$ whose preimage is defined over A. We have $D(A)/E(A) \subseteq S(A)$. $D(A)/E(A)$ is, of course, closed under definable functions $S^m \to S$ that lift to definable functions $D^m \to D$, but it is not necessarily closed under arbitrary definable functions, i.e., functions whose graph is the image of a definable subset of $D^m \times D$. For example $x \mapsto (1/n)x$ is a definable function on the value group of a model of ACVF, but if $A \leq M \models$ ACVF, $\Gamma(A)$ need not be divisible.

When $A \leq M, B \leq N$ with $M, N \models T$, a function $f : A \to B$ is called a *(partial) elementary embedding* $(A, M) \to (B, N)$ if for any definable set D of L, $f^{-1}D(B) = D(A)$. In particular, when $A = M$, $B = N$, one says that M is an *elementary submodel* of N.

By a *constructible set* over A, we mean the functor $L \mapsto \phi(L)$ on models $M \models T_A$, where $\phi = \phi(x_1, \dots, x_n, a_1, \dots, a_m)$ is a quantifier-free formula with parameters from A.

We say that T *admits quantifier elimination* if every definable set coincides with a constructible set. It follows in this case that for any A, any A-definable set is A-constructible. When T admits quantifier elimination, $f : A \to B$ is a partial elementary embedding iff it is an embedding of structures.

Theories (1)–(5) of Section 2.1 admit quantifier elimination in their natural algebraic languages (theorems of Tarski–Chevalley and Robinson; cf. [16]). The sixth admits quantifier elimination in a language that needs to be formulated with more care; see [23].

In all of this paper, except for Sections 12.1 and 12.3, we will only use structural properties of definable sets, and not explicit formulas. In this situation quantifier elimination can be assumed softly, by merely increasing the language by definition so that all definable sets become equivalent to quantifier-free ones. The above distinctions will only directly come into play in Sections 12.1 and 12.3.

If $A \leq M \models T$, L_A is the language L expanded by a constant c_a for each element a of A, so that an L_A-structure is the same as an L-structure M together with a function $A_S \to S(M)$ for each sort S. T_A is the set of L_A sentences true in M when the constant symbol c_a is interpreted as a; the models of T_A are models M of T, together with an isomorphic embedding of A as a substructure of M. In particular, M with the inclusion of A in M is an L_A-structure denoted M_A. For any subset $A_0 \subseteq M$, we write T_{A_0} for $T_{\langle A_0 \rangle}$, where $\langle A_0 \rangle$ is the substructure generated by A_0.

A definable set of T_A will also be referred to as A-definable; similarly for other notions such as those defined just below.

A *parametrically definable set* of T is by definition a T_A-definable set for some A.

An *almost definable* set is the union of classes of a definable equivalence relation with finitely many classes. An element e is called *algebraic* (respectively, definable) if the singleton set $\{e\}$ is almost definable (respectively, definable). When T is a complete theory, the set of algebraic (definable) elements of a model M of T forms a substructure that does not depend on M, up to (a unique) isomorphism.

Let $A_0 \subseteq M \models T$; the set of $e \in M$ almost definable over A_0 is called the *algebraic closure* of A_0, $\mathrm{acl}(A_0)$. If A_0 is contained in sorts $S_1, \dots, S_n$, any substructure of $\mathrm{acl}(A_0)$ containing $\mathrm{dcl}(A_0)$ is said to be *almost* $S_1, \dots, S_n$*-generated*.

Example 2.2. If a definable set D carries a definable linear ordering, then every algebraic element of D is definable. This is because the *least* element of a finite definable set F is clearly definable; the rest are contained in a smaller finite definable subset of D, so are definable by induction.

If, in addition, D has elimination of imaginaries, and Y is almost definable and definable with parameters from D, then Y is definable. Indeed, using elimination of imaginaries in D, the set Y can be defined using canonical parameters. These are algebraic elements of D, hence definable.

Two definable functions $f : X \to Y$, $f' : X \to Y'$ will be called *isogenous* if for all $x \in X$, $\mathrm{acl}(f(x)) = \mathrm{acl}(f'(x))$.

Compactness

Compactness often allows us to replace arguments in relative dimension one over a definable set, by arguments in dimension one over a different base structure. Here is an example.

Lemma 2.3. *Let $f_i : X_i \to Y$ be definable maps between definable sets of T $(i = 1, 2)$. Assume that for any $M \models T$ and $b \in Y(M)$, $X_1(b) := {f_1}^{-1}(b)$ is T_b-definably isomorphic to $X_2(b) = {f_2}^{-1}(b)$. Then X_1, X_2 are definably isomorphic.*

Proof. Let $\mathcal{F}$ be the family of pairs (U, h), where U is a definable subset of Y, and $h : {f_1}^{-1}U \to {f_2}^{-1}U$ is a definable bijection.

Claim. For any $b \in Y(M)$, $M \models T$, there exists $(U, h) \in \mathcal{F}$ with $b \in U$.

Proof. Let $b \in Y(M)$. There exists a T_b-definable bijection $X_1(b) \to X_2(b)$. This bijection can be written as $x \mapsto g(x, b)$, where g is a definable function. Let $U = \{y \in Y : (x \mapsto g(x, y)) \text{ is a bijection } X_1(y) \to X_2(y)\}$. Then $(U, g(x, f_1(x))) \in \mathcal{F}$, and $b \in U$. □

Now by compactness, there exist a finite number of definable subsets $U_1, \dots, U_k$ of Y, with $Y = \cup_i U_i$, and $(U_i, h_i) \in \mathcal{F}$ for some h_i. We define $U_i' = U_i \setminus (U_1 \cup \dots \cup U_{i-1})$ and $h = \cup_i h_i | U_i'$. Then $h : X_1 \to X_2$ is the required bijection. □

Here is another example of the use of compactness.

Example 2.4. If D is a definable set, and for any $a, b \in D$, $a \in \mathrm{acl}(b)$, then D is finite. More generally, if $a \in \mathrm{acl}(b)$ for any $b \in D$, then $a \in \mathrm{acl}(\emptyset)$.

Proof. We prove the first statement, the second being similar. For any model M, pick $a \in M$; then $D(M) \subseteq \mathrm{acl}(a)$. For $b \in \mathrm{acl}(a)$. Let ϕ_b be the formula $x \neq b \& D(x)$. Thus the set of formulas $\mathrm{Th}(M)_M \cup \{\phi_b\}$ has no common solution. By compactness, some finite subset already has no solution; this is only possible if $D(M)$ is finite. □

Transitivity, orthogonality

A definable set D is *transitive* if it has no proper, nonempty definable subsets. (The usual word is "atomic." One also says that D *generates a complete type*.) It is *(finitely) primitive* if it admits no nontrivial definable equivalence relation (with finitely many classes).

Remark 2.5. Let A be a VF-generated substructure of a model of ACVF. When A is VF-generated, we will see that an ACVF_A-definable ball b is never transitive in ACVF_A; indeed, it always contains an A-definable finite set. But b is always $\mathrm{ACVF}_{A(b)}$-definable, and quite often it is transitive; cf. Lemma 3.8.

Two definable sets D, D' are said to be *orthogonal* if any definable subset of $D^m \times D^l$ is a finite union of rectangles $E \times F$, $E \subseteq D^m$, $F \subseteq D^l$. In this case, the rectangles E, F can be taken to be almost definable. If the rectangles can actually be taken definable, we say the D, D' are *strongly orthogonal*.

Types

Let S be a product of sorts, and let $M \models T$, $a \in S(M)$. We write $\mathrm{tp}(a) = \mathrm{tp}(a; M)$ (the type of a) for the set of definable sets D with $a \in D$; when $p = \mathrm{tp}(a)$ we write $a \models p$. A *complete type* is the type of some element in some model. If $q = \mathrm{tp}(a)$, we say that a is a *realization* of q. The set $\mathcal{T}p_S$ of complete types belonging to S can be topologized: a basic open set is the set of types including a given definable set D. The *compactness theorem* of model theory implies that this is a compact topological space: if $\{D_i\}$ is any collection of definable sets with nonempty finite intersections, the compactness theorem asserts the existence of $M \models T$ with $\cap_i D_i(M) \neq \emptyset$.

The compactness theorem is often used by way of a construction called *saturated models*; cf. [9]. These are models where all types over "small" sets are realized. They enjoy excellent Galois-theoretic properties: in particular, if M is saturated, then $\mathrm{dcl}(A_0) = \mathrm{Fix}\,\mathrm{Aut}(M/A_0)$ for any finite $A_0 \subseteq M$. If D is $\mathrm{acl}(A_0)$-definable, then there exists an A_0-definable D' which is a finite union of $\mathrm{Aut}(M/A_0)$-conjugates of D.

A type p can also be identified with the functor P from models of T (under elementary embeddings) into sets; $P(M) = \{a \in M : a \models p\}$. As with definable sets, we speak as if P is simply a set. Unlike definable sets, the value of $P(M)$ at a single model does not determine P. (It could be empty, but it does determine P if M is sufficiently saturated.)

Any definable map $f : S \to S'$ induces a map $f_* : \mathcal{T}p_S \to \mathcal{T}p_{S'}$; as another consequence of the compactness theorem, f_* is continuous. We also have a restriction map from types of T_A to types of T, $\mathrm{tp}_{T(A)}(a) \mapsto \mathrm{tp}_T(a)$.

If $L \subseteq L'$ and $T \subseteq T'$, we say that T' is an *expansion* of T. In this case any T'-type p' restricts to a T-type p. If p' is the *unique* type of T' extending p, we say that p implies p'.

The simplest kind of expansion is an expansion by constants, i.e., a theory T_A (where $A \leq M \models T$). If $c \in M^n$, or more generally if $c \in M^{\mathrm{eq}}$, the type of c for M_A

is denoted $\text{tp}(c/A)$. It is rare for $\text{tp}(c)$ to imply $\text{tp}(c/A)$, but it is significant when it happens.

An instance of this is strong orthogonality: it is easy to see that strong orthogonality of two definable sets D, D' is equivalent to the following condition:

$$\text{If } A' \text{ is generated by elements of } D'\text{, then any type of elements of } D \text{ generates a complete type over } A'. \tag{$*$}$$

The asymmetry in $(*)$ is therefore only apparent.

Similarly, we have the following.

Lemma 2.6. *Let D, D' be definable sets. Then* $(1) \iff (2), (3) \iff (4)$.

(1) *Every definable function $f : D \to D'$ is piecewise constant, i.e., there exists a partition $D = \cup_{i=1}^n D_i$ of D into definable sets, with f constant on D_i.*
(2) *If $d \in D, d' \in D', d' \in \text{dcl}(d)$, then $d' \in \text{dcl}(\emptyset)$.*
(3) *If $f : E \to D$ is a definable finite-to-one map, and $g : E \to D'$ is definable, then $g(E)$ is finite.*
(4) *If $d \in D, d' \in D', d' \in \text{acl}(d)$, then $d' \in \text{acl}(\emptyset)$.*

Proof. Let us show that (3) implies (4). Let $M \models T, d \in D(M)$, and $d' \in D'(M)$, $d' \in \text{acl}(d)$. Then d' lies in some finite T_d-definable set $D'(d) \subseteq D'$. Since T_d is obtained from T by adding a constant symbol for d, there exists a formula $\phi(x, y)$ of the language of T and some m such that $M \models \phi(d, d')$ and $M \models (\exists^{\leq m} z)\phi(d, z)$. Let $X_0 = \{(x : (\exists^{\leq m} y)\phi(x, y)\}$, $E = \{(x, y) : x \in X_0, \phi(x, y)\}$, $f(x, y) = x$, $g(x, y) = y$. Then by (3), $g(E)$ is finite, but $d' \in g(E)$, so $d' \in \text{acl}(\emptyset)$.

Next, (4) implies (3): let f, E, g be as in (3) , and suppose $g(E)$ is infinite. In particular, for any finite $F \subseteq \text{acl}(\emptyset)$ there exists $d' \in g(E) \setminus F$. Thus the family consisting of $g(E)$ and the complement of all finite definable sets has nonempty intersections of finite subfamilies, so by the compactness theorem, in some $M \models T$, there exists $d' \in g(E) \setminus \text{acl}(\emptyset)$.

Let $d \in E(M)$ be such that $d' = g(d)$. Then $d' \in \text{acl}(f(d))$, but $f(d) \in D$, contradicting (4). Thus (4) implies (3).

The equivalence of (1)–(2) is similar. □

Example 2.7. Let P be a complete type, and f a definable function. Then $f(P)$ is a complete type P'. If f is injective on P, then there exist definable $D \supseteq P, D' \supseteq P'$ such that f restricts to a bijection of D with D'.

Proof. For any definable D', $f^{-1}D'$ is definable, so $P \subseteq f^{-1}D'$ or $P \cap f^{-1}D' = \emptyset$. Thus $P' \subseteq D'$ or $P' \cap D' = \emptyset$. Thus P' is complete.

Let $\{D_i\}$ be the family of definable sets containing P. Let $R_i = \{(x, y) \in D_i^2 : x \neq y, f(x) = f(y)\}$. Then $\cap_i R_i = \emptyset$. Since the family of $\{D_i\}$ is closed under finite intersections, it follows from the compactness theorem that for some i, $R_i = \emptyset$. Let $D = D_i$, $D' = f(D)$. □

Naming almost definable sets

As special case of an expansion by constants, we can move from a complete theory T to the theory T_A, where $A = \mathrm{acl}(\emptyset)$ is the set of all algebraic elements of a model M of T, including imaginaries. The effect is a theory where each class of any definable equivalence relation E with finitely many classes is definable. Since T is complete, the isomorphism type of $\mathrm{acl}(\emptyset)$ in a model M does not depend on the choice of model; so the theory T_A is determined. A definable set in this theory corresponds to an almost definable set in T.

When D is a constructible set, $T|D$ denotes the theory induced on D. If T eliminates quantifiers, the language is just the restriction to D of the relations and functions of L. If the language is countable, the countable models of D_A are of the form $D(M)$, where M is a countable model of T_A.

Stable embeddedness

A definable subset D of any product of sorts (possibly imaginary) is called *stably embedded* (in T) if for any A, any T_A-definable subset of D^m is T_B-definable for some $B \subset D$. For example, the set of open balls is not stably embedded in ACVF, since the set of open balls containing a point $a \in K$ cannot in general be defined using a finite number of balls.

Lemma 2.8. *Let D be a family of sorts of L; let $T|D$ be the theory induced on the sorts D. If D is stably embedded and $T|D$ admits elimination of imaginaries, then for any definable P and definable $S \subset P \times D^m$, viewed as a P-indexed family of subsets $S(a) \subseteq D^m$, $a \in P$, we have a definable function $f : P \to D^n$, with $f(a)$ a canonical parameter for $S(a)$.*

Proof. By stable embeddedness there exists a family $S' \subset P' \times D^m$ yielding the same family, i.e., $\{S(a) : a \in P\} = \{S'(a') : a' \in P'\}$, and with $P' \subseteq D^n$; using elimination of imaginaries we can take S' to be a canonical family; now a defines $f(a)$ to be the unique $a' \in P'$ with $S(a) = S'(a')$. □

Corollary 2.9. *If D is stably embedded and admits elimination of imaginaries, then for any substructure A,*

(1) $(T_A)|D = (T|D)_{A\cap D}$;
(2) *for $a \in A$*, $\mathrm{tp}(a/A \cap D)$ *implies* $\mathrm{tp}(a/D)$. □

Examples of definable sets of ACVF satisfying the hypotheses include the residue field $\mathbf{k}$, or the value group Γ, as well as $\mathrm{RV} \cup \Gamma$. The stable embeddedness in this case is an immediate consequence of quantifier elimination; cf. Lemma 3.30.

If M is saturated and D is stably embedded in T, then we have an exact sequence

$$1 \to \mathrm{Aut}(M/D(M)) \to \mathrm{Aut}(M) \to \mathrm{Aut}(D(M)) \to 1,$$

where $\mathrm{Aut}(M/D(M))$ is the group of automorphisms of M fixing $D(M)$ pointwise, and $\mathrm{Aut}(D(M))$ is the group of permutations of $D(M)$ preserving all definable relations. Moreover, $\mathrm{Aut}(M/D(M))$ has a good Galois theory; in particular, elements with a finite orbit are almost definable over some finite subset of D. This and some other characterizations can be found in [5, appendix].

Generic types

Let T be a complete theory with quantifier elimination. Let $\mathcal{C}$ be the category of substructures of models of T, with L-embeddings, and let $\mathcal{S}$ be the category of pairs (A, p) with $A \in \mathrm{Ob}\,\mathcal{C}$ and p a type over A. We define $\mathrm{Mor}((A, p), (B, q)) = \{f \in \mathrm{Mor}_{\mathcal{C}}(A, B) : f^*(q) = p\}$.

By a *generic type* we will mean a function p on $\mathrm{Ob}\,\mathcal{C}$, denoted $A \mapsto (p|A)$, such that $A \mapsto (A, p|A)$ is a functor $\mathcal{C} \to \mathcal{S}$. For example, when T, the theory of algebraically closed fields, is provided by any absolutely irreducible variety V: given a field F, let $p|F$ be the type of an F-generic point of V, i.e., the type of a point of $V(L)$ avoiding $U(L)$ for every proper F-subvariety U of V, where L is some extension field of F. Other examples will be given below, beginning with Example 3.3.

Lemma 2.10. *Let p be a generic type of T, and let $M \models T$, $a, b \in M$. Let $c \models p|M$.*

(1) *If $a \notin \mathrm{dcl}(\emptyset)$, then $a \notin \mathrm{dcl}(c)$.*
(2) *If $a \notin \mathrm{acl}(\emptyset)$, then $a \notin \mathrm{acl}(c)$.*
(3) *If $a \notin \mathrm{acl}(b)$, then $a \notin \mathrm{acl}(b, c)$.*

Proof.

(1) Since $a \notin \mathrm{dcl}(\emptyset)$, there exists $a' \neq a$ with $\mathrm{tp}(a) = \mathrm{tp}(a')$. Let $c' \models p|\langle\{a, a'\}\rangle$. Since $\mathrm{tp}(a) = \mathrm{tp}(a')$, there exists an isomorphism $\langle a\rangle \to \langle a'\rangle$; by functoriality of p, $\mathrm{tp}(a, c) = \mathrm{tp}(a', c)$. If $a \in \mathrm{dcl}(c)$, then a is the unique realization of $\mathrm{tp}(a/c)$, so $a = a'$; a contradiction.
(2) If $a \in \mathrm{acl}(c)$, then for some n there are at most n realizations of $\mathrm{tp}(a/c)$. Since $a \notin \mathrm{acl}(\emptyset)$, there exist distinct realizations $a_0, \ldots, a_n$ of $\mathrm{tp}(a)$. Proceed as in (1) to get a contradiction.
(3) This follows from (2) for $T_{\langle b\rangle}$. □

2.2 Grothendieck rings

We define the Grothendieck group and associated objects of a theory T; cf. [10]. $\mathrm{Def}(T)$ is the category of definable sets and functions. Let $\mathcal{C}$ be a subcategory of $\mathrm{Def}(T)$. We assume $\mathrm{Mor}(X, Y)$ is a sheaf on X: if $X_1 = X_2 \cup X_3$ are subobjects of X, and $f_i \in \mathrm{Mor}(X_i, Y)$ with $f_1|(X_2 \cap X_3) = f_2|(X_2 \cap X_3)$, then there exists $f \in \mathrm{Mor}(X_1, Y)$ with $f|X_i = f_i$. Thus the disjoint union of two constructible sets in $\mathrm{Ob}\,\mathcal{C}$ is also the category theoretic disjoint sum.

If only the objects are given, we will assume $\mathrm{Mor}\,\mathcal{C}$ is the collection of all definable bijections between them.

The *Grothendieck semigroup* $K_+(\mathcal{C})$ is defined to be the semigroup generated by the isomorphism classes $[X]$ of elements $X \in \mathrm{Ob}\,\mathcal{C}$, subject to the relation

$$[X] + [Y] = [X \cup Y] + [X \cap Y].$$

In most cases, $\mathcal{C}$ has disjoint unions; then the elements of $K_+(\mathcal{C})$ are precisely the isomorphism classes of $\mathcal{C}$.

If $\mathcal{C}$ has Cartesian products, we have a semiring structure given by

$$[X][Y] = [X \times Y].$$

In all cases we will consider the cases when products are present, the symmetry isomorphism $X \times Y \to Y \times X$ will be in the category, as well as the associativity morphisms, so that $K_+(\mathcal{C})$ is a commutative semiring.

(The assumption on Cartesian products is taken to include the presence of an object $\{p\} = X^0$ such that the bijections $X \to \{p\} \times X$, $x \mapsto (p, x)$, and $X \to X \times \{p\}$, $x \mapsto (x, p)$, are in $\mathrm{Mor}_{\mathcal{C}}$ for all $X \in \mathrm{Ob}_{\mathcal{C}}$. All such p give the same element $1 = [\{p\}] \in K(\mathcal{C})$, which serves as the identity element of the semiring.)

Let $K(\mathcal{C})$ be the Grothendieck group, the formal groupification of $K_+(\mathcal{C})$. When $\mathcal{C}$ has products, $K(\mathcal{C})$ is a commutative ring.

We will often have dimension filtrations on our categories, and hence on the semiring.

By an *semiring ideal* we mean a congruence relation, i.e., an equivalence relation on the semiring R that is a subsemiring of $R \times R$. To show that an equivalence relation E is a congruence on a commutative semiring R, it suffices to check that if $(a, b) \in E$ then $(a + c, b + c) \in E$ and $(ac, bc) \in E$.

Remark. When T is incomplete, let S be the (compact, totally disconnected) space of completions of T. Then $\{K(t) : t \in S\}$ are the fibers of a sheaf of rings over S. $K(T)$ can be identified with the ring of continuous sections of this sheaf. In this sense, Grothendieck rings reduce to the case of complete theories.

This last remark is significant even when T is complete: if one adds a constant symbol c to the language, T becomes incomplete, and so the Grothendieck ring of T in $L(c)$ is the Boolean power of $K(T_a)$, where T_a ranges over all $L(c)$-completions of T. Say c is a constant for an element of a sort S. Then an $L(c)$-definable subset of a sort S' corresponds to an L-definable subset of $S \times S'$. This allows for an inductive analysis of the Grothendieck ring of a structure, given good information about definable sets in one variable (cf. Lemma 2.3).

Groups of functions into $\mathcal{R}$

Let $\mathcal{C}(T)$ be a subcategory of the category of definable sets and bijections, defined systematically for T and for expansions by constants T. Let $\mathcal{R}(T) = K_+(\mathcal{C}(T))$ be the Grothendieck semigroup of $\mathcal{C}(T)$. When V is a definable set, we let $\mathcal{C}_V$, $\mathcal{R}_V$ denote the corresponding objects over V; the objects of $\mathcal{C}_V$ are definable sets

$X \subseteq (V \times W)$ such that for any $a \in V$, $X_a \in \mathcal{C}_a$, and similarly the morphisms. In practice, $\mathcal{R}$ will be the Grothendieck semigroup of all definable sets and definable isomorphisms satisfying some definable conditions, such as a boundedness condition on the objects, or a "measure preservation" condition on the definable bijections.

To formalize the notion of "definable function into $\mathcal{R}$" we will need to look at classes X_a of parametrically definable sets. The class of X_a makes sense only in the Grothendieck groups associated with $\mathbf{T}_a$, not $\mathbf{T}$. Moreover, the equality of such classes, say, of X_a and of X_b, begins to make sense only in Grothendieck groups of $\mathbf{T}_{(a,b)}$. Expressions like

$$[X] \underset{a,b}{=} [Y]$$

will therefore mean that X, Y are both definable in $\mathbf{T}_{a,b}$, $[X], [Y]$ denote their classes in the Grothendieck group of $\mathbf{T}_{a,b}$, and these classes are equal.

If V is a definable set, we define the *semigroup of definable functions* $V \to \mathcal{R}$, denoted $\mathrm{Fn}(V, \mathcal{R})$. An element of $\mathrm{Fn}(V, \mathcal{R})$ is represented by a definable $X \in \mathcal{C}_V$, viewed as the function $a \mapsto [X_a]$, where $[X_a]$ is a class in $\mathcal{R}_a$. X, X' represent the same function if for all a, $[X_a], [X'_a]$ are the same element of $\mathcal{R}_a$. Note that despite the name, the elements of $\mathrm{Fn}(V, \mathcal{R})$ should actually be viewed as sections $V \to \prod_{a \in V} \mathcal{R}_a$.

Addition is given by disjoint union in the image (i.e., disjoint union over X).

Usually $\mathcal{R}$ has a natural grading by dimension; in this case $\mathrm{Fn}(V, \mathcal{R})$ inherits the grading.

Assume that V is a definable group and $\mathcal{R} = K_+(T)$ is the Grothendieck semiring of all definable sets and functions of T, there is a natural convolution product on $\mathrm{Fn}(V, \mathbb{R})$. If $h_i(a) = [H_i(a)]$, $H_i \subset V \times B_i$, the convolution $h_1 * h_2$ is represented by

$$H = \{(a_1 + a_2, (a_1, a_2, y_1, y_2)) : (a_i, y_i) \in H_i\} \subseteq V \times (V^2 \times B_1 \times B_2)$$

so that $h_1 * h_2(a) = H(a) = \{(a_1, a_2, y_1, y_2) : (a_i, y_i) \in H_i, a_1 + a_2 = a\}$.

Grothendieck groups of orthogonal sets

Lemma 2.11. *Let T be a theory with two strongly orthogonal definable sets D_1, D_2, $D_{12} = D_1 \times D_2$. Let $K_+ D_i[n]$ be the Grothendieck semigroup of definable subsets of D_i^n. Then $K_+ D_{12}[n] \simeq K_+ D_1[n] \otimes K_+ D_2[n]$.*

Proof. This reduces to $n = 1$. Given definable sets $X_i \subseteq D_i^n$, it is clear that the class of $X_1 \times X_2$ in $K_+ D_{12}[n]$ depends only on the classes of X_i on $D_i[n]$. Define $[X_1] \otimes [X_2] = [X_1 \times X_2]$. This is clearly $\mathbb{Z}$-bilinear, and so extends to a homomorphism $\eta : K_+ D_1[1] \times K_+ D_2[1] \to K_+ D_{12}[1]$. By strong orthogonality, η is surjective.

To prove injectivity, note that any element of $K_+ D_1[n] \otimes K_+ D_2[n]$ can be written $\sum [X_1^i] \otimes [X_2^i]$, with $X_1^1, \dots, X_1^k$ pairwise disjoint. To see this, begin with some expression $\sum [X_1^i] \otimes [X_2^i]$; use the relation $[X' \dot{\cup} X''] \otimes [Y] = [X'] \otimes [Y] + [X''] \otimes [Y]$ to replace the X_1^i by the atoms of the Boolean algebra they generate, so that the new X_i^i are equal or disjoint; finally use the relation $[X' \otimes Y'] + [X' \otimes Y''] = [X'] \otimes [Y' \dot{\cup} Y'']$ to amalgamate the terms with equal first coordinate.

Hence it suffices to show that if $[\cup_i X_1^i \times X_2^i] = [\cup_j Y_1^i \times Y_2^i]$, with the X_1^i and the Y_1^i pairwise disjoint, then $\sum [X_1^i] \times [X_2^i] = \sum [Y_1^i] \times [Y_2^i]$. Let $F : \cup_i X_1^i \times X_2^i \to \cup_j Y_1^i \times Y_2^i$ be a definable bijection. By strong orthogonality, the graph of F is a disjoint union of rectangles. Since F is a bijection, it is easy to see that each of these rectangles has the form $f_1^k \times f_2^k$, where for $\nu = 1, 2$, $f_\nu^k : X_\nu(k) \to Y_\nu(k)$ is a bijection from a subset of $\cup_i X_\nu^i$ to a subset of $\cup_j Y_\nu^j$. The rest follows by an easy combinatorial argument; we omit the details, since a somewhat more complicated case will be needed and proved later; see Proposition 10.2. □

Integration by parts

The following will be used only in Section 9, to study the Grothendieck semiring of the valuation group.

Definition 2.12. Let us say that $Y \in \mathrm{Ob}\,\mathcal{C}$ is *treated as discrete* if for any $X \in \mathrm{Ob}\,\mathcal{C}$ and any definable $F \subset X \times Y$ such that $T \models F$ is the graph of a function, the projection map $F \to X$ is an invertible element of $\mathrm{Mor}_{\mathcal{C}}(F, X)$.

To explain the terminology, suppose each $X \in \mathrm{Ob}\,\mathcal{C}$ is endowed with a measure μ_X, and $\mathcal{C}$ is the category of measure-preserving maps. If μ_Y is the counting measure, and $\mu_{X\times Y}$ is the product measure, then for any function $f : X \to Y$, $x \mapsto (x, f(x))$ is measure preserving.

We will assume $\mathcal{C}$ is closed under products.

If Y_1, Y_2 are treated by $\mathcal{C}$ as discrete, so is $Y_1 \times Y_2$: if $F \subset X \times (Y_1 \times Y_2)$ is the graph of a function $X \to (Y_1 \times Y_2)$, then the projection to $F_1 \subset X \times Y_1$ is the graph of a function, hence the projection $F_1 \to X$ is in $\mathcal{C}$; now $F \subset (F_1 \times Y_2)$ is the graph of a function, and so $F \to F_1$ is invertibly represented, too; thus so is the composition. In particular, if Y is discretely treated, any bijection $U \to U'$ between subsets of Y^n is represented in $\mathcal{C}$.

If $\mathcal{R}$ is a Grothendieck group or semigroup, we write $[X] \underset{R}{=} [Y]$ to mean that X, Y have the same class in $\mathcal{R}$.

Lemma 2.13. *Let $f, f' \subset X \times L$ be objects of $\mathcal{C}$ such that $[f(c)] \underset{K(\mathcal{C}_c)}{=} [f'(c)]$ for any c in X. Then $[f] \underset{K(\mathcal{C})}{=} [f']$; similarly for K_+.*

Proof. By assumption, there exists $g(c)$ such that $f(c) + g(c)$, $f'(c) + g(c)$ are $\mathcal{C}_c$-isomorphic. By compactness (cf. the end of the proof of Lemma 2.3) this must be uniform (piecewise in L, and hence by glueing globally): there exists a definable $g \subset Z \times L$ and a definable isomorphism $f + g \simeq f' + g$, inducing the isomorphisms of each fiber. By the definition of $\mathcal{C}_c$, and since $\mathcal{C}$ is closed under finite glueing, $f + g$, $f' + g$ are in $\mathrm{Ob}\,\mathcal{C}$ and the isomorphism between them is in $\mathrm{Mor}\,\mathcal{C}$. □

Let L be an object of $\mathcal{C}$, treated as discrete in $\mathcal{C}$, and assume given a definable partial ordering on L.

Notation 2.14. Let $f \subset X \times L$. For $y \in L$, let $f(y) = \{x : (x, y) \in f\}$. Denote $\sum_{\gamma<\beta} f(\gamma) = [\{(x, y) : x \in f(y), y < \gamma\}]$.

Notation 2.15. Let $\phi : L \to K(X)$ be a constructible function, represented by $f \subset X \times L$; so that $\phi(y) = [f(y)]$, $f(y) = \{x : (x, y) \in f\}$. Denote $\sum_{\gamma<\beta} \phi(\gamma) = [\{(x, y) : x \in f(y), y < \gamma\}]$.

Note by Lemma 2.13 that this is well defined.

Below, we write fg for the pointwise product of two functions in $K(\mathcal{C})$; $[fg(y)] = [f(y) \times g(y)]$.

Lemma 2.16 (integration by parts). *Let* Γ *be an object of* $\mathcal{C}$*, treated as discrete in* $\mathcal{C}$*, and assume given a definable partial ordering of* Γ*. Let* $f \subset X \times \Gamma$, $F(\beta) = \sum_{\gamma<\beta} f(\gamma)$, $g \subset Y \times \Gamma$, $\mathbf{G}(\beta) = \sum_{\gamma\leq\beta} g(\gamma)$.

Then

$$F\mathbf{G}(\beta) = \sum_{\gamma<\beta} f\mathbf{G}(\gamma) + \sum_{\gamma\leq\beta} Fg(\gamma).$$

Proof. Clearly,

$$F\mathbf{G}(\beta) = \Sigma_{\gamma<\beta,\gamma'\leq\beta} f(\gamma)g(\gamma').$$

We split this into two sets, $\gamma < \gamma'$ and $\gamma' \leq \gamma$. Now

$$\Sigma_{\gamma<\gamma'\leq\beta} f(\gamma)g(\gamma') = \Sigma_{\gamma'\leq\beta} F(\gamma')g(\gamma'),$$
$$\Sigma_{\gamma'\leq\gamma<\beta} f(\gamma)g(\gamma') = \Sigma_{g<\beta} f(\gamma)\mathbf{G}(\gamma').$$

The lemma follows. □

This is particularly useful when L is treated as discrete in $\mathcal{C}$, since then, if the sets $f(\gamma)$ are disjoint, $[f] = [\cup_\gamma f_\gamma]$. Another version, with $G(\beta) = \sum_{\gamma<\beta} g(\gamma)$:

$$FG(\beta) = \sum_{\gamma<\beta} (fG + gF + fg)(\gamma).$$

3 Some C-minimal geometry

We will isolate the main properties of the theory ACVF, and work with an arbitrary theory **T** satisfying these properties. This includes the rigid analytic expansions $\mathrm{ACVF}^{\mathrm{R}}$ of [23].

The right general notion, C-minimality, has been introduced and studied in [15]. They obtain many of the results of the present section. Largely for expository reasons, we will describe a slightly less general version; it is essentially minimality with respect to an ultrametric structure in the sense of [31]. We will use notation suggestive of the case of valued fields; thus we denote the main sort by VF and a binary function by $\mathrm{val}(x - y)$. Some additional assumptions will be made explicit later on.

Let T be a theory in a language L, extending a theory T in a language L. T is said to be T-*minimal* if for any $M \models T$, any L_M-formula *in one variable* is T_M-equivalent to an L_M formula.

More generally, if D is a definable subset of T (i.e., a formula of L), we say that D is T-*minimal* if for any $M \models T$, any T_M-definable subset of D is T_M-equivalent to one defined by an L_M formula.

Strong minimality

Let $\mathsf{L} = \emptyset$. The only atomic formulas of L are thus equalities $x = y$ of two variables. T is the theory of infinite sets. T-minimality is known as *strong minimality*; see [1, 28, 29]. A theory T is strongly minimal iff for any $M \models T$, any T_M-definable subset of M is finite or cofinite. For us the primary example of a strongly minimal theory is ACF, the theory of algebraically closed fields.

Let $M \models T$. If D is strongly minimal, and X a definable subset of D^*, we define the *D-dimension* of X to be the least n such that X admits a T_M-definable map into D^n with finite fibers. In the situation we will work in, there will be more than one definable strongly minimal set up to isomorphism, and even up to definable isogeny; in particular, there will be the various sets of RES_M. However, between any of these, there exists an M-definable isogeny; so the $\mathbf{k}$ dimension agrees with the D dimension for any of them. We will call it the RES dimension. It agrees with *Morley rank*, a notion defined in greater generality, that we will not otherwise need here.

O-minimality

$\mathsf{L} = \{<\}$, $\mathsf{T} = DLO$ the theory of dense linear orders without endpoints (cf. [9]). DLO minimality is known as O-minimality, and can also be stated thusly: any T_M-definable subset of M is a finite union of points and intervals. This also forms the basis of an extensive theory; see [37].

Let D be O-minimal. Then the O-minimal dimension of a definable set $X \subseteq D^*$ is the least n such that X admits a T_M-definable map into D^n with bounded finite fibers.

The Steinitz exchange principle states that if $a \in \mathrm{acl}(B \cup \{b\})$ but $a \notin \mathrm{acl}(B)$, then $b \in \mathrm{acl}(B \cup \{a\})$.

This holds for both strongly minimal and O-minimal structures; cf. [37].

For us the relevant O-minimal theory is DOAG itself. We will occasionally use stronger facts valid for this theory. Quantifier elimination for DOAG implies the following.

Lemma 3.1.

(1) *Any parameterically definable function f of one variable is piecewise affine; there exists a finite partition of the universe into intervals and points, such that on each interval I in the partition, $f(x) = \alpha x + c$ for some rational α and some definable c.*
(2) DOAG *admits elimination of imaginaries.*

Proof.

(1) This follows from quantifier elimination for DOAG.
(2) This follows from (1) that any function definable with parameters in DOAG has a canonical code, consisting of the endpoints of the intervals of the coarsest such partition, together with a specification of the rationals α and the constants c. But from this it follows on general grounds that every definable set is coded (cf. [16, 3.2.2]). Thus DOAG admits elimination of imaginaries. □

C-minimality

Let $\mathsf{T} = \mathsf{T}_{um}$ be the theory of ultrametric spaces or, equivalently, chains of equivalence relations (cf. [31]).

In more detail, L has two sorts, VF and Γ_∞. The relations on Γ_∞ are a constant ∞ and a binary relation $<$. In addition, L has a function symbol $\mathrm{VF}^2 \to \Gamma_\infty$, written $\mathrm{val}(x-y)$. T states the following:

(1) Γ_∞ is a dense linear ordering with no least element, but with a greatest element ∞.
(2) $\mathrm{val}(x-y) = \infty$ iff $x = y$.
(3) $\mathrm{val}(x-y) \geq \alpha$ is an equivalence relation; the classes are called *closed α-balls*. Hence so is the relation $\mathrm{val}(x-y) > \alpha$, whose classes are called *open α-balls*.
(4) Let $\Gamma = \Gamma_\infty \setminus \{\infty\}$. For $\alpha \in \Gamma$, every closed α-ball contains infinitely many open α-balls.

A T_{um}-minimal theory will be said to be C-minimal. The notion considered in [15] is a little more general, but for theories T_{um} they coincide. Since we will be interested in fields, this level of generality will suffice.

A theory T extending ACVF is C-minimal iff for any $M \models T$, every T_M-definable subset of $\mathrm{VF}(M)$ is a Boolean combination of open balls, closed balls and points. If T is C-minimal, $A \leq M \models T$, and b is an A-definable ball, or an infinite intersection, let p_A^b be the collection of A-definable sets not contained in a finite union of proper subballs of b. Then by C-minimality, p_A^b is a complete type over A.

Let T be C-minimal. Then in T, Γ is O-minimal, and for any closed α-ball C, the set of open α-subballs of C is strongly minimal. Denote it $C/(1+\mathcal{M})$. (These facts are immediate from the definition.)

Assume T is C-minimal with a distinguished point 0. We define: $\mathrm{val}(x) = \mathrm{val}(x-0)$; $\mathcal{M} = \{x : \mathrm{val}(x) > 0\}$. Let $\mathfrak{B}_{\mathrm{cl}}$ be the family of all closed balls, including points. Among them are ${\mathfrak{B}_{\mathrm{cl}}}^c{}_\alpha(0) = \{x : \mathrm{val}(x) \geq \alpha\}$. Let $\mathrm{RV} = \dot{\bigcup}_{\gamma\in\Gamma} B_\gamma^c(0)/(1+\mathcal{M})$, and let $\mathrm{rv} : \mathrm{VF} \setminus \{0\} \to \mathrm{RV}$ and $\mathrm{val}_{\mathrm{rv}} : \mathrm{RV} \to \Gamma$ be the natural map. By an rv-ball we mean an open ball of the form $\mathrm{rv}^{-1}(c)$.

The T-definable fibers of $\mathrm{val}_{\mathrm{rv}}$ are referred to, collectively, as RES_T. Later we will fix a theory $\mathbf{T}$, and write RES for $\mathrm{RES}_{\mathbf{T}}$; we will also write RES_A for $\mathrm{RES}_{\mathbf{T}_A}$. The unqualified notion "definable," as well as many derived notions, will implicity refer to $\mathbf{T}$.

A certain notion of genericity plays an essential role in these theories.

Example 3.2. Let T be a strongly minimal theory. For any $A \leq M \models T$, any A-definable set is finite or has finite complement. Therefore, the collection of cofinite sets forms a complete type. A realization of this type is called a *generic* element of M, over A.

Example 3.3. Let T be an O-minimal theory. For any $A \leq M \models T$, any A-definable set contains, or is disjoint from, an infinite interval (b, ∞) for some $b \in M$. The set of A-definable sets containing such an interval is thus a complete type, the generic type of large elements of Γ. Similarly, the set of A-definable sets containing an interval $(0, a)$ with $0 < a$ is *the generic type of small positive elements*. More generally, given a subset $S \subseteq A$ $S' = \{b \in A : (\forall s \in S)(s < b)\}$; then the definable sets $x > a (a \in S), x < b (b \in S')$ generate a complete type over A, called the type of elements *just bigger than* S.

Definition 3.4. Let T be C-minimal. Let b be a T_A-definable ball, or an infinite intersection of balls. The generic type p_b of b is defined by $p_b|A' = p^b_{A'}$, for any $A \leq A' \leq M \models T$.

The completeness follows from C-minimality, since for any A'-definable subset S of b, either S is contained in a finite union of proper subballs of b, or else the complement $b \setminus S$ is contained in such a finite union.

A realization of $p_b|A'$ is said to be a *generic point of* b *over* A'. An A'-definable set is said to be b*-generic* if it contains a generic point of b over A'.

See Section 3.2 for some generalities about generic types. For our purposes it will suffice to consider generic types in one VF variable. For more information see [16, Section 2.5].

Remark 3.5. If $A = \mathrm{acl}(A)$ then any type of a field element $\mathrm{tp}(c/A)$ coincides with $p_b|A$, where b is the intersection of all A-definable balls containing c.

This is intended to include the case of closed balls of valuative radius ∞, i.e., points; these are the algebraic types $x = c$. Note also the degenerate case that c is not in any A-definable ball; then $b = \mathrm{VF}$ and $\mathrm{tp}(c/A)$ is the generic type of VF over A.

Not every generic 1-type is of the form p_b for a ball b as above. For instance, let b be an open ball, $c \in b$; then the generic type $p_b((x - c)^{-1})$ is not of this form.

For V-minimal theories (defined below) it can be shown that every generic 1-type is of the form p_b or $p_b((x - c)^{-1})$.

Let **T** be a C-minimal theory. Let b be a definable ball, or an infinite intersection of definable balls. We say that b is centered if it contains a proper definable finite union of balls. If b is open, or a properly infinite intersection of balls, we have the following:

> If b contains a proper finite union of balls, then it contains a definable closed ball (the smallest closed ball containing the finite set). $(*)$

For C-minimal fields of residue characteristic 0, $(*)$ is true of closed balls: the set of maximal open subballs of b forms an affine space over the residue field $\mathbf{k}$, where the center of mass of a finite set is well defined.

Clearly, b is centered over $\mathrm{acl}(A)$ if and only if it is centered over A. The term "centered" will be justified to some extent by the assertion of Lemma 3.39, that when A is generated by elements of $\mathrm{VF} \cup \mathrm{RV} \cup \Gamma$, any A-definable closed ball contains an A-definable point, and thus a centered ball has a definable "center."

Lemma 3.6. *b is centered over A iff b is not transitive over A.*

This is immediate from the definition, and from C-minimality, since any proper definable subset would have to be a Boolean combination of balls.

An often useful corollary of C-minimality is the following.

Lemma 3.7. *Let T be C-minimal, X a definable subset of* VF, *and Y a definable set of disjoint balls. Then for all but finitely many $b \in Y$, either $b \subseteq X$ or $b \cap X = \emptyset$.*

Proof. X is a finite Boolean combination of balls, so it suffices to prove this when X is a ball; then X is contained in at most one ball $b \in Y$; for any other $b \in Y$, either $b \subseteq X$ or $b \cap X = \emptyset$. □

Lemma 3.8. *Let $(b_t : t \in Q)$ be a definable family of pairwise disjoint balls. Then for any nonalgebraic $t \in Q$, b_t is transitive over $\langle t \rangle$.*

Proof. Consider a definable $R' \subseteq Q \times \mathrm{VF}$ with $R'(t) \subseteq b_t$. Let $Y = \cup_{t \in Q} R'(t)$. Then Y is a definable subset of VF, hence a finite combination of a finite set H of balls. The b_t are pairwise disjoint, so at most finitely many can contain an element of H, and thus no nonalgebraic b_t contains an element of H. Thus each ball in H is disjoint from, or contains, any given b_t. It follows that Y is disjoint from, or contains, any given b_t. Thus $b_t \cap Y$ cannot be a nonempty proper subset of b_t. □

Internalizing finite sets

The following lemma will be generalized later to finite sets of balls. It is of such fundamental importance in this paper that we include it separately in its simplest form. The failure of this lemma in residue characteristic $p > 0$ is the main reason for the failure of the entire theory to generalize, in its present form. Recall the definition of RV (Section 2.1).

Lemma 3.9. *Let $\mathbf{T}$ be a C-minimal theory of fields of residue characteristic* 0 *(possibly with additional structure), $A \leq M \models \mathbf{T}$. Let F be a finite $\mathbf{T}_A$-definable subset of* VF^n. *Then there exists $F' \subseteq \mathrm{RV}^m$, and a $\mathbf{T}_A$-definable bijection $h : F \to F'$.*

Proof. First consider $F = \{c_1, \dots, c_n\} \subseteq \mathrm{VF}$. Let $c = (\sum_{i=1}^n c_i)/n$ be the average; then F is $\mathbf{T}_A$-definably isomorphic to $\{c_1 - c, \dots, c_n - c\}$. Thus we may assume the average is 0. If there is no nontrivial A-definable equivalence relation on F, then $\mathrm{val}(x - y) = \alpha$ is constant on $x \neq y \in F$. In this case rv is injective on F and one can take $h = \mathrm{rv}$. Otherwise, let E be a nontrivial A-definable equivalence relation on F. By an E-symmetric polynomial, we mean a polynomial $H(x_1, \dots, x_n)$ with coefficients in A, invariant under the symmetric group on each E-class. For any such H, $H(F)$ is a $\mathbf{T}_A$-definable set with $< n$ elements. There exists H such that $H(F)$ has more than one element. By induction, there exists an injective A-definable function $h_0 : H(F) \to \mathrm{RV}^m$. Let $h_1 = h_0 \circ H$. For $d \in h_0(H(F))$, and $d' = {h_0}^{-1}d$, let $F_d = H^{-1}{h_0}^{-1}(d) = H^{-1}(d')$. By induction again, there exists an $A(d) = A(d')$-definable injective function $g_d : F_d \to \mathrm{RV}^{m'}$. (We can take the same m' for all d.) Define $h(x) = (h_1(x), g_{h_1(x)}(x))$. Then clearly h is A-definable and injective.

The case $F \subseteq \mathrm{VF}^n$ follows using a similar induction, or by finding a linear projection with $\mathbb{Q}$-coefficients $\mathrm{VF}^n \to \mathrm{VF}$ which is injective on F. □

3.1 Basic geography of C-minimal structures

Let $\mathbf{T}$ be a C-minimal theory. We begin with a rough study of the existence and nonexistence of definable maps between various regions of the structure: $\mathbf{k}$, Γ, RV, VF and VF/$\mathcal{O}$.

We will occasionally refer to *stable* definable sets.

A definable set D of a theory T is called *stable* if there is no model $M \models T$ and M-definable relation $R \subseteq D^2$ and infinite subset $J \subseteq M(D)$ such that $R \cap J^2$ is a linear ordering. This is a model-theoretic finiteness condition, greatly generalizing finite Morley rank, and in turn strong minimality (cf. [28, 29]).

It is shown in [16] that a definable subset of $\mathrm{ACVF}_A^{\mathrm{eq}}$ is stable if and only if it has finite Morley rank, if and only if it admits no parametrically definable map onto an interval of Γ; and this is if and only if it embeds, definably over $\mathrm{acl}(A)$, into a finite-dimensional $\mathbf{k}$-vector space. These vector spaces have the general form $\Lambda/\mathcal{M}\Lambda$, with $\Lambda \leq \mathrm{VF}^n$ a lattice. Within the sorts we are using here, the relevant ones are the finite products of vector spaces of RES. More generally, in a C-minimal structure with sorts VF, RV, all stable sets are definably embeddable (with parameters) into RES. We will, however, make no use of these facts, beyond justifying the terminology. Thus "X is a stable definable set" can simply be read as " there exists a definable bijection between X and a subset of RES*."

The first fact is the unrelatedness of $\mathbf{k}$ and Γ.

Lemma 3.10. *Let Y be a stable definable set. Then Y, Γ are strongly orthogonal. In particular, any definable map from Y to Γ has finite image.*

Proof. We prove the second statement first: let $M \models \mathbf{T}$. Let $f : Y \to \Gamma$ be an M-definable map. Then $f(Y)$ is stable, and linearly ordered by $<_\Gamma$; hence by the definition of stability, it is finite.

Let $\gamma = (\gamma_1, \dots, \gamma_m) \in \Gamma$. We have to show that for a Y-generated structure A, $\mathrm{tp}(\gamma)$ implies $\mathrm{tp}(\gamma/A)$. It suffices to show that for any $a, \dots, a_n \in A$, $\mathrm{tp}(\gamma_i/\langle\gamma_1, \dots, \gamma_{i-1}\rangle)$ implies $\mathrm{tp}(\gamma_i/\langle\gamma_1, \dots, \gamma_{i-1}, a_1, \dots, a_n\rangle)$, for each i. By passing to $T_{\langle\gamma_1,\dots,\gamma_{i-1}\rangle}$ we may assume $m = 1$, $\gamma \in \Gamma$. Similarly, we may assume $n = 1$; let $a = a_1 \in Y$. To show that $\mathrm{tp}(\gamma)$ implies $\mathrm{tp}(\gamma/a)$, it suffices to show that any T_a-definable subset of Γ is definable. By O-minimality, any T_a-definable subset of Γ is a finite union of intervals, so (in view of the linear ordering) it suffices to show this for intervals (c_1, c_2). But if the interval is T_a-definable then so are the endpoints, so $c_i = c_i(a)$ is a value of a definable map $Y \to \Gamma$. But such maps have finite images, so c_i lies in a finite definable set. Using the linear ordering, we see that c_i itself is definable, and hence so is the interval. □

Lemma 3.11. *There are no definable sections of* $\mathrm{val}_{\mathrm{rv}} : \mathrm{RV} \to \Gamma$ *over an infinite subset of* Γ. *In fact if* $Y \subset \mathrm{RV}^n$ *is definable and* $\mathrm{val}_{\mathrm{rv}}$ *is finite-to-one on* Y, *then* Y *is finite.*

Proof. Looking at the fibers of the projection of Y to RV^{n-1}, and using induction, we reduce the lemma to the case $n = 1$. In this case, by Lemma 3.7, every definable set is a Boolean combination of pullbacks by $\mathrm{val}_{\mathrm{rv}}$ of subsets of Γ and finite sets. □

Lemma 3.12. *Let* $M \models \mathbf{T}$ *and let* $Y \subset \mathfrak{B}_{\mathrm{cl}}{}^n$ *be an infinite definable set. Then there exists a surjective* M*-definable map of* Y *to a proper interval in* Γ.

Proof. Since Γ is O-minimal, any infinite M-definable subset contains a proper interval. Thus it suffices to find an M-definable map of Y onto an infinite subset of Γ.

If the projection of Y to $\mathfrak{B}_{\mathrm{cl}}{}^{n-1}$ as well as every fiber of this projection are finite, then Y is finite. Otherwise, replacing Y by one of the fibers or by the projection, we reduce inductively to the case $n = 1$.

Let $v(y) \in \Gamma$ be the valuative radius of the ball y. Then $v(Y)$ is an M-definable subset of Γ. If it is infinite, we are done; otherwise, we may assume all elements of Y have the same valuative radius γ.

Let $W = \cup Y$. By C-minimality, W is a Boolean combination of balls b_i (open, of valuative radius $< \gamma$, or closed, of valuative radii $\delta_i \leq \gamma$). If W contains some $W' = b_i \setminus (b_{j_1} \cup \cdots \cup b_{j_l})$, where b_{j_i} is a proper subball of b_i, and $\delta_i < \gamma$, pick a point c in W'; then for any δ with $\gamma > \delta > \delta_i$ there exists $c' \in W'$ with $\mathrm{val}(c - c') = \delta$. It follows that the balls $b_\gamma(c)$, $b_\gamma(c')$ of radius γ around c, c' are both in Y; but infinitely many such δ exist; fixing c, we obtain a map $b_\gamma(c') \mapsto \mathrm{val}(c - c')$ into an infinite subset of Γ.

Otherwise, W can only be a finite set of balls of valuative radius γ. Thus Y is finite. □

Corollary 3.13. $\mathfrak{B}_{\mathrm{cl}}{}^n$ *contains no stable definable set. In particular,* VF *contains no strongly minimal set.* □

By contrast, we have the following.

Lemma 3.14. *Any infinite definable subset of* RV^n *contains a strongly minimal* M*-definable subset.*

Proof. By Lemma 3.11, the inverse image of some point in Γ^n must be infinite. □

Lemma 3.15. *Let* $M \models \mathbf{T}$. *Let* $Y \subseteq \mathfrak{B}_{\mathrm{cl}}$ *be a definable set. Let* $\mathrm{rad}(y)$ *be the valuative radius of the ball* y. *Then either* $\mathrm{rad} : Y \to \Gamma$ *is finite-to-one, or else there exists an* M*-definable map of an* M*-definable* $Y' \subseteq Y$ *onto a strongly minimal set.*

Proof. If rad is not finite-to-one, then Y contains an infinite set Y' of balls of the same radius α. Then $\cup Y'$ contains a closed ball b of valuative radius $\beta < \alpha$. The set S of open subballs b' of b of valuative radius β forms a strongly minimal set; the map sending $y \in Y'$ to the unique $b' \in S$ containing y is surjective. □

The following lemma regarding VF/$\mathcal{O}$ will be needed for integration with an additive character (Section 11).

Lemma 3.16. *Let Y be a stable definable set, $Z \subset \mathrm{VF} \times Y$ a definable set such that for $y \in Y$, $Z(y) = \{x : (x, y) \in Z\}$ is additively $\mathcal{M}$ invariant. Then for all but finitely many $\mathcal{O}$-cosets C, $Z \cap (C \times Y)$ is a rectangle $C \times Y'$.*

Proof. For $y \in Y$, $Z(y)$ is a $\mathbf{T}_y$-definable subset of VF, hence a Boolean combination of a finite $\langle y \rangle$-definable set of balls $b_1(y), \ldots, b_k(y)$. Let $B_i(y)$ be the smallest closed ball containing $b_i(y)$. According to Lemma 3.13, since the set of closed balls occurring as $B_i(y)$ for some y is stable, it is finite:

$$\{B_i(y) : y \in Y\} = \{B_1, \ldots, B_l\}.$$

All the B_i are $\mathcal{O}$-invariant. Let R be the set of $\mathcal{O}$-cosets C that are equal to some B_i.

If $B_i(y)$ has valuative radius < 0 (i.e., it is bigger than an $\mathcal{O}$-coset), then so is $b_i(y)$, so the characteristic function of such a $b_i(y)$ is constant on any closed $\mathcal{O}$-coset C. If $C \notin R$, then it is disjoint from any B_i of valuative radius equal to (or greater than) 0, so the characteristic functions of the corresponding $b_i(y)$ are also constant on it. Thus with finitely many exceptional C, any such characteristic function is constant on C, and the claim follows. □

3.2 Generic types and orthogonality

Two generic types p, q are said to be *orthogonal* if for any base A', if $c \models p|A'$, $d \models q|A'$, then p generates a complete type over $A(d)$; equivalently, q generates a complete type over $A(c)$. We will see that generics of different kinds are orthogonal (cf. Lemma 3.19). This orthogonality of types is weaker than the orthogonality of definable sets mentioned in the introduction, and in the present case is only an indirect consequence of the orthogonality between the residue field and value group; these types do not have orthogonal definable neighborhoods.

If $\gamma \in \Gamma$ and $rk_{\mathbb{Q}}(\Gamma(C(a))/\Gamma(C)) = n$, we say that $\mathrm{tp}(\gamma/C)$ has Γ-dimension n.

Lemma 3.17. *Let p_Γ be a $\mathbf{T}_A$-type of elements of Γ^n of Γ dimension n. Let $P = \overset{-1}{\mathrm{val}}(p_\Gamma)$. Then we have the following:*

(1) $\overset{-1}{\mathrm{val}}(p_\Gamma)$ *is a complete type over A. In other words, for any A-definable set X, either* $\overset{-1}{\mathrm{val}}(p_\Gamma) \subseteq X$ *or* $\overset{-1}{\mathrm{val}}(p_\Gamma) \cap X = \emptyset$.
(2) *If D is a stable A-definable set and $d_1, \ldots, d_n \in D$, then P implies a complete type over $A(d_1, \ldots, d_n)$.*
(3) *If $c \in P$, then $D(A(c)) = D(A)$.*
(4) *P is complete over A.*

Proof.

(1) This reduces inductively to the case $n = 1$. Since $\mathrm{val}^{-1}(p_\Gamma)$ is a disjoint union of open balls, (1) for $n = 1$ follows from Lemma 3.7: an A-definable set X cannot intersect nontrivially each of an infinite family of open balls. Therefore, either X is disjoint from almost all, or X contains almost all open balls $\mathrm{val}^{-1}(c)$, $c \models p_\Gamma$; in the former case the complement of X contains $\mathrm{val}^{-1}(p_\Gamma)$, and in the latter X contains $\mathrm{val}^{-1}(p_\Gamma)$ since p_Γ is complete.
(2) By strong orthogonality, p_Γ generates a complete type q' over $A(d)$, of Γ-dimension n. By (1) over $A(d)$, $\mathrm{val}^{-1}(p_\Gamma)$ is complete over $A(d)$. But if $c \in P$ then $\mathrm{val}(c) \models p_\Gamma$ so $c \in \mathrm{val}^{-1}(p_\Gamma)$. Thus P implies a complete type over $A(d)$.
(3) follows from (2): if $d \in D(A(c))$ then there exists a formula ϕ such that $\models \phi(d, c)$ and such that $\phi(x, c)$ has a unique solution. By (2) ϕ is a consequence of $P(c) \cup \mathrm{tp}(d/A)$, and hence by compactness of a formula $\phi_1(x) \& \phi_2(c)$, where $\phi_2 \in \mathrm{tp}(d/A)$. Thus already $\phi_1(x)$ has the unique solution d, and thus $d \in D(A)$.
(4) This is immediate from (1). □

Lemma 3.18. *Let q be a $\mathbf{T}_A$-type of elements of RES^n_A of RES dimension n. Let $Q = \mathrm{rv}^{-1}(q)$. Then Q is complete over A. Moreover, if $\gamma_1, \dots, \gamma_m \in \Gamma$, then Q implies a complete type over $A(\gamma_1, \dots, \gamma_m)$.*

Proof. Again the lemma reduces inductively to the case $n = 1$, and for $n = 1$ follows from Lemma 3.7, since $\mathrm{val}^{-1}(q)$ is a union of disjoint annuli; the "moreover" also follows from orthogonality as in the proof of Lemma 3.17(2). □

Lemma 3.19 ([16, Section 2.5]).

(1) *If b is an open ball, or a properly infinite intersection of balls, and b' a closed ball, then p_b, $p_{b'}$ are orthogonal.*
(2) *Any b-definable map to $\mathbf{k}$ is constant on b away from a proper subball of b.*

Proof. We recall the proof from [16, Section 2.5]: The statement becomes stronger if the base set is enlarged. Thus we may assume that b and b' are centered; by translating we may assume both are centered at 0, and by a multiplicative renormalization that b' is the unit closed ball. Thus

$$c \models p_{b'}|A \quad \text{iff} \quad c \in \mathcal{O} \quad \text{and} \quad \mathrm{res}(c) \notin \mathrm{acl}(A). \tag{$*$}$$

On the other hand, let p_Γ be the type of elements of Γ that are just bigger than the valuative radius of b (cf. Example 3.3). Then $d \models p_b|A$ iff $\mathrm{val}(d) \models p_\Gamma$, i.e., p_b is now the type P described in Lemma 3.17. By Lemma 3.17, if $c' \in P$ then $\mathbf{k}(A(c')) = \mathbf{k}(A)$. It follows that if $c \models p_{b'}|A$, then $\mathrm{res}(c) \notin \mathrm{acl}(\mathbf{k}(A(c')))$. By $(*)$ $c \models p_{b'}|A(c')$.

For the second statement, let g be a definable map $b \to \mathbf{k}$; by Lemma 3.17(3), g is constant on the generic type of b; by compactness, g is constant on b away from some proper subball of b. □

Lemma 3.20. *Let $a = (a_1, \ldots, a_n) \in \mathrm{RV}^n$, and assume $a_i \notin \mathrm{acl}(A(a_1, \ldots, a_{i-1}))$ for $1 \leq i \leq n$. Then the formula $D(x) = \bigwedge_{i=1}^n \mathrm{rv}(x_i) = a_i$ generates a complete type over $A(a)$, and, indeed, over any $\mathrm{RV} \cup \Gamma$-generated structure A'' over A.*

In particular, if $q = \mathrm{tp}(a/A)$, any A-definable function $f : \mathrm{rv}^{-1}(q) \to \mathrm{RV} \cup \Gamma$ factors through $\mathrm{rv}(x) = (\mathrm{rv}(x_1), \ldots, \mathrm{rv}(x_n))$.

Proof. This reduces inductively to the case $n = 1$. If we replace A by a bigger set M (such that $a_i \notin \mathrm{acl}(A(a_1, \ldots, a_{i-1}))$ for $1 \leq i \leq n$), the assertion becomes stronger; so we may assume $A = M \models \mathbf{T}$. Let $\mathrm{rv}(c) = \mathrm{rv}(c') = a$. Either $\mathrm{val}(c) = \mathrm{val}(c') \notin M$, or else $\mathrm{val}(c) = \mathrm{val}(c') = \mathrm{val}(d)$ for some $d \in M$, and $\mathrm{res}(c/d) = \mathrm{res}(c'/d) \notin M$; in either case, by Lemma 3.17 or Lemma 3.18, we have $\mathrm{tp}(c/M) = \mathrm{tp}(c'/M)$. Thus $\mathrm{tp}(c, \mathrm{rv}(c)/M) = \mathrm{tp}(c', \mathrm{rv}(c')/M)$, i.e., $\mathrm{tp}(c/M(a)) = \mathrm{tp}(c'/M(a))$. This proves completeness over $A(a)$.

Let A' be a structure generated over A by finitely many elements of Γ. Then $A'(a) = A(\gamma_1, \ldots, \gamma_k, a)$, where $\gamma_i \in \Gamma$, and $\gamma_i \notin A(\gamma_1, \ldots, \gamma_{i-1}, \mathrm{val}(a))$. It follows that $\mathrm{rv}(a) \notin A(\gamma_1, \ldots, \gamma_k)$, so $D(x)$ generates a complete type over $A(\gamma_1, \ldots, \gamma_k)(a) = A'(a)$.

Let A'' be generated over $A'(a)$ by elements of stable A-definable sets. Since $D(x)$ is the (unique, and therefore) generic type of an open ball over $A'(a)$, by Lemma 3.17, it generates a complete type over A''.

Now if $A'' = A(\gamma_1, \ldots, \gamma_k, r_1, \ldots, r_n, d)$, where $\gamma_j \in \Gamma$, $r_i \in \mathrm{RV}$ and d lies in a stable set over A, let $A' = A(\gamma_1, \ldots, \gamma_k, \mathrm{val}_{\mathrm{rv}}(r_1), \ldots, \mathrm{val}_{\mathrm{rv}}(r_n))$; then A'/A is Γ-generated, and A''/A is generated by elements of stable sets (including ${\mathrm{val}_{\mathrm{rv}}}^{-1}(r_i)$). Thus the above applies.

The last statement follows by applying the first part of the lemma over $A'' = A(f(c))$: the formula $f(x) = f(c)$ must follow from the formula $D(x)$, since $D(x)$ generates a complete type over A''. □

3.3 Definable sets in group extensions

We will analyze the structure of RV in a slightly more abstract setting. In the following lemmas we assume R is a ring, and $0 \to A \to B \to C \to 0$ is a definable exact sequence of R-modules in T. This means that A, B, C are definable sets, and that one is also given definable maps $+_A : A^2 \to A$, $f_A^r : A \to A$ for each $r \in R$, and similarly for B, C; and definable maps $\iota : A \to B, \vartheta : B \to C$, such that in any $M \models T$, $A(M), B(M), C(M)$ are R-modules under the corresponding functions, and $0 \to A(M) \to_\iota B(M) \to_\vartheta C(M) \to 0$ is an exact sequence of homomorphisms of R-modules.

Lemma 3.21. *Consider a theory with a sequence $0 \to A \to B \to_\vartheta C \to 0$ of definable R-modules and homomorphisms (carrying additional structure). Assume the following:*

(1) *A, C are stably embedded and orthogonal.*
(2) *Every almost definable subgroup of A^n is defined by finitely many R-linear equations.*

(3) *("No definable quasi-sections.") If P is a definable subset of B^n whose projection to C^n is finite-to-one, then P is finite.*

Then every almost definable subset Z of B^n is a finite union of sets of the form

$$\{b : \vartheta(b) \in W, Nb \in Y\},$$

where $N \in B_{n,k}(R)$ is an $n \times k$ matrix, Y is an almost definable subset of a single coset of A^k, W is an almost definable subset of C^n.

Note the following:

(1) To verify (3), it suffices to check it for $n = 1$ but for parametrically definable P.
(2) If C is definably linearly ordered, and Z is definable, then Y, W may be taken definable.

Proof. Using a base change as in Section 2.1, we may assume almost definable sets are definable. Replacing B by B^n and R by $M_n(R)$, we may assume $n = 1$. Let Z be a definable subset of B. Given $X \subset A$, let $[X]$ denote the class of X up to translation; so $[X] = [X']$ if $X = X' + a$ for some $a \in A$. Now a definable subset U of a coset $b + A$ of A has the form $b + X$, $X \subset A$; the class $[X]$ is well defined, and we will denote $[U] = [X]$. We obtain a map

$$c \mapsto [Z \cap \vartheta^{-1}(c)].$$

In more detail, for any $b \in (\vartheta^{-1}(c) \cap Z)$, we have $(\vartheta^{-1}(c) \cap Z) - b \subseteq A$, and so by stable embeddedness of A we can write $(\vartheta^{-1}(c) \cap Z) - b = X(a)$ for some $a \in A^m$. The tuple a is not well defined; but the class of a in the definable equivalence relation

$$x \sim x' \iff (\exists t \in A)(t + X(x)) = X(x')$$

is obviously a function of c alone. By the orthogonality assumption, this map is piecewise constant. Thus we may assume it is constant and fix C_0 with $[Z \cap \vartheta^{-1}(c)] = [C_0]$. Let S be the stabilizer $S = \{a \in A : a + C_0 = C_0\}$. Then for $a \in S$, $a + (Z \cap \vartheta^{-1}(c)) = (Z \cap \vartheta^{-1}(c))$ for any $c \in C$, so that also $S = \{a \in A : a + Z = Z\}$, and S is definable.

Now $Z \cap \vartheta^{-1}(c) = C_0 + f(c)$ for some $f(c) \in \vartheta^{-1}(c)$; $f(c) + S$ is well defined.

By assumption (2), $S = \mathrm{Ker}(r_1) \cap \ldots \cap \mathrm{Ker}(r_m)$ for some $r_i \in R$. Let $I = \{r_1, \ldots, r_m\}$. For $r \in I$, $f_r(c) := rf(c)$ is a well-defined element of B, and for all $c \in \vartheta(Z)$, $r(Z \cap \vartheta^{-1}(c)) = rC_0 + f_r(c)$.

We have $\vartheta f_r(c) = rc$. If $d \in \mathrm{Ker}(r : C \to C)$, then $f_r(d + c) = rc$ also, so $f_r(d + c) - f_r(c) \in A$. By orthogonality, for fixed r, $f_r(d + c) - f_r(c)$ takes finitely many values as c, d vary in C. In other words, $\{rf(c) : c \in \vartheta(Z)\}$ is a quasi-section above $r\vartheta(Z)$. By (3), $r\vartheta(Z)$ is finite, for each $r \in I$. Let $N = (r_1, \ldots, r_m)$, $Y' = NZ$. Then $\vartheta(Y')$ is finite. It follows that Y' is contained in a finite union of cosets of A, so C, Y' are orthogonal.

Thus $\{(\vartheta(z), Nz) : z \in Z\}$ is a finite union of rectangles; upon dividing Z further, we may assume this set is a rectangle $W \times Y$. Now if $\vartheta(b) \in W$ and $Nb \in Y$ then for some $z \in Z$, $\vartheta(b) = \vartheta(z)$ and $Nb = Nz$; it follows that $b - z \in A$ and $b - z \in S$; so $b \in S + Z = Z$. Thus Z is of the required form. □

Corollary 3.22. *Let T be a complete theory in a language L satisfying the assumptions of Lemma* 3.21. *Let $L \subseteq L'$, $T \subseteq T'$, and assume (1)–(3) persist to T'. If T, T' induce the same structure on A and on C, up to constants they induce the same structure on B, i.e., every T-definable subset of B^* is parameterically T'-definable.*

Proof. Apply Lemma 3.21 to T', and note that every definable set in the normal form obtained there is already parametrically definable in T. □

We will explicitly use imaginaries in RV only rarely; but our ability to work with RV, using Γ as an auxiliary, is partly explained by the following.

Corollary 3.23. *Let $0 \to A \to B \to_\vartheta C \to 0$ be as in Lemma* 3.21, *and assume C carries a definable linear ordering. Let $\bar{V}$ be the disjoint union of the definable cosets of A in B, with structure induced from T. Let e be an imaginary element of B. Then $\langle e \rangle = \langle (a', c') \rangle$ for some pair (a', c') consisting of an imaginary of $\bar{V}$ and an imaginary of C. Thus if $\bar{V}$, C eliminate imaginaries, so does $B \cup C \cup \bar{V}$.*

Proof. Let e be an imaginary element of B; let E_0 be the set of A, $\bar{V}$-imaginaries that are algebraic over e.

By Lemma 3.21, applied to a definable set with code e in the theory T_{E_0}, there exist almost definable subsets of $\bar{V}$, C^n from which e can be defined. These are coded by imaginaries permitted in the definition of E_0. Thus e is E_0-definable. Thus $e = g(d)$ for some definable function g and some tuple d from E_0. □

Let us now show that e is equidefinable with a finite set, i.e., an imaginary of the form $(f_1, \dots, f_n)/\operatorname{Sym}(n)$. Let W be the set of elements with the same type as d over e; W is finite by the definition of E_0, and is e-definable. But $e = g(w)$ for any element $w \in W$, so e is definable from $\{W\}$.

It remains to see that every finite set of elements of E_0 is coded by imaginaries of A and C and elements of B. Since C is linearly ordered, it suffices to consider finite sets whose image in C^m consists of one point. These are subsets of some definable coset of A^m, so again by elimination of imaginaries there they are coded. □

Corollary 3.24. *The structure induced on* $\mathrm{RV} \cup \Gamma$ *from* ACVF *eliminates imaginaries.*

Proof. Γ_{E_0} eliminates imaginaries, and so does ACF (cf. [31]). Note that $\bar{V}$ is essentially a family of one-dimensional **k**-vector spaces, closed under tensor products and roots and duals. Hence by [18], $\bar{V}_{E_0}$ eliminates imaginaries, too. Our only application of this lemma will be in a situation when parameters can be freely added; in this case, it suffices to quote elimination of imaginaries in ACF. □

Corollary 3.25. *Let T be a theory as in Lemma* 3.21, *with $R = \mathbb{Z}$, and C a linearly ordered group. Then every definable subset of B^n is a disjoint union of $\mathrm{GL}_n(\mathbb{Z})$-images of products $Y \times \vartheta^{-1}(Z)$, with $|\vartheta Y| = 1$. In particular, the Grothendieck semiring $K_+(B)$ (with respect to the category of all definable sets and functions of B) is generated by the classes of elements $Y \subset B^n$ with $|\vartheta Y| = 1$, and pullbacks $\vartheta^{-1}(Z)$, $Z \subset C^m$.*

Proof. By Lemma 3.21, the Grothendieck ring is generated by classes of sets X of the form $X = \{b \in B^n : \vartheta(b) \in W, Nb \in Y\}$. After performing row and column operations on the matrix N, we may assume it is the composition of a projection $p : R^n \to R^k$ with a diagonal $k \times k$ integer matrix with nonzero determinant. The composition $\vartheta p(X)$ is finite; since C is ordered, each element of $\vartheta p(X)$ is definable, and so we may assume $\vartheta p(X)$ has one element e. Thus $W = \{(e) \times W'\}$ for some W', and $X = pX \times \vartheta^{-1}(W')$. □

Lemma 3.26. *Let T be a theory, and let $0 \to A \to B \to_\vartheta C \to 0$ be an exact sequence of definable Abelian groups and homomorphisms. If $E \leq M \models T$, we will write $E_A = A(E)$, etc. Assume the following:*

(1) *A, C are orthogonal.*
(2) *Any parametrically definable subset of B is a Boolean combination of sets Y with $\vartheta(Y)$ finite, and of full pullbacks $\vartheta^{-1}(Z)$.*
(3) *C a uniquely divisible Abelian group, and for any $E \leq M \models T$, every divisible subgroup containing E_C is algebraically closed in C over E.*
(4) *For any prime $p > 0$, $T \models (\exists x \in A)(px = 0, x \neq 0)$.*

Let $Z \subset C^n$ and $f : Z \to C$ be definable, and suppose there exists E and E-definable $X \subset B^n$ and $F : X \to B$ lifting f: $\vartheta X = Z$, $\vartheta F(x) = f(\vartheta x)$. Then there exists a partition of Z into finitely many definable sets Z_ν, such that for each ν, for some $m \in \mathbb{Z}^n$, $f(x) - \sum_{i=1}^n m_i x_i$ is constant on Z_ν.

The main point is the integrality of the coefficients m_i.

Proof. It suffices to show that for any $M \models T$ and any $c = (c_1, \ldots, c_n) \in Z(M)$, there exists $m = (m_1, \ldots, m_n) \in \mathbb{Z}$ such that $f(c) - mc \in E^0$, where $E^0 = \mathrm{dcl}(\emptyset)$ is the smallest substructure of M. For if so, there exists a formula of one variable of sort C, such that $T \models (\exists^{\leq 1} z)\psi(z)$, $M \models \psi(f(c) - mc)$. By compactness there exists a finite set F of such pairs $\nu = (m, \psi)$, such that for any $M \models T$ and $c \in Z(M)$, for some $(m, \psi) \in F$, $M \models \psi(f(c) - mc)$; the required partition is given by $X_{m,\psi} = \{z \in Z : \psi(f(z) - mz)\}$.

Fix M and $c \in Z(M)$. Let $\langle c \rangle$ be the smallest divisible subgroup of $C(M)$ containing E^0_C and $c_1, \ldots, c_n$. By (3), $\langle c \rangle$ is closed under f, so $f(c) \in \langle c \rangle$, i.e., $f(c) = \sum \alpha_i c_i + d$ for some $\alpha_i \in \mathbb{Q}$ and some $d \in E^0_C$. The only problem is to show that we can take $\alpha_i \in \mathbb{Z}$.

We will use induction on n. Let $K = \{\beta \in \mathbb{Q}^n : \beta \cdot c \in E^0_C\}$. K is a $\mathbb{Q}$-subspace of $\mathbb{Q}^n$. If $K \neq (0)$, there exists a primitive integral vector $\beta_1 \in K$. β_1 may be completed to a basis for a $\mathbb{Z}$-lattice in $\mathbb{Q}^n$. Applying a $\mathrm{GL}_n(\mathbb{Z})$ change of variables to B^n, we may assume $\beta_1 = (1, 0, \ldots, 0)$, i.e., $c_1 \in E^0_C$. But then let $f'(z_2, \ldots, z_n) = f(c_1, z_2, \ldots, z_n)$. Then f' lifts to a definable function on B^n (with parameters, of the form $F(b_1, y_2, \ldots, y_n)$) so by induction, $f(c_1, \ldots, c_n) = f'(c_2, \ldots, c_n) = \sum_{i \geq 2} m_i z_i + d'$ for some $m_2, \ldots, m_n \in \mathbb{Z}$ and $d' \in E^0_C$, as required.

Thus we can assume $K = (0)$.

We can find $m, m_i \in \mathbb{Z}$, $e \in \mathrm{dcl}(\emptyset)$ with

$$mf(c) = \sum m_i c_i + e.$$

If $m|m_i$ we are done. We will now derive a contradiction from the contrary assumption that m does not divide each m_i in such an equation, with f a liftable function. We may assume that the greatest common divisor of $m, m_1, \dots, m_n$; so there exists a prime dividing m but not (say) m_1.

Let $g(x) = f(x, c_2, \dots, c_n) - e/m - \sum_{i=2}^{n} m_i c_i/m$; then $mg(c_1) = m_1 c_1$, m does not divide m_1, g is $E = \mathrm{acl}(c_2, \dots, c_n)$-definable and liftable. Since $K = (0)$, by assumption (3), $c_1 \notin \mathrm{acl}(E)$. Let $E' \supset E$ be such that g lifts to an E'-definable function G'. Enlarging the model if necessary, let c_1' realize $\mathrm{tp}(c_1/E)$, with $c_1' \notin E'$ (cf. Example 2.4). Therefore, there exists E'' such that E'', c_1 and E', c_1' have the same type. In particular, g lifts to an E''-definable function G.

Consider any b_1 such that $\vartheta(b_1) = c_1$. Then $m\vartheta G(b_1) - m_1\vartheta(b_1) = 0$. Thus $mG(b_1) - m_1 b_1 \in A$.

Let p be prime, $p|m$ but $p \not| m_1$. Let $s, r \in \mathbb{Z}$ be such that $sp - rm_1 = 1$, and let $h(x) = sx - \frac{rm}{p} g(x)$. Then $ph(c_1) = psc_1 - rmg(c_1) = psc_1 - rm_1 c_1 = c_1$. Also h is liftable over E'': indeed, if G is E''-definable and lifts g, then $H(x) = sx - \frac{rm}{p} G(x)$ lifts h.

Thus $pH(b_1) = b_1 + d$, some $d \in A$. Let $b_2 = H(b_1)$; then $b_1 = pb_2 - d$, or

$$b_2 = H(pb_2 - d).$$

Now let $c_2 = h(c_1) = \vartheta(b_2)$. Then $pc_2 = c_1$, and so $c_2 \notin \mathrm{acl}(E'')$, since by unique divisibility $c_1 \in \mathrm{acl}(h(c_1))$. By (1), $c_2 \notin \mathrm{acl}(E''(d))$. Let $C_2 = \vartheta^{-1}c_2$. By (2), any $E''(d)$-definable set either contains C_2 or is disjoint from C_2. Hence for any $y \in C_2$, $y = H(py - d)$.

By (4) there exists $0 \neq \omega_p \in A$ with $p\omega_p = 0$. Let $b_2' = b_2 + \omega_p$. Then $b_2 \in C_2$, so $b_2' = H(pb_2' - d)$. But $pb_2' = pb_2$, so $b_2 = b_2'$ and $\omega_p = 0$, a contradiction. □

Remark 3.27.

(1) It follows from Lemma 3.26 that a definable bijection between subsets of C^n that lifts to subsets of B^n is piecewise given by an element of $\mathrm{GL}_n(\mathbb{Z}) \ltimes C^n$ (cf. Lemma 3.28).
(2) Assumption (4) on torsion does not hold in characteristic $p > 0$ for the sequence $\mathbf{k}^* \to \mathrm{RV} \to \Gamma$. In this case there is l-torsion for $l \neq p$, but no p-torsion, and the corresponding group is $\mathrm{GL}_n(\mathbb{Z}[1/p]) \ltimes C^n$.

Note as a corollary that there can be no definable sections of $B \to C$ over an infinite definable subset of C.

Lemma 3.28. *Let* $0 \to A \to B \to C \to 0$ *be as in Lemma* 3.26. *Let* $X \subset B^n$ *be definable, and let* $f : X \to B^l$ *be a definable function.* X *may be partitioned into finitely many pieces* X', *such that on each* X',

(1) $f(x) = Mx + b(x)$, *where* M *is a* $l \times n$*-integer matrix and* $\vartheta b(x)$ *is constant;*
(2) *there exists* $g \in \mathrm{GL}_n(\mathbb{Z})$ *such that* $b \circ g$ *factors through a projection* $B^n \to_\pi B^k$, *where* $\vartheta\pi(X')$ *is one point of* C^k.

Proof. We first prove (1)–(2) for complete types.

(1) This reduces to $l = 1$. Let P be a complete type of elements of X. Then on P we have $\vartheta \circ f(x) = \sum m_i \vartheta(x_i) + d$ for some constant d (Lemma 3.26).

Thus $f(x) = \sum m_i x_i + b(x)$, where $b(x) = f(x) - \sum m_i x_i$, and $\vartheta b(x) = d$ is constant.

(2) Let $\pi : B^n \to B^k$ be a projection such that $\vartheta\pi(X)$ is one point of C^k, and with k maximal. Thus $P \subset P' \times P''$, $P' \subset B^{n-k}$, $P'' \subset B^k$, and $\vartheta(P'')$ is a single point of C^k, while $\vartheta(P')$ is not contained in any proper hypersurface $\sum n_i x_i =$ constant with $n_i \in \mathbb{Z}$. Pick $b'' \in P''$. Let $\gamma = (\gamma_1, \dots, \gamma_k) \in \vartheta(P')$, γ not in any such hypersurface. Let $a = (a_1, \dots, a_k)$, $\vartheta(a_i) = \gamma_i$, and let a' be another point with $\vartheta(a') = \gamma$. Let $e = f(a, b)$. Then $\mathrm{tp}(a/b, e) = \mathrm{tp}(a'/b, e)$, so $f(a', b) = e$. Thus $f(a, b)$ depends only on $b \in P''$ and not on a (with $(a, b) \in P$).

Since (1)–(2) hold on each complete type, there exists a definable partition such that they hold on each piece. □

3.4 V-minimality

We assume from now on that **T** is a theory of C-minimal valued fields, of residue characteristic 0. When using the many-sorted language, we will still say that **T** is *a theory of valued fields* when $\mathbf{T} = \mathrm{Th}(F, \mathrm{RV}(F))$ for some valued field F, possibly with additional structure. A C-minimal **T** satisfying assumption (3) below will be said to have *centered closed balls*. If, in addition, (1)–(2) hold, we will say **T** is *V-minimal*. Expansions by the definition of the language, i.e., the addition of a relation symbol $R(x)$ to the language along with a definition $(\forall x)(R(x) \iff \phi(x))$ to the theory, do not change any of our assumptions. Thus we can assume that **T** eliminates quantifiers.

(1) *Induced structure on* RV. **T** contains ACVF(0, 0), and every parametically **T**-definable relation on RV* is parametrically definable in ACVF(0, 0).
(2) *Definable completeness.* Let $A \le M \models T$, and let $W \subset \mathfrak{B}$ be a $\mathbf{T}_A$-definable family of closed balls linearly ordered by inclusion. Then $\cap W \neq \emptyset$.
(3) *Choosing points in closed balls.* Let $M \models \mathbf{T}$, $A \subseteq \mathrm{VF}(M)$, and let b be an almost A-definable closed ball. Then b contains an almost A-definable point.

T will be called *effective* if every definable finite disjoint union of balls contains a definable set, with exactly one point in each. A substructure A of a model of **T** will be called effective if $\mathbf{T}_A$ is effective.

If every definable finite disjoint union of rv-balls contains a definable set, with exactly one point in each, we can call **T** rv-effective. However, we have the following.

Lemma 3.29. *Let* **T** *be* V*-minimal. Then* **T** *is effective iff it is* rv*-effective.*

Proof. Assume **T** is rv-effective. Let b be an algebraic ball. If b is closed, it has an algebraic point by assumption (3) of Section 3.4. If b is open, let $\bar{b}$ be the closed ball surrounding it. Then $\bar{b}$ has an algebraic point a. Let $f(x) = x - a$. Then $f(b)$ is an rv-ball, so by rv-effectivity it has an algebraic point a'. Hence $a' + a$ is an algebraic point of b. □

In general, effectivity is needed for lifting morphisms from RV to VF, not for the "integration" direction.

If $\mathbf{T}$ is V-minimal and A is a $\mathrm{VF} \cup \mathrm{RV} \cup \Gamma$-generated structure, we will see that $\mathbf{T}_A$ is V-minimal, too. The analogue for points in open balls is true but only for $\mathrm{VF} \cup \Gamma$-generated substructures; for thin annuli it is true only for VF-generated structures. For this reason the condition on closed balls is more flexible; luckily we will be able to avoid the others.

Lemma 3.30. *Let* $\mathbf{T}$ *be a* C*-minimal theory of valued fields. Then* (1) $\implies$ (2) $\implies$ (3) $\implies$ (4):

(1) $\mathbf{T}$ *admits quantifier elimination in a three-sorted language* $(\mathrm{VF}, \mathbf{k}, \Gamma)$,*such that for any basic function symbol* F *with range* VF, *the domain is a power of* VF; *and no relations on* $\mathbf{k}$, Γ *beyond the field structure on* $\mathbf{k}$ *and the ordered Abelian group structure on* Γ.
(2) *Every parametrically definable relation on* $\mathbf{k}$ *is parameterically definable in* ACF(0), *and every parametrically definable relation on* Γ *is parameterically definable in* DOAG.
(3) *Assumption* (1) *of Section* 3.4.
(4) $\mathbf{k}$, Γ, *and* RV *are stably embedded.*

Proof.

(1) $\implies$ (2) Let $\phi(a, x)$ be an atomic formula with paramaters $a = (a_1, \dots, a_n)$ from VF and $x = (x_1, \dots, x_m)$ variables for the $\mathbf{k}, \Gamma$ sorts. Then ϕ must have the form $\psi(t(a), x)$, where t is a term (composition of function symbols) $\mathrm{VF}^* \to (\mathbf{k} \cup \Gamma)$. Thus $\phi(a, x)$ defines the same set as $\psi(b, x)$ where $b = t(a)$. Since every formula is a Boolean combination of atomic ones, (2) follows.

(2) $\implies$ (3) This follows from Corollary 3.22. The assumptions of Lemma 3.26 are satisfied: (1) is automatic since by C-minimality $\mathbf{k}$ is strongly minimal and Γ is O-minimal; (2) follows from C-minimality; (3)–(4) follow from the assumptions on $\mathbf{k}, \Gamma$.

(3) $\implies$ (4) This is immediate. □

Lemma 3.31. *Let* $\mathbf{T}$ *be a theory of valued fields satisfying assumption* (1) *of Section* 3.4, *such that* res *induces a surjective map on algebraic points. Then* (1) $\implies$ (2) $\implies$ (3) $\implies$ (4):

(1) *For any* VF*-generated substructure* A *of a model* M *of* $\mathbf{T}$, *if* $\Gamma(A) \neq (0)$, *then* $\mathrm{acl}(A) \models \mathbf{T}$.
(2) *For any* VF*-generated substructure* A *of a model of* $\mathbf{T}$, *any* $\mathbf{T}_A$*-definable nonempty finite union of balls contains a nonempty* $\mathbf{T}_A$*-definable finite set.*
(3) *Assumption* (3) *of Section* 3.4 *holds.*
(4) *Let* A *be* VF*-generated, and* Y *a finite* A*-definable set of disjoint closed balls. Then there exists an* A*-definable finite set* Z *such that* $|b \cap Z| = 1$ *for each* $b \in Y$.

Proof. We first show the following.

Claim. For any VF-generated A with $\Gamma(A) = (0)$, res : VF(acl$(A)) \to \mathbf{k}($acl$(A))$ is surjective.

Proof. It suffices to prove the claim for finitely generated A. For $A = \emptyset$ this is true by assumption. Using induction on the number of generators, it suffices to show that if the claim holds for A_0 and $c \in$ VF then it holds for $A = A_0(c)$).

Since $\Gamma(A) = (0)$, res is defined and injective on VF(A). If $c \in \mathrm{acl}(A_0)$ there is nothing to prove. Otherwise, by injectivity, res$(c) \notin \mathrm{acl}(A_0)$. As a consequence of assumption (1) of Section 3.4, both dcl and acl agree with the corresponding field-theoretic notions on RV and, in particular, on the residue field.

By Lemma 3.20,

$$\mathbf{k}(A_0(c)) \subseteq \mathrm{dcl}(\mathrm{RV}(A_0), \mathrm{rv}(c)) = \mathrm{dcl}(\mathbf{k}(A_0), \mathrm{res}(c)) = \mathbf{k}(A_0)(\mathrm{res}(c)).$$

Now if $d \in \mathbf{k}(\mathrm{acl}(A))$ then $d \in \mathbf{k}$ and $d \in \mathrm{acl}(A)$, so by stable embeddedness of $\mathbf{k}$, we have $d \in \mathrm{acl}(\mathbf{k}(A))$; but $\mathrm{acl}(\mathbf{k}(A)) = \mathbf{k}(A)^{\mathrm{alg}}$ by assumption (1) of Section 3.4; so $d \in \mathbf{k}(A_0)(\mathrm{res}(c))^{\mathrm{alg}} \subseteq \mathrm{res}(A_0(c)^{\mathrm{alg}})$. □

Assume (1). If $\Gamma(\mathrm{acl}(A)) \neq (0)$, then by (1) $\mathrm{acl}(A) \models T$ and, in particular, every acl(A)-definable ball has a point in acl(A), so (2) holds. Assume therefore that $\Gamma(\mathrm{acl}(A)) = 0$. Let b be an acl(A)-definable ball. Then b must have valuative radius 0. If some element of b has valuation $\gamma < 0$ then all do, and $\gamma \in A$, a contradiction. Thus b is the (open or closed) ball of radius 0 around some $c \in \mathcal{O}$. If b is closed, then $b = \mathcal{O}$ and $0 \in b$. If b is open, then $b = \mathrm{res}^{-1}(b')$ for some element b' of the residue field $\mathbf{k}$; in this case b has an acl(A)-definable point by the claim.

(3) is included in (2), being the case of closed balls.

Assume (3). In expansions of ACVF(0, 0), the *average* of a finite subset of a ball remains within the ball. Thus if Y is a finite A-definable set of disjoint balls, by (3), there exists a finite A-definable set Z_0 including a representative of each ball in Y. Let $Z = \{\mathrm{av}(b \cap Z_0) : b \in Y\}$, where av$(u)$ denotes the average of a finite set u. □

Lemma 3.32. *When* **T** *is a complete theory, definable completeness is true as soon as T has a single* spherically complete *model M in the sense of Ribenboim and Kaplansky: every intersection of nested closed balls is nonempty.*

Proof. The proof is clear. □

Let ACVF^{an} denote any of the rigid analytic theories of [23]. For definiteness, let us assume the power series have coefficients in $\mathbb{C}((X))$. See [14] for variants living over $\mathbb{Z}_p$.

Lemma 3.33. ACVF(0, 0) *is* V*-minimal and effective. Thus is* ACVF^{an}.

Proof. C-minimality is proved in [24]. Lemma 3.30(1) for ACVF is a version of Robinson's quantifier elimination; cf. [16].

ACVF^{an} admits quantifier elimination in the sorts (VF, Γ) by [23, Theorem 3.8.2]. The residue field sort is not explicit in this language, but one can argue as follows. Let

$\mathbf{k}_1$ be a large algebraically closed field containing $\mathbb{C}$, and let $K = \cup_{n\geq 1}\mathbf{k}_1((X^{1/n}))$ be the Puiseux series ring. Then K admits a natural expansion to a model of the theory. K is not saturated, but by C-minimality the induced structure on the residue field is strongly minimal, so $\mathbf{k}_1$ is saturated. Now any automorphism of $\mathbf{k}_1$ as a field extends to an automorphism of K as a rigid analytic structure. Thus every K-definable relation on $\mathbf{k}_1$ is algebraic. (This could be repeated over a larger value group if necessary.) Lemma 3.30(2) thus holds in both cases; hence we have assumption (1) of Section 3.4.

Condition Lemma 3.31(1) is obviously true for ACVF. For ACVF^{an} it is proved in [24]. It is also evident that these theories have a spherically complete model. Thus by Lemmas 3.31 and 3.32, assumptions (3) and (2) of Section 3.4 hold, too. □

Remarks.

(1) Lemma 3.31(1)–(3) remain true for ACVF in positive residue characteristic, but (4) fails.

(2) ACVF(0, 0) also admits quantifier elimination in the two sorted language with sorts VF, RV; so assumption (1) of Section 3.4 can also be proved directly, without going through $\mathbf{k}$, Γ as in Lemma 3.30.

(3) Assumption (1) of Section 3.4 is needed for lifting definable bijections of RV to VF, Proposition 6.1, Lemma 6.3. Specifically, it implies the truth of assumptions (2) of Lemma 3.21 and (4) of Lemma 3.26. These lemmas are only needed for the injectivity of the Euler charactersitic and integration maps, not for their construction and main properties. It is also needed for the theory of differentiation and for comparing derivations in VF and RV; indeed, even for posing the question, since in general there is no notion of differentiation on RV. The theory of differentiation itself is needed neither for the Euler characteristic nor for integration of definable sets with a Γ-volume form. They are required only for the finer theory introduced here of integration of RV-volume forms.

(4) We know no examples of C-minimal fields where assumption (2) of Section 3.4 fails.

(5) Beyond effectivity of dcl(∅), assumption (3) of Section 3.4 imposes a condition on liftability of definable functions from VF to $\mathfrak{B}^{cl}$. Let $\mathbf{T}_1$ be the theory, intermediate between ACVF(0, 0) and a Lipshitz rigid analytic expansion, generated over ACVF(0, 0) by the relation

$$\mathrm{val}(f(t_0x) - y) \geq \mathrm{val}(t_1)$$

on $\mathcal{O}^2$, where t_0, t_1 are constants with $\mathrm{val}(t_1) \eth \mathrm{val}(t_0) > 0$ and f is an analytic function. It appears that balls do not necessarily remain pointed upon adding VF-points to $\mathbf{T}_1$; so assumption (3) of Section 3.4 is not redundant.

3.5 Definable completeness and functions on the value group

We assume $\mathbf{T}$ is C-minimal and definably complete. We show that the property of having centered closed balls is preserved under passage to $\mathbf{T}_A$ if A is RV, Γ, VF-generated; similarly for open balls if A is Γ, VF-generated. Also included is a lemma stating that every image of an RV-set in VF must be finite; from the point of view of content this belongs to the description of the "basic geography," but we need the lemmas on functions from Γ first.

Proposition 3.34. *Let $M \models \mathbf{T}$, $\gamma = (\gamma_1, \ldots, \gamma_m)$ a tuple of elements of $\Gamma(M)$. Any almost $A(\gamma)$-definable ball b contains an almost A-definable ball b'.*

Proof. See [16, Proposition 2.4.4]. While the proposition is stated for ACVF there, the proof uses only C-minimality and definable completeness. We review the proof in the case that $b \in A(\gamma)$, i.e., $b = f(\gamma)$ for some definable function f with domain $D \subseteq \Gamma^M$.

Let $P = \text{tp}(\gamma/A)$. Let $r(\gamma)$ be the valuative radius of $f(\gamma)$. By O-minimality, r is piecewise monotone; since P is a complete type, r is monotone, say, decreasing. For $a \in P$ let $P_a = \{b \in P : b < a\}$, and for $b \in P_a$ let $f_a(b)$ be the open ball of size $r(a)$ containing $f(b)$. By Lemma 3.15, the valuative radius map rad is finite-to-one on $f_a(P_a)$; but by definition it is constant, so $f_a(P_a)$ is finite. Using the linear ordering, $f_a(P_a)$ is constant on each complete type over a. Pick $b_1 \in P$, $\epsilon \in \Gamma$ with $\epsilon > 0$ but very small (over $A(b_1)$), and $\epsilon' \in \Gamma$ with $\epsilon' > 0$ but ϵ' very small (over $A(b_1, \epsilon)$). Let $b_2 = b_1 + \epsilon$, $a = b_2 + \epsilon'$. Then $\text{tp}(b_1, a/A) = \text{tp}(b_2, a/A)$, so $f_a(b_1) = f_a(b_2)$. Now if $f(b_1), f(b_2)$ are disjoint, let $\delta = \text{val}(x_1 - x_2)$ for (some or any) $x_i \in f(b_i)$. Then $r(b_2) > \delta$. Since ϵ' is very small, $r(a) > \delta$ also. Thus $f_a(b_1), f_a(b_2)$ are distinct, a contradiction. Thus $f(b_1) \subset f(b_2)$. Since $\text{tp}(a/A) = \text{tp}(b_2/A)$, we have $f(y) \subset f(a)$ for some $y \in P_a$. If $f(y) \subset f(a)$ for all $y \in P_a$, we are done; otherwise, let $c(a)$ be the unique smallest element such that f is monotone on $(c(a), a)$. We saw, however, that f is monotone on $(d, c(a))$ for some $d < c(a)$, hence also on (d, a), a contradiction. Thus f is monotone with respect to inclusion. By compactness, this is true on some A-definable interval, hence on some interval I containing P.

Let $U = \bigcap_{a \in I} f(a)$. By definable completeness (assumption (2) of Section 3.4), $U \neq \emptyset$. Clearly, U is a ball, and $U \subseteq b$. □

Lemma 3.35. *Let $M \models \mathbf{T}$, $\gamma = (\gamma_1, \ldots, \gamma_m)$ a tuple of elements of $\Gamma(M)$. Then any $A(\gamma)$-definable ball contains an A-definable ball. If Y is a finite $A(\gamma)$-definable set of disjoint balls, then there exists a finite A-definable set Y' of balls, such that each ball of Y contains a unique ball of Y'.*

Proof. This reduces immediately to $m = 1$. For $m = 1$, by Proposition 3.34, any almost $A(\gamma)$-definable ball b contains an almost A-definable ball b'. Thus given a finite $A(\gamma)$-definable set Y of disjoint balls, there exists a finite A-definable set Z of balls, such that any ball of Y contains a ball of Z. Given $b \in Y$, let b' be the smallest ball containing every subball c of b with $c \in Z$. Then $Y' = \{b' : b \in Y\}$ is $A(\gamma)$-definable, finite, almost A-definable, and (since b_1' is disjoint from b_2 if $b_1 \neq b_2 \in Y$) each ball of Y contains a unique ball of Y'. Using elimination of imaginaries in Γ, by Example 2.2, being $A(\gamma)$-definable and almost A-definable, Y' is A-definable. □

The following corollary of Lemma 3.35 concerning definable functions from Γ will be important for the theory of integration with an additive character in Section 11.

Corollary 3.36. *Let Y be a definable set admitting a finite-to-one map into Γ^n, and let into h be a definable map on Y into* VF *or* $\mathrm{VF}/\mathcal{O}$ *or* $\mathrm{VF}/\mathcal{M}$. *Then h has finite image.*

Proof. One can view h as a function from a subset of Γ^n into finite sets of balls. Since a ball whose radius is definable containing a definable ball is itself definable, Lemma 3.35 implies that $h(\gamma) \in \mathrm{acl}(\emptyset)$ for any $\gamma \in \Gamma^n$. By Lemma 2.6, the corollary follows. □

Corollary 3.37. *Let $Y \subseteq (\mathrm{RV} \cup \Gamma)^n$ and $Z \subseteq \mathrm{VF} \times Y$ be definable sets, with Z invariant for the action of $\mathcal{M}$ on* VF. *Then for all but finitely many $\mathcal{O}$-cosets C, $Z \cap (C \times Y)$ is a rectangle $C \times Y'$.*

Proof. Let $p : (\mathrm{RV} \cup \Gamma)^n \to \Gamma^n$ be the natural projection, and for $\gamma \in \Gamma^n$ let Z_γ be the fiber. For each γ, by Lemma 3.16, there exists a finite $F(\gamma) \subseteq \mathrm{VF}/\mathcal{O}$ such that for any $\mathcal{O}$-coset $C \notin F(\gamma)$, $Z_\gamma \cap (C \times Y)$ is $\mathcal{O}$-invariant. Now $\{(u, \gamma) : u \in F(\gamma)\}$ projects finite-to-one to Γ^n, so by Lemma 3.36, this set projects to a finite subset of $\mathrm{VF}/\mathcal{O}$. Thus there exists a finite $E \subset \mathrm{VF}/\mathcal{O}$ such that for any γ, and any $\mathcal{O}$-coset $C \notin E$, $Z_\gamma \cap (C \times Y)$ is $\mathcal{O}$-invariant. In other words, for any $C \notin E$, $Z \cap (C \times Y)$ is $\mathcal{O}$-invariant. □

Lemma 3.38. *Let $M \models$* **T**, *A a substructure of M (all imaginary elements allowed), and let $r = (r_1, \ldots, r_m)$ be a tuple of elements of* $\mathrm{RV}(M) \cup \Gamma(M)$. *Then any closed ball almost defined over $A(r)$ contains a ball almost defined over A.*

Proof. This reduces to $m = 1$, $r = r_1$; moreover, using Lemma 3.35, to the case $r \in \mathrm{RV}(M)$, $\mathrm{val}_{\mathrm{rv}}(r) = \gamma \in A$. Let $E = \{y \in \mathrm{RV} : \mathrm{val}_{\mathrm{rv}}(y) = \gamma\}$. Then E is a k^*-torsor, and so is strongly minimal within M. If c is almost defined over $A(r)$, there exists an A-definable set $W \subset E \times \mathfrak{B}_{\mathrm{cl}}$, with $W(e) = \{y : (e, y) \in W\}$ finite, and $c \in W(r)$. But then W is a finite union of strongly minimals, and hence so is the projection P of W to $\mathfrak{B}_{\mathrm{cl}}$. But any strongly minimal subset of $\mathfrak{B}_{\mathrm{cl}}$ is finite. (Otherwise, it admits a definable map onto a segment in Γ; but Γ is linearly ordered and cannot have a strongly minimal segment.) Thus $c \in P$ is almost defined over A. □

Lemma 3.39. *Let $M \models$* **T**, **T** *C-minimal with centered closed balls. Let B be substructure of* $\mathrm{VF}(M) \cup \mathrm{RV}(M) \cup \Gamma(M)$. *Then every B-definable closed ball has a B-definable point. If Y is a finite B-definable set of disjoint closed balls, there exists a finite B-definable set $Z \subset M$, meeting each ball of Y in a unique point.*

Proof. We may take B to contain a subfield K and be generated over K by finitely many points $r_1, \ldots, r_k \in RV$. Let Y be a finite B-definable set of disjoint closed balls, and let $b \in Y$. We may assume all elements of Y have the same type over B. By Lemma 3.38, there exists a closed ball b' defined almost over K and contained in b. By assumption (3) of Section 3.4, there exists a finite K-definable set Z' meeting b' in a unique point. Let $Y' = \{b'' \in Y : b'' \cap Z' \neq \emptyset\}$, and $Z = \{\mathrm{av}(Z' \cap b'') : b'' \in Y'\}$. Then Z meets each ball of Y' in a unique point, and Z, Y' are B-definable. As for $Y \setminus Y'$, it may be treated inductively. □

Corollary 3.40. *Let $M \models \mathbf{T}$, $\mathbf{T}$ C-minimal with centered closed balls, and effective. Let B be an almost Γ-generated substructure. Then $\mathbf{T}$ is effective.*

Proof. The proof is the same as the proof of Lemma 3.39, using Lemma 3.34 in place of Lemma 3.38. □

Lemma 3.41. *Let Y be a $\mathbf{T}$-definable set admitting a finite-to-one map into* RV^n. *Let $g : Y \to \mathrm{VF}^m$ be another definable map. Then $g(Y)$ is finite.*

Proof. It suffices to prove this for $\mathbf{T}_A$, where $A \models \mathbf{T}$. We may also assume $m = 1$. We will use the equivalence (3) $\iff$ (4) of Lemma 2.6. If $g(Y)$ is infinite, then by compactness there exists $a \in g(Y)$ $a \notin \mathrm{acl}(A)$. But for some b we have $a = g(b)$, so if $c = f(b)$, we have $c \in \mathrm{RV}^n$, $a \in \mathrm{acl}(c)$. Thus it suffices to show the following:

$$\text{If } a \in \mathrm{VF},\, c \in \mathrm{RV}^n \text{ and } a \in \mathrm{acl}(A(c)), \text{ then } a \in \mathrm{acl}(A). \tag{$*$}$$

This clearly reduces to the case $n = 1$, $c \in \mathrm{RV}$. Let $d = \mathrm{val}_{\mathrm{rv}}(c)$, $A' = \mathrm{acl}(A(d))$. Then c lies in an A'-definable strongly minimal set S (namely, $S = {\mathrm{val}_{\mathrm{rv}}}^{-1}(d)$). Using Lemma 2.6 in the opposite direction, since $a \in \mathrm{acl}(A'(c))$ there exists a finite-to-one map $f : S' \to S$ and a definable map $g' : S' \to \mathrm{VF}$ with $a \in g'(f^{-1}(S'))$. By Corollary 3.13, $g'(f^{-1}(S'))$ is finite. Hence $a \in \mathrm{acl}(A(d))$. But then by Lemma 3.36, $a \in \mathrm{acl}(A)$. □

In particular, there can be no definable isomorphism between an infinite subset of RV^n and one of VF^m.

Lemma 3.42. *Let $M \models \mathbf{T}$, $\mathbf{T}$ C-minimal with centered closed balls, and let A be a substructure of M. Write A_{VF} for the field elements of A, A_{RV} for the* RV*-elements of A.*

Let $c \in \mathrm{RV}(M)$, and let $A(c) = dcl(A \cup \{c\})$. Then $A(c)_{\mathrm{VF}} \subset (A_{\mathrm{VF}})^{\mathrm{alg}}$, and $\mathrm{rv}(A(c)_{\mathrm{VF}}) \cap A_{\mathrm{RV}} = \mathrm{rv}(A_{\mathrm{VF}})$.

Proof. Let $e \in A(c)_{\mathrm{VF}}$. Then $e = f(c)$ for some A-definable function $f : W \to \mathrm{VF}$, $W \subseteq \mathrm{RV}$. By Lemma 3.41, the image of f is finite, $e \in \mathrm{acl}(A)$. This proves the first point. Now if $d \in \mathrm{RV}_A$ and $\mathrm{rv}^{-1}(d)$ has a point in $A(c)$, then it has a point in $(A_{\mathrm{VF}})^{\mathrm{alg}}$, by assumption (3) of Section 3.4. □

3.6 Transitive sets in dimension one

Let b be a closed ball in a valued field. Then the set Aff(b) of maximal open subballs of b has the structure of an affine space over the residue field. We will now begin using this structure. Without it, more general transitive annuli (missing more than one ball) could exist.

Lemma 3.43. *Let $X \subseteq \mathrm{VF}$ be a transitive $\mathbf{T}_B$-definable set, where B is some set of imaginaries. Then X is a finite union of open balls of equal size, or a finite union of closed balls of equal size, or a finite union of thin annuli.*

Proof. By C-minimality, X is a finite Boolean combination of balls. There are finitely many distinct balls $b_1, \ldots, b_n$ that are almost contained in X (i.e., $b_i \setminus X$ is contained in a finite union of proper subballs of b_i) but such that no ball larger than b_i is almost contained in X. These b_i must be disjoint. If some of the b_i have different type than the others, their union (intersected with X) will be a proper B-definable subset of X. Thus they all have the same type over B; in particular, they have the same radius β.

Consider first the case where the balls b_i are open. Then $b_i \subseteq X$. Otherwise, $b_i \setminus X$ is contained in a unique smallest ball c_i. Say c_i has radius α; then $\alpha > \beta$. Let b'_i be the open ball of radius $(1/2)(\alpha + \beta)$ around c_i; then $\cup_i b'_i$ is a B-definable proper subset of X, a contradiction. Thus in the case of open balls, $X \supseteq \cup_i b_i$ and therefore $X = \cup_i b_i$.

If the balls b_i are closed, let c_{ij} be a minimal finite set of subballs of b_i needed to cover $b_i \setminus X$. The same argument shows that no c_{ij} has radius $< \beta$. Thus all c_{ij} are elements of the set V_i of open subballs of b_i of radius β. Now V_i is a $\mathbf{k}$-affine space, and if there is more than one c_{ij} then over $\operatorname{acl}(B)$, V_i admits a bijection with $\mathbf{k}$; so there is a finite B-definable set of bijections $V_i \to \mathbf{k}$; since any finite definable subset of $\mathbf{k}$ is contained in a strictly bigger one, the union of the pullbacks gives a B-definable subset of V_i properly containing the c_{ij}, leading to a proper B-definable subset of X. Thus either $b_i \subseteq X$ (and then $X = \cup_i b_i$), or else $b_i \setminus c_i \subseteq X$ for a unique maximal open subball c_i. Now $\cup c_i$ intersects X in a proper subset, which must be empty. Thus in this case $X = \cup_i (b_i \setminus c_i)$. □

Let X be a transitive B-definable set. Call $Y \subseteq X$ *potentially transitive* if there exists $B' \supset B$ such that Y is B'-definable and B'-transitive. Let $\mathcal{F}(X)$ be the collection of all proper potentially transitive subsets Y of X. Let $\mathcal{F}_{\max}(X)$ be the set of maximal elements of $\mathcal{F}(X)$.

Lemma 3.44.

(1) *If X is an open ball,* $\mathcal{F}_{\max}(X) = \emptyset$.
(2) *If X is a closed ball,* $\mathcal{F}_{\max}(X) = \{X \setminus Y : Y \in \operatorname{Aff}(X)\}$.
(3) *If X is a thin annulus $X' \setminus Y$ with X' closed, then* $\mathcal{F}_{\max}(X) = \operatorname{Aff}(X) \setminus \{Y\}$.

Proof. Any element of $\mathcal{F}(X)$ must be a ball or a thin annulus, so the lemma follows by inspection. □

Lemma 3.45. *Let b be a transitive closed ball (respectively, thin annulus). Let $Y = \operatorname{Aff}(b)$ be the set of maximal open subballs of b. Then the group of automorphisms of Y over $\mathbf{k}$ is definable, acts transitively on Y, and, in fact, contains $G_a(\mathbf{k})$ (respectively, $G_m(\mathbf{k})$).*

If b, b' are transitive definable closed balls, and $F : b \to b'$ a definable bijection, let $F_* : Y(b) \to Y(b')$ be the induced map. Then F_* is a homomorphism of affine spaces, i.e., there exists a vector space isomorphism $F_{**} : V(b) \to V(b')$ between the corresponding vector spaces, and $F_*(a + v) = F_*(a) + F_{**}(v)$. If $b = b'$ then $F_{**} = \operatorname{Id}$.

Proof. $Y = \mathrm{Aff}(b)$ is transitive, and there is a $\mathbf{k}$-affine space structure on Y (respectively, a $\mathbf{k}$-vector space structure on $V = Y' \dot{\cup} \{0\}$). Let $G = \mathrm{Aut}(Y/\mathbf{k})$ be the subgroup of the group $\mathrm{Aff} = (G_m \ltimes G_a)(\mathbf{k})$ of affine transformations of Y that preserve all definable relations. By definition, this is an intersection of definable subgroups of Aff. However, there is no infinite descending chain of definable subgroups of Aff, so G is definable.

If G is finite, then $Y \subseteq \mathrm{acl}(\mathbf{k})$, and it follows (cf. Section 2.1) that there are infinitely many algebraic points of Y, contradicting transitivity. Thus G is an infinite subgroup of $(G_m \ltimes G_a)(\mathbf{k})$ such that the set of fixed points Y^G is empty. Thus G must contain a translation, and by strong minimality it must contain $G_a(\mathbf{k})$. Similarly, in the case of the annulus, G is an infinite definable subgroup of $G_m(\mathbf{k})$, so it must equal $G_m(\mathbf{k})$.

As for the second statement, F induces a group isomorphism $\mathrm{Aut}(Y(b)/k) \to \mathrm{Aut}(Y(b')/k)$, and hence an isomorphism $G_a(k) \to G_a(k)$, which must be multiplication by some $\gamma \in \mathbf{k}^*$. Since $G_a(k)$ acts by automorphisms on $(Y(b), Y(b'))$, any definable function $Y(b) \to Y(b')$ commutes with this action and hence has the specified form. If $b = b'$ then $Y(b) = Y(b')$; now if $F_{**} \neq \mathrm{Id}$ then F_* would have a fixed point, contradicting transitivity. □

Lemma 3.46. *Let b be a transitive $\mathbf{T}_B$-definable closed (open) ball. Let F be a B-definable function, injective on b. Then $F(b)$ is a closed (open) ball.*

Proof. By Lemma 3.43, since $F(b)$ is also transitive, it is either a closed ball, or an open ball, or a thin annulus. We must rule out the possibility of a bijection between such sets of distinct types.

Consider the collection $\mathcal{F}_{\max}(b)$ defined above. Any definable bijection between b and b' clearly induces a bijection $\mathcal{F}_{\max}(b) \to \mathcal{F}_{\max}(b')$. By Lemma 3.44, the bijective image of an open ball is an open ball.

Let b be a closed ball, $b' = b'' \setminus b'''$ a closed ball minus an open ball, $A = \mathcal{F}_{\max}(b) \simeq \mathrm{Aff}(b)$, $A' = \mathcal{F}(b') \simeq \mathrm{Aff}(b'') \setminus \{b'''\}$, $G = \mathrm{Aut}(A/\mathbf{k})$, $G' = \mathrm{Aut}(A'/\mathbf{k})$. Then a definable bijection $A \to A'$ would give a definable group isomorphism $G \to G'$. But by Lemma 3.45, $G' = G_m(\mathbf{k})$ while G contains $G_a(\mathbf{k})$, so no such isomorphism is possible (say, because $G_m(\mathbf{k})$ has torsion points).

Thus the three types are distinct. □

We will see later that there can be no definable bijection between an open and a closed ball, whether transitive or not.

Lemma 3.47. *Let b be a transitive ball. Then every definable function on b into* $\mathrm{RV} \cup \Gamma$ *is constant. If b is a transitive thin annulus, every definable function on b into* $\mathbf{k} \cup \Gamma$ *is constant. More generally, this is true for definable functions into definable cosets C of* $\mathbf{k}^*$ *in* RV *that contain algebraic points.*

Proof. When a ball b is transitive, it is actually finitely primitive. For if E is a B-definable equivalence relation with finitely many classes, then exactly one of these classes is generic (i.e., is not contained in a finite union of proper subballs of b). This class is B-definable, hence must equal b.

Thus a definable function on b with finite image is constant.

Let F be a definable function on b into Γ. If F is not constant, then for some $\gamma \in \Gamma$, $F^{-1}(\gamma)$ is a proper subset of b; it follows that some finite union of proper subballs of b is γ-definable. By Lemma 3.35, it follows that some such finite union is already definable, a contradiction.

Thus it suffices to show that functions into a single coset $C = \mathrm{val_{rv}}^{-1}(\gamma)$ of $\mathbf{k}^*$ are constant on b.

Assume first that b is open, or a properly infinite intersection of balls. By Lemma 3.19 definable functions on b into C are generically constant; but then by transitivity they are constant.

Now suppose b is closed, or a thin annulus. Let Y be the set of maximal open subballs b' of b. Each $b' \in Y$ is transitive over $\mathbf{T}_{b'}$, so $F|b'$ is constant. Thus F factors through Y.

In the case of the annulus, by Lemma 3.45, $G_m(\mathbf{k})$ acts transitively on Y by automorphisms over $\mathbf{k}$. This suffices to rule out nonconstant functions into $\mathbf{k}$. More generally, if a coset C of $\mathbf{k}^*$ has algebraic points, then $\mathrm{Aut}(C/\mathbf{k})$ is finite. Since $\mathrm{Aut}(Y/\mathbf{k})$ is transitive, it follows that if $f : Y \to C$ is definable then $f(Y)$ is finite. But Y is finitely primitive, so $f(Y)$ is a point.

Assume finally that b is a closed ball. Using Lemma 3.45, we can view $G_a(\mathbf{k})$ as a subgroup of $\mathrm{Aut}(Y/\mathbf{k})$. $\mathrm{Aut}(C/\mathbf{k})$ is contained in $G_m(\mathbf{k})$. Let $S = \mathrm{Aut}(Y \times C/\mathbf{k}) \cap (G_a(\mathbf{k}) \times G_m(\mathbf{k}))$. Then S projects onto $G_a(\mathbf{k})$. By strong minimality, $S \cap (G_a(\mathbf{k}) \times (0))$ is either $G_a(\mathbf{k})$ or a finite group. In the first case, $S = G_a \times T$ for some $T \leq G_m$. In the latter, S is the graph of an finite-to-one homomorphism $G_a \to T$; but this is impossible. Thus $G_a \times (0) \leq S$ and G_a acts transitively on Y by automorphisms fixing C; it follows that F is constant. □

3.7 Resolution and finite generation

Lemma 3.48. *Let $A \leq B$ be substructures of a model of $\mathbf{T}$. Assume B is finitely generated over A. Then $\mathrm{RV}(B)$ is finitely generated over $\mathrm{RV}(A)$. Also, if $\mathrm{RV}(A) \leq C \leq \mathrm{RV}(B)$ then C is finitely generated over $\mathrm{RV}(A)$.*

Proof. Suppose $\Gamma(B)$ has infinitely many $\mathbb{Q}$-linearly independent elements, modulo $\Gamma(A)$. By Lemma 3.1, they are algebraically independent. By Lemma 3.20, they lift to algebraically independent elements of B over A, contradicting the assumption of finite generation. Thus $rk_\Gamma \Gamma(B)/\Gamma(A) < \infty$. It is thus clear that any substructure of $\Gamma(B)$ containing $\Gamma(A)$ is finitely generated over $\Gamma(A)$. Thus it suffices to show that $\mathrm{RV}(B)$ is finitely generated over $A \cup \Gamma(B)$; replacing A by $A \cup \Gamma(B)$, we may assume $\Gamma(B) = \Gamma(A)$. In this case $\mathrm{RV}(B) \subset \mathrm{RES}$. See [17, Proposition 7.3] for a proof stated for ACVF_A, but valid in the present generality. Here is a sketch. One looks at $B = A(c)$ with $c \in \mathrm{VF}$. If $c \in \mathrm{acl}(A)$ then the Galois group $\mathrm{Aut}(\mathrm{acl}(A)/A(c))$ has finite index in $\mathrm{Aut}(\mathrm{acl}(A)/A)$. Hence the same is true of their images in $\mathrm{Aut}(\mathrm{acl}(A) \cap \mathrm{RV})$, and since RV is stably embedded (by clause (1) of the definition of V-minimality) it follows that there exists a finite subset C' of $A(c) \cap \mathrm{RV}$ such that any automorphism of $\mathrm{acl}(A)$ fixing $A(C')$ fixes $A(c) \cap \mathrm{RV}$. By Galois theory for saturated structures (Section 2.1) C' generates $A(c) \cap \mathrm{RV}$ over A.

On the other hand, if $c \notin \mathrm{acl}(A)$, then $\mathrm{tp}(c/\mathrm{acl}(A))$ agrees with the generic type over A of either a closed ball, an open ball, or an infinite intersection of balls. In the latter two cases, $\mathrm{RES}(A) = \mathrm{RES}(B)$ using Lemma 3.19. In the case of a closed ball b, let b' be the unique maximal open subball of b containing c. Then $b' \in A(c)$, and $\mathrm{tp}(c/A(b'))$ is generic in the open ball b'. Thus by Lemma 3.17, $\mathrm{RES}(B) = \mathrm{RES}(A(b'))$ so it is 1-generated. □

Recall $\mathfrak{B} = \mathfrak{B}^o \cup \mathfrak{B}^{\mathrm{cl}}$ is the sort of closed and open balls.

We require a variant of a result from [17] on canonical resolutions. We state it for $\mathfrak{B}$-generated structures, but it can be generalized to arbitrary ACVF-imaginaries [16].

The proposition and corollaries will have the effect of allowing free use of the technology constructed in this paper over arbitrary base (cf. Proposition 8.3).

For this proposition, we allow $\mathfrak{B}$ (and Γ) as sorts, in addition to VF and RV, so that a structure is a subset of $\mathfrak{B}$, Γ of a model of $\mathbf{T}$, closed under definable functions.

Assume for simplicity that $\mathbf{T}$ has quantifier elimination (cf. Section 3.4).

Let us call a structure A *resolved* if any ball and any thin annulus defined over $\mathrm{acl}(A)$ has a point over $\mathrm{acl}(A)$.

Lemma 3.49. *Let* $\mathbf{T}$ *be* V*-minimal. Let* $M \models \mathbf{T}$*, and let* A *be a substructure of* M*. Then* (1) *and* (2) *are equivalent; if* $\Gamma(A) \neq (0)$*, then* (3) *is equivalent to both.*

(1) *A is effective and* $\mathrm{VF}(\mathrm{acl}(A)) \to \Gamma(A)$ *is surjective.*
(2) *A is resolved.*
(3) $\mathrm{acl}(A)$ *is an elementary submodel of* M*.*

Proof. Clearly, (3) implies (1) and (2) implies (1). To prove that (1) implies (3) it suffices to show that every definable $\phi(x)$ of $\mathbf{T}_A$ in one variable, with a solution in M, has a solution in A. If x is an RV-variable it suffices to show that $\phi(\mathrm{rv}(y))$ has a solution; so we may assume x is a VF-variable, so ϕ defines $D \subseteq \mathrm{VF}$. By C-minimality D is a finite Boolean combination of balls. D can be written as a finite union of definable sets of the form $\cup_{j=1}^m D_j \setminus E_j$, where for each j, D_j is a closed ball, and E_j a finite union of maximal open subballs of D_j, or D_j is an open ball and E_j is a proper subball of D_j, or $E_j = \emptyset$, or $D_j = K$. In the third case, by effectivity there exists a finite set meeting each D_j in a point; since $A = \mathrm{acl}(A)$, this finite set is contained in A; so $D(A) \neq \emptyset$, as required. In the first and second cases, there exists similarly a finite set Y meeting each E_j. Since $A = \mathrm{acl}(A)$, $Y \subseteq A$. By picking a point and translating by it, we may assume $0 \in E_j$ for some j. Say E_j has valuative radius α; picking a point $d \in A$ with $\mathrm{val}(d) = \alpha$ and dividing, we may assume $\alpha = 0$. Now in the open case any element of valuation 0 will be in D_j. In the closed case, the image of E_j under res is a finite subset of the residue field; pick some element $\bar{a}$ of $\mathbf{k}(A)$ outside this finite set; by effectivity, pick $a \in A$ with $\mathrm{res}(a) = \bar{a}$; then $a \in D$. In the fourth case, we use the assumption that $\Gamma(A) \neq (0)$. This proves (3).

It remains to show that (1) implies (2). Let b be a thin annulus defined over $\mathrm{acl}(\emptyset)$; so $b = b' \setminus b''$ for a unique closed ball b' and maximal open subball b''. By effectivity, b'' has an algebraic point, so translation we may assume $0 \in b''$. In this case, the assumption that $\mathrm{VF}(\mathrm{acl}(A)) \to \Gamma(A)$ is surjective gives a point of $b' \setminus b''$.□

If $\mathbf{T}_0$ is V-minimal, A is a finitely generated structure (allowing $\mathfrak{B}$, or even ACVF-imaginareis), and $T = (\mathbf{T}_0)_A$, we will call T a finitely generated extension of a V-minimal theory.

Remark 3.50. If A is effective, then A is $\mathrm{VF} \cup \Gamma$-generated. If A is resolved, then A is VF-generated.

Proposition 3.51. *Let* $\mathbf{T}$ *be* V*-minimal.*

(1) *There exists an effective structure* E_{eff} *admitting an embedding into any effective structure* E. *We have* $\mathrm{RV}(E_{\mathrm{eff}}), \Gamma(E_{\mathrm{eff}}) \subseteq \mathrm{dcl}(\emptyset)$.
(2) *There exists a resolved* E_{rslv} *embedding into any resolved structure* E. *We have* $\mathbf{k}(E_{rslv}), \Gamma(E_{rslv}) \subseteq \mathrm{dcl}(\emptyset)$. *In fact,* $C(E_{rslv}) \subseteq \mathrm{dcl}(\emptyset)$ *for any cosets* C *of* $\mathbf{k}^*$ *in* RV *that contain algebraic points.*
(3) *Let* A *be a finitely generated substructure of a model of* $\mathbf{T}$, *in the sorts* $\mathrm{VF} \cup \mathfrak{B}$. *Then* (1)–(2) *hold for* $\mathbf{T}_A$.

Proof.

(1) Let $(b_i)_{i<\lambda}$ enumerate the definable balls. Define a tower of VF-generated structures A_i, and a sequence of balls b_i, as follows. Let $A_0 = \mathrm{dcl}(\emptyset)$; if κ is a limit ordinal, let $A_\kappa = \cup_{i<\kappa} A_i$. Assume A_i has been defined. If possible, let b_i be an A_i-definable, A_i-transitive ball, not a point; and let c_i be any point of b_i. If no such ball b_i exists, the construction ends, and we let $E_{\mathrm{eff}} = A_i$ for this i.

Suppose E is any effective substructure of a model of $\mathbf{T}$. We can inductively define a tower of embeddings $f_i : A_i \to E$. At limit stages κ let $f_\kappa = \cup_{i<\kappa} f_i$. Given f_i with $A_i \neq E$, let b_i' be the image under f of b_i. By effectivity, b_i' has a point $c_i' \in E$. Since b_i is transitive over A_i, the formula $x \in b_i$ generates a complete type; so $\mathrm{tp}(c_i/A_i)$ is carried by f to $\mathrm{tp}(c_i'/A_i')$. Thus there exists an embedding $f_{i+1} : A_{i+1} \to E$ extending f_i, and with $c_i \mapsto c_i'$.

Each A_i is VF-generated; by Lemma 3.31(3) $\Longrightarrow$ (4), the process can only stop when $A_i = E_{\mathrm{eff}}$. This shows that E_{eff} embeds into E, and at the same time that the construction of E_{eff} itself must halt at some stage (of cardinality $\leq |\mathbf{T}|$).

By construction, E_{eff} is VF-generated; and hence $\mathbf{T}_{E_{\mathrm{eff}}}$ is V-minimal. Moreover, there are no E_{eff}-definable E_{eff}-transitive balls (except points). In other words all E_{eff}-definable balls are centered. By V-minimality (assumption (3) of Section 3.4) every closed ball has a definable point, so every centered ball has one. Thus E_{eff} is effective.

It remains only to show that $\mathrm{RV}(E_{\mathrm{eff}}), \Gamma(E_{\mathrm{eff}}) \subseteq \mathrm{dcl}(\emptyset)$. We show inductively that $\mathrm{RV}(A_i), \Gamma(A_i) \subseteq \mathrm{dcl}(\emptyset)$. At limit stages this is trivial, and at successor stages it follows from Lemma 3.47.

(2) The proof is identical to that of (1), but using thin annuli as well as balls. If a thin annulus is not transitive, it contains a proper nonempty finite union of balls, so by V-minimality it contains a proper nonempty finite set. Hence the construction of the A_i stops only when A_i is resolved.

(3) Let $A_0 = (A \cap (\mathrm{VF} \cup \Gamma))$. A is generated over A_0 by some $b_1, \dots, b_n \in \mathfrak{B}$ with b_i of valuative radius $\gamma_i \in A_0$. Since $\mathbf{T}_{A_0}$ is V-minimal, we may assume $\mathbf{T} = \mathbf{T}_{A_0}$ and A is generated by $b_1, \dots, b_n$, with γ_i definable.

Let J be a subset of $\{1, \dots, n\}$ of smallest size such that $\mathrm{acl}(\{b_j : j \in J\}) = \mathrm{acl}(\{b_1, \dots, b_n\})$. By minimality, no b_j is algebraic over $\{b_{j'} : j' \in J, j' \neq j\}$. Let $j \in J$, and let Y_j be the set of balls of radius γ_j; then Y_j is a definable family of disjoint balls. By Lemma 3.8 for $\mathbf{T}' = \mathbf{T}_{\langle\{b_{j'}:j' \in J, j' \neq j\}\rangle}$, b_j is transitive in $\mathbf{T}'_{b_j}$, i.e., in $\mathbf{T}_{\langle b_{j'}:j' \in J\rangle}$; hence b_j is transitive over $\mathrm{acl}(b_1, \dots, b_n) = \mathrm{acl}(A)$. Let us now show, using induction on $|J|$, that $\Pi_{j \in J} b_j$ is transitive over A. Let $c_j \in b_j$. By Lemma 2.10 the $(\{b_{j'} : j' \in J, j' \neq j\})$ remain algebraically independent over $\langle c_j \rangle$. Thus by induction, $\Pi_{j \neq j'} b_{j'}$ is transitive over $A(c_j)$; since b_j is transitive over A, $\Pi_{j \in J} b_j$ is, too. Let $A' = A(c_j : j \in J)$.

Claim. If B is a $\mathrm{VF} \cup \Gamma$-generated structure containing A, then A' embeds into B over A.

Proof. Since B is $\mathrm{VF} \cup \Gamma$-generated, every ball of $\mathbf{T}_B$ is centered; in particular, b_j has a point c'_j defined over $\mathbf{T}_B$. Let $c' = (c'_j : j \in J)$. By transitivity of $\Pi_{j \in J} b_j$, we have $\mathrm{tp}(c/A) = \mathrm{tp}(c'/A)$. Thus A' embeds into B. □

Note that A' is almost $\mathrm{VF} \cup \Gamma$-generated; indeed, since γ_i is definable, $b_i \in \mathrm{dcl}(c_i)$ so $A' \subseteq \mathrm{acl}((c_j)_{j \in J})$. Thus $\mathbf{T}_{A'}$ is V-minimal. Thus (1)–(2) applies and prove (3).□

See Lemma 3.60 for a uniqueness statement.

Corollary 3.52. *Let* $f : \mathrm{VF} \to (\mathrm{RV} \cup \Gamma)^*$ *be a definable map.*

(1) *There exists a definable* $\tilde{f} : \mathrm{RV} \to (\mathrm{RV} \cup \Gamma)^*$ *such that for any* $\mathbf{x} \in \mathrm{RV}$, *for some* $x \in \mathrm{VF}$ *with* $\mathrm{rv}(x) = \mathbf{x}$, $\tilde{f}(\mathbf{x}) = f(x)$.

(2) *Let* $\Omega = \mathrm{VF}/\mathcal{M}$. *There exists a definable map* $\tilde{f} : \Omega \to (\mathrm{RV} \cup \Gamma)^*$ *such that for any* $\mathbf{x} \in \Omega$, *for some* $x \in \mathrm{VF}$ *with* $x + \mathcal{M} = \mathbf{x}$, $\tilde{f}(\mathbf{x}) = f(x)$.

Proof.

(1) In view of Lemma 2.3, it suffices to show that for a given complete type $P \subseteq \mathrm{RV}$, there exists such a function $\tilde{f}$ on P. We fix $\mathbf{a} \in P$, and show the existence of $\mathbf{c} \in \mathrm{dcl}(\mathbf{a})$ such that for some a with $\mathrm{rv}(a) = \mathbf{a}$, $f(a) = \mathbf{c}$.

By Proposition 3.51, there exists an effective substructure A with $\mathbf{a} \in A$ and $(\mathrm{RV} \cup \Gamma)(A) = (\mathrm{RV} \cup \Gamma)(\langle \mathbf{a} \rangle)$. Thus the open ball $\mathrm{rv}^{-1}(\mathbf{a})$ has an A-definable point a. Set $\mathbf{c} = f(a)$; since $f(a) \in \mathrm{RV}(A) = \mathrm{RV}(\langle \mathbf{a} \rangle)$ we have $\mathbf{c} = \tilde{f}_P(\mathbf{a})$ for some definable function $\tilde{f}_P$. Clearly, $\tilde{f}_P$ satisfies the lemma for the input $\mathbf{a}$, hence for any input from P.

(2) The proof is identical, using Lemma 3.51(3). □

Corollary 3.53. *Let* $\mathbf{T}$ *be* V*-minimal. Assume every definable point of* Γ *lifts to an algebraic point of* RV. *Then there exists a resolved structure* E_{rslv} *such that* E_{rslv} *can be embedded into any resolved structure* E, *and* $\mathrm{RV}(E_{rslv}), \Gamma(E_{rslv}) \subseteq \mathrm{dcl}(\emptyset)$. *If* A *is a finitely generated substructure of a model of* $\mathbf{T}$, *in the sorts* $\mathrm{VF} \cup \mathfrak{B}$, *the same is true for* $\mathbf{T}_A$.

Proof. Under the assumption of the corollary, the conclusion of Proposition 3.51 implies $\mathrm{RV}(E_{rslv}) \subseteq \mathrm{dcl}(\emptyset)$. □

Remark 3.54. It is easy to see using the description of imaginaries in [16] that in a resolved structure, any definable ACVF imaginary is resolved. In other words, if A is a resolved, and $\sim$ is a definable equivalence relation on a definable set D, then $D(A) \to (D/\sim)(A)$ is surjective.

If A is only effective, then there exists $\gamma \in \Gamma(A)^n$ such that for any t with $\mathrm{val}(t) = \gamma$, $(D/\sim)(A) \subseteq \mathrm{dcl}(D(A)/\sim, t)$; this can be seen by embedding $D/\sim$ into $B_n(K)/H$ for an appropriate $H \leq B_n(\mathcal{O})$, and splitting $B_n = T_n U_n$.

3.8 Dimensions

We define the VF-*dimension* of a T_M-definable set X to be the smallest n such that for some n, X admits a T_M-definable map with finite fibers into $\mathrm{VF}^n \times (\mathrm{RV} \cup \Gamma)^*$.

By *essential bijection* $Y \to Z$ we mean a bijection $Y_0 \to Z_0$, where $\dim_{\mathrm{VF}}(Y \setminus Y_0)$, $\dim_{\mathrm{VF}}(Z \setminus Z_0) < \dim_{\mathrm{VF}}(Y) = \dim_{\mathrm{VF}}(Z)$; and where two such maps are identified if they agree away from a set of dimension $< \dim_{\mathrm{VF}}(Y)$.

We say that a map $f : X \to \mathrm{VF}^n$ has RV-fibers if there exists $g : X \to (\mathrm{RV} \cup \Gamma)^*$ with (f, g) injective.

Lemma 3.55. *Let $X \subseteq \mathrm{VF}^n \times (\mathrm{RV} \cup \Gamma)^*$ be a definable set. Then we have the following:*

(1) *X has* VF *dimension $\leq n$ iff there exists a definable map $f : X \to \mathrm{VF}^n$ with* RV*-fibers.*
(2) *If it exists, the map f is "unique up to isogeny": if $f_1, f_2 : X \to \mathrm{VF}^n$ have* RV*-fibers, then there exists a definable $h : X \to Z \subseteq \mathrm{VF}^n \times (\mathrm{RV} \cup \Gamma)^*$ and $g_1, g_2 : Z \to \mathrm{VF}^n$ with finite fibers, such that $f_i = g_i h$.*

Proof.

(1) If $f : X \to \mathrm{VF}^n$ has RV-fibers, let g be as in the definition of RV-fibers; then $(f, g) : X \to \mathrm{VF}^n \times (\mathrm{RV} \cup \Gamma)^*$ is injective, so certainly finite-to-one. If $\phi : X \to \mathrm{VF}^n \times \mathrm{RV}^*$ is finite-to-one, by Lemma 3.9, each fiber $\phi^{-1}(c)$ admits a c-definable injective map $\psi_c : {}^{-1}(c) \to \mathrm{RV}^*$. By Lemma 2.3 we can find $\theta : X \to \mathrm{VF}^n \to \mathrm{RV}^*$ that is injective on each ϕ-fiber. Let $f(x) = (\phi, \theta)$. This proves the equivalence.
(2) Now suppose $f_1, f_2 : X \to \mathrm{VF}^n$ both have RV-fibers. Let $h(x) = (f_1(x), f_2(x))$, $Z' = h(X)$, and define $g_i : Z' \to \mathrm{VF}^n$ by $g_1(x, y) = x$, $g_2(x, y) = y$. Then g_i has finite fibers. Otherwise, we can find $a \in X$ such that $f_1(a) \notin \mathrm{acl}(f_2(a))$ (or vice versa). But for any $a \in X$, we have $f_1(a) \in \mathrm{acl}(f_2(a), c)$ for some $c \in (\mathrm{RV} \cup \Gamma)^*$. By Lemma 3.41, $f_1(a) \in \mathrm{acl}(f_2(a))$, a contradiction. By Lemma 3.9 (cf. Lemma 2.3), there exists a definable bijection between Z' and a subset Z of $\mathrm{VF}^n \times \mathrm{RV}^*$. Replacing Z by Z' finishes the proof of the lemma. □

Corollary 3.56. *Let $f : X \to \mathrm{RV} \cup \Gamma$, $X_a = f^{-1}(a)$. Then $\dim(X) = \max_a \dim X_a$.*

Proof. Let $n = \max_a \dim X_a$. For each a there exist definable functions $g_a : X_a \to \mathrm{VF}^n$ and $h_a : X_a \to (\mathrm{RV} \cup \Gamma)^*$ with (g_a, f_a) injective on X_a. Thus by the compactness argument of Lemma 2.3, there exists definable functions $g : X \to \mathrm{VF}^n$ and $h : X_a \to (\mathrm{RV} \cup \Gamma)^*$ such that (g, h) is injective when restricted to each X_a. But then clearly (g, h, f) is injective, so $\dim(X) \leq n$. The other inequality is obvious.□

We continue to assume **T** is V-minimal.

Lemma 3.57. *Let* $a, b \in \mathrm{VF}$. *If* $a \in \mathrm{acl}(b) \setminus \mathrm{acl}(\emptyset)$, *then* $b \in \mathrm{acl}(a)$.

Proof. Suppose $b \notin \mathrm{acl}(a)$. Let $A_0 = \Gamma(\mathrm{acl}(a, b))$. Then by Lemma 3.36, $b \notin \mathrm{acl}(A_0(a))$.

Let C be the intersection of all $\mathrm{acl}(A_0)$-definable balls such that $b \in C$, and let C' be the union of all $\mathrm{acl}(A_0)$-definable proper subballs of C. Let $B = \cap_i \{B_i\}$ be the set of all balls defined over $\mathrm{acl}(A_0(a))$ with $b \in B_i$, and let $B' = \cup_j \{B'_j\}$ be the union of all $\mathrm{acl}(A_0(a))$-definable proper subballs of B.

Since $a \in \mathrm{acl}(b)$, we have $a \in \mathrm{acl}(b')$ for all $b' \in B \setminus B'$, outside some proper subball. It follows by compactness that for some i, j, $a \in \mathrm{acl}(b')$ for all $b' \in B_i \setminus B'_j$. Say $i = j = 1$, $B'_1 \subset B_1$. By Example 3.57, $a \in \mathrm{acl}(A_0(f_1))$, where $f_1 \in \mathfrak{B}$ codes the ball B_1.

If B_1 is a point, we are done. Otherwise, B_1 has valuative radius $\alpha_1 < \infty$ defined over A_0. It follows that if $B_1 \supseteq C$ then B_1 is $\mathrm{acl}(A_0)$-definable; but then $a \in \mathrm{acl}(A_0)$, contradicting the assumption. Since B_1 meets P nontrivially, we therefore have $B_1 \subset C$. Similarly, B_1 cannot contain any ball in C' since it is not $\mathrm{acl}(A_0)$-definable, but it cannot be contained in C' since $B_1 \cap P \neq \emptyset$. so $B_1 \cap C' = \emptyset$. Thus $B_1 \subset P$.

Let $\bar{B_1}$ be the closed ball of radius α_1 containing B_1, and let e_1 be the corresponding element of $\mathfrak{B}_{\mathrm{cl}}$. Since $\bar{B_1}$ is almost definable over $A_0(a)$, it follows from V-minimality that there exists an almost $A_0(a)$-definable point $c(a)$ in $\bar{B_1}$. Now if $a \in \mathrm{acl}(A_0(e_1))$, then $\bar{B_1}$ contains an $A_0(e_1)$-definable finite set $F_1 = F_1(e_1)$. But since B_1 is a proper subset of P, $e_1 \notin \mathrm{acl}(A_0)$, this contradicts Lemma 3.8. Thus $a \notin \mathrm{acl}(A_0(e_1))$.

Nevertheless, we have seen that $a \in \mathrm{acl}(A_0(f_1))$. Thus $B_1 \neq \bar{B_1}$, so B_1 is a maximal open subball of $\bar{B_1}$. Let b_1 be the point of $\mathrm{Aff}(\bar{B_1})$ representing B_1. Then $a \in \mathrm{acl}(b_1)$. It follows that $\mathrm{tp}(a/\mathrm{acl}(A_0(e_1)))$ is strongly minimal, contradicting Lemma 3.13. We have obtained a contradiction in all cases; so $b \in \mathrm{acl}(a)$. □

Since the lemma continues to apply over any VF-generated structure, algebraic closure is a dependence relation in the sense of Steinitz (also called a prematroid or combinatorial geometry; cf. [34]). Define the VF-transcendence degree of a finitely generated structure B to be the maximal number of elements of $\mathrm{VF}(B)$ that are algebraically independent over $\mathrm{VF}(A)$. This is the size of any maximal independent set, and also the minimal size of a subset whose algebraic closure includes all VF-points. Hence we have the following.

Corollary 3.58. *The* VF *dimension of a definable set* D *is the maximal transcendence degree of* $\langle b \rangle$. □

We can now obtain a strengthening of Lemma 3.41, and a uniqueness statement in Proposition 3.51.

Corollary 3.59. *Let Y be a $\mathbf{T}$-definable set admitting a finite-to-one map f into $\mathfrak{B}^n$. Let $g : Y \to \mathrm{VF}^m$ be a definable map. Then $g(Y)$ is finite.*

Proof. We may assume $m = 1$. We will use the equivalence (3) $\iff$ (4) of Lemma 2.6. If $g(Y)$ is infinite, then by compactness there exists $a \in g(Y)$, $a \notin \mathrm{acl}(A)$. But for some b we have $a = g(b)$, so if $c = f(b)$, we have $c \in \mathfrak{B}^n$, $a \in \mathrm{acl}(c)$. Thus it suffices to show the following:

$$\text{If } a \in \mathrm{VF}, c \in \mathfrak{B}^n \text{ and } a \in \mathrm{acl}(A(c)), \text{ then } a \in \mathrm{acl}(A). \qquad (*)$$

This clearly reduces to the case $n = 1$, $c \in \mathfrak{B}$. Let γ be the valuative radius of c. As follows from Corollary 3.36, it suffices to show that $a \in \mathrm{acl}(A(\gamma))$. Thus in $(*)$ we may assume $\gamma \in A$.

Finally, to prove $(*)$ (using again the equivalence of Lemma 2.6), we may enlarge A, so we may assume $A \models \mathbf{T}$.

Since $\gamma \in A$, $c \in \mathrm{dcl}(A(e))$ for any element e of the ball c. Thus $a \in \mathrm{acl}(A(e))$. Suppose $a \notin \mathrm{acl}(A)$; then by exchange for algebraic closure in VF, $e \in \mathrm{acl}(A(a))$. Thus any two elements of the ball c are algebraic over each other. By Example 2.4, c has finitely many points; which is absurd. This contradiction shows that $a \in \mathrm{acl}(A)$. □

Lemma 3.60 (cf. Proposition 3.51). *Let $\mathbf{T}$ be a finitely generated extension of an effective V-minimal theory. Then if E_1, E_2 are effective and both embed into any effective E, then they are finitely generated, and $E_1 \simeq E_2$.*

Proof. The finite generation is clear. Since E_1, E_2 embed into each other, they have the same VF-transcendence degree We may assume $E_1 \leq E_2$. But then by Lemma 3.58, $E_2 \subseteq \mathrm{acl}(E_1)$. By Lemma 3.9, $E_2 \subseteq \mathrm{dcl}(E_1, F)$ for some finite $F \subseteq \mathrm{RV}^* \cap \mathrm{dcl}(E_2)$. But $\mathrm{RV}(E_1) = \mathrm{RV}(E_2)$, so $F \subseteq \mathrm{dcl}(E_1)$, and thus $E_2 = E_1$.□

Remark 3.61. The analogous statement is true for resolved structures. Note that if F is a finite definable subset of RV^n, then automatically the coordinates of the points of F lie in cosets of $\mathbf{k}^*$ that have algebraic points.

Remark. The hypothesis of Lemma 3.60 can be slightly weakened to the following: $\mathbf{T}$ is finitely generated over a V-minimal theory, and there exists a finitely generated effective E.

Example 3.62. In ACVF, when $X \subseteq \mathrm{VF}^n$, the VF dimension equals the dimension of the Zariski closure of X. This is proved in [36]. The idea of the proof: the VF dimension is clearly bounded by the Zariski dimension. For the opposite inequality, in the case of dimension 0, if X is a finite A-definable subset of VF, then using quantifier elimination there exists a nonzero polynomial f with coefficients in A, such that f vanishes on X. In general, if a definable $X \subseteq \mathrm{VF}^n$ has VF dimension $< n$, one can reduce to the case where all fibers of the projection $\mathrm{pr} : X \to \mathrm{pr}\, X \subset \mathrm{VF}^{n-1}$ are finite, then X is not Zariski dense in VF^n, using the zero-dimensional case.

The RV-*dimension* of a definable set $X \subseteq \mathrm{RV}^*$ is the smallest integer n (if any) such that X admits a parametrically definable finite-to-one map into RV^n. More generally for $X \subseteq (\mathrm{RV} \cup \Gamma)^*$, $\dim_{\mathrm{RV}}(X)$ is the smallest integer n (if any) such that X admits a parametrically definable finite-to-one map into $(\mathrm{RV} \cup \Gamma)^n$.

Note that RV is one dimensional, but Γ and every fiber of $\mathrm{val}_{\mathrm{rv}}$ are also one dimensional. In this sense $\mathrm{RV} \cup \Gamma$ dimension is not additive; model-theoretically it is closer to weight than to rank. We do have $\dim(X \times Y) = \dim(X) + \dim(Y)$.

Dually, if a structure B is RV-generated over a substructure A, we can define the *weight* of B/A to be the least n such that $B \subseteq \mathrm{acl}(A, a_1, \ldots, a_n)$, with $a_i \in \mathrm{RV}$.

For subsets of RV, RV dimension can be viewed as the size of a Steinitz basis with respect to algebraic closure. One needs to note that the exchange principle holds.

Lemma 3.63 (exchange). *Let* $a, b_1, \ldots, b_n \in \mathrm{RV}$; *assume* $a \in \mathrm{acl}(A, b_1, \ldots, b_n) \setminus \mathrm{acl}(A, b_1, \ldots, b_{n-1})$. *Then* $b_n \in \mathrm{acl}(A, b_1, \ldots, b_{n-1}, a)$.

Proof. We may take $n = 1$, $b_n = b$, and $A = \mathrm{acl}(A)$. Let $\alpha = \mathrm{val}_{\mathrm{rv}}(a) \in \Gamma$, $\beta = \mathrm{val}_{\mathrm{rv}}(b)$. If $\beta \in A$ then $\Gamma(A(a, b)) = \Gamma(A(b)) = \Gamma(A)$. The first equality is true since $a \in \mathrm{acl}(A(b))$ so $A(a, b) \subset \mathrm{acl}(A(b))$, and using the stable embeddedness of Γ (Section 2.1) and the linear ordering on Γ. The second equality follows from Lemma 3.10. Thus if $\beta \in A$, then a, b lie in A-definable strongly minimal sets, cosets of $\mathbf{k}^*$, and the lemma is clear.

Assume $\beta \notin A$. If $\alpha \in A$, then $\mathrm{tp}(a/A)$ is strongly minimal, and $\mathrm{tp}(a/A)$ implies $\mathrm{tp}(a/A(b))$ by Lemma 3.10; but then $a \in \mathrm{acl}(A)$, contradicting the assumption. Thus $\alpha, \beta \notin A$; from the exchange principle in Γ, it follows that $A' := \mathrm{acl}(A, \alpha) = \mathrm{acl}(A, \beta)$. Moreover, $a \notin \mathrm{acl}(\alpha)$ by Lemma 3.11 and Lemma 2.6. By the previous case, $b \in \mathrm{acl}(A', a)$, so $b \in \mathrm{acl}(A, a)$. □

Lemma 3.64. *A definable* $X \subseteq \mathrm{RV}^n$ *has* RV *dimension* n *iff it contains an* n-*dimensional definable subset of some coset of* $\mathbf{k}^{*n}$.

Proof. Assume X has RV dimension n. Then there exists $(a_1, \ldots, a_n) \in X$ with $a_1, \ldots, a_n$ algebraically independent. Let $c \in \Gamma$; then since $a_n \notin \mathrm{acl}(a_1, \ldots, a_{n-1})$, it follows as in the proof of Lemma 3.63 that $a_n \notin \mathrm{acl}(a_1, \ldots, a_{n-1}, c)$. This applies to any index, so $a_1, \ldots, a_n$ remain algebraically independent over c; and inductively we may add to the base any finite number of elements of Γ. Let $c_i = \mathrm{val}_{\mathrm{rv}}(a_i)$, and let $A' = A(c_1, \ldots, c_n)$. Then $a_1, \ldots, a_n$ are algebraically independent over A', and they lie in $X' = X \cap \prod_{i=1}^n \mathrm{rv}^{-1}(c_i)$; thus X' is an n-dimensional definable subset of a coset of $\mathbf{k}^{*n}$. □

Definition 3.65. $\mathrm{VF}[n, \cdot]$ be the category of definable subsets of $\mathrm{VF}^* \times \mathrm{RV}^*$ of dimension $\leq n$. Morphisms are definable maps.

Let $X \in \mathrm{Ob}\,\mathrm{VF}[n, \cdot]$. By Lemma 3.55, there exists a definable $f : X \to \mathrm{VF}^n$ with RV-fibers; and the maximal RV dimension of a fiber is a well-defined quantity, depending only on the isomorphism type of X (but not on the choice of f). In particular, the subcategory of definable sets of maximal fiber dimension 0 will be denoted $\mathrm{VF}[n]$.

Definition 3.66. We define $\mathrm{RV}[n,\cdot]$ to be the category of definable pairs (U, f), with $U \subseteq \mathrm{RV}^*$, $f : U \to \mathrm{RV}^n$. If $U, U' \in \mathrm{Ob}\,\mathrm{RV}[n,\cdot]$, a morphism $h : U \to U'$ is a definable map, such that $U'' = \{(f(u), f'(h(u)) : u \in U\}$ has finite-to-one first projection to RV^n. $\mathrm{RV}[n]$ is the full subcategory of pairs (U, f) with $f : U \to \mathrm{RV}^n$ finite-to-one.

$\mathrm{RES}[n]$ is the full subcategory of $RV[n]$ whose objects are pairs $(U, f) \in \mathrm{Ob}\,\mathrm{RV}[n]$ such that $\mathrm{val}_{\mathrm{rv}}(U)$ is finite, i.e., $U \subseteq \mathrm{RES}^*$.

Remark 3.67.

(1) For $X, Y \in \mathrm{Ob}\,\mathrm{RV}[n]$, any definable bijection $X \to Y$ is in $\mathrm{Mor}_{\mathrm{RV}[n]}(X, Y)$.
(2) The forgetful map $(X, f) \mapsto X$ is an equivalence of categories between $\mathrm{RV}[n]$ and the category of all definable subsets of RV^* of RV dimension $\leq m$, with all maps between them. The presentation with f is nonetheless useful for defining $\mathbb{L}$.

By Remark 3.67, $K_+(\mathrm{RV}[m])$ is isomorphic to the Grothendieck semigroup of definable subsets of RV^* of RV dimension $\leq m$. If $\dim(X) \leq m$, let $[X]_m$ denote the class $[X]_m = [(X, f)]_m \in \mathrm{RV}[m]$, where $f : X \to \mathrm{RV}^*$ is any finite-to-one definable map.

Unlike the case of $\mathrm{VF}[n,\cdot]$ or $\mathrm{RV}[n]$, for $(U, f) \in \mathrm{Ob}\,\mathrm{RV}[n,\cdot]$ the map f cannot be reconstructed from U alone, even up to isogeny, so it must be given as part of the data. We view (U, f) as a cover of $f(U)$ with "discrete" fibers.

We denote

$$\mathrm{RV}[\leq N,\cdot] := \oplus_{0\leq n\leq N}\mathrm{RV}[n,\cdot], \qquad \mathrm{RV}[\leq N] = \oplus_{0\leq n\leq N}\mathrm{RV}[n],$$
$$\mathrm{RV}[*,\cdot] := \oplus_{0\leq n}\mathrm{RV}[n,\cdot], \qquad \mathrm{RV}[*] := \oplus_{0\leq n}\mathrm{RV}[n],$$
$$\mathrm{RES}[*] := \oplus_{0\leq n}\mathrm{RES}[n].$$

We have natural multiplication maps $K_+\,\mathrm{RV}[k,\cdot] \times K_+\,\mathrm{RV}[l,\cdot] \to K_+[k+l,\cdot]$, $([(X, f)], [(Y, g)]) \mapsto [(X \times Y, f \times g)]$. This gives a semiring structure to $K_+(\mathrm{RV}[*])$. This differs from the Grothendieck ring $K_+(\mathrm{RV})$.

Alternative description of $\mathrm{RV}[\leq N,\cdot]$

An object of $\mathrm{RV}[\leq N,\cdot]$ thus consists of a formal sum $\sum_{n=0}^{N} \mathbf{X}_n$ of objects $\mathbf{X}_n = (X_n, f_n)$ of $\mathrm{RV}[n,\cdot]$. This can be explained from another angle if one adds a formal element ∞ to RV, and extends rv to VF by $\mathrm{rv}(0) = \infty$. Define a function $f[k]$ by $f[k](x) = (f_n(x), \infty, \ldots, \infty)$ ($N-k$ times). If $\mathbf{X} = (X, f)$, let $\mathbf{X}[k] = (X, f[k])$. Then $\sum_{n=0}^{N} \mathbf{X}_n$ can be viewed as the disjoint union $\cup_{i=0}^{N} X_i \times \{\infty\}[N-i]$. The rv pullback is then a set of VF dimension N, invariant under multiplication by $1+\mathcal{M}$; the sum over dimensions $\leq N$ is necessary to ensure that any such invariant set is obtained (cf. Lemma 4.9). From this point of view, an isomorphism is a definable bijection preserving the function "number of finite coordinates." We will use $\mathrm{RV}[\leq N,\cdot]$ or $\mathrm{RV}_\infty[N,\cdot]$ interchangeably.

Lemma 3.68. *Let $X, X' \in \mathrm{Ob}\,\mathrm{RV}[n,\cdot]$, and assume a bijection $g : X' \to X$ lifts to $G : \mathbb{L}X' \to \mathbb{L}X$. Then $g \in \mathrm{Mor}_{\mathrm{RV}[n,\cdot]}(X', X)$.*

Proof. We only have to check the isogeny condition, i.e., that $f(g(a)) \in \mathrm{acl}(f'(a))$ for $a \in X'$ (and dually). By Lemma 3.42, for $x \in {\rho_{X'}}^{-1}(a)$, $G(x)_{\mathrm{VF}} \in \mathrm{acl}(x_{\mathrm{VF}})$, i.e., the VF-coordinates of $G(x)$ are algebraic over those of x. Thus $f(g(a)) \in \mathrm{acl}(x_{\mathrm{VF}})$. This is true for any $x \in {\rho_{X'}}^{-1}(a)$, so $f(g(a)) \in \mathrm{acl}(a)$. □

4 Descent to RV: Objects

We assume **T** is C-minimal with centered closed balls. We will find a very restricted set of maps that transform any definable set to a pullback from RV. This is related to Denef's cell decomposition theorem; since we work in C-minimal theories it takes a simpler form. Recall that this assumption is preserved under passage to $\mathbf{T}_A$, when A is a (VF, RV, Γ)-generated substructure of a model of **T** (Lemma 3.39).

Recall that $\mathrm{RV} = \mathrm{VF}^\times/(1+\mathcal{M})$, $\mathrm{rv} : \mathrm{VF}^\times \to \mathrm{RV}$ the quotient map. Let $\mathrm{RV}_\infty = \mathrm{RV} \cup \{\infty\}$, and define $\mathrm{rv}(0) = \infty$. We will also write rv for the induced map $\mathrm{rv}^n : (\mathrm{VF}^\times)^n \to (\mathrm{RV})^n$.

Definition 4.1. Fix n. Let $\mathcal{C}^0$ be the category whose objects are the definable subsets of $\mathrm{VF}^n \times \mathrm{RV}_\infty^*$, and whose morphisms are generated by the inclusion maps together with functions of one of the following types:

(1) Maps

$$(x_1, \dots, x_n, y_1, \dots, y_l) \mapsto (x_1, \dots, x_{i-1}, x_i + a, x_{i+1}, \dots, x_n, y_1, \dots, y_l)$$

with $a = a(x_1, \dots, x_{i-1}, y_1, \dots, y_l) : \mathrm{VF}^{i-1} \times \mathrm{RV}_\infty^l \to \mathrm{VF}$ an A-definable function of the coordinates $y, x_1, \dots, x_{i-1}$.

(2) Maps $(x_1, \dots, x_n, y_1, \dots, y_l) \mapsto (x_1, \dots, x_n, y_1, \dots, y_l, \mathrm{rv}(x_i))$.

The above functions are called *elementary admissible transformations over* A; a morphism in $\mathcal{C}_A^0$ generated by elementary admissible transformations over A will be called an *admissible transformation* over A. Taking $l = 0$, we see that all A-definable additive translations of VF^n are admissible.

Analogously, if Y is a given definable set, one defines the notion of a Y-family of admissible transformations.

If $e \in \mathrm{RV}$ and T_e is an $A(e)$-admissible transformation, then there exists an A-admissible T such that $\iota_e T_e = T \iota_e$, where $\iota_e(x_1, \dots, x_n, y_1, \dots, y_l) = (x_1, \dots, x_n, e, y_1, \dots, y_l)$. This is easy to see for each generator and follows inductively.

Informally, note that admissible maps preserve volume for any product satisfying Fubini's theorem of translation invariant measures on VF and counting measures on RV.

We will now see that any $X \subset \mathrm{VF}^n$ is a finite disjoint union of admissible transforms of pullbacks from RV. We begin with $n = 1$.

Lemma 4.2. *Let* **T** *be C-minimal with centered closed balls. Let X be a definable subset of* VF. *Then X is the disjoint union of finitely many definable sets Z_i, such that* *for some admissible transformations T_i, and definable subsets H_i of* $\mathrm{RV}_\infty^{l_i}$, $T_i Z_i = \{(x, y) : y \in H_i, \mathrm{rv}(x) = y_{l_i}\}$.

If X is bounded, H_i is bounded below; in fact, for any $h \in H_i$, $\mathrm{val}_{\mathrm{rv}}(h) \geq \mathrm{val}(x)$ *for some $x \in X$.*

Here VF will be considered a ball of valuative radius $-\infty$, and points as balls of valuative radius ∞.

Proof. We may assume X is a finite union of disjoint balls of the same valuative radius $\alpha \in \Gamma \cup \{\pm\infty\}$, each minus a finite union of proper subballs, since any definable set is a finite union of definable sets of that form.

Case 1: *X is a closed ball.* In this case, by the assumption of centered closed balls, X has a definable point a. Let $T(x) = x - a$. Then $TX \setminus \{0\}$ is the pullback of a subset of Γ, the semi-infinite interval $[\alpha, \infty)$ (where α is the valuative radius of X). Thus $TX = \mathrm{rv}^{-1}(H)$, where $H = {\mathrm{val}_{\mathrm{rv}}}^{-1}([\alpha, \infty)) \cup \{\infty\}$.

Case 2: *X is an open ball.* Let $\mathbf{X}$ be the surrounding closed ball of the same radius α, and as in Case 1 let $a \in \mathbf{X}$ be an definable point, $T(x) = x - a$. If $0 \in TX$ then $TX = \mathrm{rv}^{-1}(H)$, where $H = {\mathrm{val}_{\mathrm{rv}}}^{-1}((\alpha, \infty)) \cup \{\infty\}$. If $0 \notin TX$, then $TX = \mathrm{rv}^{-1}(H)$, where $H = \mathrm{rv}(TX)$ is a singleton of RV.

Case 3: *$X = C \setminus F$ is a ball with a single hole, the closed ball F.* Let β be the valuative radius of F. Let $a \in F$ be a definable point, $T(x) = x - a$. Then $TX = \mathrm{rv}^{-1}(H)$, $H = {\mathrm{val}_{\mathrm{rv}}}^{-1}(I)$, where I is the open interval (α, β) of Γ in case C is closed, the half-open interval $[\alpha, \beta)$ when C is open.

Case 4: *$X = C \setminus \cup_{j \in J} F_j$ is a closed ball, minus a finite union of maximal open subballs.* As in Case 1, find T_1 such that $0 \in T_1 X$. Then $T_1 X$ is the union of the maximal open subball S of radius α, with $\mathrm{rv}^{-1}(H)$, where $H = \mathrm{rv}(X \setminus S)$. S can be treated as in Case 2. Here H is a subset of ${\mathrm{val}_{\mathrm{rv}}}^{-1}(\alpha)$, consisting of ${\mathrm{val}_{\mathrm{rv}}}^{-1}(\alpha)$ minus finitely many points.

Cases 3a *and* 4a: *X is a union of m balls (perhaps with holes) of types* 1–4 *above.* Here we use induction on m; we have m balls C_j covering X. Let E be the smallest ball containing all C_j. As we may assume $m > 1$, E must be a closed ball; and each C_j is contained in some maximal open subball M_j of E. By the choice of E, not all C_j can be contained in the same maximal open ball of E. Let $a \in E$ be a definable point, $T_1(x) = x - a$. If $0 \in T_1 C_j$ for some j, the lemma is true by induction for this C_j and for the union of the others, hence also for X. Otherwise, $F = \mathrm{rv}(T_1(X))$ is a finite set, with more than one element. For $b \in F$, let $Y_b = T_1 X \cap \mathrm{rv}^{-1}(b)$. By Lemma 2.3, we can, in fact, find a definable Y whose fiber at b is Y_b. By induction again, there exists an admissible transformation T_b such that $T_b(Y)$ is a pullback of the required form. Let $T_2(x) = (x, \mathrm{rv}(x))$, $T_3((x, b)) = ((T_b(x), b))$. Then $T_3 T_2 T_1$ solves the problem.

General subsets of VF. Let $\beta \geq \alpha$ be the least size (i.e., greatest element of Γ) such that some ball of radius β contains more than one hole of X. Let $\{C_j : j \in J\}$ be the balls of radius β around the holes W of X, and let $C = \cup_{j \in J} C_j$. Then $X = (X \setminus C) \dot{\cup} (C \setminus W)$. Now $X \setminus C$ has fewer holes than X, so it can be dealt with inductively. Thus we may assume $X = C \setminus W$; and any proper subball of C of less than maximal size contains at most one hole of X. We may assume the $\{C_j\}$ form a single Galois orbit; so they each contain two or more holes of X. Since these holes are not contained in a proper subball of C_j, each C_j must be closed, and the maximal open subballs of C_j separate holes. Let $D_{j,k}$ be the maximal open subballs of C_j containing a hole $F_{j,k}$. Let $\bar{F}_{j,k}$ be the smallest closed ball containing $F_{j,k}$. Then $X = (C \setminus \cup_{j,k} D_{j,k}) \dot{\cup} \cup_{j,k} (D_{j,k} \setminus \bar{F}_{j,k}) \dot{\cup} \cup_{j,k} (\bar{F}_{j,k} \setminus F_{j,k})$. The second summand in this union falls into Case 3a, the first and third (when nonempty) into Case 4a. □

Remark. If we allow arbitrary Boolean combinations (rather than disjoint unions only), we can demand in Lemma 4.2 that the sets H_i be finite. More precisely, let X be a definable subset of VF. Then there exist definable sets Z_i, admissible transformations T_i, and finite definable subsets H_i of $\mathrm{RV}_\infty^{l_i}$ such that we have the following:

X is a Boolean combination of the sets Z_i, and $T_i Z_i$ is one of the following:

(1) VF;
(2) $(0) \times H_i$;
(4) $b_i \times H_i$, with b_i a definable ball containing 0;
(5) $\{(x, y) : y \in H_i, \mathrm{rv}(x) = f_i(y)\}$, for some definable function $f_i : H_i \to \mathrm{RV}_\infty$.

Corollary 4.3. *Let* $X \subseteq \mathrm{VF} \times \mathrm{RV}^*$ *be definable. Then there exists a definable* $\rho : X \to \mathrm{RV}^*$ *and* $c : \mathrm{RV}^* \to \mathrm{VF}$, $c' : \mathrm{RV}^* \to \mathrm{RV}_\infty$, $c'' : \mathrm{RV}^* \to \mathrm{RV}^*$ *such that every fiber* $\rho^{-1}(\alpha)$ *has the form* $(c(\alpha) + \mathrm{rv}^{-1}(c'(\alpha))) \times \{c''(a)\}$. *Moreover, c has finite image.*

Proof. The finiteness of the image of c is automatic, by Lemma 3.41. The corollary is obviously true for sets of the form $\mathbb{L}(H, h) = \{(x, u) \in \mathrm{VF} \times H : \mathrm{rv}(x) = h(u)\}$; take $\rho(x, u) = (\mathrm{rv}(x), u)$. If the statement holds for TX where T is an admissible transformation, then it holds for X. If true for two disjoint sets, it is also true for their union. (Add to ρ a map to $\{1, -1\} \subseteq \mathbf{k}^*$ whose fibers are the two sets.) Hence by Lemma 4.2 is true for all definable sets. □

Corollary 4.4. *Let* **T** *be* V*-minimal,* $X \subseteq \mathrm{VF}$ *and let* $f : X \to \mathrm{RV} \cup \Gamma$ *be a definable function. Then there exists a definable finite partition of* $X = \cup_{i=1}^m X_i$ *such that either* f *is constant on* X_i, *or else* X_i *is a finite union of balls of equal radius (possibly missing some subballs), there is a definable set* F_i *meeting each of the balls b in a single point, and for* $x \in X_i$, *letting* $n(x)$ *be the point of* F_i *nearest x, for some function* H, $f(x) = H(\mathrm{rv}(x - n(x)))$.

Proof. The conclusion is so stated that it suffices to prove it over $\mathrm{acl}(\emptyset)$, i.e., we may assume every almost definable set is definable; cf. Section 2.1. By compactness it suffices to show that for each complete type p, $f|p$ has the stated form. Let b be the intersection of all balls containing p. If b is transitive then by Lemma 3.47 $f|p$ is constant. Otherwise, by V-minimality b contains a definable point, and so we may assume $0 \in b$. It follows that $\mathrm{rv}(p)$ is infinite. Thus by Lemma 3.20, f factors through rv. □

Proposition 4.5. *Let* **T** *be C-minimal with centered closed balls, and let X be a definable subset of $\mathrm{VF}^n \times \mathrm{RV}^l$. Then X can be expressed as a finite disjoint union of A-definable sets Z, with each Z of the following form. For some A-admissible transformation T, A-definable subset H of $\mathrm{RV}_\infty^{l^*}$, and map of indices $\nu \in \{1, \dots, n\} \mapsto \nu' \in \{1, \dots, l^*\}$,*

$$TZ = \{(a, b) : b \in H, \mathrm{rv}(a_\nu) = b_{\nu'} (\nu = 1, \dots, n)\}.$$

If X projects finite-to-one to VF^n, then the projection of H to the primed coordinates $1', \dots, n'$ is finite to one.

If X is bounded, then H is bounded below in RV_∞.

Proof. By induction on n; the case $n = 0$ is trivial. Let $\mathrm{pr} : X \to \mathrm{pr}\, X$ be the projection of X to $\mathrm{VF}^{n-1} \times \mathrm{RV}^l$, so that $X \subset \mathrm{VF} \times \mathrm{pr}\, X$.

Let $\mathrm{pr}^*(Y) = \{v : (\exists y \in Y)(x, y) \in Y\}$. For any $c \in \mathrm{pr}\, X$, according to Lemma 4.2, we can write $\mathrm{pr}^*(c) = \dot{\bigcup}_{i=1}^{k} Z_i(c)$, where

$$T_i(c) Z_i(c) = \{(a, b) : b \in H_i(c), \mathrm{rv}(a) = b_{1'}\}$$

for some $A(c)$-admissible $T_i(c)$, $A(c)$-definable $Z_i(c)$, and $H_i(c) \subseteq \mathrm{RV} = \mathrm{RV}^{1'}$. We can write $Z_i(c) = \{x : (x, c) \in Z_i\}$, $H_i(c) = \{x : (x, c) \in H_i\}$ for some definable Z_i and $H_i \subset \mathrm{VF}^{n-1} \times \mathrm{RV}^{1'}$. By compactness, as in Lemma 2.3, one can assume that the $Z_i(c)$, $H_i(c)$, $T_i(c)$ are uniformly definable: there exists a partition of $\mathrm{pr}\, X$ into finitely many definable sets Y, and for each Y families Z_i, H_i, T_i over Y of definable sets and admissible transformations over Y, such that the integer k is the same for all $c \in Y$, and the $Z_i(c)$, $H_i(c)$, $T_i(c)$ are fibers over c of Z_i, H_i, T_i. In this case, $\mathrm{pr}^*(Y) = \dot{\bigcup}_{i=1}^{k} Z_i$. We can express X as a disjoint union of the various $\mathrm{pr}^*(Y)$; so we may as well assume $\mathrm{pr}\, X = Y$ and $X = Z_1$. Let T_1 be such that $\iota_c T_1(c) = T_1 \iota_c$. Then

$$T_1 X = \{(a, c, b) : (c, b) \in H_1, \mathrm{rv}(a) = b_{1'}\}.$$

Any admissible transformation is injective and so commutes with disjoint unions.

Now by induction, H_1 itself is a disjoint union $H_1 = \dot{\bigcup}_{j=1}^{k'} Z_j$, with

$$T_i' Z_i' = \{(d, b) : b \in H_i', \mathrm{rv}(d_\nu) = d_{\nu'} (\nu = 2, \dots, n)\}.$$

Notational remarks. Here $d = (d_2, \dots, d_n)$ are the VF-coordinates of c above. The $'$ depends on i but we will not represent this notationally.

Let $T_i^*(a, d, b) = (a, T_i'(d, b))$, i.e., T_i^* does not touch the first coordinate. Note that T_i^* also does not move the $1'$ coordinate, since in general admissible transformations can only add RV coordinates but not change existing ones. Let

$$Z_i = \{x : T_1(x) = (a, d, b), (d, b) \in Z_i', \mathrm{rv}(a) = b_{1'}\}.$$

Then (as one sees by applying T_1) $X = \dot{\bigcup}_{i=1}^{k} Z_i$, and if $T_i = T_i^* T_1$, we have

$$\begin{aligned} T_i Z_i &= \{(a, d', b') : (d', b') \in T_i' Z_i', \mathrm{rv}(a) = b'_{1'}\} \\ &= \{(a, d', b') : b \in H_i', \mathrm{rv}(a) = b_{1'}, \mathrm{rv}(d_\nu) = b_{\nu'}\}. \end{aligned}$$

As for the finiteness of the projection, if X admits a finite-to-one projection to VF^n, so does each Z in the statement of the proposition, and hence the isomorphic set TZ. We have $H \subset \mathrm{RV}^{n+l}$, $\pi : \mathrm{RV}^{n+l} \to \mathrm{RV}^n$, so $TZ = \{(a, b, b') : (b, b') \in H, \mathrm{rv}(a) = b'\}$. For fixed a, this yields an a-definable finite-to-one map $TZ'(a) = \{b' : (a, b, b') \in TZ\} \to \mathrm{VF}^n$. By Lemma 3.41, $TZ'(a)$ is finite. Now fix b and suppose $(b, b') \in H$ with b' not algebraic over b. Then for generic $a \in \mathrm{rv}^{-1}(b)$, b' is not algebraic over b, a. Yet $(a, b, b') \in TZ$ and so $b' \in TZ'(a)$, a contradiction.

The statement on boundedness is obvious from the proof; if $X \subseteq \{x : \mathrm{val}(x) \geq -\gamma\}^n \times \mathrm{RV}^m$, then H is bounded below by $-\gamma$ in each coordinate. □

A remark on more general base structures

Lemma 4.6. *Let* **T** *be* V*-minimal,* A *a* $\mathfrak{B}$*-generated substructure of a model of* **T**. *Let* X *be a* $\mathbf{T}_A$*-definable subset of* $\mathrm{VF}^n \times \mathrm{RV}^l$. *Then there exist* $\mathbf{T}_A$*-definable subsets* $Y_i \subset \mathrm{RV}^{m_i}$ *and (projection) maps* $f_i : Y_i \to \mathrm{RV}^n$, *a disjoint union* Z *of*

$$Z_i = Y_i \times_{f_i, \mathrm{rv}} \mathrm{VF}^n$$

and a nonempty A*-definable family* $\mathfrak{F}$ *of admissible transformations* $X \to Z$. $\mathfrak{F}$ *will have an* A'*-point for any* $\mathrm{VF} \cup \mathrm{RV} \cup \Gamma$*-generated structure containing* A.

Proof. We may assume A is finitely generated. By Proposition 3.51 there exists an almost $\mathrm{VF} \cup \Gamma$-generated $A' \supset A$ embeddable over A into any $\mathrm{VF} \cup \Gamma$-generated structure containing A, and with $\mathrm{RV}(A') = \mathrm{RV}(A)$. By Proposition 4.5, the required objects Y_i, f_i exist over A'. But since RV is stably embedded, this data is defined over $\mathrm{RV}(A') \subseteq A$. The admissible transformations $X \to Z = \dot{\bigcup}(Y_i \times_{f_i, \mathrm{rv}} \mathrm{VF}^n)$ exist over A'; so one can find a definable set D with an A'-point, and such that any element of D codes an admissible transformation $X \to Z$. □

Remark. In fact, arbitrary ACVF-imaginaries may be allowed here.

Example 4.7. $\mathfrak{F}$ need not have an A-rational point. For instance, if A consists of an element of $\mathrm{VF}/\mathcal{M}$, i.e., an open ball c, then we can take $Y = Y_1$ to be the point $0 \in \mathrm{RV}$ (since c can be transformed to $\mathcal{M}$); but there is no A-definable bijection of c with $\mathcal{M}$.

A statement in terms of Grothendieck groups

Recall Definitions 3.65 and 3.66.

Definition 4.8. Define $\mathbb{L} : \operatorname{Ob} \mathrm{RV}[n, \cdot] \to \operatorname{Ob} \mathrm{VF}[n, \cdot]$ by

$$\mathbb{L}(X, f) = (\mathrm{VF}^{\times})^n \times_{\mathrm{rv}^n, f} X \subset \mathrm{VF}^n \times \mathrm{RV}^m,$$

where $\mathrm{VF}^{\times} = \mathrm{VF} \setminus \{0\}$.

For $\mathbf{X} = \sum_i \mathbf{X}_i \in \mathrm{RV}[*]$, we let $\mathbb{L}(\mathbf{X})$ be the disjoint sum $\sum_i \mathbb{L}(\mathbf{X}_i)$ over the various components in $\mathrm{RV}[i]$.

Let ρ denote the natural map $\mathbb{L}(X, f) \to X$.

Lemma 4.9. *The image of* $\mathbb{L} : \operatorname{Ob} \mathrm{RV}[\leq n, \cdot] \to \operatorname{Ob} \mathrm{VF}[n, \cdot]$ *meets every isomorphism class of* $\mathrm{VF}[n, \cdot]$.

Proof. For $X \subseteq \mathrm{RV}^*$ and $f : X \to \mathrm{RV}_\infty$, define $\mathrm{rv}(0) = \infty$ and

$$\mathbb{L}(X, f) = \mathrm{VF}^n \times_{\mathrm{rv}^n, f} X \subset \mathrm{VF}^n \times \mathrm{RV}^m.$$

Then in the statement of Proposition 4.5, we have $TZ = \mathbb{L}(H, h)$ where h is the projection to the primed coordinates. For $x \in H$, let $s(x) = \{i : h_i(x) = \infty\}$. For $w \subseteq \{1, \ldots, n\}$, let $H_w = \{x \in H : s(x) = w\}$. Let $\bar{H}_w = (H_w, h'_w)$ where $h'_w = (h_i)_{i \notin w}$. Then $\bar{H}_w \in \mathrm{RV}[|w|, \cdot]$, and $\mathbb{L}(H_w, h|H_w) \simeq \mathbb{L}(\bar{H}_w)$. Thus $\mathbb{L}(H, h) \simeq \mathbb{L}(\sum_w \bar{H}_w)$. □

A restatement in terms of VF alone

This restatement will not be used later in the paper.

Definition 4.10. Let A be a subfield of VF. Let $\mathcal{C}^1_A(n, l)$ be the category of definable subsets of $\mathrm{VF}^n \times (\mathrm{VF}^{\times})^l$, generated by composition and restriction to subsets by maps of one of the following types:

(1) Maps

$$(x_1, \ldots, x_n, y_1, \ldots, y_l) \mapsto (x_1, \ldots, x_{i-1}, x_i + a, x_{i+1}, \ldots, x_n, y_1, \ldots, y_l)$$

with $a = a(x_1, \ldots, x_{i-1}, y_1, \ldots, y_l) : \mathrm{VF}^{i+l-1} \to \mathrm{VF}$ an A-definable function of the coordinates $y, x_1, \ldots, x_{i-1}$.

(2) Maps $(x_1, \ldots, x_n, y_1, \ldots, y_l) \mapsto (x_1, \ldots, x_n, y_1, \ldots, y_{i-1}, x_i y_i, y_{i+1}, \ldots, y_l) : X \to Y$ assuming $x_i \neq 0$ on X, and that this function takes X into Y.

Remark 4.11. The morphisms in this category are measure preserving with respect to Fubini products of invariant measures (additively for VF, multiplicatively for $\mathrm{VF}^{\times}$), viz. $dx_1 \wedge \cdots \wedge dx_n \wedge dy_1/y_1 \wedge \cdots \wedge dy_l/y_l$.

Lemma 4.12. *Let* **T** *be C-minimal with centered closed balls, X a definable subset of* VF^n. *Then X can be expressed as a disjoint union of A-definable sets Z with the following property. For some* $l \in \mathbb{N}$, *there exists an* $\mathcal{C}^1_A(n,l)$*-transformation T and a definable subset H of* $\mathrm{RV}^n_\infty \times \mathrm{RV}^l$, *such that*

$$T(Z \times ((1+\mathcal{M}))^l) = \mathrm{rv}^{-1}(H).$$

Moreover, the projection of H to RV^n_∞ *is finite-to-one.*

If $\mathrm{val}(x)$ *is bounded below, then* $\mathrm{val}(H)$ *may be taken to be bounded below in the* RV*-coordinates, and bounded in the* RV_∞*-coordinates.*

Proof. This follows from Proposition 4.5. □

5 V-minimal geometry: Continuity and differentiation

We work with a V-minimal theory.

5.1 Images of balls under definable functions

Proposition 5.1. *Let X, Y be definable subsets of* VF, *and let* $F : X \to Y$ *be a definable bijection. Then there exists a partition of X to finitely many definable equivalence classes, such that for any open ball b contained in one of the classes,* $F(b)$ *is an open ball; and dually, if* $F(b)$ *is an open ball, so is b.*

Proof. It suffices to show that such a partition exists over $\mathrm{acl}(\emptyset)$; for any finite almost definable partition has a finite definable refinement (cf. the discussion of Galois theory in Section 2.1). Thus as in Section 2.1 we may assume every almost definable set is named.

We will show that if p is a complete type, and b is an open subball of p, then $F(b)$ is an open ball; and that if b' is an open subball of $F(p)$, then b is an open ball. From this it follows by compactness that there exists a definable D_p containing p with the same property; by another use of compactness, finitely many D_p cover X; it then suffices to choose any partition, such that any class is contained in some D_p.

When p has a unique solution, the assertion is trivial. When p is the generic type of a closed ball, or of VF, or of a transitive open or ∞-definable ball, for any $\alpha \in \Gamma$, p remains complete over $\langle \alpha \rangle$. In the transitive cases, this follows from Lemma 3.47, while in the centered closed case it follows from Lemma 3.18.

Thus all open subballs b_t of p of any radius α have the same type over $\langle \alpha \rangle$; hence they are all transitive over $\langle t \rangle$, where $t \in K/\mathcal{M}_\alpha$, where $\mathcal{M}_\alpha = \{x : \mathrm{val}(x) > \alpha\}$ (Lemma 3.8, with $Q = p$). Thus by Lemma 3.46, $F(b_t)$ is an open ball.

The remaining case is that p is the generic type of a centered open or ∞-definable ball b_1. Thus b_1 contains a definable proper subball b_0. If b is an open subball of p, of radius α, then $b \cap b_0 = \emptyset$; let $\bar{b}$ be the smallest closed ball of containing b and b_0. Then b is contained in the generic type of $\bar{b}$, and so by the case of closed balls, $F(b)$ is an open ball. □

Remark 5.2. When $X \subseteq \mathrm{VF} \times \mathrm{RV}^n$, by a *ball contained in* X we will mean a subset of X of the form $b \times \{e\}$, where $b \in \mathfrak{B}$ and $e \in \mathrm{RV}^n$. With this understanding, the proposition extends immediately to such sets X. Indeed, for each $e \in \mathrm{RV}^n$, according to the proposition there is a finite partition of $X(e)$ with the required property; as in Lemma 2.3 these can be patched to form a single partition of X.

Remark 5.3. When $X \subseteq \mathrm{VF}$ there exists a finite set of points F (not necessarily A-definable) such that $F(b)$ is an open ball whenever b is an open ball disjoint from F. (This does not extend to $X \subseteq \mathrm{VF} \times \mathrm{RV}^*$.)

Indeed, by Proposition 5.1 there is a finite number of closed and open balls b_i and points, such that $F(b)$ is an open ball for any open ball b that is either contained in or is disjoint from each b_i. Now let c_i be a point of b_i. If b is an open ball and no $c_i \in b$, then b must be disjoint from, or contained in, each b_i; otherwise, b contains b_i and hence c_i.

5.2 Images of balls II

Lemma 5.4. *Let X, Y be balls, and $F : X \to Y$ a definable bijection taking open balls to open balls. Then for all $x, x' \in X$,*

$$\mathrm{val}(F(x) - F(x')) = \mathrm{val}(x - x') + v_0,$$

where v_0 is the difference of the valuative radii of X, Y.

Proof. Translating by some $a \in X$ and by $F(a) \in Y$, we may assume $0 \in X, 0 \in Y$, $F(0) = 0$; and by multiplying we may assume and both X, Y have valuative radius 0, i.e., $X = Y = \mathcal{O}$. Let $M(\alpha) = \{x : \mathrm{val}(x) < \alpha\}$. Then $F(M(\alpha)) = M(\beta)$ for some $\beta = \beta(\alpha)$. β is an increasing definable surjection from $\{\alpha \in \Gamma : \alpha > 0\}$ to itself; it must have the form $\beta(\alpha) = m\alpha$ for some rational $m > 0$. By Lemma 3.26, we have $m \in \mathbb{Z}$. Now reversing the roles of X, Y and using F^{-1} will transform m to m^{-1}, so $m^{-1} \in \mathbb{Z}$ also, i.e., $m = \pm 1$. Since $m > 0$, we have $m = 1$. □

Lemma 5.5. *Let X be a transitive open or closed ball (or infinite intersection of balls), and $F : X \to Y$ a definable bijection. Then there exists a definable $e_0 \in \mathrm{RV}$ such that for $x \neq x' \in X$, $\mathrm{rv}(F(x) - F(x')) = e_0 \,\mathrm{rv}(x - x')$.*

Proof. We first show a weaker statement.

Claim. For some definable $e_0 : \Gamma \to \mathrm{RV}$, $\mathrm{rv}(F(x) - F(x')) = e_0(\mathrm{val}(x - x'))\,\mathrm{rv}(x - x')$ for all $x \neq x' \in X$.

Proof. Fix $a \in X$. For $\delta \in \Gamma$, let $b_\delta = b_\delta(a)$, the closed ball around a of valuative radius δ. Consider those b_δ with $b_\delta \subseteq X$. As we saw in the proof of Lemma 5.1, as any $a \in X$ is generic, b_δ is transitive in $\mathbf{T}_{b_\delta}$. By Lemma 3.45, $\mathrm{rv}(F(x) - F(a)) = f_a(\delta)\,\mathrm{rv}(x - a)$, where $\mathrm{val}(x - a) = \delta$, and $f_a(\delta)$ is a function of a and δ. But then f_a is a function $\Gamma \to \mathrm{RV}$, so by Lemma 3.11 it takes finitely many values $v_1, \ldots, v_n$.

Let $Y_i = f_a{}^{-1}(v_i)$. Y_i has a canonical code $e_i \in \Gamma^*$, consisting of the endpoints of the intervals making up Y_i. Using the linear ordering on Γ, each individual e_i is definable from the set $\{e_i\}_i$, and hence from a; thus $v_i = f_a(Y_i)$ is also definable from a. Thus f_a f_a is definable from $(e_i, v_i)_i$. (This last argument could have been avoided by quoting elimination of imaginaries in $\mathrm{RV} \cup \Gamma$.) However, as X is transitive, every definable function $X \to (\mathrm{RV} \cup \Gamma)$ is constant, and so $f_a = f_b$ for any $a, b \in X$. Let $e_0(\delta) = f_a(\delta)$. □

We now have to show that the function e_0 of the claim is constant. Using the O-minimality of Γ, it suffices to show for any definable $\delta \in \mathrm{dom}(e_0)$ that

(1) if $e_0(\delta) = e$, then $e_0(\gamma) = e$ for sufficiently small $\gamma > \delta$,

and if δ is not a minimal element of $\mathrm{dom}(e_0)$), then also

(2) if $e_0(\gamma) = e$ for sufficiently large $\gamma < \delta$, then $e_0(\delta) = e$.

To determine $e_0(\delta)$, it suffices to know $\mathrm{rv}(F(x) - F(x'))$ and $\mathrm{rv}(x - x')$ for one pair x, x' with $\mathrm{val}(x - x') = \delta$. Thus in (1) we may replace X by a closed subball Y of valuative radius δ, and in (2) by any closed subball Y of X of valuative radius $< \delta$. Since such closed balls Y are transitive (over their code), we may assume X is a closed ball.

Fix $a \in X$. Pick a generic c (over a) with $\mathrm{rv}(c) = e$.

To prove (1), note that type of such c is generic in an open ball, whereas the elements of X are generic in a closed ball; these generic types are orthogonal by Lemma 3.19; so X remains transitive in $\mathbf{T}_c$. Thus we may assume (by passing to $\mathbf{T}_c$) that c is definable.

Let q_a be the generic type of the closed ball $\{x : \mathrm{val}(a - x) \geq \delta\}$. For $x \models q_a$, let $v_0 = \mathrm{val}(F(a) - F(x) - c(a - x)) - \mathrm{val}(c)$.

By the definition of e, $\mathrm{val}(F(a) - F(x) - c(a - x)) > \mathrm{val}(F(a) - F(x))$, so we have

$$\begin{aligned} v_0 + \mathrm{val}(c) &= \mathrm{val}(F(a) - F(x) - c(a - x)) > \mathrm{val}(F(a) - F(x)) \\ &= \mathrm{val}(c(a - x)) = \delta + \mathrm{val}(c). \end{aligned} \tag{5.1}$$

If $\delta < \gamma < v_0$, find $x, x' \models q_a$ with $\mathrm{val}(x - x') = \gamma$. Then $\mathrm{val}(F(x) - F(x') - c(x - x')) \geq v_0 + \mathrm{val}_{\mathrm{rv}}(e) > \gamma + \mathrm{val}_{\mathrm{rv}}(e) = \mathrm{val}(c(x - x'))$, so $\mathrm{rv}(F(x) - F(x')) = \mathrm{rv}(c(x - x'))$ showing that $e_0(\gamma) = \mathrm{rv}(c) = e$. This proves (1).

For (2), let $Q_0 = \{\gamma : \gamma < \delta\}$, Q_0^{def} the set of definable elements of Q_0, and $Q = \{\gamma \in Q_0 : (\forall y \in Q_0^{\mathrm{def}})(\gamma > y)\}$. Thus Q is a complete type of elements of Γ. For $\gamma \in Q$, according to Lemma 3.17, the formula $\mathrm{val}(x - a) = \gamma$ generates a complete type $q_{\gamma;a}(x)$. By Lemma 3.47, X is transitive over γ, so the formula $x' \in X$ generates a complete $\mathbf{T}_\gamma$-type. Thus by transitivity a complete $\mathbf{T}_\gamma$-type $q_\gamma(x, x')$ is generated by $x, x' \in X$, $\mathrm{val}(x - x') = \gamma$; namely, $(a, b) \models q_\gamma$ iff $b \models q_{\gamma;a}$.

For some definable v_0, for $(a, x) \models q_\gamma$ we have, as in (1),

$$\mathrm{val}(F(a) - F(x) - c(a - x)) = v_0(\gamma) + \mathrm{val}(c) > \gamma + \mathrm{val}(c). \tag{5.2}$$

If we show that $v_0(\gamma) > \delta$ we can finish as in (1).

Now $v_0(\gamma) = m\gamma + \gamma_0$ for some definable $\gamma_0 \in \Gamma$, and some rational m. Letting $\gamma \to \delta$ in (5.2) gives $m\delta + \gamma_0 \geq \delta$. If $m < 0$, then $v_0(\gamma) = m\gamma + \gamma_0 > m\delta + \gamma_0 \geq \delta$ so we are done; hence we may take $m \geq 0$.

By Lemma 3.47, $\mathrm{RV}(\langle\emptyset\rangle) = \mathrm{RV}(\langle a\rangle)$; by Lemma 3.20, when $x \models q_{\gamma;a}$, $\mathrm{RV}(\langle a, x\rangle)$ is generated over $\mathrm{RV}(\langle a\rangle)$ by $\mathrm{rv}(a - x)$.

In particular, on $q_{\gamma,a}$, $x \mapsto \mathrm{rv}(F(a) - F(x) - c(a - x))$ is a function of $\mathrm{rv}(a - x)$. This function lifts v_0 to a function on RV; hence by Lemma 3.26, $m \in \mathbb{Z}$. (This and $m \geq 0$ are simplifications rather than essential points.) We have

$$\mathrm{val}((F(a) - F(x) - c(a - x))(a - x)^{-m}) = \gamma_0.$$

By Lemma 3.47, $(\mathrm{RV} \cup \Gamma)(\langle a\rangle) = (\mathrm{RV} \cup \Gamma)(\langle\emptyset\rangle)$. By Lemma 3.17, then ${\mathrm{val}_{\mathrm{rv}}}^{-1}(\gamma_0) \cap \mathrm{dcl}(a, x) = {\mathrm{val}_{\mathrm{rv}}}^{-1}(\gamma_0) \cap \mathrm{dcl}(a)$. Thus ${\mathrm{val}_{\mathrm{rv}}}^{-1}(\gamma_0) \cap \mathrm{dcl}(a, x) = {\mathrm{val}_{\mathrm{rv}}}^{-1}(\gamma_0) \cap \mathrm{dcl}(\emptyset)$. Thus $\mathrm{rv}((F(a) - F(x) - c(a - x))(a - x)^{-m}) \in \mathrm{dcl}(\emptyset)$; i.e.,

$$\mathrm{rv}((F(a) - F(x) - c(a - x))(a - x)^{-m}) = e_1$$

for some definable e_1. As in (1), we may assume there exists a definable c_1 with $\mathrm{rv}(c_1) = e_1$. Thus for $(a, x) \models q_\gamma$,

$$\begin{aligned}\mathrm{val}((F(a) - F(x) - c(a - x) - c_1(a - x)^m)) &> \mathrm{val}(F(a) - F(x) - c(a - x)) \\ &= v_0(\gamma) + \mathrm{val}(c).\end{aligned} \tag{5.3}$$

Let $x' \models q_{\gamma,a}$ be generic over $\{\gamma, a, x\}$, so in particular $\mathrm{val}(x - x') = \mathrm{val}(x - a) = \mathrm{val}(a - x') = \gamma$. We have

$$\begin{aligned}\mathrm{val}((F(a) - F(x') - c(a - x') - c_1(a - x')^m)) &> \mathrm{val}(F(a) - F(x) - c(a - x)) \\ &= v_0(\gamma) + \mathrm{val}(c).\end{aligned}$$

Subtracting from (5.3), we obtain

$$\begin{aligned}\mathrm{val}((F(x') - F(x) - c(x' - x) - c_1[(a - x)^m - (a - x')^m]) &> v_0(\gamma) + \mathrm{val}(c) \\ &= \mathrm{val}(c_1(a - x')^m).\end{aligned} \tag{5.4}$$

But since $(x, x') \models q_\gamma$, by (5.3) we have

$$\begin{aligned}\mathrm{val}((F(x) - F(x') - c(x - x') - c_1(x - x')^m)) &> v_0(\gamma) + \mathrm{val}(c) \\ &= \mathrm{val}(c_1(x - x')^m).\end{aligned} \tag{5.5}$$

Comparing (5.4) and (5.5) (and subtracting $\mathrm{val}(c_1)$), we see that

$$\begin{aligned}\mathrm{val}((a - x)^m - (a - x')^m - (x' - x)^m) &> \mathrm{val}((x - x')^m) \\ &= \mathrm{val}((a - x')^m) = \mathrm{val}((a - x)^m).\end{aligned}$$

Let $u = (a - x')/(x' - x)$; then $(a - x)/(x' - x) = u + 1$, $\mathrm{val}(u) = 0 = \mathrm{val}(u + 1)$, and $\mathrm{val}((u + 1)^m - u^m - 1) > 0$. If $U = \mathrm{res}(u)$, we get $(U + 1)^m = U^m + 1$.

Since the residue characteristic is 0 this forces $m = 1$. (Note that U is generic.) Thus $v_0(\gamma) = \gamma + \gamma_0$.

From (5.2), $\gamma + \gamma_0 + \mathrm{val}(c) > \gamma + \mathrm{val}(c)$, or $\gamma_0 > 0$. But $\delta - \gamma_0 \in Q_0^{\mathrm{def}}$, so since $\gamma \in Q$ we have $\gamma > \delta - \gamma_0$, or $v_0(\gamma) = \gamma + \gamma_0 > \delta$. As noted below (5.2) this proves the lemma. □

Remark 5.6. In ACVF(p, p), the claim following Lemma 5.5 remains true, but it is possible for e_0 to take more than one value; consider $x - cx^p$ on a closed ball of valuative radius 0, where $\mathrm{val}(c) < 0$.

Lemma 5.7. *Let X be a transitive open ball, and let $F : X \to X$ be a definable bijection. Then* $\mathrm{rv}(F(x) - F(y)) = \mathrm{rv}(x - y)$ *for all $x \neq y \in X$.*

Proof. This follows from the second assertion in Lemma 3.45 and from Lemma 5.5. □

At this point, Lemma 5.1 may be improved.

Definition 5.8. Call a function G on an open ball *nice* if for some e_0, for all $x \neq x' \in \mathrm{pr}\, X$, $\mathrm{rv}(G(x) - G(x')) = e_0 \,\mathrm{rv}(x - x')$.

Proposition 5.9. *Let X, Y be definable subsets of* VF, *and let $F : X \to Y$ be a definable bijection. Then there exists a partition of X to finitely many definable classes, such that on any open ball b contained in one of the classes, $F(b)$ is an open ball, and $F|b$ is nice.*

Proof. The proof of Proposition 5.1 goes through verbatim, only quoting Lemma 5.5 along with Lemma 3.46. □

A definable translate of a ball $\mathrm{rv}^{-1}(\alpha)$ will be called a *basic* 1*-cell*. Thus Corollary 4.3 states that every fiber of ρ is a basic 1-cell. By a *basic* 2*-cell* we mean a set of the form

$$X = \{(x, y) : x \in \mathrm{pr}\, X, \mathrm{rv}(y - G(x)) = \alpha\},$$

where $\mathrm{pr}\, X$ is a basic 1-cell, and G is nice.

Corollary 5.10. *Let $X \subseteq \mathrm{VF}^2$ be definable. Then there exists a definable $\rho : X \to$ RV* such that every fiber is a basic* 2*-cell.*

Proof. Let $X(a) = \{y : (a, y) \in X\}$. By Corollary 4.3 there exist an a-definable $\rho_a : X(a) \to \mathrm{RV}^*$ and functions c, c' such that every fiber $\rho_a{}^{-1}(\alpha)$ is a basic 1-cell $\mathrm{rv}^{-1}(c'(a, \alpha)) + c(a, \alpha)$. By Lemma 2.3 we can glue these together to a function $\rho_1 : X \to \mathrm{RV}^*$ with $\rho_a(y) = \rho_1(a, y)$. Let $\rho_2(x, y) = (\rho_1(x, y), c'(x, \rho(x, y)))$. Then any fiber D of ρ_2 has the form

$$\{(x, y) : x \in \mathrm{pr}_1\, D, \mathrm{rv}(y - G_D(x)) = \alpha\},$$

where $G_D(x) = c(x, \alpha)$, α depending on the fiber D. Combining ρ_2 with a function whose fibers yield a partition as in Proposition 5.9, we may assume G takes open balls to open balls (cf. Remark 5.2). Now apply Corollary 4.3 to $\mathrm{pr}\, X$ to obtain a map $\rho' : \mathrm{pr}\, X \to \mathrm{RV}^*$ with nice fibers. □

5.3 Limits and continuity

We now assume **T** is a C-minimal theory of valued fields, satisfying assumption (1) of Section 3.4.

Let V be a VF-variety. By "almost all a" we will mean "all a away from a set of smaller VF dimension."

Lemma 5.11. *Let g be a definable function on a ball around* 0. *Then either* $\mathrm{val}\, g(x) \to -\infty$ *as* $\mathrm{val}(x) \to \infty$ *or there exists a unique $b \in$* VF *such that* $b = \lim_{x\to 0, x\neq 0} g(x)$; *i.e.,*

$$(\forall \epsilon \in \Gamma)(\exists \delta \in \Gamma)(0 \neq x \,\&\, \mathrm{val}(x)) > \delta \implies \mathrm{val}(g(x) - b) > \epsilon.$$

Proof. Let p be the generic type of an element of large valuation; so $c \models p|A$ iff $\mathrm{val}(x) > \Gamma(A)$. and let $q = \mathrm{tp}(g(c)/A)$, where $c \models p|A$. By Remark 3.5, q coincides with the generic type of P over A where P is a closed ball, an open ball, or an infinite intersection of balls, or $P = \mathrm{VF}$. The last case means that $\mathrm{val}\, g(x) \to -\infty$. The existence of g shows that p, q are nonorthogonal, so it follows from Lemma 3.19 that the first case is impossible.

We begin by reducing to the case where P is centered. Assume therefore that P is transitive. For $b \in P$, let $q_b = \mathrm{tp}(g(c')/A(b))$, where $c' \models p|A(b)$. If q_b includes a proper b-definable subball P_b of P, or a finite union of such balls, we may take them all to have the same radius $\alpha(b)$; so $\alpha(b)$ is b-definable. By Lemma 3.47, α is constant. If as b varies there are only finitely many balls P_b, then P is after all centered. If not, then there are two disjoint $P_b, P_{b'}$; but this is absurd since if $c'' \models p|A(b, b')$ then $g(c'') \in P_b \cap P_{b'}$. Thus q_b cannot include a proper subball P_b of P; so q_b is just the generic type of P, over $A(b)$. Moving from A to $A(b)$ we may thus assume that P is centered.

Thus P is a centered open or infinitely-definable ball; therefore, it has a proper definable subball b. If $y \notin b$, write $\mathrm{val}(b - y)$ for the constant value of $\mathrm{val}(c - y)$, $c \in b$. By the definition of a generic type of P, $\mathrm{val}(b - g(c)) \notin \Gamma(A)$. Now $\mathrm{val}(b - g(c)) \in \Gamma(A(c)) = \Gamma(A) \oplus \mathbb{Q}\,\mathrm{val}(c)$ (by assumption (2) of the definition of V-minimality (Section 3.4)), and $\mathrm{val}(c) > \Gamma(A)$; it follows that $\mathrm{val}(b - g(c)) < \Gamma(A)$ or $\mathrm{val}(b - g(c)) > \Gamma(A)$. The first case is again the case of $P = \mathrm{VF}$, while the second implies that P is an infinite intersection of balls P_i, whose radius is not bounded by any element of $\Gamma(A)$. In other words, $P = \{b\}$. Unwinding the definitions shows that $b = \lim_{x\to 0, x\neq 0} g(x)$. □

Remark. In reality, the transitive case considered in the proof above cannot occur.

By an (open, closed) polydisc, we mean a product of (open, closed) balls. Let B be a closed polydisc. Let $M \models T$. Let $b \in B(M)$, $a \in B(\mathrm{acl}(\emptyset))$. Write $b \to a$ if for any definable $\gamma \in \Gamma$, and each coordinate i, $\mathrm{val}(b_i - a_i) > \gamma$. Let p_0 be the type of elements of Γ greater than any given definable element. Then Lemma 5.11 can also be stated thusly: given a definable g on a ball B_0 around 0 into B, there exists $b \in \mathrm{dcl}(\emptyset)$ such that if $\mathrm{val}(t) \models p$, then $(t, g(t)) \to (0, b)$.

Stated this way, the lemma generalizes to functions defined on a finite cover of B_0.

Lemma 5.12. *Let B_0 be a ball around 0, and B a closed polydisc, both 0-definable. Let $t \in B_0$ have $\mathrm{val}(t) \models p_0$, and let $a \in \mathrm{acl}(t)$, $a \in B$. Then there exists $b \in B$, $b \in \mathrm{acl}(\emptyset)$ with $(t, a) \mapsto (0, b)$.*

Proof. The proof of Lemma 5.11 goes through. □

The following is an analogue of a result of Macintyre's for the p-adics. By the *boundary* of a set X, we mean the closure minus the interior of X.

Lemma 5.13.

(1) *Any definable $X \subseteq \mathrm{VF}^n$ of dimension n contains an open polydisc.*
(2) *Any definable function $\mathrm{VF}^n \to \mathrm{RV} \cup \Gamma$ is constant on some open polydisc.*
(3) *The boundary of any definable $X \subseteq \mathrm{VF}^n$ has dimension $< n$.*

Proof. Given (1) and (3) follows since the boundary is definable; so it suffices to prove (1)–(2). For a given n, (2) follows from (1): by Lemma 3.56, the fibers of the function cannot all have dimension $< n$.

For $n = 1$, (1) is immediate from C-minimality. Assume that (1)–(2) are true for n and let $X \subseteq \mathrm{VF} \times \mathrm{VF}^n$ be a definable set of dimension $n + 1$. For any $a \in \mathrm{VF}^n$ such that $X_a = \{b : (a, b) \in X\}$ contains an open ball, let $\gamma(a)$ be the infimum of all γ such that X_a contains an open γ-ball. By (2) for n, γ takes a constant value γ_0 on some polydisc U; pick $\gamma_1 > \gamma_0$. Let

$$X' = \{(u, z) \in X : u \in U \& (\forall z')(\mathrm{val}(z - z') > \gamma_0 \implies (u, z') \in X)\}.$$

Then $\dim(X') = n + 1$. Now consider the projection $(u, z) \mapsto z$. For some $c \in \mathrm{VF}$, the fiber $X'_c = \{u : (u, c) \in X'\}$ must have dimension n. By induction, X'_c contains a polydisc V. Now, clearly, $V \times B^o_{\gamma_1}(c) \subseteq X$. □

If $x = (x_1, \dots, x_n)$, $x' = (x'_1, \dots, x'_n)$, write $\mathrm{val}(x - x')$ for $\min \mathrm{val}(x_i - x'_i)$. Say a function F is δ-Lipschitz at x if whenever $\mathrm{val}(x - x')$ is sufficiently large, $\mathrm{val}(F(x) - F(x')) > \delta + \mathrm{val}(x - x')$. Say F is locally Lipschitz on X if for any $x \in X$, for some $\delta \in \Gamma$, F is δ-Lipschitz at x.

Lemma 5.14. *Let $F : X \subseteq \mathrm{VF}^n \to \mathrm{VF}$ be a definable function. Then F is continuous away from a subset X' of dimension $< n$. Moreover, F is locally Lipschitz on $X \setminus X'$.*

Proof. Let X' be the (definable) set of points x where F is not Lipschitz. We must show that X' has dimension $< n$. (In this case, by Lemma 5.13, the closure of X' has dimension $< n$, too.) Suppose otherwise. For $n = 1$ the lemma follows from Lemmas 5.1 and 5.4. Let $\pi_i : X' \subseteq \mathrm{VF}^n \to \mathrm{VF}^{n-1}$ be the projection along the ith coordinate axis. Let Y be the set of $b \in \mathrm{VF}^{n-1}$ such that ${\pi_i}^{-1}(b)$ is infinite or, equivalently, contains a ball; it is a definable set. For $b \in Y$, let

$$D_i(b) = \{x \in {\pi_i}^{-1}(b) : (\exists \delta \in \Gamma)(F|{\pi_i}^{-1}(b) \text{ is } \delta\text{-Lipschitz near } x)\}.$$

By the case $n = 1$, ${\pi_i}^{-1}(b) \setminus D_i(b)$ is finite. Thus if $D_i = \cup_{b \in Y} D_i(b)$, then π_i has finite fibers on $X \setminus D_i$, so $\dim(X \setminus D_i) < n$. Let $X^* = \cap_i D_i$, and for $x \in X^*$ let

$\delta(x)$ be the infinimum of all such Lipschitz constants δ (for all n projections). By Lemma 5.13, δ is constant on some open polydisc $U \subseteq X^*$. Let δ' be greater than this constant value. Then at any $x \in U$, the restriction of F to a line parallel to an axis is δ'-Lipschitz. It follows immediately (using the ultrametric inequality) that F is δ'-Lipschitz on U; but this contradicts the definition of X'. □

Remark 5.15. Via assumption (1) of Section 3.4, we used the existence of p-torsion points in the kernel of $\mathrm{RV} \to \Gamma$ for each p. In ACVF(p, p) this fails; one can still show that F is locally logarithmically Lipschitz, i.e., for some rational $\alpha > 0$, for any $x \in X \setminus X'$, for sufficiently close x', $\mathrm{val}(F(x) - F(x')) > \delta \,\mathrm{val}(x - x')$.

5.4 Differentiation in VF

Let $F : \mathrm{VF}^n \to \mathrm{VF}$ be a definable function, defined on a neighborhood of $a \in \mathrm{VF}^n$. We say that F is differentiable at a if there exists a linear map $L : \mathrm{VF}^n \to \mathrm{VF}$ such that for any $\gamma \in \Gamma$, for large enough $\delta \in \Gamma$, if $\mathrm{val}(x_i) > \delta$ for each i, $x = (x_1, \dots, x_n)$, then $\mathrm{val}(F(a + x) - F(a) - Lx) > \delta + \gamma$. If such an L exists it is unique, and we denote it dF_a.

Lemma 5.16. *Let $F : X \subseteq \mathrm{VF}^n \to \mathrm{VF}^m$ be a definable function. Then each partial derivative is defined at almost every $a \in X$.*

Proof. We may assume $n = m = 1$. Let $g(x) = (F(a + x) - F(a))/x$. By Lemma 5.4, for almost every a, for some $\delta \in \Gamma$, for all x with $\mathrm{val}(x)$ sufficiently large, $\mathrm{val}(F(a + x) - F(a)) = \delta + \mathrm{val}(x)$; so $\mathrm{val}\, g(x)$ is bounded. By Lemma 5.11, and Proposition 5.1, $g(x)$ approaches a limit $b \in \mathrm{VF}$ as $x \to 0$ (with $x \neq 0$); the lemma follows. □

Corollary 5.17. *Let $F : \mathrm{VF}^n \to \mathrm{VF}$ be a definable function. Then F is continuously differentiable away from a subset of dimension $< n$.*

Proof. F has partial derivatives almost everywhere, and these are continuous almost everywhere, so the usual proof works. □

Lemma 5.18. *Let $X \subseteq \mathrm{VF}^n \times \mathrm{RV}^m$ be definable,* $\mathrm{pr} : X \to \mathrm{VF}^n$ *the projection. Then for almost every $p \in \mathrm{VF}^n$, there exists an open neighborhood U of p and $H \subseteq \mathrm{RV}^m$ such that* $\mathrm{pr}^{-1}(U) = U \times H$. *If $h : X \to \mathrm{VF}$, then for almost all $x \in X$, h is differentiable with respect to each* VF*-coordinate.*

Proof. For $x \in \mathrm{VF}^n$, let $H(x) = \{h \in \mathrm{RV}^m : (x, h) \in X\}$. By Corollary 3.24, Lemma 2.8, there exists $H' \subseteq \mathrm{RV}^m \times \mathrm{RV}^l \times \Gamma^k$ such that for any $x \in \mathrm{VF}^n$, there exists a unique $y = f(x) \in \mathrm{RV}^l \times \Gamma^k$ with $H(x) = H'(y)$. By Lemma 5.13, f is locally constant almost everywhere. Thus for almost all x, for some neighborhood U of x, for all $x' \in U$, $H(x) = H(x')$; so $\mathrm{pr}^{-1} U = U \times H(x)$. The last assertion is immediate. □

We can now define the partial derivatives of any definable map $F : X \to \mathrm{VF}$ (almost everywhere); we just take them with respect to the VF-coordinates, ignoring the RV-coordinates.

Given $h : X \to \mathrm{VF}^n$, $h' : X' \to \mathrm{VF}^n$ with RV-fibers, and a definable map $F : X \to X'$, we define the partials of F to be those of $h' \circ F$. Then the differential dF_x exists at almost every point $x \in X$ by Corollary 5.17, and we denote the determinant by Jcb, and refer to it as usual as the Jacobian.

Definition 5.19. Let $X, X' \in \mathrm{VF}[n, \cdot]$ and let $F : X \to X'$ be a definable bijection. F is *measure preserving* if $\mathrm{rv}\,\mathrm{Jcb}(x) = 1$ for almost all $x \in X$. $\mathrm{VF}_{\mathrm{vol}}[n, \cdot]$ is the subcategory of $\mathrm{VF}[n, \cdot]$ with the same objects, and whose morphisms are the measure-preserving morphisms of $\mathrm{VF}[n, \cdot]$.

Let $\mathrm{VF}_{\mathrm{vol}}$ be the category whose objects are those of $\mathrm{VF}[n, \cdot]$, and whose morphisms $X \to Y$ are the essential bijections $f : X \to Y$ that are measure preserving.

5.5 Differentiation and Jacobians in RV

Let X, Y be definable sets, together with finite-to-one definable maps $f_X : X \to \mathrm{RV}^n$, $f_Y : Y \to \mathrm{RV}^n$. Here X, Y can be subsets of RV^* or of $\mathrm{RV}^* \times \mathrm{VF}^*$, etc.; the notion of Jacobian will not depend on the particular realization of X, Y.

Let $h : X \to Y$ be a definable map.

The notion of Jacobian will depend not only on h, X, Y but also on f_X, f_Y; to emphasize this we will write $h : (X, f_X) \to (Y, f_Y)$.

We first define smoothness. When $A = f_X(X)$, $B = f_Y(Y)$ are definable subsets of $\mathbf{k}^n$, we say that h, X, Y are smooth if A, B are Zariski open, $\{(f_X(x), f_Y(h(x))) : x \in X\} \cap (A \times B) = Z$ for some nonsingular Zariski closed set $Z \subset A \times B$, and the differentials of the projections to A and to B are isomorphisms at any point $z \in Z$. In this case, composing the inverse of one of these differentials with the other, we obtain a linear isomorphism $T_a(A) \to T_b(B)$ for any $a = f_X(x), b = f_Y(h(x))$; since $T_a(A) = k^n = T_b(B)$, this linear isomorphism is given by an invertible matrix, whose determinant is the Jacobian J.

In general, to define smoothness of X, Y at $(x, y = h(x))$, we restrict to the cosets of $(\mathbf{k}^*)^n$ containing x and y, translate multiplicatively by x and y, respectively, and pose the same condition.

Any X, Y, h are smooth outside of a set E, where $E \cap C$ has dimension $< n$ for each coset C of $(\mathbf{k}^*)^n$. Equivalently (by Lemma 3.64), E has RV-dimension $< n$.

Assume now that X, Y, h are smooth. Define

$$\mathrm{Jcb}_{\mathrm{RV}}(h)(q) = \Pi(f_X(q))^{-1}\Pi(f_Y(q'))J(1, 1) \in \mathrm{RV},$$

where $\Pi(c_1, \ldots, c_n) = c_1 \cdot \cdots \cdot c_n$.

At times it is preferable not to use a different translation at each point of a coset of $(\mathbf{k}^*)^n$. The Jacobian $\mathrm{Jcb}_{\mathrm{RV}}(h)$ of h at $q \in X$ can also be defined as follows. Let $q' = h(q)$, $\gamma = \mathrm{val}_{\mathrm{rv}}(q)$, $\gamma' = \mathrm{val}_{\mathrm{rv}}(q') \in \Gamma^n$. Pick any $c, d \in \mathrm{RV}^n$ with $\mathrm{val}_{\mathrm{rv}}(c) = \gamma$, $\mathrm{val}_{\mathrm{rv}}(d) = \gamma'$ (one can take $c = f_X(q), d = f_Y(q')$). Let

$$W(\gamma, \gamma') = \{a : f_X(a) \in \mathrm{val_{rv}}^{-1}(\gamma), f_Y(h(a)) \in \mathrm{val_{rv}}^{-1}(\gamma')\},$$
$$H' = \{(c^{-1} f_X(a), d^{-1} f_Y(h(a))) : a \in W\}.$$

Since f_X, f_Y are finite-to-one, $H' \subset (\mathbf{k}^{*n})^2$ both projections of H' to $\mathbf{k}^{*n}$ are finite-to-one, and H' is nonsingluar by the smoothness of (X, Y, h). We can thus define the Jacobian J' of H' at any point. We have

$$\mathrm{Jcb_{RV}}(h)(q) = \Pi(c)^{-1}\Pi(d)J'(qc^{-1}, q'd^{-1}) \in \mathrm{RV}.$$

We also define $\mathrm{Jcb}_\Gamma(h)(q) = \sum \gamma' - \sum \gamma \in \Gamma$ (writing Γ additively). *Note that this depends only on the value of h at q.* We have

$$\mathrm{val_{rv}}\, \mathrm{Jcb_{RV}}(h)(q) = \mathrm{Jcb}_\Gamma(h)(q).$$

Example 5.20. Jacobian of maps on Γ. If $\bar{X}, \bar{Y} \subset \Gamma^n$, we saw that a definable map $\bar{f} : \bar{X} \to \bar{Y}$ lifts to RV iff it is piecewise given by an element of $\mathrm{GL}_n(\mathbb{Z})$ composed with a translation. Assume $\bar{f}$ is given by a matrix $M \in \mathrm{GL}_n(\mathbb{Z})$, let $X = \mathrm{val_{rv}}^{-1}(\bar{X})$, $Y = \mathrm{val_{rv}}^{-1}(\bar{Y})$, and let $f : X \to Y$ be given by the same matrix, but multiplicatively. Then X, Y, f are smooth, and

$$J(f)(x) = \Pi(y)\Pi(x)^{-1} \det M,$$

where $y = f(x)$, and $\det(M) = \pm 1$.

Alternative: Γ-weighted polynomials

We have seen that the geometry on $\mathrm{val_{rv}}^{-1}(\gamma)$ $(\gamma \in \Gamma^n)$ translates to the geometry on $(\mathbf{k}^*)^n$, but this is true for the general notions and not for specific varieties; a definable subset of $C(\gamma) = \mathrm{val}^{-1}(\gamma)$ does not correspond canonically to any definable subset of $\mathrm{val}^{-1}(0)$. An invariant approach is therefore useful. Let $\Gamma_0 = \Gamma(\langle\emptyset\rangle)$. $X = (X_1, \ldots, X_n)$ be variables, $\gamma = (\gamma_1, \ldots, \gamma_n) \in \Gamma_0^n$, and let $\nu = (\nu(1), \ldots, \nu(n)) \in \mathbb{N}^n$ denote a multi-index. By a γ-weighted monomial we mean an expression $a_\nu X^\nu$ with a_ν a definable element of RV, such that $\mathrm{val_{rv}}(a_\nu) + \sum \nu(i)\gamma_i = 0 \in \Gamma$. Let $\mathrm{Mon}(\gamma, \nu)$ be the set of γ-weighted monomials of exponent ν, together with 0. Then $\mathrm{Mon}(\gamma, \nu) \setminus \{0\}$ is a copy of $\mathrm{val_{rv}}^{-1}(-(a_\nu) + \sum \nu(i)\gamma_i)$; so $\mathrm{Mon}(\gamma, \nu)$ is a one-dimensional $\mathbf{k}$-space. In particular, addition is defined in $\mathrm{Mon}(\gamma, \nu)$. We also have a natural multiplication $\mathrm{Mon}(\gamma, \nu) \times \mathrm{Mon}(\gamma, \nu') \to \mathrm{Mon}(\gamma, \nu + \nu')$. Let $R[X; \gamma] = \oplus_{\nu \in \mathbb{N}^n} \mathrm{Mon}(\gamma, \nu)$. This is a finitely generated graded $\mathbf{k}$-algebra. It may be viewed as an affine coordinate ring of $C[\gamma]$; but the ring of the product $C[\gamma, \gamma']$ is $R[X, X'; (\gamma, \gamma')]$, in general a bigger ring than $R[X, \gamma] \otimes_{\mathbf{k}} R[X', \gamma']$. Nevertheless, a Zariski closed subset of $C(\gamma)$ corresponds to a radical ideal of $R[X'; \gamma]$. In this way, notions such as smoothness may be attributed to closed or constructible subsets of any $C(\gamma)$ in an invariant way.

Definition 5.21. Let $X, Y \in \mathrm{Ob}\,\mathrm{RV}[n, \cdot]$ and let $h : X \to Y$ be a definable bijection. h is *measure preserving* if $\mathrm{Jcb}_\Gamma\, h(x) = 0$ for all $x \in X$, and $\mathrm{Jcb_{RV}}\, h(x) = 1$ for all

$x \in X$ away from a set of RV dimension $< n$. If only the first condition holds, we say h is Γ-measure preserving.

For $X, Y \in \mathrm{RV}[\leq n, \cdot]$, we say that $h : X \to Y$ is measure preserving if this is true of the $\mathrm{RV}[n]$-component of h.

$\mathrm{RV}_{\mathrm{vol}}[n, \cdot]$ (respectively, $\mathrm{RV}_{\Gamma\text{-vol}'}[n, \cdot]$) is the subcategory of $\mathrm{RV}[n, \cdot]$ with the same objects, and whose morphisms are the measure-preserving (respectively, Γ-measure-preserving) definable bijections.

$\mathrm{RV}_{\mathrm{vol}}[\leq n, \cdot] = \oplus_{k<n} \mathrm{RV}_{\Gamma\text{-vol}'}[k, \cdot] \oplus \mathrm{RV}_{\mathrm{vol}}[n, \cdot]$.

Note that when $X, Y \in \mathrm{Ob}\,\mathrm{RV}[n, \cdot]$, a bijection $h : X \to Y$ is Γ-measure preserving iff it leaves invariant the sets $S_\gamma = \{(a_1, \ldots, a_n) : \sum_{i=1}^n \mathrm{val}_{\mathrm{rv}}(a_i) = \gamma\}$.

5.6 Comparing the derivatives

Consider a definable function $F : \mathrm{VF} \to \mathrm{VF}$ lying above $f : \mathrm{RV} \to \mathrm{RV}$, i.e., $\mathrm{rv}\, F = f\, \mathrm{rv}$. The fibers of the map $\mathrm{rv} : \mathrm{VF} \to \mathrm{RV}$ above $\mathbf{k}$, for instance, are open balls of valuative radius 0, whereas the derivative is defined on the scale of balls of radius r for $r \to +\infty$. Thus the comparison between the derivatives of F and f is not tautological. Nevertheless, one obtains the expected relation almost everywhere.

While this case of the affine line would suffice (using the usual technique of partial derivatives), it is easier to place oneself in the more general context of curves. More precisely, we consider definable sets C together with finite-to-one maps $f : C \to \mathrm{RV}$. Let $\mathbb{L}C$ and $\rho : \mathbb{L}C \to C$ be as above.

In the following lemma, H', h' denote, respectively, the VF-, RV derivatives of functions H, h defined on objects of VF[1], RV[1], respectively.

Proposition 5.22. *Let $C_i \subseteq \mathrm{RV}^*$ be definable sets, $f_i : C_i \to \mathrm{RV}$ finite-to-one definable maps ($i = 1, 2$). Let $h : C_1 \to C_2$ be a definable bijection, and let $H : \mathbb{L}C_1 \to \mathbb{L}C_2$ be a lifting of h, i.e., $\rho H = h\rho$. Then we have the following:*

(1) *For all but finitely many $c \in C_1$, h is differentiable at c, H is differentiable at any $x \in \mathbb{L}c$, and* $\mathrm{rv}\, H'(x) = h'(\mathrm{rv}(x))$.
(2) *For all $c \in C_1$, H is differentiable at a generic $x \in \mathbb{L}c$, and* $\mathrm{val}\, H'(x) = (\mathrm{val}_{\mathrm{rv}} h')(x) = \mathrm{val}(f_2(h(x))) - \mathrm{val}(f_1(x))$.

Proof.

(1) Let Z' be the set of $x \in \mathbb{L}C_1$ such that H is not differentiable at x (a finite set) or that $\mathrm{rv}(H'(x)) \neq h'(\mathrm{rv}(x))$. We have to show that $\rho(Z') \subseteq C$ is finite or, equivalently, that $f_1 \circ \rho(Z')$ is finite. Otherwise, there exists $c \in \rho(Z')$ with $c \notin \mathrm{acl}(A)$. By Lemma 3.20, the formula $\mathrm{rv}(x) = f_1(c)$ generates a complete type q over $A(c)$; it defines a transitive open ball b_c over $A(c)$. Since $\rho \circ H = \rho \circ h$, we have $H(c, y) = (c, H_c(y))$ for some $A(c)$-definable bijection H_c of b_c. By Lemma 5.5, for some $e_0 \in \mathrm{RV}$, $\mathrm{rv}(H(u) - H(v)) = e_0\, \mathrm{rv}(u - v)$ for all $u, v \in b_c$; so $\mathrm{rv}((H(u) - H(v))/(u - v)) = e_0$. Since H is differentiable almost everywhere on b_c (Lemma 5.17) and b_c is transitive, it is differentiable at every point. Clearly, $\mathrm{rv}\, H'(u) = e_0$, contradicting the definition of Z'.

(2) This follows from Lemma 5.4. □

Corollary 5.23. *Let* $\mathbf{X} \in \mathrm{Ob}\,\mathrm{RV}[n]$, $F : \mathbb{L}\mathbf{X} \to \mathrm{VF}^n$ *a definable function,* $f : \mathbb{L}\mathbf{X} \to \mathrm{RV}^n$ *a definable function. Assume* $\mathrm{rv}\, F(x) = f(\mathrm{rv}(x))$. *Then Proposition* 5.22 *applies for each partial derivative of* F. *In particular,*

- *for all* $c \in X$ *away from a set of smaller dimension, for all* $x \in \mathbb{L}c$, F *is differentiable at* x, f *is differentiable at* c, *and* $\mathrm{rv}\,\mathrm{Jcb}(F)(x) = \mathrm{Jcb}_{\mathrm{RV}}(f)(x)$;
- *for all* $c \in X$, *for generic* $x \in \mathbb{L}c$, F *is differentiable at* x, *and* $\mathrm{val}\,\mathrm{Jcb}(F)(x) = (\mathrm{Jcb}_\Gamma f)(x)$. □

Corollary 5.24. *Let*

$$\mathbf{X}, \mathbf{Y} \in \mathrm{Ob}\,\mathrm{RV}[\leq n], \qquad f \in \mathrm{Mor}_{\mathrm{RV}[\leq n]}(\mathbf{X}, \mathbf{Y}), \qquad F \in \mathrm{Mor}_{\mathrm{VF}_{\mathrm{vol}}[n]}(\mathbb{L}\mathbf{X}, \mathbb{L}\mathbf{Y}).$$

Assume $\mathrm{rv}\, F(x) = f(\mathrm{rv}(x))$. *Then* $f \in \mathrm{Mor}_{\mathrm{RV}_{\mathrm{vol}}[n]}(\mathbf{X}, \mathbf{Y})$. □

Proof. The proof follows from Corollary 5.23. □

6 Lifting functions from RV to VF

Proposition 6.1. *Let* $\mathbf{T}$ *be an effective* V*-minimal theory. Let* $X \subset \mathrm{RV}^k$ *be definable and let* $\phi_1, \phi_2 : X \to \mathrm{RV}^n$ *be two definable maps with finite fibers. Then there exists a definable bijection* $F : X \times_{\phi_1,\mathrm{rv}} (\mathrm{VF}^\times)^n \to X \times_{\phi_2,\mathrm{rv}} (\mathrm{VF}^\times)^n$, *commuting with the natural projections to* X.

Proof. Let $A = \mathrm{dcl}(\emptyset) \cap (\mathrm{VF} \cup \Gamma)$. If $b \in \mathrm{dcl}(\emptyset) \cap \mathrm{RV}$, then viewed as a ball b has a point $a \in A$; since the valuative radius of b is also in A, we have $b \in \mathrm{dcl}(A)$. Thus ϕ_1, ϕ_2, X are ACVF_A-definable. Any ACVF_A-definable bijection F is a fortiori $\mathbf{T}$-definable; so the proposition for ACVF_A implies the proposition for $\mathbf{T}$. Moreover, ACVF_A is V-minimal and effective, since any algebaic ball of ACVF_A is $\mathbf{T}_A$-algebraic and hence has a point in $\mathrm{VF}(A)^{\mathrm{alg}}$. Thus we may assume $\mathbf{T} = \mathrm{ACVF}_A$.

The proof will be asymmetric, concentrating on $\phi_1 X$.

We may definably partition X, and prove the proposition on each piece.

Consider first the case where $\phi_1 : X \to U$ and $\phi_2 : X \to V$ are bijections to definable subsets $U, V \subseteq (\mathbf{k}^*)^k$. Our task is to lift the bijection $f = \phi_2 {\phi_1}^{-1}$ to VF^n. A definable subset of $\mathbf{k}^n$ (such as $\phi_i(X)$) is a disjoint union of smooth varieties. We thus consider a definable bijection $f : U \to V$ between $\mathbf{k}$ varieties $U \subset \mathbf{k}^n$ and $V \subset \mathbf{k}^n$. Induction on $\dim(U)$ will allow us to remove a subset of U of smaller dimension. Hence we may assume U is smooth, cut out by $h = (h_1, \dots, h_l)$, $TU = \mathrm{Ker}(dh)$, $f = (f_1, \dots, f_n)$, where f_i are regular on U (defined on an open subset of $\mathbf{k}^n$), and df is injective on TU at each point of U. Thus the common kernel of $dh_1, \dots, dh_l, df_1, \dots, df_n$ equals 0.

It follows that at a generic point of U (i.e., every point outside a proper subvariety), if Q is a sufficiently generic $n \times l$ matrix of elements of A (or integers) and we let $f_i' = f_i + Qh$, then the common kernel of $df_1', \dots, df_n'$ vanishes. Note that

$f_i|U = f_i'|U$. Let W be a smooth variety contained in $f(U)$ and whose complement in $f(U)$ is a constructible set of dimension smaller than $\dim(U)$. Replacing U by $f^{-1}(W)$, we may assume $f(U)$ is also a smooth variety.

Let $\tilde{U} = \mathrm{res}^{-1}(U)$. Lift each f_i' to a polynomial F_i over $\mathcal{O}$, with definable coefficients. This is possible by effectiveness of ACVF_A. Obtain a regular map F, whose Jacobian is invertible at points of $\tilde{U}$. We have $\mathrm{res} \circ F = f \circ \mathrm{res}$. Since f is 1-1 on U, the invertibility of dF implies that F is 1-1 on $\tilde{U}$. Moreover, by Hensel's lemma, $F : \mathrm{rv}^{-1}(U) \to \mathrm{rv}^{-1}(W)$ is bijective.

Next consider the case where in place of a bijection $f : U \to V$ we have a finite-to-finite correspondence $\tilde{f} \subset U \times V$ (where $U = \phi_1(X), V = \phi_2(X)$), $\tilde{f} = \{(\phi_1(x), \phi_2(x)) : x \in X\}$. We may take $\tilde{f} \subset U \times V$ to be a subvariety, unramified and quasi-finite over U and over V; and we can take U, V to be smooth varieties. As before we can lift $\tilde{f}$ to a correspondence $\widetilde{F} \subset \widetilde{U} \times \widetilde{V}$, such that $\widetilde{F} \cap \mathrm{rv}^{-1}(u) \times \mathrm{rv}^{-1}(v)$ is a bijection $\mathrm{rv}^{-1}(u) \to \mathrm{rv}^{-1}(v)$ whenever $(u, v) \in \tilde{f}$. It follows that a bijection $X \times_{\phi_1,\mathrm{rv}} (\mathrm{VF}^\times)^n \to X \times_{\phi_2,\mathrm{rv}} (\mathrm{VF}^\times)^n$ is given by $(x, y) \mapsto (x, y')$ iff $(y, y') \in \widetilde{F}$.

Let $\phi_1 : X \to U$ and $\phi_2 : X \to V$ be bijections to definable subsets U, V, each contained in a single coset of $(\mathbf{k}^*)^k$, say, $U \subseteq C(\gamma)$, $V \subseteq C(\gamma')$ for some $\gamma, \gamma' \in \Gamma^k$ (cf. Section 5.5). Let $Z = (Z_1, \ldots, Z_k)$ be variables, $R[Z; \gamma]$ be the subring of $\mathrm{VF}[Z]$ consisting of polynomials $\sum a_\nu Z^\nu$, with $\mathrm{val}(a_\nu) + \sum_{i=1}^k \nu(i)\gamma_i = 0$, and a_ν a definable element of VF. There is a natural homomorphism $R'[Z; \gamma] \to R[Z; \gamma]$, where $R[Z; \gamma]$ is the coordinate ring of $C(\gamma)$. By effectivity, this homomorphism is surjective. The proof now proceeds in exactly the same way as above.

This proves the proposition in case $\mathrm{val}_{\mathrm{rv}}\phi_i(X)$ consists of one point.

Next, assume $\mathrm{val}_{\mathrm{rv}}\phi_2$ consists of one point, and $\mathrm{val}_{\mathrm{rv}}\phi_1(X)$ is finite. Thus $\phi_1(X)$ lies in the union of finitely many cosets $(C(a) : a \in E)$, with E finite.

For $a \in E$, $A(a)$ remains almost VF, Γ-generated; since the proposition is true for ${\phi_1}^{-1}C(a)$ (definable in $\mathbf{T}_{A(a)}$), then by the one-coset case an appropriate isomorphism F exists; and the finitely many F obtained in this way can then be glued together, to yield a map defined over A.

The case of $\mathrm{val}_{\mathrm{rv}}\phi_1$, $\mathrm{val}_{\mathrm{rv}}\phi_2$ both finite, is treated similarly.

This proves the existence of a lifting in case $\mathrm{val}_{\mathrm{rv}}\phi_i(X)$ is finite. Now for the general case.

Claim. Let $P \subset X$ be a complete type. Then there exists a definable D with $P \subset D \subset X$, and definable functions θ on $\mathrm{val}_{\mathrm{rv}}(\phi_1(D))$ and θ' on $\mathrm{val}_{\mathrm{rv}}(\phi_2(D))$ such that for $x \in D$, $\theta(\mathrm{val}_{\mathrm{rv}}(\phi_1(x))) = \mathrm{val}_{\mathrm{rv}}\phi_2(x)$, $\theta'(\mathrm{val}_{\mathrm{rv}}(\phi_2(x))) = \mathrm{val}_{\mathrm{rv}}\phi_1(x)$.

Proof. Let $a \in P$, $\gamma_i = \mathrm{val}_{\mathrm{rv}}(\phi_i(a))$. Then γ_2 is definable over some points of $\phi_1^{-1}{\mathrm{val}_{\mathrm{rv}}}^{-1}(\gamma_1)$. But ${\mathrm{val}_{\mathrm{rv}}}^{-1}(\gamma_1)$ is a coset of $\mathbf{k}^*$, and ϕ_1 is finite-to-one, so $\phi_1^{-1}{\mathrm{val}_{\mathrm{rv}}}^{-1}(\gamma_1)$ is orthogonal to Γ. Thus γ_2 is algebraic over γ_1. Since Γ is linearly ordered, γ_2 is definable over γ_1; so $\gamma_2 = \theta(\gamma_1)$ for some definable θ. Similarly, $\gamma_1 = \theta'(\gamma_2)$. Clearly, θ restricts to a bijection $\mathrm{val}_{\mathrm{rv}}\phi_1 P \to \mathrm{val}_{\mathrm{rv}}\phi_2 P$, with inverse θ'. By Lemma 2.7 there exists a definable D with $\theta\phi_1 = \phi_2$, $\phi_1 = \theta'\phi_2$ on D. □

Now by compactness, there exist finitely many $(D_i, \theta_i, \theta_i')$ as in the claim with $\cup_i D_i = X$. We may cut down the D_i successively, so we may assume the union is disjoint. But in this case the proposition reduces to the case of each individual D_i, so we may assume $X = D$. Let $B_i = \mathrm{val}_{\mathrm{rv}}\phi_i(X)$. Given $b \in B_1$, let $X_b = (\mathrm{val}_{\mathrm{rv}}\phi_1)^{-1}(b)$. Then by the case already considered there exists an $A(b)$-definable $F_b : X_b \times_{\phi_1,\mathrm{rv}} (\mathrm{VF}^\times)^n \to X_b \times_{\phi_2,\mathrm{rv}} (\mathrm{VF}^\times)^n$. Let $F = \cup_{b \in B_1} F_b$. By Lemma 2.3, $F : X \times_{\phi_1,\mathrm{rv}} (\mathrm{VF}^\times)^n \to X \times_{\phi_2,\mathrm{rv}} (\mathrm{VF}^\times)^n$ is bijective (see the discussion in Section 2.1). □

We note a corollary.

Lemma 6.2. *Let* $\mathbf{T}$ *be* V*-minimal and effective, and let* A *be an almost* (VF, Γ)*-generated structure. Then* A *is effective.*

Proof. By Lemma 3.29 it suffices to show A is rv-effective. Note that if $A \subseteq \mathrm{acl}(\emptyset)$, then $\mathbf{T}$ is rv-effective iff $\mathbf{T}_A$ is rv-effective (see the proof of Lemma 3.31(2)–(3)). Thus it suffices to show that if $A_0 = \mathrm{acl}(A_0)$, $a \in \mathrm{VF} \cup \Gamma$, and $\mathbf{T}' = \mathbf{T}_{A_0}$ is effective, then so is $\mathbf{T}'(a)$. The case $a \in \Gamma$ is included in Corollary 3.40, so assume $a \in \mathrm{VF}$. Let P be the intersection of all A_0-definable balls containing a. If P is transitive over A_0, then by Lemma 3.47 we have $\mathrm{RV}(A_0(a)) = \mathrm{RV}(A_0)$, so rv-effectivity remains true trivially. Otherwise, P is centered over A_0, hence has an A_0-definable point, and by translation we may assume $0 \in P$. a is then a generic point of P over A_0. Let $c \in \mathrm{RV}(A_0(a))$; we must show that $\mathrm{rv}^{-1}(c)$ is centered over $A_0(a)$. By Lemma 3.20, if $c \in \mathrm{RV}(A_0(a))$ then $c = f(d)$ for some A_0-definable function $f : \mathrm{RV} \to \mathrm{RV}$, where $d = \mathrm{rv}(a)$. By Lemma 6.1 there exists an A_0-definable function $F : \mathrm{VF} \to \mathrm{VF}$ lifting f. Then $F(d) \in \mathrm{rv}^{-1}(c)$. □

Base change: Summary

Base change from $\mathbf{T}$ to $\mathbf{T}_A$ preserves V-minimality, effectiveness and being resolved, if A is VF-generated; V-minimality and effectiveness, if A is RV-generated; V-minimality, if A is Γ-generated. (Lemmas 6.2, 3.39, and 3.40; the resolved case follows using Lemma 3.49).

Though the notion of a morphism $g : (X_1, \phi_1) \to (X_2, \phi_2)$ does not depend on ϕ_1, ϕ_2, recall that the RV-Jacobian of g is defined with reference to these finite-to-one maps.

Lemma 6.3. *Let* $\mathbf{T}$ *be* V*-minimal and effective. Let* $X_i \subset \mathrm{RV}^{k_i}$ *be definable and let* $\phi_i : X \to \mathrm{RV}^n$ *be definable maps with finite fibers; let* $g : X_1 \to X_2$ *be a definable bijection. Assume given, in addition, a definable function* $\delta : X_1 \to \mathrm{RV}$*, such that*

(1) $\mathrm{val}_{\mathrm{rv}}\delta(x) = \mathrm{Jcb}_\Gamma\, g(x)$ *for all* $x \in X_1$;
(2) $\delta(x) = \mathrm{Jcb}_{\mathrm{RV}}\, g(x)$ *for almost all* $x \in X_1$ *(i.e., all* x *outside a set of dimension* $< n$*).*

Then there exists a definable bijection $G : X_1 \times_{\phi_1,\mathrm{rv}} (\mathrm{VF}^\times)^n \to X_2 \times_{\phi_2,\mathrm{rv}} (\mathrm{VF}^\times)^n$ *such that* $\rho_2 \circ G = g \circ \rho_1$*, where* ρ_i *are the natural projections to the* X_i*, and such that for any* $x \in X_1 \times_{\phi_1,\mathrm{rv}} (VF^*)^n$*,* G *is differentiable at* x*, and* $\mathrm{rv}(\mathrm{Jcb}(G)(x)) = \delta(x)$*.*

Proof. We follow closely the proof of Proposition 6.1. As there, we may assume $\mathbf{T} = \mathrm{ACVF}_A$, with A be an almost (VF, Γ)-generated substructure.

We first assume that $\mathrm{val}_{\mathrm{rv}}\phi_1(X_1)$ is a single point of Γ^n

Then X_1 can be definably embedded into $\mathbf{k}^N$ for some N, and it follows from the orthogonality of $\mathbf{k}$ and Γ that the image of X_1 in Γ under any definable map is finite. Thus $\phi_2 X_2$ is contained in finitely many cosets $(C(a) : a \in S)$ of $(\mathbf{k}^*)^n$; by partitioning X_1 working in $\mathbf{T}_{A(a)}$, we may assume $\phi_2 X_2$ is contained in a single coset (cf. Lemma 2.3).

As in Proposition 6.1, we may assume $\phi_i X \subseteq \mathbf{k}^n$, and, indeed, that $\phi_1 X = U, \phi_2 X = V$ are smooth varieties. If $\dim(U) = n$, then the lift constructed in Proposition 6.1 satisfies $\mathrm{rv}(\mathrm{Jcb}(G))(x) = \mathrm{Jcb}_{\mathrm{RV}}\, g(x)$ for $x \in X \times_{\phi_1,\mathrm{rv}} \mathrm{VF}^n$; thus by assumption (2), we have $\mathrm{rv}(\mathrm{Jcb}(G))(x) = \delta(x)$ for almost all x. The exceptional points have dimension $< n$, and may be partitioned into smooth varieties of dimension $< n$. Thus we are reduced to the case $\dim(U) < n$. We prove it by induction on $\dim(U)$. In this case choose any lifting G_0. We have an error term $e(x) = \mathrm{rv}(\mathrm{Jcb}(G_0))(x)^{-1}\delta(x)$. Now $A(x)$ is almost VF, Γ-generated, and so balls $\mathrm{rv}^{-1}(y)$ contain definable points; thus $e(x) = \mathrm{rv}\, E(x)$ for some definable $E : (X \times_{\phi_1,\mathrm{rv}} \mathrm{VF}^n) \to \mathrm{VF}$. Since U is a smooth subvariety of $\mathbf{k}^n$ of positive codimension, some regular h on $\mathbf{k}^n$ vanishes on V, while some partial derivative (say, h_1) vanishes only on a lower-dimensional subvariety. By induction, one may assume h_1 vanishes nowhere. Lift h to H; so H_1 lifts h_1. Compose G_0 with a map fixing all coordinates but the first, and multiplying the first coordinate by $E(x)H(y)/H_1(y)$. (Here $x = g^{-1}(y)$.) Where h vanishes, this has Jacobian $E(x)$; so the composition has RV-Jacobian $\delta(x)$ as required.

Now in general, for any $\gamma \in \Gamma^n$ let $X_1(\gamma) = \{x \in X_1 : \mathrm{val}_{\mathrm{rv}}\phi_1(x) = \gamma\}$, $X_2(\gamma) = g(X_1(\gamma))$. By the definitions of $\mathrm{Jcb}_{\mathrm{RV}}$ and Jcb_Γ, $\mathrm{Jcb}_{\mathrm{RV}}(g|X_2(\gamma)) = \mathrm{Jcb}_{\mathrm{RV}}(g)|X_2(\gamma)$ and likewise Jcb_Γ. By the case already analyzed (for the sets $X_1(\gamma), X_2(\gamma)$ defined in $\mathrm{ACVF}_{A(\gamma)}$) there exists an $A(\gamma)$-definable bijection $G_\gamma : X_1(\gamma) \times_{\phi_1,\mathrm{rv}} (VF^\times)^n \to X_2(\gamma) \times_{\phi_2,\mathrm{rv}} (VF^\times)^n$ with $\mathrm{rv}(\mathrm{Jcb}(G_\gamma)(x)) = \delta(x)$. As in Lemma 2.3 one can extend the G_γ by compactness to definable sets containing γ, cover X_1 by finitely many such definable sets, and glue together to obtain a single function G with the same property. □

Remark. Assume $\mathrm{Id}_X : (X, \phi_1) \to (X, \phi_2)$ has Jacobian 1 everywhere. Then it is possible to find F that is everywhere differentiable, of Jacobian precisely equal to 1. At the before the point where Hensel's lemma is quoted, it is possible to multiply the function by $J(F)^{-1}$ (not effecting the reduction, since $J(F) \in 1 + \mathcal{M}$). Then one obtains on each such coset a function of Jacobian 1 and therefore globally.

Example. Let $\phi_2(x) = \phi_1(x)^m$. A definable bijection

$$X \times_{\phi_1,\mathrm{rv}} (\mathrm{VF}^\times)^n \to X \times_{\phi_2,\mathrm{rv}} (\mathrm{VF}^\times)^n$$

is given by $(x, y) \mapsto (x, y^m)$. (If $\mathrm{rv}(u) = \phi(x)^m$, there exists a unique y with $\mathrm{rv}(y) = \phi(x)$ and $y^m = u$.)

Example 6.4. Proposition 6.1 need not remain valid over an RV-generated base set. Let $A = \mathrm{dcl}(c)$, c a transcendental point of $\mathbf{k}$. Let $f_1(y) = y$, $f_2(y) = 1$, $\mathbb{L}(Y, f_i) := \mathrm{VF} \times_{\mathrm{rv}, f_i} Y = \{(x, y) \in \mathrm{RV} \times Y : \mathrm{rv}(x) = y\}$. Then $\mathbb{L}(Y, f), \mathbb{L}(Y, f')$ are both open balls; over any field A' containing A, they are definably isomorphic, using a translation. But these balls are not definably isomorphic over A.

7 Special bijections and RV-blowups

We work with a V-minimal theory $\mathbf{T}$. Recall the lift $\mathbb{L} : \mathrm{RV}[\leq n, \cdot] \to \mathrm{VF}[n, \cdot]$, with $\rho_X : \mathbb{L}X \to X$. Our present goal is an intrinsic description in terms of RV of the congruence relation $\mathbb{L}X \simeq \mathbb{L}Y$.

A will denote a (VF, Γ, RV)-generated substructure of a model of $\mathbf{T}$. Note that $\mathbf{T}_A$ is also V-minimal (Corollary 3.39) so any lemma proved for $\mathbf{T}$ under our assumptions can be used for any $\mathbf{T}_A$.

The word "definable" below refers to $\mathbf{T}$. The categories VF, RV[$*$] defined below thus depend on $\mathbf{T}$; when necessary, we will denote them $\mathrm{VF}_{\mathbf{T}}$, etc. When $\mathbf{T}$ has the form $\mathbf{T} = \mathbf{T}^0_A$ for fixed $\mathbf{T}^0$ but varying A, we write VF_A, etc.

7.1 Special bijections

Let $X \subseteq \mathrm{VF}^{n+1} \times \mathrm{RV}^m$ be $\underset{\mathrm{rv}}{\sim}$-invariant. Say

$$X = \{(x, y, u) \in \mathrm{VF} \times \mathrm{VF}^n \times \mathrm{RV}^m : (\mathrm{rv}(x), \mathrm{rv}(y), u) \in \bar{X}\}.$$

(We allow x to be any of the $n + 1$ coordinates and y the others.)

Let $s(y, u)$ be a definable function into VF with $\underset{\mathrm{rv}}{\sim}$-invariant domain of definition

$$\mathrm{dom}(s) = \{(y, u) : (\mathrm{rv}(y), u) \in \bar{S}\}$$

and $\theta(u)$ a definable function on $\mathrm{pr}_u(\mathrm{dom}(s))$ into RV, such that $(s(y, u), y, u) \in X$ and $\mathrm{rv}(s(y, u)) = \theta(u)$ for $(y, u) \in \mathrm{dom}(s)$. Note that θ is uniquely defined (given s) if it exists. Let

$$X_1 = \{(x, y, u) \in X : (\mathrm{rv}(y), u) \in \bar{S}, \mathrm{rv}(x) = \theta(u)\}, \qquad X_2 = X \setminus X_1,$$
$$X_1' = \{(x, y, u) \in \mathrm{VF} \times \mathrm{dom}(s) : \mathrm{val}(x) > \mathrm{val}_{\mathrm{rv}}\theta(u)\}$$

and let $X' = X_1' \dot{\cup} X_2$. Also define $e_s : X' \to X$ to be the identity on X_2, and

$$e_s(x, y, u) = (x + s(y, u), y, u)$$

on X_1'.

Definition 7.1. $e_s : X' \to X$ is a definable bijection, called an *elementary bijection*. □

Lemma 7.2.

(1) *If X is $\underset{\mathrm{rv}}{\sim}$-invariant, so is X'. If $X \to \mathrm{VF}^{n+1}$ is finite-to-one, the same is true of X'.*
(2) *If $X_i = \mathbb{L}\bar{X}_i$, $X'_1 = \mathbb{L}\bar{X}'_1$, then $\bar{X}'_1$ is isomorphic to $(\mathrm{RV}^{>0}\dot{\cup}\{1\}) \times \bar{S}$, while $\bar{X}_1$ is isomorphic to $\bar{S}$.*
(3) *If the projection $X \to \mathrm{VF}^{n+1}$ has finite fibers, then so does the projection $\mathrm{dom}(s) \to \mathrm{VF}^n$, and also the projection $\bar{S} \to \mathrm{RV}^n$, $(y', u) \mapsto y'$.*
(4) *e_s has partial derivative matrix I everywhere, hence has Jacobian 1. Thus if $F : X \to Y$ is such that* $\mathrm{rv}\,\mathrm{Jcb}(F)$ *factors through ρ_X, then* $\mathrm{rv}\,\mathrm{Jcb}(F \circ e_s)$ *factors through $\rho_{X'}$.*

Proof. (1) and (4) are clear. The first isomorphism of (2) is obtained by dividing x by $\theta(u)$, the second is evident. For (3), note that if $(y, u) \in \mathrm{dom}(s)$ then $(s(y, u), y, u) \in X$ so by the assumption $u \in \mathrm{acl}(y, s(y, u))$. But for fixed y, $\{s(y, u) : u \in \mathrm{dom}(s)\}$ is finite, by Lemma 3.41. Thus, in fact, $(y, u) \in \mathrm{dom}(s)$ implies $u \in \mathrm{acl}(y)$. Hence $(y', u) \in \bar{S}$ implies $u \in \mathrm{acl}(y)$ for any y with $\mathrm{rv}(y) = y'$, so (fixing such a y) $\{u : (y', u) \in \bar{S}\}$ is finite for any given y'. □

A *special bijection* is a composition of elementary bijections and *auxiliary bijections* $(x_1, \ldots, x_n, u) \mapsto (x_1, \ldots, x_n, u, \mathrm{rv}(x_1), \ldots, \mathrm{rv}(x_n))$.

An elementary bijection depends on the data s of a partial section of $X \to \mathrm{VF}^n \times \mathrm{RV}^m$. Conversely, given s, if $\mathrm{rv}(s(y, u))$ depends only on u we can define $\theta(u) = \mathrm{rv}(s(y, u))$ and obtain a special bijection. If not, we can apply an auxiliary bijection to $X \subseteq \mathrm{VF}^n \times \mathrm{RV}^m$, and obtain a set $X' \subseteq \mathrm{VF} \times \mathrm{RV}^{m+n}$, such that $\mathrm{rv}(x) = \mathrm{pr}_{m+1}(u)$ for $(x, u) \in X'$. For such a set X', the condition for existence of θ is automatic and we can define an elementary bijection $X'' \to X'$ based on s, and obtain a special bijection $X'' \to X$ as the composition.

The classes of auxiliary morphisms and elementary morphisms are all closed under disjoint union with any identity morphism, and it follows that the class special morphisms is closed under disjoint unions.

7.2 Special bijections in one variable and families of RV-valued functions

We consider here special bijections in dimension 1. An elementary bijection $X' \to X$ in dimension 1 involves a finite set B of rv-balls, and a set of "centers" of these balls (i.e., a set T containing a unique point $t(b)$ of each $b \in B$), and translates each ball so as to be centered at 0 (while fixing the RV coordinates). We say that $X' \to X$ blows up the balls in B, with centers T.

Given a special bijection $h' : X' \to X$, let $\mathrm{Fn}^{\mathrm{RV}}(X; h')$ be the set of definable functions $X \to \mathrm{RV}$ of the form $H(\rho_{X'}((h')^{-1}(x)))$, where H is a definable function. This is a finitely generated set of definable functions $X \to \mathrm{RV}$. There will usually be no ambiguity in writing $\mathrm{Fn}^{\mathrm{RV}}(X, X' \to X)$ instead.

Note that while a special bijection is an isomorphism in VF, an asymmetry exists: if $e : X' \to X$ is a special bijection, then $\mathrm{Fn}^{\mathrm{RV}}(X, X)$ is usually a proper subset of $\mathrm{Fn}^{\mathrm{RV}}(X, X' \to X)$.

What is the effect on $\mathrm{Fn}^{\mathrm{RV}}$ of passing from X' to X'', where $X'' \to X'$ is a special bijection? The auxiliary bijections have no effect. Assume rv is already a coordinate function of X'. Consider an elementary bijection $e_s : X'' \to X'$. Let $B = \{(x, u) \in X' : u \in \mathrm{dom}(s)\}$. Then the characteristic function 1_B lies in $\mathrm{Fn}^{\mathrm{RV}}(X', \mathrm{Id}_{X'})$; so $1_B \circ (h')^{-1}$ lies in $\mathrm{Fn}^{\mathrm{RV}}(X, h')$. Using this, we see that $\mathrm{Fn}^{\mathrm{RV}}(X', e_s)$ is generated over $\mathrm{Fn}^{\mathrm{RV}}(X', \mathrm{Id}_{X'})$ by the function $B \to \mathrm{RV}$, $(x, u) \mapsto \mathrm{rv}(x - s(u))$ (extended by 0 outside B). Thus if $h'' = h' \circ e_s : X'' \to X$, $\mathrm{Fn}^{\mathrm{RV}}(X, h'')$ is generated over $\mathrm{Fn}^{\mathrm{RV}}(X, h')$ by the composition of the function $(x, u) \mapsto \mathrm{rv}(x - s(u))$ with $(h')^{-1}$.

Conversely, if B is a finite union of open balls whose characteristic function lies in $\mathrm{Fn}^{\mathrm{RV}}(X, h')$, and if there exists a definable set T of representatives (one point $t(b)$ in each ball b of B), and a function $\phi = (\phi_1, \ldots, \phi_n)$, $\phi_i \in \mathrm{Fn}^{\mathrm{RV}}(X, h')$, with ϕ injective on T, then one can find a special bijection $X'' \to X'$ with composition $h'' : X'' \to X$, such that $\mathrm{Fn}^{\mathrm{RV}}(X, h'')$ is generated over $\mathrm{Fn}^{\mathrm{RV}}(X, h')$ by $y \mapsto \mathrm{rv}(y - t(y))$, $y \in b \in B$. Namely, let $\mathrm{dom}(s) = \phi(T)$, and for $u \in \mathrm{dom}(s)$ set $s(u) = h'^{-1}(t)$ if $t \in T$ and $\phi(t) = u$. In this situation, we will say that the balls in B are blown up by $X'' \to X'$, with centers T. Let $\theta(u) = \mathrm{rv}(s(u))$. Because $X' \to X$ may already have blown up some of the balls in B, $\mathrm{Fn}^{\mathrm{RV}}(X, h'')$ is generated over $\mathrm{Fn}^{\mathrm{RV}}(X, h')$ by the restriction of $y \mapsto \mathrm{rv}(y - t(y))$ to some subball of b, possibly proper. Nevertheless, we have the following.

Lemma 7.3. *The function $y \mapsto \mathrm{rv}(y - t(y))$ on B lies in* $\mathrm{Fn}^{\mathrm{RV}}(X, h'')$.

Proof. This follows from the following, more general claim. □

Claim. Let $c \in \mathrm{VF}$, $b \in \mathfrak{B}$ be definable, with $c \in b$. Let b' be an rv-ball with $c \in b'$. Then the function $\mathrm{rv}(x - c)$ on b is generated by its restriction to b', rv, and the characteristic function of b.

Proof. Let $x \in b \setminus b'$. From $\mathrm{rv}(x)$ compute $\mathrm{val}(x)$. If $\mathrm{val}(x) < \mathrm{val}(c)$, $\mathrm{rv}(x - c) = \mathrm{rv}(c)$. If $\mathrm{val}(x) > \mathrm{val}(c)$, $\mathrm{rv}(x - c) = \mathrm{rv}(x)$. When $\mathrm{val}(x) = \mathrm{val}(c)$, but $x \notin b'$, $\mathrm{rv}(x - c) = \mathrm{rv}(x) - \mathrm{rv}(c)$. Recall here that $\mathrm{val}_{\mathrm{rv}}{}^{-1}(\gamma)$ is the nonzero part of a $\mathbf{k}$-vector space; subtraction, for distinct elements u, v, can therefore be defined by $u - v = u(u^{-1}v - 1)$. □

Thus any special bijection can be understood as blowing up a certain finite number of balls (in a certain sequence and with certain centers). We will say that a special bijection $X'' \to X'$ is subordinate to a given partition of X if each ball blown up by $X'' \to X'$ is contained in some class of the partition.

It will sometimes be more convenient to work with the sets of functions $\mathrm{Fn}^{\mathrm{RV}}(X, h)$ than with the special bijections h themselves.

We observe that any finite set of definable functions $X \to \mathrm{RV}$ is contained in $\mathrm{Fn}^{\mathrm{RV}}(X; h)$ for some X', h.

Lemma 7.4. *Let $X \subseteq \mathrm{VF} \times \mathrm{RV}^*$ be $\underset{\mathrm{rv}}{\sim}$-invariant, and let $f : X \to (\mathrm{RV} \cup \Gamma)$ be a definable map. Then there exists an $\underset{\mathrm{rv}}{\sim}$-invariant $X' \subseteq \mathrm{VF} \times \mathrm{RV}^*$ a special bijection $h : X' \to X$, and a definable function t such that $t \circ \rho_{X'} = f \circ h$. Moreover, if $X = \cup_{i=1}^m P_i$ is a finite partition of X into sets whose characteristic functions factor through ρ, we can find $X' \to X$ subordinate to this partition.*

Proof. Say $X \subseteq \mathrm{VF} \times \mathrm{RV}^m$; let $\pi : X \to \mathrm{VF}$, $\pi' : X \to \mathrm{RV}^m$ be the projections. Applying an auxiliary bijection, we may assume $\mathrm{rv}(\pi(x)) = \mathrm{pr}_m \pi'(x)$, i.e., $\mathrm{rv}(\pi(x))$ agrees with one of the coordinates of $\pi'(x)$. We now claim that there exists a finite $F' \subseteq \mathrm{RV}^m$, such that away from $\pi'^{-1}(F')$, f factors through π'. To prove this, it suffices to show that if p is a complete type of X and $\pi'_* p$ is nonalgebraic (i.e., not contained in a finite definable set), then $f|p$ factors through π'; this follows from Lemma 3.20.

We can thus restrict attention to $\pi'^{-1}(F')$; our special bijections will be the identity away from this. Thus we may assume $\pi'(X)$ is finite. Recall that (since an auxiliary bijection has been applied) rv is constant on each fiber of π'. In this case there is no problem relativizing to each fiber of π', and then collecting them together (Lemma 2.3), we may assume, in fact, that $\pi'(X)$ consists of a single point $\{u\}$. In this case the partition (since it is defined via ρ) will automatically be respected.

The rest of the proof is similar to Lemma 4.2. We first consider functions f with finite support F (i.e., $f(x) = 0$ for $x \notin F$) and prove the analogue of the statement of the lemma for them. If $F = \{0\} \times \{u\}$ then $F = \rho^{-1}(\{(0)\} \times \{u\})$ so the claim is clear. If $F = \{(x_0, u)\}$, let $s : \{u\} \to \mathrm{VF}$, $s(u) = x_0$. Applying e_s returns us to the previous case. If $F = F_0 \times \{u\}$ has more than one point, we use induction on the number of points. Let $s(u)$ be the average of F_0. Apply the special bijection e_s. Then the result is a situation where rv is no longer constant on the fiber. Applying an auxiliary bijection to make it constant again, the fibers of $F \to \mathrm{RV}^{m+1}$ become smaller.

The case of the characteristic function of a finite union of balls is similar (following Lemma 4.2).

Now consider a general function f. Having disposed of the case of characteristic function, it suffices to treat f on each piece of any given partition. Thus we can assume f has the form of Corollary 4.4, $f(x) = H(\mathrm{rv}(x - n(x)))$. Translating by the $n(x)$ as in the previous cases, we may assume $n(x) = 0$. But then again f factors through ρ and rv, so one additional auxiliary bijection suffices. □

Corollary 7.5. *Let $X, Y \subseteq \mathrm{VF}^n \times \mathrm{RV}^*$, and let $f : X \to Y$ be a definable bijection. Then there exists a special bijection $h : X' \to X$, and t such that $\rho_Y \circ (f \circ h) = t \circ \rho_{X'}$.*

It can be found subordinate to a given finite partition, factoring through ρ_X. □

We wish to obtain a symmetric version of Corollary 7.5. We will say that bijections $f, g : X \to Y$ *differ by special bijections* if there exist special bijections h_1, h_2 with $h_2 g = f h_1$. We show that every definable bijection between $\underset{\mathrm{rv}}{\sim}$-invariant objects differs by special bijections from an $\underset{\mathrm{rv}}{\sim}$-invariant bijection.

Lemma 7.6. *Let $X \subseteq \mathrm{VF} \times \mathrm{RV}^m$, $Y \subseteq \mathrm{VF} \times \mathrm{RV}^{m'}$ be definable, $\underset{\mathrm{rv}}{\sim}$-invariant; let $F : X \to Y$ be a definable bijection. Then there exist special bijections $h_X : X' \to X$, $h_Y : Y' \to Y$, and an $\underset{\mathrm{rv}}{\sim}$-invariant bijection $F' : X' \to Y'$ with $F = h_Y F' h_X^{-1}$; i.e., F differs from an $\underset{\mathrm{rv}}{\sim}$-invariant morphism by special bijections.*

Proof. It suffices to find h_X, h_Y such that $\mathrm{Fn}^{\mathrm{RV}}(X, h_X) = F \circ \mathrm{Fn}^{\mathrm{RV}}(Y, h_Y)$; for then we can let $F' = h_Y^{-1} F h_X$.

Let $X = \cup_{i=1}^{m} P_i$ be a partition as in Proposition 5.1. By Lemma 7.4, there exist X_0, Y_1 and special bijections $X_0 \to X, Y_1 \to Y$, such that the characteristic functions of the sets P_i (respectively, the sets $F(P_i)$) are in $\mathrm{Fn}^{\mathrm{RV}}(X, X_0 \to X)$ (respectively, $\mathrm{Fn}^{\mathrm{RV}}(Y, Y_1 \to Y)$).

By Corollary 7.5, one can find a special $X_1 \to X_0$ such that $\mathrm{Fn}^{\mathrm{RV}}(X, X_1 \to X)$ contains $F \circ \mathrm{Fn}^{\mathrm{RV}}(Y, Y_1 \to Y)$. By another application of the same, one can find a special bijection $Y_* \to Y_1$ subordinate to $\{F(P_i)\}$ such that

$$\mathrm{Fn}^{\mathrm{RV}}(Y, Y_* \to Y) \supseteq F^{-1} \circ \mathrm{Fn}^{\mathrm{RV}}(X, X_1 \to X). \tag{7.1}$$

Now Y_* is obtained by composing a sequence $Y_* = Y_m \to \cdots \to Y_1$ of elementary bijections and auxiliary bijections. We define inductively $X_m \to \cdots \to X_2 \to X_1$, such that

$$\mathrm{Fn}^{\mathrm{RV}}(Y, Y_k \to Y) \circ F \subseteq \mathrm{Fn}^{\mathrm{RV}}(X, X_k \to X). \tag{7.2}$$

Let $k \geq 1$. Y_{k+1} is obtained by blowing up a finite union of balls B of Y, with a definable set T of representatives such that some $\phi \in \mathrm{Fn}^{\mathrm{RV}}(Y, Y_k \to Y)$ is injective on T; then $\mathrm{Fn}^{\mathrm{RV}}(Y, Y_{k+1} \to Y)$ is generated over $\mathrm{Fn}^{\mathrm{RV}}(Y, Y_k \to Y)$ by ψ, where for $y \in b \in B$ $\psi(y) = \mathrm{rv}(y - t(b))$ (Lemma 7.3). By the choice of the partition $\{P_i\}$, $F^{-1}(B)$ is also a finite union of balls.

Now $F^{-1}(B)$, with center set $F^{-1}(T)$, can serve as data for a special bijection: the requirement about the characteristic function of B and the injective function on T being in $\mathrm{Fn}^{\mathrm{RV}}$ are satisfied by virtue of Lemma 7.3. We can thus define $X_{k+1} \to X_k$ so as to blow up $F^{-1}(B)$ with center set $F^{-1}(T)$. By Lemma 5.4, $\mathrm{rv}(F(x) - F(x'))$ is a function of $\mathrm{rv}(x-x')$ (and conversely) on each of these balls, so $\mathrm{Fn}^{\mathrm{RV}}(X, X_{k+1} \to X)$ is generated over $\mathrm{Fn}^{\mathrm{RV}}(X, X_k)$ by $\psi \circ F$. Hence (7.2) remains valid for $k+1$.

Now by (7.1), $\mathrm{Fn}^{\mathrm{RV}}(X, X_1 \to X) \subseteq \mathrm{Fn}^{\mathrm{RV}}(Y, Y_* \to Y) \circ F$; since the generators match at each stage, by induction on $k \leq m$,

$$\mathrm{Fn}^{\mathrm{RV}}(X, X_k \to X) \subseteq \mathrm{Fn}^{\mathrm{RV}}(Y, Y_m \to Y) \circ F. \tag{7.3}$$

By (7.2) and (7.3) for $k = m$, $\mathrm{Fn}^{\mathrm{RV}}(X, X_m \to X) = \mathrm{Fn}^{\mathrm{RV}}(Y, Y_* \to Y) \circ F$. □

For the sake of possible future refinements, we note that the proof of Lemma 7.6 also shows the following.

Lemma 7.7. *Let $X \subseteq \mathrm{VF} \times \mathrm{RV}^m$, $Y \subseteq \mathrm{VF} \times \mathrm{RV}^{m'}$ be definable, $\underset{\mathrm{rv}}{\sim}$-invariant; let $F : X \to Y$ be a definable bijection. If a Proposition* 5.1 *partition for F has characteristic functions factoring through ρ_X, ρ_Y, and if F is $\underset{\mathrm{rv}}{\sim}$-invariant, then for any special bijection $h'_X : X' \to X$, there exists a special bijection $h'_Y : Y' \to Y'$ such that $(h'_Y)^{-1} F h'_X$ is $\underset{\mathrm{rv}}{\sim}$-invariant.* □

7.3 Several variables

We will now show in general that any definable map from an $\underset{\mathrm{rv}}{\sim}$-invariant object to RV factors through the inverse of a special bijection, and the standard map ρ.

Lemma 7.8. *Let $X \subseteq \mathrm{VF}^n \times \mathrm{RV}^m$ be $\underset{\mathrm{rv}}{\sim}$-invariant, and let $\phi : X \to (\mathrm{RV} \cup \Gamma)$. Then there exists a special bijection $h : X' \to X$, and a definable function τ such that $\tau \circ \rho_{X'} = \phi \circ h$.*

Proof. By induction on n. For $n = 0$ we can take $X' = X$, since $\rho_X = \mathrm{Id}_X$.

For $n = 1$ and $X \subseteq \mathrm{VF}$, by Lemma 7.4, there exists $\mu = \mu(X, \phi) \in \mathbb{N}$ such that the lemma holds for some h that is a composition of μ elementary and auxiliary bijections. It is easy to verify the semicontinuity of μ with respect to the definable topology: if X_t is a definable family of definable sets, so that X_b is $A(b)$-definable, and $\mu(X_b, \phi|X_b) = m$, then there exists a definable set D with $b \in D$ such that if $b' \in D$, then $\mu(X_{b'}, \phi|X_{b'}) \leq m$.

Assume the lemma is known for n and suppose $X \subseteq \mathrm{VF} \times Y$, with $Y \subseteq \mathrm{VF}^n \times \mathrm{RV}^m$. For any $b \in Y$, let $X_b = \{x : (x, b) \in X\} \subseteq \mathrm{VF}$; so X_b is $A(b)$-definable.

Let $\mu = \max_b \mu(X_b, \phi|X_b)$. Consider first the case $\mu = 0$. Then $\phi|X_b = \tau_b \circ \rho|X_b$, for some $A(b)$-definable function $\tau_b : \mathrm{RV}^m \to (\mathrm{RV} \cup \Gamma)$. By stable embeddedness and elimination of imaginaries in $\mathrm{RV} \cup \Gamma$, there exists (Section 2.1) a canonical parameter $d \in (\mathrm{RV} \cup \Gamma)^l$, and an A-definable function τ, such that $\tau_b(t) = \tau(d, t)$; and d itself is definable from τ_b, so we can write $d = \delta(b)$ for some definable $\delta : Y \to (\mathrm{RV} \cup \Gamma)^l$. Using the induction hypothesis for (Y, δ) in place of (X, ϕ), we find that there exists an $\underset{\mathrm{rv}}{\sim}$-invariant $Y' \subseteq \mathrm{VF}^n \times \mathrm{RV}^*$, a special $h_Y : Y' \to Y$, and a definable τ_Y, such that $\tau_Y \circ \rho_{Y'} = \delta \circ h_Y$. Let $X' = X \times_Y Y'$, $h(x, y') = (x, h_Y(y'))$. An elementary bijection to Y determines one to X, where the function s does not make use of the first coordinate; so $h : X' \to X$ is special. In this case, the lemma is proved: $\phi \circ h(x, y') = \phi(x, h_Y(y')) = \tau(\delta(h_Y(y')), \rho(x, y)) = \tau(\tau_Y(\rho_{Y'}(y')), \rho(x, y))$.

Next suppose $\mu > 0$. Applying an auxiliary bijection, we may assume that for some definable function (in fact, projection) p, $\mathrm{rv}(x) = p(u)$ for $(x, y, u) \in X$. For each $b \in Y(M)$ (with M any model of $\mathbf{T}_A$) there exists an elementary bijection $h_b : X'_b \to X_b$, such that $\mu(X'_b, \phi|X'_b) < \mu$; h_b is determined by s_b, θ_b,with $s_b \in \mathrm{rv}(s_b) = \theta_b$, and $(s_b, \theta_b) \in X$. (The u-variables have been absorbed into b.) By compactness, one can take $s_b = s(b)$ and $\theta_b = \theta'(b)$ for some definable functions s, θ'. By the inductive hypothesis applied to (Y, θ'), as in the previous paragraph, we can assume $\theta'(y, u) = \theta(u)$ for some definable θ. Applying the special bijection with data (s, θ) now amounts to blowing up (s_b, θ_b) uniformly over each b, and thus reduces the value of μ. □

Question 7.9. Is Proposition 7.6 true in higher dimensions?

Corollary 7.10. *Let $X \subseteq \mathrm{VF}^n \times \mathrm{RV}^m$ be definable. Then every definable function $\phi : X \to \Gamma$ factors through a definable function $X \to \mathrm{RV}^*$.*

Proof. By Lemma 4.5 we may assume X is $\underset{\mathrm{rv}}{\sim}$-invariant; now the corollary follows from Lemma 7.8. □

(It is convenient to note this here, but it can also be proved with the methods of Section 3; the main point is that on the generic type of ball with center c, every function into $\mathrm{RV} \cup \Gamma$ factors through $\mathrm{rv}(x - c)$; while on a transitive ball, every function into $\mathrm{RV} \cup \Gamma$ is constant.)

Consider pairs (X', f') with X', $f' : X' \to \mathrm{VF}^n$ definable, such that f' has RV-fibers. A bijection $g : X' \to X''$ is said to be *relatively unary* (with respect to f', f'') if it commutes with $n - 1$ coordinate projections, i.e., $\mathrm{pr}_i\, f'' g = \mathrm{pr}_i\, f'$ for all but at most one value of i.

Given $X \subseteq \mathrm{VF}^n \times \mathrm{RV}^m$, we view it as a pair (X, f) with f the projection to VF^n. Thus for $X, Y \subseteq \mathrm{VF}^n \times \mathrm{RV}^*$, the notion $F : X \to Y$ is relatively unary is defined.

Note that the elementary bijections are relatively unary, as are the auxiliary bijections.

Lemma 7.11. *Let $X, Y \subseteq \mathrm{VF}^n \times \mathrm{RV}^*$, and let $F : X \to Y$ be a definable bijection. Then F can be written as the composition of relatively unary morphisms of* $\mathrm{VF}[n, \cdot]$.

Proof. We have X with two finite-to-one maps $f, g : X \to \mathrm{VF}^n$ (the projection and the composition of F with the projection $Y \to \mathrm{VF}^n$). We must decompose the identity $X \to X$ into a composition of relatively unary maps $(X, f) \to (X, g)$.

Begin with the case $n = 2$; we are given (X, f_1, f_2) and (X, g_1, g_2).

Claim. There exists a definable partition of X into sets X_{ij} such that $(f_i, g_j) : X \to \mathrm{VF}^2$ is finite-to-one.

Proof. Let $a \in X$. We wish to show that for some i, j, $a \in \mathrm{acl}(f_i(a), g_j(a))$. This follows from the exchange principle for algebraic closure in VF: if $a \in \mathrm{acl}(\emptyset)$, there is nothing to show. Otherwise, $g_j(a) \notin \mathrm{acl}(\emptyset)$ for some j; in this case either $a \in \mathrm{acl}(f_1(a), g_j(a))$ or $f_1(a) \in \mathrm{acl}(g_j(a))$, and then $a \in \mathrm{acl}(f_2(a), g_j(a))$. The claim follows by compactness. □

Let $h : X' \to X$ be a special bijection such that the characteristic functions of X_{ij} are in $\mathrm{Fn}^{\mathrm{RV}}(X, X' \to X)$. (Lemma 7.8). Since h is composition of relatively unary bijections, we may replace X by X' (and f_i, g_i by $f_i \circ h, g_i \circ h$, respectively). Thus we may assume the characteristic function of X_{ij} is in $\mathrm{Fn}^{\mathrm{RV}}(X, X)$, i.e., $X_{ij} \in VFr[n]$. But then it suffices to treat each X_{ij} separately, say, X_{11}. In this case the identity map on X takes

$$\begin{aligned}(X, f_1, f_2) &\mapsto (X, f_1, g_1) \mapsto (X, g_2, g_1) \mapsto (X, g_2, g_1 - g_2)\\ &\mapsto (X, g_1, g_1 - g_2) \mapsto (X, g_1, g_2),\end{aligned}$$

where each step is relatively unary.

When $n > 2$, we move between $(X, f_1, \dots, f_n)$ and $(X, g_1, \dots, g_n)$, by partitioning, and on a given piece replacing each f_i by some g_j, one at a time. □

7.4 RV-blowups

We now define the RV-counterparts of the special bijections, which will be called RV-blowups. These will not be bijections; the kernel of the homomorphism $\mathbb{L} : K_+[\mathrm{RV}] \to K_+[\mathrm{VF}]$ will be seen to be obtained by formally inverting RV-blowups. Let $\mathrm{RV}_\infty^{>0} = \{x \in \mathrm{RV} : \mathrm{val}(x) > 0 \cup \{\infty\}\} \subseteq \mathrm{RV}_\infty$. In the RV[$\leq 1$]-presentation, $\mathrm{RV}_\infty^{>0} = [\mathrm{RV}^{>0}]_1 + [1]_0$ (cf. Section 3.8).

Definition 7.12.

(1) Let $\mathbf{Y} = (Y, f) \in \mathrm{Ob}\,\mathrm{RV}_\infty[n, \cdot]$ be such that $f_n(y) \in \mathrm{acl}(f_1(y), \ldots, f_{n-1}(y))$, and $f_n(y) \neq \infty$. Let $Y' = Y \times \mathrm{RV}_\infty^{>0}$. For $(y, t) \in Y'$, define $f' = (f_1', \ldots, f_n')$ by $f_i'(y, t) = f_i(y)$ for $i < n$, $f_n'(y, t) = tf_n(y)$. Then $\widetilde{\mathbf{Y}} = (Y', f')$ is an *elementary blowup* of $\mathbf{Y}$. It comes with the projection map $Y' \to Y$.

(2) Let $\mathbf{X} = (X, g) \in \mathrm{Ob}\,\mathrm{RV}_\infty[n, \cdot]$, $X = X' \dot{\cup} X''$, $g' = g|X'$, $g'' = g|X''$, and let $\phi : \mathbf{Y} \to (X', g')$ be an $\mathrm{RV}_{\mathrm{vol}}$-isomorphism. Then the RV-*blowup* $\widetilde{\mathbf{X}}_\phi$ is defined to be $\widetilde{\mathbf{Y}} + (X'', g'') = (Y' \dot{\cup} X'', f' \dot{\cup} g'')$. It comes with $b : Y' \dot{\cup} X'' \to X$, defined to be the identity on X'', and the projection on Y'. X' is called the *blowup locus* of $b : \widetilde{\mathbf{X}}_\phi \to \mathbf{X}$.

An *iterated* RV-*blowup* is obtained by finitely many iterations of RV-blowups.

Since blowups in the sense of algebraic geometry will not occur in this paper, we will say "blowup" for RV-blowup.

Remark 7.13. In the definition of an elementary blowup, $\dim_{\mathrm{RV}}(Y) < n$. For such Y, $\phi : \mathbf{Y} \to (X', g')$ is an $\mathrm{RV}_{\mathrm{vol}}[\leq n, \cdot]$-isomorphism iff it is an $\mathrm{RV}_{\Gamma\text{-vol}'}$-isomorphism (Definition 5.21).

Lemma 7.14.

(1) *Let* $\mathbf{Y}'$ *be an elementary blowup of* $\mathbf{Y}$. $\mathbf{Y}'$ *is* $\mathrm{RV}_{\mathrm{vol}}[n, \cdot]$*-isomorphic to* $\mathbf{Y}'' = (Y'', f'')$*, with* $Y'' = \{(y, t) \in Y \times \mathrm{RV}_\infty : \mathrm{val}_{\mathrm{rv}}(t) > f_n(y)\}$, $f''(y, t) = (f_1(y), \ldots, f_{n-1}(y), t)$.

(2) *An elementary blowup* $\mathbf{Y}'$ *of* $\mathbf{Y}$ *is* $\mathrm{RV}_\infty[n, \cdot]$*-isomorphic to* $(Y \times \mathrm{RV}_\infty, f')$ *for any* f' *isogenous to* $(f_1, \ldots, f_n, t)$.

(3) *Up to isomorphism, the blowup depends only on the blowup locus. In other words, if* X, X', g, g' *are as in Definition* 7.12*, and* $\phi_i : \mathbf{Y}_i \to (X', g')$ $(i = 1, 2)$ *are isomorphisms, then* $\widetilde{\mathbf{X}}_{\phi_1}, \widetilde{\mathbf{X}}_{\phi_2}$ *are* $\mathbf{X}$*-isomorphic in* $\mathrm{RV}_{\mathrm{vol}}[n, \cdot]$.

Proof.

(1) The isomorphism is given by $(y, t) \mapsto (y, tf_n(y))$.

(2) The identity map on $Y \times \mathrm{RV}$ is an $\mathrm{RV}_\infty[n, \cdot]$ isomorphism.

(3) Let $\psi_0 = \phi_2{}^{-1}\phi_1$, and define $\psi_1 : Y_1 \times \mathrm{RV}_\infty^{>0} \to Y_2 \times \mathrm{RV}_\infty^{>0}$ by $\psi(y, t) = (\psi_0(y), t)$. The sum of the values of the n coordinates of $\widetilde{\mathbf{Y}}_i$ is then $(\sum_{i<n} \mathrm{val}_{\mathrm{rv}} f_i) + (\mathrm{val}_{\mathrm{rv}}(t) + \mathrm{val}_{\mathrm{rv}} f_n)$ in both cases. Since by assumption $\psi_0 : Y_1 \to Y_2$ is an $\mathrm{RV}_{\mathrm{vol}}$-isomorphism, it preserves $\sum_{i\leq n} \mathrm{val}_{\mathrm{rv}} f_i$ and so ψ_1 too is an $\mathrm{RV}_{\Gamma\text{-vol}'}$-isomorphism; thus $\mathrm{Jcb}_{\mathrm{RV}}(\psi_1) \in \mathbf{k}^*$, i.e., let $\theta Y_1 \to \mathbf{k}^*$ be a definable map such

that $\theta = \mathrm{Jcb}_{\mathrm{RV}}(\psi_1)$ almost everywhere. Define $\psi : Y_1 \times \mathrm{RV}_\infty^{>0} \to Y_2 \times \mathrm{RV}_\infty^{>0}$ by $\psi(y,t) = (\psi_0(y), t/\theta(y))$. Then one computes immediately that $\mathrm{Jcb}_{\mathrm{RV}}(\psi) = 1$, so ψ is an $\mathrm{RV}_{\mathrm{vol}}[n,\cdot]$-isomorphism, and hence so is $\psi \dot\cup \mathrm{Id}_{X''} : \widetilde{\mathbf{X}}_{\phi_1} \to \widetilde{\mathbf{X}}_{\phi_2}$. □

Here is a coordinate-free description of RV-blowups; we will not really use it in the subsequent development.

Lemma 7.15.

(1) *Let* $\mathbf{Y} = (Y, g) \in \mathrm{Ob}\,\mathrm{RV}_\infty[n,\cdot]$, *with* $\dim(g(Y)) < n$; *let* $f : Y \to \mathrm{RV}^{n-1}$ *be isogenous to* g. *Let* $h : Y \to \mathrm{RV}$ *be definable, with* $h(y) \in \mathrm{acl}(g(y))$ *for* $y \in Y$, *and with* $\sum(g) = \sum(f) + \mathrm{val}_{\mathrm{rv}}(h)$. *Let* $Y' = Y \times \mathrm{RV}_\infty^{\geq 0}$, *and* $f'(y,t) = (f(y), th(y))$. *Then* $\mathbf{Y}' = (Y', f')$ *with the projection map to* Y *is a blowup.*

(2) *Let* $\mathbf{Y}'' \to \mathbf{Y}$ *be a blowup with blowup locus* Y. *Then there exist* f, h *such that with* $\mathbf{Y}'$ *as in* (3), $\mathbf{Y}'', \mathbf{Y}'$ *are isomorphic over* $\mathbf{Y}$.

Proof.

(1) Since $\dim_{\mathrm{RV}}(g(Y)) < n$, $\mathrm{Id}_Y : (Y, (f,h)) \to (Y, g)$ is an $\mathrm{RV}_{\mathrm{vol}}$-isomorphism. Use this as ϕ in the definition of blowup.

(2) With notation as in Definition 7.12, let $h = g_n \circ \phi^{-1}$, $f = (g_1, \ldots, g_{n-1}) \circ \phi^{-1}$. □

Definition 7.16. For $\mathcal{C} = \mathrm{RV}[\leq n,\cdot]$ or $\mathcal{C} = \mathrm{RV}_{\mathrm{vol}}[\leq n,\cdot]$, let $I_{\mathrm{sp}}[\leq n]$ be the set of pairs $(\mathbf{X}_1, \mathbf{X}_2) \in \mathrm{Ob}\,\mathcal{C}$ such that there exist iterated blowups $b_i : \widetilde{\mathbf{X}}_i \to \mathbf{X}_i$ and an $\mathcal{C}$-isomorphism $F : \widetilde{\mathbf{X}}_1 \to \widetilde{\mathbf{X}}_2$.

When n is clear from the context, we will just write I_{sp}.

Definition 7.17. Let 1_0 denote the one-element object of $\mathrm{RV}[0]$. Given a definable set $X \subseteq \mathrm{RV}^n$ let $\mathbf{X}_n$ denote $(X, \mathrm{Id}_X) \in \mathrm{RV}[n]$, and $[\mathbf{X}]_n$ the class in $K_+(\mathrm{RV}[n])$. Write $[1]_1$ for $[\{1\}]_1$ (where $\{1\}$ is the singleton set of the identity element of $\mathbf{k}$).

Lemma 7.18. *Let* $\mathcal{C} = \mathrm{RV}[\leq n,\cdot]$ *or* $\mathcal{C} = \mathrm{RV}_{\mathrm{vol}}[\leq n,\cdot]$.

(1) *Let* $f : \mathbf{X}_1 \to \mathbf{X}_2$ *be a* $\mathcal{C}$*-isomorphism, and let* $b_1 : \widetilde{\mathbf{X}}_1 \to \mathbf{X}_1$ *be a blowup. Then there exists a blowup* $b_2 : \widetilde{\mathbf{X}}_2 \to \mathbf{X}_2$ *and a* $\mathcal{C}$*-isomorphism* $F : \widetilde{\mathbf{X}}_1 \to \widetilde{\mathbf{X}}_2$ *with* $b_2 F = f b_1$.

(2) *If* $b : \widetilde{\mathbf{X}} \to \mathbf{X}$ *is a blowup, then so are* $b \dot\cup \mathrm{Id} : \widetilde{\mathbf{X}} \dot\cup \mathbf{Z} \to \mathbf{X} \dot\cup \mathbf{Z}$ *and* $(b \times \mathrm{Id}) : \widetilde{\mathbf{X}} \times \mathbf{Z} \to \mathbf{X} \times \mathbf{Z}$.

(3) *Let* $b_i : \widetilde{\mathbf{X}}_{\phi_i} \to \mathbf{X}$ *be a blowup* $(i = 1, 2)$. *Then there exist blowups* $b_i' : \mathbf{Z}_i \to \widetilde{\mathbf{X}}_{\phi_i}$ *and an isomorphism* $F : \mathbf{Z}_1 \to \mathbf{Z}_2$ *such that* $b_2 b_2' F = b_1 b_1'$.

(4) *Same as* (1)–(3) *for iterated blowups.*

(5) I_{sp} *is an equivalence relation. It induces a semiring congruence on* $K_+\mathrm{RV}[*,\cdot]$, *respectively,* $K_+\mathrm{RV}_{\mathrm{vol}}[*,\cdot]$.

(6) *As a semiring congruence on* $K_+\mathrm{RV}[*,\cdot]$, I_{sp} *is generated by* $([1]_1, [\mathrm{RV}^{>0}]_1 + 1_0)$.

Proof.

(1) This reduces to the case of elementary blowups. If $\mathcal{C} = \mathrm{RV}_{\mathrm{vol}}[n,\cdot]$, then the composition $f \circ b_1$ is already a blowup. If $\mathcal{C} = \mathrm{RV}[\leq n,\cdot]$, it is also clear using Lemma 7.14(2).

(2) This follows from the definition of blowup.

(3) If b_1 is the identity, let $b'_1 = b_2, b'_2 = \mathrm{Id}$, $F = \mathrm{Id}$; similarly if b_2 is the identity. If $X = X' \dot\cup X''$ and the statement is true above X' and above X'', then by glueing it is true also above X. We thus reduce to the case that b_1, b_2 both are blowups with blowup locus equal to X. But then by Lemma 7.14(3), there exists an isomorphism $F : \widetilde{\mathbf{X}}_{\phi_1} \to \widetilde{\mathbf{X}}_{\phi_2}$ over $\mathbf{X}$. Let $b'_1 = b'_2 = \mathrm{Id}$.

(4) For (1)–(2) the induction is immediate. For (3), write k-blowup as shorthand for "an iteration of k blowups." We show by induction on k_1, k' a more precise form.

Claim. If $\mathbf{X}_1 \to \mathbf{X}$ is a k_1-blowup, and $\mathbf{X}' \to \mathbf{X}$ is a k'-blowup, then there exists an k'-blowup $\mathbf{Z}'_1 \to \mathbf{X}_1$ a k_1-blowup $\mathbf{Z}' \to \mathbf{X}$, and an $\mathrm{RV}_{\mathrm{vol}}[n,\cdot]$-isomorphism $\mathbf{Z}'_1 \to \mathbf{Z}'$ over $\mathbf{X}$.

If $k_1 = k' = 1$, this is (3). Thus say $k' > 1$. The map $\mathbf{X}' \to \mathbf{X}$ is a composition $\mathbf{X}' \to \mathbf{X}_2 \to \mathbf{X}$, where $\mathbf{X}_2 \to \mathbf{X}$ is a $k'-1$-blowup and $\mathbf{X}' \to \mathbf{X}_2$ is a blowup. By induction there is a $k'-1$-blowup $\mathbf{Z}_1 \to \mathbf{X}_1$ and a k_1-blowup $\mathbf{Z}_2 \to \mathbf{X}_2$ and an $\mathrm{RV}_{\mathrm{vol}}[n,\cdot]$-isomorphism $\mathbf{Z}_1 \to \mathbf{Z}_2$ over $\mathbf{X}$.

By induction again there is a blowup and $\mathbf{Z}'_2 \to \mathbf{Z}_2$, a k_1-blowup $\mathbf{Z}' \to \mathbf{X}'$ an $\mathrm{RV}_{\mathrm{vol}}[n,\cdot]$-isomorphism $\mathbf{Z}' \to \mathbf{Z}_2$ over $\mathbf{X}_2$. By (1) there exists a blowup $\mathbf{Z}'_1 \to \mathbf{Z}_1$ and an $\mathrm{RV}_{\mathrm{vol}}[n,\cdot]$-isomorphism $\mathbf{Z}'_1 \to \mathbf{Z}'_2$, making the $\mathbf{Z}_1, \mathbf{Z}_2, \mathbf{Z}'_1, \mathbf{Z}_2$-square commute. Thus $\mathbf{Z}_1 \to \mathbf{X}_1$ is a k'-blowup, $\mathbf{Z}' \to \mathbf{X}'$ is a k_1-blowup, and we have a composed isomorphism $\mathbf{Z}'_1 \to \mathbf{Z}'_2 \to \mathbf{Z}'$ over $\mathbf{X}$.

(5) If $(\mathbf{X}_1, \mathbf{X}_2), (\mathbf{X}_2, \mathbf{X}_3) \in \mathrm{I_{sp}}$, there are iterated blowups $\mathbf{X}'_1 \to \mathbf{X}_1, \mathbf{X}'_2 \to \mathbf{X}_2$ and an isomorphism $\mathbf{X}'_1 \to \mathbf{X}'_2$; and also $\mathbf{X}''_2 \to \mathbf{X}_2, \mathbf{X}'_3 \to \mathbf{X}_3$ and $\mathbf{X}''_2 \to \mathbf{X}_3$. Using (3) for iterated blowups, there exist iterated blowups $\widehat{\mathbf{X}_2}' \to \mathbf{X}'_2, \widehat{\mathbf{X}_2}'' \to \mathbf{X}''_2$, and an isomorphism $\widehat{\mathbf{X}_2}' \to \widehat{\mathbf{X}_2}'$. By (1), for iterated blowups there are iterated blowups $\widehat{\mathbf{X}_1} \to \mathbf{X}'_1, \widehat{\mathbf{X}_3} \to \mathbf{X}_3$ and isomorphisms $\widehat{\mathbf{X}_1} \to \widehat{\mathbf{X}_2}', \widehat{\mathbf{X}_2}'' \to \widehat{\mathbf{X}_3}$, with the natural diagrams commuting. Composing, we obtain $\widehat{\mathbf{X}_1} \to \widehat{\mathbf{X}_3}$, showing that $(\mathbf{X}_1, \mathbf{X}_3) \in \mathrm{I_{sp}}$. Hence $\mathrm{I_{sp}}$ is an equivalence relation.

Isomorphic objects are $\mathrm{I_{sp}}$-equivalent, so an equivalence relation on the semiring $K_+ \mathcal{C}$ is induced. If $(X_1, X_2) \in \mathrm{I_{sp}}$, then by (2), $(X_1 \dot\cup Z, X_2 \dot\cup Z) \in \mathrm{I_{sp}}$, and $(X_1 \times Z, X_2 \times Z) \in \mathrm{I_{sp}}$. It follows that $\mathrm{I_{sp}}$ induces a congruence on the semiring $K_+ \mathcal{C}$.

(6) We can blow up 1_1 to $\mathrm{RV}_1^{>0} + 1_0$, so $([1]_1, [\mathrm{RV}^{>0}]_1 + 1_0) \in \mathrm{I_{sp}}$. Conversely, under the conditions of Definition 7.12, let $\mathbf{Y}^- = [(Y, f_1, \dots, f_{n-1})]$; then $[\mathbf{Y}] = [(Y, f_1, \dots, f_{n-1}, 0)] = [\mathbf{Y}^-] \times [1]_1$ by Lemma 7.14, and we have

$$[\widetilde{\mathbf{X}}_{\mathbf{Y}}] = [\mathbf{Y}]_{n-1} + [\mathbf{Y}]_{n-1} \times [\mathrm{RV}^{>0}]_1 + [\mathbf{X}''] \cong_{\mathrm{I_{sp}}} [\mathbf{Y}] \times [1]_1 + [\mathbf{X}''] = [\mathbf{X}]$$

modulo the congruence generated by $([1]_1, [\mathrm{RV}^{>0}]_1 + 1_0)$. □

We now relate special bijections to blowing ups. Given $\mathbf{X} = (X, f), \mathbf{X}' = (X', f') \in \mathrm{RV}[n,\cdot]$, say, $\mathbf{X}, \mathbf{X}'$ are *strongly isomorphic* if there exists a bijection $\phi : X \to X'$ with $f' = \phi f$. Strong isomorphisms are always in $\mathrm{RV}_{\mathrm{vol}}[n,\cdot]$.

Up to strong isomorphism, an elementary blowup of (Y, f) can be put in a different form: $(\widetilde{\mathbf{Y}}) \simeq (Y'', f'')$, $Y'' = \{(z, y) : y \in Y, \mathrm{val}_{\mathrm{rv}}(z) > \mathrm{val}_{\mathrm{rv}} f_n(y)\}$, $f_i(z, y) = f_i(y)$ for $i < n$, $f_n(z, y) = z$. The strong isomorphism $Y'' \to Y'$ is given by $(z, y) \mapsto (y, z/f_n(y))$. This matches precisely the definition of special bijection, and makes evident the following lemma.

Lemma 7.19. *Let* $\mathcal{C} = \mathrm{RV}_\infty[n, \cdot]$ *or* $\mathrm{RV}_{\mathrm{vol}}[\leq n, \cdot]$.

(1) $\mathbf{X}$, $\mathbf{Y}$ *are strongly isomorphic over* RV^n *iff* $\mathbb{L}\mathbf{X}$, $\mathbb{L}\mathbf{Y}$ *are isomorphic over the projection to* VF^n.
(2) *Let* $\mathbf{X}, \mathbf{X}' \in \mathrm{RV}[\leq n, \cdot]$, *and let* $G : \mathbb{L}\mathbf{X}' \to \mathbb{L}\mathbf{X}$ *be an auxiliary special bijection. Then* $\mathbf{X}'$ *is isomorphic to* $\mathbf{X}$ *over* RV^n.
(3) *Let* $\mathbf{X}, \mathbf{X}' \in \mathrm{RV}[\leq n, \cdot]$, *and let* $G : \mathbb{L}\mathbf{X}' \to \mathbb{L}\mathbf{X}$ *be an elementary bijection. Then* $\mathbf{X}'$ *is strongly isomorphic to a blowup of* $\mathbf{X}$.
(4) *Let* $\mathbf{X}, \mathbf{X}' \in \mathrm{RV}[\leq n, \cdot]$, *and let* $G : \mathbb{L}\mathbf{X}' \to \mathbb{L}\mathbf{X}$ *be a special bijection. Then* $\mathbf{X}'$ *is strongly isomorphic to an iterated blowup of* $\mathbf{X}$.
(5) *Assume* $\mathbf{T}$ *is effective. If* $\mathbf{Y} \to \mathbf{X}$ *is an* RV*-blowup, there exists* $\mathbf{Y}'$ *strongly isomorphic to* $\mathbf{Y}$ *over* $\mathbf{X}$ *and an elementary bijection* $c : \mathbb{L}\mathbf{Y}' \to \mathbb{L}\mathbf{Y}$ *lying over* $\mathbf{Y}' \to \mathbf{Y}$.

Proof.

(1) This is clear using Lemma 3.52.
(2) This is a special case of (1).
(3) This is clear from the definitions.
(4) This is clear from (1)–(3).
(5) It suffices to consider elementary blowups; we use the notation in the definition there. Thus $f_n(x) \in \mathrm{acl}(f_1(x), \ldots, f_{n-1}(x))$ for $x \in \phi(Y)$. By effectiveness and Lemma 6.2, there exists a definable function $s(x, y_1, \ldots, y_{n-1})$ such that if $\mathrm{rv}(y_i) = f_i(x)$ for $i = 1, \ldots, n-1$, then $\mathrm{rv}\, s(x, y) = f_n(x)$. This s is the additional data needed for an elementary bijection. □

Lemma 7.20. *Let* $\mathbf{X} = (X, f), \mathbf{X}' = (X', f') \in \mathrm{RV}[\leq n, \cdot]$, *and let* $h : X \to W \subseteq \mathrm{RV}^*$, $h' : X' \to W$ *be definable maps. Let* $X_c = h^{-1}(c)$, $\mathbf{X}_c = (X_c, f|X_c)$ *and similarly* $\mathbf{X}'_c$. *If* $(\mathbf{X}_c, \mathbf{X}'_c) \in \mathrm{I}_{\mathrm{sp}}(\mathrm{RV}_c[n, \cdot])$; *then* $(\mathbf{X}, \mathbf{X}') \in \mathrm{I}_{\mathrm{sp}}$.

Proof. Lemma 2.3 applies to $\mathrm{RV}_{\Gamma\text{-vol}'}$-isomorphisms, and hence using Remark 7.13 also to blowups. It also applies to $\mathrm{RV}[\leq n, \cdot]$-isomorphisms; hence to I_{sp}-equivalence. □

Lemma 7.21. *If* $(\mathbf{X}, \mathbf{Y}) \in \mathrm{I}_{\mathrm{sp}}$ *then* $\mathbb{L}\mathbf{X} \simeq \mathbb{L}\mathbf{Y}$.

Proof. Clear, since $\mathbb{L}[1]_1$ is the unit open ball around 1, $\mathbb{L}([\mathrm{RV}^{>0}]_1$ is the punctured unit open ball around 0, and $\mathbb{L}1_0 = \{0\}$. □

7.5 The kernel of $\mathbb{L}$

Definition 7.22. $\mathrm{VFR}[k, l, \cdot]$ is the set of pairs $\mathbf{X} = (X, f)$, with $X \subseteq \mathrm{VF}^k \times \mathrm{RV}^*$, $f : X \to \mathrm{RV}^l_\infty$, such that f factors through the projection $\mathrm{pr}_{\mathrm{RV}}(X)$ of X to the RV-coordinates. $\mathrm{I_{sp}}$ is the equivalence relation on $\mathrm{VFR}[k, l, \cdot]$:

$$(X, Y) \in \mathrm{I_{sp}} \iff (X_a, Y_a) \in \mathrm{I_{sp}}(\mathbf{T}_a) \quad \text{for each } a \in \mathrm{VF}^k.$$

$K_+ \mathrm{VFR}$ is the set of equivalence classes.

By the usual compactness argument, if $(X, Y) \in \mathrm{I_{sp}}$ then there are uniform formulas demonstrating this. The relative versions of Lemmas 7.14 and 7.18 follow.

If $\mathbf{U} = (U, f) \in \mathrm{VFR}[k, l, \cdot]$, and for $u \in U$ we are uniformly given $\mathbf{V}_u = (V_u, g_u) \in \mathrm{VFR}[k', l', \cdot]$, we can define a sum $\sum_{u \in U} \mathbf{V}_u \in \mathrm{VFR}[k + k', l + l', \cdot]$: it is the set $\dot\bigcup_{u \in U} \mathrm{V}_u$, with the function $(u, v) \mapsto (f(u), g_u(v))$. When necessary, we denote this operation $\sum^{(k,l;k',l')}$. The special case $k = l = 0$ is understood as the default case.

By Proposition 7.6, the inverse of $\mathbb{L} : \mathrm{RV}[1, \cdot] \to \mathrm{VF}[1, \cdot]$ induces an isomorphism $I_1^1 : K_+ \mathrm{VF}[1, \cdot] \to K_+ \mathrm{RV}[1, \cdot]/\mathrm{I_{sp}}$:

$$I([X]) = [Y]/\mathrm{I_{sp}} \iff [\mathbb{L} Y] = [X].$$

Let J be a finite set of k elements. For $j \in J$, let $\pi^j : \mathrm{VF}^k \times \mathrm{RV}^* \to \mathrm{VF}^{J-\{j\}} \times \mathrm{RV}^*$ be the projection forgetting the jth VF coordinate. We will write VF^k, VF^{k-1} for VF^J, $\mathrm{VF}^{J-\{j\}}$, respectively, when the identity of the indices is not important.

Let $\mathbf{X} = (X, f) \in \mathrm{VFR}[k, l, \cdot]$. By assumption, f factors through π^j. We view the image $(\pi^j X, f)$ as an element of $\mathrm{VFR}[k - 1, l, \cdot]$. Note that each fiber of π^j is in $\mathrm{VF}[1, \cdot]$.

Relativizing I_1^1 to π^j, we obtain a map

$$I^j = I^j_{k,l} : \mathrm{VFR}[k, l, \cdot] \to K_+ \mathrm{VFR}[k - 1, l + 1, \cdot]/\mathrm{I_{sp}}.$$

Lemma 7.23. *Let* $\mathbf{X} = (X, f), \mathbf{X}' = (X', f') \in \mathrm{VFR}[k, l, \cdot]$.

(1) I^j *commutes with maps into* RV*: if* $h : \mathbf{X} \to W \subseteq \mathrm{RV}^*$ *is definable,* $\mathbf{X}_c = h^{-1}(c)$, *then* $I^j(\mathbf{X}) = \sum_{c \in W} I^j(\mathbf{X}_c)$.
(2) *If* $([\mathbf{X}], [\mathbf{X}']) \in \mathrm{I_{sp}}$ *then* $(I^j(\mathbf{X}), I^j(\mathbf{X}')) \in \mathrm{I_{sp}}$.
(3) I^j *induces a map* $K_+ \mathrm{VFR}[k, l, \cdot]/\mathrm{I_{sp}} \to K_+ \mathrm{VFR}[k - 1, l + 1, \cdot]/\mathrm{I_{sp}}$.

Proof.

(1) This reduces to the case of I_1^1, where it is an immediate consequence of the uniqueness, and the fact that $\mathbb{L}$ commutes with maps into RV in the same sense.
(2) All equivalences here are relative to the $k - 1$ coordinates of VF other than j, so we may assume $k = 1$. For $a \in \mathrm{VF}$, $([\mathbf{X}_a], [\mathbf{X}'_a]) \in \mathrm{I_{sp}}(\mathbf{T}_a)$. By stable embeddedness of RV, there exists $\alpha = \alpha(a) \in \mathrm{RV}^*$ such that $\mathbf{X}_a, \mathbf{X}'_a$ are $\mathbf{T}_\alpha$-definable and $([\mathbf{X}]_a, [\mathbf{X}']_a) \in \mathrm{I_{sp}}(\mathbf{T}_\alpha)$. Fibering over the map α we may assume by (1) and Lemma 7.20 that α is constant; so for some $W \in \mathrm{VF}[1]$, $\mathbf{Y}, \mathbf{Y}' \in \mathrm{RV}[l, \cdot]$, we have $\mathbf{X} = W \times \mathbf{Y}, \mathbf{X}' = W \times \mathbf{Y}'$, and $([\mathbf{Y}], [\mathbf{Y}']) \in \mathrm{I_{sp}}$. Then $I^j(\mathbf{X}) = I^j(W) \times \mathbf{Y}, I^j(\mathbf{X}') = I^j(W) \times \mathbf{Y}'$, and the conclusion is clear.

(3) This follows from (2). □

Lemma 7.24. *Let* $\mathbf{X} = (X, f)$, $X \subseteq \mathrm{VF}^J \times \mathrm{RV}^\infty$, $f : X \to \mathrm{RV}^l$. *If* $j \neq j' \in J$, *then* $I^j I^{j'} = I^{j'} I^j : K_+ \mathrm{VFR}[k, l, \cdot]/\mathrm{I_{sp}} \to K_+ \mathrm{VFR}[k-2, l+2, \cdot]/\mathrm{I_{sp}}$.

Proof. We may assume $S = \{1, 2\}$, $j = 1$, $j' = 2$, since all is relative to $\mathrm{VF}^{S \setminus \{j, j'\}}$. By Lemma 7.23(1) it suffices to prove the statement for each fiber of a given definable map into RV.

Hence we may assume $X \subseteq \mathrm{VF}^2$ and f is constant; and by Lemma 5.10, we can assume X is a basic 2-cell:

$$X = \{(x, y) : x \in X_1, \mathrm{rv}(y - G(x)) = \alpha_1\}, \quad X_1 = \mathrm{rv}^{-1}(\delta_1) + c_1.$$

The case where G is constant is easy since then X is a finite union of rectangles. Otherwise, G is invertible, and by the niceness of G we can also write

$$X = \{(x, y) : y \in X_2, \mathrm{rv}(x - G^{-1}(y)) = \beta\}, \quad X_2 = \mathrm{rv}^{-1}(\delta_2) + c_2.$$

We immediately compute

$$I_2 I_1(X) = (\delta_1, \alpha_1), \qquad I_1 I_2(X) = (\alpha_2, \delta_2).$$

Clearly, $[(\delta_1, \alpha_1)]_2 = [(\alpha_2, \delta_2)]_2$. □

Proposition 7.25. *Let* $\mathbf{X}, \mathbf{Y} \in \mathrm{RV}[\leq n, \cdot]$. *If* $\mathbb{L}\mathbf{X}$, $\mathbb{L}\mathbf{Y}$ *are isomorphic, then* $([X], [Y]) \in \mathrm{I_{sp}}$.

Proof. Define $I = I_1 \dots I_n : \mathrm{VF}[n, \cdot] = \mathrm{VFR}[n, 0, \cdot] \to \mathrm{VFR}[0, n, \cdot] = \mathrm{RV}[\leq n, \cdot]$. Let $V \in \mathrm{VF}[n, \cdot]$.

Claim 1. If $\sigma \in \mathrm{Sym}(n)$ then $I = I_{\sigma(1)} \dots I_{\sigma(n)}$.

Proof. We may assume σ just permutes two adjacent coordinates, say, 2, 3 out of 1, 2, 3, 4. Then $I = I_1 I_2 I_3 I_4 = I_1 I_3 I_2 I_4$ by Lemma 7.24. □

Claim 2. When $F : V \to F(V)$ is a relatively unary bijection, we have $I(V) = I(F(V))$.

Proof. By Claim 1 we may assume F is relatively unary with respect to pr^n. Thus $F(V_a) = F(V)_a$, where V_a, $F(V)_a$ are the pr^n-fibers. By the definition of I_1^1, we have $I_1^1(V_a) = I_1^1(F(V)_a) \in \mathrm{RV}[1, \cdot](\mathbf{T}_a)$; but by the definition of I^n, $I_n(V)_a = I_1^1(V_a)$. Thus $I^n(V) = I^n(F(V))$ and thus $I(V) = I(F(V))$. □

Claim 3. When $F : V \to F(V)$ is any definable bijection, $I(V) = I(F(V))$.

Proof. The proof is immediate from Claim 2 and Lemma 7.11. □

Now turning to the statement of the proposition, assume $\mathbb{L}\mathbf{X}$, $\mathbb{L}\mathbf{Y}$ are isomorphic. We compute inductively that $\mathbb{L}(\mathbf{X}) = [\mathbf{X}]$. By Claim 3, $[\mathbf{X}] = I(\mathbb{L}\mathbf{X}) = I(\mathbb{L}\mathbf{Y}) = [\mathbf{Y}]$. □

Notation 7.26. Let $\mathbb{L}^* : K_+(\mathrm{VF}) \to K_+(\mathrm{RV}[*])/Isp$ be the inverse map to $\mathbb{L}$.

Remark 7.27. When **T** is rv-effective, one can restate the conclusion of Proposition 7.25 as follows: if $X, Y \in \mathrm{VF}[n, \cdot]$ are $\underset{\mathrm{rv}}{\sim}$-invariant and $F : X \to Y$ is a definable bijection, then there exist special bijections $X' \to X$ and $Y' \to Y$ and an $\underset{\mathrm{rv}}{\sim}$-invariant-definable bijection $G : X' \to Y'$. (This follows from Propositions 7.25 and 6.1 and Lemmas 7.18 and 7.19.) The effectiveness hypothesis is actually unnecessary here, as will be seen in the proof of Proposition 8.26. Perhaps Question 7.9 can be answered simply by tracing the connection between F and G through the proof.

8 Definable sets over VF and RV: The main theorems

In stating the theorems, we restrict attention to $\mathrm{VF}[n]$, i.e., to definable subsets of varieties, though the proof was given more generally for $\mathrm{VF}[n, \cdot]$ (definable subsets of $\mathrm{VF}^n \times \mathrm{RV}^*$).

8.1 Definable subsets of varieties

Let **T** be V-minimal. We will look at the category of definable subsets of varieties, and definable maps between them. The results will be stated for $\mathrm{VF}[n]$; analogous statements for $\mathrm{VF}[n, \cdot]$ are true with the same proofs.

We define three variants of the sets of objects. $\mathrm{VF}''[n]$ is the category of $\leq n$-dimensional definable sets over VF, i.e., of definable subsets of n-dimensional varieties. Let $\mathrm{VF}[n]$ be the category of definable subsets $X \subseteq \mathrm{VF}^n \times \mathrm{RV}^*$ such that the projection $X \to \mathrm{VF}^n$ has finite fibers. $\mathrm{VF}'[n]$ is the category of definable subsets X of $V \times \mathrm{RV}^*$, where V ranges over all $\mathrm{VF}(A)$-definable sets of dimension $n, m \in \mathbb{N}$, such that the projection $X \to V$ is finite-to-one. VF, VF′, VF″ are the unions over all n. In all cases, the morphisms $\mathrm{Mor}(X, Y)$ are the definable functions $X \to Y$.

Lemma 8.1. *The natural inclusion of* $\mathrm{VF}[n]$ *in* $\mathrm{VF}'[n]$ *is an equivalence. If* **T** *is effective, so is the inclusion of* $\mathrm{VF}''[n]$ *in* $\mathrm{VF}'[n]$.

Proof. We will omit the index $\leq n$. The inclusion is fully faithful by definition, and we have to show that it hits every VF′-isomorphism type; in other words, that any definable $X \subseteq (V \times \mathrm{RV}^m)$ is definably isomorphic to some $X' \subseteq \mathrm{VF}^n \times \mathrm{RV}^{m+l}$ for some l (with $n = \dim(V)$). Definable isomorphisms can be glued on pieces, so we may assume V is affine, and admits a finite-to-one map $h : V \to \mathrm{VF}^m$. By Lemma 3.9, each fiber $h^{-1}(a)$ is $A(a)$-definably isomorphic to some $F(a) \subseteq \mathrm{RV}^l$. By compactness, F can be chosen uniformly definable, $F(a) = \{y \in \mathrm{RV}^l : (a, y) \in F\}$ for some definable $F \subseteq \mathrm{VF}^m \times \mathrm{RV}^l$; and there exists a definable isomorphism $\beta : V \to F$, over VF^m. Let $\alpha(v, t) = (\beta(v), t)$, $X' = \alpha(X)$.

Now assume **T** is effective. Let $X \in \mathrm{Ob}\,\mathrm{VF}'$; $X \subseteq V \times \mathrm{RV}^m$, $V \subseteq \mathrm{VF}^n$, such that the projection $X \to V$ has finite fibers. Then by effectivity, for any $v \in V$ (over any extension field), if $(v, c_1, \ldots, c_m) \in X$ then each c_i, viewed as a ball, has a point defined over $A(v)$. Hence the partial map $V \times \mathrm{VF}^m \to X$, $(v, x_1, \dots, x_m) \mapsto (v, \mathrm{rv}(x_1), \dots, \mathrm{rv}(x_m))$ has an A-definable section; the image of this section is a subset S of $V \times \mathrm{VF}^m$, definably isomorphic to X; and the Zariski closure V' of S in $V \times \mathrm{VF}^m$ has dimension $\leq \dim(V)$. □

The following definition and proposition apply both to the category of definable sets, and to the definable sets with volume forms.

Definition 8.2. X, Y are *effectively isomorphic* if

for any effective A, X, Y are definably isomorphic in $\mathbf{T}_A$. If $K_+^{\mathrm{eff}}(\mathrm{VF})$ is the semiring of effective isomorphic classes of definable sets. $K(\mathrm{VF})$ is the corresponding ring; similarly $K_+^{\mathrm{eff}}(\mathrm{VF}[n])$, etc.

Over an effective base, in particular, if $\mathbf{T}$ is effective over any field-generated base, effectively isomorphic is the same as isomorphic. But Example 4.7 shows that this is not so in general.

Proposition 8.3. *Let T be* V*-minimal, or a finitely generated extension of a* V*-minimal theory. The following conditions are equivalent (let $X, Y \in \mathrm{VF}[n]$):*

(1) $[\mathbb{L}^* X] = [\mathbb{L}^* Y]$ *in* $K_+(\mathrm{RV}[\leq n])/\mathrm{I}_{\mathrm{sp}}[\leq n]$.
(2) *There exists a definable family $\mathfrak{F}$ of definable bijections $X \to Y$ such that for any effective structure A, $F(A) \neq \emptyset$.*
(3) *X, Y are effectively isomorphic.*
(4) *X, Y are definably isomorphic over any A such that $\mathrm{VF}^*(A) \to \mathrm{RV}(A)$ is surjective.*
(5) *For some finite $A_0 \subseteq \mathrm{RV}(\langle\emptyset\rangle)$, X, Y are definably isomorphic over any A such that $A_0 \subseteq \mathrm{rv}(\mathrm{VF}^*(A))$.*

Proof.

(1) implies (5): By Proposition 6.1 (Proposition 6.3 in the measured case), the given isomorphism $[\mathbb{L}^* X] \to [\mathbb{L}^* Y]$ lifts to an isomorphism $\mathbb{L}\mathbb{L}^* X \to \mathbb{L}\mathbb{L}^* Y$; since $\mathbf{T}_A \supseteq \mathrm{ACVF}_A$, this is also a $\mathbf{T}_A$ isomorphism; it can be composed with the isomorphisms $X \to \mathbb{L}\mathbb{L}^* X$, $Y \to \mathbb{L}\mathbb{L}^* Y$.

(2) implies (3), (5) implies (4) implies (3), trivially.

(3) implies (1)–(2): Let E_{eff} be as in Proposition 3.51. By (3), X, Y are E_{eff}-isomorphic. By Proposition 7.25, $[\mathbb{L}^* X] = [\mathbb{L}^* Y]$ in $K_+(\mathrm{RV}_{E_{\mathrm{eff}}}[*])/\mathrm{I}_{\mathrm{sp}}$. But $\mathrm{RV}(E_{\mathrm{eff}}), \Gamma(E_{\mathrm{eff}}) \subseteq \mathrm{dcl}(\emptyset)$, so every E_{eff}-definable relation on RV is definable; i.e., $\mathrm{RV}_{E_{\mathrm{eff}}}$, RV are the same structure. Thus (1) holds.

Now by assumption, there exists an E_{eff}-definable bijection $f' : X \to Y$. f' is an E_{eff}-definable element of a definable family $\mathfrak{F}$ of definable bijections $X \to Y$. Since this family has an E_{eff}-point, and E_{eff} embeds into any effective B, it has a B point, too. Thus (3) implies (2). □

8.2 Invariants of all definable maps

Let $[X]$ denote the class of X in $K_+^{\text{eff}}(\text{VF}[n])$.

Proposition 8.4. *Let* **T** *be* V*-minimal. There exists a canonical isomorphism of Grothendieck semigroups*

$$\oint : K_+^{\text{eff}}(\text{VF}[n]) \to K_+(\text{RV}[\leq n])/\text{I}_{\text{sp}}[\leq n]$$

satisfying

$$\oint [X] = W/\text{I}_{\text{sp}}[\leq n] \iff [X] = [\mathbb{L}W] \in K_+^{\text{eff}}(\text{VF}[n]).$$

Proof. Recall Definition 4.8. Given $\mathbf{X} = (X, f) \in \text{Ob}\,\text{RV}[k]$ we have $\mathbb{L}\mathbf{X} \in \text{Ob}\,\text{VF}[k] \subseteq \text{Ob}\,\text{VF}[n]$. If $\mathbf{X}, \mathbf{X}'$ are isomorphic,then by Proposition 6.1, $\mathbb{L}\mathbf{X}, \mathbb{L}\mathbf{X}'$ are effectively isomorphic. Direct sums are clearly respected, so we have a semigroup homomorphism $\mathbb{L} : K_+(\text{RV}[\leq n]) \to K_+^{\text{eff}}(\text{VF}[n])$. It is surjective by Proposition 4.5. By Proposition 8.3, the kernel is precisely $\text{I}_{\text{sp}}[\leq n]$. Inverting, we obtain $\oint$. □

Definition 8.5. Let $K_+\,\text{VF}[n]/(\dim < n)$ be the Grothendieck ring of the category of definable subsets of n-dimensional varieties, and essential bijections between them. Let $\text{I}_{\text{sp}}'[n]$ be the congruence on $RV[n]$ generated by pairs $(X, X \times \text{RV}^{>0})$ (where $X \subseteq \text{RV}^*$ is definable, of dimension $< n$).

Corollary 8.6. $\oint$ *induces an isomorphism*

$$K_+^{\text{eff}}(\text{VF}[n])/(\dim < n) \to \text{RV}[n]/\text{I}_{\text{sp}}'[n].$$ □

Corollary 8.7. *Let* $A, B \in \text{RV}[\leq n]$*. Let* $n' > n$*, and let* A_N, B_N *be their images in* $\text{RV}[\leq N]$*. If* $(A_N, B_N) \in \text{I}_{\text{sp}}[\leq N]$*, then* $(A, B) \in \text{I}_{\text{sp}}[\leq n]$.

Proof. By Proposition 8.4, $(A, B) \in \text{I}_{\text{sp}}[\leq n]$ iff $\mathbb{L}A, \mathbb{L}B$ are definably isomorphic; this latter condition does not depend on n. □

Putting Proposition 8.4 together for all n, we obtain the following.

Theorem 8.8. *Let* **T** *be* V*-minimal. There exists a canonical isomorphism of filtered semirings*

$$\oint : K_+(\text{VF}) \to K_+(\text{RV}[*])/\text{I}_{\text{sp}}.$$

Let $[X]$ *denote the class of* X *in* $K_+(\text{VF})$*. Then*

$$\oint [X] = \frac{W}{\text{I}_{\text{sp}}} \iff [X] = [\mathbb{L}W] \in K_+^{\text{eff}}(\text{VF}).$$ □

On the other hand, using the Grothendieck group isomorphisms of Proposition 8.4 and passing to the limit, we have the following.

Corollary 8.9. *Let* $\mathbf{T}$ *be* V*-minimal. The isomorphisms of Proposition* 8.4 *induce an isomorphism of Grothendieck groups:*

$$\int^K : K^{\text{eff}}(\text{VF}[n]) \to K(\text{RV}[n]).$$

The isomorphism $\oint$ *of Theorem* 8.8 *induces an injective ring homomorphism*

$$\int^K : K^{\text{eff}}(\text{VF}) \to K(\text{RV})[J^{-1}],$$

where $J = \{1\}_1 - [\text{RV}^{>0}]_1 \in K(\text{RV})$.

Proof. We may work over an effective base. With subtraction allowed, the generating relation of I_{sp} can be read as $[\{1\}]_0 = \{1\}_1 - [\text{RV}^{>0}]_1 := J$, so that the groupification of $K_+(\text{RV}[\leq n])/\text{I}_{\text{sp}}[\leq n]$ is isomorphic to $K(\text{RV}[n])$, via the embedding of $K_+(\text{RV}[n])$ as a direct factor in $K_+(\text{RV}[\leq n])$. Thus the groupification of the homomorphism of Theorem 8.8 is a homomorphism

$$\int^K : K(\text{VF}) \to \lim_{n\to\infty} K(\text{RV}[n]),$$

where the direct limit system maps are given by $[X]_d \mapsto ([X]_{d+1} - ([X]_d \times (\text{RV}^{>0}))) = [X_d]J$. This direct limit embeds into $K(\text{RV})[J^{-1}]$ by mapping $X \in K(\text{RV}[n])$ to XJ^{-n}. □

8.3 Definable volume forms: VF

We will now define the category $\mu\text{VF}[n]$ of "n-dimensional $\mathbf{T}_A$-definable sets with definable volume forms, up to RV-equivalence" and the same up to Γ-equivalence. We will represent the forms as functions to RV, that transform in the way volume forms do.

By way of motivation, in a local field with an absolute value, a top differential form ω induces a measure $|d\omega|$. For a regular isomorphism $f : V \to V'$, we have $\omega = hf^*\omega'$ for a unique h, and f is measure preserving between $(V, |\omega|)$ and $(V', |\omega'|)$ iff $|h| = 1$.

We do not work with an absolute value into the reals, but instead define the analogue using the map rv or, a coarser version, the map val into Γ. When $\Gamma = \mathbb{Z}$, the latter is the usual practice in Denef-style motivic integration. Using rv leaves room for considering an absolute value on the residue field, and iterating the integration functorially when places are composed, for instance, $\mathbb{C}((x))((y)) \to \mathbb{C}((x)) \to \mathbb{C}$. This functoriality will be described in a future work.

In the definition below, the words "almost every $y \in Y$" will mean for all y outside a set of VF dimension $< \dim_{\text{VF}}(Y)$.

Definition 8.10. Ob μVF$[n, \cdot]$ consists of pairs (Y, ω), where Y is a definable subset of $\mathrm{VF}^n \times \mathrm{RV}^*$, and $\omega : Y \to \mathrm{RV}$ is a definable map. A morphism $(Y, \omega) \to (Y', \omega')$ is a definable essential bijection F such that for almost every $y \in Y$,

$$\omega(y) = \omega'(F(y)) \cdot \mathrm{rv}(\mathrm{Jcb}\, F(y)).$$

(We will say "$F : (Y, \omega) \to (Y', \omega')$ is measure preserving.")

μ_ΓVF$[n, \cdot]$ is the category of pairs (Y, ω) with $\omega : Y \to \Gamma$ a definable function A morphism $(Y, \omega) \to (Y', \omega')$ is a definable essential bijection $F : Y \to Y'$ such that for almost every $y \in Y$,

$$\omega(y) = \omega'(F(y)) + \mathrm{val}(\mathrm{Jcb}\, F(y)).$$

("$F : (Y, \omega) \to (Y', \omega')$ is Γ-measure preserving.")

μVF$[n]$, μ_ΓVF$[n]$ are the full subcategories of μVF$[n, \cdot]$, μ_ΓVF$[n, \cdot]$ (respectively) whose objects admit a finite-to-one map to VF^n.

In this definition, let $t_1(y), \dots, t_n(y)$ be the VF-coordinates of $y \in Y$. One can think of the form as $\omega(y) dt_1 \cdot \dots \cdot dt_n$.

Note that $\mathrm{VF}_{\mathrm{vol}}$ of Definition 5.19 is isomorphic to the full subcategory of μVF whose objects are pairs $(Y, 1)$.

Remark 8.11. When **T** is V-minimal and effective, the data ω of an object (Y, ω) of μVF$[n]$ can be written as $\mathrm{rv} \circ \Phi$ for some $\Phi : Y \to \mathrm{VF}$. (Write $\omega = \bar{\omega} \circ \mathrm{rv} \circ F$ for some F, and use Proposition 6.1 to lift $\bar{\omega}$ to some G, so that $\omega = \mathrm{rv} \circ G \circ F$.) It is thus possible to view ω as the RV-image (respectively, Γ-image) of a definable volume form on Y. One could equivalently take ω to be a definable section of $\Lambda^n TY/(1+\mathcal{M})$, where TY is the (appropriately defined) tangent bundle, Λ^n the nth exterior power with $n = \dim(Y)$.

For VF_Γ the category we take is slightly more flexible than taking varieties with absolute values of volume forms, even if **T** is V-minimal and effective, in that expressions such as $\int |\sqrt{x}| dx$ are allowed.

In either of these categories, one could restrict the objects to bounded ones.

Definition 8.12. Let $\mu\mathrm{VF}_{\mathrm{bdd}}[n]$ be the full subcategory of μVF$[n]$ whose objects are bounded definable sets, with bounded definable forms ω. Similarly, one defines $\mu\mathrm{VF}_{\Gamma;\mathrm{bdd}}$.

Here *bounded* means that there is a lower bound on the valuation of any coordinate of any element of the set. A similar definition applies in RV and μRV.

Note that if an object of μVF$[n]$ is μVF$[n]$-isomorphic to an object of $\mu\mathrm{VF}_{\mathrm{bdd}}[n]$, it must lie in $\mu\mathrm{VF}_{\mathrm{bdd}}[n]$.

8.4 Definable volume forms: RV

We will define a category $\mu\mathrm{RV}[n]$ of definable subsets of $(\mathrm{RV})^m$, with additional data that can be viewed as a volume form. Unlike $\mu\mathrm{VF}[n]$, in $\mu\mathrm{RV}[n]$ subsets of dimension $< n$ are *not* ignored: a point of RV^n corresponds to an open polydisc of VF^n, with nonzero n-dimensional volume.

In particular, the Jacobian of a morphism needs to be defined at every point, not just away from a lower-dimensional set. However, in accord with Lemma 6.3, it may be modified by $\mathbf{k}^*$-multiplication on a lower-dimensional set.

Definition 8.13. The objects of $\mu\mathrm{RV}[n]$ are definable triples (X, f, ω), $X \subseteq \mathrm{RV}^{n+m}$, $f : X \to \mathrm{RV}^n$ finite-to-one, and $\omega : X \to \mathrm{RV}$.

We define a multiplication $\mu\mathrm{RV}[n] \times \mu\mathrm{RV}[n'] \to \mu\mathrm{RV}[n+n']$ by $(X, f, \omega) \times (X', f', \omega') = (X \times X', f \times f', \omega \cdot \omega')$. Here $\omega \cdot \omega'(x, x') = \omega(x)\omega'(x')$.

Given $\mathbf{X} = (X, f, \omega)$, we define an object $\mathbb{L}\mathbf{X}$ of $\mathrm{VF}[n]$; namely, $(\mathbb{L}X, \mathbb{L}f, \mathbb{L}\omega)$, where $\mathbb{L}X = X \times_{f,\mathrm{rv}} (\mathrm{VF}^\times)^n$, $\mathbb{L}f(a, b) = f(a, \mathrm{rv}(b))$, $\mathbb{L}\omega(a, b) = \omega(a, \mathrm{rv}(b))$. (Sometimes we will write f, ω for $\mathbb{L}f, \mathbb{L}\omega$.)

A morphism $\alpha : \mathbf{X} = (X, f, \omega) \to \mathbf{X}' = (X', f', \omega')$ is a definable bijection $\alpha : X \to X'$ such that

$$\omega(y) = \omega'(\alpha(y)) \cdot \mathrm{rv}(\mathrm{Jcb}_{\mathrm{RV}}(\alpha)(y)) \quad \text{for almost all } y,$$

where "almost all" means "away from a set Y with $\dim_{\mathrm{RV}}(f(Y)) < n$"; and

$$\mathrm{val}_{\mathrm{rv}}\omega(y) + \sum_{i=1}^{n} \mathrm{val}_{\mathrm{rv}} f_i(y) = \mathrm{val}_{\mathrm{rv}}\omega'(\alpha(y)) + \sum_{i=1}^{n} \mathrm{val}_{\mathrm{rv}} f_i'(\alpha y) \quad \text{for all } y.$$

The objects of $\mu_\Gamma\mathrm{RV}[n]$ are triples (X, f, ω), with $f : X \to \mathrm{RV}^n$, $\omega : X \to \Gamma$. A morphism $\alpha : (X, f, \omega) \to (X', f', \omega')$ is a definable bijection $\alpha : X \to X'$ such that $\mathrm{val}_{\mathrm{rv}}\omega(y) + \sum_{i=1}^{n} \mathrm{val}_{\mathrm{rv}} f_i(y) = \mathrm{val}_{\mathrm{rv}}\omega'(\alpha(y)) + \sum_{i=1}^{n} \mathrm{val}_{\mathrm{rv}} f_i'(\alpha y)$ for all y. Disjoint sums and products are defined as for $\mu\mathrm{RV}$.

$\mu_\Gamma\mathrm{RES}[n]$ is the full subcategory of $\mu_\Gamma\mathrm{RV}[n]$ with objects (X, f, ω), such that $\mathrm{val}_{\mathrm{rv}}(X)$ is finite. In this case, ω takes finitely many values, too.

$K_+^{\mathrm{eff}}\, \mu\mathrm{RV}[n]$ is the Grothendieck semigroup of $\mu\mathrm{RV}[n]$ with respect to effective isomorphism. $K_+^{\mathrm{eff}}\, \mu\mathrm{RV}$ is the direct sum $\bigoplus_n K_+^{\mathrm{eff}}\, \mu\mathrm{RV}[n]$; it clearly inherits a semiring structure from Cartesian multplication, $(X, f, \omega) \times (X', f', \omega') = (X \times X', (f, f'), \omega \cdot \omega')$.

The morphisms of $\mu_\Gamma\mathrm{RV}[n]$ are called Γ-measure preserving.

The category $\mathrm{RV}_{\mathrm{vol}}[n, \cdot]$ of Definition 5.21 is isomorphic to the full subcategory whose objects have $\omega = 1$.

Remark. The semiring $K_+^{\mathrm{eff}}\, \mathrm{RV}_{\mathrm{vol}}$ is naturally a subsemiring of $K_+^{\mathrm{eff}}\, \mu\mathrm{RV}$. The latter is obtained by inverting $[\{a\}]_1$ for $a \in \mathrm{RV}$ and taking the zeroth graded component. This process is needed in order to identify integrals of functions in n variables with volumes in $n+1$ variables. Thus as semirings they are closely related. But if the dimension grading is taken into account, the subsemiring of RV-volumes contains finer information connected to integrability of forms.

8.5 The kernel of $\mathbb{L}$ in the measured case

The description of the kernel of $\mathbb{L}$ on the semigroups of definable sets with volume forms is essentially the same as for definable sets. We will now run through the proof, indicating the modficiations. The principal change is the introduction of a category with fewer morphisms, defined not only with reference to RV but also to VF. For effective bases, the category is identical to μRV, so it will be invisible in the statements of the main theorems; but during the induction in the proof, bases will not in general be effective and the mixed category introduced here has better properties.

Both the introduction of the various intermediate categories and the repetition of the proof would be unnecessary if we had a positive answer to Question 7.9. In this case the proof of Lemma 8.23 would immediately lift to higher dimensions. Indeed, the characterization of the kernel of the map $\mathbb{L}$ on Grothendieck groups would be uniformized not only for the categories we consider, but for a range of categories carrying more structure.

The integer n will be fixed in this subsection.

Lemma 8.14. *Let* $(X, \omega) \in \mathrm{Ob}\, \mu\mathrm{VF}[n, \cdot]$, $Y \in \mathrm{Ob}\,\mathrm{VF}[n, \cdot]$, *and let* $F : Y \to X$ *be a definable bijection.*

(1) *There exists* $\psi : Y \to \mathrm{RV}$ *such that* $F : (Y, \psi) \to (X, \omega)$ *is measure preserving.*
(2) ψ *is essentially unique in the sense that if* ψ' *meets the same condition, then* ψ, ψ' *are equal away from a subset of* X *of lower dimension.*
(3) *Dually, given* F, X, Y, ψ, *there exists an essentially unique* ω *such that* $F : (Y, \psi) \to (X, \omega)$ *is measure preserving.*
(4) *Lemma* 7.11 *applies to* $\mu\mathrm{VF}[n, \cdot]$ *and to* $\mu\mathrm{VF}[n]$.

Proof.

(1)–(2) Let $\psi(y) = \omega(\alpha(y)) \cdot \mathrm{rv}(\mathrm{Jcb}_{\mathrm{RV}}(\alpha)(y))$. By the definition of μVF this works, and is the only choice "almost everywhere."
(3) This follows from the case of F^{-1}.
(4) Now let $\mathbf{X}, \mathbf{Y} \in \mathrm{Ob}\,\mu\mathrm{VF}[n]$ and let $F \in \mathrm{Mor}_{\mu\mathrm{VF}[n]}(X, Y)$. We have $\mathbf{X} = (X, \omega_X), \mathbf{Y} = (Y, \omega_Y)$ for some $X, Y \in \mathrm{Ob}\,\mathrm{VF}[n]$ and $\omega_X : X \to \mathrm{RV}, \omega_Y : Y \to \mathrm{RV}$. By Lemma 7.11 there exist $X = X_1, \ldots, X_n = Y \in \mathrm{Ob}\,\mathrm{VF}[n]$ and essentially unary $F_i : X_i \to X_{i+1}$ with $F = F_{n-1} \circ \cdots \circ F_1$. Let $\omega_1 = \omega_X$, and inductively let ω_{i+1} be such that $F_i \in \mathrm{Mor}_{\mu\mathrm{VF}[n]}((X_i, \omega_i), (X_{i+1}, \omega_{i+1}))$. Then $F \in \mathrm{Mor}_{\mu\mathrm{VF}[n]}((X, \omega), (Y, \omega_n))$. By uniqueness it follows that ω_Y, ω_n are essentially equal. □

Definition 8.15. Given $\mathbf{X}, \mathbf{Y} \in \mathrm{Ob}\,\mu\mathrm{RV}[n, \cdot]$ call a definable bijection $h : X \to Y$ *liftable* if there exists $F \in \mathrm{Mor}_{\mu\mathrm{VF}[n,\cdot]}(\mathbb{L}X, \mathbb{L}Y)$ with $\rho_Y F = h\rho_X$.

Let $\mathcal{C} = \mu_l\mathrm{RV}[n, \cdot]$ be the subcategory of $\mu\mathrm{RV}[n, \cdot]$ consisting of all objects and liftable morphisms.

By Proposition 5.22, liftable morphisms must preserve the volume forms, so $\mathcal{C}$ is a subcategory of $\mu\mathrm{RV}[n, \cdot]$.

Over an effective base, $\mathcal{C} = \mu\mathrm{RV}[n,\cdot]$ (Lemma 6.3), and the condition of existence of s in Definition 8.16(1) below is equivalent to $f_n(y) \in \mathrm{acl}(f_1(y),\ldots,f_{n-1}(y))$.

Definition 8.16.

(1) Let $\mathbf{Y} = (Y, f, \omega) \in \mathrm{Ob}\,\mu\mathrm{RV}[n,\cdot]$ be such that there exists $s : \mathbf{Y} \times_{f_1,\ldots,f_{n-1}} \mathrm{VF}^{n-1} \to \mathrm{VF}$ with $\mathrm{rv}(s(y,u_1,\ldots,u_{n-1})) = f_n(y)$. Let $Y' = Y \times \mathrm{RV}^{>0}$. For $(y,t) \in Y'$, define $f' = (f'_1,\ldots,f'_n)$ by $f'_i(y,t) = f_i(y)$ for $i < n$, $f'_n(y,t) = tf_n(y)$. Let $\omega'(y,t) = \omega(y)$. Then $\widetilde{\mathbf{Y}} = (Y', f', \omega')$ is an *elementary blowup* of $\mathbf{Y}$. It comes with the projection map $Y' \to Y$.

(2) Let $\mathbf{X} = (X, g, \omega) \in \mathrm{Ob}\,\mu\mathrm{RV}[n,\cdot]$, $X = X' \dot\cup X''$, $g' = g|X'$, $g'' = g|X''$, $\omega' = \omega|X'$, $\omega'' = \omega|X''$, and let $\phi : \mathbf{Y} \to (X', g', \omega')$ be a $\mu_l\mathrm{RV}[n,\cdot]$-isomorphism. Then the RV-*blowup* $\widetilde{\mathbf{X}}_\phi$ is defined to be $\widetilde{\mathbf{Y}} + (X'', g'', \omega'') = (Y' \dot\cup X'', f' \dot\cup g'', \omega' \dot\cup \omega'')$. It comes with $b : Y' \dot\cup X'' \to X$, defined to be the identity on X'', and the projection on Y'. X' is called the *blowup locus* of $b : \widetilde{\mathbf{X}}_\phi \to \mathbf{X}$.

An *iterated* RV-*blowup* is obtained by finitely many iterations of RV-blowups.

Definition 8.17. Let $\mathrm{I}^{\mu}_{\mathrm{sp}}[n]$ be the set of pairs $(\mathbf{X}_1, \mathbf{X}_2) \in \mathrm{Ob}\,\mu\mathrm{RV}[n,\cdot]$ such that there exist iterated blowups $b_i : \widetilde{\mathbf{X}}_i \to \mathbf{X}_i$ and a $\mu_l\mathrm{RV}[n,\cdot]$-isomorphism $F : \widetilde{\mathbf{X}}_1 \to \widetilde{\mathbf{X}}_2$.

When n is fixed, we will simply write $\mathrm{I}^{\mu}_{\mathrm{sp}}$. On the other hand, we will need to make explicit the dependence on the theory; we write $\mathrm{I}^{\mu}_{\mathrm{sp}}(A)$ for the congruence $\mathrm{I}^{\mu}_{\mathrm{sp}}$ of the theory $\mathbf{T}_A$.

When $\mathbf{X} = (X, f, \omega) \in \mathrm{Ob}\,\mu\mathrm{RV}[n,\cdot]$, $h : X \to W$ is a definable map, and $c \in W$, define $\mathbf{X}_c = (h^{-1}(c), f|h^{-1}(c), \omega|h^{-1}(c))$.

Let $X_1, X_2 \in \mathrm{Ob}\,\mu\mathrm{RV}[n,\cdot]$, and let $f_i : X_i \to Y$ be a definable map, with $Y \subseteq \mathrm{RV}^*$. In this situation the existence of $\mu\mathrm{RV}[n,\cdot](\langle a\rangle)$-isomorphisms between each pair of fibers $X_1(a)$, $X_2(a)$ $(a \in Y)$ does not necessarily imply that $X_1 \simeq_{\mu\mathrm{RV}[\le n,\cdot]} X_2$, because of the explicit reference to dimension in the definition of morphisms; the dimension of the allowed exceptional sets may accumulate over Y. The definition of morphisms for $\mu\mathrm{VF}[n]$ also allows a lower-dimensional exceptional set; but this does not create a problem when fibered over $W \subseteq \mathrm{RV}^*$, since by Lemma 3.56 $\max_{c\in W} \dim_{\mathrm{VF}}(Z_c) = \dim_{\mathrm{VF}}(Z)$. Thus an RV-disjoint union of $\mu\mathrm{VF}[n]$-isomorphisms is again a $\mu\mathrm{VF}[n]$-isomorphism, and it follows that the same is true for $\mu_l\mathrm{RV}[n,\cdot]$. We thus have the following.

Lemma 8.18. *Let* $\mathbf{X} = (X, f, \omega)$, $\mathbf{X}' = (X', f', \omega) \in \mu\mathrm{RV}[n,\cdot]$, *and let* $h : X \to W \subseteq \mathrm{RV}^*$, $h' : X' \to W$ *be definable maps. If for each* $c \in W$, $(\mathbf{X}_c, \mathbf{X}'_c) \in \mathrm{I}^{\mu}_{\mathrm{sp}}(\langle c\rangle)$, *then* $(\mathbf{X}, \mathbf{X}') \in \mathrm{I}^{\mu}_{\mathrm{sp}}$.

Proof. Lemma 2.3 applies to $\mathrm{RV}_{\mathrm{vol}}$-isomorphisms, and hence using Remark 7.13, also to blowups. It also applies to $\mu_l\mathrm{RV}[n,\cdot]$-isomorphisms by the discussion above, and hence to $\mathrm{I}^{\mu}_{\mathrm{sp}}$-equivalence. □

In other words, there exists a well-defined direct sum operation on $\mu\mathrm{RV}[n,\cdot]/\mathrm{I}^{\mu}_{\mathrm{sp}}$, with respect to RV-indexed systems.

Lemma 8.19.

(1) *Let* $\mathbf{Y}'$ *be an elementary blowup of* $\mathbf{Y}$. $\mathbf{Y}'$ *is* $\mathcal{C}$*-isomorphic to* $\mathbf{Y}'' = (Y'', f'', \omega')$, *with*

$$Y'' = \{(y, t) \in Y \times \mathrm{RV}_\infty : \mathrm{val}_{\mathrm{rv}}(t) > f_n(y)\},$$
$$f''(y, t) = (f_1(y), \dots, f_{n-1}(y), t), \quad \omega'(y, t) = \omega(y).$$

(2) *Up to isomorphism, the blowup depends only on the blowup locus. In other words, if* X, X', g, g', ω, ω' *are as in Definition* 8.16, *and* $\phi_i : \mathbf{Y}_i \to (X', g', \omega')$ $(i = 1, 2)$ *are* $\mu_l\mathrm{RV}[n, \cdot]$*-isomorphisms, then* $\widetilde{\mathbf{X}}_{\phi_1}$, $\widetilde{\mathbf{X}}_{\phi_2}$ *are* $\mathbf{X}$*-isomorphic in* $\mu_l\mathrm{RV}[n, \cdot]$.

Proof.

(1) The isomorphism is given by $h((y, t)) = (y, t f_n(y))$; since f_n always lifts to a function $F_n : \mathbb{L}Y \to \mathrm{VF}$ (a coordinate projection), h can be lifted to H defined by $H((y, t)) = (y, t F_n(y))$.
(2) By assumption, ϕ_1, ϕ_2 lift to measure-preserving maps $\Phi_i : \mathbb{L}\mathbf{Y}_i \to \mathbb{L}\mathbf{X}'$. On the other hand, by the assumption on existence of a section s of f_n, we have measure-preserving isomorphisms $\alpha_1 : \mathbb{L}\mathbf{Y}_1 \to \mathbb{L}\widetilde{\mathbf{Y}}_1$, $(y, u_1, \dots, u_n) \mapsto (y, u_1, \dots, u_{n-1}, (u_n - s)/s)$. Similarly, we have $\alpha_2 : \mathbb{L}\mathbf{Y}_2 \to \mathbb{L}\widetilde{\mathbf{Y}}_2$. Composing, we obtain $\alpha_2 {\Phi_2}^{-1} \Phi_1 {\alpha_1}^{-1} : \mathbb{L}\widetilde{\mathbf{Y}_1} \to \mathbb{L}\widetilde{\mathbf{Y}}_2$; it is easy to check that this is $\underset{\mathrm{rv}}{\sim}$-invariant and shows that $\mathbb{L}\widetilde{\mathbf{Y}}_1, \mathbb{L}\widetilde{\mathbf{Y}}_2$ are $\mathbf{Y}$-isomorphic in $\mu_l\mathrm{RV}[n, \cdot]$. Taking the disjoint sum with the complement X'' of X', we obtain the result. □

Remark. There is also a parallel of Lemma 7.15: Let $\mathbf{Y} = (Y, g) \in \mathrm{Ob}\,\mathrm{RV}_\infty[n, \cdot]$, with $\dim(g(Y)) < n$; let $f : Y \to \mathrm{RV}^{n-1}$ be isogenous to g. Let $h : Y \to \mathrm{RV}$ be definable, with $h(y) \in \mathrm{acl}(g(y))$ for $y \in Y$, and with $\sum(g) = \sum(f) + \mathrm{val}_{\mathrm{rv}}(h)$. Let $Y' = Y \times \mathrm{RV}_\infty^{\geq 0}$, and $f'(y, t) = (f(y), t h(y))$. Then for appropriate ω', $\mathbf{Y}' = (Y', f', \omega')$ with the projection map to Y is a blowup. This follows from Lemma 7.15 and Lemma 8.14(3).

Notation. For $X \in \mathrm{RV}[n, \cdot]$, $[X] = [(X, 1)]$ denotes the corresponding object of $\mu\mathrm{RV}[n, \cdot]$ with form 1.

Lemma 8.20. *Lemma* 7.18(1)–(5) *holds for* $\mu_l\mathrm{RV}[n, \cdot]$. *We also have the following:*

(6) *As a semiring congruence on* $K_+ \mu_l\mathrm{RV}[n, \cdot]$, $\mathrm{I}_{\mathrm{sp}}^{\mu}$ *is generated by* $([[1_{\mathbf{k}}]_1], [[\mathrm{RV}^{>0}]_1])$ (*with the forms* 1).

Proof. (1)–(5) go through with the same proof. For (6), Let $\sim$ be the congruence generated by this element. By blowing up a point one sees immediately that $([[1]_1], [[\mathrm{RV}^{>0}]_1]) \in \mathrm{I}_{\mathrm{sp}}^{\mu}$, so $\sim \leq \mathrm{I}_{\mathrm{sp}}^{\mu}$. For the converse direction we have to show that $(\widetilde{\mathbf{Y}}, \mathbf{Y}) \in \sim$ whenever $\widetilde{\mathbf{Y}}$ is a blowup of $\mathbf{Y}$; the elementary case suffices, since the $\mu_l\mathrm{RV}[n, \cdot]$-isomorphisms of Definition 8.16(2) are already accounted for in the semigroup $K_+ \mu_l\mathrm{RV}[n, \cdot]$. Now $\mathbf{Y} = (Y, f, \omega)$ with $f_n(y) \in \mathrm{RV}$. Since $\dim(Y) < n$, we have $\mathbf{Y} \simeq (Y, f', \omega')$ where $f'_i = f_i$ for $i < n$, $f'_n = 1$, and $\omega' = f_n\omega$. Thus we may assume $f_n = 1$. In this case, as in the proof of Lemma 7.18(6), $(\widetilde{\mathbf{Y}}, \mathbf{Y}) \in \sim$. □

Definition 8.21. Let J be a k-element set of natural numbers. $\mathrm{VFR}_\mu[J,l,\cdot]$ is the set of triples $\mathbf{X}=(X,f,\omega)$, with $X \subseteq \mathrm{VF}^J \times \mathrm{RV}^*$, $f: X \to \mathrm{RV}^l_\infty$, $\omega: X \to \mathrm{RV}$, and such that f and ω factor through the projection $\mathrm{pr}_{\mathrm{RV}}(X)$ of X to the RV-coordinates. $\mathrm{I}^\mu_{\mathrm{sp}}$ is the equivalence relation on $\mathrm{VFR}_\mu[J,l,\cdot]$:

$$(X,Y) \in \mathrm{I}^\mu_{\mathrm{sp}} \iff (X_a,Y_a) \in \mathrm{I}^\mu_{\mathrm{sp}}(\langle a\rangle) \quad \text{for each } a \in \mathrm{VF}^J.$$

$K_+\,\mathrm{VFR}_\mu$ is the set of equivalence classes.

For $j \in J$, let $\pi^j: \mathrm{VF}^k \times \mathrm{RV}^* \to \mathrm{VF}^{J-\{j\}} \times \mathrm{RV}^*$ be the projection forgetting the jth VF coordinate. We will write $\mathrm{VFR}_\mu[k,l,\cdot]$, VF^k, VF^{k-1} for $\mathrm{VFR}_\mu[J,l,\cdot]$ VF^J, $\mathrm{VF}^{J-\{j\}}$, respectively, when the identity of the indices is not important.

The map $\mathbb{L}: \mathrm{Ob}\,\mu\mathrm{RV}[n,\cdot] \to \mathrm{Ob}\,\mu\mathrm{VF}[n]$ induces, by Lemma 6.3, a homomorphism $\mathbb{L}: K_+\,\mu\mathrm{RV}[n,\cdot] \to K_+\,\mu\mathrm{VF}[n]$. By Proposition 4.5 it is surjective.

Lemma 8.22. *Let $\mathbf{X},\mathbf{X}' \in \mu\mathrm{RV}[n,\cdot]$, and let $G: \mathbb{L}\mathbf{X}' \to \mathbb{L}\mathbf{X}$ be a special bijection. Then $\mathbf{X}'$ is isomorphic to an iterated blowup of $\mathbf{X}$.*

Proof. The proof is clear from Lemma 7.19 since strong isomorphisms are also $\mu_l\mathrm{RV}[n,\cdot]$-isomorphisms. □

Lemma 8.23. *The homomorphism $\mathbb{L}: K_+\,\mu\mathrm{RV}[1,\cdot] \to K_+\,\mu\mathrm{VF}[1,\cdot]$ is surjective, with kernel equal to $\mathrm{I}^\mu_{\mathrm{sp}}[1]$. The image of $K_+\,\mathrm{RV}_{\mathrm{vol}}[1,\cdot]$ is $K_+\,\mathrm{VF}_{\mathrm{vol}}[1,\cdot]$*

Proof. Let $\mathbf{X},\mathbf{Y} \in \mu\mathrm{RV}[1,\cdot]$, and let $F: \mathbb{L}\mathbf{X} \to \mathbb{L}\mathbf{Y}$ be a definable measure-preserving bijection. We have $\mathbf{X}=(X,f,\omega)$, $\mathbf{Y}=(Y,g,\omega)$ with $(X,f),(Y,g) \in \mathrm{RV}[1,\cdot]$. By Lemma 7.6 there exist special bijections $b_X: \mathbb{L}\mathbf{X}' \to \mathbb{L}\mathbf{X}$, $b_Y: \mathbb{L}\mathbf{Y}' \to \mathbb{L}\mathbf{Y}$ and an $\underset{\mathrm{rv}}{\sim}$-invariant definable bijection $F': \mathbb{L}\mathbf{X}' \to \mathbb{L}\mathbf{Y}'$ such that $b_YF' = Fb_X$. We used here that any $\underset{\mathrm{rv}}{\sim}$-invariant object can be written as $\mathbb{L}\mathbf{X}'$ for some $\mathbf{X}'$. Since F, b_X, b_Y are measure-preserving bijections, so is F'. By Lemma 8.22, $\mathbf{X}' \to \mathbf{X}$ and $\mathbf{Y}' \to \mathbf{Y}$ are blowups; and F' descends to a definable bijection between them. This bijection is measure preserving by Lemma 5.22. Hence by definition $(\mathbf{X},\mathbf{Y}) \in \mathrm{I}^\mu_{\mathrm{sp}}$.□

By Proposition 8.23, the inverse of $\mathbb{L}: \mathrm{RV}[1,\cdot] \to \mathrm{VF}[1,\cdot]$ induces an isomorphism $I^{\mathrm{vol}}_1: K_+\,\mathrm{VF}_{\mathrm{vol}}[1,\cdot] \to K_+\,\mathrm{RV}_{\mathrm{vol}}[1,\cdot]/\mathrm{I}^\mu_{\mathrm{sp}}$.

$$I^{\mathrm{vol}}_1([X]) = [Y]/\mathrm{I}^\mu_{\mathrm{sp}} \iff [\mathbb{L}Y] = [X].$$

Let $\mathbf{X}=(X,f,\omega) \in \mathrm{VFR}_\mu[k,l,\cdot]$. By assumption, f,ω factor through π^j, so that they can be viewed as functions on π^jX. We view the image (π^jX,f,ω) as an element of $\mathrm{VFR}_\mu[k-1,l,\cdot]$. Each fiber of π^j is a subset of VF; it can be viewed as an element of $\mathrm{VF}_{\mathrm{vol}}[1] \subseteq \mu\mathrm{VF}[1] \subseteq \mu\mathrm{VF}[1,\cdot]$.

Claim. Relative $\mathrm{I}^\mu_{\mathrm{sp}}$-equivalence implies $\mathrm{I}^\mu_{\mathrm{sp}}$-equivalence, in the following sense. Let $X_i \subseteq \mathrm{RV}^*$ $(i=1,2)$; $h_i: X_i \to W \subseteq \mathrm{RV}^*$; $f_W: W \to \mathrm{RV}^l$, $\omega: W \to \mathrm{RV}$, and $f_i: X \to \mathrm{RV}^k$ be definable sets and functions. Let $\mathbf{X}_i = (X_i,(f_W \circ h_i, f_i), \omega \circ h_i)$. Let $\mathbf{X}_i(w) = (X_i(w), f_i|X_i(w), \omega \circ h_i|X_i(w))$, where $X_i(w) = {h_i}^{-1}(w)$. If $\mathbf{X}_1(w),\mathbf{X}_2(w) \in \mathrm{I}_{\mathrm{sp}}(\langle w\rangle)$ for each $w \in W$, then $(\mathbf{X}_1,\mathbf{X}_2) \in \mathrm{I}^\mu_{\mathrm{sp}}$.

Proof. The proof is clear using Lemma 8.18. □

The claim allows us to relativize I_1^{vol} to π^j. We obtain a map

$$I^j = I^j_{k,l} : \mathrm{VFR}_\mu[k, l, \cdot] \to K_+ \mathrm{VFR}_\mu[k-1, l+1, \cdot]/\mathrm{I}^\mu_{\mathrm{sp}}.$$

Lemma 8.24. *Let* $\mathbf{X} = (X, f, \omega), \mathbf{X}' = (X', f', \omega') \in \mathrm{VFR}_\mu[k, l, \cdot]$.

(1) I^j *commutes with maps into* RV*: if* $h : \mathbf{X} \to W \subseteq \mathrm{RV}^*$ *is definable,* $\mathbf{X}_c = h^{-1}(c)$*, then* $I^j(\mathbf{X}) = \sum_{c \in W} I^j(\mathbf{X}_c)$.
(2) *If* $([\mathbf{X}], [\mathbf{X}']) \in \mathrm{I}^\mu_{\mathrm{sp}}$*, then* $(I^j(\mathbf{X}), I^j(\mathbf{X}')) \in \mathrm{I}^\mu_{\mathrm{sp}}$.
(3) I^j *induces a map* $K_+ \mathrm{VFR}_\mu[k, l, \cdot]/\mathrm{I}^\mu_{\mathrm{sp}} \to K_+ \mathrm{VFR}_\mu[k-1, l+1, \cdot]/\mathrm{I}^\mu_{\mathrm{sp}}$.

Proof.

(1) This reduces to the case of I_1^{vol}, where it is an immediate consequence of uniqueness, and the fact that $\mathbb{L}$ commutes with maps into RV in the same sense.
(2) All equivalences here are relative to the $k-1$ coordinates of VF other than j, so we may assume $k = 1$, and write I for I^j. For $a \in \mathrm{VF}$, $([\mathbf{X}_a], [\mathbf{X}'_a]) \in \mathrm{I}^\mu_{\mathrm{sp}}(\langle a \rangle)$. By stable embeddedness of RV, there exists $\alpha = \alpha(a) \in \mathrm{RV}^*$ such that $\mathbf{X}_a, \mathbf{X}'_a$ are $\langle \alpha \rangle$-definable there are $\langle \alpha \rangle$-definable blowups $\widetilde{\mathbf{X}}_a, \widetilde{\mathbf{X}'}_a$ and an $\langle \alpha \rangle$-definable isomorphism between them, lifting to an a-definable isomorphism. Using (1) and Lemma 8.18 we may assume that α is constant. Thus for some $W \in \mathrm{Ob\,VF}[1]$, $\mathbf{Y}, \mathbf{Y}' \in \mu\mathrm{RV}[l+1, \cdot]$, we have $\mathbf{X} = W \times \mathbf{Y}$, $\mathbf{X}' = W \times \mathbf{Y}'$, $\widetilde{\mathbf{Y}}, \widetilde{\mathbf{Y}}'$ are blowups of $\mathbf{Y}, \mathbf{Y}'$, respectively, $\phi : \mathbf{Y} \to \mathbf{Y}'$ is a bijection, and for any $w \in W$ there exists a measure-preserving $F_w : \mathbb{L}\widetilde{\mathbf{Y}} \to \mathbb{L}\widetilde{\mathbf{Y}}$ lifting ϕ. Then $I(\mathbf{X}) = I(W) \times \mathbf{Y}$, $I(\mathbf{X}') = I(W) \times \mathbf{Y}'$ and the bijection $\mathrm{Id}_{I(W)} \times \phi$ is lifted by the measure-preserving bijection $(w, y) \mapsto (w, F_w(y))$.
(3) This follows by (2). □

Lemma 8.25. *Let* $\mathbf{X} = (X, f, \omega) \in \mathrm{Ob\,VFR}_\mu[J, l, \cdot]$*. If* $j \neq j' \in J$*, then* $I^j I^{j'} = I^{j'} I^j : K_+ \mathrm{VFR}_\mu[J, l, \cdot]/\mathrm{I}^\mu_{\mathrm{sp}} \to K_+ \mathrm{VFR}_\mu[J \setminus \{j, j'\}, l+2, \cdot]/\mathrm{I}^\mu_{\mathrm{sp}}$.

Proof. We may assume $S = \{1, 2\}$, $j = 1$, $j' = 2$, since all is relative to $\mathrm{VF}^{S \setminus \{j, j'\}}$. By Lemma 7.23(1) and Lemma 8.18 it suffices to prove the statement for each fiber of a given map into $\mathrm{RV}[l]$. Hence we may assume $X \subseteq \mathrm{VF}^2$ so that f is constant; and by Lemma 5.10, we can assume X is a basic 2-cell:

$$X = \{(x, y) : x \in X_1, \mathrm{rv}(y - G(x)) = \alpha_1\}, \quad X_1 = \mathrm{rv}^{-1}(\delta_1) + c_1.$$

The case where G is constant is easy since then X is a finite union of rectangles. Otherwise, G is invertible, and by the niceness of G we can also write

$$X = \{(x, y) : y \in X_2, \mathrm{rv}(x - G^{-1}(y)) = \beta\}, \quad X_2 = \mathrm{rv}^{-1}(\delta_2) + c_2.$$

We immediately compute

$$I_2 I_1(X) = (\delta_1, \alpha_1), \qquad I_1 I_2(X) = (\alpha_2, \delta_2)$$

and necessarily $\mathrm{val}_{\mathrm{rv}}\delta_1 + \mathrm{val}_{\mathrm{rv}}\alpha_1 = \mathrm{val}_{\mathrm{rv}}\alpha_2 + \mathrm{val}_{\mathrm{rv}}\delta_2$ (Lemma 5.4). We have bijections $F_j : X \to \mathbb{L}I_j(X)$. The map $F_1 F_2 F_1{}^{-1} F_2{}^{-1} : \mathbb{L}I_2 I_1(X) \to \mathbb{L}I_1 I_2(X)$ lifts the unique bijection between the singleton sets $\{(\delta_1, \alpha_1)\}$, $\{(\alpha_2, \delta_2)\}$, and shows that $[(\delta_1, \alpha_1)]_2 = [(\alpha_2, \delta_2)]_2$. □

Proposition 8.26. *Let* $\mathbf{X}, \mathbf{Y} \in \mu\mathrm{RV}[\leq n, \cdot]$. *If* $\mathbb{L}\mathbf{X}$, $\mathbb{L}\mathbf{Y}$ *are isomorphic, then* $([X], [Y]) \in \mathrm{I}_{\mathrm{sp}}^{\mu}$.

Proof. The proof is identical to the proof of Proposition 7.25, only quoting Lemma 8.25 in place of Lemma 7.24, and Lemma 8.14 to enable using Lemma 7.11. □

Proposition 8.27. *Proposition* 8.3 *is valid for* $\mu\mathrm{VF}[n]$, $\mu\mathrm{RV}[n]$, $\mathrm{I}_{\mathrm{sp}}^{\mu}[n]$.

Proof. The proof is the same as that of Proposition 8.3, but using Proposition 6.3 in place of 6.1 and Proposition 8.26 in place of Proposition 7.25. □

8.6 Invariants of measure-preserving maps, and some induced isomorphisms

Theorem 8.28. *Let* $\mathbf{T}$ *be* V*-minimal. There exists a canonical isomorphism of Grothendieck semigroups*

$$\oint : K_+^{\mathrm{eff}}\ \mu\mathrm{VF}[n, \cdot] \to K_+(\mu\mathrm{RV}[n, \cdot])/\mathrm{I}_{\mathrm{sp}}^{\mu}[n].$$

Let $[X]$ *denote the class of* X *in* $K_+^{\mathrm{eff}}(\mu\mathrm{VF}[n])$. *Then*

$$\oint [X] = W/\mathrm{I}_{\mathrm{sp}}^{\mu}[n] \iff [X] = [\mathbb{L}W] \in K_+^{\mathrm{eff}}(\mu\mathrm{VF}[n]).$$

Proof. Given $\mathbf{X} = (X, f, \omega) \in \mathrm{Ob}\,\mu\mathrm{RV}[n]$ we have $\mathbb{L}\mathbf{X} \in \mathrm{Ob}\,\mu\mathrm{VF}[n]$. If $\mathbf{X}, \mathbf{X}'$ are isomorphic, then by Lemma 6.3, $\mathbb{L}\mathbf{X}$, $\mathbb{L}\mathbf{X}'$ are effectively isomorphic. Direct sums are clearly respected, so we have a semigroup homomorphism $\mathbb{L} : K_+(\mu\mathrm{RV}[n]) \to K_+^{\mathrm{eff}}(\mu\mathrm{VF}[n])$. It is surjective by Proposition 4.5 and injective by Proposition 8.3. Inverting, we obtain I. □

Let $\mathrm{I}_{\mathrm{sp}}^{\mu\,\prime}$ be the semigroup congruence on $\mathrm{RV}_{\mathrm{vol}}[n]$ generated by $((Y, f), (Y \times \mathrm{RV}^{>0}, f'))$, where Y, f, f' are as in Definition 7.12. Let $\mu_\Gamma \mathrm{I}_{\mathrm{sp}}$ be the congruence on $K_+\,\mu_\Gamma\mathrm{RV}[n]$ generated by $([[1_{\mathbf{k}}]_1], [[\mathrm{RV}^{>0}]_1])$, with the constant Γ-form $0 \in \Gamma$.

Assume given a distinguished subgroup N_1 of the multiplicative group of the residue field $\mathbf{k}$. For example, N_1 may be the group of elements of norm one, with respect to some absolute value $|,|$ on $\mathbf{k}$. With this example in mind, write $|x| = 1$ for $x \in N_1$. Let $|\mu|\mathrm{VF}[n]$ be the subcategory of $\mathrm{VF}[n]$ with the same objects, and such that $F \in \mathrm{Mor}_{|\mu|\mathrm{VF}[n]}$ iff $F \in \mathrm{Mor}_{\mu_\Gamma \mathrm{VF}[n]}$ and $|JRV(F)| = 1$ almost everywhere. Define $|\dot{\mu}|\mathrm{RV}[n]$ similarly.

Theorem 8.29. *The isomorphism* $\oint$ *of Theorem* 8.28 *induces isomorphisms:*

$$K_+^{\mathrm{eff}}\ \mathrm{VF}_{\mathrm{vol}}[n] \to K_+\,\mathrm{RV}_{\mathrm{vol}}[n]/\mathrm{I}_{\mathrm{sp}}^{\mu\,\prime}[n], \tag{8.1}$$

$$K_+^{\mathrm{eff}}\ \mathrm{VF}^{\mathrm{bdd}}_{\mathrm{vol}}[n] \to K_+\,\mathrm{RV}^{\mathrm{bdd}}_{\mathrm{vol}}[n]/\mathrm{I}^{\mu}_{\mathrm{sp}}[n], \tag{8.2}$$

$$K_+^{\mathrm{eff}}\ \mu\mathrm{VF}^{\mathrm{bdd}}[n] \to K_+\,\mu\mathrm{RV}^{\mathrm{bdd}}[n]/\mathrm{I}^{\mu}_{\mathrm{sp}}[n], \tag{8.3}$$

$$K_+^{\mathrm{eff}}\ |\mu|\mathrm{VF}[n] \to K_+\,|\dot{\mu}|\mathrm{RV}[n]/\mathrm{I}^{\mu}_{\mathrm{sp}}[n], \tag{8.4}$$

$$K_+^{\mathrm{eff}}\ \mu_\Gamma\mathrm{VF}[n] \to K_+\,\mu_\Gamma\mathrm{RV}[n]/\mu_\Gamma\mathrm{I}_{\mathrm{sp}}[n]. \tag{8.5}$$

Proof. Since Proposition 4.5 uses measure-preserving maps, Proposition 6.1 does not go out of the subcategory $\mathrm{VF}_{\mathrm{vol}}$, and $\mathrm{RV}_{\mathrm{vol}}[n]$ is a full subcategory of $\mu\mathrm{RV}[n]$, we have (8.1). It is similarly easy to see that "dimension $< n$" and boundedness are preserved, hence (8.2)–(8.3).

We have $K_+^{\mathrm{eff}}\ |\mu|\mathrm{VF} = K_+^{\mathrm{eff}}\ \mu\mathrm{VF}/N_{\mathrm{VF}}$, where $N_{\mathrm{VF}} = \{([X,\omega],[X,g\omega]) : g : X \to \mathrm{RV}, |g| = 1\}$; similarly for $K_+^{\mathrm{eff}}\ |\dot{\mu}|\mathrm{RV}$. Thus for (8.4) it suffices to show that $(\not\!\int(\mathbf{X}), \not\!\int(\mathbf{Y})) \in N_{\mathrm{RV}} \iff (\mathbf{X},\mathbf{Y}) \in N_{\mathrm{VF}}$. For $X \in \mathrm{Ob}\,\mu\mathrm{VF}[n]$ or $X \in \mathrm{Ob}\,\mu\mathrm{RV}[n]$ with RV-volume form ω, given $g : X \to \mathrm{RV}$, let ${}^gX$ denote the same object but with volume form $g\omega$. In one direction, we have to show that $(\mathbb{L}X, \mathbb{L}Y) \in N_{\mathrm{VF}}$ if $(X,Y) \in N_{\mathrm{RV}}$. This is clear since $\mathbb{L}({}^gX) =^g (\mathbb{L}X)$. Conversely we have to show that $(\not\!\int[{}^gZ], \not\!\int[Z]) \in N_{\mathrm{RV}}$. Since $\not\!\int$ commutes with RV-sums, we may assume g is constant, with value a. But then $\mathbb{L}({}^aX) =^a (\mathbb{L}X)$ implies $\not\!\int({}^aZ) =^a \not\!\int Z$ as required. This gives (8.4); (8.5) is a special case. □

9 The Grothendieck semirings of Γ

Let $T = \mathrm{DOAG}_A$ be the theory of divisible ordered Abelian groups Γ, with distinguished constants for elements of a subgroup A. Let $\mathrm{DOAG}_A[*]$ be the category of all DOAG_A definable sets and bijections. Our primary concern is not with DOAG_A, but rather a proper subcategory $\Gamma[*]$, having the same objects but only piecewise integral morphisms (Definition 9.1). Our interest in $\Gamma[*]$ derives from this: the morphisms of $\Gamma[*]$ are precisely those that lift to morphisms of $\mathrm{RV}[*]$, and it is $K_+[\Gamma[*]]$ that forms a part of $K_+[\mathrm{RV}[*]]$ (cf. Section 3.3). This category depends on A, but will nevertheless be denoted $\Gamma[*]$ when A is fixed and understood.

We will first describe $K(\Gamma^{\mathrm{fin}}[*])$, the subring of classes of finite definable sets. Next, we will analyze $K(\mathrm{DOAG}_A)$, obtaining two Euler characteristics. This repeats earlier work by Maříková. We retain our proofs as they give a rapid path to the Euler characteristics, but [26] includes a complete analysis of the semiring $K(\mathrm{DOAG}_A)$, that may well be useful in future applications.

At the level of Grothendieck rings, the categories $\Gamma[*]_A$ and DOAG_A may be rather close; see Lemma 9.8 and Question 9.9. But the semiring homomorphism $K_+(\Gamma[*]_A) \to K(\mathrm{DOAG}_A)$ is far from being an isomorphism, and it remains important to give a good description of $K_+(\Gamma[*]_A)$. We believe that further invariants can be found by mapping $K_+[\Gamma[*]]$ into the Grothendieck semirings of other completions of the universal theory of ordered Abelian groups over A, as well as DOAG, in the manner of Proposition 9.2; it is possible that all invariants appear in this way.

A description of $K_+(\Gamma[*]_A)$ would include information about the Grothendieck group of subcategories, such as the category of bounded definable sets. We will only sample one bit of the information available there, in the form of a "volume" map on bounded subsets of $K_+[\Gamma[*]]$ into the rationals, and a discrete analogue.

Definition 9.1. An object of $\Gamma[n]$ is a finite disjoint union of subsets of Γ^n defined by linear equalities and inequalities with $\mathbb{Z}$-coefficients and parameters in A. Given $X, Y \in \mathrm{Ob}\,\Gamma[n]$, $f \in \mathrm{Mor}_\Gamma(X, Y)$ iff f is a bijection, and there exists a partition $X = \cup_{i=1}^n X_i$, $M_i \in \mathrm{GL}_n(\mathbb{Z})$, $a_i \in A^n$, such that for $x \in X_i$,

$$f(x) = M_i x + a_i.$$

$\Gamma[*]$ is the category of definable subsets of Γ^n for any n, with the same morphisms. Since there are no morphisms between different dimensions, it is simply the direct sum of the categories $\Gamma[n]$, and the Grothendieck semiring $K_+[\Gamma]$ of $\Gamma[*]$ is the graded direct sum of the semigroups $K_+(\Gamma[n])$. We will write $K[\Gamma]$ for the corresponding group.

Let $\Gamma^{\mathrm{bdd}}[*]$ be the full subcategory of $\Gamma[*]$ consisting of bounded sets, i.e., an element of $\mathrm{Ob}\,\Gamma^{\mathrm{bdd}}[n]$ is a definable subset of $[-\gamma, \gamma]^n$ for some $\gamma \in \Gamma$.

Γ_A is a subcategory of $\Gamma_{\mathbb{Q}\otimes A}$ (a category with the same objects, but more morphisms, generated by additional translations) and this in turn is a subcategory of $\mathrm{DOAG}_{\mathbb{Q}\otimes A}$.

There is therefore always a natural morphism from $K_+(\Gamma_A[*])$ to the simpler semigroup $K_+(\mathrm{DOAG}_{\mathbb{Q}\otimes A})$. We will exhibit two independent Euler characteristics on $\mathrm{DOAG}_{\mathbb{Q}\otimes A}$ and show that they define an isomorphism $K(\mathrm{DOAG}_{\mathbb{Q}\otimes A}) \to \mathbb{Z}^2$. Taking the dimension grading into account, this will give rise to two families of Euler characteristics on $K(\Gamma_A)$, with $\mathbb{Z}[T]$-coefficients.

9.1 Finite sets

Let $\Gamma^{\mathrm{fin}}[n]$ be the full subcategory of $\Gamma_A[n]$ consisting of finite sets. The Grothendieck semiring of $\Gamma^{\mathrm{fin}}[*]$ embeds into the semirings of both Γ_A and RES, within the Grothendieck semiring of RV_A, and we will see that $K_+(\mathrm{RV}_A)$ is freely generated by them over $K_+(\Gamma^{\mathrm{fin}}[*])$. We proceed to analyze $K_+(\Gamma^{\mathrm{fin}}[*])$ in detail.

Let $\boldsymbol{\tau} = [0]_1 \in K_+(\Gamma^{\mathrm{fin}}[1])$ be the class of the singleton $\{0\}$.

The unit element of $K(\Gamma)$ is the class of Γ^0. Note that the bijection between $\boldsymbol{\tau}$ and Γ^0 is not a morphism in $\Gamma[*]$; in fact $1, \boldsymbol{\tau}, \boldsymbol{\tau}^2, \ldots$ are distinct and $\mathbb{Q}$-linearly independent in $K(\Gamma)$. The motivation for this choice of category becomes clear if one thinks of the lift to RV: the inverse image of $\boldsymbol{\tau}^n$ in RV (also denoted $\boldsymbol{\tau}^n$) has dimension n, and cannot be a union of isomorphic copies of $\boldsymbol{\tau}^m$ for smaller m.

Let $K(\Gamma^{\mathrm{fin}})[\boldsymbol{\tau}^{-1}]$ be the localization. This ring is a naturally $\mathbb{Z}$-graded ring; let H_{fin} be the zero-dimensional component.

Let Ξ_A be the space of subgroups of $(\mathbb{Q} \otimes A)/A$ or, equivalently, of subgroups of $\mathbb{Q} \otimes A$ containing A. View it as a closed subspace of the Tychonoff space $2^{(\mathbb{Q}\otimes A)/A}$, via the characteristic function 1_s of a subgroup $s \in \Xi_A$. Let $C(\Xi_A, \mathbb{Z})$ be the ring of continuous functions $\Xi_A \to \mathbb{Z}$ (where $\mathbb{Z}$ is discrete).

A *cancellation* semigroup is a semigroup where $a + b = a + c$ implies $b = c$; in other words, a subsemigroup of an Abelian group.

Proposition 9.2. *$K_+(\Gamma^{\mathrm{fin}}[n])$ is a cancellation semigroup. As a semiring, $K_+(\Gamma^{\mathrm{fin}}[*])$ is generated by $K_+(\Gamma^{\mathrm{fin}}[1])$. We have*

$$K(\Gamma^{\mathrm{fin}})[\boldsymbol{\tau}^{-1}] = H_{\mathrm{fin}}[\boldsymbol{\tau}, \boldsymbol{\tau}^{-1}],$$
$$H_{\mathrm{fin}} \simeq C(\Xi_A, \mathbb{Z}).$$

Proof. Since Γ is ordered, any finite definable subset of Γ^n is a union of definable singletons. Thus the semigroup $K_+(\Gamma^{\mathrm{fin}}[n])$ is freely generated by the isomorphism classes of singletons $a \in \Gamma^n$ and, in particular, is a cancellation semigroup. The displayed equality is thus clear; we proceed to prove the isomorphism.

A definable singleton of Γ^n has the form $(a_1, \dots, a_n)$, where for some $N \in \mathbb{N}$, $Na_1, \dots, Na_n \in A$. Thus $[(a_1, \dots, a_n)] = [(a_1)] \cdots [(a_n)]$.

For any commutative ring R, let $\mathrm{Idem}(R)$ be the Boolean algebra of idempotent elements of a commutative ring R with the operations $1, 0, xy, x + y - xy$. Note that the elements $[(a_1, \dots, a_n)]\boldsymbol{\tau}^{-n} \in H_{\mathrm{fin}}$ belong to $\mathrm{Idem}(H_{\mathrm{fin}})$: in $K_+(\Gamma^{\mathrm{fin}})$: for any $a \in \Gamma$ we have the relation $[a]^2 = [a]\boldsymbol{\tau}$. Let B be the Boolean subalgebra of $\mathrm{Idem}(H_{\mathrm{fin}})$ generated by the elements $[(a_1, \dots, a_n)]\boldsymbol{\tau}^{-n}$. For a maximal ideal M of B, let I_M be the ideal of H_{fin} generated by M. Note $H_{\mathrm{fin}} = \mathbb{Z}B$. Hence we have to show the following:

(1) The Stone space of B is Ξ_A.
(2) For any maximal ideal M of B, $H_{\mathrm{fin}}/I_M \simeq \mathbb{Z}$ naturally.

For any commutative ring R, a finitely generated Boolean ideal of $\mathrm{Idem}(R)$ is generated by a single element b; if $b \neq 1$, then $bR \neq R$ since $b(1-b) = 0$. Thus if M is a proper ideal of $\mathrm{Idem}(R)$, then MR is a proper ideal of R. Applying this to B, viewed as a Boolean subalgebra of $\mathrm{Idem}(\mathbb{Q} \otimes H_{\mathrm{fin}})$, we see that $I_M \cap \mathbb{Z} = (0)$ for any maximal ideal M of B. Thus the composition $\mathbb{Z} \to H_{\mathrm{fin}} \to H_{\mathrm{fin}}/I_M$ is injective. On the other hand, H_{fin} is generated over $\mathbb{Z}$ by the elements $[a]/\boldsymbol{\tau}$, and each of them equals 0 or 1 modulo I_M, so the map is surjective, too. This proves the second point.

To prove the first, we define a map $\Phi : \Xi_A \to \mathrm{Stone}(B)$.

Let $t = T/A$, $T \leq \mathbb{Q} \otimes A$. If $[(a_1, \dots, a_n)] = [(b_1, \dots, b_n)]$, then some element of $\mathrm{GL}_n(\mathbb{Z}) \ltimes A^n$ takes $(a_1, \dots, a_n)$ to $(b_1, \dots, b_n)$; in this case, if $a_i \in T$ for each i then $b_i \in T$ for each i; so $\Pi_{i=1}^n 1_t(a_i + A) = \Pi_{i=1}^n 1_t(b_i + A)$. Thus, given $t \in \Xi_A$, we can define a homomorphism $h_t : H_{\mathrm{fin}} \to \mathbb{Z}$ by

$$[(a_1, \dots, a_n)]/\boldsymbol{\tau}^n \mapsto \Pi_{i=1}^n 1_t(a_i + A).$$

Let $M(t) = \ker(h_t) \cap B$.

The map $\Phi : t \mapsto M(t)$ is clearly continuous. If t, t' are distinct subgroups, let $a \in t, a \notin t'$ (say); then $[a]/\boldsymbol{\tau} \in M(t)$, $[a]/\boldsymbol{\tau} \notin M(t')$. Thus Φ is injective. If P is a maximal filter of B, let $t_P = \{a + A : [a]/\boldsymbol{\tau} \in P\}$.

Claim. t_P is a subgroup.

Proof. Suppose $a + A, b + A \in t_P$ and let $c = a + b$. Then we have the relation

$$[a][b]\boldsymbol{\tau} = [a][b][c]$$

in $K_+(\Gamma^{\text{fin}})$, arising from the map

$$(x, y, z) \mapsto (x, y, xyz).$$

Thus $([a]/\boldsymbol{\tau})([b]/\boldsymbol{\tau})(1-[c]/\boldsymbol{\tau}) = 0$. As $([a]/\boldsymbol{\tau}), ([b]/\boldsymbol{\tau}) \in P$ we have $(1-[c]/\boldsymbol{\tau}) \notin P$, so $[c]/\boldsymbol{\tau} \in P$. □

Clearly, $P = M(t_P)$. Thus Φ is surjective, and so a homeomorphism. □

Example. We always have a homomorphism $K(\Gamma^{\text{fin}}) \to \mathbb{Z}$ (by counting points of a finite set in the divisible hull); when A is divisible, this identifies $K(\Gamma^{\text{fin}})$ with $\mathbb{Z}[\tau]$. In general, we have the surjective morphism $K(\Gamma^{\text{fin}})) \to K(\Gamma^{\text{fin}}_{\mathbb{Q}\otimes A}) = \mathbb{Z}[\tau]$.

Lemma 9.3. *Let Y be an A-definable subset of Γ^n, of dimension $< n$. Then Y is a finite union of $\mathrm{GL}_n(\mathbb{Z})$-conjugates of sets $Y_i \subseteq \{c_i\} \times \Gamma^{n-1}$, with $c_i \in \mathbb{Q} \otimes A$.*

Proof. Y can be divided into finitely many A-definable pieces, each contained in some A-definable hyperplane of Γ^n. Thus we may assume Y itself is contained in some such hyperplane, i.e., $\sum r_i y_i = c$ for some $c \in \mathbb{Q} \otimes \mathrm{val}_{\mathrm{rv}}(A)$. We may assume $r_i \in \mathbb{Z}$ and $(r_1, \ldots, r_n)$ have no common divisor. In this case $\mathbb{Z}^n/\mathbb{Z}(r_1, \ldots, r_n)$ is torsion free, hence free, so $\mathbb{Z}(r_1, \ldots, r_n)$ is a direct summand of $\mathbb{Z}^n$. Thus after effecting a transformation of $\mathrm{GL}_n(\mathbb{Z})$, we may assume $(r_1, \ldots, r_n) = (1, 0, \ldots, 0)$, i.e., Y lies in the hyperplane $y_1 = c$. Let Z be the projection of Y to the coordinates $(2, \ldots, n)$. Then $Y = \{c\} \times Z$. □

9.2 Euler characteristics of DOAG

We describe two independent Euler characteristics on A-definable subsets of Γ, i.e., additive, multiplicative $\mathbb{Z}[\boldsymbol{\tau}]$-valued functions invariant under all definable bijections. The values are in $\mathbb{Z}[\boldsymbol{\tau}]$ rather than $\mathbb{Z}$ because $\Gamma[*] = \oplus_n \Gamma[n]$ is graded by ambient dimension. Proposition 9.4–Lemma 9.6 were obtained earlier in [26], and independently in [20].

In fact, these two Euler characteristics come from Euler characteristics of $\mathrm{DOAG}_{\mathbb{Q}\otimes A}$. Ther they are the only ones.

Proposition 9.4. *Let A be a divisible ordered Abelian group. Then $K(\mathrm{DOAG}_A) \simeq \mathbb{Z}^2$.*

Proof. We begin by noting that there are at most two possibilities.

In DOAG, all definable singletons are isomorphic. The identity element of the ring $K(\mathrm{DOAG})$ is the class of any singleton. Thus the image of $K(\Gamma^{\text{fin}}[*])$ in $K(\mathrm{DOAG}_A)$ is isomorphic to $\mathbb{Z}$.

Claim. The image of $K(\Gamma^{\text{bdd}}_A)$ in $K(\mathrm{DOAG}_A)$ equals the image of $K(\Gamma^{\text{fin}}[*])$ there.

Translation by a gives an equality of classes in $K(\Gamma)$, $[(0,\infty)] = [(a,\infty)]$, so

$$[(0,a)] + [\{pt\}] = [(0,a] = 0.$$

Thus bounded segments are equivalent to linear combinations of points. This can be seen directly by induction on dimension and on ambient dimension: consider the class of a bounded set $Y \subset \Gamma^{n+1}$. Y is a Boolean combination of sets of the form $\{(x,y) : x \in X, f(x) < y < g(x)\}$. This is DOAG_A-isomorphic to $Y' = \{(x,y) : x \in X, 0 < y < h(x)\}$, where $h(x) = g(x) - f(x)$. Let $Z = \{(x,y) : x \in X, y > 0\}$, $Z' = \{(x,y) : x \in X, y > h(x)\}$. Then the map $(x,y) \mapsto (x, y + h(x))$ shows that $[Z] = [Z']$. On the other hand, Z' is the disjoint union of Z, Y and a lower-dimensional set W. Thus $[Z] = [Z'] = [Z'] + [Y] + [W]$ so $[Y] = -[W]$, and by induction $[Y]$ lies in the image of $K(\Gamma^{\mathrm{fin}}[*])$.

Now consider $t = [(0,\infty)] \in K(\Gamma_A)$. We have a homomorphism $K(\Gamma_A^{\mathrm{bdd}})[t] \to K(\Gamma)$. To see that it is surjective, again by induction it suffices to look at sets such as $\{(x,y) : x \in X, f(x) < y\}$ or $\{(x,y) : x \in X, f(x) < y < g(x)\}$. The latter is equivalent to a lower-dimensional set, by induction, as above. The former is equivalent to $\{(x,y) : x \in X, 0 < y\}$ so that it has the class $[X] \times t$ and thus is in the image of $K(\Gamma_A^{\mathrm{bdd}})[t]$.

Let $T = \{(x,y) : 0 < y \leq x\}$. The map $(x,y) \mapsto (x, y + x)$ takes T to $\{(x,y) : 0 < x < y \leq 2x\}$, so $2[T] = [\{(x,y) : 0 < y \leq 2x\}]$. The same map shows that $t^2 - [T] = t^2 - 2[T]$ so $[T] = 0$. But then $[\{(x,y) : 0 < x \leq y\}] = 0$, and adding we obtain $0 + 0 = t^2 + [\{(x,x) : 0 < x\}] = t^2 + t$. Thus $K(\mathrm{DOAG}_A)$ is a homomorphic image of $\mathbb{Z}[t]/(t^2+t) \simeq \mathbb{Z}^2$. To see that the homomorphism is bijective, it remains to exhibit a homomorphism $K(\mathrm{DOAG}_A) \to \mathbb{Z}$ with $t \mapsto 0$ and another with $t \mapsto -1$. The two lemmas below show this, in a form suitable also for a dimension-graded version. □

Lemma 9.5. *There exists a ring homomorphism* $\chi_O : K(\Gamma) \to \mathbb{Z}[\boldsymbol{\tau}]$, *such that* $\chi_O((0,\infty)) = \boldsymbol{\tau}$. *It is invariant under* $\mathrm{GL}_n(\mathbb{Q})$ *acting on* Γ^n.

Proof. Let RCF be the theory of real closed fields. See [37] for the existence and definability of an Euler characteristic map $\chi : K(\mathrm{RCF}) \to \mathbb{Z}$. For any definable $X, P, f : X \to P$ of RCF, there exists $m \in \mathbb{N}$ and a definable partition $P = \bigcup_{-m \leq i \leq m} P_i$, such that for any i, any $M \models \mathrm{RCF}$ and $b \in P_i(M)$, $\chi(X_b) = i$. Here $X_b = f^{-1}(b)$, and $\chi(X_b) = i$ iff there exists an M-definable partition of X_b into definable cells C_j, with $\sum_j (-1)^{\dim(C_j)} = i$.

The language of Γ (the language of ordered Abelian groups) is contained in the language of RCF. Thus if $X, P, f : X \to P$ are definable in the language of ordered Abelian groups, they are RCF-definable. Therefore, the above result specializes, and we obtain an Euler characteristic map $\chi : K(\Gamma_A[n]) \to \mathbb{Z}$, valid for any divisible group A. This χ is invariant under all definable bijections (not only the morphisms of $\Gamma[*]$), and is additive and multiplicative. We have $\chi_O(\{0\}) = 1$, $\chi_O((a,b)) = -1$ for $a < b$, and $\chi_O(0,\infty) = -1$, too (though $(0,1)$ and $(0,\infty)$ are not definably isomorphic in the linear structure). Now let $\chi_O(X) = \chi(X)\boldsymbol{\tau}^n$ for $X \subseteq \Gamma^n$, and extend to $\Gamma[*]$ by additivity. □

Remark. The Euler characteristic constructed in this proof appears to depend on an embedding of A into the additive group of a model of RCF. But by the uniqueness shown above, it does not. In fact, as pointed out to us by Van den Dries, Ealy and Maříková, an Euler characteristic with the requisite properties is defined in [37] directly for any O-minimal structure; moreover, the use of RCF in the lemma below can also be replaced by a direct inductive argument, and some simple facts about Fourier–Motzkin elimination.

Another Euler characteristic can be obtained as follows: given a definable set $Y \subset \Gamma^n$, let

$$\chi'(Y) = \lim_{r \to \infty} \chi(Y \cap C_r),$$

where C_r is the bounded closed cube $[-r, r]^n$. By O-minimality, the value of $\chi(Y \cap C_r)$ is eventually constant.

Note that χ' is not invariant under semialgebraic bijections, since the bounded and unbounded open intervals are given different measures. Still,

Lemma 9.6. *χ' induces a group homomorphism $K(\Gamma[n]) \to \mathbb{Z}$; and yields a ring homomorphism $K(\Gamma[*]) \to \mathbb{Z}[\boldsymbol{\tau}]$. Moreover, χ' is invariant under piecewise* $\mathrm{GL}_n(\mathbb{Q})$*-transformations.*

Proof. χ' is clearly additive and multiplicative. Isomorphism invariance can be checked as follows: First, we make the following claim.

Claim. If $X \neq \emptyset$ is defined by a finite number of weak ($\leq$) affine equalities and inequalities, then $\chi'(X) = 1$.

Proof. It suffices to show that this is true in $(\mathbb{R}, +)$; since then it is true in any model of the theory of divisible ordered Abelian groups. Now we may compute the Euler characteristic χ of the bounded sets $X \cap C_r$ in $(\mathbb{R}, +, \cdot)$. Let $p \in X$. For large enough r, $p \in X \cap C_r$ there is a definable retraction of the closed bounded set $X \cap C_r$ to p (along lines through p). Thus $X \cap C_r$ has the same homology groups as a point, and so Euler characteristic 1. □

To prove the lemma we must show that if $\phi : X \to Y$ is a definable bijection, $X, Y \subseteq \Gamma^n$, then $\chi'(X) = \chi'(Y)$. We use induction on $\dim(X)$. By additivity, if X is a Boolean combination of finitely many pieces, it suffices to prove the lemma for each piece. We may therefore assume that ϕ is linear (rather than only piecewise linear) on X. Let ϕ' be a linear automorphism extending ϕ. Expressing X as a union of basic pieces, we may assume X is defined by some inequalities $\sum \alpha_i x_i \leq c$, as well as some equalities and strict inequalities. Thus X is convex. We have to show that $\chi'(X) = \chi'(\phi' X)$. Let $\bar{X}$ be the closure of X (defined by the corresponding weak inequalities). Then $\bar{X} \setminus X$ has dimension $< \dim(X)$, so by induction $\chi'(\phi'(\bar{X} \setminus X) = \chi'(\bar{X} \setminus X)$. But $\bar{X}$ is closed and convex, so $\chi_{O'}(\bar{X}) = 1 = \chi_{O'}(\phi' \bar{X})$. Subtracting, $\chi'(\phi'(\bar{X})) = \chi'(\bar{X})$.

Once again, using the ambient dimension grading, we can define $\chi'_O : \Gamma[*] \to \mathbb{Z}[\boldsymbol{\tau}]$ with $\chi'_O(x) = \chi'(x)\boldsymbol{\tau}^n$ for $x \in \Gamma[n]$. □

In the following lemma, all classes are taken in $K(\Gamma_A)[*]$. Let e_a be the class in $K(\Gamma_A)[1]$ of the singleton $\{a\}$, and τ_a the class of the segment $(0, a)$.

Lemma 9.7. *Let $a \in \mathbb{Q} \otimes A$, $b \in A$.*

(1) $\tau_a = \tau_{a+b}$, $e_a = e_{a+b}$.
(2) *If $b < c \in A$ then* $[(b, c)] = -e_0$.
(3) $e_a e_0 = e_a^2$.
(4) $\tau_a(\tau_a + e_0) = 0$.
(5) *If $2a \in A$ then $2\tau_a + e_a = -e_0$, and $e_0(e_a - e_0) = 0$.*

Proof.

(1) $\tau_a = [(0, a)] = [(0, \infty)] - [(a, \infty)] - e_a$, and similarly τ_{a+b}. The map $x \mapsto x+b$ shows that $[(a, \infty)] = [(a + b, \infty)]$ and $e_a = e_{a+b}$, hence also $\tau_a = \tau_{a+b}$.
(2) $[(b, c)] = [(b, \infty)] - [(c, \infty)] - e_c = -e_0$ by (1), since $c - b \in A$.
(3) The map $(x, y) \mapsto (x, y+x)$ is an $SL_2(\mathbb{Z})$-bijection between $\{(a, 0)\}$ and (a, a).
(4) Let

$$\begin{aligned} D &= \{(x, y) : 0 < x < a, 0 < y \le x\}, \\ D' &= \{(x, y) : 0 < y < a, 0 < x \le y\}, \\ D_1 &= \{(x, y) : 0 < x < a, y > 0\}, \\ T(x, y) &= (x, y + x). \end{aligned}$$

Then $T(D_1) = D_1 \setminus D$. Since $[T(D_1)] = [D_1]$, $[D] = 0$. Similarly, $[D'] = 0$. Note also

$$T((0, a) \times \{0\}) = \{(x, x) : 0 < x < a\}.$$

Thus

$$0 = [D] + [D'] = [(0, a)^2] + [\{(x, x) : 0 < x < a\}] = \tau_a^2 + \tau_a e_0.$$

(5) Let $0 < 2a \in A$. Then $[(0, a)] = [(a, 2a)]$ using the map $x \mapsto 2a - x$. Thus $2\tau_a + e_a = [(0, a) \cup \{a\} \cup (a, 2a)] = [(0, 2a)] = -e_0$ (by (2)). Therefore, $(-e_0 - e_a)(e_0 - e_a) = (2\tau_a)(2\tau_a + 2e_0) = 0$ by (1). Thus $e_a e_0 = e_a^2 = e_0^2$. □

The next lemma will not be used, except as a partial indication towards the question that follows, regarding the difference at the level of Grothendieck groups between $\mathrm{GL}_n(\mathbb{Z})$ and $\mathrm{GL}_n(\mathbb{Q})$ transformations. Let $\mathrm{Ann}(e_0)$ be the annihilator ideal of e_0; it is a graded ideal. Let $R = K(\Gamma_A)[*]/\mathrm{Ann}(e_0)$, the image of $K(\Gamma_A)[*]$ in the localization $K(\Gamma_A)[*](e_0{}^{-1})$. In the next lemma, the classes of definable sets are taken in R, viewed as a subring of $K(\Gamma_A)[*](e_0{}^{-1})$. Let $\mathbf{e}_a = e_a/e_0$, $t_a = \tau_a/e_0$.

Lemma 9.8. *Let $A' = \{a \in \mathbb{Q} \otimes A : \mathbf{e}_a = 1\}$.*

(1) *If $X \subseteq \Gamma^n$ is definable by linear inequalities over A, and $T \in \mathrm{GL}_n(\mathbb{Z}) \ltimes (A')^n$, then $[TX] = [X] \in R$.*
(2) *A' is a subgroup of $\mathbb{Q} \otimes A$.*

(3) $\mathbf{e}_a^2 = \mathbf{e}_a$, $t_a(t_a + 1) = 0$.
(4) *A' is 2-divisible.*

Proof.

(1) It suffices to show this when T is a translation by an element $a \in (A')^n$. The map $(x, y) \mapsto (x+y, y)$ is in $SL_{2n}(\mathbb{Z})$, hence $[X \times \{a\}] = [TX \times \{a\}]$ in $K(\Gamma_A)[2n]$. Since $a \in (A')^n$, $[a] = e_0^n$. Thus $[X]e_0^n = [TX]e_0^n$, and upon dividing by e_0^n the statement follows.
(2) This is clear from (1). For the following clauses, note that by (1)–(2), Lemma 9.7 applies with A replaced by A'.
(3) This follows from Lemma 9.7(3)–(4) divided by e_0^2.
(4) By Lemma 9.7(5) applied to A', if $2a \in A'$ then $e_0(e_a - e_0) = 0$; so $\mathbf{e}_a - 1 = 0$, i.e., $a \in A'$. □

Question 9.9. Is it true that $K(\Gamma_A[*])/\operatorname{Ann}(e_0) = K(\mathrm{DOAG}_A[*])/\operatorname{Ann}(e_0)$?

A positive answer would follow from an extension of (4) to odd primes, over arbitrary A; by an inductive argument, or by integration by parts.

9.3 Bounded sets: Volume homomorphism

Let $\bar{A} = \mathbb{Q} \otimes A$. Recall that $\Gamma^{\mathrm{bdd}}[n]$ is the category of bounded A-definable subsets of Γ^n, with piecewise $\mathrm{GL}_n(\mathbb{Z}) \ltimes A$-bijections for morphisms. Let $\mathrm{Sym}(\bar{A})$ be the symmetric algebra on A.

Proposition 9.10. *There exists a natural "volume" ring homomorphism $K(\Gamma^{\mathrm{bdd}}[*]) \to \mathrm{Sym}(\bar{A})$.*

Proof. We first work with DOAG without parameters, defining a polynomial associated with a family of definable sets.

Let $C(x, u) = C(x_1, \ldots, x_n; u_1, \ldots, u_m)$ be a formula of DOAG. Write $C_b = \{x : C(x, b)\}$; this is a definable family of definable sets. Assume the sets C_b are uniformly bounded: equivalently, as one easily sees, for some $q \in \mathbb{N}$, for each i, $C(x, u)$ implies $|x_i| \leq q \sum_j |u_j|$. For $b \in \mathbb{R}^m$, let $v(b) = \mathrm{vol} C_b(\mathbb{R}^n)$. Here vol is the Lebesgue measure.

By a *constructible function into* $\mathbb{Q}$, we mean a $\mathbb{Q}$-linear combination of characteristic functions of definable sets of DOAG. Let R be the $\mathbb{Q}$-algebra of constructible functions into $\mathbb{Q}$.

Claim 1. There exists a polynomial $P_C(u) \in R[u]$ such that for all $b \in \mathbb{R}^m$, $\mathrm{vol} C_b(\mathbb{R}^n) = P_C(b)$.

In other words, the volume of a rational polytope is piecewise polynomial in the parameters, with linear pieces. The proof of the claim is standard, using iterated integration. For each C, fix such a polynomial P_C.

At this point we reintroduce A. Any A-definable bounded subset of Γ^n has the form C_b for some C as above and some $b \in \bar{A}^m$.

Claim 2. If $C_b = C'_{b'}$ then $P_C(b) = P_{C'}(b')$.

Proof. (See also below for a more algebraic proof). Fix the formulas C, C'. Write $b = Ne, b' = N'e$ where $e \in \bar{A}^l$ is a vector of $\mathbb{Q}$-linearly independent elements of $\bar{A}$, and N, N' are rational matrices. Write $P_C = \sum a_\nu(u)u^\nu$ where a_ν is a constructible function into $\mathbb{Q}$; similarly for $P_{C'}$.

Now note that any formula $\psi(x_1, \dots, x_l)$ of DOAG of dimension l has a solution in $\mathbb{R}^l$ whose entries are algebraically independent. Use this to find algebraically independent $\tilde{e} \in \mathbb{R}^l$ such that $C_{N\tilde{e}} = C'_{N'\tilde{e}}$, and $a_\nu(N\tilde{e}) = a_\nu(b), a_\nu(N'\tilde{e}) = a'_\nu(b')$ for each multi-index ν of degree d.

By the definition of P_C we have $P_C(N\tilde{e}) = P_{C'}(N'\tilde{e})$. Thus $\sum a_\nu(b)(N\tilde{e})^\nu = \sum a'_\nu(b')(N'\tilde{e})^\nu$. By algebraic independence, $\sum a_\nu(b)(Nv)^\nu = \sum a'_\nu(b')(N'v)^\nu$ as $\mathbb{Q}$-polynomials. Therefore, $P_C(Ne) = P_{C'}(N'e)$. □

Thus we can define: $v(C_b) = P_C(b)$. Let us show that v defines a ring homomorphism.

Given C, C' one can find C'' such that $C''_{b,b'} = C_b \cup C_{b'}$, and similarly C''' with $C'''_{b,b'} = C_b \cap C_{b'}$. Then $P_C + P_{C'} = P_{C''} + P_{C'''}$. It follows that v is additive. Similarly, v is multiplicative, and translation invariant. Since $|\det(M)| = 1$ for $M \in \mathrm{GL}_n(\mathbb{Z})$, if $\phi^M(x, u) = \phi(Mx, u)$ then $P_{\phi^M} = P_\phi$. □

Van den Dries, Ealy, and Maříková pointed out that Claim 2 can also be reduced to the following statement: if $Q \in R[u]$, B is any 0-definable set of , and Q vanishes on $B(\mathbb{R})$, then Q vanishes on $B(\Gamma)$. They prove it as follows: let $\bar{B}$ be the Zariski closure of B; $\bar{B}$ is clearly a finite union of linear subspaces, and by intersecting B with each of these, we may assume $\bar{B}$ is linear, so it is cut out by homogeneous linear polynomials $Q_1, \dots, Q_m$. Each Q_i vanishes on $B(\mathbb{R})$ and hence on $B(\Gamma)$. Thus Q lies in the (radical) ideal generated by $Q_1, \dots, Q_m$, hence vanishes on $B(\Gamma)$.

The counting homomorphism in the discrete case

Suppose A has a least positive element 1, and assume given a homomorphism $h_p : A \to \mathbb{Z}_p$ for each p. Then A embeds into a $\mathbb{Z}$-group $\widetilde{A}$, i.e., an ordered Abelian group whose theory is the theory $\mathrm{Th}(\mathbb{Z})$ of $(\mathbb{Z}, <, +)$. (We have $\widetilde{A} \cap (\mathbb{Q} \otimes A) = \{a/n \in \mathbb{Q} \otimes A : (\forall p)(n | h_p(a)).\}$) We have a homomorphism $[X] \mapsto [X(\widetilde{A})]$ from $K_+(\Gamma[*])$ to $K_+(\mathrm{Th}(\mathbb{Z})_A)$. On the other hand, the polynomial formula for the number of integral points in a polytope defined by linear equations over $\mathbb{Z}$ yields a homomorphism $K(\mathrm{Th}(\mathbb{Z})^{\mathrm{bdd}}[*]) \to \mathbb{Q}[A]$. By composing we obtain a homomorphism $K(\Gamma^{\mathrm{bdd}}[*]) \to \mathbb{Q}[A]$.

Remark. Using integration by parts, one can see that the homomorphism

$$K(\mathrm{Th}(\mathbb{Z})^{\mathrm{bdd}}[*]) \to \mathbb{Q}[A]$$

above is actually an isomorphism.

9.4 The measured case

We repeat the definition of $\mu\Gamma$ from the introduction, along with two related categories. The category $\mathrm{vol}\,\Gamma$ corresponds to integrable volume forms, i.e., those that can be transformed by a definable change of variable to the standard form on a definable subsets of affine n-space. By Lemma 3.26, the liftability condition in (2) is equivalent to being piecewise in $\mathrm{GL}_n(\mathbb{Z}) \ltimes A^n$, A^n being the group of definable points.

Definition 9.11.

(1) For $c = (c_1, \ldots, c_n) \in \Gamma^n$, let $\sum(c) = \sum_{i=1}^n c_i$.
(2) For $n \geq 0$, let $\mu\Gamma[n]$ be the category whose objects are pairs (X, ω), with $X \in \mathrm{Ob}\,\Gamma[n]$ and $\omega : X \to \Gamma$ a definable map. A morphism $(X, \omega) \to (X', \omega')$ is a definable bijection $f : X \to X'$ liftable to a definable bijection $\mathrm{val}_{\mathrm{rv}}{}^{-1}X \to \mathrm{val}_{\mathrm{rv}}{}^{-1}X'$, such that $\sum(x) + \omega(x) = \sum(x') + \omega'(x')$ for $x \in X$, $x' = f(x)$.
(4) Let $\mu\Gamma^{\mathrm{bdd}}[n]$ be the full subcategory of $\mu\Gamma[n]$ with objects $X \subseteq [\gamma, \infty)^n$ for some $\gamma \in \Gamma$.
(3) Let $\mathrm{Ob}\,\mathrm{vol}\,\Gamma[n]$ be the set of finite disjoint unions of definable subsets of Γ^n. Given $X, Y \in \mathrm{Ob}\,\mathrm{vol}\,\Gamma[n]$, $f \in \mathrm{Mor}_{\mathrm{vol}\,\Gamma[n]}(X, Y)$ iff $f \in \mathrm{Mor}_{\Gamma[n]}$ and $\sum(x) = \sum(f(x))$ for $x \in X$.
(5) $\mu\Gamma[*]$ is the direct sum of the $\mu\Gamma[n]$, and similarly for the related categories.

Recall the Grothendieck rings of functions from Section 2.2. $\mathrm{Fn}(\Gamma, K_+(\Gamma))$ is a semigroup with pointwise addition. We also have a convolution product: if f is represented by a definable $F \subseteq \Gamma \times \Gamma^m$, in the sense that $f(\gamma) = [F(\gamma)]$, and g by a definable $G \subseteq \Gamma \times \Gamma^n$, let

$$f * g(\gamma) = [\{(\alpha, b, c) : \alpha \in \Gamma, b \in F(\alpha), c \in G(\gamma - \alpha)\}].$$

The coordinate α in the definition is needed in order to make the union disjoint. In general, it yields an element represented by a subset of $\Gamma \times \Gamma^{m+n+1}$ rather than $m + n$. But let $\mathrm{Fn}(\Gamma, K_+(\Gamma))[n]$ be the set of $[F] \in \mathrm{Fn}(\Gamma, K_+(\Gamma[n]))$ such that $\dim(F(a)) < n$ for all but finitely many $a \in \Gamma$. If $f \in \mathrm{Fn}(\Gamma, K_+(\Gamma))[m]$ and $g \in \mathrm{Fn}(\Gamma, K_+(\Gamma))[n]$, then $f * g \in \mathrm{Fn}(\Gamma, K_+(\Gamma))[m+n]$. Let $\mathrm{Fn}(\Gamma, K_+(\Gamma))[*] = \oplus_m \mathrm{Fn}(\Gamma, K_+(\Gamma))[m]$, a graded semiring.

Lemma 9.12.

(1) $K_+(\mu\Gamma)[n] \simeq \mathrm{Fn}(\Gamma, K_+(\Gamma))[n]$.
(2) $K_+\,\mu\Gamma^{\mathrm{bdd}}[n] \simeq \{f \in \mathrm{Fn}(\Gamma, K_+(\Gamma^{\mathrm{bdd}}))[n] : (\exists\gamma_0)(\forall\gamma < \gamma_0)(f(\gamma) = 0)\}$.
(3) $K_+\,\mathrm{vol}\,\Gamma[n] \simeq \mathrm{Fn}(\Gamma, K_+(\Gamma[n-1]))$.

Proof.
(1) Let $(X, \omega) \in \mathrm{Ob}\,\mu\Gamma[n]$, with $X \subseteq \Gamma^n$ and $\omega : X \to \Gamma$. Let $d(x) = \omega(x) + \sum(x)$. For $a \in \Gamma$, let $X_a = \{x \in X : d(x) = a\}$. This determines an element $F(X, \omega) \in \mathrm{Fn}(\Gamma, K_+(\Gamma[n]))$, namely, $a \mapsto [X_a]$. It is clear from additivity of dimension that $\dim(X_a) < n$ for all but finitely many a; so

$F(X,\omega) \in \mathrm{Fn}(\Gamma, K_+(\Gamma))[n]$. If $h \in \mathrm{Mor}_{\mu\Gamma[n]}(X,Y)$, then by the definition of $\mu\Gamma$ we have $h(X_a) = Y_a$; so $[X_a] = [Y_a]$ in $K_+(\Gamma)[n]$. Conversely if for all $a \in \Gamma$ we have $[X_a] = [Y_a]$ in $K_+(\Gamma)[n]$, then ${\mathrm{val}_{\mathrm{rv}}}^{-1}(X_a)$, ${\mathrm{val}_{\mathrm{rv}}}^{-1}(Y_a)$ are a-definably isomorphic. By Lemma 2.3 there exists a definable $H : {\mathrm{val}_{\mathrm{rv}}}^{-1}(X) \to {\mathrm{val}_{\mathrm{rv}}}^{-1}(Y)$ such that for any $x \in {\mathrm{val}_{\mathrm{rv}}}^{-1}(X)$, $H(x) = h_a(x)$, where $a = \sum \mathrm{val}_{\mathrm{rv}}(x)$. Clearly, H descends to $\bar{H} : X \to Y$; by construction $\bar{H}$ lifts to RV, and preserves $\sum + \omega$, so $\bar{H} \in \mathrm{Mor}_{\mu\Gamma[n]}(X,Y)$. We have thus shown that $[X] \mapsto [F(X)]$ is injective. It is clearly a semiring homomorphism.

For surjectivity, let $g \in \mathrm{Fn}(\Gamma, K_+(\Gamma))[n]$ be represented by $G \subseteq \Gamma \times \Gamma^n$. It suffices to consider either g with singleton support $\{\gamma_0\}$, or g such that $\dim(G(a)) < n$ for all $a \in \Gamma$. In the first case, $g = F(X,\omega)$ where $X = G(\gamma_0)$ and $\omega(x) = \gamma_0 - \sum(x)$. In the second: after effecting a partition and a permuation of the variables, we may assume $G(a) \subseteq \Gamma^{n-1} \times \{\psi(a)\}$ for some definable function $\psi(a)$. With another partition of Γ, we may assume g is supported on $S \subseteq \Gamma$, i.e., $g(x) = 0$ for $x \notin S$, and ψ is either injective or constant on S. In fact, we may assume ψ is injective on S: if ψ is constant on S, let $G' = \{(a, (b_1, \ldots, b_{n-1}, b_n + a)) : (a, (b_1, \ldots, b_n)) \in G, a \in S\}$. Then G' also represents g, and for G' the function ψ is injective. Now let $X = \cup_{a \in S} G(a)$, and let $\omega(x) = -\sum(x) + \psi^{-1}(x_n)$. Then $F(X,\omega) = g$.

(2) This follows from (1) by restricting the isomorphism.

(3) This is proved in a similar manner to (1) though more simply and we omit the details. The key point is that $\mathrm{GL}_n(\mathbb{Z})$ acts transitively on $\mathbb{P}^n(\mathbb{Q})$; this can be seen as a consequence of the fact that finitely generated torsion free Abelian groups are free. More specifically, the covector $(1, \ldots, 1)$ is $\mathrm{GL}_n(\mathbb{Z})$-conjugate to $(1, 0, \ldots, 0)$. Thus the catgegory $\mathrm{vol}\,\Gamma[n]$ is equivalent to the one defined using the weighting x_1 in place of $\sum(x_i)$. For this category the assertion is clear. □

This lemma reduces the study of $K_+(\mu\Gamma)$ to that of $K_+(\Gamma)$.

10 The Grothendieck semirings of RV

10.1 Decomposition to Γ, RES

Recall that RV is a structure with an exact sequence

$$0 \to \mathbf{k}^* \to RV \underset{\mathrm{val}_{\mathrm{rv}}}{\to} \Gamma \to 0.$$

We study here the Grothendieck semiring of RV in a theory $\mathbf{T}_{\mathrm{RV}}$ satisfying the assumptions of Lemma 3.26. The intended case is the structure induced from ACVF_A for some RV, Γ-generated base structure A.

We show that the Grothendieck ring of RV decomposes into a tensor product of those of RES, and of Γ.

The category $\Gamma[*]$ was described in Section 9. We used $\mathrm{GL}_n(\mathbb{Z})$ rather than $\mathrm{GL}_n(\mathbb{Q})$ morphisms. The reason is given by the following.

Lemma 10.1. *The morphisms of* $\Gamma[n]$ *are precisely those definable maps that lift to morphisms of* $\mathrm{RV}[n]$. *The map* $X \mapsto {\mathrm{val}_{\mathrm{rv}}}^{-1}(X)$ *therefore induces a functor* $\Gamma[n] \to \mathrm{RV}[n]$, *yielding an embedding of Grothendieck semirings* $K_+[\Gamma[n]] \to K_+[\mathrm{RV}[n]]$.

Proof. Any morphism of $\Gamma[*]$ obviously lifts to RV, since $\mathrm{GL}_n(\mathbb{Z})$ acts on C^n for any group C. The converse is a consequence of Lemma 3.28. □

We also have an inclusion morphism $K_+(\mathrm{RES}) \to K_+(\mathrm{RV})$.

Observe that $K_+(\Gamma^{\mathrm{fin}})$ forms a part of both $K_+(\mathrm{RES}[*])$ and $K_+(\Gamma[*])$: the embedding of $K_+(\Gamma[*])$ into $K_+(\mathrm{RV}[*])$ takes $K_+(\Gamma^{\mathrm{fin}})$ to a subring of $K_+(\mathrm{RES}[*])$, namely, the subring generated by the pullbacks $\mathrm{val}_{\mathrm{rv}}(\gamma)$, $\gamma \in \Gamma$ a definable point.

Given two semirings R_1, R_2 and a homomorphism $f_i : S \to R_i$, define $R_1 \otimes_S R_2$ by the universal property for triples (R, h_1, h_2), with R a semiring and $h_i : R_i \to R$ a semiring homomorphism, satisfying $h_1 f_1 = h_2 f_2$.

We have a natural map $K_+(\mathrm{RES}) \otimes K_+(\Gamma[*]) \to K_+(\mathrm{RV})$, $[X] \otimes [Y] \mapsto [X \times {\mathrm{val}_{\mathrm{rv}}}^{-1}(Y)]$. By the universal property it induces a map on $K_+(\mathrm{RES}) \otimes_{K_+(\Gamma^{\mathrm{fin}})} K_+(\Gamma[*])$. A typical element of the image is represented by a definable set of the form $\dot{\cup}(X_i \times {\mathrm{val}_{\mathrm{rv}}}^{-1}(Y_i))$, with $X_i \subseteq \mathrm{RES}^*$, $Y_i \subseteq \Gamma^*$.

Proposition 10.2. *The natural map* $K_+(\mathrm{RES}) \otimes_{K_+(\Gamma^{\mathrm{fin}})} K_+(\Gamma[*]) \to K_+(\mathrm{RV})$ *is an isomorphism.*

Proof. Surjectivity is Corollary 3.25. We will prove injectivity. In this proof, $X \otimes Y$ will always denote an element of $K_+(\mathrm{RES}) \otimes_{K_+(\Gamma^{\mathrm{fin}})} K_+(\Gamma[*])$.

Claim 1. Any element of $K_+(\Gamma[*])$ can be expressed as $\sum_{j=1}^{l} [Y_j] \times \{p_j\}$, for some $Y_j \subseteq \Gamma^{m_j}$, $\dim(Y_j) = m_j$, and $p_j \in \Gamma^{l_j}$.

Proof. Let $Y \subseteq \Gamma^m$ be definable. If $\dim(Y) < m$, then Y can be partitioned into finitely many sets Y_j, each of which lies in some definable affine hypersurface $\sum_{i=1}^{m} \alpha_i x_i = c$, with $\alpha_i \in \mathbb{Q}$, not all 0. In other words $x \mapsto \alpha \cdot x$ is constant on Y_j, where $\alpha = (\alpha_1, \dots, \alpha_m)$. We may assume that each $\alpha_i \in \mathbb{Z}$ and that they are relatively prime. Then (α) is the first row of a matrix $M \in \mathrm{GL}_m(\mathbb{Z})$. The map $x \mapsto Mx$ takes Y_j to a set of the form $Y_j' \times \{c\}$, $Y_j' \subseteq \Gamma^{m-1}$. Since $[MY_j] = [Y_j]$ in $K_+(\Gamma[*])$, the claim follows by induction. □

Claim 2. Any element of $K_+(\mathrm{RES}) \otimes_{K_+(\Gamma^{\mathrm{fin}})} K_+(\Gamma[*])$ can be represented as

$$\sum_{i=1}^{k} X_i \otimes {\mathrm{val}_{\mathrm{rv}}}^{-1} Y_i,$$

where $X_i \subseteq \mathrm{RES}^{n_i}$ and $Y_i \subseteq \Gamma^{m_i}$ are definable sets, and $m_i = \dim Y_i$.

Proof. By the definition of $K_+(\mathrm{RES}) \otimes_{K_+(\Gamma^{\mathrm{fin}})} K_+(\Gamma[*])$ and by Claim 1, any element is a sum of tensors $X \otimes {\mathrm{val}_{\mathrm{rv}}}^{-1}(Y \times \{p\})$; using the $\otimes_{K_+(\Gamma^{\mathrm{fin}})}$-relation, $X \otimes {\mathrm{val}_{\mathrm{rv}}}^{-1}(Y \times \{p\}) = (X \times {\mathrm{val}_{\mathrm{rv}}}^{-1}(p)) \otimes Y$. □

Now let $X_i, X_i' \subseteq \mathrm{RES}^*$, $Y_i, Y_i' \subseteq \Gamma^*$ be definable sets, and let

$$F : \dot{\cup}(X_i \times \mathrm{val_{rv}}^{-1}(Y_i)) \to \dot{\cup}(X_{i'}' \times \mathrm{val_{rv}}^{-1}(Y_{i'}'))$$

be a definable isomorphism. Let m be the maximal dimension m of any Y_i or $Y_{i'}'$. Assume the following (by Claim 2):

$$\text{For each } i',\ Y_{i'}' \subseteq \Gamma^{\dim(Y_{i'}')} \text{ and similarly for the } Y_i. \qquad (*)$$

Claim 3. Let P be a complete type of Y_i of dimension m, and Q a complete type of X_i. Then $F(Q \times \mathrm{val_{rv}}^{-1} P) = Q' \times \mathrm{val_{rv}}^{-1} P'$, where Q' is a complete type of some $X_{i'}'$, and P' a complete type type of $Y_{i'}'$.

Moreover, there exist definable sets $\widetilde{P}$, $\widetilde{Q}$, $\widetilde{P}'$, $\widetilde{Q}'$ containing P, Q, P', Q', respectively, such that

(1) F restricts to a bijection $\widetilde{Q} \times \mathrm{val_{rv}}^{-1} \widetilde{P} \to \widetilde{Q}' \times \mathrm{val_{rv}}^{-1} \widetilde{P}'$;
(2) there exist definable bijections $f : \widetilde{P} \to \widetilde{P}'$ and $g : \widetilde{Q} \to \widetilde{Q}'$;
(3) For any $x \in \widetilde{Q}$, $y \in \widetilde{P}$, F restricts to a bijection $\{x\} \times \mathrm{val_{rv}}^{-1}(y) \to \{f(x)\} \times \mathrm{val_{rv}}^{-1}(g(y))$.

Proof. By Lemma 3.17, $\mathrm{val_{rv}}^{-1}(P)$ is a complete type; by the same lemma, $Q \times \mathrm{val_{rv}}^{-1}(P)$ is complete; hence so is $F(Q \times \mathrm{val_{rv}}^{-1}(P))$. We have $F(Q \times \mathrm{val_{rv}}^{-1}(P)) \subseteq (X_{i'}' \times \mathrm{val_{rv}}^{-1}(Y_{i'}'))$ for some i'. Let $Q' = \mathrm{pr}_1(F(Q \times \mathrm{val_{rv}}^{-1}(P)))$, $V' = \mathrm{pr}_2(F(Q \times \mathrm{val_{rv}}^{-1}(P)))$, $P' = \mathrm{val_{rv}}(V') \subseteq Y_{i'}'$. where $\mathrm{pr}_1 : X_{i'}' \times \mathrm{val_{rv}}^{-1}(Y_{i'}') \to X_i \subseteq \mathrm{RES}$, $\mathrm{pr}_2 : X_{i'}' \times \mathrm{val_{rv}}^{-1}(Y_{i'}') \to \mathrm{val_{rv}}^{-1}(Y_{i'}')$ are the projections. Then Q', V', P' are complete types. We have $m = \dim(P') \geq \dim(Y_{i'}')$, so by maximality of m, equality holds. We thus have $P' \subseteq \Gamma^{\dim(P')}$. By Lemma 3.17, $Q' \times \mathrm{val_{rv}}^{-1}(P')$ is also complete type. Thus $F(Q \times P) = Q' \times \mathrm{val_{rv}}^{-1} P'$.

By one more use of Lemma 3.17, the function $f_y : x \mapsto \mathrm{pr}_1 F(x, y)$, whose graph is a subset of the stable set $Q \times Q'$, cannot depend on $y \in P$. Thus $f_y = f$, i.e., $F(x, y) = (f(x), \mathrm{pr}_2 F(x, y))$.

Since $Q \times \mathrm{val_{rv}}^{-1}(y)$ is stable, $\mathrm{val_{rv}}\, \mathrm{pr}_2 F$ must be constant on it; so $\mathrm{val_{rv}}\, \mathrm{pr}_2 F(x, y) = g(y)$ on $P \times Q$. This shows that (3) of the "moreover" holds on $P \times Q$. By compactness, it holds on some definable $\widetilde{Q} \times \widetilde{P}$ (and we may take f injective on $\widetilde{Q}$, and g on $\widetilde{P}$). Let $\widetilde{Q}' = f(\widetilde{Q})$, $\widetilde{P}' = g(\widetilde{P})$. Then (1)–(2) hold also. □

Claim 4. Assume $(*)$ holds. Then there exist finitely many definable Y_i^j ($j = 0, \ldots, N_i$) and X_i^j such that $\dim(Y_i^0) < m$, and the conclusion of Claim 3 holds on each $X_i^j \times \mathrm{val_{rv}}^{-1} Y_i^j$ for $j \geq 1$. Moreover, we may take the Y_i^j, X_i^j pairwise disjoint.

Proof. This follows from Claim 3 by compactness; the disjointness can be achieved by noting that if Claim 3(3) holds for $\widetilde{P}$, $\widetilde{Q}$, then it holds for their definable subsets, too. □

We now show that if $\dot\cup(X_i \times \mathrm{val_{rv}}^{-1}(Y_i))$ and$\dot\cup(X'_{i'} \times \mathrm{val_{rv}}^{-1}(Y'_{i'}))$ are definably isomorphic then $\sum_{i'}[X_{i'}] \otimes [Y_{i'}] = \sum_i [X_i \otimes Y_i]$. We use induction on the maximal dimension m of any Y_i or $Y'_{i'}$, and also on the number of indices i such that $\dim(Y_i) = m$. Say $\dim(Y_1) = m$.

By Claim 2, without changing $\sum_{i'} X'_{i'} \otimes \mathrm{val_{rv}}^{-1}(Y'_{i'}))$ as an element of

$$K_+(\mathrm{RES}) \otimes_{K_+(\Gamma^{\mathrm{fin}})} K_+(\Gamma[*]),$$

we can arrange that $\dim(Y_{i'}) = m_{i'}$, i.e., $(*)$ holds. Thus Claims 3 and 4 apply.

The Y_1^j for $j \geq 1$ may be removed from Y_1, if their images are correspondingly excised from the appropriate Y'_j, since $[\widetilde{Q}] \otimes_{K_+(\Gamma^{\mathrm{fin}})} [\widetilde{P}] = [f(\widetilde{Q})] \otimes_{K_+(\Gamma^{\mathrm{fin}})} [g(\widetilde{P})]$. What is left in Y_1 has Γ dimension $< m$, and so by induction the classes are equal.

The injectivity and the proposition follow. □

For applications to VF, we need a version of Proposition 10.2 keeping track of dimensions. Below, the tensor product is in the category of graded semirings.

Corollary 10.3. *The natural map* $K_+(\mathrm{RES}[*]) \otimes_{K_+(\Gamma^{\mathrm{fin}})} K_+(\Gamma[*]) \to K_+(\mathrm{RV}[*])$ *is an isomorphism.*

Proof. For each n we have a surjective homomorphism

$$\oplus_{k=1}^{n} K_+(\mathrm{RES}[k]) \otimes K_+(\Gamma[n-k]) \to K_+(\mathrm{RV}[n]).$$

$K_+\,\mathrm{RV}[n]$ can be identified with a subset of the semiring $K_+\,\mathrm{RV}$, namely, $\{[X] : \dim(X) \leq n\}$. The proof of Proposition 10.2 shows that the kernel is generated by relations of the form

$$(X \times \mathrm{val_{rv}}^{-1}(Y)) \otimes Z = X \otimes (Y \otimes Z)$$

when $Y \in K_+(\Gamma^{\mathrm{fin}})$ and $\dim(X) + \dim(\mathrm{val_{rv}}^{-1}(Y)) + \dim(\mathrm{val_{rv}}^{-1}(Z)) = n$. These relations are taken into account in the ring $K_+(\mathrm{RES}[*]) \otimes_{K_+(\Gamma^{\mathrm{fin}})} K_+(\Gamma[*])$, so that the natural map $K_+(\mathrm{RES}[*]) \otimes_{K_+(\Gamma^{\mathrm{fin}})} K_+(\Gamma[*]) \to K_+(\mathrm{RV}[*])$ is injective and hence an isomorphism. □

Recall the classes $e_a = [\{a\}]_1$ in $K(\Gamma[1])$, defined for $a \in \Gamma(\langle\emptyset\rangle)$. They are in $K_+(\Gamma^{\mathrm{fin}})$, hence identified with classes in $K(\mathrm{RES}[1])$, namely, $e_a = [\mathrm{val_{rv}}^{-1}(a)]$. When denoting classes of varieties V over the residue field, we will write $[V]$ for $[V(\mathbf{k})]$, when no confusion can arise.

Definition 10.4. Let $I!$ be the ideal of $K(\mathrm{RES}[*])$ generated by all differences $e_a - e_0$, where $a \in \Gamma(\langle\emptyset\rangle)$. Let $!\mathrm{K}(\mathrm{RES}[*]) = K(\mathrm{RES}[*])/I!$.

By Lemma 9.7(3), the natural homomorphism $K(\mathrm{RES}[*])$ into the localization of $K(\mathrm{RV}[*])$ by all classes e_a factors through $!\mathrm{K}(\mathrm{RES}[*])$.

Since $I!$ is a homogeneous ideal, $!\mathrm{K}(\mathrm{RES}[*])$ is a graded ring.

The theorem that follows, when combined with the canonical isomorphisms $K(\mathrm{VF}[n]) \to K(\mathrm{RV}[\leq n])/\mathrm{I_{sp}}$ and $K(\mathrm{VF}) \to K(\mathrm{RV}[*])/\mathrm{I_{sp}}$,

$$\int\!\!\!\!\!\!\mathrm{R} : K(\mathrm{VF}) \to !\mathrm{K}(\mathrm{RES})[[\mathbb{A}_1(\mathbf{k})]^{-1}],$$

$$\int\!\!\!\!\!\!\mathrm{R}' : K(\mathrm{VF}) \to !\mathrm{K}(\mathrm{RES}).$$

Theorem 10.5.

(1) *There exists a group homomorphism*

$$\mathcal{E}_n : K(\mathrm{RV}[\leq n])/\mathrm{I_{sp}} \to !\mathrm{K}(\mathrm{RES}[n])$$

with

$$[\mathrm{RV}^{>0}]_1 \mapsto -[\mathbb{A}^{n-1} \times G_m]_n$$

and

$$[X]_k \mapsto [X \times \mathbb{A}^{n-k}]_n$$

for $X \in \mathrm{RES}[k]$.

(2) *There exists a ring homomorphism* $\mathcal{E} : K(\mathrm{RV}[*])/\mathrm{I_{sp}} \to !\mathrm{K}(\mathrm{RES})[[\mathbb{A}_1]^{-1}]$ *with* $\mathcal{E}([X]_k) = [X]_k/\mathbb{A}^k$ *for* $X \in \mathrm{RES}[k]$.

(3) *There exists a group homomorphism*

$$\mathcal{E}'_n : K(\mathrm{RV}[\leq n])/\mathrm{I_{sp}} \to !\mathrm{K}(\mathrm{RES}[n])$$

with $[\mathrm{RV}^{>0}]_1 \mapsto 0$ *and* $[X]_k \mapsto [X]_n$ *for* $X \in \mathrm{RES}[k]$.

(4) *There exists a ring homomorphism* $\mathcal{E}' : K(\mathrm{RV}[*])/\mathrm{I_{sp}} \to !\mathrm{K}(\mathrm{RES})$ *with* $\mathcal{E}([X]_k) = [X]_k$ *for* $X \in \mathrm{RES}[k]$.

Proof.

(1) We first define a homomorphism $\chi[m] : K(\mathrm{RV}[m]) \to !\mathrm{K}(\mathrm{RES}[m])$. By Corollary 10.3,

$$K(\mathrm{RV}[m]) = \oplus_{l=1}^{m} K(\mathrm{RES}[m-l]) \otimes_{K_+(\Gamma^{\mathrm{fin}})} K(\Gamma[l]).$$

Let $\chi_0 = \mathrm{Id}_{K\,\mathrm{RES}[m]}$. For $l \geq 1$, recall the homomorphism $\chi : K(\Gamma[l]) \to \mathbb{Z}$ of Lemma 9.5. It induces $\chi_l : K(\mathrm{RES}[k]) \otimes_{K_+(\Gamma^{\mathrm{fin}})} K(\Gamma[l]) \to !\mathrm{K}(\mathrm{RES}[k])$ by $a \otimes b \mapsto \chi(b) \cdot [G_m]^l \cdot a$.

Define a group homomorphism

$$\chi[m] : K(\mathrm{RV}[m]) \to K(\mathrm{RES}[m]), \quad \chi[m] = \oplus_l \chi_l.$$

We have

$$\chi[m_1 + m_2](ab) = \chi[m_1](a)\chi[m_2](b)$$

when $a \in K(\mathrm{RV}[m_1])$, $b \in K(\mathrm{RV}[m_2])$. This can be checked on homogeneous elements with respect to the grading $\oplus_l K_+(\mathrm{RES}[m-l]) \otimes K_+(\Gamma[l])$.

We compute $\chi[1]([\mathrm{RV}^{>0}]_1) = \chi_1(1 \otimes [\Gamma^{>0}]_1) = -[G_m] \in K(\mathrm{RES}[1])$.

Next, define a group homomorphism $\beta_m : !\mathrm{K}(\mathrm{RES}[m]) \to !\mathrm{K}(\mathrm{RES}[n])$ by $\beta_m([X]) = [X \times \mathbb{A}^{n-m}]$. Define $\gamma : \oplus_{m \le n} K(\mathrm{RV}[m]) \to !\mathrm{K}(\mathrm{RES}[n])$ by $\gamma = \sum_m \beta_m \circ \chi[m]$. Then γ is a group homomorphism, and $\gamma(a)\gamma(b) = \gamma(ab) \times [\mathbb{A}^n]$ for $a \in K(\mathrm{RV}[m_1])$, $b \in K(\mathrm{RV}[m_2])$, $m_1 + m_2 \le n$. Again this is easy to verify on homogeneous elements.

Finally, we compute γ on the standard generator $J = [\mathrm{RV}^{>0}]_1 + [1]_0 - [1]_1$ of I_{sp}. Since $\chi[1]([\mathrm{RV}^{>0}]_1) = -[G_m]$, we have

$$\gamma([\mathrm{RV}^{>0}]_1) = \beta_1(-[G_m]) = -[G_m \times \mathbb{A}^{n-1}]_1$$

On the other hand,

$$\gamma([1]_0) = \beta_0([1]_0) = [\mathbb{A}^n]_n,$$
$$\gamma([1]_1) = \beta_1([1]_1) = [\mathbb{A}^{n-1}]_n.$$

Thus $\gamma(J) = [\mathbb{A}^{n-1}]_{n-1} \times (-[G_m]_1 + [\mathbb{A}^1]_1 - [1]_1) = 0$. A homomorphism $K(\mathrm{RV}[\le n])/\mathrm{I}_{\mathrm{sp}} \to K(\mathrm{RES}[n])$ is thus induced.

(2) For $a \in K(\mathrm{RV}[m])$, let $\mathcal{E}(a) = \beta_m(a)/[\mathbb{A}^m]$. For any large enough n, we have $\mathcal{E}(a) = \mathcal{E}_n(a)/[\mathbb{A}^n]$. The formulas in (1) prove that $\mathcal{E}$ is a ring homomorphism.

(3)–(4) The proof is similar, using χ' from Lemma 9.6 in place of χ of Lemma 9.5, and the identity in place of β_m. □

Corollary 10.6. *The natural morphism* $K(\mathrm{RES}[n]) \to K(\mathrm{RV}[\le n])/\mathrm{I}_{\mathrm{sp}}$ *has the kernel contained in* $!I$. □

Lemma 10.7. *Let* $\mathbf{T} = \mathrm{ACVF}_{F((t))}$ *or* $\mathbf{T} = \mathrm{ACVF}^{\mathrm{R}}{}_{F((t))}$, *$F$ a field of characteristic 0, with* $\mathrm{val}(F) = (0)$, $\mathrm{val}(F((t))) = \mathbb{Z}$, *and* $\mathrm{val}(t) = 1 \in \mathbb{Z}$. *Then there exists a retraction* $\rho_t : K_+(\mathrm{RES}) \to K_+(\mathrm{Var}_F)$. *It induces a retraction* $!\mathrm{K}(\mathrm{RES}) \to K(\mathrm{Var}_F)$.

Proof. Let $t_n \in F((t))^{\mathrm{alg}}$ be such that $t_1 = t$ and $t_{nm}^n = t_m$. For $\alpha = m/n \in \mathbb{Q}$, with $m \in \mathbb{Z}$, $n \in \mathbb{N}$, let $t_\alpha = t_n^m$. Thus $\alpha \to t_a$ is a homomorphism $\mathbb{Q} \to G_m(F((t))^{\mathrm{alg}})$.

Let $V(\alpha) = \mathrm{val}_{\mathrm{rv}}{}^{-1}(\alpha)$. Let $\mathbf{t}_\alpha = \mathrm{rv}(t_\alpha)$. Then $\mathbf{t}_a \in V(\alpha)$.

Let $X \in \mathrm{RES}[n]$. Then for some $\alpha_1, \dots, \alpha_n \in \mathbb{Q}$, we have $X \subseteq \Pi_{i=1}^n V(\alpha_i)$, where $V(\alpha_i) = \mathrm{val}_{\mathrm{rv}}{}^{-1}(\alpha_i)$. Define $f(x_1, \dots, x_n) = (x_1/\mathbf{t}_{\alpha_1}, \dots, x_n/\mathbf{t}_{\alpha_n})$. Then f is $F((t^{1/m}))$-definable for some m, but not in general. Nevertheless, $F(X) =: Y$ is definable. This is because the Galois group $G = \mathrm{Aut}(F^a((t^{1/m}))/F^a((t)))$ extends to a group of valued field automorphisms $\mathrm{Aut}(\mathbf{k}((t^{1/m}))/\mathbf{k}((t)))$ fixing the entire residue field $\mathbf{k}$; while $Y \subseteq \mathbf{k}$; thus G fixes Y pointwise and hence setwise.

The map $X \mapsto Y$ of definable sets described above clearly respects disjoint unions. It also respects definable bijections: if $h : X \to X'$ is a definable bijection, $Y = f(X)$, $Y' = F(Y')$, then fhf^{-1} is an $F((t^{1/\infty}))$-definable bijection $Y \to Y'$; by the Galois argument above, it is, in fact, definable.

The definable subsets of $\mathbf{k}$ are just the F-constructible sets. Thus we have an induced homomorphism $\rho_t : K_+(\mathrm{RES}) \to K_+(\mathrm{Var}_F)$; it is clearly the identity on $K_+(\mathrm{RES})$. It induces a homomorphism $K(\mathrm{RES}) \to K(\mathrm{Var}_F)$.

Finally $\rho_t(\mathrm{val}_{\mathrm{rv}}{}^{-1}(\alpha)) = [G_m]$ for any $\alpha \in \mathbb{Q}$; so a homomorphism on $!\mathrm{K}(\mathrm{RES})$ is induced. □

This example can be generalized as follows. Let L be a valued field with residue field F of characteristic 0, $\mathbf{T} = \mathrm{ACVF}_L$ or $\mathrm{ACVF^R}_L$. Let $A = \mathrm{res}(L)$, $\mathbf{A} = \mathbb{Q} \otimes A$, and let $t : \mathbf{A} \to G_m(L^a)$ be a monomorphism, with $t(A) \subseteq G_m(L)$. Then there exists a retraction $\rho_t : K_+(\mathrm{RES}) \to K_+(\mathrm{Var}_F)$.

From Theorem 10.5 and Lemma 10.7, we obtain the example discussed in the introduction.

Proposition 10.8. *Let* $\mathbf{T} = \mathrm{ACVF^R}_{F((t))}$, *$F$ a field of characteristic* 0, *with* $\mathrm{val}(F) = (0)$ *and* $\mathrm{val}(t) = 1 \in \mathbb{Z}$. *Then there exists a ring homomorphism* $\mathcal{E}_t : K(\mathrm{VF}) \to K(\mathrm{Var}_F)[[\mathbb{A}^1]^{-1}]$, *with* $[\mathcal{M}] \mapsto -[G_m]/[G_a]$, $\mathbb{L}([X]_k) \mapsto [X]_k/[\mathbb{A}^k]$ *for* $X \in \mathrm{Var}_F[k]$. *There is also a ring homomorphism* $\mathcal{E}'_t : K(\mathrm{VF}) \to K(\mathrm{Var}_F)$ *with* $\mathbb{L}([X]_k) \mapsto [X]_k$.

10.2 Decomposition of μRV

An analogous decomposition is valid for the measured Grothendieck semiring $\mu_\Gamma\mathrm{RV}$ (Definition 8.13).

Lemma 10.9. *There exists a homomorphism* $K_+\, \mu\Gamma[n] \to K_+\, \mu_\Gamma\mathrm{RV}[n]$ *with* $[(X, \omega)] \mapsto [(\mathrm{val_{rv}}^{-1}(X), \mathrm{Id}, \omega \circ \mathrm{val_{rv}})]$.

Proof. We have to show that a $\mu\Gamma[n]$-isomorphism $X \to Y$ lifts to a $\mu_\Gamma\mathrm{RV}[n]$-isomorphism. This follows immediately from the definitions. □

Recall $\mu_\Gamma\mathrm{RES}$ from Definition 8.13. Along the lines of Lemma 9.12, we can also describe $K_+\, \mu_\Gamma\mathrm{RES}[n]$ as the semigroup of functions with finite support $\Gamma \to K_+(\mathrm{RES}[n])$. We also have the inclusion $K_+\, \mu_\Gamma\mathrm{RES}[n] \to K_+\, \mu_\Gamma\mathrm{RV}[n]$, $[(X, f)] \mapsto [(X, f, 1)]$.

Let $\mu\Gamma^{\mathrm{fin}}[n]$ be full subcategory of $\mu\Gamma[n]$ whose objects are finite. We have a homomorphism $K_+(\mu\Gamma^{\mathrm{fin}})[n] \to \mu_\Gamma\mathrm{RES}[n]$, $(X, \omega) \mapsto (\mathrm{val_{rv}}^{-1}(X), \mathrm{Id}, \omega \circ \mathrm{val_{rv}})$. As before, we obtain a homomorphism $K_+\, \mu_\Gamma\mathrm{RES}[*] \otimes_{K_+(\mu\Gamma^{\mathrm{fin}})} K_+(\mu\Gamma[*]) \to K_+(\mathrm{RV}[*])$.

Let $\mathrm{RES}_{\Gamma\text{-vol}'}$ be the full subcategory of $\mathrm{RV}_{\Gamma\text{-vol}'}$ whose objects are in RES; this is the same as RV except that morphisms must respect $\sum \mathrm{val_{rv}}$. Let $\mathrm{vol}\,\Gamma^{\mathrm{fin}}$ be the subcategory of finite objects of $\mathrm{vol}\,\Gamma$.

Proposition 10.10.

(1) *The natural map* $K_+(\mu_\Gamma\mathrm{RES}[*]) \otimes_{K_+(\mu\Gamma^{\mathrm{fin}})} K_+(\mu\Gamma[*]) \to K_+(\mu_\Gamma\mathrm{RV}[*])$ *is an isomorphism.*
(2) *So is* $K_+(\mathrm{RES}_{\Gamma\text{-vol}'}[*]) \otimes_{K_+(\mathrm{vol}\,\Gamma^{\mathrm{fin}}[*])} K_+(\mathrm{vol}\,\Gamma[*]) \to K_+(\mathrm{RV}_{\Gamma\text{-vol}'}[*])$.
(3) *The decompositions of this section preserve the subsemirings of bounded sets.*

Proof. We first prove surjectivity in (1). By the surjectivity in Corollary 10.3, it suffices to consider a class $c = [(X \times \mathrm{val_{rv}}^{-1}(Y), f, \omega)]$ with $X \in \mathrm{RES}[k]$, $Y \subseteq \Gamma^l$, $f(x, y) = (f_0(x), y)$, and $\omega : X \times (\mathrm{val}\, r^{-1}(Y)) \to \mathrm{RV}$. In fact, as in Proposition 10.2 we may take $\dim(Y) = l$, and inductively we may assume that any class

$[(X' \times Y', f', \omega')]$ with $\dim(Y') < l$ is in the image. Since we may remove a subset of Y of smaller dimension, applying Lemma 3.17 to $\omega : X \times \mathrm{val}_{\mathrm{rv}}{}^{-1}(Y) \to \Gamma$, we may assume $\omega(x, y) = \omega'(\gamma)$ when $\mathrm{val}_{\mathrm{rv}}(y) = \gamma$. Now $c = [(X, f_0, 1)] \otimes [(Y, \omega')]$.

The proof of surjectivity in (2) is similar.

The proof of injectivity in (1)–(2) is the same as of Proposition 10.2 and Corollary 10.3. (3) is clear by inspection of the homomorphisms. □

We now deduce Theorem 1.3. For a finite extension L of $\mathbb{Q}_p$, write $\mathrm{vol}_L(U)$ for $\mathrm{vol}_L(U(L))$. Let r be the ramification degree, i.e., $\mathrm{val}(L^*) = (1/r)\mathbb{Z}$. Let $Q = q^r$. The normalization is such that $\mathcal{M}$ has volume 1; so an open ball of valuative radius γ has volume $q^{r\gamma} = Q^\gamma$. Thus the volume of $\mathrm{val}_{\mathrm{rv}}{}^{-1}(\gamma)$ is $(q-1)Q^\gamma$. Also the norm satisfies $|y| = Q^{\mathrm{val}(y)}$.

Proof of Theorem 1.3. For $a \in \Gamma^k$, let $Z(a) = \{x \in \mathcal{O}_L^n : \mathrm{val}(f_1(x)) = a_1 \ldots \mathrm{val}(f_k(x)) = a_k\}$. Then

$$\int_{\mathcal{O}_L^n} |f|^s = \sum_{a \in (\Gamma^{\geq 0})^k} Q^{s \cdot a} \mathrm{vol}_L(Z(a)).$$

According to Propositions 4.5 and 10.10, we can write

$$Z(a) \sim \dot{\bigcup_{i=1}^{\nu}} \mathbb{L}\mathbf{X}_i \times \mathbb{L}\Delta_i(a),$$

where Δ_i is a definable subset of $\Gamma^{k+n_2(i)}$, $h^i : \Delta_i \to \Gamma^k$ the projection to the first k coordinates, $\Delta_i(a) = \{d \in \Gamma^{n_2(i)} : h^i(d) = a\}$, $\mathbf{X}_i = (X_i, f_i) \in \mathrm{RES}[n_1(i)]$, and $\sim$ denotes equivalence up to an admissible transformation. Thus

$$\mathrm{vol}_L(Z(a)) = \mathrm{vol}_L\left(\dot{\bigcup_{i=1}^{\nu}} \mathbb{L}\mathbf{X}_i \times \mathbb{L}\Delta_i(a)\right) = \sum_{i=1}^{\nu} \mathrm{vol}_L(\mathbb{L}\mathbf{X}_i)\mathrm{vol}_L(\mathbb{L}\Delta_i(a)).$$

If $b = (b_1, \ldots, b_{k+n_2(i)}) \in \Delta_i$, let $h_0^i(b)$ be the sum of the last $n_2(i)$ coordinates.

Since $\mathrm{val}_{\mathrm{rv}}$ takes only finitely many values on a definable subset of RES, we may assume $\sum \mathrm{val}_{\mathrm{rv}}(f(x)) = \gamma(i)$ is constant on $x \in X_i$. Then $\mathrm{vol}_L(\mathbb{L}X_i(L)) = Q^{\gamma(i)}|X_i(L)|$. Thus

$$\int_{\mathcal{O}_L^n} |f|^s = \sum_i |X_i(L)| Q^{\gamma(i)} \sum_{a \in (\Gamma^{\geq 0})^k} Q^{s \cdot a} \mathrm{vol}_L(\mathbb{L}\Delta_i(a)). \tag{10.1}$$

Now $\mathrm{vol}_L(\mathbb{L}\Delta_i(a)) = \sum_{b \in \Delta_i, h(b) = a} (q-1)^{n_2(i)} Q^{h_0(b)}$. Thus

$$\begin{aligned}\sum_{a \in (\Gamma^{\geq 0})^k} Q^{s \cdot a} \mathrm{vol}_L(\mathbb{L}\Delta_i(a)) &= \sum_{b \in \Delta_i} Q^{s_1 h_1^i(b) + \cdots + s_k h_k^i(b)} (q-1)^{n_2(i)} Q^{h_0(b)} \\ &= (q-1)^{n_2(i)} \mathrm{ev}_{h^i, s, Q}(\Delta_i).\end{aligned} \tag{10.2}$$

The theorem follows from equations (10.1)–(10.2). □

Let A be the set of definable points of Γ. Recall that for $X \subseteq \mathrm{RV}$, $[X]_1$ denotes the class $[(X, \mathrm{Id}_X)] \in \mathrm{RV}[1]$ of X with the identity map to RV, and the constant form 1. For $a \in A$, let $\tilde{e}_a = [(\mathrm{val_{rv}}^{-1}(0), \mathrm{Id}, a)] \in \mathrm{RES}[1]$, $f_a = [\{1\}_{\mathbf{k}}, \mathrm{Id}, a] \in \mathrm{RES}[1]$ where a in the third coordinate is the constant function with value a. If a lifts to a definable point d of RV, multiplication by d shows that $\tilde{e}_a = [\mathrm{val_{rv}}^{-1}(a), \mathrm{Id}, 0]$, $f_a = [\{d\}, \mathrm{Id}, 0]$. Note $\tilde{e}_a\tilde{e}_b = \tilde{e}_{a+b}\tilde{e}_0$; and $\tilde{e}_0 = [G_m]$. Let $\tau_a \in \mathrm{RES}[1]$ be the class of $(\mathrm{val_{rv}}^{-1}((a, \infty)), \mathrm{Id}, 0)$. The generating relation of $\mu_\Gamma \mathrm{I_{sp}}$ is thus (τ_0, f_0) (Lemma 8.20(6)). Let $\mathfrak{h}$ be the class of $[(\mathrm{RV}^{>0}, \mathrm{Id}, x^{-1})]$.

Let $!I^0_\mu$ be the ideal of $K(\mu_\Gamma\mathrm{RES}[*])$ generated by the relations $\tilde{e}_{a+b} = [(\mathrm{val_{rv}}^{-1}(a), \mathrm{Id}, b)]$, where $a, b \in A$, b denoting the constant function b. Let $!I_\mu$ be the ideal generated by $!I^0_\mu$ as well as the element $[\mathbb{A}_1]_1$.

Theorem 10.11. *There exist two graded ring homomorphisms*

$$\oint, \oint' : K^{\mathrm{eff}}(\mu_\Gamma\mathrm{VF}[*]) = K(\mu_\Gamma\mathrm{RV}[*])/\mathrm{I}^\mu_{\mathrm{sp}} \to K(\mu_\Gamma\mathrm{RES}[*])/!I_\mu$$

such that the composition

$$K(\mu_\Gamma\mathrm{RES}[*]) \to K(\mu_\Gamma\mathrm{RV}[*])/\mathrm{I}^\mu_{\mathrm{sp}} \to K(\mu_\Gamma\mathrm{RES}[*])/!I_\mu$$

equals the natural projection

$$\pi : K(\mu_\Gamma\mathrm{RES}[*]) \to K(\mu_\Gamma\mathrm{RES}[*])/!I_\mu,$$

with

$$\oint \mathfrak{h} = -[\{0_{\mathbf{k}}\}]_1, \qquad \oint' \mathfrak{h} = 0.$$

Proof. The identification $K^{\mathrm{eff}}(\mu_\Gamma\mathrm{VF}[*]) = K(\mu_\Gamma\mathrm{RV}[*])/\mathrm{I}^\mu_{\mathrm{sp}}$ is given by Theorem 8.28, and we work with $K(\mu_\Gamma\mathrm{RV}[*])/\mathrm{I}^\mu_{\mathrm{sp}}$.

According to Proposition 10.10, we can identify

$$K(\mu_\Gamma\mathrm{RV}[*]) = K(\mu_\Gamma\mathrm{RES}[*]) \otimes_{K_+(\mu\Gamma^{\mathrm{fin}})} K(\mu\Gamma[*]).$$

We first construct two homomorphisms of graded rings $R, R' : K(\mu_\Gamma\mathrm{RV}[*]) \to K(\mu_\Gamma\mathrm{RES}[*])/!I_\mu$. This amounts to finding graded ring homomorphisms $K(\mu\Gamma[*]) \to K(\mu_\Gamma\mathrm{RES}[*])/!I_\mu$, agreeing with π on the graded ring $K_+(\mu\Gamma^{\mathrm{fin}})$. It will be simpler to work with R, R' together, i.e., construct

$$R'' = (R, R') : K(\mu\Gamma[n]) \to (K(\mu_\Gamma\mathrm{RES}[n])/!I_\mu)^2.$$

Recall from Lemma 9.12 the isomorphism

$$\phi : K(\mu\Gamma[n]) \to \mathrm{Fn}(\Gamma, K(\Gamma))[n].$$

Let $\chi'' : K(\Gamma[n]) \to \mathbb{Z}^2$ be the Euler characteristic of Proposition 9.4; so that $\chi'' = (\chi, \chi')$; cf. Lemmas 9.5 and 9.6. We obtain by composition a map $E''_n =$

$(E_n, E'_n) : \mathrm{Fn}(\Gamma, K(\Gamma[n])) \to \mathrm{Fn}(\Gamma, \mathbb{Z})^2$. Here $\mathrm{Fn}(\Gamma, \mathbb{Z})$ is the group of functions $g : \Gamma \to \mathbb{Z}$ such that $g(\Gamma)$ is finite and $g^{-1}(z)$ is a definable subset of Γ (a finite union of definable intervals and points). Thus $\mathrm{Fn}(\Gamma, \mathbb{Z})$ is freely generated as an Abelian group by $\{p_a, q_a, r\}$, where r is the constant function 1, and for $a \in A$, p_a,q_a are the characteristic functions of $\{a\}$, $\{(a, \infty)\}$, respectively. Define $\psi_n : \mathrm{Fn}(\Gamma, \mathbb{Z}) \to K(\mu_\Gamma \mathrm{RES}[*])$:

$$\psi_m(p_a) = [G_m]^{n-1}\tilde{e}_a = [G_m]^n f_a, \qquad \psi_n(q_a) = -[G_m]^n f_a, \qquad \psi_n(r) = 0.$$

For $u \in K(\mu\Gamma[n])$, let $R''(u) = \psi_n(E''_n(\phi(u)))$.

Claim. $R'' : K(\mu\Gamma[*]) \to K(\mu_\Gamma \mathrm{RES}[m])^2$ is a graded ring homomorphism.

Proof. We have already seen that ϕ is a ring homorphism, so it remains to show this for $\psi_* \circ E''_*$. Now by Proposition 9.4, $\chi''(Y) = \chi''(Y')$ iff $[Y] = [Y']$ in the Grothendieck group of DOAG. Hence given families Y_t, $Y_{t'}$ of pairwise disjoint sets with $\chi''(Y_t) = \chi''(Y'_t)$, by Lemma 2.3 we have $\chi''(\cup_t Y_t) = \chi''(\cup_t Y'_t)$. From this and the definition of multiplication in $\mathrm{Fn}(\Gamma, K(\Gamma))[*]$, and the multiplicativity of E''_n, it follows that if $E''_n(f) = E''_n(f')$ and $E''_m(g) = E''_m(g')$ then $E''_{n+m}(fg) = E''_{n+m}(fg)$. In other words, E''_* is a graded homomorphism from into $(\mathrm{Fn}(\Gamma, \mathbb{Z})^2, \star)$ for some uniquely determined multplication $\star$ on $\mathrm{Fn}(\Gamma, \mathbb{Z})^2$. Clearly, $(a, b) \star (c, d) = (a *_1 c, b *_2 d)$ for two operations $*_1$, $*_2$ on $\mathrm{Fn}(\Gamma, \mathbb{Z})$.

Now we can compute these operations explicitly on the generators:

$$p_a * p_b = p_{a+b}, \qquad p_a * q_b = q_{a+b}, \qquad q_a * q_b = -q_{a+b}$$

for both $*_1$ and $*_2$, and

$$r *_1 \tilde{e}_a = r, \qquad r *_1 q_a = -r, \qquad r *_1 r = r,$$
$$r *_2 \tilde{e}_a = -r, \qquad r *_2 q_a = 0, \qquad r *_2 r = -r.$$

Composing with ψ, we see that R'' is, indeed, a graded ring homomorphism. □

Let R, R' be the components of R''.

Claim. R, R', π agree on $K_+(\mu\Gamma^{\mathrm{fin}})$. $R(\tau_0) = R'(\tau_0) = -\tilde{e}_0$.

This is a direct computation. It follows that R, R' induce homomorphisms $K(\mu_\Gamma \mathrm{RV}[*]) \to K(\mu_\Gamma \mathrm{RES}[*])$. Since $\tilde{e}_0 + f_0 = [(\mathbb{A}_1, \mathrm{Id}, 0)]$, modulo $!I_\mu$ both R, R' equalize $\mu_\Gamma \mathrm{I_{sp}}$, and hence induce homomorphisms on $K(\mu_\Gamma \mathrm{RV}[*])/\mu_\Gamma \mathrm{I_{sp}} \to K(\mu_\Gamma \mathrm{RES}[*])/!I_\mu$. □

Remark. The construction is heavily, perhaps completely constrained. The value of $\psi_m(p_a)$ is determined by the tensor relation over $K_+(\mu\Gamma^{\mathrm{fin}})$. The value of $\psi_m(q_a)$ is determined by the relation $\mathrm{I_{sp}}$. The choice $\psi(r) = 0$ is not forced, but the multiplicative relation shows that either r or $-r$ is idempotent, so one has a product of two rings, with $\psi(r) = 0$ and with $\psi(r) = \pm 1$. In the latter case we obtain the isomorphisms of Theorem 10.5. Thus the only choice involved is to factor the fibers of an element of $\mathrm{Fn}(\Gamma, K(\Gamma))[n]$ through χ'', i.e., through $K(\mathrm{DOAG})$. It is possible that $K(\Gamma[n]) = K(\mathrm{DOAG}[n])$ (cf. Question 9.9). In this case, $\oint, \oint', \not\oint, \not\oint'$ are injective as a quadruple, and determine $K(\mu\mathrm{VF}[*])$ completely, at least when localized by the volume of a unit ball.

11 Integration with an additive character

Let $\Omega = \mathrm{VF}/\mathcal{M}$. Let $\psi : \mathrm{VF} \to \Omega$ be the canonical map.

Motivation. For any p, $\Omega(\mathbb{Q}_p)$ can be identified with the pth power roots of unity via an additive character on $\mathbb{Q}_p$. For other local fields, the universal ψ we use is tantamount to integration with respect to all additive characters of conductor $\mathcal{M}$ at once. Thus Ω is our motivic analogue of the roots of unity, and the natural map $\mathrm{VF} \to \mathrm{VF}/\mathcal{M}$, an analogue of a generic additive character.

Throughout this paper, we have been able to avoid subtractions and work with semigroups, but here it appears to be essential to work with a group or at least a cancellation semigroup. The reason is that we will introduce, as the essential feature of integration with an additive character, an identification of the integral of a function f with $f + g$ if g is $\mathcal{O}$-invariant. This corresponds to the rule that the sum over a subgroup of a nontrivial character vanishes. Now for any $h : \Omega \to K_+(\mu\mathrm{VF})$, it is easy to construct $h' : \Omega \to K_+(\mu\mathrm{VF})$ such that $h + h'$ is $\mathcal{O}$-invariant. Thus if $f + h = f' + h$ for some h, then $f = f + h + h' = f' + h + h' = f'$. Thus cancellation appears to come by itself.

If we allow all definable sets and volume forms, a great deal of collapsing is caused by the cancellation rule. We thus use the classical remedy and work with bounded sets and volume forms. The setting is flexible and can be compatible with stricter notions of boundedness. This is only a partial remedy in the case of higher-dimensional local fields; cf. Example 12.12.

The theory can be carried out for any of the settings we considered. Let $\mathcal{R}$ be one of these groups or rings, with $\mathcal{D}$ the corresponding data. For instance, $\mathcal{D}$ is the set of pairs (X, ϕ) with X a bounded definable subset of $\mathrm{VF}^n \times \mathrm{RV}^*$, and $\phi : X \to \mathrm{RV}$ is a bounded definable function; $\mathcal{R}$ is the corresponding Grothendieck ring. Similarly, we can take Γ-volumes, or pure isomorphism invariants without volume forms. In this last case there is no point restricting to bounded sets. As we saw, two Euler characteristics into the Grothendieck group of varieties over RES do survive.

In each case, we think of $\mathcal{R}$ as a Grothendieck ring of associated RV-data, modulo a canonical ideal.

Everything can be graded by dimension, but for the moment we have no need to keep track of it, so in the volume case we can take the direct sum over all n or fix one n and omit it from the notation.

The corresponding group for the theory $\mathbf{T}_A$ or $\mathbf{T}_{\langle a \rangle}$ will be denoted $\mathcal{R}_A, \mathcal{R}_a$, etc. When V is a definable set, we let $\mathcal{D}_V$, $\mathcal{R}_V$ denote the corresponding objects over V. For instance, in the case of bounded RV-volumes, $\mathcal{D}_V$ is the set of pairs $(X \subseteq V \times W, \phi : X \to \mathrm{RV}^*)$ such that for any $a \in V$, $(X_a, \phi|X_a)$ with X_a bounded.

If $\mathcal{R}$ is our definable analogue of the real numbers (as recipients of values of p-adic integration), the group ring $\mathcal{C} = \mathcal{R}[\Omega]$ will take the role of the complex numbers. We have a canonical group homomorphism $(\mathrm{VF}, +) \to \Omega \subseteq G_m(\mathcal{C})$, corresponding to a generic additive character.

Integration with an additive character can be presented in two ways: in terms of definable functions $f : X \to \Omega$ (Riemann style), where we wish to evaluate expressions such as $\int_X f(x)\phi(x)$; classically f usually has the form $\psi(h(x))$, where

h is a regular function and ψ is the additive character. Or we can treat definable functions $F : \Omega \to \mathcal{R}$ (Lebesgue style), and evaluation $\int_{\omega \in \Omega} F(\omega)$. We will work with the latter. Given this, to reconstruct a Riemann style integral, given $f : X \to \Omega$, and an $\mathcal{R}$-valued volume form ϕ on X, let

$$F(\omega) = \int_{f^{-1}(\omega)} \phi(x).$$

Then we can define

$$\int_X f(x)\phi(x) = \int_{\omega \in \Omega} \omega F(\omega).$$

It thus suffices to define the integral of a definable function on Ω. Such a function can be interpreted as an $\mathcal{M}$-invariant function on VF. We impose one rule (cancellation): the integral of a function that is constant on each $\mathcal{O}$-class equals zero. The integral is a homomorphism on the group of $\mathcal{M}$-invariant functions $\mathrm{VF} \to \mathcal{R}$, vanishing on the $\mathcal{O}$-invariant ones. We give a full description of the quotient group, showing that the universal homomorphism of this type factors through a similar group on the residue field.

Recall the group $\mathrm{Fn}(V, \mathcal{R})$ of Section 2.2. We will not need to refer to the dimension grading explicitly.

If V is a definable group, V acts on on $\mathrm{Fn}(V, \mathcal{R})$ by translation.

Definition 11.1. For a definable subgroup W of V, let $\mathrm{Fn}(V, \mathcal{R})^W$ be the set of W-invariant elements of $\mathrm{Fn}(V, \mathcal{R})$: they are represented by a definable X, such that if $t \in W$ and $a \in V$ then $X[a]$, $X[a+t]$ represent the same class in $K(\mu\mathrm{VF}_{a,t})[n]$.

Lemma 11.2. *An element of* $\mathrm{Fn}(\mathrm{VF}, \mathcal{R})^{\mathcal{M}}$ *can be represented by an* $\mathcal{M}$*-invariant* $X \subseteq (\mathrm{VF} \times *)$.

Proof. Let $Y \in \mathcal{D}_{\mathrm{VF}}^{RV}$ represent an element of $\mathrm{Fn}(\mathrm{VF}, \mathcal{R})^{\mathcal{M}}$. Thus each fiber $Y_a \in \mathcal{D}^{RV}$. By Lemma 3.52, for $\mathbf{a} \in \mathrm{VF}/\mathcal{M}$ one can find $Y'_{\mathbf{a}} \in \mathcal{D}^{RV}$ such that for some $a \in \mathrm{VF}$ with $a + \mathcal{M} = \mathbf{a}$, $Y_a = Y'_{\mathbf{a}}$. As in Lemma 2.3 there exists $Y' \in \mathcal{R}_{\mathrm{VF}/\mathcal{M}}$ such that $Y'_{\mathbf{a}}$ is the fiber of Y' over $\mathbf{a}$. Pulling back to VF gives the required $\mathcal{M}$-invariant representative. □

Since the equivalence is defined in terms of effective isomorphism, Definition 8.2, it is clear that two elements of $\mathcal{D}_\Omega$ are equivalent iff the corresponding pullbacks to $\mathrm{Fn}(\mathrm{VF}, \mathcal{R})^{\mathcal{M}}$ are equivalent.

The groups $\mathrm{Fn}(\mathrm{VF}, \mathcal{R})^{\mathcal{M}}$ and $\mathrm{Fn}(\mathrm{VF}/\mathcal{M}, \mathcal{R})$ can thus be identified.

Note that the effective isomorphism agrees with pointwise isomorphism for $\mathrm{Fn}(\mathrm{VF}, \mathcal{R})^{\mathcal{M}}$, but not for $\mathrm{Fn}(\mathrm{VF}/\mathcal{M}, \mathcal{R})$.

The group we seek to describe is $\mathcal{A} = \mathcal{A}_{\mathbf{T}} = \mathrm{Fn}(\mathrm{VF}, \mathcal{R})^{\mathcal{M}} / \mathrm{Fn}(\mathrm{VF}, \mathcal{R})^{\mathcal{O}}$. The quotient corresponds to the cancellation rule discussed earlier.

Let $\mathrm{Fn}(\mathbf{k}, \mathcal{R})$ be the Grothendieck group of functions $\mathbf{k} \to \mathcal{R}$, with addition induced from $\mathcal{R}$.

Let $\mathcal{C} = \mathcal{R}[\Omega]$ be the ring of definable functions $\Omega \to \mathcal{R}$ with finite support, convolution product.

Remark. $\mathcal{C}$ embeds into the Galois-invariant elements of the abstract group ring $\mathcal{R}_{\widetilde{T}}[\Omega_{\widetilde{T}}]$, where $\widetilde{T} = \mathbf{T}_{\mathrm{acl}(\emptyset)}$.

The additive group $\mathbf{k} = \mathcal{O}/\mathcal{M}$ is a subgroup of $\Omega = \mathrm{VF}/\mathcal{M}$, and so acts on Ω by translation. It also acts naturally on $\mathrm{Fn}(\mathbf{k}, \mathcal{R})$. This gives two actions on $\mathrm{Fn}(\mathbf{k}, \mathcal{C}) = \mathrm{Fn}(\mathbf{k}, \mathcal{R})[\Omega]$. Let $\mathrm{Fn}(\mathbf{k}, \mathcal{C})_{\mathbf{k}}$ denote the coinvariants with respect to the anti-diagonal action, i.e., the largest quotient on which the two actions coincide.

In general, the upper index denotes invariants, the lower index coinvariants.

$\mathrm{Fn}(\mathrm{VF}, \mathcal{R})$ is the ring of definable functions from VF to $\mathcal{R}$. $\mathrm{Fn}(\mathbf{k}, \mathcal{R})$ is the ring of definable functions from $\mathbf{k}$ to $\mathcal{R}$. $\mathrm{Fn}(\mathbf{k}, \mathcal{C})$ is the ring of definable functions from $\mathbf{k}$ to $\mathcal{C}$; equivalently, it is the set of Galois-invariant elements of the group ring $\mathrm{Fn}(\mathbf{k}, \mathcal{R})[\Omega]$.

The action of $\mathbf{k}$ on $\mathrm{Fn}(\mathbf{k}, \mathcal{C})$ is by translation on $\mathbf{k}$, and negative translation on Ω and hence on $\mathcal{C}$. The term (Const) refers to the image of the constant functions of $\mathrm{Fn}(\mathbf{k}, \mathcal{C})$ in $\mathrm{Fn}(\mathbf{k}, \mathcal{C})_{\mathbf{k}}$. (It is isomorphic to $(\mathcal{C}/\mathbf{k})$.)

Theorem 11.3. *There exists a canonical isomorphism* $\mathrm{Fn}(\mathbf{k}, \mathcal{C})_{\mathbf{k}}/(\mathrm{Const}) \overset{\rightarrow}{\cong} \mathrm{Fn}(\mathrm{VF}, \mathcal{R})^{\mathcal{M}} / \mathrm{Fn}(\mathrm{VF}, \mathcal{R})^{\mathcal{O}}$.

Proof. Let $\mathcal{A}_{\mathrm{fin}}$ be the subring of $\mathrm{Fn}(\mathrm{VF}, \mathcal{R})^{\mathcal{M}}$ consisting of functions represented by elements of $\mathrm{Fn}(\mathrm{VF}, \mathcal{D})^{\mathcal{M}}$ whose support projects to a finite subset of $\mathrm{VF}/\mathcal{O}$.

A definable function on $\mathbf{k}$ can be viewed as an $\mathcal{M}$-invariant function on $\mathcal{O}$; this gives

$$\mathrm{Fn}(\mathbf{k}, \mathcal{R}) \overset{\rightarrow}{\cong} \mathrm{Fn}(\mathcal{O}, \mathcal{R})^{\mathcal{M}}. \tag{11.1}$$

On the other hand, we can define a homomorphism

$$\mathrm{Fn}(\mathcal{O}, \mathcal{R})^{\mathcal{M}}[\Omega] \to \mathcal{A}_{\mathrm{fin}} : \sum_{\omega \in W} a(\omega)\omega \mapsto \sum_{\omega \in W} a(\omega)_{\omega}, \tag{11.2}$$

where W is a finite A-definable subset of Ω, $a : W \to \mathrm{Fn}(\mathcal{O}, \mathcal{R})^{\mathcal{M}}$ is an A-definable function, (so that $\sum_{a \in W} a(\omega)\omega$ is a typical element of the group ring $\mathrm{Fn}(\mathcal{O}, \mathcal{R})^{\mathcal{M}}[\Omega]$), and b_{ω} is the translation of b by ω, i.e., $b_{\omega}(x) = b(x - \omega)$.

(11.2) is surjective: Let $f \in \mathcal{A}_{\mathrm{fin}}$ be represented by F, with support Z, a finite union of translates of $\mathcal{O}$. By Lemma 3.39 there exists a finite definable set W, meeting each ball of Z in a unique point. Define $a : W \to \mathrm{Fn}(\mathcal{O}, \mathcal{R})^{\mathcal{M}}$ by

$$a(\omega) = (f | \omega + \mathcal{O})_{-\omega}.$$

Then (11.2) maps $\sum a(\omega)\omega$ to f.

The kernel of (11.2) is the equalizer of the two actions of $\mathbf{k}$. Composing with (11.1), we obtain an isomorphism $(\mathrm{Fn}(\mathbf{k}, \mathcal{R})[\Omega])_{\mathbf{k}} \overset{\rightarrow}{\cong} \mathcal{A}_{\mathrm{fin}}$ or, equivalently,

$$\mathrm{Fn}(\mathbf{k}, \mathcal{C})_{\mathbf{k}} \overset{\rightarrow}{\cong} \mathcal{A}_{\mathrm{fin}}. \tag{11.3}$$

The last ingredient is the homomorphism

$$\mathcal{A}_{\text{fin}} \to \mathcal{A}. \tag{11.4}$$

We need to show that it is surjective, and to describe the kernel.

Using the representation $\mathcal{D}$ of elements of $\mathcal{R}$ by RV-data, an element of $\mathcal{A}$ is represented by an $\mathcal{M}$-invariant definable $W \subset \text{VF} \times \text{RV}^*$.

By Lemma 3.37, for each coset C of $\mathcal{O}$ in VF apart from a finite number, $W \cap (C \times \text{RV}^{n+l})$ is invariant under translation of the first coordinate by elements of $\mathcal{O}$. Thus W is the disjoint sum of an $\mathcal{O}$-invariant set W' and a set $W'' \subset \text{VF} \times \text{RV}^*$ projecting to a finite union Z of cosets of $\mathcal{O}$ in VF, i.e., representing a function in $\mathcal{A}_{\text{fin}}$.

Clearly, $W' \times_{\text{RV}^n} \text{VF}^n$ lies in $\text{Fn}(\text{VF}, \mathcal{R})^{\mathcal{O}}$.

Thus (11.4) is surjective; the kernel is $\mathcal{A}_{\text{fin}}^{\mathcal{O}}$. Composing (11.3),(11.4) we obtain an isomorphism

$$\mathcal{A} \overset{\rightarrow}{\cong} (\text{Fn}(\mathbf{k}, \mathcal{R})[\Omega])_{\mathbf{k}} / (\text{Const}).$$

Using the identification $\text{Fn}(\mathbf{k}, \mathcal{R})[\Omega] \simeq \text{Fn}(\mathbf{k}, \mathcal{C})$, the theorem follows. □

Note that $\text{Fn}(\mathbf{k}, \mathcal{C})^{\mathbf{k}} \simeq \mathcal{C}$, via $\text{Fn}(\mathbf{k}, \mathcal{C}) \simeq \text{Fn}(\mathbf{k} \times \Omega, \mathcal{R})_{\text{fin}}$.

11.1 Definable distributions

$\mathcal{R}$ is graded by dimension (VF-presentation) or ambient dimension (RV-presentation). Write $\mathcal{R} = \oplus_{n \ge 0} \mathcal{R}[n]$.

Let $\mathcal{R}_{df}$ be the dimension-free version: first form the localization $\mathcal{R}[[0]_1^{-1}]$, where $[0]_1$ is the class of the point $1 \in \text{RV}$, as an element of $\text{RV}[1]$. Equivalently, $[0]_1^n$ is the volume of the open n-dimensional polydisc $\mathcal{O}^n$. Let $\mathcal{R}_{df}$ be the zero-dimensional component of this localization. Similarly, define $\mathcal{C}_{df}$ so that $\mathcal{C}_{df} = \mathcal{R}_{df}[\Omega]$. We can also define $K_+(\mathcal{D})_{df}$, and check that the groupification is $\mathcal{R}_{df}$.

Given $a = (a_1, \ldots, a_n) \in \text{VF}^n$ and $\gamma = (\gamma_1, \ldots, \gamma_n) \in \Gamma^n$, let $B(a, \gamma) = \Pi_{i=1}^n B(a_i, \gamma_i)$, where $B(a_i, \gamma_i) = \{c \in \text{VF} : \text{val}(c - a_i) > \gamma_i\}$. Call $B(a, \gamma)$ an open polydisc of dimension γ. If $\gamma \in \Gamma$, let $B(a, \gamma) = B(a, (\gamma, \ldots, \gamma))$ (the open cube of side γ).

Note that $[B(0, \gamma)]$ is invertible in $\mathcal{R}_{df}$, in each dimension. In particular, in dimension 1, $[B(0, \gamma)][B(0, -\gamma)] = [0]_1^2$. Note also that $[B(a, \gamma)] \underset{a}{=} [B(0, \gamma)]$.

We proceed to define integrals of definable functions.

Let U be a bounded definable subset of VF^n. A definable function $f : U \to K_+(\mathcal{D})_{df}$ has the form $[0]_1^{-m} F$, where $F : U \to K_+ \mathcal{D}[m]$ is a definable function, represented by some $\bar{F} \in \mathcal{D}[m + n]_U$. In case $\bar{F}$ can be taken bounded, define

$$\int_U f = [0]_1^{-m+n} [F]_{n+m}.$$

We say that f is *boundedly represented* in this case.

In particular, $\text{vol}(U) = \int_U 1 = [0]_1^{-m} [U]_m$ is treated as a pure number now, without dimension units. (Check the independence of the choices.)

This extends by linearity to $\int_U f$ for $f : U \to \mathcal{R}_{df}$, provided f can be expressed as the difference of two boundedly represented functions $U \to K_+(\mathcal{D})_{df}$.

We now note that averaging twice, with appropriate weighting, is the same as doing it once. The function γ' in the lemmas below corresponds to a partition of U into cubes; $\gamma'(u)$ is the side of the cube around $u \in U$.

Lemma 11.4. *Let U be a bounded open subset of* VF^n, *f a boundedly represented function on U. Let $\gamma' : U \to \Gamma$ be a definable function such that if $u \in U$ and $u' \in B(u, \gamma'(u))$ then $u' \in U$ and $\gamma'(u') = \gamma'(u)$. Then*

$$\int_U f = \int_U \left[\mathrm{vol}(B(u, \gamma'(u)))^{-1} \int_{B(u,\gamma'(u))} f \right].$$

Proof. Let $f = [0]_1^{-m} F$, where $F : U \to K_+ \mathcal{D}[m]$ is bounded. We have $\mathrm{vol}(B(u, \gamma')) = [0]_1^{-n}[\gamma'(u)]^n$ so

$$\mathrm{vol}(B(u, \gamma'))^{-1} = [0]_1^n[\gamma'(u)]^{-n} = [0]_1^{-n}[-\gamma'(u)]^n.$$

Thus, multiplying by $[0]_1^{3n+m}$, we have to show

$$[0]_1^{2n}[F] = [-\gamma'(u)]^n[\{(u, u', z) : u \in U, u' \in B(u, \gamma'(u)), (u', z) \in F\}].$$

Now $u' \in B(u, \gamma'(u))$ iff $u \in B(u', \gamma'(u'))$. Applying the measure-preserving bijection $(u, u', z) \mapsto (u - u', u', z')$ we see that the $[\{(u, u', z) : u \in U, u' \in B(u, \gamma'(u)), (u', z) \in F\}] = [\gamma]_1^n[\{(u', z) : (u', z) \in F]$, so the equality is clear. □

We now define the integral of definable functions into $\mathcal{C}_{df}$. By definition, such a function is a finite sum of products fg, with $f \in \mathrm{Fn}(U, \mathcal{R}_{df})$ and $g \in \mathrm{Fn}(U, \Omega)$. Define

$$\int_U fg = \int_{\omega \in \Omega} \omega \int_{g^{-1}(\omega)} f$$

and extend by linearity.

Note that this is defined as soon as g is boundedly represented. (Again, check the independence of the choices.)

Definition 11.5. A definable distribution on an open $U \subseteq \mathrm{VF}^n$ is a definable function $\mathfrak{d} : U \times \Gamma \to \mathcal{C}_{df}$, such that $\mathfrak{d}(a, \gamma) = \mathfrak{d}(a', \gamma)$ if $B(a, \gamma) = B(a', \gamma)$, and whenever $\gamma' > \gamma$ in each coordinate,

$$\mathfrak{d}(b, \gamma) = \int_{u \in B(b,\gamma)} \mathrm{vol}(B(0, \gamma'))^{-1} \mathfrak{d}(u, \gamma').$$

As in Lemma 11.2, the invariance condition means that $\mathfrak{d}$ can be viewed as a function on open polydiscs, and we will view it this way below.

If $\mathfrak{d}$ takes values in $\mathcal{R}_{df}$, we say it is $\mathcal{R}_{df}$-valued. By definition, $\mathfrak{d}$ can be written as a finite sum $\sum \omega_i \mathfrak{d}_i$, where $\mathfrak{d}_i$ is an $\mathcal{R}_{df}$-valued function; in fact, $\mathfrak{d}_i$ is an $\mathcal{R}_{df}$-valued distribution.

We wish to strengthen the definition of a distribution so as to apply to subpolydiscs of variable size. For this we need a preliminary lemma.

Lemma 11.6. *Let $U = B(a, \gamma)$ be a polydisc. Let $\gamma' : B(a, \gamma) \to \Gamma$ be a definable function such that $\gamma'(u') = \gamma'(u)$ for $u' \in B(u, \gamma'(u))$. Then γ' is bounded on U.*

Proof. Suppose for contradiction that γ' is not bounded on $B(a, \gamma)$; i.e.,

$$(\forall \delta \in \Gamma)(\exists u \in B(a, \gamma))(\gamma'(u) > \delta).$$

This will not change if we add a generic element of Γ to the base, so we may assume $\Gamma(\mathrm{dcl}(\emptyset)) \neq (0)$. By Lemma 3.51, there exists a resolved structure with the same RV-part as $\langle \emptyset \rangle$; hence we may assume **T** is resolved. By Section 6 any VF-generated structure is resolved. By Lemma 3.49, for any $M \models \mathbf{T}$ and $c \in \mathrm{VF}(M)$, $\mathrm{acl}(c)$ is an elementary submodel of M. Consider c with $\mathrm{val}(c) \models p_0$, where p_0 is the generic type at ∞ of elements of Γ, i.e., $p_0|A = \{x > \delta : \delta \in \Gamma(A)\}$. Since

$$\mathrm{acl}(c) \models (\forall \delta \in \Gamma)(\exists u \in B(a, \gamma))(\gamma'(u) > \delta)$$

there exists $e \in \mathrm{acl}(c)$ with $e \in B(a, \gamma)$ and $\gamma'(e) > \mathrm{val}(c)$. By Lemma 5.12, there exists $e_0 \in \mathrm{acl}(\emptyset)$ such that $(c, e) \to (0, e_0)$. In particular, $e_0 \in B(a, \gamma)$. But then since $e \to e_0$ and $\gamma'(e_0) \in \Gamma(\mathrm{acl}(\emptyset))$, we have $e \in B(e_0, \gamma'(e_0))$. Thus $\gamma'(e) = \gamma'(e_0)$. But then $\gamma'(e_0) > \mathrm{val}(c)$, contradicting the choice of c. □

Lemma 11.7.

(1) *Let $\mathfrak{d} : U \times \Gamma \to \mathfrak{C}_{df}$ be a definable distribution. Let $\gamma' : U \to \Gamma$ be a definable function with $\gamma'(u) > \gamma$, such that $\gamma'(u') = \gamma'(u)$ for $u' \in B(u, \gamma'(u))$. Then*

$$\mathfrak{d}(b, \gamma) = \int_{u \in B(b,\gamma)} \mathrm{vol}(B(0, \gamma'(u))^{-1} \mathfrak{d}(u, \gamma'(u)). \tag{11.5}$$

(2) *Let $\mathfrak{d}_1$, dd_2 be definable distributions on U such that for any $x \in U$, for all large enough $\gamma \in \Gamma$, for any $y \in B(x, \gamma)$ and any $\gamma' > \gamma$, $\mathfrak{d}_1(B(y, \gamma')) = \mathfrak{d}_2(B(y, \gamma'))$. Then $\mathfrak{d}_1 = \mathfrak{d}_2$.*

Proof.

(1) To prove (11.5), fix b, γ. We may assume $U = B(b, \gamma)$. Using Lemma 11.6, pick a constant γ'' with $\gamma'' > \gamma'(u)$ for all $u \in B(b, \gamma)$. Use the definition of a distribution with respect to γ'' to compute both $\mathfrak{d}(B(b, \gamma))$ and for each u $\mathfrak{d}(u, \gamma'(u))$, and compare the integrals using Lemma 11.4.

(1) Define $\gamma'(u)$ to be the smallest γ' such that for all $\gamma'' > \gamma'$ and all $y \in B(u, \gamma)$, $\mathfrak{d}_1(B(y, \gamma'')) = \mathfrak{d}_2(B(y, \gamma''))$. It is clear that $\gamma'(u') = \gamma'(u)$ for $u' \in B(u, \gamma'(u))$. (11.5) gives the same integral formula for $\mathfrak{d}_1(b, \gamma)$ and $\mathfrak{d}_2(b, \gamma)$. □

Let $\mathfrak{d}$ be a definable distribution, and U an arbitrary bounded open set. We can define $\mathfrak{d}(U)$ as follows. For any $x \in U$, let $\rho(x, U)$ be the smallest $\rho \in \Gamma$ such that $B(x, \rho) \subseteq U$. Let $B(x, U) = B(x, \rho(x, U))$; this is the largest open cube around x contained in U. Note that two such cubes $B(x, U)$, $B(x', U)$ are disjoint or equal. Define

$$\mathfrak{d}(U) = \int_{x \in U} \mathrm{vol}(B(x, U))^{-1} \mathfrak{d}(x, \rho(x, U)).$$

More generally, if h is a locally constant function on VF^n into $\mathcal{R}_{df}$ with bounded support, we can define

$$\mathfrak{d}(h) = \int_{x \in \mathrm{VF}^n} h(x)[B(x, h)]^{-1} \mathfrak{d}(x, \rho(x, U)), \tag{11.6}$$

where now $B(x, h) = B(x, \rho(x, U))$ is the largest open cube around x on which h is constant.

Proposition 11.8. *Let $\mathfrak{d}$ be a definable distribution. Then there exists a definable open set $G \subseteq \mathrm{VF}^n$ whose complement Z has dimension $< n$, and a definable function $g : G \to \mathcal{C}_{df}$ such that for any polydisc $U \subseteq G$*

$$\mathfrak{d}(U) = \int_U g.$$

Proof. Since $\mathfrak{d}$ is a finite sum of $\mathcal{R}_{df}$-valued distributions, we may assume it is $\mathcal{R}_{df}$-valued. Given $a \in \mathrm{VF}^n$, we have a function $\alpha_a : \Gamma \to \mathcal{R}_{df}$ defined by $\alpha_a(\rho) = \mathfrak{d}(B(a, \rho))$. Using the RV-description of $\mathcal{R}$, and the stable embeddedness of $\mathrm{RV} \cup \Gamma$, we see that α_a has a canonical code $c(a) \in (\mathrm{RV} \cup \Gamma)^*$.

Let G be the union of all polydiscs W such that c is constant on W. Let $Z = \mathrm{VF}^n \setminus G$. By Lemma 5.13, $\dim(Z) < n$.

Claim. Let W be a polydisc such that c is constant on W. Then for some $r \in \mathcal{R}_{df}$, for any polydisc $U = B(a, \rho) \subseteq W$, $\mathfrak{d}(a, \rho) = r\mathrm{vol}(U)$.

Proof. Since c is constant on W, for some function δ, all ρ and all $b \in W$ with $B(w, \rho) \subseteq W$, we have $\mathfrak{d}(B(w, \rho)) = \delta(\rho)$. By the definition of a distribution we have, for any $a \in W$,

$$\delta(\rho)\mathrm{vol}(B(a, \rho')) \underset{a}{=} \mathrm{vol} B(a, \rho)\delta(\rho').$$

Now $\mathrm{vol}(B(a, \rho)) \underset{a}{=} \mathrm{vol} B(0, \rho)$. Thus $\delta(\rho)\mathrm{vol}(B(0, \rho')) \underset{a}{=} \mathrm{vol} B(0, \rho)\delta(\rho')$. Since this holds for any $a \in W$, by Proposition 3.51 we have

$$\delta(\rho)\mathrm{vol}(B(0, \rho'))) = \mathrm{vol} B(0, \rho)\delta(\rho').$$

Thus $\delta(\rho)/\mathrm{vol} B(0, \rho) = r$ is constant. The claim follows. □

The proposition also follows using Lemma 11.7. □

11.2 Fourier transform

Let ψ be the tautological projection $K \to K/\mathcal{M} = \Omega$.

Let $g : \mathrm{VF}^n \to \mathcal{C}_{df}$ be a definable function, bounded on bounded subsets of VF^n. Define a function $\mathcal{F}(g)$ by

$$\mathcal{F}(g)(U) = \int_{y \in \mathrm{VF}} g(y)(\int_{x \in U} \psi(x \cdot y)).$$

This makes sense since for a given U, $(\int_{x \in U} \psi(x \cdot y))$ vanishes for y outside a certain polydisc (with sides inverse to U). Moreover, we have the following.

Lemma 11.9. *$\mathcal{F}(g)$ is a definable distribution.*

Proof. This follows from Fubini, Lemma 11.4, and chasing the definitions. □

Corollary 11.10. *Fix integers n, d. For all local fields L of sufficiently large residue characteristic, for any polynomial $G \in L[X_1, \dots, X_n]$ of degree $\leq d$, there exists a proper variety V_G of L^n such that $\mathcal{F}(|G|)$ agrees with a locally constant function outside V_G.*

Proof. The proof follows from Lemmas 11.9 and 11.8. □

See [4] for the real case.

12 Expansions and rational points over Henselian fields

We have worked everywhere with the geometry of algebraically closed valued fields, or more generally of **T**, but at a geometric level; all objects and morphisms can be lifted to the algebraic closure, and the quantifiers are interpreted there.

For many purposes, we believe this is the right framework. It includes, for instance, Igusa integrals $\int_{x \in X(F)} |f(x)|^s$, and we will show in a future work how to interpret in it some constructions of representation theory. See also [21].

In other situations, however, one wishes to integrate definable sets over Henselian fields rather than only constructible sets; and to have a change of variable formula for definable maps, as obtained by Denef–Loeser and Cluckers-Loeser (cf. [7]). It turns out that our formalism lends itself immediately to this generalization; we explain in this section how to recover it. The point is that an arbitrary definable set is an RV-union of constructible ones, and the integration theory commutes with RV-unions.

We will consider F that admits quantifier elimination in a language $\mathbb{L}^+$ obtained from the language of **T** *by adding relations to* RV *only*. For example, if $F = \mathrm{Th}(\mathbb{C}((X)))$, F has quantifier elimination in a language expanded with names D_n for subgroups of Γ (with $D_n(F) = n\Gamma(F)$).

There are two steps in moving from F^{alg} to F. We will try to clarify the situation by taking them one at a time. The two steps are to restrict the points to a smaller set (the F-rational points), and they enlarge the language to a larger one (with enough relation symbols for F-quantifier elimination). We will take these steps in the reverse order. In Section 12.1 we show how to extend the results of this paper to expansions of the language in the RV sorts, and in Section 12.3 how to pass to sets of rational points over a Hensel field.

The reader who wishes to restrict attention to constructible integrals (still taking rational points) may skip Section 12.1, taking $\mathbf{T}^+ = \mathbf{T}$ in Section 12.3. In this case one still has a change of variable formula for a constructible change of variable, but not for a definable change of variable. An advantage is that the target ring correspondingly involves the Grothendieck group of constructible sets and maps rather than definable ones, which sometimes has more faithful information; cf. Example 12.12.

12.1 Expansions of the RV sort

Let $\mathbf{T}$ be V-minimal.

Let $\mathbf{T}^+$ be an expansion of $\mathbf{T}$ obtained by adding relations to RV. We assume that every $M \models \mathbf{T}$ embeds into the restriction to the language of $\mathbf{T}$ of some $N \models \mathbf{T}^+$. (As $\mathbf{T}$ is complete, this is actually automatic.) By adding some more basic relations, without changing the class of definable relations, we may assume $\mathbf{T}^+$ eliminates RV-quantifiers. As $\mathbf{T}$ eliminates field quantifiers, and $\mathbf{T}^+$ has no new atomic formulas with VF variables, $\mathbf{T}^+$ eliminates VF-quantifiers, too, and hence all quantifiers.

For instance, $\mathbf{T}+$ may include a name for a subfield of the residue field (say, pseudofinite) or the angular coefficients the Denef–Pas language (where RV is split). Write $+$-definable for $\mathbf{T}^+$-definable; similarly, tp_+ will denote the type in $\mathbf{T}^+$, etc. The unqualified words formula, type, and definable closure will refer to quantifier-free formulas of $\mathbf{T}$.

Lemma 12.1. *Let $M \models \mathbf{T}^+$. Let A be a substructure of M, $c \in M$, $B = A(c) \cap \mathrm{RV}$.*

(1) $\mathrm{tp}(c/A \cup B) \cup {\mathbf{T}^+}_{A \cup B}$ *implies* $\mathrm{tp}_+(c/A \cup B)$.
(2) *Assume c is ${\mathbf{T}^+}_A$-definable. Then $c \in \mathrm{dcl}(A, b)$ for some $b \in A(c) \cap \mathrm{RV}$.*

Proof.

(1) This follows immediately from the quantifier elimination for $\mathbf{T}^+$. Indeed, let $\phi(x) \in \mathrm{tp}_+(c/A \cup B)$. Then ϕ is a Boolean combination of atomic formulas, and it is sufficient to consider the case of ϕ atomic, or the negation of an atomic formula. Now since any basic function $\mathrm{VF}^n \to \mathrm{VF}$ is already in the language of $\mathbf{T}$, every basic function of the language of $\mathbf{T}^+$ denoting a function $VF^n \to \mathrm{RV}$ factors through a $\mathbf{T}$-definable function into RV. Hence the same is true for all terms (compositions of basic functions). And any basic relation is either the equality relation on VF, or else a relation between variables of RV. If ϕ is an equality or inequality between $f(x), g(x)$, it is already in $\mathrm{tp}(c/A)$. Now suppose ϕ is a relation $R(f_1(x), \ldots, f_n(x))$ between elements of RV. Since $B(c) \cap \mathrm{RV} \subseteq B$, the formula $f_i(x) = b_i$ lies in $\mathrm{tp}(c/A \cup B)$ for some $b_i \in B$. On the other hand, $R(b_1, \ldots, b_n)$ is part of ${\mathbf{T}^+}_B$. These formulas together imply $R(f_1(x), \ldots, f_n(x))$.

(2) We must show that $c \in \mathrm{dcl}(A \cup B)$. Let $p = \mathrm{tp}(c/A \cup B)$. By (1), p generates a complete type of $\mathbf{T}^{+}{}_{A\cup B}$. Since this is the type of c and c is $\mathbf{T}^{+}{}_{A}$-definable, and since any model of $\mathbf{T}$ embeds into a model of $\mathbf{T}^{+}$, p has a unique solution solution in any model of $\mathbf{T}$. Thus $c \in \mathrm{dcl}(A \cup B)$. □

We will now see that any $\mathbf{T}^{+}$-definable bijection decomposes into $\mathbf{T}$-bijections, and bijections of the form $x \mapsto (x, j(g(x)))$ where g is a $\mathbf{T}$-definable map into RV^m and j is a $\mathbf{T}^{+}$-definable map on RV.

Corollary 12.2.

(1) *Let P be a $\mathbf{T}^{+}$-definable set. There exist $\mathbf{T}$-definable $f : \widetilde{P} \to \mathrm{RV}^*$ and a $\mathbf{T}^{+}$-definable $Q \subseteq \mathrm{RV}^*$ such that $P = f^{-1}Q$.*

(2) *Let P_1, P_2 be $\mathbf{T}^{+}$-definable sets, and let $F : P_1 \to P_2$ be a $\mathbf{T}^{+}$-definable bijection. Then there exist $g_i : \widetilde{P_i} \to R_i \subseteq \mathrm{RV}^m$, $R \subseteq \mathrm{RV}^m$, $h_i : R \to R_i$, and a bijection $H : \widetilde{P_1} \times_{g_1,h_1} R \to \widetilde{P_2} \times_{g_2,h_2} R$ over R, all $\mathbf{T}$-definable, and $\mathbf{T}^{+}$-definable $Q_i \subseteq R_i$, $Q \subseteq R$, and $j_i : Q_i \to Q$ such that $P_i = {g_i}^{-1}Q_i$, $h_i j_i = \mathrm{Id}_{Q_i}$, and for $x \in P_1$,*

$$j_1 g_1(x) = j_2 g_2(F(x)) =: j(x) \quad \text{and} \quad H(x, j(x)) = (F(x), j(x)). \qquad (\diamond)$$

Moreover, if $P_i \subseteq \mathrm{VF}^n \times \mathrm{RV}^m$ projects finite-to-one to VF^n, then $R \to R_i$ is finite-to-one.

Proof.

(1) Let $\mathcal{F}$ be the family of all $\mathbf{T}$-definable functions $f : W \to \mathrm{RV}^m$, where W is a definable set.

Claim. If $\mathrm{tp}(c) = \mathrm{tp}(d)$ and $f(c) = f(d)$ for all $f \in \mathcal{F}$ with $c, d \in \mathrm{dom}(f)$, then $c \in P \iff d \in P$.

Proof. We have $\mathrm{tp}(c, f(c)) = \mathrm{tp}(d, f(d)) = \mathrm{tp}(d, f(c))$, so $\mathrm{tp}(c/f(c)) = \mathrm{tp}(d/f(c))$ for all $f \in \mathcal{F}$ with $c \in \mathrm{dom}(f)$, and thus $\mathrm{tp}(c/B) = \mathrm{tp}(d/B)$, where $B = A(c) \cap \mathrm{RV}$. It follows that $\mathrm{tp}_{+}(c) = \mathrm{tp}_{+}(d)$ and, in particular, $c \in P \iff d \in P$. □

By compactness, there are $(f_i, W_i)_{i=1}^m \in \mathcal{F}$ such that if $c \in W_i \iff d \in W_i$ and $f_i(c) = f_i(d)$ whenever $c, d \in W_i$, then $c \in P \iff d \in P$. Let $\widetilde{P} = \cup_i W_i$, and extend f_i to $\widetilde{P}$ by $f_i(x) = \infty$ if $x \notin W_i$. Let $f(x) = (f_1(x), \ldots, f_m(x))$. Letting $\widetilde{P} = \cup_i W_i$ and $Q = f(P)$, (1) follows.

(2) Consider first a $\mathbf{T}^{+}$-type $p = \mathrm{tp}_{+}(c_1)$, $c_1 \in P_1$. Let $c_2 = F(c_1)$. Using Lemma 3.48, there exists $g_i^p \in \mathcal{F}$ such that $e_i = g_i^p(c_i)$ generates $\mathrm{dcl}(c_i) \cap \mathrm{RV}$. It follows as in Lemma 12.1(1) that e_i generates $\mathrm{dcl}_{+}(c_i) \cap \mathrm{RV}$. Let e generate $\mathrm{dcl}(c_1, c_2) \cap \mathrm{RV}$; we have $e_i = h_i^p(e)$ for appropriate $\mathbf{T}$-definable h_i^p. Note $\mathrm{dcl}_{+}(c_1) = \mathrm{dcl}_{+}(c_2)$, and so $e \in \mathrm{dcl}_{+}(c_i)$. Now quantifier elimination for $\mathbf{T}^{+}$ implies the stable embeddedness of RV, in the same way as for ACVF (cf. Section 2.1). By Lemma 2.9 $\mathrm{tp}_{+}(c_i/e_i)$ implies $\mathrm{tp}_{+}(c_i/\mathrm{RV})$; in particular, since $e \in \mathrm{dcl}_{+}(c_i)$ $e = j_i^p(e_i)$ for some $\mathbf{T}^{+}$-definable j_i^p. By Lemma 12.1(2) over $\mathrm{dcl}(c_1)$, $c_2 \in \mathrm{dcl}(c_1, e)$; similarly, $c_1 \in \mathrm{dcl}(c_2, e)$. Thus there exists a $\mathbf{T}$-definable invertible

H^p with $H^p(c_1, e) = (c_2, e)$. Equations ($\diamond$) have been shown to hold on p. Now g_i extends to a **T**-definable function $g_i : \widetilde{P_i} \to R_i$. By compactness ($\diamond$) holds on some definable neighborhood of p; and by (1) this neighborhood can be taken to have the form ${g_1}^{-1}Q_1$ for some Q_1. Finitely many such neighborhoods cover P_1, and the data can be sewed together as in (1). We thus find $\widetilde{P_1}$, R, R_1, R_2, g_1, g_2, h_1, h_2, H, Q_1, j_1, j_2 such that $h_i j_i(x) = x$ and ($\diamond$) holds on ${g_1}^{-1}Q_1 = P_1$. Let $Q_2 = h_2 j_1 Q_1$; it follows that $P_2 = F(P_1) = {g_2}^{-1}Q_2$.

To prove the last point, since $c_2 \in \mathrm{dcl}(c_1, e)$ we have (Lemma 3.41) $c_2 \in \mathrm{acl}(c_1)$. But $e \in \mathrm{dcl}(c_1, c_2)$ so $e \in \mathrm{acl}(\mathrm{dcl}(c_1))$; and as $e \in \mathrm{RV}^m$ for some m, $e \in \mathrm{acl}(\mathrm{dcl}(e_1))$.

Let VF^+ be the category of $+$-definable subsets of varieties over $\mathrm{VF} \cap \mathrm{dcl}(\emptyset)$, and $+$-definable maps. Define effective isomorphism as in Definition 8.2; let ${K_+}^{\mathrm{eff}}$ denote the Grothendieck group of effective isomorphism classes, and let $[X]$ be the class of X.

Let $\mathrm{RV}^+[*]$ be the category of pairs (Y, f), where Y is a $+$-definable subset of X for some $(X, f) \in \mathrm{Ob}\,\mathrm{RV}[*]$ (Definition 3.66). A morphism $(Y, f) \to (Y', f')$ is a definable bijection $h : Y \to Y'$ such that $f'(h(y)) \in \mathrm{acl}(f(y))$ for $y \in Y$.

Let $K_+(\mathrm{RV}^+[*])$ be the Grothendieck semigroup of isomrphism classes of RV^+; let $\mathrm{I_{sp}}$ be the congruence generated by $(J, 1_1)$, where $J = \{1\}_0 + [\mathrm{RV}^{>0}]_1$.

Proposition 12.3. *There exists a canonical surjective homomorphism of Grothendieck semigroups*

$$\oint : K_+(\mathrm{VF}^+[*]) \to K_+(\mathrm{RV}^+[*])/\mathrm{I_{sp}}$$

determined by

$$\oint [X] = [W]/\mathrm{I_{sp}} \iff [X] = [\mathbb{L}W].$$

Proof. We have to show the following:

(i) Any element of $K_+(\mathrm{VF}^+)$ is effectively isomorphic to one of the form $[\mathbb{L}W]$.
(ii) If $[\mathbb{L}W_1] = [\mathbb{L}W_2]$ then $([W_1], [W_2]) \in \mathrm{I_{sp}}$.

(i) By Corollary 12.2(1), a typical element of $K_+(\mathrm{VF}^+)$ is represented by $P = f^{-1}Q$, where $Q \subseteq \mathrm{RV}^*$is $\mathbf{T}^+$-definable, $f : \widetilde{P} \to \mathrm{RV}^*$ is **T**-definable. For any $a \in \mathrm{RV}^*$, $f^{-1}(a)$ is $\mathbf{T}_a$-definable, and $[f^{-1}(a)] = [\mathbb{L}C_a]$ where $[C_a] = [\int f^{-1}(a)]$. Since $\mathbb{L}$ commutes with RV-disjoint unions, it follows that $[P] = [\mathbb{L}W]$ where $W = \dot{\bigcup}_{a \in Q} C_a$.

(ii) Assume $[\mathbb{L}W_1] = [\mathbb{L}W_2]$. By Proposition 3.51, the base can be enlarged so as to be made effective, without change to RV; thus to show that $([W_1], [W_2]) \in \mathrm{I_{sp}}$ we may assume $\mathbb{L}W_1, \mathbb{L}W_2$ are isomorphic. Let $f : \mathbb{L}W_1 \to \mathbb{L}W_2$ be an isomorphism. Let $P_i = \mathbb{L}W_i$ and let $\widetilde{P_i}$, R_i, g_i, h_i, R, H, Q, Q_i, j_i be as in Corollary 12.2(2).

Since $P_i = {g_i}^{-1}Q_i = \mathbb{L}W_i$, the maximal $\underset{\mathrm{rv}}{\sim}$-invariant subset of $\widetilde{P_i}$ contains P_i, so we may assume $\widetilde{P_i}$ is $\underset{\mathrm{rv}}{\sim}$-invariant; in other words, $\widetilde{P_i} = \mathbb{L}\widetilde{W_i}$ for some **T**-definable $\widetilde{W_i} \in \mathrm{RV}[*, \cdot]$ containing W_i.

By Lemma 7.8, there exists a special bijection $\sigma : \mathbb{L}\widetilde{W}_i^* \to \mathbb{L}\widetilde{W}_i$ such that $g_i \circ \sigma$ factors through ρ, i.e., for some $e_i : \widetilde{W}_i^* \to R_i$ we have $g_i \circ \sigma = e_i \circ \rho$ on $\mathbb{L}\widetilde{W}_i$. Let $W_i{}^*$ be the pullback of W_i to $\widetilde{W}_i^*$, so that $\sigma(\mathbb{L}W_i^*) = \mathbb{L}W_i = P_i$. Then $([W_i], [W_i{}^*]) \in \mathrm{I_{sp}}$, so it suffices to show that $(W_1^*, W_2^*) \in \mathrm{I_{sp}}$. Since $P_i = g_i{}^{-1}Q_i$, we have $W_i^* = e_i{}^{-1}Q_i$.

For $c \in R$, let $\widetilde{P}_i(c) = \sigma^{-1}g_i{}^{-1}(h_i(c))$, $\widetilde{W}_i(c) = e_i{}^{-1}(h_i(c))$. Then $\widetilde{P}_i(c) = \mathbb{L}\widetilde{W}_i(c)$. Now H induces a bijection $\widetilde{P}_1(c) \to \widetilde{P}_2(c)$. Thus by Proposition 7.25, $(\widetilde{W}_1(c), \widetilde{W}_2(c)) \in \mathrm{I_{sp}}$. In particular, this is true for $c \in Q$; now $h_i : Q \to Q_i$ is a bijection, and $W_i^* = \dot{\bigcup}_{c\in Q} \widetilde{W}_i(c)$. Thus $([W_1^*], [W_2^*]) \in \mathrm{I_{sp}}$. □

Remark. Since the structure on RV in $\mathbf{T}^+$ is arbitrary, we cannot expect the homomorphism of Corollary 12.3 to be injective. We could make it so tautologically by modifying the category RV^+, taking only *liftable* morphisms, i.e., those that lift to VF; we then obtain an isomorphism. In specific cases it may be possible to check that all morphisms are liftable.

12.2 Transitivity

Motivation. Consider a tower of valued fields, such as $\mathbb{C} \leq \mathcal{C}((s)) \leq \mathcal{C}((s))((t))$. Given a definable set over $\mathcal{C}((s))((t))$, we can integrate with respect to the t-valuation, obtaining data over $\mathcal{C}((s))$ and the value group. The $\mathcal{C}((s))$ can then be integrated with respect to the s-valuation. On the other hand, we can consider directly the $\mathbb{Z}^2$-valued valuation of $\mathcal{C}((s))((t))$, and integrate so as to obtain an answer involving the Grothendieck group of varieties over $\mathbb{C}$. Below we develop the language for comparing these answers, and show that they coincide.

For simplicity we accept here a Denef–Pas splitting, i.e., we expand RV so as to split the sequence $\mathbf{k}^* \to \mathrm{RV}^* \to \Gamma$. Then rv splits into two maps, $\mathrm{ac} : \mathrm{VF}^* \to \mathbf{k}^*$ and $\mathrm{val} : \mathrm{VF}^* \to \Gamma$. This expansion of ACVF(0, 0) is denoted $\mathrm{ACVF^{DP}}$. Note that this falls under the framework of Section 12.1, as will the further expansions below.

Consider two expansions of $\mathrm{ACVF^{DP}}$: (1) Expand the residue field to have the structure of a valued field (itself a model of $\mathrm{ACVF^{DP}}$). (2) Expand the value group to be a lexicographically ordered product of two ordered Abelian groups. Then (1)–(2) yield bi-interpretable theories. In more detail, we have the following:

First expansion. Rename the VF sort as VF_{21}, the residue field as VF_1, and the value group Γ_1. VF_1 carries a field structure; expand it to a model of $\mathrm{ACVF^{DP}}$, with residue field F_0 and value group Γ_0. Let ac_{21}, val_{21} have their natural meanings.

Second expansion. Rename the VF-sort as VF_{20}, the residue field as F_0 and the value group as Γ_{20}. Add a predicate Γ_0 for a proper convex subgroup of Γ_{20}, and a predicate Γ_1 for a complementary subgroup, so that Γ_{20} is identified with the lexicographically ordered $\Gamma_0 \times \Gamma_1$.

Lemma 12.4. *The two theories described above are bi-interpretable. A model of* (1) *can canonically be made into a model of* (2) *with the same class of definable relations, and vice versa.*

Proof. Given (1), let $\mathrm{VF}_{20} = \mathrm{VF}_{21}$ as fields. Define

$$\mathrm{ac}_{20} = \mathrm{ac}_{10} \circ \mathrm{ac}_{21}\,. \tag{12.1}$$

Let $\Gamma_{20} = \Gamma_1 \times \Gamma_0$, and define $\mathrm{val}_{20} : \mathrm{VF}_{21}^* \to \Gamma_{20}$ by

$$\mathrm{val}_{20}(x) = (\mathrm{val}_{21}(x), \mathrm{val}_{10}(\mathrm{ac}_{21}(x))). \tag{12.2}$$

Conversely, given (2), let $\mathrm{VF}_{21} = \mathrm{VF}_{20}$ as fields;

$$\begin{aligned} \mathcal{O}_{21} &= \{x \in \mathrm{VF}_{21} : (\exists t \in \Gamma_0)(\mathrm{val}_{20}(x) \geq t)\}, \\ \mathcal{M}_{21} &= \{x \in \mathrm{VF}_{21} : (\forall t \in \Gamma_0)(\mathrm{val}_{20}(x) > t)\}, \\ \mathrm{VF}_1 &= \mathcal{O}_{21}/\mathcal{M}_{21}. \end{aligned}$$

Let VF_{21} have the valued field structure with residue field VF_1; note that the value group $\mathrm{VF}_{21}^*/\mathcal{O}_{21}^*$ can be identified with Γ_1. Note that $\ker \mathrm{ac}_{20} \supset 1 + \mathcal{M}_{21}$, so that factors through VF_1^*, and define ac_{10}, ac_{21} so as to make (12.1) hold. Then define val_{21}, val_{10} so that (12.2) holds. □

Let $\mathrm{VF}^+[*]$ denote the category of definable subsets of VF_{21}, equivalently, VF_{20}, in the expansions (1) or (2). According to Proposition 12.3 and Lemma 2.11, we have canonical maps $K_+(\mathrm{VF}^+[*]) \to K_+(\mathrm{RV}_1^+[*])/\mathrm{I_{sp}}$ and $K_+(\mathrm{VF}^+[*]) \to K_+(\mathrm{RV}_2^+[*])/\mathrm{I_{sp}}$, where $\mathrm{RV}_i^+[*]$ denotes the expansion of RV according to (1)–(2), respectively.

By Proposition 8.4 we have canonical maps

$$\begin{aligned} K_+(\mathrm{VF}^+[*]) &\to K_+(\mathrm{VF}_1[*]) \otimes K_+(\Gamma_{21}[*])/\mathrm{I_{sp}} \\ &\to (K_+(F_0) \otimes K_+(\Gamma_{10})) \otimes K_+(\Gamma_{21})/\mathrm{I_{sp\,1}} \end{aligned} \tag{12.3}$$

for a certain congruence $\mathrm{I_{sp\,1}}$. And, on the other hand,

$$\begin{aligned} K_+(\mathrm{VF}^+[*]) &\to K_+(F_0[*]) \otimes K_+(\Gamma_{20}[*])/\mathrm{I_{sp}} \\ &= K_+(F_0[*]) \otimes (K_+(\Gamma_{10}[*]) \otimes K_+(\Gamma_{21}[*]))/\mathrm{I_{sp\,2}}. \end{aligned} \tag{12.4}$$

For an appropriate $\mathrm{I_{sp\,2}}$. The tensor products here are over $\mathbb{Z}$, in each dimension separately.

Using transitivity of the tensor product we identify $(K_+(F_0) \otimes K_+(\Gamma_{10})) \otimes K_+(\Gamma_{21})$ with $K_+(F_0[*]) \otimes (K_+(\Gamma_{10}[*]) \otimes K_+(\Gamma_{21}[*]))$. Then

Theorem 12.5. $\mathrm{I_{sp\,1}}$, $\mathrm{I_{sp\,2}}$ *are equal and the maps of* (12.3),(12.4) *coincide.*

Proof. It suffices to show in the opposite direction that the compositions of maps induced by $\mathbb{L}$

$$\begin{aligned} (K_+(F_0[*]) \otimes K_+(\Gamma_{10}[*]) \otimes K_+(\Gamma_{21}[*]) &\to K_+(\mathrm{VF}_1[*]) \otimes K_+(\Gamma_{21}[*]) \\ &\to K_+(\mathrm{VF}^+[*]), \end{aligned} \tag{12.5}$$

$$\begin{aligned} K_+(F_0[*]) \otimes K_+(\Gamma_{10}[*]) \otimes K_+(\Gamma_{21}[*]) &\to K_+(F_0[*]) \otimes K_+(\Gamma_{20}[*]) \\ &\to K_+(\mathrm{VF}^+[*]) \end{aligned} \tag{12.6}$$

coincide. But this reduces by RV-additivity to the case of points, and by multplicativity to the individual factors F_0, Γ_{21}, Γ_{10}, yielding to an obvious computation in each case. □

12.3 Rational points over a Henselian subfield: Constructible sets and morphisms

Let $\mathbf{T}$ be V-minimal, and $\mathbf{T}^+$ an expansion of $\mathbf{T}$ in the RV sorts.

Let F be an effective substructure of a model of $\mathbf{T}$. Thus $F = (F_{\mathrm{VF}}, F_{\mathrm{RV}})$, with F_{VF} a field, and $\mathrm{rv}(F_{\mathrm{VF}}) = F_{\mathrm{RV}}$; and F is closed under definable functions of $\mathbf{T}$. For example, if $\mathbf{T} = \mathbf{T}^+ = \mathrm{ACVF}(0,0)$, this is the case iff F_{VF} is a Henselian field and $F_{\mathrm{RV}} = F/\mathcal{M}(F)$; any Hensel field of residue characteristic 0 can be viewed in this way. See Example 12.8.

By a $+$-*constructible* subset of F^n, we mean a set of the form $X(F) = X \cap F^n$, with X a quantifier-free formula of $\mathbf{T}^+$. Let $\mathrm{VF}^+(F)$ be the category of such sets, and $+$-constructible functions between them. The Grothendieck semiring $K_+\,\mathrm{VF}^+(F)$ is thus the quotient of $K_+\,\mathrm{VF}$ by the semiring congruence

$$I_F = \{([X],[Y]) : X, Y \in \mathrm{Ob}\,\mathrm{VF}^+, X(F) = Y(F)\}.$$

(One can verify this is an ideal; in fact, if $X(F) = Y(F)$ and $X \simeq X'$, then there exists $Y' \simeq Y$ with $X'(F) = Y'(F)$.)

Similarly, we can define I_F^{RV} and form $K_+\,\mathrm{RV}(F) \simeq K_+(\mathrm{RV})/I_F^{RV}$. As usual, let $\mathrm{I_{sp}}$ denote the congruence generated by $([1]_0 + [\mathrm{RV}^{>0}]_1, [1]_1)$, and $I_F^{RV} + \mathrm{I_{sp}}$ their sum.

Claim. If $([X],[X']) \in I_F$ then $(\not{b}[X], \not{b}[X']) \in I_F^{RV} + \mathrm{I_{sp}}$.

Proof. We may assume, changing X within the VF-isomorphism class $[X]$, that $X(F) = X'(F)$. Then $X(F) = (X \cup X')(F) = X'(F)$, and it suffices to show that $(\not{b}[X], \not{b}[X \cup X']), (\not{b}[X'], \not{b}[X \cup X']) \in I_F^{RV}$. Thus we may assume $X \subseteq X'$. Let $Z = X' \setminus X$. Then $Z(F) = \emptyset$, and it suffices to show that $(\not{b}(Z), \emptyset) \in I_F^{RV}$. Now $\not{b}(Z) = [Y]$ for some Y with Z definably isomorphic to $\mathbb{L}Y$. Thus $\mathbb{L}Y(F) = \emptyset$; hence $Y(F) = \emptyset$. Thus $([Y], \emptyset) \in I_F^{RV}$, as required. □

As an immediate consequence, we have the following.

Proposition 12.6. *Assume $F \leq M \models \mathbf{T}$, with F closed under definable functions of $\mathbf{T}$. The homomorphism $\not{b}$ of Theorem 8.8 induces a homomorphism*

$$\int_F : K_+\,\mathrm{VF}^+(F) \to K_+\,\mathrm{RV}^+(F)/\mathrm{I_{sp}}. \qquad \square$$

12.4 Quantifier elimination for Hensel fields

Let $\mathbf{T}$ be a V-minimal theory in a language $L_{\mathbf{T}}$, with sorts (VF, RV) (cf. Section 2.1). Assume $\mathbf{T}$ admits quantifier elimination and, moreover, that any definable function is given by a basic function symbol. This can be achieved by an expansion-by-definition of the language.

Let $\mathbf{T}_h = (\mathbf{T})_\forall \cup \{(\forall y \in \mathrm{RV})(\exists x \in \mathrm{VF})(\mathrm{rv}(x) = y)\}$.

A model of $\mathbf{T}_h$ is thus the same as a substructure A of a model of $\mathbf{T}$, such that $\mathrm{RV}(A) = \mathrm{rv}(\mathrm{VF}(A))$.

Lemma 12.7. *Any formula of $L_{\mathbf{T}}$ is* $\mathbf{T}$*-equivalent to a Boolean combination of formulas in* VF*-variables alone, and formula $\psi(t(x), u)$ where t is a sequence of terms for functions* $\mathrm{VF}^n \to \mathrm{RV}$*, u is a sequence of* RV*-variables, and ψ is a formula of* RV *variables only.*

Proof. This follows from stable embeddedness of RV, Corollary 3.24, Lemma 2.8 and the fact (Lemma 7.10) that definable functions into Γ factor through definable functions into RV. □

Example 12.8. If $\mathbf{T} = \mathrm{ACVF}(0, 0)$, then $\mathbf{T}_h$ is an expansion-by-definition of the theory of Hensel fields of residue characteristic zero.

Proof. We must show that a Henselian valued field is definably closed in its algebraic closure, in the two sorts VF, RV.

Let $K \models T_{Hensel}$, $K \le M \models \mathrm{ACVF}$. Let $X \subseteq \mathrm{VF}^k \times \mathrm{RV}^l$, $Y \subseteq \mathrm{VF}^{k'} \times \mathrm{RV}^{l'}$ be ACVF_K-definable sets, and $F : X \to Y$ an ACVF_K-definable bijection. We have to show that $F(X \cap K^k \times \mathrm{RV}(K)^l) = Y \cap K^{k'} \times \mathrm{RV}(K)^{l'}$.

K^{alg} is an elementary submodel of M; we may assume $K^{\mathrm{alg}} = M$. By one of the characterizations of Henselianity, the valuation on K extends uniquely to K^{alg}. Hence every field automorphism of M over K is a valued field automorphism. Thus K is the fixed field of $\mathrm{Aut}(M/K)$ (in the sense of valued fields), and hence $K = \mathrm{dcl}(K)$. Since ACVF_K is effective, any definable point of RV lifts to a definable point of VF; so $\mathrm{dcl}(K) \cap \mathrm{RV} = \mathrm{RV}_K$. Thus K is definably closed in M in both sorts. □

Let $L \supset L_{\mathbf{T}}$; assume $L \setminus L_{\mathbf{T}}$ consists of relations and functions on RV only. If $A \le M \models \mathbf{T}$, let $L_{\mathbf{T}}(A)$ be the languages enriched with constants for each element of A; let $\mathbf{T}_h(A) = \mathbf{T}_A \cup \mathbf{T}_h$, where $\mathbf{T}_A$ is the set of quantifier-free valued field formulas true of A.

Proposition 12.9. $\mathbf{T}_h$ *admits elimination of field quantifiers.*

Proof. Let A be as above. Let Φ_A be the set of $L(A)$-formulas with no VF-quantifiers.

Claim. Let $\phi(x, y) \in \Phi_A$ with x a free VF-variable. Then $(\exists x)\phi(x, y)$ is $\mathbf{T}_h(A)$-equivalent to a formula in Φ_A.

Proof. By the usual methods of compactness and absorbing the y-variables into A, it suffices to prove this when x is the only variable. Assume first that $\phi(x)$ is an $L_{\mathbf{T}}(A)$-formula. By Lemma 4.2, there exists an ACVF-definable bijection between the definable set defined by $\phi(x)$, and a definable set of the form $\mathbb{L}\phi'(x', u)$, where ϕ' is an $L_{\mathbf{T}}(A)$-formula in RV-variables only (including a distinguished variable x' on which $\mathbb{L}$ acts.) By the definition of $\mathbf{T}_h$, in any model of $\mathbf{T}_h$, ϕ has a solution iff $\mathbb{L}\phi'(x', u)$ has a solution. But clearly $\mathbb{L}\phi'(x', u)$ has a solution iff $\phi'(x', u)$ does. Thus $\mathbf{T}_h(A) \models (\exists x)\phi(x) \iff (\exists x', u)\phi'(x', u)$.

Now let $\phi(x)$ be an arbitrary Φ_A formula. Let Ψ be the set of formulas of $L(A)$ involving RV-variables only. Let Θ be the set of conjunctions of formulas of $L_{\mathbf{T}}(A)$ in VF-variables only, and of formulas of the form $\psi(t(x))$, where $\psi \in \Psi$ and t

is a term of $L_{\mathbf{T}}(A)$. The set of disjunctions of formulas in Θ is then closed under Boolean combinations, and under existential RV-quantification. By Lemma 12.7 it includes all $L_{\mathbf{T}}$-formulas, up to equivalence; and also all formulas in RV-variables only. Thus $\phi(x)$ is a disjunction of formulas in Θ, and we may assume $\phi(x) \in \Theta$. Say $\phi = \phi_0(x) \wedge \psi(t(x))$, with $\phi_0 \in L_{\mathbf{T}}(A)$ and $\psi \in \Psi$. By the claim, for some formula $\rho(y)$ of Φ_A, we have $T_h(A) \models \rho(y) \iff (\exists x)(t(x) = y \& \phi_0(x))$. Hence $(\exists x)\phi(x) \iff (\exists y)(\psi(y) \& \rho(y))$. □

Quantifier elimination now follows by induction. □

Remark. Since only field quantifiers are mentioned, this immediately extends to expansions in the field sort.

In particular, one can split the sequence $0 \to \mathbf{k}^* \to \mathrm{RV} \to \Gamma \to 0$ if one wishes. This yields the quantifier elimination [30] in the Denef–Pas language.

The results of Ax-Kochen and Ershov, and the large literature that developed around them, appeared to require methods of "quasi-convergent sequences." It is thus curious that they can also be obtained directly from Robinson's earlier and purely "algebraic" quantifier elimination for ACVF. Note that in the case of ACVF, there is no need to expand the language to obtain QE; and then Lemma 12.7 requires no proof beyond inspection of the language.

12.5 Rational points: Definable sets and morphisms

In this subsection we will work with completions T of $\mathbf{T}_h \cup \{(\exists x \in \Gamma)(x > 0)\}$. These are theories of valued fields of residue characteristic 0, possibly expanded, not necessarily algebraically closed. The language of T is thus the language of $\mathbf{T}^+$. The words formula, type, definable closure will refer to quantifier-free formulas of $\mathbf{T}^+$. Definable closure, types with respect to T are referred to explicitly as dcl^T, $\mathbf{T}\,\mathrm{tp}$, etc.

Let $F \models T$. Since $F \models \mathbf{T}_\forall$, F embeds into a model M' of $\mathbf{T}^+$. Since $\Gamma(F) \neq (0)$, by Proposition 3.51 and Lemma 3.49, there exists $F' \subseteq M'$ containing F, with $\Gamma(F') = \Gamma(F)$, and $M = \mathrm{acl}(F')$ an elementary submodel of M'. Hence F embeds into a model M of $\mathbf{T}^+$ with $\Gamma(F)$ cofinal in $\Gamma(M)$.

Lemma 12.10. *Let $F \models T$, $F \leq M \models \mathbf{T}^+$, $\Gamma(F)$ cofinal in $\Gamma(M)$. Let A be a substructure of M, $c \in F$, $B = A(c) \cap \mathrm{RV} \cap F$,*

(1) $\mathrm{tp}(c/B) \cup T_B$ *implies* $\mathbf{T}\,\mathrm{tp}(c/B)$.
(2) *Assume c is T_A-definable. Then $c \in \mathrm{dcl}(A, b)$ for some $b \in B$.*

Proof.

(1) This follows immediately from the quantifier elimination for T and from Lemma 12.1(1).
(2) We have $B \subseteq \mathrm{dcl}^T(A) \cap \mathrm{RV}$. We must show that $c \in \mathrm{dcl}(A \cup B)$. Let $p = \mathrm{tp}(c/A \cup B)$. By (1), p generates a complete type of $T_{A \cup B}$. Since this is the type of c and c is T_A-definable, some formula P in the language of $T_{A \cup B}$ with $P \in p$

has a unique solution in F. Now the values of F are cofinal in the value group of F^a; so P cannot contain any ball around c. (Any such ball would have an additional point of F, obtained by adding to c some element of large valuation.) Let P' be the set of isolated elements of P; then P' is finite (as is the case for every definable P), $\mathbf{T}_A$-definable, and $c \in P'$. By Lemma 3.9, there exists an $\mathbf{T}_A$-definable bijection $f : P' \to Q$ with $Q \subseteq \mathrm{RV}^n$. Then $f(c) \in \mathrm{dcl}_T(A) = B$, and $c = f^{-1}(f(c)) \in \mathrm{dcl}(A \cup B)$. □

Corollary 12.11. *Two definably isomorphic definable subsets of F have the same class in $K_+ \mathrm{VF}^+(F)$.*

Proof. T-definable bijections are restrictions of $\mathbf{T}^+$-definable bijections. Hence Corollary 12.2 is true with T replacing $\mathbf{T}^+$. □

Thus Proposition 12.6 includes a change-of-variable formalism for definable bijections.

12.6 Some specializations

Tim Mellor's Euler characteristic

Consider the theory RCVF of real closed valued fields. Let $\mathrm{RV}_{\mathrm{RCVF}}$, $\mathrm{RES}_{\mathrm{RCVF}}$, $\mathrm{VAL}_{\mathrm{RCVF}}$ denote the categories of definable sets and maps that lift to bijections of RCVF (on RV and on the residue field, value group, respectively; we do not need to use the sorts of RES other than the residue field here, say, all structures A of interest have Γ_A divisible). From Proposition 12.6 and Corollary 12.11, we obtain an isomorphism: $K(\mathrm{RCVF}) \to K(\mathrm{RV}_{\mathrm{RCVF}})/([0]_1 - [\mathrm{RV}^{>0}]_1 - [0]_0)$.

The residue field is a model of the theory RCF of real closed fields; $K(\mathrm{RCF}) = \mathbb{Z}$ via the Euler characteristic (cf. [37]). Since the ambient dimension grading is respected here, $K(\mathrm{RES}_{\mathrm{RCVF}}) = \mathbb{Z}[t]$.

The value group is a model of DOAG, and moreover, any definable bijection on $\Gamma[n]$ for fixed n lifts to RV and, indeed, to RCVF. This is because the multiplicative group of positive elements is uniquely divisible, and so $SL_n(\mathbb{Q})$ acts on the nth power of this group. By Proposition 9.4, $K(\mathrm{DOAG})[n] = \mathbb{Z}^2$ for each $n \geq 1$, and $K(\mathrm{VAL}_{\mathrm{RCVF}}) = \mathbb{Z}[s]^{(2)} := \{(f, g) \in \mathbb{Z}[s] : f(0) = g(0)\}$.

Thus $K(\mathrm{RV}_{\mathrm{RCVF}}) = \mathbb{Z}[t] \otimes \mathbb{Z}[s]^{(2)} \leq \mathbb{Z}[t, s]^2$; and J is identified with the class $(1, 1) - (0, -s) - (t, t)$. Thus we obtain two homomorphisms $K(\mathrm{RV}_{\mathrm{RCVF}})/J \to \mathbb{Z}[s]$ (one mapping $t \mapsto 1$, the other with $t \mapsto 1 - s$; and as a pair they are injective).

Equivalently, we have found two ring homomorphisms $\chi, \chi' : K(\mathrm{RCVF}) \to \mathbb{Z}[t]$. One of these was found in [27].

Cluckers–Haskell

Take the theory of the p-adics. By Proposition 12.6 and Corollary 12.11 we obtain an isomorphism: $K(\mathrm{pCF}) \to K(\mathrm{RV}_{\mathrm{pCF}})/\mathrm{I}_{\mathrm{sp}}$. However, $\mathrm{RV}_{\mathrm{pCF}}$ is a finite extension of $\mathbb{Z}$, and evidently $K(\mathbb{Z}) = 0$, since $[[0, \infty)] = [[1, \infty)]$. Thus $K(\mathrm{pCF}) = 0$.

12.7 Higher-dimensional local fields

We have seen that the Grothendieck group of definable sets with volume forms loses a great deal of information compared to the semigroup. Over fields with discrete value groups, restricting to bounded sets is helpful; in this way the Grothendieck group retains information about volumes. In the case of higher-dimensional local fields, with value group $A = \mathbb{Z}^n$, simple boundedness is insufficient to save it from collapse. We show that using a simple-minded notion of boundedness is only partly helpful, and loses much of the volume information (all but one $\mathbb{Z}$ factor).

Example 12.12. Let $K_\mu{}^{\mathrm{bdd}}(\mathrm{Th}(\mathbb{C}((s_1))((s_2)))[n])$ be the Grothendieck ring of definable bounded sets and measure-preserving maps in $\mathbb{C}((s_1))((s_2))$ (with $\mathrm{val}(s_1) \ll \mathrm{val}(s_2)$). Let Q^t denote the class of the thin annulus of radius t. In particular, Q^0 is the volume of the units of the valuation ring. Then in $K_\mu{}^{\mathrm{bdd}}(\mathrm{Th}(\mathbb{C}((s_1))((s_2)))[2])$, we have, for example, $(Q^0)^2 = 0$. To see this directly, let

$$\begin{aligned} Y &= \{(x, y) : \mathrm{val}(x) = 0, \mathrm{val}(y) = 0\}, \\ X &= \{(x, y) : 0 < 2\,\mathrm{val}(x) < \mathrm{val}(s_2), \mathrm{val}(x) + \mathrm{val}(y) = 0\}. \end{aligned}$$

Then X is bounded. Let $f(x, y) = (x/s_1, s_1 y)$. Then f is a measure-preserving bijection $X \to X' = \{(x, y) : 0 < 2(\mathrm{val}(x) + \mathrm{val}(s_1)) < \mathrm{val}(s_2), \mathrm{val}(x) + \mathrm{val}(y) = 0\}$. But in $\mathbb{C}((s_1))((s_2))$, $2\,\mathrm{val}(x) < \mathrm{val}(s_2)$ iff $2(\mathrm{val}(x) + \mathrm{val}(s_1)) < \mathrm{val}(s_2)$, so $X'(\mathbb{C}((s_1))((s_2))) = X(\mathbb{C}((s_1))((s_2)) \cup Y(\mathbb{C}((s_1))((s_2)))$.

Remark 12.13. $(2[[0, \gamma/2]]-[[0, \gamma]])(2[[0, \gamma/2)]-[[0, \gamma)])$, is a class of the Grothendieck group of Γ that vanishes identically in the $\mathbb{Z}$-evaluation, but not in the $\mathbb{Z}^2$-evaluation.

13 The Grothendieck group of algebraic varieties

Let X, Y be smooth nonsingular curves in $\mathbb{P}^3$, or in some other smooth projective variety Z, and assume $Z \setminus X$, $Z \setminus Y$ are biregularly isomorphic. Say X, Y, Z are defined over $\mathbb{Q}$. Then for almost all p, $|X(\mathbb{F}_p)| = |Y(\mathbb{F}_p)|$, as one may see by counting points of Z, $Z \setminus X$ and subtracting. It follows from Weil's Riemann hypothesis for curves that X, Y have the same genus, from Faltings that X, Y are isomorphic if the genus is 2 or more, and from Tate that X, Y are isogenous if the genus is one. It was this observation that led Kontsevich and Gromov to ask if X, Y must actually be isomorphic. We show that this is the case below.[2]

Theorem 13.1. *Let X, Y be two smooth d-dimensional subvarieties of a smooth projective n-dimensional variety V, and assume $V \setminus X$, $V \setminus Y$ are biregularly isomorphic. Then X, Y are stably birational, i.e., $X \times \mathbb{A}^{n-d}$, $Y \times \mathbb{A}^{n-d}$ are birationally equivalent. If X, Y contain no rational curves, then X, Y are birationally equivalent.*

[2] This already follows from [22], who use different methods.

While we do not obtain a complete characterization in dimensions > 1, the results and method of proof do show that the answer lies in synthetic geometry and is not cohomological in nature.

Let Var_K be the category of algebraic varieties over a field K of characteristic 0.

Let $[X]$ denote the class of a variety X in the Grothendieck semigroup $K_+(\mathrm{Var}_K)$. We allow varieties to be disconnected. As all varieties will be over the same field K, we will write Var for Var_K. Let $K_+ \mathrm{Var}_n$ be the Grothendieck semigroup of varieties of dimension $\leq n$.

For the proof, we view K as a trivially valued subfield of a model of ACVF(0, 0). We work with the theory ACVF_K, so that "definable" means K-definable with quantifier-free ACVF-formulas.

Note that $\mathrm{RES} = \mathbf{k}^*$ in ACVF_K; the only definable point of Γ is 0, so the only definable coset of $\mathbf{k}^*$ is $\mathbf{k}^*$ itself.

The residue map is an isomorphism on K onto a subfield K_{RES} of the residue field $\mathbf{k}$. In particular, any smooth variety V over K lifts canonically to a smooth scheme $V_{\mathcal{O}} = V \otimes_K \mathcal{O}$ over $\mathcal{O}$, with generic fiber $V_{\mathrm{VF}} = V_{\mathcal{O}} \otimes_{\mathcal{O}} \mathrm{VF}$ and special fiber $V_{\mathcal{O}} \otimes_{\mathcal{O}} \mathbf{k} = V \otimes_K \mathbf{k}$. We have a reduction homomorphism $\rho_V : V(\mathcal{O}) \to V(\mathbf{k})$. We will write $V(\mathcal{O})$, $V(\mathrm{VF})$ for $V_{\mathcal{O}}(\mathcal{O})$, $V_{\mathrm{VF}}(\mathrm{VF})$.

Given $k \leq n$ and a definable subset X of RV^* of dimension $\leq k$, let $[X]_k$ be the class of X in $K_+ \mathrm{RV}[k] \subseteq K_+ \mathrm{RV}[\leq n]$. Thus if $\dim(X) = d$ we have $n - d + 1$ classes $[X]_k$, $d \leq k \leq n$, in different direct factors of $K_+ \mathrm{RV}[\leq n]$. We also use $[X]_k$ to denote the image of this class in $K_+ \mathrm{RV}[\leq n]/\mathrm{I_{sp}}$. This abuse of notation is not excessive since for $n \leq N$, $K_+ \mathrm{RV}[\leq n]/\mathrm{I_{sp}}$ embeds in $K_+ \mathrm{RV}[\leq N]/\mathrm{I_{sp}}$ (Lemma 8.7).

Let SD_d be the image of $K_+ \mathrm{RV}[\leq d]$ in $K_+ \mathrm{RV}[\leq N]/\mathrm{I_{sp}}$. Let WD^n_d be the subsemigroup of $\mathrm{RV}[n]$ generated by $\{[X] : \dim(X) \leq d\}$, and use the same letter to denote the image in $\mathrm{RV}[\leq N]/\mathrm{I_{sp}}$. Let $FD^n = SD_{n-1} + WD^n_{n-1}$. We write $a \sim b(FD^n_d)$ for $(\exists u, v \in FD^n_d)(a + u = b + v)$. More generally, for any subsemigroup S' of a semigroup S, write $a \sim b(S')$ for $(\exists u, v \in S)(a + u = b + v)$.

We write $K(\mathrm{RV}[\leq n])/\mathrm{I_{sp}}$ for the groupification of $K_+(\mathrm{RV}[\leq n])/\mathrm{I_{sp}}$.

Lemma 13.2. *Let V be a smooth projective* $\mathbf{k}$*-variety of dimension n, X a definable subset of $V(\mathbf{k})$. Then*

$$\oint [\rho_V{}^{-1}(X)] = [X]_n.$$

Proof. Let $\mathbf{X} = (X, f)$ where $f : X \to \mathrm{RV}^n$ is a finite-to-one map. We have to show that $[\mathbb{L}\mathbf{X}] = [\rho_V{}^{-1}(X)]$ in $K_+(\mathrm{VF}[n])$, i.e., that $\mathbb{L}\mathbf{X}$, $\rho_V{}^{-1}(X)$ are definably isomorphic. By Lemma 2.3 this reduces to the case that X is a point p. Find an open affine neighborhood U of V such that $\rho_V{}^{-1}(p) \subseteq U(\mathcal{O})$, and U admits an étale map $g : V \to \mathbb{A}^n$ over $\mathbf{k}$. Now $U(\mathcal{O}) \simeq \mathcal{O}^n \times_{\mathrm{res},g} U(\mathbf{k})$. This reduces the lemma to the case of affine space, where it follows from the definition of $\mathbb{L}$. □

Lemma 13.3. *Let X be a K-variety of dimension $\leq d$.*

(1) $\oint(X(\mathrm{VF})) \in SD_d = K_+(\mathrm{RV}[\leq d])/\mathrm{I_{sp}}$.

(2) *If X is a smooth complete variety of dimension d, then* $\oint X(\mathrm{VF}) = [X]_d$.

(3) *If X is a variety of dimension d, then* $\oint X(\mathrm{VF}) \sim [X]_d(FD^d)$.

Proof.

(1) This is obvious, since $\dim(X(\mathrm{VF})) \leq d$.
(2) By Grothendieck's valuative criterion for properness, $X(\mathrm{VF}) = X(\mathcal{O})$. We thus have a map $\rho_V : X(\mathrm{VF}) = X(\mathcal{O}) \to X(\mathbf{k})$. For $\alpha \in X(\mathbf{k})$, let $X_\alpha(\mathrm{VF}) = {\rho_V}^{-1}(\alpha)$. Since X is smooth of dimension d it is covered by Zariski open neighborhoods U admitting an étale map $f_U : U \to \mathbb{A}^d$, defined over K; let $\mathcal{S}$ be a finite family of such pairs (U, f_U), with $\cup_{(U,f_U)\in\mathcal{S}} U = X$. We may choose a definable finite-to-one $f : X \to \mathbb{A}^d$, defined over K, such that for any $x \in X$, for some pair $(U, f_U) \in \mathcal{S}$, $f(x) = f_U(x)$. We have $\mathbb{L}([X]_d) = \mathbb{L}(X, f) = \mathrm{VF}^d \times_{\mathrm{rv},f} X(\mathbf{k})$. We have to show that $\mathbb{L}(X, f)$ is definably isomorphic to $X(\mathrm{VF})$. By Lemma 2.3 it suffices to show that for each $\alpha \in X(\mathbf{k})$, $\mathrm{VF}^d \times_{\mathrm{rv},f} \{a\}$ is α-definably isomorphic to $X_\alpha(\mathrm{VF})$. Now $\mathrm{VF}^d \times_{\mathrm{rv},f} \{a\} \underset{\alpha}{=} \mathrm{rv}^{-1}(f(\alpha))$. We have $f(\alpha) = f_U(\alpha)$ for some $(U, f) \in \mathcal{S}$ with $\alpha \in U$. Since f_U is étale, it induces a bijective map $U_\alpha(\mathrm{VF}) \to \mathrm{rv}^{-1}(f(\alpha))$. But $X_\alpha(\mathrm{VF}) = U_\alpha(\mathrm{VF})$, so the required isomorphism is proved.
(3) If X, Y are birationally equivalent, then $[X]_d \sim [Y]_d(WD^d_{<d})$, while $X(\mathrm{VF})$, $Y(\mathrm{VF})$ differ by VF-definable sets of dimension $< d$, so

$$\oint(X(\mathrm{VF})) \sim \oint(Y(\mathrm{VF}))(SD_d).$$

Using the resolution of singularities in the following form: every variety is birationally equivalent to a smooth nonsingular one; we are done by (2). With a more complicated induction we should be able to dispense with this use of Hironaka's theorem. □

Lemma 13.4. *Let V be a smooth projective K-variety, X, Y closed subvarieties, Let $F : V \setminus X \to V \setminus Y$ a biregular isomorphism. Let $V_\mathcal{O}$, V_{VF}, $V_\mathbf{k}$, F_{VF}, etc., be the objects obtained by base change. Then F_{VF} induces a bijection $V(\mathrm{VF}) \setminus X(\mathrm{VF}) \to V(\mathrm{VF}) \setminus Y(\mathrm{VF})$, and*

$$F_{\mathrm{VF}}({\rho_V}^{-1}(X) \setminus X(\mathrm{VF})) = {\rho_V}^{-1}(Y) \setminus Y(\mathrm{VF}).$$

Proof. The first statement follows from the Lefschetz principle since VF is algebraically closed.

Since V is projective, $V(\mathrm{VF}) = V(\mathcal{O})$, and one can define for $v \in V$ the valuative distance $d(v, X)$, namely, the greatest $\alpha \in \Gamma$ such that the image of x in $V(\mathcal{O}/\alpha)$ lies in $X(\mathcal{O}/\alpha)$.

Let $\mathbf{F}$ be the Zariski closure in V^2 of the graph of F. Then $\mathbf{F} \cap (V \setminus X) \times (V \setminus Y)$ is the graph of F. In fact, in any algebraically closed field L, we have

$$\text{if } a \in V(L) \setminus X(L) \quad \text{and} \quad (a, b) \in \mathbf{F}(L), \quad \text{then } b \in V(L) \setminus Y(L), \tag{13.1}$$

and conversely.

Suppose for the sake of contradiction that in some $M \models \mathrm{ACVF}_K$ there exist $a \in \rho_V{}^{-1}(X)$, $b \notin \rho_V{}^{-1}(Y)$, $(a, b) \in \mathbf{F}$. Thus $d(a, X) = \alpha > 0$, $d(b, Y) = 0$. Let

$$C = \{\gamma \in \Gamma : (\forall n \in \mathbb{N}) n\gamma < \alpha\}.$$

We may assume by compactness that $C(M) \neq \emptyset$. Let

$$I = \{y \in \mathcal{O}(M) : \mathrm{val}(y) \notin C\}$$

so that I is a prime ideal of $\mathcal{O}(M)$. Let L be the field of fractions of $\mathcal{O}(M)/I$. Let $\bar{a}, \bar{b}$ be the images of a, b in L. Then $(\bar{a}, \bar{b}) \in \bar{F}$, and $\bar{a} \in X, \bar{b} \notin Y$; contradicting (13.1). □

Proof of Theorem 13.1. By Lemma 13.4, there exists a definable bijection $\rho_V{}^{-1}(X) \setminus X \to \rho_V{}^{-1}(Y) \setminus Y$. Applying $\not{\int} : K(\mathrm{VF}[n]) \to K(\mathrm{RV}[\leq n])/\mathrm{I}_{\mathrm{sp}}$, and using Lemmas 13.2 and 13.3, we have $[X]_n - [X]_d = [Y]_n - [Y]_d$. Applying the first retraction $K(\mathrm{RV}[\leq n])/\mathrm{I}_{\mathrm{sp}} \to K(\mathrm{RES}[n])$ of Theorem 10.5, we obtain

$$[X_n] - [X \times \mathbb{A}^{n-d}]_n = [Y]_n - [Y \times \mathbb{A}^{n-d}]_n$$

in $!\, K(\mathrm{RES}[n]) = K(\mathrm{Var}_n)$. Thus

$$[X \times \mathbb{A}^{n-d} \dot{\cup} Y]_n + [Z] = [Y \times \mathbb{A}^{n-d} \dot{\cup} X]_n + [Z]$$

for some Z with $\dim(Z) \leq n$, where now the equality is of classes in $K_+ \mathrm{Var}_n$. Counting birational equivalence classes of varieties of dimension n, we see that $X \times \mathbb{A}^{n-d}$, $Y \times \mathbb{A}^{n-d}$ must be birationally equivalent. The last sentence follows from the lemma below. □

Lemma 13.5. *Let X, Y be varieties containing no rational curve. Let U be a variety such that there exists a surjective morphism $\mathbb{A}^m \to U$. If $X \times U$, $Y \times U$ are birationally equivalent, then so are X, Y.*

Proof. For any variety W, let $\mathcal{F}(W)$ be the set of all rational maps $g : \mathbb{A}^1 \to W$. Write $\mathrm{dom}(g)$ for the maximal subset of $\mathbb{A}^1$ where g is regular; so $\mathrm{dom}(g)$ is cofinite in $\mathbb{A}^1$. Let $R_W = \{(g(t), g(t')) \in W^2 : g \in \mathcal{F}(W), t, t' \in \mathrm{dom}(g)\}$. Let E_W be the equivalence relation generated by R_W, on points in the algebraic closure. R_W, E_W may not be constructible in general, but in the case we are concerned with, they are as follows.

Claim. Let $W \subseteq X \times U$ be a Zariski dense open set. Let $\pi : W \to X$ be the projection. Then $\pi(w) = \pi(w')$ iff $(w, w') \in E_W$ iff $(w, w') \in R_W$.

Proof. If $g \in \mathcal{F}(U)$, then $\pi \circ g : \mathrm{dom}(g) \to X$ is a regular map; hence by assumption on X it is constant. It follows that if $(w, w') \in R_U$ then $\pi(w) = \pi(w')$, and hence if $(w, w') \in E_U$ then $\pi(w) = \pi(w')$. Conversely, assume $w', w'' \in W$ and $\pi(w') = \pi(w'')$; then $w' = (x, u')$, $w' = (x, u'')$ for some $x \in X$, $u', u'' \in U$. Let $U_x = \{u \in U : (x, u) \in W\}$. Since W is open, U_x is open in U. Let $h : \mathbb{A}^m \to U$ be a surjective morphism; let $h(v') = u'$, $h(v'') = u''$. The line through v', v'' intersects $h^{-1}(U_x)$ in a nonempty open set. This gives a regular map f from the affine lines, minus finitely many points, into U, passing through u', u''. Thus $t \mapsto (x, f(t))$ gives a rational map from $\mathbb{A}^1$ to W, passing through (w', w''); and so $(w', w'') \in R_U$ and certainly in E_U. □

Using the claim, we prove the lemma. Let $W_X \subseteq X \times U$, $W_Y \subseteq Y \times U$ be Zariski dense open, and $F : W_X \to W_Y$ a biregular isomorphism. Then F takes E_{W_X} to E_{W_Y}. Moving now to the category of constructible sets and maps, quotients by constructible equivalence relations exist, and W_X/E_{W_X} is isomorphic as a constructible set to W_Y/E_{W_Y}. Let $\pi_X : W_X \to X$, $\pi_Y : W_Y \to Y$ be the projections. By the claim, $W_X/E_{W_X} = \pi_X(W_X) =: X'$. Similarly, $W_Y/E_{W_Y} = \pi_Y(W_Y) =: Y'$. Now since W_X, W_Y are Zariski dense, so are X', Y'. Thus X, Y contain isomorphic Zariski dense constructible sets, so they are birationally equivalent. □

Remark. The condition on X, Y may be weakened to the statement that they contain no rational curve through a generic point; i.e., that there exist proper subvarieties $(X_i : i \in I)$ defined over K, such that for any field $L \supset K$, any rational curve on $X \times_K L$ is contained in some $X_i \times_K L$.

Acknowledgments. Thanks to Aviv Tatarsky and Moshe Kaminsky, and to Lou Van den Dries, Clifton Ealy, and Jana Maříková for useful comments and corrections.

References

[1] J. T. Baldwin and A. H. Lachlan, On strongly minimal sets, *J. Symbolic Logic*, **36** (1971), 79–96.

[2] Ş. A. Basarab and F. V. Kuhlmann, An isomorphism theorem for Henselian algebraic extensions of valued fields, *Manuscripta Math.*, **77**-2–3 (1992), 113–126.

[3] V. V. Batyrev, Birational Calabi-Yau n-folds have equal Betti numbers, in *New Trends in Algebraic Geometry* (*Warwick*, 1996) London Mathematical Society Lecture Note Series, Vol. 264, Cambridge University Press, Cambridge, UK, 1999, 1–11.

[4] I. N. Bernstein, Analytic continuation of generalized functions with respect to a parameter, *Funk. Anal. Priložen.*, **6**-4 (1972), 26–40.

[5] Z. Chatzidakis and E. Hrushovski, Model theory of difference fields, *Trans. Math. Soc. Amer.*, **351**-8 (1999), 2997–3071.

[6] R. Cluckers and D. Haskell, Grothendieck rings of Z-valued fields, *Bull. Symbolic Logic*, **7**-2 (2001), 262–226.

[7] R. Cluckers and F. Loeser, Fonctions constructibles et intégration motivique I, II, math.AG/0403350 and math.AG/0403349, 2004; also available online from http://www.dma.ens.fr/~loeser/.

[8] R. Cluckers and F. Loeser, Fonctions constructibles exponentiel les, transformation de Fourier motivique et principe de transfert, math.NT/0509723, 2005; *C. R. Acad. Sci. Paris Sér.* I *Math.*, to appear.

[9] C. C. Chang and H. J. Keisler, *Model Theory*, 3rd ed., Studies in Logic and the Foundations of Mathematics, Vol. 73, North-Holland, Amsterdam, 1990.

[10] J. Denef and F. Loeser, Definable sets, motives and p-adic integrals, *J. Amer. Math. Soc.*, **14** (2001), 429–469.

[11] H. B. Enderton, *A Mathematical Introduction to Logic* 2nd ed., Harcourt/Academic Press, Burlington, MA, 2001.

[12] I. Fesenko and M. Kurihara, eds., *Invitation to Higher Local Fields*, Geometry and Topology Monographs, Vol. 3, Mathematics Institute, University of Warwick, Coventry, UK, 2000.

[13] M. Gromov, Endomorphisms of symbolic algebraic varieties, *J. European Math. Soc.*, **1**-2 (1999), 109–197.

[14] R. Cluckers, L. Lipshitz, and Z. Robinson, Analytic cell decomposition and analytic motivic integration, math.AG/0503722, 2005; *Ann. Sci. École Norm. Sup.*, to appear.

[15] D. Haskell, and D. Macpherson, Cell decompositions of C-minimal structures, *Ann. Pure Appl. Logic*, **66**-2 (1994), 113–162.

[16] D. Haskell, E. Hrushovski, and H. D. Macpherson, Definable sets in algebraically closed valued fields, Part I: Elimination of imaginaries, preprint, 2002; *Crelle*, to appear.

[17] D. Haskell, E. Hrushovski, and H. D. Macpherson, Stable domination and independence in algebraically closed valued fields, math.LO/0511310, 2005.

[18] E. Hrushovski, Elimination of imaginaries for valued fields, preprint.

[19] P. T. Johnstone, *Notes on Logic and Set Theory*, Cambridge Mathematical Textbooks, Cambridge University Press, Cambridge, UK, 1987.

[20] M. Kageyama and M. Fujita, Grothendieck rings of o-minimal expansions of ordered Abelian groups, math.LO/0505331, 2005.

[21] D. Kazhdan, An algebraic integration, in *Mathematics: Frontiers and Perspectives*, American Mathematical Society, Providence, RI, 2000, 93–115.

[22] M. Larsen and V. A. Lunts, Motivic measures and stable birational geometry, *Moscow Math. J.*, **3**-1 (2003), 85–95.

[23] L. Lipshitz, Rigid subanalytic sets, *Amer. J. Math.*, **115**-1 (1993), 77–108.

[24] L. Lipshitz and Z. Robinson, One-dimensional fibers of rigid subanalytic sets, *J. Symbolic Logic*, **63**-1 (1998), 83–88.

[25] F. Loeser and J. Sebag, Motivic integration on smooth rigid varieties and invariants of degenerations, *Duke Math. J.*, **119**-2 (2003), 315–344.

[26] J. Maříková, *Geometric Properties of Semilinear and Semibounded Sets*, M.A. thesis, Charles University, Prague, 2003; preprint.

[27] T. Mellor, talk, Isaac Newton Institute for Mathematical Sciences, Cambridge, UK, 2005.

[28] A. Pillay, *An Introduction to Stability Theory*, Oxford Logic Guides, Oxford University Press, Oxford, UK, 1983.

[29] A. Pillay, *Geometric Stability Theory*, Oxford Logic Guides, Oxford University Press, Oxford, UK, 1996.

[30] J. Pas, Uniform p-adic cell decomposition and local zeta functions, *J. Reine Angew. Math.*, **399** (1989), 137–172.

[31] B. Poizat, *Cours de Théorie des modeles: Nur al mantiq wal ma'arifah* (*A Course in Model Theory: An Introduction to Contemporary Mathematical Logic*), Universitext, Springer-Verlag, New York, 2000 (translated from the French by M. Klein and revised by the author).

[32] B. Poizat, Une théorie de Galois imaginaire, *J. Symbolic Logic*, **48**-4 (1983), 1151–1170.

[33] A. Robinson, *Complete Theories*, North-Holland, Amsterdam, 1956.

[34] H. H. Crapo and G.-C. Rota, *On the Foundations of Combinatorial Theory: Combinatorial Geometries*, preliminary ed., M.I.T. Press, Cambridge, MA, London, 1970.

[35] S. Shelah, *Classification Theory and the Number of Nonisomorphic Models*, 2nd ed., Studies in Logic and the Foundations of Mathematics 92, North-Holland, Amsterdam, 1990.

[36] L. van den Dries, Dimension of definable sets, algebraic boundedness and Henselian fields, *Ann. Pure Appl. Logic*, **45**-2 (1989), 189–209.

[37] L. van den Dries, *Tame Topology and o-Minimal Structures*, London Mathematical Society Lecture Note Series, Vol. 248, Cambridge University Press, Cambridge, UK, 1998.

On the Euler–Kronecker constants of global fields and primes with small norms

Yasutaka Ihara

Department of Mathematics
Graduate School of Science and Engineering
Chuo University
Kasuga 1-13-27
Bunkyo-ku, Tokyo 112-8551
Japan
ihara@math.chuo-u.ac.jp, ihara@kurims.kyoto-u.ac.jp

Dedicated to V. Drinfeld.

Subject Classifications: Primary 11R42. Secondary 11R47, 11R58.

0 Introduction

Let K be a global field, i.e., either an algebraic number field of finite degree (abbreviated NF), or an algebraic function field of one variable over a finite field (FF). Let $\zeta_K(s)$ be the Dedekind zeta function of K, with the Laurent expansion at $s = 1$:

$$\zeta_K(s) = c_{-1}(s-1)^{-1} + c_0 + c_1(s-1) + \cdots \quad (c_{-1} \neq 0). \tag{0.1}$$

In this paper, we shall present a systematic study of the real number

$$\gamma_K = c_0/c_{-1} \tag{0.2}$$

attached to each K, which we call the *Euler–Kronecker constant* (or *invariant*) of K. When $K = \mathbb{Q}$ (the rational number field), it is nothing but the Euler–Mascheroni constant

$$\gamma_{\mathbb{Q}} = \lim_{n\to\infty}\left(1 + \frac{1}{2} + \cdots + \frac{1}{n} - \log n\right) = 0.57721566\ldots,$$

and when K is imaginary quadratic, the well-known Kronecker limit formula expresses γ_K in terms of special values of the Dedekind η function. This constant γ_K appears here and there in several articles in analytic number theory, but as far as the author knows, it has not played a main role nor has it been systematically studied. We shall consider γ_K more as an *invariant* of K.

Before explaining our motivation for systematic study, let us briefly look at the FF case. When K is the function field of a curve X over a finite field $\mathbb{F}_q$ of genus g, so that $\zeta_K(s)$ is a rational function of $u = q^{-s}$ of the form

$$\zeta_K(s) = \frac{\prod_{\nu=1}^{g}(1-\pi_\nu u)(1-\bar{\pi}_\nu u)}{(1-u)(1-qu)}, \quad \pi_\nu\bar{\pi}_\nu = q \quad (1 \le \nu \le g), \tag{0.3}$$

then γ_K is closely related to the *harmonic* mean of the g positive real numbers

$$(1-\pi_\nu)(1-\bar{\pi}_\nu) \quad (1 \le \nu \le g), \tag{0.4}$$

in contrast to the facts that their arithmetic (respectively, geometric) means are related to the number of $\mathbb{F}_q$-rational points of X (respectively, its Jacobian J_X). More explicitly,

$$\begin{aligned} \frac{\gamma_K}{\log q} &= (q-1)\sum_{\nu=1}^{g}\frac{1}{(1-\pi_\nu)(1-\bar{\pi}_\nu)} - (g-1) - \frac{q+1}{2(q-1)} \\ &= \sum_{m=1}^{\infty}\left(\frac{q^m+1-N_m}{q^m}\right) + 1 - \frac{q+1}{2(q-1)}, \end{aligned} \tag{0.5}$$

where N_m denotes the number of $\mathbb{F}_{q^m}$-rational points of X (see Section 1.4). The first expression shows that γ_K is a rational multiple of $\log q$, while the second shows that when X has many $\mathbb{F}_{q^m}$-rational points for small m (especially $m = 1$), γ_K tends to be *negative*.

Our first basic observation is that, including the NF case, γ_K can sometimes be "conspicuously negative," and that this occurs when K has "many primes with small norms." In the FF case, there are known interesting towers of curves over $\mathbb{F}_q$ with many rational points, and we ask how negative γ_K can be, in general and for such a tower. In the NF case, there is no notion of rational points, but those K having many primes with small norms would be equally interesting for applications (to coding theory, etc.). Moreover, the related problems often have their own arithmetic significance (e.g., the fields K_p described below). We wish to know how negative γ_K can be also in the NF case. A careful comparison of the two cases is very interesting. Thus we are led to studying γ_K in both cases under a unified treatment, basically assuming the generalized Riemann hypothesis (GRH)in the NF case. We shall give a method for systematic computation of γ_K, give some general upper and lower bounds, and study three special cases more closely, including that of curves with many rational points, for comparisons and applications.

In Section 1, after basic preliminaries, we shall give some *explicit* estimations of γ_K , and also discuss possibilities of improvements when we specialize to smaller families of K (see Section 1.6). Among them, Theorem 1 gives a general upper bound for γ_K. The main term of this upper bound is

$$\begin{cases} 2\log\log\sqrt{|d|} & \text{(NF, under GRH)}, \\ 2\log((g-1)\log q) + \log q & \text{(FF)}, \end{cases} \tag{0.6}$$

$d = d_K$ being the discriminant. The lower bound is, as we shall see, *necessarily* much weaker. First, the main term of our general lower bound (Proposition 3) reads as

$$\begin{cases} -\log\sqrt{|d|} & \text{(NF, unconditionally)}, \\ -(g-1)\log q & \text{(FF)}. \end{cases} \tag{0.7}$$

Secondly, when we *fix* q, the latter will be improved to be

$$-\frac{1}{\sqrt{q}+1}(g-1)\log q \quad \text{(FF)} \tag{0.8}$$

(Theorem 2). In other words,

$$C(q) := \liminf \frac{\gamma_K}{(g_K-1)\log q} \geq -\frac{1}{\sqrt{q}+1}. \tag{0.9}$$

This is based on a result of Tsfasman [Ts_1] and is somewhat stronger than what we can prove only by using the Drinfeld–Vlăduţ asymptotic bound [D-V] for N_1 . We shall, moreover, see that the equality holds in (0.9) when q is a square (see below). In the NF case, our attention will be focused on the absolute constant

$$C = \liminf \frac{\gamma_K}{\log\sqrt{|d_K|}}. \tag{0.10}$$

Clearly, (0.7) gives $C \geq -1$ (unconditionally), but quite recently, Tsfasman proved, as a beautiful application of [T-V], that

$$C \geq -0.26049\ldots \quad \text{(under GRH)} \tag{0.11}$$

(see [Ts_2] in this volume). The estimation of $C(q)$ or C from *above* is related to finding a sequence of K having many primes with small norms. As for $C(q)$, see below. As for C, the author obtained $C \leq -0.1635$ (under GRH; see Section 1.6), but [Ts_2] contains a sharper unconditional estimation. At any rate, in each of the FF and NF cases, we see that the general (negative) lower bound for γ_K cannot be so close to 0 as the (positive) upper bound.

Thirdly, when the degree N of K over $\mathbb{Q}$, respectively, $\mathbb{F}_q(t)$ is fixed ($N > 1$), or grows slowly enough, (0.7) will be improved to be

$$\begin{cases} -2(N-1)\log\left(\dfrac{\log\sqrt{|d|}}{N-1}\right) & \text{(NF, under GRH)}, \\ -2(N-1)\log\left(\dfrac{(g-1)\log q}{N-1}\right) & \text{(FF)} \end{cases} \tag{0.12}$$

(Theorem 3), which is nearly as strong as the upper bound, and exactly so (with opposite signs) when $N = 2$. Granville–Stark [G-S, Section 3.1] gave an equivalent statement when $N = 2$ (NF case), and our Theorem 3 was inspired by this work. The bound (0.12) is quite sharp. In fact, some families of K having many primes with small norms imply that (0.12) cannot be replaced by its quotient even by $\log\log N$. To be precise, it cannot be replaced by its quotient by any such $f(N)$ (NF) (respectively, $f_q(N)$ for a fixed $q > 2$ (FF)) as satisfying $f(N) \to \infty$ (respectively, $f_q(N) \to \infty$).

Section 1.7 is for supplementary remarks related to computations of γ_K.

In Section 2, we shall study some special cases. First, let q be any fixed prime power. Then, as an application of a result in [E], we obtain

$$C(q) \le -c_0 \frac{\log q}{q-1} \tag{0.13}$$

(Section 2.1), where c_0 is a certain positive absolute constant. Then we treat the case where K is the function field over $\mathbb{F}_q$ of a Shimura curve, with q a square, and $g_K \gg q$ (Section 2.1). In this case, as a reflection of the fact that such a curve has an abundance of $\mathbb{F}_q$-rational points, we can prove

$$\gamma_K \le -\frac{1}{\sqrt{q}+1}(g_K - 1)\log q + \varepsilon. \tag{0.14}$$

Therefore, combining this with (0.9), we obtain

$$C(q) = -\frac{1}{\sqrt{q}+1} \quad (q \text{ a square}). \tag{0.15}$$

Secondly, when K is imaginary quadratic, we combine our *upper* bound for γ_K with the Kronecker limit formula, to give a lower bound for its class number h_K;

$$\frac{h_K \log|d_K|}{\sqrt{|d_K|}} > \frac{\pi}{3} - \varepsilon, \tag{0.16}$$

with an *explicit* description of the ε part (under GRH) (Theorem 5 in Section 2.2). As an *asymptotic* formula, this is weaker than Littlewood's [Li] and almost equivalent to Granville–Stark's [G-S] (both conditional) formulas; its merit is explicitness.

Thirdly, we consider the case where $K = K_p$ is the "first layer" of the cyclotomic $\mathbb{Z}_p$-extension over $\mathbb{Q}$ (Section 2.3). It is the unique cyclic extension over $\mathbb{Q}$ of degree p contained in the field of p^2th roots of unity. By class-field theory, a prime ℓ decomposes completely in K_p if and only if

$$\ell^{\,p-1} \equiv 1 \pmod{p^2}. \tag{0.17}$$

We shall apply our estimations of γ_K to this case $K = K_p$ (Theorem 6 and its corollaries). Among them, Corollary 7 gives information on small ℓs satisfying (0.17) for a fixed large p, while Corollary 9 relates the question of the existence of "many" p satisfying (0.17) for a fixed ℓ to that of $\liminf(\gamma_{K_p}/p)$. (Incidentally, $\lim(\gamma_{K_p}/\log\sqrt{d_{K_p}}) = \lim(\gamma_{K_p}/(p-1)\log p) = 0$ under GRH.) From Table 1, notice how the existence of a *very* small ℓ satisfying (0.17) pushes the value of γ_{K_p} drastically towards the left on the negative real axis. For example, (0.17) is satisfied for $\ell = 2$ and $p = 1093$, and accordingly, $\gamma_{K_{1093}}$ is as negative as about -747, while for several neighboring primes p, the absolute values of γ_{K_p} are at most 10. Finally, in Sections 2.4 and 2.5, we shall give some application to the "field index" of K_p.

Our main tool is "the explicit formula" for the prime counting function

$$\Phi_K(x) = \frac{1}{x-1} \sum_{N(P)^k \le x} \left(\frac{x}{N(P)^k} - 1 \right) \log N(P) \quad (x > 1), \tag{0.18}$$

where (P, k) runs over the pairs of (non-archimedean) primes P of K and positive integers k such that $N(P)^k \le x$ (Sections 2.2 and 2.3). This function $\Phi_K(x)$ is quite close to $\log x$ when x is large, and the connection with our constant γ_K is

$$\lim_{x\to\infty} (\log x - \Phi_K(x)) = \gamma_K + 1 \qquad \text{(NF, unconditionally)}, \tag{0.19}$$

$$\lim_{\substack{x \in q^{\mathbb{Z}} \\ x\to\infty}} (\log x - \Phi_K(x)) = \gamma_K + \frac{q+1}{2(q-1)} \log q \quad \text{(FF)}. \tag{0.20}$$

It is a simple combination of two well-known prime counting functions, but two characteristic features of $\Phi_K(x)$ are (i) it is *continuous* and (ii) the oscillating term in the explicit formula for $\Phi_K(x)$ has the form

$$-\frac{1}{2(x-1)} \lim_{T\to\infty} \sum_{|\rho|<T} \frac{(x^\rho - 1)(x^{1-\rho} - 1)}{\rho(1-\rho)}, \tag{0.21}$$

where ρ runs over the nontrivial zeros of $\zeta_K(s)$, which, under GRH, is very easy to evaluate. In fact, it is then sandwiched between two multiples by

$$((\sqrt{x}+1)/(\sqrt{x}-1))^{\pm 1}$$

of the negative real constant

$$-\frac{1}{2} \sum_{\rho} \frac{1}{\rho(1-\rho)}. \tag{0.22}$$

And $-\gamma_K$ is a translate of (0.22) by a more elementary constant associated to K. This is why (under GRH in the NF case) we can obtain results always *with explicit error terms*, and using only *simple elementary* arguments. Usually, one uses the "truncated explicit formula" where the summation over ρ is restricted to $|\rho| < T$ and instead contains an error term $R(x, T)$ which is not easy to evaluate systematically.

We add three more observations here.

(i) In some sense, the quantity on the RHS of (0.19)–(0.20) may be more canonical than γ_K as an invariant of K. Note that (0.20) with $q = 1$ "corresponds to" (0.19), and that (0.5) will be simplified if we use the RHS of (0.20) instead of γ_K itself (see Section 1.4).

(ii) One can of course generalize the definition of γ_K to the case of L-functions, although then they will not usually be real numbers. Multiplicative relations among the L-functions give rise to *additive* relations among these constants. In particular, when H runs over the subgroups of a given finite group G, any linear relation among those characters of G induced from the trivial character of H gives rise to the corresponding linear relation among the γ_K, where K runs over the intermediate extensions of a given G-extension.

(iii) When K is either the *cyclotomic field* $\mathbb{Q}(\boldsymbol{\mu}_m)$ or its *maximal real subfield* $\mathbb{Q}(\boldsymbol{\mu}_m)^+$, it seems fairly likely that γ_K is always *positive*! The author has computed γ_K in both cases up to $m = 600$, and Mahoro Shimura more recently checked the first case $K = \mathbb{Q}(\boldsymbol{\mu}_m)$ for m as far as up to 8000, and we have found no counterexamples. On the other hand, their difference, "the relative" γ_K, seems to take both signs "almost equally."

Studies of γ_K for various families of global fields K including these cases will be left to future publications. Some open problems and numerical data can be found in my article in the (informal) "Proceedings of the 2004 Workshop on Cryptography and Related Mathematics" (Chuo University, 2005). The 2003 workshop proceedings contains a short summary of the present paper.

1 The "explicit formula" for $\Phi_K(x)$, and estimations of γ_K

1.1 The function $\Phi_K(x)$

Let K be a global field. We denote by P any (non-archimedean) prime divisor of K, and by $N(P)$ its norm. As mentioned in the introduction, we shall consider the prime counting function

$$\Phi_K(x) = \frac{1}{x-1} \sum_{N(P)^k \le x} \left(\frac{x}{N(P)^k} - 1 \right) \log N(P) \quad (x > 1). \tag{1.1.1}$$

Here (P, k) runs over all pairs with $k \ge 1$ and $N(P)^k \le x$ (or what amounts to the same thing, $N(P)^k < x$). Call a point on the real axis *critical* if it is of the form $N(P)^k$. Then $\Phi_K(x)$ remains 0 until the first critical point, then is monotone increasing, and is everywhere continuous. In fact, at each critical point $\Phi_K(x)$ acquires new summands but their values are 0 at this point, so the visible increase at each critical point is that of the slope. The slope of $\Phi_K(x)$ between two adjacent critical points $a < b$ is $c(x-1)^{-2}$, where

$$c = \sum_{N(P)^k \le a} \left(1 - \frac{1}{N(P)^k} \right) \log N(P) > 0.$$

So the slope near x is close to

$$\left(\sum_{N(P)^k < x} \log N(P) \right) x^{-2} \sim x^{-1}.$$

Thus $\Phi_K(x)$ is an arithmetic approximation of $\log x$. If the field K has many primes P with small $N(P)$, then $\Phi_K(x)$ increases faster than $\log x$, at least for awhile. The difference $\log x - \Phi_K(x)$ "at infinity" is closely related to γ_K, as we shall see later.

1.2 The explicit formula for $\Phi_K(x)$

From Weil's general explicit formula [W1, W2], we obtain, as will be indicated in Section 1.3, the following formula for $\Phi_K(x)$:

$$\Phi_K(x) = \log x + (\alpha_K + \beta_K) + \ell_K(x) + r_K(x) \quad (x > 1). \tag{1.2.1}$$

Here

$$\begin{aligned} \alpha_K &= \frac{1}{2}\log|d| && \text{(NF)}, \\ &= (g-1)\log q && \text{(FF)} \end{aligned} \tag{1.2.2}$$

($d = d_K$ is the discriminant; $g = g_K$ is the genus; $\mathbb{F}_q$ is the exact constant field),

$$\begin{aligned} \beta_K &= -\left\{\frac{r_1}{2}(\gamma + \log 4\pi) + r_2(\gamma + \log 2\pi)\right\} && \text{(NF)}, \\ &= 0 && \text{(FF)} \end{aligned} \tag{1.2.3}$$

(r_1, r_2 is the number of real, imaginary places of K, respectively; $\gamma = \gamma_{\mathbb{Q}}$ is the Euler–Mascheroni constant $= 0.57721566\ldots$),

$$\begin{aligned} \ell_K(x) &= \frac{r_1}{2}\left(\log\frac{x+1}{x-1} + \frac{2}{x-1}\log\frac{x+1}{2}\right) \\ &\quad + r_2\left(\log\frac{x}{x-1} + \frac{1}{x-1}\log x\right) && \text{(NF)}, \\ &= \phi(q, x) && \text{(FF)}, \end{aligned} \tag{1.2.4}$$

where $\phi(q, x)$ is a certain continuous function of x parametrized by q, satisfying

$$\begin{aligned} &0 \le \phi(q, x) < \log q, \\ &\phi(q, x) = 0 \longleftrightarrow x = q^m \quad \text{with some } m \in \mathbb{N} \end{aligned} \tag{1.2.5}$$

(see below). Finally,

$$r_K(x) = -\frac{1}{2(x-1)}\sum_{\rho}\frac{(x^{\rho}-1)(x^{1-\rho}-1)}{\rho(1-\rho)}, \tag{1.2.6}$$

where ρ runs over all nontrivial zeros of $\zeta_K(s)$, counted with multiplicities, and

$$\sum_{\rho} = \lim_{T\to\infty}\sum_{|\rho|<T}. \tag{1.2.7}$$

By the functional equation for $\zeta_K(s)$, if ρ is a nontrivial zero of $\zeta_K(s)$, then so is $1-\rho$, with the same multiplicity.

In the FF case, when $x = q^m$ $(m \in \mathbb{N})$,

$$\begin{cases} \Phi_K(x)/\log q = \dfrac{1}{q^m - 1} \displaystyle\sum_{k \deg P \le m} (q^{m-k\deg P} - 1)\deg P, \\ \log x/\log q = m, \\ \alpha_K/\log q = g - 1, \\ \beta_K/\log q = \ell_K(x)/\log q = 0, \\ r_K(x)/\log q = -\left(\dfrac{q-1}{q^m-1}\right) \displaystyle\sum_{\nu=1}^{g} \dfrac{(\pi_\nu^m - 1)(\bar{\pi}_\nu^m - 1)}{(\pi_\nu - 1)(\bar{\pi}_\nu - 1)}, \end{cases} \tag{1.2.8}$$

where

$$\zeta_K(s) = \frac{\prod_{\nu=1}^{g}(1-\pi_\nu u)(1-\bar{\pi}_\nu u)}{(1-u)(1-qu)}, \quad u = q^{-s}, \quad \pi_\nu\bar{\pi}_\nu = q \quad (1 \le \nu \le g). \tag{1.2.9}$$

(To derive the last formula for $r_K(q^m)/\log q$ from the definition (1.2.6) of $r_K(x)$, take any $\alpha \in \mathbb{C}^\times$ and $q > 1$, and substitute $e^z = \alpha^{-1}q^s$ in the partial fraction expansion formula

$$(e^z - 1)^{-1} + \frac{1}{2} = \lim_{T\to\infty} \sum_{n=-T}^{T} (z - 2\pi i n)^{-1}, \tag{1.2.10}$$

which gives

$$\frac{\log q}{\alpha^{-1}q^s - 1} + \frac{\log q}{2} = \lim_{T\to\infty} \sum_{\substack{q^\rho = \alpha \\ |\rho| \le T}} (s-\rho)^{-1}. \tag{1.2.11}$$

Now let $q = \alpha\bar{\alpha}$, $s = 0$ and take the real part of (1.2.11) to obtain

$$\frac{q-1}{(\alpha-1)(\bar{\alpha}-1)} \log q = \lim_{T\to\infty} \sum_{\substack{q^\rho = \alpha \\ |\rho| \le T}} \left(\frac{1}{\rho} + \frac{1}{\bar{\rho}}\right). \tag{1.2.12}$$

The desired formula follows immediately from this.)

Note that each reciprocal zero π_ν (respectively, $\bar{\pi}_\nu$) of $\zeta_K(s)$ in $u = q^{-s}$ corresponds to infinitely many zeros in s, which are translations of one of them by $2\pi i n/\log q$ $(n \in \mathbb{Z})$. It also has poles at all translations of 0, 1 by $2\pi i n/\log q$ $(n \in \mathbb{Z})$. The function $\phi(q, x)$ arises from the poles $\theta \ne 0, 1$;

$$\phi(q, x) = \frac{1}{2(x-1)} \sum_{\text{poles } \theta \ne 0,1} \frac{(x^\theta - 1)(x^{1-\theta} - 1)}{\theta(1-\theta)}, \tag{1.2.13}$$

where

$$\sum_\theta = \lim_{T\to\infty} \sum_{|\theta| < T}. \tag{1.2.14}$$

Since either $q^\theta = 1$ or $q^{1-\theta} = 1$, it is clear that $\phi(q, x) = 0$ when $x = q^m$ ($m \in \mathbb{N}$). In a finite form,

$$\phi(q, x) = \log\left(\frac{q^m}{x}\right) - \frac{(q^{m-1} - 1)(q^m - x)}{(x-1)(q^m - q^{m-1})}\log q \tag{1.2.15}$$

for $q^{m-1} \le x \le q^m$ ($m \in \mathbb{Z}$, $m \ge 1$, $x \ne 1$). This follows immediately from the following.

Proposition 1.

(i) *The functions* $\ell_K(x)$ *and* $r_K(x)$ *are continuous.*
(ii) *(FF):* $(x-1)(\ell_K(x) + \log x)$ *and* $(x-1)r_K(x)$ *are linear on each interval* $q^{m-1} \le x \le q^m$ $(m \ge 1)$.

Proof.

(NF) $\ell_K(x)$ is continuous by definition. Since $\Phi_K(x)$ and $\ell_K(x)$ are both continuous, $r_K(x)$ is also continuous by (1.2.1).

(FF) In this case, $\ell_K(x) = \phi(q, x)$ is a function of x determined only by q. By (1.2.1) applied to the case $g = 0$, we have

$$\phi(q, x) = \Phi_{\mathbb{F}_q(t)}(x) - \log x + \log q; \tag{1.2.16}$$

hence $\phi(q, x)$ is continuous. Now, when $q^{m-1} \le x \le q^m$,

$$(x-1)\Phi_K(x) = \sum_{N(P)^k \le q^{m-1}} \left(\frac{x}{N(P)^k} - 1\right)\log N(P) \tag{1.2.17}$$

is linear. Hence by (1.2.16), $(x-1)(\phi(q, x) + \log x)$ is also linear on this interval. Moreover, the function

$$(x-1)r_K(x) = (x-1)\Phi_K(x) - (x-1)\left(\phi(q, x) + \log x\right) - (x-1)(\alpha_K + \beta_K)$$

is also linear in the same interval. □

Remarks.

(i) In the NF case, β_K and $\ell_K(x)$ both come from the archimedean places. Among them, β_K is the value at $s = 1$ of the logarithmic derivative of the "standard Γ-factor" of $\zeta_K(s)$ (see Section 1.3 below), and $\ell_K(x)$ comes from the trivial zeros of $\zeta_K(s)$. Thus $\ell_K(s)$ for the (FF) and the (NF) cases have quite different origins—poles $\ne 0, 1$, vs. trivial zeros. We have given them the same name here only to save notation.

(ii) In the NF case, $\beta_K + \ell_K(x)$ is the term coming from the archimedean places, and our separation into β_K and $\ell_K(x)$ can also be characterized by

$$\lim_{x \to \infty} \ell_K(x) = 0$$

(cf. Lemma 1 below (Section 1.5)).

(iii) We note also that

$$\ell_K(x) \ge 0 \qquad (x > 1)$$

in both cases (cf. Lemma 1 in Section 1.5).

1.3 The explicit formula for $\Phi_K(x)$ (continued)

The above explicit formula (1.2.1) for $\Phi_K(x)$, at least in the NF case, is a special case of Weil's general explicit formula. To be precise, use t for x of [W$_1$], keeping x for our x, and put

$$F(t) = \begin{cases} \frac{1}{x-1}(xe^{-t/2} - e^{t/2}), & 0 < t < \log x, \\ \frac{1}{2}, & t = 0, \\ 0, & \text{otherwise}, \end{cases}$$

in [W$_1$, formula (11)]. Then we obtain (1.2.1) by straightforward computations. The FF case is not fully treated in [W$_1$] (nor in [W$_2$] except when t is an integral multiple of $\log q$), but this case is easier.

In this Section 1.3, we shall give a brief account of some basic materials for, and a sketch of, the proof of (1.2.1) valid in both cases, which hopefully is enough for the readers to convince themselves of the validity also in the FF case, and to see why the term $\phi(q, x)$ should appear. The formula (1.3.11) obtained in this process will anyway be needed later. The advanced readers can skip this section.

The explicit formula itself, and its connection with γ_K, both follow from the partial fraction decomposition of the logarithmic derivative of $\zeta_K(s)$. Put

$$Z_K(s) = -\frac{\zeta_K'(s)}{\zeta_K(s)}. \tag{1.3.1}$$

Then from the Euler product expansion

$$\prod_P (1 - N(P)^{-s})^{-1} \quad (\mathrm{Re}(s) > 1) \tag{1.3.2}$$

of $\zeta_K(s)$ follows the Dirichlet series expansion

$$Z_K(s) = \sum_{P, k \geq 1} \frac{\log N(P)}{N(P)^{ks}} \quad (\mathrm{Re}(s) > 1) \tag{1.3.3}$$

for $Z_K(s)$. In terms of $Z_K(s)$, the Euler–Kronecker constant γ_K has the expression

$$\gamma_K = -\lim_{s \to 1} \left(Z_K(s) - \frac{1}{s-1} \right). \tag{1.3.4}$$

This $Z_K(s)$ has the following partial fraction expansion ("Stark's lemma"):

$$Z_K(s) = \frac{1}{s} + \frac{1}{s-1} - \sum_\rho \frac{1}{s-\rho} + \alpha_K + \beta_K + \xi_K(s), \tag{1.3.5}$$

with

$$
\begin{aligned}
\xi_K(s) &= \frac{r_1}{2}\left(g\left(\frac{s}{2}\right) - g\left(\frac{1}{2}\right)\right) + r_2(g(s) - g(1)) \\
&= -r_1\left(\frac{1-s}{s} + \sum_{n=1}^{\infty}\left(\frac{1}{s+2n} - \frac{1}{1+2n}\right)\right) \\
&\quad - r_2\left(\frac{1-s}{s} + \sum_{n=1}^{\infty}\left(\frac{1}{s+n} - \frac{1}{1+n}\right)\right) \qquad \text{(NF)} \\
&= \sum_{\theta \neq 0,1} \frac{1}{s-\theta} \qquad \text{(FF)},
\end{aligned}
\tag{1.3.6}
$$

where ρ runs over the nontrivial zeros of $\zeta_K(s)$, θ runs over all poles $\neq 0, 1$ of $\zeta_K(s)$ (FF case),

$$
\sum_{\rho} = \lim_{T\to\infty} \sum_{|\rho|<T}, \qquad \sum_{\theta} = \lim_{T\to\infty} \sum_{|\theta|<T}, \tag{1.3.7}
$$

and

$$
g(s) = \frac{\Gamma'(s)}{\Gamma(s)}. \tag{1.3.8}
$$

(Note that $g(1) = -\gamma_{\mathbb{Q}}$, $g(\frac{1}{2}) = -\gamma_{\mathbb{Q}} - \log 4$.)

In the NF case, (1.3.5) is Stark's lemma [St, (9)] itself. The FF case follows directly from the rational expression

$$
\zeta_K(s) = \prod_{\alpha \in A} (1 - \alpha q^{-s})^{\lambda_\alpha} \quad (A : \text{a finite subset of } \mathbb{C}^\times,\ \lambda_\alpha = \pm 1) \tag{1.3.9}
$$

of $\zeta_K(s)$;

$$
\begin{aligned}
Z_K(s) &= \frac{-d\log\zeta_K(s)}{ds} = \sum_{\alpha\in A} \lambda_\alpha \frac{\log q}{1-\alpha^{-1}q^s} \\
&= \sum_{\alpha\in A} \lambda_\alpha \left(\frac{\log q}{2} - \sum_{q^\beta=\alpha} \frac{1}{s-\beta}\right) \qquad \text{(by (1.2.11))} \\
&= (g-1)\log q - \sum_{\rho} \frac{1}{s-\rho} + \left(\frac{1}{s} + \frac{1}{s-1} + \sum_{\substack{\theta\neq 0,1 \\ \text{poles}}} \frac{1}{s-\theta}\right).
\end{aligned}
\tag{1.3.10}
$$

Now by combining (1.3.4) with (1.3.5), we easily obtain

$$
\begin{aligned}
\gamma_K &= \sum_{\rho} \frac{1}{\rho} - \alpha_K - \beta_K - c_K \\
&= \frac{1}{2}\sum_{\rho} \frac{1}{\rho(1-\rho)} - \alpha_K - \beta_K - c_K,
\end{aligned}
\tag{1.3.11}
$$

where α_K, β_K are as defined by (1.2.2), (1.2.3), respectively, and

$$\begin{aligned} c_K &= 1 && \text{(NF)}, \\ &= c_q = \frac{q+1}{2(q-1)} \log q && \text{(FF)}. \end{aligned} \tag{1.3.12}$$

The last formula for c_K in the FF case follows directly from (1.2.11) for $s = \alpha = 1$, because

$$\xi_K(1) + 1 = \sum_{\substack{q^\theta = 1, q \\ \theta \neq 1}} (1-\theta)^{-1} = \sum_{q^\theta = 1} (1-\theta)^{-1}.$$

Remark. If we define c_q for each $q \in \mathbb{R}$, $q > 1$ by (1.3.12), then $c_q > 1$ and $\lim_{q \to 1} c_q = 1$. This matches with the well-known belief that "the constant field of a number field should be $\mathbb{F}_1$."

The explicit formula (1.2.1) follows from the evaluation of the integral

$$\Phi_K^{(\mu)}(x) = \frac{1}{2\pi i} \int_{c-i\infty}^{c+i\infty} \frac{x^{s-\mu}}{s-\mu} Z_K(s) ds \quad (c \gg 0) \tag{1.3.13}$$

for $\mu = 0$ and 1 in two ways, based on the classical formula

$$\frac{1}{2\pi i} \int_{c-i\infty}^{c+i\infty} \frac{y^s}{s} ds = \begin{cases} 0, & 0 < y < 1, \\ \frac{1}{2}, & y = 1, \\ 1, & 1 < y. \end{cases}$$

The Dirichlet series expansion (1.3.3) of $Z_K(s)$ gives the connection

$$x\Phi_K^{(1)}(x) - \Phi_K^{(0)}(x) = (x-1)\Phi_K(x), \tag{1.3.14}$$

while the partial fraction decomposition (1.3.5) of $Z_K(s)$ gives, via standard residue calculations,

$$x\Phi_K^{(1)}(x) - \Phi_K^{(0)}(x) = (x-1)\{\log x + (\alpha_K + \beta_K) + \ell_K(x) + r_K(x)\}. \tag{1.3.15}$$

The terms $\log x$, $\ell_K(x)$, and $r_K(x)$ inside { } correspond to

$$\frac{1}{s} + \frac{1}{s-1}, \quad \xi_K(s), \quad \text{and} \quad -\sum_\rho \frac{1}{s-\rho}$$

in (1.3.5), respectively.

Remarks.

(i) A word about the constant β_K (NF case). If

$$\Gamma_{\mathbb{R}}(s) = \pi^{-\frac{s}{2}} \Gamma\left(\frac{s}{2}\right), \quad \text{respectively,} \quad \Gamma_{\mathbb{C}}(s) = (2\pi)^{-s}\Gamma(s) \tag{1.3.16}$$

denote the standard Γ-factors at the real (respectively, imaginary) places, so that

$$\Lambda_K(s) = \Gamma_{\mathbb{R}}(s)^{r_1}\Gamma_{\mathbb{C}}(s)^{r_2}\zeta_K(s) \tag{1.3.17}$$

satisfies the functional equation

$$\Lambda_K(s) = (\sqrt{|d_K|})^{1-2s}\Lambda_K(1-s), \tag{1.3.18}$$

then one has

$$\left(\frac{d\log\Gamma_{\mathbb{R}}(s)}{ds}\right)_{s=1} = -\frac{1}{2}(\gamma_{\mathbb{Q}} + \log 4\pi), \tag{1.3.19}$$

$$\left(\frac{d\log\Gamma_{\mathbb{C}}(s)}{ds}\right)_{s=1} = -(\gamma_{\mathbb{Q}} + \log 2\pi). \tag{1.3.20}$$

Therefore,

$$\lim_{s\to 1}\left(\left(\frac{d\log\Lambda_K(s)}{ds}\right) + \frac{1}{s-1}\right) = \gamma_K + \beta_K. \tag{1.3.21}$$

Thus β_K is the "archimedean counterpart" of γ_K.

(ii) Incidentally, the functional equation in the function field case for $\Lambda_K(s) = \zeta_K(s)$ is

$$\Lambda_K(s) = (q^{g-1})^{1-2s}\Lambda_K(1-s), \tag{1.3.22}$$

and the comparison of (1.3.18) and (1.3.22) leads to our common recognition that the FF analogue of $\frac{1}{2}\log|d|$ should be $(g-1)\log q$. In both cases, the constant term in the partial fraction decomposition of $Z_K(s)$ is determined from the functional equation.

1.4 Some elementary formulas related to γ_K

We shall give a few more remarks related to the quantity

$$\gamma_K = \sum_\rho \frac{1}{\rho} - \alpha_K - \beta_K - c_K. \tag{1.4.1}$$

When α_K is large, each of $\sum_\rho \rho^{-1}$ and $\alpha_K + \beta_K$ is *usually* much larger than the absolute value of γ_K. (Only for some special families of K do they have the same order of magnitude; see Section 1.6.) So, γ_K is a *finer* object for study than $\sum_\rho \rho^{-1}$.

In the FF case, in terms of the reciprocal roots $\pi_\nu, \bar{\pi}_\nu$ $(1 \le \nu \le g)$ of $\zeta_K(s)$ in $u = q^{-s}$, we have (as is obvious by (1.2.12))

$$\sum_\rho \frac{1}{\rho} = (q-1)\sum_{\nu=1}^{g}\frac{1}{(\pi_\nu - 1)(\bar{\pi}_\nu - 1)}\log q; \tag{1.4.2}$$

hence

$$\begin{aligned}\gamma_K &= \left\{(q-1)\sum_{\nu=1}^{g}\frac{1}{(\pi_\nu-1)(\bar{\pi}_\nu-1)}-(g-1)\right\}\log q-c_q\\ &=\sum_{\nu=1}^{g}\left(\frac{1}{\pi_\nu-1}+\frac{1}{\bar{\pi}_\nu-1}\right)\log q+(\log q-c_q)\\ &=\sum_{m=1}^{\infty}\sum_{\nu=1}^{g}(\pi_\nu^m+\bar{\pi}_\nu^m)q^{-m}\log q+(\log q-c_q)\\ &=\sum_{m=1}^{\infty}(q^m+1-N_m)q^{-m}\log q+(\log q-c_q).\end{aligned}\tag{1.4.3}$$

Consider the arithmetic, geometric, and harmonic means of g positive real numbers

$$(\pi_\nu-1)(\bar{\pi}_\nu-1)\quad(1\le\nu\le g).\tag{1.4.4}$$

Then if X denotes the proper smooth curve over $\mathbb{F}_q$ corresponding to K, and J its Jacobian, the above three means of (1.4.4) are given, respectively, by

$$\begin{aligned}a.m. &= \frac{1}{g}\#X(\mathbb{F}_q)+\left(1-\frac{1}{g}\right)(q+1)\\ &\quad\vee\!\mid\\ g.m. &= (\#J(\mathbb{F}_q))^{1/g}\\ &\quad\vee\!\mid\\ h.m. &= \frac{g(q-1)\log q}{\gamma_K+\alpha_K+c_q}.\end{aligned}\tag{1.4.5}$$

The three properties $\#X(\mathbb{F}_q)$ large, $\#J(\mathbb{F}_q)$ large, and $-\gamma_K$ large, are different but correlated, and are in a sense in the same direction. (The denominator of $h.m.$ is $\sum_\rho \rho^{-1}>0$.)

By the Riemann hypothesis for curves, we have

$$(\sqrt{q}-1)^2\le(\pi_\nu-1)(\bar{\pi}_\nu-1)\le(\sqrt{q}+1)^2;$$

hence by (1.4.3), we obtain immediately

$$\left(\frac{-2g}{\sqrt{q}+1}+1\right)\log q\le\gamma_K+c_q\le\left(\frac{2g}{\sqrt{q}-1}+1\right)\log q.\tag{1.4.6}$$

Later, we shall obtain much better bounds (Section 1.6). In particular, when g is fixed and $q\to\infty$ (e.g., the constant field extensions), we have the limit formula

$$\lim_{q\to\infty}\frac{\gamma_K}{\log q}=\frac{1}{2}.\tag{1.4.7}$$

When $g=0$, γ_K is given by the equality

$$\gamma_K+c_q=\log q.\tag{1.4.8}$$

(See Remark (i) below.)

Remarks.

(i) We defined γ_K as a natural generalization of the Euler–Mascheroni constant $\gamma_{\mathbb{Q}}$. But, in a sense, the quantity $\gamma_K + c_K$ may be more canonical, as some of the preceding formulas indicate! The quantities obtained by (further) adding β_K (NF) or $-\log q$ (FF) can sometimes be better.

(ii) In the FF case, if we use other poles of $\zeta_K(s)$, instead of $s = 1$, to define γ_K similarly, then what we obtain is still γ_K if the pole is congruent to 1 modulo $2\pi i/\log q$, and is $-\alpha_K - \gamma_K$ if the pole is $0 \mod (2\pi i/\log q)$.

In the NF case, the order of zero at $s = 0$ of $\zeta_K(s)$ is $r_1 + r_2 - 1$, and

$$\gamma_K = -\gamma_K' - \log|d_K| + [K : \mathbb{Q}](\gamma_{\mathbb{Q}} + \log(2\pi)), \tag{1.4.9}$$

where γ_K' is the coefficient of $s^{r_1+r_2}$ divided by that of $s^{r_1+r_2-1}$ in the Taylor expansion of $\zeta_K(s)$ at $s = 0$.

1.5 Estimations of $r_K(x)$ and $\ell_K(x)$

Now we return to the explicit formula (1.2.1). By (1.3.11), one may rewrite it as

$$\begin{aligned}\log x - \Phi_K(x) &= -(\alpha_K + \beta_K) - r_K(x) - \ell_K(x)\\ &= \gamma_K + c_K - (r_K(x) + \sum_\rho \rho^{-1}) - \ell_K(x)\\ &= \gamma_K + c_K - \left(r_K(x) + \frac{1}{2}\sum_\rho \frac{1}{\rho(1-\rho)}\right) - \ell_K(x).\end{aligned} \tag{1.5.1}$$

We are going to estimate the nonconstant terms on the right side of (1.5.1). In most of what follows, we shall assume GRH (which is satisfied in the FF case).

Main Lemma (FF; and NF under GRH). *For any $x > 1$, we have*

$$\frac{\sqrt{x}-1}{\sqrt{x}+1}\left(\frac{1}{2}\sum_\rho \frac{1}{\rho(1-\rho)}\right) \le -r_K(x) \le \frac{\sqrt{x}+1}{\sqrt{x}-1}\left(\frac{1}{2}\sum_\rho \frac{1}{\rho(1-\rho)}\right). \tag{1.5.2}$$

Proof. Since

$$-r_K(x) = \frac{1}{2}\sum_\rho \left\{\frac{(x^\rho - 1)(x^{1-\rho} - 1)}{(x-1)} \cdot \frac{1}{\rho(1-\rho)}\right\} \tag{1.5.3}$$

and $\rho = \frac{1}{2} + i\gamma$ $(\gamma \in \mathbb{R})$, it follows that $\rho(1-\rho) = \frac{1}{4} + \gamma^2 > 0$, and that

$$(x^\rho - 1)(x^{1-\rho} - 1) = x + 1 - 2\sqrt{x}\cos(\gamma \log x) \tag{1.5.4}$$

lies in between $(\sqrt{x}-1)^2$ and $(\sqrt{x}+1)^2$, whence our inequalities. □

The graphs of the three functions of x appearing in (1.5.2) in the Main Lemma, for the cases $K = \mathbb{Q}, \mathbb{Q}(\sqrt{481})$, are as shown in Figures 1(a)–1(b). (As (1.5.3) indicates, each ρ with small $|\gamma|$ contributes to a "high wave calm on the surface," whereas a larger $|\gamma|$ to a lower "ripple." The effect of the first few ρ is not particularly large, but sometimes determines the main shape of the graph (for x not too large). Thus these graphs seem to indicate that the smallest $|\gamma|$ for $\mathbb{Q}(\sqrt{481})$ would be much smaller than that of $K = \mathbb{Q}$ (i.e., 14.1347 . . .).)

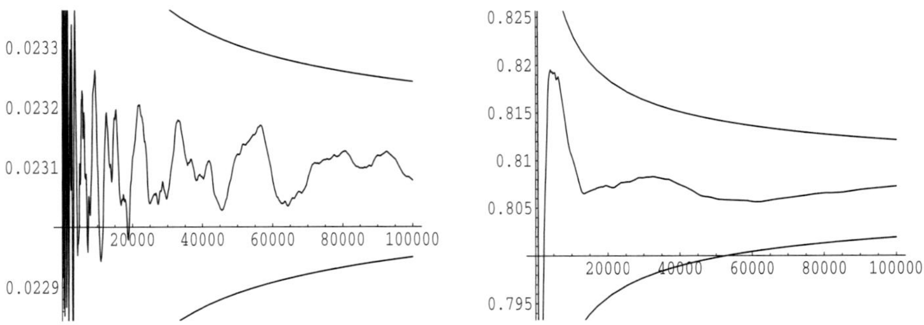

(a) $K = \mathbb{Q}$ (b) $K = \mathbb{Q}(\sqrt{481})$

Fig. 1.

By this lemma and (1.3.11), we obtain

$$\begin{aligned} \frac{-2}{\sqrt{x}+1}(\gamma_K + \alpha_K + \beta_K + c_K) &\leq -r_K(x) - \frac{1}{2}\sum_{\rho} \frac{1}{\rho(1-\rho)} \\ &\leq \frac{2}{\sqrt{x}-1}(\gamma_K + \alpha_K + \beta_K + c_K), \end{aligned} \tag{1.5.5}$$

and hence by (1.5.1),

$$\begin{aligned} &\frac{\sqrt{x}-1}{\sqrt{x}+1}(\gamma_K + c_K) - \frac{2}{\sqrt{x}+1}(\alpha_K + \beta_K) - \ell_K(x) \\ &\quad \leq \log x - \Phi_K(x) \\ &\quad \leq \frac{\sqrt{x}+1}{\sqrt{x}-1}(\gamma_K + c_K) + \frac{2}{\sqrt{x}-1}(\alpha_K + \beta_K) - \ell_K(x) \end{aligned} \tag{1.5.6}$$

(under GRH).

As for $\ell_K(x)$, we have the following.

Lemma 1.

(i) (NF case): *$\ell_K(x)$ is monotone decreasing,*

$$\lim_{x\to 1} \ell_K(x) = +\infty, \qquad \lim_{x\to\infty} \ell_K(x) = 0,$$

and

$$0 < \ell_K(x) < [K:\mathbb{Q}]\left(\frac{\log x + 1}{x-1}\right) \quad (x > 1).$$

(ii) (FF case): $\ell_K(x) = 0$ if and only if $x = q^m$ $(m \in \mathbb{N})$; *for other* x, $\ell_K(x)$ *always remains within the open interval* $(0, \log q)$ *but does* not *tend to* 0 *as* $x \to \infty$.

Proof.

(i) *NF case.* In this case,

$$\begin{cases} \ell_K(x) = \dfrac{r_1}{2}F_1(x) + r_2F_2(x) \quad \text{with} \\ F_1(x) = \log\dfrac{x+1}{x-1} + \dfrac{2}{x-1}\log\dfrac{x+1}{2}, \\ F_2(x) = \log\dfrac{x}{x-1} + \dfrac{1}{x-1}\log x. \end{cases} \tag{1.5.7}$$

First, since $F_1'(x) = -2(x-1)^{-2}\log((x+1)/2) < 0$, $F_1(x)$ is monotone decreasing. Second, since $F_2(x) = F_1(2x-1)$, $F_2(x)$ is also monotone decreasing and $F_2(x) < F_1(x)$. Third, since $\log(\frac{x+1}{x-1}) < 2(x-1)^{-1}$ and $\log(\frac{x+1}{2}) < \log x$, we obtain

$$F_1(x) < 2(\log x + 1)(x-1)^{-1}, \tag{1.5.8}$$

and it is clear that $F_2(x) > 0$. The desired inequalities follow immediately from these. The assertions for the limits at $x \to 1, \infty$ of $\ell_K(x)$ are also obvious.

The following inequality will be used later (Section 2.4):

$$\frac{1}{2}(x-1)F_1(x) = \log(x+1) + \log\left[\frac{1}{2}\left(1+\frac{2}{x-1}\right)^{\frac{x-1}{2}}\right] \geq \log(x+1) \quad (x \geq 3). \tag{1.5.9}$$

(ii) *FF case.* We already know that $\phi(q,x) = 0$ if $x = q^m$ $(m \in \mathbb{N})$. So, put $x = q^{m-1+y}$, with $m \geq 1$, $0 < y < 1$. Then by (1.2.15),

$$\begin{aligned} \phi(q,x) &= \left(1 - y - \frac{(q^{m-1}-1)(q-q^y)}{(q^{m-1+y}-1)(q-1)}\right)\log q \\ &= \left(\frac{(q^m-1)(q^y-1)}{(q^{m-1+y}-1)(q-1)} - y\right)\log q. \end{aligned} \tag{1.5.10}$$

It is easy to see that if we fix y, then this is monotone decreasing as a function of m, and tends uniformly to

$$s_q(y) = \left(\frac{1-q^{-y}}{1-q^{-1}} - y\right)\log q \quad (> 0) \tag{1.5.11}$$

as $m \to \infty$. Therefore,

$$0 < \frac{1-q^{-y}}{1-q^{-1}} - y < \frac{\phi(q,x)}{\log q} \leq 1 - y < 1, \tag{1.5.12}$$

which proves all the assertions stated in Lemma 1(ii). □

Remark. Note that $s_q(0) = s_q(1) = 0$, $s_q(y) > 0$ for $0 < y < 1$. The maximal value of $s_q(y)$ for $0 < y < 1$ is

$$\frac{\log q}{1 - q^{-1}} - (\log\log q - \log(1 - q^{-1}) + 1), \tag{1.5.13}$$

which is attained at

$$y = \frac{\log\log q - \log(1 - q^{-1})}{\log q}. \tag{1.5.14}$$

The graphs of $F_1(x) > F_2(x)$ will be shown in Figure 2(a), and that of $\phi(q, q^z)$ for $q = 5$, in Figure 2(b). The horizontal line in the latter gives the value of (1.5.13) for $q = 5$.

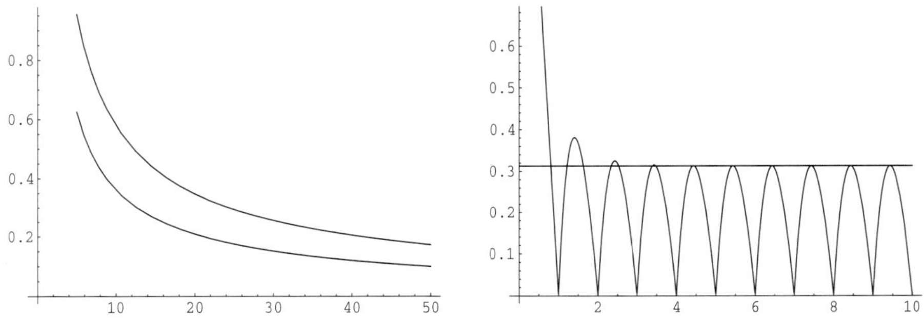

(a) (b)

Fig. 2.

1.6 Estimations related to γ_K

From (1.5.6) we immediately obtain the following.

Proposition 2 (under GRH in the NF case). *For any* $x > 1$, *we have*

$$\gamma_K \leq \frac{\sqrt{x}+1}{\sqrt{x}-1}(\log x - \Phi_K(x) + \ell_K(x)) + \frac{2}{\sqrt{x}-1}(\alpha_K + \beta_K) - c_K, \tag{i}$$

$$\gamma_K \geq \frac{\sqrt{x}-1}{\sqrt{x}+1}(\log x - \Phi_K(x) + \ell_K(x)) - \frac{2}{\sqrt{x}+1}(\alpha_K + \beta_K) - c_K. \tag{ii}$$

Since by (1.5.6) the difference between the upper and the lower bounds tends to 0 as $x \to \infty$, this gives a method for *computing the constant* γ_K (under GRH) to as much accuracy as one desires. Although the convergence is slow, one can usually determine the approximate size of γ_K (e.g., its sign) even by hand calculations.

Figures 3(a)–3(b) show two examples for the graphs of the upper and the lower bounds given by Proposition 2 (denoted, respectively, as $\mathrm{upp}_K(x)$, $\mathrm{low}_K(x)$). The horizontal lines indicate the expected values of γ_K.

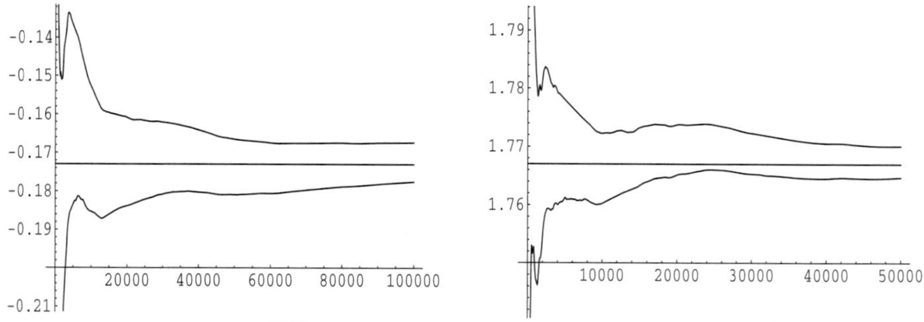

(a) $K = \mathbb{Q}(\sqrt{481})$ (b) $K = \mathbb{Q}(\cos \frac{2\pi}{9})$

Fig. 3.

Examples (by computer)

Let $K = \mathbb{Q}(\sqrt{-1})$, and take $x = 50000$. Then the upper and the lower bounds for γ_K given by Proposition 2(i)–(ii) are 0.8239498, 0.8221413, respectively. The value of γ_K computed by using the Kronecker formula (cf. Section 2.2) is 0.82282525. Incidentally, in this case, the value of $\log x - \Phi_K(x) - 1$ is 0.82280515, which is close to the actual value, and lies in between the above upper and lower bounds. But in general, $\log x - \Phi_K(x) - 1$ need not lie in between the two bounds of Proposition 2 (see Remark (ii) below).

For other imaginary quadratic fields, $0 < \gamma_K < 1$ holds for $|d_K| \leq 43$, but $\gamma_K < 0$ for $d_K = -47, -56, \dots$. For example,

$$-0.072 < \gamma_{\mathbb{Q}(\sqrt{-47})} < -0.053.$$

For real quadratic fields, $0 < \gamma_K < 2$ for $d_K < 100$, but

$$-0.181 < \gamma_{\mathbb{Q}(\sqrt{481})} < -0.167.$$

These are, of course, under (GRH).

Some other examples will be given in Sections 1.7 and 2.3.

We shall give some applications. First, by letting $x \to \infty$ in (1.5.6), we obtain the following.

Corollary 1.

$$\gamma_K = \lim_{x \to \infty} (\log x - \Phi_K(x) - 1) \qquad (NF), \tag{1.6.1}$$

$$\gamma_K = \lim_{\substack{m \to \infty \\ m \in \mathbb{N}}} (\log(q^m) - \Phi_K(q^m) - c_q) \quad (FF). \tag{1.6.2}$$

A formula equivalent to (1.6.1) can also be found in a recent preprint [HIKW, Theorem B].

Remarks.

(i) For (1.6.1), GRH is unnecessary. In fact, by using a standard zero-free region for $\zeta_K(s)$, one can show, unconditionally, that

$$\lim_{x\to\infty}\left\{r_K(x)+\frac{1}{2}\sum_{\rho}\frac{1}{\rho(1-\rho)}\right\}=0, \tag{1.6.3}$$

from which (1.6.1) follows directly by (1.5.1) and Lemma 1. The proof of (1.6.3) runs as follows. As is well known (cf., e.g., [L-O, Lemma 8.1]), there exists a positive constant c (depending on K) such that if $\rho=\beta+i\gamma$ is a nontrivial zero of $\zeta_K(s)$ with $|\gamma|$ sufficiently large, then

$$\beta<1-c(\log|\gamma|)^{-1}. \tag{1.6.4}$$

We claim that

$$\lim_{x\to\infty}\left(\sum_{\rho}\frac{x^{\beta-1}}{\gamma^2}\right)=0, \tag{1.6.5}$$

where ρ runs over all imaginary zeros of $\zeta_K(s)$. To show this, since $\beta<1$, we may exclude finitely many ρs and assume that (1.6.4) is satisfied. Then for $x>1$,

$$\sum_{\rho}\frac{x^{\beta-1}}{\gamma^2}<\sum_{\rho}\frac{x^{-c(\log|\gamma|)^{-1}}}{\gamma^2}=\sum_{\log|\gamma|<T}+\sum_{\log|\gamma|\geq T},$$

where we choose $T=\sqrt{\log x}$. Then

$$\sum_{\log|\gamma|<T}\leq\left(\sum_{\log|\gamma|<T}\frac{1}{\gamma^2}\right)x^{-cT^{-1}}\leq\left(\sum_{\rho}\frac{1}{\gamma^2}\right)\exp(-c\sqrt{\log x})\to 0,$$

and

$$\sum_{\log|\gamma|\geq T}\leq\sum_{\log|\gamma|\geq\sqrt{\log x}}\frac{1}{\gamma^2}\to 0,$$

whence (1.6.5). But since

$$r_K(x)+\frac{1}{2}\sum_{\rho}\frac{1}{\rho(1-\rho)}=\frac{1}{2}\sum_{\rho}\frac{1}{\rho(1-\rho)}\left(\frac{x^{\rho}+x^{1-\rho}-2}{x-1}\right), \tag{1.6.6}$$

and each term tends to 0 as $x\to\infty$, by (1.6.5), we obtain

$$\lim_{x\to\infty}\left|r_K(x)+\frac{1}{2}\sum_{\rho}\frac{1}{\rho(1-\rho)}\right|\leq\lim_{x\to\infty}\sum_{\substack{\rho,\\ \gamma\neq 0}}\frac{1}{\gamma^2}\frac{x^{\beta}}{x-1}=0.$$ □

(ii) Some readers may be interested in the comparison between $\mathrm{A}_K(x) := \log x - \Phi_K(x) - 1$ and the two bounds $\mathrm{upp}_K(x)$, $\mathrm{low}_K(x)$ of Proposition 2. From (1.2.1) we obtain easily

$$\frac{\sqrt{x}-1}{2}(\mathrm{upp}_K(x) - \mathrm{A}_K(x)) = -r_K(x) + \frac{\sqrt{x}-1}{2}\ell_K(x),$$
$$\frac{\sqrt{x}+1}{2}(\mathrm{A}_K(x) - \mathrm{low}_K(x)) = -r_K(x) - \frac{\sqrt{x}+1}{2}\ell_K(x),$$

and, moreover, we have $-r_K(x) > 0$ under GRH. Therefore, $\mathrm{A}_K(x) < \mathrm{upp}_K(x)$ always holds under GRH. But $\mathrm{A}_K(x) > \mathrm{low}_K(x)$ need not hold in general, a counterexample being given by $K = \mathbb{Q}(\sqrt{-1})$, $x =$ (say) 1800.

Examples of graphs of $\mathbf{A}_K(x) = \log x - \Phi_K(x) - 1$

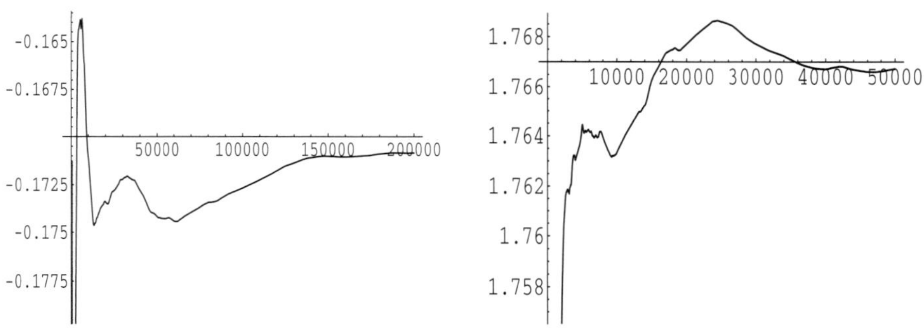

(a) $K = \mathbb{Q}(\sqrt{481})$ (b) $K = \mathbb{Q}(\cos \frac{2\pi}{9})$

Fig. 4.

Upper bounds

The second application of Proposition 2 is to the problem of finding a reasonably good *general* upper bound for γ_K in terms of more elementary invariants of K. It can be obtained from Proposition 2(i) by the substitution of a *suitable* value of x. Since we do not know a priori the local behavior of $\Phi_K(x)$, except that $\Phi_K(x) \geq 0$, what we do is try to minimize

$$\frac{\sqrt{x}+1}{\sqrt{x}-1}(\log x + \ell_K(x)) + \frac{2}{\sqrt{x}-1}(\alpha_K + \beta_K) - c_K. \tag{1.6.7}$$

We leave the discussion of this delicate question until a little later (after the proof of Theorem 1), and first see what we can obtain by choosing the value x_0 of x which minimizes

$$\log x + \frac{2\alpha_K}{\sqrt{x}},$$

i.e., $x_0 = \alpha_K^2$. We then obtain the following.

Theorem 1 (under GRH in the NF case).

$$\begin{aligned}\gamma_K &< \left(\frac{\alpha_K + 1}{\alpha_K - 1}\right)(2\log\alpha_K + a - \Phi_K(\alpha_K^2)) \\ &\le \left(\frac{\alpha_K + 1}{\alpha_K - 1}\right)(2\log\alpha_K + a),\end{aligned} \tag{1.6.8}$$

provided that

$$\begin{aligned} &g > 2, \quad or \quad g = 2 \quad and \quad q > 2 \qquad (FF) \\ &n > 2, \quad or \quad n = 2 \quad and \quad |d_K| > 8 \qquad (NF). \end{aligned} \tag{1.6.9}$$

Here $a = 1$ *and* $n = [K : \mathbb{Q}]$ *(NF),* $a = 1 + \log q$ *(FF).*

Proof. The right-hand side of Proposition 2(i) for $x = \alpha_K^2$ can be rewritten as

$$\begin{cases} \dfrac{\alpha_K + 1}{\alpha_K - 1}\Big(2\log\alpha_K - \Phi_K(\alpha_K^2) + 1\Big) + \dfrac{1}{\alpha_K - 1}\Big((\alpha_K + 1)\ell_K(\alpha_K^2) + 2\beta_K\Big) & \text{(NF)}, \\ \dfrac{\alpha_K + 1}{\alpha_K - 1}\Big(2\log\alpha_K - \Phi_K(\alpha_K^2) + 1 + \phi(q, \alpha_K^2)\Big) + (1 - c_q) & \text{(FF)}. \end{cases} \tag{1.6.10}$$

In the FF case, (1.6.9) implies $\alpha_K > 1$. And since $\phi(q, \alpha_K^2) < \log q$ and $1 - c_q < 0$, we are done. In the NF case, we have the inequalities

$$\begin{cases} f_1(x) := (x+1)\left\{\log\left(\dfrac{x^2+1}{x^2-1}\right) + \dfrac{2}{x^2-1}\log\left(\dfrac{x^2+1}{2}\right)\right\} - 2(\gamma_{\mathbb{Q}} + \log 4\pi) < 0 \\ \qquad \text{for } x > 1.16, \\ f_2(x) := (x+1)\left\{\log\left(\dfrac{x^2}{x^2-1}\right) + \dfrac{1}{x^2-1}\log(x^2)\right\} - 2(\gamma_{\mathbb{Q}} + \log 2\pi) < 0 \\ \qquad \text{for } x > 1.16. \end{cases} \tag{1.6.11}$$

They hold because $f_1(x)$, $f_2(x)$ are both monotone decreasing for $x > 1$, and their values at 1.16 are both negative (being $-0.08762\ldots$, $-0.03882\ldots$, respectively). Therefore,

$$(\alpha_K + 1)\ell_K(\alpha_K^2) + 2\beta_K < 0 \quad \text{for } \alpha_K > 1.16. \tag{1.6.12}$$

But even the Minkowski lower bound for $|d_K|$ shows that $\alpha_K > 1.16$ holds for $n > 2$ and for $n = 2$ with $|d_K| > 10.2$ (which is actually the same as $|d_K| > 8$). □

Discussions on minimizing (1.6.7)

Write $x = t^2$ $(t > 1)$, and put

$$s = s_K = \alpha_K + \beta_K. \tag{1.6.13}$$

Then (1.6.7) can be expressed as

$$g(t) = g_1(t) + g_2(t), \tag{1.6.14}$$

with

$$g_1(t) := \frac{2}{t-1}((t+1)\log t + s) - c_K, \tag{1.6.15}$$

$$\begin{aligned} g_2(t) &:= \frac{t+1}{t-1}\left\{\frac{r_1}{2}F_1(t^2) + r_2F_2(t^2)\right\} \quad (>0) \quad \text{(NF)} \\ &= \frac{t+1}{t-1}\phi(q,t^2) \quad (\geq 0) \quad \text{(FF)} \end{aligned} \tag{1.6.16}$$

(cf. (1.5.7)). These are continuous functions of $t > 1$ parametrized by $s(\in \mathbb{R})$; $r_1, r_2 \geq 0$, $r_1 + r_2 > 0$ (NF), or q (FF). We shall exclude the trivial case of genus 0 (FF). Then $g(t)$ always achieves its minimal value at some θ $(1 < \theta < \infty)$, because it is continuous and tends to $+\infty$ at both ends, i.e., $t \to 1$ and $t \to \infty$.

NF case

In this case, θ is unique, as $(t-1)^2 g'(t)$ is monotone increasing. Indeed,

$$(t-1)^2 g_1'(t) = 2(t - 2\log t - t^{-1} - s) \tag{1.6.17}$$

is monotone increasing, and so are the r_1 and r_2 components of $(t-1)^2 g_2'(t)$. When $s > 1$, θ is close to

$$s + 2\left(1 + \frac{r_1 + r_2}{s}\right)\log s. \tag{1.6.18}$$

We have included the $(r_1 + r_2)s^{-1}$ term, because when $s_K \to \infty$, the quantity $(r_1 + r_2)s_K^{-1}$, though bounded (by a standard unconditional lower bound for $|d_K|$), does not tend to 0.

FF case

The local differential structure of $g(t)$ is, in a sense, opposite to the NF case. The graph of $g(t)$ looks like a bouncing ball, bouncing at each integral power of $\sqrt{q}$, first coming down a slope and then going up another forever. (The slopes correspond to the graph of $g_1(t)$.) Indeed, $(t-1)^2 g'(t)$ is a negative constant $-2g\log q$ (g: the genus) on $1 < t^2 < q$, and is monotone decreasing on every open interval $q^{m-1} < t^2 < q^m$ $(m > 1)$. The derivative of $(t-1)^2 g'(t)$ on this interval is $-2a(t-1)^{-2}$, where

$$a = (q^{m-1} - 1)(q^m - 1)(q^m - q^{m-1})^{-1}\log q.$$

Therefore, $g''(t) < 0$ wherever $g'(t) = 0$. Therefore, $g(t)$ can acquire its minimal value only at the bouncing points $t \in (\sqrt{q})^{\mathbb{Z}}$. Note that $\phi(q, t^2) = 0$ at bouncing points. Moreover, by (1.6.17), we conclude that θ must be one of the (at most two) integral powers of $\sqrt{q}$ which are adjacent to the unique root of the equation

$$t - 2\log t - t^{-1} = s. \tag{1.6.19}$$

Thus, again, θ is close to

$$s + 2\log s$$

as long as s is large compared with q. If, at the other extreme,

$$g < \frac{1}{\log q}\left(\sqrt{q} - \frac{1}{\sqrt{q}}\right), \tag{1.6.20}$$

so that the root of (1.6.19) is smaller than $\sqrt{q}$, then θ is always equal to $\sqrt{q}$, and this gives rise only to the trivial general upper bound (1.4.6) for γ_K.

Each of Figures 5(a)–5(b) gives the graphs of two functions

$$g(t) \geq g_1(t)$$

when $K = \mathbb{Q}(\sqrt{-5003})$ (Figure 5(a)), and $q = 2$, $g = 5$ for $t^2 = q^y$ (the horizontal axis is for y) (Figure 5(b)).

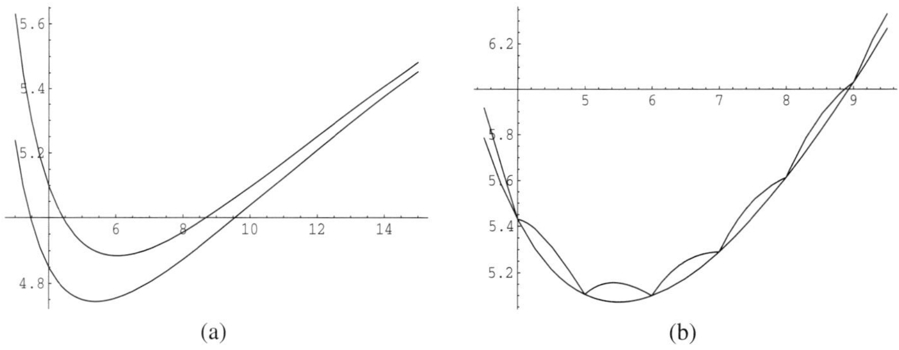

Fig. 5.

Thus the best possible approximation of θ would depend on the specific family of K and purpose of applications. For example, the β_K part of s_K in the NF case should not be neglected if $(r_1 + r_2)s_K^{-1}$ is not small. But here we are satisfied with having given a basic result expressed simply in terms of α_K only, together with some indications for possible improvements. We note that choices of other θ can improve only minor terms in Theorem 1, unless we restrict ourselves to some special families of K. Further studies of the upper bounds of γ_K for various families of K will be left to future publications.

The author has as yet no idea about the minimal possible size of $\Phi_K(\theta^2)$ for each given r_1, r_2 and given approximate range of s. This is of course related to the question of how sharp an upper bound our method can possibly give.

Lower bounds

A general, unconditional (and trivial) lower bound is as follows.

Proposition 3.

$$\gamma_K > -\alpha_K - \beta_K - c_K.$$

This follows immediately from

$$\gamma_K + \alpha_K + \beta_K + c_K = \sum_{\rho} \frac{1}{\rho} = \frac{1}{2} \sum_{\rho} \left(\frac{1}{\rho} + \frac{1}{\bar{\rho}} \right) > 0. \tag{1.6.21}$$

(Under (GRH), one can deduce this also from Proposition 2(ii) by letting $x \to 1$.)

We note that the absolute value of the negative lower bound given above is much larger than that of the upper bound given in Theorem 1. For example, in the NF case, the former is $\sim \frac{1}{2} \log |d|$ while the latter is $\sim 2 \log \log |d|$. This is not just because we have not assumed GRH for the above lower bound as we did for the upper bound. Indeed, in the FF case, γ_K can be of the order of

$$\left(-\frac{1}{\sqrt{q}+1} + \varepsilon \right) (g_K - 1) \log q \tag{1.6.22}$$

for Shimura curves over $\mathbb{F}_q$ (q a square, $g_K \to \infty$); see Section 2.1. But we can show the following.

Theorem 2 (FF). *Fix q. Then*

$$\liminf \frac{\gamma_K}{(g_K - 1) \log q} \geq -\frac{1}{\sqrt{q}+1}. \tag{1.6.23}$$

Proof. This is obtained by using Tsfasman's [Ts$_1$, Corollary 1]. (If we combine the inequality $h.m \leq a.m$ in (1.4.5) with the Drinfeld–Vlăduţ asymptotic upper bound [D-V] for N_1, then what we obtain is a somewhat weaker statement, where the denominator $\sqrt{q}+1$ on the RHS of (1.6.23) is replaced by $\sqrt{q}$.)

First, let us recall the basic materials from [Ts$_1$] that will be needed. By a curve over $\mathbb{F}_q$, we shall always mean a complete, smooth, geometrically irreducible algebraic curve over $\mathbb{F}_q$. For a curve C over $\mathbb{F}_q$, let $g = g(C)$ denote the genus, and $B_m = B_m(C)$ denote the number of prime divisors (i.e., scheme theoretic closed points) of C with degree m over $\mathbb{F}_q$. Thus

$$N_m(C) = \sum_{d|m} d B_d(C)$$

is the number of $\mathbb{F}_{q^m}$-rational points of C. If $\{C_\alpha\}$ is a family of curves over $\mathbb{F}_q$ (q fixed) with growing genus such that

$$\beta_m = \lim_{\alpha} \frac{B_m(C_\alpha)}{g(C_\alpha)}$$

exists for all $m \geq 1$, we call the family $\{C_\alpha\}$ *asymptotically exact*. Each sequence of curves over $\mathbb{F}_q$ with growing genus contains a subsequence which is asymptotically exact [Ts$_1$, p. 182]. Moreover, for any asymptotically exact family of curves over $\mathbb{F}_q$, one has

$$\sum_{m=1}^{\infty} \frac{m\beta_m}{q^{m/2}-1} \leq 1 \tag{1.6.24}$$

[Ts$_1$, Corollary 1]. Now, to prove Theorem 2, let us write $\gamma(C) = \gamma_K$, where K is the function field of C. Put

$$\lambda = \liminf \frac{\gamma(C)}{(g(C)-1)\log q} = \liminf \frac{\gamma(C)}{g(C)\log q}.$$

Then there exists a family of curves C over $\mathbb{F}_q$ with $g(C) \to \infty$ such that

$$\lim_C \frac{\gamma(C)}{g(C)\log q} = \lambda,$$

and we may assume that this family is asymptotically exact. Let C run over such a family. Then by (1.4.3),

$$\frac{\gamma(C)}{g(C)\log q} = \sum_{m=1}^{\infty} \frac{q^m+1-N_m(C)}{q^m g(C)} + \frac{1}{g(C)}\left(1-\frac{c_q}{\log q}\right). \tag{1.6.25}$$

Since the summand on the right-hand side of (1.6.25) has absolute value at most equal to $2q^{-m/2}$, the sum is *uniformly* convergent w.r.t. C. We thus obtain

$$\begin{aligned}
\lambda &= \sum_{m=1}^{\infty} \lim_C \left(\frac{q^m+1-N_m(C)}{q^m g(C)}\right) \\
&= -\sum_{m=1}^{\infty} q^{-m}\left(\sum_{d|m} d\beta_d\right) = -\sum_{d=1}^{\infty} \frac{d\beta_d}{q^d-1} \\
&\geq -\frac{1}{\sqrt{q}+1}\left(\sum_{d=1}^{\infty} \frac{d\beta_d}{q^{d/2}-1}\right) \geq -\frac{1}{\sqrt{q}+1}
\end{aligned} \tag{1.6.26}$$

by (1.6.24), as desired. □

Remark. The above proof shows also that the *equality* $\lambda = -\frac{1}{\sqrt{q}+1}$ holds if and only if $\beta_1 = \sqrt{q}-1$ holds, i.e., if and only if the Drinfeld–Vlăduţ asymptotic upper bound for $\frac{N_1(C)}{g(C)}$ is attained by this family.

In Section 2.1, we shall show that when q is a square, then the equality holds for (1.6.23) (see (2.1.11)).

It is also a very interesting problem to find out the precise value of the quantity

$$C = \liminf \frac{\gamma_K}{\alpha_K} \tag{1.6.27}$$

in the NF case. As for this, what the author obtained are the following:

(i) By Proposition 3 (Section 1.6), we have

$$C \geq -1 \quad \text{(unconditionally)}. \tag{1.6.28}$$

(ii) If there exists an *infinite* unramified Galois extension M/k over a number field k in which some prime ideals $\mathfrak{p}_1, \ldots, \mathfrak{p}_m$ of k decompose completely, then, under (GRH), we can show easily that

$$C \leq -\frac{1}{\alpha_k}\left(\sum_{i=1}^{m} \frac{\log N(\mathfrak{p}_i)}{N(\mathfrak{p}_i)-1}\right), \tag{1.6.29}$$

by applying Theorem 1 (the first inequality in (1.6.8)) to finite intermediate extensions of M/k. For example, one may choose the examples given in [T-V, Corollaries 9.3–9.5], among which ([T-V, Corollary 9.5] is an old example due to the present author, but) [T-V, Corollary 9.4] gives the best result. In this case,

$$k = \mathbb{Q}(\sqrt{d}), \quad d = -d_1 \times 73 \times 79,$$

where d_1 is the product of all prime numbers q with $13 \leq q \leq 61$. It has an infinite 2-class-field tower in which ten primes above 2, 3, 5, 7, 11 split completely. By taking $\mathfrak{p}_1, \ldots, \mathfrak{p}_m$ to be these ten primes of k, we obtain, by (1.6.29),

$$C \leq -0.16352\ldots \quad \text{(under GRH)}. \tag{1.6.30}$$

Now, after the present work was submitted to this volume, Tsfasman kindly informed me that he can prove better results, namely,

$$-0.26049\cdots \leq C \leq -0.17849\ldots \tag{1.6.31}$$

(the LHS inequality under GRH) by using [T-V] for the LHS inequality, and an *unconditional* (1.6.29) [Ts_2, Theorem 5] with a better class-field tower, for the RHS inequality (see [Ts_2]). The LHS inequality was surprising to the author who had considered it plausible that $C = -\frac{1}{2}$ (the value obtained by putting $q = 1$ on the RHS of (1.6.23)).

On the other hand,

$$\limsup \frac{\gamma_K}{\alpha_K} = 0 \quad \text{(under GRH)} \tag{1.6.32}$$

by Theorems 1 and 3 below (for, say, $N = 2$).

When the degree of K over $\mathbb{Q}$ (NF), or over a rational subfield of K (FF), is relatively small, there is a much better lower bound for γ_K, as follows.

Theorem 3. *Put $k = \mathbb{Q}$ (NF), $= \mathbb{F}_q(t)$ (FF), and let K be an extension of k of degree $N > 1$. Put*

$$\alpha_K^* = \frac{\alpha_K}{N-1} = \frac{\log\sqrt{|d|}}{N-1} \quad (NF), \qquad = \frac{(g-1)\log q}{N-1} \quad (FF), \tag{1.6.33}$$

and assume $\alpha_K^* > 1$. *Then*

$$\frac{\alpha_K^*+1}{\alpha_K^*-1}(\gamma_K + c_K) > -2(N-1)(\log\alpha_K^* + 1) \qquad \textit{(NF, under GRH)}$$

$$> -2(N-1)\Big(\log\alpha_K^* + \frac{\alpha_K^*}{\alpha_K^*-1}\Big) \qquad (FF). \tag{1.6.34}$$

Remarks.

(i) Granville–Stark [G-S, Section 3.1] gave an equivalent statement when $[K:\mathbb{Q}] = 2$ (in fact, $L'(1,\chi)/L(1,\chi) = \gamma_K - \gamma_{\mathbb{Q}}$), whose argument applies also to abelian extensions over $\mathbb{Q}$. Our theorem was motivated by [G-S].

(ii) The bound given by Theorem 3 is sharp in the following sense. The RHS of (1.6.34) cannot be replaced by its quotient by such an $f(N)$ (NF) (respectively, $f_q(N)$ (FF, for a fixed $q > 2$)) that tends to ∞ as $N \to \infty$. This can be proved easily by using a family of K satisfying (1.6.29) (NF), respectively, (2.1.8) (Section 2.1) (FF). The point is that, in each case, one can find a subsequence of K such that $\alpha_K \to \infty$ and that the following (finite) limits $\lim \alpha_K^* > 1$ and $\lim \frac{\gamma_K}{\alpha_K} < 0$ exist.

Proof. This will be based on the Main Lemma and the four inequalities

$$\Phi_K(x) \le N \cdot \Phi_k(x), \tag{1.6.35}$$

$$\Phi_k(x) < \log x, \tag{1.6.36}$$

$$\ell_K(x) \ge 0, \tag{1.6.37}$$

each for all $x > 1$, and

$$\beta_K < -[K:\mathbb{Q}]. \tag{1.6.38}$$

Among them, (1.6.35) is trivial, (1.6.36)(FF) and (1.6.37) both follow directly from Lemma 1 of Section 1.5, and (1.6.38) holds because $\gamma + \log 2\pi = 2.415\cdots > 2$. The inequality $\Phi_{\mathbb{Q}}(x) < \log x$ can be proved easily as follows. Since $\Phi_{\mathbb{Q}}(x)$, $\log x$ are both monotone increasing, it is enough to show $\Phi_{\mathbb{Q}}(x) \le \log(x-1)$ for integers $x = n \ge 2$ (note the shift $x \to x-1$ on the right side). But by the prime factorization of $n!$, we have

$$\log n! = \sum_{p^k \le n}\left[\frac{n}{p^k}\right]\log p \ge \sum_{p^k \le n}\left(\frac{n+1}{p^k} - 1\right)\log p$$
$$= n\Phi_{\mathbb{Q}}(n+1); \tag{1.6.39}$$

hence

$$\Phi_{\mathbb{Q}}(n+1) \le \frac{1}{n}\log(n!) \le \log n \tag{1.6.40}$$

for all $n \ge 1$.

Now we proceed to the proof of Theorem 3. By (1.6.35), (1.6.36), and then by (1.4.1), (1.5.2), (1.6.37), we obtain

$$
\begin{aligned}
N \log x > \Phi_K(x) &= \log x + \alpha_K + \beta_K + r_K(x) + \ell_K(x) \\
&\geq \log x + \alpha_K + \beta_K - \left(\frac{\sqrt{x}+1}{\sqrt{x}-1}\right)(\gamma_K + \alpha_K + \beta_K + c_K).
\end{aligned}
$$

Therefore,

$$
(N-1)\log x > -\frac{2}{\sqrt{x}-1}(\alpha_K + \beta_K) - \frac{\sqrt{x}+1}{\sqrt{x}-1}(\gamma_K + c_K).
$$

Putting $x = (\alpha_K^*)^2$, we obtain

$$
\frac{\alpha_K^*+1}{\alpha_K^*-1}(\gamma_K + c_K) > -2(N-1)\log \alpha_K^* - \frac{2}{\alpha_K^*-1}(\alpha_K + \beta_K).
$$

The rest follows directly by (1.6.38) in the NF case. □

Corollary 2. *If N and q (in the FF case) are fixed and $\alpha_K^* \to \infty$, then*

$$
\begin{aligned}
\gamma_K &> -2(N-1+\varepsilon)\log(\log|d|) && \text{(NF, under GRH)} \\
&> -2(N-1+\varepsilon)\log((g-1)\log q) && \text{(FF)}.
\end{aligned}
\tag{1.6.41}
$$

1.7 Supplementary remarks related to computations of γ_K

An accurate computation of γ_K *for each individual* K is not the main issue of this paper. Still, it should probably be pointed out that there are other ways of computing γ_K that are better at least microscopically.

One is the classical Landau formula generalizing the well-known Euler formula for $K = \mathbb{Q}$. Let K be a number field, let $\zeta_K(s) = \sum_{n=1}^{\infty} a_n n^{-s}$ be the Dirichlet series expansion of $\zeta_K(s)$ on $\mathrm{Re}(s) > 1$, and put $S_K(x) = \sum_{n \leq x} a_n$, so that

$$
\zeta_K(s) = s \int_1^{\infty} S_K(x) x^{-s-1} dx \quad (\mathrm{Re}(s) > 1). \tag{1.7.1}
$$

Further, put $\kappa_K = \lim_{s \to 1}(s-1)\zeta_K(s) = \lim_{x \to \infty}(S_K(x)/x)$, and

$$
B_K(x) = \kappa_K^{-1} \sum_{n \leq x} a_n n^{-1} - \log x \quad (x \geq 1). \tag{1.7.2}
$$

Then the Landau formula asserts that

$$
\gamma_K = \lim_{x \to \infty} B_K(x). \tag{1.7.3}
$$

(Compare this with (1.6.1), and note that $\log(x)$ appears with opposite signs!)

Now, there exist positive constants ϵ, C and $x_0 \geq 1$ (all depending on K) such that

$$
|S_K(x) - \kappa_K x| \leq C x^{1-\epsilon} \quad (x \geq x_0). \tag{1.7.4}
$$

So we can change the order of integration and passage to the limit $s \to 1$, on the RHS of (1.7.1) *with* $S_K(x) - \kappa_K x$ *in place of* $S_K(x)$. This gives an expression for γ_K in terms of the definite integral of $(S_K(x) - \kappa_K x)x^{-2}dx$ from 1 to ∞. But since that from 1 to x is nothing but $\kappa_K B_K(x) - S_K(x)x^{-1}$, we obtain directly

$$|\gamma_K - B_K(x)| \leq \kappa_K^{-1} C(1+\epsilon^{-1})x^{-\epsilon} \quad (x \geq x_0). \tag{1.7.5}$$

As for (1.7.4), although certainly not the best possible bound (as the well-known case of $K = \mathbb{Q}(\sqrt{-1})$—counting lattice points in circles—indicates), a general method (cf., e.g., [La, VI]) shows that one can take $\epsilon = [K : \mathbb{Q}]^{-1}$ and can compute C by using the geometry of numbers. For $[K : \mathbb{Q}] = 2$, the exponent $= -\frac{1}{2}$ of x in (1.7.5) is as strong as our conditional estimate given in Section 1.6. The constant C computed by following the method of loc. cit. is generally large compared with α_K, but the actual convergence seems considerably faster than what we expect from such a bound.

The second is for the case of quadratic fields. We have to rely on the notation of Section 2.2 below. When K is imaginary, we have the Kronecker formula (2.2.1). When K is real, there is also Hecke's formula [H], which gives

$$\gamma_K = -\frac{1}{2}\log(d_K) + 2\gamma_{\mathbb{Q}} + \frac{1}{h_K}\sum_C i_C. \tag{1.7.6}$$

Here C runs over the narrow ideal classes of K, h_K is the narrow class number, and i_C is defined as follows. Pick any ideal from C^{-1} with a $\mathbb{Z}$-basis α_1,α_2 satisfying $\alpha_1\alpha_2'$-$\alpha_1'\alpha_2 > 0$ (α_i' is the conjugate of α_i), and put

$$\omega(y) = \frac{\alpha_2 y + i\alpha_2'}{\alpha_1 y + i\alpha_1'} \quad (i = \sqrt{-1},\ 0 < y < \infty). \tag{1.7.7}$$

(Its image is a semicircle in the complex upper half-plane orthogonal to the real axis.) Then

$$i_C = -\frac{1}{\log\epsilon}\int_{\epsilon^{-1}}^{\epsilon} \log(|\eta(\omega)|^2 \operatorname{Im}(\omega)^{1/2})dy/y, \tag{1.7.8}$$

where ϵ is the fundamental unit > 1 of K. (There is a small error in [H]. The formula for m [H, two lines below (4)] is actually that for $4m$; the next two formulas will give $\frac{1}{2}$ of the residue and of the constant term, if $\log 4$ in [H, (5)] is replaced by $\log 2$.)

For example, when $K = \mathbb{Q}(\sqrt{10})$, where $\epsilon = 3+\sqrt{10}$ and $h = 2$, (1.7.6) (rewritten as the average of $-\frac{1}{2}\log(d_K) + 2\gamma_{\mathbb{Q}} + i_C$) gives

$$\gamma_K = \frac{1}{2}(0.868877 + 0.402405) = 0.635641.$$

On the other hand, $B_K(10^5) = 0.635861$, and the GRH bounds given by Proposition 2 in Section 1.6 for $x = 10^5$ give $0.634696 < \gamma_K < 0.639418$. Finally, $A_K(10^5) = 0.636813$, for

$$A_K(x) = \log(x) - \Phi_K(x) - 1. \tag{1.7.9}$$

2 Some special families of K

2.1 Curves over $\mathbb{F}_q$ with many rational points

When K corresponds to such a curve, γ_K tends to be negative with a large absolute value. In fact, as a direct application of Theorem 1, we obtain the following.

Theorem 4. (*FF*): *Fix any prime power q and $\varepsilon > 0$. Then*

$$\frac{\gamma_K}{(g_K - 1)\log q} < \varepsilon - \frac{N_1(K)}{(q-1)(g_K - 1)} \tag{2.1.1}$$

holds as long as the exact constant field of K is $\mathbb{F}_q$ and the genus g_K of K is sufficiently large. Here $N_1(K)$ is the number of $\mathbb{F}_q$-rational points of the curve corresponding to K.

Proof. Let $\alpha_K = (g_K - 1)\log q > 0$. Then Theorem 1 gives

$$\gamma_K < \left(\frac{\alpha_K + 1}{\alpha_K - 1}\right)\left(2\log\alpha_K + 1 + \log q - \Phi_K(\alpha_K^2)\right). \tag{2.1.2}$$

Let g_K be so large that $\alpha_K > q$, and take $m \in \mathbb{N}$ such that

$$q^2 \le q^m \le \alpha_K^2 < q^{m+1}. \tag{2.1.3}$$

Then by the definition of $\Phi_K(x)$,

$$\Phi_K(\alpha_K^2) \ge \frac{N_1(K)}{\alpha_K^2 - 1}\left(\alpha_K^2\left(\frac{1}{q} + \cdots + \frac{1}{q^m}\right) - m\right)\log q;$$

hence

$$-\left(\frac{\alpha_K + 1}{\alpha_K - 1}\right)\Phi_K(\alpha_K^2) \le -\frac{N_1(K)}{(\alpha_K - 1)^2}\left\{\frac{\alpha_K^2(1 - q^{-m})}{q - 1} - m\right\}\log q.$$

Now let g_K be so large that

$$\alpha_K > (q-1)\left(\frac{\log\alpha_K}{\log q} + \frac{1}{2}\right) + 1;$$

hence

$$\alpha_K > \frac{1}{2}(q-1)(m+1) + 1. \tag{2.1.4}$$

Then by (2.1.3), (2.1.4), we obtain $\alpha_K^2 q^{-m} + mq < (m+1)q < 2\alpha_K + m - 1$; hence

$$\alpha_K^2(1 - q^{-m}) - m(q-1) > (\alpha_K - 1)^2. \tag{2.1.5}$$

Therefore,

$$-\left(\frac{\alpha_K+1}{\alpha_K-1}\right)\Phi_K(\alpha_K^2) < -\frac{N_1(K)}{q-1}\log q; \tag{2.1.6}$$

hence by (2.1.2), we obtain

$$\frac{\gamma_K}{\alpha_K} < \left(\frac{\alpha_K+1}{\alpha_K-1}\right)\left(\frac{2\log\alpha_K}{\alpha_K}+\frac{1+\log q}{\alpha_K}\right)-\frac{N_1(K)}{(q-1)(g_K-1)}. \tag{2.1.7}$$

Therefore, if g_K is so large that the first term on the RHS is $< \varepsilon$, we have

$$\frac{\gamma_K}{\alpha_K} < \varepsilon - \frac{N_1(K)}{(q-1)(g_K-1)}. \qquad \square$$

We shall combine this with two typical results on curves with many rational points. First, we refer to the following.

Theorem (Elkies–Howe–Kresch–Poonen–Wetherell–Zieve [E, Section 3.2]). *There exists a positive absolute constant c_0 such that for any prime power q, and any $g \geq 1$, there exists a curve X over $\mathbb{F}_q$ of genus g such that*

$$\#X(\mathbb{F}_q) \geq c_0(\log q)(g-1). \tag{2.1.8}$$

Combining this with Theorem 4, we obtain the following.

Corollary 3. *For any fixed prime power q, we have*

$$C(q) \leq -c_0\frac{\log q}{q-1}. \tag{2.1.9}$$

Corollary 4. *Fix any prime power q. Then for any sufficiently large g, there exists K over $\mathbb{F}_q$ with genus g such that $\gamma_K < 0$.*

Secondly, let us recall the following.

Theorem ([I$_1$, I$_2$, TVZ]). *When q is a square, there exist Shimura curves X over $\mathbb{F}_q$ of growing genus g such that*

$$\#X(\mathbb{F}_q) \geq (\sqrt{q}-1)(g-1). \tag{2.1.10}$$

Therefore, by Theorems 2 and 4, we obtain the following.

Corollary 5. *Let q be a square. Then*

$$C(q) = -\frac{1}{\sqrt{q}+1}. \tag{2.1.11}$$

Corollary 6. *Let q be a square, and K be the function field of a Shimura curve over $\mathbb{F}_q$ corresponding to a $(\infty \times p)$-adic discrete subgroup Γ in the sense of* [I$_1$, I$_2$]. *Suppose that Γ is torsion-free, and*

$$\begin{cases} \sqrt{q} > 3, \\ g-1 > 3(q+1)/2(\sqrt{q}-3). \end{cases} \tag{2.1.12}$$

Then $\gamma_K < 0$.

Proof. In Proposition 2(i), take $x = q^2$. Then

$$\begin{aligned}\gamma_K &\leq \frac{q+1}{q-1}\Big(2\log q - \Phi_K(q^2)\Big) + \frac{2\alpha_K}{q-1} - c_q \\ &\leq \frac{q+1}{q-1}\left(2\log q - \frac{N_1(K)}{q^2-1}(q-1)\log q\right) + \frac{2\alpha_K}{q-1} - c_q.\end{aligned} \tag{2.1.13}$$

But $N_1(K) \geq (\sqrt{q}-1)(g_K-1)$ when K corresponds to such a Shimura curve. Therefore,

$$\begin{aligned}\frac{\gamma_K}{\alpha_K} &\leq \frac{q+1}{q-1}\left(\frac{2}{g_K-1} - \frac{\sqrt{q}-1}{q+1}\right) + \frac{2}{q-1} - \frac{c_q}{\alpha_K} \\ &= \frac{1}{q-1}\left\{(3-\sqrt{q}) + \frac{3}{2}\frac{q+1}{g_K-1}\right\},\end{aligned} \tag{2.1.14}$$

from which the desired assertion follows at once. □

2.2 Imaginary quadratic fields

Let K be an imaginary quadratic field with discriminant $d(<0)$. Then the Kronecker limit formula, averaged over all the ideal classes of K, gives the following.

Theorem (Kronecker).

$$\gamma_K = -\frac{1}{2}\log|d| + 2\gamma_{\mathbb{Q}} - \log 2 + \frac{1}{h}\sum_C t_C. \tag{2.2.1}$$

Here $h = h_K$ is the class number of K, C runs over all ideal classes of K,

$$t_C = -2\log(|\eta(\omega_C)|^2 \cdot \mathrm{Im}(\omega_C)^{1/2}), \tag{2.2.2}$$

$$\eta(\tau) = q^{\frac{1}{24}}\prod_{n=1}^{\infty}(1-q^n), \quad q = e^{2\pi i\tau} \quad (\mathrm{Im}(\tau) > 0) \tag{2.2.3}$$

is the Dedekind η-function, and ω_C is defined as follows. Pick any ideal from C^{-1}, with a $\mathbb{Z}$-basis $[\omega_1, \omega_2]$ such that $\mathrm{Im}(\omega_2/\omega_1) > 0$. Then we put $\omega_C = \omega_2/\omega_1$. Since the function

$$|\eta(\tau)|^2(\mathrm{Im}(\tau))^{\frac{1}{2}} \tag{2.2.4}$$

is $\mathrm{SL}_2(\mathbb{Z})$-invariant, t_C is well defined.

Remark. Each t_C is positive. Indeed, the maximal value M of (2.2.4) on the upper half-plane is attained at $\tau = \frac{1}{2}(1+\sqrt{3}i)$, and

$$M = 0.596450134\ldots, \qquad \log M = -0.516759638\ldots. \tag{2.2.5}$$

Since $-t_C \leq 2\log M < 0$, we see that t_C is always positive.

Now, one notes that the contribution of the principal ideal class C_0 to the formula (2.2.1) is *extraordinarily large*. Indeed,

$$t_{C_0} \sim \frac{1}{6}\pi\sqrt{|d|}, \tag{2.2.6}$$

which is "too large" compared with our upper bound $\sim \log\log|d|$ for γ_K under GRH. So, this "too outstanding a contribution of the principal class" should be "pulled down" by averaging over a large number of nonprincipal classes. We thus obtain, by combining with Theorem 1, the following.

Theorem 5 (under GRH). *If* $\alpha_K = \frac{1}{2}\log|d_K| > 1.16$. *(i.e.,* $|d_K| \geq 11$*), then*

$$h_K > \frac{\frac{\pi}{6}\sqrt{|d_K|} - \alpha_K + b_1}{\alpha_K + 2\log\alpha_K + b_2 + c(\alpha_K)}, \tag{2.2.7}$$

with small b_1, b_2, $c(\alpha_K)$ *given by*

$$\begin{cases} b_1 = 2\log M + \log 2 - 4q_0 = -0.34037 - 4q_0, & q_0 = e^{-\pi\sqrt{|d_K|}}, \\ b_2 = 2\log M - 2\gamma_{\mathbb{Q}} + \log 2 + 1 = -0.49480\ldots, \\ c(t) = \dfrac{4\log t + 2}{t-1}. \end{cases} \tag{2.2.8}$$

Proof. Write

$$\begin{cases} \gamma_K = \dfrac{1}{h}\left(\displaystyle\sum_C t_C\right) - \xi, \\ \xi = \alpha_K - 2\gamma_{\mathbb{Q}} + \log 2. \end{cases} \tag{2.2.9}$$

Take $t_0 > t_1 > 0$ such that

$$t_0 \leq t_{C_0}, \quad t_1 \leq t_C \quad (\text{all } C), \tag{2.2.10}$$

and a majorant U for γ_K, viz.,

$$\gamma_K \leq U. \tag{2.2.11}$$

Then

$$\frac{t_0 - t_1}{h} + t_1 \leq \frac{1}{h}\left(\sum_C t_C\right) \leq U + \xi; \tag{2.2.12}$$

hence $U + \xi - t_1$ is positive and

$$h \geq \frac{t_0 - t_1}{U + \xi - t_1}. \tag{2.2.13}$$

For t_1, we choose $t_1 = -2\log M$; and for t_0, we may choose

$$t_0 = \frac{\pi\sqrt{|d_K|}}{6} - 4q_0 - \log\frac{\sqrt{|d_K|}}{2} \quad (q_0 = e^{-\pi\sqrt{|d_K|}}) \tag{2.2.14}$$

(see below). Finally, choose U as in Theorem 1,

$$U = \left(\frac{\alpha_K + 1}{\alpha_K - 1}\right)(2\log\alpha_K + 1), \tag{2.2.15}$$

and we obtain the theorem by (2.2.13). Here it remains to check that t_0 given by (2.2.14) satisfies both

$$\text{(i)}\ t_0 \le t_{C_0} \quad \text{and} \quad \text{(ii)}\ \ -2\log M < t_0.$$

(i) For ω_{C_0}, we may choose $(\sqrt{|d_K|}i)/2$, respectively, $(1+\sqrt{|d_K|}i)/2$, according to whether $d_K \equiv 0 \pmod 4$, respectively, $d_K \equiv 1 \pmod 4$; hence

$$t_{C_0} = -\log\frac{\sqrt{|d_K|}}{2} + \frac{\pi\sqrt{|d_K|}}{6} - 4\log\prod_{n=1}^{\infty}(1-(\varepsilon q_0)^n), \tag{2.2.16}$$

with $\varepsilon = 1$, respectively, -1. But $\prod_{n=1}^{\infty}(1-(\varepsilon q_0)^n) < 1$, respectively, $< 1 + q_0$; hence

$$t_{C_0} > -\log\frac{\sqrt{|d_K|}}{2} + \frac{\pi\sqrt{|d_K|}}{6} - \begin{cases} 0, & d_K \equiv 0 \pmod 4, \\ 4q_0, & d_K \equiv 1 \pmod 4, \end{cases} \tag{2.2.17}$$

which settles (i).

(ii) For $\alpha_K > 1.16$, we have $t_0 > 1.2032 > 1.0335\cdots = -2\log M$. □

Remark. As the above proof shows, the term $-4q_0$ in the formula for b_1 in Theorem 5 is unnecessary when $d_K \equiv 0 \pmod 4$.

A similar result has already been obtained by Granville–Stark [G-S, Theorem 1] (note that the unconditional [G-S, Theorem 2] still contains $L'(1,\chi)/L(1,\chi) = \gamma_K - \gamma_{\mathbb{Q}}$).

S. Louboutin kindly informed the author that, as an asymptotic formula, there is an essentially stronger (and much older) result due to Littlewood [Li];

$$\frac{h\log\log|d|}{\sqrt{|d|}} > \frac{\pi.\exp(-\gamma_{\mathbb{Q}})}{12} - o(1) \quad \text{(under GRH)}. \tag{2.2.18}$$

(As asymptotic formula, this is better than Theorem 5 when $\log|d|/\log\log|d| \ge 4\exp(\gamma_{\mathbb{Q}})$, and hence when $|d|$ has 10 or more digits.)

2.3 The field K_p

For each odd prime p, let K_p denote the unique cyclic extension of degree p over $\mathbb{Q}$ contained in the field $\mathbb{Q}(\boldsymbol{\mu}_{p^2})$ of p^2th roots of unity. It is totally real, with discriminant $d = d_p = p^{2p-2}$, whence

$$\log\sqrt{d} = (p-1)\log p. \tag{2.3.1}$$

Let ℓ be any prime number $\neq p$. Then, by class-field theory,

$$\ell \text{ decomposes completely in } K_p \longleftrightarrow \ell^{\,p-1} \equiv 1 \pmod{p^2}.$$

We shall study the invariant

$$\gamma_p = \gamma_{K_p} \tag{2.3.2}$$

in connection with the following set of primes

$$W(p) = \{\ell;\ \text{primes} < p,\ \ell^{p-1} \equiv 1 \ (\mathrm{mod}\ p^2)\}. \tag{2.3.3}$$

For example, the list of nonempty $W(p)$ with $p < 100$ is

$$\begin{array}{lll} W(11) = \{3\}, & W(43) = \{19\}, & W(59) = \{53\}, \\ W(71) = \{11\}, & W(79) = \{31\}, & W(97) = \{53\}. \end{array}$$

Among the 14 primes p with $900 < p < 1000$, only three primes p satisfy $W(p) \neq \emptyset$ (namely, $p = 907, 919, 983$). The known primes p such that $W(p)$ contains 2 (respectively, 3) are $p = 1093, 3511$ (respectively, $p = 11, 1006003$).

Theorem 6. *Under GRH for K_p, we have*

(i)

$$\gamma_p < i'_p\{2\log(p-1) + 2\log\log p + 1\} - p\left(\sum_{\substack{\ell^k < p \\ \ell \in W(p)}} \frac{\log \ell}{\ell^k}\right).$$

(ii)

$$\gamma_p > -i_p\{2(p-1)(\log\log p + 1)\} - 1.$$

Here

$$i'_p = 1 + \frac{2}{(p-1)\log p - 1}, \qquad i_p = 1 - \frac{2}{\log p + 1}. \tag{2.3.4}$$

The first inequality is a direct consequence of Theorem 1 in Section 1.6. Indeed, each $\ell \in W(p)$ has p distinct primes of K_p above ℓ, so if we write $\alpha_p = \alpha_{K_p} = (p-1)\log p$, then (since $\alpha_p^2 - p \geq (\alpha_p - 1)^2$ for $p \geq 3$) we have

$$\frac{\alpha_p + 1}{\alpha_p - 1}\Phi_K(\alpha_p^2) \geq p \sum_{\substack{\ell^k < p \\ \ell \in W(p)}} \frac{\alpha_p^2 - \ell^k}{(\alpha_p - 1)^2}\frac{\log \ell}{\ell^k} \geq p \sum_{\substack{\ell^k < p \\ \ell \in W(p)}} \frac{\log \ell}{\ell^k}. \tag{2.3.5}$$

The second inequality is a special case of Theorem 3. Note that $\alpha_K^* = \log p$.

Remark. One may replace the sum

$$\sum_{\substack{\ell^k < p \\ \ell \in W(p)}} \frac{\log \ell}{\ell^k}$$

in Theorem 6(i) and the following Corollaries 7 and 8 by a somewhat larger sum

$$\sum_{\ell,k} \frac{\log \ell}{\ell^k},$$

where the primes ℓ satisfy $\ell^{p-1} \equiv 1 \pmod{p^2}$ and k satisfies $\ell^k < 2\alpha_p - 1$ (instead of $\ell^k < p$). The latter sum is also interesting, but in order not to blur the present focus, we just give it as a remark instead of incorporating it in Theorem 6.

Combining (i) and (ii) of Theorem 6, we immediately obtain the following.

Corollary 7 (under GRH). *For any $\varepsilon > 0$, there is an effectively computable bound N_ε such that if $p > N_\varepsilon$, then*

$$\sum_{\substack{\ell^k < p \\ \ell \in W(p)}} \frac{\log \ell}{\ell^k} < 2 \log\log p + 2 + \varepsilon. \tag{2.3.6}$$

Since $\sum_{\ell^k < x} (\log \ell)/\ell^k \sim \log x$, this is in agreement with a result of Lenstra [Le, Theorem 3] which asserts (unconditionally!) that there exists some prime $\ell \notin W(p)$ with $\ell < 4(\log p)^2$.

Remark. Lenstra also gives an asymptotic bound

$$(4e^{-2} + \varepsilon)(\log p)^2.$$

The agreement would have been perfect if the second term on the right side of (2.3.6) had been $\log 4 - 2$ instead of 2. But we have not been able to make this replacement.

Corollary 8 (under GRH).

(i)

$$\lim \left(\frac{\gamma_p}{(p-1)\log p} \right) = 0.$$

(ii)

$$\limsup \left(\frac{\gamma_p}{p} \right) \le \limsup \left(\frac{\gamma_p}{p} + \sum_{\substack{\ell^k < p \\ \ell \in W(p)}} \frac{\log \ell}{\ell^k} \right) \le 0.$$

(iii)

$$-2 \le \liminf \left(\frac{\gamma_p}{p \log\log p} \right).$$

By Theorem 6(i), if $W(p)$ contains small primes ℓ, then γ_p tends to be negative. For example, $\gamma_3, \gamma_5, \gamma_7$ are positive, but γ_{11} is negative, reflecting $W(11) \ni 3$. Also, γ_{1093} is "very negative," reflecting $W(1093) \ni 2$ (see Table 1 below). It is an interesting problem to investigate the asymptotic behaviors of $\gamma_p, \gamma_p/p$, etc. In particular, the determination of the value of

$$\liminf \frac{\gamma_p}{p} \quad (\leq 0)$$

will have the following implications.

Corollary 9 (under GRH).

(i) *If* $\liminf \frac{\gamma_p}{p} = 0$, *then for each prime* ℓ, *there exist at most finitely many primes* p *that satisfy*

$$\ell^{p-1} \equiv 1 \pmod{p^2}.$$

(ii) *If for each prime* ℓ, *all but finitely many primes* p *satisfy*

$$\ell^{p-1} \equiv 1 \pmod{p^2},$$

then $\liminf \frac{\gamma_p}{p} = -\infty$.

Proof.

(i) If for some ℓ there exist infinitely many p such that $\ell^{p-1} \equiv 1 \pmod{p^2}$, then by Theorem 6(i),

$$\liminf \frac{\gamma_p}{p} \leq -\frac{\log \ell}{\ell - 1} < 0.$$

(ii) Under this assumption, by Theorem 6(i),

$$\liminf \frac{\gamma_p}{p} \leq -\sum_{\ell} \frac{\log \ell}{\ell - 1} = -\infty. \qquad \square$$

Table 1 shows the approximate values of γ_p for $p < 110$, and for several primes around $p = 1093, 3511$ (the two known p such that $W(p)$ contains 2) , under GRH. Let ℓ_p (respectively, u_p) denote the lower (respectively, upper) bound for γ_p given by Proposition 2 in Section 1.6 for $x = x_0 = 5 \times 10^4$, and put

$$\gamma_p' = \frac{1}{2}(\ell_p + u_p),$$
$$\gamma_p'' = \log x_0 - \Phi_{K_p}(x_0) - 1,$$
$$\varepsilon_p = \frac{1}{2}(u_p - \ell_p),$$

so that (under GRH) γ_p should lie in between $\gamma_p' \pm \varepsilon_p$, and γ_p'' should also be close to γ_p.

Note how "conspicuously negative" the values of γ_p', γ_p'' are when $W(p)$ contains small primes!

Table 1.

p	γ''_p	γ'_p	ε_p
3	1.76673	1.76741	0.00270354
5	1.6981	1.69927	0.0122214
7	1.84553	1.84723	0.032591
11	−1.43302	−1.43032	0.0577191
13	0.468641	0.472016	0.107757
17	3.5781	3.58283	0.210134
19	4.53435	4.53974	0.25948
23	4.47064	4.47731	0.346256
29	2.32308	2.33163	0.46998
31	4.61964	4.62896	0.540857
37	5.6061	5.6175	0.70755
41	4.2761	4.28883	0.805977
43	−0.929757	−0.916538	0.81594
47	−2.6783	−2.66375	0.91587
53	6.05396	6.071	1.17309
59	0.428977	0.447956	1.30809
61	4.62301	4.64288	1.40864
67	6.03706	6.05918	1.6139
71	−12.8724	−12.8496	1.57591
73	5.99832	6.02267	1.81104
79	−3.85765	−3.83146	1.92486
83	1.21387	1.24177	2.10718
89	7.51911	7.54953	2.37227
97	−5.02725	−4.99428	2.54395
101	2.75934	2.79415	2.75782
103	−2.22423	−2.18885	2.7859
107	5.75378	5.79103	3.00361
109	5.59505	5.63306	3.07587
1069	−4.10435	−3.63507	51.7394
1087	−5.5176	−5.03975	52.7617
1091	−3.11201	−2.63214	53.0135
1093	−748.191	−747.74	46.4644
1097	3.54759	4.03061	53.4188
1103	7.84455	8.33062	53.8033
1109	−0.666736	−0.178118	54.0736
3499	9.81761	11.521	206.78
3511	−2423.07	−2421.45	185.836
3517	7.66195	9.37476	207.986

2.4 The field index of K_p

We shall give some applications to the *field index* of K_p in the sense of [G]. In general, let K be a number field and O_K be the ring of integers of K. For each $\xi \in O_K$, consider the discriminant $D(\xi) = I(\xi)^2 \cdot d_K$ of ξ. Thus $I(\xi) = (O_K : \mathbb{Z}[\xi])$. The greatest common divisor I_K of the $I(\xi)$ is called the field index of K. Clearly, $I_K = 1$ if O_K is generated by a single element. When $K = K_p$, write $I_K = I_p$. Then

$$\ell | I_p \longleftrightarrow \ell \in W(p) = \{\ell;\ \text{primes} < p,\ \ell^{p-1} \equiv 1 \ (\text{mod}\ p^2)\}.$$

This is obvious by the Dirichlet pigeonhole principle (p pigeons are the conjugates of ξ, and ℓ holes are the residue classes modulo a fixed prime factor of ℓ) and the Chinese remainder theorem. In particular, $I_p > 1$ if and only if $W(p) \neq \emptyset$. The exponent of $\ell \in W(p)$ in I_p can be expressed explicitly as

$$\text{ord}_\ell I_p = \sum_{\ell^k < p} C(p, \ell^k), \tag{2.4.1}$$

where $C(m, n)$ denotes the following combinatorial number.

Consider finite sets M, N of orders m, n, respectively, with $m > n$. For each map $f : M \to N$, let n_f be the number of unordered pairs (μ, μ') of distinct elements of M such that $f(\mu) = f(\mu')$. Define $C(m, n) = \text{Min}_f(n_f)$ (> 0). Clearly, n_f attains the minimal value $C(m, n)$ if and only if the maximal difference among $\# f^{-1}(\mu)$ $(\mu \in N)$ is at most 1. (So if we write $M = \{1, 2, \ldots, m\}$, the mod ℓ^k map $f_k : M \to \mathbb{Z}/\ell^k$ satisfies $n_{f_k} = C(m, \ell^k)$ for all k such that $\ell^k < m$. This explains the remaining key point underlying the equality (2.4.1).) Explicitly,

$$C(m, n) = \left[\frac{m}{n}\right]\left(m - \frac{1}{2}n - \frac{1}{2}\left[\frac{m}{n}\right]n\right). \tag{2.4.2}$$

We have

$$\frac{m}{2n}(m-n) \le C(m, n) \le \frac{m}{2n}(m-n) + \frac{n}{8}. \tag{2.4.3}$$

(In fact, the left and right sides of (2.4.3) are given, respectively, by

$$C(m, n) - \frac{1}{2n}k(n-k), \qquad C(m, n) + \frac{1}{8n}(n-2k)^2,$$

where k is defined by $m = [\frac{m}{n}]n + k$ $(0 \le k < n)$.) Since

$$\begin{aligned} \log I_p &= \sum_{\ell \in W(p)} \text{ord}_\ell(I_p) \log \ell \\ &= \sum_{\substack{\ell \in W(p) \\ \ell^k < p}} C(p, \ell^k) \log \ell, \end{aligned} \tag{2.4.4}$$

we see that $\log I_p$ is fairly close to, and is bounded from below, by

$$\frac{p}{2} \sum_{\substack{\ell \in W(p) \\ \ell^k < p}} \left(\frac{p}{\ell^k} - 1\right) \log \ell = \frac{1}{2}(p-1)\Phi_{K_p}(p). \tag{2.4.5}$$

This was the initial motivation for our study of $\Phi_K(x)$. Combining with our previous bounds for γ_p, we obtain some estimations of $\log I_p$, as follows.

Proposition 4 (under GRH).

(i) *For each* $\varepsilon > 0$, *if* $p \geq N_\varepsilon$, *then*

$$\log I_p < (1+\varepsilon)p^2 \log\log p.$$

(ii) *If* $\gamma_p < -2\sqrt{p}\log p$, *then* $W(p) \neq \emptyset$, *and*

$$\log I_p > \frac{1}{2}(\gamma_{\mathbb{Q}} + \log 4\pi)p\sqrt{p}.$$

Sketch of proof.

(i) By (2.4.3), (2.4.4), we have

$$\begin{aligned}\log I_p &\leq \frac{p-1}{2}\Phi_{K_p}(p) + \frac{1}{8}\sum_{\substack{\ell \in W(p)\\ \ell^k < p}} \ell^k \log \ell \\ &< \frac{p-1}{2}\Phi_{K_p}(p) + \frac{p}{8}\left(\sum_{\ell^k < p} \log \ell\right).\end{aligned} \tag{2.4.6}$$

By using the explicit formula for $\Phi_{K_p}(p)$, together with the upper bound

$$r_{K_p}(p) \leq -\frac{\sqrt{p}-1}{\sqrt{p}+1}(\alpha_{K_p} + \beta_{K_p} + \gamma_p + 1) \tag{2.4.7}$$

(the Main Lemma and (1.3.11)), we obtain

$$\log I_p < -\frac{p-1}{2}\gamma_p + O(p^2). \tag{2.4.8}$$

Therefore, by Theorem 6(ii), we obtain

$$\frac{\log I_p}{p^2 \log\log p} < 1 + o(1). \tag{2.4.9}$$

(ii) We have

$$\log I_p \geq \frac{p-1}{2}\Phi_{K_p}(p). \tag{2.4.10}$$

By using the explicit formula for $\Phi_{K_p}(p)$, the lower bound for $r_{K_p}(p)$ ((1.5.2), (1.4.1)), and the inequality $\ell_{K_p}(p) > \log p$ (by (1.5.9)), we obtain

$$\Phi_{K_p}(p) > -2\sqrt{p}\log p + \sqrt{p}(\gamma_{\mathbb{Q}} + \log 4\pi) - \frac{\sqrt{p}+1}{\sqrt{p}-1}(\gamma_p + 1), \tag{2.4.11}$$

from which the desired inequality follows directly. □

2.5 The field index of K_p (continued)

Finally, we shall show that the above conditional upper bound (Proposition 4(i)) for $\log I_p$ is essentially stronger than what one can obtain by the "easier" method, i.e., by using the index of a standard generator of K_p. Put

$$\begin{cases} \Delta_p = \{\delta \in (\mathbb{Z}/p^2)^\times;\ \delta^{p-1} = 1\}, \\ \eta_p = \sum_{\delta \in \Delta_p} \zeta^\delta, \end{cases} \tag{2.5.1}$$

where $\zeta = exp(2\pi\sqrt{-1}/p^2)$. Then $K_p = \mathbb{Q}(\eta_p)$, and the ring of integers O_p is spanned over $\mathbb{Z}$ by 1 and the conjugates of η_p. (The trace of η_p is 0). As I_p divides $I(\eta_p)$, any estimation of the latter from above gives rise to that of the former. Recall that

$$D(\eta_p) = I(\eta_p)^2 p^{2p-2}. \tag{2.5.2}$$

By Lemma 2 below, we obtain

$$D(\eta_p) < p^p \cdot (p-1)^{p-1} \cdots 2^2 \cdot 1^1. \tag{2.5.3}$$

This gives

$$\begin{aligned} \log D(\eta_p) &< \sum_{i=1}^{p} i(\log i) \\ &< \frac{1}{2}p^2 \log p - \frac{1}{4}p^2 + p \log p + 1 - 2\log 2, \end{aligned} \tag{2.5.4}$$

and hence we have the following.

Proposition 5 (unconditional).

$$\log I_p < \left(\frac{1}{4} + \varepsilon\right) p^2 \log p. \tag{2.5.5}$$

Note the difference "log log vs. log," between Propositions 4(i) and 5.

Remarks.

(i) Mahoro Shimura has shown by numerical computations that the upper bound for $\log D(\eta_p)$ provided by (2.5.3) is quite close to the actual values (the ratio of their log is $0.98\ldots$ for $p \doteqdot 400$).

(ii) Each prime factor ℓ of $I(\eta_p)$ must satisfy $\ell^{p-1} \equiv 1 \pmod{p^2}$, but not necessarily $\ell < p$. Shimura has also shown that $I(\eta_p)$ is often divisible by much larger primes ℓ. This phenomenon appears already at $p = 11$, where $I_{11} = 3^{17}$ but $I(\eta_{11}) = 3^{22} \times 457$, and continues on.

We conclude this section by stating the lemma in question. Take $n(\geq 2)$ real numbers $x_1, \ldots, x_n$, and put

$$\begin{cases} nS = x_1 + \cdots + x_n, \\ nT = x_1^2 + \cdots + x_n^2. \end{cases} \tag{2.5.6}$$

Then $S^2 \leq T$, with equality only if $x_1 = \cdots = x_n$. Now fix $S, T \in \mathbb{R}$ satisfying $S^2 < T$, and let $x_1, \ldots, x_n$ vary under the restrictions (2.5.6). Consider the maximal value of the discriminant

$$D = \prod_{1 \leq i < j \leq n} (x_i - x_j)^2 \tag{2.5.7}$$

under (2.5.6).

Lemma 2 (Schur). *The maximal value of D is attained at the unique (unordered) n-ple*

$$(x_1^0, \ldots, x_n^0)$$

determined as follows. Put

$$f(x) = \prod_{i=1}^{n} (x - x_i^0). \tag{2.5.8}$$

Then this monic polynomial of degree n is determined uniquely by the differential equation

$$f''(x) + (a - bx) f'(x) + nbf(x) = 0, \tag{2.5.9}$$

where

$$b = \frac{n-1}{T - S^2}, \quad a = bS. \tag{2.5.10}$$

The maximal value of D is given by

$$D = \frac{n^n (n-1)^{n-1} \cdots 2^2 1^1}{b^{\frac{1}{2}n(n-1)}}. \tag{2.5.11}$$

Remark. This "another Schur's lemma" was kindly pointed out to me by J-P.Serre. We may assume $S = 0$ by translation,and then the maximal value problem will be the same whether we impose $S = 0$ or do not fix S (see [Sc, Section 2]).

Now let us take $n = p$ and $\{x_i\}$ to be the conjugates of η_p. Then $S = 0, T = p-1$, hence $a = 0$, $b = 1$. Hence (2.5.3) follows. Comparison of exponents of p shows that the two sides of (2.5.3) can never be equal.

2.6 Concluding remarks

This paper consists mostly of *inequalities*, functional and numerical, under the generalized Riemann hypothesis in the number field case. Computational data, including graphical ones related to $\Phi_K(x)$, $\log x$, γ_K and the upper and the lower bounds given by Proposition 2, impressively fit with the conditional results.

However, a more interesting problem related to γ_K is its total behavior when we consider a natural family of K, for example, a family of curves over $\mathbb{F}_p$ arising from a two-dimensional scheme over $\mathbb{Z}$ or $\mathbb{F}_p$. We hope to be able to report on this, too, in the near future.

Acknowledgments. I wish to express my deep gratitude to Professor M. A. Tsfasman for valuable comments related to the NF analogue of Theorem 2 and for providing a related text to this volume [Ts_2], to Professor J.-P. Serre for helpful information, to Dr. M. Shimura for all his technical assistance including those related to numerical computations, and to the referee for valuable suggestions for the improvement of the presentation.

References

[D-V] V. G. Drinfeld and S. G. Vlăduţ, On the number of points of algebraic curves, *Functional Anal.*, **17** (1983), 68–69 (inRussian).

[E] N. Elkies, E. Howe, A. Kresch, B. Poonen, J. Wetherell, and M. Zieve, Curves of every genus with many points II: Asymptotically good families, *Duke Math. J.*, **122** (2004), 399–422.

[G] K. Győry, Discriminant form and index form equations, in F. Halter-Koch and R. F. Tichy, eds., *Algebraic Number Theory and Diophantine Analysis*, Walter De Gruyter, Berlin, New York, 2000, 191–214.

[G-S] A. Granville and H. M. Stark, ABC implies no "Siegel zeros" for L-functions of characters with negative discriminant, *Invent. Math.*, **139** (2000), 509–523.

[H] E. Hecke, Über die Kroneckersche Grenzformel für reelle Quadratische Körper und die Klassenzahl relativ-abelscher Körper, *Verh. Naturf. Ges. Basel*, **28** (1917), 363–372.; also in *Mathematische Werke*, 2nd ed., Vandenhoeck and Ruprecht, Göttingen, the Netherlands, 1970, 198–207.

[HIKW] Y. Hashimoto, Y. Iijima, N. Kurokawa, and M. Wakayama, Euler's constants for the Selberg and the Dedekind zeta functions, *Bull. Belgian Math. Soc.*, **11**-4 (2004), 493–516.

[I_1] Y. Ihara, The congruence monodromy problems, *J. Math. Soc. Japan*, **20** (1968), 107–121.

[I_2] Y. Ihara, Shimura curves over finite fields and their rational points, *Contemp. Math.*, **245** (1999), 15–23.

[L-O] J. C. Lagarias and A. M. Odlyzko, Effective versions of the Chebotarev density theorem, in A. Fröhlich, ed., *Algebraic Number Fields: Proceedings of the* 1975 *Durham Symposium*, Academic Press, London, New York, 1977, 409–464.

[La] S. Lang, *Algebraic Number Theory*, Addison–Wesley, Reading, MA, 1970.

[Le] H. W. Lenstra, Jr., Miller's primality test, *Inform. Process. Lett.*, **8** (1979), 86–88.

[Li] J. E. Littlewood, On the class-number of the corpus $P(\sqrt{-k})$, *Proc. London Math. Soc. Ser.* 2, **27** (1928), 358–372; also in *Collected Papers*, Vol. II, Oxford University Press, Oxford, UK, 920–934.

[Sc] I. Schur, Über die Verteilung der Wurzeln bei gewissen algebraischen Gleichungen mit ganzzahligen Koeffizienten, *Math. Z.*, **1** (1918), 377–402; also in *Gesammelte Abhandlungen*, Vol. II, Springer-Verlag, Berlin, 1973, number 32, 213–238.

[St] H. M. Stark, Some effective cases of the Brauer–Siegel theorem, *Invent. Math.*, **23** (1974), 135–152.

[Ts_1] M. A. Tsfasman, Some remarks on the asymptotic number of points, in *Coding Theory and Algebraic Geometry* (*Luminy*, 1991), Lecture Notes in Mathematics, Vol. 1518, Springer-Verlag, Berlin, 1992, 178–192.

[Ts_2] M. A. Tsfasman, Asymptotic behavior of the Euler–Kronecker constant, in V. Ginzburg, ed., *Algebraic Geometry and Number Theory: In Honor of Vladimir Drinfeld's 50th Birthday*, Progress in Mathematics, Vol. 850, Birkhäuser Boston, Cambridge, MA, 2006, 453–458 (this volume).

[T-V] M. A. Tsfasman and S. G. Vlăduţ, Infinite global fields and the generalized Brauer–Siegel theorem, *Moscow Math. J.*, **2** (2002), 329–402.

[TVZ] M. A. Tsfasman, S. G. Vlăduţ, and Th. Zink, Modular curves, Shimura curves, and Goppa codes better than the Varshamov–Gilbert bound, *Math. Nachr.*, **109** (1982), 21–28.

[W_1] A. Weil, Sur les "formules explicites" de la théorie des nombres premiers, *Comm. Sém. Math. Lund*, **Suppl. vol.** (1952), 252–265; also in *Collected Works*, Vol. 2, 48–62.

[W_2] A. Weil, Sur les formules explicites de la théorie des nombres, *Izv. Akad. Nauk SSSR Ser. Mat.*, **36** (1972), 3–18; also in *Collected Works*, Vol. 3, 249–264 (in French).

Asymptotic behaviour of the Euler–Kronecker constant

M. A. Tsfasman

Poncelet Laboratory (UMI 2615 of CNRS and the Independent University of Moscow)
11 Bolshoi Vlasievskii per.
Moscow 119002
Russia

and

Institut de Mathématiques de Luminy
Luminy Case 907
Marseille 13288
France
tsfasman@iml.univ-mrs.fr

To Volodya Drinfeld with friendship and admiration.

Subject Classifications: Primary 11R42. Secondary 11R47.

Summary. This appendix to the beautiful paper [1] of Ihara puts it in the context of infinite global fields of our papers [2] and [3]. We study the behaviour of Euler–Kronecker constant γ_K when the discriminant (genus in the function field case) tends to infinity. Results of [2] easily give us good lower bounds on the ratio $\gamma_K / \log\sqrt{|d_K|}$. In particular, for number fields, under the generalized Riemann hypothesis we prove

$$\liminf \frac{\gamma_K}{\log\sqrt{|d_K|}} \geq -0.26049\dots.$$

Then we produce examples of class-field towers, showing that

$$\liminf \frac{\gamma_K}{\log\sqrt{|d_K|}} \leq -0.17849\dots.$$

1 Introduction

Let K be a global field, i.e., a finite algebraic extension either of the field $\mathbb{Q}$ of rational numbers, or of the field of rational functions in one variable over a finite field of constants. Let $\zeta_K(s)$ be its zeta-function. Consider its Laurent expansion at $s = 1$,

$$\zeta_K(s) = c_{-1}(s-1)^{-1} + c_0 + c_1(s-1) + \cdots.$$

In [1] Yasutaka Ihara introduces and studies the constant

$$\gamma_K = c_0/c_{-1}.$$

There are several reasons to study it:

- It generalizes the classical Euler constant $\gamma = \gamma_{\mathbb{Q}}$.
- For imaginary quadratic fields, it is expressed by a beautiful Kronecker limit formula.
- For fields with large discriminants, its absolute value is at most of the order of const $\times \log\sqrt{|d_K|}$, while the residue c_{-1} itself may happen to be exponential in $\log\sqrt{|d_K|}$; see [2].

In this appendix, we study the asymptotic behaviour of this constant when the discriminant (genus in the function field case) of the field tends to infinity. It is but natural to compare Ihara's results [1] with the methods of infinite zeta-functions developed in [2].

Let $\alpha_K = \log\sqrt{|d_K|}$ in the number field case and $\alpha_K = (g_K - 1)\log q$ in the function field case over $\mathbb{F}_q$. In the number field case, Ihara shows that

$$0 \geq \limsup_K \frac{\gamma_K}{\alpha_K} \geq \liminf_K \frac{\gamma_K}{\alpha_K} \geq -1.$$

We improve the lower bound to the following.

Theorem 1. *Assuming the generalized Riemann hypothesis, we have*

$$\begin{aligned}
&\liminf_K \frac{\gamma_K}{\alpha_K}\\
&\geq -\frac{\log 2 + \frac{1}{2}\log 3 + \frac{1}{4}\log 5 + \frac{1}{6}\log 7}{\frac{1}{\sqrt{2}-1}\log 2 + \frac{1}{\sqrt{3}-1}\log 3 + \frac{1}{\sqrt{5}-1}\log 5 + \frac{1}{\sqrt{7}-1}\log 7 + \frac{1}{2}(\gamma + \log 8\pi)}\\
&= -0.26049\ldots.
\end{aligned}$$

Remarks. Unconditionally we get $\liminf \gamma_K/\alpha_K \geq -0.52227\ldots$.

In the function field case, using the same method, we get $0 \geq \limsup \gamma_K/\alpha_K \geq \liminf \gamma_K/\alpha_K \geq -(\sqrt{q}+1)^{-1}$, which, of course, coincides with Ihara's result [1, Theorem 2].

Let us remark that the upper bound 0 is attained for any asymptotically bad family of global fields, and that the lower bound in the function field case is attained for any asymptotically *optimal* family (such that the ratio of the number of $\mathbb{F}_q$-points to the genus tends to $\sqrt{q}-1$), which we know to exist whenever q is a square. Hence $\limsup \gamma_K/\alpha_K = 0$ and in the function field case with a square q, we have $\liminf \gamma_K/\alpha_K = -(\sqrt{q}+1)^{-1}$.

In Section 3 we construct examples of class-field towers proving (unconditionally) the following.

Theorem 2.

$$\liminf_K \gamma_K/\alpha_K \le -\frac{2\log 2 + \log 3}{\log\sqrt{5\cdot 7\cdot 11\cdot 13\cdot 17\cdot 19\cdot 23\cdot 29\cdot 31\cdot 37}} = -0.17849\ldots.$$

This slightly improves the examples given by Ihara in [1].

In the number field case set $\beta_K = -(\frac{r_1}{2}(\gamma + \log 4\pi) + r_2(\gamma + \log 2\pi))$. If we complete γ_K by archimedean terms, we get the following.

Theorem 3. *Let $\tilde{\gamma}_K = \gamma_K + \beta_K$. Then under the generalized Riemann hypothesis, we have*

$$\liminf_K \frac{\tilde{\gamma}_K}{\alpha_K} \ge -\frac{\gamma + \log(2\pi)}{\gamma + \log(8\pi)} = -0.6353\ldots.$$

It is much easier to see that $\limsup \tilde{\gamma}_K/\alpha_K \le 0$, and that 0 is attained for any *asymptotically bad* family (i.e., such that all ϕs vanish; see the definitions below).

The best example we know gives (unconditionally) the following.

Theorem 4.

$$\liminf_K \tilde{\gamma}_K/\alpha_K \le -0.5478\ldots.$$

2 Bounds

Let us consider the asymptotic behaviour of γ_K. We treat the number field case. (The same argument in the function field case leads to [1, Theorem 2].) Let $|d_K|$ tend to infinity. By [2, Lemma 2.2], any family of fields contains an *asymptotically exact subfamily*, i.e., such that for any q there exists the limit ϕ_q of the ratio of the number $\Phi_q(K)$ of prime ideals of norm q to the "genus" α_K, and also the limits $\phi_{\mathbb{R}}$ and $\phi_{\mathbb{C}}$ of the ratios of r_1 and r_2 to α_K. To find $\liminf \gamma_K/\alpha_K$ and $\liminf \tilde{\gamma}_K/\alpha_K$, it is enough to find corresponding limits for a given asymptotically exact family, and then to look for their minimal values. In what follows, we consider only asymptotically exact families.

Theorem 5. *For an asymptotically exact family $\{K\}$, we have*

$$\lim_K \frac{\gamma_K}{\alpha_K} = -\sum \frac{\phi_q \log q}{q-1},$$

where q runs over all prime powers.

Proof. The right-hand side equals $\xi_\phi^0(1)$, where $\xi_\phi^0(s)$ is the log-derivative of the infinite zeta-function $\zeta_\phi(s)$ of [2]. The corresponding series converges for $\operatorname{Re} s \ge 1$ [2, Proposition 4.2]. We know [1, (1.3.3) and (1.3.4)] that

$$\gamma_K = -\lim_{s\to 1}\left(Z_K(s) - \frac{1}{s-1}\right),$$

where for $\mathrm{Re}(s) > 1$,

$$Z_K(s) = -\frac{\zeta_K'}{\zeta_K}(s) = \sum_{P,k\geq 1} \frac{\log N(P)}{N(P)^{ks}} = \sum_q \Phi_q(K)\frac{\log q}{q^s - 1}.$$

By the same [2, Proposition 4.2], $\frac{\zeta_K'}{\zeta_K}(s) \to \xi_\phi^0(s)$, and hence $\gamma_K/\alpha_K \to \xi_\phi^0(1)$. □

Proof of Theorem 1. We have to maximize $\sum \frac{\phi_q \log q}{q-1}$ under the following conditions:

- $\phi_q \geq 0$.
- For any prime p we have $\sum_{m=1}^{\infty} m\phi_{p^m} \leq \phi_{\mathbb{R}} + 2\phi_{\mathbb{C}}$.
- $\sum_q \frac{\phi_q \log q}{\sqrt{q}-1} + \phi_{\mathbb{R}}(\log 2\sqrt{2\pi} + \frac{\pi}{4} + \frac{\gamma}{2}) + \phi_{\mathbb{C}}(\log 8\pi + \gamma) \leq 1$ (the Basic Inequality, [2, GRH-Theorem 3.1]).

If we put

$$a_0 = \log\sqrt{8\pi} + \frac{\pi}{4} + \frac{\gamma}{2}, \qquad a_1 = \log 8\pi + \gamma, \qquad a_q = \frac{\log q}{\sqrt{q}-1},$$

$$b_0 = b_1 = 0, \qquad b_q = \frac{\log q}{q-1},$$

we are under [2, Section 8, conditions (1)–(4) and (i)–(iv)].

Theorem 1 is now straightforward from [2, Proposition 8.3]. Indeed, the maximum is attained for $\phi_{p^m} = 0$ for $m > 1$, $\phi_{\mathbb{R}} = 0$, and $\phi_2 = \phi_3 = \phi_5 = \phi_7 = 2\phi_{\mathbb{C}}$. (Calculation shows that starting from $p' = 11$, the last inequality of [2, Proposition 8.3] is violated.) □

Proof of Theorem 3. This proof is much easier. Since in this case all coefficients are positive and the ratio of the coefficient of the function we maximize to the corresponding coefficient of the Basic Inequality is maximal for $\phi_{\mathbb{C}}$, the maximum is attained when all ϕs vanish except for $\phi_{\mathbb{C}}$. □

Remarks. If we want unconditional results, then instead of the Basic Inequality we have to use [2, Proposition 3.1]:

$$2\sum_q \phi_q \log q \sum_{m=1}^{\infty} \frac{1}{q^m+1} + \phi_{\mathbb{R}}\left(\frac{\gamma}{2} + \frac{1}{2} + \log 2\sqrt{\pi}\right) + \phi_{\mathbb{C}}(\gamma + \log 4\pi) \leq 1.$$

For $\tilde{\gamma}_K/\alpha_K$, one easily gets

$$\liminf \frac{\tilde{\gamma}_K}{\alpha_K} \geq -\frac{\gamma + \log(2\pi)}{\gamma + \log(4\pi)} = -0.7770\ldots.$$

The calculation for γ_K/α_K is trickier since the last condition of [2, Proposition 8.3] is not violated until very large primes. Changing the coefficients by the first term $(q+1)^{-1}$, Zykin [5] gets

$$\liminf \frac{\gamma_K}{\alpha_K} \geq -0.52227\ldots.$$

Note that (for an asymptotically exact family) $1 + \tilde{\gamma}_K/\alpha_K$ is just the value at 1 of the log-derivative $\xi(s)$ of the *completed infinite zeta-function* $\tilde{\zeta}(s)$ of [2].

3 Examples

Let us bound lim inf γ_K/α_K from above. To do this, one should provide some examples of families. The easiest is, just as in [2, Section 9], to produce quadratic fields having infinite class-field towers with prescribed splitting. The proof of Theorem 1 suggests that we should look for towers of totally complex fields, where 2, 3, 5, and 7 are totally split. This is, however, imprecise, because the sum of [2, Proposition 8.3] varies only slightly when we change p_0. Therefore, I also look at the cases when 2, 3, 5, 7, and 11 are split, and when only 2, 3, and 5 are split, or even only 2 and 3. This leads to a slight improvement on [1, (1.6.30)].

Each of the following fields has an infinite 2-class-field tower with prescribed splitting (just apply [2, Theorem 9.1]), and Theorem 5 gives the following list.

- For $\mathbb{Q}(\sqrt{11 \cdot 13 \cdot 17 \cdot 19 \cdot 23 \cdot 29 \cdot 31 \cdot 37 \cdot 41 \cdot 43 \cdot 47 \cdot 53 \cdot 59 \cdot 61 \cdot 67})$ (the example of [2, Theorem 9.4]) $\mathbb{R}$, 2, 3, 5, 7 totally split, we get $\liminf \gamma_K/\alpha_K \leq -0.1515\dots$.
- For $\mathbb{Q}(\sqrt{-13 \cdot 17 \cdot 19 \cdot 23 \cdot 29 \cdot 31 \cdot 37 \cdot 41 \cdot 43 \cdot 47 \cdot 53 \cdot 59 \cdot 61 \cdot 73 \cdot 79})$ (the example of [2, Theorem 9.5]) with 2, 3, 5, 7, and 11 split, we get $-0.1635\dots$.
- For $\mathbb{Q}(\sqrt{-7 \cdot 11 \cdot 13 \cdot 17 \cdot 19 \cdot 23 \cdot 29 \cdot 31 \cdot 37 \cdot 41 \cdot 43 \cdot 79})$ with 2, 3, 5 split, we get $-0.1727\dots$.
- For $\mathbb{Q}(\sqrt{-7 \cdot 11 \cdot 13 \cdot 17 \cdot 19 \cdot 23 \cdot 29 \cdot 31 \cdot 37 \cdot 41 \cdot 47 \cdot 59})$ with 2, 3, 5 split, we get $-0.1737\dots$.
- An even better example is found by Zykin [5]:

$$\mathbb{Q}(\sqrt{-5 \cdot 7 \cdot 11 \cdot 13 \cdot 17 \cdot 19 \cdot 23 \cdot 29 \cdot 31 \cdot 37})$$

  with 2 and 3 split gives us $-0.17849\dots$. This proves Theorem 2.

For $\liminf \tilde{\gamma}_K/\alpha_K$, the Martinet field $\mathbb{Q}(\cos\frac{2\pi}{11}, \sqrt{2}, \sqrt{-23})$ (see [2, Theorem 9.2]) gives $-0.5336\dots$ The best Hajir–Maire example (see [4, Section 3.2]) gives $\liminf \tilde{\gamma}_K/\alpha_K \leq -0.5478\dots$ This proves Theorem 4.

Acknowledgments. I am grateful to Professor Yasutaka Ihara for letting me know his results prior to publication and for a useful e-mail discussion, to my student Alexei Zykin for many fruitful discussions and for computer verification of my calculations, and to Don Zagier for several valuable remarks on the text.

This research was supported in part by the RFBR grants 02-01-01041 and 02-01-22005.

References

[1] Y. Ihara, On the Euler–Kronecker constants of global fields and primes with small norms, in V. Ginzburg, ed., *Algebraic Geometry and Number Theory: In Honor of Vladimir Drinfeld's* 50*th Birthday*, Progress in Mathematics, Vol. 850, Birkhäuser Boston, Cambridge, MA, 2006, 407–452 (this volume).

[2] M. A. Tsfasman and S. G. Vlăduţ, Infinite global fields and the generalized Brauer–Siegel theorem, *Moscow Math. J.*, **2**-2 (2002), 329–402.

[3] M. A. Tsfasman and S. G. Vlăduţ, Asymptotic properties of zeta-functions, *J. Math. Sci. (N.Y.)*, **84**-5 (1997), 1445–1467.
[4] F. Hajir and C. Maire, Tamely ramified towers and discriminant bounds for number fields II, *J. Symbol. Comput.*, **33**-4 (2002), 415–423.
[5] A. Zykin, Private communication.

Crystalline representations and F-crystals

Mark Kisin

Department of Mathematics
University of Chicago
5734 S. University Avenue
Chicago, IL 60637
USA
kisin@math.uchicago.edu

To Vladimir Drinfeld on his 50th birthday.

Summary. Following ideas of Berger and Breuil, we give a new classification of crystalline representations. The objects involved may be viewed as local, characteristic 0 analogues of the "shtukas" introduced by Drinfeld. We apply our results to give a classification of p-divisible groups and finite flat group schemes, conjectured by Breuil, and to show that a crystalline representation with Hodge–Tate weights 0, 1 arises from a p-divisible group, a result conjectured by Fontaine.

Subject Classifications: Primary 11S20. Secondary 14F30.

Introduction

Let k be a perfect field of characteristic $p > 0$, $W = W(k)$ its ring of Witt vectors, $K_0 = W(k)[1/p]$, and K/K_0 a finite totally ramified extension. In [Br 4] Breuil proposed a new classification of p-divisible groups and finite flat group schemes over the ring of integers $\mathcal{O}_K$ of K. For p-divisible groups and $p > 2$, this classification was established in [Ki], where we also used a variant of Breuil's theory to describe flat deformation rings, and thereby establish a modularity lifting theorem for potentially Barsotti–Tate Galois representations.

In this paper we generalize Breuil's theory to describe crystalline representations of higher weight or, equivalently, their associated weakly admissible modules. To explain our main theorem, fix a uniformiser $\pi \in K$ with Eisenstein polynomial $E(u)$, and write $\mathfrak{S} = W[\![u]\!]$. We equip $\mathfrak{S}$ with the endomorphism φ, which acts via the Frobenius on W, and sends u to u^p. Let $\mathrm{Mod}^{\varphi}_{/\mathfrak{S}}$ denote the category of finite free $\mathfrak{S}$-modules $\mathfrak{M}$ equipped with a map $\varphi^*(\mathfrak{M}) \to \mathfrak{M}$ whose cokernel is killed by some power of $E(u)$.

Theorem 0.1. *The category of crystalline representations with all Hodge–Tate weights $\leqslant 0$ admits a fully faithful embedding into the isogeny category $\mathrm{Mod}^{\varphi}_{/\mathfrak{S}} \otimes \mathbb{Q}_p$ of $\mathrm{Mod}^{\varphi}_{/\mathfrak{S}}$.*

Unfortunately the embedding of the theorem is not essentially surjective. In this sense the situation is not as good as for p-divisible groups. However, we do give an explicit description of the image of the functor. To explain it, let $\mathcal{O}$ denote the ring of rigid analytic functions on the open unit u-disk. Then $\mathfrak{S}[1/p]$ corresponds to the bounded functions in $\mathcal{O}$. It turns out that the module $\mathcal{M} = \mathfrak{M} \otimes_{\mathfrak{S}} \mathcal{O}$ is equipped with a canonical connection which has poles at a sequence of points corresponding to the ideals $\varphi^n(E(u))\mathcal{O} \subset \mathcal{O}$. A module $\mathfrak{M}$ is in the image of our functor if and only if these poles are logarithmic (see Corollary 1.3.15 below).

In fact the theorem we prove is slightly more general than the above, and includes the case of semistable representations. We refer to the reader to the body of the text for the more general statement.

To prove the theorem we adapt the techniques of Berger [Be 1]. One can view his results as relating the weakly admissible module attached to a semistable representation and the (φ, Γ)-module attached to the same representation [Fo 1]. (φ, Γ)-modules are constructed using norm fields for the cyclotomic extension. We develop an analogue of Berger's theory in a setting where the cyclotomic extension has been replaced by the Kummer extension $K_\infty = \cup_{n\geq 1} K(\sqrt[p^n]{\pi})$ (cf. [Br 1]). As in Berger's case, a crucial role in the construction is played by Kedlaya's theory of slopes [Ke 1]. In particular, we again make use of Berger's beautiful observation that the notion of weak admissibility for filtered (φ, N)-modules is intimately related to that of a Frobenius module over the Robba ring being of slope 0 in the sense of [Ke 1]. For $K = K_0$, the analogue of the theorem in the setting of the cyclotomic extension is proved in [Be 2, Theorem 2].

Let us mention some applications of our results. Fix an algebraic closure $\bar{K}$ of K, and write $G_K = \mathrm{Gal}(\bar{K}/K)$ and $G_{K_\infty} = \mathrm{Gal}(\bar{K}/K_\infty)$. The following result was conjectured by Breuil [Br 1], and proved by him for representations of G_K arising from p-divisible groups [Br 3, 3.4.3].

Theorem 0.2. *The functor from crystalline representations of G_K to p-adic G_{K_∞}-representations, obtained by restricting the action of G_K to G_{K_∞}, is fully faithful.*

We also obtain a proof of Fontaine's conjecture that weakly admissible modules are admissible (see Proposition 2.1.5). This is at least the fourth proof, following those of Colmez–Fontaine [CF], Colmez [Co], and Berger [Be 1]. Of course our proof is related to the one of Berger.

As alluded to above, for crystalline representations with all Hodge–Tate weights equal to 0 or -1, there is a refinement of Theorem 0.1. Namely the category of such representations is equivalent to $\mathrm{BT}^{\varphi}_{/\mathfrak{S}} \otimes \mathbb{Q}_p$, where $\mathrm{BT}^{\varphi}_{/\mathfrak{S}}$ denotes the full subcategory of $\mathrm{Mod}^{\varphi}_{/\mathfrak{S}}$ consisting of objects $\mathfrak{M}$ such that the cokernel of $\varphi^*(\mathfrak{M}) \to \mathfrak{M}$ is killed by $E(u)$. On the other hand there is a functor from $\mathrm{BT}^{\varphi}_{/\mathfrak{S}}$ to the category of p-divisible groups. This functor was first constructed for $p > 2$ in [Br 4] using the theory of

[Br 2], and it was conjectured to exist and be an equivalence for all p [Br 4, 2.1.2]. Here we construct it for all p using Grothendieck–Messing theory. As a consequence, we establish the following two results.

Theorem 0.3. *Any crystalline representation with all Hodge–Tate weights equal to* 0 *or* 1 *arises from a p-divisible group.*

Theorem 0.4. *There is a functor from* $\mathrm{BT}^{\varphi}_{/\mathfrak{S}}$ *to p-divisible groups. If* $p > 2$ *this functor is an equivalence. For* $p = 2$ *it induces an equivalence on the associated isogeny categories.*

Theorem 0.3 was conjectured by Fontaine [Fo 3, 5.2.5], and proved by Laffaille for ramification degree $e(K/K_0) \leqslant p - 1$ [La, Section 2], and by Breuil for $p > 2$, and k finite [Br 2, Theorem 1.4].

For $p > 2$, Theorem 0.4 was proved in [Ki] by a completely different method. Finally, it was pointed out by Beilinson that, using Theorem 0.4, one can deduce a classification of finite flat group schemes over $\mathcal{O}_K$ when $p > 2$. A special case of this had been conjectured by Breuil [Br 4, 2.1.1]. To explain this result we denote by $(\mathrm{Mod}\,/\mathfrak{S})$ the category of finite $\mathfrak{S}$-modules $\mathfrak{M}$ which are killed by some power of p, have projective dimension 1 (i.e., $\mathfrak{M}$ has a two-term resolution by finite free $\mathfrak{S}$-modules) and are equipped with a map $\varphi^*(\mathfrak{M}) \to \mathfrak{M}$ whose cokernel is killed by $E(u)$. Then we have

Theorem 0.5. *For* $p > 2$*, the category* $(\mathrm{Mod}\,/\mathfrak{S})$ *is equivalent to the category of finite flat group schemes over* $\mathcal{O}_K$.

During the writing of this paper, I learned from V. Lafforgue that, with Genestier, he had recently developed a theory remarkably parallel to ours in the function field case [GL]. The characteristic p analogues of modules in $\mathrm{Mod}^{\varphi}_{/\mathfrak{S}}$ are a sort of local version of a "shtuka" in the sense of Drinfeld [Ka]. Drinfeld introduced these objects, with stunning success, in order to study the arithmetic of function fields. Lafforgue pointed out to us that the modules in our theory could be regarded as analogues of local shtukas in the case of mixed characteristic. The connection with shtukas gives a first hint that our theory, and related constructions using norm fields, which have no known geometric interpretation, may have some deeper meaning. The question of whether there is a global analogue of a shtuka for number fields is extremely tantalizing, and suggests that Drinfeld's ideas, which revolutionized the study of automorphic forms over function fields, may yet find an application in this case. It is a pleasure to dedicate this article to him.

1 F-crystals and weakly admissible modules

1.1 Preliminaries

Throughout the paper we will fix a uniformiser $\pi \in K$, and we denote by $E(u) \in K_0[u]$ the Eisenstein polynomial of π. We also fix an algebraic closure $\bar{K}$ of K, and a sequence of elements $\pi_n \in \bar{K}$, for n a nonnegative integer, such that $\pi_0 = \pi$, and $\pi_{n+1}^p = \pi_n$. We write $K_{n+1} = K(\pi_n)$.

1.1.1. Let $\mathfrak{S} = W[\![u]\!]$. We denote by $\widehat{\mathfrak{S}}_n$ the completion of $K_{n+1} \otimes_W \mathfrak{S}$ at the maximal ideal $(u - \pi_n)$. The ring $\widehat{\mathfrak{S}}_n$ is equipped with its $(u - \pi_n)$-adic filtration, and this extends to a filtration on the quotient field $\widehat{\mathfrak{S}}_n[1/(u - \pi_n)]$.

Denote by $D[0, 1)$ the open rigid analytic disk of radius 1 with co-ordinate u. Thus the $\bar{K}$-points of $D([0, 1))$ correspond to $x \in \bar{K}$ such that $|x| < 1$. Suppose that $I \subset [0, 1)$ is a subinterval. We denote by $D(I) \subset D[0, 1)$ the admissible open subspace whose $\bar{K}$-points correspond to $x \in \bar{K}$ with $|x| \in I$. We set $\mathcal{O}_I = \Gamma(D(I), \mathcal{O}_{D(I)})$, and $\mathcal{O} = \mathcal{O}_{[0,1)}$. If $I = (a, b)$ we will write $D(a, b)$ rather than $D((a, b))$, and similarly for half open and closed intervals.

Note that for any n we have natural maps $\mathfrak{S}[1/p] \to \mathcal{O} \to \widehat{\mathfrak{S}}_n$ given by sending u to u, where the first map has dense image. On $\mathfrak{S}$ we have the Frobenius φ which sends u to u^p, and acts as the natural Frobenius on W. We will write $\varphi_W : \mathfrak{S} \to \mathfrak{S}$ for the $\mathbb{Z}_p[\![u]\!]$-linear map which acts on W via the Frobenius, and by $\varphi_{\mathfrak{S}/W} : \mathfrak{S} \to \mathfrak{S}$ the W-linear map which sends u to u^p. For any $I \subset [0, 1)$, φ_W induces a map $\varphi_W : \mathcal{O}_I \to \mathcal{O}_I$, while $\varphi_{\mathfrak{S}/W}$ induces a map $\varphi_{\mathfrak{S}/W} : \mathcal{O}_I \to \mathcal{O}_{p^{-1}I}$, where $p^{-1}I = \{r : r^p \in I\}$. We will write $\varphi = \varphi_W \circ \varphi_{\mathfrak{S}/W} : \mathcal{O}_I \to \mathcal{O}_{p^{-1}I}$.

Let $c_0 = E(0) \in K_0$. Set

$$\lambda = \prod_{n=0}^{\infty} \varphi^n(E(u)/c_0) \in \mathcal{O}.$$

Thinking of functions in $\mathcal{O}$ as convergent power series in u, we define a derivation $N_\nabla := -u\lambda\frac{d}{du} : \mathcal{O} \to \mathcal{O}$. We denote by the same symbol the induced derivation $\mathcal{O}_I \to \mathcal{O}_I$, for each $I \subset [0, 1)$.

We adjoin a formal variable ℓ_u to $\mathcal{O}$ which acts formally like $\log u$. We extend the natural maps $\mathcal{O} \to \widehat{\mathfrak{S}}_n$ to $\mathcal{O}[\ell_u]$ by sending ℓ_u to

$$\log\left[\left(\frac{u - \pi_n}{\pi_n}\right) + 1\right] := \sum_{i=1}^{\infty} (-1)^{i-1} i^{-1} \left(\frac{u - \pi_n}{\pi_n}\right)^i \in \widehat{\mathfrak{S}}_n.$$

We extend φ to $\mathcal{O}[\ell_u]$ by setting $\varphi(\ell_u) = p\ell_u$, and we extend N_∇ to a derivation on $\mathcal{O}[\ell_u]$ by setting $N_\nabla(\ell_u) = -\lambda$. Finally, we write N for the derivation on $\mathcal{O}[\ell_u]$ which acts as differentiation of the formal variable ℓ_u. These satisfy the relations

$$N\varphi = p\varphi N \quad \text{and} \quad N_\nabla\varphi = (p/c_0)E(u)\varphi N_\nabla. \tag{1.1.2}$$

Finally, we remark that N and N_∇ commute.

1.1.3. Recall [Fo 2] that a φ-module is a finite-dimensional K_0-vector space D together with a bijective, Frobenius semilinear map $\varphi : D \to D$. A (φ, N)-module is a φ-module D, together with a linear (necessarily nilpotent) map $N : D \to D$ which satisfies $N\varphi = p\varphi N$. (φ, N)-modules (respectively, φ-modules) form a Tannakian category.

If D is a one-dimensional (φ, N)-module, and $v \in D$ is a basis vector, then $\varphi(v) = \alpha v$ for some $\alpha \in K_0$, and we write $t_N(D)$ for the p-adic valuation of α. If D has dimension $d \in \mathbb{N}^+$, then we write $t_N(D) = t_N(\bigwedge^d D)$.

A *filtered* (φ, N)-module (respectively, φ-module) is a (φ, N)-module (respectively, φ-module) D equipped with a decreasing, separated, exhaustive filtration on $D_K = D \otimes_{K_0} K$. These again form a Tannakian category. Given a one-dimensional filtered (φ, N)-module D, we denote by $t_H(D)$ the unique integer i such that $\mathrm{gr}^i D_K$ is nonzero. In general, if D has dimension d, we set $t_H(D) = t_H(\bigwedge^d D)$. A filtered (φ, N)-module D is called *weakly admissible* if $t_H(D) = t_N(D)$ and for any (φ, N)-submodule $D' \subset D$, $t_H(D') \leqslant t_N(D')$, where $D'_K \subset D_K$ is equipped with the induced filtration.

We will call a filtered (φ, N)-module *effective* if $\mathrm{Fil}^0 D = D$.

1.1.4. By a φ-module over $\mathcal{O}$ we mean a finite free $\mathcal{O}$-module $\mathcal{M}$, equipped with a φ-semilinear, injective map $\varphi : \mathcal{M} \to \mathcal{M}$. A (φ, N_∇)-module over $\mathcal{O}$ is a φ-module $\mathcal{M}$ over $\mathcal{O}$, together with a differential operator $N_\nabla^{\mathcal{M}}$ over N_∇. That is, for $f \in \mathcal{O}$, and $m \in \mathcal{M}$, we have

$$N_\nabla^{\mathcal{M}}(fm) = N_\nabla(f)m + fN_\nabla^{\mathcal{M}}(m).$$

φ and N_∇ are required to satisfy the relation $N_\nabla^{\mathcal{M}}\varphi = (p/c_0)E(u)\varphi N_\nabla^{\mathcal{M}}$. We will usually write N_∇ for $N_\nabla^{\mathcal{M}}$ since this will cause no confusion. The category of (φ, N_∇)-modules over $\mathcal{O}$ has a natural structure of a Tannakian category.

It will often be convenient to think of $\mathcal{M}$ as a coherent sheaf on $D[0, 1)$. Then we may speak of $\mathcal{M}$ or $1 \otimes \varphi : \varphi^*(\mathcal{M}) \to \mathcal{M}$ having some property (e.g., being an isomorphism) in the neighbourhood of a point of $D[0, 1)$, or over some admissible open subset. We will need the following.

Lemma 1.1.5. *Let $I \subset [0, 1)$ be an interval, $\mathcal{M}$ a finite free $\mathcal{O}_I$-module, and $\mathcal{N} \subset \mathcal{M}$ an $\mathcal{O}_I$-submodule. Then the following conditions are equivalent:*

(1) *$\mathcal{N} \subset \mathcal{M}$ is closed.*
(2) *$\mathcal{N}$ is finitely generated.*
(3) *$\mathcal{N}$ is finite free.*

Proof. We obviously have (3) $\implies$ (2). If $\mathcal{N}$ is finitely generated, then it is free of rank at most that of $\mathcal{M}$ by [Be 3, 4.13], so (2) $\implies$ (3). Moreover, in this case, $\mathcal{N}$ is the image of a map $\mathcal{M} \to \mathcal{M}$, hence by [Be 3, 4.12(5)], we may choose an isomorphism $\mathcal{M} \xrightarrow{\sim} \mathcal{O}_I^d$ under which $\mathcal{N}$ maps onto $\sum_{i=1}^d f_i\mathcal{O}_I$ for some $f_i \in \mathcal{O}_I$. Since $f_i\mathcal{O}_I \subset \mathcal{O}_I$ is a closed ideal by [Laz, 8.11], it follows that $\mathcal{N}$ is closed in $\mathcal{M}$.

Finally, suppose that $\mathcal{N} \subset \mathcal{M}$ is closed. We will show that $\mathcal{N}$ is free by induction on the $\mathcal{O}_I$-rank of $\mathcal{M}$. If $\mathcal{M}$ has rank 1, then this follows from [Laz, 7.3]. In general choose a nonzero element $n \in \mathcal{N}$. Let $\mathcal{M}' = (\mathcal{M} \cap n \cdot \mathcal{O}_I) \otimes_{\mathcal{O}_I} \mathrm{Fr}\,\mathcal{O}_I \subset \mathcal{M} \otimes_{\mathcal{O}_I} \mathrm{Fr}\,\mathcal{O}_I$, where $\mathrm{Fr}\,\mathcal{O}_I$ denotes the field of fractions of $\mathcal{O}_I$. Write $\mathcal{N}' = \mathcal{N} \cap \mathcal{M}'$. By Lazard's results and [Ke 1, Lemma 2.4], $\mathcal{M}' \subset \mathcal{M}$ is a direct summand and is free of rank 1 over $\mathcal{O}_I$. Since $\mathcal{N}'$ is closed in $\mathcal{M}'$, and $\mathcal{N}/\mathcal{N}'$ is closed in $\mathcal{M}/\mathcal{M}'$ by the open mapping theorem, we deduce by induction that both $\mathcal{N}'$ and $\mathcal{N}/\mathcal{N}'$ are finite free over $\mathcal{O}_I$, whence the same holds for $\mathcal{N}$. □

1.2 Filtered (φ, N)-modules and (φ, N_∇)-modules

Let D be an effective filtered (φ, N)-module. We define a (φ, N_∇)-module over $\mathcal{O}$, as follows: For each nonnegative integer n, write ι_n for the composite

$$\mathcal{O}[\ell_u] \otimes_{K_0} D \overset{\varphi_W^{-n} \otimes \varphi^{-n}}{\longrightarrow} \mathcal{O}[\ell_u] \otimes_{K_0} D \to \widehat{\mathfrak{S}}_n \otimes_{K_0} D = \widehat{\mathfrak{S}}_n \otimes_K D_K,$$

where the second map is deduced from the map $\mathcal{O}[\ell_u] \to \widehat{\mathfrak{S}}_n$ defined in (1.1.2). We may extend this to a map

$$\iota_n : \mathcal{O}[\ell_u, 1/\lambda] \otimes_{K_0} D \to \widehat{\mathfrak{S}}_n[1/(u - \pi_n)] \otimes_K D_K.$$

Set

$$\mathcal{M}(D) = \{x \in (\mathcal{O}[\ell_u, 1/\lambda] \otimes_{K_0} D)^{N=0} : \iota_n(x) \in \mathrm{Fil}^0(\widehat{\mathfrak{S}}_n[1/(u - \pi_n)] \otimes_K D_K), \\ n \geq 0\}.$$

Note that $(\mathcal{O}[\ell_u, 1/\lambda] \otimes_{K_0} D)^{N=0}$ is an $\mathcal{O}$-module with a φ-semilinear Frobenius given by those on D and $\mathcal{O}[\ell_u, 1/\lambda]$, where the latter ring is equipped with a Frobenius, because $\varphi(1/\lambda) = E(u)/(c_0\lambda)$. It is equipped with a differential operator N_∇, induced by the operator on $N_\nabla \otimes 1$ on $\mathcal{O}[\ell_u, 1/\lambda] \otimes_{K_0} D$.

Lemma 1.2.1. *If we regard $\widehat{\mathfrak{S}}_n$ as an $\mathcal{O}$-module via φ_W^{-n}, then*

(1) *The map*
$$\widehat{\mathfrak{S}}_n \otimes_{\mathcal{O}} (\mathcal{O}[\ell_u] \otimes_{K_0} D)^{N=0} \to \widehat{\mathfrak{S}}_n \otimes_K D_K$$
induced by ι_n is an isomorphism.

(2) *We have*
$$\widehat{\mathfrak{S}}_n \otimes_{\mathcal{O}} \mathcal{M}(D) \overset{\sim}{\longrightarrow} \sum_{j \geq 0} (u - \pi_n)^{-j} \widehat{\mathfrak{S}}_n \otimes_K \mathrm{Fil}^j D_K$$
$$= \sum_{j \geq 0} \varphi^n_{\mathfrak{S}/W}(E(u))^{-j} \widehat{\mathfrak{S}}_n \otimes_K \mathrm{Fil}^j D_K.$$

Proof. Since both sides in (1) are easily seen to be free $\widehat{\mathfrak{S}}_n$-modules of the same rank, it suffices to show that the map obtained by reducing modulo $u - \pi_n$ is an isomorphism. The latter map is $(K_{n+1}[\ell_u] \otimes_{K_0} D)^{N=0} \overset{\ell_u \mapsto 0}{\to} K_{n+1} \otimes_K D_K$ (where N acts on $K_{n+1}[\ell_u]$ K_{n+1}-linearly), and this is easily seen to be an isomorphism. This establishes (1), and (2) follows easily. □

Lemma 1.2.2. *Suppose that D is effective. Then the operators φ and N_∇ on $(\mathcal{O}[\ell_u, 1/\lambda] \otimes_{K_0} D)^{N=0}$ induce on $\mathcal{M}(D)$ the structure of a (φ, N_∇)-module over $\mathcal{O}$. Moreover, there is an isomorphism of $\mathcal{O}$-modules*

$$\mathrm{coker}(1 \otimes \varphi : \varphi^*\mathcal{M}(D) \to \mathcal{M}(D)) \overset{\sim}{\longrightarrow} \oplus_{i \geq 0} (\mathcal{O}/E(u)^i)^{h_i}$$

where $h_i = \dim_K \mathrm{gr}^i D_K$.

Proof. First, we check that $\mathcal{M}(D)$ is finite free over $\mathcal{O}$. Let r be a nonnegative integer such that $\mathrm{Fil}^{r+1} D = 0$. Then $\mathcal{M}(D) \subset \lambda^{-r}(\mathcal{O}[\ell_u] \otimes_{K_0} D)^{N=0}$, and the right-hand side is a finite free $\mathcal{O}$-module. Since the maps ι_n are continuous, and the filtration on $\widehat{\mathfrak{S}}_n[1/(u-\pi_n)]$ is by closed K-subspaces, this submodule is closed, and hence finite free by Lemma 1.1.5.

Now let $\mathcal{D}_0 = (\mathcal{O}[\ell_u] \otimes_{K_0} D)^{N=0}$. To prove the rest of the lemma, we have to check that the natural map $\varphi^*(\lambda^{-r}\mathcal{D}_0) \to \lambda^{-r}\mathcal{D}_0$, induced by the isomorphism $1 \otimes \varphi_{\mathcal{D}_0} : \varphi^*\mathcal{D}_0 \xrightarrow{\sim} \mathcal{D}_0$, takes $\varphi^*(\mathcal{M}(\mathcal{D}))$ into $\mathcal{M}(D)$, and that the cokernel of $\varphi^*(\mathcal{M}(D)) \to \mathcal{M}(D)$ is as claimed. For this it will be convenient to think of finite $\mathcal{O}$-modules as coherent sheaves on $D[0,1)$.

At any point of $D[0,1)$ not corresponding to a maximal ideal of the form $\varphi^n(E(u))$ for some $n \geq 0$, $\mathcal{M}(D)$ is isomorphic to $\mathcal{D}_0$, and so $1 \otimes \varphi_{\mathcal{D}_0}$ induces an isomorphism $\varphi^*\mathcal{M}(D) \xrightarrow{\sim} \mathcal{M}(D)$ at such a point. Now for any $n \geq 1$, the map $\varphi_{\mathfrak{S}/W}$ on $\mathfrak{S}$ induces a map of K_{n+1}-algebras $\varphi_{\mathfrak{S}/W} : \widehat{\mathfrak{S}}_n \overset{u \mapsto u^p}{\longrightarrow} \widehat{\mathfrak{S}}_{n+1}$, and we have a commutative diagram

$$\begin{array}{ccc} \lambda^{-r}\mathcal{D}_0 & \xrightarrow{\iota_n} & (u-\pi_n)^{-r}\widehat{\mathfrak{S}}_n \otimes_K D_K \\ \Big\downarrow{\scriptstyle \varphi} & & \Big\downarrow{\scriptstyle \varphi_{\mathfrak{S}/W}\otimes 1} \\ \lambda^{-r}\mathcal{D}_0 & \xrightarrow{\iota_{n+1}} & (u-\pi_{n+1})^{-r}\widehat{\mathfrak{S}}_{n+1} \otimes_K D_K . \end{array}$$

If we regard $\widehat{\mathfrak{S}}_n$ as an $\mathcal{O}$-module via φ_W^{-n}, then $\varphi_{\mathfrak{S}/W}$ becomes a φ-semilinear map, and the induced $\mathcal{O}$-linear map

$$1 \otimes \varphi_{\mathfrak{S}/W} : \varphi^*\widehat{\mathfrak{S}}_n[1/(u-\pi_n)] \to \widehat{\mathfrak{S}}_{n+1}[1/(u-\pi_{n+1})] \tag{1.2.3}$$

is an isomorphism, which takes $\varphi^*(u-\pi_n)^s\widehat{\mathfrak{S}}_n$ onto $(u-\pi_{n+1})^s\widehat{\mathfrak{S}}_{n+1}$ for each integer s. Now let

$$\mathcal{M}_n(D) = \{x \in \mathcal{D}_0[1/\lambda] : \iota_n(x) \in \mathrm{Fil}^0(\widehat{\mathfrak{S}}_n[1/(u-\pi_n)] \otimes_K D_K)\}.$$

Then $\mathcal{M}(D) \subset \mathcal{M}_n(D)$ and this inclusion is an isomorphism at the point $x_n \in D[0,1)$ corresponding to the ideal $(\varphi^n(E(u))) \subset \mathcal{O}$.

By Lemma 1.2.1, the map

$$\begin{aligned} \mathcal{D}_0/\varphi_W^n(E(u))\mathcal{D}_0 &= (\mathcal{O}/\varphi_W^n(E(u))\mathcal{O}[\ell_u] \otimes_{K_0} D)^{N=0} \\ &\to \widehat{\mathfrak{S}}_n \otimes_K D_K/(u-\pi_n)\widehat{\mathfrak{S}}_n \otimes_K D_K \xrightarrow{\sim} K_{n+1} \otimes_K D_K \end{aligned}$$

induced by ι_n is a bijection. Hence we have an exact sequence

$$\begin{aligned} &0 \to \mathcal{M}_n(D) \to \lambda^{-r}\mathcal{D}_0 \\ &\to ((u-\pi_n)^{-r}\widehat{\mathfrak{S}}_n \otimes_K D_K)/\mathrm{Fil}^0(\widehat{\mathfrak{S}}_n[1/(u-\pi_n)] \otimes_K D_K) \to 0 \end{aligned}$$

Denote by Q_n the term on the right of this exact sequence. Then its pullback by the flat map $\varphi : \mathcal{O} \to \mathcal{O}$ sits in a commutative diagram with exact rows

$$\begin{array}{ccccccccc} 0 & \longrightarrow & \varphi^*\mathcal{M}_n(D) & \longrightarrow & \varphi^*(\lambda^{-r}\mathcal{D}_0) & \longrightarrow & \varphi^*(Q_n) & \longrightarrow & 0 \\ & & \downarrow & & \downarrow & & \downarrow & & \\ 0 & \longrightarrow & \mathcal{M}_{n+1}(D) & \longrightarrow & \lambda^{-r}\mathcal{D}_0 & \longrightarrow & Q_{n+1} & \longrightarrow & 0. \end{array}$$

Here the map on the right is induced by the map $1 \otimes \varphi_{\mathfrak{S}/W}$ of (1.2.3), and the remarks above show that it is a bijection. The map in the middle has image $E(u)^r \lambda^{-r}\mathcal{D}_0$. In particular, we may fill in the left hand map $\varphi^*(\mathcal{M}_n(D)) \to \mathcal{M}_n(D)$, as shown, and we see that its cokernel is contained in $\lambda^{-r}\mathcal{D}_0/(E(u)^r\lambda^{-r}\mathcal{D}_0)$. Since the inclusions $\varphi^*(\mathcal{M}(D)) \subset \varphi^*(\mathcal{M}_n(D))$ and $\mathcal{M}(D) \subset \mathcal{M}_{n+1}(D)$ are isomorphisms at x_{n+1}, this shows that $1 \otimes \varphi_{\mathcal{D}_0}$ induces an isomorphism $\varphi^*(\mathcal{M}(D)) \xrightarrow{\sim} \mathcal{M}(D)$ at x_{n+1}.

Finally, since $\varphi(x_0) \neq x_n$ for any $n \geq 0$, the inclusion $\mathcal{D}_0 \subset \mathcal{M}(D)$ gives rise to an inclusion $\varphi^*\mathcal{D}_0 \subset \varphi^*(\mathcal{M}(D))$ which is an isomorphism at x_0. Since $1 \otimes \varphi_{\mathcal{D}_0}$ maps $\varphi^*(\mathcal{D}_0)$ isomorphically onto $\mathcal{D}_0 \subset \mathcal{M}(D)$, it induces a map $\varphi^*(\mathcal{M}(D)) \to \mathcal{M}(D)$ whose cokernel is supported on x_0. Moreover, we have a commutative diagram with exact rows

$$\begin{array}{ccccccccc} 0 & \longrightarrow & \mathcal{D}_0 & \longrightarrow & \lambda^{-r}\mathcal{D}_0 & \longrightarrow & ((u-\pi)^{-r}\widehat{\mathfrak{S}}_0/\widehat{\mathfrak{S}}_0) \otimes_K D_K & \longrightarrow & 0 \\ & & \downarrow & & \| & & \downarrow & & \\ 0 & \longrightarrow & \mathcal{M}_0(D) & \longrightarrow & \lambda^{-r}\mathcal{D}_0 & \longrightarrow & Q_0 & \longrightarrow & 0. \end{array}$$

Hence

$$\begin{aligned} \operatorname{coker}(\varphi^*(\mathcal{M}(D)) \to \mathcal{M}(D)) &\xrightarrow{\sim} \mathcal{M}_0(D)/\mathcal{D}_0 \\ &\xrightarrow{\sim} \operatorname{Fil}^0(\widehat{\mathfrak{S}}_0[1/(u-\pi)] \otimes_K D_K)/(\widehat{\mathfrak{S}}_0 \otimes_K D_K) \end{aligned}$$

and the lemma follows. □

1.2.4. We will say that a φ-module $\mathcal{M}$ over $\mathcal{O}$ is of finite E-height if the cokernel of the $\mathcal{O}$-linear map $\varphi^*\mathcal{M} \to \mathcal{M}$ is killed by some power of $E(u)$, that is, if this cokernel is supported on $x_0 \in D[0,1)$. A (φ, N_∇)-module over $\mathcal{O}$ is of finite E-height if it is of finite E-height as a φ-module. We denote by $\mathrm{Mod}^{\varphi,N_\nabla}_{/\mathcal{O}}$ (respectively, $\mathrm{Mod}^{\varphi}_{/\mathcal{O}}$) the category of (φ, N_∇)-modules (respectively, φ-modules) over $\mathcal{O}$ of finite E-height. Both these categories are stable under $\otimes$-products.

1.2.5. Suppose that $\mathcal{M}$ is in $\mathrm{Mod}^{\varphi}_{/\mathcal{O}}$. We define a filtered φ-module $D(\mathcal{M})$ as follows: The underlying K_0-vector space of $D(\mathcal{M})$ is $\mathcal{M}/u\mathcal{M}$, and the operator φ is induced by φ on $\mathcal{M}$.

To construct the filtration on $D(\mathcal{M})_K$, it will be convenient to adopt the following notation: If $J \subset I \subset [0,1)$ are intervals, and $\mathcal{M}$ is a finite $\mathcal{O}_I$-module, we will write $\mathcal{M}_J = \mathcal{M} \otimes_{\mathcal{O}_I} \mathcal{O}_J$. If we think of $\mathcal{M}$ as a coherent sheaf on $D(I)$, then $\mathcal{M}_J$ corresponds to the restriction of $\mathcal{M}$ to $D(J)$. Similarly, if $\xi : \mathcal{M} \to \mathcal{M}'$ is a map of finite $\mathcal{O}_I$-modules we denote by $\xi_J : \mathcal{M}_J \to \mathcal{M}'_J$ the induced map. We will need the following

Lemma 1.2.6. *Let $\mathcal{M}$ be a φ-module over $\mathcal{O}$. There is a unique $\mathcal{O}$-linear, φ-equivariant morphism*

$$\xi : D(\mathcal{M}) \otimes_{K_0} \mathcal{O} \to \mathcal{M}$$

whose reduction modulo u induces the identity on $D(\mathcal{M})$. ξ is injective, and its cokernel is killed by a finite power of λ. If $r \in (|\pi|, |\pi|^{1/p})$, then the image of the map $\xi_{[0,r)}$ induced by ξ coincides with the image of $1 \otimes \varphi : (\varphi^*\mathcal{M})_{[0,r)} \to \mathcal{M}_{[0,r)}$.

Proof. Recall that $\mathcal{O}$ is a Fréchet space, with its topology defined by the norms $|\cdot|_r$ for $r \in (0, 1)$, given by $|f|_r = \sup_{x \in D[0,r]} |f(x)|$. Since $\mathcal{M}$ is free we may identify it with $\mathcal{O}^d$, where $d = \mathrm{rk}_{\mathcal{O}} \mathcal{M}$, and we will again denote by $|\cdot|_r$ the norm on $\mathcal{M}$ obtained by taking the maximum of $|\cdot|_r$ applied to the co-ordinates of an element $m \in \mathcal{M} = \mathcal{O}^d$. For a subset $\Sigma \subset \mathcal{M}$ we set $|\Sigma|_r = \sup_{x \in \Sigma} |x|_r$.

Now choose any K_0-linear map $s_0 : D(\mathcal{M}) \to \mathcal{M}$ whose reduction modulo u is the identity. We define a new map $s : D(\mathcal{M}) \to \mathcal{M}$ by

$$s = s_0 + \sum_{i=1}^{\infty} (\varphi^i \circ s_0 \circ \varphi^{-i} - \varphi^{i-1} \circ s_0 \circ \varphi^{1-i})$$

To check that the right-hand side converges to a well defined map, fix an $r \in (0, 1)$, and let $L \subset D(\mathcal{M})$ be a $\mathcal{O}_{K_0}$-lattice. Then $\varphi^{-1}(L) \subset p^{-j} L$ for some nonnegative integer j. After increasing j, we may also assume that $|\varphi(m)|_r \leqslant |p^{-j} m|_r$ for all $m \in \mathcal{M}$. Since $\varphi \circ s_0 \circ \varphi^{-1} - s_0 \in u\mathcal{M}$, we have $\tilde{L} := u^{-1}(\varphi \circ s_0 \circ \varphi^{-1} - s_0)(L) \subset \mathcal{M}$ so that

$$|(\varphi^{i+1} \circ s_0 \circ \varphi^{-i-1} - \varphi^i \circ s_0 \circ \varphi^{-i})(L)|_r \leqslant |p^{-ij} u^{p^i} \varphi^i(\tilde{L})|_r \leqslant p^{2ij} r^{p^i} |\tilde{L}|_r.$$

Since $|\tilde{L}|_r$ is finite, and $p^{2ij} r^{p^i} \to 0$ as $i \to \infty$, for any $j \geq 0$ and $r \in (0, 1)$, the map s is well defined. Once checks immediately that $\varphi \circ s = s \circ \varphi$.

Given any other such map s', the difference $s - s'$ sends $D(\mathcal{M})$ into $u\mathcal{M}$. But since φ is a bijection on $D(\mathcal{M})$, and $\varphi^j \circ (s - s') = (s - s') \circ \varphi^j$, for $j \geq 1$, we see that $(s - s')(D(\mathcal{M})) \subset u^{p^j} \mathcal{M}$, so that $s - s' = 0$. It follows that s is the unique such map. Extending s to $D(\mathcal{M}) \otimes_{K_0} \mathcal{O}$ by $\mathcal{O}$-linearity yields the required map ξ, and the uniqueness of s implies the that of ξ.

To establish the claim regarding the image of ξ, note that ξ is an isomorphism modulo u, so for some sufficiently large positive integer i, $\xi_{[0,r^{p^i})}$ is an isomorphism. Since ξ commutes with φ, we have a commutative diagram

$$\begin{array}{ccc}
\varphi^*(D(\mathcal{M}) \otimes_{K_0} \mathcal{O}) & \xrightarrow{\varphi^*\xi} & \varphi^*\mathcal{M} \\
\downarrow \wr & & \downarrow {\scriptstyle 1\otimes\varphi} \\
D(\mathcal{M}) \otimes_{K_0} \mathcal{O} & \xrightarrow{\xi} & \mathcal{M}.
\end{array}$$

If $i > 1$, then the restriction of the right vertical map to $[0, r^{p^{i-1}})$ is an isomorphism, so that $\xi_{[0,r^{p^{i-1}})}$ is also. Repeating this argument, we find that $\xi_{[0,r^p)}$ is an isomorphism, and making use of the above commutative diagram once more, we find that the image of $\xi_{[0,r)}$ coincides with $(1 \otimes \varphi)_{[0,r)}$.

Finally, we have seen that $\xi_{[0,r)}$ is injective with cokernel killed by a finite power $E(u)^s$ of $E(u)$. It follows from the same commutative diagram above that ξ is injective with cokernel killed by λ^s. □

1.2.7. Now define a decreasing filtration on $\varphi^*\mathcal{M}$ by

$$\mathrm{Fil}^i\, \varphi^*\mathcal{M} = \{x \in \varphi^*\mathcal{M} : 1 \otimes \varphi(x) \in E(u)^i \mathcal{M}\}.$$

This is a filtration on $\varphi^*\mathcal{M}$ by finite free $\mathcal{O}$-modules (for example, using Lemma 1.1.5), whose successive graded pieces are $E(u)$-torsion modules. By transport of structure, this defines a filtration on $(1 \otimes \varphi)(\varphi^*\mathcal{M})$, and hence on $(1 \otimes \varphi)(\varphi^*\mathcal{M})_{[0,r)}$, where r is as in Lemma 1.2.2. Using the map $\xi_{[0,r)}$ of Lemma 1.2.6, we obtain a filtration on $(D(\mathcal{M}) \otimes_{K_0} \mathcal{O})_{[0,r)}$. The required filtration on $D(\mathcal{M})_K$ is defined to be the image filtration under the composite

$$(D(\mathcal{M}) \otimes_{K_0} \mathcal{O})_{[0,r)} \to D(\mathcal{M}) \otimes_{K_0} \mathcal{O}/E(u)\mathcal{O} \xrightarrow{\sim} D(\mathcal{M}) \otimes_{K_0} K = D(\mathcal{M})_K.$$

Finally, if $\mathcal{M}$ is a (φ, N_∇)-module over $\mathcal{O}$ of finite E-height, then we equip $D(\mathcal{M})$ with a K_0-linear operator N, by reducing the operator N_∇ on $\mathcal{M}$ modulo u. This gives $D(\mathcal{M})$ the structure of a filtered (φ, N)-module.

We will show that the functors D and $\mathcal{M}$ induce quasi-inverse equivalences of categories.

Proposition 1.2.8. *Let D be an effective filtered (φ, N)-module. There is a natural isomorphism of filtered (φ, N)-modules $D(\mathcal{M}(D)) \xrightarrow{\sim} D$.*

Proof. As in Lemma 1.2.2, we set $\mathcal{D}_0 = (\mathcal{O}[\ell_u] \otimes_{K_0} D)^{N=0}$. The natural inclusion $\mathcal{D}_0 \subset \mathcal{M}(D)$ is an isomorphism at $u = 0$, so that

$$D(\mathcal{M}(D)) = \mathcal{M}(D) \otimes_{\mathcal{O}} \mathcal{O}/u\mathcal{O} \xrightarrow{\sim} (K_0[\ell_u] \otimes_{K_0} D)^{N=0} \tag{1.2.9}$$

We claim that the composite map

$$\eta : (K_0[\ell_u] \otimes_{K_0} D)^{N=0} \subset K_0[\ell_u] \otimes_{K_0} D \overset{\ell_u \mapsto 0}{\to} D. \tag{1.2.10}$$

is an isomorphism of filtered (φ, N)-modules, where on the left-hand side N acts by $-N \otimes 1$. This is the operator induced by reducing the operator $N_\nabla \otimes 1$ on $\mathcal{O}[\ell_u] \otimes_{K_0} D$ modulo u. First, one checks easily that η is an injection, and that both sides have the same dimension. Hence η is a bijection. Since both maps in (1.2.10) are evidently compatible with φ, so is the composite. Finally, suppose that $d = \sum_{j \geq 0} d_j \ell_u^j \in (K_0[\ell_u] \otimes_{K_0} D)^{N=0}$, with $d_j \in D$. Since $N(d) = 0$, we see that $N(d_0) + d_1 = 0$. Hence

$$\eta(N_\nabla \otimes 1(d)) = -d_1 = N(d_0) = N(\eta(d)),$$

so η is compatible with N.

It remains to check that η is strictly compatible with filtrations. As remarked in the proof of Lemma 1.2.2, the submodule $\mathcal{D}_0 \subset \mathcal{M}(D)$, is contained in $(1 \otimes \varphi)(\varphi^*\mathcal{M})$, and this containment is an isomorphism at x_0. By definition of $\mathcal{M}$, an element $d \in \mathcal{D}_0$ is in $E(u)^i\mathcal{M}$ if and only if $\iota_0(d) \in \mathrm{Fil}^i(\widehat{\mathfrak{S}}_0 \otimes_K D_K)$. Hence, using Lemma 1.2.1, one sees that under the isomorphisms

$$\begin{aligned} D(\mathcal{M}(D))_K &= (K_0[\ell_u] \otimes_{K_0} D)^{N=0} \otimes_{K_0} \mathcal{O}/E(u)\mathcal{O} = \mathcal{D}_0/E(u)\mathcal{D}_0 \\ &\underset{\iota_0}{\xrightarrow{\sim}} \widehat{\mathfrak{S}}_0 \otimes_K D_K/(u-\pi)\widehat{\mathfrak{S}}_0 \otimes_K D_K = D_K, \end{aligned} \tag{1.2.11}$$

the filtration on $D(\mathcal{M}(D))_K$ is identified with the given filtration on D_K.

Thus, to show that η is strictly compatible with filtrations, we have to check that the composite

$$D \overset{\eta^{-1}}{\to} D(\mathcal{M}(D)) \hookrightarrow D(\mathcal{M}(D))_K \overset{(1.2.11)}{\to} D_K.$$

is the natural inclusion. However, this is clear because both η and (1.2.11) send an element $\sum_{i\geq 0} \ell_u^i d_i \in (K_0[\ell_u] \otimes_{K_0} D)^{N=0}$ to d_0. □

Lemma 1.2.12. *Let $\mathcal{M}$ be in $\mathrm{Mod}_{/\mathcal{O}}^{\varphi,N_\nabla}$. Then we have*

(1) *The $\mathcal{O}$-submodule $(1 \otimes \varphi)\varphi^*\mathcal{M} \subset \mathcal{M}$ is stable under N_∇.*
(2) *For $i \geq 0$, $N_\nabla(E(u)^i\mathcal{M}) \subset E(u)^i\mathcal{M}$. In particular, if we identify $\varphi^*\mathcal{M}$ with $(1 \otimes \varphi)\varphi^*\mathcal{M}$ via $1 \otimes \varphi$, then N_∇ respects the filtration on $\varphi^*\mathcal{M}$ defined in Section* 1.2.7.
(3) *The map*

$$\begin{aligned} (\mathcal{O}[\ell_u] \otimes_{K_0} D(\mathcal{M}))^{N=0} &= (K_0[\ell_u] \otimes_{K_0} D(\mathcal{M}))^{N=0} \otimes_{K_0} \mathcal{O} \\ &\overset{\eta\otimes 1}{\to} D(\mathcal{M}) \otimes_{K_0} \mathcal{O} \overset{\xi}{\to} \mathcal{M} \end{aligned}$$

is compatible with N_∇. Here η is the isomorphism of (1.2.10), *and N_∇ acts on the left via its action on $\mathcal{O}[\ell_u]$.*
(4) *For $i \geq 1$, applying $\widehat{\mathfrak{S}}_0 \otimes_{\mathcal{O}}$ to the map of* (3) *and using the isomorphism of Lemma* 1.2.1(1) *induces an isomorphism*

$$\sum_{j\geq 0} E(u)^j \widehat{\mathfrak{S}}_0 \otimes_K \mathrm{Fil}^{i-j} D(\mathcal{M})_K \underset{\xi\circ(\eta\otimes 1)}{\xrightarrow{\sim}} \widehat{\mathfrak{S}}_0 \otimes_{\mathcal{O}} (1 \otimes \varphi)(\mathrm{Fil}^i \varphi^*\mathcal{M}).$$

Proof. (1) follows from the relation $N_\nabla\varphi = E(u)\varphi N_\nabla$, while (2) follows from the Leibniz rule for N_∇, and the fact that $N_\nabla(E(u)) = -u\lambda E'(u)iE(u)^{i-1}$, since $E(u)$ divides λ in $\mathcal{O}$.

For (3) let $\sigma = N_\nabla \circ (\xi \circ \eta) - (\xi \circ \eta) \circ N_\nabla$, and write $D_0(\mathcal{M}) = (K_0[\ell_u] \otimes_{K_0} D(\mathcal{M}))^{N=0}$. Then σ is $\mathcal{O}$-linear, and it suffices to show that $\sigma(D_0(\mathcal{M})) = 0$. Since the map η of (1.2.10) is compatible with N, and ξ reduces to the identity modulo u, we have $\sigma(D_0(\mathcal{M})) \subset u\mathcal{M}$. On the other hand, $\xi \circ \eta$ is compatible with φ, so that $\sigma \circ \varphi = pE(u)/c_0\varphi \circ \sigma$, and for $i \geq 1$

$$\begin{aligned}\sigma(D_0(\mathcal{M})) &= \sigma\circ\varphi^i(D_0(\mathcal{M}))\\ &= p^i E(u)/c_0\varphi(E(u)/c_0)\dots\varphi^{i-1}(E(u)/c_0)\varphi^i\circ\sigma(D_0(\mathcal{M}))\\ &\subset \mathcal{O}\cdot\varphi^i(u\mathcal{M})\subset u^{p^i}\mathcal{M}.\end{aligned}$$

It follows that $\sigma = 0$, which proves (3).

Finally, for (4) it will be convenient to again denote by N_∇ the operator $-u\lambda\frac{d}{du}$ on $\widehat{\mathfrak{S}}_0$, and to extend $N_\nabla^{\mathcal{M}}$ to a differential operator on $\widehat{\mathfrak{S}}_0\otimes_{\mathcal{O}}\mathcal{M}$, which we again denote by N_∇. By (2), N_∇ leaves $\widehat{\mathfrak{S}}_0\otimes_{\mathcal{O}}(1\otimes\varphi)(\mathrm{Fil}^i\,\varphi^*\mathcal{M})$ stable.

Set $M_i = \widehat{\mathfrak{S}}_0\otimes_{\mathcal{O}}(E(u)(1\otimes\varphi)\varphi^*\mathcal{M}\cap E(u)^i\mathcal{M})$ for $i\geq 1$. We have a commutative diagram with exact rows

$$\begin{array}{ccccccccc} 0 & \longrightarrow & M_i & \longrightarrow & M_{i-1} & \longrightarrow & M_{i-1}/M_i & \longrightarrow & 0\\ & & \big\downarrow{\scriptstyle N_\nabla|_{M_i}} & & \big\downarrow{\scriptstyle N_\nabla|_{M_{i-1}}} & & \big\downarrow & & \\ 0 & \longrightarrow & M_i & \longrightarrow & M_{i-1} & \longrightarrow & M_{i-1}/M_i & \longrightarrow & 0,\end{array}$$

where the vertical maps are induced by N_∇. We claim that $N_\nabla|_{M_i}$ is a bijection for $i\geq 0$.

By Lemmas 1.2.1(1) and 1.2.6 and (3) above, we have an N_∇-compatible isomorphism

$$\widehat{\mathfrak{S}}_0\otimes_{K_0} D(\mathcal{M})\xrightarrow{\sim}\widehat{\mathfrak{S}}_0\otimes_K D(\mathcal{M})_K\xrightarrow{\sim}\widehat{\mathfrak{S}}_0\otimes_{\mathcal{O}}\varphi^*\mathcal{M},$$

where N_∇ acts on the left via $N_\nabla\otimes 1$. Since N_∇ induces a bijection on $E(u)\widehat{\mathfrak{S}}_0$, our claim holds for $i=0$. For $i\geq 1$, we may assume by induction that $N_\nabla|_{M_{i-1}}$ is a bijection. Hence $N_\nabla|_{M_{i-1}/M_i}$ is surjective, and it is therefore injective as M_{i-1}/M_i is a finite-dimensional K-vector space. Finally, it follows from the snake lemma that $N_\nabla|_{M_i}$ is surjective. In particular, we see that $N_\nabla|_{M_0/M_i}$ is bijective for all i.

To prove (4), we proceed by induction on i. For $i=0$, this follows from Lemma 1.2.6. For $i\geq 1$, the induction hypothesis implies that

$$(u-\pi)\,\mathrm{Fil}^{i-1}(\widehat{\mathfrak{S}}_0\otimes_K D(\mathcal{M})_K) = (u-\pi)\widehat{\mathfrak{S}}_0\otimes_{\mathcal{O}}(1\otimes\varphi)(\mathrm{Fil}^{i-1}\,\varphi^*\mathcal{M}).$$

Since the filtrations on $\varphi^*\mathcal{M}$ and on $\widehat{\mathfrak{S}}_0\otimes_K D(\mathcal{M})_K$ both induce the same filtration on their common quotient $D(\mathcal{M})_K$, it suffices to show that

$$\xi\circ(\eta\otimes 1)(\mathrm{Fil}^i\,D(\mathcal{M})_K)\subset\widehat{\mathfrak{S}}_0\otimes_{\mathcal{O}}(1\otimes\varphi)(\mathrm{Fil}^i\,\varphi^*\mathcal{M}).$$

Let $d\in\xi\circ(\eta\otimes 1)(\mathrm{Fil}^i\,D(\mathcal{M})_K)$. We may write $d = d_0+d_1$, with $d_0\in\widehat{\mathfrak{S}}_0\otimes_{\mathcal{O}}\mathrm{Fil}^i\,\varphi^*\mathcal{M}$, and $d_1\in E(u)\widehat{\mathfrak{S}}_0\otimes_{\mathcal{O}}\varphi^*\mathcal{M} = M_0$. Since $N_\nabla(d)=0$,

$$N_\nabla(d_1) = -N_\nabla(d_0)\in\widehat{\mathfrak{S}}_0\otimes_{\mathcal{O}}\mathrm{Fil}^i\,\varphi^*\mathcal{M}\cap E(u)\widehat{\mathfrak{S}}_0\otimes_{\mathcal{O}}\varphi^*\mathcal{M} = M_i.$$

Hence, by what we saw above, we must have $d_1\in M_i\subset\widehat{\mathfrak{S}}_0\otimes_{\mathcal{O}}\mathrm{Fil}^i\,\varphi^*\mathcal{M}$. □

Proposition 1.2.13. *Let $\mathcal{M}$ be in $\mathrm{Mod}_{/\mathcal{O}}^{\varphi,N_\nabla}$. There is a canonical isomorphism $\mathcal{M}(D(\mathcal{M}))\xrightarrow{\sim}\mathcal{M}$.*

Proof. Let $\mathcal{M}' = \mathcal{M}(D(\mathcal{M}))$. We will write $\mathcal{D}_0(\mathcal{M}) = (\mathcal{O}[\ell_u] \otimes_{K_0} D(\mathcal{M}))^{N=0}$. By construction $\mathcal{M}' \subset \mathcal{D}_0(\mathcal{M})[1/\lambda]$. On the other hand, if we identify $\mathcal{D}_0(\mathcal{M})$ with an $\mathcal{O}$-submodule of $\mathcal{M}$ via the map $\xi \circ (\eta \otimes 1)$ of Lemma 1.2.12(3), then $\mathcal{M} \subset \mathcal{D}_0(\mathcal{M})[1/\lambda]$, by Lemma 1.2.6. Since both these inclusions are compatible with N_∇ and φ, it suffices to check that $\mathcal{M}' = \mathcal{M}$.

It is enough to check that $\mathcal{M}'_{[0,r)} = \mathcal{M}_{[0,r)}$ where $r \in (|\pi|, |\pi|^{1/p})$, for then pulling back by $(\varphi^*)^i$, and using the fact that $\mathcal{M}$ and $\mathcal{M}'$ are both of finite E-height, we find that $\mathcal{M}'_{[0,r^{1/p^i})} = \mathcal{M}_{[0,r^{1/p^i})}$, and hence that $\mathcal{M} = \mathcal{M}'$.

Now at any point of $D[0,r)$ other than x_0, we have $\mathcal{M} = \mathcal{D}_0(\mathcal{M}) = \mathcal{M}'$, so we have to check that

$$\widehat{\mathfrak{S}}_0 \otimes_{\mathcal{O}} \mathcal{M} = \widehat{\mathfrak{S}}_0 \otimes_{\mathcal{O}} \mathcal{M}' \subset \widehat{\mathfrak{S}}_0 \otimes_{\mathcal{O}} \mathcal{D}_0(\mathcal{M})[1/\lambda] \underset{1.2.1}{\xrightarrow{\sim}} \widehat{\mathfrak{S}}_0[1/(u-\pi)] \otimes_K D(\mathcal{M})_K$$

For this it suffices to check that an element $x \in \widehat{\mathfrak{S}}_0 \otimes_{\mathcal{O}} \mathcal{D}_0(\mathcal{M})$ is divisible by $E(u)^i$ in $\mathcal{M}$ for some $i \geq 0$ if and only if it is divisible by $E(u)^i$ in $\mathcal{M}'$. Now by Lemma 1.2.6, and the observations made in the proof of Lemma 1.2.2, we have

$$\widehat{\mathfrak{S}}_0 \otimes_{\mathcal{O}} (1 \otimes \varphi)\varphi^*\mathcal{M} = \widehat{\mathfrak{S}}_0 \otimes_{\mathcal{O}} \mathcal{D}_0(\mathcal{M}) = \widehat{\mathfrak{S}}_0 \otimes_{\mathcal{O}} (1 \otimes \varphi)\varphi^*\mathcal{M}', \qquad (1.2.14)$$

so it is enough to show that the filtrations on the left- and right-hand sides of (1.2.14), defined in Section 1.2.7, coincide. This follows by comparing Lemma 1.2.1(2) with Lemma 1.2.12(4). □

Theorem 1.2.15. *The functors D and $\mathcal{M}$ induce exact, quasi-inverse equivalences of $\otimes$-categories between effective filtered (φ, N)-modules and the category* $\mathrm{Mod}_{/\mathcal{O}}^{\varphi,N_\nabla}$.

Proof. By Propositions 1.2.8 and 1.2.13 we know that D and $\mathcal{M}$ induce quasi-inverse equivalences of categories. It remains to check that they they are exact and compatible with tensor products.

Consider a sequence of filtered (φ, N)-modules

$$D^\bullet : 0 \to D'' \to D \to D' \to 0$$

and denote by $\mathcal{M}(D^\bullet)$ the corresponding sequence of (φ, N_∇)-modules over $\mathcal{O}$. If $D^\bullet$ is exact then, thinking of $\mathcal{M}(D^\bullet)$ as a sequence of coherent sheaves on $D[0,1)$, we see that it is evidently exact outside the set of points $\{x_n\}_{n\geq 0}$, and the exactness at x_n follows from Lemma 1.2.1(2). Conversely, if $\mathcal{M}(D^\bullet)$ is exact then Lemma 1.2.1(2) implies that $D^\bullet$ is exact. Thus $\mathcal{M}$ and D are exact functors.

Suppose we are given filtered (φ, N)-modules D_1 and D_2. There is an obvious morphism of (φ, N_∇)-modules over $\mathcal{O}$, $\mathcal{M}(D_1) \otimes_{\mathcal{O}} \mathcal{M}(D_2) \to \mathcal{M}(D_1 \otimes_{K_0} D_2)$, which is an isomorphism outside the points $\{x_n\}_{n\geq 0}$. That it is an isomorphism at x_n follows from Lemma 1.2.1(2). Hence $\mathcal{M}$ commutes with tensor products.

Finally, suppose that $\mathcal{M}_1$ and $\mathcal{M}_2$ are (φ, N_∇)-modules over $\mathcal{O}$. From the definitions, one sees that there is an isomorphism $D(\mathcal{M}_1) \otimes_{K_0} D(\mathcal{M}_2) \xrightarrow{\sim} D(\mathcal{M}_1 \otimes_{\mathcal{O}} \mathcal{M}_2)$ compatible with the action of φ, and that the map of K-vector spaces obtained by tensoring both sides by $\otimes_{K_0} K$ is compatible with filtrations. That it is strictly compatible with filtrations may be deduced from the strict compatibility with filtrations of the map

$$\varphi^*\mathcal{M}_1 \otimes_{\mathcal{O}} \varphi^*\mathcal{M}_2 \to \varphi^*(\mathcal{M}_1 \otimes_{\mathcal{O}} \mathcal{M}_2). \qquad \square$$

1.3 Weakly admissible modules and F-crystals

In this section we show how to produce (φ, N_∇)-modules over $\mathcal{O}$ using φ-modules over $\mathfrak{S}$ of finite E-height.

1.3.1. We begin by reviewing the results of Kedlaya [Ke 1], [Ke 2]. Recall that the Robba ring $\mathcal{R}$ is defined by

$$\mathcal{R} = \lim_{r \to 1^-} \mathcal{O}_{(r,1)}$$

$\mathcal{R}$ is equipped with a Frobenius φ induced by the maps $\varphi : \mathcal{O}_{(r,1)} \to \mathcal{O}_{(r^{1/p},1)}$. We denote by $\mathrm{Mod}^{\varphi}_{/\mathcal{R}}$ the category of finite free $\mathcal{R}$-modules $\mathcal{M}$ equipped with an isomorphism $\varphi^*\mathcal{M} \xrightarrow{\sim} \mathcal{M}$. This has a natural structure of a Tannakian category.

We also have the bounded Robba ring $\mathcal{R}^b$, defined by

$$\mathcal{R} = \lim_{r \to 1^-} \mathcal{O}^b_{(r,1)}$$

where $\mathcal{O}^b_{(r,1)} \subset \mathcal{O}_{(r,1)}$ denotes the functions on $D(r,1)$ which are bounded. The ring $\mathcal{R}^b$ is a discrete valuation field, with a valuation $v_{\mathcal{R}^b}$ given by

$$v_{\mathcal{R}^b}(f) = -\log_p \lim_{r \to 1^-} \sup_{x \in D(r,1)} |f(x)|$$

The Frobenius φ on $\mathcal{R}$ induces a Frobenius φ on $\mathcal{R}^b$. We denote by $\mathrm{Mod}^{\varphi}_{/\mathcal{R}^b}$ the category of finite-dimensional $\mathcal{R}^b$-vector spaces $\mathcal{M}$ equipped with an isomorphism $\varphi^*\mathcal{M} \xrightarrow{\sim} \mathcal{M}$.

Kedlaya defines an $\mathcal{R}$-algebra $\mathcal{R}^{\mathrm{alg}}$ (denoted by $\Gamma^{\mathrm{alg}}_{\mathrm{an,con}}$ in [Ke 1]), which contains a copy of $W(\bar{k})$, where $\bar{k}$ denotes an algebraic closure of k, is equipped with a lifting φ of the Frobenius on $\mathcal{R}$, and such that for any $\mathcal{M}$ in $\mathrm{Mod}^{\varphi}_{/\mathcal{R}}$, there exists a finite extension E of $W(\bar{k})[1/p]$ such that $\mathcal{M} \otimes_{\mathcal{R}} \mathcal{R}^{\mathrm{alg}} \otimes_{W(\bar{k})[1/p]} E$ admits a basis of φ-eigenvectors $\mathbf{v}_1, \dots \mathbf{v}_n$ such that $\varphi(\mathbf{v}_i) = \alpha_i \mathbf{v}_i$ for some $\alpha_i \in E$. The set of p-adic valuations of $\alpha_1, \dots, \alpha_n$ is uniquely determined by $\mathcal{M}$, and called the set of slopes of $\mathcal{M}$ [Ke 1, Theorem 4.16]. If these are all equal to some $s \in \mathbb{Q}$, then $\mathcal{M}$ is called pure of slope s. We denote by $\mathrm{Mod}^{\varphi,s}_{/\mathcal{R}}$ the full subcategory of $\mathrm{Mod}^{\varphi}_{/\mathcal{R}}$ consisting of modules which are pure of slope s. We write $\mathrm{Mod}^{\varphi,s}_{/\mathcal{R}^b}$ for the full subcategory of $\mathrm{Mod}^{\varphi}_{/\mathcal{R}^b}$ consisting of modules which are pure of slope s (as φ-modules over a discretely valued field).

Theorem 1.3.2. (1) *The functor $\mathcal{M} \mapsto \mathcal{M} \otimes_{\mathcal{R}^b} \mathcal{R}$ induces an equivalence*

$$\mathrm{Mod}^{\varphi,s}_{/\mathcal{R}^b} \xrightarrow{\sim} \mathrm{Mod}^{\varphi,s}_{/\mathcal{R}}.$$

(2) *For any* $\mathcal{M}$ *in* $\mathrm{Mod}^{\varphi}_{/\mathcal{R}}$*, there exists a canonical filtration—called the slope filtration—*$0 = \mathcal{M}_0 \subset \mathcal{M}_1 \subset \cdots \subset \mathcal{M}_r = \mathcal{M}$ *by* φ*-stable submodules such that* $\mathcal{M}_i/\mathcal{M}_{i-1}$ *is finite free over* $\mathcal{R}$ *and pure of slope* s_i*, and* $s_1 < s_2 < \cdots < s_r$.

Proof. The first part is [Ke 2, Theorem 6.3.3], while the second follows from [Ke 1, Theorem 6.10]. □

1.3.3. We want to show that if $\mathcal{M}$ in $\mathrm{Mod}^{\varphi}_{/\mathcal{R}}$ arises from a module $\mathcal{M}_{\mathcal{O}}$ in $\mathrm{Mod}^{\varphi,N_\nabla}_{/\mathcal{O}}$, then the slope filtration of (4) is induced by a filtration on $\mathcal{M}_{\mathcal{O}}$.

We denote by N_∇ the operator $-u\lambda\frac{d}{du}$ on $\mathcal{R}$ and we write $\mathrm{Mod}^{\varphi,N_\nabla}_{/\mathcal{R}}$ for the category whose objects consist of a module $\mathcal{M}$ in $\mathrm{Mod}^{\varphi}_{/\mathcal{R}}$ equipped with a differential operator $N_\nabla = N^{\mathcal{M}}_\nabla$ over the operator N_∇ on $\mathcal{R}$, such that $N_\nabla\varphi = (pE(u)/c_0)\varphi N_\nabla$.

For $\mathcal{M}$ a finite free $\mathcal{R}$-module (respectively, an $\mathcal{O}_I$-module for some interval $I \subset [0,1)$,) we say that an $\mathcal{R}$ (respectively, $\mathcal{O}_I$) submodule $\mathcal{N} \subset \mathcal{M}$ is saturated if it is finitely generated and if $\mathcal{M}/\mathcal{N}$ is torsion-free or, equivalently, free over $\mathcal{R}$ (respectively, $\mathcal{O}_I$). If $\mathcal{N} \subset \mathcal{M}$ is any submodule, then there is a smallest submodule $\mathcal{N}' \subset \mathcal{M}$ containing $\mathcal{N}$ which is saturated, and we call this the saturation of $\mathcal{N}$.

Lemma 1.3.4. *Let* $\mathcal{M}$ *be a finite free* $\mathcal{O}$*-module equipped with a* φ*-semilinear map* $\varphi : \mathcal{M} \to \mathcal{M}$ *such that the induced map* $\varphi^*\mathcal{M} \to \mathcal{M}$ *is an injection. Let* $\mathcal{N}_{\mathcal{R}} \subset \mathcal{M}_{\mathcal{R}} := \mathcal{M} \otimes_{\mathcal{O}} \mathcal{R}$ *be a saturated submodule which is stable under* φ*. Then there is a unique saturated submodule* $\mathcal{N}_{(0,1)} \subset \mathcal{M}_{(0,1)}$ *such that* $\mathcal{N}_{(0,1)} \otimes_{\mathcal{O}_{(0,1)}} \mathcal{R} = \mathcal{N}_{\mathcal{R}}$*.* $\mathcal{N}_{(0,1)}$ *is* φ*-stable.*

Proof. Since $\mathcal{N}_{\mathcal{R}}$ is finitely generated, there exists $r \in (0,1)$ and a saturated $\mathcal{O}_{(r,1)}$-submodule $\mathcal{N}_{(r,1)} \subset \mathcal{M}_{(r,1)}$ such that $\mathcal{N}_{(r,1)} \otimes_{\mathcal{O}_{(r,1)}} \mathcal{R} = \mathcal{N}$. Since $\mathcal{N}_{(r,1)}$ is clearly the unique such saturated submodule of $\mathcal{M}_{(r,1)}$, $1 \otimes \varphi$ induces a map $\varphi^*\mathcal{N}_{(r,1)} \to \mathcal{N}_{(r^{1/p},1)}$.

Set $\mathcal{N}_{(r^p,1)} = \mathcal{M}_{(r^p,1)} \cap \mathcal{N}_{(r,1)}$. Since $\mathcal{N}_{(r^p,1)}$ is clearly a closed $\mathcal{O}_{(r^p,1)}$-submodule, it is finitely generated, and one sees immediately that it is saturated. We claim that its rank is equal to $h = \mathrm{rk}_{\mathcal{O}_{(r,1)}} \mathcal{N}_{(r,1)}$. It suffices to show that $\varphi^*\mathcal{N}_{(r^p,1)}$ has $\mathcal{O}_{(r,1)}$-rank h. Since the map $\varphi : \mathcal{O}_{(r,1)} \to \mathcal{O}_{(r^p,1)}$ is finite flat, we have a commutative diagram with exact rows

$$\begin{array}{ccccccc} 0 \longrightarrow & \varphi^*\mathcal{N}_{(r^p,1)} & \longrightarrow & \varphi^*\mathcal{M}_{(r^p,1)} \oplus \varphi^*\mathcal{N}_{(r,1)} & \longrightarrow & \varphi^*\mathcal{M}_{(r,1)} \\ & \downarrow & & \downarrow & & \downarrow \\ 0 \longrightarrow & \mathcal{N}_{(r,1)} & \longrightarrow & \mathcal{M}_{(r,1)} \oplus \mathcal{N}_{(r^{1/p},1)} & \longrightarrow & \mathcal{M}_{(r^{1/p},1)}. \end{array}$$

Since the central and right vertical maps are injective, and the cokernel of the central vertical map is a torsion $\mathcal{O}_{(r,1)}$-module, we see that $\mathrm{rk}_{\mathcal{O}_{(r,1)}} \varphi^*\mathcal{N}_{(r^p,1)} = \mathrm{rk}_{\mathcal{O}_{(r,1)}} \mathcal{N}_{(r,1)} = h$.

Since $\mathcal{N}_{(r^p,1)} \otimes_{\mathcal{O}_{(r^p,1)}} \mathcal{O}_{(r,1)} \subset \mathcal{N}_{(r,1)}$ and both modules are saturated $\mathcal{O}_{(r,1)}$-submodules of $\mathcal{M}_{(r,1)}$ of the same rank, this inclusion must be an equality. Repeating the argument, we obtain for each $i \geq 0$ a saturated $\mathcal{O}_{(r^{p^i},1)}$-submodule $\mathcal{N}_{(r^{p^i},1)}$ of

$\mathcal{M}_{(r^{p^i},1)}$ such that the restriction of $\mathcal{N}_{(r^{p^i},1)}$ to $D(r^{p^{i-1}}, 1)$ is $\mathcal{N}_{(r^{p^{i-1}},1)}$. These modules glue to a coherent sheaf $\underline{\mathcal{N}}_{(0,1)}$ on $D(0, 1)$. Write $\mathcal{N}_{(0,1)}$ for the global sections of $\underline{\mathcal{N}}_{(0,1)}$. Then $\mathcal{N}_{(0,1)}$ is a closed $\mathcal{O}_{(0,1)}$-submodule of $\mathcal{M}_{(0,1)}$, and hence finitely generated, and $\underline{\mathcal{N}}_{(0,1)}$ is the coherent sheaf corresponding to $\mathcal{N}_{(0,1)}$. In particular, we see that $\mathcal{N}_{(0,1)} \subset \mathcal{M}_{(0,1)}$ is saturated and that $\mathcal{N}_{(0,1)} \otimes_{\mathcal{O}_{(0,1)}} \mathcal{R} = \mathcal{N}_{\mathcal{R}}$. Since $\mathcal{N}_{(0,1)} = \mathcal{M}_{(0,1)} \cap \mathcal{N}_{\mathcal{R}}$ is the unique saturated submodule with this property, we see that $\mathcal{N}_{(0,1)}$ is stable under φ. □

Lemma 1.3.5. *Let $\mathcal{M}$ be a finite free $\mathcal{O}$-module equipped with a differential operator ∂ over $-u\frac{d}{du}$, and suppose that the operator $N : \mathcal{M}/u\mathcal{M} \to \mathcal{M}/u\mathcal{M}$ induced by ∂ is nilpotent. If $\mathcal{N}_{(0,1)} \subset \mathcal{M}_{(0,1)}$ is a saturated $\mathcal{O}_{(0,1)}$-submodule which is stable under ∂, then $\mathcal{N}_{(0,1)}$ extends uniquely to a saturated, ∂-stable $\mathcal{O}$-submodule $\mathcal{N} \subset \mathcal{M}$.*

Proof. This is part of the theory of connections with regular singular points. In fact one can even suppress the assumption on the nilpotence of N (cf. [De, Proposition 5.4]). Since we could not find a good reference, and for the convenience of the reader, we give a proof here. Closely related arguments may be found in the literature—see, for example, [Ba] and [An].

We equip $\mathcal{M}$ with the connection given by $\nabla(m) = -u^{-1}\partial(m)du$, and $\mathcal{M}' := \mathcal{M}/u\mathcal{M} \otimes_{K_0} \mathcal{O}$ with the logarithmic connection given by $\nabla(m \otimes f) = -N(m)/u \otimes f du + m \otimes df$. Then $\mathrm{Hom}_{\mathcal{O}}(\mathcal{M}', \mathcal{M})$ is naturally equipped with a logarithmic connection, given by $\nabla(f)(m') = -f(\nabla(m')) + \nabla(f(m'))$. Let $s_0 : \mathcal{M}/u\mathcal{M} \to \mathcal{M}$ be any K_0-linear map lifting the identity on $\mathcal{M}/u\mathcal{M}$. We define a new section by

$$s = \sum_{i=0}^{\infty} \nabla\left(\frac{d}{du}\right)^i (s_0)(-u)^i/i!.$$

Note that since s_0 lifts the identity section, $\nabla(\frac{d}{du})^i(s_0)(-u)^i/i!$ sends $\mathcal{M}/u\mathcal{M}$ into $u\mathcal{M}$ for each $i \geq 1$. Moreover, since N is nilpotent, this summand sends $\mathcal{M}/u\mathcal{M}$ into $u^{[i/d]}\mathcal{M}$, where d denotes the rank of $\mathcal{M}$. Using this one sees easily that there is a positive integer n such that the formula for s gives a well defined section $s : \mathcal{M}/u\mathcal{M} \to \mathcal{M}$ over $D[0, p^{-n})$. After replacing u by u/p^n, we may assume that s gives a well defined section over $D[0, 1)$, so that $\mathcal{M} \xrightarrow{\sim} \mathcal{M}'$ as $\mathcal{O}$-modules with logarithmic connection. In particular, $\mathcal{M}^{\partial^d=0} \subset \mathcal{M}$ is a K_0-vector space of dimension d, which spans $\mathcal{M}$.

Now let $\mathcal{L}$ be any finite free $\mathcal{O}_{(0,1)}$-module equipped with a connection ∇, and define $\partial = \nabla(-u\frac{d}{du})$. We claim that the natural map $\mathcal{L}^{\partial^d=0} \otimes_{K_0} \mathcal{O}_{(0,1)} \to \mathcal{L}$ is injective. To see this we remark that we may replace $\mathcal{L}$ by the image of the above map, and assume that $\mathcal{L}^{\partial^d=0}$ spans $\mathcal{L}$. Note also that, if $\mathcal{L}'$ and $\mathcal{L}''$ are two finite free $\mathcal{O}_{(0,1)}$-modules with connection, and $\mathcal{L}$ is an extension of $\mathcal{L}'$ by $\mathcal{L}''$, then it suffices to prove the claim for $\mathcal{L}'$ and $\mathcal{L}''$. Furthermore if $\mathcal{L}' \subset \mathcal{L}$ is any $\mathcal{O}_{(0,1)}$-submodule which is stable by ∇, then $\mathcal{L}/\mathcal{L}'$ is equipped with a connection, and is hence $\mathcal{O}_{(0,1)}$-free [Kat 2, Proposition 8.9]. Applying these remarks with $\mathcal{L}' = a \cdot \mathcal{O}$ where $a \in \mathcal{L}^{\partial=0}$ is nonzero, and using induction on the rank of $\mathcal{L}$, it suffices to consider the case where $\mathcal{L} = \mathcal{O}_{(0,1)}$ with the trivial connection. In this case the result is clear since any

$f \in \mathcal{O}_{(0,1)}$ can be written as a convergent sum $f = \sum_{i\in\mathbb{Z}} a_i u^i$, so that $\partial^d(f) = 0$ if and only if f is constant.

Now the natural map $\mathcal{M}^{\partial^d=0} \otimes_{K_0} \mathcal{O} \to \mathcal{M}$ is evidently an isomorphism. Hence, applying the above remarks with $\mathcal{L} = \mathcal{N}_{(0,1)}$ and $\mathcal{M}_{(0,1)}/\mathcal{N}_{(0,1)}$, and using the snake lemma, we see that the map $\mathcal{N}_{(0,1)}^{\partial^d=0} \otimes_{K_0} \mathcal{O}_{(0,1)} \to \mathcal{N}_{(0,1)}$ is an isomorphism. In particular, $\mathcal{N}_{(0,1)}$ extends to $\mathcal{N} = \mathcal{N}_{(0,1)}^{\partial^d=0} \otimes_{K_0} \mathcal{O}$. □

Lemma 1.3.6. *Let $\mathcal{R}\xi$ denote a free $\mathcal{R}$-module of rank 1 with a generator ξ. We think of $\mathcal{L} := \mathcal{R} \oplus \mathcal{R}\xi$ as a right $\mathcal{R}$-module by setting $\xi \cdot a = a\xi + N_\nabla(a)$, and letting $\mathcal{R}$ act on itself in the natural way. This makes $\mathcal{L}$ into an $(\mathcal{R}, \mathcal{R})$-bimodule.*

Let $\mathcal{M}$ be in $\mathrm{Mod}_{/\mathcal{R}}^{\varphi,N_\nabla}$. Then $\mathcal{L} \otimes_{\mathcal{R}} \mathcal{M}$ has a natural structure of an object of $\mathrm{Mod}_{/\mathcal{R}}^{\varphi}$ given by

$$\varphi(a \otimes n + b\xi \otimes m) = \varphi(a)\varphi(n) + \varphi(b)(pE(u)/c_0)^{-1}\xi \otimes \varphi(m),$$

and the set of slopes of $\mathcal{L} \otimes_{\mathcal{R}} \mathcal{M}$ is equal to those of $\mathcal{M}$. More precisely, if s is a slope of $\mathcal{M}$ which appears with multiplicity h, then s appears with multiplicity $2h$ in $\mathcal{L} \otimes_{\mathcal{O}} \mathcal{M}$.

Proof. First, observe that the formula giving φ defines a well defined Frobenius because

$$\begin{aligned}\varphi(\xi \otimes bm) &= (pE(u)/c_0)^{-1}\xi \otimes \varphi(bm)\\ &= (pE(u)/c_0)^{-1}\varphi(b)\xi \otimes \varphi(m) + (pE(u)/c_0)^{-1}N_\nabla(\varphi(b))\varphi(m)\\ &= \varphi(b\xi \otimes m + N_\nabla(b) \otimes m).\end{aligned}$$

To prove the second claim, we may reduce by dévissage to the case where $\mathcal{M}$ is irreducible and of pure slope $s \in \mathbb{Q}$. Then we have an exact sequence in $\mathrm{Mod}_{/\mathcal{R}}^{\varphi}$

$$0 \to \mathcal{M} \to \mathcal{L} \otimes_{\mathcal{R}} \mathcal{M} \to \mathcal{M}(1) \to 0.$$

Here $\mathcal{M}(1)$ denotes the object of $\mathrm{Mod}_{/\mathcal{R}}^{\varphi}$ whose underlying $\mathcal{R}$-module is equal to $\mathcal{M}$, but whose Frobenius is the Frobenius on $\mathcal{M}$ multiplied by $(pE(u)/c_0)^{-1}$. The first map is given by $m \mapsto m \oplus 0$, while the second sends $a + b\xi \otimes m$ to bm. It follows by [Ke 1, Proposition 4.5] that $\mathcal{L} \otimes_{\mathcal{R}} \mathcal{M}$ is pure of slope s. □

Proposition 1.3.7. *Let $\mathcal{M}$ be in $\mathrm{Mod}_{/\mathcal{O}}^{\varphi,N_\nabla}$ and set $\mathcal{M}_{\mathcal{R}} = \mathcal{M} \otimes_{\mathcal{O}} \mathcal{R}$. If*

$$0 = \mathcal{M}_{0,\mathcal{R}} \subset \mathcal{M}_{1,\mathcal{R}} \subset \cdots \subset \mathcal{M}_{r,\mathcal{R}} = \mathcal{M}_{\mathcal{R}}$$

denotes the slope filtration of $\mathcal{M}_{\mathcal{R}}$ then for $i = 0, 1, \ldots, r$, $\mathcal{M}_{i,\mathcal{R}}$ extends uniquely to a saturated $\mathcal{O}$-submodule $\mathcal{M}_i \subset \mathcal{M}$ which is stable by φ and N_∇.

Proof. For any interval $I \subset [0, 1)$, $\mathcal{M}_I = \mathcal{M} \otimes_{\mathcal{O}} \mathcal{O}_I$ is equipped with a differential operator induced by N_∇ on $\mathcal{M}$ and $-u\lambda\frac{d}{du}$ on $\mathcal{O}_I$. Passing to the limit we also get a differential operator on $\mathcal{M}_{\mathcal{R}}$. We again denote these differential operators by N_∇,

By Lemma 1.3.4 $\mathcal{M}_{i,\mathcal{R}}$ extends to a saturated φ-stable submodule $\mathcal{M}_{i,(0,1)} \subset \mathcal{M}_{(0,1)}$. We claim that $\mathcal{M}_{i,(0,1)}$ is stable by N_∇. Since $\mathcal{M}_{i,(0,1)} = \mathcal{M}_{i,\mathcal{R}} \cap \mathcal{M}_{(0,1)}$, it suffices to show that $\mathcal{M}_{i,\mathcal{R}}$ is stable by N_∇. For this we use the notation of Lemma 1.3.6. Consider the map of $\mathcal{R}$-modules

$$\delta : \mathcal{L} \otimes_{\mathcal{R}} \mathcal{M} \to \mathcal{M}; \qquad (a + b\xi) \otimes m \mapsto am + bN_\nabla(m).$$

A simple calculation shows that δ respects the action of φ. Let $\mathcal{M}'_{i,\mathcal{R}} = \delta(\mathcal{L} \otimes_{\mathcal{R}} \mathcal{M}_{i,\mathcal{R}})$. We obviously have $\mathcal{M}_{i,\mathcal{R}} \subset \mathcal{M}'_{i,\mathcal{R}}$. To show this inclusion is an equality we proceed by induction on i. Let s_i denote the slope of $\mathcal{M}_i/\mathcal{M}_{i-1}$. When $i = 0$ there is nothing to prove. If $\mathcal{M}_{i-1,\mathcal{R}} = \mathcal{M}'_{i-1,\mathcal{R}}$, then we have surjections

$$\mathcal{L} \otimes_{\mathcal{R}} \mathcal{M}_{i,\mathcal{R}}/\mathcal{M}_{i-1,\mathcal{R}} \xrightarrow{\delta} \mathcal{M}'_{i,\mathcal{R}}/\mathcal{M}'_{i-1,\mathcal{R}} = \mathcal{M}'_{i,\mathcal{R}}/\mathcal{M}_{i-1,\mathcal{R}} \to \mathcal{M}'_{i.\mathcal{R}}/\mathcal{M}_{i,\mathcal{R}}.$$

Since the $\mathcal{R}$-submodule $\mathcal{M}'_{i,\mathcal{R}}/\mathcal{M}_{i,\mathcal{R}} \subset \mathcal{M}/\mathcal{M}_{i,\mathcal{R}}$ is finitely generated, it is finite free over $\mathcal{R}$ by Lemma 1.1.5. By Lemma 1.3.6, $\mathcal{L}\otimes_{\mathcal{R}}\mathcal{M}_{i,\mathcal{R}}/\mathcal{M}_{i-1,\mathcal{R}}$ is pure of slope s_i, so if $\mathcal{M}'_{i,\mathcal{R}}/\mathcal{M}_{i,\mathcal{R}}$ is nonzero, its smallest slope is $\leqslant s_i$ by [Ke 1, Lemma 4.1]. But then the smallest slope of $\mathcal{M}/\mathcal{M}_{i,\mathcal{R}}$ is $\leqslant s_i$, which is a contradiction as all the slopes of this module are $\geq s_{i+1} > s_i$.

Finally, N_∇ induces a differential operator $\partial = \lambda^{-1} N_\nabla$ over $-u\frac{d}{du}$ on $\mathcal{M}_{[0,p^{-2})}$. By Lemma 1.3.5, $\mathcal{M}_{i,(0,p^{-2})}$ extends to a unique ∂-stable saturated $\mathcal{O}_{[0,p^{-2})}$-submodule $\mathcal{M}_{i,[0,p^{-2})} \subset \mathcal{M}_{i,[0,p^{-2})}$. Hence $\mathcal{M}_{i,(0,1)}$ extends to a unique, N_∇-stable, saturated $\mathcal{O}$-submodule $\mathcal{M}_i \subset \mathcal{M}$. Since $\mathcal{M}_i = \mathcal{M} \cap \mathcal{M}_{i,\mathcal{R}}$ it is stable by φ. □

Theorem 1.3.8. *Let D be an effective filtered (φ, N)-module. Then D is weakly admissible if and only if $\mathcal{M}(D)$ is pure of slope* 0.

Proof. Suppose first that D has rank 1, and choose a basis $\boldsymbol{e} \in D$. Write $\varphi(e) = \alpha \boldsymbol{e}$ for some $\alpha \in K_0$. Set $\mathcal{D}_0 = (\mathcal{O}[\ell_u] \otimes_{K_0} D)^{N=0}$. Then the definition of $\mathcal{M}$ shows that $\mathcal{M}(D) = \lambda^{-t_H(D)}\mathcal{D}_0$, so that

$$\varphi(\lambda^{-t_H(D)}\boldsymbol{e}) = (E(u)/c_0)^{t_H(D)}\alpha\lambda^{-t_H(D)}\boldsymbol{e}.$$

Hence $\mathcal{M}(D)$ has slope $t_N(D) - t_H(D)$. This proves the theorem for rank 1 (φ, N)-modules.

Suppose that D is weakly admissible. By Proposition 1.3.7, the slope filtration on $\mathcal{M}(D)_{\mathcal{R}}$ is induced by a filtration of $\mathcal{M}(D)$ by saturated $\mathcal{O}$-submodules, stable by φ and N_∇

$$0 = \mathcal{M}_0 \subset \mathcal{M}_1 \subset \cdots \subset \mathcal{M}_r = \mathcal{M}.$$

Write s_i for the unique slope of $\mathcal{M}_{i,\mathcal{R}}/\mathcal{M}_{i-1,\mathcal{R}}$, and d_i for its $\mathcal{R}$-rank. By Theorem 1.2.15, $\mathcal{M}_1 = \mathcal{M}(D_1)$ for some filtered (φ, N) submodule $D_1 \subset D$. Then $\bigwedge^{d_1} \mathcal{M}_1$ has slope $d_1 s_1$ [Ke 1, Proposition 5.13], and the compatibility with tensor products in Theorem 1.2.15 and the rank 1 case considered above imply that this slope is $t_N(D_1) - t_H(D_1)$. Hence the weak admissibility of D implies that $s_1 \geq 0$. Since s_1 is the smallest slope this implies that $s_i \geq 0$ for all i. On the other hand, applying the rank 1 case as above, $\sum_{i=1}^r d_i s_i = t_N(D) - t_H(D) = 0$, so that $r = 1$ and $s_1 = 0$.

Conversely, suppose that $\mathcal{M}(D)$ is pure of slope 0. We have already seen that this implies $t_N(D) = t_H(D)$. If $D' \subset D$ is a (φ, N)-submodule, then $\mathcal{M}(D') \subset \mathcal{M}(D)$ has all slopes ≥ 0 by [Ke 1, Proposition 4.4]. In particular, the slope of the top exterior product of $\mathcal{M}(D')$ is ≥ 0, so we have $t_N(D') - t_H(D') \geq 0$. □

1.3.9. A (φ, N)-module over $\mathcal{O}$ is a φ-module $\mathcal{M}$ together with a K_0-linear map $N : \mathcal{M}/u\mathcal{M} \to \mathcal{M}/u\mathcal{M}$ which satisfies $N\varphi = p\varphi N$, where we have written φ for the endomorphism of $\mathcal{M}/u\mathcal{M}$ obtained by reducing $\varphi : \mathcal{M} \to \mathcal{M}$ modulo u. We say that $\mathcal{M}$ is pure of slope 0 if $\mathcal{M} \otimes_{\mathcal{O}} \mathcal{R}$ is. As usual, $\mathcal{M}$ is said to be of finite E-height if it has this property as a φ-module over $\mathcal{O}$. We denote by $\mathrm{Mod}_{/\mathcal{O}}^{\varphi,N}$ the category of (φ, N)-modules over $\mathcal{O}$ of finite E-height, and by $\mathrm{Mod}_{/\mathcal{O}}^{\varphi,N,0}$ the full subcategory consisting of modules which are pure of slope 0. Each of these categories has a natural structure of a Tannakian category.

Given a module $\mathcal{M}$ in $\mathrm{Mod}_{/\mathcal{O}}^{\varphi,N_\nabla}$ we obtain a module $\tilde{\mathcal{M}}$ in $\mathrm{Mod}_{/\mathcal{O}}^{\varphi,N}$ by taking $\tilde{\mathcal{M}} = \mathcal{M}$ equipped with the operator φ, and taking N to be the reduction of N_∇ modulo u.

Lemma 1.3.10. *Let $\mathcal{M}$ be in $\mathrm{Mod}_{/\mathcal{O}}^{\varphi,N}$. Then*

(1) *$\mathcal{M}[1/\lambda]$ is canonically equipped with an operator N_∇ such that $N_\nabla\varphi = (p/c_0)E(u)\varphi N_\nabla$ and $N_\nabla|_{u=0} = N$.*
(2) *The functor $\mathcal{N} \mapsto \tilde{\mathcal{N}}$ is fully faithful, and a module $\mathcal{M}$ is in its image if and only if it is stable under the operator N_∇ on $\mathcal{M}[1/\lambda]$.*
(3) *Any $\mathcal{M}$ in $\mathrm{Mod}_{/\mathcal{O}}^{\varphi,N}$ which has $\mathcal{O}$-rank 1 is in the image of the functor in (2).*

Proof. The construction of Section 1.2.5 shows that given an $\mathcal{M}$ in $\mathrm{Mod}_{/\mathcal{O}}^{\varphi,N}$, we obtain a filtered (φ, N)-module $D(\mathcal{M})$, and that for $\mathcal{N}$ in $\mathrm{Mod}_{/\mathcal{O}}^{\varphi,N_\nabla}$ we have $D(\tilde{\mathcal{N}}) = D(\mathcal{N})$.

Now, given $\mathcal{M}$ we set $\mathcal{D}_0 = (\mathcal{O}[\ell_u] \otimes_{K_0} D(\mathcal{M}))^{N=0}$, equipped with an operator N_∇ induced by the corresponding operator on $\mathcal{O}[\ell_u]$. As in Lemma 1.2.12, we may consider the composite

$$\mathcal{D}_0 = (K_0[\ell_u] \otimes_{K_0} D(\mathcal{M}))^{N=0} \otimes_{K_0} \mathcal{O} \stackrel{\eta\otimes 1}{\to} D(\mathcal{M}) \otimes_{K_0} \mathcal{O} \stackrel{\xi}{\to} \mathcal{M} \qquad (1.3.11)$$

where η is a bijection, and ξ has cokernel killed by some power of λ by Lemma 1.2.6. Using (1.3.11), we obtain an isomorphism $\mathcal{D}_0[1/\lambda] \xrightarrow{\sim} \mathcal{M}[1/\lambda]$, which is compatible with the action of φ, and with N after applying $\otimes_{\mathcal{O}}\mathcal{O}/u\mathcal{O}$. From the definition of $\mathcal{M}(D)$, we have an isomorphism $\mathcal{D}_0[1/\lambda] \xrightarrow{\sim} \mathcal{M}(D(\mathcal{M}))[1/\lambda]$, compatible with φ and N_∇.

This proves (1). Moreover, by Theorem 1.2.15 $\mathcal{M}$ is in the image of $\mathcal{N} \mapsto \tilde{\mathcal{N}}$ if and only if $\mathcal{M} \xrightarrow{\sim} \mathcal{M}(D(\mathcal{M}))$ in $\mathcal{D}_0[1/\lambda]$, and this is equivalent to $\mathcal{M}$ being stable under N_∇. This also shows the claim regarding full faithfulness.

Finally, suppose $\mathcal{M}$ in $\mathrm{Mod}_{/\mathcal{O}}^{\varphi,N}$ has $\mathcal{O}$-rank 1. The above discussion shows that $\xi(D(\mathcal{M})) \subset \mathcal{M}$ is killed by N_∇. If $\boldsymbol{e}$ is a K_0-basis vector for $D(\mathcal{M})$, then there exists $f \in \mathcal{O}$ such that $\mathcal{M} = f^{-1}\mathcal{O}\boldsymbol{e}$, and

$$N_\nabla(f^{-1}\boldsymbol{e}) = -u\lambda\frac{df^{-1}}{du}\boldsymbol{e} = u\lambda\frac{df}{du}f^{-1}(f^{-1}\boldsymbol{e}).$$

So it suffices to show that $\lambda\frac{df}{du}f^{-1} \in \mathcal{O}$. Since $\mathcal{M} \subset \mathcal{D}_0[1/\lambda]$ the set of zeroes of f is contained in the set of zeroes of λ. Since $\frac{df}{du}f^{-1}$ has at most a simple pole at each such zero, this completes the proof of (3). □

1.3.12. A (φ, N)-module over $\mathfrak{S}$ is a finite free $\mathfrak{S}$-module $\mathfrak{M}$, equipped with a φ-semilinear Frobenius $\varphi : \mathfrak{M} \to \mathfrak{M}$, and a linear endomorphism $N : \mathfrak{M}/u\mathfrak{M} \otimes_{\mathbb{Z}_p} \mathbb{Q}_p \to \mathfrak{M}/u\mathfrak{M} \otimes_{\mathbb{Z}_p} \mathbb{Q}_p$ such that $N\varphi = p\varphi N$ on $\mathfrak{M}/u\mathfrak{M} \otimes_{\mathbb{Z}_p} \mathbb{Q}_p$. We say that $\mathfrak{M}$ is of finite E-height if the cokernel of $\varphi^*\mathfrak{M} \to \mathfrak{M}$ is killed by some power of $E(u)$. We denote by $\mathrm{Mod}^{\varphi,N}_{/\mathfrak{S}}$ the category (φ, N)-modules over $\mathfrak{S}$ of finite E-height, and by $\mathrm{Mod}^{\varphi,N}_{/\mathfrak{S}} \otimes \mathbb{Q}_p$ the associated isogeny category.

The reader may wonder why we do not insist that the operator N takes $\mathfrak{M}/u\mathfrak{M}$ to itself. The reason is that with this definition we could not prove Lemma 1.3.13 below. We do not know if the two definitions give rise to the same isogeny category.

Lemma 1.3.13. *The functor*

$$\Theta : \mathrm{Mod}^{\varphi,N}_{/\mathfrak{S}} \otimes \mathbb{Q}_p \xrightarrow{\sim} \mathrm{Mod}^{\varphi,N,0}_{/\mathcal{O}}; \qquad \mathfrak{M} \mapsto \mathfrak{M} \otimes_{\mathfrak{S}} \mathcal{O} \tag{1.3.14}$$

is an equivalence of Tannakian categories.

Proof. Let $\mathcal{M}$ be in $\mathrm{Mod}^{\varphi,N,0}_{/\mathcal{O}}$. Then $\mathcal{M}_{\mathcal{R}} := \mathcal{M} \otimes_{\mathcal{O}} \mathcal{R}$ is in $\mathrm{Mod}^{\varphi,0}_{/\mathcal{R}}$, and hence Theorem 1.3.2 implies that $\mathcal{M}$ is of the form $\mathcal{M}_{\mathcal{R}^b} \otimes_{\mathcal{R}^b} \mathcal{R}$ for some $\mathcal{M}_{\mathcal{R}^b}$ in $\mathrm{Mod}^{\varphi,0}_{/\mathcal{R}^b}$, whose construction is functorial in $\mathcal{M}_{\mathcal{R}}$. Thus we have

$$\mathcal{M}_{\mathcal{R}^b} \otimes_{\mathcal{R}^b} \mathcal{R} \xrightarrow{\sim} \mathcal{M}_{\mathcal{R}} \xrightarrow{\sim} \mathcal{M} \otimes_{\mathcal{O}} \mathcal{R}.$$

Choose an $\mathcal{R}^b$-basis for $\mathcal{M}_{\mathcal{R}^b}$, and an $\mathcal{O}$-basis for $\mathcal{M}$. The composite of the above isomorphisms is then given by a matrix with values in $\mathcal{R}$. By [Ke 1, Proposition 6.5], after modifying the chosen bases, we may assume that this matrix is the identity. In other words $\mathcal{M}_{\mathcal{R}^b}$ and $\mathcal{M}$ are spanned by a common basis. Let $\mathcal{M}^b$ denote the $\mathfrak{S}[1/p]$-span of this basis. Since $\mathfrak{S}[1/p] = \mathcal{O} \cap \mathcal{R}^b \subset \mathcal{R}$, we have $\mathcal{M}^b = \mathcal{M}_{\mathcal{R}^b} \cap \mathcal{M} \subset \mathcal{M}_{\mathcal{R}}$.

Hence $\mathcal{M}^b$ is stable by φ, and of finite E-height, since $\mathcal{M}$ is. This already shows that Θ is fully faithful, since given any $\mathcal{N}$ in $\mathrm{Mod}^{\varphi,N}_{/\mathfrak{S}}$, $\mathcal{N} \otimes \mathbb{Q}_p$ can be recovered as $\Theta(\mathcal{N})_{\mathcal{R}^b} \cap \Theta(\mathcal{N})$. To show that it is essentially surjective, we have to check that $\mathcal{M}^b$ arises from an object of $\mathrm{Mod}^{\varphi,N}_{/\mathfrak{S}}$.

Let $\mathcal{O}_{\mathcal{R}^b}$ denote the valuation ring of $\mathcal{R}^b$. Since $\mathcal{M}_{\mathcal{R}^b}$ has slope 0, there exists a φ-stable $\mathcal{O}_{\mathcal{R}^b}$-lattice $\mathcal{L}$ in $\mathcal{M}_{\mathcal{R}^b}$. Let $\mathfrak{M}' = \mathcal{M}^b \cap \mathcal{L}$, and set

$$\mathfrak{M} = \mathcal{O}_{\mathcal{R}^b} \otimes_{\mathfrak{S}} \mathfrak{M}' \cap \mathfrak{M}'[1/p] \subset \mathcal{M}_{\mathcal{R}^b}.$$

Then $\mathfrak{M} \subset \mathcal{M}_{\mathcal{R}^b}$ is a finite, φ-stable $\mathfrak{S}$-submodule. Moreover, the structure theory of finite $\mathfrak{S}$-modules shows that there exists an inclusion $\mathfrak{M}' \subset F$ of $\mathfrak{M}'$ into a finite free $\mathfrak{S}$-module F, such that $F/\mathfrak{M}'$ has finite length. This implies that $\mathfrak{M}$ may be identified with F. Thus $\mathfrak{M}$ is free over $\mathfrak{S}$.

To check that $\mathfrak{M}$ is in $\mathrm{Mod}^{\varphi,N}_{/\mathfrak{S}}$, we have to check that the cokernel of $\varphi^*\mathfrak{M} \to \mathfrak{M}$ is killed by a power of $E(u)$. Let d be the $\mathfrak{S}$-rank of $\mathfrak{M}$. Then φ on $\bigwedge^d_{\mathfrak{S}} \mathfrak{M}$ with respect to some choice of basis vector is given by $p^r E(u)^s w$ where $r, s \geq 0$, and $w \in \mathfrak{S}^\times$. Since $\mathcal{M}_{\mathcal{R}^b} = \mathfrak{M} \otimes_{\mathfrak{S}} \mathcal{R}^b$, and $\wedge^d_{\mathcal{R}^b}\mathcal{M}_{\mathcal{R}^b}$ is pure of slope 0, we must have that $r = 0$. □

Corollary 1.3.15. *There exists a fully faithful $\otimes$-functor from the category of effective weakly admissible filtered (φ, N)-modules to* $\mathrm{Mod}^{\varphi,N}_{/\mathfrak{S}} \otimes \mathbb{Q}_p$.

If $\mathfrak{M}$ is in $\mathrm{Mod}^{\varphi,N}_{/\mathfrak{S}}$, *and $\mathcal{M} = \mathfrak{M} \otimes_{\mathfrak{S}} \mathcal{O}$, then $\mathcal{M}[1/\lambda]$ is canonically equipped with a connection ∇ such that $\varphi \circ \nabla = \nabla \circ \varphi$. The module $\mathfrak{M}$ is in the image of the functor above if and only if ∇ induces a singular connection on $\mathcal{M}$ with only logarithmic singularities.*

Proof. By Theorems 1.3.8 and 1.2.15, $D \mapsto \mathcal{M}(D)$ is an equivalence between the category of effective weakly admissible filtered (φ, N)-modules, and $\mathrm{Mod}^{\varphi,N_\nabla,0}_{/\mathcal{O}}$. By Lemma 1.3.10, the latter category is a full subcategory of $\mathrm{Mod}^{\varphi,N,0}_{/\mathcal{O}}$ which is equivalent to $\mathrm{Mod}^{\varphi,N}_{/\mathfrak{S}} \otimes \mathbb{Q}_p$ by Lemma 1.3.13. This proves the first claim, and the second follows from Lemma 1.3.10(2), the connection on $\mathcal{M}$ being given by $\nabla(m) = -\lambda^{-1} N_\nabla(m)\frac{du}{u}$. □

2 Galois representations and p-divisible groups

2.1 G_K-representations and G_{K_∞}-representations

In this section we will use the theory of the previous section to compare constructions of crystalline representations, and representations of finite E-height. We show that the functor from crystalline G_K-representations to G_{K_∞}-representations is fully faithful.

2.1.1. Let $\mathcal{O}_{\bar{K}}$ denote the ring of integers of $\bar{K}$. Let $R = \varprojlim \mathcal{O}_{\bar{K}}/p$ where the transition maps are given by Frobenius. There is a unique surjective map $\theta : W(R) \to \widehat{\mathcal{O}_{\bar{K}}}$ to the p-adic completion $\widehat{\mathcal{O}_{\bar{K}}}$ of $\mathcal{O}_{\bar{K}}$, which lifts the projection $R \to \mathcal{O}_{\bar{K}}/p$ onto the first factor in the inverse limit.

Write $\underline{\pi} = (\pi_n)_{n\geq 0} \in R$, where $\pi_n \in \bar{K}$ are the elements introduced in Section 1.1.1. Let $[\underline{\pi}] \in W(R)$ be the Teichmüller representative. We embed the W-algebra $W[u]$ into $W(R)$ by $u \mapsto [\underline{\pi}]$. Since $\theta([\underline{\pi}]) = \pi$ this embedding extends to an embedding $\mathfrak{S} \hookrightarrow W(R)$, and $\theta|_{\mathfrak{S}}$ is the map $\mathfrak{S} \to \mathcal{O}_K$ sending u to π. This embedding is compatible with Frobenius endomorphisms.

We denote by A_{cris} the p-adic completion of the divided power envelope of $W(R)$ with respect to $\ker(\theta)$. As usual, we write $B^+_{\mathrm{cris}} = A_{\mathrm{cris}}[1/p]$, we denote by B^+_{st} the ring obtained by formally adjoining the element "$\log[\underline{\pi}]$" to B^+_{cris}, and by B^+_{dR} the $\ker(\theta)$-adic completion of $W(R)[1/p]$.

Let $\mathcal{O}_{\mathcal{E}}$ be the p-adic completion of $\mathfrak{S}[1/u]$. Then $\mathcal{O}_{\mathcal{E}}$ is a discrete valuation ring with residue field the field of Laurent series $k((u))$. We write $\mathcal{E}$ for the field of fractions of $\mathcal{O}_{\mathcal{E}}$. If Fr R denotes the field of fractions of R, then the inclusion $\mathfrak{S} \hookrightarrow W(R)$ extends to an inclusion $\mathcal{E} \hookrightarrow W(\mathrm{Fr}\, R)$. Let $\mathcal{E}^{\mathrm{ur}} \subset W(\mathrm{Fr}\, R)[1/p]$ denote the maximal unramified extension of $\mathcal{E}$ contained in $W(\mathrm{Fr}\, R)[1/p]$, and $\mathcal{O}_{\mathcal{E}^{\mathrm{ur}}}$ its ring of integers. Since Fr R is algebraically closed [Fo 1, A.3.1.6], the residue field $\mathcal{O}_{\mathcal{E}^{\mathrm{ur}}}/p\mathcal{O}_{\mathcal{E}^{\mathrm{ur}}}$ is a separable closure of $k((u))$. We denote by $\widehat{\mathcal{E}^{\mathrm{ur}}}$ the p-adic completion of $\mathcal{E}^{\mathrm{ur}}$, and by $\mathcal{O}_{\widehat{\mathcal{E}^{\mathrm{ur}}}}$ its ring of integers. $\widehat{\mathcal{E}^{\mathrm{ur}}}$ is also equal to the closure of $\mathcal{E}^{\mathrm{ur}}$ in $W(\mathrm{Fr}\, R)$. We write $\mathfrak{S}^{\mathrm{ur}} = \mathcal{O}_{\widehat{\mathcal{E}^{\mathrm{ur}}}} \cap W(R) \subset W(\mathrm{Fr}\, R)$. We regard all these rings as subrings of $W(R)$.

Let $K_\infty = \cup_{n\geq 0} K_n$, and write $G_{K_\infty} = \mathrm{Gal}(\bar{K}/K_\infty)$. Since G_{K_∞} fixes the subring $\mathfrak{S} \subset W(R)$, it acts on $\mathfrak{S}^{\mathrm{ur}}$ and $\mathcal{E}^{\mathrm{ur}}$.

Lemma 2.1.2. *Let $\mathfrak{M}$ be a finitely generated $\mathfrak{S}$-module equipped with an $\mathfrak{S}$-linear map $\varphi^*\mathfrak{M} \to \mathfrak{M}$. Suppose that $\mathfrak{M}$ is isomorphic as an $\mathfrak{S}$-module to a finite direct sum $\oplus_{i\in I}\mathfrak{S}/p^{n_i}\mathfrak{S}$ where $n_i \in \mathbb{N}^+$ and that* $\mathrm{coker}(1\otimes\varphi)$ *is killed by some power of $E(u)$. Then*

(1) *The association $\mathfrak{M} \mapsto \mathrm{Hom}_{\mathfrak{S},\varphi}(\mathfrak{M}, \mathfrak{S}^{\mathrm{ur}}[1/p]/\mathfrak{S}^{\mathrm{ur}})$ is an exact functor in $\mathfrak{M}$.*
(2) *The natural map*

$$\mathrm{Hom}_{\mathfrak{S},\varphi}(\mathfrak{M}, \mathfrak{S}^{\mathrm{ur}}[1/p]/\mathfrak{S}^{\mathrm{ur}}) \to \mathrm{Hom}_{\mathfrak{S},\varphi}(\mathfrak{M}, \mathcal{E}^{\mathrm{ur}}/\mathcal{O}_{\mathcal{E}^{\mathrm{ur}}})$$

is an isomorphism, and both sides are isomorphic to $\oplus_{i\in I}\mathbb{Z}_p/p^{n_i}\mathbb{Z}_p$ as $\mathbb{Z}_p$-modules.

Proof. The first part of (2) follows from [Fo 1, B.1.8.4]. The rest of the lemma then follows from [Fo 1, Section A.1.2]. □

2.1.3. We denote by $\mathrm{Mod}^{\varphi}_{/\mathfrak{S}}$ the category of finite free $\mathfrak{S}$-modules equipped with an $\mathfrak{S}$-linear map $1\otimes\varphi : \varphi^*\mathfrak{M} \to \mathfrak{M}$ whose cokernel is killed by some power of $E(u)$. We may regard $\mathrm{Mod}^{\varphi}_{/\mathfrak{S}}$ as a full subcategory of $\mathrm{Mod}^{\varphi,N}_{/\mathfrak{S}}$ by taking the operator N to be 0 on an object of $\mathrm{Mod}^{\varphi}_{/\mathfrak{S}}$.

Corollary 2.1.4. *Let $\mathfrak{M}$ be in $\mathrm{Mod}^{\varphi}_{/\mathfrak{S}}$. Then*

$$V_{\mathfrak{S}}(\mathfrak{M}) := \mathrm{Hom}_{\mathfrak{S},\varphi}(\mathfrak{M}, \mathfrak{S}^{\mathrm{ur}})$$

is a free $\mathbb{Z}_p$-module of rank $r = \mathrm{rk}_{\mathfrak{S}}\,\mathfrak{M}$, and the functor $\mathfrak{M} \mapsto V_{\mathfrak{S}}(\mathfrak{M})$ is exact in $\mathfrak{M}$. Moreover, the natural map

$$\mathrm{Hom}_{\mathfrak{S},\varphi}(\mathfrak{M}, \mathfrak{S}^{\mathrm{ur}}) \to \mathrm{Hom}_{\mathcal{O}_{\mathcal{E}},\varphi}(\mathcal{O}_{\mathcal{E}} \otimes_{\mathfrak{S}} \mathfrak{M}, \mathcal{O}_{\widehat{\mathcal{E}^{\mathrm{ur}}}})$$

is a bijection.

Proof. This follows immediately from Lemma 2.1.2. □

Proposition 2.1.5. *Let D be an effective, weakly admissible filtered (φ, N)-module, and $\mathfrak{M}$ in $\mathrm{Mod}^{\varphi,N}_{/\mathfrak{S}}$ a module whose image in $\mathrm{Mod}^{\varphi,N}_{/\mathfrak{S}} \otimes \mathbb{Q}_p$ is equal to the image of D under the functor of Corollary* 1.3.15.

Then there exists a canonical bijection

$$\mathrm{Hom}_{\mathfrak{S},\varphi}(\mathfrak{M}, \mathfrak{S}^{\mathrm{ur}}) \otimes_{\mathbb{Z}_p} \mathbb{Q}_p \xrightarrow{\sim} \mathrm{Hom}_{\mathrm{Fil},\varphi,N}(D, B^+_{\mathrm{st}}),$$

which is compatible with the action of G_{K_∞} on the two sides. In particular, both sides have dimension $\dim_{K_0} D$, and D is admissible.

Proof. Set $\mathcal{M} = \mathfrak{M} \otimes_{\mathfrak{S}} \mathcal{O}$. Using Proposition 1.2.8, we may identify D with $D(\mathcal{M})$. The inclusion $\mathfrak{S} \subset B^+_{\mathrm{cris}}$ admits a unique continuous extension to $\mathcal{O}$, and we will regard B^+_{cris} as an $\mathcal{O}$-algebra in this way. Since the inclusion of $\mathcal{O}$ in B^+_{cris} sends $E(u)$ to $E([\underline{\pi}]) \in \mathrm{Fil}^1 B^+_{\mathrm{dR}}$, it extends to an inclusion of $\widehat{\mathfrak{S}}_0$ into B^+_{dR}. Recall that the $\mathcal{O}$-module $\varphi^*\mathcal{M}$ is equipped with a decreasing filtration as in Section 1.2.7, while the ring $B^+_{\mathrm{cris}} \otimes_{K_0} K$ is equipped with a filtration via its inclusion into the discrete valuation ring B^+_{dR}.

Observe that we have natural maps

$$\mathrm{Hom}_{\mathfrak{S},\varphi}(\mathfrak{M}, \mathfrak{S}^{\mathrm{ur}}) \to \mathrm{Hom}_{\mathcal{O},\varphi}(\mathcal{M}, B^+_{\mathrm{cris}}) \to \mathrm{Hom}_{\mathcal{O},\mathrm{Fil},\varphi}(\varphi^*\mathcal{M}, B^+_{\mathrm{cris}}). \quad (2.1.6)$$

Here the term on the right means $\mathcal{O}$-linear, φ-compatible maps which induce a filtered map $\widehat{\mathfrak{S}}_0 \otimes_{\mathcal{O}} \varphi^*\mathcal{M} \to B^+_{\mathrm{dR}}$, and the second map is obtained by composing morphisms with the inclusion $1 \otimes \varphi : \varphi^*\mathcal{M} \to \mathcal{M}$. It follows from the definition of the filtration on $\varphi^*\mathcal{M}$ (and the fact that $E([\underline{\pi}]) \in \mathrm{Fil}^1 B^+_{\mathrm{dR}}$)) that any such composed morphism respects filtrations. Note that both maps in (2.1.6) are injective. This is clear for the first map, and for the second it follows from the fact that the cokernel of $1 \otimes \varphi$ is a killed by some power of λ, while B^+_{cris} is a domain.

Next, we set $\mathcal{D}_0 = (\mathcal{O}[\ell_u] \otimes_{K_0} D)^{N=0}$. By Lemma 1.2.1 (and since we are identifying D and $D(\mathcal{M})$), we have an isomorphism $\widehat{\mathfrak{S}}_0 \otimes_{\mathcal{O}} \mathcal{D}_0 \xrightarrow{\sim} \widehat{\mathfrak{S}}_0 \otimes_K D_K$, and we regard the left-hand side of this isomorphism as equipped with the filtration induced by that on the right-hand side. By Lemma 1.2.12, $\xi \circ (\eta \otimes 1)$ induces a map

$$\mathrm{Hom}_{\mathcal{O},\mathrm{Fil},\varphi}(\varphi^*\mathcal{M}, B^+_{\mathrm{cris}}) \to \mathrm{Hom}_{\mathcal{O},\mathrm{Fil},\varphi}(\mathcal{D}_0, B^+_{\mathrm{cris}}), \quad (2.1.7)$$

where the term on the right means φ-compatible maps, which induce a map $\widehat{\mathfrak{S}}_0 \otimes_{\mathcal{O}} \mathcal{D}_0 \to B^+_{\mathrm{dR}}$ that is compatible with filtrations. Since the map of Lemma 1.2.12(3) is an isomorphism at the point x_0, (2.1.7) is an injection.

Finally, note that multiplication in the ring $\mathcal{O}[\ell_u]$ induces a natural isomorphism $\mathcal{O}[\ell_u] \otimes_{\mathcal{O}} \mathcal{D}_0 \underset{\mu\otimes 1}{\xrightarrow{\sim}} \mathcal{O}[\ell_u] \otimes_{K_0} D$ which is compatible with φ and N. Hence given any map f in the right-hand side of (2.1.7), we may form the composite

$$D \hookrightarrow \mathcal{O}[\ell_u] \otimes_{K_0} D \underset{(\mu\otimes 1)^{-1}}{\xrightarrow{\sim}} \mathcal{O}[\ell_u] \otimes_{\mathcal{O}} \mathcal{D}_0 \overset{1\otimes f}{\to} \mathcal{O}[\ell_u] \otimes_{\mathcal{O}} B^+_{\mathrm{cris}} \underset{\ell_u \mapsto \log[\underline{\pi}]}{\xrightarrow{\sim}} B^+_{\mathrm{st}}.$$

It follows from the definition of the filtration on $\widehat{\mathfrak{S}}_0 \otimes_{\mathcal{O}} \mathcal{D}_0$ that any such composed map respects filtrations after tensoring by $K \otimes_{K_0}$. Hence we obtain an injective map

$$\mathrm{Hom}_{\mathcal{O},\mathrm{Fil},\varphi}(\mathcal{D}_0, B^+_{\mathrm{cris}}) \to \mathrm{Hom}_{\mathrm{Fil},\varphi,N}(D, B^+_{\mathrm{st}}). \tag{2.1.8}$$

Combining (2.1.6)–(2.1.8), we obtain a G_{K_∞}-equivariant inclusion

$$\mathrm{Hom}_{\mathfrak{S},\varphi}(\mathfrak{M}, \mathfrak{S}^{\mathrm{ur}}) \otimes_{\mathbb{Z}_p} \mathbb{Q}_p \to \mathrm{Hom}_{\mathrm{Fil},\varphi,N}(D, B^+_{\mathrm{st}}),$$

Since the left-hand side has $\mathbb{Q}_p$-dimension $d = \dim_{K_0} D$ by Corollary 2.1.4, the dimension of the right-hand side is $\geq d$. But now an elementary argument [CF, Proposition 4.5] shows that the right-hand side has dimension d, and D is admissible. Hence our map is a bijection, as required. □

Lemma 2.1.9. *Let $h : \mathfrak{M} \to \mathfrak{M}'$ be a morphism in $\mathrm{Mod}^{\varphi}_{/\mathfrak{S}}$ which becomes an isomorphism after tensoring by $\mathcal{O}_{\mathcal{E}}$. Then h is an isomorphism.*

Proof. Since h is a morphism of free $\mathfrak{S}$-modules of the same rank, it is an isomorphism if the induced map on determinants is. Hence we may assume that $\mathfrak{M}$ and $\mathfrak{M}'$ are free of rank 1 over $\mathfrak{S}$.

Let $\mathcal{M} = \mathfrak{M} \otimes_{\mathfrak{S}} \mathcal{O}$ and $\mathcal{M}' = \mathfrak{M}' \otimes_{\mathfrak{S}} \mathcal{O}$. By Lemmas 1.3.13 and 1.3.10(3), $\mathcal{M}$ and $\mathcal{M}'$ may be regarded as objects of $\mathrm{Mod}^{\varphi,N_\nabla}_{/\mathcal{O}}$. Let $D = D(\mathcal{M})$ and $D' = D(\mathcal{M}')$. By Lemma 1.3.10(2) and Theorem 1.2.15 the map $D \to D'$ induced by h is nonzero, and hence is an isomorphism of filtered (φ, N)-modules. Hence h becomes an isomorphism after inverting p by Corollary 1.3.15. This means that in a suitable choice of bases h is given by multiplication by p^i for some nonnegative integer i. Since h becomes an isomorphism after tensoring by $\mathcal{O}_{\mathcal{E}}$, we must have $i = 0$. □

Lemma 2.1.10. *Let $\mathfrak{M}$ be in $\mathrm{Mod}^{\varphi}_{/\mathfrak{S}}$, and let $V_{\mathfrak{S}}(\mathfrak{M})$ be as in Corollary* 2.1.4. *Then $\mathfrak{M}' = \mathrm{Hom}_{\mathbb{Z}_p[G_{K_\infty}]}(V_{\mathfrak{S}}(\mathfrak{M}), \mathfrak{S}^{\mathrm{ur}})$ is a free $\mathfrak{S}$-module of rank $d = \mathrm{rk}_{\mathfrak{S}}\,\mathfrak{M}$, and the natural map*

$$\mathfrak{M} \to \mathrm{Hom}_{\mathbb{Z}_p[G_{K_\infty}]}(\mathrm{Hom}_{\mathfrak{S},\varphi}(\mathfrak{M}, \mathfrak{S}^{\mathrm{ur}}), \mathfrak{S}^{\mathrm{ur}}) = \mathfrak{M}'$$

is an injection.

Proof. Set $\mathcal{M} = \mathfrak{M} \otimes_{\mathfrak{S}} \mathcal{O}_{\mathcal{E}}$. The natural map $\mathfrak{S}^{\mathrm{ur}}/p\mathfrak{S}^{\mathrm{ur}} \to \mathcal{O}_{\mathcal{E}^{\mathrm{ur}}}/p\mathcal{O}_{\mathcal{E}^{\mathrm{ur}}}$ is an injection, so that we have an injection

$$\mathrm{Hom}_{\mathbb{Z}_p[G_{K_\infty}]}(V_{\mathfrak{S}}(\mathfrak{M}), \mathfrak{S}^{\mathrm{ur}}/p\mathfrak{S}^{\mathrm{ur}}) \hookrightarrow \mathrm{Hom}_{\mathbb{Z}_p[G_{K_\infty}]}(V_{\mathfrak{S}}(\mathfrak{M}), \widehat{\mathcal{O}_{\mathcal{E}^{\mathrm{ur}}}}/p\mathcal{O}_{\mathcal{E}^{\mathrm{ur}}}).$$

By [Fo 1, A.1.2.7], the right-hand side is a $\mathcal{O}_{\mathcal{E}}/p\mathcal{O}_{\mathcal{E}} = k((u))$-vector space of dimension $d = \mathrm{rk}_{\mathbb{Z}_p} V_{\mathfrak{S}}(\mathfrak{M})$. The left-hand side is clearly a u-adically separated, torsion-free $k[\![u]\!]$-module. Hence it is a free $k[\![u]\!]$-module of rank at most d.

Now $\mathfrak{M}'$ is a p-adically separated torsion-free $\mathfrak{S}$-module. Moreover, we have an injection

$$\mathfrak{M}'/p\mathfrak{M}' \hookrightarrow \mathrm{Hom}_{\mathbb{Z}_p[G_{K_\infty}]}(V_{\mathfrak{S}}(\mathfrak{M}), \mathfrak{S}^{\mathrm{ur}}/p\mathfrak{S}^{\mathrm{ur}}).$$

Hence $\mathfrak{M}'$ is a quotient of $\mathfrak{S}^d$. On the other hand, the natural map $\mathfrak{M} \to \mathfrak{M}'$ is an injection because the map $\mathcal{O}_{\mathcal{E}} \otimes_{\mathfrak{S}} \mathfrak{M} \to \mathrm{Hom}_{\mathbb{Z}_p[G_{K_\infty}]}(V_{\mathfrak{S}}(\mathfrak{M}), \mathcal{O}_{\widehat{\mathcal{E}^{\mathrm{ur}}}})$ is an isomorphism by [Fo 1, A.1.2.7]. Thus $\mathfrak{M}'$ must be a free $\mathfrak{S}$-module of rank $d = \mathrm{rk}_{\mathfrak{S}}\,\mathfrak{M}$ by Corollary 2.1.4. □

2.1.11. Denote by $\mathrm{Mod}^{\varphi}_{/\mathcal{O}_{\mathcal{E}}}$ the category of finite free $\mathcal{O}_{\mathcal{E}}$-modules $\mathcal{M}$ equipped with an isomorphism $\varphi^*\mathcal{M} \to \mathcal{M}$.

Proposition 2.1.12. *The functor*

$$\mathrm{Mod}^{\varphi}_{/\mathfrak{S}} \to \mathrm{Mod}^{\varphi}_{/\mathcal{O}_{\mathcal{E}}}; \qquad \mathfrak{M} \mapsto \mathfrak{M} \otimes_{\mathfrak{S}} \mathcal{O}_{\mathcal{E}}$$

is fully faithful.

Proof. Let $\mathcal{M}$ be in $\mathrm{Mod}^{\varphi}_{/\mathcal{O}_{\mathcal{E}}}$. If $\mathfrak{M} \subset \mathcal{M}$ is any finitely generated $\mathfrak{S}$-module which is stable under φ, and is such that $\mathfrak{M}/\varphi^*(\mathfrak{M})$ is killed by some power of $E(u)$, then we set $F(\mathfrak{M}) = \mathcal{O}_{\mathcal{E}} \otimes_{\mathfrak{S}} \mathfrak{M} \cap \mathfrak{M}[1/p]$. As in the proof of Lemma 1.3.13, $F(\mathfrak{M})$ is a finite free $\mathfrak{S}$-module, and is naturally a submodule of $\mathcal{M}$, which contains $\mathfrak{M}$, is stable by φ, and such that $F(\mathfrak{M})/\varphi^*(F(\mathfrak{M}))$ is killed by some power of $E(u)$. In particular, $F(\mathfrak{M})$ is an object of $\mathrm{Mod}^{\varphi}_{/\mathfrak{S}}$.

Now suppose that $\mathfrak{M}_1$ and $\mathfrak{M}_2$ are in $\mathrm{Mod}^{\varphi}_{/\mathfrak{S}}$, and write $\mathcal{M}_1 = \mathfrak{M}_1 \otimes_{\mathfrak{S}} \mathcal{O}_{\mathcal{E}}$ and $\mathcal{M}_2 = \mathfrak{M}_2 \otimes_{\mathfrak{S}} \mathcal{O}_{\mathcal{E}}$. Suppose we are given a morphism $h : \mathcal{M}_1 \to \mathcal{M}_2$ in $\mathrm{Mod}^{\varphi}_{/\mathcal{O}_{\mathcal{E}}}$. We have to show this induces a map $\mathfrak{M}_1 \to \mathfrak{M}_2$.

Suppose first that h is the identity morphism. By Corollary 2.1.4, we have $V_{\mathfrak{S}}(\mathfrak{M}_1) = V_{\mathfrak{S}}(\mathfrak{M}_2)$, so both $\mathfrak{M}_1$ and $\mathfrak{M}_2$ are contained in $\mathrm{Hom}_{\mathbb{Z}_p[G_{K_\infty}]}(V_{\mathfrak{S}}(\mathfrak{M}_1), \mathfrak{S}^{\mathrm{ur}})$, which is a finite $\mathfrak{S}$-module of rank $d = \mathrm{rk}_{\mathcal{O}_{\mathcal{E}}} \mathcal{M}_1$, by (2.1.10). In particular, $\mathfrak{M}_3 = \mathfrak{M}_1 + \mathfrak{M}_2 \subset \mathcal{M}_1$ is a finite $\mathfrak{S}$-module of rank d, which is stable under the action of φ, and $\mathfrak{M}_3/\varphi^*(\mathfrak{M}_3)$ is killed by a power of $E(u)$. Hence the morphism $\mathfrak{M}_1 \to F(\mathfrak{M}_3)$ is an isomorphism by Lemma 2.1.9, and similarly $\mathfrak{M}_2 = F(\mathfrak{M}_3) = \mathfrak{M}_1$.

Now consider the case of any map h. Let $\mathcal{M}_3 = h(\mathcal{M}_1)$, $\mathfrak{M}_3 = h(\mathfrak{M}_1)$, and $\mathfrak{M}'_3 = \mathcal{M}_3 \cap \mathfrak{M}_2$. Then $\mathcal{M}_3$ is in $\mathrm{Mod}^{\varphi}_{/\mathcal{O}_{\mathcal{E}}}$, and $\mathfrak{M}_3$ and $\mathfrak{M}'_3$ are finitely generated, φ-stable $\mathfrak{S}$-modules, such that $\mathfrak{M}_3/\varphi^*(\mathfrak{M}_3)$ and $\mathfrak{M}'_3/\varphi^*(\mathfrak{M}'_3)$ are killed by some power of $E(u)$. To see this for $\mathfrak{M}'$ note that we have an exact sequence

$$0 \to \mathfrak{M}'_3 \to \mathcal{M}_3 \oplus \mathfrak{M}_2 \to \mathcal{M}_2$$

and that the map $1 \otimes \varphi$ is injective on all the terms of this sequence. Thus the cokernel of $1 \otimes \varphi$ on $\mathfrak{M}'_3$ may be identified with an $\mathfrak{S}$-submodule of the cokernel of $1 \otimes \varphi$ on $\mathfrak{M}_2$. By what we have seen above, we must have $F(\mathfrak{M}_3) = F(\mathfrak{M}'_3) \subset \mathcal{M}_3$, so h induces the composite map

$$\mathfrak{M}_1 \to F(\mathfrak{M}_3) = F(\mathfrak{M}'_3) \to F(\mathfrak{M}_2) = \mathfrak{M}_2. \qquad \square$$

2.1.13. Denote by $\mathrm{Rep}_{G_{K_\infty}}$ the category of continuous representations of G_{K_∞} on finite-dimensional $\mathbb{Q}_p$-vector spaces. Similarly, we denote by $\mathrm{Rep}^{\mathrm{cris}}_{G_K}$ the category of crystalline representations of $G_K = \mathrm{Gal}(\bar{K}/K)$. The following result had been conjectured by Breuil [Br 1, p. 202].

Corollary 2.1.14. *The functor* $\mathrm{Rep}^{\mathrm{cris}}_{G_K} \to \mathrm{Rep}_{G_{K_\infty}}$ *obtained by restricting the action of a G_K-representation to G_{K_∞} is fully faithful.*

Proof. It suffices to prove the corollary for the full subcategory $\mathrm{Rep}^{\mathrm{cris},+}_{G_K} \subset \mathrm{Rep}^{\mathrm{cris}}_{G_K}$ consisting of crystalline representations with nonnegative Hodge–Tate weights.

Consider the diagram of functors

$$\begin{array}{ccc} \mathrm{Rep}^{\mathrm{cris},+}_{G_K} & \longrightarrow & \mathrm{Rep}_{G_{K_\infty}} \\ \downarrow & & \uparrow \\ \mathrm{Mod}^{\varphi}_{/\mathfrak{S}} \otimes \mathbb{Q}_p & \longrightarrow & \mathrm{Mod}^{\varphi}_{/\mathcal{O}_{\mathcal{E}}} \otimes \mathbb{Q}_p. \end{array}$$

Here $\otimes\mathbb{Q}_p$ means that we have passed to the associated isogeny category. The map on the left is given by composing the (contravariant) functor from crystalline representations to weakly admissible modules with the fully faithful functor of Corollary 1.3.15. The map on the bottom is given by Proposition 2.1.12, and hence is fully faithful, while the map on the right is given by sending $\mathcal{M}$ in $\mathrm{Mod}^{\varphi}_{/\mathcal{O}_{\mathcal{E}}}$ to $\mathrm{Hom}_{\mathcal{O}_{\mathcal{E}}}(\mathcal{M}, \widehat{\mathcal{E}^{\mathrm{ur}}})$, and this functor is an equivalence by [Fo 1, A.1.2.7]. That the square commutes (up to a natural equivalence) follows from Proposition 2.1.5. It follows that the top functor is also fully faithful. □

Lemma 2.1.15. *Let $\mathfrak{M}$ be in $\mathrm{Mod}^{\varphi}_{/\mathfrak{S}}$ and set $V = V_{\mathfrak{S}}(\mathfrak{M}) \otimes_{\mathbb{Z}_p} \mathbb{Q}_p$, and $\mathcal{M} = \mathcal{E} \otimes_{\mathfrak{S}} \mathfrak{M}$. Then the map $\mathfrak{N} \mapsto \mathrm{Hom}_{\mathfrak{S},\varphi}(\mathfrak{N}, \mathfrak{S}^{\mathrm{ur}})$ is a bijection between finite free, φ-stable $\mathfrak{S}$-submodules $\mathfrak{N} \subset \mathcal{M}$ such that $\mathcal{E} \otimes_{\mathfrak{S}} \mathfrak{N} \xrightarrow{\sim} \mathcal{M}$ and $\mathfrak{N}/\varphi^*(\mathfrak{N})$ is killed by a power of $E(u)$, and G_{K_∞}-stable $\mathbb{Z}_p$-lattices $L \subset V$.*

Proof. By [Fo 1, A.1.2.7] the set of G_{K_∞}-stable lattices $L \subset V$ is in bijection with the set of finite free, φ-stable $\mathcal{O}_{\mathcal{E}}$-lattices $\mathcal{N} \subset \mathcal{M}$ such that the map $\varphi^*\mathcal{N} \to \mathcal{N}$ is an isomorphism.

Given $\mathfrak{N}$, $\mathrm{Hom}_{\mathfrak{S},\varphi}(\mathfrak{N}, \mathfrak{S}^{\mathrm{ur}})$ is a G_{K_∞}-stable lattice in V by Corollary 2.1.4. Moreover, the above remarks together with Corollary 2.1.4 and Lemma 2.1.9 show that the map of the lemma is an injection. Suppose we are given a G_{K_∞}-stable lattice $L \subset V$, and let $\mathcal{N} = \mathrm{Hom}_{\mathbb{Z}_p[G_{K_\infty}]}(L, \mathcal{O}_{\widehat{\mathcal{E}^{\mathrm{ur}}}})$ be the corresponding finite free $\mathcal{O}_{\mathcal{E}}$-module. Let $\mathfrak{N} = \mathcal{N} \cap \mathfrak{M}[1/p] \subset \mathcal{M}$. As in the proof of Lemma 1.3.13 $\mathfrak{N}$ is a finite free $\mathfrak{S}$-module such that $\mathfrak{N}/\varphi^*(\mathfrak{N})$ is killed by some power of $E(u)$, and $\mathcal{O}_{\mathcal{E}} \otimes_{\mathfrak{S}} \mathfrak{N} = \mathcal{N}$. Hence $\mathfrak{N}$ maps to L by Corollary 2.1.4. □

2.2 Applications to p-divisible groups

In this section we apply the theory of Section 1 to the special case of p-divisible groups. We give a classification of p-divisible groups (up to isogeny when $p = 2$) using $\mathfrak{S}$-modules, and we show Fontaine's conjecture that a crystalline representation with Hodge–Tate weights 0 and 1 arises from a p-divisible group.

2.2.1. We will denote by $\mathrm{BT}^{\varphi}_{/\mathfrak{S}}$ the full subcategory of $\mathrm{Mod}^{\varphi}_{/\mathfrak{S}}$ consisting of objects $\mathfrak{M}$ such that $\mathfrak{M}/\varphi^*(\mathfrak{M})$ is killed by $E(u)$ (not just some power). Similarly we denote by $\mathrm{BT}^{\varphi,N_\nabla}_{/\mathcal{O}}$ (respectively, $\mathrm{BT}^{\varphi}_{/\mathcal{O}}$) the full subcategory of $\mathrm{Mod}^{\varphi,N_\nabla}_{/\mathcal{O}}$ (respectively, $\mathrm{Mod}^{\varphi,N}_{/\mathcal{O}}$)

consisting of objects $\mathcal{M}$ such that $N_\nabla = 0$ modulo u, (respectively, $N = 0$) and $\mathcal{M}/\varphi^*(\mathcal{M})$ is killed by $E(u)$.

We say a weakly admissible module D is of Barsotti–Tate type if $\mathrm{gr}^i D_K = 0$ for $i \neq 0, 1$.

Proposition 2.2.2. *The functor of Corollary* 1.3.15 *induces an exact equivalence between the category of weakly admissible modules of Barsotti–Tate type and* $\mathrm{BT}^{\varphi}_{/\mathfrak{S}} \otimes \mathbb{Q}_p$.

Proof. Let $\mathfrak{M}$ be in $\mathrm{BT}^{\varphi}_{/\mathfrak{S}}$ and $\tilde{\mathcal{M}} = \mathfrak{M} \otimes_{\mathfrak{S}} \mathcal{O}$ the corresponding object of $\mathrm{Mod}^{\varphi,N}_{/\mathcal{O}}$. Then $\tilde{\mathcal{M}}$ is evidently in $\mathrm{BT}^{\varphi}_{/\mathcal{O}}$. As in (1.3.11) we obtain a map $D(\tilde{\mathcal{M}}) \otimes_{K_0} \mathcal{O} \to \tilde{\mathcal{M}}$, which lifts the isomorphism $D(\tilde{\mathcal{M}}) \xrightarrow{\sim} \tilde{\mathcal{M}}/u\tilde{\mathcal{M}}$, and is compatible with φ and N_∇. Here N_∇ acts on $D(\tilde{\mathcal{M}}) \otimes_{K_0} \mathcal{O}$ as $1 \otimes -u\lambda\frac{d}{du}$. Now since $\tilde{\mathcal{M}}/\varphi^*(\tilde{\mathcal{M}})$ is killed by $E(u)$, one sees easily using Lemma 1.2.6 that $\tilde{\mathcal{M}}/(D(\tilde{\mathcal{M}}) \otimes_{K_0} \mathcal{O})$ is killed by λ. Let $m \in \tilde{\mathcal{M}}$, and write $m = \sum_{i=1}^r d_i \otimes \lambda^{-1} f_i$, where $d_i \in D(\tilde{\mathcal{M}})$ and $f_i \in \mathcal{O}$. Then

$$N_\nabla(m) = -u\lambda \sum_{i=1}^{r} d_i \otimes (-\lambda^{-2}\frac{d\lambda}{du} f_i + \lambda^{-1}\frac{df_i}{du}) = u\frac{d\lambda}{du}m - u\sum_{i=1}^{r} d_i \otimes \frac{df_i}{du} \in \tilde{\mathcal{M}}.$$

Hence, by Lemma 1.3.10, $\tilde{\mathcal{M}}$ arises from a module $\mathcal{M}$ in $\mathrm{Mod}^{\varphi,N_\nabla}_{/\mathcal{O}}$, and $D(\tilde{\mathcal{M}}) = D(\mathcal{M})$ is weakly admissible by Theorems 1.3.8 and 1.2.15. By construction, the functor in Corollary 1.3.15 takes $D(\mathcal{M})$ to (an object isomorphic to) $\mathfrak{M}$.

It remains to remark that if D is an effective weakly admissible module, and $\mathfrak{M}$ in $\mathrm{Mod}^{\varphi,N}_{/\mathfrak{S}}$ is the image of D under the functor of Corollary 1.3.15, then $\mathfrak{M}$ is in $\mathrm{BT}^{\varphi}_{/\mathfrak{S}}$ if and only if D is of Barsotti–Tate type. This follows from Lemma 1.2.2. □

2.2.3. We will use the notation introduced in the appendix. Given a module $\mathfrak{M}$ in $\mathrm{BT}^{\varphi}_{/\mathfrak{S}}$, $\mathcal{M} = S \otimes_{\varphi,\mathfrak{S}} \mathfrak{M}$ has a natural structure of an object of $\mathrm{BT}^{\varphi}_{/S}$, where this is the category introduced in Section A.5. Here the tensor product is taken with respect to the map $\mathfrak{S} \to S$ sending u to u^p. Following [Br 4], we set

$$\mathrm{Fil}^1 \mathcal{M} = \{m \in \mathcal{M} : 1 \otimes \varphi(m) \in \mathrm{Fil}^1 S \otimes_{\mathfrak{S}} \mathfrak{M} \subset S \otimes_{\mathfrak{S}} \mathfrak{M}\},$$

and we define the map φ_1 as the composite

$$\varphi_1 : \mathrm{Fil}^1 \mathcal{M} \overset{1\otimes\varphi}{\to} \mathrm{Fil}^1 S \otimes_{\mathfrak{S}} \mathfrak{M} \overset{\varphi_1\otimes 1}{\to} S \otimes_{\varphi,\mathfrak{S}} \mathfrak{M} = \mathcal{M}$$

By Lemma A.2, given a p-divisible group G over $\mathcal{O}_K$, $\mathcal{M}(G) := \mathbb{D}(G)(S)$ is naturally an object of $\mathrm{BT}^{\varphi}_{/S}$. By Proposition A.6 the functor $G \mapsto \mathcal{M}(G)$ is an equivalence between $\mathrm{BT}^{\varphi}_{/S}$ and the category of p-divisible groups if $p > 2$. If $p = 2$ it induces an equivalence between the corresponding isogeny categories.

Given a p-divisible group G over $\mathcal{O}_K$, we will denote by $T_p(G)$ its Tate module.

Lemma 2.2.4. *Let G be a p-divisible group over $\mathcal{O}_K$. If we regard the ring A_{cris} of Section 2.1.1 as an S-algebra via $u \mapsto [\underline{\pi}]$, then there is a canonical injection of G_{K_∞}-modules*

$$T_p(G) \hookrightarrow \mathrm{Hom}_{S,\mathrm{Fil},\varphi}(\mathcal{M}(G), A_{\mathrm{cris}}).$$

This map is an isomorphism if $p > 2$, and has cokernel killed by p when $p = 2$.

Proof. An element of $T_p(G)$ is a map of p-divisible groups over $\mathcal{O}_{\bar{K}}$, $\mathbb{Q}_p/\mathbb{Z}_p \to G \otimes_{\mathcal{O}_K} \mathcal{O}_{\bar{K}}$. Since A_{cris} is a divided power thickening of $\widehat{\mathcal{O}_{\bar{K}}}$, we can pull this map back to $\widehat{\mathcal{O}_{\bar{K}}}$, and then evaluate the corresponding crystals on A_{cris} (see the appendix). This gives rise to a map $\mathcal{M} \otimes_S A_{\mathrm{cris}} \to A_{\mathrm{cris}}$ compatible with filtrations and Frobenius. That the resulting map is injective, an isomorphism when $p > 2$, and has cokernel killed by p when $p = 2$, follows from [Fa, Theorem 7]. □

2.2.5. We remark that the fact that the map of Lemma 2.2.4 is an isomorphism when $p > 2$ also follows from the calculations of [Br 2, Section 5.3]; however, Faltings' argument is quite direct and does not rely on reduction to calculations with finite flat group schemes.

The following result had been conjectured by Fontaine [Fo 3, 5.2.5]

Corollary 2.2.6. *Let V be a crystalline representation of G_K with all Hodge–Tate weights equal to 0 or 1. Then there exists a p-divisible group G such that $V \xrightarrow{\sim} T_p(G) \otimes_{\mathbb{Z}_p} \mathbb{Q}_p$.*

Proof. Let $D = \mathrm{Hom}_{\mathbb{Z}_p[G_K]}(V, B^+_{\mathrm{cris}})$ denote the admissible filtered (φ, N)-module attached to V, and let $\mathfrak{M}$ in $\mathrm{BT}^{\varphi}_{/\mathfrak{S}} \otimes \mathbb{Q}_p$ be the module associated to D by the functor of Proposition 2.2.2. We again denote by $\mathfrak{M}$ the object of $\mathrm{BT}^{\varphi}_{/\mathfrak{S}}$ underlying $\mathfrak{M}$. Write $\mathcal{M} = S \otimes_{\varphi,\mathfrak{S}} \mathfrak{M}$ for the associated object of $\mathrm{BT}^{\varphi}_{/S}$. Then $\mathcal{M}$ is associated to a p-divisible group G as above, and by Lemma 2.2.4 we have an isomorphism of $\mathbb{Q}_p$-vector spaces with G_{K_∞}-action

$$T_p(G) \otimes_{\mathbb{Z}_p} \mathbb{Q}_p \xrightarrow{\sim} \mathrm{Hom}_{S,\mathrm{Fil},\varphi}(\mathcal{M}, A_{\mathrm{cris}}) \otimes_{\mathbb{Z}_p} \mathbb{Q}_p \xrightarrow{\sim} \mathrm{Hom}_{S,\mathrm{Fil},\varphi}(D, B^+_{\mathrm{cris}}) = V.$$

Here the final isomorphism follows from the fact that, by [Br 2, 5.1.3], we have a canonical isomorphism $\mathcal{M} \otimes_{\mathbb{Z}_p} \mathbb{Q}_p \xrightarrow{\sim} D \otimes_W S$, compatible with φ and filtrations. This fact is also easily deduced from Lemma 1.2.6. That this map is actually compatible with the action of G_K follows from Corollary 2.1.14. □

Theorem 2.2.7. *There exists an exact functor between $\mathrm{BT}^{\varphi}_{/\mathfrak{S}}$ and the category of p-divisible groups over $\mathcal{O}_K$. When $p > 2$, this functor is an equivalence, and when $p = 2$ it induces an equivalence between the corresponding isogeny categories.*

Proof. Let $\mathfrak{M}$ be in $\mathrm{BT}^{\varphi}_{/\mathfrak{S}}$ and $\mathcal{M} = S \otimes_{\varphi,\mathfrak{S}} \mathfrak{M}$ the corresponding module in $\mathrm{BT}^{\varphi}_{/S}$. We have natural maps

$$\mathrm{Hom}_{\mathfrak{S},\varphi}(\mathfrak{M}, \mathfrak{S}^{\mathrm{ur}}) \to \mathrm{Hom}_{S,\mathrm{Fil},\varphi}(\mathcal{M}, A_{\mathrm{cris}}) \tag{2.2.8}$$

obtained by composing maps $\mathfrak{M} \to \mathfrak{S}^{\mathrm{ur}}$ with the inclusion $\mathfrak{S}^{\mathrm{ur}} \xrightarrow{\varphi} A_{\mathrm{cris}}$, and extending the resulting map to $\mathcal{M}$ by S-linearity. By Lemma 2.2.4 and Corollary 2.1.4, both sides of (2.2.8) are finite free $\mathbb{Z}_p$-modules of the same rank, and (2.2.8) is clearly injective. Hence it becomes an isomorphism after inverting p. In particular, any map in the right-hand side induces a map $\mathfrak{M} \to \mathfrak{S}^{\mathrm{ur}}[1/p]$. It follows that (2.2.8) is an isomorphism provided that any map $\mathfrak{M} \to \mathfrak{S}^{\mathrm{ur}}$ in the left-hand side whose composite with $\mathfrak{S}^{\mathrm{ur}} \xrightarrow{\varphi} A_{\mathrm{cris}}$ factors through pA_{cris} actually factors through $p\mathfrak{S}^{\mathrm{ur}}$. That this is the case for $p > 2$, was observed in the proof of [Br 3, 3.3.2].

Now given $\mathfrak{M}$ in $\mathrm{BT}^{\varphi}_{/\mathfrak{S}}$, the construction of Section 2.2.3 produces a p-divisible group $G(\mathfrak{M})$. Conversely, given a p-divisible group G, its Tate module $T_p(G)$ is a lattice in the Barsotti–Tate representation $V_p(G) = T_p(G) \otimes_{\mathbb{Z}_p} \mathbb{Q}_p$. By Proposition 2.2.2 and Lemma 2.1.15, there is an $\mathfrak{M}$ in $\mathrm{BT}^{\varphi}_{/\mathfrak{S}}$, determined up to canonical isomorphism, such that $T_p(G) \xrightarrow{\sim} \mathrm{Hom}_{\mathfrak{S},\varphi}(\mathfrak{M}, \mathfrak{S}^{\mathrm{ur}})$, and it follows from Proposition 2.1.12 that the assignment $G \mapsto \mathfrak{M} = \mathfrak{M}(G)$ is functorial.

Now suppose that $p > 2$. Then Lemma 2.2.4 and the fact that (2.2.8) is an isomorphism imply that for $\mathfrak{M}$ in $\mathrm{Mod}^{\varphi}_{/\mathfrak{S}}$ there is a natural isomorphism $\mathfrak{M}(G(\mathfrak{M})) \xrightarrow{\sim} \mathfrak{M}$. On the other hand, if G is a p-divisible group over $\mathcal{O}_K$, then we have natural isomorphisms,

$$T_p(G(\mathfrak{M}(G))) \xrightarrow{\sim} \mathrm{Hom}_{\mathfrak{S},\varphi}(\mathfrak{M}(G), \mathfrak{S}^{\mathrm{ur}}) \xrightarrow{\sim} T_p(G),$$

and hence a natural isomorphism $G(\mathfrak{M}(G)) \xrightarrow{\sim} G$ by Tate's theorem.

For $p = 2$, the same arguments show that the functors G and $\mathfrak{M}$ induce equivalences on the associated isogeny categories. We could also have deduced the theorem in this case directly from Proposition 2.2.2 and Corollary 2.2.6. □

2.2.9. In [Ki, 2.2.22] we gave a different proof of the above theorem when $p > 2$, which, in particular, made no use of Tate's theorem. One can recover Tate's result from Theorem 2.2.7 by using the full faithfulness of Proposition 2.1.12 together with Lemma 2.2.4 and the isomorphism (2.2.8).

2.3 Classification of finite flat group schemes

In this final subsection of the paper, we use Theorem 2.2.7 to give a classification of finite flat group schemes over $\mathcal{O}_K$. The idea that one could do this is due to A. Beilinson, and we are grateful to him for allowing us to include his argument here. The final result was conjectured by Breuil [Br 4, 2.1.1]

2.3.1. Following the notation of [Ki] we denote by $'(\mathrm{Mod}\,/\mathfrak{S})$ the category consisting of $\mathfrak{S}$-modules $\mathfrak{M}$ equipped with a Frobenius semilinear map φ, such that the cokernel of $\varphi^*(\mathfrak{M}) \to \mathfrak{M}$ is killed by $E(u)$. We denote by $(\mathrm{Mod}\,/\mathfrak{S})$ the full subcategory of $'(\mathrm{Mod}\,/\mathfrak{S})$ consisting of modules $\mathfrak{M}$ such that $\mathfrak{M}$ has projective dimension 1 as an $\mathfrak{S}$-module and is killed by some power of p.

Later we will need the full subcategory $(\mathrm{ModFI}\,/\mathfrak{S})$ of $(\mathrm{Mod}\,/\mathfrak{S})$ consisting of modules $\mathfrak{M}$ which are of the form $\oplus_{i\in I}\mathfrak{S}/p^{n_i}\mathfrak{S}$, where I is a finite set and $n_i \in \mathbb{N}^+$.

Lemma 2.3.2. *A module* $\mathfrak{M}$ *in* $'(\mathrm{Mod}\,/\mathfrak{S})$ *is in* $(\mathrm{Mod}\,/\mathfrak{S})$ *if and only if* $\mathfrak{M}$ *is an extension in* $'(\mathrm{Mod}\,/\mathfrak{S})$ *of objects which are finite free* $\mathfrak{S}/p\mathfrak{S}$*-modules.*

Proof. We remark that since $\mathfrak{S}$ is a regular ring of dimension 2, the Auslander-Buchsbaum theorem implies that a finitely generated torsion $\mathfrak{S}$-module $\mathfrak{M}$ has projective dimension 1 if and only if it has depth 1. The latter condition holds if and only if the associated primes of $\mathfrak{M}$ are all of height 1 or, equivalently, if $\mathfrak{M}$ has no section supported on the closed point of Spec $\mathfrak{S}$.

Thus, if $\mathfrak{M}$ is in $(\mathrm{Mod}\,/\mathfrak{S})$ then the quotients $\mathfrak{M}[p^i]/\mathfrak{M}[p^{i-1}]$ for $i = 0, 1, 2\dots$ are easily seen to be free $\mathfrak{S}/p\mathfrak{S}$-modules, and one sees by descending induction on i that $\varphi^*(\mathfrak{M}[p^i]/\mathfrak{M}[p^{i-1}]) \to \mathfrak{M}[p^i]/\mathfrak{M}[p^{i-1}]$ has kernel killed by $E(u)$, and is therefore injective [Ki, 1.1.9]. Hence $\mathfrak{M}$ is an extension of objects which are free over $\mathfrak{S}/p\mathfrak{S}$.

Conversely, any such extension has projective dimension 1, and is killed by some power of p. □

2.3.3. Let $D^b(\mathrm{BT}^{\varphi}_{/\mathfrak{S}})$ denote the bounded derived category of the exact category $\mathrm{BT}^{\varphi}_{/\mathfrak{S}}$. We write $(\mathrm{Mod}\,/\mathfrak{S})^{\bullet}$ for the full subcategory of $D^b(\mathrm{BT}^{\varphi}_{/\mathfrak{S}})$ consisting of two-term complexes $\mathfrak{M}^{\bullet} = \mathfrak{M}_1 \to \mathfrak{M}_2$ concentrated in degrees 0 and -1, such that $H^{-1}(\mathfrak{M}^{\bullet}) = 0$, and $H^0(\mathfrak{M}^{\bullet})$ is killed by a power of p. This is equivalent to asking that the map $\mathfrak{M}_1 \to \mathfrak{M}_2$ becomes an isomorphism in the isogeny category of $\mathrm{BT}^{\varphi}_{/\mathfrak{S}}$. Concretely, $(\mathrm{Mod}\,/\mathfrak{S})^{\bullet}$ is obtained by taking the category of two-term complexes $\mathfrak{M}^{\bullet}$, as above, dividing by homotopy equivalences—that is, by morphisms $\mathfrak{N}^{\bullet} \to \mathfrak{M}^{\bullet}$ of the form $(h \circ d, d \circ h)$, where $h : \mathfrak{N}_2 \to \mathfrak{M}_1$ is a morphism in $\mathrm{BT}^{\varphi}_{/\mathfrak{S}}$—and inverting quasi-isomorphisms.

Lemma 2.3.4. *The functor* $\mathfrak{M}^{\bullet} \mapsto H^0(\mathfrak{M}^{\bullet})$ *induces an equivalence between* $(\mathrm{Mod}\,/\mathfrak{S})^{\bullet}$ *and* $(\mathrm{Mod}\,/\mathfrak{S})$.

Proof. It is easy to check that the functor is fully faithful. To check essential surjectivity it suffices, given $\mathfrak{M}$ in $(\mathrm{Mod}\,/\mathfrak{S})$, to find $\tilde{\mathfrak{M}}$ in $\mathrm{BT}^{\varphi}_{/\mathfrak{S}}$ and a surjection $\tilde{\mathfrak{M}} \to \mathfrak{M}$ compatible with φ. Indeed, the kernel of any such surjection is automatically a finite free module, and since $\varphi^*(\mathfrak{M}) \to \mathfrak{M}$ is injective, this cokernel is in $\mathrm{BT}^{\varphi}_{/\mathfrak{S}}$.

Let $L = \mathfrak{M}/(1 \otimes \varphi)(\varphi^*\mathfrak{M})$. Then L is a finite $\mathcal{O}_K$-module (via $\mathfrak{S} \overset{u \mapsto \pi}{\to} \mathcal{O}_K$). Let $\tilde{L}$ be a free $\mathcal{O}_K$-module, and $\tilde{L} \to L$ a surjection. Choose a free $\mathfrak{S}$-module $\tilde{\mathfrak{M}}$ and surjections of $\mathfrak{S}$-modules $\tilde{\mathfrak{M}} \to \tilde{L}$ and $\tilde{\mathfrak{M}} \to \mathfrak{M}$ compatible with the projections of $\tilde{L}$ and $\mathfrak{M}$ to L.

Since we may always replace $\tilde{\mathfrak{M}}$ with $\tilde{\mathfrak{M}} \oplus \mathfrak{S}^r$ for $r \in \mathbb{N}^+$, and map the second factor to 0 in $\tilde{L}$ and arbitrarily to $\mathfrak{N} = \ker(\mathfrak{M} \to L)$, we may assume that $\tilde{\mathfrak{N}} := \ker(\tilde{\mathfrak{M}} \to \tilde{L})$ surjects onto $\mathfrak{N}$. Finally, we may write $\tilde{\mathfrak{N}} = \tilde{\mathfrak{N}}_0 \oplus \tilde{\mathfrak{N}}_1$, where $\tilde{\mathfrak{N}}_1$ maps to 0 in $\mathfrak{N}$, and $\tilde{\mathfrak{N}}_0 \otimes_{\mathfrak{S}} k \overset{\sim}{\longrightarrow} \mathfrak{N} \otimes_{\mathfrak{S}} k$. Since $\varphi^*(\tilde{\mathfrak{M}})$ is a free $\mathfrak{S}$-module, the composite

$$\varphi^*(\tilde{\mathfrak{M}}) \to \varphi^*(\mathfrak{M}) \overset{\sim}{\longrightarrow} \mathfrak{N} \subset \mathfrak{M}$$

lifts to $\tilde{\mathfrak{N}}_0$, and any such lift is automatically a surjection. We may then lift this further to a surjection $\varphi^*(\tilde{\mathfrak{M}}) \to \tilde{\mathfrak{N}}$. Any such lift is an isomorphism, since both sides are free $\mathfrak{S}$-modules of the same rank.

The induced map $\varphi^*(\tilde{\mathfrak{M}}) \to \tilde{\mathfrak{M}}$ has cokernel $\tilde{L}$, and hence gives $\tilde{\mathfrak{M}}$ the structure of a module in $\mathrm{BT}^{\varphi}_{/\mathfrak{S}}$, which surjects onto $\mathfrak{M}$. □

Theorem 2.3.5. *If $p > 2$, there is an exact anti-equivalence between* $(\mathrm{Mod}\,/\mathfrak{S})$ *and the category* $(p\text{-}Gr\,/\mathcal{O}_K)$ *of finite flat group schemes over* $\mathcal{O}_K$.

Proof. Let $D^b(p\text{-div}\,/\mathcal{O}_K)$ denote the bounded derived category of the exact category of p-divisible groups over $\mathcal{O}_K$. We write $(p\text{-Gr}\,/\mathcal{O}_K)^\bullet$ for the full subcategory of $D^b(p\text{-div}\,/\mathcal{O}_K)$ consisting of isogenies of p-divisible groups $G_1 \to G_2$. This category has an explicit description analogous to the one given for $(\mathrm{Mod}\,/\mathfrak{S})^\bullet$ in Section 2.3.3.

The kernel of any isogeny is a finite flat group scheme, and conversely given any finite flat group scheme G there exists an embedding of G into a p-divisible group G_1 [BBM, 3.1.1]. The quotient G_1/G (taken, for example, in the category of fppf sheaves) is a p-divisible group. Hence one sees easily that the functor $(p\text{-Gr}\,/\mathcal{O}_K)^\bullet \to (p\text{-Gr}\,/\mathcal{O}_K)$ given by sending an isogeny to its kernel is an equivalence of categories.

On the other hand, $(p\text{-Gr}\,/\mathcal{O}_K)^\bullet$ is anti-equivalent to $(\mathrm{Mod}\,/\mathfrak{S})^\bullet$ by Theorem 2.2.7, and the theorem follows, since $(\mathrm{Mod}\,/\mathfrak{S})^\bullet$ is equivalent to $(\mathrm{Mod}\,/\mathfrak{S})$ by Lemma 2.3.4. □

Corollary 2.3.6. *If $p > 2$, the category* $(\mathrm{ModFI}\,/\mathfrak{S})$ *is anti-equivalent to the category of finite flat group schemes G over $\mathcal{O}_K$ such that $G[p^n]$ is finite flat for $n \geq 1$.*

Proof. This can be deduced by formal arguments from Theorem 2.3.5 in the same way that [Br 2, 4.2.2.5] is deduced from [Br 2, 4.2.1.6]. □

Appendix A: Crystals and p-divisible groups

A.1

Let T be a W-scheme on which p is locally nilpotent, and denote by $(T/W)_{\mathrm{cris}}$ the crystalline site of T over W, corresponding to embeddings of W-schemes $T \hookrightarrow T'$, defined by a sheaf of ideal J on T', which is equipped with divided powers, and such that the local sections of J are nilpotent.

Let G be a p-divisible group on T. Recall [MM, II Section 9] that there is a contravariant functor $G \mapsto \mathbb{D}(G)$ from the category of p-divisible groups over T to the category of crystals on $(T/W)_{\mathrm{cris}}$. The functor is defined using the Lie algebra of the universal vector extension of the dual p-divisible group G^*.

The formation of $\mathbb{D}(G)$ is compatible with arbitrary base change. In particular, if $p = 0$ on T, then we can pull G back by the Frobenius φ on T. The relative Frobenius on G, gives a map $G \to \varphi^*(G)$, and hence a map of crystals

$$\varphi^*(\mathbb{D}(G)) \xrightarrow{\sim} \mathbb{D}(\varphi^*(G)) \to \mathbb{D}(G).$$

Suppose now that T_0 is a W-scheme with $p = 0$ on T_0, and G_0 is a p-divisible group over T_0. Let $T_0 \hookrightarrow T$ be an object of $(T_0/W)_{\text{cris}}$ on which p is locally nilpotent, and G a lifting of G_0 to T. By construction of $\mathbb{D}$, we have an isomorphism $\mathbb{D}(G_0)(T) \xrightarrow{\sim} \mathbb{D}(G)(T)$. Moreover, the $\mathcal{O}_T$-module $\mathbb{D}(G)(T)$ sits in an exact sequence

$$0 \to (\operatorname{Lie} G)^* \to \mathbb{D}(G)(T) \to \operatorname{Lie} G^* \to 0$$

where $(\operatorname{Lie} G)^*$ denotes the $\mathcal{O}_T$-dual of $\operatorname{Lie} G$. Hence specifying G equips $\mathbb{D}(G_0)(T)$ with an $\mathcal{O}_T$-submodule L such that $\mathbb{D}(G_0)(T)/L$ is a free $\mathcal{O}_T$-module.

The main result of [Me] asserts that if the divided powers on the ideal defining $T_0 \hookrightarrow T$ are nilpotent, then G is determined by L, and that, conversely, given a submodule $L \subset \mathbb{D}(G_0)(T)$ such that $\mathbb{D}(G_0)(T)/L$ is $\mathcal{O}_T$-free, and $L \otimes_{\mathcal{O}_T} \mathcal{O}_{T_0} \subset \mathbb{D}(G_0)(T_0)$ coincides with $(\operatorname{Lie} G_0)^*$, there is a p-divisible group G over T with $L = (\operatorname{Lie} G)^* \subset \mathbb{D}(G)(T) = \mathbb{D}(G_0)(T)$. (Strictly speaking, the result in [Me] applies when, locally on T_0, G_0 admits some lift to T, but this condition is always satisfied [MM, II Section 9]).

If $T = \operatorname{Spec} A$ is affine we will write $\mathbb{D}(G)(A)$ for $\mathbb{D}(G)(\operatorname{Spec} A)$.

Lemma A.2. *Let $A \to A_0$ be a surjection of p-adically complete and separated, local $\mathbb{Z}_p$-algebras with residue field k, whose kernel* $\operatorname{Fil}^1 A$ *is equipped with divided powers. Suppose that*

(1) *A is p-torsion-free, and equipped with an endomorphism $\varphi : A \to A$ lifting the Frobenius on A/pA.*
(2) *The induced map $\varphi^*(\operatorname{Fil}^1 A) \xrightarrow{1\otimes\varphi/p} A$ is surjective.*

If G is a p-divisible group over A_0, write $\operatorname{Fil}^1 \mathbb{D}(G)(A) \subset \mathbb{D}(G)(A)$ *for the preimage of* $(\operatorname{Lie} G)^* \subset \mathbb{D}(G)(A_0)$. *Then the restriction of $\varphi : \mathbb{D}(G)(A) \to \mathbb{D}(G)(A)$ to* $\operatorname{Fil}^1 \mathbb{D}(G)(A)$ *is divisible by p, and the induced map*

$$\varphi^* \operatorname{Fil}^1 \mathbb{D}(G)(A) \xrightarrow{1\otimes\varphi/p} \mathbb{D}(G)(A)$$

is a surjection.

Proof. Let $\mathcal{M} = \mathbb{D}(G)(A)$. Let $\tilde{G}$ be a lifting of G to A, and set $\tilde{G}_0 = G \otimes_A A/pA$. Note that φ induces the zero endomorphism of $(\operatorname{Lie} \tilde{G}_0)^*$, and that φ restricted to $\operatorname{Fil}^1 A = \ker(A \to A_0)$ is divisible by p, since this ideal is equipped with divided powers. In particular, the map of (2) makes sense. Since

$$\operatorname{Fil}^1 \mathcal{M} = (\operatorname{Lie} \tilde{G})^* + \operatorname{Fil}^1 A \cdot \mathcal{M},$$

we see that $\varphi(\operatorname{Fil}^1 \mathcal{M}) \subset p\mathcal{M}$, so we may define a map

$$\varphi_1 = \varphi/p : \operatorname{Fil}^1 \mathcal{M} \to \mathcal{M}.$$

We have to check that the image of this map generates the A-module $\mathcal{M}$. The hypothesis (2) implies that $\varphi(\mathcal{M}) = \varphi_1(\operatorname{Fil}^1 A)A\varphi(\mathcal{M}) \subset \varphi_1(\operatorname{Fil}^1 \mathcal{M})A$. Hence it suffices to show that the map

$$\varphi^*(\mathrm{Fil}^1 \mathcal{M} + p\mathcal{M}) \stackrel{1\otimes\varphi/p}{\longrightarrow} \mathcal{M} \tag{A.2.1}$$

is surjective.

There is a unique map $A \to W(k)$ which lifts the projection $A \to k$ and is compatible with the action of Frobenius. Write $H = \tilde{G} \otimes_A W(k)$ and $H_0 = H \otimes_{W(k)} k$. By [MM, II Section 15] $\mathbb{D}(H)(W(k))$ is naturally isomorphic to the Dieudonné module of H, and this isomorphism is compatible with the action of Frobenius. Hence if V denotes the Verschiebung, then we have

$$(\mathrm{Lie}\, H)^* = V(F/p)(\mathrm{Lie}\, H^*) \subset V\mathbb{D}(H)(W(k)).$$

Hence $(\mathrm{Lie}\, H_0)^* \subset V\mathbb{D}(H_0)(k)$, and this inclusion must be an equality since both sides have the same k-dimension. (They may both be identified with the quotient $\mathbb{D}(H_0)(k)/F\mathbb{D}(H_0)(k)$.) Hence $(\mathrm{Lie}\, H)^* + p\mathbb{D}(H)(W(k)) = V\mathbb{D}(H)(W(k))$, and since $(F/p)V = 1$, we see that F/p induces a surjection of $(\mathrm{Lie}\, H)^* + p\mathbb{D}(H)(W(k))$ onto $\mathbb{D}(H)(W(k))$. Hence (A.2.1) is also a surjection. □

A.3

By a *special* ring we shall mean a p-adically complete, separated, p-torsion-free, local $\mathbb{Z}_p$-algebra A with residue field k, equipped with an endomorphism φ lifting the Frobenius on A/pA.

For such an A, we denote by $\mathcal{C}_A$ the category of finite free A-modules $\mathcal{M}$, equipped with a Frobenius semilinear map $\varphi : \mathcal{M} \to \mathcal{M}$ and an A-submodule $\mathcal{M}_1 \subset \mathcal{M}$ such that $\varphi(\mathcal{M}_1) \subset p\mathcal{M}$ and the map $1 \otimes \varphi/p : \varphi^*(\mathcal{M}_1) \to \mathcal{M}$ is surjective.

Given a map of special rings $A \to B$, (that is a map of $\mathbb{Z}_p$-algebras compatible with φ) and $\mathcal{M}$ in $\mathcal{C}_A$, we give $\mathcal{M} \otimes_A B$ the structure of an object in $\mathcal{C}_B$, by giving it the induced Frobenius, and setting $(\mathcal{M} \otimes_A B)_1$ equal to the image of $\mathcal{M}_1 \otimes_A B$ in $\mathcal{M} \otimes_A B$.

Lemma A.4. *Let $h : A \to B$ be a surjection of special rings with kernel J. Suppose that for $i \geq 1$, $\varphi^i(J) \subset p^{i+j_i} J$, where $\{j_i\}_{i\geq 1}$ is a sequence of integers such that $\varinjlim_i j_i = \infty$.*

Let $\mathcal{M}$ and $\mathcal{M}'$ be in $\mathcal{C}_A$, and $\theta_B : \mathcal{M} \otimes_A B \xrightarrow{\sim} \mathcal{M}' \otimes_A B$ an isomorphism in $\mathcal{C}_B$ Then there exists a unique isomorphism of A-modules $\theta_A : \mathcal{M} \xrightarrow{\sim} \mathcal{M}'$ lifting θ_B, and compatible with φ.

Proof. Let $\theta_0 : \mathcal{M} \to \mathcal{M}'$ be any map of A-modules lifting θ_B. Since $\varphi(J) \subset pA$ the truth of the proposition is unaffected if we replace $\mathcal{M}_1$ and $\mathcal{M}'_1$ by $\mathcal{M}_1 + J\mathcal{M}$ and $\mathcal{M}'_1 + J\mathcal{M}'$ respectively. In particular, we may assume that $\theta_0(\mathcal{M}_1) \subset \mathcal{M}'_1$.

We claim that the composite

$$\varphi^*(\mathcal{M}_1) \stackrel{\varphi^*(\theta_0|\mathcal{M}_1)}{\longrightarrow} \varphi^*(\mathcal{M}'_1) \stackrel{1\otimes\varphi/p}{\longrightarrow} \mathcal{M}' \tag{A.4.1}$$

factors through $\mathcal{M}$ via the map $1 \otimes \varphi/p$. To see this note that the map $\varphi^*\mathcal{M} \to \mathcal{M}$ is injective because, after inverting p, it becomes a surjection of finite free $A[1/p]$-modules of the same rank, and hence an isomorphism. Hence if $x \in \varphi^*(\mathcal{M}_1)$ is in the kernel of $1 \otimes \varphi/p$, then the image of x in $\varphi^*(\mathcal{M})$ is 0, and hence so is $\varphi^*(\theta_0)(x) \in \varphi^*(\mathcal{M}')$. It follows that (A.4.1) maps px and hence also x to 0.

Let $\theta_1 : \mathcal{M} \to \mathcal{M}'$ be the map induced by (A.4.1). Then for $x \in \mathcal{M}_1$ we have $\theta_1 \circ \varphi/p(x) = \varphi/p \circ \theta_0(x)$. Repeating the construction, we obtain a sequence of maps $\theta_0, \theta_1, \dots$ lifting θ_B, and such that $\theta_i \circ \varphi/p(x) = \varphi/p \circ \theta_{i-1}(x)$ for $x \in \mathcal{M}_1$. In particular, we have $(\theta_{i+1} - \theta_i) \circ \varphi/p = \varphi/p \circ (\theta_i - \theta_{i-1})$ on $\mathcal{M}_1$, and since $\varphi/p(\mathcal{M}_1)$ generates $\mathcal{M}$ as an A-module, we see that $(\theta_{i+1} - \theta_i)(\mathcal{M}) \subset (\varphi/p)^i(J)\mathcal{M}' \subset p^{j_i}\mathcal{M}'$. Hence the θ_i converge to a well defined map $\theta_A : \mathcal{M} \to \mathcal{M}'$, which commutes with φ and lifts θ_B.

If θ_A and θ'_A are two such maps, then as above, we obtain that $(\theta_A - \theta'_A)(\mathcal{M}) \subset (\varphi/p)^i(J)\mathcal{M}'$ for each i so that $\theta_A = \theta'_A$. □

A.5

We will apply Lemma A.4 in the following situation: J is equipped with divided powers, and there exist a finite set of elements $x_1, \dots x_n \in J$ such that J is topologically (for the p-adic topology) generated by the x_i and their divided powers, and $\varphi(x_i) = x_i^p$. The integers j_i may then be taken to be $v_p((p^i - 1)!) - i$.

Denote by S the p-adic completion of the divided power envelope of $W[u]$ with respect to the ideal $E(u)$. The ring S is equipped with an endomorphism φ given by the Frobenius on W, and $\varphi(u) = u^p$. We denote by $\mathrm{Fil}^1 S \subset S$ the closure of the ideal generated by $E(u)$ and its divided powers. Note that $\varphi(\mathrm{Fil}^1 S) \subset pS$. We set $\varphi_1 = \varphi/p|_{\mathrm{Fil}^1 S}$.

We will denote by $\mathrm{BT}^{\varphi}_{/S}$ the category of finite free S-modules $\mathcal{M}$ equipped with an S-submodule $\mathrm{Fil}^1 \mathcal{M}$ and a φ-semilinear map $\varphi_1 : \mathrm{Fil}^1 \mathcal{M} \to \mathcal{M}$ such that

(1) $\mathrm{Fil}^1 S \cdot \mathcal{M} \subset \mathrm{Fil}^1 \mathcal{M}$, and the quotient $\mathcal{M}/\mathrm{Fil}^1 \mathcal{M}$ is a free $\mathcal{O}_K$-module.
(2) The map $\varphi^*(\mathrm{Fil}^1 \mathcal{M}) \stackrel{1\otimes\varphi_1}{\to} \mathcal{M}$ is surjective.

Any $\mathcal{M}$ in $\mathrm{BT}^{\varphi}_{/S}$ is equipped with a Frobenius semilinear map $\varphi : \mathcal{M} \to \mathcal{M}$ defined by $\varphi(x) = \varphi_1(E(u))^{-1}\varphi_1(E(u)x)$.

Proposition A.6. *There is an exact contravariant functor $G \mapsto \mathbb{D}(G)(S)$ from the category of p-divisible groups over $\mathcal{O}_K$ to $\mathrm{BT}^{\varphi}_{/S}$. If $p > 2$ this functor is an anti-equivalence, and if $p = 2$ it induces an anti-equivalence of the corresponding isogeny categories.*

Proof. Given a p-divisible group G over $\mathcal{O}_K$, the S-module $\mathcal{M}(G) := \mathbb{D}(G)(S)$ has a natural structure of an object of $\mathrm{BT}^{\varphi}_{/S}$ by (A.2). This gives a functor from p-divisible groups over $\mathcal{O}_K$ to $\mathrm{BT}^{\varphi}_{/S}$. We will construct a quasi-inverse (up to isogeny if $p = 2$).

Let $\mathcal{M}$ be in $\mathrm{BT}^{\varphi}_{/S}$. We begin by constructing from $\mathcal{M}$ a p-divisible group G_i over $\mathcal{O}_K/\pi^i$ for $i = 1, 2, \dots, e$. More precisely, for any such i let $R_i = W[u]/u^i$. It is equipped with a Frobenius endomorphism φ given by the usual Frobenius on W and $u \mapsto u^p$. We regard $\mathcal{O}_K/\pi^i$ as an R_i-algebra via $u \mapsto \pi$. This is a surjection with kernel pR_i, so R_i is a divided power thickening of $\mathcal{O}_K/\pi^i$ and given any p-divisible group G_i over $\mathcal{O}_K/\pi^i$ we may form $\mathbb{D}(G_i)(R_i)$. As in (A.2), we denote by $\mathrm{Fil}^1\,\mathbb{D}(G_i)(R_i)$ the preimage of $(\mathrm{Lie}\,G_i)^* \subset \mathbb{D}(G_i)(\mathcal{O}_K/\pi^i)$ in $\mathbb{D}(G_i)(R_i)$. On the other hand, we have a φ-compatible map $S \to R_i$, sending u to u, and $u^{ej}/j!$ to 0 for $j \geq 1$. Write I_i for the kernel of this map. We equip $\mathcal{M}_i = R_i \otimes_S \mathcal{M}$ with the induced Frobenius φ, and we set $\mathrm{Fil}^1\,\mathcal{M}_i \subset \mathcal{M}_i$ equal to the image of $\mathrm{Fil}^1\,\mathcal{M}$ in $\mathcal{M}_i$. Note that $1 \otimes \varphi_1 : \varphi^*(\mathrm{Fil}^1\,\mathcal{M}) \to \mathcal{M}$ induces a surjective map $\varphi^*(\mathrm{Fil}^1\,\mathcal{M}_i) \to \mathcal{M}_i$. We will construct a p-divisible group G_i together with a canonical isomorphism $\mathbb{D}(G_i)(R_i) \xrightarrow{\sim} \mathcal{M}_i$ compatible with φ and filtrations.

Denote by $F : \mathcal{M}_1 \to \mathcal{M}_1$ the map induced by $\varphi : \mathcal{M} \to \mathcal{M}$. A simple computation shows that both sides of the surjective map $\varphi^*(\mathrm{Fil}^1\,\mathcal{M}_1) \to \mathcal{M}_1$, are free W-modules of the same rank, hence this map is an isomorphism. Composing the inverse of this isomorphism with the composite

$$\varphi^*(\mathrm{Fil}^1\,\mathcal{M}_1) \to \varphi^*(\mathcal{M}_1) \xrightarrow{\sim} \mathcal{M}_1,$$

where the first map is induced by the inclusion $\mathrm{Fil}^1\,\mathcal{M} \subset \mathcal{M}$, while the second is given by $a \otimes m \mapsto \varphi^{-1}(a)m$, gives a φ^{-1} semilinear map $V : \mathcal{M}_1 \to \mathcal{M}_1$, such that $FV = VF = p$. Denote by G_1 to be the p-divisible group associated (contravariantly) to this Dieudonné module. The tautological isomorphism $\mathbb{D}(G_1)(W) \xrightarrow{\sim} \mathcal{M}_1$ is compatible with Frobenius, and it is compatible with filtrations because $\mathrm{Fil}^1\,\mathbb{D}(G_1)$ may be identified with $V\mathbb{D}(G_1)$, as explained at the end of the proof of Lemma A.2.

Now suppose that $i \in [2, e]$ is an integer and that we have constructed G_{i-1} such that $\mathbb{D}(G_{i-1})(R_{i-1}) \xrightarrow{\sim} \mathcal{M}_{i-1}$ is compatible with Frobenius and filtrations. Note that the kernel of $R_i \to \mathcal{O}_K/\pi^{i-1}$ is equal to (u^{i-1}, p) which admits divided powers, so we may evaluate $\mathbb{D}(G_{i-1})$ on R_i. By Lemma A.2 and what we have already seen $\mathbb{D}(G_{i-1})(R_i)$, and $\mathcal{M}_i$ both have the structure of objects of $\mathcal{C}_{R_i}$, and the above isomorphism is an isomorphism in $\mathcal{C}_{R_{i-1}}$. Hence by Lemma A.4 applied to the surjection $R_i \to R_{i-1}$, it lifts to a unique φ-compatible isomorphism $\mathbb{D}(G_{i-1})(R_i) \xrightarrow{\sim} \mathcal{M}_i$. By the main result of [Me] there is a unique p-divisible group G_i over $\mathcal{O}_K/\pi^i$ which lifts G_{i-1}, and such that $(\mathrm{Lie}\,G_i)^* \subset \mathbb{D}(G_{i-1})(\mathcal{O}_K/\pi^i)$ is equal to the image of $\mathrm{Fil}^1\,\mathcal{M}_i$ under the composite

$$\mathrm{Fil}^1\,\mathcal{M}_i \subset \mathcal{M}_i \xrightarrow{\sim} \mathbb{D}(G_{i-1})(R_i) \to \mathbb{D}(G_{i-1})(\mathcal{O}_K/\pi^i).$$

By construction we have $\mathbb{D}(G_i)(R_i) \xrightarrow{\sim} \mathcal{M}_i$ compatible with φ and filtrations, which completes the induction.

We now apply Lemma A.4 to the surjection $S \to R_e$, and the modules $\mathcal{M}$ and $\mathbb{D}(G_e)(S)$ in $\mathcal{C}_S$. Note that the kernel of $S \to \mathcal{O}_K/\pi^e = \mathcal{O}_K/p$ admits divided powers, so we may evaluate $\mathbb{D}(G_e)$ on S, and the result is in $\mathcal{C}_S$ by Lemma A.2. Since $\mathcal{M}_e \xrightarrow{\sim} \mathbb{D}(G_e)(R_e)$ in $\mathcal{C}_{R_e}$, we have a canonical φ-compatible isomorphism $\mathcal{M} \xrightarrow{\sim} \mathbb{D}(G_e)(S)$ by Lemma A.4.

Suppose that $p > 2$. Then the divided powers on the kernel of $\mathcal{O}_K \to \mathcal{O}_K/p$ are nilpotent, and we may take $G = G(\mathcal{M})$ to be the unique lift of G_e to $\mathcal{O}_K$ such that

$(\mathrm{Lie}\, G)^* \subset \mathbb{D}(G_e)(\mathcal{O}_K)$ is equal to the image of $\mathrm{Fil}^1 \mathcal{M}$ under the composite of the above isomorphism and the projection $\mathbb{D}(G_e)(S) \to \mathbb{D}(G_e)(\mathcal{O}_K)$. Strictly speaking what Grothendieck–Messing theory produces is a sequence of p-divisible groups over $\mathcal{O}_K/p^i$ for $i = 1, 2, \dots$ which are compatible under the maps $\mathcal{O}_K/p^i \to \mathcal{O}_K/p^{i-1}$. However, this data corresponds to a unique p-divisible group over $\mathcal{O}_K$ [deJ, 2.4.4].

From the construction we clearly have $\mathcal{M} \xrightarrow{\sim} \mathcal{M}(G(\mathcal{M}))$. On the other hand using the uniqueness at every stage of the construction, one sees by induction on i that for $i = 1, 2, \dots, e$ and any p-divisible group G over $\mathcal{O}_K$, $G_i(\mathcal{M}(G))$ is isomorphic to $G \otimes_{\mathcal{O}_K} \mathcal{O}_K/\pi^i$, and then that $G \xrightarrow{\sim} G(\mathcal{M}(G))$.

Now suppose that $p = 2$. We may regard the kernel of $\mathcal{O}_K/p^2 \to \mathcal{O}_K/p$ as being equipped with divided powers by taking the divided powers $p^{[i]}$ to be 0 for $i \geq 2$. We denote by G_{2e} the unique lift of G_e to $\mathcal{O}_K/p^2$, such that the image of the composite

$$\mathrm{Fil}^1 \mathcal{M} \to \mathcal{M} \xrightarrow{\sim} \mathbb{D}(G_e)(S) \to \mathbb{D}(G_e)(\mathcal{O}_K/p^2)$$

is equal to $(\mathrm{Lie}\, G_{2e})^*$. Finally, as for the case $p = 2$, we set G equal to the unique lift of G_{2e} to $\mathcal{O}_K$, such that the image of $\mathrm{Fil}^1 \mathcal{M}$ in $\mathbb{D}(G_{2e})(\mathcal{O}_K) = \mathbb{D}(G_e)(\mathcal{O}_K)$ is equal to $(\mathrm{Lie}\, G)^*$.

As for $p > 2$, we still have a natural isomorphism $\mathcal{M}(G(\mathcal{M})) \xrightarrow{\sim} \mathcal{M}$. Given a p-divisible group G over $\mathcal{O}_K$, we also obtain, as before, an isomorphism $G_e(\mathcal{M}(G)) \xrightarrow{\sim} G \otimes_{\mathcal{O}_K} \mathcal{O}_K/p$. In general, G_{2e} need not be isomorphic to $G'_{2e} := G \otimes_{\mathcal{O}_K} \mathcal{O}_K/p^2$, because the divided power structure on $(p) \subset \mathcal{O}_K/p^2$, its not compatible with the divided powers on $(p) \subset S$. However, since both these p-divisible groups lift G_e, there exist maps $G_{2e} \leftrightarrows G'_{2e}$, lifting multiplication by p^2 on G_e [Kat, 1.1.3]. Since G and $G(\mathcal{M}(G))$ are obtained from G'_{2e} and G_{2e} as the unique lifts corresponding to the image of $\mathrm{Fil}^1 \mathcal{M}$ in

$$\mathbb{D}(G_{2e})(\mathcal{O}_K) \xrightarrow{\sim} \mathbb{D}(G_e)(\mathcal{O}_K) \xrightarrow{\sim} \mathbb{D}(G'_{2e})(\mathcal{O}_K),$$

these maps lift to maps $G(\mathcal{M}(G)) \leftrightarrows G$ whose composite in either order is multiplication by p^4. □

Acknowledgments. I would like to thank A. Beilinson, L. Berger, P. Berthelot, C. Breuil, J.-M. Fontaine, K. Kedlaya, and V. Lafforgue for extremely useful conversations and correspondence. I would especially like to thank Beilinson for allowing me to include the proof of Theorem 2.3.5.

The author is partially supported by NSF grant DMS-0400666 and a Sloan Research Fellowship.

References

[An] Y. André, Différentielles non commutatives et théorie de Galois différentielle ou aux différences, *Ann. Sci. École Norm. Sup.*, **34** (2001), 685–739.

[Ba] F. Baldassarri, Comparaison entre la cohomologie algébrique et la cohomologie p-adique rigide à coefficients dans un module différentiel II: Cas des singularités régulières à plusieurs variables, *Math. Ann.*, **280** (1998), 417–439.

[BBM] P. Berthelot, L. Breen, and W. Messing, *Théorie de Dieudonné cristalline* II, Lecture Notes in Mathematics, Vol. 930, Springer-Verlag, Berlin, New York, Heidelberg, 1982.

[Be 1] L. Berger, Équations différentielles p-adiques et (φ, N)-modules filtrés, preprint, 2004.

[Be 2] L. Berger, Limites des représentations cristallines, *Compositio Math.*, **140** (2004), 1473–1498.

[Be 3] L. Berger, Représentations p-adiques et équations différentielles, *Invent. Math.*, **148**-2 (2002), 219–284.

[Br 1] C. Breuil, Une application du corps des normes, *Compositio Math.*, **117** (1999), 189–203.

[Br 2] C. Breuil, Groupes p-divisibles, groupes finis et modules filtrés, *Ann. Math.*, **152** (2000), 489–549.

[Br 3] C. Breuil, Integral p-adic Hodge theory, in S. Usui, ed., *Algebraic Geometry* 2000, *Azumino*, Advanced Studies in Pure Mathematics, Vol. 36, Mathematical Society of Japan, Tokyo, 2002, 51–80.

[Br 4] C. Breuil, Schémas en groupes et corps des normes, unpublished manuscript, 1998.

[CF] P. Colmez and J. M. Fontaine, Construction des représentations p-adiques semi-stables, *Invent. Math.*, **140**-1 (2000), 1–43.

[Co] P. Colmez, Espaces de Banach de dimension finie, *J. Inst. Math. Jussieu*, **1** (2002), 331–439.

[deJ] A. de Jong, Crystalline Dieudonné module theory via formal and rigid geometry, *Inst. Études Sci. Publ. Math.*, **82** (1995), 5–96.

[De] P. Deligne, *Equations Différentielles à points singuliers réguliers*, Lecture Notes in Mathematics, Vol. 163, Springer-Verlag, Berlin, Heidelberg, New York, 1970.

[Fa] G. Faltings, Integral crystalline cohomology over very ramified valuation rings, *J. Amer. Math. Soc.*, **12** (1999), 117–144.

[Fo 1] J.-M. Fontaine, Représentations p-adiques des corps locaux, in P. Carier, ed., *Grothendieck Festschrift*, Vol. 2, Progress in Mathematics, Vol. 87, Birkhäuser, Basel, 1991, 249–309.

[Fo 2] J. M. Fontaine, Représentations p-adiques semi-stables, in *Périodes p-adiques*, Astérisque, Vol. 223, Société Mathématique de France, Paris, 1994, 113–184.

[Fo 3] J.-M. Fontaine, Modules galoisiens, modules filtrés et anneaux de Barsotti-Tate, in *Journées de Géométrie Algébrique de Rennes* (*Orsay* 1978), Astérisque, Vol. 65, Société Mathématique de France, Paris, 1979, 3–80.

[GL] A. Genestier and V. Lafforgue, Théories de Fontaine et Fontaine-Laffaille en égale caractéristique, preprint, 2004.

[La] G. Laffaille, Groupes p-divisibles et modules filtrés: Le cas peu ramifié, *Bull. Soc. Math. France*, **108** (1980), 187–206.

[Laz] M. Lazard, Les zéros des fonctions analytiques d'une variable sur un corps valué complet, *Inst. Hautes Études Sci. Publ. Math.*, **14** (1962), 47–75.

[Ka] D. Kazhdan, An introduction to Drinfeld's "Shtuka," in A. Borel and W. Casselman, eds., *Automorphic Forms, Representations and L-Functions* II, Proceedings of Symposia in Pure Mathematics, Vol. 33, American Mathematical Society, Providence, RI, 1979, 347–356.

[Kat] N. Katz, Serre-Tate local moduli, in J. Giraud, ed., *Surfaces Algébriques: Séminaire de Géométrie Algébriques d'Orsay* 1976–1978, Lecture Notes in Mathematics, Vol. 868, Springer-Verlag, Berlin, 1981, 128–202.

[Kat 2] N. Katz, Nilpotent connections and the monodromy theorem: Applications of a result of Turrittin, *Inst. Hautes Études Sci. Publ. Math.*, **39** (1970), 175–232.

[Ke 1] K. S. Kedlaya, A p-adic local monodromy theorem, *Ann. Math.*, **160**-1 (2004), 93–184.

[Ke 2] K. S. Kedlaya, Slope filtrations revisited, math.NT/0504204, 2005; *Documenta Math.*, to appear.

[Ki] M. Kisin, Moduli of finite flat group schemes and modularity, preprint, 2004.

[Me] W. Messing, *The Crystals Associated to Barsotti-Tate Groups: With Applications to Abelian Schemes*, Lecture Notes in Mathematics, Vol. 264, Springer-Verlag, Berlin, New York, Heidelberg, 1972.

[MM] B. Mazur and W. Messing, *Universal Extensions and One Dimensional Crystalline Cohomology*, Lecture Notes in Mathematics, Vol. 370, Springer, Berlin, New York, Heidelberg, 1974.

Integrable linear equations and the Riemann–Schottky problem

I. Krichever

Department of Mathematics
Columbia University
2990 Broadway
509 Mathematics Building
Mail Code: 4406
New York, NY 10027
USA
krichev@math.columbia.edu

and

Landau Institute for Theoretical Physics
Kosygina Street 2
119334 Moscow
Russia
kricheve@itp.ac.ru

Summary. We prove that an indecomposable principally polarized abelian variety X is the Jacobain of a curve if and only if there exist vectors $U \neq 0$, V such that the roots $x_i(y)$ of the theta-functional equation $\theta(Ux + Vy + Z) = 0$ satisfy the equations of motion of the *formal infinite-dimensional Calogero–Moser system.*

Subject Classifications: Primary 14H70. Secondary 14H40, 14K05, 37K20, 14H42.

1 Introduction

The Riemann–Schottky problem on the characterization of the Jacobians of curves among abelian varieties is more than 120 years old. Quite a few geometrical characterizations of Jacobians have been found. None of them provides an explicit system of equations for the image of the Jacobian locus in the projective space under the level-two theta imbedding.

The first effective solution of the Riemann–Schottky problem was obtained by T. Shiota [1], who proved the famous Novikov conjecture:

An indecomposable principally polarized abelian variety (X, θ) is the Jacobian of a curve of a genus g if and only if there exist g-dimensional vectors $U \neq 0$, V, W such that the function

$$u(x,y,t) = -2\partial_x^2 \ln \theta(Ux + Vy + Wt + Z) \tag{1.1}$$

is a solution of the Kadomtsev–Petviashvili (*KP*) *equation*

$$3u_{yy} = (4u_t + 6uu_x - u_{xxx})_x. \tag{1.2}$$

Here $\theta(Z) = \theta(Z|B)$ is the Riemann theta-function,

$$\theta(z) = \sum_{m\in\mathbb{Z}^g} e^{2\pi i(z,m)+\pi i(Bm,m)}, \qquad (z,m) = m_1 z_1 + \cdots + m_g z_g, \tag{1.3}$$

where B is the corresponding symmetric matrix with positive definite imaginary part.

It is easy to show [2] that the KP equation with u of the form (1.1) is in fact equivalent to the following system of algebraic equations for the fourth-order derivatives of the level-two theta constants:

$$\partial_U^4 \Theta[\varepsilon,0] - \partial_U \partial_W \Theta[\varepsilon,0] + \partial_V^2 \Theta[\varepsilon,0] + c\Theta[\varepsilon,0] = 0, \qquad c = \text{const}. \tag{1.4}$$

Here $\Theta[\varepsilon,0] = \Theta[\varepsilon,0](0)$, where $\Theta[\varepsilon,0](z) = \theta[\varepsilon,0](2z|2B)$ are level-two theta-functions with half-integer characteristics $\varepsilon \in \frac{1}{2}\mathbb{Z}_2^g$.

The KP equation admits the so-called zero-curvature representation [3, 4], which is the compatibility condition for the following over-determined system of linear equations:

$$(\partial_y - \partial_x^2 + u)\psi = 0, \tag{1.5}$$

$$\left(\partial_t - \partial_x^3 + \frac{3}{2}\partial_x + w\right)\psi = 0. \tag{1.6}$$

The main goal of the present paper is to show that the KP equation contains excessive information and that the Jacobians can be characterized in terms of *only the first* of its auxiliary linear equations.

Theorem 1.1. *An indecomposable principally polarized abelian variety* (X,θ) *is the Jacobian of a curve of genus* g *if and only if there exist* g*-dimensional vectors* $U \neq 0$, V, A *such that equation* (1.5) *is satisfied for*

$$u = -2\partial_x^2 \ln \theta(Ux + Vy + Z) \tag{1.7}$$

and

$$\psi = \frac{\theta(A + Ux + Vy + Z)}{\theta(Ux + Vy + Z)} e^{px+Ey}, \tag{1.8}$$

where p, E *are constants.*

The "if" part of this statement follows from the exact theta-functional expression for the Baker–Akhiezer function [5, 6].

The addition formula for the Riemann theta-function directly implies that equation (1.5) with u and ψ of the form (1.7) and (1.8) is equivalent to the system of equations

$$(\partial_V - \partial_U^2 - 2p\partial_U + (E - p^2))\Theta[\varepsilon, 0](A/2) = 0, \quad \varepsilon \in \frac{1}{2}\mathbb{Z}_2^g. \tag{1.9}$$

Recently Theorem 1.1 was proved by E. Arbarello and G. Marini and the author [7] under the additional assumption that the closure $\langle A \rangle$ of the subgroup of X generated by A is irreducible. The geometric interpretation of Theorem 1.1 is equivalent to the characterization of Jacobians via flexes of Kummer varieties (see details in [7]), which is a particular case of the so-called *trisecant conjecture*, first formulated in [8].

Theorem 1.1 is not the strongest form of our main result. What we really prove is that the Jacobian locus in the space of principally polarized abelian varieties is characterized by a system of equations which formally can be seen as the equations of motion of the *infinite-dimensional* Calogero–Moser system.

Let $\tau(x, y)$ be an entire function of the complex variable x smoothly depending on a parameter y. Consider the equation

$$\operatorname{res}_x(\partial_y^2 \ln \tau + 2(\partial_x^2 \ln \tau)^2) = 0, \tag{1.10}$$

which means that the meromorphic function given by the left-hand side of (1.10) has no *residues* in the x variable. If $x_i(y)$ is a simple zero of τ, i.e., $\tau(x_i(y), y) = 0$, $\partial_x \tau(x_i(y), y) \neq 0$, then (1.10) implies

$$\ddot{x}_i = 2w_i, \tag{1.11}$$

where "dots" stands for the y-derivatives and w_i is the third coefficient of the Laurent expansion of $u(x, y) = -2\partial_x^2 \tau(x, y)$ at x_i, i.e.,

$$u(x, y) = \frac{2}{(x - x_i(y))^2} + v_i(y) + w_i(y)(x - x_i(y)) + \cdots. \tag{1.12}$$

Formally, if we represent τ as an infinite product,

$$\tau(x, y) = c(y) \prod_i (x - x_i(y)), \tag{1.13}$$

then equation (1.10) can be written as the infinite system of equations

$$\ddot{x}_i = -4 \sum_{j \neq i} \frac{1}{(x_i - x_j)^3}. \tag{1.14}$$

Equations (1.14) are purely formal because, even if τ has simple zeros at $y = 0$, in the general case there is no nontrivial interval in y where the zeros stay simple. For the moment, the only reason for representing (1.11) in the form (1.14) is to show that in the case when τ is a rational, trigonometric or elliptic polynomial the system (1.11) coincides with the equations of motion for the rational, trigonometrical or elliptic Calogero–Moser systems, respectively.

Equations (1.11) for the zeros of the function $\tau = \theta(Ux + Vy + Z)$ were derived in [7] as a direct corollary of the assumptions of Theorem 1.1. Simple expansion of θ at the points of its divisor $z \in \Theta : \theta(z) = 0$ gives the equation

$$[(\partial_2\theta)^2 - (\partial_1^2\theta)^2]\partial_1^2\theta + 2[\partial_1^2\theta\partial_1^3\theta - \partial_2\theta\partial_1\partial_2\theta]\partial_1\theta + [\partial_2^2\theta - \partial_1^4\theta](\partial_1\theta)^2 \\ = 0 \ (\mathrm{mod}\,\theta) \tag{1.15}$$

which is valid on Θ. Here and below Θ is the divisor on X defined by the equation $\theta(Z) = 0$ and ∂_1 and ∂_2 are constant vector fields on $\mathbb{C}^g$ corresponding to the vectors U and V.

It would be very interesting to understand if any reasonable general theory of equation (1.10) exists. The following form of our main result shows that in any case such a theory has to be interesting and nontrivial.

Let Θ_1 be defined by the equations $\Theta_1 = \{Z : \theta(Z) = \partial_1\theta(Z) = 0\}$. The ∂_1-invariant subset Σ of Θ_1 will be called the *singular locus*.

Theorem 1.2. *An indecomposable principally polarized abelian variety* (X, θ) *is the Jacobian of a curve of genus* g *if and only if there exist* g*-dimensional vectors* $U \neq 0$, V, *such that for each* $Z \in \mathbb{C}^g \setminus \Sigma$ *equation* (1.10) *for the function* $\tau(x, y) = \theta(Ux + Vy + Z)$ *is satisfied, i.e., equation* (1.15) *is valid on* Θ.

The main idea of Shiota's proof of the Novikov conjecture is to show that if u is as in (1.1) and satisfies the KP equation, then it can be extended to a τ-function of the KP *hierarchy*, as a *global* holomorphic function of the infinite number of variables $t = \{t_i\}$, $t_1 = x$, $t_2 = y$, $t_3 = t$. Local existence of τ directly follows from the KP equation. The global existence of the τ-function is crucial. The rest is a corollary of the KP theory and the theory of commuting ordinary differential operators developed by Burchnall–Chaundy [9, 10] and the author [5, 6].

The core of the problem is that there is a homological obstruction for the global existence of τ. It is controlled by the cohomology group $H^1(\mathbb{C}^g \setminus \Sigma, \mathcal{V})$, where $\mathcal{V}$ is the sheaf of ∂_1-*invariant* meromorphic functions on $\mathbb{C}^g \setminus \Sigma$ with poles along Θ (see details in [11]). The hardest part of Shiota's work (clarified in [11]) is the proof that the locus Σ is empty. That ensures the vanishing of $H^1(\mathbb{C}^g, \mathcal{V})$. Analogous obstructions have occurred in all the other attempts to apply the theory of soliton equations to various characterization problems in the theory of abelian varieties. None of them has been completely successful. Only partial results were obtained. (Note that Theorem 1.1 in one of its equivalent forms was proved earlier in [12] under the additional assumption that Θ_1 does not contain a ∂_1-invariant line.)

Strictly speaking, the KP equation and the KP hierarchy are not used in the present paper. But our main construction of *the formal wave solutions* of (1.5) is reminiscent of the construction of the τ-function. All its difficulties can be traced back to those in Shiota's work. The wave solution of (1.5) is a solution of the form

$$\psi(x, y, k) = e^{kx+(k^2+b)y}\left(1 + \sum_{s=1}^{\infty} \xi_s(x, y)k^{-s}\right). \tag{1.16}$$

At the beginning of the next section, we show that the assumptions of Theorem 1.2 are necessary and sufficient conditions for the *local* existence of the wave solutions such that

$$\xi_s = \frac{\tau_s(Ux + Vy + Z, y)}{\theta(Ux + Vy + Z)}, \qquad Z \notin \Sigma, \tag{1.17}$$

where $\tau_s(Z, y)$, as a function of Z, is holomorphic in some open domain in $\mathbb{C}^g$. The functions ξ_s are defined recursively by the equation $2\partial_1\xi_{s+1} = \partial_y\xi_s - \partial_1^2\xi_s + u\xi_s$. Therefore, the global existence of ξ_s is controlled by the same cohomology group $H^1(\mathbb{C} \setminus \Sigma, \mathcal{V})$ as above. At the local level the main problem is to find a translational invariant normalization of ξ_s which defines wave solutions uniquely up to a ∂_1-invariant factor.

In the case of periodic potentials $u(x + T, y) = u(x)$ the normalization problem for the wave functions was solved by D. Phong and the author in [13]. It was shown that the condition that ξ_s *is periodic* completely determines the y-dependence of the integration constants and the corresponding wave solutions are related by an x-independent factor. In general, the potential $u = -2\partial_x^2\theta(Ux+Vy+Z)$ is only quasi-periodic in x. In that case the solution of the normalization problem is technically more involved but mainly goes along the same lines as in the periodic case. The corresponding wave solutions are called λ-periodic.

In the last section, we showed that for each $Z \notin \Sigma$ a local λ-periodic wave solution is the common eigenfunction of a commutative ring $\mathcal{A}^Z$ of ordinary differential operators. The coefficients of these operators are independent of ambiguities in the construction of ψ. For generic Z the ring A^Z is maximal and the corresponding spectral curve Γ is Z-independent. The correspondence $j : Z \longmapsto \mathcal{A}^Z$ allows us to take the next crucial step and prove the global existence of the wave function. Namely, on $X \setminus \Sigma$ the wave function can be globally defined as the preimage $j^*\psi_{BA}$ under j of the Baker–Akhiezer function on Γ and then can be extended to X by the usual Hartog-type arguments. The global existence of the wave function implies that X contains an orbit of the KP hierarchy, as an abelian subvariety. The orbit is isomorphic to the generalized Jacobian $J(\Gamma) = \mathrm{Pic}^0(\Gamma)$ of the spectral curve [1]. Therefore, the generalized Jacobian is compact. The compactness of $\mathrm{Pic}^0(\Gamma)$ implies that the spectral curve is smooth and the correspondence j extends by linearity and defines an isomorphism $j : X \to J(\Gamma)$.

2 λ-periodic wave solutions

As was mentioned above, the formal Calogero–Moser equations (1.11) were derived in [7] as a necessary condition for the existence of a meromorphic solution to equation (1.5).

Let $\tau(x, y)$ be a holomorphic function of the variable x in some open domain $D \in \mathbb{C}$ smoothly depending on a parameter y. Suppose that for each y the zeros of τ are simple,

$$\tau(x_i(y), y) = 0, \qquad \tau_x(x_i(y), y) \neq 0. \tag{2.1}$$

Lemma 2.1 ([7]). *If equation* (1.5) *with the potential* $u = -2\partial_x^2 \ln \tau(x, y)$ *has a meromorphic in* D *solution* $\psi_0(x, y)$, *then equations* (1.11) *hold.*

Proof. Consider the Laurent expansions of ψ_0 and u in the neighborhood of one of the zeros x_i of τ:

$$u = \frac{2}{(x - x_i)^2} + v_i + w_i(x - x_i) + \cdots, \tag{2.2}$$

$$\psi_0 = \frac{\alpha_i}{x - x_i} + \beta_i + \gamma_i(x - x_i) + \delta_i(x - x_i)^2 + \cdots, \tag{2.3}$$

(All coefficients in these expansions are smooth functions of the variable y.) Substitution of (2.2), (2.3) in (1.5) gives a system of equations. The first three of them are

$$\alpha_i \dot{x}_i + 2\beta_i = 0, \tag{2.4}$$

$$\dot{\alpha}_i + \alpha_i v_i + 2\gamma_i = 0, \tag{2.5}$$

$$\dot{\beta}_i + v_i\beta_i - \gamma_i \dot{x}_i + \alpha_i w_i = 0. \tag{2.6}$$

Taking the y-derivative of the first equation and the using other two, we get (1.11).

Let us show that equations (1.11) are sufficient for the existence of meromorphic wave solutions. □

Lemma 2.2. *Suppose that equations* (1.11) *for the zeros of* $\tau(x, y)$ *hold. Then there exist meromorphic wave solutions of equation* (1.5) *that have simple poles at* x_i *and are holomorphic everywhere else.*

Proof. Substitution of (1.16) into (1.5) gives a recurrent system of equations

$$2\xi'_{s+1} = \partial_y \xi_s + u\xi_s - \xi''_s. \tag{2.7}$$

We are going to prove by induction that this system has meromorphic solutions with simple poles at all the zeros x_i of τ.

Let us expand ξ_s at x_i:

$$\xi_s = \frac{r_s}{x - x_i} + r_{s0} + r_{s1}(x - x_i), \tag{2.8}$$

where for brevity we omit the index i in the notation for the coefficients of this expansion. Suppose that ξ_s are defined and equation (2.7) has a meromorphic solution. Then the right-hand side of (2.7) has zero residue at $x = x_i$, i.e.,

$$\text{res}_{x_i}(\partial_y \xi_s + u\xi_s - \xi''_s) = \dot{r}_s + v_i r_s + 2r_{s1} = 0. \tag{2.9}$$

We need to show that the residue of the next equation also vanishes. From (2.7) it follows that the coefficients of the Laurent expansion for ξ_{s+1} are equal to

$$r_{s+1} = -\dot{x}_i r_s - 2r_{s0}, \tag{2.10}$$

$$2r_{s+1,1} = \dot{r}_{s0} - r_{s1} + w_i r_s + v_i r_{s0}. \tag{2.11}$$

These equations imply

$$\dot{r}_{s+1} + v_i r_{s+1} + 2r_{s+1,1} = -r_s(\ddot{x}_i - 2w_i) - \dot{x}_i(\dot{r}_s - v_i r_s s + 2r_{s1}) = 0, \tag{2.12}$$

and the lemma is proved. □

Our next goal is to fix a *translation-invariant* normalization of ξ_s which defines wave functions uniquely up to an x-independent factor. It is instructive to consider first the case of the periodic potentials $u(x+1, y) = u(x, y)$ (see details in [13]).

Equations (2.7) are solved recursively by the formulae

$$\xi_{s+1}(x, y) = c_{s+1}(y) + \xi^0_{s+1}(x, y), \tag{2.13}$$

$$\xi^0_{s+1}(x, y) = \frac{1}{2}\int_{x_0}^{x} (\partial_y \xi_s - \xi_s'' + u\xi_s)dx, \tag{2.14}$$

where $c_s(y)$ are *arbitrary* functions of the variable y. Let us show that the periodicity condition $\xi_s(x+1, y) = \xi_s(x, y)$ defines the functions $c_s(y)$ uniquely up to an additive constant. Assume that ξ_{s-1} is known and satisfies the condition that the corresponding function ξ^0_s is periodic. The choice of the function $c_s(y)$ does not affect the periodicity property of ξ_s, but it does affect the periodicity in x of the function $\xi^0_{s+1}(x, y)$. In order to make $\xi^0_{s+1}(x, y)$ periodic, the function $c_s(y)$ should satisfy the linear differential equation

$$\partial_y c_s(y) + B(y)c_s(y) + \int_{x_0}^{x_0+1} (\partial_y \xi^0_s(x, y) + u(x, y)\xi^0_s(x, y))dx, \tag{2.15}$$

where $B(y) = \int_{x_0}^{x_0+1} u dx$. This defines c_s uniquely up to a constant.

In the general case, when u is quasi-periodic, the normalization of the wave functions is defined along the same lines.

Let $Y_U = \langle Ux\rangle$ be the closure of the group Ux in X. Shifting Y_U if needed, we may assume, without loss of generality, that Y_U is not in the singular locus, $Y_U \notin \Sigma$. Then for a sufficiently small y, we have $Y_U + Vy \notin \Sigma$ as well. Consider the restriction of the theta-function onto the affine subspace $\mathbb{C}^d + Vy$, where $\mathbb{C}^d = \pi^{-1}(Y_U)$, and $\pi : \mathbb{C}^g \to X = \mathbb{C}^g/\Lambda$ is the universal cover of X:

$$\tau(z, y) = \theta(z + Vy), \quad z \in \mathbb{C}^d. \tag{2.16}$$

The function $u(z, y) = -2\partial_1^2 \ln \tau$ is periodic with respect to the lattice $\Lambda_U = \Lambda \cap \mathbb{C}^d$ and, for fixed y, has a double pole along the divisor $\Theta^U(y) = (\Theta - Vy) \cap \mathbb{C}^d$.

Lemma 2.3. *Let equation* (1.10) *for* $\tau(Ux + z, y)$ *hold and let* λ *be a vector of the sublattice* $\Lambda_U = \Lambda \cap \mathbb{C}^d \subset \mathbb{C}^g$. *Then*

(i) *equation* (1.5) *with the potential* $u(Ux + z, y)$ *has a wave solution of the form* $\psi = e^{kx+k^2y}\phi(Ux + z, y, k)$ *such that the coefficients* $\xi_s(z, y)$ *of the formal series*

$$\phi(z, y, k) = e^{by}\left(1 + \sum_{s=1}^{\infty} \xi_s(z, y)k^{-s}\right) \tag{2.17}$$

are λ*-periodic meromorphic functions of the variable* $z \in \mathbb{C}^d$ *with a simple pole along the divisor* $\Theta^U(y)$,

$$\xi_s(z + \lambda, y) = \xi_s(z, y) = \frac{\tau_s(z, y)}{\tau(z, y)}; \tag{2.18}$$

(ii) $\phi(z, y, k)$ *is unique up to a factor* $\rho(z, k)$ *that is* ∂_1*-invariant and holomorphic in* z,

$$\phi_1(z, y, k) = \phi(z, y, k)\rho(z, k), \quad \partial_1\rho = 0. \tag{2.19}$$

Proof. The functions $\xi_s(z)$ are defined recursively by the equations

$$2\partial_1\xi_{s+1} = \partial_y\xi_s + (u + b)\xi_s - \partial_1^2\xi_s. \tag{2.20}$$

A particular solution of the first equation $2\partial_1\xi_1 = u + b$ is given by the formula

$$2\xi_1^0 = -2\partial_1 \ln \tau + (l, z)b, \tag{2.21}$$

where (l, z) is a linear form on $\mathbb{C}^d$ given by the scalar product of z with a vector $l \in \mathbb{C}^d$ such that $(l, U) = 1$, and $(l, \lambda) \neq 0$. The periodicity condition for ξ_1^0 defines the constant b,

$$(l, \lambda)b = (2\partial_1 \ln \tau(z + \lambda, y) - 2\partial_1 \ln \tau(z, y)), \tag{2.22}$$

which depends only on a choice of the lattice vector λ. A change of the potential by an additive constant does not affect the results of the previous lemma. Therefore, equations (1.11) are sufficient for the local solvability of (2.20) in any domain, where $\tau(z + Ux, y)$ has simple zeros, i.e., outside of the set $\Theta_1^U(y) = (\Theta_1 - Vy) \cap \mathbb{C}^d$. Recall that $\Theta_1 = \Theta \cap \partial_1\Theta$. This set does not contain a ∂_1-invariant line because any such line is dense in Y_U. Therefore, the sheaf $\mathcal{V}_0$ of ∂_1-invariant meromorphic functions on $\mathbb{C}^d \setminus \Theta_1^U(y)$ with poles along the divisor $\Theta^U(y)$ coincides with the sheaf of holomorphic ∂_1-invariant functions. That implies the vanishing of $H^1(C^d \setminus \Theta_1^U(y), \mathcal{V}_0)$ and the existence of global meromorphic solutions ξ_s^0 of (2.20) which have a simple pole along the divisor $\Theta^U(y)$ (see details in [1, 11]). If ξ_s^0 are fixed, then the general global meromorphic solutions are given by the formula $\xi_s = \xi_s^0 + c_s$, where the constant of integration $c_s(z, y)$ is a holomorphic ∂_1-invariant function of the variable z.

Let us assume, as in the example above, that a λ-periodic solution ξ_{s-1} is known and that it satisfies the condition that there exists a periodic solution ξ_s^0 of the next equation. Let ξ_{s+1}^* be a solution of (2.20) for fixed ξ_s^0. Then it is easy to see that the function

$$\xi_{s+1}^0(z, y) = \xi_{s+1}^*(z, y) + c_s(z, y)\xi_1^0(z, y) + \frac{(l, z)}{2}\partial_y c_s(z, y) \tag{2.23}$$

is a solution of (2.20) for $\xi_s = \xi_s^0 + c_s$. A choice of a λ-periodic ∂_1-invariant function $c_s(z, y)$ does not affect the periodicity property of ξ_s, but it does affect the periodicity of the function ξ_{s+1}^0. In order to make ξ_{s+1}^0 periodic, the function $c_s(z, y)$ should satisfy the linear differential equation

$$(l, \lambda)\partial_y c_s(z, y) = 2\xi_{s+1}^*(z + \lambda, y) - 2\xi_{s+1}^*(z, y). \tag{2.24}$$

This equation, together with an initial condition $c_s(z) = c_s(z, 0)$ uniquely defines $c_s(x, y)$. The induction step is then completed. We have shown that the ratio of two periodic formal series ϕ_1 and ϕ is y-independent. Therefore, equation (2.19), where $\rho(z, k)$ is defined by the evaluation of the two sides at $y = 0$, holds. The lemma is thus proved. □

Corollary 2.1. *Let* $\lambda_1, \dots, \lambda_d$ *be a set of linear independent vectors of the lattice* Λ_U *and let* z_0 *be a point of* $\mathbb{C}^d$. *Then, under the assumptions of the previous lemma, there is a unique wave solution of equation* (1.5) *such that the corresponding formal series* $\phi(z, y, k; z_0)$ *is quasi-periodic with respect to* Λ_U, *i.e., for* $\lambda \in \Lambda_U$

$$\phi(z + \lambda, y, k; z_0) = \phi(z, y, k; z_0)\mu_\lambda(k) \tag{2.25}$$

and satisfies the normalization conditions

$$\mu_{\lambda_i}(k) = 1, \qquad \phi(z_0, 0, k; z_0) = 1. \tag{2.26}$$

The proof is identical to that of [1, Lemma 12, part (b)]. Let us briefly present its main steps. As shown above, there exist wave solutions corresponding to ϕ which are λ_1-periodic. Moreover, from statement (ii) above it follows that for any $\lambda' \in \Lambda_U$,

$$\phi(z + \lambda, y, k) = \phi(z, y, k)\rho_\lambda(z, k), \tag{2.27}$$

where the coefficients of ρ_λ are ∂_1-invariant holomorphic functions. Then the same arguments as in [1] show that there exists a ∂_1-invariant series $f(z, k)$ with holomorphic in z coefficients and formal series $\mu_\lambda^0(k)$ with constant coefficients such that the equation

$$f(z + \lambda, k)\rho_\lambda(z, k) = f(z, k)\mu_\lambda(k) \tag{2.28}$$

holds. The ambiguity in the choice of f and μ corresponds to the multiplication by the exponent of a linear form in z vanishing on U, i.e.,

$$f'(z, k) = f(z, k)e^{(b(k), z)}, \qquad \mu'_\lambda(k) = \mu_\lambda(k)e^{(b(k), \lambda)}, \quad (b(k), U) = 0, \tag{2.29}$$

where $b(k) = \sum_s b_s k^{-s}$ is a formal series with vector-coefficients that are orthogonal to U. The vector U is in general position with respect to the lattice. Therefore, the ambiguity can be uniquely fixed by imposing $(d - 1)$ normalizing conditions $\mu_{\lambda_i}(k) = 1, i > 1$. (Recall that $\mu_{\lambda_1}(k) = 1$ by construction.)

The formal series $f\phi$ is quasi-periodic and its multiplicators satisfy (2.26). Then, by these properties it is defined uniquely up to a factor which is constant in z and y. Therefore, for the unique definition of ϕ_0, it is enough to fix its evaluation at z_0 and $y = 0$. The corollary is proved.

3 The spectral curve

In this section, we show that λ-periodic wave solutions of equation (1.5), with u as in (1.7), are common eigenfunctions of rings of commuting operators and identify X with the Jacobian of the spectral curve of these rings.

Note that a simple shift $z \to z + Z$, where $Z \notin \Sigma$, gives λ-periodic wave solutions with meromorphic coefficients along the affine subspaces $Z + \mathbb{C}^d$. These λ-periodic wave solutions are related to each other by a ∂_1-invariant factor. Therefore, choosing, in the neighborhood of any $Z \notin \Sigma$, a hyperplane orthogonal to the vector U and fixing initial data on this hyperplane at $y = 0$, we define the corresponding series $\phi(z + Z, y, k)$ as a *local* meromorphic function of Z and *global* meromorphic function of z.

Lemma 3.1. *Let the assumptions of Theorem* 1.2 *hold. Then there is a unique pseudodifferential operator*

$$\mathcal{L}(Z, \partial_x) = \partial_x + \sum_{s=1}^{\infty} w_s(Z)\partial_x^{-s} \tag{3.1}$$

such that

$$\mathcal{L}(Ux + Vy + Z, \partial_x)\psi = k\psi, \tag{3.2}$$

where $\psi = e^{kx+k^2y}\phi(Ux+Z, y, k)$ *is a* λ*-periodic solution of* (1.5). *The coefficients* $w_s(Z)$ *of* $\mathcal{L}$ *are meromorphic functions on the abelian variety* X *with poles along the divisor* Θ.

Proof. The construction of $\mathcal{L}$ is standard for the KP theory. First we define $\mathcal{L}$ as a pseudodifferential operator with coefficients $w_s(Z, y)$, which are functions of Z and y.

Let ψ be a λ-periodic wave solution. The substitution of (2.17) in (3.2) gives a system of equations that recursively define $w_s(Z, y)$ as differential polynomials in $\xi_s(Z, y)$. The coefficients of ψ are local meromorphic functions of Z, but the coefficients of $\mathcal{L}$ are well-defined *global meromorphic functions* on $\mathbb{C}^g \setminus \Sigma$, because different λ-periodic wave solutions are related to each other by a ∂_1-invariant factor, which does not affect $\mathcal{L}$. The singular locus is of codimension ≥ 2. Then Hartog's holomorphic extension theorem implies that $w_s(Z, y)$ can be extended to a global meromorphic function on $\mathbb{C}^g$.

The translational invariance of u implies the translational invariance of the λ-periodic wave solutions. Indeed, for any constant s the series $\phi(Vs + Z, y - s, k)$ and $\phi(Z, y, k)$ correspond to λ-periodic solutions of the same equation. Therefore, they coincide up to a ∂_1-invariant factor. This factor does not affect $\mathcal{L}$. Hence $w_s(Z, y) = w_s(Vy + Z)$.

The λ-periodic wave functions corresponding to Z and $Z + \lambda'$ for any $\lambda' \in \Lambda$ are also related to each other by a ∂_1-invariant factor:

$$\partial_1(\phi_1(Z + \lambda', y, k)\phi^{-1}(Z, y, k)) = 0. \tag{3.3}$$

Hence w_s are periodic with respect to Λ and therefore are meromorphic functions on the abelian variety X. The lemma is proved. □

Consider now the differential parts of the pseudodifferential operators $\mathcal{L}^m$. Let $\mathcal{L}_+^m$ be the differential operator such that $\mathcal{L}_-^m = \mathcal{L}^m - \mathcal{L}_+^m = F_m\partial^{-1} + O(\partial^{-2})$. The leading coefficient F_m of $\mathcal{L}_-^m$ is the residue of $\mathcal{L}^m$:

$$F_m = \operatorname{res}_\partial \mathcal{L}^m. \tag{3.4}$$

From the construction of $\mathcal{L}$ it follows that $[\partial_y - \partial_x^2 + u, \mathcal{L}^n] = 0$. Hence

$$[\partial_y - \partial_x^2 + u, \mathcal{L}_+^m] = -[\partial_y - \partial_x^2 + u, \mathcal{L}_-^m] = 2\partial_x F_m. \tag{3.5}$$

The functions F_m are differential polynomials in the coefficients w_s of $\mathcal{L}$. Hence $F_m(Z)$ are meromorphic functions on X. The next statement is crucial for the proof of the existence of commuting differential operators associated with u.

Lemma 3.2. *The abelian functions F_m have at most a second-order pole along the divisor Θ.*

Proof. We need a few more standard constructions from the KP theory. If ψ is as in Lemma 3.1, then there exists a unique pseudodifferential operator Φ such that

$$\psi = \Phi e^{kx+k^2y}, \qquad \Phi = 1 + \sum_{s=1}^{\infty} \varphi_s(Ux + Z, y)\partial_x^{-s}. \tag{3.6}$$

The coefficients of Φ are universal differential polynomials on ξ_s. Therefore, $\varphi_s(z + Z, y)$ is a global meromorphic function of $z \in \mathbb{C}^d$ and a local meromorphic function of $Z \notin \Sigma$. Note that $\mathcal{L} = \Phi(\partial_x)\Phi^{-1}$.

Consider the dual wave function defined by the left action of the operator Φ^{-1}: $\psi^+ = (e^{-kx-k^2y})\Phi^{-1}$. Recall that the left action of a pseudodifferential operator is the formal adjoint action under which the left action of ∂_x on a function f is $(f\partial_x) = -\partial_x f$. If ψ is a formal wave solution of (3.5), then ψ^+ is a solution of the adjoint equation

$$(-\partial_y - \partial_x^2 + u)\psi^+ = 0. \tag{3.7}$$

The same arguments, as before, prove that if equations (1.11) for poles of u hold then ξ_s^+ have simple poles at the poles of u. Therefore, if ψ as in Lemma 2.3, then the dual wave solution is of the form $\psi^+ = e^{-kx-k^2y}\phi^+(Ux + Z, y, k)$, where the coefficients $\xi_s^+(z + Z, y)$ of the formal series

$$\phi^+(z + Z, y, k) = e^{-by}\left(1 + \sum_{s=1}^{\infty} \xi_s^+(z + Z, y)k^{-s}\right) \tag{3.8}$$

are λ-periodic meromorphic functions of the variable $z \in \mathbb{C}^d$ with a simple pole along the divisor $\Theta^U(y)$.

The ambiguity in the definition of ψ does not affect the product

$$\psi^+\psi = (e^{-kx-k^2y}\Phi^{-1})(\Phi e^{kx+k^2y}). \tag{3.9}$$

Therefore, although each factor is only a local meromorphic function on $\mathbb{C}^g \setminus \Sigma$, the coefficients J_s of the product

$$\psi^+\psi = \phi^+(Z, y, k)\phi(Z, y, k) = 1 + \sum_{s=2}^{\infty} J_s(Z, y)k^{-s}. \tag{3.10}$$

are *global meromorphic functions* of Z. Moreover, the translational invariance of u implies that they have the form $J_s(Z, y) = J_s(Z + Vy)$. Each of the factors in the left-hand side of (3.10) has a simple pole along $\Theta - Vy$. Hence $J_s(Z)$ is a meromorphic function on X with a second-order pole along Θ.

From the definition of $\mathcal{L}$, it follows that

$$\mathrm{res}_k(\psi^+(\mathcal{L}^n\psi)) = \mathrm{res}_k(\psi^+k^n\psi) = J_{n+1}. \tag{3.11}$$

On the other hand, using the identity

$$\operatorname{res}_k(e^{-kx}\mathcal{D}_1)(\mathcal{D}_2 e^{kx}) = \operatorname{res}_\partial(\mathcal{D}_2\mathcal{D}_1), \tag{3.12}$$

which holds for any two pseudodifferential operators [14], we get

$$\operatorname{res}_k(\psi^+\mathcal{L}^n\psi) = \operatorname{res}_k(e^{-kx}\Phi^{-1})(\mathcal{L}^n\Phi e^{kx}) = \operatorname{res}_\partial \mathcal{L}^n = F_n. \tag{3.13}$$

Therefore, $F_n = J_{n+1}$ and the lemma is proved. □

Let $\hat{\mathbf{F}}$ be a linear space generated by $\{F_m, m = 0, 1, \dots\}$, where we set $F_0 = 1$. It is a subspace of the 2^g-dimensional space of the abelian functions that have at most second-order pole along Θ. Therefore, for all but $\hat{g} = \dim \hat{\mathbf{F}}$ positive integers n, there exist constants $c_{i,n}$ such that

$$F_n(Z) + \sum_{i=0}^{n-1} c_{i,n} F_i(Z) = 0. \tag{3.14}$$

Let I denote the subset of integers n for which there are no such constants. We call this subset the gap sequence.

Lemma 3.3. *Let $\mathcal{L}$ be the pseudodifferential operator corresponding to a λ-periodic wave function ψ constructed above. Then for the differential operators*

$$L_n = \mathcal{L}_+^n + \sum_{i=0}^{n-1} c_{i,n}\mathcal{L}_+^{n-i} = 0, \quad n \notin I, \tag{3.15}$$

the equations

$$L_n\psi = a_n(k)\psi, \quad a_n(k) = k^n + \sum_{s=1}^{\infty} a_{s,n}k^{n-s}, \tag{3.16}$$

where $a_{s,n}$ are constants, hold.

Proof. First, note that from (3.5), it follows that

$$[\partial_y - \partial_x^2 + u, L_n] = 0. \tag{3.17}$$

Hence if ψ is a λ-periodic wave solution of (1.5) corresponding to $Z \notin \Sigma$, then $L_n\psi$ is also a formal solution of the same equation. This implies the equation $L_n\psi = a_n(Z, k)\psi$, where a is ∂_1-invariant. The ambiguity in the definition of ψ does not affect a_n. Therefore, the coefficients of a_n are well-defined *global* meromorphic functions on $\mathbb{C}^g \setminus \Sigma$. The ∂_1-invariance of a_n implies that a_n, as a function of Z, is holomorphic outside of the locus. Hence it has an extension to a holomorphic function on $\mathbb{C}^g$. Equations (3.3) imply that a_n is periodic with respect to the lattice Λ. Hence a_n is Z-independent. Note that $a_{s,n} = c_{s,n}, s \leq n$. The lemma is proved.□

The operator L_m can be regarded as a $Z \notin \Sigma$-parametric family of ordinary differential operators L_m^Z whose coefficients have the form

$$L_m^Z = \partial_x^n + \sum_{i=1}^{m} u_{i,m}(Ux + Z)\partial_x^{m-i}, \quad m \notin I. \tag{3.18}$$

Corollary 3.1. *The operators L_m^Z commute with each other,*

$$[L_n^Z, L_m^Z] = 0, \quad Z \notin \Sigma. \tag{3.19}$$

From (3.16) it follows that $[L_n^Z, L_m^Z]\psi = 0$. The commutator is an ordinary differential operator. Hence the last equation implies (3.19).

Lemma 3.4. *Let $\mathcal{A}^Z$, $Z \notin \Sigma$, be a commutative ring of ordinary differential operators spanned by the operators L_n^Z. Then there is an irreducible algebraic curve Γ of arithmetic genus $\hat{g} = \dim \hat{\mathbf{F}}$ such that $\mathcal{A}^Z$ is isomorphic to the ring $A(\Gamma, P_0)$ of meromorphic functions on Γ with the only pole at a smooth point P_0. The correspondence $Z \to \mathcal{A}^Z$ defines a holomorphic imbedding of $X \setminus \Sigma$ into the space of torsion-free rank-1 sheaves $\mathcal{F}$ on Γ,*

$$j : X\backslash\Sigma \longmapsto \overline{\mathrm{Pic}}(\Gamma). \tag{3.20}$$

Proof. It is a fundamental fact of the theory of commuting linear ordinary differential operators [5, 6, 9, 10, 15] that there is a natural correspondence

$$\mathcal{A} \longleftrightarrow \{\Gamma, P_0, [k^{-1}]_1, \mathcal{F}\} \tag{3.21}$$

between *regular* at $x = 0$ commutative rings $\mathcal{A}$ of ordinary linear differential operators containing a pair of monic operators of coprime orders, and sets of algebraic-geometrical data $\{\Gamma, P_0, [k^{-1}]_1, \mathcal{F}\}$, where Γ is an algebraic curve with a fixed first jet $[k^{-1}]_1$ of a local coordinate k^{-1} in the neighborhood of a smooth point $P_0 \in \Gamma$ and $\mathcal{F}$ is a torsion-free rank-1 sheaf on Γ such that

$$H^0(\Gamma, \mathcal{F}) = H^1(\Gamma, \mathcal{F}) = 0. \tag{3.22}$$

The correspondence becomes one-to-one if the rings $\mathcal{A}$ are considered modulo conjugation, $\mathcal{A}' = g(x)\mathcal{A}g^{-1}(x)$.

Note that in [5, 6, 9, 10] the main attention was paid to the generic case of commutative rings corresponding to smooth algebraic curves. The invariant formulation of the correspondence given above is due to Mumford [15].

The algebraic curve Γ is called the spectral curve of $\mathcal{A}$. The ring $\mathcal{A}$ is isomorphic to the ring $A(\Gamma, P_0)$ of meromorphic functions on Γ with the only pole at the puncture P_0. The isomorphism is defined by the equation

$$L_a\psi_0 = a\psi_0, \quad L_a \in \mathcal{A}, \quad a \in A(\Gamma, P_0). \tag{3.23}$$

Here ψ_0 is a common eigenfunction of the commuting operators. At $x = 0$, it is a section of the sheaf $\mathcal{F} \otimes \mathcal{O}(-P_0)$.

Important Remark. The construction of the correspondence (3.21) depends on a choice of initial point $x_0 = 0$. The spectral curve and the sheaf $\mathcal{F}$ are defined by the evaluations of the coefficients of generators of $\mathcal{A}$ and a finite number of their derivatives at the initial point. In fact, the spectral curve is independent of the choice of x_0, but the sheaf does depend on it, i.e., $\mathcal{F} = \mathcal{F}_{x_0}$.

Using the shift of the initial point it is easy to show that the correspondence (3.21) extends to commutative rings of operators whose coefficients are *meromorphic* functions of x at $x = 0$. The rings of operators having poles at $x = 0$ correspond to sheaves for which the condition (3.22) is violated.

Let Γ^Z be the spectral curve corresponding to $\mathcal{A}^Z$. Note that, due to the remark above, it is well defined for all $Z \notin \Sigma$. The eigenvalues $a_n(k)$ of the operators L_n^Z defined in (3.16) coincide with the Laurent expansions at P_0 of the meromorphic functions $a_n \in A(\Gamma^Z, P_0)$. They are Z-independent. Hence the spectral curve is Z-independent as well, $\Gamma = \Gamma^Z$. The first statement of the lemma is thus proved. □

The construction of the correspondence (3.21) implies that if the coefficients of the operators $\mathcal{A}$ holomorphically depend on parameters then the algebraic-geometrical spectral data are also holomorphic functions of the parameters. Hence j is holomorphic away from Θ. Then using the shift of the initial point and the fact, that $\mathcal{F}_{x_0}$ holomorphically depends on x_0, we get that j holomorphically extends over $\Theta \setminus \Sigma$, as well. The lemma is proved.

Recall that a commutative ring $\mathcal{A}$ of linear ordinary differential operators is called maximal if it is not contained in any bigger commutative ring. Let us show that for a generic Z the ring $\mathcal{A}^Z$ is maximal. Suppose that it is not. Then there exists $\alpha \in I$, where I is the gap sequence defined above, such that for each $Z \notin \Sigma$ there exists an operator L_α^Z of order α which commutes with $L_n^Z, n \notin I$. Therefore, it commutes with $\mathcal{L}$. A differential operator commuting with $\mathcal{L}$ up to order $O(1)$ can be represented in the form $L_\alpha = \sum_{m<\alpha} c_{i,\alpha}(Z)\mathcal{L}_+^i$, where $c_{i,\alpha}(Z)$ are ∂_1-invariant functions of Z. It commutes with $\mathcal{L}$ if and only if

$$F_\alpha(Z) + \sum_{i=0}^{n-1} c_{i,\alpha}(Z)F_i(Z) = 0, \quad \partial_1 c_{i,\alpha} = 0. \tag{3.24}$$

Note the difference between (3.14) and (3.24). In the first equation the coefficients $c_{i,n}$ are constants. The λ-periodic wave solution of equation (1.5) is a common eigenfunction of all commuting operators, i.e., $L_\alpha\psi = a_\alpha(Z,k)\psi$, where $a_\alpha = k^\alpha + \sum_{s=1}^\infty a_{s,\alpha}(Z)k^{\alpha-s}$ is ∂_1-invariant. The same arguments as those used in the proof of equation (3.16) show that the eigenvalue a_α is Z-independent. We have $a_{s,\alpha} = c_{s,\alpha}$, $s \le \alpha$. Therefore, the coefficients in (3.24) are Z-independent. This contradicts the assumption that $\alpha \notin I$.

Our next goal is to finally prove the global existence of the wave function.

Lemma 3.5. *Let the assumptions of Theorem* 1.2 *hold. Then there exists a common eigenfunction of the corresponding commuting operators* L_n^Z *of the form* $\psi = e^{kx}\phi(Ux + Z, k)$ *such that the coefficients of the formal series*

$$\phi(Z, k) = 1 + \sum_{s=1}^{\infty} \xi_s(Z) k^{-s} \tag{3.25}$$

are global meromorphic functions with a simple pole along Θ.

Proof. It is instructive to consider first the case when the spectral curve Γ of the rings $\mathcal{A}^Z$ is smooth. Then as shown in [5, 6], the corresponding common eigenfunction of the commuting differential operators (the Baker–Akhiezer function), normalized by the condition $\psi_0|_{x=0} = 1$, is of the form [5, 6]

$$\hat{\psi}_0 = \frac{\hat{\theta}(\hat{A}(P) + \hat{U}x + \hat{Z})\hat{\theta}(\hat{Z})}{\hat{\theta}(\hat{U}x + \hat{Z})\hat{\theta}(\hat{A}(P) + \hat{Z})} e^{x\Omega(P)}. \tag{3.26}$$

Here $\hat{\theta}(\hat{Z})$ is the Riemann theta-function constructed with the help of the matrix of b-periods of normalized holomorphic differentials on Γ; $\hat{A} : \Gamma \to J(\Gamma)$ is the Abel map; Ω is the abelian integral corresponding to $d\Omega$; $d\Omega$ is the meromorphic differential of the second kind and has the only pole at the puncture P_0, where its singularity is of the form dk; and $2\pi i \hat{U}$ is the vector of its b-periods.

Remark. Let us emphasize, that the formula (3.26) is not the result of solution of some differential equations. It is a direct corollary of analytic properties of the Baker–Akhiezer function $\hat{\psi}_0(x, P)$ on the spectral curve:

(i) $\hat{\psi}_0$ *is a meromorphic function of* $P \in \Gamma \setminus P_0$*; its pole divisor is of degree* $\tilde{g}$ *and is* x*-independent. It is nonspecial if the operators are regular at the normalization point* $x = 0$.
(ii) *In the neighborhood of* P_0 *the function* $\hat{\psi}_0$ *has the form* (1.16) (*with* $y = 0$).

From the Riemann–Roch theorem, it follows that, if $\hat{\psi}_0$ exists, then it is unique. It is easy to check that the function $\hat{\psi}_0$ given by (3.26) is single-valued on Γ and has all the desired properties.

The last factors in the numerator and the denominator of (3.26) are x-independent. Therefore, the function

$$\hat{\psi}_{BA} = \frac{\hat{\theta}(\hat{A}(P) + \hat{U}x + \hat{Z})}{\hat{\theta}(Ux + \hat{Z})} e^{x\Omega(P)} \tag{3.27}$$

is also a common eigenfunction of the commuting operators.

In the neighborhood of P_0 the function $\hat{\psi}_{BA}$ has the form

$$\hat{\psi}_{BA} = e^{kx} \left(1 + \sum_{s=1}^{\infty} \frac{\tau_s(\hat{Z} + \hat{U}x)}{\hat{\theta}(\hat{U}x + \hat{Z})} k^{-s} \right), \quad k = \Omega, \tag{3.28}$$

where $\tau_s(\hat{Z})$ are global holomorphic functions.

According to Lemma 3.4, we have a holomorphic imbedding $\hat{Z} = j(Z)$ of $X \setminus \Sigma$ into $J(\Gamma)$. Consider the formal series $\psi = j^* \hat{\psi}_{BA}$. It is globally well defined away from Σ. If $Z \notin \Theta$, then $j(Z) \notin \hat{\Theta}$ (which is the divisor on which the condition (3.22) is violated). Hence the coefficients of ψ are regular away from Θ. The singular locus is at least of codimension 2. Hence, once again using Hartog-type arguments, we can extend ψ on X.

If the spectral curve is singular, we can proceed along the same lines using the generalization of (3.27) given by the theory of the Sato τ-function [16]. Namely, a set of algebraic-geometrical data (3.21) defines a point of the Sato Grassmannian, and therefore the corresponding τ-function: $\tau(t; \mathcal{F})$. It is a holomorphic function of the variables $t = (t_1, t_2, \dots)$, and is a section of a holomorphic line bundle on $\overline{\mathrm{Pic}}(\Gamma)$.

The variable x is identified with the first time of the KP-hierarchy, $x = t_1$. Therefore, the formula for the Baker–Akhiezer function corresponding to a point of the Grassmannian [16] implies that the function $\hat{\psi}_{BA}$ given by the formula

$$\hat{\psi}_{BA} = \frac{\tau(x-k, -\frac{1}{2}k^2, -\frac{1}{3}k^3, \dots; \mathcal{F})}{\tau(x, 0, 0, \dots; \mathcal{F})} e^{kx} \tag{3.29}$$

is a common eigenfunction of the commuting operators defined by $\mathcal{F}$. The rest of the arguments proving the lemma are the same as in the smooth case. □

Lemma 3.6. *The linear space* $\hat{\mathbf{F}}$ *generated by the abelian functions* $\{F_0 = 1, F_m = \mathrm{res}_\partial \mathcal{L}^m\}$, *is a subspace of the space* $\mathbf{H}$ *generated by* F_0 *and by the abelian functions* $H_i = \partial_1 \partial_{z_i} \ln \theta(Z)$.

Proof. Recall that the functions F_n are abelian functions with at most second-order poles on Θ. Hence a priori $\hat{g} = \dim \hat{\mathbf{F}} \leq 2^g$. In order to prove the statement of the lemma, it is enough to show that $F_n = \partial_1 Q_n$, where Q_n is a meromorphic function with a pole along Θ. Indeed, if Q_n exists, then, for any vector λ in the period lattice, we have $Q_n(Z+\lambda) = Q_n(Z) + c_{n,\lambda}$. There is no abelian function with a simple pole on Θ. Hence there exists a constant q_n and two g-dimensional vectors l_n, l'_n, such that $Q_n = q_n + (l_n, Z) + (l'_n, h(Z))$, where $h(Z)$ is a vector with the coordinates $h_i = \partial_{z_i} \ln \theta$. Therefore, $F_n = (l_n, U) + (l'_n, H(Z))$.

Let $\psi(x, Z, k)$ be the formal Baker–Akhiezer function defined in the previous lemma. Then the coefficients $\varphi_s(Z)$ of the corresponding wave operator Φ (3.6) are global meromorphic functions with poles along Θ.

The left and right actions of pseudodifferential operators are formally adjoint, i.e., for any two operators the equality $(e^{-kx}\mathcal{D}_1)(\mathcal{D}_2 e^{kx}) = e^{-kx}(\mathcal{D}_1\mathcal{D}_2 e^{kx}) + \partial_x(e^{-kx}(\mathcal{D}_3 e^{kx}))$ holds. Here $\mathcal{D}_3$ is a pseudodifferential operator whose coefficients are differential polynomials in the coefficients of $\mathcal{D}_1$ and $\mathcal{D}_2$. Therefore, from (3.9)–(3.13) it follows that

$$\psi^+\psi = 1 + \sum_{s=2}^{\infty} F_{s-1} k^{-s} = 1 + \partial_x \left(\sum_{s=2}^{\infty} Q_s k^{-s} \right). \tag{3.30}$$

The coefficients of the series Q are differential polynomials in the coefficients φ_s of the wave operator. Therefore, they are global meromorphic functions of Z with poles along Θ. The lemma is proved. □

In order to complete the proof of our main result, we need one more standard fact of the KP theory: flows of the KP hierarchy define deformations of the commutative rings $\mathcal{A}$ of ordinary linear differential operators. The spectral curve is invariant under these flows. For a given spectral curve Γ the orbits of the KP hierarchy are isomorphic to the generalized Jacobian $J(\Gamma) = \mathrm{Pic}^0(\Gamma)$, which is the set of equivalence classes of zero degree divisors on the spectral curve (see details in [1, 5, 6, 16]).

The KP hierarchy in the Sato form is a system of commuting differential equation for a pseudodifferential operator $\mathcal{L}$,

$$\partial_{t_n}\mathcal{L} = [\mathcal{L}^n_+, \mathcal{L}]. \tag{3.31}$$

If the operator $\mathcal{L}$ is as above, i.e., if it is defined by λ-periodic wave solutions of equation (1.5), then equations (3.31) are equivalent to the equations

$$\partial_{t_n} u = \partial_x F_n. \tag{3.32}$$

The first two times of the hierarchy are identified with the variables $t_1 = x$, $t_2 = y$.

Equations (3.32) identify the space $\hat{\mathbf{F}}_1$ generated by the functions $\partial_1 F_n$ with the tangent space of the KP orbit at $\mathcal{A}^Z$. Then from Lemma 3.6, it follows that this tangent space is a subspace of the tangent space of the abelian variety X. Hence for any $Z \notin \Sigma$, the orbit of the KP flows of the ring $\mathcal{A}^Z$ is in X, i.e., it defines a holomorphic imbedding:

$$i_Z : J(\Gamma) \longmapsto X. \tag{3.33}$$

From (3.33), it follows that $J(\Gamma)$ is *compact*.

The generalized Jacobian of an algebraic curve is compact if and only if the curve is *smooth* [17]. On a smooth algebraic curve a torsion-free rank-1 sheaf is a line bundle, i.e., $\overline{\mathrm{Pic}}(\Gamma) = J(\Gamma)$. Then (3.20) implies that i_Z is an isomorphism. Note that for the Jacobians of smooth algebraic curves the bad locus Σ is empty [1], i.e., the imbedding j in (3.20) is defined everywhere on X and is inverse to i_Z. Theorem 1.2 is proved.

Acknowledgments. This research was supported in part by National Science Foundation grant DMS-04-05519.

The author would like to thank Enrico Arbarello for very useful conversations and help during the preparation of this paper. The author is greatful to Takahiro Shiota, whose remarks helped the author to clarify some missing arguments in the first version of the paper.

References

[1] T. Shiota, Characterization of Jacobian varieties in terms of soliton equations, *Invent. Math.*, **83**-2 (1986), 333–382.

[2] B. A. Dubrovin, Kadomtsev-Petviashvili equation and relations for the matrix of periods on Riemann surfaces, *Izv. Akad. Nauk SSSR Ser. Mat.*, **45** (1981), 1015–1028.

[3] V. E. Zakharov and A. B. Shabat, Integration method of nonlinear equations of mathematical physics with the help of the inverse scattering problem, *Funk. Anal Pril.*, **8**-3 (1974), 43–53.

[4] V. S. Druma, On analytic solution of the two-dimensional Korteweg-de Vries equation, *JETP Lett.*, **19**-12 (1974), 219–225.

[5] I. M. Krichever, Integration of non-linear equations by methods of algebraic geometry, *Functional Anal. Appl.*, **11**-1 (1977), 12–26.

[6] I. M. Krichever, Methods of algebraic geometry in the theory of non-linear equations, *Russian Math. Surveys*, **32**-6 (1977), 185–213.

[7] E. Arbarello, I. Krichever, and G. Marini, Characterizing Jacobians via flexes of the Kummer variety, 2005, math.AG/0502138.

[8] G. E. Welters, A criterion for Jacobi varieties, *Ann. Math.*, **120**-3 (1984), 497–504.

[9] J. L. Burchnall and T. W. Chaundy, Commutative ordinary differential operators I, *Proc. London Math Soc.*, **21** (1922), 420–440.

[10] J. L. Burchnall and T. W. Chaundy, Commutative ordinary differential operators II, *Proc. Roy. Soc. London*, **118** (1928), 557–583.

[11] E. Arbarello and C. De Concini, Another proof of a conjecture of S. P. Novikov on periods of abelian integrals on Riemann surfaces, *Duke Math. J.*, **54** (1987), 163–178.

[12] G. Marini, A geometrical proof of Shiota's theorem on a conjecture of S. P. Novikov, *Comp. Math.*, **111** (1998), 305–322.

[13] I. Krichever and D. H. Phong, Symplectic forms in the theory of solitons, in C. L. Terng and K. Uhlenbeck, eds., *Surveys in Differential Geometry* IV, International Press, Cambridge, MA, 1998, 239–313.

[14] L. A. Dickey, *Soliton Equations and Hamiltonian Systems*, Advanced Series in Mathematical Physics, Vol. 12, World Scientific, Singapore, 1991.

[15] D. Mumford, An algebro-geometric construction of commuting operators and of solutions to the Toda lattice equation, Korteweg-de Vries equation and related non-linear equations, in *Proceedings of the International Symposium on Algebraic Geometry, Kyoto*, 1977, Kinokuniya Book Store, Kyoto, 1978, 115–153.

[16] G. Sigal and G. Wilson, Loop groups and equations of KdV type, *IHES Publ. Math.*, **61** (1985), 5–65.

[17] P. Deligne and D. Mumford, The irreducibility of the space of curves of given genus, *IHES Publ. Math.*, **36** (1969), 75–109.

Fibres de Springer et jacobiennes compactifiées

Gérard Laumon

CNRS et Université Paris-Sud
UMR 8628
Mathématique, Bâtiment 425
F-91405 Orsay Cedex
France
gerard.laumon@math.u-psud.fr

Résumé. L'objet de ce travail est d'identifier, à homéomorphisme près, les fibres de Springer affines pour $\mathrm{GL}(n)$ sur un corps local d'égales caractéristiques à des revêtements de jacobiennes compactifiées de courbes projectives singulières. Ce lien permet de démontrer certaines propriétés géométriques de ces fibres de Springer, dont une propriété d'irréductibilité, et aussi d'en construire des déformations à homéomorphismes près.

Subject Classifications: Primary 14G20. Secondary 14F20, 14K30.

0 Introduction

Soient F un corps local non archimédien d'égales caractéristiques, $\mathcal{O}_F$ son anneau des entiers, k son corps résiduel et E un F-espace vectoriel de dimension finie.

La grassmannienne affine pour le F-schéma en groupes $\mathrm{Aut}_F(E)$ des automorphismes de E est le ind-k-schéma des $\mathcal{O}_F$-réseaux M dans E. Pour tout endomorphisme régulier semi-simple et topologiquement nilpotent γ de E, on peut considérer le fermé réduit X_γ de la grassmannienne affine formé des réseaux $M \subset E$ tels que $\gamma(M) \subset M$. Ce fermé est appelé la *fibre de Springer affine en* γ par analogie avec les fibres de Springer classiques dans les variétés de drapeaux.

D'après Kazhdan et Lusztig [K-L], X_γ est un vrai schéma, localement de type fini sur k, qui est muni d'une action libre naturelle d'un groupe abélien libre de type fini Λ_γ, et le quotient $Z_\gamma = X_\gamma/\Lambda_\gamma$ est un k-schéma projectif.

Dans ce travail, nous attachons à γ une courbe intègre et projective C_γ sur k, qui n'a au plus qu'un point singulier, point en lequel l'anneau local complété de la courbe n'est autre que $\mathcal{O}_F[\gamma] \subset F[\gamma] \subset \mathrm{Au}_F(E)$. Puis, nous relions la fibre de Springer affine X_γ et son quotient Z_γ à la jacobienne compactifiée de C_γ.

Nous déduisons alors des résultats d'Altman et Kleiman sur les jacobiennes compactifiées, un énoncé d'irréductibilité pour Z_γ (Corollaire 2.3.1) et la possibilité de déformer à homéomorphisme près Z_γ (et aussi dans une certaine mesure X_γ) en faisant varier γ (cf. Chapitre 4).

Nous énonçons une variante de la conjecture de pureté de Goresky, Kottwitz, et MacPherson pour les jacobiennes compactifiées (cf. Chapitre 3), que nous démontrons dans le cas homogène par un argument similaire à celui utilisé par Springer dans pour les fibres de Springer classiques (cf. Section 4.3).

Dans un long appendice, nous rappelons quelques résultats de Teissier d'une part, et de Diaz et Harris d'autre part, sur les déformations des singularités isolées de courbes planes.

1 Fibres de Springer

1.1 Les données

On fixe un corps parfait k. Dans ce travail, on appelle simplement *corps local* tout corps K contenant k et muni d'une valuation discrète $v_K : K \to \mathbb{Z} \cup \{\infty\}$ pour laquelle K est complet et de corps résiduel k. Pour un tel corps local, on note $\mathcal{O}_K = \{x \in K \mid v_K(x) \geq 0\} \subset K$ l'anneau des entiers de K et $\mathfrak{p}_K = \{x \in K \mid v_K(x) > 0\}$ l'idéal maximal de $\mathcal{O}_K$. Le choix d'une uniformisante ϖ_K de K identifie $\mathfrak{p}_K \subset \mathcal{O}_K \subset K$ à $\varpi_K k[[\varpi_K]] \subset k[[\varpi_K]] \subset k((\varpi_K))$.

On fixe un corps local F et une famille finie $(E_i)_{i\in I}$ non vide d'extensions finies, séparables et totalement ramifiées de F. Pour chaque $i \in I$, on note n_i le degré de E_i sur F et on se donne un élément γ_i de $\mathfrak{p}_{E_i} \subset \mathcal{O}_{E_i} \subset E_i$ qui engendre E_i sur F, de sorte que $E_i \cong F[T]/(p_i(T))$ où $p_i(T) \in \mathcal{O}_F[T]$ est le polynôme minimal de γ_i sur F.

On suppose que les polynômes $p_i(T)$ unitaires et irréductibles dans $F[T]$ sont deux à deux distincts.

On note A_i la k-algèbre intègre

$$A_i = \mathcal{O}_F[\gamma_i] \subset \mathcal{O}_{E_i}.$$

Elle est locale d'idéal maximal $\mathfrak{m}_i = \mathfrak{p}_{E_i} \cap A_i$, son corps des fractions est E_i, et sa normalisation $\widetilde{A}_i \subset E_i$ n'est autre que $\mathcal{O}_{E_i}$.

On note $E_I = \prod_{i\in I} E_i$, $n_I = \sum_{i\in I} n_i$ la dimension de cet espace vectoriel sur F, $\gamma_I = (\gamma_i)_{i\in I} \in \prod_{i\in I} A_i$ et

$$A_I = \mathcal{O}_F[\gamma_I] \subset \prod_{i\in I} A_i.$$

La k-algèbre A_I est locale d'idéal maximal $\mathfrak{m}_I = \left(\prod_{i\in I} \mathfrak{m}_i\right) \cap A_I$, son anneau total des fractions est E_I et sa normalisation est égale à $\widetilde{A}_I := \prod_{i\in I} \widetilde{A}_i = \prod_{i\in I} \mathcal{O}_{E_i} =: \mathcal{O}_{E_I} \subset E_I$.

Comme

$$A_I \cong \mathcal{O}_F[T]/(p_I(T)) = k[[\varpi_F]][T]/(p_I(T)) \cong k[[\varpi_F, T]]/(p_I(T))$$

où $p_I(T) = \prod_{i\in I} p_i(T)$, la k-algèbre locale (intègre et de dimension 1) A_I est de Gorenstein et son module dualisant ω_{A_I} est égal à

$$\omega_{A_I} = \{y \in \Omega^1_{E_I/k} \mid \mathrm{Res}_I(xy) = 0,\ \forall x \in A_I\} \supset \Omega^1_{\widetilde{A}_I/k},$$

où l'application k-linéaire $\mathrm{Res}_I : \Omega^1_{E_I/k} \to k$ est la somme des applications résidus $\mathrm{Res}_i : \Omega^1_{E_i/k} \to k$.

Proposition 1.1.1 (Rosenlicht). *L'accouplement $(\widetilde{A}_I/A_I) \times (\omega_{A_I}/\Omega^1_{\widetilde{A}_I/k}) \to k$ qui envoie $(x + A_I, y + \Omega^1_{\widetilde{A}_I/k})$ sur $\mathrm{Res}_I(xy)$ est un accouplement parfait.*

Démonstration. Voir [A-K 1, VIII, Proposition 1.16]. □

On pose

$$\delta_I = \dim_k(\widetilde{A}_I/A_I)$$

et on note $\mathfrak{a}_I$ le conducteur de $\widetilde{A}_I$ dans A_I, c'est-à-dire l'idéal de $\widetilde{A}_I$ formé des $x \in \widetilde{A}_I$ tels que $x\widetilde{A}_I \subset A_I$. Cet idéal est contenu dans A_I et il résulte de la proposition ci-dessus que

$$\dim_k(A_I/\mathfrak{a}_I) = \delta_I.$$

Pour chaque $i \in I$, on pose

$$\delta_i = \dim_k(\widetilde{A}_i/A_i)$$

et, pour chaque $i \neq j$ dans I, on note

$$r_{ij} = v_{E_i}(p_j(\gamma_i)) = v_{E_j}(p_i(\gamma_j)) = r_{ji}$$

la valuation du résultant dans F des polynômes $p_i(T)$ et $p_j(T)$. On vérifie que

$$\delta_I = \sum_{i\in I}\delta_i + \frac{1}{2}\sum_{\substack{i,j\in I\\ i\neq j}} r_{ij}$$

et que

$$\mathfrak{a}_I = \prod_{i\in I}\mathfrak{p}_{E_i}^{2\delta_i + \sum_{j\in I\setminus\{i\}} r_{ij}} \subset \prod_{i\in I}\mathcal{O}_{E_i} = \mathcal{O}_{E_I}.$$

1.2 La grassmannienne affine

Rappelons qu'un $\mathcal{O}_F$-réseau dans un F-espace vectoriel de dimension finie E est un sous-$\mathcal{O}_F$-module de E de rang égal à la dimension de E. Si M et N sont deux tels $\mathcal{O}_F$-réseaux, l'indice de M relativement à N est l'entier

$$[M : N] = \dim_k(M/P) - \dim_k(N/P)$$

où P est n'importe quel $\mathcal{O}_F$-réseau de E contenu à la fois dans M et dans N. Par exemple, A_I et $\widetilde{A}_I$ sont des $\mathcal{O}_F$-réseaux dans E_I et $[\widetilde{A}_I : A_I] = \delta_I$.

Soient $N \geq 0$ et d des entiers et soit

$$M \subset \varpi_F^{-N} A_I \subset E_I$$

un $\mathcal{O}_F$-réseau qui est d'indice d relativement au $\mathcal{O}_F$-réseau particulier A_I (bien entendu, pour qu'il existe un tel M, il faut que $d \leq n_I N$). La multiplication par ϖ_F induit un endomorphisme nilpotent du k-espace vectoriel $\varpi_F^{-N} A_I / M$ de dimension $n_I N - d$. Par suite, M contient automatiquement $\varpi_F^{(n_I-1)N-d} A_I$ et la donnée de M équivaut à la donnée du sous-espace vectoriel

$$M / \varpi_F^{(n_I-1)N-d} A_I \subset \varpi_F^{-N} A_I / \varpi_F^{(n_I-1)N-d} A_I$$

stable par l'endomorphisme nilpotent induit par la multiplication par ϖ_F.

Pour toute extension k' de k, on note $F' = k' \widehat{\otimes}_k F = k'((\varpi_F))$, $E_i' = k' \widehat{\otimes}_k E_i = k'((\varpi_{E_i}))$, etc. les complétés ϖ_F-adiques de $k' \otimes_k F$, $k' \otimes_k E_i$, etc. Ce que l'on vient de dire vaut encore après que l'on ait remplacé k par k', F par F', etc.

Pour k' variable, les $\mathcal{O}_{F'}$-réseaux dans E_I' qui sont d'indice d relativement au $\mathcal{O}_{F'}$-réseau particulier A_I' et qui sont contenus dans $\varpi_F^{-N} A_I'$ sont donc naturellement les k'-points d'un k-schéma projectif réduit $R_{I,N}^d$, à savoir le fermé (réduit) de la grassmannienne des $(n_I-1)(n_I N-d)$-plans dans le k-espace vectoriel $\varpi_F^{-N} A_I / \varpi_F^{(n_I-1)N-d} A_I$ de dimension $n_I(n_I N - d)$, formé des plans qui sont stables par l'endomorphisme nilpotent induit par la multiplication par ϖ_F.

Pour d fixé, les k-schémas projectifs $R_{I,N}^d$ s'organisent en un système inductif d'immersions fermées

$$\cdots \hookrightarrow R_{I,N}^d \hookrightarrow R_{I,N+1}^d \hookrightarrow \cdots$$

et on note R_I^d le ind-k-schéma «limite».

La *grassmannienne affine* ou *ind-k-schéma des $\mathcal{O}_F$-réseaux de E_I* est par définition la somme disjointe

$$R_I = \coprod_{d \in \mathbb{Z}} R_I^d.$$

1.3 Fibres de Springer

Toujours pour k' variable, les $\mathcal{O}_{F'}$-réseaux $(M \subset E_I') \in R_{I,N}^d(k')$ tels que

$$\gamma_I M \subset M$$

sont les k'-points d'un sous-k-schéma fermé réduit $X_{I,N}^d$ de $R_{I,N}^d$. Pour d fixé, les $X_{I,N}^d \subset R_{I,N}^d$ s'organisent en un système inductif et on note $X_I^d \subset R_I^d$ le sous-ind-k-schéma fermé réduit «limite».

Définition 1.3.1. La *fibre de Springer* en γ_I est le sous-ind-k-schéma fermé réduit

$$X_I = \coprod_{d \in \mathbb{Z}} X_I^d \subset \coprod_{d \in \mathbb{Z}} R_I^d = R_I$$

des $\mathcal{O}_F$-réseaux de E_I stabilisé par γ_I.

Bien entendu, chaque $X_I^d = X_I \cap R_I^d$ est une partie ouverte et fermée de X_I.

La remarque évidente suivante est essentielle pour la suite :

On peut identifier les k'-points de X_I aux sous-A'_I-modules $M \subset E'_I$ qui sont de rang 1 en chaque point générique de $\mathrm{Spec}(A'_I)$.

Pour k' variable, le groupe $E_I'^{\times}/A_I'^{\times}$ est de manière naturelle le groupe des k'-points d'un k-schéma en groupes commutatifs G_I qui est lisse et de dimension δ_I sur k. Plus précisément, le groupe des composantes connexes de G_I est le quotient $E_I^{\times}/\mathcal{O}_{E_I}^{\times} = \prod_{i\in I}(E_i^{\times}/\mathcal{O}_{E_i}^{\times})$, qui est canoniquement isomorphe à $\Lambda_I := \mathbb{Z}^I$; la composante neutre G_I^0 de G_I, qui admet $\mathcal{O}_{E'_I}^{\times}/A_I'^{\times}$ pour groupe des k'-points, est une extension d'un tore T_I (le quotient de $\mathbb{G}_{\mathrm{m},k}^I$ par $\mathbb{G}_{\mathrm{m},k}$ plongé diagonalement) par un schéma en groupes unipotents U_I de type fini dont le groupe des k'-points est

$$U_I(k') = \big(\textstyle\prod_{i\in I}(1+\mathfrak{p}_{E'_I})\big)/(1+\mathfrak{m}'_I)$$

où $\mathfrak{m}'_I$ est l'idéal maximal de A'_I.

L'action par homothéties de $E_I'^{\times}/A_I'^{\times}$ sur les réseaux $M \in X_I(k')$ provient d'une action algébrique naturelle de G_I sur X_I. Cette action permute les composantes X_I^d de X_I suivant la règle

$$g \cdot X_I^d = X_I^{d+|\lambda(g)|}$$

où $\lambda(g)$ est l'image de $g \in G_I$ dans Λ_I et où on a posé $|\lambda| = \sum_{\iota\in I}\lambda_i$ pour chaque $\lambda \in \Lambda_I$.

Le k-schéma X_I contient le k-point particulier $M = A_I$. Le fixateur dans G_I de ce point particulier est réduit à l'élément neutre et son orbite $X_I^{\circ} = G_I \cdot A_I$ est donc une partie de X_I isomorphe à G_I.

Lemme 1.3.1. *La G_I-orbite X_I° est l'ouvert de X_I dont les k'-points sont les M qui sont libres de rang 1 en tant que A'_I-modules.* □

Tout scindage $\sigma : \Lambda_I^0 \hookrightarrow G_I$ de l'extension

$$1 \to G_I^0 \to G_I \to \Lambda_I \to 0$$

au-dessus du sous-groupe

$$\Lambda_I^0 := \{\lambda \in \Lambda_I \mid |\lambda| = 0\} \subset \Lambda_I$$

définit une action libre de Λ_I^0 sur X_I qui préserve les composantes X_I^d. On notera $Z_I = X_I/\sigma(\Lambda_I^0)$ le k-espace quotient correspondant et $Z_I^d = X_I^d/\sigma(\Lambda_I^0)$ ses composantes.

Théorème 1.3.1 (Kazhdan–Lusztig). *La fibre de Springer X_I est en fait un k-schéma localement de type fini et de dimension finie, dont les composantes connexes sont exactement les X_I^d, $d \in \mathbb{Z}$.*

Pour tout scindage $\sigma : \Lambda_I^0 \hookrightarrow G_I$ comme ci-dessus, les k-espaces quotients Z_I^d correspondants sont des k-schémas projectifs, Z_I est le k-schéma somme disjointe des Z_I^d et l'application quotient $X_I \to Z_I$ est un revêtement étale galoisien de groupe de Galois Λ_I^0.

Démonstration. Voir [K-L, Section 3]. □

2 Les fibres de Springer comme revêtements de jacobiennes compactifiées

L'objet de ce chapitre est d'identifier, à homéomorphisme près, le quotient $Z_I = X_I/\sigma(\Lambda_I^0)$ de la fibre de Springer X_I à la jacobienne compactifiée d'une courbe projective, et d'identifier, toujours à homéomorphisme près, le revêtement $X_I \to Z_I$ à un revêtement étale galoisien de cette jacobienne compactifiée.

La clé de cette identification est la remarque très simple suivante : un $\mathcal{O}_F$-réseau $M \subset E_I$ tel que $\gamma_I M \subset M$ n'est rien d'autre qu'un A_I-module de type fini sans torsion M, de rang 1 en tout point générique de $\mathrm{Spec}(A_I)$, muni d'un isomorphisme E_I-linéaire $E_I \otimes_{A_I} M \xrightarrow{\sim} E_I$.

Le revêtement de la jacobienne compactifiée qui intervient peut être construit en utilisant une formule d'adjonction de Grothendieck explicitée par Raynaud (cf. la Section 2.6) et un résultat d'auto-dualité partielle pour les jacobiennes compactifiées démontré par Estève, Gagné, et Kleiman, résultat que nous rappelons dans la Section 2.5.

2.1 La courbe C_I

Dans la situation de la Section 1.1, le schéma formel $\mathrm{Spf}(A_I)$ est un germe formel de courbe plane dont la famille des branches irréductibles est $(\mathrm{Spf}(A_i))_{i\in I}$ et dont le normalisé est le schéma formel semi-local $\mathrm{Spf}(\widetilde{A}_I) = \coprod_{i\in I} \mathrm{Spf}(\mathcal{O}_{E_i})$. *On suppose dans la suite que le nombre d'éléments du corps k est au moins égal au nombre d'éléments de l'ensemble fini I.*

Proposition 2.1.1. *Il existe une courbe projective et géométriquement intègre C_I sur k, munie de deux k-points distincts c_I et ∞_I, ayant les propriétés suivantes :*

1. *C_I est lisse sur k en dehors de c_I,*
2. *le complété de l'anneau local de C_I en c_I est isomorphe à A_I,*
3. *la normalisée $\widetilde{C}_I$ de C_I est isomorphe à la droite projective standard $\mathbb{P}_k^1$ sur k par un isomorphisme qui envoie le point à l'infini ∞ de $\mathbb{P}_k^1$ sur le point ∞_I de C_I.*

Pour une telle courbe C_I, son morphisme de normalisation $\pi_I : \widetilde{C}_I \to C_I$ est un isomorphisme au-dessus de $C_I \setminus \{c_I\}$, et $\pi_I^{-1}(c_I) \subset \widetilde{C}_I$ est l'ensemble des branches de $\mathrm{Spf}(A_I)$; pour chaque $i \in I$, on notera $\widetilde{c}_i$ le point de $\pi_I^{-1}(c_I)$ correspondant à la branche $\mathrm{Spf}(A_i)$.

Démonstration. On fixe arbitrairement une injection $\iota : I \hookrightarrow k$ et, pour chaque $i \in I$, on fixe arbitrairement une uniformisante ϖ_{E_i} de $\mathcal{O}_{E_i}$. On plonge $k[x]$ dans $\mathcal{O}_{E_i}$ en envoyant x sur $\iota(i) + \varpi_{E_i}$. On en déduit un plongement de k-algèbres de $k[x]$ dans $\mathcal{O}_{E_I} = \widetilde{A}_I$ qui identifie $\widetilde{A}_I$ au complété de l'anneau semi-local de la droite affine $\mathbb{A}_k^1 = \mathrm{Spec}(k[x])$ en l'ensemble fini de points $\iota(I)$.

Considérons alors la k-algèbre B_I définie par le carré cartésien

$$\begin{array}{ccc} B_I & \subset & A_I \\ \cap & \square & \cap \\ k[x] & \subset & \widetilde{A}_I. \end{array}$$

Elle est intègre, de type fini sur k et de dimension 1, l'inclusion $B_I \hookrightarrow A_I$ induit un isomorphisme du complété de B_I le long de son idéal maximal $\mathfrak{m}_I \cap B_I$ sur A_I et l'inclusion $B_I \hookrightarrow k[x]$ fait de $k[x]$ une B_I-algèbre finie. En effet, d'une part on a

$$\widetilde{A}_I = k[x] + \mathfrak{a}_I^{n+1}, \ \forall n \in \mathbb{N},$$

et $\mathfrak{a}_I \subset A_I \subset \widetilde{A}_I$, de sorte que

$$A_I = B_I + \mathfrak{a}_I^{n+1}, \ \forall n \in \mathbb{N},$$

et d'autre part

$$k[x] \cap \mathfrak{a}_I \subset B_I \subset k[x]$$

est l'idéal principal engendré par

$$\prod_{i \in I} (x - \iota(i))^{2\delta_i + \sum_{j \in I \setminus \{i\}} r_{ij}}.$$

On peut donc effectuer la «somme amalgamée» de $\mathbb{A}_k^1$ et de $\mathrm{Spf}(A_I)$ le long de $\mathrm{Spf}(\widetilde{A}_I)$; c'est par définition le k-schéma affine $\mathrm{Spec}(B_I)$. Bien sûr, le morphisme fini $\mathbb{A}_k^1 \to \mathrm{Spec}(B_I)$, induit par l'inclusion $B_I \subset k[x]$, envoie le sous-ensemble fini $\iota(I) \subset \mathbb{A}_k^1$ sur un unique k-point c_I de $\mathrm{Spec}(B_I)$ et il induit un isomorphisme de $\mathbb{A}_k^1 \setminus \iota(I)$ sur $\mathrm{Spec}(B_I) \setminus \{c_I\}$.

On définit la courbe géométriquement intègre et projective C_I sur k en recollant $\mathrm{Spec}(B_I)$ et $\mathbb{P}_k^1 \setminus \iota(I)$ le long de leur ouvert commun $\mathrm{Spec}(B_I) \setminus \{c_I\} \cong \mathbb{A}_k^1 \setminus \iota(I)$. □

2.2 Schémas de Picard compactifiés

Soit C une courbe réduite, projective et géométriquement irréductible sur k. Comme k est parfait, C est géométriquement réduite sur k et son lieu singulier C^{sing} est donc fini. On suppose que toutes les singularités de C sont *planes*, c'est-à-dire que pour tout $c \in C^{\mathrm{sing}}$, le complété de l'anneau local de C en c est isomorphe à $\kappa(c)[[x,y]]/(f)$ pour une série formelle $f \in \kappa(c)[[x,y]]$. On note $\widetilde{C} \to C$ le morphisme de normalisation de C. La courbe $\widetilde{C}$ est donc géométriquement connexe, projective et lisse sur k. On note $g(C) = \dim_k H^1(\widetilde{C}, \mathcal{O}_{\widetilde{C}})$ son genre, c'est-à-dire par définition le genre géométrique de C.

On a la suite exacte

$$0 \to \mathcal{O}_C \to \pi_*\mathcal{O}_{\widetilde{C}} \to \bigoplus_{c \in C^{\mathrm{sing}}} (\pi_*\mathcal{O}_{\widetilde{C}}/\mathcal{O}_C)_c \to 0$$

et on pose

$$\delta(C) = \sum_{c \in C^{\mathrm{sing}}} [\kappa(c) : k]\delta_c(C)$$

où, pour chaque $c \in C^{\mathrm{sing}}$, $\delta_c(C)$ est la dimension du $\kappa(c)$-espace vectoriel $(\pi_* \mathcal{O}_{\widetilde{C}}/\mathcal{O}_C)_c$.

Le genre arithmétique $\dim_k H^1(C, \mathcal{O}_C) = 1 - \chi(C, \mathcal{O}_C)$ de C est égal à $g(C) + \delta(C)$. Pour tout $\mathcal{O}_C$-Module cohérent $\mathcal{M}$, on pose

$$\deg(\mathcal{M}) = \chi(C, \mathcal{M}) - \mathrm{rang}(\mathcal{M})\chi(C, \mathcal{O}_C)$$

où $\mathrm{rang}(\mathcal{M})$ est le rang générique de $\mathcal{M}$. Pour tout $\mathcal{O}_C$-Module inversible $\mathcal{L}$, $\pi^*\mathcal{L}$ est un $\mathcal{O}_{\widetilde{C}}$-Module inversible et on a

$$\deg(\mathcal{L}) = \deg(\pi^*\mathcal{L})$$

et

$$\deg(\mathcal{L} \otimes_{\mathcal{O}_C} \mathcal{M}) = \mathrm{rang}(\mathcal{M}) \deg(\mathcal{L}) + \deg(\mathcal{M}).$$

Soit $P(C) = \mathrm{Pic}_{C/k}$ le *k-schéma en groupes de Picard de C*. Pour toute extension k' de k, ses k'-points sont les classes d'isomorphie de $\mathcal{O}_{k' \otimes_k C}$-Modules inversibles (avec la multiplication définie par le produit tensoriel). Ce k-schéma est lisse de dimension $\delta(C)$. Ses composantes connexes sont les sous-k-schémas $P^d(C)$, $d \in \mathbb{Z}$, découpés par le degré du Module inversible universel, et elles sont en fait géométriquement connexes. La composante neutre $P^0(C)$ de $P(C)$ est quasi-projective.

Soit $\overline{P}(C) = \overline{\mathrm{Pic}}_{C/k}$ le *k-schéma de Picard compactifié de C* défini par Mayer et Mumford (cf. [A-K 2]) dont les k'-points sont les classes d'isomorphie de $\mathcal{O}_{k' \otimes_k C}$-Modules cohérents sans torsion de rang générique 1. Par définition, $P(C)$ est un ouvert de $\overline{P}(C)$ et l'action par translation de $P(C)$ sur lui-même se prolonge en une action de $P(C)$ sur $\overline{P}(C)$ (encore définie par produit tensoriel). On a aussi un découpage en parties ouvertes et fermées

$$\overline{P}(C) = \coprod_{d \in \mathbb{Z}} \overline{P}^d(C)$$

par le degré du Module sans torsion universel, avec bien entendu

$$P^d(C) = P(C) \cap \overline{P}^d(C)$$

et

$$P^d(C) \cdot \overline{P}^e(C) = \overline{P}^{d+e}(C)$$

quels que soient les entiers d, e. D'après Mayer et Mumford (cf. [A-K 2] et [A-K 3]), chaque composante $\overline{P}^d(C)$ est un k-schéma projectif.

Théorème 2.2.1 (Altman, Iarrobino, Kleiman [A-I-K, Corollary 7], Rego [Re, Theorem A]). *Chaque composante $\overline{P}^d(C)$ de $\overline{P}(C)$ est géométriquement intègre et localement d'intersection complète de dimension $\delta(C)$.* □

La composante connexe $\overline{P}^0(C)$ est aussi appelée la jacobienne compactifiée de C puisqu'elle compactifie la jacobienne $P^0(C)$ de C.

2.3 Lien entre fibres de Springer et schémas de Picard compactifiés

On note dans la suite $P_I = P(C_I)$, $\overline{P}_I = \overline{P}(C_I)$, etc.

Faisons le lien entre les k-schémas X_I et $\overline{P}_I$. On a la suite exacte

$$\begin{aligned}1 &\to H^0(C_I, \mathbb{G}_{\mathrm{m}}) \to H^0(\widetilde{C}_I, \mathbb{G}_{\mathrm{m}}) \to H^0(C_I, \pi_{I*}\mathbb{G}_{\mathrm{m}}/\mathbb{G}_{\mathrm{m}})\\ &\to H^1(C_I, \mathbb{G}_{\mathrm{m}}) \to H^1(\widetilde{C}_I, \mathbb{G}_{\mathrm{m}}) \to 1\end{aligned}$$

dont la flèche de co-bord identifie les k-schémas en groupes

$$G_I^0 = H^0(C_I, \pi_{I*}\mathbb{G}_{\mathrm{m}}/\mathbb{G}_{\mathrm{m}}) \quad \text{et} \quad P_I^0 = \mathrm{Ker}(H^1(C_I, \mathbb{G}_{\mathrm{m}}) \to H^1(\widetilde{C}_I, \mathbb{G}_{\mathrm{m}}))$$

puisque $\widetilde{C}_I$ est une droite projective sur k. On prolonge cette identification en un k-épimorphisme de k-schémas en groupes

$$G_I \twoheadrightarrow P_I$$

en envoyant $x \in E_I^\times/A_I^\times$ sur le $\mathcal{O}_{C_I}$-Module inversible $\mathcal{L}$ obtenu en recollant $\mathcal{O}_{C_I\setminus\{c_I\}}$ et A_I le long de $\mathrm{Spec}(E_I) = (C_I \setminus \{c_I\}) \times_{C_I} \mathrm{Spec}(A_I)$ à l'aide de la multiplication par x. Cet épimorphisme n'est autre que la flèche $H^1_{\{c_I\}}(C_I, \mathbb{G}_{\mathrm{m}}) \to H^1(C_I, \mathbb{G}_{\mathrm{m}})$ qui s'insère dans la suite exacte longue

$$\begin{aligned}(1) = H^0_{\{c_I\}}(C_I, \mathbb{G}_{\mathrm{m}}) &\to H^0(C_I, \mathbb{G}_{\mathrm{m}}) \to H^0(C_I \setminus \{c_I\}, \mathbb{G}_{\mathrm{m}})\\ \to H^1_{\{c_I\}}(C_I, \mathbb{G}_{\mathrm{m}}) &\to H^1(C_I, \mathbb{G}_{\mathrm{m}}) \to H^1(C_I \setminus \{c_I\}, \mathbb{G}_{\mathrm{m}}) = (1)\end{aligned}$$

et son noyau est donc le groupe discret

$$\begin{array}{ccc} H^0(C_I \setminus \{c_I\}, \mathbb{G}_{\mathrm{m}})/H^0(C_I, \mathbb{G}_{\mathrm{m}}) & \subset & E_I^\times/A_I^\times \\ \| & & \downarrow \\ H^0(\widetilde{C}_I \setminus \pi_I^{-1}(c_I), \mathbb{G}_{\mathrm{m}})/H^0(\widetilde{C}_I, \mathbb{G}_{\mathrm{m}}) & \subset & E_I^\times/\mathcal{O}_{E_I}^\times \end{array}$$

des diviseurs de degré 0 sur $\widetilde{C}_I$ qui sont supportés par le fermé réduit $\pi_I^{-1}(c_I)_{\mathrm{red}} = \{\widetilde{c}_i \mid i \in I\}$, groupe que l'on identifie à Λ_I^0 par la flèche $\lambda \mapsto \sum_{i\in I} \lambda_i[\widetilde{c}_i]$. Compte tenu de cette identification, le plongement

$$H^0(C_I \setminus \{c_I\}, \mathbb{G}_{\mathrm{m}})/H^0(C_I, \mathbb{G}_{\mathrm{m}}) \hookrightarrow E_I^\times/A_I^\times = G_I(k)$$

définit un scindage $\sigma : \Lambda_I^0 \to G_I$ de l'extension $1 \to G_I^0 \to G_I \to \Lambda_I \to 0$ au-dessus de $\Lambda_I^0 \subset \Lambda_I$.

On prolonge $G_I \twoheadrightarrow P_I$ en le morphisme de k-schémas

$$X_I \to \overline{P}_I$$

qui envoie $M \subset E$ sur le $\mathcal{O}_{C_I}$-Module sans torsion $\mathcal{M}$ de rang générique 1 obtenu en recollant $\mathcal{O}_{C_I\setminus\{c_I\}}$ et M le long de $\mathrm{Spec}(E_I) = (C_I \setminus \{c_I\}) \times_{C_I} \mathrm{Spec}(A_I)$. Pour chaque entier d, ce morphisme envoie la composante connexe X_I^d dans la composante connexe $\overline{P}_I^d$. Il est G_I-équivariant pour l'action naturelle de G_I sur X_I et l'action de

G_I sur P_I induite par celle de P_I sur $\overline{P}_I$ et l'épimorphisme $G_I \twoheadrightarrow P_I$ ci-dessus, et il passe au quotient en un morphisme de k-schémas projectifs

$$Z_I = X_I/\sigma(\Lambda_I^0) \to \overline{P}_I$$

qui envoie, pour chaque entier d, la composante connexe Z_I^d dans la composante connexe $\overline{P}_I^d$. Le morphisme $Z_I \to \overline{P}_I$ est birationnel puisqu'il induit un isomorphisme de l'ouvert $Z_I^\circ = X_I^\circ/\Lambda_I^0 \subset Z_I$ sur l'ouvert $P_I \subset \overline{P}_I$.

Proposition 2.3.1. *Le k-morphisme birationnel $Z_I \to \overline{P}_I$ ci-dessus est un homéomorphisme universel, c'est-à-dire est fini, radiciel et surjectif.*

Démonstration. Il suffit de voir que les fibres géométriques du morphisme $Z_I \to \overline{P}_I$ sont toutes réduites à un point, avec éventuellement des nilpotents. Or tout $\mathcal{O}_{C_I}$-Module $\mathcal{M}$ sans torsion de rang générique 1 s'obtient par recollement de $\mathcal{O}_{C_I \setminus \{c_I\}}$ et d'un A_I-réseau $M \subset E_I$, le couple formé de M et de la donnée de recollement étant uniquement déterminé modulo l'action de

$$\sigma(\Lambda_I^0) = H^0(C_I \setminus \{c_I\}, \mathbb{G}_{\mathrm{m}})/H^0(C_I, \mathbb{G}_{\mathrm{m}}). \qquad \square$$

Compte tenu du Théorème 2.2.1, on déduit de cette proposition le suivant.

Corollaire 2.3.1. *Le k-schéma Z_I est irréductible et la G_I-orbite X_I° de $M = A_I$ est dense dans X_I. En particulier on a* $\dim_k(Z_I) = \dim_k(X_I) = \delta_I$. □

Bezrukavnikov [Be] a donné une formule très générale pour la dimension des fibres de Springer, formule qui contient bien entendu l'égalité $\dim(X_I) = \delta_I$. Par contre, sa méthode ne permet pas de démontrer que X_I n'admet pas de composantes irréductibles de dimension $< \delta_I$.

Corollaire 2.3.2. *Le revêtement étale galoisien $X_I \to Z_I$ de groupe de Galois $\Lambda_I^0 \cong \sigma(\Lambda_I^0)$ provient par le changement de base $Z_I \to \overline{P}_I$ d'un revêtement étale galoisien*

$$\overline{\varphi}_I : \overline{P}_I^\natural \to \overline{P}_I$$

dont la description au niveau des k'-points est la suivantes : $\overline{P}_I^\natural(k')$ est l'ensemble des couples $(\mathcal{M}, \iota)$, où $\mathcal{M}$ est un $\mathcal{O}_{k'\otimes_k C_I}$-Module sans torsion de rang générique 1 *et $\iota : \mathcal{M}_{|k'\otimes_k(C_I\setminus\{c_I\})} \xrightarrow{\sim} \mathcal{O}_{k'\otimes_k(C_I\setminus\{c_I\})}$ est une trivialisation de la restriction de $\mathcal{M}$ à $k' \otimes_k (C_I \setminus \{c_I\})$, et $\overline{\varphi}_I$ est le morphisme d'oubli de ι.* □

On notera encore $P_I^\natural = G_I$ et $\varphi_I : P_I^\natural \twoheadrightarrow P_I$ le k-épimorphisme défini plus haut.

Exemples 2.3.1. Pour $|I| = 1$ et $p_I(T) = T^2 - \varpi_F^3$, C_I est la cubique n'ayant pour seule singularité qu'un cusp ordinaire et l'homéomorphisme $Z_I^0 \to \overline{P}_I^0$ est naturellement isomorphe au morphisme de normalisation $\pi_I : \widetilde{C}_I \to C_I$.

Pour $I = \{1, 2\}$, $p_1(T) = T - \varpi_F$ et $p_2(T) = T + \varpi_F$, C_I est la cubique n'ayant pour seule singularité qu'un point double ordinaire, $\overline{\varphi}_I : Z_I \to \overline{P}_I$ est un isomorphisme et la composante de degré 0 du revêtement $\overline{\varphi}_I : \overline{P}_I^\natural \to \overline{P}_I$ est naturellement isomorphe au revêtement $C_I^\natural \to C_I$ de groupe de Galois $\mathbb{Z}$ dont l'espace total est la chaîne de droites projective indexée par $\mathbb{Z}$ obtenue en prenant $\mathbb{Z}$ copies de la droite projective standard sur k et en identifiant le point à l'infini de la n-ème copie à l'origine de la $(n + 1)$-ème. □

2.4 Généralisation

On peut généraliser la proposition 2.3.1 comme suit. Soit C une courbe géométriquement intègre et projective sur k n'ayant que des singularités planes. Soit $\{c_j \mid j \in J\}$ l'ensemble fini des points singuliers de C et, pour chaque $j \in J$, soit $\pi_C^{-1}(c_j) = \{\widetilde{c}_i \mid i \in I_j\}$ l'ensemble des branches de C en c_j. Pour chaque $j \in J$, notons A_{I_j} le complété de l'anneau local de C en c_j.

Pour simplifier, on suppose que tous les points c_j de C et tous les points $\widetilde{c}_i$ de $\widetilde{C}$ sont rationnels sur k.

Pour chaque $j \in J$ on a la fibre de Springer affine X_{I_j} qui paramètre les idéaux fractionnaires de A_{I_j}, c'est-à-dire les sous-A_{I_j}-modules M de l'anneau total des fractions E_{I_j} de A_{I_j} tels que $E_{I_j}M = E_{I_j}$. Cette fibre de Springer est munie d'une action du k-schéma en groupes G_{I_j} «défini» par $G_{I_j}(k) = E_{I_j}^{\times}/A_{I_j}^{\times}$, action dont on a vu qu'elle admet une orbite dense.

On note

$$X(C) = \prod_{j\in J} X_{I_j},\ G(C) = \prod_{j\in J} G_{I_j},\ I = \coprod_{j\in J} I_j \text{ et } \Lambda(C) = \mathbb{Z}^I.$$

On a un dévissage

$$1 \to G^0(C) \to G(C) \to \Lambda(C) \to 0$$

qui au niveau des k-points n'est autre que le dévissage

$$1 \to \prod_{j\in J} \mathcal{O}_{E_{I_j}}^{\times}/A_{I_j}^{\times} \to \prod_{j\in J} E_{I_j}^{\times}/A_{I_j}^{\times} \to \prod_{j\in J} \mathbb{Z}^{I_j} \to 0.$$

On a de plus une suite exacte de schémas en groupes en groupes commutatifs connexes et de type fini sur k

$$1 \to G^0(C) \to P^0(C) \to P^0(\widetilde{C}) \to 1$$

où $G^0(C)$ est affine, et donc produit d'un tore par un groupe unipotent, et où $P^0(\widetilde{C})$ est un k-schéma abélien de dimension $g(C)$.

On forme le k-schéma de Picard compactifié $\overline{P}(C)$ de C et on considère le k-morphisme

$$X(C) \to \overline{P}(C)$$

qui envoie $(M_j \subset E_{I_j})_{j\in J}$ sur le $\mathcal{O}_C$-Module $\mathcal{M}$ sans torsion de rang générique 1 obtenu en recollant $\mathcal{O}_{C\setminus\{c_j \mid j\in J\}}$ et les M_j. Compte tenu de l'action de $P^0(C) \subset P(C)$ sur $\overline{P}(C)$ on en déduit un morphisme $P^0(C) \times_k X(C) \to \overline{P}(C)$ qui passe au quotient en un morphisme

$$[P^0(C) \times_k X(C)]/G^0(C) \to \overline{P}(C)$$

où $G^0(C)$ agit librement sur $P^0(C) \times_k X(C)$ par $g \cdot (p, x) = (pg^{-1}, g \cdot x)$. On remarque que l'on a par construction une «fibration»

$$[P^0(C) \times_k X(C)]/G^0(C) \to P^0(\widetilde{C})$$

de fibre type $X(C)$.

Considérons le groupe discret

$$H^0(C \setminus \{c_j \mid j \in J\}, \mathbb{G}_{\mathrm{m}})/H^0(C, \mathbb{G}_{\mathrm{m}}) \subset \prod_{j \in J} E_{I_j}^\times / A_{I_j}^\times = G(C)(k)$$

qui est l'image d'une section σ de $G(C) \twoheadrightarrow \Lambda(C)$ au-dessus de $\Lambda^0(C) = \mathrm{Ker}(\mathbb{Z}^I \to \mathbb{Z})$. Ce groupe discret agit librement sur $X(C)$ et le k-morphisme $[P^0(C) \times_k X(C)]/G^0(C) \to \overline{P}(C)$ passe au quotient en un k-morphisme

$$[P^0(C) \times_k (X(C)/\sigma(\Lambda^0(C)))]/G^0(C) \to \overline{P}(C).$$

On remarque que l'on a encore une «fibration»

$$[P^0(C) \times_k (X(C)/\sigma(\Lambda^0(C)))]/G^0(C) \to P^0(\widetilde{C})$$

de fibre type $Z(C) = X(C)/\sigma(\Lambda^0(C))$.

Proposition 2.4.1. *Le k-morphisme $[P^0(C) \times_k (X(C)/\sigma(\Lambda^0(C)))]/G^0(C) \to \overline{P}(C)$ défini ci-dessus est un homéomorphisme universel.* □

Corollaire 2.4.1. *Le revêtement étale galoisien*

$$[P^0(C) \times_k X(C)]/G^0(C) \to [P^0(C) \times_k (X(C)/\sigma(\Lambda^0(C)))]/G^0(C)$$

de groupe de Galois $\Lambda^0(C) \cong \sigma(\Lambda^0(C))$ provient par le changement de base $[P^0(C) \times_k (X(C)/\sigma(\Lambda^0(C)))]/G^0(C) \to \overline{P}(C)$ d'un revêtement étale galoisien

$$\overline{\varphi} : \overline{P}^\natural(C) \to \overline{P}(C).$$ □

2.5 Auto-dualité des jacobiennes compactifiées d'après Esteves, Gagné, et Kleiman

Dans [E-G-K], Esteves, Gagné et Kleiman ont démontré un théorème d'auto-dualité pour les jacobiennes compactifiées des courbes projectives et intègres dont toutes les singularités sont planes et de multiplicité 2. Ce théorème généralise l'énoncé classique d'auto-dualité des jacobiennes des courbes projectives et lisses.

Nous n'aurons besoin que de la partie «facile» de ce théorème, partie qui vaut en fait sans l'hypothèse restrictive de multiplicité 2 et que nous allons rappeler maintenant.

On se place dans la situation naturelle pour ce résultat. Soient donc S un schéma nœthérien et $C \to S$ un morphisme projectif et plat, dont toutes les fibres géométriques sont intègres et de dimension 1. Pour simplifier, on supposera qu'il existe une section globale de C sur S dont l'image est contenue dans le lieu de lissité de C sur S et on fixera une fois pour toute une telle section $\infty : S \to C$. On notera $[\infty]$ le diviseur de Cartier relatif sur C image de cette section.

Pour chaque entier d, soit $P^d = \mathrm{Pic}^d_{C/S}$ la composante du S-schéma de Picard de la courbe C/S qui paramètre les classes d'isomorphie de Modules inversibles sur

C qui sont de degré d fibre à fibre du morphisme $C \to S$, et soit $\overline{P}^d = \overline{\mathrm{Pic}}^d_{C/S}$ la compactification relative de P^d qui paramètre les Modules cohérents sur C qui sont plats sur S et, fibre à fibre, sans torsion, de rang générique 1 et de degré d. La torsion $\mathcal{M} \mapsto \mathcal{M}(-(d+1)[\infty])$ par le diviseur de Cartier $-(d+1)[\infty]$ identifie P^d et $\overline{P}^d$ à P^{-1} et $\overline{P}^{-1}$ respectivement.

On supposera que, pour un entier d ou ce qui revient au même pour tout entier d, le foncteur de Picard de $\overline{P}^d/S$ est représentable par un S-schéma

$$\mathrm{Pic}_{\overline{P}^d/S}$$

qui est une réunion disjointe de S-schémas quasi-projectifs. C'est le cas d'après Grothendieck (cf. [Gr 1, Théorème 3.1] et aussi [B-L-R, Section 8.2, Theorem 1]) si toutes les fibres de C/S ont au pire des singularités planes, car le S-schéma $\overline{P}^d$ est alors projectif, plat et à fibres géométriques intègres d'après Altman, Iarrobino et Kleiman, et Rego (cf. [A-I-K, Re] et notre Section 2.1). C'est aussi le cas si S est le spectre d'un corps d'après Murre et Oort (cf. [B-L-R, Section 8.2, Theorem 3]).

Si S est le spectre d'un corps k, le k-schéma en groupes $\mathrm{Pic}_{\overline{P}^d/k}$ admet une composante neutre $\mathrm{Pic}^0_{\overline{P}^d/k}$ et on définit

$$\mathrm{Pic}^{\tau}_{\overline{P}^d/k} = \bigcup_{n>0} [n]^{-1} \,\mathrm{Pic}^0_{\overline{P}^d/k},$$

où $[n] : \mathrm{Pic}_{\overline{P}^d/k} \to \mathrm{Pic}_{\overline{P}^d/k}$ est la multiplication par l'entier n. Pour S arbitraire, on note $\mathrm{Pic}^0_{\overline{P}^d/S}$ (resp., $\mathrm{Pic}^{\tau}_{\overline{P}^d/S}$) le sous-foncteur de $\mathrm{Pic}_{\overline{P}^d/S}$ formé des classes d'isomorphie de Modules inversibles sur $\overline{P}^d$ dont la restriction à chaque fibre $\overline{P}^d_s$ de $\overline{P}^d \to S$ est dans $\mathrm{Pic}^0_{\overline{P}^d_s/\kappa(s)}$ (resp., $\mathrm{Pic}^{\tau}_{\overline{P}^d_s/\kappa(s)}$). Le sous-foncteur $\mathrm{Pic}^{\tau}_{\overline{P}^d/S}$ est représentable par une partie ouverte et fermée de $\mathrm{Pic}_{\overline{P}^d/S}$ (cf. [B-L-R, Section 8.4, Theorem 4]) ; par contre, le sous-foncteur $\mathrm{Pic}^0_{\overline{P}^d/S}$ n'est pas représentable en général.

On a l'application d'Abel

$$A_{-1} : C \to \overline{P}^{-1}$$

définie par l'Idéal de la diagonale $C \subset C \times_S C$: la première projection $C \times_S C \to C$ est un changement de base de $C \to S$ et cet Idéal est un $\mathcal{O}_{C\times_S C}$-Module cohérent qui est plat sur C et, fibre à fibre, sans torsion, de rang générique 1 et de degré -1. Pour tout entier d, on définit

$$A_d : C \to \overline{P}^d$$

comme le composé de A_{-1} et de l'isomorphisme $\overline{P}^{-1} \xrightarrow{\sim} \overline{P}^d$ de torsion par $(d+1)[\infty]$. Le morphisme A_d induit un homomorphisme de S-schémas en groupes

$$A_d^* : \mathrm{Pic}_{\overline{P}^d/S} \to \mathrm{Pic}_{C/S}\,.$$

Dans [E-G-K], Esteves, Gagné et Kleiman construisent un inverse à droite de l'homomorphisme A_d^* sur $P^0 = \mathrm{Pic}^0_{C/S} \subset \mathrm{Pic}_{C/S}$ en utilisant le déterminant de la cohomologie.

Plus précisément, notons

$$\begin{array}{ccccc} C\times_S \overline{P} & \xleftarrow{\mathrm{pr}_{12}} & C\times_S \overline{P}\times_S P^0 & \xrightarrow{\mathrm{pr}_{13}} & C\times_S P^0 \\ \downarrow{\scriptstyle \mathrm{pr}_2} & & \downarrow{\scriptstyle \mathrm{pr}_{23}} & & \downarrow{\scriptstyle \mathrm{pr}_2} \\ \overline{P} & \xleftarrow[\mathrm{pr}_1]{} & \overline{P}\times_S P^0 & \xrightarrow[\mathrm{pr}_2]{} & P^0 \end{array}$$

les projections canoniques et $\mathcal{L}^{\mathrm{univ}}$ et $\mathcal{M}^{\mathrm{univ}}$ les Modules universels sur $C\times_S P^0$ et $C\times_S \overline{P}$ respectivement, rigidifiés le long des sections $P^0\to C\times_S P^0$ et $\overline{P}\to C\times_S\overline{P}$ induites par la section $\infty : S\to C$. Alors, on peut former le $\mathcal{O}_{\overline{P}\times_S P^0}$-Module inversible

$$(\det R\,\mathrm{pr}_{23,*}(\mathrm{pr}_{12}^*\,\mathcal{M}^{\mathrm{univ}}\otimes \mathrm{pr}_{13}^*\,\mathcal{L}^{\mathrm{univ}}))^{\otimes -1}\otimes \det R\,\mathrm{pr}_{23,*}\,\mathrm{pr}_{12}^*\,\mathcal{M}^{\mathrm{univ}}$$

sur $\overline{P}\times_S P^0$, Module inversible qui définit un morphisme

$$\beta=\prod_{d\in\mathbb{Z}}\beta_d : P^0\to \mathrm{Pic}_{\overline{P}/S}=\prod_{d\in\mathbb{Z}}\mathrm{Pic}_{\overline{P}^d/S}\,.$$

Proposition 2.5.1 (Esteves, Gagné, et Kleiman [E-G-K, Proposition 2.2]). *Pour chaque entier d, le morphisme $\beta_d : P^0\to \mathrm{Pic}_{\overline{P}^d/S}$ est un homomorphisme de S-schémas en groupes dont l'image ensembliste est contenue dans le sous-foncteur $\mathrm{Pic}^0_{\overline{P}^d/S}$, et donc dans l'ouvert et fermé $\mathrm{Pic}^{\tau}_{\overline{P}^d/S}$, et la formation de β_d commute à tout changement de base $S'\to S$. De plus, le composé $A_d^*\circ\beta_d$ est l'identité de P^0.* □

Remarque 2.5.1. Si on note

$$\mu : \overline{P}\times_S P^0\to\overline{P},\ (\mathcal{M},\mathcal{L})\mapsto \mathcal{M}\otimes_{\mathcal{O}_C}\mathcal{L},$$

l'action naturelle de P^0 sur $\overline{P}$, on a le carré cartésien

$$\begin{array}{ccc} C\times_S\overline{P}\times_S P^0 & \xrightarrow{\mathrm{Id}_C\times\mu} & C\times_S\overline{P} \\ \downarrow{\scriptstyle \mathrm{pr}_{23}} & \square & \downarrow{\scriptstyle \mathrm{pr}_2} \\ \overline{P}\times_S P^0 & \xrightarrow[\mu]{} & \overline{P} \end{array}$$

et on a un isomorphisme canonique

$$\mathrm{pr}_{12}^*\,\mathcal{M}^{\mathrm{univ}}\otimes\mathrm{pr}_{13}^*\,\mathcal{L}^{\mathrm{univ}}\cong(\mathrm{Id}_C\times\mu)^*\mathcal{M}^{\mathrm{univ}}.$$

Le théorème de changement de base assure alors que

$$R\,\mathrm{pr}_{23,*}(\mathrm{pr}_{12}^*\,\mathcal{M}^{\mathrm{univ}}\otimes\mathrm{pr}_{13}^*\,\mathcal{L}^{\mathrm{univ}})\cong\mu^*R\,\mathrm{pr}_{2,*}\,\mathcal{M}^{\mathrm{univ}}$$

et le morphisme β d'Esteves, Gagné et Kleiman est encore défini par le Module inversible

$$(\mu^*\det R\,\mathrm{pr}_{2,*}\,\mathcal{M}^{\mathrm{univ}})^{\otimes -1}\otimes\det R\,\mathrm{pr}_{2,*}\,\mathcal{M}^{\mathrm{univ}}.$$ □

2.6 Schémas de Picard et dualité pour les tores

Nous aurons besoin dans la section suivante d'un résultat général de Grothendieck qui nous a été communiqué par Raynaud.

Soient $f : Z \to S$ un morphisme propre, plat et de présentation finie de schémas tel que $f_*\mathcal{O}_Z = \mathcal{O}_S$, et T un S-tore (plat et de présentation finie) de faisceau des caractères $X^*(T) = \mathcal{H}om_{S\text{-sch.gr.}}(T, \mathbb{G}_{\mathrm{m},S})$. Pour chaque changement de base

$$\begin{array}{ccc} Z' & \longrightarrow & Z \\ {\scriptstyle f'}\downarrow & \square & \downarrow{\scriptstyle f} \\ S' & \longrightarrow & S \end{array},$$

chaque section globale t' du S'-tore $T' = S' \times_S T$ et chaque $f'^*X^*(T')$-torseur Y' sur Z', on note $\langle Y', t'\rangle$ le $\mathbb{G}_{\mathrm{m},Z'}$-torseur sur Z' obtenu en poussant Y' par le caractère image réciproque par f' du caractère $X^*(T') \to \mathbb{G}_{\mathrm{m},S'}$, $\chi' \mapsto \chi'(t')$.

Proposition 2.6.1 (cf. [Ra, Proposition 6.2.1]). *Notons*

$$\mathrm{Pic}_{Z/S} = R^1 f_*\mathbb{G}_{\mathrm{m},Z}$$

le foncteur de Picard relatif. Alors, l'homomorphisme de faisceaux étales sur S

$$R^1 f_* f^* X^*(T) \to \mathcal{H}om_{S\text{-sch.gr.}}(T, \mathrm{Pic}_{Z/S})$$

qui, quel que soit le S-schéma S', envoie la classe d'un $f'^*X^*(T')$-torseur $Y' \to Z'$ sur l'homomorphisme $\langle Y', \cdot\rangle : T' \to \mathrm{Pic}_{Z'/S'}$, est un isomorphisme.

Démonstration. On raisonne comme le fait Raynaud pour prouver [Ra, Proposition 6.2.1]. Pour tout faisceau fppf en groupes commutatifs $\mathcal{F}$ sur S, on considère le foncteur

$$\mathcal{G} \mapsto H(\mathcal{G}) = f_*\mathcal{H}om(f^*\mathcal{F}, \mathcal{G})$$

sur la catégorie des faisceaux fppf en groupes commutatifs sur Z. On a deux suites spectrales

$$E_2^{pq} = R^q f_*\,\mathcal{E}xt^p(f^*\mathcal{F}, \mathcal{G}) \Rightarrow R^{p+q}H(\mathcal{G})$$

et

$$E_2^{pq} = \mathcal{E}xt^q(\mathcal{F}, R^p f_*\mathcal{G}) \Rightarrow R^{p+q}H(\mathcal{G})$$

et donc deux suites exactes courtes des termes de bas degrés, qui s'écrivent pour $\mathcal{F} = T$ et $\mathcal{G} = \mathbb{G}_{\mathrm{m},Z}$,

$$0 \to R^1 f_* f^* X^*(T) \to R^1 H(\mathbb{G}_{\mathrm{m},Z}) \to f_*\,\mathcal{E}xt^1(f^*T, \mathbb{G}_{\mathrm{m},Z})$$

et

$$\begin{aligned} 0 \to \mathcal{E}xt^1(T, f_*\mathbb{G}_{\mathrm{m},Z}) &\to R^1 H(\mathbb{G}_{\mathrm{m},Z}) \to \mathcal{H}om(T, R^1 f_*\mathbb{G}_{\mathrm{m},Z}) \\ &\to \mathcal{E}xt^2(T, f_*\mathbb{G}_{\mathrm{m},Z}) \to R^2 H(\mathbb{G}_{\mathrm{m},Z}). \end{aligned}$$

L'homomorphisme de la proposition est alors l'homomorphisme composé

$$R^1 f_* f^* X^*(T) \to R^1 H(\mathbb{G}_{\mathrm{m},Z}) \to \mathcal{H}om(T, R^1 f_* \mathbb{G}_{\mathrm{m},Z}).$$

On a $f_*\mathbb{G}_{\mathrm{m},Z} = \mathbb{G}_{\mathrm{m},S}$ par hypothèse et on sait que $\mathcal{E}xt^1(f^*T, \mathbb{G}_{\mathrm{m},Z}) = (0)$ et $\mathcal{E}xt^1(T, \mathbb{G}_{\mathrm{m},S}) = (0)$. Par suite, l'homomorphisme composé ci-dessus est injectif et, pour démontrer qu'il est bijectif, il suffit de vérifier que l'application $\mathcal{E}xt^2(T, f_*\mathbb{G}_{\mathrm{m},Z}) \to R^2 H(\mathbb{G}_{\mathrm{m},Z})$ est injective. Comme cette propriété d'injectivité est locale pour la topologie fppf sur S, on peut supposer que f admet une section globale $z : S \to Z$, et donc que l'on a une flèche $\iota : f_*\mathcal{G} \to z^*\mathcal{G}$ fonctorielle en $\mathcal{G}$. Or, la flèche ι induit un morphisme $R^2 H(\mathcal{G}) \to \mathcal{E}xt^2(\mathcal{F}, z^*\mathcal{G})$ tel que l'homomorphisme composé $\mathcal{E}xt^2(\mathcal{F}, f_*\mathcal{G}) \to R^2 H(\mathcal{G}) \to \mathcal{E}xt^2(\mathcal{F}, z^*\mathcal{G})$ soit égal à $\mathcal{E}xt^2(\mathcal{F}, \iota)$, et elle est un isomorphisme pour $\mathcal{G} = \mathbb{G}_{\mathrm{m},Z}$ par hypothèse. Notre assertion d'injectivité est donc vérifiée et la proposition est démontrée. □

2.7 Auto-dualité des jacobiennes compactifiées et fibres de Springer

Revenons maintenant à la situation de la section 2.1. Nous avons donc la courbe intègre et projective $C = C_I$ sur k, de normalisée $\pi = \pi_I : \widetilde{C} = \widetilde{C}_I \to C$ une droite projective, avec son unique point singulier $c = c_I$ en lequel la singularité est plane. Le complété de l'anneau local de C en c est notre anneau $A = A_I$ de la section 1.1 et l'ensemble des branches $\pi^{-1}(c) = \{\widetilde{c}_i \mid i \in I\}$ est indexé par I.

On considère le k-schéma de Picard compactifié $\overline{P} = \overline{P}_I$ de C et son revêtement étale galoisien

$$\overline{P}^\natural \to \overline{P},$$

de groupe de Galois

$$\begin{aligned}\Lambda^0 &= H^0(C \setminus \{c\}, \mathbb{G}_{\mathrm{m}})/H^0(C, \mathbb{G}_{\mathrm{m}}) \\ &= H^0(\widetilde{C} \setminus \{\widetilde{c}_i \mid i \in I\}, \mathbb{G}_{\mathrm{m}})/H^0(\widetilde{C}, \mathbb{G}_{\mathrm{m}}) = \mathrm{Ker}(\mathbb{Z}^I \to \mathbb{Z}),\end{aligned}$$

défini dans la section 2.3. Rappelons que les k-points de $\overline{P}^\natural$ sont les couples $(\mathcal{M}, \iota)$ où $\mathcal{M}$ est un $\mathcal{O}_C$-Module sans torsion de rang générique 1 et $\iota : \mathcal{M}_{|C\setminus\{c\}} \xrightarrow{\sim} \mathcal{O}_{C\setminus\{c\}}$ est une rigidification de $\mathcal{M}$ sur le lieu de lissité $C \setminus \{c\}$ de C, et que la flèche du revêtement est l'oubli de la rigidification.

Par construction, la restriction du revêtement $\overline{P}^\natural \to \overline{P}$ par le morphisme radiciel

$$Z = Z_I \to \overline{P}$$

défini dans la section 2.3 est le revêtement

$$X \to X/\Lambda^0 = Z$$

où $X = X_I$ est la fibre de Springer de la première partie.

Nous allons utiliser les résultats généraux des sections 2.3 et 2.6 pour donner une autre définition du revêtement $\overline{P}^\natural \to \overline{P}$.

Soit T le tore maximal de la composante neutre P^0 du k-schéma de Picard de C. On a $T = \mathbb{G}_{\mathrm{m},k}^I/\mathbb{G}_{\mathrm{m},k}$ et le groupe des caractères de T est naturellement isomorphe au noyau

$$\mathrm{Ker}(\mathbb{Z}^I \to \mathbb{Z})$$

de l'homomorphisme somme, avec pour accouplement

$$\mathrm{Ker}(\mathbb{Z}^I \to \mathbb{Z}) \times T \to \mathbb{G}_{\mathrm{m},k}, \ (\lambda, t) \mapsto t^\lambda = \prod_{i\in I} t_i^{\lambda_i}.$$

L'homomorphisme de k-schémas en groupes

$$T \to \mathrm{Pic}_{\overline{P}/k}$$

obtenu en composant l'inclusion $T \subset P^0$ et l'homomorphisme canonique β d'Esteves, Gagné, et Kleiman (cf. Section 2.3), définit donc d'après la Proposition 2.6.1 un Λ^0-torseur $\overline{P}^\dagger \to \overline{P}$.

Proposition 2.7.1. *Le Λ^0-torseur $\overline{P}^\dagger \to \overline{P}$ n'est autre que le revêtement $\overline{P}^\natural \to \overline{P}$.*

Démonstration. Il s'agit de démontrer que, pour chaque entier d, la restriction à Z^d du Λ^0-torseur $\overline{P}^\dagger \to \overline{P}$ est isomorphe à $X^d \to Z^d = X^d/\Lambda^0$ et, bien sûr, il suffit de le faire pour $d = 0$.

Si l'on interprète le k-schéma de Picard $\mathrm{Pic}_{Z^0/k}$ de Z^0 comme le k-schéma de Picard Λ^0-équivariant $\mathrm{Pic}_{X^0/k}^{\Lambda^0}$ de X^0, le Λ^0-torseur $X^0 \to Z^0$ correspond de manière tautologique par la Proposition 2.6.1 au k-homomorphisme

$$T \to \mathrm{Pic}_{X^0/k}^{\Lambda^0} = \mathrm{Pic}_{Z^0/k}$$

qui envoie $t \in T(k)$ sur le $\mathcal{O}_{X^0}$-Module inversible trivial muni de l'action de Λ^0 donnée par le caractère

$$\chi_t : \Lambda^0 \to \mathbb{G}_{\mathrm{m},k}, \ \lambda \mapsto t^\lambda,$$

et on veut montrer que ce k-homomorphisme coïncide avec le k-homomorphisme composé

$$T \subset P^0 \xrightarrow{\beta_0} \mathrm{Pic}_{\overline{P}^0/k} \to \mathrm{Pic}_{Z^0/k}$$

où la dernière flèche est la flèche de restriction par le morphisme radiciel $Z^0 \to \overline{P}^0$. Il suffit donc de démontrer que, pour tout $t \in T(k)$, la restriction à Z^0 du $\mathcal{O}_{\overline{P}^0}$-Module inversible $\beta_0(t)$ est le $\mathcal{O}_{X^0}$-Module inversible trivial muni de l'action de Λ^0 donnée par le caractère χ_t.

Rappelons que l'anneau total des fractions de A est $E = E_I = \prod_{i\in I} E_i$, que $\widetilde{C}$ est la droite projective standard $\mathbb{P}_k^1$, que le point ∞ de $\widetilde{C}$ est distinct des $\widetilde{c}_i$, que pour chaque $i \in I$, on peut prendre pour uniformisante $\varpi_i = \varpi_{E_i}$ de E_i la fonction $x - \widetilde{c}_i$ où x est une coordonnée affine sur $\widetilde{C} \setminus \{\infty\} \cong \mathbb{A}_k^1$, que $\widetilde{A} = \prod_{i\in I} k[[\varpi_i]] \subset \prod_{i\in I} k((\varpi_i)) \cong E_I$ et que Λ^0 est l'ensemble des fractions rationnelles de la forme $\prod_{i\in I}(x - \widetilde{c}_i)^{\lambda_i}$ avec $\lambda \in \mathbb{Z}^I$ et $\sum_{i\in I} \lambda_i = 0$.

Rappelons de plus que X^0 est le k-schéma des A-réseaux $M \subset E$ d'indice 0 relativement à $A \subset E$, que l'image $\mathcal{M} \in \overline{P}^0(k)$ de $M \in X^0(k)$ est le $\mathcal{O}_C$-Module obtenu en recollant $\mathcal{O}_{C\setminus\{c\}}$ et M, et que l'action du groupe de Galois Λ^0 est donnée par

$$(\lambda, M) \mapsto \lambda \cdot M = (\varpi_i^{\lambda_i})_{i\in I} M.$$

Rappelons enfin que le k-schéma X^0 est localement de type fini, réunion d'une suite croissante

$$X_0^0 \subset \cdots \subset X_n^0 \subset X_{n+1}^0 \subset \cdots \subset X^0$$

de fermés de type fini stables sous l'action par translation de $P^0(k) = \widetilde{A}^\times / A^\times$; par exemple, on peut prendre pour X_n^0 le fermé formé des M qui sont contenus dans $(\varpi_i^{-n})_{i\in I} \widetilde{A} \subset E$.

Il suffit donc de construire un système compatible de trivialisations des restrictions du Module inversible $\beta_0(t)$ aux X_n^0 et de montrer que, pour tout $M \in X^0(k)$ et tout $\lambda \in \Lambda^0$, l'isomorphisme de la fibre en M de la restriction de $\beta_0(t)$ à X^0 sur celle en $\lambda \cdot M$ donné par l'action de λ s'exprime dans ce système de trivialisations comme la multiplication par t^λ.

La fibre de $\beta_0(t)$ en l'image $\mathcal{M} \in \overline{P}^0(k)$ de $M \in X_n^0(k)$ peut se calculer comme suit. Soit $R = H^0(C \setminus \{c\}, \mathcal{O}_C) \subset E$ la k-algèbre des fonctions rationnelles sur $\widetilde{C}$ qui sont régulières en dehors de $\{\widetilde{c}_i \mid i \in I\}$. On a une suite exacte

$$0 \to H^0(C, \mathcal{M}) \to R \to E/M \to H^1(C, \mathcal{M}) \to 0$$

et il existe $N \geq n$, indépendant de $M \in X_n^0(k)$, tel que la restriction de la flèche surjective $E/M \twoheadrightarrow H^1(C, \mathcal{M})$ à $(\varpi_i^{-N})_{i\in I} \widetilde{A}/M \subset E/M$ soit encore surjective. La suite induite

$$0 \to H^0(C, \mathcal{M}) \to R \cap (\varpi_i^{-N})_{i\in I} \widetilde{A} \to (\varpi_i^{-N})_{i\in I} \widetilde{A}/M \to H^1(C, \mathcal{M}) \to 0.$$

est donc aussi exacte.

La classe d'isomorphie fixée $t \in T(k) \subset P^0(k)$ est représentée par le $\mathcal{O}_C$-Module inversible $\mathcal{L}$ obtenu en recollant $\mathcal{O}_{C\setminus\{c\}}$ et le réseau $L = (t_i)_{i\in I} A \subset E$, et on peut remplacer dans les suites exactes ci-dessus $\mathcal{M}$ par $\mathcal{L} \otimes_{\mathcal{O}_C} \mathcal{M}$ et M par $t \cdot M = (t_i)_{i\in I} M \subset (\varpi_i^{-n})_{i\in I} \widetilde{A} \subset E$. Par conséquent, la fibre de $\beta_0(t)$ en $\mathcal{M}$ s'identifie canoniquement à la k-droite

$$\left(\bigwedge^{\max} (\varpi_i^{-N})_{i\in I} \widetilde{A}/t \cdot M\right) \otimes_k \left(\bigwedge^{\max} (\varpi_i^{-N})_{i\in I} \widetilde{A}/M\right)^{\otimes -1},$$

ou ce qui revient au même à la k-droite

$$\left(\bigwedge^{\max} (\varpi_i^{-n})_{i\in I} \widetilde{A}/t \cdot M\right) \otimes_k \left(\bigwedge^{\max} (\varpi_i^{-n})_{i\in I} \widetilde{A}/M\right)^{\otimes -1}.$$

Le déterminant de l'isomorphisme $(\varpi_i^{-n})_{i\in I} \widetilde{A}/M \xrightarrow{\sim} (\varpi_i^{-n})_{i\in I} \widetilde{A}/t \cdot M$ induit par la multiplication par $(t_i)_{i\in I}$ définit un vecteur de base $e_{n,M}$ de cette dernière

k-droite. La section $M \mapsto e_M = \left(\prod_{i\in I} t_i^{-n}\right) e_{n,M}$ de la restriction de $\beta_0(t)$ à X_n^0 est la trivialisation cherchée : elle est «indépendant de n» puisque le déterminant de l'automorphisme de multiplication par $(t_i)_{i\in I}$ sur $(\varpi_i^{-n-1})_{i\in I}\widetilde{A}/(\varpi_i^{-n})_{i\in I}\widetilde{A}$ est égal à $\prod_{i\in I} t_i$.

Maintenant, si M et $\lambda\cdot M$ sont dans $X_n^0(k)$, la multiplication par $(\varpi_i^{-\lambda_i})_{i\in I}$ induit un isomorphisme de la k-droite

$$\left(\bigwedge^{\max}(\varpi_i^{-n})_{i\in I}\widetilde{A}/t\cdot(\lambda\cdot M)\right)\otimes_k\left(\bigwedge^{\max}(\varpi_i^{-n})_{i\in I}\widetilde{A}/\lambda\cdot M\right)^{\otimes -1}$$

sur la k-droite

$$\left(\bigwedge^{\max}(\varpi_i^{-\lambda_i-n})_{i\in I}\widetilde{A}/t\cdot M\right)\otimes_k\left(\bigwedge^{\max}(\varpi_i^{-\lambda_i-n})_{i\in I}\widetilde{A}/M\right)^{\otimes -1}$$

qui envoie le vecteur de base $e_{n,M}$ sur le déterminant e' de l'isomorphisme

$$(\varpi_i^{-\lambda_i-n})_{i\in I}\widetilde{A}/M \xrightarrow{\sim} (\varpi_i^{-\lambda_i-n})_{i\in I}\widetilde{A}/t\cdot M$$

induit par la multiplication par $(t_i)_{i\in I}$. Or cette dernière k-droite est canoniquement isomorphe à la k-droite

$$\left(\bigwedge^{\max}(\varpi_i^{-n})_{i\in I}\widetilde{A}/t\cdot M\right)\otimes_k\left(\bigwedge^{\max}(\varpi_i^{-n})_{i\in I}\widetilde{A}/M\right)^{\otimes -1}$$

par un isomorphisme qui envoie e' sur $t^{-\lambda}e_{n,M}$ (choisir arbitrairement un entier $m \geq \lambda_i + n$ quel que soit $i\in I$ et utiliser les inclusions $(\varpi_i^{-\lambda_i-n})_{i\in I}\widetilde{A}\subset(\varpi_i^{-m})_{i\in I}\widetilde{A}\supset(\varpi_i^{-n})_{i\in I}\widetilde{A}$). On en déduit donc que l'action de $-\lambda$ envoie $e_{\lambda\cdot M}$ sur $t^{-\lambda}e_M$, ce que l'on voulait démontrer. □

Remarque 2.7.1. Pour tout $\widetilde{a}\in\widetilde{A}^\times/A^\times$, la multiplication par $\widetilde{a}$ induit un isomorphisme de $(\varpi_i^{-n})_{i\in I}\widetilde{A}/M$ sur $(\varpi_i^{-n})_{i\in I}\widetilde{A}/\widetilde{a}M$ dont le déterminant ne dépend que du «terme constant» $\widetilde{a}(0)$ de $\widetilde{a}$, c'est-à-dire de l'image de $\widetilde{a}$ par la projection canonique $\widetilde{A}^\times/A^\times \twoheadrightarrow (k^\times)^I/k^\times$. Les mêmes calculs de déterminant de la cohomologie que ceux de la démonstration de la proposition montrent donc que la flèche composée

$$P^0 \xrightarrow{\beta} \mathrm{Pic}_{\overline{P}} \to \mathrm{Pic}_Z$$

est triviale sur la composante unipotente de P^0. □

Remarque 2.7.2. On laisse le soin au lecteur d'étendre la construction ci-dessus pour le revêtement $\overline{\varphi}:\overline{P}^\natural(C)\to\overline{P}(C)$ du corollaire 2.4.1. □

3 La conjecture de pureté

Dans tout ce chapitre, k est un corps fini à q éléments. On fixe une clôture algébrique $\overline{k}$ de k, un nombre premier ℓ distinct de la caractéristique p de k et une clôture algébrique $\overline{\mathbb{Q}}_\ell$ de $\mathbb{Q}_\ell$. Pour tout k-schéma de type fini S on note simplement

$$H_{\mathrm{c}}^{\bullet}(S) := H_{\mathrm{c}}^{\bullet}(\overline{k} \otimes_k S, \overline{\mathbb{Q}}_\ell)$$

la cohomologie ℓ-adique à supports compacts de $\overline{k} \otimes_k S$. Bien entendu, si S est propre sur k, on peut ôter l'indice c.

Pour chaque entier m, l'élément de Frobenius géométrique $\mathrm{Frob}_k \in \mathrm{Gal}(\overline{k}/k)$ (l'élévation à la puissance q^{-1} dans $\overline{k}$) agit sur $H_{\mathrm{c}}^m(S)$ et d'après la forme forte de la conjecture de Weil prouvée par Deligne (cf. [De]), pour chaque valeur propre $\alpha \in \overline{\mathbb{Q}}_\ell$ pour cette action il existe un entier $w(\alpha) \leq m$ (appelé le poids de α) tel que $|\iota(\alpha)| = q^{\frac{w(\alpha)}{2}}$ pour tout plongement $\iota : \overline{\mathbb{Q}}_\ell \hookrightarrow \mathbb{C}$. Deligne a défini la filtration par le poids $W_\bullet H_{\mathrm{c}}^{\bullet}(S)$: pour tous entiers m et w, $W_w H_{\mathrm{c}}^m(S)$ est le sous-$\overline{\mathbb{Q}}_\ell$-espace vectoriel de $H_{\mathrm{c}}^m(S)$ somme des espaces propres généralisés pour les valeurs propres de Frob_k de poids $\leq w$. Pour tout fermé $T \subset S$, la flèche de restriction $H_{\mathrm{c}}^m(S) \to H_{\mathrm{c}}^m(T)$ est strictement compatibles aux filtrations par le poids.

On dit que la cohomologie ℓ-adique de S est pure si, pour tout entier m, les valeurs propres de Frob_k sur $H_{\mathrm{c}}^m(S)$ sont toutes de poids m, ou ce qui revient au même si

$$(0) = W_{m-1} H_{\mathrm{c}}^m(S) \subset W_m H_{\mathrm{c}}^m(S) = H_{\mathrm{c}}^m(S).$$

Toujours d'après Deligne (cf. loc. cit.), c'est automatiquement le cas si S est propre et lisse sur k.

3.1 La conjecture de Goresky, Kottwitz, et MacPherson

La composante connexe X_I^0 de la fibre de Springer affine X_I du chaptire 1 est un k-schéma connexe, localement de type fini, de dimension finie et réunion croissante de sous-schémas fermés $X_{I,N}^0$ qui sont projectifs sur k. On définit la cohomologie ℓ-adique $H^{\bullet}(X_I^0)$ comme la limite projective des cohomologies ℓ-adiques des $X_{I,N}^0$ relativement aux flèches de restriction $H^{\bullet}(X_{I,N+1}^0) \to H^{\bullet}(X_{I,N}^0)$. On a $H^m(X_I^0) = (0)$ pour $m > 2\delta_I$ où on rappelle que δ_I est la dimension de X_I.

On définit la filtration par le poids $W_\bullet H^{\bullet}(X_I^0)$ par

$$W_w H^m(X_I^0) = \varprojlim_N W_w H^m(X_{I,N}^0).$$

On dit que la cohomologie ℓ-adique de X_I^0 est pure si, pour chaque entier m, on a

$$(0) = W_{m-1} H^m(X_I^0) \subset W_m H^m(X_I^0) = H^m(X_I^0).$$

La cohomologie ℓ-adique des fibres de Springer classiques est pure (cf. [Sp]). Ce résultat est une des motivations pour la conjecture suivante :

Conjecture (Goresky, Kottwitz, et MacPherson). La cohomologie ℓ-adique de X_I^0 est pure.

Le premier cas de cette conjecture a été obtenu (avant la formulation de cette dernière) par Lusztig et Smelt (cf. [Lu-Sm]). Il s'agit du cas où I est réduit à un élément, $E = F[\varpi_E]/(\varphi_E^n - \varpi_F)$ et $\gamma = \varpi_E^v$ pour des entiers n et v premiers entre eux et premiers à p. Lusztig et Smelt montre plus précisément que dans ce cas, X_I^0 est pavé par des espaces affines standard. Nous donnerons à la section 4.3 une autre démonstration de ce résultat inspirée par l'argument de Springer dans [Sp].

Plus généralement, Goresky, Kottwitz, et MacPherson ont pavé X_I^0 par des fibrés en espaces affines standard sur des variétés projectives et lisses—et ont donc démontré la conjecture de pureté ci-dessus—dans le cas d'égales valuations. Il s'agit du cas où toutes les extensions E_i sont égales à une même extension finie $E = F[\varpi_E]/(\varphi_E^n - \varpi_F)$ de F pour un entier n premier à p et où $v_E(\gamma_i)$ ne dépend pas de i.

3.2 Une variante pour les jacobiennes compactifiées

Soit C une courbe projective et géométriquement intègre sur k dont toutes les singularités sont planes. Quitte à étendre les scalaires à une extension finie de k, on peut supposer et on supposera dans la suite que toutes les singularités de C sont rationnelles sur k, ainsi que toutes leurs branches formelles.

On a vu que la composante connexe $\overline{P}^0(C)$ du k-schéma de Picard compactifié de C qui paramètre les classes d'isomorphie de $\mathcal{O}_C$-Modules cohérents sans torsion de rang générique 1 et de degré 0 est un k-schéma projectif, géométriquement intègre et localement d'intersection complète de dimension $g(C) + \delta(C)$.

On a construit dans les sections 2.4 et 2.7 un revêtement étale galoisien $\overline{P}^{\natural,0} \to \overline{P}^0(C) = J(C)$ de groupe de Galois $\mathbb{Z}^{b-1}$ où b est le nombre total des branches des singularités de C.

Comme $\overline{P}^{\natural,0}$ est homéomorphe au k-schéma $[P^0(C) \times_k X^0(C)]/G^0(C)$ qui est l'espace total d'une fibration au dessus de la variété abélienne $P^0(\widetilde{C})$, de fibre type $X(C)$, on voit facilement que la conjecture de pureté de Goresky, Kottwitz et MacPherson est équivalente à la conjecture suivante :

Conjecture. La cohomologie ℓ-adique de $\overline{P}^{\natural,0}$ est pure.

En particulier, on a :

Conjecture. Soit C une courbe projective et géométriquement intègre sur k dont toutes les singularités sont planes et géométriquement unibranches, alors la cohomologie ℓ-adique de sa jacobienne compactifiée $\overline{J}(C)$ est pure.

4 «Déformations» des fibres de Springer

4.1 Déformations de la courbe C

Revenons à la situation de la première partie. On rappelle (cf. Section 2.1) qu'on a introduit une courbe intègre, projective $C = C_I$ sur k qui n'a qu'un seul point singulier

$c = c_I$ pour lequel le complété $\widehat{\mathcal{O}}_{C,c}$ de l'anneau local de C en c est isomorphe à

$$A = A_I = \mathcal{O}_F[\gamma_I] \cong k[[\varpi_F, T]]/(p_I(T)).$$

Par hypothèse de séparabilité de $p_I(T)$, on sait que l'idéal

$$(\partial_T p_I(T), p_I(T)) \subset k[[\varpi_F]][T]$$

est de codimension finie. Il en est donc de même des idéaux

$$(\partial_T p_I(T), p_I(T)) \text{ et } (\partial_{\varpi_F} p_I(T), \partial_T p_I(T), p_I(T)) \subset k[[\varpi_F, T]].$$

On supposera dans la suite que l'idéal

$$(\partial_{\varpi_F} p_I(T), \partial_T p_I(T)) \subset k[[\varpi_F, T]]$$

est lui aussi de codimension finie (cette hypothèse est automatiquement vérifiée si k est de caractéristique nulle ; cf. [Te 1, 1.1]).

Dans la suite on note

$$x = \varpi_F,\ y = T \text{ et } f(x, x) = p_I(T)$$

de sorte que $A = k[[x, y]]/(f)$.

Par construction la courbe C est muni d'un point ∞ dans son lieu de lissité et la normalisée $\widetilde{C}$ de C est identifiée à la droite projective $\mathbb{P}^1_k$ de telle sorte que ∞ soit le point à l'infini.

Lemme 4.1.1. *Supposons de plus que C n'est pas lisse en c et que la caractéristique de k est $> |I|$. Alors, le k-schéma en groupes* $\mathrm{Aut}_k(C, \infty)$ *des automorphismes (C, ∞) sur k est soit fini, soit isomorphe à $\mathbb{G}_{\mathrm{m},k}$.*

Démonstration. Tout automorphisme g de (C, ∞) induit un automorphisme de $\widetilde{C} = \mathbb{P}^1_k$ qui fixe le point à l'infini et $\{\widetilde{c}_i \mid i \in I\} \subset \widetilde{C}$ dans son ensemble. Par suite, $\mathrm{Aut}_k(C, \infty)$ est un sous-k-schéma en groupes fermé du sous-groupe de Borel des matrices triangulaires supérieures dans PGL(2). De plus, la puissance $g^{|I|!}$ de g fixe chacun des $\widetilde{c}_i$. Par suite, si $|I| \geq 2$, $g^{|I|!}$ est nécessairement l'identité et $\mathrm{Aut}_k(C, \infty)$ est un k-schéma en groupes fini d'ordre premier à la caractéristique de k, et si $|I| = 1$, $\mathrm{Aut}_k(C, \infty)$ est un sous-k-schéma en groupes du tore maximal diagonal $\mathbb{G}^2_{\mathrm{m},k}/\mathbb{G}_{\mathrm{m},k} \subset \mathrm{PGL}(2)$. □

Comme C est de dimension 1 et localement d'intersection complète, on a

$$H^2(C, \mathcal{H}om_{\mathcal{O}_C}(\Omega^1_{C/k}(\infty), \mathcal{O}_C)) = (0)$$

et il n'y a pas d'obstruction à déformer le couple (C, ∞). Par suite la base $\mathcal{R}$ de la déformation miniverselle de (C, ∞) est une k-algèbre de séries formelles en $\tau(C, \infty)$ variables, où $\tau(C, \infty)$ est la dimension de l'espace tangent de Zariski

$$\mathrm{Ext}^1_{\mathcal{O}_C}(\Omega^1_{C/k}(\infty), \mathcal{O}_C).$$

De plus, le k-espace vectoriel sous-jacent à l'algèbre de Lie de $\mathrm{Aut}_k(C,\infty)$ est égal à $\mathrm{Hom}_{\mathcal{O}_C}(\Omega^1_{C/k}(\infty),\mathcal{O}_C)$.

De même, le foncteur des déformations topologiques de $A = k[[x,y]]/(f)$ est formellement lisse avec pour espace tangent

$$\mathrm{Def}_A^{\mathrm{top}}(k[\varepsilon]) = \mathrm{Ext}^1_A(\Omega_A^{1,\mathrm{top}}, A) \cong k[[x,y]]/(f,\partial_x f,\partial_y f)$$

($\Omega_A^{1,\mathrm{top}}$ est le conoyau de la flèche

$$A \hookrightarrow A\mathrm{d}x \oplus A\mathrm{d}y$$

qui envoie 1 sur $\mathrm{d}f = (\partial_x f)\mathrm{d}x + (\partial_y f)\mathrm{d}y$). En fait on obtient une déformation miniverselle topologique

$$\mathrm{Spf}(k[[x,y,z_1,\ldots,z_{\tau(A))}]]/(\widetilde{f})) \to \mathrm{Spf}(k[[z_1,\ldots,z_{\tau_c(C,\infty)}]])$$

du germe formel de C_s en c en choisissant arbitrairement des série formelles

$$g_1,\ldots,g_{\tau(A)} \in k[[x,y]]$$

dont les réductions modulo l'idéal $(f,\partial_x f,\partial_y f)$ forment une base de $\mathrm{Def}_A^{\mathrm{top}}(k[\varepsilon])$, et en posant

$$\widetilde{f} = \widetilde{f}(x,y,z) = f(x,y) + \sum_{t=1}^{\tau(A)} z_t g_t(x,y).$$

Bien sûr, on peut prendre $g_{\tau(A)} = 1$, de sorte que $k[[x,y,z_1,\ldots,z_{\tau(A)}]]/(\widetilde{f})$ est isomorphe à la k-algèbre de séries formelles $k[[x,y,z_1,\ldots,z_{\tau(A)-1}]]$.

Soient $S = \mathrm{Spec}(\mathcal{R})$ et $s = \mathrm{Spec}(k)$ l'unique point fermé de S.

On note dorénavant $C_s = C_{I,s_I}$, $\infty_s = \infty$ la courbe C et son point ∞, ce qui libère les notation C et ∞ que nous allons pouvoir utiliser pour les déformations de ces mêmes objets. On conserve la notation c pour l'unique point singulier de C_s.

Théorème 4.1.1. *La déformation formelle miniverselle de (C_s,∞_s) est algébrisable : il existe un morphisme de schémas $C \to S$ propre et plat le complété formel pour la topologie $\mathfrak{m}_{\mathcal{R}}$-adique est cette déformation formelle miniversel. Ce morphisme est en fait projectif et à fibres géométriquement intègres.*

Démonstration. On applique le théorème d'algébrisation de Grothendieck [Gr 2, Théorème 5.4.5] et [Gr 3, Théorème 12.2.4]. □

Proposition 4.1.1. *Le schéma C est formellement lisse sur k, le morphisme $C \to S$ est localement d'intersection complète et la fibre générique du morphisme $C \to S$ est lisse.*

Démonstration. Les deux premières assertions sont locales au point singulier $c \in C_s$ et résultent immédiatement des écritures

$$\mathrm{Spf}(A) \text{ et } \mathrm{Spf}(k[[x,y,z_1,\ldots,z_{\tau(A)}]]/(\widetilde{f}))$$

pour les complétés formels de C_s et C en c, avec

$$\widetilde{f} = \widetilde{f}(x, y, z) = f(x, y) + \sum_{t=1}^{\tau(A)} z_t g_t(x, y).$$

et $g_{\tau(A)} = 1$.

Passons à la dernière assertion. On sait déjà que la fibre C_η de $C \to S$ au point générique η de S est géométriquement intègre et donc génériquement lisse.

Maintenant, si C_η admettait une singularité en un point fermé c_η, on pourrait localiser la situation au voisinage du point fermé de C_s qui spécialise c_η et qui serait nécessairement c. Pour conclure, il suffit donc de vérifier que, pour $\widetilde{f}$ comme ci-dessus, le lieu critique de la projection canonique

$$(*) \qquad \mathrm{Spf}\,(k[[x, y, z_1, \ldots, z_{\tau(A)}]]/(\widetilde{f})) \to \mathrm{Spf}\,(k[[z_1, \ldots, z_{\tau(A)}]]) = \mathcal{S}$$

est fini sur $\mathcal{S}$ et son discriminant, c'est-à-dire l'image dans $\mathcal{S}$ du lieu critique, est un fermé strict de $\mathcal{S}$.

Or, le lieu critique de $(*)$ est le fermé de $\mathrm{Spf}\,(k[[x, y, z_1, \ldots, z_{\tau(A)}]])$ d'équations

$$\widetilde{f} = \partial_x \widetilde{f} = \partial_y \widetilde{f} = 0.$$

Il est donc bien fini sur $\mathcal{S}$ puisque $k[[x, y]]/(f, \partial_x f, \partial_y f)$ est de dimension finie sur k, et le discriminant de $(*)$ est le fermé défini par le 0-ème idéal de Fitting du $k[[z_1, \ldots, z_{\tau(A)}]]$-module de type fini $k[[x, y, z_1, \ldots, z_{\tau(A)}]]/(\widetilde{f}, \partial_x \widetilde{f}, \partial_y \widetilde{f})$.

Si le discriminant de $(*)$ était $\mathcal{S}$ tout entier, il contiendrait en particulier tout l'axe des z_1. Or la formation de ce discriminant commute à tout changement de base (cf. [Te 2, Section 5]). Sa restriction à l'axe des z_1 est donc le discriminant du morphisme

$$\mathrm{Spf}\,(k[[x, y]]) \to \mathrm{Spf}\,(k[[z_1]])$$

qui envoie (x, y) sur $z_1 = -f(x, y)$. Mais, il est facile de voir (cf. [Te 2, Section 2.6]) que ce dernier discriminant est le fermé de $\mathrm{Spf}\,(k[[z_1]])$ défini par l'équation $z_1^{\mu(f)}$ où

$$\mu(f) = \dim_k(k[[x, y]]/(\partial_x f, \partial_y f)),$$

d'où la conclusion. □

4.2 Déformations de jacobiennes compactifiées

On s'intéresse de nouveau à la fibre de Springer $X = X_I$ et à son quotient $Z = Z_I$ par le réseau $\Lambda^0 = \Lambda_I^0$.

On a vu d'une part que Z est naturellement homéomorphe au k-schéma de Picard compactifié $\overline{P} = \overline{P}_I$ et que le revêtement $X \to Z$ provient d'un revêtement $\overline{P}^\natural = \overline{P}_I^\natural \to \overline{P}$. On a vu d'autre part comment déformer la courbe $C = C_I$. On se propose maintenant d'utiliser les déformations de C pour déformer $\overline{P}$, et aussi d'une certaine manière $\overline{P}^\natural$.

Comme on l'a déjà fait dans la section précédente on note dorénavant les objets ci-dessus par $X_s = X_{I,s_I}$, $Z_s = Z_{I,s_I}$, $\Lambda_s^0 = \Lambda_{I,s_I}^0$, $P_s = P_{I,s_I}$, $\overline{P}_s = \overline{P}_{I,s_I}, \ldots$, *ce qui libère les notation* X, Z, Λ, P, $\overline{P}, \ldots$ *que nous allons pouvoir utiliser pour les déformations de ces mêmes objets.*

Soient $P = P_I \subset \overline{P}_I = \overline{P}$ les schémas de Picard et de Picard compactifié de la déformation algébrique $(C, \infty)/S$ de (C_s, ∞_s) étudiée dans la section précédente.

On rappelle que $\overline{P}$ paramètre les classes de $\mathcal{O}_C$-Modules plats sur S qui sont fibre à fibre sans torsion et de rang générique 1, et que P est l'ouvert de $\overline{P}$ formé des classes de Modules inversibles.

Les schémas P et $\overline{P}$ sont purement de dimension relative $\delta = \delta_I$ sur S, réunions disjointes de composantes connexes $P^d = P_I^d \subset \overline{P}_I^d = \overline{P}^d$, respectivement quasi-projectives et projectives sur S. Les fibres P_t^d et $\overline{P}_t^d$ de $P^d \to S$ et $\overline{P}^d \to S$ en tout point géométrique t de S sont les composantes de degré d des schémas de Picard et de Picard compactifié de la fibre C_t de $C \to S$ en t ; chaque $\overline{P}_t^d$ est intègre et localement d'intersection complète (cf. Théorème 2.2.1).

Théorème 4.2.1 (Fantechi, Göttsche, et van Straten [F-G-S, Corollary B.2]). *Le schéma* $\overline{P}$ *est régulier (formellement lisse sur* k*) et la fibre spéciale* $\overline{P}_s$ *de* $\overline{P} \to S$ *est localement d'intersection complète dans* $\overline{P}$.

Pour prouver ce théorème, Fantechi, Göttsche et van Straten considèrent le foncteur des déformations

$$\mathrm{Def}_{C_s,\mathcal{M}_s} : \mathrm{Art}_k \to \mathrm{Ens}$$

du couple $(C_s, \mathcal{M}_s)$ où $\mathcal{M}_s$ est un $\mathcal{O}_{C_s}$-Module cohérent sans torsion de rang générique 1, et sa variante locale

$$\mathrm{Def}_{A,M}^{\mathrm{top}} : \mathrm{Art}_k \to \mathrm{Ens}$$

en l'unique point singulier c de C_s, où $A = \widehat{\mathcal{O}}_{C_s,c}$ et $M = \widehat{\mathcal{M}_{s,c}}$ est un A-module de type fini, sans torsion et de rang générique 1. On a le carré commutatif de morphismes naturels de foncteurs

$$\begin{array}{ccc} \mathrm{Def}_{C_s,\mathcal{M}_s} & \longrightarrow & \mathrm{Def}_{A,M}^{\mathrm{top}} \\ \downarrow & + & \downarrow \\ \mathrm{Def}_{C_s} & \longrightarrow & \mathrm{Def}_{A}^{\mathrm{top}} \end{array}$$

et il est facile de vérifier :

Lemme 4.2.1. *Le morphisme de foncteurs*

$$\mathrm{Def}_{C_s,\mathcal{M}_s} \to \mathrm{Def}_{C_s} \times_{\mathrm{Def}_M^{\mathrm{top}}} \mathrm{Def}_{A,M}^{\mathrm{top}}$$

induit par le diagramme ci-dessus est formellement lisse. □

Le théorème résulte donc de la proposition suivante.

Proposition 4.2.1. *Le foncteur des déformations* $\mathrm{Def}_{A,M}^{\mathrm{top}}$ *est formellement lisse sur* k.

Démonstration. D'après le lemme A.3.3, M admet, en tant que $k[[x,y]]$-module, une résolution

$$0 \to k[[x,y]]^n \xrightarrow{F} k[[x,y]]^n \to M \to 0$$

où $n \in \{1, \ldots, \delta(A)+1\}$ et F est une matrice carré de taille $n \times n$ à coefficients dans $k[[x,y]]$. En raisonnant comme dans la démonstration du Théorème A.3.1 avec la matrice des co-facteurs de F, on peut exiger de plus que le déterminant $\det F$ est égal à f.

Toute déformation topologique (R-plate) M_R du $k[[x,y]]$-module M sur $R \in$ ob Art peut-être obtenue en déformant la matrice F en une matrice carrée F_R de taille $n \times n$ à coefficients dans $R[[x,y]] = R \otimes_k k[[x,y]]$, de telle sorte que M_R admette la présentation

$$0 \to R[[x,y]]^n \xrightarrow{F_R} R[[x,y]]^n \to M_R \to 0.$$

Pour une telle déformation topologique M_R de M en tant que $k[[x,y]]$-module, la R-algèbre quotient $A_R = R[[x,y]]/(\det F_R)$ est une déformation topologique de A sur R et M_R est de manière évidente un A_R-module sans torsion. Si maintenant $B_R = R[[x,y]]/(g_R)$ est une autre déformation plate de A sur R telle que $g_R M_R = (0)$, alors on a $g_R = h_R \det F_R$ où $h_R \in R[[x,y]]$ est congru modulo l'idéal maximal de R à un élément inversible de $k[[x,y]]$ et est donc inversible dans $R[[x,y]]$, de sorte que $B_R = A_R$.

Si $\mathrm{Def}_F^{\mathrm{top}} : \mathrm{Art}_k \to \mathrm{Ens}$ est le foncteur des déformations de la matrice F ci-dessus, on a donc construit un morphisme formellement lisse de foncteurs

$$\mathrm{Def}_F^{\mathrm{top}} \to \mathrm{Def}_{A,M}^{\mathrm{top}},\ F_R \mapsto (R[[x,y]]/(\det F_R), \mathrm{Coker}(F_R)).$$

Comme le foncteur $\mathrm{Def}_F^{\mathrm{top}}$ est trivialement formellement lisse sur k, la proposition s'en suit. □

Dans la situation de la proposition ci-dessus, on note $V(A)$ le sous-k-espace vectoriel

$$V(A) = \mathfrak{a}/(\partial_x f, \partial_y f) \subset A/(\partial_x f, \partial_y f) = \mathrm{Def}_A^{\mathrm{top}}(k[\varepsilon]) \tag{4.2.4}$$

de l'espace tangent de $\mathrm{Def}_A^{\mathrm{top}}$, où $\mathfrak{a} \subset A$ est le conducteur du normalisé $\widetilde{A}$ de A dans A.

Le résultat suivant généralise le théorème 4.2.1.

Théorème 4.2.2 (Fantechi, Göttsche, van Straten ; [F-G-S] Corollary B.3). *Soit $C_T \to T$ une courbe relative de base $T \cong \mathrm{Spec}(k[[t_1,\ldots,t_m]])$ qui provient de la courbe universelle $C \to S$ par un changement de base local $T \to S$. Alors, le T-schéma*

$$\overline{P}_T = T \times_S \overline{P}$$

de Picard compactifié de C_T/T est régulier (formellement lisse sur k) si et seulement si l'espace tangent à l'origine de T est transverse au sous-espace $V(C_s) \subset T_s S$ de l'espace tangent à l'origine de S obtenu par image inverse du sous-k-espace vectoriel

$$V(A) \subset \mathrm{Def}_A^{\mathrm{top}}(k[\varepsilon])$$

par l'épimorphisme naturel $T_s S \twoheadrightarrow \mathrm{Def}_A^{\mathrm{top}}(k[\varepsilon])$.

Démonstration. On a un morphisme d'oubli

$$\mathrm{Def}_{A,M}^{\mathrm{top}} \to \mathrm{Def}_A^{\mathrm{top}}$$

et il suffit de démontrer que l'image de l'application tangente à ce morphisme contient le sous-espace $V(A)$ de l'espace tangent à $\mathrm{Def}_A^{\mathrm{top}}$.

L'image de cette application tangente est le quotient $I/(f, \partial_x f, \partial_y f)$, où $I \subset k[[x, y]]$ est l'idéal engendrée par les entrées de la matrice F^* des co-facteurs de F puisque l'on a

$$\det(F + \varepsilon E^{ij}) = \det F + \varepsilon F_{ji}^*$$

quels que soient $0 \leq i, j \leq n$, où E^{ij} est la matrice élémentaire dont l'entrée (i, j) est égale à 1 et dont toutes les autres entrées sont nulles.

Or, le A-module $M = \mathrm{Coker}(F)$ admet la résolution périodique, de période 2,

$$\cdots \xrightarrow{F} A^n \xrightarrow{F^*} A^n \xrightarrow{F} A^n \to M \to 0,$$

et M est encore égal à

$$M = \mathrm{Ker}(F) = \mathrm{Im}(F^*).$$

Comme on a $\mathrm{Ext}_A^1(N, A) = (0)$ pour tout A-module sans torsion N, le morphisme $\mathrm{Hom}_A(F^*, A)$ admet la factorisation canonique

$$\mathrm{Hom}_A(A^n, A) \twoheadrightarrow \mathrm{Hom}_A(M, A) \hookrightarrow \mathrm{Hom}_A(A^n, A)$$

et l'image $I/(f)$ de I dans A est égale à l'idéal engendré par les $\varphi(m)$ pour φ parcourant $\mathrm{Hom}_A(M, A)$ et m parcourant M.

Mais, à isomorphisme près, on peut supposer que $A \subset M \subset \widetilde{A}$ (cf. la démonstration du Lemme A.3.3), et alors il est clair que

$$\{\varphi(m) \mid \varphi \in \mathrm{Hom}_A(M, A),\ m \in M\} \supset \mathfrak{a},$$

ce qui termine la démonstration du théorème. □

4.3 Application à la pureté dans le cas homogène

Soient $m > n \geq 1$ premiers entre eux. On suppose que $m!$ est inversible dans k et que k contient toutes les racines m-ème de l'unité. On considère la fibre de Springer X associée à l'extension totalement ramifiée $F \subset E$ de degré n définie par $\varpi_F = \varpi_E^n$ et à l'élément $\gamma = \varpi_E^m$, ou encore celle associée à l'extension totalement ramifiée $F \subset E$ de degré m définie par $\varpi_F = \varpi_E^m$ et à l'élément $\gamma = \varpi_E^n$. Dans les deux cas, le germe formel de courbe plane correspondant est le même, à savoir $\mathrm{Spf}(A)$ où $A = k[[x, y]]/(x^m - y^n)$. Comme ce germe n'a qu'une seule branche, on a $\Lambda = \mathbb{Z}$ et $X = Z = Z^0 \times \mathbb{Z}$.

Rappelons que, si C est n'importe quelle courbe intègre, projective sur k, de normalisée isomorphe à la droite projective $\mathbb{P}^1_k$, et qui est munie d'un k-point c en dehors duquel elle est lisse sur k et en lequel le germe formel de C est isomorphe à $\mathrm{Spf}(A)$, alors le k-schéma Z^0 est homéomorphe à la composante $\overline{\mathrm{Pic}}^0_{C/k}$ de degré 0 du k-schéma de Picard compactifié de C.

On peut construire une telle courbe C de la façon suivante. Soit

$$\mathbb{P} = \mathrm{Proj}(k[X, Y, Z])$$

le plan projectif pondéré où $\deg X = n$, $\deg Y = m$ et $\deg Z = 1$, et soit $C \subset \mathbb{P}$ le fermé défini par l'équation homogène

$$F(X, Y, Z) = X^m - Y^n = 0$$

de degré mn. L'intersection de C avec la carte affine $\{Z \neq 0\} = \mathrm{Spec}(k[x, y])$ de $\mathbb{P}$ est la courbe d'équation

$$f = x^m - y^n = 0;$$

elle est donc lisse en dehors de l'origine $(0, 0)$ et admet le germe formel voulu en $(0, 0)$. L'intersection de C avec le diviseur à l'infini $\{Z = 0\}$ de $\mathbb{P}$ est réduite au point de coordonnées homogènes $(1; 1; 0)$. De plus le germe formel de C en ce point est isomorphe à celui en $(x = 1, z = 0)$ de la courbe d'équation

$$x^m - 1 = 0$$

dans le plan affine $\mathrm{Spec}(k[x, z])$. (La carte affine $\{Y \neq 0\}$ de $\mathbb{P}$ est le quotient de ce plan affine par le groupe fini des racines m-ème de l'unité dans k pour l'action définie par $(\zeta, (x, z)) \mapsto (\zeta^n x, \zeta z)$ et cette action est libre en dehors de l'origine.) Par suite, C est lisse au point $(1; 1; 0)$. Enfin, la normalisation de C est donnée par

$$\mathbb{P}^1_k \to C,\ (T; U) \mapsto (T^n; T^m; U).$$

L'espace tangent au foncteur des déformations plates du germe formel de courbe $\mathrm{Spf}(k[[x, y]]/(f))$ est égal à

$$k[[x, y]]/(f, \partial_x f, \partial_y f) = k[[x, y]]/(x^{m-1}, y^{n-1})$$

et est donc de dimension

$$\mu = (m - 1)(n - 1).$$

Une déformation miniverselle de f est donnée par

$$\mathrm{Spf}(k[[\ldots, a_{ij}, \ldots, x, y]]/(\widetilde{f}))$$

où les a_{ij} sont des indéterminées sur le corps de base k et

$$\widetilde{f}(x, y) = x^m - y^n + \sum_{\substack{0 \leq i \leq m-2 \\ 0 \leq j \leq n-2}} a_{ij} x^i y^j.$$

Le conducteur $\mathfrak{a}$ de $A = k[[t^n, t^m]]$ dans son normalisé $\widetilde{A} = k[[t]]$ est égal à $t^{(m-1)(n-1)}k[[t]]$, soit encore à

$$\mathfrak{a} = \{ \sum_{i,j\geq 0} a_{ij}x^i y^j \in k[[x,y]] \mid a_{ij} = 0,\ \forall i,j \text{ tels que } in+jm < (m-1)(n-1)\}/(f).$$

En particulier, le quotient

$$V(A) = \mathfrak{a}/(\partial_x f, \partial_y f) \subset A/(\partial_x f, \partial_y f) = k[[x,y]]/(x^{m-1}, y^{n-1})$$

admet pour base les classes des monômes $x^i y^j$ pour lesquels $0 \leq i \leq m-2$, $0 \leq j \leq n-2$ et $in + jm \geq (m-1)(n-1)$.

Considérons l'espace affine $S = \mathrm{Spec}(k[\ldots, a_{ij}, \ldots])$, son fermé $T \subset S$ défini par les équations $a_{ij} = 0$, $\forall i, j$ tels que $in + jm \geq (m-1)(n-1)$, le plan projectif pondéré

$$\mathbb{P}_T = \mathrm{Proj}(\mathcal{O}_T[X, Y, Z])$$

où $\deg X = n$, $\deg Y = m$ et $\deg Z = 1$, et la courbe projective et plate relative

$$\begin{array}{ccc} C_T & \hookrightarrow & \mathbb{P}_T \\ \downarrow & \swarrow & \\ T & & \end{array}$$

définie par l'équation homogène

$$\widetilde{F}(X,Y,Z) = X^m - Y^n + \sum_{\substack{0\leq i\leq m-2\\ 0\leq j\leq n-2\\ in+jm<(m-1)(n-1)}} a_{ij}X^iY^jZ^{mn-in-jm} = 0$$

de degré mn. Pour chaque $t \in T$, l'intersection de la fibre C_t de $C_T \to T$ en t avec le diviseur à l'infini $Z = 0$ du plan projectif pondéré $\mathbb{P}_t$ est réduite au point $(1; 1; 0)$ et on voit comme ci-dessus que C_t est lisse en ce point.

On a des actions compatibles du groupe multiplicatif $\mathbb{G}_{\mathrm{m},k}$ sur T et C_T données par

$$\lambda \cdot (\ldots, a_{ij}, \ldots) = (\ldots, \lambda^{mn-in-jm} a_{ij}, \ldots)$$

et

$$\lambda \cdot (\ldots, a_{ij}, \ldots, X; Y; Z)) = (\ldots, \lambda^{mn-in-jm} a_{ij}, \ldots, \lambda^n X, \lambda^m Y, Z).$$

Par définition du fermé T de S, l'action sur T est contractante.

Considérons la composante $\overline{\mathrm{Pic}}^0_{C_T/T}$ de degré 0 du schéma de Picard compactifié relatif de C_T sur T. L'action de $\mathbb{G}_{\mathrm{m},k}$ sur C_T induit une action de $\mathbb{G}_{\mathrm{m},k}$ sur $\overline{\mathrm{Pic}}^0_{C_T/T}$ qui relève celle sur T.

L'espace tangent au fermé $T \subset S$ est un supplémentaire du sous-espace $V(A)$ de $A/(\partial_x f, \partial_y f) = T_{(0,0)}S$. Il s'en suit, d'après le théorème 4.2.2, que le schéma $\overline{\mathrm{Pic}}_{C_T/T}$ est lisse sur le corps de base k le long de sa fibre $\overline{\mathrm{Pic}}_{C/k}$ à l'origine 0 de T. Compte tenu de l'action de $\mathbb{G}_{\mathrm{m},k}$, $\overline{\mathrm{Pic}}_{C_T/T}$ est partout lisse sur k.

Soient ℓ un nombre premier distinct de la caractéristique de k et $K \in \mathrm{ob}\, D_c^b(T, \mathbb{Q}_\ell)$ l'image directe du faisceau constant $\mathbb{Q}_\ell$ par la projection $\overline{\mathrm{Pic}}^0_{C_T/T} \to T$. L'action de $\mathbb{G}_{\mathrm{m},k}$ sur T se relève en une «action» sur K et on peut appliquer [Sp, Proposition 1] qui assure que

$$H^i(T, K) = \begin{cases} K_0 & \text{si } i = 0, \\ (0) & \text{sinon.} \end{cases}$$

Or $R\Gamma(T, K)$ n'est autre que la cohomologie ℓ-adique $R\Gamma(\overline{\mathrm{Pic}}^0_{C_T/T}, \mathbb{Q}_\ell)$ du k-schéma lisse et quasi-projectif $\overline{\mathrm{Pic}}^0_{C_T/T}$, et le théorème de changement de base propre implique que la fibre K_0 de K à l'origine de T n'est autre que la cohomologie ℓ-adique $R\Gamma(\overline{\mathrm{Pic}}^0_{C/k}, \mathbb{Q}_\ell)$ du k-schéma projectif $\overline{\mathrm{Pic}}^0_{C/k}$, soit encore la cohomologie ℓ-adique $R\Gamma(X^0, \mathbb{Q}_\ell)$ de la composante de degré 0 de la fibre de Springer puisque cette composante $X^0 = Z^0$ est universellement homéomorphe à $\overline{\mathrm{Pic}}^0_{C/k}$.

Si k est de caractéristique $p > 0$, toute la situation ci-dessus est définie sur $\mathbb{F}_p$. Appliquant alors la forme forte de la conjecture de Weil prouvée par Deligne comme l'a fait Springer dans [Sp], on déduit de qui précède :

Proposition 4.3.1. *Chaque groupe de cohomologie $H^i(X^0, \mathbb{Q}_\ell)$ est pur de poids i.* □

Bien sûr, cette proposition résulte aussi de fait que X^0 peut être pavé par des espaces affines (cf. [Lu-Sm]).

4.4 Construction de déformations de fibres de Springer affine

Revenons à la situation générale de la Section 4.2. Nous avons donc la courbe intègre, projective $C_s = C_{I,s_I}$ sur k, de normalisée $\pi_s : \widetilde{C}_s \to C_s$ une droite projective, avec son unique point singulier c en lequel la singularité est plane et l'ensemble des branches $\pi_s^{-1}(c) = \{\widetilde{c}_i \mid i \in I\}$ est indexé par I. Nous allons utiliser les résultats de la section 2.7 pour «déformer» le revêtement

$$X_s \to Z_s$$

de groupe de Galois

$$\Lambda_s^0 = H^0(\widetilde{C}_s \setminus \pi_s^{-1}(c), \mathbb{G}_{\mathrm{m}})/H^0(\widetilde{C}_s, \mathbb{G}_{\mathrm{m}}) = \mathrm{Ker}(\mathbb{Z}^I \to \mathbb{Z}),$$

de la section 2.3. Plus exactement, nous allons déformer le revêtement

$$\overline{P}_s^\natural \to \overline{P}_s$$

qui lui est homéomorphe.

Considérons l'algébrisation $(C = C_I, \infty) \to S_I = S$ de la déformation miniverselle de (C_s, ∞_s) introduite dans la section 4.2 et la strate à δ constant $S^\delta \subset S$ (cf. la Section A.4 de l'appendice). Considérons aussi le morphisme de normalisation $\pi_{S^\delta} : \widetilde{S^\delta} \to S^\delta$ de cette strate et la courbe

$$C_{\widetilde{S^\delta}} = C_{I,\widetilde{S^\delta}} \to \widetilde{S^\delta}$$

déduite de $C \to S$ par le changement de base

$$\widetilde{S^\delta} \twoheadrightarrow S^\delta \hookrightarrow S.$$

D'après le Corollaire A.2.1, cette courbe relative admet une normalisation en famille $\widetilde{C_{\widetilde{S^\delta}}} \to C_{\widetilde{S^\delta}}$, et on a en fait

$$\widetilde{C_{\widetilde{S^\delta}}} \cong \mathbb{P}^1_{\widetilde{S^\delta}}$$

puisque la normalisée de C_s est une droite projective sur k. On peut choisir l'isomorphisme ci-dessus de telle sorte que la section de $\widetilde{S^\delta} \to \widetilde{C_{\widetilde{S^\delta}}}$ induite par la section $\infty : S \to C$ corresponde à la section à l'infini de $\mathbb{P}^1_{\widetilde{S^\delta}}$ sur $\widetilde{S^\delta}$.

Au moins pour k de caractéristique nulle (cf. Théorème A.4.2), le morphisme $C_{\widetilde{S^\delta}} \to \widetilde{S^\delta}$ a les propriétés suivantes :

(1) $\widetilde{S^\delta}$ est un schéma strictement local régulier (formellement lisse sur k) de dimension $\tau(C_s, \infty_s) - \delta$,

(3) toutes les fibres géométriques de $\widetilde{C_{\widetilde{S^\delta}}} \to \widetilde{S^\delta}$ sont des courbes intègres à singularités planes,

(2) la fibre générique géométrique de $\widetilde{C_{\widetilde{S^\delta}}} \to \widetilde{S^\delta}$ n'a comme seules singularités que δ points doubles ordinaires.

Hypothèses et Notations 4.4.1. *Dans la suite, nous supposerons que les propriétés* (1) *à* (3) *sont vérifiées sur notre corps k.*

De plus, comme la déformation totale $C \to S$ n'interviendra plus, nous noterons simplement $C \to S$ la courbe relative $C_{\widetilde{S^\delta}} \to \widetilde{S^\delta}$ pour alléger l'exposition.

Nous avons donc un schéma strictement local S formellement lisse sur k, de dimension $\tau(C_s, \infty_s) - \delta$, et une courbe relative C sur S, munie d'une section globale ∞ le long de laquelle C est lisse sur S, de fibre spéciale C_s et qui admet une normalisation en famille $\pi_C : \widetilde{C} \to C$ dont l'espace total $\widetilde{C}$ est une droite projective sur S. Le morphisme structural $f : C \to S$ est lisse en dehors du sous-schéma $D \subset C$ fini sur S qui est défini par l'Idéal conducteur $\mathfrak{a}$ de $\pi_* \mathcal{O}_{\widetilde{C}}$ dans $\mathcal{O}_C$ et

$$\widetilde{f} = f \circ \pi_C : \widetilde{C} \to S$$

est identifié à la droite projective standard $\mathbb{P}^1_S \to S$ de telle sorte que la section à l'infini corresponde à la section $\infty : S \to C$, et donc ne rencontre pas $\widetilde{D} = \pi_C^{-1}(D)$.

Considérons les S-schémas de Picard P et de Picard compactifié $\overline{P}$ relatifs de C sur S. Le S-schéma en groupes P est isomorphe à $P^0 \times \mathbb{Z}$ puisque le lieu de lissité de $C \to S$ admet une section. Pour chaque entier d, la composante connexe $\overline{P}^d$ contient $P^d = P^0 \times \{d\}$ comme ouvert dense.

La composante neutre P^0 du schéma de Picard est par définition le S-schéma affine et lisse qui représente le faisceau fppf

$$\widetilde{f}_*\mathbb{G}_{\mathrm{m},\widetilde{C}}/f_*\mathbb{G}_{\mathrm{m},C} = f_*(\pi_{C,*}\mathbb{G}_{\mathrm{m},\widetilde{C}}/\mathbb{G}_{\mathrm{m},C}).$$

Notons f_D (resp., $\widetilde{f}_D$) les restrictions de f et $\widetilde{f}$ aux fermés $D \subset C$ (resp., $\widetilde{D} \subset \widetilde{C}$). On a le S-schéma en groupes affine et lisse

$$\mathrm{Res}_{D/S}\,\mathbb{G}_{\mathrm{m}} \text{ (resp., } \mathrm{Res}_{\widetilde{D}/S}\,\mathbb{G}_{\mathrm{m}})$$

restriction à la Weil du groupe multiplicatif de D à S (resp., de $\widetilde{D}$ à S), qui représente le faisceau fppf $f_{D,*}\mathbb{G}_{\mathrm{m},D}$ (resp., $\widetilde{f}_{D,*}\mathbb{G}_{\mathrm{m},\widetilde{D}}$), et on a la flèche d'adjonction

$$\alpha : \pi_{C,*}\mathbb{G}_{\mathrm{m},\widetilde{C}}/\mathbb{G}_{\mathrm{m},C} \to i_*(\pi_{D,*}\mathbb{G}_{\mathrm{m},\widetilde{D}}/\mathbb{G}_{\mathrm{m},D})$$

où $i : D \hookrightarrow C$ est l'inclusion et $\pi_D : \widetilde{D} \to D$ est la restriction de π.

Lemme 4.4.1. *La flèche d'adjonction α est un isomorphisme et induit un isomorphisme de S-schémas en groupes*

$$P^0 \xrightarrow{\sim} \mathrm{Res}_{\widetilde{D}/S}\,\mathbb{G}_{\mathrm{m}}/\,\mathrm{Res}_{D/S}\,\mathbb{G}_{\mathrm{m}}.$$

Démonstration. Soit $\mathrm{Spec}(A) \to C$ une carte affine de C. Au dessus cette carte, $\widetilde{C}$ est égal à $\mathrm{Spec}(B)$ pour une A-algèbre finie B, la trace de D est définie par l'idéal conducteur I de B dans A et α est donnée par la flèche naturelle

$$B^\times/A^\times \to (B/I)^\times/(A/I)^\times$$

(on rappelle que $I \subset A \subset B$ est à la fois un idéal de A et un idéal de B). Comme cette dernière flèche est trivialement injective, l'injectivité de α est démontrée.

Maintenant, si $b, b' \in B$ sont tels que $bb' = 1 + a$ avec $a \in I \subset A$, quitte à remplacer $\mathrm{Spec}(A)$ par le voisinage ouvert $\mathrm{Spec}(A[(1+a)^{-1}])$ du fermé $\mathrm{Spec}(A/I) \subset \mathrm{Spec}(A)$, on voit que b est inversible dans B, ce qui démontre la surjectivité de α. □

La fibre spéciale $\mathrm{Res}_{D_s/s}\,\mathbb{G}_{\mathrm{m}}$ du S-schéma en groupes $\mathrm{Res}_{D/S}\,\mathbb{G}_{\mathrm{m}}$ admet pour tore maximal le tore $\mathbb{G}_{\mathrm{m},k}$ puisque $(D_s)_{\mathrm{red}}$ est réduit au point $c \in D_s \subset C_s$. De même, $\mathrm{Res}_{\widetilde{D}_s/s}\,\mathbb{G}_{\mathrm{m}}$ admet pour tore maximal le tore $\mathbb{G}_{\mathrm{m},k}^I$ puisque $(\widetilde{D}_s)_{\mathrm{red}} = \{\widetilde{c}_i \mid i \in I\}$. Le tore maximal T_s de P_s^0 est donc canoniquement isomorphe à $\mathbb{G}_{\mathrm{m},k}^I/\mathbb{G}_{\mathrm{m},k}$ (quotient pour le plongement diagonal).

Le tore maximal $\mathbb{G}_{\mathrm{m},k}$ de $\mathrm{Res}_{D_s/s}\,\mathbb{G}_{\mathrm{m}}$ est la fibre spéciale du tore canonique $\mathbb{G}_{\mathrm{m},S} \subset \mathrm{Res}_{D/S}\,\mathbb{G}_{\mathrm{m}}$ défini par la flèche d'adjonction $\mathrm{id} \to f_{D,*}f_D^*$. Comme S est strictement hensélien, $\widetilde{D}$ se casse en autant de composantes connexes qu'il y a de points dans $(\widetilde{D}_s)_{\mathrm{red}}$ et le tore maximal $\mathbb{G}_{\mathrm{m},k}^I$ de $\mathrm{Res}_{\widetilde{D}_s/s}\,\mathbb{G}_{\mathrm{m}}$ est aussi la fibre spéciale du tore canonique $\mathbb{G}_{\mathrm{m},S}^I \subset \mathrm{Res}_{\widetilde{D}/S}\,\mathbb{G}_{\mathrm{m}}$ défini par la flèche d'adjonction $\mathrm{id} \to \widetilde{f}_{D,*}\widetilde{f}_D^*$, composante connexe par composante connexe de $\widetilde{D}$.

Par suite, le tore maximal T_s de P_s^0 est la fibre spéciale d'un tore canonique

$$T = \mathbb{G}_{\mathrm{m},S}^I/\mathbb{G}_{\mathrm{m},S} \subset P^0.$$

Plus généralement, pour chaque point géométrique t de S, la fibre T_t de T en t est contenue dans le tore maximal de

$$P_t^0 \xrightarrow{\sim} \mathrm{Res}_{\widetilde{D}_t/t}\,\mathbb{G}_{\mathrm{m}}/\,\mathrm{Res}_{D_t/t}\,\mathbb{G}_{\mathrm{m}},$$

tore maximal qui est isomorphe à $\prod_{j\in J_t}(\mathbb{G}_{\mathrm{m},\kappa(t)}^{I_{t,j}}/\mathbb{G}_{\mathrm{m},t})$ où $\{c_{j,t}\mid j\in J_t\}$ est l'ensemble des points singuliers de C_t et, pour chaque $j\in J_t$, $\{\widetilde{c}_{t,i}\mid i\in I_{t,j}\}$ est l'ensemble des branches du germe formel de C_t en son point singulier $c_{t,j}$.

On identifie le groupe des caractères de T_s, ou ce qui revient au même celui de T, à $\Lambda_s^0 := \Lambda^0$ comme dans la section 2.7. Pour chaque point géométrique t de S, Λ^0 est donc un quotient du groupe des caractères

$$\Lambda_t^0 = \prod_{j\in J}\mathrm{Ker}(\mathbb{Z}^{I_{t,j}}\to\mathbb{Z})$$

du tore maximal de P_t^0, quotient que l'on peut voir concrètement en considérant la manière dont les points singuliers $c_{t,j}$ et leurs branches $\widetilde{c}_{t,i}$ confluent vers le point singulier c et ses branches $\widetilde{c}_i$.

Nous sommes maintenant en mesure de construire la déformation de $\overline{P}_s^\natural\to\overline{P}_s$.

Pour chaque entier d, la restriction de l'homomorphisme β d'Esteves, Gagné et Kleiman à T est un homomorphisme $T\to\mathrm{Pic}_{\overline{P}/S}$ qui définit d'après la proposition 2.7.1 un Λ^0-torseur

$$Q^d\to\overline{P}^d$$

dont la formation commute à tout changement de base $S'\to S$. Notons

$$Q\to\overline{P}$$

la somme disjointe de ces torseurs. C'est la déformation cherchée.

En effet, la fibre spéciale est par définition le revêtement $\overline{P}_s^\natural\to\overline{P}_s$. Plus généralement, pour chaque point géométrique t de S, $Q_t\to\overline{P}_t$ est le quotient du revêtement $\overline{P}_t^\natural\to\overline{P}_t$ dans le Corollaire 2.3.2, quotient qui correspond au quotient $\Lambda_t^0\twoheadrightarrow\Lambda^0$ entre les groupes de Galois.

Appendice A Rappels sur les déformations de courbes

A.1 Déformations miniverselles : généralités

Soit k un corps. On note Art_k la catégorie des k-algèbres locales artiniennes de corps résiduel k. Tous les k-schémas considérés dans la suite sont supposés séparés et localement nœthériens.

Soit X_k un k-schéma. On a le foncteur des *déformations*

$$\mathrm{Def}_{X_k}:\mathrm{Art}_k\to\mathrm{Ens}$$

qui associe à une k-algèbre locale artinienne R l'ensemble des classes d'isomorphie de R-schémas plats X_R munis d'un isomorphisme de k-schémas

$$\iota : X_k \xrightarrow{\sim} k \otimes_R X_R.$$

Si $X_k = \mathrm{Spec}(A_k)$ est affine, toute déformation X_R de X_k sur $R \in \mathrm{ob}\,\mathrm{Art}_k$ est aussi un schéma affine $X_R = \mathrm{Spec}(A_R)$ où A_R est une R-algèbre plate munie d'un isomorphisme de k-algèbres $k \otimes_R A_R \xrightarrow{\sim} A_k$. On identifiera donc Def_{X_k} au foncteur Def_{A_k} qui envoie R sur l'ensemble des classes d'isomorphie des A_R.

On définit aussi le foncteur des déformations

$$\mathrm{Def}_{X_k}^{\mathrm{top}} = \mathrm{Def}_{A_k}^{\mathrm{top}}$$

d'un k-schéma formel affine $X_k = \mathrm{Spf}(A_k)$ en tenant compte de la topologie de A_k.

Si U_k est un ouvert de X_k et si $x \in U_k$, on a des morphismes de foncteurs évidents

$$\mathrm{Def}_{X_k} \to \mathrm{Def}_{U_k} \to \mathrm{Def}_{\mathcal{O}_{X_k,x}} \to \mathrm{Def}_{\widehat{\mathcal{O}}_{X_k,x}}^{\mathrm{top}}.$$

Une *déformation formelle* $(\mathcal{R}, \mathcal{X})$ de X_k est un couple formé d'une k-algèbre locale complète nœthérienne $\mathcal{R}$ de corps résiduel k et d'un couple $\mathcal{X} = (X_\bullet, \iota_\bullet)$ où $X_\bullet$ est une suite de déformations X_n de X_k sur $R_n = \mathcal{R}/\mathfrak{m}_{\mathcal{R}}^{n+1}$ avec $X_0 = X_k$ et où $\iota_\bullet$ est une suite d'isomorphismes

$$\iota_n : X_n \xrightarrow{\sim} R_n \otimes_{R_{n+1}} X_{n+1},\ n \in \mathbb{N}.$$

On voit dans la suite $\mathcal{X}$ comme un $\mathcal{S}$-schéma formel où $\mathcal{S} = \mathrm{Spf}(\mathcal{R})$.

Une telle déformation définit un morphisme de foncteurs

$$\mathcal{S} \to \mathrm{Def}_{X_k}$$

qui envoie le R-point $\varphi : \mathcal{R} \to R$ de $\mathcal{S}$ sur $(R \otimes_{\varphi,\mathcal{R}} \mathcal{X}, \iota)$ où on a posé

$$R \otimes_{\varphi,\mathcal{R}} \mathcal{X} = R \otimes_{\varphi_n, R_n} X_n$$

quel que soit l'entier n assez grand pour que φ se factorise en $\mathcal{R} \twoheadrightarrow R_n \xrightarrow{\varphi_n} R$.

Définition A.1.1. Une déformation formelle $(\mathcal{R}, \mathcal{X})$ de X_k est dite *miniverselle* si le morphisme de foncteurs $\mathrm{Hom}_{k-\mathrm{alg}}(\mathcal{R}, \cdot) \to \mathrm{Def}_{X_k}$ associé est formellement lisse et si l'application entre les espaces tangents

$$\mathrm{Hom}_k(\mathfrak{m}_{\mathcal{R}}/\mathfrak{m}_{\mathcal{R}}^2, k) \xleftarrow{\sim} \mathcal{S}(k[\varepsilon]) \to \mathrm{Def}_{X_k}(k[\varepsilon])$$

est bijective.

On définit de même une déformation formelle miniverselle d'une k-algèbre A_k et une déformation formelle miniverselle topologique d'une k-algèbre topologique $\mathcal{A}_k$. Une déformation formelle miniverselle de X_k ou A_k (resp., une déformation formelle miniverselle topologique A_k) est unique à isomorphisme près.

Théorème A.1.1 (Schlessinger ; cf. [Ri, Theorem 4.5]). *Soit X_k un schéma séparé et de type fini sur k. Supposons de plus que*

– *soit X_k est propre sur k,*

– *soit X_k est affine avec un lieu singulier X_k^{sing} fini sur k.*

Alors, X_k admet une déformation formelle miniverselle $(\mathcal{R}, \mathcal{X})$. □

Théorème A.1.2 (Rim [Ri, Theorem 4.11, Corollary 4.13]). *Soit X_k un schéma séparé et de type fini sur k dont le lieu singulier X_k^{sing} est fini sur k. Pour chaque $x \in X_k^{\text{sing}}$, soit $U_{k,x}$ un ouvert affine de X_k qui ne rencontre X_k^{sing} qu'en x. On a alors les propriétés suivantes.*

1. *Si $H^2(X_k, \mathcal{H}om_{\mathcal{O}_{X_k}}(\Omega^1_{X_k/k}, \mathcal{O}_{X_k})) = (0)$, le morphisme de foncteurs*

$$\mathrm{Def}_{X_k} \to \prod_{x \in X_k^{\text{sing}}} \mathrm{Def}_{U_{k,x}}$$

est formellement lisse.

2. *Pour chaque $x \in X_k^{\text{sing}}$, le morphisme de foncteurs*

$$\mathrm{Def}_{U_{k,x}} \to \mathrm{Def}^{\text{top}}_{\widehat{\mathcal{O}}_{X,x}}$$

est formellement lisse et induit une bijection entre les espaces tangents de Zariski

$$\mathrm{Def}_{U_{k,x}}(k[\varepsilon]) \xrightarrow{\sim} \mathrm{Def}^{\text{top}}_{\widehat{\mathcal{O}}_{X,x}}(k[\varepsilon]);$$

en particulier, si $(\mathcal{R}, \mathcal{U}_x)$ est une déformation formelle miniverselle du k-schéma affine $U_{k,x}$, alors $(\mathcal{R}, \widehat{\mathcal{O}}_{\mathcal{U}_x,x})$ est une déformation formelle miniverselle topologique de $\widehat{\mathcal{O}}_{X,x}$.

3. *Si $H^2(X_k, \mathcal{H}om_{\mathcal{O}_{X_k}}(\Omega^1_{X_k/k}, \mathcal{O}_{X_k})) = (0)$ et si X_k est localement d'intersection complète, tous les foncteurs de déformations considérés dans* (i) *et* (ii) *sont formellement lisses sur k ; si de plus le lieu de lissité est dense dans X_k, la suite exacte entre les espaces tangents associée au morphisme de foncteurs*

$$\mathrm{Def}_{X_k} \to \prod_{x \in X_k^{\text{sing}}} \mathrm{Def}^{\text{top}}_{\widehat{\mathcal{O}}_{X_k,x}}$$

n'est autre que la suite exacte courte

$$\begin{aligned} 0 \to H^1(X_k, \mathcal{H}om_{\mathcal{O}_{X_k}}(\Omega^1_{X_k/k}, \mathcal{O}_{X_k})) &\to \mathrm{Ext}^1_{\mathcal{O}_{X_k}}(\Omega^1_{X_k/k}, \mathcal{O}_{X_k}) \\ &\to H^0(X_k, \mathcal{E}xt^1_{\mathcal{O}_{X_k}}(\Omega^1_{X_k/k}, \mathcal{O}_{X_k})) \to 0. \end{aligned}$$

□

A.2 Normalisation en famille et constance de l'invariant δ, d'après Teissier

Nous noterons dans la suite $\pi_X : \widetilde{X} \to X$ le morphisme de normalisation de tout schéma intègre X.

Pour toute courbe C_κ géométriquement intègre sur un corps κ, le lieu singulier C_κ^{sing} est fini et on définit

$$\delta(C_\kappa) = \sum_{c \in C_\kappa^{\mathrm{sing}}} [\kappa(c) : \kappa] \delta_c(C_\kappa)$$

comme dans la section 2.2.

Soient S un schéma local complet, intègre et nœthérien, de point fermé s, $\varphi : C \to S$ une courbe relative plate à fibres géométriquement réduites et $D \subset C$ un diviseur effectif fini et plat sur S tel φ soit lisse en dehors du support de D.

Soit $\mathfrak{a} \subset \mathcal{O}_C$ l'Idéal annulateur du $\mathcal{O}_C$-Module cohérent $\pi_{C,*}\mathcal{O}_{\widetilde{C}}/\mathcal{O}_C$ et, pour chaque $t \in S$, soit $\mathfrak{a}_t \subset \mathcal{O}_{C_t}$ l'Idéal annulateur du $\mathcal{O}_{C_t}$-Module cohérent $\pi_{C_t,*}\mathcal{O}_{\widetilde{C}_t}/\mathcal{O}_{C_t}$, où bien entendu C_t est la fibre de φ en t. Il n'est pas vrai en général que $\mathfrak{a}_t$ soit la restriction de $\mathfrak{a}$ à C_t.

Lemme A.2.1. *On peut trouver un diviseur effectif $D' \subset C$, fini et plat sur S, tel que l'on ait les inclusions*

$$\mathcal{O}_{C_t}(-D'_t) \subset \mathfrak{a}_t \subset \mathcal{O}_{C_t}$$

quel que soit $t \in S$.

Démonstration. Par hypothèse, on a un diviseur effectif $D \subset C$ fini et plat sur S tel que φ soit lisse en dehors du support de D.

Pour chaque point t de S, D induit un diviseur $D_t \subset C_t$ et, comme π_{C_t} est un isomorphisme en dehors du support de D_t, le support du fermé de C_t défini par $\mathfrak{a}_t$ est contenu dans celui de D_t. Il existe donc un entier $n \geq 0$ tel que $\mathcal{O}_{C_t}(-nD_t) \subset \mathfrak{a}_t \subset \mathcal{O}_{C_t}$; notons n_t le plus petit entier $n \geq 0$ ayant cette propriété.

La fonction $S \to \mathbb{N}$, $t \mapsto n_t$, est constructible. En effet, par induction nœthérienne, il suffit de montrer que cette fonction est localement constante sur un ouvert dense de S. Mais cela résulte de l'existence d'un ouvert dense normal U de S tel que, pour tout $t \in U$, π_{C_t} soit la fibre en t de π_C et $\mathfrak{a}_t$ soit la restriction de $\mathfrak{a}$ à C_t.

On peut donc choisir un entier n tel que $n \geq n_t$ quel que soit $t \in S$ et le diviseur $D' = nD$ répond à la question. □

Proposition A.2.1 (Teissier [Te 3, I, 1.3.2]). *La fonction $S \to \mathbb{Z}$, $t \mapsto \delta(C_t)$, est semi-continue supérieurement. Elle est même constante si l'on suppose de plus que le morphisme composé $\widetilde{\varphi} = \varphi \circ \pi_C : \widetilde{C} \to S$ est lisse.*

Inversement, si S est normal *et si la fonction $S \to \mathbb{Z}$, $t \mapsto \delta(C_t)$, est constante sur S, le morphisme $\widetilde{\varphi}$ est lisse.*

Chaque fois que le morphisme composé $\varphi \circ \pi_C : \widetilde{C} \to S$ sera lisse, on verra le morphisme de normalisation $\pi_C : \widetilde{C} \to C$ comme une *normalisation en famille* de la courbe relative $\varphi : C \to S$.

Teissier démontre dans un premier temps le cas particulier plus précis suivant.

Lemme A.2.2. *Supposons de plus que $S =$ Spec(V) soit un trait (V est donc un anneau de valuation discrète), de point générique η. Alors, le morphisme composé $\widetilde{\varphi} = \varphi \circ \pi_C : \widetilde{C} \to S$ est plat et on a la relation*

$$\delta(C_s) - \delta(C_\eta) = \delta((\widetilde{C})_s) \geq 0$$

où $(\widetilde{C})_s$ *est la fibre spéciale de* $\widetilde{\varphi}$.

Démonstration du Lemme A.2.2. Il résulte du lemme A.2.1 que l'image réciproque de la restriction de $\mathfrak{a}$ à $C_s \subset C$ par le morphisme de normalisation $\pi_{C_s} : \widetilde{C_s} \to C_s$ est non nulle. La propriété universelle de la normalisation $\pi_C : \widetilde{C} \to C$ nous donne alors une factorisation

$$\widetilde{C_s} \longrightarrow \widetilde{C} \xrightarrow{\pi_C} C$$

de $\widetilde{C_s} \xrightarrow{\pi_{C_s}} C_s \subset C$, ou ce qui revient au même une factorisation

$$\widetilde{C_s} \xrightarrow{\rho} (\widetilde{C})_s \xrightarrow{\pi_{C,s}} C_s$$

de π_{C_s} par la fibre spéciale $\pi_{C,s}$ de π_C.

Fixons une uniformisante v de V. On vérifie que

$$(v \cdot \pi_{C,*}\mathcal{O}_{\widetilde{C}}) \cap \mathcal{O}_C = v \cdot \mathcal{O}_C,$$

et donc que le $\mathcal{O}_S$-Module cohérent $\pi_{C,*}\mathcal{O}_{\widetilde{C}}/\mathcal{O}_C$ est sans v-torsion et que les homomorphismes de k-Algèbres

$$\mathcal{O}_{C_s} \to (\pi_{C,s})_*\mathcal{O}_{(\widetilde{C})_s} \to \pi_{C_s,*}\mathcal{O}_{\widetilde{C_s}}$$

correspondant à la factorisation ci-dessus sont injectifs. On vérifie aussi que la fibre générique $\pi_{C,\eta} : (\widetilde{C})_\eta \to C_\eta$ de π_C est la normalisation de C_η. Par suite, le $\mathcal{O}_S$-Module $\varphi_*(\pi_{C,*}\mathcal{O}_{\widetilde{C}}/\mathcal{O}_C)$ est libre de rang fini égal à

$$\begin{aligned}
\dim_{\kappa(s)} \varphi_{s,*}((\pi_{C,s})_*\mathcal{O}_{(\widetilde{C})_s}/\mathcal{O}_{C_s}) &= \dim_{\kappa(s)}(\varphi_*(\pi_{C,*}\mathcal{O}_{\widetilde{C}}/\mathcal{O}_C))_s \\
&= \dim_{\kappa(\eta)}(\varphi_*(\pi_{C,*}\mathcal{O}_{\widetilde{C}}/\mathcal{O}_C))_\eta \\
&= \dim_{\kappa(\eta)} \varphi_{\eta,*}((\pi_{C,\eta})_*\mathcal{O}_{(\widetilde{C})_\eta}/\mathcal{O}_{C_\eta}) = \delta(C_\eta),
\end{aligned}$$

le morphisme $\widetilde{\varphi}$ est plat, sa fibre spéciale $(\widetilde{C})_s$ est intègre, ρ est le morphisme de normalisation de $(\widetilde{C})_s$ et on a bien

$$\delta(C_s) = \delta((\widetilde{C})_s) + \dim_k \varphi_{s,*}((\pi_{C,s})_*\mathcal{O}_{(\widetilde{C})_s}/\mathcal{O}_{C_s}) = \delta((\widetilde{C})_s) + \delta(C_\eta). \qquad \square$$

Démonstration de la Proposition A.2.1. La semi-continuité de la fonction $t \mapsto \delta(C_t)$ dans le cas général résulte de sa constructibilité laissée au lecteur (voir la démonstration du lemme A.2.1) et du cas déjà traité où S est un trait.

Si $\widetilde{\varphi}$ est lisse, il résulte du lemme A.2.2 que

$$\delta(C_s) - \delta(C_\eta) = \delta((\widetilde{C})_s) = 0$$

dans le cas où S est un trait. On en déduit immédiatement que la fonction $t \mapsto \delta(C_t)$ est constante pour S arbitraire.

Inversement, supposons que $t \mapsto \delta(C_t)$ est constante de valeur δ et fixons un diviseur effectif $D' \subset S$ comme dans le Lemme A.2.1. Le $\mathcal{O}_S$-Module $\varphi_*(\mathcal{O}_C(D')/\mathcal{O}_C)$

est localement libre de rang fini $d \geq \delta$. Soit $G \to S$ la grassmannienne des quotients localement libres de rang $d-\delta$ de ce $\mathcal{O}_S$-Module. On définit une section *ensembliste* $\sigma : S \to G$ de G sur S en envoyant le point t sur le conoyau de l'inclusion

$$\Gamma(C_t, \pi_{C_t,*}\mathcal{O}_{\widetilde{C}_t}/\mathcal{O}_{C_t}) \subset \Gamma(C_t, \mathcal{O}_{C_t}(D'_t)/\mathcal{O}_{C_t})$$

(on a $\mathcal{O}_{C_t}(-D'_t) \cdot \pi_{C_t,*}\mathcal{O}_{\widetilde{C}_t} \subset \mathfrak{a}_t \cdot \pi_{C_t,*}\mathcal{O}_{\widetilde{C}_t} \subset \mathcal{O}_{C_t}$ et donc $\pi_{C_t,*}\mathcal{O}_{\widetilde{C}_t} \subset \mathcal{O}_{C_t}(D'_t)$).

Soit $Z \subset G$ l'ensemble des points $\sigma(t)$ pour t parcourant S. On vérifie par induction nœthérienne (comme dans la démonstration du lemme A.2.1) que l'ensemble Z est constructible. On vérifie aussi que les constructions de $\mathcal{F}$, G et Z commutent aux changements de bases $S' \to S$. On vérifie enfin, à l'aide du lemme A.2.2, que dans le cas où S est un trait, Z est l'image ensembliste d'une section algébrique de G sur S.

On déduit de ces propriétés que Z est l'ensemble des points d'un fermé réduit de G, noté encore Z, et que la restriction de la projection canonique $G \to S$ à ce fermé est un homéomorphisme de Z sur S.

Supposons de plus que S soit normal. Cet homéomorphisme est alors nécessairement un isomorphisme et σ est en fait une section algébrique de $G \to S$. En d'autres termes, les espaces vectoriels $\Gamma(C_t, \pi_{C_t,*}\mathcal{O}_{\widetilde{C}_t}/\mathcal{O}_{C_t})$ pour $t \in S$ sont les fibres d'un fibré vectoriel $\mathcal{E}$ de rang δ sur S qui est un sous-$\mathcal{O}_S$-Module localement facteur direct de $\varphi_*(\mathcal{O}_C(D')/\mathcal{O}_C)$.

Considérons maintenant le $\mathcal{O}_C$-Module cohérent

$$\mathcal{M} = (\mathcal{O}_C \oplus \mathcal{O}_C(D'))/\mathcal{O}_C$$

où $\mathcal{O}_C$ est plongé diagonalement dans $\mathcal{O}_C \oplus \mathcal{O}_C(D')$. On a d'une part une suite exacte évidente

$$0 \to \mathcal{O}_C \to \mathcal{M} \to \mathcal{O}_C(D')/\mathcal{O}_C \to 0.$$

On a d'autre part le plongement

$$\mathcal{M} \hookrightarrow \mathcal{O}_C(D'),\ (f \oplus f') + \mathcal{O}_C \mapsto f - f',$$

que l'on peut composer avec le plongement canonique de $\mathcal{O}_C(D')$ dans l'Anneau total des fractions $\mathcal{K}_C$ de $\mathcal{O}_C$.

Notons $\mathcal{F} \subset \mathcal{M}$ l'image réciproque de $\mathcal{E}$ par la surjection $\mathcal{M} \twoheadrightarrow \mathcal{O}_C(D')/\mathcal{O}_C$. On peut voir $\mathcal{F} \subset \mathcal{M}$ comme un sous-$\mathcal{O}_C$-Module de $\mathcal{K}_C$. Comme $\mathcal{O}_{C_t}(-D'_t) \subset \mathfrak{a}_t$, on vérifie facilement que la restriction de $\mathcal{F}$ à C_t est une sous-$\mathcal{O}_{C_t}$-Algèbre de l'Anneau total des fractions $\mathcal{K}_{C_t}$ de $\mathcal{O}_{C_t}$ quel que soit $t \in S$. Il s'en suit que $\mathcal{F}$ est une sous-$\mathcal{O}_C$-Algèbre de $\mathcal{K}_C$, et donc que

$$\mathcal{O}_C \subset \mathcal{F} \subset \pi_{C,*}\mathcal{O}_{\widetilde{C}} \subset \mathcal{K}_C$$

puisque $\pi_{C,*}\mathcal{O}_{\widetilde{C}}$ est la clôture intégrale de $\mathcal{O}_C$ dans $\mathcal{K}_C$ et que $\mathcal{F}$ est évidemment un $\mathcal{O}_C$-Module cohérent.

Comme on l'a vu au cours de la démonstration du Lemme A.2.2, pour chaque $t \in S$, on a les inclusions

$$\mathcal{O}_{C_t} \subset (\pi_{C,t})_*\mathcal{O}_{(\widetilde{C})_t} \subset \pi_{C_t,*}\mathcal{O}_{\widetilde{C}_t} \subset \mathcal{K}_{C_t}$$

et donc $(\pi_{C,t})_*\mathcal{O}_{(\widetilde{C})_t}$ est contenu dans la restriction de $\mathcal{F}$ à C_t. Il s'en suit que $\pi_{C,*}\mathcal{O}_{\widetilde{C}} \subset \mathcal{F} \subset \mathcal{K}_C$ et donc que $\mathcal{F} = \pi_{C,*}\mathcal{O}_{\widetilde{C}}$.

Comme $\mathcal{O}_C$ et $\mathcal{O}_C(D')/\mathcal{O}_C$ sont S-plats, il en est de même de $\mathcal{M}$. De plus, comme $\mathcal{E}$ est aussi S-plat, il en est de même de $\mathcal{F} = \pi_{C,*}\mathcal{O}_{\widetilde{C}}$. On a donc bien montré que $\widetilde{\varphi}$ est plat et qu'en outre

$$(\widetilde{C})_t = \widetilde{C}_t$$

pour tout $t \in S$. □

Corollaire A.2.1. *Les points $t \in S$ pour lesquels $\delta(C_t) = \delta(C_s)$ sont les points d'un fermé réduit $S^\delta \subset S$,* la strate à δ-constant.

La courbe plate relative $C_{\widetilde{S^\delta}} = \widetilde{S^\delta} \times_S C \to \widetilde{S^\delta}$ déduite de $\varphi : C \to S$ par le changement de base par le morphisme

$$\widetilde{S^\delta} \twoheadrightarrow S^\delta \hookrightarrow S,$$

composé du morphisme de normalisation $\pi_{\widetilde{S^\delta}}$ de S^δ et de l'inclusion de S^δ dans S, admet une normalisation en famille $\pi_{C_{\widetilde{S^\delta}}} : \widetilde{C}_{\widetilde{S^\delta}} \to C_{\widetilde{S^\delta}}$. □

A.3 Le foncteur des déformations de $A \hookrightarrow \widetilde{A}$

Soit $A = k[[x, y]]/(f)$ l'anneau formel d'un germe de courbe plane à singularité isolée, c'est-à-dire tel que le k-espace vectoriel $k[[x, y]]/(\partial_x f, \partial_y f)$ est de dimension finie. Comme on l'a rappelé dans la section 4.1, si k est de caractéristique nulle, il revient au même de demander que $k[[x, y]]/(f, \partial_x f, \partial_y f)$ soit de dimension finie (cf. [Te 1, 1.1]).

On note K l'anneau total des fractions de A et $\widetilde{A} \subset K$ la normalisation de A dans K. On peut décomposer K en un produit fini de corps $K = \prod_{i \in I} K_i$ et $\widetilde{A}$ en le produit correspondant d'anneaux intègres $\widetilde{A} = \prod_{i \in I} \widetilde{A}_i$. On pose $\delta(A) = \dim_k(\widetilde{A}/A)$.

On fixe dans la suite une uniformisante t_i de $\widetilde{A}_i$ pour chaque $i \in I$, de sorte que $\widetilde{A}_i = k[[t_i]]$. Le plongement $A \hookrightarrow \widetilde{A}$ est donné par une famille de couples $(x_i(t_i), y_i(t_i)) \in k[[t_i]]^2$ indexée par $i \in I$.

On rappelle (cf. [A-K 1, Chapter 8]) que :

- le module dualisant ω_A est le A-module libre de rang 1 défini par

$$\omega_A = \operatorname{Ext}^1_{k[[x,y]]}(A, \omega_{k[[x,y]]})$$

où $\omega_{k[[x,y]]} = \Omega^2_{k[[x,y]]/k}$ est un $k[[x, y]]$-module libre de rang 1,
- pour tout A-module M et tout entier i, on a un isomorphisme canonique de A-modules

$$\operatorname{Ext}^i_A(M, \omega_A) \cong \operatorname{Ext}^{i+1}_{k[[x,y]]}(M, \omega_{k[[x,y]]})$$

où M est vu comme un $k[[x, y]]$-module via l'épimorphisme canonique $k[[x, y]] \twoheadrightarrow A$,

– le module dualisant ω_A est donné concrètement par

$$\omega_A = A\tfrac{\mathrm{d}x\wedge \mathrm{d}y}{\mathrm{d}f} = A\tfrac{\mathrm{d}x}{\partial_y f} = A\big(-\tfrac{\mathrm{d}y}{\partial_x f}\big) \subset \Omega^1_{K/k},$$

et aussi par

$$\omega_A = \{\alpha \in \Omega^1_{K/k} \mid \mathrm{Res}(A\alpha) = (0)\}$$

où

$$\mathrm{Res} = \sum_{i\in I} \mathrm{Res}_i : \Omega^1_{K/k} = \bigoplus_{i\in I} \Omega^1_{K_i/k} \to k,\ \oplus_{i\in I} a_i(t_i)\frac{\mathrm{d}t_i}{t_i} \to \sum_{i\in I} a_i(0),$$

est la somme des homomorphismes résidus,

– on a

$$\omega_{\widetilde{A}} = \Omega^1_{\widetilde{A}/k} \subset \omega_A$$

et l'accouplement

$$(\widetilde{A}/A) \times (\omega_A/\omega_{\widetilde{A}}) \to k, \qquad (\widetilde{a} + A, \alpha + \omega_{\widetilde{A}}) \to \mathrm{Res}(\widetilde{a}\alpha),$$

est un accouplement parfait entre deux k-espaces vectoriels de dimension $\delta(A)$,

– le conducteur

$$\mathfrak{a} = \{a \in A \mid a\widetilde{A} \subset A\}$$

est aussi le conducteur

$$\mathfrak{a} = \{a \in A \mid a\omega_A \subset \omega_{\widetilde{A}}\}.$$

En particulier, on a :

Lemme A.3.1. *L'idéal de A engendré par les classes de $\partial_x f$ et $\partial_y f$ modulo (f) est contenu dans le conducteur $\mathfrak{a}$.* □

On considère maintenant le foncteur de déformations

$$\mathrm{Def}^{\mathrm{top}}_A : \mathrm{Art}_k \to \mathrm{Ens}$$

(cf. la section A.1 de l'appendice) et on s'intéresse plus particulièrement aux déformations de A à δ constant. Pour cela, on va étudier les déformations de l'homomorphisme de k-algèbres $A \hookrightarrow \widetilde{A}$. On considère donc le foncteur

$$\mathrm{Def}^{\mathrm{top}}_{A\hookrightarrow\widetilde{A}} : \mathrm{Art}_k \to \mathrm{Ens}$$

qui envoie $R \in \mathrm{ob}\,\mathrm{Art}_k$ sur l'ensemble des classes d'isomorphie d'homomorphismes de R-algèbres

$$A_R \to \widetilde{A}_R$$

dont la réduction modulo $\mathfrak{m}_R$ est l'inclusion $A \subset \widetilde{A}$, où A_R est une déformation plate sur R de A et $\widetilde{A}_R$ est une déformation plate sur R de $\widetilde{A}$. On a bien sûr un morphisme d'oubli

$$\mathrm{Def}^{\mathrm{top}}_{A\hookrightarrow\widetilde{A}} \to \mathrm{Def}^{\mathrm{top}}_A .$$

Lemme A.3.2. *Tout homomorphisme de R-algèbres* $(A_R \to \widetilde{A}_R) \in \mathrm{Def}^{\mathrm{top}}_{A \hookrightarrow \widetilde{A}}(R)$ *est nécessairement injectif et son conoyau est automatiquement R-plat.*

Démonstration. Notons N_R, I_R et C_R les noyau, image et conoyau de l'homomorphisme $A_R \to \widetilde{A}_R$. On a les suites exactes

$$0 \to \mathrm{Tor}_1^R(C_R, k) \to I_R \otimes_R k \to \widetilde{A}_R \otimes_R k \to C_R \otimes_R k \to 0$$

et

$$0 \to \mathrm{Tor}_1^R(I_R, k) \to N_R \otimes_R k \to A_R \otimes_R k \to I_R \otimes_R k \to 0$$

et l'homomorphisme composé

$$A_R \otimes_R k \twoheadrightarrow I_R \otimes_R k \to \widetilde{A}_R \otimes_R k$$

est par hypothèse l'inclusion $A \subset \widetilde{A}$; par suite, la surjection $A_R \otimes_R k \twoheadrightarrow I_R \otimes_R k$ est nécessairement un isomorphisme, la flèche $I_R \otimes_R k \to \widetilde{A}_R \otimes_R k$ est injective et $\mathrm{Tor}_1^R(C_R, k) = (0)$, de sorte que C_R est bien R-plat ; mais alors I_R est aussi R-plat puisque $\widetilde{A}_R$ l'est, et on a $N_R \otimes_R k = (0)$. Il s'en suit que $N_R = (0)$ et le lemme est démontré. □

Lemme A.3.3. *Soit M un A-module de type fini sans torsion. On a*

$$\mathrm{Ext}^i_A(M, A) = (0), \ \forall i \neq 0,$$

et il existe un entier $n \geq 0$ *tel que M, vu comme* $k[[x, y]]$*-module via l'épimorphisme canonique* $k[[x, y]] \twoheadrightarrow A$*, admette une résolution*

$$0 \to k[[x, y]]^n \to k[[x, y]]^n \to M \to 0.$$

Si on suppose de plus que M est de rang générique 1*, il existe même un tel entier* $n \leq \delta(A) + 1$.

Démonstration. Comme le A-module ω_A est libre de rang 1, pour tout A-module M on a

$$\mathrm{Ext}^i_A(M, A) \cong \mathrm{Ext}^i_A(M, \omega_A) \cong \mathrm{Ext}^{i+1}_{k[[x,y]]}(M, \omega_{k[[x,y]]})$$

et, comme la k-algèbre $k[[x, y]]$ est régulière de dimension 2, il s'en suit que

$$\mathrm{Ext}^i_A(M, A) = (0)$$

quel que soit $i \neq 0, 1$. Si M est sans torsion, on a de plus la suite exacte

$$\mathrm{Ext}^1_A(K \otimes_A M, A) \to \mathrm{Ext}^1_A(M, A) \to \mathrm{Ext}^2_A(K \otimes_A M/M, A)$$

où

$$\mathrm{Ext}^1_A(K \otimes_A M, A) \cong \mathrm{Ext}^1_K(K \otimes_A M, K) = (0)$$

et $\mathrm{Ext}^2_A(K \otimes_A M/M, A) = (0)$, et donc on a aussi $\mathrm{Ext}^1_A(M, A) = (0)$ et a fortiori

$$\mathrm{Ext}^2_{k[[x,y]]}(M, k[[x, y]]) \cong \mathrm{Ext}^2_{k[[x,y]]}(M, \omega_{k[[x,y]]}) = (0).$$

Par suite, si M est de type fini et sans torsion, le noyau de tout épimorphisme de $k[[x,y]]$-modules $k[[x,y]]^n \twoheadrightarrow M$ est nécessairement libre de rang fini et donc non canoniquement isomorphe à $k[[x,y]]^n$ puisqu'en tant que $k[[x,y]]$-module, M est de rang générique 0.

Montrons enfin que, si M un A-module de type fini, sans torsion et de rang générique 1, M peut être engendré par $\delta(A)+1$ éléments. Comme $\widetilde{A}\otimes_A M$ est un $\widetilde{A}$-module libre de rang 1 et que l'homomorphisme canonique $M \to \widetilde{A}\otimes_A M$ est injectif, on peut supposer $M \subset \widetilde{A}$ et que $\widetilde{A}M = \widetilde{A}$. Mais alors, M contient au moins un élément inversible m de $\widetilde{A}$ et, à isomorphisme près, on peut supposer que $m = 1$, c'est-à-dire que

$$A \subset M \subset \widetilde{A}.$$

Sous ces conditions, on a $\dim_k(M/A) \leq \delta(A)$ et on conclut en remarquant que M est engendré sur A par $\{1, m_1, \ldots, m_n\}$ où $\{m_1, \ldots, m_n\} \subset M$ représente une base de M/A sur k. □

Considérons maintenant les foncteurs

$$\mathrm{Def}^{\mathrm{top}}_{k[[x,y]]\twoheadrightarrow A},\ \mathrm{Def}^{\mathrm{top}}_{k[[x,y]]\to \widetilde{A}} : \mathrm{Art}_k \to \mathrm{Ens}$$

qui envoie $R \in \mathrm{ob}\,\mathrm{Art}_k$ sur l'ensemble des classes d'isomorphie d'homomorphismes de R-algèbres

$$R[[x,y]] \to A_R \text{ et } R[[x,y]] \to \widetilde{A}_R$$

dont les réductions modulo $\mathfrak{m}_R$ sont l'épimorphisme canonique $k[[x,y]] \twoheadrightarrow A$ et l'homomorphisme composé $k[[x,y]] \twoheadrightarrow A \hookrightarrow \widetilde{A}$, où bien entendu A_R et $\widetilde{A}_R$ sont des déformations plates sur R de A et $\widetilde{A}$ respectivement. On remarque que $R[[x,y]] \to A_R$ est automatiquement un épimorphisme d'après le lemme de Nakayama. On a les morphismes de foncteurs évidents

$$\begin{array}{ccc} & & \mathrm{Def}^{\mathrm{top}}_{A\hookrightarrow \widetilde{A}} \\ & & \downarrow \\ \mathrm{Def}^{\mathrm{top}}_{k[[x,y]]\twoheadrightarrow A} & \longrightarrow & \mathrm{Def}^{\mathrm{top}}_{A} \end{array}$$

et on pose

$$\mathrm{Def}^{\mathrm{top}}_{k[[x,y]]\twoheadrightarrow A\hookrightarrow \widetilde{A}} = \mathrm{Def}^{\mathrm{top}}_{k[[x,y]]\twoheadrightarrow A} \times_{\mathrm{Def}^{\mathrm{top}}_A} \mathrm{Def}^{\mathrm{top}}_{A\hookrightarrow \widetilde{A}}.$$

Le foncteur $\mathrm{Def}^{\mathrm{top}}_A$, le morphisme de foncteurs $\mathrm{Def}^{\mathrm{top}}_{k[[x,y]]\twoheadrightarrow A} \to \mathrm{Def}^{\mathrm{top}}_A$ et donc aussi le foncteur $\mathrm{Def}^{\mathrm{top}}_{k[[x,y]]\twoheadrightarrow A}$, sont formellement lisses.

Théorème A.3.1. *Le morphisme de composition* $\mathrm{Def}^{\mathrm{top}}_{k[[x,y]]\twoheadrightarrow A\hookrightarrow \widetilde{A}} \to \mathrm{Def}^{\mathrm{top}}_{k[[x,y]]\to \widetilde{A}}$ *est un isomorphisme.*

Démonstration. Nous allons expliciter un inverse pour ce morphisme de composition. Suivant une suggestion de J.-B. Bost, nous utiliserons pour cela un cas très simple de la construction «Div» de Mumford (cf. [M-F, Chapter 5, Section 3]).

Soit $R[[x,y]] \to \widetilde{A}_R$ une déformation de $k[[x,y]] \to \widetilde{A}$. On voit $\widetilde{A}$ et $\widetilde{A}_R$ comme des modules sur $k[[x,y]]$ et $R[[x,y]]$ à l'aide de ces homomorphismes. Fixons arbitrairement une résolution

$$0 \to k[[x,y]]^n \xrightarrow{F} k[[x,y]]^n \to \widetilde{A} \to 0$$

de $\widetilde{A}$ en tant que $k[[x,y]]$-module (il en existe d'après le lemme précédent). Si F^* est la matrice des co-facteurs de F, on a $F^*F = FF^* = \det F$ et $\det F$ annule $\widetilde{A}$. Inversement si une fonction $g \in k[[x,y]]$ annule $\widetilde{A}$, c'est-à-dire est telle que $gk[[x,y]]^n \subset Fk[[x,y]]^n$, il existe une matrice carrée G de taille $n \times n$ à coefficient dans $k[[x,y]]^n$ telle que $g = FG$ et $g^n = \det F \cdot \det G$. Comme $k[[x,n]]$ est réduit, il s'en suit que $\det F$ et f engendre le même idéal de $k[[x,y]]$ et que l'on peut demander de plus que $\det F = f$.

Comme $\widetilde{A}_R$ est plat sur R, on peut relever la résolution ci-dessus en une résolution

$$0 \to R[[x,y]]^n \xrightarrow{F_R} R[[x,y]]^n \to \widetilde{A}_R \to 0.$$

En considérant la matrice des co-facteurs de F_R, on montre comme ci-dessus que $f_R = \det F_R$ dans le noyau de l'homomorphisme $R[[x,y]] \to \widetilde{A}_R$, ou ce qui revient au même que l'on a une factorisation

$$R[[x,y]] \twoheadrightarrow R[[x,y]]/(f_R) \to \widetilde{A}_R$$

qui relève la factorisation $k[[x,y]] \twoheadrightarrow A \hookrightarrow \widetilde{A}$. Alors, $A_R := R[[x,y]]/(f_R)$ est un relèvement plat sur R de A et la flèche $A_R \to \widetilde{A}_R$ est nécessairement injective (à conoyau R-plat) d'après le Lemme A.3.2. La factorisation $R[[x,y]] \twoheadrightarrow A_R \hookrightarrow \widetilde{A}_R$ est donc la factorisation canonique par l'image et la construction de Mumford produit bien un inverse au morphisme de foncteurs $\mathrm{Def}^{\mathrm{top}}_{k[[x,y]]\twoheadrightarrow A\hookrightarrow \widetilde{A}} \to \mathrm{Def}^{\mathrm{top}}_{k[[x,y]]\to \widetilde{A}}$. □

Si J est un idéal de carré nul dans $R \in \mathrm{ob}\,\mathrm{Art}_k$ et si $\overline{R} = R/J$, il n'y a pas d'obstruction à relever un homomorphisme de $\overline{R}$-algèbres $\overline{R}[[x,y]] \to \widetilde{A}_{\overline{R}}$ en un homomorphisme de $\overline{R}$-algèbres $R[[x,y]] \to \widetilde{A}_R$. En particulier, le foncteur $\mathrm{Def}^{\mathrm{top}}_{k[[x,y]]\to \widetilde{A}}$ est formellement lisse et son espace tangent est le k-espace vectoriel $\widetilde{A} \oplus \widetilde{A}$. Le théorème admet donc le corollaire suivant :

Corollaire A.3.1. *Le foncteur* $\mathrm{Def}^{\mathrm{top}}_{k[[x,y]]\to A\hookrightarrow \widetilde{A}}$*, et donc aussi le foncteur* $\mathrm{Def}^{\mathrm{top}}_{A\hookrightarrow \widetilde{A}}$*, est formellement lisse.* □

Remarque A.3.1. En évaluant l'isomorphisme du théorème sur $R = k[\varepsilon]$ avec $\varepsilon^2 = 0$, on trouve que, pour tout $(\dot{x}(t_i), \dot{y}(t_i))_{i\in I} \in \prod_{i\in I}(k[[t_i]] \times k[[t_i]]) = \widetilde{A} \oplus \widetilde{A}$, il existe $\dot{f}(x,y) \in k[[x,y]]$ tel que

$$(f + \varepsilon \dot{f})(x(t_i) + \varepsilon \dot{x}(t_i), y(t_i) + \varepsilon \dot{y}(t_i)) \equiv 0, \ \forall i \in I,$$

c'est-à-dire tel que

$$\dot{f}(x(t_i), y(t_i)) + \partial_x f(x(t_i), y(t_i))\dot{x}(t_i) + \partial_y f(x(t_i), y(t_i))\dot{y}(t_i) \equiv 0, \ \forall i \in I,$$

ce qui est une reformulation du Lemme A.3.1. □

Nous allons maintenant déterminer l'espace tangent au foncteur $\mathrm{Def}^{\mathrm{top}}_{A\hookrightarrow\widetilde{A}}$. Plus généralement étudions le problème de relèvement d'une déformation $A_{\overline{R}}\to\widetilde{A}_{\overline{R}}$ de $A\hookrightarrow\widetilde{A}$ sur $\overline{R}=R/J$, où J est un idéal de carré nul dans R, en une déformation $A_R\to\widetilde{A}_R$ de $A\hookrightarrow\widetilde{A}$ sur R.

Commençons par fixer un relèvement R-plat $\widetilde{A}_R$ de $\widetilde{A}_{\overline{R}}$ sur R. On a donc un diagramme

$$\begin{array}{ccc} R & \twoheadrightarrow & \overline{R} \\ & & \downarrow \\ & & A_{\overline{R}} \\ & & \downarrow \\ \widetilde{A}_R & \longrightarrow & \widetilde{A}_{\overline{R}} \end{array}$$

que l'on cherche à compléter en un diagramme

$$\begin{array}{ccc} R & \twoheadrightarrow & \overline{R} \\ \downarrow & & \downarrow \\ A_R & \twoheadrightarrow & A_{\overline{R}} \\ \downarrow & & \downarrow \\ \widetilde{A}_R & \twoheadrightarrow & \widetilde{A}_{\overline{R}} \end{array}$$

où A_R est une déformation plate de $A_{\overline{R}}$ sur R. D'après Illusie [Il, Chapitre III, Section 2.3], ce problème de relèvement est contrôlé par un complexe T qui s'insère dans un triangle distingué

$$T\to\mathrm{RHom}_{A_{\overline{R}}}(L_{A_{\overline{R}}/\overline{R}},J\otimes_{\overline{R}}A_{\overline{R}})\to\mathrm{RHom}_{\widetilde{A}_{\overline{R}}}(\widetilde{A}_{\overline{R}}\otimes_{A_{\overline{R}}}L_{A_{\overline{R}}/\overline{R}},J\otimes_{\overline{R}}\widetilde{A}_{\overline{R}})\to$$

où $L_{A_{\overline{R}}/\overline{R}}$ est le complexe cotangent de la $\overline{R}$-algèbre $A_{\overline{R}}$. Plus précisément, il y a une obstruction à relever dans $H^2(T)$ et, si cette obstruction est nulle, l'ensemble des classes d'isomorphie de relèvements est un torseur sous $H^1(T)$ et le groupe des automorphismes d'un relèvement donné s'identifie canoniquement à $H^0(T)$. Comme la déformation plate $A_{\overline{R}}$ de A sur $\overline{R}$ est nécessairement de la forme $A_{\overline{R}}=\overline{R}[[x,y]]/(f_{\overline{R}})$ pour $f_{\overline{R}}\in\overline{R}[[x,y]]$, le complexe cotangent

$$L_{A_{\overline{R}}/\overline{R}}=[A_{\overline{R}}\to A_{\overline{R}}\mathrm{d}x\oplus A_{\overline{R}}\mathrm{d}y]$$

est concentré en degrés $[-1,0]$ et son unique différentielle non triviale envoie 1 sur $df_{\overline{R}}=(\partial_x f_{\overline{R}})\mathrm{d}x\oplus(\partial_y f_{\overline{R}})\mathrm{d}y$. Comme de plus la flèche $A_{\overline{R}}\to\widetilde{A}_{\overline{R}}$ est nécessairement injective à conoyau $\overline{R}$-plat d'après le Lemme A.3.2, le complexe

$$T=[J\otimes_{\overline{R}}(\widetilde{A}_{\overline{R}}/A_{\overline{R}})\partial_x\oplus J\otimes_{\overline{R}}(\widetilde{A}_{\overline{R}}/A_{\overline{R}})\partial_y\to J\otimes_{\overline{R}}(\widetilde{A}_{\overline{R}}/A_{\overline{R}})]$$

est concentré en degrés $[1,2]$ et son unique différentielle envoie $\widetilde{a}\partial_x\oplus\widetilde{b}\partial_y$ sur $\widetilde{a}\partial_x f_{\overline{R}}+\widetilde{b}\partial_y f_{\overline{R}}$. On a donc $H^0(T)=(0)$ et la suite exacte

$$0\to H^1(T)\to J\otimes_{\overline{R}}(\widetilde{A}_{\overline{R}}/A_{\overline{R}})\partial_x\oplus J\otimes_{\overline{R}}(\widetilde{A}_{\overline{R}}/A_{\overline{R}})\partial_y$$

$$\to J\otimes_{\overline{R}}(\widetilde{A}_{\overline{R}}/A_{\overline{R}})\to H^2(T)\to 0.$$

L'obstruction au relèvement se calcule comme suit. On se donne des relèvements arbitraires $f_R\in R[[x,y]]$ de $f_{\overline{R}}$ et $R[[x,y]]\to\widetilde{A}_R$ de l'homomorphisme de $\overline{R}$-algèbres composé $\overline{R}[[x,y]]\twoheadrightarrow A_{\overline{R}}\hookrightarrow\widetilde{A}_{\overline{R}}$. Alors l'image de f_R par l'homomorphisme de R-algèbres $R[[x,y]]\to\widetilde{A}_R$ est dans $J\widetilde{A}_R\subset\widetilde{A}_R$. L'obstruction cherchée est l'image de l'élément ainsi construit de $J\widetilde{A}_R$ par l'épimorphisme composé

$$J\widetilde{A}_R\cong J\otimes_{\overline{R}}\widetilde{A}_{\overline{R}}\twoheadrightarrow J\otimes_{\overline{R}}(\widetilde{A}_{\overline{R}}/A_{\overline{R}})\twoheadrightarrow H^2(T).$$

Bien entendu, cette obstruction ne dépend pas des choix faits et elle est nulle d'après le Corollaire A.3.1.

En particulier, l'espace tangent relatif au morphisme de foncteurs $\mathrm{Def}^{\mathrm{top}}_{\widetilde{A}}\to\mathrm{Def}^{\mathrm{top}}_{A\hookrightarrow\widetilde{A}}$ est le noyau de l'application k-linéaire

$$(\widetilde{A}/A)\partial_x\oplus(\widetilde{A}/A)\partial_y\to(\widetilde{A}/A),\ \widetilde{a}\partial_x\oplus\widetilde{b}\partial_y\mapsto\widetilde{a}\partial_x f+\widetilde{b}\partial_y f,$$

application qui est identiquement nulle d'après le Lemme A.3.1. Cet espace tangent est donc égal au k-espace vectoriel $(\widetilde{A}/A)\partial_x\oplus(\widetilde{A}/A)\partial_y$ de dimension $2\delta(A)$.

Comme $\widetilde{A}=\prod_{i\in I}k[[t_i]]$, le foncteur $\mathrm{Def}^{\mathrm{top}}_{\widetilde{A}}$ est trivial. Plus précisément, toute déformation plate de $\widetilde{A}$ sur $R\in\mathrm{ob}\,\mathrm{Art}_k$ est isomorphe à $\prod_{i\in I}R[[t_i]]$ et, en termes de relèvements, pour tout idéal J de carré nul dans R et pour $\overline{R}=R/J$, il n'y a pas d'obstruction à relever $\widetilde{A}_{\overline{R}}=\prod_{i\in I}\overline{R}[[t_i]]$ à R, il n'y a qu'une seule classe d'isomorphie de tels relèvements, à savoir celle de $\widetilde{A}_R=\prod_{i\in I}R[[t_i]]$, et le groupe des automorphismes d'un relèvement arbitraire dans cette classe est le groupe

$$\mathrm{Hom}_{\widetilde{A}_{\overline{R}}}(\Omega^1_{\widetilde{A}_{\overline{R}}/\overline{R}},J\otimes_{\overline{R}}\widetilde{A}_{\overline{R}})=\prod_{i\in I}(J\otimes_{\overline{R}}\overline{R}[[t_i]])\partial_{t_i}.$$

On a donc montré :

Proposition A.3.1. *Soient $R\in\mathrm{ob}\,\mathrm{Art}_k$, $J\subset R$ un idéal de carré nul et $\overline{R}=R/J$.*

1. *Tout objet $A_R\hookrightarrow\widetilde{A}_R$ de $\mathrm{Def}^{\mathrm{top}}_{A\hookrightarrow\widetilde{A}}(R)$ est de la forme*

$$R[[x,y]]/(f_R)\hookrightarrow\prod_{i\in I}R[[t_i]],\ x\mapsto(x_{R,i}(t_i))_{i\in I},\ y\mapsto(y_{R,i}(t_i))_{i\in I},$$

pour des séries $f_R\in R[[x,y]]$ et $x_{R,i}(t_i),y_{R,i}(t_i)\in R[[t_i]]$, qui relèvent les séries f et $x_i(t),y_i(t)$, et qui vérifient bien sûr

$$f_R(x_{R,i}(t_i),y_{R,i}(t_i))\equiv 0,\ \forall i\in I.$$

2. *Soit $(A_{\overline{R}}\hookrightarrow\widetilde{A}_{\overline{R}})\in\mathrm{Def}^{\mathrm{top}}_{A\hookrightarrow\widetilde{A}}(\overline{R})$ isomorphe à*

$$\overline{R}[[x,y]]/(f_{\overline{R}})\hookrightarrow\prod_{i\in I}\overline{R}[[t_i]],\ x\mapsto(x_{\overline{R},i}(t_i))_{i\in I},\ y\mapsto(y_{\overline{R},i}(t_i))_{i\in I}.$$

Il n'y a pas d'obstruction à relever cet objet à R, l'ensemble des classes d'isomorphie des relèvements est le conoyau de la flèche

$$\bigoplus_{i\in I}(J\otimes_{\overline{R}}\overline{R}[[t_i]])\partial_{t_i}\to \operatorname{Ker}(J\otimes_{\overline{R}}(\widetilde{A}_{\overline{R}}/A_{\overline{R}})\partial_x\oplus J\otimes_{\overline{R}}(\widetilde{A}_{\overline{R}}/A_{\overline{R}})\partial_y$$
$$\to J\otimes_{\overline{R}}(\widetilde{A}_{\overline{R}}/A_{\overline{R}}))$$

qui envoie $\oplus_{i\in I}a_i(t_i)\partial_{t_i}$ *sur*

$$\big((a_i(t_i)(\partial_{t_i}x_{\overline{R},i})(t_i))_{i\in I}+J\otimes_{\overline{R}}A_{\overline{R}}\big)\partial_x\oplus\big((a_i(t_i)(\partial_{t_i}y_{\overline{R},i})(t_i))_{i\in I}+J\otimes_{\overline{R}}A_{\overline{R}}\big)\partial_y$$

et que le groupe des automorphismes d'un relèvement arbitraire dans cette classe est son noyau. □

Corollaire A.3.2. *L'espace tangent au foncteur formellement lisse* $\mathrm{Def}^{\mathrm{top}}_{A\hookrightarrow\widetilde{A}}$ *est le conoyau de l'application k-linéaire*

$$\bigoplus_{i\in I}k[[t_i]]\partial_{t_i}\to(\widetilde{A}/A)\partial_x\oplus(\widetilde{A}/A)\partial_y$$

qui envoie $\oplus_{i\in I}a_i(t_i)\partial_{t_i}$ *sur*

$$\big((a_i(t_i)(\partial_{t_i}x_i)(t_i))_{i\in I}+A\big)\partial_x\oplus\big((a_i(t_i)(\partial_{t_i}y_i)(t_i))_{i\in I}+A\big)\partial_y.$$

De plus, l'application tangente au morphisme de foncteur $\mathrm{Def}^{\mathrm{top}}_{A\hookrightarrow\widetilde{A}}\to\mathrm{Def}^{\mathrm{top}}_A$ *est induite par l'application k-linéaire*

$$(\widetilde{A}/A)\partial_x\oplus(\widetilde{A}/A)\partial_y\to A/(\partial_xf,\partial_yf)$$

qui envoie la classe d'un élément $\widetilde{a}\partial_x\oplus\widetilde{b}\partial_y\in\widetilde{A}\partial_x\oplus\widetilde{A}\partial_y$ *sur la classe de l'élément* $-(\widetilde{a}\partial_xf+\widetilde{b}\partial_yf)\in A\subset\widetilde{A}$. □

A.4 La strate à δ constant, d'après Diaz et Harris

Les résultats de cette section devraient être vérifiés pour k de caractéristique arbitraire. Faute de référence nous nous limiterons à la caractéristique nulle.

Supposons donc de plus que k est de caractéristique nulle.

Théorème A.4.1 (Diaz et Harris [D-H, Proposition 4.17 et Theorem 4.15]). *Le morphisme de k-schémas formels* $\mathrm{Def}^{\mathrm{top}}_{A\hookrightarrow\widetilde{A}}\to\mathrm{Def}^{\mathrm{top}}_A$ *est fini, son image schématique*

$$\mathrm{Def}^{\mathrm{top},\delta}_A\subset\mathrm{Def}^{\mathrm{top}}_A$$

est une fermé intègre de codimension $\delta(A)$ *et le morphisme canonique* $\mathrm{Def}^{\mathrm{top}}_{A\hookrightarrow\widetilde{A}}\to\mathrm{Def}^{\mathrm{top},\delta}_A$ *est le morphisme de normalisation.*

De plus, le cône tangent de $\mathrm{Def}^{\mathrm{top},\delta}_A$ *est le sous-k-espace vectoriel*

$$V(A)=\mathfrak{a}/(\partial_xf,\partial_yf)\subset A/(\partial_xf,\partial_yf)=\mathrm{Def}^{\mathrm{top}}_A(k[\varepsilon])$$

de l'espace tangent de $\mathrm{Def}^{\mathrm{top}}_A$*, où* $\mathfrak{a}\subset A$ *est le conducteur du normalisé* $\widetilde{A}$ *de A dans A (cf. Lemme* A.3.1.) □

Soient maintenant C_k une courbe projective, intègre et à singularités planes isolées sur k, et $\varphi : C \to S$ une algébrisation d'une déformation formelle miniverselle de C_k comme dans la section 4.1. Toutes les fibres de φ sont géométriquement intègre et le lieu singulier $\bigcup_{t\in S} C_t^{\mathrm{sing}}$ est fini sur S. On peut donc appliquer le Corollaire A.2.1 à cette courbe et on obtient la strate à δ constant

$$S^\delta \subset S.$$

Pour chaque point singulier c de C_k, on peut aussi considérer une déformation miniverselle $\mathcal{C}_c = \mathrm{Spf}(\mathcal{A}_c) \to \mathcal{S}_c = \mathrm{Spf}(\mathcal{R}_c)$ de $\mathcal{C}_{c,s} = \mathrm{Spf}(\widehat{\mathcal{O}}_{C_k,c})$. Il résulte du théorème A.1.3 que l'on a un morphisme canonique formellement lisse

$$\mathcal{S} \to \prod_{c\in C_k^{\mathrm{sing}}} \mathcal{S}_c$$

où $\mathcal{S} = \mathrm{Spf}(\mathcal{O}_{S,s})$ est le complété de S en son point fermé s.

Pour chaque $c \in C_k^{\mathrm{sing}}$, soit

$$\mathcal{S}_c^\delta \subset \mathcal{S}_c$$

la strate à δ constant.

Lemme A.4.1. *Le complété de S^δ au point fermé s est le fermé de $\mathcal{S}$ image réciproque du fermé $\prod_{c\in C_k^{\mathrm{sing}}} \mathcal{S}_c^\delta$ de $\prod_{c\in C_k^{\mathrm{sing}}} \mathcal{S}_c$ par le morphisme canonique ci-dessus.* □

L'énoncé suivant est une variante globale du théorème A.4.1. La dernière assertion est bien connue mais ne semble être démontrée nulle part ; elle résulte du théorème (1.3) de [D-H] dans le cas où C_k est une courbe plane.

Théorème A.4.2. *La strate S^δ est irréductible de codimension dans S égale à $\delta(C_k)$. Le schéma normalisé de S^δ est formellement lisse sur k.*

De plus, le cône tangent à l'origine de S^δ est le sous-k-espace vectoriel

$$V(C_k) \subset T_s S$$

de l'espace tangent à l'origine s de S obtenu par image inverse du sous-k-espace vectoriel

$$\bigoplus_{c\in C_k^{\mathrm{sing}}} V(\widehat{\mathcal{O}}_{C_s,c}) \subset \bigoplus_{c\in C_k^{\mathrm{sing}}} \mathrm{Def}^{\mathrm{top}}_{\widehat{\mathcal{O}}_{C_s,c}}(k[\varepsilon])$$

par l'épimorphisme naturel $T_s S \twoheadrightarrow \bigoplus_{c\in C_k^{\mathrm{sing}}} \mathrm{Def}^{\mathrm{top}}_{\widehat{\mathcal{O}}_{C_s,c}}(k[\varepsilon])$.

En outre, la fibre de $C \to S$ en tout point géométrique générique de $S^\delta \subset S$ est une courbe n'ayant comme seules singularités que $\delta(C_k)$ points doubles ordinaires. □

Remerciements. Je remercie J.-B. Bost, E. Esteves, L. Göttsche, L. Illusie, S. Kleiman, Y. Laszlo, F. Loeser et M. Raynaud pour l'aide qu'ils m'ont apportée durant la préparation de ce travail.

Litérature

[A-I-K] A. Altman, A. Iarrobino, and S. Kleiman, Irreducibility of the compactified Jacobian, in P. Holm, ed., *Real and Complex Singularities : Proceedings, Oslo* 1976, Sijthoff and Nordhoff, Alpen van Rijn, The Netherlands, 1977, 1–12.

[A-K 1] A. Altman and S. Kleiman, *Introduction to Grothendieck Duality Theory*, Lecture Notes in Mathematics, Vol. 146, Springer-Verlag, Berlin, New York, Heidelberg, 1970.

[A-K 2] A. Altman and S. Kleiman, Compactifying the Jacobian, *Bull. Amer. Math. Soc.*, **82** (1976), 947–949.

[A-K 3] A. Altman and S. Kleiman, Compactifying the Picard scheme, *Adv. Math.*, **35** (1980), 50–112.

[B-L-R] AUTHOR: PLEASE SUPPLY INFORMATION FOR REFERENCE [B-L-R].

[Be] R. Bezrukavnikov, The dimension of the fixed point set on affine flag manifolds, *Math. Res. Lett.*, **3** (1996), 185–189.

[De] P. Deligne, La conjecture de Weil II, *Publ. Math. IHÉS*, **52** (1980), 313–428.

[D-H] S. Diaz and J. Harris, Ideals associated to deformations of singular plane curves, *Trans. Amer. Math. Soc.*, **309** (1988), 433–468.

[E-G-K] E. Esteves, M. Gagné, and S. Kleiman, Autoduality of the compactified Jacobian, math.AG/9911071, 1999.

[F-G-S] D. Fantechi, L. Göttsche, and D. Van Straten, Euler number of the compactified Jacobian and multiplicity of rational curves, *J. Algebraic Geom.*, **8** (1999), 115–133.

[Gr 1] A. Grothendieck, *Fondements de la géométrie algébrique*, Séminaire Bourbaki, Vol. 232, W. A. Benjamin, New York, 1966.

[Gr 2] A. Grothendieck, Éléments de géométrie algébrique III : Étude cohomologique des faisceaux cohérents, *Publ. Math. IHÉS*, **11** (1961).

[Gr 3] A. Grothendieck, Éléments de géométrie algébrique IV : Étude locale des schémas et des morphismes de schémas (troisième partie), *Publ. Math. IHÉS*, **28** (1966).

[Il] L. Illusie, *Complexe Cotangent et Déformations* I, Lecture Notes in Mathematics, Vol. 239, Springer-Verlag, Berlin, 1971.

[K-L] D. Kazhdan and G. Lusztig, Fixed point varieties on affine flag manifolds, *Israel J. Math.*, **62** (1988), 129–168.

[Lu-Sm] G. Lusztig and J. M. Smelt, Fixed point varieties on the space of lattices, *Bull. London Math. Soc.*, **23** (1991), 213–218.

[M-F] D. Mumford and J. Fogarty, *Geometric Invariant Theory*, 2nd enlarged ed., Springer-Verlag, Berlin, New York, Heidelberg, 1982.

[Ra] M. Raynaud, Spécialisation du foncteur de Picard, *Publ. Math. IHÉS*, **38** (1970), 27–76.

[Re] C. J. Rego, The compactified Jacobian, *Ann. Sci. École Norm. Sup.*, **13** (1980), 211–223.

[Ri] D. S. Rin, Formal deformation theory, in *SGA* 7 I : *Groupe de Monodromie en Géométrie Algébrique, dirigé par A. Grothendieck*, Lecture Notes in Mathematics, Vol. 288, Springer-Verlag, Berlin, 1972, 32–132.

[Sp] T. A. Springer, A purity result for fixed point varieties in flag manifolds, *J. Fac. Sci. Univ. Tokyo*, **31** (1984), 271–282.

[Te 1] B. Teissier, Cycles évanescents, sections hyperplanes et condition de Whitney, in *Singularités à Cargèse*, Astérisque, Vols. 7–8, Société Mathématique de France, Paris, 1973, 285–362.

[Te 2] B. Teissier, The hunting of invariants in the geometry of discriminants, in P. Holm, ed., *Real and Complex Singularities : Proceedings, Oslo* 1976, Sijthoff and Nordhoff, Alpen van Rijn, The Netherlands, 1977, 565–677.

[Te 3] B. Teissier, Résolution simultanée I, II, in M. Demazure, H. Pinkham, and B. Teissier, eds., *Séminaire sur les Singularités des Surfaces, Palaiseau, France* 1976–1977, Lecture Notes in Mathematics, Vol. 777, Springer-Verlag, Berlin, New York, Heidelberg, 1980, 71–146.

Iterated integrals of modular forms and noncommutative modular symbols

Yuri I. Manin

Max-Planck Institut für Mathematik
Vivatsgasse 7
D-53111 Bonn
Germany
manin@mpim-bonn.mpg.de
and
Department of Mathematics
Northwestern University
2033 Sheridan Road
Evanston, IL 60208-2730
USA

To Volodya Drinfeld, cordially and friendly.

Summary. The main goal of this paper is to study properties of the iterated integrals of modular forms in the upper half-plane, possibly multiplied by z^{s-1}, along geodesics connecting two cusps. This setting generalizes simultaneously the theory of modular symbols and that of multiple zeta values.

Subject Classifications: Primary 11F67, 11M41. Secondary 11G55.

0 Introduction and summary

This paper was inspired by two sources: the theory of multiple zeta values on the one hand (see [Za2]) and the theory of modular symbols and periods of cusp forms on the other [Ma1, Ma2, Sh1, Sh2, Sh3, Me]. Roughly speaking, it extends the theory of periods of modular forms, replacing integration along geodesics in the upper complex half-plane by iterated integration. Here are some details.

0.1 Multiple zeta values

They are the numbers given by the k-multiple Dirichlet series

$$\zeta(m_1, \ldots, m_k) = \sum_{0<n_1<\cdots<n_k} \frac{1}{n_1^{m_1} \cdots n_k^{m_k}}, \tag{0.1}$$

which converge for all integer $m_i \geq 1$ and $m_k > 1$, or equivalently by the m-multiple iterated integrals, $m = m_1 + \cdots + m_k$,

$$\zeta(m_1, \ldots, m_k) = \int_0^1 \frac{dz_1}{z_1} \int_0^{z_1} \frac{dz_2}{z_2} \int_0^{z_2} \cdots \int_0^{z_{m_{k-1}}} \frac{dz_{m_k}}{1 - z_{m_k}} \cdots \tag{0.2}$$

where the sequence of differential forms in the iterated integral consists of consecutive subsequences of the form $\frac{dz}{z}, \ldots, \frac{dz}{z}, \frac{dz}{1-z}$ of lengths $m_k, m_{k-1}, \ldots, m_1$.

Easy combinatorial considerations allow one to express in two different ways products $\zeta(l_1, \ldots, l_j) \cdot \zeta(m_1, \ldots, m_k)$ as linear combinations of multiple zeta values.

If one uses for this the integral representation (0.2), one gets a sum over shuffles which enumerate the simplices of highest dimension occurring in the natural simplicial decomposition of the product of two integration simplices.

If one uses instead (0.1), one gets sums over shuffles with repetitions which enumerate some simplices of lower dimension as well.

These relations and their consequences are called double shuffle relations. Both types of relations can be succinctly written down in terms of formal series on free noncommutating generators. One can include in these relations regularized multiple zeta values for arguments where the convergence of (0.1), (0.2) fails.

For a very clear and systematic exposition of these results, see [De, Ra1, Ra2].

In fact, the formal generating series for (regularized) iterated integrals (0.2) appeared in the celebrated Drinfeld paper [Dr2], essentially as *the Drinfeld associator*, and more relations for multiple zeta values were implicitly deduced there. The question about interdependence of (double) shuffle and associator relations does not seem to be settled at the moment of writing this; cf. [Ra3]. The problem of completeness of these systems of relations is equivalent to some difficult transcendence questions.

Multiple zeta values are interesting, because they and their generalizations appear in many different contexts involving mixed Tate motives [DeGo, T], deformation quantization [Kon], knot invariants, etc.

0.2 Modular symbols and periods of modular forms

Let Γ be a congruence subgroup of the modular group acting upon the union $\overline{H}$ of the upper complex half-plane H and the set of cusps $\mathbf{P}^1(\mathbf{Q})$.

The quotient $\Gamma \setminus \overline{H}$ is the modular curve X_Γ. Differentials of the first kind on X_Γ lift to cusp forms of weight 2 on H (multiplied by dz).

The modular symbols $\{\alpha, \beta\}_\Gamma \in H_1(X_\Gamma, \mathbf{Q})$, where $\alpha, \beta \in \mathbf{P}^1(\mathbf{Q})$, were introduced in [Ma1] as linear functionals on the space of differentials of the first kind obtained by lifting and integrating. The fact that one lands in $H_1(X_\Gamma, \mathbf{Q})$ and not just $H_1(X_\Gamma, \mathbf{R})$ is not obvious. It was proved in [Dr1] by refining a weaker argument given in [Ma1]. This is equivalent to the statement that the difference of any two cusps in Γ has finite order in the Jacobian, or else that the mixed Hodge structure on $H^1(X_\Gamma \setminus \{\text{cusps}\}, \mathbf{Q})$ is split (cf. [El]). One of the basic new insights of [Ma1] consisted in the realization that studying the action of Hecke operators on modular symbols one gets new arithmetic facts about periods and Fourier coefficients of cusp forms of weight 2.

The further generalizations of modular symbols proceeded, in particular, in the following directions:

(a) In [Ma2] it was demonstrated that the same technique applies to the integrals of cusp forms of higher weight, possibly multiplied by polynomials in z, producing similar information about their periods and Fourier coefficients. In principle, such integrals cannot be pushed down to X_Γ, but they can be pushed down to the appropriate Kuga–Sato varieties over X_Γ, that is, relative Cartesian powers of the universal elliptic curve. In this way, modular symbols of higher weight can be interpreted as rational homology classes of middle dimension of Kuga–Sato varieties; cf. [Sh1, Sh2, Sh3].

(b) Pushing down an oriented geodesic connecting two cusps in $\overline{H}$ to X_Γ, we get a singular chain with boundary in cusps, which is a relative cycle modulo cusps with integral coefficients. This is the viewpoint of [Me]. Hence it is more natural to consider the relative/noncompact version of modular symbols, and allow integration of the Eisenstein series, that is, differential forms of the third kind with poles at cusps as well. The same remark applies to the modular symbols of higher weight.
This refinement appears as well in the study of the "noncommutative boundary" of the modular space, that is, the (tower of) space(s) $\Gamma \setminus \mathbf{P}^1(\mathbf{R})$; cf. [MaMar]. Namely, it turns out that the relative 1-homology modulo cusps (and additional groups of similar nature) can be interpreted as (sub)groups of the K-theory of the noncommutative boundary.
In this paper I suggest a generalization in the third direction, namely

(c) The study of iterated integrals of cusp forms and Eisenstein series, possibly multiplied by a power of z, along geodesics connecting two cusps. Some of these integrals can be pushed down to X_Γ and thus produce a de Rham version of modular symbols which assigns iterated (possibly regularized) periods to the elements of the fundamental groupoid of $(X_\Gamma, \{\text{cusps}\})$ instead of its 1-homology group. One may call them *noncommutative modular symbols.*

Other integrals can only be pushed down to the Kuga–Sato varieties, or preferably, to some (covers of the) moduli spaces $\overline{M}_{1,n}$, in the same vein as was done for multiple zeta values and $\overline{M}_{0,n}$ in [GoMa]. The related geometry deserves further study, both for integrands related to cusp forms and to Eisenstein series.

Notice in conclusion that the discussion above implicitly referred only to the case of SL_2-modular symbols. It would be quite interesting to extend it to groups of higher rank, along the lines of [AB] and [AR].

0.3 Summary of this paper

I recall the basic properties of iterated integrals of holomorphic 1-forms on a simply connected Riemann surface in Section 1. The shuffle relations for the iterated integrals are reflected directly in terms of a generating function J stating that it is a grouplike element with respect to a comultiplication; cf. Proposition 1.4.1.

Then I turn to the main object of study. In Section 2 I define 1-forms of modular and cusp modular type, introduce and study the iterated and total Mellin transform for families of such forms. The functional equation for the total Mellin transform is deduced which extends the classical functional equation for L-series.

Using only critical values of these Mellin transforms, I introduce in Section 2.5 an iterated modular symbol as a certain noncommutative 1-cohomology class of the relevant subgroup of the modular group.

In Section 3, I study the representation of such Mellin transforms at integer values of their Mellin arguments in terms of multiple Dirichlet series. The results differ from the classical ones expressed by the identity (0.1) = (0.2) in two essential respects. First, iterated integrals are only linear combinations of certain multiple Dirichlet series. Second, the latter are *not* of the usual type

$$\sum_{0<n_1<\cdots<n_k} \frac{a_{1,n_1}\cdots a_{n,n_k}}{n_1^{m_1}\cdots n_k^{m_k}},$$

in fact, their coefficients depend on pairwise differences $n_j - n_i$.

In Section 4, the properties of the multiple Dirichlet series which emerged in Section 3 are axiomatized, and the shuffle relations for them are deduced. This requires, however, a considerable extension of the initial supply of series; the system of those coming from 1-forms of modular type is not closed.

Section 5 is dedicated to the iterated analogues of the so-called Eichler–Shimura and Manin relations for periods of cusp forms. Whereas the relations of the first type are quite straightforward, the relations of the second type, involving Hecke operators, are not obvious. The results presented here (Theorem 5.3) are preliminary; they clearly afford generalizations and deserve further study.

Finally, in Section 6 I return to the formalism of Section 1 and extend it by allowing our integrands to have logarithmic singularities at the boundary. A version of the regularization procedure I use here is the same as in Drinfeld's paper [Dr2]. It exploits complex analyticity in place of Boutet de Monvel's technique of [De] and [Ra2].

Using the Manin–Drinfeld theorem on cusps, I suggest a generalization of Drinfeld's associator and extend to this case a part of the identities satisfied by the latter. This list includes the grouplike property, the duality, and the hexagonal relation, which turn out to have the same source as the Shimura–Eichler relations for the periods of cusp forms. In contrast, the pentagonal relation seems to be specific for the original Drinfeld's associator.

1 Iterated integrals of holomorphic 1-forms

1.1 Setup

Let X be a connected Riemann surface, not necessarily compact, $\mathcal{O}_X$ its structure sheaf of holomorphic functions, Ω^1_X the sheaf of holomorphic 1-forms. If ω is a (local) 1-form, $z \in X$ a point, $\omega(z)$ denotes the value of ω at z, i.e., the respective cotangent vector.

Let V be a finite set which will be used as a set indexing various families. Consider the completed unital semigroup ring freely generated by V. We will write it as the ring of associative formal series $\mathbf{C}\langle\langle A_V\rangle\rangle$, where $A_V := (A_v | v \in V)$ are noncommuting free formal variables.

More generally, we may consider the ring $\mathcal{O}_X(U)\langle\langle A_V\rangle\rangle$, where $\mathcal{O}_X(U)$ is the ring of holomorphic functions on an open subset $U \subset X$ (A_v commute with $\mathcal{O}_X(U)$), and the bimodule $\Omega^1_X(U)\langle\langle A_V\rangle\rangle$ over this ring, connected by the differential d such that $dA_v = 0$ for all $v \in V$. Varying U, we will get two presheaves; the sheaves associated with these presheaves are denoted $\mathcal{O}_X\langle\langle A_V\rangle\rangle$, respectively, $\Omega^1_X\langle\langle A_V\rangle\rangle$, and d extends to them, so that $\operatorname{Ker} d$ is the constant sheaf $\mathbf{C}\langle\langle A_V\rangle\rangle$.

Let $\omega_V := (\omega_v | v \in V)$ be a family of 1-forms holomorphic in U and indexed by V. Put

$$\Omega := \sum_{v \in V} A_v \omega_v. \tag{1.1}$$

The total iterated integral of this form along a piecewise smooth path $\gamma : [0, 1] \to U$ is denoted $J_\gamma(\Omega)$ or $J_\gamma(\omega_V)$ and is defined by the formula

$$J_\gamma(\Omega) := 1 + \sum_{n=1}^{\infty} \int_0^1 \gamma^*(\Omega)(t_1) \int_0^{t_1} \gamma^*(\Omega)(t_2) \cdots \int_0^{t_{n-1}} \gamma^*(\Omega)(t_n) \in \mathbf{C}\langle\langle A_V\rangle\rangle, \tag{1.2}$$

where the integration is taken over the simplex $0 < t_n < \cdots < t_1 < 1$. If γ, γ' with the same ends are homotopic, $J_\gamma(\Omega) = J_{\gamma'}(\Omega)$.

Putting $z_i = \gamma(t_i) \in X$, $a = \gamma(0)$, $z = \gamma(1)$, and considering the whole integral as a function of a variable z we will also write (1.2) in the form

$$J_a^z(\Omega) = J_a^z(\omega_V) = 1 + \sum_{n=1}^{\infty} \int_a^z \Omega(z_1) \int_a^{z_1} \Omega(z_2) \cdots \int_a^{z_{n-1}} \Omega(z_n). \tag{1.3}$$

If U is connected and simply connected, this expression is an unambiguously defined element of $\mathcal{O}_X(U)\langle\langle A_V\rangle\rangle$. Otherwise it is a multivalued function of z in this domain.

The following result is classical.

Proposition 1.2.

(i) *$J_a^z(\Omega)$ as a function of z satisfies the equation*

$$dJ_a^z(\Omega) = \Omega(z) J_a^z(\Omega). \tag{1.4}$$

In other words, $J_a^z(\Omega)$ is a horizontal (multi)section of the flat connection $\nabla_\Omega := d - l_\Omega$ on $\mathcal{O}_X\langle\langle A_V\rangle\rangle$, where l_Ω is the operator of left multiplication by Ω.

(ii) *If U is a simply connected neighborhood of a, $J_a^z(\Omega)$ is the only horizontal section with initial condition $J_a^a = 1$. Any other horizontal section K^z can be uniquely written in the form $J_a^z(\Omega)C$, $C \in \mathbf{C}\langle\langle A_V\rangle\rangle$. In particular, for any $b \in U$,*

$$J_b^z(\Omega) = J_a^z(\Omega) J_b^a(\Omega). \tag{1.5}$$

Proof. (i) follows directly from (1.3). Since $J_a^z(\Omega)$ is an invertible element of the ring $\mathcal{O}_X(U)\langle\langle A_V\rangle\rangle$, we can form $J_a^z(\Omega)^{-1}K^z$ and then directly check that $d(J_a^z(\Omega)^{-1}K^z)=0$. Hence this element belongs to $\mathbf{C}\langle\langle A_V\rangle\rangle$, and, moreover, equals its value at $z=a$, that is, K^a. Choosing $K^z=J_b^z(\Omega)$, we get (1.5). □

1.3 $J_a^z(\Omega)$ as a generating series

Clearly, we have

$$J_a^z(\omega_V)=J_a^z(\Omega)=1+\sum_{n=1}^{\infty}\sum_{(v_1,\dots,v_n)\in V^n}A_{v_1}\cdots A_{v_n}I_a^z(\omega_{v_1},\dots,\omega_{v_n}), \tag{1.6}$$

where

$$I_a^z(\omega_{v_1},\dots,\omega_{v_n})=\int_a^z\omega_{v_1}(z_1)\int_a^{z_1}\omega_{v_2}(z_2)\cdots\int_a^{z_{n-1}}\omega_{v_n}(z_n) \tag{1.7}$$

are the usual iterated integrals.

In the remaining part of this section, and in the main body of the paper, we will encode various (infinite families of) relations among the iterated integrals (1.7) in the form of relations between the generating functions $J_a^z(\omega_V)$. Generally, our relations between the generating functions will be (noncommutative) polynomial ones. They may also involve different families (ω_V), different integration paths, and some linear transformations of the formal variables A_v; cf. especially Theorem 2.2.1, Proposition 5.1.1, Theorem 5.3, and Section 6.5 (in the context requiring a regularization).

1.4 Basic relations between total iterated integrals

There are three types of basic relations, which we will call *grouplike property*, *cyclicity*, and *functoriality*, respectively.

Proposition 1.4.1. *Consider the comultiplication*

$$\Delta:\mathbf{C}\langle\langle A_V\rangle\rangle\to\mathbf{C}\langle\langle A_V\rangle\rangle\,\widehat{\otimes}_{\mathbf{C}}\,\mathbf{C}\langle\langle A_V\rangle\rangle,\quad \Delta(A_v)=A_v\otimes 1+1\otimes A_v$$

and extend it to the series with coefficients $\mathcal{O}_X$ and Ω_X^1. Then

$$\Delta(J_a^z(\omega_V))=J_a^z(\omega_V)\otimes_{\mathcal{O}_X}J_a^z(\omega_V). \tag{1.8}$$

Proof. Both sides of (1.8) satisfy the equation $dJ=\Delta(\Omega)J$ and have the initial value 1 at $z=a$. □

N.B. Coefficientwise, (1.8) is a compact version of the shuffle relations for the iterated integrals (1.7).

1.4.2 Cyclicity

Let γ be a closed oriented contractible contour in U, $a_1, \ldots, a_n$ points along this contour (cyclically) ordered compatibly with orientation. Then

$$J_{a_2}^{a_1}(\Omega) J_{a_3}^{a_2}(\Omega) \cdots J_{a_n}^{a_{n-1}}(\Omega) J_{a_1}^{a_n}(\Omega) = 1. \tag{1.9}$$

This follows from (1.5) by induction.

1.4.3 Functoriality

Consider an automorphism $g : X \to X$ such that g^* maps into itself the linear space spanned by ω_v. In particular, there is a constant matrix $G = (g_{vu})$ with rows and columns labeled by V such that $g^*(\omega_v) = \sum_u g_{vu}\omega_u$. Define the automorphism g_* of any of the rings/modules of formal series $\mathbf{C}\langle\langle A_V \rangle\rangle$, $\mathbf{C}(X)\langle\langle A_V \rangle\rangle$, $\Omega^1(X)\langle\langle A_V \rangle\rangle$ by the formula $g_*(A_u) = \sum_v A_v g_{vu}$. On coefficients g_* acts as the identity.

Claim 1.4.4. *We have*

$$J_{ga}^{gz}(\omega_V) = g_*(J_a^z(\omega_V)). \tag{1.10}$$

Proof. In fact, both sides coincide with $J_a^z(g^*(\omega_V))$. We will give below a calculation which proves a slightly more general statement. □

1.5 A variant: Multiple lower integration limits

Somewhat more generally, in the simply connected case we can consider a family of points $(a_\bullet) := (a_{i,v})$ in X indexed by pairs $i = 1, 2, 3, \ldots$, $v \in V$.

Given such a family and ω_V, we can construct the following formal series in $\mathbf{C}(X)\langle\langle A_V \rangle\rangle$ with constant term 1:

$$J_{(a_\bullet)}^z(\omega_V) := \sum_{n=0}^{\infty} \sum_{(v_1,\ldots,v_n)\in V^n} A_{v_1} \cdots A_{v_n} I_{a_{1,v_1},\ldots,a_{n,v_n}}^z(\omega_{v_1}, \ldots, \omega_{v_n}), \tag{1.11}$$

where

$$I_{a_{1,v_1},\ldots,a_{n,v_n}}^z(\omega_{v_1}, \ldots, \omega_{v_n}) := \int_{a_{1,v_1}}^z \omega_{v_1}(z_1) \int_{a_{2,v_2}}^{z_1} \omega_{v_2}(z_2) \cdots \int_{a_{n,v_n}}^{z_{n-1}} \omega_{v_n}(z_n). \tag{1.12}$$

As above, $z \in X$ denotes a variable point, the argument of our functions. Then we have

$$dJ_{(a_\bullet)}^z(\omega_V) = \Omega J_{(a_\bullet)}^z(\omega_V) \tag{1.13}$$

and

$$J_{(a_\bullet)}^z(\omega_V) = J_a^z(\omega_V) J_{(a_\bullet)}^a(\omega_V). \tag{1.14}$$

The series (1.11) satisfies the following functoriality relation generalizing (1.10).

Claim 1.5.1. *We have*

$$J^{gz}_{(ga_\bullet)}(\omega_V) = g_*(J^z_{(a_\bullet)}(\omega_V)). \tag{1.15}$$

Proof. We will check that both sides coincide with $J^z_{(a_\bullet)}(g^*(\omega_V))$. In fact, $\int_{gu}^{gv} \nu(z) = \int_u^v \nu(gz)$ so that, removing g step by step from the integration limits, we get

$$I^{gz}_{ga_{1,v_1},\dots,ga_{n,v_n}}(\omega_{v_1},\dots,\omega_{v_n}) = I^z_{a_{1,v_1},\dots,a_{n,v_n}}(g^*(\omega_{v_1}),\dots,g^*(\omega_{v_n})).$$

Multiplying the left-hand side by $A_{v_1}\cdots A_{v_n}$ and summing, we get the left-hand side of (1.15).

On the other hand,

$$\sum_{v_1,\dots,v_n\in V^n} A_{v_1}\cdots A_{v_n} I^z_{a_{1,v_1},\dots,a_{n,v_n}}(g^*(\omega_{v_1}),\dots,g^*(\omega_{v_n}))$$

$$= \sum_{v_1,\dots,v_n\in V^n} A_{v_1}\cdots A_{v_n} I^z_{a_{1,v_1},\dots,a_{n,v_n}}\left(\sum_{u_1\in V} g_{v_1,u_1}\omega_{u_1},\dots,\sum_{u_n\in V} g_{v_n,u_n}\omega_{u_n}\right)$$

$$= \sum_{\substack{v_1,\dots,v_n\in V^n\\ u_1,\dots,u_n\in V^n}} A_{v_1} g_{v_1,u_1}\cdots A_{v_n} g_{v_n,u_n} I^z_{a_{1,v_1},\dots,a_{n,v_n}}(\omega_{v_1},\dots,\omega_{v_n})$$

$$= g_*\left(\sum_{v_1,\dots,v_n\in V^n} A_{v_1}\cdots A_{v_n} I^z_{a_{1,v_1},\dots,a_{n,v_n}}(\omega_{v_1},\dots,\omega_{v_n})\right).$$

Summation over n produces the right-hand side of (1.15), proving the lemma. □

1.6 A variant: Nonlinear Ω

Now let $\Omega \in \Omega^1_X(U)\langle\langle A_V\rangle\rangle$ be an arbitrary form without constant term in A_v:

$$\Omega = \sum_{n=1}^{\infty} \sum_{(v_1,\dots,v_n)\in V^n} A_{v_1}\cdots A_{v_n}\Omega_{v_1,\dots,v_n}, \tag{1.16}$$

where $\Omega_{v_1,\dots,v_n} \in \Omega^1_X(U)$.

The total iterated integrals $J_\gamma(\Omega)$ and $J^z_a(\Omega)$ are defined by exactly the same formulas (1.2) and (1.3). It is not true anymore that the coefficients of this series are the usual iterated integrals. However, an analogue of Proposition 1.2 and the cyclic identity remain true.

Proposition 1.6.1. *$J^z_a(\Omega)$ as a function of z satisfies the equation*

$$dJ^z_a(\Omega) = \Omega(z)J^z_a(\Omega). \tag{1.17}$$

If U is a simply connected neighborhood of a, $J^z_a(\Omega)$ is the only horizontal section with initial condition $J^a_a = 1$. Any other horizontal section K^z can be uniquely written in the form $J^z_a(\Omega)C$, $C \in \mathbf{C}\langle\langle A_V\rangle\rangle$. In particular, for any $b \in U$,

$$J^z_b(\Omega) = J^z_a(\Omega)J^a_b(\Omega). \tag{1.18}$$

Corollary 1.6.2. *Let γ be a closed oriented contractible contour in U, $a_1, \dots, a_n$ points along this contour (cyclically) ordered compatibly with orientation. Then*

$$J_{a_2}^{a_1}(\Omega) J_{a_3}^{a_2}(\Omega) \cdots J_{a_n}^{a_{n-1}}(\Omega) J_{a_1}^{a_n}(\Omega) = 1. \tag{1.19}$$

Notice in conclusion that the integral formula (0.2) for the multiple zeta values is not quite covered by the formalism reviewed so far because the integrands in (0.2) have logarithmic poles at the boundary. We will return to this situation in Section 6, to which some readers may prefer to turn right away. However, for applications to the integration of cusp forms in Sections 2–5 the regular case treated here suffices.

2 1-forms of modular type, iterated Mellin transform, and noncommutative modular symbols

2.1 Setup

In this section, X will be the upper half-plane H and z the standard complex coordinate. H is endowed with the metric of constant curvature -1: $ds^2 = |dz|^2/(\mathrm{Im}\, z)^2$.

The limits of integration in our iterated integrals generally lie in H, but may be "improper" as well, that is, belong to the set of cusps $\mathbf{Q} \cup \{i\infty\}$. If this is the case, we always assume that the respective integration path in some neighborhood of the cusp coincides with a segment of a geodesic curve.

Our 1-forms generally will have the following structure.

Definition 2.1.1.

(i) A 1-form ω on H is called a form of modular type if it can be represented as $f(z)z^{s-1}dz$, where s is a complex number and $f(z)$ is a modular form of some weight with respect to a congruence subgroup of the modular group.
The modular form $f(z)$ is then well defined and called the associated modular form (to ω), and the number s is called the Mellin argument of ω.

(ii) ω is called a form of cusp modular type if the associated $f(z)$ is a cusp form.

To fix notation, we will recall below some classical facts.

2.1.2 Action of automorphisms

Any matrix $\gamma \in \mathrm{GL}_2^+(\mathbf{R})$ defines a holomorphic isometry of H, namely $z \mapsto [\gamma]z$, where $[\gamma]$ is the fractional linear transformation corresponding to γ. We will denote this automorphism also by γ. It induces the inverse image maps on the sheaves $(\Omega_H^1)^{\otimes r}$ of holomorphic tensor differentials of degree r:

$$\gamma^*(f(z)(dz)^r) = f([\gamma]z)(d[\gamma]z)^r = (\det \gamma)^r f([\gamma]z) \frac{(dz)^r}{(c_\gamma z + d_\gamma)^{2r}}, \tag{2.1}$$

where (c_γ, d_γ) is the lower row of γ.

If one identifies $(\Omega^1_H)^{\otimes r}$ with $\mathcal{O}_H$ by sending $(dz)^r$ to 1, (2.1) turns into the action of weight $2r$ on functions, which is traditionally written as a right action:

$$f\|[\gamma]_{2r}(z) := (\det\gamma)^r f([\gamma]z)(c_\gamma z + d_\gamma)^{-2r}. \tag{2.2}$$

Assume that $f(z)(dz)^r$ is invariant with respect to γ. Then, writing $f(z)z^{s-1}dz = f(z)(dz)^r \cdot z^{s-1}(dz)^{1-r}$, we see that

$$\gamma^*(f(z)z^{s-1}dz) = (\det\gamma)^{1-r} f(z)(a_\gamma z + b_\gamma)^{s-1}(c_\gamma z + d_\gamma)^{2r-1-s}dz, \tag{2.3}$$

where (a_γ, b_γ) is the upper row of γ. In particular, if $2r \geq 2$ is an integer, γ^* maps into itself the space of 1-forms spanned by

$$f(z)z^{s-1}dz, \quad 1 \leq s \leq 2r-1, \quad s \in \mathbf{Z}. \tag{2.4}$$

More generally, if

$$\gamma^*(f(z)(dz)^r) = \chi(\gamma)f(z)(dz)^r \tag{2.5}$$

for some $\chi(\gamma) \in \mathbf{C}$, then

$$\gamma^*(f(z)z^{s-1}dz) = (\det\gamma)^{1-r} f(z)\chi(\gamma)(a_\gamma z + b_\gamma)^{s-1}(c_\gamma z + d_\gamma)^{2r-1-s}dz, \tag{2.6}$$

and the space (2.4) will still remain invariant.

We can apply this formalism to the spaces of modular forms of weight $2r$ with respect to a congruence subgroup Γ of $\mathrm{SL}_2(\mathbf{Z})$, i.e., to the functions f in $(dz)^{-r}((\Omega^1_H)^{\otimes r})^\Gamma$. Two special cases will be of particular interest:

(i) For any such f, the space of 1-forms spanned by (2.4) is Γ-invariant.
(ii) Assume that $\Gamma = \Gamma_0(N)$. This group is normalized by the involution

$$g = g_N := \begin{pmatrix} 0 & -1 \\ N & 0 \end{pmatrix}. \tag{2.7}$$

Therefore, this involution maps into itself the space of $\Gamma_0(N)$-modular forms, and the latter has a basis consisting of forms with

$$g_N^*(f(z)(dz)^r) = \varepsilon_f f(z)(dz)^r, \quad \varepsilon_f = \pm 1. \tag{2.8}$$

Applying (2.6) with $\gamma = g_N$ we get, for any complex s,

$$g_N^*(f(z)z^{s-1}dz) = \varepsilon_f N^{r-s} f(z)z^{2r-1-s}dz. \tag{2.9}$$

2.1.3 Geodesics and cusp forms

The geodesic from 0 to $i\infty$ is the upper half of the pure imaginary line. The unoriented distance of a point iy on it to i is $|\log y|$. The exponential of this distance is thus y, if $y > 1$, and y^{-1}, if $y < 1$. If we replace i by another reference point, even outside of the imaginary axis, the exponential of the distance will behave like $e^{O(1)}y$ (respectively, $e^{O(1)}y^{-1}$) as $y \to \infty$ (respectively, $y \to 0$).

Let $f(z)$ be a cusp form of weight $2r$ for a congruence subgroup. Then it can be represented by a Fourier series $f(z) = \sum_{n=1}^{\infty} c_n e^{2\pi i n z/N}$ for some $N \in \mathbf{Z}_+$, whose coefficients are polynomially bounded: $c_n = O(n^C)$ for some $C > 0$. Therefore, we have $|f(iy)| = O(e^{-ay})$ for some $a > 0$ as $y \to \infty$. From the previous analysis it follows that more generally, for any cusp form and any geodesic connecting two cusps, $|f(z)| = O(e^{-ay(z)})$ for some $a > 0$ as z tends along the geodesics to one of its ends, where this time $y(z)$ means the exponentiated geodesic distance from z to any reference point in H, fixed once and for all.

Now let $\omega(z) = f(z)z^{s-1}dz$ be a 1-form of cusp modular type. Then the estimates above show that the following expected properties indeed hold:

(a) As $z_0 \to i\infty$ along the imaginary axis, the family $\int_{z_0}^{z} \omega$ of holomorphic functions of z in any bounded domain H converges absolutely and uniformly to a holomorphic function of z, which is denoted $\int_{i\infty}^{z} \omega$. The same remains true if one replaces $i\infty$ by 0.

These integrals are holomorphic functions of the Mellin argument s of ω as well.

(b) The sum $(\int_{i\infty}^{z} + \int_{z}^{0})\omega$ does not depend on z in H and is denoted $\int_{i\infty}^{0} \omega$. As a function of s, it is called *the classical Mellin transform of* ω.

Denote this classical transform by $\Lambda(f; s)$. Assume that f satisfies (2.8). Then we have the classical functional equation

$$\Lambda(f; s) = -\varepsilon_f N^{r-s} \Lambda(f; 2r - s), \tag{2.10}$$

because in view of (2.9)

$$\begin{aligned}\int_{i\infty}^{0} \omega &= -\int_{0}^{i\infty} \omega = -\int_{g_N(i\infty)}^{g_N(0)} \omega = -\int_{i\infty}^{0} g_N^*(\omega)\\ &= -\varepsilon_f N^{r-s} \int_{i\infty}^{0} f(z) z^{2r-1-s} dz.\end{aligned}$$

Another identity in the same vein uses the fact that $i/\sqrt{N}$ is the fixed point of g_N, so that $\Lambda(f; s)$ can be written as

$$\Lambda(f; s) = \int_{i\infty}^{i/\sqrt{N}} \omega - \int_{i\infty}^{i/\sqrt{N}} g_N^*(\omega). \tag{2.11}$$

This allows one to use the Fourier expansions of $f(z)$ and $f\|[g_N]_{2r}(z)$ in order to deduce series expansions for $\Lambda(f; s)$. (Notice that the Fourier expansions cannot be termwise integrated near $z = 0$ because the formal integration produces a divergent series.)

Now we can finally write down the analogues of these definitions and results for iterated integrals.

Definition 2.2.

(i) Let $f_1, \dots, f_k$ be a finite sequence of cusp forms with respect to a congruence subgroup, $\omega_j(z) := f_j(z)z^{s_j-1}dz$. The iterated Mellin transform of (f_j) is, by definition,

$$\begin{aligned} M(f_1, \dots, f_k; s_1, \dots, s_k) &:= I_{i\infty}^0(\omega_1, \dots, \omega_k) \\ &= \int_{i\infty}^0 \omega_1(z_1) \int_{i\infty}^{z_1} \omega_2(z_2) \cdots \int_{i\infty}^{z_{n-1}} \omega_n(z_n). \end{aligned} \tag{2.12}$$

(ii) Let $f_V = (f_v | v \in V)$ be a finite family of cusp forms with respect to a congruence subgroup, $s_V = (s_v | v \in V)$ a finite family of complex numbers, $\omega_V = (\omega_v)$, where $\omega_v(z) := f_v(z)z^{s_v-1}dz$. The total Mellin transform of f_V is, by definition,

$$\begin{aligned} \mathrm{TM}(f_V; s_V) &:= J_{i\infty}^0(\omega_V) \\ &= \sum_{n=0}^{\infty} \sum_{(v_1,\dots,v_n)\in V^n} A_{v_1} \cdots A_{v_n} M(f_{v_1}, \dots, f_{v_n}; s_{v_1}, \dots, s_{v_n}) \end{aligned} \tag{2.13}$$

(cf. (1.3)).

Below we will assume that the space spanned by all ω_v is stable with respect to some g_N^*. Then, as in Section 1.4.3, denote by $G = (g_{vu})$ the matrix of this action on (ω_v), and by g_{N*} the action of the transposed matrix on the formal variables (A_v).

For example, if (ω_v) and $(f_v(dz)^{r_v})$, respectively, can be represented as a union of pairs of forms, corresponding to the left- and right-hand sides of (2.9), the matrix G consists of two-by-two antidiagonal blocks, each of which after the classical Mellin transform produces a functional equation of the form (2.10).

Theorem 2.2.1.

(i) *If the space spanned by all ω_v is stable with respect to some g_N^*, we have the functional equation*

$$J_{i\infty}^0(\omega_V) = g_{N*}(J_{i\infty}^0(\omega_V))^{-1}. \tag{2.14}$$

(ii) *Under the assumptions of Definition* 2.2(ii), *denote the weight of f_v by $2r_v$ and assume that f_v is an eigenvector for g_N^* with eigenvalue ε_v. Then the total Mellin transform* (2.13) *satisfies*

$$\mathrm{TM}(f_V; s_V) = g_*(\mathrm{TM}(f_V; 2r_V - s_V))^{-1}, \tag{2.15}$$

where g_* multiplies each A_v by $\varepsilon_v N^{r_v - s_v}$.

Proof. This is a straightforward corollary of the definitions and formulas (1.9) and (1.10) as soon as one has checked that the latter formulas are applicable to the improper iterated integrals of the 1-forms of cusp modular type.

This check is a routine matter, since at each step of an iterated integration we multiply the result of the previous step by a holomorphic function of the type $f(z)z^{s-1}$ which is bounded by $O(e^{-ay(z)})$ as in Section 2.1.3 above as z tends to 0 or $i\infty$.

Notice in conclusion that no analogue of the functional equation (2.11) can be written for the individual Mellin transforms (2.12), because applying g_N to the integration limits in them we get an expression which is not a Mellin transform in our sense. Only putting them all together produces the necessary environment for replacing the overall minus sign on the right-hand side of (2.10) by the overall exponent -1 on the right-hand side of (2.15). □

A similar reasoning establishes the iterated analogue of (2.11).

Proposition 2.3. *We have*

$$\mathrm{TM}(f_V; s_V) = (g_{N*} J_{i\infty}^{i/\sqrt{N}}(\omega_V))^{-1} J_{i\infty}^{i/\sqrt{N}}(\omega_V). \tag{2.16}$$

2.4 Pushing down iterated integrals

Let ω be a 1-form of modular type whose associated modular form has weight 2 with respect to a subgroup Γ of the modular group, and whose Mellin argument is 1. In this case ω is Γ-invariant so that it can be pushed down to a 1-form ν on $X_\Gamma^\circ := \Gamma \setminus H$. Instead of integrating ω along a path in H, we can integrate ν along the push-down of this path to X_Γ°. If all ω_v have this property, all relevant iterated integrals can be pushed down to X_Γ°.

This argument admits a partial generalization to higher weights. Assume that the modular form associated with ω has weight $2r > 2$, whereas its Mellin argument is an integer belonging to the critical strip (2.4), $1 \le s \le 2r - 1$. In this case the relevant simple integral along, say, $\{i\infty, 0\}$ can be pushed down to the Kuga–Sato variety $X_\Gamma^{(2r-2)}$ which is the $(2r-2)$ fibered power of the universal elliptic curve over X_Γ, or rather its compactified smooth model. However, on $X_\Gamma^{(2r-2)}$ we obtain an integral of a holomorphic form $\widehat{\omega}$ of degree $2r-1$ over a relative cycle of the same dimension, which is > 1. Therefore, iterated "line" integrals of such forms on H cannot be directly translated into integrals of the same type on $X_\Gamma^{(2r-2)}$.

On the other hand, one can generally define Chen's iterated integrals of forms of arbitrary degree, say, $\widehat{\omega}_v$ on $X_\Gamma^{(2r-2)}$, which take values in the space of differential forms on the path space $PX_\Gamma^{(2r-2)}$ and not just $\mathbf{C}$; cf. the papers [Ch] and [Ha], as well as references therein. Studying properties of such iterated integrals in the modular case presents an interesting challenge.

Here I will only give a formula for $\widehat{\omega}$ and show that its periods coincide with integrals of ω along geodesics. It turns out that $\widehat{\omega}$ depends only on the modular form associated with ω, whereas geodesic integrals of 1-forms ωz^k for integral critical values of k become periods of $\widehat{\omega}$ along cycles depending on k. For more details, see [Sh1, Sh2] and especially [Sh3].

Denote by $\Gamma^{(r)}$ the semidirect product $\Gamma \ltimes (\mathbf{Z}^{2r-2} \times \mathbf{Z}^{2r-2})$ acting upon $H \times \mathbf{C}^{2r-2}$ via

$$(\gamma; n, m)(z, \zeta) := ([\gamma]z; (c_\gamma z + d_\gamma)^{-1}(\zeta + zn + m)).$$

Here $n = (n_1, \dots, n_{2r-2})$, $m = (m_1, \dots, m_{2r-2})$, $\zeta = (\zeta_1, \dots, \zeta_{2r-2})$, and $nz = (n_1 z, \dots, n_{2r-2}z)$.

If $f(z)$ is a holomorphic modular form of weight $2r$, then $f(z)dz \wedge d\zeta_1 \wedge \dots \wedge d\zeta_{2r-2}$ is a $\Gamma^{(r)}$-invariant holomorphic volume form on $H \times \mathbf{C}^{2r-2}$. Hence one can push it down to (a Zariski open subset of) the quotient $\Gamma^{(r)} \setminus (H \times \mathbf{C}^{2r-2})$ which is a Zariski open subset of the respective Kuga–Sato variety. Denote by $\widehat{\omega}$ the image of this form. It is common for all 1-forms of modular type $\omega = f(z)z^{s-1}dz$ with different Mellin arguments s.

A detailed analysis of singularities performed in [Sh2, Sh3] shows that the map $f \mapsto \widehat{\omega}$ induces an isomorphism of the space of cusp forms of weight $2r$ with the space of holomorphic volume forms on an appropriate smooth projective Kuga–Sato variety. (As I have already remarked in the introduction, it would be useful to replace it by the base extension $(\overline{M}_{1,2r-2})_{X_\Gamma}$.)

The dependence of the period of ω on the integration path and on the Mellin argument is reflected in the choice of the relative cycle over which we integrate $\widehat{\omega}$.

More precisely, let $\alpha, \beta \in \mathbf{P}^1(\mathbf{Q})$ be two cusps in $\overline{H}$ and let p be a geodesic joining α to β. Fix (n_i) and (m_i) as above. Construct a cubic singular cell $p \times (0, 1)^{2r-2} \to H \times \mathbf{C}^{2r-2}$: $(z, (t_i)) \mapsto (z, (t_i(zn_i + m_i)))$. Take the S_{2r-2}-symmetrization of this cell and push down the result to the Kuga–Sato variety. We will get a relative cycle whose homology class is Shokurov's higher modular symbol $\{\alpha, \beta; n, m\}_\Gamma$. From this construction, it is almost obvious that

$$\int_\alpha^\beta f(z) \prod_{i=1}^{2r-2} (n_i z + m_i) dz = \int_{\{\alpha,\beta;n,m\}_\Gamma} \widehat{\omega}.$$

In particular, if k of the coordinates n_i are 1, and the rest are zero, whereas $m_i = 1 - n_i$, the left-hand side equals $\int_\alpha^\beta f(z)z^k dz$.

The singular cube $(0, 1)^{2r-2}$ may also be replaced by an evident singular simplex. This can be useful for transposing the results of [GoMa] to the genus one moduli spaces.

2.5 Noncommutative modular symbols and continued fractions

I will define in this subsection a generalization of modular symbols involving iterated integrals and allowing a mixture of forms of different weights with respect to the same subgroup Γ of $\mathrm{SL}(2, \mathbf{Z})$.

Let (ω_v) be a family of linearly independent 1-forms of cusp modular type whose Mellin arguments are integers lying in the respective critical strip as in (2.4). Let Γ be a subgroup of modular group acting on the space spanned by (ω_v) as in (2.3). Denote by Π the multiplicative group of power series in (A_v) with constant term 1. Clearly, the map $J \mapsto g_* J$ (see Section 1.4.3) defines a left action of Γ on Π.

Proposition–Definition 2.5.1.

(i) *For each* $a \in \mathbf{P}^1(\mathbf{Q})$*, the map* $\Gamma \to \Pi : \gamma \mapsto J^a_{\gamma a}(\Omega)$ *is a noncommutative 1-cocycle* ζ_a *in* $Z^1(\Gamma, \Pi)$.

(ii) *The cohomology class of* ζ_a *in* $H^1(\Gamma, \Pi)$ *does not depend on the choice of* a *and is called the noncommutative modular symbol.*

Proof. We have, omitting Ω for brevity, and using (1.9), (1.10),

$$J^a_{\gamma\beta a} = J^a_{\gamma a} J^{\gamma a}_{\gamma\beta a} = J^a_{\gamma a} \gamma_*(J^a_{\beta a}),$$

which means that ζ_a is a 1-cocycle. Moreover, if b is another cusp,

$$J^a_{\gamma a} = J^a_b J^b_{\gamma b} J^{\gamma b}_{\gamma a} = J^a_b J^b_{\gamma b} (\gamma_*(J^a_b))^{-1},$$

that is, ζ_a and ζ_b are homologous. □

Remark. Assume that the cusp forms associated with (ω_v) span the sum of all spaces of cusp forms of certain weights, and for each weight and each cusp form all admissible Mellin arguments actually occur. Then the linear in A_v term of ζ_a encodes all periods of the involved cusp forms along all classical modular symbols corresponding to loops in $X_\Gamma(\mathbf{C})$ starting and ending at the cusp Γa.

2.5.2 Iterated integrals between arbitrary cusps

The group Γ generally does not act transitively on cusps, so that the components of cocycles ζ_a do not contain iterated integrals along all geodesics connecting two cusps. One can use the technique of continued fractions as in [Ma1, Ma2] in order to express all such integrals through a finite number of them.

Namely, choose a set of representatives C of the left cosets $\Gamma \setminus \mathrm{SL}_2(\mathbf{Z})$. Call the iterated integrals of the form $(J^{g(0)}_{g(i\infty)})^{\pm 1}$, $g \in C$, *primitive* ones. Notice that when $g \notin \Gamma$ the space spanned by (ω_v) is not generally g^*-stable, so that we cannot define g_*.

Proposition 2.5.3. *Each* J^a_b *can be expressed as a noncommutative monomial in* $\gamma_*(J^c_d)$*, where* γ *runs over* Γ *and* J^c_d *runs over primitive integrals.*

Proof. First, we can write $J^a_b = (J^{i\infty}_a)^{-1} J^{i\infty}_b$. So it remains to find a required expression for $J^a_{i\infty}$. Assume that $a > 0$; the case $a < 0$ can be treated similarly. Consider the consequent convergents to a:

$$a = \frac{p_n}{q_n}, \quad \frac{p_{n-1}}{q_{n-1}}, \quad \dots, \quad \frac{p_0}{q_0} = \frac{p_0}{1}, \quad \frac{p_{-1}}{q_{-1}} := \frac{1}{0}.$$

Put

$$g_k := \begin{pmatrix} p_k & (-1)^{k-1} p_{k-1} \\ q_k & (-1)^{k-1} q_{k-1} \end{pmatrix}, \quad k = 0, \dots, n.$$

We have $g_k = g_k(a) \in \mathrm{SL}_2(2, \mathbf{Z})$. Put $g_k = \gamma_k c_k$, where $\gamma_k \in \Gamma$ and $c_k \in C$ are two sequences of matrices depending on a. Then from cyclicity we get

$$J^a_{i\infty} = \prod_{k=n}^{0} J^{p_k/q_k}_{p_{k-1}/q_{k-1}},$$

and from functoriality we obtain

$$J^{p_k/q_k}_{p_{k-1}/q_{k-1}} = \gamma_{k*}(J^{c_k(i\infty)}_{c_k(0)}). \qquad \square$$

3 Values of iterated Mellin transforms at integer points and multiple Dirichlet series

In this section, we collect some formulas expressing iterated Mellin transforms (2.12) at integer values of their Mellin arguments as linear combinations of "multiple Dirichlet series."

3.1 Notation

Consider a family of 1-forms ω_V, $v \in V$, satisfying the following conditions. First,

$$\omega_v(z) = \sum_{n=1}^{\infty} c_{v,n} e^{2\pi i n z} z^{m_v - 1} dz, \quad c_{v,n} \in \mathbf{C}, \quad m_v \in \mathbf{Z}, \quad m_v \geq 1. \tag{3.1}$$

Moreover, assume that $c_{v,n} = O(n^C)$ for some C and each v.

Until a problem of analytic continuation arises, we do not have to assume modularity. The notation m_v, replacing the former s_v, is chosen to emphasize that these Mellin arguments are natural numbers.

We start with introducing some notation.

3.1.1 Functions $L(z; \omega_{v_k}, \ldots, \omega_{v_1}; j_k, \ldots, j_1)$

Choose $k \geq 1$; $v_k, \ldots, v_1 \in V$, and nonnegative integers $j_k, \ldots, j_1$; it is convenient to add $j_0 = 0$. In our applications, j_a will satisfy the following restrictions:

$$j_a \leq m_{v_a} - 1 + j_{a-1}. \tag{3.2}$$

Now put

$$L(z; \omega_{v_k}, \ldots, \omega_{v_1}; j_k, \ldots, j_1)$$

$$:= (2\pi i z)^{j_k} \sum_{n_1, \ldots, n_k \geq 1} \frac{c_{v_1,n_1} \cdots c_{v_k,n_k} e^{2\pi i (n_1 + \cdots + n_k) z}}{n_1^{m_{v_1} + j_0 - j_1} (n_1 + n_2)^{m_{v_2} + j_1 - j_2} \cdots (n_1 + \cdots + n_k)^{m_{v_k} + j_{k-1} - j_k}}. \tag{3.3}$$

Thanks to the presence of exponential terms in (3.2), this series converges absolutely for any z with $\mathrm{Im}\, z > 0$ and defines a holomorphic function in H.

Notice that the enumeration of arguments of L is reversed in order to get a more natural enumeration of factors in the summands of (3.3).

3.1.2 Numbers $L(0; \omega_{v_k}, \dots, \omega_{v_1}; j_k, j_{k-1}, \dots, j_1)$

If we formally put $z = 0$ in the expansion for

$$(2\pi i z)^{-j_k} L(z; \omega_{v_k}, \dots, \omega_{v_1}; j_k, \dots, j_1),$$

we will get the formal series

$$\sum_{n_1,\dots,n_k \geq 1} \frac{c_{v_1,n_1} \cdots c_{v_k,n_k}}{n_1^{m_{v_1}+j_0-j_1} (n_1+n_2)^{m_{v_2}+j_1-j_2} \dots (n_1+\dots+n_k)^{m_{v_k}+j_{k-1}-j_k}}. \tag{3.4}$$

We have

$$c_{v_1,n_1} \cdots c_{v_k,n_k} = O((n_1 n_2 \dots n_k)^C) = O((n_1+\dots+n_k)^{kC}). \tag{3.5}$$

Assume that (3.2) holds. Then the general term of (3.4) is bounded by

$$\frac{1}{n_1(n_1+n_2)\cdots(n_1+\dots+n_{k-1})(n_1+\dots+n_k)^{m_{v_k}+j_{k-1}-j_k-1-kC}}.$$

Hence (3.4) converges absolutely as long as

$$m_{v_k} + j_{k-1} - j_k > 1 + kC.$$

Summarizing, we get three alternatives, describing the possible behavior of

$$L(z; \omega_{v_k}, \dots, \omega_{v_1}; j_k, j_{k-1}, \dots, j_1)$$

as $z \to 0$. We will later identify the respective limit as a component of the total Mellin transform of (ω_V).

Case 1: $j_k = 0$ *and* $m_{v_k} + j_{k-1} > 1 + kC$. Then the limit exists, and is equal to the "multiple Dirichlet series"

$$L(0; \omega_{v_k}, \dots, \omega_{v_1}; 0, j_{k-1}, \dots, j_1) = \sum_{n_1,\dots,n_k \geq 1} \frac{c_{v_1,n_1} \cdots c_{v_k,n_k}}{n_1^{m_{v_1}+j_0-j_1} (n_1+n_2)^{m_{v_2}+j_1-j_2} \dots (n_1+\dots+n_k)^{m_{v_k}+j_{k-1}-j_k}}. \tag{3.6}$$

($j_0 = j_k = 0$ appear on the right-hand side only for uniformity of notation.)

Case 2: $j_k > 0$ *and* $m_{v_k} + j_{k-1} - j_k > 1 + kC$. Then the limit exists, and vanishes thanks to the factor $(2\pi i z)^{j_k}$:

$$L(0; \omega_{v_k}, \dots, \omega_{v_1}; j_k, j_{k-1}, \dots, j_1) = 0 \tag{3.7}$$

Case 3: $j_k > 0$ *and* $m_{v_k} + j_{k-1} - j_k \leq 1 + kC$. In this case additional study is needed.

We can now formulate the first main result of this section.

Theorem 3.2. *For any $k \geq 1$, $(v_1, \ldots, v_k) \in V^k$, and* $\operatorname{Im} z > 0$, *we have*

$$(2\pi i)^{m_{v_1}+\cdots+m_{v_k}} I^z_{i\infty}(\omega_{v_k}, \ldots, \omega_{v_1})$$
$$= (-1)^{\sum_{i=1}^k (m_{v_i}-1)} \sum_{j_1=0}^{m_{v_1}-1} \sum_{j_2=0}^{m_{v_2}-1+j_1} \cdots \sum_{j_k=0}^{m_{v_k}-1+j_{k-1}} (-1)^{j_k}$$
$$\times \frac{(m_{v_1}-1)!(m_{v_2}-1+j_1)!\cdots(m_{v_k}-1+j_{k-1})!}{j_1!j_2!\cdots j_k!} L(z; \omega_{v_k}, \ldots, \omega_{v_1}; j_k, \ldots, j_1). \tag{3.8}$$

The proof requires an auxiliary construction.

3.3 Auxiliary polynomials $D^{n_1,\ldots,n_k}_{m_{v_1},\ldots,m_{v_k}}(t)$

Choose now $k \geq 1$; $v_k, \ldots, v_1 \in V$, and positive integers $n_k, \ldots, n_1$. It is convenient to agree that for $k = 0$ the respective families are empty.

Define inductively polynomials

$$D^{n_1,\ldots,n_k}_{m_{v_1},\ldots,m_{v_k}}(t) \in \mathbf{Q}[t]$$

putting $D^{\emptyset}_{\emptyset} = 1$, and

$$D^{n_1,\ldots,n_{k+1}}_{m_{v_1},\ldots,m_{v_{k+1}}}(t) = (1+\partial_t)^{-1}\left(D^{n_1,\ldots,n_k}_{m_{v_1},\ldots,m_{v_k}}\left(\frac{n_1+\cdots+n_k}{n_1+\cdots+n_{k+1}}t\right)\cdot t^{m_{v_{k+1}}-1}\right), \tag{3.9}$$

where

$$(1+\partial_t)^{-1} := \sum_{k\geq 0}(-1)^k \partial_t^k$$

as a linear operator on polynomials.

For example,

$$D^{n_1}_{m_{v_1}}(t) = (-1)^{m_{v_1}-1}(m_{v_1}-1)! \sum_{j_1=0}^{m_{v_1}-1} \frac{(-1)^{j_1} t^{j_1}}{j_1!}. \tag{3.10}$$

In particular, $D^{n_1}_{m_{v_1}}(0) = (-1)^{m_{v_1}-1}(m_{v_1}-1)!$. Furthermore,

$$D^{n_1,n_2}_{m_{v_1},m_{v_2}}(t)$$
$$= (-1)^{m_{v_1}-1}(m_{v_1}-1)! \sum_{j_1=0}^{m_{v_1}-1} \frac{(-1)^{j_1}}{j_1!}(1+\partial_t)^{-1}\frac{n_1^{j_1} t^{j_1+m_{v_2}-1}}{(n_1+n_2)^{j_1}}$$
$$= (-1)^{m_{v_1}-1+m_{v_2}-1}(m_{v_1}-1)!(m_{v_2}-1)! \sum_{j_1=0}^{m_{v_1}-1} \frac{(-1)^{j_1} n_1^{j_1}}{j_1!(n_1+n_2)^{j_1}} \sum_{j_2=0}^{m_{v_2}-1+j_1} \frac{(-1)^{j_2} t^{j_2}}{j_2!}. \tag{3.11}$$

In particular,

$$D^{n_1,n_2}_{m_{v_1},m_{v_2}}(0) = (-1)^{m_{v_1}-1+m_{v_2}-1}(m_{v_1}-1)!(m_{v_2}-1)! \sum_{j_1=0}^{m_{v_1}-1} \frac{(-1)^{j_1} n_1^{j_1}}{j_1!(n_1+n_2)^{j_1}}.$$

The general formula looks as follows.

Proposition 3.3.1. *We have for $k \geq 1$*

$$\begin{aligned} &D^{n_1,\ldots,n_k}_{m_{v_1},\ldots,m_{v_k}}(t) \\ &= (-1)^{\sum_{i=1}^k (m_{v_i}-1)} \sum_{j_1=0}^{m_{v_1}-1} \sum_{j_2=0}^{m_{v_2}-1+j_1} \cdots \sum_{j_k=0}^{m_{v_k}-1+j_{k-1}} (-1)^{j_k} \\ &\quad \times \frac{(m_{v_1}-1)!(m_{v_2}-1+j_1)!\cdots(m_{v_k}-1+j_{k-1})!}{j_1!j_2!\cdots j_k!} \\ &\quad \times \frac{1}{n_1^{-j_1}(n_1+n_2)^{j_1-j_2}\cdots(n_1+\cdots+n_{k-1})^{j_{k-2}-j_{k-1}}(n_1+\cdots+n_k)^{j_{k-1}}} t^{j_k}. \end{aligned} \tag{3.12}$$

Proof. We argue by induction on k. Assume that (3.12) holds for k and apply the operator at the right-hand side of (3.9) to the right-hand side of (3.12). Looking for brevity only at the last line of (3.11), we get

$$\begin{aligned} &\frac{1}{n_1^{-j_1}(n_1+n_2)^{j_1-j_2}\cdots(n_1+\cdots+n_k)^{j_{k-1}-j_k}(n_1+\cdots+n_{k+1})^{j_k}} \\ &\qquad \cdot (1+\partial_t)^{-1} t^{m_{v_{k+1}}-1+j_k} \\ &\quad = \frac{1}{n_1^{-j_1}(n_1+n_2)^{j_1-j_2}\cdots(n_1+\cdots+n_k)^{j_{k-1}-j_k}(n_1+\cdots+n_{k+1})^{j_k}} \\ &\qquad \times \sum_{j_{k+1}=0}^{m_{v_{k+1}}-1+j_k} \frac{(-1)^{j_{k+1}}(-1)^{m_{v_{k+1}}-1+j_k}(m_{v_{k+1}}-1+j_k)!}{j_{k+1}!} t^{j_{k+1}}. \end{aligned}$$

Combining this with (3.12) for k, we get (3.12) for $k+1$.

3.4 Proof of Theorem 3.2

By induction on k we will prove the formula

$$\begin{aligned} &(2\pi i)^{m_{v_1}+\cdots+m_{v_k}} I^z_{i\infty}(\omega_{v_k},\ldots,\omega_{v_1}) \\ &\quad = \sum_{n_1,\ldots,n_k \geq 1} \frac{c_{v_1,n_1}\cdots c_{v_k,n_k} e^{2\pi i(n_1+\cdots+n_k)z}}{n_1^{m_{v_2}}(n_1+n_2)^{m_{v_1}}\cdots(n_1+\cdots+n_k)^{m_{v_k}}} \\ &\qquad\qquad \cdot D^{n_1,\ldots,n_k}_{m_{v_1},\ldots,m_{v_k}}(2\pi i(n_1+\cdots+n_k)z). \end{aligned} \tag{3.13}$$

Combining it with (3.12) and (3.3), we will get (3.8).

For $k = 1$ we check (3.13) directly:

$$(2\pi i)^{m_{v_1}} I_{i\infty}^{z}(\omega_{v_1}) = (2\pi i)^{m_{v_1}} \int_{i\infty}^{z} \omega_{v_1}(z_1)$$
$$= (2\pi i)^{m_{v_1}} \sum_{n_1=1}^{\infty} c_{v_1,n_1} \int_{i\infty}^{z} e^{2\pi i n_1 z_1} z_1^{m_{v_1}-1} dz_1.$$

Putting in the n_1th summand $t = 2\pi i n_1 z_1$, we can rewrite this as

$$\sum_{n_1=1}^{\infty} \frac{c_{v_1,n_1}}{n_1^{m_{v_1}}} \int_{-\infty}^{2\pi i n_1 z} e^t t^{m_{v_1}-1} dt. \tag{3.14}$$

Since

$$\int e^t P(t) dt = e^t (1 + \partial_t)^{-1} P(t) + \text{const}, \tag{3.15}$$

this is equivalent to (3.13).

The inductive step from k to $k+1$ is similar: Assuming $(3.13)_k$, we have

$$\begin{aligned}
&(2\pi i)^{m_{v_1}+\cdots+m_{v_{k+1}}} I_{i\infty}^{z}(\omega_{v_{k+1}}, \ldots, \omega_{v_1}) \\
&= (2\pi i)^{m_{v_{k+1}}} \sum_{n_{k+1}=1}^{\infty} c_{v_{k+1},n_{k+1}} \int_{i\infty}^{z} e^{2\pi i n_{k+1} z_k} z_k^{m_{v_{k+1}}-1} \\
&\quad \times \sum_{n_1,\ldots,n_k \ge 1} \frac{c_{v_1,n_1} \cdots c_{v_k,n_k} e^{2\pi i (n_1+\cdots+n_k) z_k}}{n_1^{m_{v_1}} (n_1+n_2)^{m_{v_1}} \cdots (n_1+\cdots+n_k)^{m_{v_k}}} \\
&\qquad \cdot D^{n_1,\ldots,n_k}_{m_{v_1},\ldots,m_{v_k}} (2\pi i (n_1+\cdots+n_k) z_k) dz_k \\
&= (2\pi i)^{m_{v_{k+1}}} \sum_{n_1,\ldots,n_{k+1} \ge 1} \frac{c_{v_1,n_1} \cdots c_{v_k,n_k} c_{v_{k+1},n_{k+1}}}{n_1^{m_{v_1}} (n_1+n_2)^{m_{v_1}} \cdots (n_1+\cdots+n_k)^{m_{v_k}}} \\
&\quad \times \int_{i\infty}^{z} e^{2\pi i (n_1+\cdots+n_k+n_{k+1}) z_k} D^{n_1,\ldots,n_k}_{m_{v_1},\ldots,m_{v_k}} (2\pi i (n_1+\cdots+n_k) z_k) z_k^{m_{v_{k+1}}-1} dz_k.
\end{aligned} \tag{3.16}$$

Putting here $t = 2\pi i (n_1 + \cdots + n_{k+1}) z_k$, we can rewrite the last integral as

$$\frac{1}{(2\pi i)^{m_{v_{k+1}}}} \frac{1}{(n_1+\cdots+n_{k+1})^{m_{v_{k+1}}}} \cdot \int_{i\infty}^{z} e^t D^{n_1,\ldots,n_k}_{m_{v_1},\ldots,m_{v_k}} \left(\frac{n_1+\cdots+n_k}{n_1+\cdots+n_{k+1}} t \right) t^{m_{v_{k+1}}-1} dt,$$

that is,

$$\frac{1}{(2\pi i)^{m_{v_{k+1}}}} \frac{1}{(n_1+\cdots+n_k)^{m_{v_{k+1}}}} e^{2\pi i (n_1+\cdots+n_{k+1}) z} \tag{3.17}$$

$$\times (1+\partial_t)^{-1}\left(D^{n_1,\ldots,n_k}_{m_{v_1},\ldots,m_{v_k}} \left(\frac{n_1+\cdots+n_k}{n_1+\cdots+n_{k+1}} t\right) t^{m_{v_{k+1}}-1}\right)\Bigg|_{t=2\pi i(n_1+\cdots+n_{k+1})z} . \tag{3.18}$$

Substituting this into (3.16), we finally obtain (3.13) and (3.12). □

3.5 The limit $z \to 0$

We can try to get an expression for

$$I^0_{i\infty}(\omega_{v_k}, \ldots, \omega_{v_1})$$

as a (linear combination of) multiple Dirichlet series, by formally putting $z = 0$ in the right-hand side of (3.8). However, we will find out that this cannot be done automatically for a certain range of values of (j_k, j_{k-1}), namely, for $j_k \geq m_{v_k} + j_{k-1} - 1 - kC$; cf. Case 3 at the end of Section 3.1.2.

To solve this problem, we will have for the first time to assume that $z^{1-m_v}\omega_v(z)$ are of cusp modular type, say, for the group $\Gamma_0(N)$, or for any modular subgroup which is normalized by the involution $z \mapsto gz$,

$$g = g_N := \begin{pmatrix} 0 & -1 \\ N & 0 \end{pmatrix}.$$

We can then apply Proposition 2.3 which we reproduce and slightly augment:

Proposition 3.5.1. *Assume that ω_V as above is a basis of a space of 1-forms invariant with respect to g_N. Then*

$$J^0_{i\infty}(\omega_V) = (g_{N*}(J^{\frac{i}{\sqrt{N}}}_{i\infty}(\omega_V)))^{-1} J^{\frac{i}{\sqrt{N}}}_{i\infty}(\omega_V). \tag{3.19}$$

Replacing the coefficients of the formal series on the right-hand side of (3.19) *by their (convergent) representations via multiple Dirichlet series* (3.8)*, we get such representations for $I^0_{i\infty}(\omega_{v_k}, \ldots, \omega_{v_1})$.*

3.5.2 Main application

If we fix a modular subgroup normalized by g_N, this proposition becomes applicable to any family of cusp forms of the type $\omega_i(z)z^{m-1}$, $m \geq 1$, where $\omega_i(z)$ runs over a basis of the space of forms of a fixed weight $2r$, and m runs over $[1, 2r-1]$ (cf. (2.4)). Moreover, we can mix different weights, that is, take a finite union of such families.

Passing to a different basis of such a space, we may even assume that (ω_v) consists of eigenforms for g_N: $g_N^*(\omega_v) = \varepsilon_v\omega_v$, $\varepsilon_v = \pm 1$, for all $s \in V$.

Coefficients of $J^{\frac{i}{\sqrt{N}}}_{i\infty}$ are the series (3.3) at $z = \frac{i}{\sqrt{N}}$.

4 Shuffle relations between multiple Dirichlet series

4.1 Notation

In this section, we will consider (formal) multiple Dirichlet series of a special form generalizing expressions (3.4), and deduce bilinear relations between them generalizing the well-known *harmonic* shuffle relations involving shuffles with repetitions.

Each such series will depend on a set of *coefficients data* C and several complex or formal arguments s_i. Here are the precise definitions. Let $k \geq 1$ be a natural number.

Definition 4.1.1. (i) Coefficients data C of depth k is a family of numbers $c_{n,m}^{(j,i)}$ indexed by two pairs of integers satisfying $j > i \geq 0$, $j \leq k$, and $n > m \geq 0$.
(ii) The multiple Dirichlet series associated with C and arguments $s_1, \dots, s_k$ is

$$L_C(s_1, \dots, s_k) := \sum_{0=u_0<u_1<\dots<u_k\in\mathbf{Z}} \frac{\prod_{k\geq j>i\geq 0} c_{u_j,u_i}^{(j,i)}}{u_1^{s_1}u_2^{s_2}\cdots u_k^{s_k}}. \tag{4.1}$$

4.1.2 Examples

(a) Assume that $c_{n,m}^{(j,i)} = 1$ if $m > 0$ or $i > 0$ and put $c_{n,0}^{(j,0)} = a_n^{(j)}$. Then

$$L_C(s_1, \dots, s_k) = \sum_{0<u_1<\dots<u_k\in\mathbf{Z}} \frac{a_{u_1}^{(1)}a_{u_2}^{(2)}\cdots a_{u_k}^{(k)}}{u_1^{s_1}u_2^{s_2}\cdots u_k^{s_k}} \tag{4.2}$$

is an ordinary multiple Dirichlet series.

(b) Define $c_{v,n}$ as in Section 3.1, and choose $v_1, \dots, v_k \in V$ as in Section 3.1.1. Construct the coefficients data C putting

$$c_{n,m}^{(j,j-1)} := c_{v_j,n-m}, \tag{4.3}$$

and $c_{n,m}^{(j,i)} = 1$ otherwise. Then $L_C(m_{v_1} + j_0 - j_1, \dots, m_{v_k} + j_{k-1} - j_k)$ becomes the formal series (3.4) if we redenote $u_j = n_1 + \dots + n_j$.

4.1.3 Shuffles and a composition of the coefficients data

Let $C = (c_{n,m}^{(j,i)})$ and $D = (d_{n,m}^{(j,i)})$ be two data of depths k and l, respectively. A *(k, l, p)-shuffle with repetitions* is a pair of strictly increasing maps $\sigma = (\sigma_1, \sigma_2)$,

$$\sigma_1 : [0, k] \to [0, p], \quad \sigma_2 : [0, l] \to [0, p]$$

satisfying the conditions

$$\sigma_1(0) = \sigma_2(0) = 0, \qquad \sigma_1([0, k]) \cup \sigma_2([0, l]) = [0, p]. \tag{4.4}$$

It follows that $\max(k, l) \leq p \leq k+l$. We will say that *the σ-multiplicity of $j \in [0, p]$ is* 1 if $j \notin \sigma_1([0, k]) \cap \sigma_2([0, l])$. Otherwise, the σ-multiplicity of $j \in [0, p]$ is 2. In particular, the σ-multiplicity of 0 is 2.

Given such C, D and $\sigma = (\sigma_1, \sigma_2)$, we will define the third coefficients data $E = (e_{n,m}^{(j,i)})$ of depth p, which we denote $E = C *_\sigma D$. Choose j, i with $p \geq j > i \geq 0$. We have the following set of mutually exclusive and exhaustive alternatives $(\mathrm{A})_1$, $(\mathrm{A})_2$, (B), and (C):

(A) Assume that both j and i have multiplicity 1.

$(\mathrm{A})_1$ Both j and i belong to the image of one and the same σ_a with $a = 1$ or $a = 2$. Then we put

$$e_{n,m}^{(j,i)} := c_{n,m}^{(\sigma_1^{-1}(j), \sigma_1^{-1}(i))} \quad \text{for } a = 1$$

and

$$e_{n,m}^{(j,i)} := d_{n,m}^{(\sigma_2^{-1}(j), \sigma_2^{-1}(i))} \quad \text{for } a = 2.$$

$(\mathrm{A})_2$ Assume that j, respectively, i, belongs to the image of σ_a, respectively, σ_b, with $a \neq b$. Then we put

$$e_{n,m}^{(j,i)} := 1.$$

(B) Assume that exactly one of j, i has multiplicity 2. Then there exists only one value $a = 1$ or 2 such that j and i belong to the image of σ_a. We put then as in the case $(\mathrm{A})_1$

$$e_{n,m}^{(j,i)} := c_{n,m}^{(\sigma_1^{-1}(j), \sigma_1^{-1}(i))} \quad \text{for } a = 1$$

and

$$e_{n,m}^{(j,i)} := d_{n,m}^{(\sigma_2^{-1}(j), \sigma_2^{-1}(i))} \quad \text{for } a = 2.$$

(C) Assume that both i and j have multiplicity 2. Then we put

$$e_{n,m}^{(j,i)} := c_{n,m}^{(\sigma_1^{-1}(j), \sigma_1^{-1}(i))} d_{n,m}^{(\sigma_2^{-1}(j), \sigma_2^{-1}(i))}.$$

4.1.4 Shuffles and a composition of the arguments

Let $s := (s_1, \ldots, s_k)$ and $t := (t_1, \ldots, t_l)$ be arguments for the data C and D as above, and σ a (k, l, p)-shuffle as above. We define $s +_\sigma t := (r_1, \ldots, r_p)$ as follows.

If i has multiplicity one and is covered by σ_1, respectively, σ_2, then $r_i := s_{\sigma_1^{-1}(i)}$, respectively, $r_i := t_{\sigma_2^{-1}(i)}$.

If i has multiplicity 2, then $r_i := s_{\sigma_1^{-1}(i)} + t_{\sigma_2^{-1}(i)}$.

We can now state the main result of this section.

Theorem 4.2. *Let C, respectively, D, be some coefficients data of depths k, respectively, l, as above. Then we have*

$$L_C(s) \cdot L_D(t) = \sum_\sigma L_{C *_\sigma D}(s +_\sigma t), \tag{4.5}$$

where the summation is taken over all (k, l, p)-shuffles with repetitions.

Proof. Consider a term of the series (4.1) corresponding to $(u_0 = 0, u_1, u_2, \dots, u_k)$ and a term of the series $L_D(t)$ corresponding to, say, $(w_0 = 0, w_1, w_2, \dots, w_l)$. This pair of terms determines a unique (k, l, p)-shuffle (σ_1, σ_2), where p is the cardinality of the union of sets

$$\{u_1, u_2, \dots, u_k\} \cup \{w_1, w_2, \dots, w_l\} := \{q_1, \dots, q_p\}.$$

Namely, we may and will assume that $q_0 = 0 < q_1 < \dots < q_p$. Then $\sigma_1(i) = j$ if $u_i = q_j$, and $\sigma_2(i) = j$ if $w_i = q_j$.

Group together all pairwise products corresponding to one and the same shuffle, and denote the resulting sum by L_σ.

The denominator of one such a product will obviously be $q_1^{r_1} \cdots q_p^{r_p}$, where $r = s +_\sigma t$. Moreover, knowing such a denominator, we uniquely reconstruct the two terms from $L_C(s)$ and $L_D(t)$, from which it was produced, at least if s, t take generic values so that in the family $\{s_a, t_b, s_a + t_b\}$ all terms are pairwise distinct. Finally, all possible sequences $q_0 = 0 < q_1 < \dots < q_p$ will occur.

To prove that $L_\sigma = L_{C *_\sigma D}(s +_\sigma t)$, it remains to check that the numerator of such a product will be as predicted by (4.5), in other words, that

$$\prod_{p \geq j > i \geq 0} e_{q_j, q_i}^{(j,i)} \overset{(?)}{=} \prod_{k \geq j > i \geq 0} c_{u_j, u_i}^{(j,i)} \prod_{l \geq j > i \geq 0} d_{w_j, w_i}^{(j,i)}$$

if $e_{n,m}^{(j,i)}$ are defined as in Section 4.1.3.

This is straightforward, although somewhat tedious. □

4.3 Concluding remarks

It would be interesting to describe some nontrivial spaces of Dirichlet series containing periods of cusp forms, closed with respect to the series shuffle relations, and consisting entirely of periods in the sense of [KonZa].

Regarding shuffle relations themselves, motivic philosophy predicts that they should be obtainable by standard manipulations with integrals. For the harmonic shuffle relations between multiple zeta values, A. Goncharov established this in [Go6, Chapter 9] (for convergent integrals), and in [Go6, 7.5] elaborating the last page of [Go4] (for regularized integrals). Conversely, integral shuffle relations can be deduced from harmonic ones: see [Go5, Chapter 2].

5 Iterated Eichler–Shimura and Hecke relations

5.1 Eichler–Shimura relations for iterated integrals

In this subsection, we take for X the upper half-plane H and for (ω_V) a family of 1-forms of cusp modular type (see 2.1.1) spanning a finite-dimensional linear space stable with respect to the modular transformations

$$\sigma = \begin{pmatrix} 0 & -1 \\ 1 & 0 \end{pmatrix}, \qquad \tau = \begin{pmatrix} 1 & -1 \\ 1 & 0 \end{pmatrix}.$$

Proposition 5.1.1. *With these assumptions, we have*

$$\sigma_*(J_0^{i\infty}(\omega_V)) \cdot J_0^{i\infty}(\omega_V) = 1, \tag{5.1}$$

$$\tau_*^2(J_0^{i\infty}(\omega_V)) \cdot \tau_*(J_0^{i\infty}(\omega_V)) \cdot J_0^{i\infty}(\omega_V) = 1. \tag{5.2}$$

Proof. From (1.15) we get

$$\sigma_*(J_0^{i\infty}(\omega_V)) = J_{\sigma(0)}^{\sigma(i\infty)}(\omega_V) = J_{i\infty}^{0}(\omega_V) = J_0^{i\infty}(\omega_V)^{-1}.$$

which shows (5.1).

Similarly, τ transforms $(0, i\infty)$ into $(i\infty, 1)$, then to $(1, 0)$, and according to (1.9)

$$J_1^0(\omega_V) \cdot J_{i\infty}^1(\omega_V) \cdot J_0^{i\infty}(\omega_V) = 1, \tag{5.3}$$

which is the same as (5.2). □

Notice that some care is needed in establishing (5.3): the geodesic triangle with vertices $0, i\infty, 1$ should first be replaced by a sequence of geodesic hexagons lying entirely in H and cutting the corners of the triangle, and then it must be checked that in the limit the hexagon relation replacing (5.3) tends to (5.3). This is routine for cusp modular 1-forms; cf. Section 2.1.3.

We now pass to the relations involving Hecke operators.

5.2 Hecke operators

In this subsection, V denotes a finite set, $\omega_v = f_v(z)dz$ a family of modular forms *of weight* 2, and p_v a family of primes, both indexed by $v \in V$. Moreover, we assume that $T_{p_v}\omega_v = \lambda_v\omega_v$, $\lambda_v \in \mathbf{C}$, where T_{p_v} is the Hecke operator

$$T_{p_v} := \begin{pmatrix} p_v & 0 \\ 0 & 1 \end{pmatrix} + \sum_{b=0}^{p_v-1} \begin{pmatrix} 1 & b \\ 0 & p_v \end{pmatrix} = \mathbf{p_v} + \sum_{b=0}^{p_v-1} h(p_v, b). \tag{5.4}$$

Put $U := V \coprod V'$, where V' is another copy of the indexing set V, and for $v' \in V'$ corresponding to $v \in V$ put $\omega_{v'} := (\mathbf{p_v})^*(\omega_v)$. Let ω_U be the family consisting of all ω_v and $\omega_{v'}$. When we consider formal series of the type $J_a^z(\omega_U)$ as in Section 1, we denote the variables corresponding to V, respectively, V', by A_v, respectively, $A_{v'}$.

Denote by W the set of pairs $w = (v, b)$, where $v \in V$ and $b \in [0, p_v - 1]$. Let ω_W be the family consisting of $\omega_{(v,b)} := h(p_v, b)^*(\omega_v)$. When we consider formal series of the type $J_a^z(\omega_W)$, we denote the variables corresponding to w, by B_w or $B_{(v,b)}$.

Define the following two continuous homomorphisms of rings of formal series:

$$l : \mathbf{C}\langle\langle A_U\rangle\rangle \to \mathbf{C}\langle\langle A_V\rangle\rangle : l(A_v) := \lambda_v A_v, \quad l(A_{v'}) := -A_v, \tag{5.5}$$

$$r : \mathbf{C}\langle\langle B_W\rangle\rangle \to \mathbf{C}\langle\langle A_V\rangle\rangle : r(B_{(v,b)}) := A_v. \tag{5.6}$$

Theorem 5.3. *We have*

$$l(J^0_{i\infty}(\omega_U)) = r(J^0_{i\infty}(\omega_W)). \tag{5.7}$$

Proof. We will check that

$$l(J^0_{i\infty}(\omega_U)) = J^0_{i\infty}((\lambda_v\omega_v - \omega_{v'})) \tag{5.8}$$

whereas

$$r(J^0_{i\infty}(\omega_W)) = J^0_{i\infty}((\sum_{b=0}^{p_v-1} \omega_{(v,b)})), \tag{5.9}$$

where on the right-hand sides we consider both families as indexed by V. Since from (5.4) and the above definitions we obtain for each $v \in V$

$$\lambda_v\omega_v - \omega_{v'} = \sum_{b=0}^{p_v-1} \omega_{(v,b)}, \tag{5.10}$$

this will prove the theorem. □

We have

$$J^0_{i\infty}(\omega_U) = \sum_{n=0}^{\infty} \sum_{(u_1,\dots,u_n)\in U^n} A_{u_n}\cdots A_{u_1} I^0_{i\infty}(\omega_{u_n},\dots,\omega_{u_1}). \tag{5.11}$$

Consider one summand in (5.11). In the sequence $(u_1,\dots,u_n)$ there are several, say $0 \le k \le n$, elements $v_i \in V$ and the remaining $n-k$ elements $v'_j \in V'$. Application of l eliminates all primes in the subscripts of $A_{u_n}\cdots A_{u_1}$ and produces a monomial in A_v; besides, it multiplies this monomial by $(-1)^{n-k}\prod\lambda_{v_i}$. Hence the coefficient at any monomial $A_{v_n}\cdots A_{v_1}$ in $l(J^0_{i\infty}(\omega_U))$ can be written as

$$\sum_{S\subset[1,\dots n]} (-1)^{n-|S|}\left(\prod_{i\in S}\lambda_{v_i}\right) I^0_{i\infty}(\pi_S(\omega_{v_n},\dots,\omega_{v_1})), \tag{5.12}$$

where the operator π_S replaces v_j by v'_j whenever $j\notin S$.

On the other hand, the similar coefficient in $J^0_{i\infty}((\lambda_v\omega_v - \omega_{v'}))$ is

$$I^0_{i\infty}(\lambda_{v_n}\omega_{v_n} - \omega_{v'_n},\dots,\lambda_{v_1}\omega_{v_1} - \omega_{v'_1}),$$

which obviously coincides with (5.12) because the iterated integrals are polylinear in ω. This proves (5.8).

The check of (5.9) is similar. We have

$$J^0_{i\infty}(\omega_W) = \sum_{((v_n,b_n),\dots,(v_1,b_1))\in W^n} B_{(v_n,b_n)}\cdots B_{(v_1,b_1)} I^0_{i\infty}(\omega_{(v_n,b_n)},\dots,\omega_{(u_1,b_1)}).$$

Application of r produces a series in (A_v) whose coefficient of $A_{v_n}\cdots A_{v_1}$ equals

$$\sum_{(b_1,\dots,b_n)} I^0_{i\infty}(\omega_{(v_n,b_n)},\dots,\omega_{(v_1,b_1)}).$$

This is the same as the respective coefficient on the right-hand side of (5.9).

6 Differentials of the third kind and generalized associators

6.1 Normalized horizontal sections

In this subsection, following [Dr2], we will define and study solutions of the differential equation (1.4), $dJ^z(\Omega) = \Omega(z)J^z(\Omega)$ in the case when $\Omega = \sum A_v\omega_v$ may have a logarithmic singularity at a point a, so that it cannot be normalized by the condition $J^a(\Omega) = 1$ and in fact cannot be defined by the series (1.3).

We start with a local situation. Put

$$r_{v,a} := \mathrm{res}_a\, \omega_v, \qquad R_a := \mathrm{res}_a\, \Omega = \sum_v r_{v,a} A_v. \tag{6.1}$$

The normalized solution will depend on the choice of a local parameter t_a at a, and a branch of the logarithm $\log t_a$.

Let U be a disc around a uniformized by t_a. Denote by $\log t_a$ the branch of the logarithm in U which is real on $\mathrm{Im}\, t_a = 0$, $\mathrm{Re}\, t_a > 0$. Delete from U a cut from a to the boundary which does not intersect the latter interval and denote by U' the remaining domain. Write $t_a^{R_a}$ for $e^{R_a \log t_a}$. It is a formal series in A_v with coefficients which are holomorphic functions near a in U'. Assume that, outside of a, all ω_v are regular in U.

Definition 6.1.1. A ∇_Ω-horizontal section J in U' is called normalized at a (with respect to a choice of t_a) if it is of the form $J = K \cdot t_a^{R_a}$, where K can be extended to a holomorphic section in some neighborhood of a in U which takes the value 1 at a.

We will see that this definition produces a version of $J_a^z(\Omega)$. In fact, we get precisely $J_a^z(\Omega)$ if $R_a = 0$, so that $t_a^{R_a} = 1$.

Proposition 6.1.2. *For any a and t_a as above, there exists a unique local section holomorphic in U' normalized at a.*

Proof. In the course of proof, we will be considering only one point a, so we will omit it in the notation for brevity and write R, t, r_v, etc., in place of the former R_a, t_a, $r_{v,a}$.

The equation $\nabla_\Omega(K \cdot t^R) = 0$ is equivalent to

$$dK = \Omega' K + t^{-1}[R, K]dt, \tag{6.2}$$

where

$$\Omega' := \Omega - R\frac{dt}{t} = \sum_v r_v \nu_v A_v.$$

We look for a solution to (6.2) of the form

$$K = 1 + \sum_{n=1}^{\infty} \sum_{(v_1,\dots,v_n)\in V^n} f_{v_1,\dots,v_n} A_{v_1} \cdots A_{v_n},$$

where $f_{v_1,\dots,v_n}$ must be holomorphic functions defined in some common neighborhood of a in U and vanishing at a. From (6.2) we find $df_v = \nu_v$, so that the only choice is $f_v(z) := \int_a^z \nu_v$. Notice that all ν_v are regular in a common neighborhood of a uniformized by t_a. We will see that in this neighborhood all other coefficients can be defined as well.

If $f_{v_1,\dots,v_{n-1}}$ with the required properties are defined for some $n-1 \geq 1$, we find from (6.2)

$$df_{v_1,\dots,v_n} = \nu_{v_1} f_{v_2,\dots,v_n} + t^{-1}(r_{v_1} f_{v_2,\dots,v_n} - f_{v_2,\dots,v_{n-1}} r_{v_n})dt.$$

By the inductive assumption, the right-hand side is well defined and regular at a, so integrating it from a to z, we get $f_{v_1,\dots,v_n}(z)$. □

6.2 Scattering operators

Now let a, b be two points where Ω may have logarithmic singularities. Choose t_a, t_b as above, construct the neighborhoods U_a, U_b and neighborhoods with deleted cuts U_a', U_b' in which we have the holomorphic normalized horizontal sections J_a, J_b. Now embed U_a', U_b' into a connected simply connected domain W to which both J_a and J_b can be analytically extended. Clearly, they are invertible elements of $\mathcal{O}_x\langle\langle A_V\rangle\rangle$ at almost all points $x \in W$. Put

$$\widetilde{J}_b^a = J_a^{-1} J_b. \tag{6.3}$$

As in the proof of Proposition 1.2, one sees that $\widetilde{J}_a^b \in \mathbf{C}\langle\langle A_V\rangle\rangle$. Borrowing the physics terminology, we can call this transition element *the scattering operator.*

I added twiddle in the notation in order to remind the reader that $\widetilde{J}_a^b$, as well as Ω, depends on t_a and t_b as well, if at least one of the residues R_a, R_b does not vanish. This dependence however is pretty mild. Let $J_a' = K'(t_a')^{R_a}$ be another horizontal section normalized with respect to some t_a'. Denote by $\tau_a \in \mathbf{C}^*$ the value of t_a'/t_a at a, and let $t_a' = T_a \cdot t_a \tau_a$, $T_a(a) = 1$.

Proposition 6.2.1.

(i) *We have*

$$J_a' = J_a \cdot \tau_a^{R_a}, \qquad K = K' \cdot T^{R_a}. \tag{6.4}$$

(ii) *Therefore, after replacing the two uniformizing parameters t_a, t_b by t_a', t_b', we get*

$$\widetilde{J}_b'^a = \tau_a^{-R_a} \widetilde{J}_b^a \tau_b^{R_b}. \tag{6.5}$$

Proof.

(i) We have

$$J_a' = K'(t_a')^{R_a} = K' \cdot T_a^{R_a} \cdot (t_a)^{R_a} \tau^{R_a}$$

Since $\tau^{R_a} \in \mathbf{C}\langle\langle A_V\rangle\rangle$, and this section is invertible, $K' \cdot T_a^{R_a} \cdot (t_a)^{R_a}$ is ∇_Ω-horizontal as well, and since $K' \cdot T_a^{R_a}$ at a equals 1, it is normalized, so $K' \cdot T_a^{R_a} \cdot (t_a)^{R_a} = J_a$, which proves (6.4).

(ii) Similarly, from $J_b' = J_a' \widetilde{J}_b'^a$ and (6.4), we get

$$J_b = J_a \tau_a^{R_a} \widetilde{J}_b'^a \tau_b^{-R_b},$$

which together with (6.3) proves (6.5). □

6.3 Example: Drinfeld's associator

Let $X = \mathbf{P}^1(\mathbf{C})$, $V = \{0, 1\}$,

$$\omega_0 = \frac{1}{2\pi i}\frac{dz}{z}, \qquad \omega_1 = \frac{1}{2\pi i}\frac{dz}{z-1}.$$

Then

$$\Omega = A_0\omega_0 + A_1\omega_1$$

has poles at 0, 1, ∞ with residues $A_0/2\pi i$, $A_1/2\pi i$, $-(A_0 + A_1)/2\pi i$, respectively. Put $t_0 = z, t_1 = 1-z$. Then $\widetilde{J}_0^1$ in our notation is the Drinfeld associator $\phi_{KZ}(A_0, A_1)$ from [Dr2, Section 2].

6.4 Generalized associators

An essential feature of the the last example is that ω_v are global logarithmic differentials on the compact Riemann surface $\mathbf{P}^1$.

Generally, let X be such a surface, and (a_i) a finite set of N points on it. The dimension of the space of global logarithmic differentials with poles in this set, $\sum c_i d \log f_i$, where $c_i \in \mathbf{C}$, f_i are meromorphic on X, is bounded by $N - 1$. It achieves the maximum value $N - 1$ iff the difference of any two points $a_i - a_j$ is torsion in the divisor class group. I will call such a set (a_i) *logarithmic*. The supply of logarithmic sets depends on the genus of X.

(a) *Genus zero*. Any finite set of points is logarithmic. The respective iterated integrals in this case include the multiple polylogarithms introduced in [Go1]. In fact, as Goncharov remarked, general iterated integrals can be reduced to multiple polylogarithms.

(b) *Genus one*. In this case we can take any finite set of points of finite order, for example, the subgroup of all points of a given order M.
For a general subset of points, Goncharov found a Feynman integral presentation (in the sense of the last section of [Go2]) of the respective real periods. His formulas involve the generalized Kronecker–Eisenstein series and can be considered as an elliptic version of multiple zeta values. A similar problem is addressed in the recent work of A. Levin and G. Racinet.

(c) *Higher genera*. If the genus of X is > 1, the order of such a logarithmic set is bounded. The most interesting explicitly known examples are modular curves and cusps on them; cf. [Dr1, El].
Notice that the initial Drinfeld's setting is modular as well: $\mathbf{P}^1$ with three marked points "is" the modular curve $\Gamma_0(4) \setminus \overline{H}$ together with its cusps.

According to Deligne and Elkik [El], a set of points is logarithmic iff the mixed Hodge structure on $H^1(X^\circ, \mathbf{Q})$ (where X° is the complement to the set of points) is split, that is, the direct sum of pure Hodge structures.

Definition 6.4.1. Let X be a compact Riemannian surface, and (a_i) a logarithmic set of points on it. Then any scattering operator of the form $\widetilde{J}_{a_i}^{a_j}$ is called a generalized associator.

6.5 Relations between scattering operators

Three types of relations established for $J_a^z(\omega_V)$ in Section 1 are extended below to the case of the general scattering operators.

6.5.1 Grouplike property

We have

$$\Delta(\widetilde{J}_a^b) = \widetilde{J}_a^b \otimes \widetilde{J}_a^b \tag{6.6}$$

where Δ is defined in 1.4.1.

To see this, notice that $\Delta(J_a)$, by definition, is the series which is obtained from J_a by replacing each A_v with $B_v := A_v \otimes 1 + 1 \otimes A_v$. Hence $\Delta(R_a) = R_a \otimes 1 + 1 \otimes R_a$ is the residue of $\Delta(\Omega)$ at a. Therefore, $\Delta(J_a)$ is the normalized $\nabla_{\Delta(\Omega)}$-horizontal section in the ring of formal series in B_v, and $\Delta(\widetilde{J}_a^b)$ is the respective scattering operator.

On the other hand, $J_a \otimes J_a$ satisfies the same equation $d(J_a \otimes J_a) = (\Omega \otimes 1 + 1 \otimes \Omega)(J_a \otimes J_a)$ and clearly is as well normalized at t_a. Hence the passage from $J_a \otimes J_a$ to $J_b \otimes J_b$ is governed by the same scattering operator, this time represented as the right-hand side of (6.6).

6.5.2 Cycle identities

Let γ be a closed oriented contractible contour in U, inside which there are no singularities of Ω. Let $a_1, \dots, a_n$ be points along this contour (cyclically) ordered compatibly with orientation. Then

$$\widetilde{J}_{a_2}^{a_1} \widetilde{J}_{a_3}^{a_2} \cdots \widetilde{J}_{a_n}^{a_{n-1}} \widetilde{J}_{a_1}^{a_n} = 1. \tag{6.7}$$

Of course, in this statement we assume that at each point a_i one and the same local parameter t_{a_i} and one and the same branch of its logarithm are used for the definition of the two relevant normalized sections corresponding to the incoming and outcoming segments of the contour. Otherwise the relevant τ^R factors as in (6.5) must be inserted.

6.5.3 Functoriality

Let g be an automorphism of X acting compatibly upon all the relevant objects, with the possible exception of the parameters t_a, and transforming the space spanned by ω_v into itself. Define g_* as in Section 1.4. Then

$$\widetilde{J}_{ga}^{gb} = \tau_a^{-R_a} g_*(\widetilde{J}_a^b)\tau_b^{R_b}. \tag{6.8}$$

where the τ^R factors account for the passage from t_a, respectively, t_b, to $g^*(t_a)$, respectively, $g^*(t_a)$.

The proof is essentially the same as in Sections 1.4 and 1.5.

6.6 Example: Drinfeld associator revisited

If we treat Drinfeld's setup as the $\Gamma_0(4)$-modular curve, lift it to H and apply to the respective family of scattering operators the Eichler–Shimura relations (5.1), (5.2) (following from the cycle identities of lengths 2 and 3, and functoriality), we will get the duality and the hexagonal relations. Of course, this is how they were deduced in the first place, with σ, τ pushed down to $\mathbf{P}^1$ rather than everything else lifted to H.

It seems very probable that the somewhat mysterious relationship between the double logarithms at roots of unity and the modular complex discovered in [Go3] can be explained in the same way.

Acknowledgments. I am grateful to A. Levin and A. Goncharov, who have read the first draft of this paper and made a number of useful suggestions which are incorporated in the text.

References

[AB] A. Ash and A. Borel, Generalized modular symbols, in J.-P. Labesse and J. Schwermer, eds., *Cohomology of Arithmetic Groups and Automorphic Forms (Luminy*, 1989, Lecture Notes in Mathematics, Vol. 1447, Springer-Verlag, New York, 1990, 57–75.

[AR] A. Ash and L. Rudolph, The modular symbol and continued fractions in higher dimensions, *Invent. Math.*, **55** (1979), 241–250.

[Ch] K.-T. Chen, Iterated path integrals, *Bull. Amer. Math. Soc.*, **83** (1977), 831–879.

[De] P. Deligne, *Multizeta Values*, notes d'exposés, Institute for Advanced Study, Princeton, NJ, 2001.

[DeGo] P. Deligne and A. Goncharov, Groupes fondamentaux motiviques de Tate mixte, *Ann. Sci. École Norm. Sup* (4), **38**-1 (2005), 1–56.

[Dr1] V. G. Drinfeld, Two theorems on modular curves, *Functional Anal. Appl.*, **7**-2 (1973), 155–156.

[Dr2] V. G. Drinfeld, On quasi-triangular quasi-Hopf algebras and some groups closely associated with Gal $\overline{\mathbf{Q}}/\mathbf{Q}$) *Algebra Anal.*, **2**-4 (1990); *Leningrad Math. J.*, **2**-4 (1991), 829–860.

[El] R. Elkik, *Le théorème de Manin–Drinfeld*, Astérisque, Vol. 183, Société Mathématique de France, Paris, 1990, 59–67.

[Go1] A. Goncharov, Polylogarithms in arithmetic and geometry, in *Proceedings of the International Congress of Mathematicians* (*ICM*'94, *Zurich, August* 3–11, 1994, Vol. 1, Birkhäuser, Basel, 1995, 374–387.

[Go2] A. Goncharov, Multiple ζ-values, Galois groups and geometry of modular varieties, in *Proceedings of the Third European Congress of Mathematicians*, Progress in Mathematics, Vol. 201, Birkhäuser Boston, Cambridge, MA, 2001, 361–392.

[Go3] A. Goncharov, The double logarithm and Manin's complex for modular curves, *Math. Res. Lett.*, **4** (1997), 617–636.

[Go4] A. Goncharov, Multiple polylogarithms, cyclotomy, and modular complexes, *Math. Res. Lett.*, **5** (1998), 497–516.

[Go5] A. Goncharov, Multiple polylogarithms and mixed Tate motives, math.AG/0103059, 2001.

[Go6] A. Goncharov, Periods and mixed motives, math.AG/0202154, 2002.

[GoMa] A. Goncharov and Yu. Manin, Multiple zeta-motives and moduli spaces $\overline{M}_{0,n}$, *Composito Math.*, **140**-1 (2004), 1–14.

[Ha] R. Hain, Iterated integrals and algebraic cycles: Examples and prospects, in *Contemporary Trends in Algebraic Geometry and Algebraic Topology* (*Tianjin*, 2000), Nankai Tracts in Mathematics, Vol. 5, World Scientific, River Edge, NJ, 2002, 55–118.

[Kon] M. Kontsevich, Operads and motives in deformation quantization, *Lett. Math. Phys.*, **48** (1999), 35–72.

[KonZa] M. Kontsevich and D. Zagier, Periods, in B. Engquist and W. Schmid, eds., *Mathematics Unlimited:* 2001 *and Beyond*, Springer-Verlag, Berlin, 2001, 771–808.

[Ma1] Yu. Manin, Parabolic points and zeta-functions of modular curves, *Izv. Akad. Nauk SSSR Ser. Mat.*, **36**-1 (1972), 19–66 (in Russian); *Math. USSR Izv.*, **6**-1 (1972), 19–64 (in English); in *Selected Papers*, World Scientific, Singapore, 1996, 202–247.

[Ma2] Yu. Manin, Periods of parabolic forms and p–adic Hecke series, *Mat. Sb.*, **92**-3 (1973), 378–401 (in Russian); *Math. USSR Sb.*, **21**-3 (1973), 371–393 (in English); in *Selected Papers*, World Scientific, Singapore, 1996, 268–290.

[MaMar] Yu. Manin and M. Marcolli, Continued fractions, modular symbols, and non-commutative geometry, *Selecta Math.* (*N.S.*), **8** (2002), 475–521.

[Me] L. Merel, *Quelques aspects arithmétiques et géométriques de la théorie des symboles modulaires*, thèse de doctorat, Université Paris VI, Paris, 1993.

[Ra1] G. Racinet, *Séries génératrices non-commutatives de polyzêtas et associateurs de Drinfeld*, Thèse de doctorat, Universitde Picardie-Jules-Verne, Amiens, France, 2000.

[Ra2] G. Racinet, Doubles mélanges des polylogarithms multiples aux racines de l'unité, *Publ. Math. IHÉS*, **95** (2002), 185–231.

[Ra3] G. Racinet, *Summary of Algebraic Relations between Multiple Zeta Values*, lecture notes, Max-Planck Institut für Mathematik, Bonn, 17 August, 2004.

[Sh1] V. Shokurov, Modular symbols of arbitrary weight, *Functional Anal. Appl.*, **10**-1 (1976), 85–86.

[Sh2] V. Shokurov, The study of the homology of Kuga varieties, *Math. USSR Izv.*, **16**-2 (1981), 399–418.

[Sh3] V. Shokurov, Shimura integrals of cusp forms, *Math. USSR Izv.*, **16**-3 (1981), 603–646.

[T] T. Terasoma, Mixed Tate motives and multiple zeta values, *Invent. Math.*, **149** (2002), 339–369.

[Za1] D. Zagier, Hecke operators and periods of modular forms, in S. Gelbart, R. Howe, and P. Sarnak, eds., *Festschrift in honor of I. I. Piatetski-Shapiro on the Occasion of his Sixtieth Birthday*, Israel Mathematics Conference Proceedings, Vol. 3, Weizmann Science Press, Jerusalem, 1990, 321–336.

[Za2] D. Zagier, Values of zeta functions and their applications, in *Proceedings of the First European Congress of Mathematics*, Part II, Progress in Mathematics, Vol. 120, Birkhäuser, Basel, 1994, 497–512.

[Za3] D. Zagier, Periods of modular forms and Jacobi theta functions, *Invent. Math.*, **104** (1991), 449–465.

Structures membranaires

Vadim Schechtman

Laboratoire Émile Picard
Université Paul Sabatier
31062 Toulouse
France
schechtman@picard.ups-tlse.fr

À Volodya Drinfeld en témoignage d'amitié à l'occasion de son 50*ième anniversaire.*

0 Introduction

Dans cette article on généralise quelques formules de [GMS] au cas de la troisième classe de Chern–Simons.

0.1

Soit X un schéma lisse sur un anneau commutatif $k \supset \mathbb{Q}$. La théorie de classes de Chern "style de Rham" associe à chaque fibré vectoriel E sur X des classes de cohomologie

$$c_i^{\mathrm{DR}}(E) \in H^{2i}(X, \Omega^{\cdot}_X), \quad i \geq 1,$$

qui satisfont aux propriétés usuelles; cf. [Gr, Section 6]. Ici $\Omega^{\cdot}_X$ est le complexe de de Rham de X sur k. On sait (cf., par exemple, [BD, 2.8]) que ces classes admettent une version plus fine: les classes *de Chern–Simons*

$$c_i^{\mathrm{CS}}(E) \in H^i(X, \Omega_X^{[i,2i-1\rangle}).$$

Ici

$$\Omega_X^{[i,2i-1\rangle} = \sigma_{\geq i}\tau_{\leq 2i-1}\Omega_X[i] : \Omega_X^i \longrightarrow \ldots \longrightarrow \Omega_X^{2i-1,\mathrm{fer}},$$

où $\Omega_X^{j,\mathrm{fer}} \subset \Omega_X^j$ est le sous-faisceau des formes fermées. L'image de $c_i^{\mathrm{CS}}(E)$ par le morphisme canonique

$$H^i(X, \Omega_X^{[i,2i-1\rangle}) \longrightarrow H^{2i}(X, \Omega^{\cdot}_X)$$

est égale à $c_i^{\mathrm{DR}}(E)$.

Les images des $c_i^{\mathrm{CS}}(E)$ par les morphismes évidents

$$H^i(X, \Omega_X^{[i,2i-1\rangle}) \longrightarrow H^i(X, \Omega_X^i)$$

sont les classes de Chern "style Hodge" $c_i^{\mathrm{Hdg}}(E)$; cf. [Gr, Section 6].

Il est très facile de décrire explicitement la première classe de Chern–Simons $c_1^{\mathrm{CS}}(E) \in H^1(X, \Omega_X^{1,\mathrm{fer}})$. En effet, si $r = \mathrm{rk}(E)$, choisissons un 1-cocycle de Čech définissant E, $\phi = \{\phi_{ij}\} \in \check{Z}^1(\mathfrak{U}, \mathrm{GL}_r(\mathcal{O}_X))$ sur un recouvrement ouvert $\mathfrak{U} = \{U_i\}$ de X convenable; $\phi_{ij} \in \Gamma(U_{ij}, \mathrm{GL}_r(\mathcal{O}_X))$, $U_{ij} = U_i \cap U_j$. Alors $c_1^{\mathrm{CS}}(E)$ est la classe de cohomologie du 1-cocycle de Čech

$$z_1^{\mathrm{CS}}(E) = \{\mathrm{tr}(\phi_{ij}^{-1} d\phi_{ij})\} \in \check{Z}^1(\mathfrak{U}, \Omega_X^{1,\mathrm{fer}}).$$

Il est clair que $z_1^{\mathrm{CS}}(E) = z_1^{\mathrm{CS}}(\det(E))$, où $\det(E) = \Lambda_{\mathcal{O}_X}^r(E)$ est le fibré en droites des puissances extérieures maximals de E.

Soit L un fibré en droites. Considérons le faisceau $\mathcal{C}onn\mathcal{I}nt(L)$ de connexions intégrables sur L; ceci est un torseur sous $\Omega_X^{1,\mathrm{fer}}$, dont la classe caractéristique $c(\mathcal{C}onn\mathcal{I}nt(L))$ (l'obstruction à l'existence d'une section globale) est égale à $c_1^{\mathrm{CS}}(L)$.

Donc, pour E arbitraire, $c_1^{\mathrm{CS}}(E)$ est la classe du $\Omega_X^{1,\mathrm{fer}}$-torseur $\mathcal{C}onn\mathcal{I}nt(\det(E))$. On peut donc dire que ce torseur est à l'origine de l'existence de la première classe de Chern–Simons.

Désormais nous ne nous intéresserons qu'au cas $E = \mathcal{T}_X$, où $\mathcal{T}_X$ est le fibré tangent de X. Au lieu des classes de Chern c_i on va considerer les "caractères de Chern" ch_i, donnés par les polynômes de Newton usuels, i.e., $\mathrm{ch}_1 = c_1$, $\mathrm{ch}_2 = c_1^2 - c_2/2$, etc. D'ailleurs, dans cette note on ne discutera que les cas $i = 1, 2, 3$.

Dans le Section 3 on décrit explicitement des cocycles qui représentent les caractères $\mathrm{ch}_i^{\mathrm{CS}}(\mathcal{T}_X)$ pour $i = 2, 3$.

Il est facile de voir (cf., par exemple, [GMS, Section 11]) que la donnée d'une connexion intégrable sur $\det(\mathcal{T}_X)$ équivaut à celle d'une application k-linéaire $c : \mathcal{T}_X \longrightarrow \mathcal{O}_X$ vérifiant deux propriétés

$$c(a\tau) = ac(\tau) + \tau(a) \tag{CY 1}$$

et

$$c([\tau, \tau']) = \tau c(\tau') - \tau' c(\tau) \tag{CY 2}$$

($a \in \mathcal{O}_X$, $\tau, \tau' \in \mathcal{T}_X$). Ces axiomes entraînent d'ailleurs que c est un opérateur différentiel (d'ordre 1). On appelle un tel opérateur *une structure de Calabi–Yau* sur $\mathcal{T}_X$.

0.2

Les structures semblables, mais correspondant au *deuxième* caractère de Chern–Simons, sont liées aux *algèbres vertex*. Une algèbre vertex est l'algèbre des symétries fondamentale de la *théorie des cordes quantiques*. Dans [GMS] on a étudié une classe particulière de ces algèbres: les algèbres vertex des opérateurs differentiels (VDO).

Définir un faisceau de VDO sur X revient à définir *une structure vertex* sur $\mathcal{T}_X$, c'est-à-dire: trois opérateurs différentiels

$$\gamma : \mathcal{O}_X \otimes_k \mathcal{T}_X \longrightarrow \Omega^1_X, \qquad \langle , \rangle : S^2\mathcal{T}_X \longrightarrow \mathcal{O}_X, \quad \text{et} \quad c : \Lambda^2\mathcal{T}_X \longrightarrow \Omega^1_X,$$

vérifiants cinq axiomes analogues aux (CY 1), (CY 2); cf. op. cit. 1.4 où 7.2 ci-dessous. Le théorème principal de op. cit. dit que toutes les structures vertex sur $\mathcal{T}_X$ forment une *gerbe* sous le complexe $\Omega_X^{[1,2\rangle}$, dont la classe caractéristique est égale à $\mathrm{ch}_2^{\mathrm{CS}}(\mathcal{T}_X)$. Cette gerbe est donc à l'origine de l'existence de la classe $\mathrm{ch}_2^{\mathrm{CS}}(\mathcal{T}_X)$.

Les axiomes d'une structure vertex ont l'air assez mystérieux; dans [GMS] ils se déduisent des axiomes de Borcherds pour une algèbre vertex. Par contre, dans [S] il est donné une interprétation de ces axiomes qui ne fait aucune référence aux algèbres vertex. Il y est introduit un complexe naturel de faisceaux sur X, $\mathcal{HKR}(2)^{\cdot}_X$, dit le deuxième complexe de *Hochschild–Koszul–de Rham*, muni d'une inclusion de complexes

$$\sigma_{\geq 2}\sigma_{\leq 5}\Omega_X[2] \hookrightarrow \mathcal{HKR}(2)^{\cdot}_X \tag{0.2.1}$$

et d'un 2-cocycle canonique

$$\epsilon_X^{(2)} \in Z^2\Gamma(X, \mathcal{HKR}(2)^{\cdot}_X).$$

Pour expliquer la structure de $\mathcal{HKR}(2)^{\cdot}_X$ on utilise la notion d'un *tricomplexe tordu* (de k-modules). Un tel objet est un k-module $\mathbb{Z}^3$-gradué $C^{\cdots} = \{C^{pqr}\}_{p,q,r\in\mathbb{Z}}$ munie d'une collection d'endomorphismes $\{d_{ij}\}_{i,j\in\mathbb{Z}}$ où d_{ij} est de degré $(i, j, -i - j + 1)$, c'est à dire que d_{ij} est une collection $d_{ij} = \{d_{ij}^{pqr}\}_{p,q,r\in\mathbb{Z}}$ où $d_{ij}^{pqr} \in \mathrm{Hom}(C^{pqr}, C^{p+i,q+j,r-i-j+1})$. On exige les deux propriétés (a) et (b) si-dessous.

(a) (Finitude.) Pour chaque p, q, r et chaque $x \in C^{pqr}$ il n'y a qu'un nombre fini de $d_{ij}(x)$ qui sont différents de 0.

Définissons un k-module $\mathbb{Z}$-gradué $C^{\cdot}$ par

$$C^i = \oplus_{p+q+r=i} C^{pqr}.$$

Posons $d = \sum_{ij} d_{ij}$; ceci est un endomorphisme de degré 1 de $C^{\cdot}$, bien défini grâce à (a).

(b) $d^2 = 0$.

(Cette définition est plus générale que celle donnée dans la paragraphe 1.)

Le couple $(C^{\cdot}, d)$ est appelé *le complexe simple associé à* $C^{\cdots}$.

Par exemple, si que $d' = d_{10}, d'' = d_{01}$ et $d''' = d_{00}$ sont différents de zéro, on retrouve la notion usuelle d'un complexe triple.

Notre complexe $\mathcal{HKR}(2)^{\cdot}_X$ est le complexe simple associé à un fasceau de tricomplexes tordus $\{\mathcal{HKR}(2)_X^{pqr}\}$, qui est une structure de dimension 3, p étant la dimension "de de Rham," q étant la dimension "de Koszul" et r étant la dimension "de Hochschild." Les seules composantes d_{ij} non triviales sont: d_{10} ("de de Rham"), d_{01} ("de Koszul"), d_{00} ("de Hochschild") et deux flèches complémentaires, $d_{2,-1}$ et $d_{1,-1}$.

Ce complexe tordu apparaît de manière essentiellement unique, lorsque l'on ajoute une dimension hochschildienne au complexe de de Rham. La dimension koszulienne et les flèches complémentaires $d_{2,-1}$ et $d_{1,-1}$ viennent alors naturellement. La construction est reproduite (suivant [S]) dans le la paragraphe 6.

Par exemple, la composante $\mathcal{HKR}(2)^1_X$ est égale à

$$\mathcal{H}om_k(\mathcal{O}_X \otimes_k \mathcal{T}_X, \Omega^1_X) \oplus \mathcal{H}om_k(S^2\mathcal{T}_X, \mathcal{O}_X) \oplus \mathcal{H}om_k(\Lambda^2\mathcal{T}_X, \Omega^1_X).$$

Une structure vertex (au-dessus de X) est un élément

$$v_X = (\gamma_X, \langle , \rangle_X, c_X) \in \Gamma(X, \mathcal{HKR}(2)^1_X)$$

tel que $d_{\mathcal{HKR}}(v_X) = \epsilon^{(2)}_X$, où $d_{\mathcal{HKR}}$ désigne la différentielle dans $\mathcal{HKR}(2)^{\cdot}_X$.

Par exemple, soit $\mathfrak{b} = \{\tau_p\} \subset \Gamma(X, \mathcal{T}_X)$ *une base abélienne*, c'est-à-dire que $\Gamma(X, \mathcal{T}_X)$ est un $\Gamma(X, \mathcal{O}_X)$-module libre de base $\mathfrak{b}$ et tous les vecteurs de cette base commutent. Grâce à la lissité de X, de telles bases existent Zariski localement. À une telle base $\mathfrak{b}$ on peut associer une structure vertex $v_{\mathfrak{b}}$. De plus, si $\mathfrak{b}, \mathfrak{b}'$ sont deux bases abéliennes, on peut définir un élément $h_{\mathfrak{b},\mathfrak{b}'} \in \Gamma(X, \mathcal{HKR}(2)^0_X)$ tel que

$$d_{\mathcal{HKR}} h_{\mathfrak{b},\mathfrak{b}'} = v_{\mathfrak{b}'} - v_{\mathfrak{b}}.$$

Soit $\mathfrak{U} = \{U_i\}$ un recouvrement ouvert de X et pour chaque i fixons une base abélienne $\mathfrak{b}_i \subset \Gamma(U_i, \mathcal{T}_X)$, d'où les collections

$$v_{\mathfrak{U},\mathfrak{b}} = \{v_{\mathfrak{b}_i}\} \in \prod_i \Gamma(U_i, \mathcal{HKR}(2)^1_X)$$

et

$$h_{\mathfrak{U},\mathfrak{b}} = \{h_{\mathfrak{b}_i\mathfrak{b}_j}\} \in \prod_{ij} \Gamma(U_i \cap U_j, \mathcal{HKR}(2)^0_X).$$

On peut considérer le couple $\hat{v} = (v_{\mathfrak{U},\mathfrak{b}}, h_{\mathfrak{U},\mathfrak{b}})$ comme une 1-cochaîne du complexe $\check{C}^{\cdot}(\mathfrak{U}; \mathcal{HKR}(2)_X)$ (complexe simple associé au complexe double $\check{C}^{\cdot}(\mathfrak{U}; \mathcal{HKR}(2)^{\cdot}_X)$).

L'inclusion (0.2.1) induit l'inclusion de complexes

$$\mu : \check{C}^{\cdot}(\mathfrak{U}; \Omega^{[2,3\rangle}_X) \hookrightarrow \check{C}^{\cdot}(\mathfrak{U}; \mathcal{HKR}(2)_X).$$

D'autre part, on a le morphisme évident de complexes

$$\delta : \Gamma(X, \mathcal{HKR}(2)^{\cdot}_X) \longrightarrow \check{C}^{\cdot}(\mathfrak{U}; \mathcal{HKR}(2)_X).$$

La cochaîne $\hat{v}$ est construite dans la paragraphe 7, dont résultat principal dit que $D\hat{v} = \mu(\beta^{(2)}) + \delta(\epsilon^{(2)})$, où

$$\beta^{(2)} \in \check{C}^2(\mathfrak{U}, \Omega^{[2,3\rangle}_X)$$

est un cocycle représentant le deuxième caractère de Chern–Simons du fibré tangent et D désigne la différentielle totale dans $\check{C}^{\cdot}(\mathfrak{U}; \mathcal{HKR}(2)_X)$. Cela est une variante du le calcul principal de [GMS].

Les structures de Calabi–Yau admettent une interprétation tout à fait parallèle; cf. [S] et les paragraphes 4, 5 ci-dessous.

0.3

On peut appeler *i-branaires* les structures de la Géométrie Différentielle sous-jacentes au $i+1$-ième caractère de Chern–Simons.

Pour passer à dimension 3, il faudra définir le troisième complexe de Hochschild–Koszul–de Rham $\mathcal{HKR}(3)^{\cdot}_X$ et procéder comme ci-dessus. Dans cette note on fait une partie de travail.

(a) La partie “Hochschild–Koszul.” On construit dans la paragraphe 8 un complexe naturel $\mathcal{HK}(3)^{\cdot}_X$ qui est le complexe simple associé au complexe double. Il est muni d'une inclusion

$$\Omega^3_X \hookrightarrow \mathcal{HK}(3)^{\cdot}_X$$

et d'un cocycle canonique

$$\epsilon^{(3)}_{X,\mathcal{HK}} \in Z^3\Gamma(X, \mathcal{HK}(3)^{\cdot}_X).$$

On appellera “structure prémembranaire” un élément $\pi \in \Gamma(X, \mathcal{HK}(3)^2_X)$ tel que $d_{\mathcal{HK}}\pi = \epsilon_X$. Cette notion est analogue à celle d'une préalgébroïde vertex; cf. [GMS, 4.1]. Les structures prémembranaires forment une “2-gerbe” sous Ω^3_X dont la classe caractéristique est égale au troisième caractère de Chern style Hodge $\mathrm{ch}_3^{\mathrm{Hdg}}(\mathcal{T}_X) \in H^3(X, \Omega^3_X)$; cf. Section 9.

(b) La partie “Koszul–de Rham.” On construit dans Section 10 un complexe naturel $\mathcal{KR}(3)^{\cdot}_X$ qui est le complexe simple associé à un “bicomplexe tordu”; cf. Section 1. Il est muni d'inclusion de complexes

$$\sigma_{\geq 3}\sigma_{\leq 6}\Omega^{\cdot}_X[3] \hookrightarrow \mathcal{KR}(3)^{\cdot}_X.$$

Étant donné un recouvrement $\mathfrak{U}$ de X, cette inclusion induit l'inclusion de complexes

$$\mu : \check{C}^{\cdot}(\mathfrak{U}; \Omega^{[3,5\rangle}_X) \hookrightarrow \check{C}^{\cdot}(\mathfrak{U}; \mathcal{KR}(3)_X).$$

Si l'on choisit une collection de bases abéliennes $\mathfrak{b}$ comme ci-dessus, on définira dans Section 11 une cochaîne $\hat{m} \in \check{C}^2(\mathfrak{U}; \mathcal{KR}(3)_X)$ telle que $D\hat{m} = \mu(\hat{\beta}^{(3)})$, où D est la différentielle dans $\check{C}^{\cdot}(\mathfrak{U}; \mathcal{KR}(3)_X)$ et

$$\beta^{(3)} \in \check{Z}^3(\mathfrak{U}, \Omega^{[3,5\rangle}_X)$$

représente la classe $\mathrm{ch}_3^{\mathrm{CS}}(\mathcal{T}_X)$. Ceci est le résultat principal de cette note, analogue, en dimension 3, au calcul principal de [GMS]. En effet, la formule pour $\beta^{(3)}$ écrite dans Section 3 est la conséquence du calcul du paragraphe 11.

Les théorèmes principaux de cette note sont: 8.4.1, 9.4.1, 10.7 et 11.4.1. On ne décrit que la stratégie de leurs démonstrations; la vérification est tout à fait directe et exige *plus laboris quam artis*.

Quant à l'idée de base, elle est simple: tout le contenu de cette note est obtenu par “bootstrap.” D'abord, tous les complexes se déduisent du complexe de de Rham. Ensuite, le rôle fondamental est joué par l'opérateur $\{,\ldots,\}_{\mathfrak{b}} : S^n\mathcal{T}_X \longrightarrow \mathcal{O}_X$ défini (pour une base abélienne $\mathfrak{b} = \{\tau_i\} \subset \Gamma(X, \mathcal{T}_X)$) par la formule

$$\{a_1\tau_{i_1}, \ldots, a_n\tau_{i_n}\}_{\mathfrak{b}} = \frac{1}{n!} \operatorname{Sym}_{1\ldots n} \tau_{i_n}(a_1)\tau_{i_1}(a_2)\ldots\tau_{i_{n-1}}(a_n) \tag{0.3.1}$$

($a_j \in \mathcal{O}_X$, $\tau_{i_j} \in \mathfrak{b}$). Tous les cocycles se déduisent de manière essentiellement unique de (0.3.1). Ce phénomène d'unicité pour nous est un mystère.

1 Complexes tordus

1.1

Soit $\mathcal{C}$ une catégorie abélienne. Suivant l'usage, un bicomplexe dans $\mathcal{C}$ est un triple $(C^{\cdot\cdot}, d', d'')$, $C^{\cdot\cdot} = \{C^{ij}, i, j \in \mathbb{Z}\}$ étant une collection d'objets de $\mathcal{C}$ et $d' = \{d'^{ij}, i, j \in \mathbb{Z}\}$, $d'' = \{d''^{ij}, i, j \in \mathbb{Z}\}$, étant des collections de morphismes, où d' a le degré $(1, 0)$, i.e.

$$d'^{ij} : C^{ij} \longrightarrow C^{i+1,j}$$

et d'' a le degré $(0, 1)$, c'est-à-dire,

$$d''^{ij} : C^{ij} \longrightarrow C^{i,j+1},$$

qui vérifient les relations

$$d'^2 = 0, \qquad d''^2 = 0, \qquad d'd'' = d''d'.$$

Le complexe simple associé $\mathrm{Tot}(C^{\cdot\cdot})$ est défini par

$$\mathrm{Tot}(C^{\cdot\cdot})^i = \oplus_{p+q=i} C^{pq}, \qquad d_{\mathrm{Tot}}(x) = d'x + (-1)^p d''x, \quad x \in C^{pq}.$$

Dans nos applications les bicomplexes seront limités inférieurement, i.e., $C^{ij} = 0$ pour $i < i_0$ ou $j < j_0$, donc les sommes directes seront finies.

Exemple typique. Si X est un espace topologique, $\mathcal{F}^{\cdot}$ est un complexe limité inférieurement de faisceaux de groupes abéliens sur X, $\mathfrak{U} = \{U_p\}$ un recouvrement ouvert de X, alors les cochaînes de Čech $\check{C}^j(\mathfrak{U}, \mathcal{F}^i)$ forment un bicomplexe.

1.2

Un *bicomplexe tordu* est une collection $C^{\cdot\cdot}$ comme ci-dessus munie de trois endomorphismes d', d'', R, où d' a le degré $(1, 0)$, d'' a le degré $(0, 1)$ et $R = \{R^{ij}\}$ a le degré $(2, -1)$, i.e.,

$$R^{ij} : C^{ij} \longrightarrow C^{i+2,j-1}$$

On exige les relations suivantes:

$$d''^2 = 0, \qquad d'd'' = d''d', \qquad d'^2 = Rd'' + d''R, \qquad Rd' = d'R, \qquad R^2 = 0 \tag{1.2.1}$$

Le complexe simple associé est défini par

$$\begin{aligned} &\mathrm{Tot}(C^{\cdot\cdot})^i = \oplus_{p+q=i} C^{pq}, \\ &d_{\mathrm{Tot}}(x) = d'x + (-1)^p d''x + (-1)^{p+1} Rx, \quad x \in C^{pq}. \end{aligned} \tag{1.2.2}$$

On vérifie aussitôt que $d_{\mathrm{Tot}}^2 = 0$.

1.3

De même, un tricomplexe est une collection d'objets $C^{\cdots} = \{C^{pqr},\ p, q, r \in \mathbb{Z}\}$ munie de trois endomorphismes d', d'', d''' de degrés $(1, 0, 0)$, $(0, 1, 0)$, et $(0, 0, 1)$, respectivement, tels que

$$d'^2 = d''^2 = d'''^2 = 0, \qquad d'd'' = d''d', \qquad d'd''' = d'''d', \qquad d''d''' = d'''d''.$$

Le complexe simple associé est défini par

$$\begin{aligned} \mathrm{Tot}(C^{\cdots})^i &= \oplus_{p+q+r=i} C^{pqr}, \\ d_{\mathrm{Tot}}(x) &= d'x + (-1)^p d''x + (-1)^{p+q} d'''x, \quad x \in C^{pqr}. \end{aligned}$$

Définition équivalente: on prend d'abord le complexe simple par rapport aux deux premiers degrés, on obtient le bicomplexe $\mathrm{Tot}_{12}(C^{\cdots})$ et on prend le complexe simple associé à ce bicomplexe $\mathrm{Tot}(\mathrm{Tot}_{12}(C^{\cdots}))$.

1.4

Un *tricomplexe tordu* est une collection $C^{\cdots}$ comme ci-dessus munie de 5 endomorphismes d', d'', d''', R, et M de degrés

$$\deg d' = (1, 0, 0), \qquad \deg d'' = (0, 1, 0), \deg d''' = (0, 0, 1),$$
$$\deg R = (2, -1, 0), \qquad \deg M = (1, -1, 1).$$

qui vérifient les relations (a), (b), et (c) ci-dessous.

(a) Pour chaque p la collection $(C^{p''}, d'', d''')$ est un bicomplexe.
(b) Pour chaque r la collection $(C^{\cdot\cdot r}, d', d'', R)$ est un bicomplexe tordu.
(c) Les conditions sur M:

$$\begin{aligned} M^2 &= 0, \\ d'''d' &= d'd''' + Md'' + d''M, \\ d'''R &= Rd''' + Md' + d'M, \\ d'''M + Md''' &= 0, \qquad RM + MR = 0. \end{aligned}$$

Le complexe simple associé est défini par

$$\begin{aligned} \mathrm{Tot}(C^{\cdots})^i &= \oplus_{p+q+r=i} C^{pqr}, \\ d_{\mathrm{Tot}}(x) &= d'x + (-1)^p d''x + (-1)^{p+q} d'''x + (-1)^{p+1} Rx + (-1)^{q+1} Mx, \end{aligned} \tag{1.4.1}$$

$x \in C^{pqr}$. On vérifie que $d_{\mathrm{Tot}}^2 = 0$.

1.5

Chaque objet de $\mathcal{C}$ sera considéré comme un complexe concentré en degré 0.

Rappelons les notations usuelles pour les complexes. Soit C un complexe. Alors sa translation $C[a] (a \in \mathbb{Z})$ est défini par $C[a]^i = C^{a+i}$, $d_{C[a]} = (-1)^a d_C$.

Tronquations. “Bêtes”:

$$\sigma_{\leq a} C : \cdots \longrightarrow C^{a-1} \longrightarrow C^a \longrightarrow 0,$$
$$\sigma_{\geq a} : 0 \longrightarrow C^a \longrightarrow C^{a+1} \longrightarrow \cdots .$$

“Intelligentes”:

$$\tau_{\leq a} : \cdots \longrightarrow C^{a-2} \longrightarrow C^{a-1} \longrightarrow \operatorname{Ker} d^a \longrightarrow 0.$$

Si $f : C \longrightarrow D$ est un morphisme de complexes, alors Cône(f) est le complexe simple associé au bicomplexe $C^{\cdot} \xrightarrow{f} D^{\cdot}$, $D^{\cdot}$ ayant le premier degré 0.

2 Rappels sur le complexe de de Rham

2.1

Désormais on fixe un anneau commutatif de base k et une k-algèbre commutative A. Chaque A-module est un k-module par restriction de scalaires.

On va utiliser les notations suivantes.

k-Mod: la catégorie de k-modules; $\otimes := \otimes_k$, Hom $:=$ Hom$_k$. Si $M_1, \ldots, M_n$, $N \in k$-Mod on identifiera $\mathrm{Hom}(M_1 \otimes \cdots \otimes M_n, N)$ avec l'ensemble d'applications k-multilinéaires $M_1 \times \cdots \times M_n \longrightarrow N$. $M^{\otimes n}$ désignera la n-ième puissance tensorielle sur k d'un k-module M.

De même, $\mathrm{Hom}(\Lambda^n M, N)$ (resp., $\mathrm{Hom}(S^n M, N)$) désignera le k-module des fonctions $f : M^n \longrightarrow N$ k-multilinéaires alternées (resp., symétriques) (ceci peut être considéré comme une définition de la puissance extérieure (resp., symétrique). Par définition, $\Lambda^0 M = S^0 M = k$.

Plus généralement, $\mathrm{Hom}(\Lambda^n M \otimes S^m M', N)$ désignera le k-module des fonctions $f : M^n \times M'^m \longrightarrow N$ k-multilinéaires, alternées par rapport aux premiers n arguments et symétriques par rapport aux derniers m-arguments, etc.

Si $M_1, \ldots, M_n, N \in A$-Mod, $\mathrm{Hom}_A(\Lambda^n_A M, N)$ (resp., $\mathrm{Hom}_A(S^n_A M, N)$) désignera le A-module des fonctions $f : M^n \longrightarrow N$ A-multilinéaires alternées (resp., symétriques).

2.2

Soit $\mathfrak{g}$ une k-algèbre de Lie. $\mathfrak{g}$-Mod va désigner la catégorie de $\mathfrak{g}$-modules.

Si $M, N \in \mathfrak{g}$-Mod, alors $M \otimes N$, $M^{\otimes n}$, $\Lambda^n M$, $S^n M$ sont des $\mathfrak{g}$-modules sur lequels $\mathfrak{g}$ agit par la règle de Leibniz, par exemple,

$$\tau(x_1 \otimes \cdots \otimes x_n) = \sum_{i=1}^{n} x_1 \otimes \cdots \otimes \tau x_i \otimes \cdots \otimes x_n, \qquad (2.2.1)$$

etc.

De plus, $\mathrm{Hom}(M, N)$ est aussi un $\mathfrak{g}$-module; ici $\mathfrak{g}$ agit par la règle

$$(\tau(f))(x) = \tau(f(x)) - f(\tau(x)), \quad f \in \mathrm{Hom}(M, N). \qquad (2.2.2)$$

On utilisera aussi les notations τf et $\mathrm{Lie}_\tau f$ pour $\tau(f)$.

Il en découle que si $M_1, \ldots, M_n, N$ sont de $\mathfrak{g}$-modules, alors $\mathrm{Hom}(M_1 \otimes \cdots \otimes M_n, N)$ sera un $\mathfrak{g}$-module, etc.

Le *complexe de cochaînes de Chevalley* de $\mathfrak{g}$ à coefficients dans M, $C^{\cdot}(\mathfrak{g}, M)$, est défini par $C^n(\mathfrak{g}, M) = \mathrm{Hom}(\Lambda^n \mathfrak{g}, M)$, $n \geq 0$, la différentielle de Chevalley $d_{\mathrm{Ch}} : C^{n-1}(\mathfrak{g}, M) \longrightarrow C^n(\mathfrak{g}, M)$ agit par la formule

$$\begin{aligned} d_{\mathrm{Ch}} f(\tau_1, \ldots, \tau_n) = &\sum_{1 \leq i < j \leq n} (-1)^{i+j} f([\tau_i, \tau_j], \tau_1, \ldots, \hat{\tau}_i, \ldots, \hat{\tau}_j, \ldots, \tau_n) \\ &+ \sum_{1 \leq i \leq n} (-1)^{i+1} \tau_i f(\tau_1, \ldots, \hat{\tau}_i, \ldots, \tau_n) \end{aligned} \qquad (2.2.3)$$

Attention: $\tau_i f(\tau_1, \ldots, \hat{\tau}_i, \ldots, \tau_n)$ désigne $\tau_i(f(\tau_1, \ldots, \hat{\tau}_i, \ldots, \tau_n))$!

2.3

Rappelons qu'une *A-algébroïde de Lie* est une k-algèbre de Lie T agissant sur A par dérivations k-linéaires, munie d'une structure de A-module à gauche, telle que

$$[\tau, a\tau'] = a[\tau, \tau'] + \tau(a)\tau'$$

$(a \in A,\ \tau, \tau' \in T)$.

Exemple typique. $T = \mathrm{Der}_k(A)$ (algèbre de Lie des dérivations k-linéaires $\tau : A \longrightarrow A$).

Désormais nous fixons une A-algébroïde Lie T. On désigne par T^{Lie} le même T considérée comme une algèbre de Lie, avec la structure d'une algébroïde de Lie oubliée.

On pose $\Omega = \mathrm{Hom}_A(T, A)$; ceci est un A-module. Les éléments de T (resp., de Ω) seront notés τ, τ' (resp., ω), etc.; les éléments de A seront notés a, b, c.

On désigne par $\langle , \rangle : T \times \Omega \longrightarrow A$ l'accouplement A-bilinéaire canonique. On a la A-dérivation canonique $d : A \longrightarrow \Omega$ définie par

$$\langle \tau, da \rangle = \tau(a).$$

De plus, T^{Lie} opère sur Ω par la règle

$$\langle \tau', \tau(\omega) \rangle = \tau(\langle \tau', \omega \rangle) - \langle [\tau, \tau'], \omega \rangle$$

$(\tau, \tau' \in T,\ \omega \in \Omega)$. Cette action vérifie les propriétés

$$\tau(a\omega) = \tau(a)\omega + a\tau(\omega),$$
$$(a\tau)(\omega) = a\tau(\omega) + \langle \tau, \omega \rangle da.$$

La flèche d est un morphisme de T^{Lie}-modules.

2.4

Complexe de Chevalley–de Rham. Définissons les A-modules $\Omega^n := \mathrm{Hom}_A(\Lambda^n_A T, A)$ $(n \geq 0)$. En particulier, $\Omega^0 = A$, $\Omega^1 = \Omega$.

Les opérations fondamentales suivantes agissent sur ces modules.

Convolution avec un champ vectoriel. Pour $\tau \in T$, l'opérateur $i_\tau = \langle \tau, ? \rangle : \Omega^{n+1} \longrightarrow \Omega^n$ est défini par

$$i_\tau \omega(\tau_1, \ldots, \tau_n) = \omega(\tau, \tau_1, \ldots, \tau_n).$$

Il s'en suit que pour $\omega \in \Omega^n$

$$\omega(\tau_1, \ldots, \tau_n) = i_{\tau_n} i_{\tau_{n-1}} \ldots i_{\tau_1} \omega.$$

Évidemment,

$$i_\tau i_{\tau'} = -i_{\tau'} i_\tau, \quad \text{ou bien} \quad [i_\tau, i_{\tau'}] = 0; \quad i_\tau^2 = 0.$$

Dérivée de Lie. $\mathrm{Lie}_\tau : \Omega^n \longrightarrow \Omega^n$ est définie par

$$\mathrm{Lie}_\tau\, \omega(\tau_1, \ldots, \tau_n) = \tau(\omega(\tau_1, \ldots, \tau_n)) - \sum_i \omega(\tau_1, \ldots, [\tau, \tau_i], \ldots, \tau_n).$$

En d'autres termes, T^{Lie} agit sur $C^n(T, A) = \mathrm{Hom}(\Lambda^n T, A)$ par les règles (2.2.1), (2.2.2) et cette action respecte le sous-module Ω^n.

On désignera $Lie_\tau \omega$ aussi par $\tau(\omega)$ ou simplement par $\tau\omega$.

On a

$$\mathrm{Lie}_\tau \circ i_{\tau'} = i_{[\tau,\tau']} + i_{\tau'} \circ \mathrm{Lie}_\tau, \quad \text{ou bien} \quad [\mathrm{Lie}_\tau, i_{\tau'}] = i_{[\tau,\tau']}.$$

Différentielle de Chevalley–de Rham. Par recurrence sur n on établit sans peine qu'il existe une unique collection d'opérateurs $d : \Omega^n \longrightarrow \Omega^{n+1}$ qui satisfont à l'identité

$$i_\tau \circ d + d \circ i_\tau = \mathrm{Lie}_\tau, \quad \text{ou bien} \quad [i_\tau, d] = \mathrm{Lie}_\tau .$$

Pour $n = 0$, d a été déjà défini dans 2.3. On a $d^2 = 0$.

La formule explicite sera

$$d\omega(\tau_1, \ldots, \tau_n) = \sum_{i<j} (-1)^{i+j} \omega([\tau_i, \tau_j], \tau_1, \ldots, \hat{\tau}_i, \ldots, \hat{\tau}_j, \ldots)$$

$$+ \sum_i (-1)^{i+1} \tau_i \omega(\tau_1, \ldots, \hat{\tau}_i, \ldots).$$

Autrement dit, la différentielle de Chevalley (2.2.3) respecte les sous-modules $\Omega^n \subset C^n(T, A)$, donc le complexe de de Rham $(\Omega^{\cdot}, d)$ est un sous-somplexe du complexe de Chevalley $C^{\cdot}(T, A)$. On utilise la notation $\Omega^{p,\text{fer}} := \text{Ker}(d : \Omega^p \longrightarrow \Omega^{p+1})$.

On a $[d, \text{Lie}_\tau] = 0$.

2.5

Multiplication. Il existe une unique multiplication $\Omega^{\cdot} \times \Omega^{\cdot} \longrightarrow \Omega^{\cdot}$, $(\omega, \omega') \mapsto \omega\omega'$, $\Omega^p \Omega^q \subset \Omega^{p+q}$ qui vérifie la propriété

$$i_\tau(\omega\omega') = i_\tau(\omega)\omega' + (-1)^p \omega i_\tau(\omega') \quad (\omega \in \Omega^p).$$

En effet, cette règle implique la formule explicite "shuffle" évidente. Soient $\omega \in \Omega^p$, $\omega' \in \Omega^q$; désignons par $\mathcal{P}_{pq}$ l'ensemble des sous-ensembles $P \subset [1, p+q] := \{1, \ldots, p+q\}$ de cardinalité p. Pour $P = \{i_1, \ldots, i_p\}$, $i_1 < \cdots < i_p$, posons $P' = [1, p+q] - P = \{j_1, \ldots, j_q\}$. On désigne par $\text{sgn}(P) \in \mathbb{Z}/2\mathbb{Z}$ le signe de la permutation $\{i_1, \ldots, i_p, j_1, \ldots, j_q\}$. Dans ces notations

$$\omega\omega'(\tau_1, \ldots, \tau_{p+q}) = \sum_{P \in \mathcal{P}_{pq}} (-1)^{\text{sgn}(P)} \omega(\tau_P)\omega'(\tau_{P'}),$$

où $\omega(\tau_P)$ désigne $\omega(\tau_{i_1}, \ldots, \tau_{i_p})$.

La composante de degré $(0, p)$ correspond à la structure d'un A-module sur Ω^p et $(\Omega^{\cdot}, d)$ devient une algèbre différentielle graduée associative et commutative.

3 Formes de Chern–Simons

3.1

Fixons un entier $n \geq 1$. Pour chaque algèbre B on désigne par $\text{Mat}_n(B)$ l'algèbre de matrices $n \times n$ à coéfficients dans B.

Par exemple, considérons $\text{Mat}_n(\Omega^{\cdot})$. Ceci est une algèbre différentielle graduée associative (cf. 2.5); la graduation est définie par $\text{Mat}_n(\Omega^{\cdot})^i = \text{Mat}_n(\Omega^i)$. Pour $P \in \text{Mat}_n(\Omega^i)$ on pose $|P| := i$.

Notre algèbre est munie de la fonction trace $\text{tr} : \text{Mat}_n(\Omega^{\cdot}) \longrightarrow \Omega^{\cdot}$ qui commute avec la différentielle et satisfait à la propriété fondamentale

$$\text{tr}(PQ) = (-1)^{|P||Q|} \text{tr}(QP). \tag{3.1.1}$$

Par contre, on utilisera parfois la notation usuelle pour le commutateur, $[P, Q] = PQ - (-1)^{|P||Q|} QP$, d'où (3.1.1) se recrit comme $\text{tr}([P, Q]) = 0$.

On remarque aussi que

$$\mathrm{tr}(P^2) = 0 \quad \text{si } |P| \text{ est impair.} \tag{3.1.2}$$

Bien sûr, ceci est une conséquence de (3.1.1) ci $1/2 \in k$, mais c'est vrai toujours.

Le groupe $\mathrm{GL}_n(\Omega^\cdot)$ opère sur $\mathrm{Mat}_n(\Omega^\cdot)$ par conjugaison; on n'utilisera que l'action du sous-groupe $\mathrm{GL}_n(A)$, avec la notation

$$P^\phi = \phi^{-1} P \phi \quad (\phi \in \mathrm{GL}_n(A)).$$

Évidemment, $\mathrm{tr}(P^\phi) = \mathrm{tr}(P)$.

3.2

Pour $\phi \in \mathrm{GL}_n(A)$ on a

$$d(\phi^{-1}) = -\phi^{-1} d\phi, \quad \phi^{-1} \in \mathrm{Mat}_n(\Omega^1).$$

On introduit la notation

$$\ell(\phi) := \phi^{-1} d\phi \in \mathrm{Mat}_n(\Omega^1). \tag{3.2.1}$$

On a

$$d\ell(\phi) = -\ell(\phi)^2.$$

Donc la forme $\mathrm{tr}\{\ell(\phi)\}$ est fermée,

$$d\,\mathrm{tr}\{\ell(\phi)\} = -\,\mathrm{tr}\{\ell(\phi)^2\} = 0 \tag{3.2.2}$$

grâce à (3.1.2).

Ensuite,

$$d(P^\phi) = -[\ell(\phi), P] + (dP)^\phi \quad (P \in \mathrm{Mat}_n(\Omega^\cdot)).$$

3.3

D'un autre côté,

$$\ell(\psi\phi) = \ell(\psi)^\phi + \ell(\phi) \tag{3.3.1}$$

pour $\psi, \phi \in \mathrm{GL}_n(A)$, d'où

$$\mathrm{tr}\{\ell(\psi\phi)\} = \mathrm{tr}\{\ell(\psi)\} + \mathrm{tr}\{\ell(\phi)\}. \tag{3.3.2}$$

3.4

Bicomplexe de Čech–de Rham. Supposons qu'un groupe G opère à droit sur un groupe abélien M; on utilise la notation exponentielle: $x^g, x \in M, g \in G$. Rappelons que le complexe de cochaînes $C^\cdot(G, M)$ est défini par $C^n(G, M) = \mathrm{Hom}_{\mathcal{E}ns}(G^n, M)$, avec la différentielle, dit *de Čech*,

$$\begin{aligned} &d_c f(g_1, \ldots, g_{n+1}) \\ &\quad = f(g_2, \ldots, g_{n+1}) \\ &\quad\quad + \sum_{i=1}^{n} (-1)^i f(g_1, \ldots, g_i g_{i+1}, \ldots, g_{n+1}) + (-1)^{n+1} f(g_1, \ldots, g_n)^{g_{n+1}}. \end{aligned}$$

Par exemple, (3.3.1) signifie que ℓ est un 1-cocycle de $\mathrm{GL}_n(A)$ à coefficients dans $\mathrm{Mat}_n(\Omega^1)$ et (3.3.2) et (3.2.1) signifient que

$$\mathrm{tr}\{\ell\} \in Z^1(\mathrm{GL}_n(A), \Omega^{1,\mathrm{fer}}) = H^1(\mathrm{GL}_n(A), \Omega^{1,\mathrm{fer}}) = \mathrm{Hom}_{\mathrm{Groupes}}(\mathrm{GL}_n(A), \Omega^{1,\mathrm{fer}}).$$

(L'action de $\mathrm{GL}_n(A)$ sur $\Omega^{1,\mathrm{fer}}$ étant triviale.)

Considérons le complexe de de Rham $\Omega^{\cdot}$ comme muni de l'action triviale de $\mathrm{GL}_n(A)$. Le complexe de cochaînes $C^{\cdot}(\mathrm{GL}_n, \Omega^{\cdot})$ devient un bicomplexe avec les *colonnes* $C^{\cdot}(\mathrm{GL}_n(A), \Omega^i)$, donc la première differentielle sera de de Rham et la seconde celle de Čech.

3.5

Complexes de Chern–Simons. Pour $i \geq 1$ considérons les complexes

$$\Omega^{[i,2i-1\rangle} := \tau_{\leq 2i-1}\sigma_{\geq i}\Omega^{\cdot}[i] : \Omega^i \longrightarrow \ldots \longrightarrow \Omega^{2i-1,\mathrm{fer}}.$$

Le i-ième bicomplexe de Chern–Simons $\mathrm{CS}(i)^{\cdot\cdot}$ est le bicomplexe $C^{\cdot}(\mathrm{GL}_n(A), \Omega^{[i,2i-1\rangle})$ dont le première degré est le degré dans $\Omega^{[i,2i-1\rangle}$.

Le i-ième complexe de Chern–Simons $\mathrm{CS}(i)$ est le complexe simple associé

$$\mathrm{CS}(i) = \mathrm{Tot}\,\mathrm{CS}(i)^{\cdot\cdot}.$$

Par exemple, $\mathrm{CS}(1) = C^{\cdot}(\mathrm{GL}_n(A), \Omega^{1,\mathrm{fer}})$. On a $\mathrm{tr}\{\ell\} \in \mathrm{CS}(1)^{01}$ et $d_c(\mathrm{tr}\{\ell\}) = 0$, i.e.,

$$\mathrm{tr}\{\ell\} \in Z^1(\mathrm{CS}(1)). \tag{3.5.1}$$

On appelle l'élément (3.5.1) *la première forme de Chern–Simons* et l'on désigne par β_1.

On définira ci-dessous des cocycles analogues $\beta_i \in Z^i(\mathrm{CS}(i))$ pour $i = 2, 3$.

Deuxième forme

3.6

Posons, pour abréger la notation, $a := \ell(\phi)$. On vérifie par recurrence que

$$d(a^{2i-1}) = -a^{2i}; \quad d(a^{2i}) = 0.$$

Supposons que 2 est inversible dans k. Alors $a^{2i} = \frac{1}{2}[a, a^{2i-1}]$, d'où $\mathrm{tr}\{a^{2i}\} = 0$. (D'ailleurs, ceci est vrai sans hypothèse que $1/2 \in k$; nous n'aurons pas besoin de cela.) Donc

$$\operatorname{tr}\{a^{2i-1}\} \in \Omega^{2i-1,\mathrm{fer}}.$$

Soit $\psi \in \mathrm{GL}_n(A)$. Posons $b := \ell(\psi)^\phi$. On a

$$db = -[a, b] - b^2 = -(ab + ba + b^2),$$

d'où

$$d(ba) = -aba - b^2 a$$

et

$$d \operatorname{tr}\{ba\} = -\operatorname{tr}\{a^2 b + ab^2\}.$$

3.7

D'autre part, rappelons que $d_c f(\psi, \phi) = f(\phi) - f(\psi\phi) + f(\psi)^\phi$, donc

$$\begin{aligned} d_c a^3(\psi, \phi) &= a^3 - (a+b)^3 + b^3 \\ &= -(a^2 b + aba + ab^2 + ba^2 + bab + b^2 a), \end{aligned}$$

d'où

$$d_c \operatorname{tr}(a^3) = -3 \operatorname{tr}\{a^2 b + ab^2\} = 3d \operatorname{tr}(ba).$$

Enfin, soit $\chi \in CL_n(A)$. Posons $c := \ell(\chi)^{\psi\phi}$. Alors on aura

$$d_c(ba)(\chi, \psi, \phi) = ba - (b + c)a + c(a + b) - cb = 0$$

3.8

Supposons que 6 est inversible dans k. Définissons les formes: $\beta^{11} \in C^1(G, \Omega^{3,\mathrm{fer}}) = \mathrm{CS}(2)^{11}$ par

$$\beta^{11}(\phi) = \frac{1}{6} \operatorname{tr}\{\ell(\phi)^3\}$$

et $\beta^{02} \in C^2(G, \Omega^2) = \mathrm{CS}(2)^{02}$ par

$$\beta^{02}(\psi, \phi) = \frac{1}{2} \operatorname{tr}\{\ell(\psi)^\phi \ell(\phi)\}.$$

Alors on a montré que $d_c\beta^{02} = 0$ et $d\beta^{02} = d_c\beta^{11}$. Donc l'élément $\beta_2 = (\beta^{02}, \beta^{11})$ est un 2-cocycle dans $\mathrm{CS}(2)$. Nous l'appelons *la deuxième forme de Chern–Simons*.

Troisième forme

3.9

Ici nous supposons que 30 est inversible dans k. On a $\operatorname{tr}\{\ell(\phi)^5\} = 0$. Considérons la forme $\operatorname{tr}\{\ell^5\} \in \operatorname{Hom}(\mathrm{GL}_n(A), \Omega^{5,\mathrm{fer}}) = \mathrm{CS}(3)^{21}$. On a

$$\begin{aligned} d_c \operatorname{tr}\{\ell^5\}(\psi, \phi) &= \operatorname{tr}\{b^5 - (a+b)^5 + a^5\} \\ &= -5 \operatorname{tr}\{ba^4 + b^2a^3 + b^4a + b^3a^2 + baba^2 + b^2aba\}. \end{aligned}$$

On définit

$$\beta^{21} := -\frac{1}{60} \operatorname{tr}\{\ell^5\} \in \operatorname{Hom}(\mathrm{GL}_n(A), \Omega^{5,\mathrm{fer}}) = \mathrm{CS}(3)^{21}.$$

3.10

D'un autre côté, on introduit une forme $\beta^{12} \in \operatorname{Hom}(\mathrm{GL}_n(A)^2, \Omega^4) = \mathrm{CS}(3)^{12}$ par

$$\begin{aligned} \beta^{12}(\psi, \phi) &:= -\frac{1}{12} \operatorname{tr}\{ba^3\} - \frac{1}{12} \operatorname{tr}\{b^3a\} - \frac{1}{24} \operatorname{tr}\{baba\} + \frac{1}{12} \operatorname{tr}\{b^2a^2\} \\ &= \beta' + \beta'' + \beta''' + \beta''''. \end{aligned}$$

Alors on a

$$\begin{aligned} d\beta' &= \frac{1}{12} \operatorname{tr}\{ba^4 + b^2a^3\}, \\ d\beta'' &= \frac{1}{12} \operatorname{tr}\{b^4a + b^3a^2\}, \\ d\beta''' Z &= \frac{1}{12} \operatorname{tr}\{baba^2 + b^2aba\}, \end{aligned}$$

et $d\beta'''' = 0$. En ajoutant,

$$d\beta^{12}(\psi, \phi) = \frac{1}{12} \operatorname{tr}\{ba^4 + b^2a^3 + b^4a + b^3a^2 + baba^2 + b^2aba\} = -d_c\beta^{21}.$$

3.11

Enfin, on définit la forme $\beta^{03} \in \operatorname{Hom}(\mathrm{GL}_n(A)^3, \Omega^3) = \mathrm{CS}(3)^{03}$ par

$$\beta^{03}(\chi, \psi, \phi) = \frac{1}{6} \operatorname{tr}\{\ell(\chi)^{\psi\phi}\ell(\psi)^{\phi}\ell(\phi)\} = \frac{1}{6} \operatorname{tr}\{cba\}.$$

On voit aussitôt que $d_c(cba) = 0$, donc $d_c\beta^{03} = 0$.

D'autre part,

$$dc = -ac - bc - c^2 - cb - ca,$$

d'où

$$d(cba) = -(a+b+c)cba$$

Par contre, calculons

$$d_c\beta^{12}(\chi,\psi,\phi)=\beta^{12}(\psi,\phi)-\beta^{12}(\chi\psi,\phi)+\beta^{12}(\chi,\psi\phi)-\beta^{12}(\chi,\psi).$$

On trouve

$$\begin{aligned}d_c\beta' &= -\frac{1}{12}\operatorname{tr}\{cb^2a+cab^2+cba^2+ca^2b+caba+cbab\},\\ d_c\beta'' &= \frac{1}{12}\operatorname{tr}\{c^2ba-c^2ab+cb^2a+cab^2+cbca-cbab\},\\ d_c\beta''' &= \frac{1}{12}\operatorname{tr}\{caba-cacb\},\end{aligned}$$

et

$$d_c\beta''' = \frac{1}{12}\operatorname{tr}\{-cba^2-bca^2+c^2ba+c^2ab\}.$$

En ajoutant, on obtient

$$\begin{aligned}d_c\beta^{12}(\chi,\psi,\phi) &= \frac{1}{6}\operatorname{tr}\{c^2ba-cbab-cba^2\}=\frac{1}{6}\operatorname{tr}\{(a+b+c)cba\}\\ &= -d\beta^{03}(\chi,\psi,\phi).\end{aligned}$$

Il s'en suit:

Théorème 3.11.1. *L'èlément* $\beta_3=(\beta^{03},\beta^{12},\beta^{21})\in \mathrm{CS}(3)^3$ *est un cocycle.*

On l'appele *la troisième forme de Chern–Simons*.

4 Bicomplexe de Hochschild–de Rham

4.1

Soit M, N deux A-modules. On peut plonger le module $\mathrm{Hom}_A(M,N)$ dans le module de homomorphismes k-linéaires $\mathrm{Hom}(M,N)$; évidemment,

$$\mathrm{Hom}_A(M,N)=\mathrm{Ker}(d_H:\mathrm{Hom}(M,N)\longrightarrow \mathrm{Hom}(A\otimes M,N)),$$

où la flèche d_H est définie par

$$d_Hf(a,x)=af(x)-f(ax)$$

$(x\in M, a\in A)$.

Plus généralement, définissons le complexe de Hochschild $C_H^{\cdot}(M,N)$ par $C_H^n(M,N)=\mathrm{Hom}(A^{\otimes n-1}\otimes M,N)$ $(n\geq 0)$, la différentielle $d_H: C_H^n(M,N)\longrightarrow C_H^{n+1}(M,N)$ agit par la formule

$$\begin{aligned}&d_Hf(a_1,\dots,a_n;x)\\ &\quad=a_1f(a_2,\dots,a_n;x)\\ &\qquad+\sum_{i=1}^{n-1}(-1)^i f(a_1,\dots,a_ia_{i+1},\dots,a_n;x)+(-1)^n f(a_2,\dots,a_{n-1};a_nx).\end{aligned}$$

4.2

Définissons un complexe augmenté de A-modules

$$B^{\cdot}_A(M) \xrightarrow{e} M \tag{4.2.1}$$

où $B_A^{-n}(M) = A^{\otimes n+1} \otimes M$ $(n \geq 0)$, la structure d'un A-module sur $B_A^{-n}(M)$ étant définie par

$$a(a_1 \otimes \cdots \otimes a_{n+1} \otimes x) = aa_1 \otimes \cdots \otimes a_{n+1} \otimes x,$$

la différentielle $d_H : B_A^{-n}(M) \longrightarrow B_A^{-n+1}(M)$ agissant par

$$\begin{aligned} d_H(a_1 \otimes \cdots \otimes a_{n+1} \otimes x) = &\sum_{i=1}^{n} (-1)^{i+1} a_1 \otimes \ldots a_i a_{i+1} \otimes \cdots \otimes a_{n+1} \otimes x \\ &+ (-1)^n a_1 \otimes \cdots \otimes a_n \otimes a_{n+1} x \end{aligned}$$

et l'augmentation e étant définie par $e(a \otimes x) = ax$.

Il est facile de voir que (4.2.1) est une résolution (on construit sans peine une homotopie).

Il découle des définitions que

$$C^{\cdot}_H(M, N) = \mathrm{Hom}_A(B^{\cdot}_A(M), N).$$

En pratique (sous les hypothèses faibles sur A) les A-modules $B_A^n(M)$ seront projectifs, donc

$$H^n(C^{\cdot}(M, N)) = Ext_A^n(M, N) \quad (n \geq 0).$$

Donc si M est projectif sur A, $C^{\cdot}(M, N)$ sera une résolution de $\mathrm{Hom}_A(M, N)$.

4.3

Maintenant on applique cette construction à $\Omega = \mathrm{Hom}_A(T, A)$. On obtient le complexe de Hochschild

$$C^{\cdot}_H(T, A) : \mathrm{Hom}(T, A) \longrightarrow \mathrm{Hom}(A \otimes T, A) \longrightarrow \cdots .$$

On va noter ce complexe $\mathcal{HR}(1)^{0\cdot}$, et cela sera la 0-ième colonne du *bicomplexe de Hochschild–de Rham.*

Plus généralement, définissons pour $i \geq 1$ les complexes analogues $\mathcal{HR}(1)^{i\cdot}$. Posons $\mathcal{HR}(1)^{i0} = \mathrm{Hom}(\Lambda^{i+1} T, A)$. On remarque que l'on a l'inclusion évidente

$$\Omega^{i+1} = \mathrm{Hom}_A(\Lambda_A^{i+1} T, A) \hookrightarrow \mathcal{HR}(1)^{i0}.$$

Ensuite, posons

$$\mathcal{HR}(1)^{ij} = \mathrm{Hom}(A^{\otimes j} \otimes T \otimes \Lambda^i T, A) \quad (j \geq 1);$$

les différentielles de Hochschild seront definies par la formule 4.1 *par rapport au premier argument*:

$$\begin{aligned} d_H f(a_1, \dots, a_j; \tau_1; \tau_2, \dots, \tau_i) \\ = a_1 f(a_2, \dots, a_j; \tau_1; \tau_2, \dots, \tau_i) \\ + \sum_{p=1}^{j-1} (-1)^p f(a_1, \dots, a_p a_{p+1}, \dots, a_j; \tau_1; \tau_2, \dots, \tau_i) \\ + (-1)^j f(a_1, \dots, a_{j-1}; a_j \tau_1; \tau_2, \dots, \tau_i). \end{aligned}$$

Le complexe $\mathcal{HR}(1)^{i\cdot}$ sera la i-ième colonne de notre bicomplexe de Hochschild–de Rham.

4.4

Il nous reste à définir les flèches horizontales $d_{\mathrm{DR}} : \mathcal{HR}(1)^{i\cdot} \longrightarrow \mathcal{HR}(1)^{i+1\cdot}$.

D'abord la 0-ième ligne

$$\mathcal{HR}(1)^{\cdot 0} : \mathrm{Hom}(T, A) \longrightarrow \mathrm{Hom}(\Lambda^2 T, A) \longrightarrow \cdots .$$

La différentielle ici sera $-d_{\mathrm{Ch}}$, où d_{Ch} est la différentielle de Chevalley dans $C^\cdot(T^{\mathrm{Lie}}, A)$, (2.2.3).

Par contre, pour $j \geq 1$ la j-ième ligne sera

$$\begin{aligned} \mathcal{HR}(1)^{\cdot j} : \mathrm{Hom}(A^{\otimes j} \otimes T, A) &\longrightarrow \mathrm{Hom}(A^{\otimes j} \otimes T \otimes T, A) \\ &\longrightarrow \mathrm{Hom}(A^{\otimes j} \otimes T \otimes \Lambda^2 T, A) \longrightarrow \cdots . \end{aligned}$$

Identifions $\mathrm{Hom}(A^{\otimes j} \otimes T \otimes \Lambda^i T, A)$ avec $\mathrm{Hom}(\Lambda^i T, \mathrm{Hom}(A^{\otimes j} \otimes T, A))$ et définissons la différentielle de de Rham comme la différentielle de Chevalley dans $C^\cdot(T^{\mathrm{Lie}}, \mathrm{Hom}(A^{\otimes j} \otimes T, A))$, où l'action de T^{Lie} sur $\mathrm{Hom}(A^{\otimes j} \otimes T, A)$ est définie en accord avec la règle usuelle (cf. 2.2):

$$\begin{aligned} (\tau f)(a_1, \dots, a_j; \tau') \\ = \tau f(a_1, \dots, a_j; \tau') - \sum_p f(a_1, \dots, \tau(a_p), \dots; \tau') - f(a_1, \dots, a_j; [\tau, \tau']). \end{aligned}$$

Donc

$$\begin{aligned} & d_{\mathrm{DR}} f(a_1, \dots, a_j; \tau; \tau_1, \dots) \\ & = \sum_p (-1)^{p+1} \{\tau_p f(a_1, \dots; \tau; \dots, \hat{\tau}_p, \dots) \\ & \quad - \sum_r f(a_1, \dots, \tau_p(a_r), \dots; \tau; \dots, \hat{\tau}_p, \dots) - f(a_1, \dots; [\tau_p, \tau]; \dots, \hat{\tau}_p, \dots)\} \\ & \quad + \sum_{p<q} (-1)^{p+q} f(a_1, \dots; \tau; \dots, \hat{\tau}_p, \dots, \hat{\tau}_q, \dots). \end{aligned}$$

On vérifie par un calcul direct que $d_H^0 d_{\mathrm{Ch}} = -d_{\mathrm{Ch}} d_H^0$ (sic!), d'où $d_H^0 d_{\mathrm{DR}} = d_{\mathrm{DR}} d_H^0$.
D'autre part, on vérifie:

Lemme 4.4.1. *Les Hochschilds*

$$d_H : \mathrm{Hom}(A^{\otimes j} \otimes T, A) \longrightarrow \mathrm{Hom}(A^{\otimes j+1} \otimes T, A)$$

sont des morphismes de T^{Lie}-modules.

4.5

Il s'en suit que $d_H^j d_{\mathrm{DR}} = d_{\mathrm{DR}} d_H^j$ pour $j \geq 1$. Ceci donne le bicomplexe de Hochschild–de Rham $\mathcal{HR}(1)^{\cdot\cdot}$ promis. Le complexe de Hochschild–de Rham est le complexe simple associé: $\mathcal{HR}(1) = \mathrm{Tot}\,\mathcal{HR}(1)^{\cdot\cdot}$.

On a l'inclusion canonique

$$\sigma_{\geq 1}\Omega^{\cdot}[1] \hookrightarrow \mathcal{HR}(1). \tag{4.5.1}$$

Évidemment, $\Omega^1 = \mathrm{Ker}(d_H^0 : \mathcal{HR}(1)^{00} \longrightarrow \mathcal{HR}(1)^{01})$, donc (4.5.1) induit un isomorphisme sur H^0:

$$H^0(\mathcal{HR}(1)) = \Omega^{1,\mathrm{fer}}.$$

5 Structures de Calabi–Yau

5.1

Cocycle ϵ_1. Considérons l'élément "evaluation" $e \in \mathrm{Hom}(A \otimes T, A) = \mathcal{HR}(1)^{01}$ défini par

$$e(a; \tau) = \tau(a).$$

Lemme. $d_H e = d_{\mathrm{DR}} e = 0$.

En effet,

$$\begin{aligned} d_H e(a, b; \tau) &= ae(b; \tau) - e(ab; \tau) + e(a; b\tau) \\ &= a\tau(b) - \tau(ab) + b\tau(a) = 0. \end{aligned}$$

D'autre part,

$$\begin{aligned} d_{\mathrm{DR}} e(a; \tau, \tau') &= (\tau' e)(a; \tau) = \tau' e(a; \tau) - e(\tau'(a); \tau) - e(a; [\tau', \tau]) \\ &= \tau'\tau(a) - \tau\tau'(a) - [\tau', \tau](a) = 0. \end{aligned}$$

(Autrement dit, l'opérateur e est invariant.)

Il s'en suit que $\epsilon_1 = (-e, 0) \in \mathcal{HR}(1)^{01} \oplus \mathcal{HR}(1)^{10} = \mathcal{HR}(1)^1$ est un cocycle.

5.2

Une structure de Calabi–Yau sur T. est un élément $c \in \mathcal{HR}(1)^0 = \mathrm{Hom}(T, A)$ tel que $d_{\mathcal{HR}} c = \epsilon_1$.

Ceci signifie que

$$d_H c = -e, \quad \text{c'est-à-dire,} \qquad c(a\tau) - ac(\tau) = \tau(a), \tag{CY1}$$

et

$$d_{\mathrm{DR}} c = 0, \qquad \text{c'est-à-dire,} \quad c([\tau, \tau']) - \tau c(\tau') + \tau' c(\tau) = 0. \tag{CY2}$$

Par exemple, si A est lisse sur k et $T = Der_k(A)$, alors une structure de CY sur T est la même chose qu'une structure d'un $\mathcal{D}$-module *à droite* sur A (définie par la règle $1 \cdot \tau = -c(\tau)$) (cf. [GMS, Section 11]), et ceci est la même chose qu'une connexion intégrable sur det T, d'où le nom.

On désigne par $\mathcal{CY}$ l'ensemble de structures de CY sur T. Évidemment, celui-ci est un torseur sous le groupe abélien $H^0(\mathcal{HR}(1)) = \Omega^{1,\mathrm{fer}}$. On va calculer sa classe.

5.3

Supposons que deux conditions (a) et (b) sont vérifiées.

(a) T est un A-module libre de rang fini n.
Une base $\mathfrak{b} = \{\tau_1, \ldots, \tau_n\}$ comme un A-module est appelée *abélienne* si $[\tau_i, \tau_j] = 0$ pour tous i, j.

(b) T admet une base abélienne.
En effet, l'hypothèse (b) est superflue: sans doute tous les résultats du présent papier restent vrais sans supposer (b). Par contre, elle simplifie énormement les calculs et est suffisante pour les applications: en pratique elle est toujours vérifiée. Donc nous supposons désormais que (b) est vérifiée.

On désigne par $\mathfrak{B}$ l'ensemble de bases abéliennes de T.

5.4

Formules utiles. Soit $\mathfrak{b} = \{\tau_i\}, \mathfrak{b}' = \{\tau_i'\} \in \mathfrak{B}$ où $\tau_i' = \phi^{ij}\tau_j, \phi^{ij} \in A$ (la sommation par les indices répétes est sous-entendue). On ecrit $\mathfrak{b}' = \phi\mathfrak{b}, \phi = (\phi^{ij}) \in \mathrm{GL}_n(A)$.

On aura

$$\begin{aligned} 0 = [\tau_i', \tau_j'] &= [\phi^{ip}\tau_p, \phi^{jq}\tau_q] = \phi^{ip}\tau_p(\phi^{jq})\tau_q - \phi^{jq}\tau_q(\phi^{ip})\tau_p \\ &= \tau_i'(\phi^{jq})\tau_q - \tau_j'(\phi^{ip})\tau_p, \end{aligned}$$

d'où

$$\tau_i'(\phi^{jp}) = \tau_j'(\phi^{ip})$$

pour tous i, j, p. Donc

$$\tau_a(\phi^{bc}) = \phi^{-1ap}\tau'_p(\phi^{bc}) = \phi^{-1ap}\tau'_b(\phi^{pc}) = [\phi^{-1}\tau'_b(\phi)]^{ac}$$

On utilisera les notations $\ell'_i(\phi) = \phi^{-1}\tau'_i(\phi)$, etc.:

$$\tau_a(\phi^{bc}) = \ell'_b(\phi)^{ac}, \tag{5.4.1}$$

$$\tau_a(\phi^{ba}) = \mathrm{tr}\{\ell'_b(\phi)\}. \tag{5.4.2}$$

5.5

Pour un $\mathfrak{b} = \{\tau_i\} \in \mathfrak{B}$ il existe une seule structure de CY $c_{\mathfrak{b}}$ telle que $c_{\mathfrak{b}}(\tau_i) = 0$ pour chaque i. On a

$$c_{\mathfrak{b}}(a\tau_i) = \tau_i(a).$$

Soit $\mathfrak{b}' = \{\tau'_i\}$ une autre base abélienne, $\mathfrak{b}' = \phi\mathfrak{b}$. Calculons la différence $c_{\mathfrak{b}'} - c_{\mathfrak{b}} \in \Omega^{1,\mathrm{fer}} \subset \mathrm{Hom}(T, A)$. On a

$$c_{\mathfrak{b}'}(\tau'_i) - c_{\mathfrak{b}}(\tau'_i) = -c_{\mathfrak{b}}(\tau'_i) = -c_{\mathfrak{b}}(\phi^{ip}\tau_p) = -\tau_p(\phi^{ip}) = -\,\mathrm{tr}\{\ell'_i(\phi)\} = -\langle\tau'_i, \ell(\phi)\rangle,$$

i.e.,

$$c_{\mathfrak{b}'} - c_{\mathfrak{b}} = -\ell(\phi).$$

5.6

Complexe de Čech–Hochschild–de Rham. Soit X un groupe abélien. On définit le complexe de Čech à coefficients dans X, $\check{C}^{\cdot}(\mathfrak{B}; X)$ par $\check{C}^i(\mathfrak{B}; X) := \mathrm{Hom}_{\mathcal{E}ns}(\mathfrak{B}^{i+1}, X)$ $(i \geq 0)$. Des éléments de $\check{C}^i(\mathfrak{B}; X)$ seront notés $f = \{f_{\mathfrak{b}_0\mathfrak{b}_1\ldots\mathfrak{b}_i}\}$. La différentielle agit comme

$$(d_c f)_{\mathfrak{b}_0\ldots\mathfrak{b}_{i+1}} = \sum_{p=0}^{i+1}(-1)^p f_{\mathfrak{b}_0\ldots\hat{\mathfrak{b}}_p\ldots\mathfrak{b}_{i+1}}.$$

Le morphisme d'augmentation $\delta : X \longrightarrow \check{C}^0(\mathfrak{B}; X)$ est défini par $(\delta x)_{\mathfrak{b}} = x$ $(x \in X)$.

Ensuite, on a le morphisme canonique

$$\nu : C^{\cdot}(\mathrm{GL}_n(A), X) \longrightarrow \check{C}^{\cdot}(\mathfrak{B}; X) \tag{5.6.1}$$

(où $\mathrm{GL}_n(A)$ agit trivialement sur X) défini par

$$(\nu f)_{\mathfrak{b}_0\mathfrak{b}_1\ldots\mathfrak{b}_i} = f(g_i, g_{i-1}, \ldots, g_1), \quad \text{où} \quad \mathfrak{b}_i = g_i\mathfrak{b}_{i-1}.$$

Considérons le bicomplexe de Hochschild–de Rham $\mathcal{HR}(1)^{\cdot\cdot}$. En appliquant $\check{C}^{\cdot}(\mathfrak{B}; ?)$ on obtient: le tricomplexe $\check{C}^{\cdot}(\mathfrak{B}; \mathcal{HR}(1)^{\cdot\cdot})$ (avec les différentielles: la première, de de Rham; la seconde, de Hochschild; et la troisième, de Čech), le bicomplexe $\check{C}^{\cdot}(\mathfrak{B}; \mathcal{HR}(1)^{\cdot})$ et le complexe simple noté $\check{C}^{\cdot}(\mathfrak{B}; \mathcal{HR}(1))$, donc par définition

$$\check{C}^i(\mathfrak{B}; \mathcal{HR}(1)) = \oplus_{s+r=i}\check{C}^r(\mathfrak{B}; \mathcal{HR}(1)^s) = \oplus_{p+q+r=i}\check{C}^r(\mathfrak{B}; \mathcal{HR}(1)^{pq}).$$

On a le morphisme d'augmentation

$$\delta : \mathcal{HR}(1) \longrightarrow \check{C}^{\cdot}(\mathfrak{B}; \mathcal{HR}(1)).$$

D'autre part, les inclusions $\Omega^{1,\mathrm{fer}} \hookrightarrow \Omega^1 \hookrightarrow \mathrm{Hom}(T, A) = \mathcal{HR}(1)^{00}$ fournissent le morphisme de complexes

$$\Omega^{1,\mathrm{fer}} \longrightarrow \mathcal{HR}(1)$$

qui induit, à l'aide de ν (5.6.1), le morphisme de complexes

$$\mu : \mathrm{CS}(1) = C^{\cdot}(\mathrm{GL}_n(A), \Omega^{1,\mathrm{fer}}) \longrightarrow \check{C}^{\cdot}(\mathfrak{B}; \mathcal{HR}(1)),$$

d'où la flèche

$$(\mu, \delta) : \mathrm{CS}(1) \oplus \mathcal{HR}(1) \longrightarrow \check{C}^{\cdot}(\mathfrak{B}; \mathcal{HR}(1)).$$

On appele *le premier complexe de Chern–Simons étendu* le cône

$$\hat{\mathrm{CS}}(1) := \mathrm{Cône}(\mu, \delta)[-1]. \tag{5.6.2}$$

Évidemment, on a la projection canonique

$$\pi : \hat{\mathrm{CS}}(1) \longrightarrow \mathrm{CS}(1)$$

Les calculs précédents sont résumés en:

Théorème 5.6.1. *Considérons la collection $c_1 = \{c_{\mathfrak{b}}\}_{\mathfrak{b}\in\mathfrak{B}}$ comme un élément de $\check{C}^0(\mathfrak{B}; \mathcal{HR}(1)^{00})$. Alors $d_H c_1 = \epsilon_1$ et $d_c c_1 = -\beta_1$.*

Donc l'élément $\hat{\beta}_1 = (\beta_1, -\epsilon_1, -c_1)$ est un 1-cocycle de $\hat{\mathrm{CS}}(1)$ tel que $\pi(\hat{\beta}_1) = \beta_1$.

On appelera cet élément le premier cocycle de Chern–Simons raffiné.

6 Deuxième Hochschild–Koszul–de Rham

6.1

Dans cette section on réproduit la construction de [S], Caput Secundum. Pour les détails de calculs, voir op. cit.

Considérons Ω^2 comme un sous-module de $\mathrm{Hom}(T, \Omega)$, en associant à $\omega \in \Omega^2$ le morphisme $\iota\omega : T \longrightarrow \Omega$ défini par $\iota\omega(\tau) = i_\tau\omega$. Alors l'image de Ω^2 est caractérisée par deux propriétés: (a) le morphisme $\iota\omega$ est A-linéaire; (b) l'expression $\langle\tau', \iota\omega(\tau)\rangle$ est antisymétrique en τ, τ'.

Donc on peut réprésenter Ω^2 comme l'intersection de deux noyaux

$$\Omega^2 = \mathrm{Ker}(d_H : \mathrm{Hom}(T, \Omega) \longrightarrow \mathrm{Hom}(A \otimes T, \Omega))$$
$$\cap\, \mathrm{Ker}(Q : \mathrm{Hom}(T, \Omega) \longrightarrow \mathrm{Hom}(S^2T, A))$$

où le Hochschild d_H est défini par la formule usuelle $d_H f(a,\tau) = af(\tau) - f(a\tau)$, et la “différentielle de Koszul” Q est définie par

$$Qf(\tau,\tau') = \mathrm{Sym}_{\tau,\tau'}\langle \tau, f(\tau')\rangle. \tag{6.1.1}$$

Dans cette section on va définir un tricomplexe tordu $\mathcal{HKR}(2)^{\cdot\cdot\cdot} = \{\mathcal{HKR}(2)^{ijk}\}$ avec $\mathcal{HKR}(2)^{000} = \mathrm{Hom}(T,\Omega)$, $\mathcal{HKR}(2)^{010} = \mathrm{Hom}(S^2T, A)$, et $\mathcal{HKR}(2)^{001} = \mathrm{Hom}(A\otimes T,\Omega)$.

Dans ce tricomplexe i sera le degré de de Rham, j sera le degré de Koszul et k sera le degré de Hochschild. Il sera concentré en degrés $0 \le i+j+k \le 3$, $j = 0, 1$.

Le complexe simple associé $\mathcal{HKR}(2) = \mathrm{Tot}\,\mathcal{HKR}(2)^{\cdot\cdot\cdot}$ sera muni d'un plongement de complexes canonique

$$\iota : \Omega^{[2,5]} := \sigma_{\ge 2}\sigma_{\le 5}\Omega^{\cdot}[2] \hookrightarrow \mathcal{HKR}(2). \tag{6.1.2}$$

Voici ces termes non nuls.

Rez-de-chaussée. On aura $\mathcal{HKR}(2)^{000} = \mathrm{Hom}(T,\Omega)$; les lignes

$$\begin{aligned}\mathcal{HKR}(2)^{\cdot 00} : \mathrm{Hom}(T,\Omega) &\longrightarrow \mathrm{Hom}(\Lambda^2T,\Omega) \longrightarrow \mathrm{Hom}(\Lambda^3T,\Omega)\\ &\longrightarrow \mathrm{Hom}(\Lambda^4T,\Omega),\\ \mathcal{HKR}(2)^{\cdot 10} : \mathrm{Hom}(S^2T,A) &\longrightarrow \mathrm{Hom}(S^2T\otimes T,A) \longrightarrow \mathrm{Hom}(S^2T\otimes\Lambda^2T,A).\end{aligned}$$

Premier étage:

$$\begin{aligned}\mathcal{HKR}(2)^{\cdot 01} : \mathrm{Hom}(A\otimes T,\Omega) &\longrightarrow \mathrm{Hom}(A\otimes T\otimes T,\Omega)\\ &\longrightarrow \mathrm{Hom}(A\otimes T\otimes\Lambda^2T,\Omega),\\ \mathcal{HKR}(2)^{\cdot 11} : \mathrm{Hom}(A\otimes T^{\otimes 2},A) &\longrightarrow \mathrm{Hom}(A\otimes T^{\otimes 3},A).\end{aligned}$$

Deuxième étage:

$$\begin{aligned}\mathcal{HKR}(2)^{\cdot 02} : \mathrm{Hom}(A^{\otimes 2}\otimes T,\Omega) &\longrightarrow \mathrm{Hom}(A^{\otimes 2}\otimes T^{\otimes 2},\Omega),\\ \mathcal{HKR}(2)^{012} &= \mathrm{Hom}(A^{\otimes 2}\otimes T^{\otimes 2},A).\end{aligned}$$

Troisième étage:

$$\mathcal{HKR}(2)^{003} = \mathrm{Hom}(A^{\otimes 3}\otimes T,\Omega).$$

Les différentielles agiront: celle de de Rham $d_{\mathrm{DR}}^{ijk} : \mathcal{HKR}(2)^{ijk} \longrightarrow \mathcal{HKR}(2)^{i+1,jk}$, celle de Koszul $Q^{i0k} : \mathcal{HKR}(2)^{i0k} \longrightarrow \mathcal{HKR}(2)^{i1k}$ et le Hochschild $d_H^{ijk} : \mathcal{HKR}(2)^{ijk} \longrightarrow \mathcal{HKR}(2)^{ij,k+1}$.

De plus, on aura 3 opérateurs

$$R^{ijk} : \mathcal{HKR}(2)^{ijk} \longrightarrow \mathcal{HKR}(2)^{i+2,j-1,k}$$

pour $(ijk) = (010)$, (110) ou (011), et 2 opérateurs

$$M^{ijk} : \mathcal{HKR}(2)^{ijk} \longrightarrow \mathcal{HKR}(2)^{i+1,j-1,k+1}$$

pour $(ijk) = (010)$ ou (011).

Rez-de-chaussée

6.2

Définissons les plongements

$$\iota_n : \Omega^n \longrightarrow \mathrm{Hom}(\Lambda^{n-1}T, \Omega) \tag{6.2.1}$$

par

$$\iota\omega(\tau_1, \ldots, \tau_{n-1}) = i_{\tau_{n-1}} i_{\tau_{n-2}} \ldots i_{\tau_1}\omega. \tag{6.2.2}$$

Alors, en employant la formule de Cartan et la formule $i_\tau \operatorname{Lie}_{\tau'} = \operatorname{Lie}_{\tau'} i_\tau - i_{[\tau',\tau]}$, on établit sans peine que

$$\iota_{n+1} d\omega = d_{\mathrm{Ch}}\iota_n\omega + E'\omega,$$

où $d_{\mathrm{Ch}} : \mathrm{Hom}(\Lambda^{n-1}T, \Omega) \longrightarrow \mathrm{Hom}(\Lambda^n T, \Omega)$ est la différentielle de Chevalley dans le complexe $C^{\cdot}(T, \Omega)$, i.e.,

$$\begin{aligned} d_{\mathrm{Ch}} f(\tau_1, \ldots, \tau_n) = &\sum_{p<q} (-1)^{p+q} f([\tau_p, \tau_q], \ldots, \hat{\tau}_p, \ldots, \hat{\tau}_q, \ldots) \\ &+ \sum_p (-1)^{p+1} \tau_p f(\tau_1, \ldots, \hat{\tau}_p, \ldots) \end{aligned} \tag{6.2.3}$$

et

$$E'\omega(\tau_1, \ldots, \tau_n) = (-1)^n d\langle \tau_n, \iota_n\omega(\tau_1, \ldots, \tau_{n-1})\rangle$$

Supposons que n est inversible dans A. Alors on peut récrire cela sous une forme plus symétrique

$$E'\omega(\tau_1, \ldots, \tau_n) = \frac{1}{n}\sum_{p=1}^{n} (-1)^p d\langle \tau_p, \iota_n\omega(\tau_1, \ldots, \hat{\tau}_p, \ldots)\rangle.$$

Ceci entraîne:

Lemme 6.2.1. *Supposons que n est inversible dans A. Définissons la flèche E : $\mathrm{Hom}(\Lambda^{n-1}T, \Omega) \longrightarrow \mathrm{Hom}(\Lambda^n T, \Omega)$ par*

$$Ef(\tau_1, \ldots, \tau_n) = \frac{1}{n}\sum_{p=1}^{n} (-1)^p d\langle \tau_p, f(\tau_1, \ldots, \hat{\tau}_p, \ldots)\rangle \tag{6.2.4}$$

et posons $d_{\mathrm{DR}} = d_{\mathrm{Ch}} + E$. Alors $d_{\mathrm{DR}}\iota_n = \iota_{n+1} d$.

6.3

Ligne principale. Ceci justifie la définition: les différentielles de de Rham d_{DR} dans la ligne

$$\mathcal{HKR}(2)^{\cdot 00} : \mathrm{Hom}(T, \Omega) \longrightarrow \mathrm{Hom}(\Lambda^2 T, \Omega) \longrightarrow \mathrm{Hom}(\Lambda^3 T, \Omega) \\ \longrightarrow \mathrm{Hom}(\Lambda^4 T, \Omega)$$

sont définies par $d_{\mathrm{DR}} = d_{\mathrm{Ch}} + E$, où d_{Ch} est donnée par (6.2.3) et E est donnée par (6.2.4). Ici on suppose que 6 est inversible dans A. On utilisera la notation d_{DR}^{i00} pour la flèche $\mathcal{HKR}(2)^{i00} \longrightarrow \mathcal{HKR}(2)^{i+1,00}$.

Les inclusions ι_n (6.2.1) donnent lieu à l'inclusion

$$\iota : \Omega^{[2,5]} \hookrightarrow \mathcal{HKR}(2)^{\cdot 00}. \tag{6.3.1}$$

Par contre, $\mathcal{HKR}(2)^{\cdot 00}$ n'est pas un complexe: tandis que $d_{\mathrm{Ch}}^2 = 0$, à cause du terme E, $d_{\mathrm{DR}}^2 \neq 0$.

6.4

Par exemple, on calcule facilement que

$$d_{\mathrm{DR}}^{100} d_{\mathrm{DR}}^{000} f(\tau_1, \tau_2, \tau_3) = -\frac{1}{6} \mathrm{Cycle}_{123}\{d\langle[\tau_1, \tau_2], f(\tau_3)\rangle + d\langle\tau_3, f([\tau_1, \tau_2])\rangle\},$$

d'où le:

Lemme 6.4.1. *Définissons le morphisme*

$$R^{010} : \mathcal{HKR}(2)^{010} = \mathrm{Hom}(S^2 T, A) \longrightarrow \mathrm{Hom}(\Lambda^3 T, \Omega) = \mathcal{HKR}(2)^{200}$$

par

$$R^{010} f(\tau_1, \tau_2, \tau_3) = -\frac{1}{6} \mathrm{Cycle}_{123}\, df([\tau_1, \tau_2], \tau_3).$$

Alors $d_{\mathrm{DR}}^{100} d_{\mathrm{DR}}^{000} = R^{010} Q^{000}$, *où* Q^{000} *est défini par* (6.1.1).

6.5

Plus généralement, définissons les morphismes de Koszul

$$Q^{i00} : \mathcal{HKR}(2)^{i00} = \mathrm{Hom}(\Lambda^{i+1} T, \Omega) \longrightarrow \mathrm{Hom}(S^2 T \otimes \Lambda^i T, A) = \mathcal{HKR}(2)^{i10}$$

$(i = 0, 1, 2)$ par

$$Q^{i00} f(\tau_1, \tau_2; \tau_3, \dots) = \mathrm{Sym}_{12}\langle\tau_1, f(\tau_2, \tau_3, \dots)\rangle.$$

Lemme 6.5.1. *On a* $d_{\mathrm{DR}}^{200} d_{\mathrm{DR}}^{100} = R^{110} Q^{100}$, *où*

$$R^{110} : \mathcal{HKR}(2)^{110} = \mathrm{Hom}(S^2 T \otimes T, A) \longrightarrow \mathrm{Hom}(\Lambda^4 T, \Omega) = \mathcal{HKR}(2)^{300}$$

et défini par

$$R^{110} f(\tau_1, \tau_2, \tau_3, \tau_4) = -\frac{1}{24} \mathrm{Alt}_{1234}\, df([\tau_1, \tau_2], \tau_3; \tau_4).$$

6.6

Les deux différentielles de de Rham dans la ligne

$$\mathcal{HKR}(2)^{\cdot 10} : \mathrm{Hom}(S^2T, A) \longrightarrow \mathrm{Hom}(S^2T \otimes T, A) \longrightarrow \mathrm{Hom}(S^2T \otimes \Lambda^2 T, A)$$

sont trouvées grace à condition de commutativité de deux carrés: $d_{\mathrm{DR}}^{010} Q^{000} = Q^{100} d_{\mathrm{DR}}^{000}$ et $d_{\mathrm{DR}}^{110} Q^{100} = Q^{200} d_{\mathrm{DR}}^{100}$. On arrive aux reponses suivantes.

Les opérateurs

$$d_{\mathrm{DR}}^{i10} : \mathrm{Hom}(S^2T \otimes \Lambda^i T, A) \longrightarrow \mathrm{Hom}(S^2T \otimes \Lambda^{i+1} T, A) \quad (i = 0, 1)$$

sont définis par $d_{\mathrm{DR}}^{i10} = -d_{\mathrm{Ch}}^{i10} - E^{i10}$.

Ici d_{Ch}^{i10} est la différentielle de Chevalley venant de l'identification

$$\mathrm{Hom}(S^2T \otimes \Lambda^i T, A) = \mathrm{Hom}(\Lambda^i T, \mathrm{Hom}(S^2T, A)) = C^i(T^{\mathrm{Lie}}, \mathrm{Hom}(S^2T, A)),$$

l'action de T^{Lie} sur $\mathrm{Hom}(S^2T, A)$ étant définie suivant l'usage, $(\tau f)(\tau', \tau'') = \tau f(\tau', \tau'') - f([\tau, \tau'], \tau'') - f(\tau', [\tau, \tau''])$.

Les opérateurs E sont définis par

$$E^{010} f(\tau, \tau'; \tau_1) = -\frac{1}{2} \mathrm{Sym}_{\tau, \tau'} \tau f(\tau', \tau_1)$$

et

$$E^{110} f(\tau, \tau'; \tau_1, \tau_2) = -\frac{1}{3} \mathrm{Sym}_{\tau, \tau'} \{\tau f(\tau', \tau_1; \tau_2) - \tau f(\tau', \tau_2; \tau_1)\}.$$

Lemme 6.6.1. *On a les identités* $d_{\mathrm{DR}}^{110} d_{\mathrm{DR}}^{010} = Q^{200} R^{010}$ *et* $R^{110} d_{\mathrm{DR}}^{010} = d_{\mathrm{DR}}^{200} R^{010}$.

6.7

Cela termine la définition de la partie $\mathcal{HKR}(2)^{\cdot\cdot 0}$; les relations écrites signifient que celui-ci est un bicomplexe tordu.

Premier étage

6.8

Les Hochschilds (par rapport au premier argument)

$$d_H^{i00} : \mathcal{HKR}(2)^{i00} = \mathrm{Hom}(\Lambda^{i+1} T, \Omega) \longrightarrow \mathrm{Hom}(A \otimes T \otimes \Lambda^i T, \Omega) = \mathcal{HKR}(2)^{i01}$$

$(i = 0, 1, 2)$ seront définis par les formules usuelles

$$d_H^{i00} f(a, \tau; \tau_1, \ldots) = a f(\tau, \tau_1, \ldots) - f(a\tau, \tau_1, \ldots)$$

Pour $f \in \mathrm{Hom}(T, A) = \mathcal{HKR}(2)^{000}$ on calcule:

$$d_H^{100} d_{\mathrm{DR}}^{000} f(a,\tau,\tau') = -\tau' d_H f(a,\tau) + d_H f(\tau'(a),\tau) + d_H f(a,[\tau',\tau]) + \frac{1}{2} d\langle \tau', d_H f(a,\tau)\rangle - \frac{1}{2} da Q f(\tau,\tau').$$

Ceci justifie les définitions suivantes: le de Rham $d_{\mathrm{DR}}^{001} := -d_{\mathrm{Ch}}^{001} - E^{001}$, où

$$d_{\mathrm{Ch}}^{001} f(a,\tau,\tau') = \tau' f(a,\tau) - f(\tau'(a),\tau) - f(a,[\tau',\tau])$$

et

$$E^{001} f(a,\tau,\tau') = -\frac{1}{2} d\langle \tau', f(a,\tau)\rangle.$$

D'autre côté, un opérateur exotique

$$M^{010} : \mathcal{HKR}(2)^{010} = \mathrm{Hom}(S^2 T, A) \longrightarrow \mathrm{Hom}(A \otimes T \otimes T, \Omega) = \mathcal{HKR}(2)^{101}$$

sera défini par

$$M^{010} f(a,\tau,\tau') = -\frac{1}{2} da f(\tau,\tau').$$

Alors notre calcul signifie que

$$d_H^{100} d_{\mathrm{DR}}^{000} = d_{\mathrm{DR}}^{001} d_H^{000} + M^{010} Q^{000}.$$

6.9

De même, le Chevalley

$$d_{\mathrm{Ch}}^{101} : \mathrm{Hom}(A \otimes T \otimes T, \Omega) \longrightarrow \mathrm{Hom}(A \otimes T \otimes \Lambda^2 T, \Omega)$$

vient de l'identification $\mathrm{Hom}(A \otimes T \otimes \Lambda^i T, \Omega) = \mathrm{Hom}(\Lambda^i T, \mathrm{Hom}(A \otimes T, \Omega)) = C^i(T^{\mathrm{Lie}}, \mathrm{Hom}(A \otimes T, \Omega))$, explicitement,

$$d_{\mathrm{Ch}}^{101} f(a;\tau;\tau_1,\tau_2) = -f(a;\tau;[\tau_1,\tau_2]) + \mathrm{Alt}_{12}\{\tau_1 f(a;\tau;\tau_2) - f(\tau_1(a);\tau;\tau_2) - f(a;[\tau_1,\tau];\tau_2)\}.$$

On pose

$$E^{101} f(a;\tau;\tau_1,\tau_2) = -\frac{1}{3} \mathrm{Alt}_{12}\, d\langle \tau_1, f(a;\tau;\tau_2)\rangle,$$

et l'on définit $d_{\mathrm{DR}}^{101} := -d_{\mathrm{Ch}}^{101} - E^{101}$.

D'un autre côté, on introduit

$$M^{110} : \mathrm{Hom}(S^2 T \otimes T, A) \longrightarrow \mathrm{Hom}(A \otimes T \otimes \Lambda^2 T, \Omega)$$

par

$$M^{110} f(a;\tau;\tau_1,\tau_2) = -\frac{1}{3} da\, \mathrm{Alt}_{12}\, f(\tau,\tau_1;\tau_2).$$

Alors nous aurons

$$d_H^{200} d_{\mathrm{DR}}^{100} = d_{\mathrm{DR}}^{101} d_H^{100} + M^{110} Q^{100}.$$

6.10

Les Hochschilds

$$d_H^{i10} : \mathrm{Hom}(S^2T \otimes \Lambda^i T, A) \longrightarrow \mathrm{Hom}(A \otimes T \otimes T \otimes \Lambda^i T, A)$$

$(i = 0, 1)$ sont définis par

$$d_H f(a; \tau; \tau'; \dots) = af(\tau, \tau'; \dots) - f(a\tau, \tau'; \dots).$$

Les Koszuls

$$Q^{i01} : \mathrm{Hom}(A \otimes T \otimes \Lambda^i T, \Omega) \longrightarrow \mathrm{Hom}(A \otimes T \otimes T \otimes \Lambda^i T, A)$$

$(i = 0, 1)$ seront définis à partir de la commutativité $Qd_H = d_H Q$, ce qui donne

$$Q^{i01} f(a; \tau; \tau'; \dots) = \langle \tau', f(a; \tau; \dots) \rangle.$$

6.11

Le de Rham

$$d_{\mathrm{DR}}^{011} : \mathrm{Hom}(A \otimes T^{\otimes 2}, A) \longrightarrow \mathrm{Hom}(A \otimes T^{\otimes 2} \otimes T, A)$$

est défini par $d_{\mathrm{DR}}^{011} = -d_{\mathrm{Ch}}^{011} - E^{011}$, où

$$\begin{aligned} &d_{\mathrm{Ch}}^{011} f(a; \tau, \tau'; \tau'') \\ &\quad = \tau'' f(a; \tau, \tau') - f(\tau''(a); \tau, \tau') - f(a; [\tau'', \tau], \tau') - f(a; \tau, [\tau'', \tau']) \end{aligned}$$

et

$$E^{011} f(a; \tau, \tau'; \tau'') = -\frac{1}{2}\tau' f(a; \tau', \tau'').$$

Ces formules sont déduites soit de la condition (a), soit de la condition (b) du lemme suivant.

Lemme 6.11.1. (a) $Q^{101} d_{\mathrm{DR}}^{001} = d_{\mathrm{DR}}^{011} Q^{001}$.
(b) $d_H^{110} d_{\mathrm{DR}}^{010} = d_{\mathrm{DR}}^{011} d_H^{010} + Q^{101} M^{010}$.

Enfin, l'opérateur R du premier étage sera défini dans le lemme ci-dessous.

Lemme 6.11.2. *Si l'on définit la flèche*

$$R^{011} : \mathrm{Hom}(A \otimes T \otimes T, A) \longrightarrow \mathrm{Hom}(A \otimes T \otimes \Lambda^2 T, \Omega)$$

par

$$\begin{aligned} &R^{011} f(a; \tau; \tau_1, \tau_2) \\ &\quad = -\frac{1}{6} d[f(a; \tau; [\tau_1, \tau_2]) - \mathrm{Alt}_{12}\{f(\tau_1(a); \tau; \tau_2) + f(a; [\tau_1, \tau]; \tau_2)\}] \end{aligned}$$

alors

(a) $d_{\mathrm{DR}}^{101} d_{\mathrm{DR}}^{001} = R^{011} Q^{001}$

et

(b) $d_H^{200} R^{010} = R^{011} d_H^{010} + M^{110} d_{\mathrm{DR}}^{010} + d_{\mathrm{DR}}^{101} M^{010}$.

Ceci termine la définition de la partie $\{\mathcal{HKR}(2)^{\cdot\cdot i}\}_{0 \le i \le 1}$ *du notre tricomplexe tordu.*

Deuxième étage

6.12

Les formules sont tout à fait pareilles à celles du premier étage.

On définit les Hochschilds:

$$d_H^{001} : \mathrm{Hom}(A \otimes T, \Omega) \longrightarrow \mathrm{Hom}(A^{\otimes 2} \otimes T, \Omega)$$

par

$$d_H^{001} f(a, b; \tau) = af(b; \tau) - f(ab; \tau) + f(a; b\tau)$$

et

$$d_H^{101} : \mathrm{Hom}(A \otimes T^{\otimes 2}, \Omega) \longrightarrow \mathrm{Hom}(A^{\otimes 2} \otimes T^{\otimes 2}, \Omega)$$

par

$$d_H^{101} f(a, b; \tau, \tau') = af(b; \tau, \tau') - f(ab; \tau, \tau') + f(a; b\tau, \tau')$$

Ensuite,

$$M^{011} : \mathrm{Hom}(A \otimes T^{\otimes 2}, A) \longrightarrow \mathrm{Hom}(A^{\otimes 2} \otimes T^{\otimes 2}, \Omega)$$

est défini par

$$M^{011} f(a, b; \tau, \tau') = -\frac{1}{2} daf(b; \tau, \tau').$$

Le de Rham

$$d_{\mathrm{DR}}^{002} : \mathrm{Hom}(A^{\otimes 2} \otimes T, \Omega) \longrightarrow \mathrm{Hom}(A^{\otimes 2} \otimes T^{\otimes 2}, \Omega)$$

sera défini par $d_{\mathrm{DR}}^{002} = -d_{\mathrm{Ch}}^{002} - E^{002}$, où

$$d_{\mathrm{Ch}}^{002} f(a, b; \tau, \tau') = \tau' f(a, b; \tau) - f(\tau'(a), b; \tau) - f(a, \tau'(b); \tau) - f(a, b; [\tau', \tau])$$

et

$$E^{002} f(a, b; \tau, \tau') = -\frac{1}{2} d\langle \tau', f(a, b; \tau)\rangle.$$

Alors on aura:

Lemme 6.12.1.

(a) $d_H^{101} d_{\mathrm{DR}}^{001} = d_{\mathrm{DR}}^{002} d_H^{001} + M^{011} Q^{001}$.
(b) $d_H^{101} M^{010} = -M^{011} d_H^{010}$.

6.13

On définit le Hochschild

$$d_H^{011} : \mathrm{Hom}(A \otimes T^{\otimes 2}, A) \longrightarrow \mathrm{Hom}(A^{\otimes 2} \otimes T^{\otimes 2}, A)$$

par

$$d_H^{011} f(a, b; \tau, \tau') = af(b; \tau, \tau') - f(ab; \tau, \tau') + f(a; b\tau, \tau')$$

et le Koszul

$$Q^{002} : \mathrm{Hom}(A^{\otimes 2} \otimes T, \Omega) \longrightarrow \mathrm{Hom}(A^{\otimes 2} \otimes T^{\otimes 2}, \Omega)$$

par

$$Q^{002} f(a, b; \tau, \tau') = \langle \tau', f(a, b; \tau) \rangle.$$

Alors on aura $Q^{002} d_H^{001} = d_H^{011} Q^{001}$.

Troisième étage

6.14

Enfin, le Hochschild

$$d_H^{002} : \mathrm{Hom}(A^{\otimes 2} \otimes T, \Omega) \longrightarrow \mathrm{Hom}(A^{\otimes 3} \otimes T, \Omega)$$

est défini par la formule usuelle

$$d_H^{002} f(a, b, c; \tau) = af(b, c; \tau) - f(ab, c; \tau) + f(a, bc; \tau) - f(a, b; c\tau).$$

Pour tous les Hochschilds, on a évidemment $d_H^2 = 0$.

Ceci termine la définition du tricomplexe tordu $\mathcal{HKR}(2)^{\cdots}$, avec $d' = d_{\mathrm{DR}}$, $d'' = Q$ et $d''' = d_H$.

Attention: Notre de Rham est $-d_{\mathrm{DR}}$ de [S], nos $d_H^{\cdot\cdot 0}$ sont $-d_H$ de [S], et notre M^{011} et $-M$ de [S]; les autres morphismes sont les mêmes.

Le plongement ι (6.1.2) est induit par (6.3.1). Son image

$$\mathrm{Im}\, \iota = \mathrm{Ker}\, d_H^{\cdot 00} \cap \mathrm{Ker}\, Q^{\cdot 00}.$$

7 Structures vertex

Dans 7.1–7.2 on réproduit les considérations de [S], Finale. Pour les détails de calculs, voir op. cit.

7.1

Cocycle ϵ_2. Définissons des éléments suivants: $\epsilon_2^{002} \in \mathcal{HKR}(2)^{002} = \mathrm{Hom}(A^{\otimes 2} \otimes T, \Omega)$ par

$$\epsilon_2^{002}(a, b; \tau) = -\tau(a)db - \tau(b)da;$$

$\epsilon_2^{011} \in \mathcal{HKR}(2)^{011} = \mathrm{Hom}(A \otimes T^{\otimes 2}, A)$ par

$$\epsilon_2^{011}(a; \tau, \tau') = \tau\tau'(a)$$

et $\epsilon_2^{101} \in \mathcal{HKR}(2)^{101} = \mathrm{Hom}(A \otimes T^{\otimes 2}, \Omega)$ par

$$\epsilon_2^{101}(a; \tau, \tau') = \frac{1}{2} d\tau\tau'(a).$$

Posons $\epsilon_2 = (\epsilon_2^{002}, \epsilon_2^{011}, \epsilon_2^{101}) \in \mathcal{HKR}(2)^2$.

Théorème 7.1.1. $d_{\mathcal{HKR}}\epsilon_2 = 0$. *Cette assertion est équivalente aux cinq identités:*

$$d_H \epsilon_2^{002} = 0, \tag{B1}$$

$$Q\epsilon_2^{002} - d_H \epsilon_2^{011} = 0, \tag{B2}$$

$$d_{\mathrm{DR}}\epsilon_2^{002} - d_H \epsilon_2^{101} + M\epsilon_2^{011} = 0, \tag{B3}$$

$$d_{\mathrm{DR}}\epsilon_2^{011} - Q\epsilon_2^{101} = 0, \tag{B4}$$

et

$$d_{\mathrm{DR}}\epsilon_2^{101} - R\epsilon_2^{011} = 0, \tag{B5}$$

qui se vérifient très facilement.

7.2

Une structure vertex sur T est un cochaîne $v \in \mathcal{HKR}(2)^1$ telle que $d_{\mathcal{HKR}}v = \epsilon_2$.

En composantes, $v = (v^{001}, v^{010}, v^{100})$. Introduisons les notations $\gamma = v^{001} \in \mathrm{Hom}(A \otimes T, \Omega)$, $\langle , \rangle = -v^{010} \in \mathrm{Hom}(S^2T, A)$ et $c = v^{100} \in \mathrm{Hom}(\Lambda^2 T, \Omega)$. Alors l'axiome de structure vertex est équivalent à cinq identités (A1)–(A5) ci-dessous.

$$d_H \gamma = \epsilon_2^{002}, \tag{A1}$$

$$d_H \langle , \rangle + Q\gamma = \epsilon_2^{011}, \tag{A2}$$

$$-d_H c - M\langle , \rangle + d_{\mathrm{DR}}\gamma = \epsilon_2^{101}, \tag{A3}$$

$$d_{\mathrm{DR}}\langle , \rangle + Qc = 0, \tag{A4}$$

et

$$d_{\mathrm{DR}}c + R\langle , \rangle = 0. \tag{A5}$$

On voit aussitôt que ces axiomes coïncident avec [GMS, 1.4, les axiomes (A1)–(A5)].

De même, *un morphisme* $h : v \longrightarrow v'$ de deux structures vertex est une cochaîne $h \in \mathcal{HKR}(2)^0 = \mathrm{Hom}(T, \Omega)$ telle que $d_{\mathcal{HKR}}h = v' - v$. On voit que ceci coïncide avec la notion introduite dans op. cit. 3.5 (avec g_A, g_T, g_Ω étant les morphismes identiques).

7.3

Fixons une base abélienne $\mathfrak{b} = \{\tau_i\}$ de T. Définissons $\langle , \rangle_{\mathfrak{b}} \in \mathrm{Hom}(S^2T, A)$ par

$$\langle a\tau_i, b\tau_j\rangle_{\mathfrak{b}} = -\tau_j(a)\tau_i(b)$$

Il existe un unique $\gamma_{\mathfrak{b}} \in \mathrm{Hom}(A \otimes T, \Omega)$ quie vérifie l'axiome (A2) avec ce $\langle , \rangle_{\mathfrak{b}}$. En effet, (A2) s'ecrit explicitement comme

$$\langle \gamma_{\mathfrak{b}}(a, \tau), \tau'\rangle = \langle a\tau, \tau'\rangle_{\mathfrak{b}} - a\langle \tau, \tau'\rangle_{\mathfrak{b}} + \tau\tau'(a), \tag{7.3.1}$$

qui nous donne

$$\gamma_{\mathfrak{b}}(b, a\tau_i) = ad\tau_i(b).$$

De même, il existe un unique $c_{\mathfrak{b}} \in \mathrm{Hom}(\Lambda^2T, \Omega)$ tel que $c_{\mathfrak{b}}(\tau_i, \tau_j) = 0$ pour tous i, j et satisfait à l'axiome (A3). Il est donné par la formule

$$c_{\mathfrak{b}}(a\tau_i, b\tau_j) = \frac{1}{2}\,\mathrm{Alt}_{a\tau_i, b\tau_j}\, \tau_i(b)d\tau_j(a).$$

Théorème 7.3.1. *Le triple $v_{\mathfrak{b}} = (\gamma_{\mathfrak{b}}, -\langle , \rangle_{\mathfrak{b}}, c_{\mathfrak{b}})$ est une structure vertex.*

On remarque que la structure vertex $v_{\mathfrak{b}}$ diffère de celle utilisée dans [GMS], 5.7; notre $v_{\mathfrak{b}}$ est plus simple. Nous laissons la preuve de 7.3.1 au lecteur.

7.4

Soient $\mathfrak{b} = \{\tau_i\}$, $\mathfrak{b}' = \{\tau_i'\}$, $\mathfrak{b}'' = \{\tau_j''\}$ trois bases abéliennes, $\mathfrak{b}' = \phi\mathfrak{b}$, $\mathfrak{b}'' = \psi\mathfrak{b}'$, $\phi, \psi \in \mathrm{GL}_n(A)$.

On définit un élément $h_{\mathfrak{b}\mathfrak{b}'} \in \mathrm{Hom}(T, \Omega) = \mathcal{HKR}(2)^0$ par

$$h_{\mathfrak{b}\mathfrak{b}'}(a\tau_i') = \frac{1}{2}\,\mathrm{tr}\{\phi^{-1}a\tau_i'(\phi)\phi^{-1}d\phi\} + [d\phi\phi^{-1}]^{ip}\tau_p'(a).$$

Rappelons les composantes de la deuxième forme de Chern–Simons (cf. 3.8):

$$\beta^{11}(\phi) = \frac{1}{6}\,\mathrm{tr}\{\ell(\phi)^3\}; \qquad \beta^{02}(\psi, \phi) = \frac{1}{2}\,\mathrm{tr}\{\ell(\psi)^{\phi}\ell(\phi)\},$$

où $\ell(\phi) = \phi^{-1}d\phi$, $\ell(\psi)^{\phi} = \psi^{-1}\ell(\phi)\psi$.

Le théorème ci-dessous réproduit le résultat principal de [GMS, Théorème 5.14].

Théorème 7.4.1.

(a) $d_{\mathrm{DR}}h_{\mathfrak{b}\mathfrak{b}'}(a\tau_i', b\tau_j') = c_{\mathfrak{b}'}(a\tau_i', b\tau_j') - c_{\mathfrak{b}}(a\tau_i', b\tau_j') - \langle b\tau_j', \langle a\tau_i', \beta^{11}(\phi)\rangle\rangle$.
(b) $\langle a\tau_i', h_{\mathfrak{b}\mathfrak{b}'}(b\tau_j')\rangle + \langle b\tau_j', h_{\mathfrak{b}\mathfrak{b}'}(a\tau_i')\rangle = \langle a\tau_i', b\tau_j'\rangle_{\mathfrak{b}} - \langle a\tau_i', b\tau_j'\rangle_{\mathfrak{b}'}$.
(c) $ah_{\mathfrak{b}\mathfrak{b}'}(b\tau_j') - h_{\mathfrak{b}\mathfrak{b}'}(ab\tau_j') = \gamma_{\mathfrak{b}'}(a, b\tau_j) - \gamma_{\mathfrak{b}}(a, b\tau_j)$.
(d) $h_{\mathfrak{b}',\mathfrak{b}''}(a\tau_i'') - h_{\mathfrak{b},\mathfrak{b}''}(a\tau_i'') + h_{\mathfrak{b}\mathfrak{b}'}(a\tau_i'') = -\langle a\tau_i'', \beta^{02}(\psi, \phi)\rangle$.

On peut démontrer ce théorème par la méthode de [GMS], ou bien faire la vérification directe. Nous laissons les détails comme un exercice au lecteur.

7.5

Considérons le bicomplexe de Čech $\check{C}^{\cdot}(\mathfrak{B}, \mathcal{HKR}(2)^{\cdot})$ où le degré de Čech est le premier et le complexe simple associé $\check{C}^{\cdot}(\mathfrak{B}, \mathcal{HKR}(2))$.

Le composé

$$\Omega^{[2,3\rangle} \hookrightarrow \Omega^{[2,5]} \xrightarrow{\iota} \mathcal{HKR}(2)$$

induit le morphisme canonique

$$\mu : \mathrm{CS}(2) = C^{\cdot}(\mathrm{GL}_n(A), \Omega^{[2,3\rangle}) \longrightarrow \check{C}^{\cdot}(\mathfrak{B}, \mathcal{HKR}(2)).$$

Posons $v = \{v_{\mathfrak{b}}\} \in \check{C}^0(\mathfrak{B}; \mathcal{HKR}(2)^1)$; $h = \{h_{\mathfrak{b}\mathfrak{b}'}\} \in \check{C}^1(\mathfrak{B}; \mathcal{HKR}(2)^0)$.

Alors le théorème précédent se réécrit:

Théorème 7.5.1. (a) $d_{\mathcal{HKR}}h - d_c v = -\mu(\beta^{11})$.
(b) $d_c h = -\mu(\beta^{02})$.

De plus, on a comme d'habitude le morphisme d'augmentation $\delta : \mathcal{HKR}(2) \longrightarrow \check{C}^{\cdot}(\mathfrak{B}; \mathcal{HKR}(2))$, et 7.3.1 signifie que $d_{\mathcal{HKR}}v = \delta(\epsilon_2)$.

Définissons le complexe

$$\hat{\mathrm{CS}}(2) = \mathrm{Cône}\{(\mu, \delta) : \mathrm{CS}(2) \oplus \mathcal{HKR}(2) \longrightarrow \check{C}^{\cdot}(\mathfrak{B}; \mathcal{HKR}(2))[-1]\};$$

cf. (5.6.2). On a le morphisme canonique de complexes $\pi : \hat{\mathrm{CS}}(2) \longrightarrow \mathrm{CS}(2)$.

On définit

$$\hat{v} = (v, h) \in \check{C}^1(\mathfrak{B}; \mathcal{HKR}(2)).$$

Toute l'information précédente est rassemblée dans:

Théorème 7.5.2. *La cochaîne* $\hat{\beta}_2 = (\beta_2, -\epsilon_2, \hat{v}) \in \hat{\mathrm{CS}}(2)^2$ *est un cocycle tel que* $\pi(\hat{\beta}_2) = \beta_2$.

8 Structures prémembranaires

8.1

Complexes de Koszul. Le n-ième complexe de Koszul $\mathcal{K}(n) = \{\mathcal{K}(n)^i\}$ est concentré en degrés $0 \leq i \leq n-1$

$$\mathcal{K}(n)^{\cdot} : \mathrm{Hom}(T, \Omega^{n-1}) \longrightarrow \mathrm{Hom}(S^2T, \Omega^{n-2}) \longrightarrow \ldots \longrightarrow \mathrm{Hom}(S^nT, A),$$

i.e., $\mathcal{K}(n)^i = \mathrm{Hom}(S^{i-1}T, \Omega^{n-i+1})$. Les différentielles $Q : \mathcal{K}(n)^i \longrightarrow \mathcal{K}(n)^{i+1}$ sont définis par

$$Qf(\tau_1, \ldots, \tau_i) = \sum_{p=1}^{i} \langle \tau_p, f(\tau_1, \ldots, \hat{\tau}_p, \ldots) \rangle.$$

Il est clair que $Q^2 = 0$.

8.2

Complexes de Hochschild–Koszul. Le n-ième bicomplexe de Hochschild–Koszul $\mathcal{HK}(n)^{\cdot\cdot} = \{\mathcal{HK}(n)^{ij}\}$ habite en degrés $0 \leq i \leq n-1$, $j \geq 0$. Par définition, la 0-ième ligne $\mathcal{HK}(n)^{i0}$ coïncide avec $\mathcal{K}(n)$.

Les colonnes sont les complexes de Hochschild par rapport au premier argument:

$$\begin{aligned}\mathcal{HK}(n)^{i\cdot} : \mathrm{Hom}(S^{i-1}T, \Omega^{n-i+1}) &\longrightarrow \mathrm{Hom}(A \otimes T \otimes S^{i-2}T, \Omega^{n-i+1}) \longrightarrow \cdots \\ &\longrightarrow \mathrm{Hom}(A^{\otimes j} \otimes T \otimes S^{i-2}T, \Omega^{n-i+1}) \cdots,\end{aligned}$$

les différentielles verticales étant les Hochschilds usuels:

$$\begin{aligned} d_H f(a_1, \ldots, a_j; \tau; \ldots) &= a_1 f(a_2, \ldots, a_j; \tau; \ldots) \\ &+ \sum_{p=1}^{j-1} (-1)^p f(\ldots, a_p a_{p+1}, \ldots; \tau; \ldots) \\ &+ (-1)^j f(a_1, \ldots, a_{j-1}; a_j \tau; \ldots). \end{aligned}$$

Par contre, les différentielles horisontales dans la j-ième ligne $\mathcal{HK}(n)^{\cdot j}$, $j > 0$, sont les Koszuls suivants:

$$Qf(a_1, \ldots, a_j; \tau; \tau_1, \ldots, \tau_i) = \sum_{p=1}^{i} \langle \tau_p, f(a_1, \ldots, a_j; \tau; \tau_1, \ldots, \hat{\tau}_p, \ldots) \rangle.$$

On vérifie sans peine que $d_H Q = Q d_H$. Donc le premier degré est de Koszul, et le deuxième est de Hochschild.

Le complexe simple associé sera désigné par $\mathcal{HK}(n) = \{\mathcal{HK}(n)^i\}$.

En effet, désormais on ne sera interessé qu'au *troisième* complexe $\mathcal{HK}(3)$. On suppose dans cette section et dans la section suivante que 6 est inversible dans l'anneau de base k.

8.3

Cocycle $\epsilon_3^{\mathcal{HK}}$. Définissons les éléments: $\epsilon \in \mathrm{Hom}(A \otimes T \otimes S^2 T, A) = \mathcal{HK}(3)^{21}$ par

$$\epsilon(a; \tau; \tau', \tau'') = \frac{1}{2} \mathrm{Sym}_{\tau', \tau''} \tau\tau'\tau''(a),$$

$\epsilon' \in \mathrm{Hom}(A^{\otimes 2} \otimes T^{\otimes 2}, \Omega) = \mathcal{HK}(3)^{12}$ par

$$\epsilon'(a, b; \tau; \tau') = -\mathrm{Sym}_{a,b}\{\frac{1}{2}\tau(a) d\tau'(b) + \tau\tau'(a) db\}$$

et $\epsilon'' = \mathrm{Hom}(A^{\otimes 3} \otimes T, \Omega^2) = \mathcal{HK}(3)^{03}$ par

$$\epsilon''(a, b, c; \tau) = \tau(b) da dc + \frac{1}{2}\tau(c) da db + \frac{1}{2}\tau(a) db dc.$$

On les rassemble en

$$\epsilon_3^{\mathcal{HK}} = (-\epsilon'', -\epsilon', \epsilon) \in \mathcal{HK}(3)^3.$$

Théorème 8.3.1. $d_{\mathcal{HK}}(\epsilon_3^{\mathcal{HK}}) = 0$.

En effet, on commence par l'élément ϵ "bien naturel." Après on trouve ϵ' à partir de la condition

$$d_H\epsilon = Q\epsilon'. \tag{8.3.1}$$

Ensuite, on trouve ϵ'' de la condition

$$d_H\epsilon' = Q\epsilon''. \tag{8.3.2}$$

Enfin, on vérifie que

$$d_H\epsilon'' = 0. \tag{8.3.3}$$

Les trois équations (8.3.1)–(8.3.3) sont équivalentes à l'assertion du théorème. Pour les détails, voir 8.7–8.9.

Une structure prémembranaire sur T est une cochaîne $m = (m^{02}, m^{11}, m^{20}) \in \mathcal{HK}(3)^2$ telle que $d_{\mathcal{HK}}m = \epsilon_3^{\mathcal{HK}}$.

8.4

Fixons une base abélienne $\mathfrak{b} = \{\tau_i\}$ de T. Considérons un élément $\rho_{\mathfrak{b}} \in \mathrm{Hom}(T^{\otimes 3}, A)$:

$$\rho_{\mathfrak{b}}(a\tau_i, b\tau_j, c\tau_k) = \tau_k(a)\tau_i(b)\tau_j(c). \tag{8.4.1}$$

Manifestement, il possède la symétrie cyclique: $\rho_{\mathfrak{b}}(\tau, \tau', \tau'') = \rho_{\mathfrak{b}}(\tau', \tau'', \tau)$. Donc, si l'on définit $\{, , \}_{\mathfrak{b}}$ par

$$\{\tau, \tau', \tau''\}_{\mathfrak{b}} = -\frac{1}{2}\,\mathrm{Sym}_{\tau', \tau''}\,\rho_{\mathfrak{b}}(\tau, \tau', \tau''), \tag{8.4.2}$$

cet élément appartiendra à $\mathrm{Hom}(S^3T, A) = \mathcal{HK}(3)^{20}$.

De plus, on introduit les deux éléments suivants: $\gamma_{\mathfrak{b}} \in \mathrm{Hom}(A \otimes T^{\otimes 2}, \Omega) = \mathcal{HK}(3)^{11}$,

$$\gamma_{\mathfrak{b}}(a; b\tau_j, c\tau_k) = -\frac{1}{2}[bd\tau_j\{c\tau_k(a)\} + b\tau_j(c)d\tau_k(a)] \tag{8.4.3}$$

et $\gamma'_{\mathfrak{b}} \in \mathrm{Hom}(A^{\otimes 2} \otimes T, \Omega^2) = \mathcal{HK}(3)^{02}$,

$$\gamma'_{\mathfrak{b}}(a, b; c\tau_k) = \frac{1}{2}\,\mathrm{Sym}_{a,b}\,dacd\tau_k(b). \tag{8.4.4}$$

On les rassemble en

$$m_{\mathfrak{b}} = (-\gamma'_{\mathfrak{b}}, -\gamma_{\mathfrak{b}}, \{, , \}_{\mathfrak{b}}) \in \mathcal{HK}(3)^2.$$

Théorème 8.4.1. *$m_{\mathfrak{b}}$ est une structure prémembranaire.*

En effet, on commence par l'élément $\{,,\}_{\mathfrak{b}}$ "bien naturel." Après on trouve $\gamma_{\mathfrak{b}}$ à partir de l'équation

$$d_H\{,,\}_{\mathfrak{b}} = \epsilon + Q\gamma_{\mathfrak{b}}. \tag{8.4.5}$$

Ensuite, on trouve $\gamma'_{\mathfrak{b}}$ à partir de la condition

$$d_H\gamma_{\mathfrak{b}} = -\epsilon' + Q\gamma'_{\mathfrak{b}}. \tag{8.4.6}$$

Enfin, on vérifie que

$$d_H\gamma'_{\mathfrak{b}} = \epsilon''. \tag{8.4.7}$$

Les trois équations ci-dessus sont équivalentes à 8.4.1.

9 Troisième cocycle de Chern–Hodge raffiné

9.1

Maintenant considérons le tricomplexe $\check{C}^{\cdot}(\mathfrak{B}; \mathcal{HK}(3)^{\cdot\cdot})$, le degré de Koszul étant le premier, le degré de Hochschild étant le deuxième et le degré de Čech étant le troisième. On désigne par $\check{C}^{\cdot}(\mathfrak{B}; \mathcal{HK}(3))$ le complexe simple associé.

On a, comme d'habitude, le morphisme d'augmentation

$$\delta : \mathcal{HK}(3) \longrightarrow \check{C}^{\cdot}(\mathfrak{B}; \mathcal{HK}(3)).$$

De plus, on considère Ω^3 comme un sous-module de $\mathrm{Hom}(T, \Omega^2) = \mathcal{HK}(3)^{00}$ par la règle usuelle: on associe à une forme $\omega \in \Omega^3$ la flèche $\iota\omega : T \longrightarrow \Omega^2$, $\iota\omega(\tau) = i_\tau\omega$, dù l'inclusion

$$\iota : \Omega^3 \hookrightarrow \mathcal{HK}(3),$$

d'où le morphisme

$$\mu : C^{\cdot}(\mathrm{GL}_n(A), \Omega^3) \longrightarrow \check{C}^{\cdot}(\mathfrak{B}; \mathcal{HK}(3)).$$

Rappelons la forme $\beta_3^{\mathrm{CH}} := \beta^{03} \in Z^3(\mathrm{GL}_n(A), \Omega^3)$,

$$\beta_3^{\mathrm{CH}}(\chi, \psi, \phi) = \frac{1}{6}\,\mathrm{tr}\{\ell(\chi)^{\psi\phi}\ell(\psi)^{\phi}\ell(\phi)\};$$

cf. 3.11. On l'appele *le troisième cocycle de Chern–Hodge*, puisqu'il définit le troisième caractère de Chern style Hodge.

9.2

On considère la collection $m = \{m_{\mathfrak{b}}\}$ construite dans 8.4 ci-dessus comme une cochaîne $\hat{m}^{\cdot\cdot 0} \in \check{C}^0(\mathfrak{B}; \mathcal{HK}(3)^{\cdot\cdot})$, avec les composantes $(\hat{m}^{020}, \hat{m}^{110}, \hat{m}^{200}) = (-\gamma', -\gamma, \{,,\})$.

9.3

Soient $\mathfrak{b} = \{\tau_i\}$, $\mathfrak{b}' = \{\tau_i'\} \in \mathfrak{B}$, $\mathfrak{b}' = \phi\mathfrak{b}$. On définit l'élément $-h_{\mathfrak{b}\mathfrak{b}'} = \hat{m}^{101}_{\mathfrak{b}\mathfrak{b}'} \in \mathrm{Hom}(S^2T, \Omega) = \mathcal{HK}(3)^{10}$ par

$$\begin{aligned}-h_{\mathfrak{b}\mathfrak{b}'}(a\tau_i', b\tau_j') &= \hat{m}^{101}_{\mathfrak{b}\mathfrak{b}'}(a\tau_i', b\tau_j')\\ &= -\operatorname{Sym}_{a\tau_i', b\tau_j'}\left[\frac{1}{6}\operatorname{tr}\{\phi^{-1}a\tau_i'(\phi)\phi^{-1}b\tau_j'(\phi)\phi^{-1}d\phi\}\right.\\ &\quad\left.+\frac{1}{2}[d\phi\phi^{-1}a\tau_i'(\phi)\phi^{-1}]^{jp}\tau_p'(b)+\frac{1}{2}[d\phi\phi^{-1}]^{jp}\tau_p'(a)\tau_i'(b)\right].\end{aligned} \tag{9.3.1}$$

Ensuite, on définit l'élément $\Gamma_{\mathfrak{b}\mathfrak{b}'} = \hat{m}^{011}_{\mathfrak{b}\mathfrak{b}'} \in \mathrm{Hom}(A \otimes T, \Omega^2) = \mathcal{HK}(3)^{01}$ par

$$\Gamma_{\mathfrak{b}\mathfrak{b}'}(a; b\tau_j') = \hat{m}^{011}_{\mathfrak{b}\mathfrak{b}'}(a; b\tau_j') = \frac{1}{2}bd\phi^{jp}d\tau_p(a). \tag{9.3.2}$$

Cela fournit une cochaîne $\hat{m}^{\cdot\cdot 1} = (\hat{m}^{011}_{\mathfrak{b}\mathfrak{b}'}, \hat{m}^{101}_{\mathfrak{b}\mathfrak{b}'}) \in \check{C}^1(\mathfrak{B}; \mathcal{HK}(3)^{\cdot\cdot})$.

9.4

Soit $\mathfrak{b}'' = \{\tau_i''\} \in \mathfrak{B}$, $\mathfrak{b}'' = \psi\mathfrak{b}'$. Définissons un élément $-H_{\mathfrak{b}\mathfrak{b}'\mathfrak{b}''} = \hat{m}^{002}_{\mathfrak{b}\mathfrak{b}'\mathfrak{b}''} \in \mathrm{Hom}(T, \Omega^2) = \mathcal{HK}(3)^{00}$ par

$$\begin{aligned}-H_{\mathfrak{b}\mathfrak{b}'\mathfrak{b}''}(a\tau_i'') &= \hat{m}^{002}_{\mathfrak{b}\mathfrak{b}'\mathfrak{b}''}(a\tau_i'')\\ &= -\frac{1}{6}a\operatorname{tr}\{\ell_i''(\psi)^{\phi}\ell(\psi)^{\phi}\ell(\phi) - \ell_i''(\psi)^{\phi}\ell(\phi)\ell(\psi)^{\phi}\\ &\quad + \ell_i''(\phi)\ell(\psi)^{\phi}\ell(\phi) - \ell_i''(\psi)^{\phi}\ell(\phi)\ell(\phi)\} - \frac{1}{2}[d\psi d\phi]^{ip}\tau_p(a).\end{aligned} \tag{9.4.1}$$

Rappelons que $\ell(\phi) := \phi^{-1}d\phi$, $\ell_i''(\psi) := \psi^{-1}\tau_i''(\psi)$, et $x^{\phi} := \phi^{-1}x\phi$.

Cela fournit une cochaîne $\hat{m}^{\cdot\cdot 2} = (\hat{m}^{002}_{\mathfrak{b}\mathfrak{b}'\mathfrak{b}''}) \in \check{C}^2(\mathfrak{B}; \mathcal{HK}(3)^{\cdot\cdot})$.

En rassemblant, on définit la 2-cochaîne

$$\hat{m}^{\mathcal{HK}} = (m^{\cdot\cdot 0}, m^{\cdot\cdot 1}, m^{\cdot\cdot 2}) \in \check{C}^2(\mathfrak{B}; \mathcal{HK}(3)).$$

Théorème 9.4.1. *Si l'on désigne par $d_{\mathcal{CHK}}$ la différentielle totale dans $\check{C}^{\cdot}(\mathfrak{B}; \mathcal{HK}(3))$, on aura $d_{\mathcal{CHK}}(\hat{m}^{\mathcal{HK}}) = \delta(\epsilon_3^{\mathcal{HK}}) + \mu(\beta_3^{\mathrm{CH}})$.*

En effet, on trouve $h_{\mathfrak{b}\mathfrak{b}'}$ à partir de la condition

$$d_c\{,\,\}_{\mathfrak{b}\mathfrak{b}'} = Qh_{\mathfrak{b}\mathfrak{b}'}. \tag{9.4.2}$$

Après on trouve $H_{\mathfrak{b}\mathfrak{b}'\mathfrak{b}''}$ à partir de la condition

$$d_c h_{\mathfrak{b}\mathfrak{b}'\mathfrak{b}''} = QH_{\mathfrak{b}\mathfrak{b}'\mathfrak{b}''}. \tag{9.4.3}$$

Ensuite, on vérifie que

$$d_c H_{\mathfrak{b}\mathfrak{b}'\mathfrak{b}''\mathfrak{b}'''} = -\iota\beta_3^{\mathrm{CH}}(\chi, \psi, \phi) \tag{9.4.4}$$

(où $\mathfrak{b}''' = \chi\mathfrak{b}''$). Cela fournit la partie Koszul–Čech de notre cocycle.

Ensuite, on trouve $\Gamma_{\mathfrak{b}\mathfrak{b}'}$ à partir de la condition

$$d_c\gamma'_{\mathfrak{b}\mathfrak{b}'} = d_H\Gamma_{\mathfrak{b}\mathfrak{b}'}. \tag{9.4.5}$$

Enfin, on vérifie que

$$d_c\Gamma_{\mathfrak{b}\mathfrak{b}'\mathfrak{b}''} = -d_H H_{\mathfrak{b}\mathfrak{b}'\mathfrak{b}''} \tag{9.4.6}$$

et

$$-d_c\gamma_{\mathfrak{b}\mathfrak{b}'} + Q\Gamma_{\mathfrak{b}\mathfrak{b}'} + d_H h_{\mathfrak{b}\mathfrak{b}'} = 0. \tag{9.4.7}$$

Les relations (9.4.2)–(9.4.7), combinées avec 8.4.1, sont équivalentes à notre théorème.

Cela est analogue, en dimension 3, de [GMS, 5.10–5.13]. En langage "catégorique," les structures prémembranaires forment une "2-gerbe" lié par Ω^3, dont la classe est representée par la forme β_3^{CH}.

10 Troisième Koszul–de Rham

10.1

Dans cette section on va introduire un bicomplexe tordu $\mathcal{KR}(3)^{\cdot\cdot}$, dit *le troisième Koszul–de Rham.*

Définissons les plongements

$$\iota_n : \Omega^n \longrightarrow \mathrm{Hom}(\Lambda^{n-2}T, \Omega^2)$$

par la règle usuelle

$$\iota_n\omega(\tau_1, \ldots, \tau_{n-2}) = i_{\tau_{n-2}} i_{\tau_{n-3}} \ldots i_{\tau_1}\omega.$$

Supposons que p est inversible dans k. On définit le morphisme

$$d_{\mathrm{DR}} : \mathrm{Hom}(\Lambda^{p-1}, \Omega^2) \longrightarrow \mathrm{Hom}(\Lambda^p, \Omega^2)$$

par $d_{\mathrm{DR}} = -d_{\mathrm{Ch}} - E$, où

$$\begin{aligned} d_{\mathrm{Ch}} f(\tau_1, \ldots) = & \sum_{i<j} (-1)^{i+j} f([\tau_i, \tau_j], \ldots, \hat{\tau}_i, \ldots, \hat{\tau}_j, \ldots) \\ & + \sum_i (-1)^{i+1} \tau_i f(\ldots, \hat{\tau}_i, \ldots) \end{aligned}$$

et

$$Ef(\tau_1, \ldots, \tau_p) = \frac{1}{p}\sum_{i=1}^{p}(-1)^i d\langle \tau_i, f(\ldots, \hat{\tau}_i, \ldots)\rangle.$$

On utilisera la notation $d_{\mathrm{DR}}^{(0)}$ pour les opérateurs d_{DR} définis ci-dessus; ils nous serviront comme les différentielles dans la 0-ième ligne du notre Koszul–de Rham.

Lemme 10.1.1. $d_{\mathrm{DR}}^{(0)}\iota_{p+1} = -\iota_{p+2}d.$

Donc les morphismes ι_n *donnent lieu à l'inclusion (si* $k \supset \mathbb{Q}$*) du complexe de de Rham tronqué et decalé*

$$\Omega^{[3} := \sigma_{\geq 3}\Omega^{\cdot}[3] : 0 \longrightarrow \Omega^3 \longrightarrow \Omega^4 \longrightarrow$$

dans la ligne

$$0 \longrightarrow \mathrm{Hom}(T, \Omega^2) \xrightarrow{d_{\mathrm{DR}}^{(0)}} \mathrm{Hom}(\Lambda^2 T, \Omega^2) \xrightarrow{d_{\mathrm{DR}}^{(0)}} \cdots. \tag{10.1.1}$$

Par contre, cette ligne n'est pas un complexe, car $d_{\mathrm{DR}}^{(0)2} \neq 0$.

10.2

Suivant l'usage, on définit les opérateurs de Koszul

$$Q : \mathrm{Hom}(\Lambda^n T, \Omega^2) \longrightarrow \mathrm{Hom}(S^2 T \otimes \Lambda^{n-1} T, \Omega)$$

par

$$Qf(\tau_1, \tau_2; \ldots) = \mathrm{Sym}_{12}\langle \tau_1, f(\tau_2, \ldots)\rangle.$$

10.3

On introduit l'opérateur (en supposant que $n(n-1)$ est inversible dans k)

$$R : \mathrm{Hom}(S^2 T \otimes \Lambda^{n-3} T, \Omega) \longrightarrow \mathrm{Hom}(\Lambda^n T, \Omega^2)$$

par

$$\begin{aligned} &Rf(\tau_1, \ldots, \tau_n) \\ &\quad = -\frac{1}{n(n-1)} \\ &\qquad \cdot \sum_{i<j<k} (-1)^{i+j+k}\, \mathrm{Cycle}_{ijk}\, df([\tau_i, \tau_j], \tau_k; \ldots, \hat{\tau}_i, \ldots, \hat{\tau}_j, \ldots, \hat{\tau}_k, \ldots). \end{aligned}$$

Lemme 10.3.1. $d_{\mathrm{DR}}^{(0)2} = RQ.$

10.4

Première ligne. On introduit les opérateurs

$$d^{(1)}_{\mathrm{DR}} : \mathrm{Hom}(S^2T \otimes \Lambda^{n-2}T, \Omega) \longrightarrow \mathrm{Hom}(S^2T \otimes \Lambda^{n-1}T, \Omega)$$

(en supposant que n est inversible dans k) par $d^{(1)}_{\mathrm{DR}} = d_{\mathrm{Ch}} + E$.

Ici d_{Ch} est la différentielle de Chevalley, après l'identification

$$\mathrm{Hom}(S^2T \otimes \Lambda^{\cdot}T, \Omega) = \mathrm{Hom}(\Lambda^{\cdot}T, \mathrm{Hom}(S^2T, \Omega)) = C^{\cdot}(T^{\mathrm{Lie}}, \mathrm{Hom}(S^2T, \Omega))$$

Explicitement,

$$\begin{aligned} d_{\mathrm{Ch}} f(\tau', \tau''; \dots) = & \sum_{i<j} (-1)^{i+j} f(\tau', \tau''; [\tau_i, \tau_j], \dots, \hat{\tau}_i, \dots, \hat{\tau}_j, \dots) \\ & + \sum_i (-1)^{i+1} \tau_i f(\tau', \tau''; \dots, \hat{\tau}_i, \dots) \\ & + \sum_i (-1)^i \,\mathrm{Sym}_{\tau', \tau''}\, c([\tau_i, \tau'], \tau''; \dots, \hat{\tau}_i, \dots). \end{aligned}$$

D'un autre côté, $E = E' + E''$,

$$E' f(\tau', \tau''; \tau_1, \dots, \tau_{n-1}) = \frac{1}{n} \sum_i (-1)^i \,\mathrm{Sym}_{\tau', \tau''}\, \tau' f(\tau'', \tau_i; \dots, \hat{\tau}_i, \dots)$$

et

$$E'' f(\tau', \tau''; \tau_1, \dots, \tau_{n-1}) = \frac{1}{n} \sum_i (-1)^i d\langle \tau_i, f(\tau', \tau''; \dots, \hat{\tau}_i, \dots)\rangle.$$

Lemme 10.4.1. $Q d^{(0)}_{\mathrm{DR}} = d^{(1)}_{\mathrm{DR}} Q$.

Lemme 10.4.2. *Définissons l'opérateur de Koszul*

$$Q : \mathrm{Hom}(S^2T, \Omega) \longrightarrow \mathrm{Hom}(S^3T, A)$$

par

$$Qf(\tau, \tau', \tau'') = \mathrm{Cycle}_{\tau, \tau', \tau''} \langle \tau, f(\tau', \tau'')\rangle. \tag{10.4.1}$$

Cet opérateur est un morphisme de T^{Lie}-modules.

Par définition, les opérateurs de Koszul

$$Q : \mathrm{Hom}(S^2T \otimes \Lambda^n T, \Omega) \longrightarrow \mathrm{Hom}(S^3T \otimes \Lambda^n T, A)$$

seront induits par (10.4.1), i.e.,

$$Qf(\tau, \tau', \tau''; \dots) = \mathrm{Cycle}_{\tau, \tau', \tau''} \langle \tau, f(\tau', \tau''; \dots)\rangle. \tag{10.4.2}$$

10.5

On va définir le Koszul–de Rham $\mathcal{KR}(3)^{\cdot\cdot} = \{\mathcal{KR}(3)^{ij}\}$. Il sera concentré dans la domaine $0 \leq i + j \leq 3, 0 \leq j \leq 2$, le premier degré i étant le degré de de Rham et le deuxième degré j étant le degré de Koszul.

La 0-ième ligne sera (10.1.1) tronquée:

$$\mathcal{KR}(3)^{\cdot 0} : \mathrm{Hom}(T, \Omega^2) \xrightarrow{d_{\mathrm{DR}}^{(0)}} \cdots \xrightarrow{d_{\mathrm{DR}}^{(0)}} \mathrm{Hom}(\Lambda^4 T, \Omega^2).$$

La première ligne sera

$$\mathcal{KR}(3)^{\cdot 1} : \mathrm{Hom}(S^2 T, \Omega) \xrightarrow{d_{\mathrm{DR}}^{(1)}} \mathrm{Hom}(S^2 T \otimes T, \Omega) \xrightarrow{d_{\mathrm{DR}}^{(1)}} \mathrm{Hom}(S^2 T \otimes \Lambda^2 T, \Omega).$$

La deuxième ligne sera

$$\mathcal{KR}(3)^{\cdot 2} : \mathrm{Hom}(S^3 T, A) \xrightarrow{d_{\mathrm{DR}}^{(2)}} \mathrm{Hom}(S^3 T \otimes T, A)$$

avec $d_{\mathrm{DR}}^{(2)} = d_{\mathrm{Ch}} + E$,

$$d_{\mathrm{Ch}} f(\tau', \tau'', \tau'''; \tau) = \tau f(\tau', \tau'', \tau''') - \mathrm{Cycle}_{\tau,\tau',\tau''} f([\tau, \tau'], \tau'', \tau''')$$

et

$$E f(\tau', \tau'', \tau'''; \tau) = -\frac{1}{2} \mathrm{Cycle}_{\tau,\tau',\tau''} \tau' f(\tau'', \tau'''; \tau).$$

Les flèches Q ont été définies dans 10.2 et 10.4.2. En utilisant Lemme 10.4.2 on vérifie:

Lemme 10.5.1. $Q d_{\mathrm{DR}}^{(1)} = d_{\mathrm{DR}}^{(2)} Q$.

Les flèches

$$R^{i1} : \mathcal{KR}(3)^{i1} = \mathrm{Hom}(S^2 T \otimes \Lambda^i T, \Omega) \longrightarrow \mathrm{Hom}(\Lambda^{i+3} T, \Omega^2)$$

$(i = 0, 1)$ ont été définies dans 10.3.

10.6

Enfin, on définit la flèche

$$R^{02} : \mathcal{KR}(3)^{i1} = \mathrm{Hom}(S^3 T, A) \longrightarrow \mathrm{Hom}(S^2 T \otimes \Lambda^2 T, \Omega)$$

par

$$\begin{aligned} &R^{02} f(\tau', \tau''; \tau_1, \tau_2) \\ &\quad = -\frac{1}{6}[df(\tau', \tau''; [\tau_1, \tau_2]) + \mathrm{Sym}_{\tau',\tau''} \mathrm{Alt}_{12}\, df(\tau', [\tau'', \tau_1], \tau_2)]. \end{aligned}$$

Théorème 10.6.1. *On a les relations:* $d_{\mathrm{DR}}^{20} R^{01} = R^{11} d_{\mathrm{DR}}^{01}$ *et* $d_{\mathrm{DR}}^{11} d_{\mathrm{DR}}^{01} = Q R^{01} + R^{02} Q$.

Donc, les opérateurs d_{DR}, Q *et* R *définissent sur* $\mathcal{KR}(3)^{\cdot\cdot}$ *une structure d'un bicomplexe tordu. On désigne par* $(\mathcal{KR}(3), d_{\mathcal{KR}})$ *le complexe simple associé.*

L'inclusion décrite après 10.1.1 *induit l'inclusion de complexes*

$$\iota : \Omega^{[3,6]} := \sigma_{\leq 6} \sigma_{\geq 3} \Omega^{\cdot}[3] \hookrightarrow \mathcal{KR}(3). \tag{10.6.1}$$

10.7

Soit $\mathfrak{b} = \{\tau_i\}$ une base abélienne. On a déjà vu un élément intéressant $\{,,\}_{\mathfrak{b}} \in \mathrm{Hom}(S^3T, A) = \mathcal{KR}(3)^{02}$,

$$\begin{aligned}\{a\tau_i, b\tau_j, c\tau_k\}_{\mathfrak{b}} &= -\frac{1}{2}\,\mathrm{Sym}_{a\tau_i, b\tau_j}\, \tau_k(a)\tau_i(b)\tau_j(c)\\ &= -\frac{1}{6}\,\mathrm{Sym}_{a\tau_i, b\tau_j, c\tau_k}\, \tau_k(a)\tau_i(b)\tau_j(c);\end{aligned}$$

cf. 8.4.

Disons qu'une cochaîne $x = (x^{02}, x^{11}, x^{20}) \in \mathcal{KR}(3)^2$ est *un cocycle intéressant* si $x^{02} = \{,,\}_{\mathfrak{b}}$ (pour une $\mathfrak{b} \in \mathfrak{B}$) et $d_{\mathcal{KR}}x = 0$.

Théorème 10.7.1. *Introduisons les éléments* $\langle,;\rangle_{\mathfrak{b}} \in \mathrm{Hom}(S^2T \otimes T, \Omega) = \mathcal{KR}(3)^{11}$,

$$\langle a\tau_i, b\tau_j; c\tau_k\rangle_{\mathfrak{b}} = -\frac{1}{2}\,\mathrm{Sym}_{a\tau_i, b\tau_j}\, \tau_k(a)\tau_i(b)d\tau_j(c) - \frac{1}{2}d\{a\tau_i, b\tau_j, c\tau_k\}_{\mathfrak{b}}$$

et $c_{\mathfrak{b}} \in \mathrm{Hom}(\Lambda^3 T, \Omega^2) = \mathcal{KR}(3)^{20}$,

$$c_{\mathfrak{b}}(a\tau_i, b\tau_j, c\tau_k) = \frac{1}{6}\,\mathrm{Alt}_{a\tau_i, b\tau_j, c\tau_k}\, \tau_k(a)d\tau_i(b)d\tau_j(c).$$

Alors $m_{\mathfrak{b}}^{\mathcal{KR}} := (\{,,\}_{\mathfrak{b}}, \langle,;\rangle_{\mathfrak{b}}, c_{\mathfrak{b}})$ *est un cocycle intéressant.*

En composantes, cela signifie que

$$d_{\mathrm{DR}}\{,,\}_{\mathfrak{b}} - Q\langle,;\rangle_{\mathfrak{b}} = 0, \tag{10.7.1}$$

$$d_{\mathrm{DR}}\langle,;\rangle_{\mathfrak{b}} - R\{,,\}_{\mathfrak{b}} + Qc_{\mathfrak{b}} = 0, \tag{10.7.2}$$

$$d_{\mathrm{DR}}c_{\mathfrak{b}} + R\langle,;\rangle_{\mathfrak{b}} = 0. \tag{10.7.3}$$

En effet, on trouve $\langle,;\rangle_{\mathfrak{b}}$ grace à la condition (10.7.1), après on trouve $c_{\mathfrak{b}}$ grace à la condition (10.7.2), et enfin on vérifie (10.7.3).

Le cocycle $m_{\mathfrak{b}}^{\mathcal{KR}}$ est analogue, en dimension 3, de la partie Koszul–de Rhamienne $(-\langle,\rangle_{\mathfrak{b}}, c_{\mathfrak{b}})$ de la structure vertex $v_{\mathfrak{b}}$, 7.3, 7.3.1.

11 Troisième cocycle de Chern–Simons raffiné

11.1

Maintenant passons au bicomplexe $\check{C}^{\cdot}(\mathfrak{B}; \mathcal{KR}(3)^{\cdot})$, le degré de Čech étant comme d'habitude le deuxième. On désigne par $(\check{C}^{\cdot}(\mathfrak{B}; \mathcal{KR}(3)), d_{C\mathcal{KR}})$ le complexe simple associé. On suppose que 30 est inversible dans k.

L'inclusion ι, (10.11.1), induit le morphisme de complexes

$$\mu : \mathrm{CS}(3) = C^{\cdot}(\mathrm{GL}_n(A), \Omega^{[3,5\rangle}) \longrightarrow \check{C}^{\cdot}(\mathfrak{B}; \mathcal{KR}(3))$$

Rappelons la troisième forme de Chern–Simons (cf. 3.11.1) $\beta_3 = (\beta^{03}, \beta^{12}, \beta^{21}) \in \mathrm{CS}(3)^3$, où $\beta^{03} \in \mathrm{Hom}(\mathrm{GL}_n(A)^3, \Omega^3)$,

$$\beta^{03}(\chi, \psi, \phi) = \frac{1}{6}\,\mathrm{tr}\{\phi^{-1}\psi^{-1}\chi^{-1}d\chi d\psi d\phi\},$$

$\beta^{12} \in \mathrm{Hom}(\mathrm{GL}_n(A)^2, \Omega^4)$ est défini dans 3.10, et $\beta^{21} \in \mathrm{Hom}(\mathrm{GL}_n(A), \Omega^{5,\mathrm{fer}})$,

$$\beta^{21}(\phi) = -\frac{1}{60}\,\mathrm{tr}\{(\phi^{-1}d\phi)^5\}.$$

11.2

Introduisons une cochaîne $\hat{m}_{\mathcal{KR}} = (\hat{m}_{\mathcal{KR}}^{ijk}) \in \check{\mathrm{C}}^2(\mathfrak{B}; \mathcal{KR}(3))$.

Les composantes de degré de Čech 0. On pose

$$\hat{m}_{\mathcal{KR}}^{\cdot\cdot 0} = (\hat{m}_{\mathcal{KR}}^{020}, \hat{m}_{\mathcal{KR}}^{110}, \hat{m}_{\mathcal{KR}}^{200}) = (\{,\,\}_{\mathfrak{b}}, \langle,;\rangle_{\mathfrak{b}}, c_{\mathfrak{b}}) \in \check{\mathrm{C}}^0(\mathfrak{B}; \mathcal{KR}(3)^{\cdot\cdot});$$

cf. 10.6.1, 10.7.

11.3

Les composantes de degré de Čech 1. Soient $\mathfrak{b} = \{\tau_i\}$, $\mathfrak{b}' = \{\tau_i'\} \in \mathfrak{B}$, $\mathfrak{b}' = \phi\mathfrak{b}$. Rappelons la forme $-h_{\mathfrak{b}\mathfrak{b}'} \in \mathrm{Hom}(S^2T, \Omega) = \mathcal{KR}(3)^{01}$ définie dans 9.3.

D'un autre côté, définissons une forme $h'_{\mathfrak{b}\mathfrak{b}'} \in \mathrm{Hom}(\Lambda^2T, \Omega^2) = \mathcal{KR}(3)^{10}$ par l'expression suivante:

$$\begin{aligned}
&-h'_{\mathfrak{b}\mathfrak{b}'}(b\tau_j', c\tau_k')\\
&\quad= \frac{1}{4}\,\mathrm{tr}\{\phi^{-1}c\tau_k'(\phi)\phi^{-1}d\phi\phi^{-1}b\tau_j'(\phi)\phi^{-1}d\phi\}\\
&\qquad+ \frac{1}{12}\,\mathrm{Alt}_{b\tau_j', c\tau_k'}\,\mathrm{tr}\{\phi^{-1}c\tau_k'(\phi)\phi^{-1}b\tau_j'(\phi)(\phi^{-1}d\phi)^2\}\\
&\qquad+ \frac{1}{12}\,\mathrm{Alt}_{b\tau_j', c\tau_k'}\,\mathrm{tr}\{dc[\phi^{-1}\tau_k'(\phi)\phi^{-1}b\tau_j'(\phi) + \phi^{-1}b\tau_j'(\phi)\phi^{-1}\tau_k'(\phi)]\phi^{-1}d\phi\}\\
&\qquad+ \frac{1}{12}\,\mathrm{Alt}_{b\tau_j', c\tau_k'}\,\mathrm{tr}\{[\phi^{-1}cd\tau_k'(\phi)\phi^{-1}b\tau_j'(\phi) + \phi^{-1}b\tau_j'(\phi)\phi^{-1}cd\tau_k'(\phi)]\phi^{-1}d\phi\}\\
&\qquad- \frac{1}{4}\,\mathrm{Alt}_{b\tau_j', c\tau_k'}[d\phi\phi^{-1}cd\tau_k'(\phi)\phi^{-1}]^{jp}\tau_p'(b)\\
&\qquad+ \frac{1}{4}\,\mathrm{Alt}_{b\tau_j', c\tau_k'}[d\phi\phi^{-1}d\phi\phi^{-1}c\tau_k'(\phi)\phi^{-1}]^{jp}\tau_p'(b)\\
&\qquad- \frac{1}{4}\,\mathrm{Alt}_{b\tau_j', c\tau_k'}[d\phi\phi^{-1}b\tau_j'(\phi)\phi^{-1}]^{kp}d\tau_p'(c)\\
&\qquad+ \frac{1}{4}\,\mathrm{Alt}_{b\tau_j', c\tau_k'}[d\phi\phi^{-1}b\tau_j'(\phi)\phi^{-1}d\phi\phi^{-1}]^{ku}\tau_u'(c)\\
&\qquad+ \frac{1}{4}\,\mathrm{Alt}_{b\tau_j', c\tau_k'}[d\phi\phi^{-1}d\phi\phi^{-1}]^{jp}\tau_p'(c)\tau_k'(b)
\end{aligned}$$

$$
\begin{aligned}
&-\frac{1}{2}[d\phi\phi^{-1}]^{ju}\tau'_u(c)[d\phi\phi^{-1}]^{kv}\tau'_v(b)\\
&-\frac{1}{4}\operatorname{Alt}_{b\tau'_j,c\tau'_k}[d\phi\phi^{-1}dc\tau'_k(\phi)\phi^{-1}]^{jp}\tau'_p(b)\\
&-\frac{1}{4}\operatorname{Alt}_{b\tau'_j,c\tau'_k}[d\phi\phi^{-1}]^{ju}d\tau'_u(c)\tau'_k(b)-\frac{1}{4}\operatorname{Alt}_{b\tau'_j,c\tau'_k}[d\phi\phi^{-1}]^{kw}\tau'_w(b)d\tau'_j(c).
\end{aligned}
$$

On pose

$$\hat{m}^{\cdot\cdot 1}_{\mathcal{KR}}=(\hat{m}^{011}_{\mathcal{KR}},\hat{m}^{101}_{\mathcal{KR}})=(-h_{\mathfrak{b}\mathfrak{b}'},h'_{\mathfrak{b}\mathfrak{b}'})\in\check{C}^1(\mathfrak{B};\mathcal{KR}(3)^{\cdot\cdot}).$$

11.4

La composante de degré de Čech 2.

$$\hat{m}^{002}_{\mathcal{KR}}=(-H_{\mathfrak{b}\mathfrak{b}'\mathfrak{b}''})\in\check{C}^2(\mathfrak{B};\mathcal{KR}(3)^{00}),$$

où $-H_{\mathfrak{b}\mathfrak{b}'\mathfrak{b}''}\in\operatorname{Hom}(T,\Omega^2)=\mathcal{KR}(3)^{00}$ est écrite dans 9.4.

En rassemblant,

$$\hat{m}_{\mathcal{KR}}=(\{,,\}_{\mathfrak{b}},\langle,;\rangle_{\mathfrak{b}},c_{\mathfrak{b}};-h_{\mathfrak{b}\mathfrak{b}'},h'_{\mathfrak{b}\mathfrak{b}'};-H_{\mathfrak{b}\mathfrak{b}'\mathfrak{b}''}).$$

Théorème 11.4.1. $d_{\mathcal{CKR}}(\hat{m}_{\mathcal{KR}})=\mu(\beta_3)$.

En composantes, c'est exprimé par 9 relations:

$$-d_cH_{\mathfrak{b}\mathfrak{b}'\mathfrak{b}''\mathfrak{b}'''}=\iota\beta^{03}(\chi,\psi,\phi)\qquad(11.4.1)$$

si $\mathfrak{b}'''=\chi\mathfrak{b}''$, $\mathfrak{b}''=\psi\mathfrak{b}'$ et $\mathfrak{b}'=\phi\mathfrak{b}$; ceci est une partie du Théorème 9.4.1;

$$
\begin{aligned}
d_{\mathrm{DR}}H_{\mathfrak{b}\mathfrak{b}'\mathfrak{b}''}+d_ch'_{\mathfrak{b}\mathfrak{b}'\mathfrak{b}''}&=\iota\beta^{12}(\psi,\phi),&(11.4.2)\\
d_{\mathrm{DR}}h'_{\mathfrak{b}\mathfrak{b}'}+Rh_{\mathfrak{b}\mathfrak{b}'}+d_cc_{\mathfrak{b}\mathfrak{b}'}&=\iota\beta^{21}(\phi),&(11.4.3)\\
-d_{\mathrm{DR}}h_{\mathfrak{b}\mathfrak{b}'}-Qh'_{\mathfrak{b}\mathfrak{b}'}+d_c\langle,;\rangle_{\mathfrak{b}\mathfrak{b}'}&=0.&(11.4.4)
\end{aligned}
$$

Encore deux relations qui font partie du 9.4.1:

$$QH_{\mathfrak{b}\mathfrak{b}'\mathfrak{b}''}-d_ch_{\mathfrak{b}\mathfrak{b}'\mathfrak{b}''}=0\qquad(11.4.5)$$

et

$$-Qh_{\mathfrak{b}\mathfrak{b}'}+d_c\{,,\}_{\mathfrak{b}\mathfrak{b}'}=0\qquad(11.4.6)$$

et enfin, les 3 relations du Théorème 10.7.1.

En effet, on dérive la formule compliquée de 11.3 pour $h'_{\mathfrak{b}\mathfrak{b}'}$ de l'équation (11.4.4). Ensuite on obtient la forme β^{12} de l'équation (11.4.2) (sic!). Enfin, on vérifie que la relation (11.4.3) est satisfaite.

References

[BD] A. Beilinson and V. Drinfeld, *Chiral Algebras*, AMS Colloquium Publications, Vol. 51, American Mathematical Society, Providence, RI, 2004.

[GMS] V. Gorbounov, F. Malikov, and V. Schechtman, Gerbes of chiral differential operators II: Vertex algebroids, *Invent. Math.*, **155** (2004), 605–680.

[Gr] A. Grothendieck, Classes de Chern et représentations linéaires des groupes discrets, in *Dix exposés sur la cohomologie des schémas*, North-Holland and Masson, Amsterdam, Paris, 1968, 215–305.

[S] V. Schechtman, *Definitio nova algebroidis verticiani*, in J. Bernstein, V. Hinich, and A. Melnikov, eds., *Studies in Lie Theory: Dedicated to A. Joseph on his Sixtieth Birthday*, Progress in Mathematics, Vol. 243, Birkhäuser Boston, Cambridge, MA, 2006.

PRINTED IN THE UNITED STATES OF AMERICA